D1134617

# Human Biochemistry

# Human Biochemistry

**Gerald Litwack, Ph.D.**
Los Angeles, CA, United States

**ACADEMIC PRESS**
An imprint of Elsevier

Academic Press is an imprint of Elsevier
125 London Wall, London EC2Y 5AS, United Kingdom
525 B Street, Suite 1800, San Diego, CA 92101-4495, United States
50 Hampshire Street, 5th Floor, Cambridge, MA 02139, United States
The Boulevard, Langford Lane, Kidlington, Oxford OX5 1GB, United Kingdom

**Notices**

Knowledge and best practice in this field are constantly changing. As new research and experience broaden our understanding, changes in research methods, professional practices, or medical treatment may become necessary.

Practitioners and researchers must always rely on their own experience and knowledge in evaluating and using any information, methods, compounds, or experiments described herein. In using such information or methods they should be mindful of their own safety and the safety of others, including parties for whom they have a professional responsibility.

**British Library Cataloguing-in-Publication Data**
A catalogue record for this book is available from the British Library

**Library of Congress Cataloging-in-Publication Data**
A catalog record for this book is available from the Library of Congress

ISBN: 978-0-12-383864-3

For Information on all Academic Press publications
visit our website at https://www.elsevier.com/books-and-journals

Working together
to grow libraries in
developing countries

www.elsevier.com • www.bookaid.org

*Publisher*: Sara Tenney
*Acquisition Editor*: Linda Versteeg-Buschman
*Editorial Project Manager*: Fenton Coulthurst
*Production Project Manager*: Poulouse Joseph
*Cover Designer*: Alan Studholme

Typeset by MPS Limited, Chennai, India

# Dedication

For my family, especially for the younger ones: Kate, Geoff, Suzie and Deb.

# Contents

## 20. Vitamins and Nutrition

## 21. Blood and Lymphatic System

Additional teaching materials to support this textbook are available to instructors at: http://textbooks.elsevier.com/web/Manuals.aspx?isbn=9780123838643

# Preface

This book is designed for senior undergraduates, graduate students, and, especially, medical students. It should also be useful for teachers, researchers, and clinicians for designing lectures and as a desk reference. Although the book has considerable depth, the instructor can make use of various illustrations in the book to present a course at the level he/she wishes. The more extensive information can be gained by the interested student and the list of references following each chapter will be useful for further individual research.

Many medical students question the reason that they have to study biochemistry when their focus is thought to be exclusively on medicine. They often ask: "Why do I have to learn biochemistry?" In this volume, each chapter opens with a disease or clinical condition that is presented in clinical terms and subsequently dissected to the molecular biochemical level, demonstrating the principles of the chapter. This approach explains the basis of the disease and frequently discusses diagnosis and treatment, cementing the relationship of biochemistry to medicine and disease. Thus the question of relevance is answered.

Undergraduate and graduate students often have little exposure to medicine and disease. *Human Biochemistry* should serve to introduce these students to clinical examples in molecular terms that will broaden their education.

A clinically related example at the beginning of each chapter was first employed in *Human Biochemistry & Disease* by myself, published in 2008. *Human Biochemistry* presents many modifications. In particular, the only three-dimensional X-ray structures of molecules used here are those conveying some property or activity that is obvious to the reader. The figures used here all have been redrawn with a consistent font and white or very light backgrounds. Most figures are in color. As further teaching aids, many figures are converted to slides and other teaching aids by Academic Press/Elsevier, available to the instructor. Multiple-choice questions at the end of each chapter emphasize the principles of the chapter and prepare the student for future examinations.

The first three chapters are introductory. The first review chapter deals with the gross anatomy of the major organs in the human body, less important for the medical student if he/she is taking a class in gross anatomy concurrently with a course in biochemistry. However, this review may well be useful for nonmedical students. The first two chapters end with a summary, suggested reading list involving journal papers and books, and multiple-choice review questions. The third chapter and all the subsequent chapters close with a summary, suggested reading list, multiple-choice review questions, and a case-based problem. The case-based problem is most effectively used in small group teaching where a clinician and scientist lead the discussion. The objective of the case-based problem is to, step-by-step, reach a diagnosis and treatment.

Early on, Janice Audet of Elsevier, now at Harvard University Press, aided my work. Subsequently, Fenton Coulthurst of Elsevier, Oxford, United Kingdom, was instrumental in the progress and publication of the book.

**Gerald Litwack, PhD**
*Los Angeles, CA, United States*

# Introduction

Enough is known at present about the biochemistry of disease to securely link this science to medicine. Biochemistry and physiology are foundations of medicine. Pharmacology, the practice of which generates medicines, can be defined as the biochemistry of drug action. As most medical students, who encounter biochemistry, cell biology, molecular biology, and genetics in the first year of medical school, thirst for training and experience in medicine, it behooves instructors of biochemistry in this setting to establish firmly the interconnection between biochemistry and disease. Otherwise, students tend to separate this science from medicine and often wonder why they are taking a course that would seem remote from medicine and disease. This book, which introduces the basics of biochemistry, emphasizes the connection between biochemistry and disease so that the student is aware of the basic information supporting our current knowledge. Accordingly, relevant diseases are introduced in each chapter that relate to the principles explored in the chapter and extend the understanding of the disease often to the level of molecules.

The biochemical information is up to date. At the end of each chapter, a summary appears. In addition, each chapter has supporting online materials, including a set of review questions in the form of USLME (United States Medical Licensing Examination) examinations serving to familiarize the student with this type of testing. These questions cover the major points of the chapter and emphasize the central principles. Furthermore, the online support supplements each chapter with a case presentation that would incorporate the principles of the chapter in a clinical context. This is becoming a favored mechanism for the deductive diagnosis of a set of symptoms by a small group under the direction of a mentor. The experiential nature of this exercise serves to incorporate basic information into the thinking of the student so that he/she may retain useful information for a long period compared to rote memorization, recapitulation during testing and forgetting the information after the test is over. This technique facilitates the capability of a student to solve problems and introduces the technology for gathering relevant information, a prescription for the lifelong learner.

In all, this book contains 19 chapters that cover the essential information in a basic course in medical biochemistry. Importantly, the stress is on the integration of biochemistry, disease and medicine. The chapters are ordered so that a discussion of proteins and enzymes comes first. In my view, all succeeding information depends upon knowledge of enzymes that allow chemical reactions to occur under bodily conditions. I realize that all instructors may not agree with this order and might prefer to introduce the subject of nucleic acids at the outset. In that case, one can begin with Chapter 8, Glycolysis and Gluconeogenesis, leaving Chapters 4 and 5 for later introduction.

There are many figures in this book. They are used to provide pathways and overall views of mechanisms of action. There are some figures showing three-dimensional structures. These are used only when they shed light upon a mechanism or interaction of a macromolecule with a ligand, to show protein—protein interactions or to graphically demonstrate an enzymatic action. The use of such figures is limited to increasing the understanding of a particular mechanism.

Clinical case-based exercises are developed by individuals familiar with this technique. Likewise, the USMLE type questions are developed by separate experts. General references are included for further reading at the end of each chapter.

The first three chapters are introductory. In Chapter 1, Organ Systems and Tissues are discussed the organ systems and tissues, including aspects of tissue development. Chapter 2, The Cell focuses down to the cell and its composition with particular reference to organelles and subcellular particles. In Chapter 3, Water, pH, Buffers and Introduction to the General Features of Receptors, Channels, and Pumps, there is a discussion of water, pH, buffer systems, and general features of receptors and channels. These are basic concepts that a student should learn at the outset. The chapters introduce physiology and cell biology as larger contexts of biochemistry which focuses on molecules but has become miscible with cell—and molecular—biology.

At the end of the book is a Glossary, explaining specific names and abbreviations, an in depth index and Appendices giving the names of amino acids, their abbreviations, and some characteristics, the genetic code, and weights and measures.

Studying the basic information in this book should provide a format for lectures, if they are used, and as well a source of information for small group study and reference during the exercise of case- or problem-based learning.

**Gerald Litwack**

Chapter 1

# Organ Systems and Tissues

## TREATMENT OF THE INJURED KNEE: USE OF STEM CELLS TO REPLACE DAMAGED CARTILAGE

In the human the knee, as it supports the entire body weight is susceptible, not only to acute injury, but, in particular, to the development of chronic osteoarthritis. Osteoarthritis can be defined as the degeneration of joint cartilage and the bone beneath it and it occurs in any joint, especially middle age and older, in the hip, knee, and thumb. The most common type of osteoarthritis is in the knee and leads to the deterioration of the articular cartilage. The **articular cartilage** is that which covers the bone and the **hyaline cartilage** is located within the joints. Arthroscopic surgery has been most widely used to treat this condition but with the advent of new research on stem cells, this condition is being ameliorated without surgery by the injection of stem cells into the site. Normal and osteoarthritic cartilage is shown in Fig. 1.1.

Under the microscope healthy and osteoarthritic knee hyaline cartilage can be differentiated as shown in Fig. 1.2.

The zones of the articular cartilage down to the bone, including the tidemark, is shown in Fig. 1.3.

In attempting to repair cartilage of an osteoarthritic knee, chondrocytes are extracted arthroscopically from normal cartilage in the nonload-bearing intercondylar notch or the upper ridge of femoral condyles. These cells are then grown in tissue culture for 4–6 weeks (Fig. 1.4) under conditions where no contamination of any sort can take place ("good practices"). When sufficient numbers of cells have grown up, they are injected into the damaged area, usually in combination with a matrix structure (Fig. 1.5).

The implanted cells, provided with growth factors grow in the native environment and form new cartilage tissue. In most cases, there is a definite improvement in which there is a reduction of pain and increased ability to use the affected knee, even in competitive sports. While this sort of success in treatment of knee cartilage problems has been reported in several clinical trials there is, as yet, no uniform consensus on the effectiveness of this procedure, indicating the need for further clinical trials. This clinical example serves to highlight some of the features of this chapter, particularly with reference to development in which certain layers of cells are equivalent to stem cells that can, under the appropriate conditions, develop into any tissue in the body.

## DEVELOPMENT OF ORGANS

Development begins with fertilization of the ovum by a sperm. The early embryo undergoes a cell division generating the two-cell stage, each daughter cell containing the genome of both parents. The division is of a cleavage type (without cell growth) indicating that the size of the cells becomes progressively smaller with each division. Within 5 days cell division continues through the four-cell stage, the eight-cell stage and the sixteen-cell stage, the **morula** (by Day 3) and forming the **blastocyst** which contains an outer and inner cell mass. At the 16-cell stage, there are eight large cells outside and eight smaller cells inside the structure. By the 5th day the **trophoblast** is distinguishable with a large blastocyst cavity (the **blastocoel**) and a distinct inner and outer cell mass. By Day 7 the blastocyst becomes enlarged. This progression is depicted in Fig. 1.6.

## STEM CELLS

Cells of the blastocyst can form all the tissues of all the organs in the body. The only cells that are totipotent (can develop into every cell type in the body including placental cells) are the **embryonic (stem) cells** existing after the first and second cell divisions following fertilization. Similar in potency are pluripotent stem cells (embryonic stem cells capable of generating all cell types in the body) derived from the inner cell mass of the blastocyst. These cells, in the

Human Biochemistry. DOI: http://dx.doi.org/10.1016/B978-0-12-383864-3.00001-6

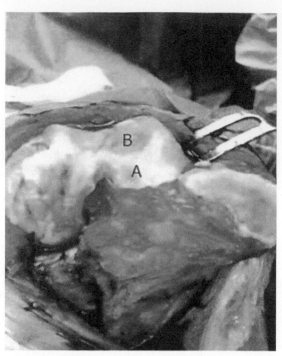

**FIGURE 1.1** Macroscopic signs of osteoarthritis knee hyaline cartilage. (A) Healthy cartilage. (B) Osteoarthritis cartilage. *Reproduced from: http://www.ncbi.nlm.nih.gov/pmc/articles/PMC4017310/figure/F1/.*

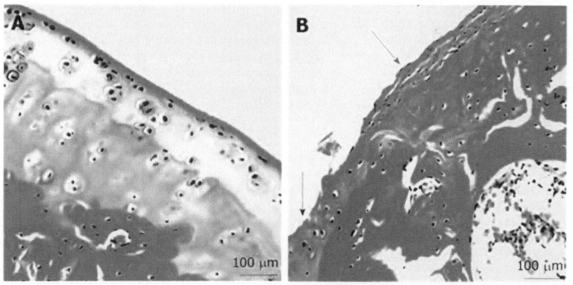

**FIGURE 1.2** Microscopic signs: (A) microscopic signs of healthy knee hyaline cartilage. The histological (HE, hemotoxylin-eosin staining) analysis of cartilage form normal donor showed a preserved morphological structure with no sign of cartilage degradation. Moreover, the surface of healthy hyaline cartilage appears white, shiny, elastic, and firm. Magnification 20×; scale bars: 100 μm. (B) Microscopic signs of osteoarthritis (OA) knee hyaline cartilage. The histological staining (HE staining) analysis of cartilage from OA donor. The donor demonstrated joint swelling and edema, horizontal cleavage tears or flaps, the surface became dull and irregular and had minimal healing capacity. Magnification 20×; scale bars: 100 μm. Moderate OA cartilage (*black arrow*); the structural alterations included a reduction of cartilage thickness of the superficial and middle zones. The structure of the collagen network is damaged, which leads to reduced thickness of the cartilage. The chondrocytes are unable to maintain their repair activity with subsequent loss of the cartilage tissue. Severe OA cartilage (*blue arrow*), demonstrated deep surface clefts, disappearance of cells from the tangential zone, cloning, and a lack of cells in the intermediate and radial zone, which are not arranged in columns. The tidemark is no longer intact and the subchondral bone shows fibrillation. *Reproduced from: http://www.ncbi.nlm.nih.gov/pmc/articles/PMC4017310/figure/F2/.*

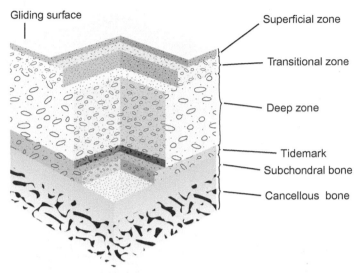

Gliding surface

Superficial zone

Transitional zone

Deep zone

Tidemark

Subchondral bone

Cancellous bone

**FIGURE 1.3** The articular (hyaline) cartilage. The articular cartilage is one of five forms of cartilage: the hyaline or articular cartilage, the fibroblastic cartilage comprising the meniscus, the fibrocartilage located at the tendon and the ligament insertion into bone, the elastic cartilage of the trachea and the physeal cartilage of the growth plate (*physeal*, area of bone separating metaphysis and epiphysis where cartilage grows). *Reproduced from: http://www.orthobullets.com/basic-science/9017/articular-cartilage.*

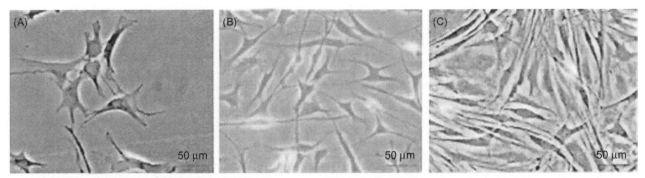

**FIGURE 1.4** Development of mesenchymal stem cells. (A) First day of culture; (B) 3rd day of culture; (C) 1 week of culture. Magnification 40 × ; scale bars: 50 μm. *Reproduced from: http://www.ncbi.nlm.nih.gov/pmc/articles/PMC4017310/figure/F4/.*

appropriate environment, can differentiate into every cell type present in the organism. What is more, this differentiation can take place either *in vivo* or *in vitro*. Stem cells also can be obtained from adults (generally multipotent stem cells that can develop into more than one cell type) as most tissues in the body have stores of stem cells (caches) that are used to replace the cells in the tissue that have died. Some caches are more plentiful than others and a great deal of research has centered on the stem cells in caches stored in joints (mesenchymal stem cells), for example. Adult stem cells, in addition to embryonic stem cells, can be transformed into many types of tissue cells under the appropriate environmental conditions. The various stem cells and tissue progenitor cells have different properties reflected in differences in their metabolism as listed in Table 1.1.

Mesenchymal stem cells have been proliferated *in vitro* and subsequently used to regenerate cartilage in knee joints that have suffered degeneration, for example. Many such clinical interventions, however, may not be permanent and the process has to be repeated after a time. There is some concern that stem cells injected into patients could generate into cancer cells although the evidence that this is an inherent danger in using them for clinical treatments, so far, has not borne out. Since each tissue in the body may have a cache of stem cells, the aging process could be the result of insufficient stem cells to replace tissue cells that become damaged and die out. Consequently the generation of such stem cells *in vitro* constitutes the practice and hope for further clinical treatments to replace tissue cells and organs. Stem cells generated *in vitro* must be safe from any contaminating substances or organisms and all are generated under good manufacturing practices (GMP) promulgated by the FDA (Food and Drug Administration). A newer development under

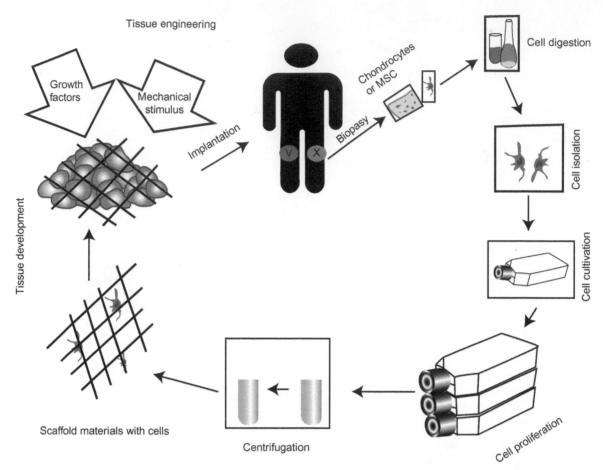

**FIGURE 1.5** Graphic representation of cartilage tissue engineering. Staring with "X" in upper center, chondrocytes, or mesenchymal stem cells are removed from healthy cartilage; the mesenchymal stem cells are isolated and cultured in tissue culture media for about 7 days under strict conditions so that no impurities can contaminate the culture. The cells are concentrated by centrifugation, combined with a matrix substance and various growth factors and injected back into the damaged knee ("V"). The damaged knee, by itself, cannot produce enough stem cells to repair the damage. *Reproduced from: http://www.ncbi.nlm.nih.gov/pmc/articles/PMC4017310/figure/F3/.*

the heading of "therapeutic cloning" involves the use of pluripotent stem cells (without destruction of a human embryo) that are genetically identical to the donor to produce pluripotent stem cells that can be used to treat disease. In this case, adult skin cells from a donor can be used to extract his/her DNA and this DNA can be fused electrically with donated adult human eggs whose DNA has been removed. The resulting embryo at the level of the blastocyst is the source of pluripotent stem cells. The resulting pluripotent cells should be useful in treating many human diseases, including Parkinson's disease, type 1 diabetes, heart disease, and others.

The next stage from the blastula is the **gastrula** developed by the invasion (**invagination**) of the bottom layer of cells of the blastula up into the blastocoel forming two layers with a pore (**blastopore**) on the bottom where the invasion began as shown in Fig. 1.7. Now, two germ layers are determined, the outer layer of **ectoderm** and the inner layer of **endoderm**.

Further development reveals the internal **mesoderm** as shown in Fig. 1.8.

These three layers of cells: the ectoderm, endoderm, and mesoderm lead to the development of specific tissues and organs. From the endoderm is derived the **primitive gut** that, in turn, generates the lungs, liver, pancreas, and digestive tubes. The **ectoderm** leads to the development of the epidermis, the forerunner of the skin, hair, and mammary glands. It also leads to the epidermal placodes and then to the lens of the eye and the inner ear. Placodes are thickenings of the ectoderm. Apparently, there are a number of different groups of placodes but they all have a similar developmental history and they lead to different aspects of the nervous system. Importantly, the ectoderm is the precursor of the neural tubes and neural tissue leading to the development of the brain and spinal cord (central nervous system, CNS) and the peripheral nervous system (PNS). The **mesoderm** gives rise to the axial skeleton (bones of the head and trunk), skeletal muscle as well as the connective tissue of the skin. It also gives rise to other organs: the oviducts and uterus, kidney,

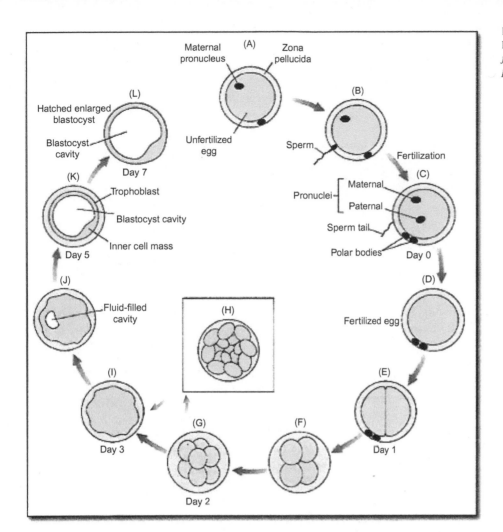

**FIGURE 1.6** The stages of early human development. *Reproduced from: http://openlearn.open.ac.uk/file. php/1638/formats/print.htm.*

Labels in figure:
- (A) Maternal pronucleus, Zona pellucida, Unfertilized egg
- (B) Sperm, Fertilization
- (C) Pronuclei — Maternal, Paternal; Sperm tail; Polar bodies; Day 0
- (D) Fertilized egg
- (E) Day 1
- (F)
- (G) Day 3
- (H)
- (I)
- (J) Fluid-filled cavity
- (K) Trophoblast, Blastocyst cavity, Inner cell mass, Day 5
- (L) Hatched enlarged blastocyst, Blastocyst cavity, Day 7
- Day 2

## TABLE 1.1 Summary of Metabolism in Respective Stem and Progenitor Cells

| Mammalian Cell Type | Active Metabolic Pathways |
| --- | --- |
| Totipotent stem cells/blastomeres | Pyruvate oxidation, bicarbonate fixation |
| Pluripotent stem cells/embryonic stem cells | Anabolic glycolysis, PPP, Thr-Gly metabolism |
| Differentiating embryonic stem cells | OxPhos, ROS, eicosanoids |
| Long-term hematopoietic stem cells | Catabolic glycolysis, fatty acid oxidation |
| Hematopoietic progenitors | Anabolic glycolysis, OxPhos, ROS, eicosanoids |
| Neural stem cells | Low glycolysis |
| Neural progenitors | High glycolysis, OxPhos, ROS, eicosanoids, fatty acid synthesis |
| Mesenchymal stem cells | Low glycolysis |
| Chondroblasts | High glycolysis |
| Osteoblasts | OxPhos |
| Preadipocytes | OxPhos, ROS |
| Myoblasts | Anabolic glycolysis, PPP |
| Myotubes | Glycolysis, OxPhos |
| Cardiomyocyte progenitors | Lactate oxidation |
| Intestinal stem cells *(Drosophila)* | OxPhos |

*OxPhos*, oxidative phosphorylation; *PPP*, pentose phosphate pathway; *ROS*, reactive oxygen species.
Source: Reproduced from: Shyh-Chang, N., Daley, G.Q., Cantley, L.C., 2013. Stem cell metabolism in tissue development and aging. *Development*, 140, 2535–2547.

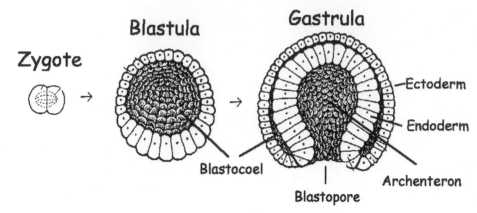

**FIGURE 1.7** Formation of the gastrula from the blastula. *Reproduced from: http://chsweb.lr.k12.nj.us/mstanley/outlines/animals/antax/image51.gif.*

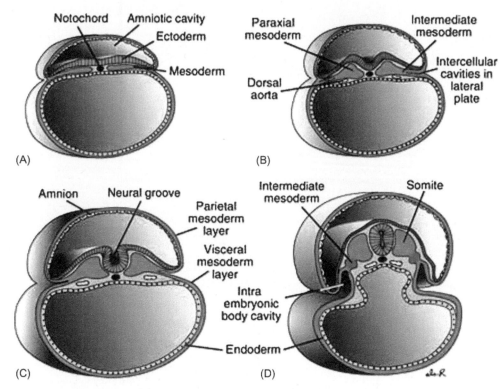

**FIGURE 1.8** Further development of the gastrula. *Reproduced from: http://www.carolguze.com/images/embryos/gastrulation.jpg.*

ovary, and testes, the connective tissue of the body wall and limbs, the mesenteries, heart, and blood vessels. The **endoderm** is the precursor for the remaining bodily systems. The summary of the further development of these three embryonic layers of cells is shown in Fig. 1.9.

## GROSS STRUCTURES AND FUNCTIONS OF ORGAN SYSTEMS

Most of this information is already obvious to you. However, the organ systems will be described briefly. Organs usually contain more than one type of tissue. Tissues will be discussed in the next section.

## SKIN

The **skin** is the largest organ in the body. It is shown in cross-section in Fig. 1.10.

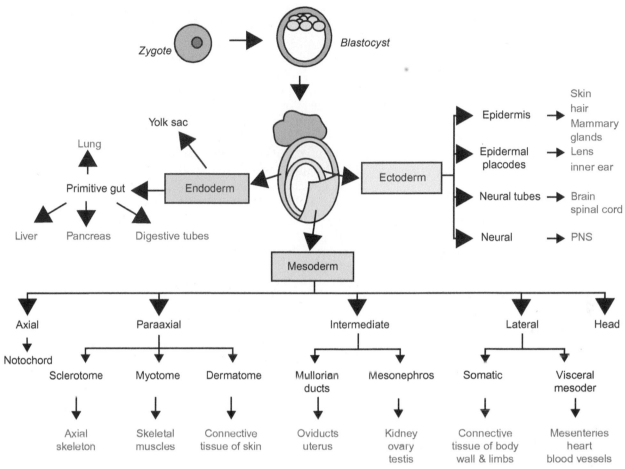

**FIGURE 1.9** Summary of the three embryonic layers of cells and their fates in the human body. *Reproduced from: http://www.unige.ch/cyberdocuments/theses2004/HeQ/images/Fig.3C.jpg.*

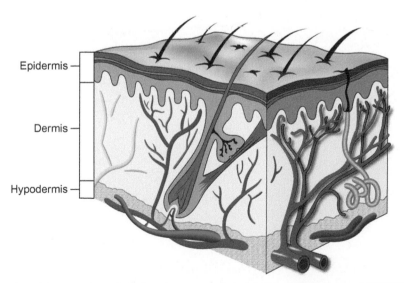

**FIGURE 1.10** Cross-section of the skin. *Reproduced from: http://content.revolutionhealth.com/contentimages/n5551176.jpg.*

From the outside extending internally, the skin consists of **epidermis, dermis** and a **subcutaneous layer** or **hypodermis** (Fig. 1.10). The three layers are derived from ectoderm. The skin functions in various ways providing insulation, regulation of temperature, and sensation. It is the site of the synthesis of **vitamin D** (cholecalciferol) from 7-dehydrocholesterol. 7-Dehydrocholesterol must be activated in cells of the skin by the UV radiation in sunlight in order for it to proceed to form the active form of vitamin D in the body via liver and kidney enzymes. Theoretically, exposure to sunlight for 10 min each day would satisfy the requirement for vitamin D. Vitamin D is considered as a vitamin because the content in the body is inadequate and it must be made available through the effect of sunlight and vitamin D in foods. The activated form of vitamin D in the body acts like a steroid hormone. Cholecalciferol must undergo further hydroxylations in the body to be generated as the active hormonal form, as will be discussed in the chapter on steroid hormones, because the active form of vitamin D is a ligand for the vitamin D receptor, a member of the steroid nuclear gene family.

Pigmentation of the skin is due to **melanin** and the extent of pigmentation varies among populations. Skin color is controlled by genes and by the environment. In hotter environments the body hair is decreased along with an increase in sweat glands to cool the body through evaporation of perspiration. Strong sunlight can cause skin damage and destroy folic acid and also can lead to basal cell carcinoma and sometimes to melanoma. Humans living in these environments have increased melanin that protects against damaging radiation from sunlight. On the other hand, humans living in colder and darker climates have lighter skin to optimize the penetration of UV sunlight for the production of vitamin D. Children who are raised in a dark environment with a small number of sunny days can develop colon cancer in adulthood, owing to a deficiency in this vitamin. Vitamin D is known for its beneficial effects on the immune surveillance system. However, in certain regions, such as Alaska, the natives have dark skin in the face of a cold and dark environment. In this case the diet of these natives is very rich in vitamin D and the skin pigmentation protects them against the strong radiation reflected from surrounding ice and snow. Skin pigmentation is under genetic control and

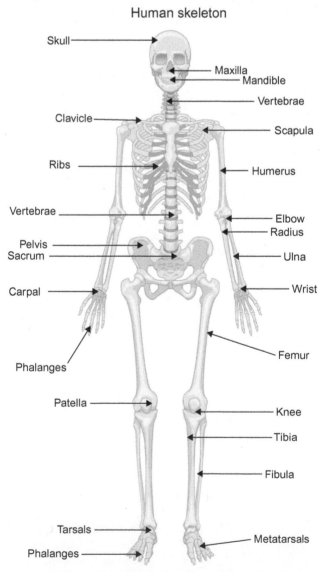

Human skeleton

FIGURE 1.11 Gross structure of the human skeleton. *Reproduced from: http://www.enchantedlearning.com/subjects/anatomy/skeleton/Skelprintout.shtml.*

involves the regulation of genes involved in the synthesis of melanin and other factors. For example, stress can affect skin color because products of the activated gene for proopiomelanocortin include ACTH and melanocyte stimulating hormone that stimulate the formation and release of cortisol and the formation of melanin. Also, the skin is the site of as many as one thousand species of bacteria. Many of the mechanisms involved are discussed in later chapters.

## THE SKELETON

The **skeletal system** is shown in Fig. 1.11. The major bones are labeled in the figure and there are 206 bones in the mature human skeleton. There are many more separate bones in early development but there are fusions in the growing child. The thigh bone (femur) is the longest bone in the body and the smallest bone resides in the inner ear (stirrup). Males have longer legs and arms than females and females have a wider pelvis. Importantly, bones are the sites of manufacture of blood cells and the storage sites for minerals. Calcium is a major component of the bone structure. Associated organs are tendons, ligaments, and cartilage.

The **muscle system** is shown in Fig. 1.12. The more than six hundred skeletal muscles (striated muscle) attach to bones and connect to joints. Muscles can work in pairs for motion and also control the movement of substances through some organs (circulation, heart, stomach, and intestines). There are also **smooth muscles** that are involuntary and are responsible for the contraction of hollow organs, such as the gastrointestinal tract, blood vessels, airways, bladder, and uterus. The other muscle type is the **striated muscle**, a voluntary muscle that is connected to bone at one or both ends. It has dark and light bands with repeating sarcomeres (the basic mechanical unit of muscle consisting of thin filaments each of which contains two strands of actin, a single strand of a regulatory protein and two thick filaments of myosin).

The **circulatory system** transports oxygen and $CO_2$, nutrients, hormones, and waste products throughout the body. It involves the heart, blood vessels, and blood. A diagram of the circulatory system is shown in Fig. 1.13.

The nervous system is made up of the **central nervous system (CNS)** and the **peripheral nervous system (PNS)**. The function of the nervous system is to send electrical signals throughout the body that direct movement and behavior.

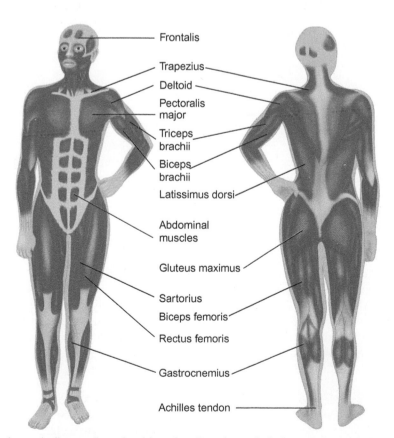

**FIGURE 1.12** The human body muscle diagram. *Reproduced from: http://www.human-body-facts.com/human-body-muscle-diagram.html.*

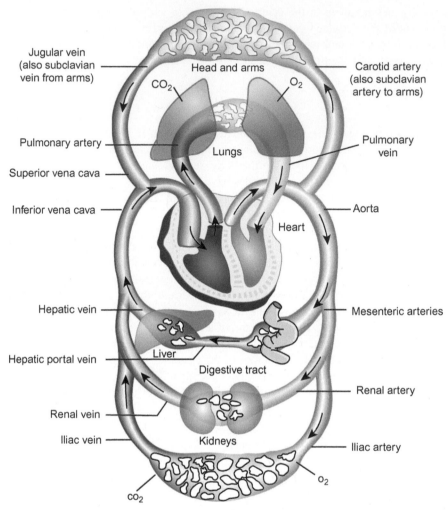

**FIGURE 1.13**   Diagram of the circulatory system. *Reproduced from: http://image.tutorvista.com/content/circulation-animals/human-blood-circulatory-system.jpeg.*

It is a major part of the control of physiological processes including the circulation and digestion. The central nervous system consists of the brain and spinal column; the rest are peripheral nerves. A diagram of the nervous system is shown in Fig. 1.14.

The **respiratory system** consists of the nose, trachea, and lungs as shown in Fig. 1.15. Oxygen is taken in from the outside atmosphere and $CO_2$ is expelled. This system exchanges oxygen and $CO_2$ between the blood and the organs.

The **digestive system** consists of the mouth, esophagus, stomach, small intestine, and large intestine. The liver produces detergents and the pancreas produces digestive enzymes and both are transported to the intestine for digestion. The overall functions are to take in nutrients, break them down in the intestinal tract and absorb the products into the bodily circulation. The transport mechanisms for moving the products of food digestion into the bloodstream and the tissues are presented in later chapters. A diagram of the digestive system is shown in Fig. 1.16.

There are many glands in the body that secrete hormones. The hypothalamus, pituitary (anterior, posterior, and cells of the intermediate pituitary), thyroid, pancreas, and adrenals are some of the major glands in the **endocrine system**. Hormones are chemical messages acting on organs distant from the secreting gland that control many functions at the cellular level. Hormones activate their signals through specific receptors. Fig. 1.17 shows glands that constitute the endocrine system.

In a stress situation, for example, many endocrine glands come into play. A stress event in the outside (the body) environment is filtered through a mechanism in the brain and the appropriate cells in the hypothalamus are signaled to secrete the corticotropin releasing hormone (CRH). This hormone is secreted into a closed portal system connecting the hypothalamus to the pituitary. CRH binds to membrane receptors on one cell type in the anterior pituitary, the

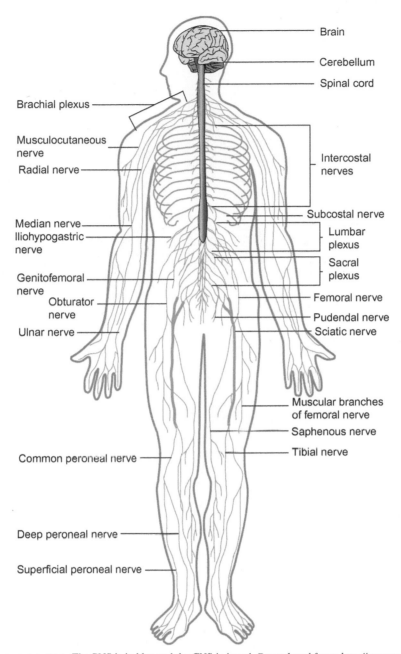

**FIGURE 1.14** The human nervous system. The PNS is in blue and the CNS is in red. *Reproduced from: http://www.nationmaster.com/encyclopedia/Peripheral-nervous-system.*

corticotrope. This receptor activates a biochemical cascade of signals that result in the secretion of adrenocorticotrophic hormone (ACTH; corticotropin) that enters the general circulation through thin membranes (fenestrations). ACTH circulates in the blood until it reaches its cognate receptor on the membranes of the middle layer of cells in the adrenal cortex. The signaling from the activated ACTH receptor results in the hydrolysis of cholesteryl esters, stored in lipid droplets, to liberate free cholesterol that, with the aid of a specific carrier, enters the mitochondria for the production of corticosteroids (cortisol and some aldosterone) that cross the cell membrane into the general circulation. Cortisol is carried in the circulation by specific proteins but about 10% of the circulating cortisol in is the free form. Unbound cortisol enters virtually every cell in the body and is retained by cells that have the glucocorticoid receptor to which it binds (virtually all cells of the body, except the cells in the space between the posterior pituitary and the anterior pituitary and the hepatobiliary cells, contain some glucocorticoid receptors). If there is more cortisol than there is receptor molecules

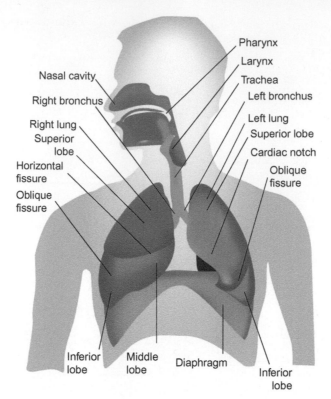

**FIGURE 1.15**   Organs of the respiratory system. *Reproduced from: http://www.byronsmith.ca/everest2000/gfx/ehb_respiratorysystem.gif.*

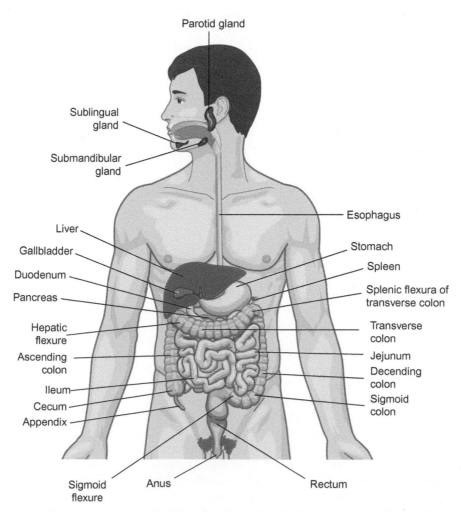

**FIGURE 1.16**   Diagram of the digestive system. *Reproduced from: http://eatwellgetwell.files.wordpress.com/2006/05/digestion_good2.jpg.*

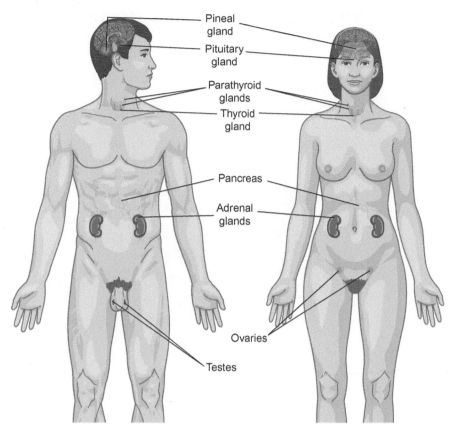

**FIGURE 1.17** Diagram showing major glands that constitute the endocrine system. *Reproduced from: http://cwx.prenhall.com/bookhind/pubbooks/ morris5/medialib/images/F02_17.jpg.*

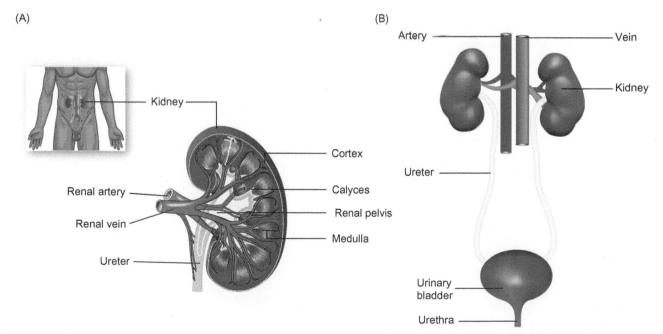

**FIGURE 1.18** The components of the excretory system. (A) A transverse section of the kidney is shown; (B) the components of the system are shown. *Reproduced from: http://sinquefield.com/id6.html.*

to bind it, the residual cortisol passes back out of the cell by free diffusion into the bloodstream. In certain tissues, such as the liver and kidney, there are large amounts of the receptor (as many as 50,000 receptor molecules in the liver cell). Not all of these activated receptor molecules are required to generate a transcriptional response. The activated receptors are then carried through the nuclear pore into the cell nucleus (as homodimers) where they interact with many genes that display a specific glucocorticoid responsive element (GRE) resulting in many genes that are transcribed into specific messenger RNAs (or, in some cases, suppressed) and subsequently translated in the cell cytoplasm into specific proteins used metabolically to adapt to stress. Thus after a stress event, virtually every cell type in the body is affected by cortisol (excepting the two cell types mentioned) so that most tissues are changed in their protein populations and the changes in the tissues summate into the bodily adaptation to the stress event. An individual cannot survive without

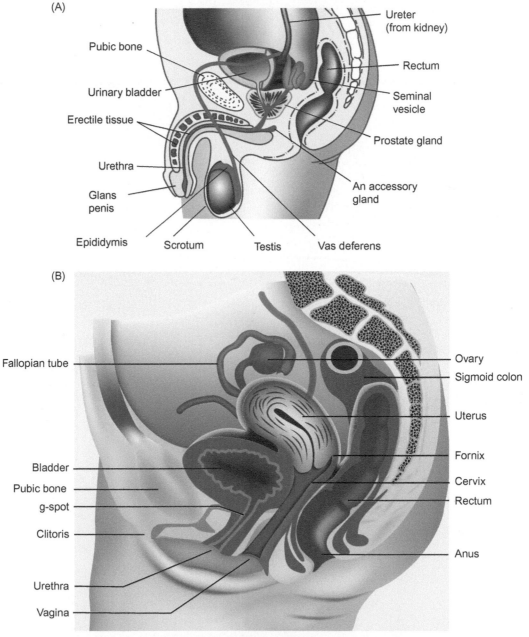

**FIGURE 1.19** (A) The reproductive system of the human male. (B) The reproductive system of the human female. The g spot has not been proven to be a distinct organ and there is controversial discussion concerning its existence. *(A) Reproduced from: http://www.cartage.org.1b/en/themes/ Sciences/LifeScience/GeneralBiology/Physiology/ReproductiveSystem/HumanReproduction/malerepro_2.gif. (B) Reproduced from: http://upload.wiki-media.org/wikipedia/commons/7/7a/Female_reproductive_system_lateral.png*

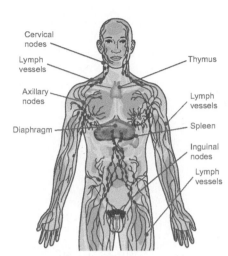

**FIGURE 1.20**   A diagram of the lymphatic system showing the major vessels, nodes, and organs (thymus, spleen, and diaphragm). *Reproduced from: http://www.cancervic.org.au/images/CISS/cancer-types/lymphatic.gif.*

cortisol and its receptor and when this system is somehow damaged (e.g., by trauma to the head, causing a break in the transport system between the hypothalamus and the anterior pituitary) replacement cortisol can be given orally but it is problematic that the patient can predict a major stress event requiring increased doses of the steroid, such as an automobile crash and survival can be threatened by inability to adapt. Cortisol is essential to life. This system is discussed in more detail in the chapter on steroid hormones.

The **excretory system** filters out toxins, cellular wastes and excess nutrients or water from the circulation. This system includes the kidneys, ureters, bladder, and urethra as shown in Fig. 1.18. An aspect of the kidney in transverse section is also shown.

The reproductive systems of the male and female are diagramed in Fig. 1.19. Part (A) shows the components of the male reproductive system and part (B) shows them for the female.

The **lymphatic system** is an elaborate network of lymph nodes with lymph vessels connecting the nodes. As shown in Fig. 1.20 the vessels distribute to every part of the body except the CNS. Important nodes are located in the neck, chest, armpits, abdomen, the pelvis, and the groin. Lymphatic fluid, containing lymphocytes and **antibodies** (as part of the immune system), flows within the lymph vessels. The nodes act as filters for the removal of bacteria and other unwanted particulate substances.

## CELLULAR COMPOSITION OF TISSUES

There are four main types of cells in tissues. They are epithelial, muscular, connective tissue, and nervous tissue type cells. Also added here are adipose cells and stem cells. These are shown in composite in Fig. 1.21. The complexity of human cell types that constitute variants of the four main types is much greater than suggested by these four major types. There are probably more than two hundred different cell types in the human body.

## SUMMARY

After fertilization of the human egg, there are a series of cell cleavages generating 2 cells, 4 cells, 8 cells, and 16 cells. By Day 3 the morula is formed and the blastocyst. In the next stage the gastrula is formed. At this stage the cells are still pluripotent (stem cells) and two primordial layers, the ectoderm and endoderm are apparent. With further development of the gastrula the internal endoderm is formed. From these three layers of cells, all the tissues and organs of the body are formed. The ectoderm forms mainly nervous tissue, the endoderm forms internal organs through the primitive gut and the mesoderm forms the majority of the internal organs.

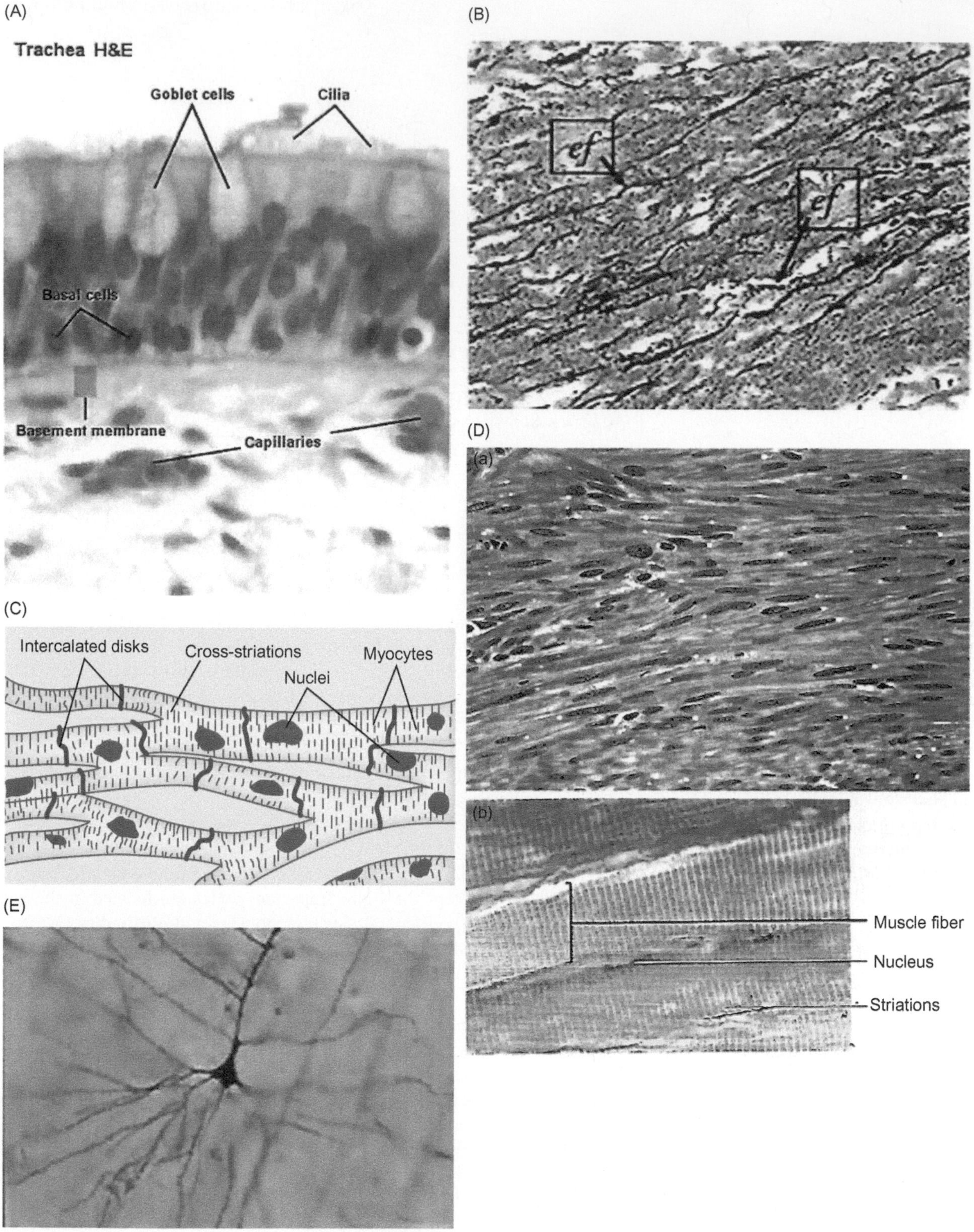

**FIGURE 1.21** Composite of the major human cell types in tissues. (A) Human tracheal epithelial cells: those pictured are pseudostratified columnar epithelium lining the trachea. The nuclei belong to cells contacting the basement membrane. In this case the epithelial cells are goblet cells (they can alternatively, be ciliated cells). The basal cells regenerate other cell types of the epithelium. (B) Connective tissue of human aorta (*ef* = elastin fibers). (C) Cardiac muscle cells (cardiomyocytes; labeled as myocytes in the figure). (D) (a) Human smooth muscle, a second type of muscle cell and (b) striated muscle. (E) Human nervous tissue cell. (F) Human adipose tissue. Note the presence of fat globules inside the adipose cells. (G) Human stem cells, pluripotent blastocyst cells. *(A) Reproduced from: http://www.lab.anhb.uwa.edu.au/mb140/CorePages/Epithelia/images/trachea041he.jpg. (B) Reproduced from: http://www. austincc.edu/histologyhelp/tissues/images/tn400.jpg. (C) Reproduced from: http://www.medical-look.com/systems_images/Cardiac_Muscle.gif. (D) (a) Reproduced from: http://images.encarta.msn.com/xrefmedia/sharemed/targets/images/pho/t790/T790539A.jpg. (b) Reproduced from: http:// images.tutorvista.com/cms/images/123/striated-muscle-fibre.jpeg. (E) Reproduced from: http://images.absoluteastronomy.com/images/encyclopediai-mages/g/go/golgistainedpyramidalcell.jpg. (F) Reproduced from: http://biology.nebrwesleyan.edu/courses/Labs/Biology_of_Animals/Images/Lab% 20Images/images%20for%20the%20Web/Adipose_Human_400X.jpg. (G) Reproduced from: http://www.sarahwray.com/USERIMAGES/blastocyst.jpg.*

(F)  (G)

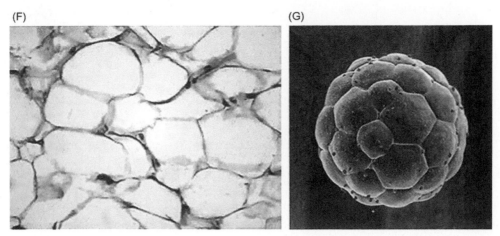

FIGURE 1.21   (Continued).

The organ systems consist of the skin, skeleton, muscle system, circulation, nervous system (central nervous system and peripheral nervous system), respiratory system, digestive system, endocrine system, excretory system, reproductive systems of the male and female, and the lymphatic system.

The organs are composed of tissues that are represented by four major cell types: epithelial cells, muscle cells (striated and smooth), nervous cells (neurons and others), and connective tissue. In addition, there are adipose cells and stem cells (typified by the cells of the blastocyst). There are more than two hundred variations of cells of the four major types.

## SUGGESTED READING

### Literature

Berry, D.C., Stenesen, D., Zeve, D., Graff, J.M., 2013. The developmental origins of adipose tissue. Development. 140, 3939−3949.

El-Osta, A., Wolffe, A.P., 2000. DNA methylation and histone deacetylation in the control of gene expression: basic biochemistry to human development and disease. Gene Exp. 9, 63−75.

Fagotto, F., 2014. The cellular basis of tissue separation. Development. 141, 3303−3318.

Forster, R., et al., 2014. Human intestinal tissue with adult stem cell properties derived from pleuripotent stem cells. Stem Cell Rep. 2, 838−852.

Horie, M., et al., 2012. Implantation of allogenic synovial stem cells promotes meniscal regeneration in a rabbit meniscal defect model. J. Bone Joint Surg. Am. 18, 701−712.

Kenneth, K.B., et al., 2014. Characterization of fetal keratinocytes, showing enhanced stem cell-like properties: a potential source of cells for skin reconstruction. Stem Cell Rep. 3, 324−338.

Nakamura, T., et al., 2012. Arthroscopic, histological and MRI analyses of cartilage repair after a minimally invasive method of transplantation of allogenic synovial mesenchymal stromal cells into vartilage defects in pigs. Cytotherapy. 14, 327−338.

Scheiner, Z.S., Talib, S., Feigal, E.G., 2014. The potential for immunogenicity of autologous induced pleuripotential stem cell-derived therapies. J. Biol. Chem. 289, 4571−4577.

Shyh-Chang, N., Daley, G.Q., Cantley, L.C., 2013. Stem cell metabolism in tissue development and aging. Development. 140, 2535−2547.

Wong, K.L., et al., 2013. Injectable cultured bone marrow-derived mesenchymal stem cells in various knees with cartilage defects undergoing high tibial osteotomy: a prospective, randomized controlled clinical trial with 2 years' follow-up. Arthroscopy. 29, 2020−2028.

Yabut, O., Bernstein, H.S., 2011. The promise of human embryonic stem cells in aging-associated diseases. Aging. 3, 494−508.

Zentner, G.E., Schacheri, P.C., 2012. The chromatin fingerprint of gene enhancer elements. J. Biol. Chem. 287, 30888−30896.

### Books

Applegate, E., 2006. The Anatomy and Physiology Learning System. 2nd ed. W.B. Saunders, Philadelphia.

Boron, W.F., Boulpaep, E.L., 2009. Medical Physiology. 2nd ed. Saunders.

Koeppen, B.M., Stanton, B.A., 2008. Berne and Levy, Physiology. 6th ed. Elsevier, Amsterdam.

Litwack, G. (Ed.), 2011. Stem Cell Regulators, volume 87 of Vitamins and Hormones. Academic Press/Elsevier, Amsterdam.

Sherwood, L., 2008. Human Physiology: From Cells to Systems. Brooks/Cole, Belmont, CA.

# Chapter 2

# The Cell

## CELLULAR TRAFFICKING IN ALZHEIMER'S DISEASE

According to the Alzheimer's Association, it is estimated in 2014 that about 5.2 million Americans have Alzheimer's disease (AD). Of these, about 200,000 are below the age of 65. It is projected that by 2050, unless there is a future effective treatment, there may be about 16 million individuals who are 65 or older who have AD. The annual death rate from AD in the United States is about 500,000. In the year 2014, the estimated cost of AD in the United States is about $214 billion. Because the pathology of AD at the cellular level involves trafficking among different compartments within the affected cell (neuron), AD represents a clinical problem that is illustrative of cellular structure.

Alzheimer's disease, a disease characterized by progressive loss of memory and the loss of cognition (thinking), is related to the excessive degradation by brain **secretases** (protease enzymes) of the **amyloid precursor protein** (**APP**) to **amyloid** β (Aβ) peptides that are neurotoxic. Thus, a disease can be caused by an endogenous normally functioning protein which, when modified to an abnormal shape, size, or folding, becomes pathologic similar to other neurodegenerative diseases like Prion disease (Creutzfeldt–Jakob disease in the human; see Chapter 4: Proteins). Aggregation and spread of product of a [alpha]-secretase action that damage the brain. This mechanism is pictured in Fig. 2.1.

Deposition of amyloid (amorphous parenchymal deposits) takes the form of ordered proteinaceous β-sheets (see Chapter 4: Proteins). Amyloid deposition of this type not only occurs in AD, but also occurs in Lewy body (abnormal protein aggregates in neurons) dementia, vascular dementia, and Down's syndrome, all of which are considered to be age-related neurogenerative diseases.

The amyloid precursor protein (APP) and its breakdown products (CTFs and AICD) are collected in multivesicular bodies within the cell and in secreted exosomes. This trafficking within the cell involves the early endosome that generates multivesicular bodies (multivesicular endosomes) that can either fuse with lysosomes for degradation of the ingredients or be secreted (exosomes) to the extracellular space. Movements of intracellular particles and secretory contents are captured in Fig. 2.2.

In the nonamyloidogenic pathway, the intracellular soluble APPα product of a [alpha]-secretase action is not further degraded into products of the actions of β- and γ-secretase that ultimately generate Aβ products leading to their pathological aggregation.

The discovery of a sorting receptor called **SORLA** reveals that it acts to prevent the amyloid protein precursor from sorting to the late endosomes where the breakdown product, Aβ, is generated and leads to amyloid deposition. *Normal SORLA activity insures that the reactions leading to Alzheimer's disease do not occur, whereas the mutations in the gene for SORLA to make the sorting receptor less functional or lower in activity can lead to reactions involving the late endosome and the production of Aβ for amyloid deposition.* In Fig. 2.3 are shown the VPS10 (vacuolar **p**rotein sorting-10) domain receptors of which one is the SORLA receptor as well as the overall trafficking of SORLA.

There are two models for the actions of the sorting receptor SORLA that can take trafficking paths to avoid Alzheimer's disease. In one version, SORLA retains APP in the transGolgi network (TGN) preventing the formation of APP homodimers that are the preferred substrates for secretase. In the other model, SORLA binds APP and shuttles between the TGN and the early endosomes, thus reducing APP from amyloidogenic processing in the endosomes. However, the further processing by the secretases (β and γ) finally avails the aggregatable form, Aβ. These models are shown in Fig. 2.4.

In addition to the deposition of aggregates of Aβ, a second lesion, known as **intraneuronal neurofibrillary tangles** (**NFT**), is involved in the development of AD. This lesion is an interneuronal aggregation of microtubule-associated **Tau proteins** (CNS neuronal protein stabilizers of microtubules) that are abnormally modified. AD progresses by virtue of a synergistic relationship between these two types of lesions (Aβ and NFT). A comparison of a normal vs an Alzheimer neuron is shown in Fig. 2.5.

**Human Biochemistry.** DOI: http://dx.doi.org/10.1016/B978-0-12-383864-3.00002-8

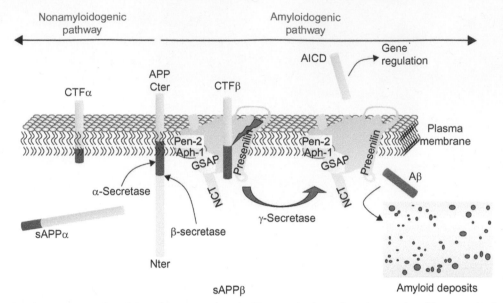

**FIGURE 2.1** APP structure and metabolism. Schematic representation of APP processing by α-, β-, and γ-secretases. The processing is divided into nonamyloidogenic pathway (left) and the amyloidogenic pathway (right). α- and β-secretases cleave APP in its extracellular domain to release a soluble fragment sAPPα or sAPPβ in the extracellular space and generate carboxy terminal fragments CTFα (83 amino acids long) or CTFβ. These CTFs can subsequently be processed by γ-secretase complex to generate AICD (**amyloid precursor protein intracellular domain**) and Aβ. Aβ is a small peptide of 39–43 amino acids. Aggregation of Aβ causes amyloidosis, apparently at the root of neurodegenerative disease. The γ-secretase complex is composed of presenilin, nicastrin (NCT), γ-secretase activating protein (GSAP), pen-2, and aph-1. *Presenilin*, transmembrane protein part of γ-secretase intramembrane protease complex; *nicastrin*, protein constituent of γ-secretase complex; *pen-2*, **p**resenilin **en**hancer of a regulatory protein in γ-secretase complex; *aph-1*, **a**nterior **ph**alanx-**d**efective protein 1, a subunit of γ-secretase complex. *Reproduced from http://journal.frontiersin.org/Journal/10.3389/fphys.2012.00229/full.*

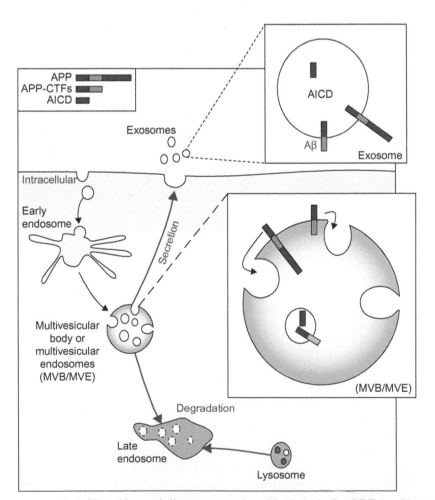

**FIGURE 2.2** Amyloid precursor protein (APP) and its metabolites are present in multivesicular bodies (MVBs) and in exosomes. APP and APP-C-terminal fragments (APP-CTFs) are internalized and directed into the internal vesicles of multivesicular bodies (MVBs). At this point APP and its metabolites can either be degraded after the fusion of MVB with lysosomes or can be released in the extracellular space in association with exosomes consecutively to the fusion of MVB with the plasma membrane. *APP, Amyloid protein precursor; CTF, C-terminal fragment; AICD*, **a**myloid precursor intracellular **d**omain. *Reproduced from http://www.frontiersin.org/Articles/18560/fphys-03-00229-r2/Image.m/fphys-03-00229-g004.jpg.*

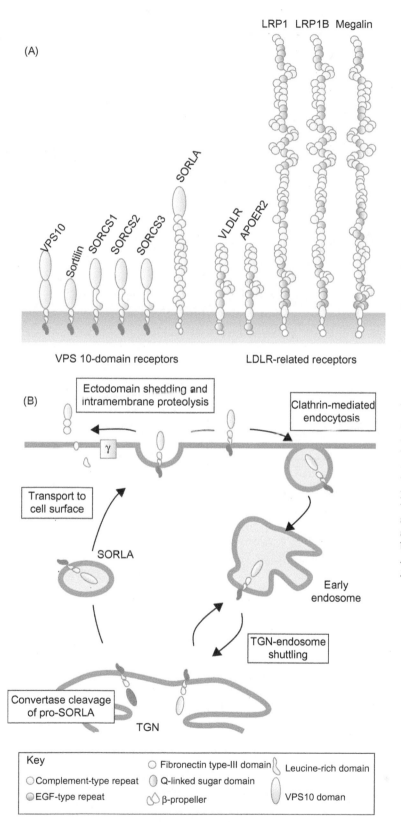

**FIGURE 2.3** Structural organization and trafficking path of SORLA. (A) SORLA is a member of the VPS10 domain receptor family, a group of sorting receptors characterized by a VPS10 domain (the name "VPS10" derives from the yeast carboxykinase Y sorting receptor [Vps10 protein]). This domain adopts the structure of a large tunnel that is involved in the binding of peptide ligands. In contrast to all other VPS10 domain receptors, SORLA contains complement-type repeats and a β-propeller, structural elements that are found in LRPs (low-density lipoprotein (LDL) receptor-related proteins). The cluster of complement-type repeats is also a site that interacts with ligands. The β-propeller is required for pH-dependent ligands in endosomes. SORLA is produced in the cell as a pro-receptor with a 53 amino acid pro-peptide that folds back on the VPS10 domain to block binding of ligands that target this receptor domain. Cleavage of the pro-peptide by convertases in the TGN (trans-Golgi network) produces the mature receptor, which is able to interact with its target proteins. All known members of the VPS10 domain receptor family are shown to include the yeast receptor VPS10 and the vertebrate proteins sortilin, SORLA, as well as SORCS1, SORCS2, and SORCS3 (*SORC*, sortilin-related VPS10 domain-containing receptor). For LRPs, the only receptors depicted are those that have been shown to interact with APP. (B) Newly synthesized pro-SORLA is activated in the Golgi by convertase cleavage. From the TGN (transGolgi network), nascent SORLA is directed to the plasma membrane through constitutive secretory vesicles. At the cell surface, some receptor molecules are subject to ectodomain shedding and subsequent intramembrane proteolysis by γ-secretase (γ), resulting in soluble fragments of the extracellular domain and the intracellular tail. Most SORLA molecules at the cell surface remain intact and undergo clathrin-mediated endocytosis. Clathrin is a major protein involved in the formation of coated vesicles. From the early endosomes, internalized receptors, and probably some of their cargo, are returned to the TGN to continue anterograde (forwardly directed) and retrograde shuttling between the secretory and early endosomal compartments. *Reproduced from http://jcs.biologists.org/content/126/13/2751/F2.expansion.html.*

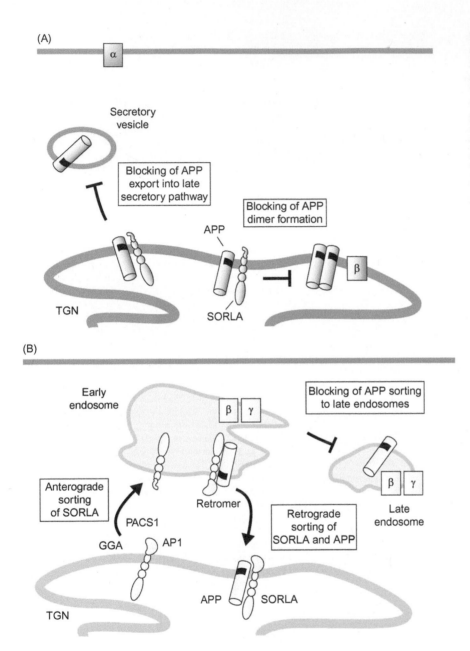

**FIGURE 2.4** Sorting receptor SORLA operates in a trafficking pathway that avoids Alzheimer's disease. (A) Newly synthesized amyloid protein precursor (APP) molecules traverse the Golgi and the transGolgi network (TGN) to the plasma membrane where most precursor molecules are cleaved by α-secretase (α). Nonprocessed precursors internalize from the cell surface through clathrin-mediated endocytosis, which is guided by the interaction between the cytoplasmic tail of APP and the clathrin adapter AP2 (adapter protein 2). From early endosomes, APP moves to the late endosomal−lysosomal compartments or backward to the TGN. Amyloidogenic processing of internalized APP through sequential cleavage by β- and γ-secretases (β, γ) is believed to proceed in endosomes and in the TGN. (B) Internalization of APP is controlled by LRPs (low density lipoprotein [LDL] receptor-related proteins), a group of endocytic receptors expressed in neurons and many other cell types. Fe65 (multidomain adapter protein) mediated association of APP with LRP1 on the cell surface facilitates its endocytic uptake and intracellular processing to Aβ. By contrast, binding to the slow-endocytosing receptors apolipoprotein E receptor 2 (APOER2, also known as LRP8) and LRP1B delays endocytosis but promotes cleavage to sAPPα. Binding of APP to APOER2 is mediated through Fe65 and F-spondin (SPON1; floor plate and thrombospondin homology or VSGP, vascular smooth muscle growth-promoting factor). The mode of interaction between LRP1B and APP might also involve Fe65 or yet unknown adaptors. *Reproduced from http://jcs.biologists.org/contents/126/13/2751/F1.expansion.html.*

It is unclear whether **cortical atrophy** (Fig. 2.5) always occurs in AD because it usually occurs in a different part of the brain (posterior brain cortex). Posterior cortical atrophy is often associated with **Lewy body dementia** or with **Creutzfeld−Jakob disease**.

## CELL MEMBRANE

As seen in **Fig. 1.19**, the cells in the body display many different shapes. Therefore, in this chapter, we will deal with an ideal model of a typical cell. An example is shown in cross section in Fig. 2.6. The higher **eukaryotic cell** that contains a distinct membrane-bound nucleus, shown here, contrasts with a prokaryotic cell, a bacterium or a cyanobacterium, that do not contain membrane-bound organelles and do not have their DNAs in the form of chromosomes. It also contrasts with the eukaryotic yeast cell that contains a rigid cell wall outside the plasma membrane. The cell membrane or the plasma membrane is the outermost layer surrounding the cell (Fig. 2.6) consisting of a double layer of lipids with polar head groups facing the exterior and interior. The lipid bilayer membrane is penetrated with transmembrane proteins with extensions to the outside of the cell and inside to contact the cytoplasm (Fig. 2.7). From the outside layer

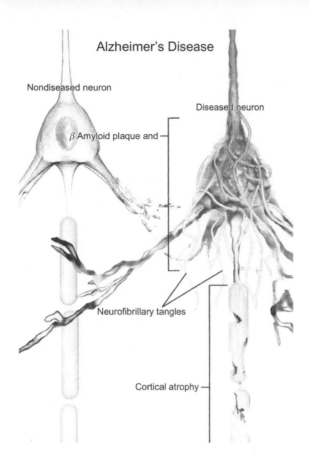

Alzheimer's Disease

Nondiseased neuron

Diseased neuron

β Amyloid plaque and

Neurofibrillary tangles

Cortical atrophy

**FIGURE 2.5** Neuronal degeneration associated with Alzheimer's disease. The figure of the diseased neuron shows β amyloid plaque and neurofibrillary tangles. *Reproduced from http://upload.wikimedia.org/wikipedia/commons/7/77/ Blausen_0017_AlzheimersDisease.png.*

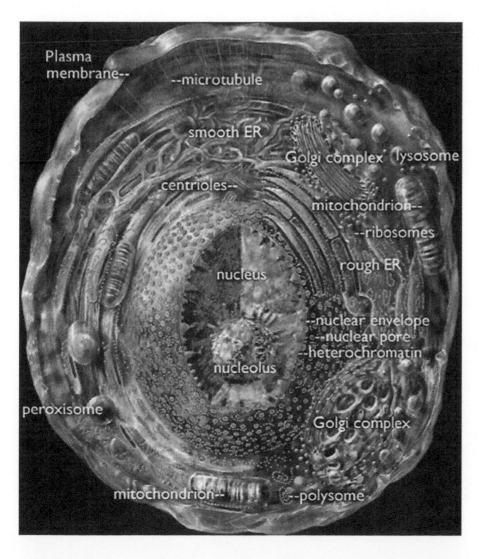

Plasma membrane--

--microtubule

smooth ER

Golgi complex   lysosome

centrioles--

mitochondrion--

--ribosomes

rough ER

nucleus

--nuclear envelope
--nuclear pore
--heterochromatin

nucleolus

peroxisome

Golgi complex

mitochondrion--   --polysome

**FIGURE 2.6** Cross-sectional model of a typical eukaryotic cell. *Reproduced from http://www.rkm.com.au/CELL/cellimages/ animal-cell-label.jpg.*

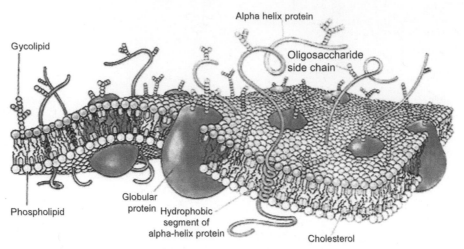

**FIGURE 2.7** Drawing of a section of the plasma membrane showing typical constituents. *Reproduced from http://www.molecularstation.com/molecular-biology-images/data/504/CellMembraneDrawing.jpg.*

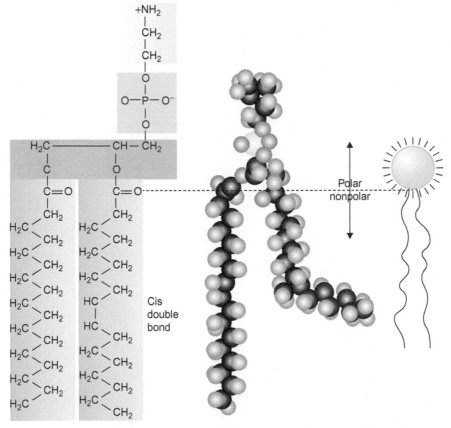

**FIGURE 2.8** Diagram of the structure of one membrane leaflet showing a polar group at the top (outside) with the nonpolar hydrocarbon structure extending to the center of the cell membrane bilayer. An opposing leaflet (mirror image approximating a similar structure) creates the double layer membrane as shown in Fig. 2.7. *Reproduced from Fig. 1.6 of Litwack, G., 2008. Human Biochemistry and Disease, Academic Press/Elsevier, p. 8.*

inward, the polar head groups are **ammonium** ($NH_4^+$), **hydrocarbon chain** of the fatty acid, and a **phosphate group** (substituent of glycerol, for one) on the interior side.

The nonpolar stretches of the glycerol connected fatty acids meet in the middle of the two layers forming the nonpolar region. Each portion of the membrane consisting of a polar group on either end connected to nonpolar stretches in the middle is referred to as a **leaflet** (Fig. 2.8).

**TABLE 2.1 Functions of the Cell Membrane**

| Function | Activity |
|---|---|
| Structure | Confines the cellular organelles and cytoplasm; separates cell from the external environment |
| Function | Mediates physical and chemical signals to the cell; regulates exchange of molecules between exterior and cellular interior; maintains charge controlling some exchange functions with the outside; involved is secretion (**exocytosis**) and uptake (**endocytosis**) of molecules |
| Composition | in order to maintain its physical state under various conditions composition of phospholipids changes; phospholipids may move to different positions in the same leaflet; the balance between saturated and unsaturated fatty acids may change; spatial distribution of components changes; membrane selectively controls traffic of water, ions and various molecules |
| Specialized functions | Some cells are capable of engulfment of large particles (**phagocytosis**); site of interaction with many **xenobiotics** (natural substances that are foreign to the body), including toxins, bacteria, viruses; some cells have specific functions involving structural variations of the membrane, e.g., neuronal axons and nerve endings |

**Cholesterol**, located between the hydrocarbon components (fatty acid chains), is a critical ingredient of the membrane and is present in about a one-to-one ratio with the phospholipid (Fig. 2.7). A phospholipid consists of 1 molecule of glycerol substituted by 2 molecules of fatty acids, a phosphate group, and a polar molecule; examples are phosphatidylcholine, phosphatidylethanolamine, or phosphatidic acid. The polar group attracts water and faces the interior aqueous cytoplasm while the nonpolar (hydrophobic; repels water) fatty acid tail faces away from the aqueous cytoplasm (see Chapter 9: Lipids). Cholesterol provides some rigidity to the otherwise flexible semipermeable membrane. Cholesterol also enhances the nonpolar solubility of the membrane for entry of nonpolar substances that can dissolve in and permeate the membrane, essentially by free diffusion. There are various groups protruding from the outside surface of the membrane. **Glycolipids**, for example, may be involved in the specific binding of extracellular proteins to the cell surface. **Glycoproteins** in the membrane have carbohydrate portions available on the outer cell membrane, each substituent having sugar molecules up to as many as 15 but usually a smaller number (Fig. 2.7). These groups provide the cell's ability to **distinguish self from non-self** and are critical for the specific identity of **blood types** as will be described in later chapters. The cell membrane is a dynamic structure containing many groupings on the outside surface including extracellular portions of **receptors** (for subsequent signaling) and **channels** (serving transporting functions) of all kinds for the recognition of micromolecules and macromolecules. *The cell (plasma) membrane permits **free diffusion** of gases, such as oxygen and carbon dioxide and small nonpolar substances but it excludes water, charged ions, and sugars for which there exist specific channels in appropriate cells.* Thus, the interior composition of the cell is protected from the external environment. Table 2.1 summarizes many of the functions of the cell membrane.

## NUCLEUS AND CELL DIVISION

Within the cytoplasm, encapsulated by the cell membrane, are found all of the particulate structures that function inside of the cell. The nucleus is the central structure in the cell (about 5 μm or 0.005 mm in diameter) that contains the genetic information of the organism (46 chromosomes in the human, containing information for as many as 32,000 genes involving up to 3 billion base pairs). A model of a typical nucleus is shown in Fig. 2.9.

The nucleus is the site of gene expression and gene repression under the control of many **transcription factors**. The nuclear genes express information for all of the cell's proteins through messenger RNAs except for a small number of proteins encoded in mitochondrial DNA. Some of the proteins encoded by the nucleus ultimately are completed in the cytoplasm and are transported into the mitochondrion.

The nuclear membrane is similar to the plasma membrane and is covered with **nuclear pores** that are the entries for molecules, such as receptors and other transcription factors. Some molecules are large enough to require transporters or **nucleoporins** to escort these molecules to the nuclear pores; they may ferry the macromolecules through the cytoplasm to docking sites on the nuclear pore complex (a megadalton translocase complex integrated into the nuclear envelope). These pores are also used for egress from the nucleus to the cytoplasm (e.g., export of mRNAs). The nucleus contains cytoplasm, called **nucleoplasm**. The nucleus contains the **chromatin** and **nucleolus** (there may be more than one; Fig. 2.9). Chromatin is constituted by the genes. **Heterochromatin** is the tightly packed or condensed chromatin,

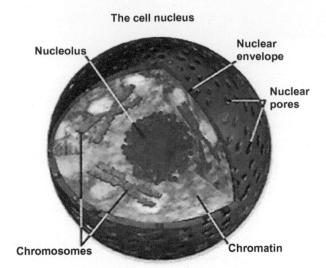

**FIGURE 2.9** A model of a typical nucleus. *Reproduced from http://www.cartage.org1b/en/themes/Sciences/Zoology/AnimalPhysiology/Anatomy/AnimalCellStructure/Nucleus/cellnucleus.jpg.*

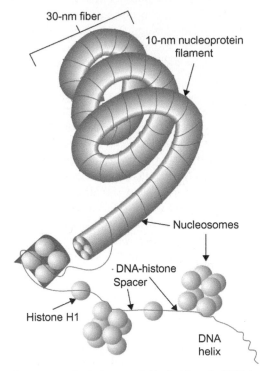

**FIGURE 2.10** Double-stranded DNA in the chromosome. At the bottom of this drawing, the DNA is uncoiled to show the attached histone proteins. **H1 histone** is bound in the linker region between nucleosomes. Each nucleosome contains eight histones (histone octamer) comprised of two copies each of histones **H2A, H2B, H3,** and **H4**. *Reproduced from Litwack, G., 2008. Human Biochemistry and Disease, Academic Press/Elsevier, p. 12.*

whereas **euchromatin** is the opened form, is less dense, and stringy in appearance. Chromatin needs to be opened to receive messages from transcription factors and to actively transcribe genetic information. **Histone proteins** are bound to chromosomes, primarily in the **nucleosomes** that occur periodically on the DNA as shown in Fig. 2.10.

A nucleus may contain one to four **nucleoli** (Fig. 2.9). The nucleolus is the main site of **ribosomal RNA (rRNA)** synthesis. It surrounds chromosomes that contain genes for rRNA (there may be hundreds of rRNA genes located in **nucleolar-organizing regions** at various positions on chromosomes) and the nucleolus does not have a visible membrane. Ribosomal RNA is synthesized in the nucleolus, translocated to the nucleoplasm, and then through nucleopores to the cytoplasm where it is assembled into ribosomes as part of the protein synthetic machinery.

The **cell cycle** dictates the point at which **cell division** occurs. A diagram of the cell cycle is shown in Fig. 2.11.

After cell division (mitosis, comprising 2% of the cycle), the smaller daughter cells proceed to the $G_1$ interphase (comprising 40% of the cycle), the first growth phase, where the majority of the cells are found. Cells in the $G_1$ phase may enter the $G_0$ phase where they may await further cell differentiation or may be programmed for cell death (apoptosis). Also during this phase, cells store **adenosine triphosphate (ATP)** and increase in size. Cells entering the S phase (comprising 39% of the cycle) replicate their DNA and then move into the second growth phase, $G_2$ (comprising 19% of the cycle). DNA synthesis results in the duplication of chromosomes. The cell is now ready to divide again in mitosis bestowing one set of chromosomes in the daughter cell and retaining one set in the parent. The daughter splits from the parent to form two equal and smaller cells.

The mitotic phase consists of four phases. In prophase, the chromosomes become prominent, the nucleolus and nuclear envelope disappear, and the **mitotic spindle** forms. In metaphase, the chromosomes are condensed and highly coiled and are located equidistant from the two poles (**metaphase plate**) and become attached to the newly formed **spindle**. In anaphase, the sister chromatids are now separated. The daughter chromosomes become stationed at opposite poles of the cell. A **chromatid** is a replicated chromosome having two daughter strands joined by a single **centromere** (the two strands separate during cell division to become individual chromosomes). A drawing of a chromatid is shown in Fig. 2.12. In the final telophase, sets of chromosomes are assembled at opposite poles.

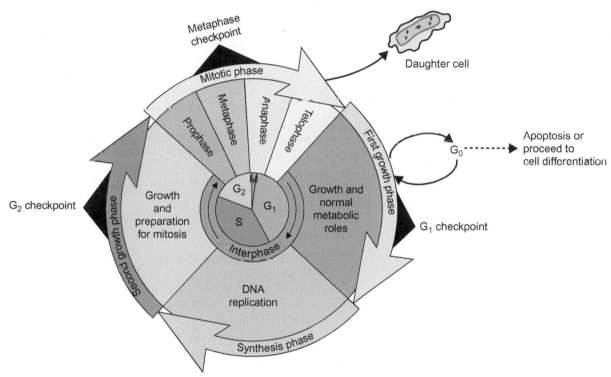

**FIGURE 2.11** A diagram of the cell cycle. There are four major phases: M, **mitotic phase** containing **prophase, metaphase, anaphase,** and **telophase** resulting in two daughter cells each with half the number of chromosomes of the parental cell, prior to entering mitosis; $G_1$, the first growth phase from which a cell may go into $G_0$ phase or **growth arrest** (the growth-arrested cell can reside in $G_0$ or, with an appropriate stimulus, reenter the $G_1$ phase or, with a specific signal, enter a cell suicide program (**programmed cell death** or **apoptosis**)); **DNA replication** in which DNA is synthesized; and, finally, $G_2$, the second growth phase. The **checkpoints** at three sites along the cycle are indicated. The phases of the cycle occupy: $G_1$ (40%), S (39%), $G_2$ (19%), and M (2%). *Reproduced in part from http://www.biologycorner.com/resouces/cell_cycle.jpg.*

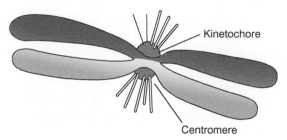

**FIGURE 2.12** Drawing of a chromatid. *Reproduced from http://www.emc.maricopa.edu/faculty/farabee/BIOBK/chromosome1.gif.*

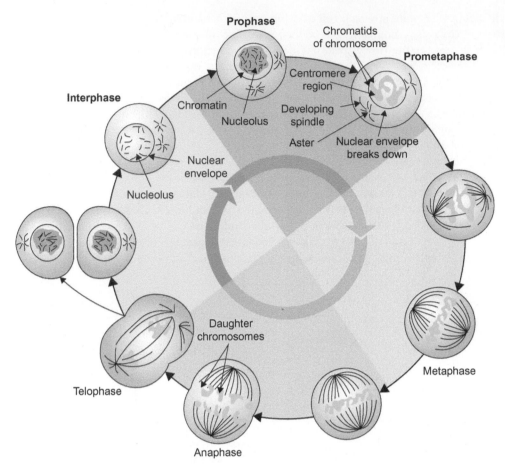

**FIGURE 2.13**  Events in the process of cell division, starting from the top, center of the figure. *Reproduced from Fig. 1.16 of Litwack, G., 2008. Human Biochemistry and Disease, Academic Press/Elsevier, p. 16.*

A nuclear envelope again forms around each set and division of the cytoplasm (**cytokinesis**) follows. After telophase, the cells enter the first growth phase, $G_1$, as described earlier. There are three **checkpoints** at specific places in the cell cycle (Fig. 2.11). At these points, the cell monitors to insure that the functions of the specific phase have been fulfilled. At the **metaphase checkpoint**, all chromosomes must be attached to the mitotic spindle in order for the cell to continue through the cycle. At the $G_1$ **checkpoint**, the cell size, availability of nutrients, and the presence of growth factors from other cells are assessed. When the appropriate growth factors from other cells arrive, they generate an increase in the concentration of **cyclins** (proteins that control progression of cells through the cell cycle) activating **cyclin-dependent kinase (CdK)**. CdK phosphorylates and activates S phase proteins leading to **cell division** (in cells that are not cycling, the transcription factor, **E2F**, is normally bound to another protein, the tumor suppressor **retinoblastoma protein (Rb)**; the complex inhibits the cell cycle. Arrival of **growth factors** from other cells activates the **cyclin−CdK complex** by phosphorylating Rb. As a result, E2F is released and stimulates the production of S phase proteins). At the $G_2$ **checkpoint**, adequate cell size and the successful completion of chromosome replication are evaluated. The cell completes the cycle when all of these measures have been attained successfully. For clarity, cell division is recapitulated from the point of view of the changes occurring within the cell at phases of the cycle in Fig. 2.13.

There are 23 **chromosomes** from each parent in the human. X and Y are the sex-determining chromosomes. The remaining chromosomes from each set are **autosomes**. The chromosomes in metaphase and a diagram of the human chromosomes are shown in Fig. 2.14.

The **nuclear membrane** resembles the plasma membrane in that it has two layers. **Ribosomes** are attached to the cytoplasmic side of the membrane. The two membranes are interrupted at each **nucleopore** (**nuclear pore**; Fig. 2.9). Molecules of molecular weight up to 44,000 can diffuse into the pore although larger molecules up to 60,000 Da diffuse more slowly through the pore. Molecules above 60,000 Da and 10 nM in diameter, exceeding the pore's diameter, require energized transport to negotiate the pore. **Nuclear transporting factors** attach to proteins needing energized

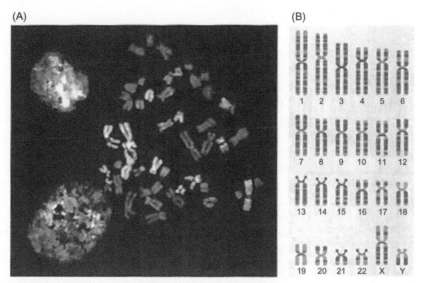

**FIGURE 2.14** (A) Metaphase chromosomes as viewed under the electron microscope. (B) Twenty-three human chromosomes from each parent. *Reproduced from Fig. 1.14 of Litwack, G., 2008. Human Biochemistry and Disease, Academic Press/Elsevier, p. 15.*

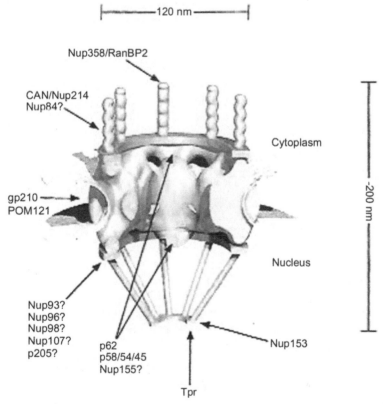

**FIGURE 2.15** Diagram of a nucleopore with names of the nucleoporins and other protein constituents. *Reproduced from Dellaire, G., Farall, R, Bickmore, W.A., 2003. The nuclear protein database (NPD): subnuclear localization and functional annotation of the nuclear proteome. Nucleic Acids Res., 31, 328–330.*

transport and hydrolyze ATP to provide the needed energy. **Nucleoporins (Nups)** are a family of proteins that are involved in the transport of molecules through the nuclear pore and they are part of the **nuclear pore complex**. There are multiple copies of about 30 different nucleoporins comprising the nuclear pore complex. Fig. 2.15 is a schematic representation of the nuclear pore complex and its constituents.

# CYTOPLASM AND CYTOSKELETON

The **cytoplasm** is the material inside the plasma membrane containing the soluble medium (**cytosol**) and the organelles including the nucleus, skeleton, mitochondria, endoplasmic reticula (microsomal fraction), Golgi apparatus, lysosomes, peroxisomes, vesicles, lipid droplets (in appropriate cells), secretion granules, and others. Importantly, many metabolic reactions (and those starting with glucose) take place in the cytoplasm, notably **glycolysis**, whose primary product, **pyruvate**, feeds into the **tricarboxylic acid cycle** in the mitochondria for the production of ATP energy, water, and carbon dioxide. Organelles and other structures (e.g., platforms and cytoplasmic parts of channels) are positioned by components of the **cytoskeleton**.

The cytoskeleton consists of three major structures: **microtubules**, that are hollow tubes constructed of **tubulin**, **microfilaments** composed of **actin**, and **intermediate filaments** composed of **fibrous proteins** and microtubules. Microtubules are hollow tubes of the globular protein, tubulin, and are about 25 nm in diameter. Usually, they are composed of 13 protofilaments which are polymers of α- and β-tubulin. They bind **GTP** and are organized by the **centrosome**. GTP hydrolysis powers the process of **depolymerization** (similarly to the function of ATP in the depolymerization of microfilaments). Microtubules maintain the shape of the cell, anchor organelles, and provide a track for **motor proteins** (specific proteins that can move along an appropriate surface whose movement is powered by hydrolysis of ATP). With **dyneins** and **kinesins**, they can transport mitochondria, vesicles, or the mitotic spindle within the cell. Dyneins and kinesins move cellular organelles on microtubules either toward the nucleus (dyneins) or away from the nucleus toward the plasma membrane (kinesins). **Microfilaments** are composed of actin proteins and have a diameter of about 7 nm. Actin is globular with an ATP-binding site at the center of the molecule. ATP is hydrolyzed just after the molecule is constructed into a filament and the resulting ADP is trapped in the binding site. ATP is required to depolymerize the filament and the ADP, previously bound, is trapped until the filament depolymerizes. When ATP is bound (unpolymerized form of actin), it is called **G-actin**. In the filament form, it is called **F-actin**. In the presence of ATP, $Mg^{2+}$, and $K^+$ and a lower than critical concentration of G-actin, the filament will depolymerize. Concentrations of G-actin above the critical concentration will produce polymerization and form a filament. Microfilaments are located just beneath the cell membrane and are involved in cell-to-cell and cell-to-matrix interactions and in the transduction of signals. **Intermediate filaments** are more strongly bound than actin filaments. They are about 8−11 nm diameter and are heterogeneous constituents of the cytoskeleton. They are components of the nuclear envelope and are involved in the organization of the three-dimensional structure of the cell and in some cell−cell and cell−matrix functions. The components of the cytoskeleton are pictured in Fig. 2.16.

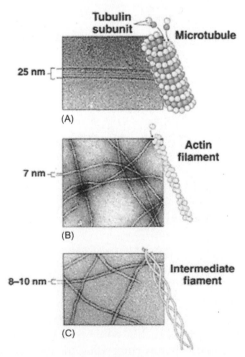

FIGURE 2.16   Components of the cytoskeleton. *Reproduced from http://migration.files.wordpress.com/2007/07/cytoskeleton02.jpg.*

## ENDOPLASMIC RETICULUM

The endoplasmic reticulum (ER) exists in the cell in two forms, the **smooth ER (SER)** and the **rough ER (RER)**. It is a network of interconnected membrane-bound and flattened sacs throughout the cell cytoplasm. The RER is similar to the SER except that the RER displays **ribosomes** attached to its surface facing outward to the cytoplasm (away from the nucleus) as shown in Fig. 2.17.

The **cisternal spaces** (**cisternae**) occur by the arrangement of the continuous convoluted membrane and are identified in the figure. The cisternal space or the lumen occupies more than 10% of the volume of the cell. The ER is connected to the nuclear envelope providing a direct connection as well as to the cytoplasm. Proteins synthesized on ribosomes of the RER (not the SER) can be directed to the interior of the RER by a **chaperone** protein, **BiP** (an Ig-binding protein), that recognizes a **retention signal sequence** (of amino acids) at the N-terminal of the growing peptide chain. Newly formed proteins may be directed to the cytoplasm or for export out of the cell as will be explained later. Some of these proteins will be destined for the **Golgi apparatus** for further processing and modification. Much of this information will be elaborated in Chapter 4, Proteins. Several activities reside in the SER including **cholesterol metabolism**, **membrane synthesis**, **detoxification**, and the major intracellular **storage site of calcium ions**.

## GOLGI APPARATUS

This saclike structure resides between the RER and the plasma membrane and is in an intimate relationship to the RER (Fig. 2.18). It modifies proteins that have been synthesized in the RER. The proteins in the Golgi system coming from the RER may form part of the plasma membrane, may be directed to lysosomes, may be destined for other parts of the cell, or secreted by a process of **exocytosis**. This process directs the contents of **secretory vesicles** to the extracellular space through the cell membrane. By this mechanism, soluble proteins are secreted from the cell and the membrane components of the secretory cell become components of the cell membrane. The two types of exocytosis are $Ca^{2+}$-activated nonconstitutive and $Ca^{2+}$-activated constitutive; most cells have the capacity for the latter type. There are five sequential activities of each vesicle undergoing exocytosis: trafficking, tethering, docking, priming, and fusion with the inner plasma membrane. The last step is driven by **SNARE** (*s*oluble *N*-ethylmaleimide-sensitive factor *a*ttachment protein *r*eceptor) proteins. Vesicles recognize its target by **vesicle SNAREs** on the surface of the vesicular surface and by **target SNAREs** on the surface of the target membrane.

One of the main functions of the Golgi apparatus is **glycosylation** in which process sugar molecules are added to proteins to form **glycoproteins** (some are called mucopolysaccharides or proteoglycans). With some proteins that are too small to merit a separate mRNA, a large precursor protein is the product of an mRNA and this large precursor is cut in the Golgi to the constituent smaller proteins. A case in point is **proopiomelanocortin** (265 amino acids) that is broken down into ACTH, MSH, β-endorphin, and other peptides (see Chapter 15: Polypeptide Hormones). Protein

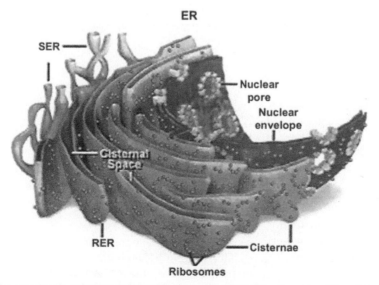

**FIGURE 2.17**  A drawing of the endoplasmic reticulum showing the rough and smooth ERs. *Reproduced from http://www.microscopy.fsu.edu/cells/ endoplasmicreticulum/images/endoplasmicreticulumfigure1.jpg.*

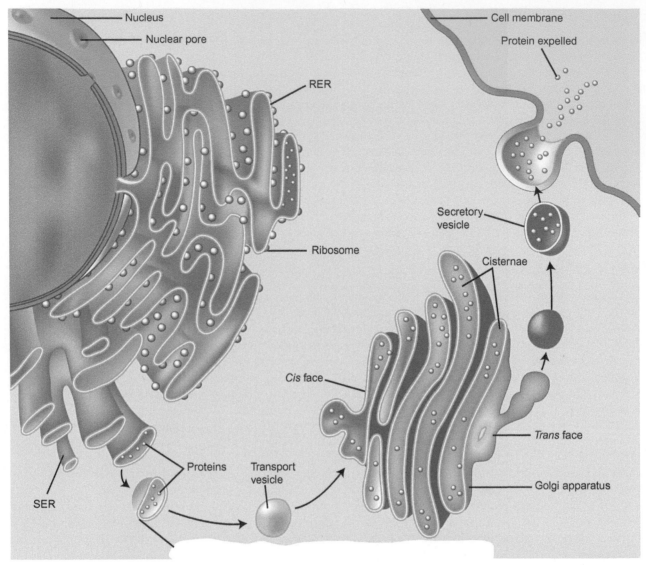

**FIGURE 2.18**  The Golgi apparatus in a portion of the cell. The Golgi has a close relationship to the RER where proteins are synthesized. Transport vesicles move newly synthesized proteins to the Golgi in transport vesicles and the proteins are released into the Golgi for further modification and export to other cellular sites including secretion from the cell (shown here). This occurs with the aid of secretory vesicles that fuse with the cell membrane to release their contents. *Reproduced from http://employees.csbsju.edu/HJAKUBOWSKI/classes/ch331/cho/ergolgi.jpeg.*

sorting that takes place in the Golgi is possibly directed by specific amino acid sequences (motifs) in precursor forms that signal subsequent cellular trafficking. Proteins can flow through the interconnected cisternae of the Golgi apparatus and for this movement do not require the vesicular system described earlier for transport to the plasma membrane. The Golgi disintegrates at the beginning of mitosis but reappears at telophase.

## MITOCHONDRIA

**Mitochondria** are the organelles that provide energy for the growth and functioning of the cell. Presumably, bacteria were engulfed by an early mammalian cell progenitor to form a symbiotic relationship. This event possibly occurred 1.5 billion years ago. This scenario would explain the presence of a **mitochondrial genome** (16,569 base pairs long), a closed circular structure (5–10 per mitochondrion), which, over time, has degenerated partially so that the mitochondrion also is dependent on the cellular nucleus. **Mitochondrial DNA (mtDNA)** encodes information for 13 protein components of **oxidative phosphorylation**. It also encodes the **ribosomal RNA (rRNA) genes** (12S and 16S) as well as 22 **transfer RNA (tRNA) genes** for protein synthesis in the mitochondrion. The proteins encoded in the mitochondrial genome are subunits of enzymes located in the inner membrane including NADH dehydrogenase, cytochrome c oxidase, and ATP synthase. Mitochondrial protein subunits that are not encoded by the mtDNA are encoded by the nuclear DNA, which, after translation in the cytoplasm, are translocated to the mitochondria to complete the mitochondrial enzymes. Interestingly, mtDNA is transferred continuously into the nucleus and some of it occurs in the nucleus in a nonfunctional form. Furthermore, there are some differences between the nuclear genetic code and the code in the mitochondrion: UGA in the nucleus means "stop" but in the mitochondrion, it is read as tryptophan; AGA and AGG are read as arginine in the nuclear code but read as "stop" in the mitochondrial code; AUA is read as isoleucine in the nuclear code and as methionine in the mitochondrion; and AUA and AUU can be **initiation codons** in the mitochondrion, whereas the initiation signal in the nuclear code is AUG (this precedes a discussion of the genetic code that will occur later in Chapter 10, Nucleic Acids and Molecular Genetics; in the meantime, consult the Appendix for the table representing the nuclear genetic code).

Mitochondria appear throughout the cellular cytoplasm as shown in Fig. 2.6. A diagram of an individual mitochondrion is shown in Fig. 2.19.

A mitochondrion approximates the size of a bacterial cell and has two membranes, made up of phospholipids and proteins, enclosing a rodlike structure. Between the two membranes (**intermembrane space**) is the mitochondrial cytoplasm similar to the cellular cytoplasm outside the mitochondrion (although the nature of the proteins differs). The **outer membrane** is smooth. It contains **porins** that facilitate the passage through this membrane of molecules that are less than 5000 in molecular weight and it allows small molecules (including ATP, ADP, ions, pyruvate, etc.) to pass through it by free diffusion. The **inner membrane** contains the **electron transport chain**, including the **ATP synthase complex**. The permeability of this membrane is restricted to the passage of $O_2$, $CO_2$, and $H_2O$. Infoldings of this membrane are called **cristae** that increase the surface area. The matrix is comprised of enzymes involved in the production of ATP as well as **mitochondrial ribosomes**, mtDNA and tRNAs. Mitochondria can multiply and they do so by a process of fission similarly to the process in bacteria. When a mitochondrion enlarges it can divide into two daughter mitochondria similarly to the cell nucleus. The need for energy production in a tissue dictates the number of mitochondria in a given cell. Thus, a skeletal muscle cell will have more mitochondria than a skin cell, for example.

Following the **glycolysis** pathway, in the cytoplasm, which converts glucose to pyruvic acid, pyruvate enters the mitochondrion and is metabolized to ATP energy (from ADP), water ("metabolic water"), and carbon dioxide through the **tricarboxylic acid cycle** (also known as the **Kreb's cycle** or the **citric acid cycle**). This is the central aerobic (requiring oxygen) oxidative machinery for obtaining energy from food. Many amino acids and fatty acids, through degradation or metabolic reactions, can be converted to compounds that become intermediates of the cycle to produce energy (to be discussed in detail in later chapters), thus, fatty acids, for example, are degraded by the process of β-oxidation whereby, the final product is acetylCoA that enters the tricarboxylic acid cycle (see Chapter 9: Lipids).

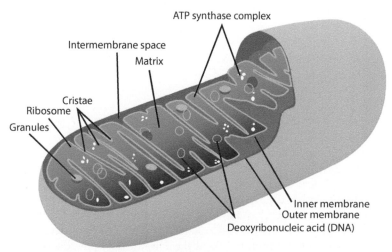

**FIGURE 2.19** Drawing of a mitochondrion showing the two membranes, cristae and matrix, as well as key intramitochondrial components. *Reproduced from http://en.wikipedia.org/wiki/File:Animal_mitochondrion_diagram_en_(edit).svg.*

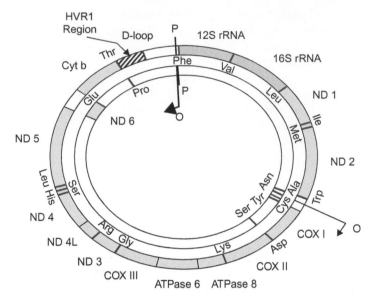

**FIGURE 2.20** The human mitochondrial genome is a circular DNA encoding 13 subunits of respiratory chain complexes; seven subunits (ND1-6 and ND4L) of complex I, cytochrome b of complex III, the COX I–III subunits of cytochrome oxidase or complex IV, and the ATPase 6 and 8 subunits of $F_oF_1$ ATP synthase. Mitochondrial DNA also encodes 12S and 16S rRNA genes and 22 tRNA genes. The abbreviated amino acid names indicate the corresponding amino acid tRNA genes. The outer strand is heavy-chain DNA and the inner one light-chain DNA. $O_H$ and $O_L$ are the replication origins of the light and heavy chains, respectively, while $P_H$ and $P_L$ indicate the transcription sites. *Reproduced from http://herkules.oulu.fi/isbn9514255674/html/x287.html.*

The mammalian egg contains about 100,000 molecules of mtDNA, whereas the sperm contains far fewer molecules (about 100−1500 mtDNAs). Although the male mtDNA appears at the early blastocyst stage of development, this component appears to be lost entirely during embryogenesis. Consequently, *mtDNA* (Fig. 2.20) *is inherited maternally*. Sequencing the **HVR1 region** (D-loop region just to the left of the region encoding 12S rRNA at the top of the circle) of mtDNA derives ancestral information for the maternal line. The HVR1 region is a 400 base pair sequence from 16,000 to 16,400 of mtDNA that provides **ancestry analysis** through the maternal line for over 100,000 years. More than 28 **haplogroups** (a haplogroup contains a **single nucleotide polymorphism, SNP**; mutation of one base in a nucleotide) of mtDNA have been identified that contain unique sets of mutations associated with each haplogrouping. The data indicate geographical location and ethnic group of ancestors. Because of the direct line of descent of mtDNA from mother to child the sequence of mtDNA of the offspring always will be identical to the mother's.

An apparent lack of a DNA repair mechanism combined with a lack of protective **histone** proteins subjects mtDNA to mutation at a rate more than 10 times that of nuclear DNA. The location of mtDNA in the inner membrane subjects it to **oxygen radicals** that are generated by the **oxidative phosphorylation** system and to products of abnormal mitochondrial metabolism. These mutations may be causative of various diseases, such as **Parkinson's disease, Alzheimer's disease (AD)**, and other neurological diseases. Presumably, a decline in mitochondrial metabolism and energy production contributes to the aging process.

## PEROXISOME

A peroxisome, like other particles in the cell, is classified as a **microbody**. It is generally spherical and is bounded by a single membrane with a diameter of 0.1−1 μm (Fig. 2.21). About 50 enzymes are contained in a peroxisome, including catalase, urate oxidase, and D-amino acid oxidase. These enzymes degrade **uric acid** and amino acids. **Hydrogen peroxide** ($H_2O_2$) is a product of the oxidative degradation of many of the organic molecules. $H_2O_2$ is further degraded to oxygen and water by **catalase**. Other enzymes degrade long chain fatty acids by β-**oxidation**, an important function of this microbody. The peroxisome synthesizes **bile acids**, **cholesterol**, and **plasmalogen**, and metabolizes amino acids and purines. Although there has been some confusion about the formation of peroxisomes, it appears that they bud off from the endoplasmic reticulum and mature. Peroxisomes are plentiful in cells, such as liver cells active in lipid biochemistry. A protein destined to enter a peroxisome contains a **peroxisome targeting signal (PTS1)** that is a tripeptide

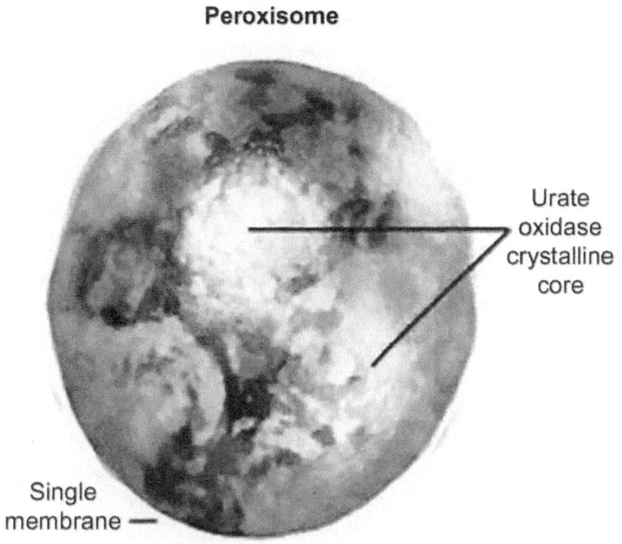

**Peroxisome**

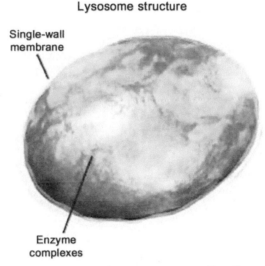

Urate
oxidase
crystalline
core

Single
membrane —

**FIGURE 2.21**  Depiction of a peroxisome. *Reproduced from http://www.palaeos.com/Eikarya/Lists/EuGlossary/EuGlossaryP.htm.*

**Lysosome structure**

Single-wall
membrane

Enzyme
complexes

**FIGURE 2.22**  Depiction of a lysosome. *Reproduced from http://www.cartage.org.lb/en/themes/sciences/zoology/AnimalPhysiology/Anatomy/AnimalCellStructure/Lysosomes/lysosome.jpg.*

(Ser−Lys−Leu or others) at its C-terminus with other sequences somewhat upstream. The receptor (**PEX5**) identifies the C-terminal motifs and traps the incoming protein through remodeling of its structure.

## LYSOSOME

The appearance and dimensions of a lysosome are similar to a peroxisome (Fig. 2.22). Lysosomes are formed by budding off from the Golgi apparatus. The pathway is from the **RER** to the **Golgi** and budding off to the cytoplasm (Fig. 2.6). The interior of the lysosome is acidic at pH 5 (whereas the cytoplasm is at pH about 7.2) and many hydrolase enzymes (probably 40 or so) are located in the lysosome. They are **proteases**, **lipases**, **nucleases**, and **polysacchari-dases**. These enzymes have pH optima for activity near the pH of the lysosome interior and would be considerably less active if released by a mishap into the cytosol. In order to maintain this acidic pH, lysosomes contain an **ion pump** that pumps in **hydrogen ions** ($H^+$, protons) with energy provided by the release of the energetic terminal phosphate group of ATP (to form ADP). Products of the **digestion of macromolecules** inside the lysosome can diffuse out of the particle into the cytoplasm for use by the cell. In order to enter the lysosome, an enzyme precursor from the RER, bearing a specific four amino acid signal (KDEL or −lysyl−aspartyl−glutamyl−leucyl−) within its C-terminus, is inactivated by phosphorylation and then binds to a receptor (the **mannose-6-phosphate receptor, M6PR**) that requires the phosphorylated form of the enzyme in order to bind; the complex traverses the Golgi apparatus and becomes pinched off into a **transport vesicle** that travels from the Golgi to the lysosome and fuses with it, emptying the receptor of the phosphorylated enzyme. The inactive enzyme, previously attached to the M6PR, is then dephosphorylated to become enzymatically active and a component of the lysosomal interior. The empty transport vesicle is returned to the Golgi apparatus along with the empty M6PR. By these means, the hydrolytic enzyme component of the lysosome is established.

A number of **storage diseases** are associated with the lysosome. Characteristic of these diseases, there is an **accumulation of macromolecules** in the lysosome owing to a loss of ability of the cell to form enzymes needed for the degradation of these macromolecules, such as proteins, lipids, polysaccharides, and others. Usually, such a disease involves two defective alleles of a gene encoding one of the hydrolytic enzymes. In **I (inclusion)-cell disease**, phosphorylation of the hydrolytic enzymes, required for binding to the M6PR, is aberrant and these unphosphorylated enzymes are secreted from the cell to the extracellular space instead of taking the normal transporting route from the Golgi apparatus to the lysosome. Failure to produce an enzyme needed to break down **sphingolipids** (structural and protective components of the cell membrane) can result in **sphingolipidosis diseases**, such as **Tay−Sachs disease** or **Gaucher's disease** and others. Failure to synthesize α-L-iduronidase, an enzyme that can degrade **proteoglycans**, such as **heparin sulfate** can cause **mucopolysaccharidosis 1** storage disease. In these storage diseases, macromolecules normally destined for degradation in the lysosome remain undegraded and form **inclusion bodies** in the cell.

In certain cases, specialized cells can release a component of the lysosome by **exocytosis** (a $Ca^{2+}$-dependent process needed to reseal damage, e.g., from wound to the cell membrane; this process is the opposite of endocytosis). Exocytosis of a lysosomal component also occurs when **cytotoxic T cells** secrete **perforin** (a cytolytic protein); when **mast cells** secrete **inflammatory mediators** or when **melanocytes** secrete **melanin**.

Other inclusion bodies in the cell, e.g., lipid droplet, secretion granules, etc. will be described in the context of specific cellular functions in later chapters.

## SUMMARY

Alzheimer's disease (AD) is used as an example of a clinical subject that exemplifies the principles exhibited in this chapter. The mechanism by which this disease occurs involves the intracellular trafficking of various products of the amyloid precursor protein (APP). The cell surface, the Golgi apparatus, the transGolgi network (TGN), early and late endosomes are all trafficking destinations that are required for the development of either the normal or the disease-producing pathways. Thus, when the digestion of the amyloid precursor protein occurs so that the α-, β-, and γ-secretases are involved, the amyloidβ (Aβ) product leads to the deposition of aggregates. Together with the development of neurofibrillary tangles (NFT) these elements form the basis of this disease. The normal situation where this disease does not develop is a function of a sorting receptor called SORLA whose activity is illustrated. It appears that mutations in the gene for SORLA that reduce its functioning may be the primary cause of Alzheimer's disease.

This chapter describes the major organelles in the cytoplasm of the cell in addition to the cell membrane. The cell membrane, or plasma membrane, consists of two layers opposing each other. Polar groups appear on the outer and inner surfaces of the membrane. Transmembrane proteins, such as receptors and channels, have groupings on the exterior and on the interior. Some nonpolar substances can dissolve in and diffuse through the membrane. Membrane glycolipids on the cell surface are involved in binding of proteins while glycoproteins differentiate self from non-self on the surface of the cell. Oxygen, carbon dioxide, and small nonpolar compounds freely diffuse through the membrane while charged ions and sugars require specific channels to enter the cell. The nucleus houses the genome with all the genetic information of the organism (46 chromosomes). Transcription factors control gene expression. The nuclear membrane, resembling the plasma membrane, contains nucleopores that permit the import of macromolecules from the cytoplasm and export from the nucleoplasm to the cytoplasm. Nucleoli inside the nucleus synthesize ribosomal RNA. As cells grow, they divide and daughter cells contain the same genetic information. The point of cell division is dictated by the cell cycle. There are four phases to the cell cycle plus an optional phase of growth arrest in which cells may reenter the cycle or may die by apoptosis if an appropriate signal occurs. Generally, cell growth reflects a buildup of energy in the form of ATP. There are three checkpoints in different phases of the cell cycle in which the cell determines whether progress through the cycle is normal. Specific factors are involved in the progress through the checkpoints. The cytoplasm is structured by the cytoskeleton. The major structures are comprised of microtubules, microfilaments, and intermediate filaments, and they involve various functions including intracellular transportation. The endoplasmic reticulum exists in two forms, the smooth endoplasmic reticulum and the rough. The latter has ribosomes on its surface where protein synthesis occurs. The Golgi apparatus has an intimate relationship with the endoplasmic reticulum and proteins are transferred from the rough endoplasmic reticulum to the Golgi through which the destination of the protein within or outside of the cell is determined. The mitochondria are the energy producers of the cell in terms of ATP production from ADP and contain the tricarboxylic acid cycle of enzymes. Mitochondria have their own genome that encodes subunits of needed enzymes that are completed by subunits encoded by the cell nucleus. The DNA of mitochondria is less stable than that of the nucleus and mutations occur more frequently than in the nucleus. Some of these mutations lead to diseases. Other organelles include the peroxisome and the lysosome. The peroxisome degrades uric acid and amino acids and, importantly, hydrogen peroxide. The lysosome contains hydrolytic enzymes and functions to breakdown macromolecules.

## SUGGESTED READING

### Literature

Dellaire, G., Farall, R., Bickmore, W.A., 2003. The nuclear protein database (NPD): subnuclear localization and functional annotation of the nuclear proteome. Nucleic Acids Res. 31, 328–330.

Garg, S., et al., 2011. Lysosomal trafficking, antigen presentation, and microbial killing are controlled by the Arf-like GTPase Arl8b. Immunity. 35, 182–193.

Halliday, M., Radford, H., Mallucci, G.R., 2014. Prions: generation and spread versus neurotoxicity. J. Biol. Chem. 289, 19862–19868.

Jahn, R., Lang, T., Sudhof, T.C., 2003. Membrane fusion. Cell. 112, 519–533.

Kohler, A., Hurt, E., 2010. Gene regulation by nucleoporins and links to cancer. Mol. Cell. 38, 6–15.

Lewis, B.A., Hanover, J.A., 2014. Glycobiology and extracellular matrices—DNA and chromosomes. J. Biol. Chem. 289, 34440–34448.

Rao, M., Gottesfeld, J.M., 2014. Development of human therapeutics based on induced pleuripotent stem cell (iPSC) technology. J. Biol. Chem. 289, 4553–4554.

Rogaeva, E., et al., 2007. The neuronal sortilin-related receptor SORL1 is genetically associated with Alzheimer disease. Nature Genet. 39, 168–177.

Solnaz, S.R., Chauhan, R., Blobel, G., Melcak, I., 2011. Molecular architecture of the transport channel of the nuclear pore complex. Cell. 147, 590–602.

Song, C., et al., 2014. Analytic 3D imaging of mammalian nucleus at nanoscale using coherent X-rays and optical fluorescence microscopy. Biophys. J. 107, 1074–1081.

Tang, D., Wang, Y., 2013. Cell cycle regulation of Golgi membrane dynamics. Trends Cell Biol. 23, 296–304.

Tokumaru, H., et al., 2001. SNARE complex oligomerization by synaphin/complexin is essential for synaptic vesicle exocytosis. Cell. 104, 421–432.

Vingtdeux, V., Sergeant, N., Buee, L., 2012. Potential contribution of exosomes to the prion-like propagation of lesions in Alzheimer's disease. Front. Physiol. 3, 229–253.

Willnow, T.E., Anderson, O.M., 2013. Sorting receptor SORLA—a trafficking path to avoid Alzheimer disease. J. Cell Sci. 126, 2751–2760.

Willnow, T.E., Carlo, A.S., Rohe, A.S., Schmidt, V., 2010. SORLA/SORL1, a neuronal sorting receptor implicated in Alzheimer's disease. Rev. Neurosci. 21, 315–329.

### Books

Alberts, B., Bray, D., Hopkin, K., Johnson, A., Lewis, J., Raff, M., et al., 2009. Essential Cell Biology. Garland Science, New York, NY.

Chapter 3

# Introductory Discussion on Water, pH, Buffers and General Features of Receptors, Channels and Pumps

Knowledge concerning water, pH, and buffers is basic for the study of biological systems. Some general features of receptors, channels, and pumps are critical to the understanding of cellular biochemistry because the human body is made up of separate organs and tissues, and for it to function, these parts of the body are interdependent and need to be able to communicate. Intercellular communication from one organ to another is usually accomplished by the release of a signaling molecule from specific cells in the originating organ for transport in the blood to a remote target cell (in other target organs) in which the signal is generated by a receptor as involved in the clinical example introducing this chapter. Molecules and ions are transported through channels, and some ions are passaged by an enzymatic pump. All of these aspects are covered in more detail in succeeding chapters but are useful by way of introduction in this chapter along with the more basic information on water, pH, and buffers.

But, first, a clinical example that reflects the economy of water in the body is the condition known as diabetes *insipidus*.

## DIABETES *INSIPIDUS*

Failure of the kidneys to reabsorb water and concentrate the urine results in greatly diluted urine and substantial loss of water from the body. There are two causes for this condition, undersecretion of the posterior pituitary hormone, **arginine vasopressin** (**AVP** or ADH, **a**ntidiuretic **h**ormone) or failure of the distal tubule of the kidney to respond to AVP. Failure of the kidney to respond to AVP can be a function of loss of function of the **AVP receptors** or loss of function of the **water channels [aquaporins (AQPs)]** required for the reabsorption of water from the urine. Undersecretion of arginine vasopressin (AVP) from the posterior pituitary is called **central diabetes** *insipidus*, and failure of the kidney to respond to AVP is called **nephrogenic diabetes** *insipidus*. Central diabetes *insipidus* is rare accounting for one person in 25,000. Nephrogenic diabetes *insipidus* can occur with the abuse of alcohol or some types of drugs and with kidney disease, sometimes associated with the aging process. Substantially, smaller amounts of cellular proteins are synthesized in cells during the aging process and, consequently, there may be some loss of the receptor for ADH in the kidney. Moreover, there may be loss of function mutations of the genes for the **AVP precursor** (**preprovasopressin**), the genes for the AVP receptor or the genes for aquaporins. Excessive urination and loss of water also occur in **untreated diabetes** *mellitus* (due to loss of insulin by destruction or underactivity of the beta cells of the pancreas or due to insulin resistance). In this case, the loss of water is related to high levels of glucose in the blood and urine. Glucose carries bound water with it (Fig. 3.1) in the process of excretion, a completely different causation than the factors in diabetes *insipidus*.

Water is a dipolar molecule wherein the oxygen has a large nucleus that attracts electrons causing the oxygen atom to be negatively charged, whereas the hydrogens are positively charged. This dipolar character allows water molecules to interact with each other through hydrogen bonding (Fig. 3.2).

Sugar hydroxyl groups are attracted to the oxygens of water to form hydrogen bonds and maintain the sugar in solution. Anions (negatively charged atoms or molecules that would be attracted to the positively charged *anode* during electrophoresis) will have strong interactions with water. The same is true for cations (atoms or molecules with a positive charge that would be attracted to the negatively charged *cathode* during electrophoresis). For example, a sodium ion (cation) will attract six molecules of water as shown in Fig. 3.3.

Human Biochemistry. DOI: http://dx.doi.org/10.1016/B978-0-12-383864-3.00003-X

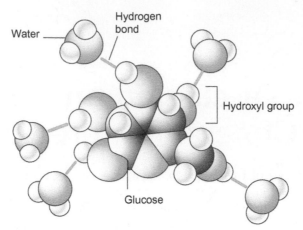

**FIGURE 3.1**   Illustration showing the interaction of a sugar in solution with water molecules. Sugar molecules are hydrophilic (water loving). In order to remain in solution, the sugar is surrounded by water molecules shown interacting with its polar hydroxyl groups. *Reproduced from http://edutech.csun.edu/eduwiki/ULimages/a/a3/Sugarinwater1.jpg.*

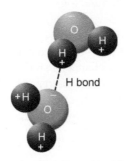

**FIGURE 3.2**   Interaction of water molecules through hydrogen bonding. *Reproduced from http://www.ces.fau.edu/nasa/module-3/why-does-temperature-vary/land-and-water.php.*

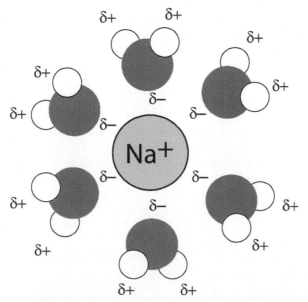

**FIGURE 3.3**   A sodium ion solvated by water molecules. The assumption here is that the $\delta$ on hydrogen is half the $\delta$ on oxygen. $\delta$, delta; the process of association of molecules of solvent and molecules of solute. From http://www.memidex.com/solvated + change.

The determination of the number of water molecules interacting with a cation ($Na^+$) is complex and depends on a number of variants solved by the formula, $[M(H_2O)_n]^{z+}$, where M is the metal ion, $z^+$ is the electrical charge, and $n$ is the number of interacting water molecules. An important factor in the determination of the number of interacting water molecules is the radius of the ion where the interaction is greater with a smaller radius ($Na^+$ has a small radius).

## THIRST AND ARGININE VASOPRESSIN (AVP)

Diabetes *insipidus* is also characterized by intense **thirst**. *Both thirst and the mechanism to release AVP are interrelated through neuronal connections in the hypothalamus.* AVP is cut out of its precursor, **preprovasopressin** (Fig. 3.4).

AVP is synthesized in vasopressinergic (releasing AVP) neurons located in the supraoptic and paraventricular nuclei of the anterior hypothalamus. AVP is transported in combination with a **neurophysin (NPII)** protein down the long axon to the neuronal terminal located in the posterior pituitary. It is secreted in response to **increased osmolality** (an increase in the solute particles in plasma but usually an increase in sodium ion concentration) from the neuronal terminal by a process of exocytosis activated by increased calcium ion concentration. AVP is secreted when the osmolality of the plasma is increased by only 1%−2%. It appears that the **TRPV channel** (transient receptor potential channel) family of cation channels may be involved as **osmomechanical receptors**. Possibly, they could mediate neuronal responses to changes in tonicity involving the changes primarily in $Na^+$ concentrations in plasma. For example, the **TRPV4 channel** is a $Ca^{2+}$-permeable cation channel that is involved in the regulation of systemic osmotic pressure. AVP has a relatively short half-life of about 15 minutes as it is converted in the liver and kidney to inactive products.

AVP has the following amino acid sequence:

C-Y-F-Q-N-C-P-**R**-G or Cys-Tyr-Phe-Gln-Asn-Cys-Pro-Arg-Gly

with a **disulfide bridge** connecting the two cysteine (C) residues. This hormone is similar to oxytocin, another posterior pituitary hormone. Oxytocin is also a nine amino acid peptide having the same sequence except that amino acid 3 (phenylalanine, F) is replaced by isoleucine (I) and amino acid 8 (arginine, R) is replaced by leucine. Both hormones have identical disulfide bridges connecting to the cysteines.

There are four major stimuli leading to **thirst**. As already described, **hypertonicity**, particularly an increase in the $Na^+$ concentration in plasma, operates through an **osmoreceptor** in the hypothalamus. Decreased volume of water, **hypovolemia**, is sensed by **low-pressure baroreceptors** located in the large veins and in the right atrium of the heart. Lowered blood pressure, **hypotension**, is sensed through **high-pressure baroreceptors** located in the carotid sinus and aorta. In decreased water volume, the consequently increased osmolality in the extracellular fluid will lead to the **decreased secretion of saliva** that, in turn, leads to **dry mouth**, creating the sensation of thirst. Some cases of **migraine headache** may be the result of inadequate hydration. Finally, the hormone, **angiotensin II (ANGII)**, is produced through the action of the enzyme **renin** secreted from the kidney in response to **renal hypotension**. ANGII binds and activates specific **osmoreceptors** located in the subfornical organ (SFO; in the lower fornix) and in the *organum vasculosum* of the lamina *terminalis* (OVLT). These are connected to the median **preoptic nucleus** of the hypothalamus (Fig. 3.5). The action of AII also leads to thirst.

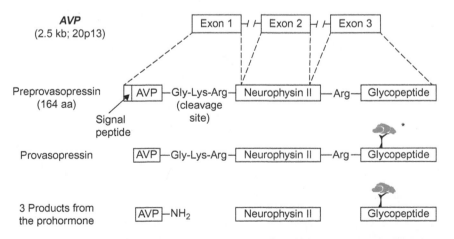

**FIGURE 3.4** Structure of the human vasopressin (AVP) gene and preprovasopressin. The cleavage sites are shown leading to the maturation of AVP, neurophysin II, and glycopeptide. The gene is first transcribed to form the messenger RNA (mRNA) that is translated in the cytoplasm to form the preprovasopressin peptide. Provasopressin is generated from its precursor by the action of an endopeptidase. The C-terminal portion of provasopressin, shown as glycopeptide, is also referred to as **copeptin**. Provasopressin is converted to the mature products by the actions of three enzymes: exopeptidase, monooxidase, and lyase. *Reproduced from http://www.smw.ch/fileadmin/smw/images/SMW-13613-Fig-04.jpg.*

*Addition of a carbohydrate chain

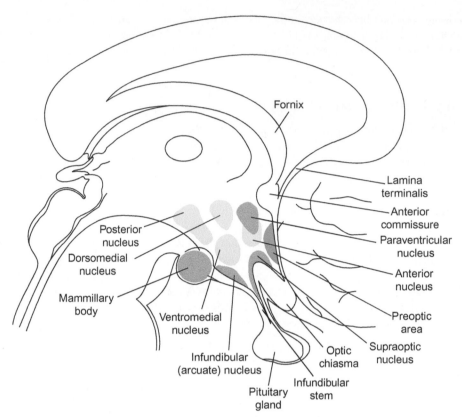

FIGURE 3.5   Illustration of the regions of the hypothalamus. The subfornical region is located in the lower fornix. *Reproduced from http://what-when-how.com/wp-content/uploads/2012/04/tmp3627.jpg.*

ANGII acts as a neurotransmitter in the neuronal path from subfornical organ (SFO) to the hypothalamus. In the same area of the hypothalamus, there are ascending pathways from low- and high-pressure baroreceptors that signal thirst development. It is believed that *the* **osmoreceptor** *stimulating* **release of AVP** *and the receptors stimulating* **thirst** *are located in the same region of the hypothalamus.* The overall picture for development of thirst and the roles of angiotensin II are summarized in Fig. 3.6.

## ACTION OF ARGININE VASOPRESSIN ON THE DISTAL KIDNEY TUBULE

The molecular actions of AVP in the kidney are discussed in the chapter on polypeptide hormones. Here, only a brief description will be given. After the AVP-neurophysin II, complex is released from the posterior pituitary into the general circulation, it reaches the vicinity of its receptor (AVPR). AVPR is located in the membranes of cells of the distal convoluted tubule and the collecting ducts of the kidney. In the vicinity of the AVP receptor, neurophysin II is dissociated from the complex so that free AVP can bind and activate its receptor. AVPR is a G-protein-coupled receptor and, through its action, cyclic AMP (cAMP) is formed from ATP. cAMP activates protein kinase A that phosphorylates **microtubular subunits**. The phosphorylated microtubular subunits form **aquaporin channels** that insert into the apical cell membrane. These channels allow the entry of water so that the water in vesicles can be transported across the cell to the basolateral membrane where the water is released into the extracellular space and finally into the bloodstream to increase its volume.

## WATER AND BIOLOGICAL ROLES

Water is a key substance for life. While the human body, depending on its condition and many other factors, can live for weeks, even months without food, one will usually die if deprived of water for 3 days. The human body is composed of about 72%−78% water. Of this, about 70% is intracellular and 30% is extracellular. The water molecule is unique; it has polarity with oxygen bearing a negative charge at one end and two hydrogens bearing positive charges at the other end (Fig. 3.7).

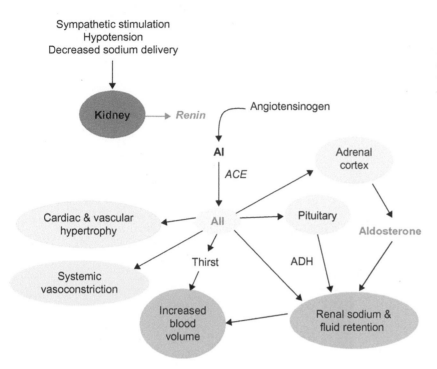

FIGURE 3.6  Relationships between hypotension, the synthesis of angiotensin II and its actions, including the sensation of thirst. *AI*, angiotensin I; *AII*, angiotensin II; *ACE*, angiotensin converting enzyme; *ADH*, antidiuretic hormone or *AVP*, arginine vasopressin.

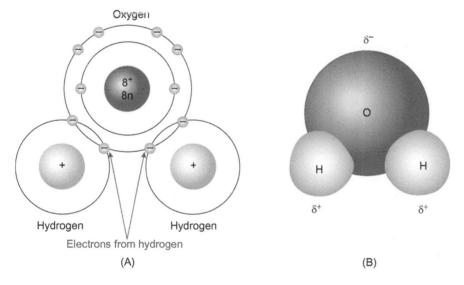

FIGURE 3.7  Structure of the water molecule. Water is a polar molecule with one end positively charged and the other, negatively charged. The hydrogen atoms are at 105 degrees from the center of the oxygen atom. (A) Electron shells in a water molecule and (B) distribution of partial charges in a water molecule. *Reproduced from http://rst.gsfc.nasa.gov/Sect20/water_molecules_con_c_la_784.jpg.*

Water is the major solvent in the body for all kinds of nutrients, salts, intermediate compounds, and others. It is critical for the activity of proteins (Fig. 3.8).

## WATER CHANNELS: AQUAPORINS (AQPs)

Water is not able to cross the nonpolar cell membrane because water is insoluble in lipids, whereas lipid molecules often can dissolve in the lipophilic cell membrane and freely diffuse into the cell. The special mechanism by which water enters a cell is a **water channel**, called an **aquaporin**, discovered in the 1980s. Aquaporins occur in most cell types including the red blood cell. Because of the nature of the water reabsorption process in the kidney, it has been revealed that there are at least four different aquaporins there. There are two apical aquaporins (aquaporin 1 and aquaporin 2) for water reabsorption, the second operating in response to **anti-diuretic hormone** (**ADH**; or AVP, **arginine vasopressin**). The other two aquaporins (aquaporin 3 and aquaporin 4) in the kidney operate basolaterally for water reabsorption. In the human, at least 13 different aquaporin protein variants have been identified starting with **aquaporin 0** (**AQP0** through **aquaporin 12**

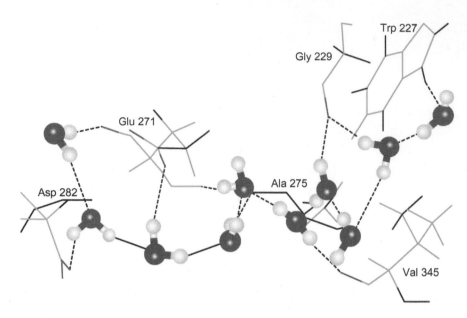

Trp 227
Gly 229
Glu 271
Asp 282
Ala 275
Val 345

**FIGURE 3.8** Water molecules involved in the hydration of a protein. Shown is a chain of 10 water molecules linking the end of 1 α-helix to the middle of another. This information is derived from X-ray diffraction data. *Reproduced from http://www1.lsbu.ac.uk/water/images/hydrprot.gif.*

**TABLE 3.1  The Mammalian AQPs and Their Characteristics**

| Aquaporin | Size | Permeability | Distribution |
|---|---|---|---|
| AQP0 | 26 kDa, 263 aa | Water (low) | Lens epithelium |
| AQP1 | 28 kDa, 269 aa | Water | Kidney capillary endothelia (except brain) red blood cells, cornea, choroid plexus |
| AQP2 | 29 kDa, 271 aa | Water | Kidney collecting duct cells (intracellular and apical membranes) |
| AQP3 | 31 kDa, 292 aa | Water, glycerol, urea | Kidney, colon |
| AQP4 | M1: 32 kDa, 301 aa M23: 34 kDa 323 aa | Water (Hg2* in-sensitive) | CNS, skeletal muscle, lung, kidney, inner ear, gastric parietal cells |
| AQP5 | 29 kDa, 265 aa | Water | Lung, salivary glands, lacrimal glands, trachea, cornea |
| AQP6 | 28−30 kDa, 276 aa | Water (low), anions ($HNO_3^-$, high) | Intracellular vesicles in kidney intercalated cells, proximal tubules |
| AQP7 | 26 kDa, 269 aa | Water, glycerol | Adipose tissue, testis, kidney |
| AQP8 | 27 kDa, 269 aa | Water | Testis, liver, pancreas |
| AQP9 | 32 kDa, 342 aa | Water, glycerol, urea | Liver, testis, brain |
| AQP10 | 28 kDa, 301 aa | Water (low), glycerol, urea | Small intestine |
| AQP11 | | | Kidney proximal tubule, liver, testis |
| AQP12 | | | Pancreas |

aa, amino acid; Aqp, aquaporin; CNS, central nervous system; M1/M23, two different isoforms of Aqp4
Source: Reproduced from http://flipper.diff.org/app/pathways/Aquaporins.

or **AQP12**). **Aquaporin 1 (AQP1)** from the red blood cell is the best studied (Table 3.1). AQP4 is highly expressed at the **blood−brain barrier** in the end feet where it regulates water homeostasis and the formation of cerebrospinal fluid. When water accumulates in the brain (**edema**), it appears that AQP9 is involved. Experimental observations suggest that AQP4 may play a role in fluid elimination rather than edema formation after brain injury.

A simplified drawing of an aquaporin water channel is shown in Fig. 3.9. It has been predicted that the aquaporin 1 channel could facilitate the transport of water at a theoretical rate of about 3 billion water molecules per second (http://www.bio.miami.edu/~cmallery/150/memb/water.channels.htm). One group of aquaporins is impermeable to charged molecules or protons but able to admit some other small molecules, such as glycerol (**aquaglyceroporins**).

Water channel

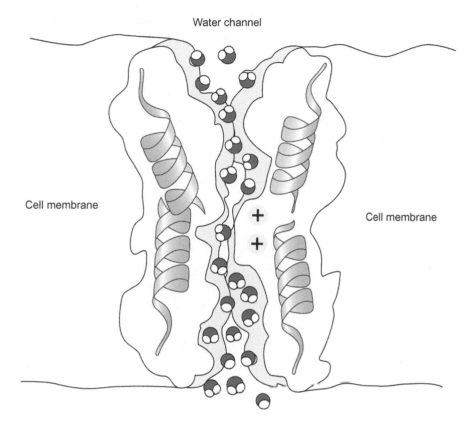

Cell membrane

Cell membrane

**FIGURE 3.9** A simplified drawing of an aquaporin water channel. *Reproduced from http://www.bio.miami.edu/~cmallery/150/ memb/water_channel_lgl.jpg.*

Fig. 3.9 shows that the water molecules on the apical side of the cell enter the channel with the oxygen atom facing downward and the two hydrogen atoms facing upward. As water proceeds downward, single file, midstream through the channel, the water molecule inverts so that the water molecule has the two hydrogen atoms facing downward and the oxygen atom facing upward. This controlled inversion is mediated by the side chains of amino acids of the channel as the water molecule proceeds through it. **Asparagines**, hydrogen bonding to the oxygen of water apparently make this inversion possible (some publications question this exact version of the flow of water molecules and also question the exclusion of proton flow through the aquaporin channel). Aquaporins can conduct water in both directions, in and out of the cell. The inward action takes place at the apical membrane and the outward action at the basolateral membrane (facing the extracellular space and ultimately the bloodstream). When water leaves a cell through an aquaporin channel, it leaves with the oxygen atom facing upward. In a kidney cell, e.g., as will be discussed later on, the assembly of water channels in a membrane is under the hormonal control of arginine vasopressin (AVP or antidiuretic hormone, ADH). The hormone signals from the activated AVP receptor, through a G protein that causes the elevation of cAMP and the consequent activation of protein kinase (PKA) to phosphorylate aquaporin subunits in the cytoplasm. The **phosphorylated subunits** migrate to the apical cell membrane, where they form a **tetrameric channel** and the process of water uptake is initiated (by itself, the subunit has the property of taking up water but four of them function in a channel). Conversely, the channels can be rendered nonfunctional by **dephosphorylation** of the subunits. In the kidney glomerulus, about 70% of water in primary urine is reabsorbed by AQP1 and passed into the blood. Ten percent more water is reabsorbed by AQP2 at the end of the glomerulus. It is here that **ADH/vasopressin** acts on **AQP2** to increase the reabsorption of water from urine. In humans, a deficiency of ADH leads to excessive urination (**diabetes *insipidus***), as discussed earlier, in which there can be a urinary output of 10−15 L/day. Conversely, **water retention** is caused by hypersecretion of **ADH**, certain drugs, high sodium ion intake, or other conditions. A virtual image of an aquaporin channel in a cell membrane is shown in Fig. 3.10.

Fig. 3.11 shows a view from the top of the aquaporin from red blood cells (AQP1).

In Fig. 3.12, an aquaglyceroporin subunit is shown.

The many aquaporins are distributed to different cell types. **AQP1** is present mostly in **red blood cells**, kidneys, and the **choroid plexus (CP)** (the CP is a structure lining the ventricular system in the brain where **cerebrospinal fluid** is produced). The CP can be an entrance point for some molecules or particles (e.g., viruses) forming a means to bypass the **blood−brain barrier**. **AQP4** is located in the brain and **AQP7** and **AQP9** are located in **adipocytes**. **AQP1** and **AQP2** are located in the apical and basolateral glomerulus and **AQP2** is in the apical cortical collecting duct of the

**FIGURE 3.10** Virtual image of an aquaporin channel. The channel is in the right center and shows the water molecules coursing through the channel single file. *Reproduced from http://www.ks.uiuc.edu/Gallery/Science/Structure/membrane-waterchannels-print-300pi_st.jpg.*

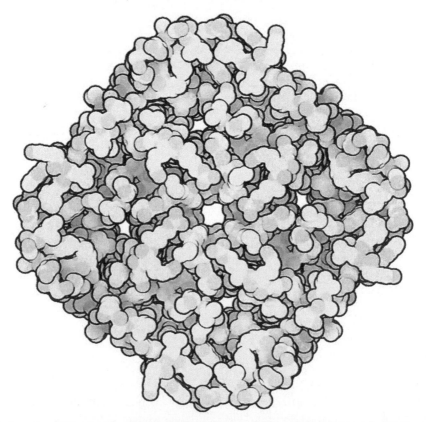

**FIGURE 3.11** Aquaporin from the red blood cell is viewed from the top. There are four identical chains each with its own channel at the center of the subunit. The larger hole at the center of the four chains is lined with carbon-rich amino acids and is plugged by membrane lipids in the cell. Two charged amino acids are located at the entry of the water channel are important, ensuring that only water passes through the channel in a single queue, eliminating hydronium or hydroxyl ions from passing through the channel. *Reproduced from ftp://resources.rcsb.org/motm/tiff/173-Aquaporin_1fqy.tif.*

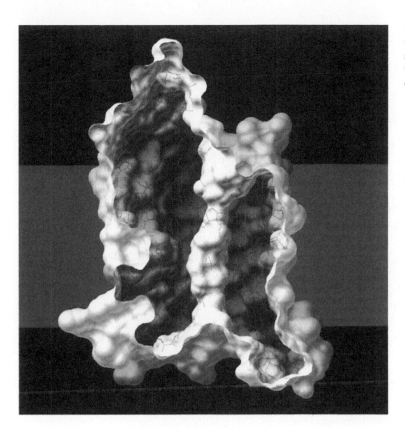

**FIGURE 3.12** A subunit of aquaglyceroporin is shown which passes both water, glycerol, and urea. Only a single subunit is shown, and it is clipped to show the tunnel passage through the membrane. *Reproduced from ftp://resources.rcsb. org/motm/tiff/173-Aquaporin_1fx8_pmv.tif.*

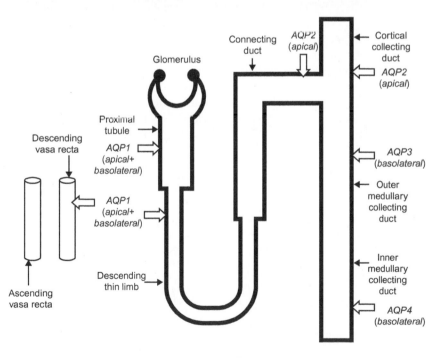

**FIGURE 3.13** The locations of specific aquaporins in the kidney. *Reproduced from http://flipper. diff.org/app/pathways/Aquaporins.*

kidney. **AQP3** is in the basolateral outer medullary-collecting duct, and **AQP4** is located in the basolateral inner medullary-collecting duct (Fig. 3.13).

**Polycystic kidneys** are produced by the disruption of **aquaporin 11** following vacuolization of the proximal tubule. This condition is often inherited and can lead to cysts in other organs and can result in kidney disease, loss of kidney function, and even, ultimately, in death, although there are gradations of this condition. AQP11 and AQP12 are different from the major two classes of AQPs: those that transport water only and the **aquaglyceroporins** that transport glycerol, urea, and other small molecules in addition to water. AQP11 is 30 kDa and has been localized to the kidney, testes, liver, and brain. AQP11 is most similar to AQP12 and least similar to AQP4 and AQP7. AQP11 and AQP12 may be the forerunners of a new family of aquaporins.

## THE ROLE OF WATER IN PROTEIN FOLDING

The details of the protein folding mechanism from the point of translation of the "linear" polypeptide chain emerging from the ribosome have been obscure until a more recent appreciation of the role of water molecules in this process has been realized. The protein-folding pattern is determined by the burying internally of the hydrophobic amino acids that repel water molecules. The hydrophilic (charged) amino acids in the polypeptide chain interact with water molecules. The hydrophobic amino acids repel water and aggregate with each other through nonpolar bonding and form the center of the molecule by retreating from the polar amino acids bound with water. The hydrophilic amino acids are in random conformations but attached to water molecules. At elevated temperatures, the water molecules associate through hydrogen bonding and dissociate from the hydrophilic amino acids allowing for new configurations to occur as the polar amino acids associate with one another via opposite charges. At various stages of this process, the water molecules reassociate with the altered configurations of the hydrophilic amino acids, and this process continues until the polypeptide chain is fully folded, encapsulating the hydrophobic core, and has reached its lowest energy stable configuration at a reduced temperature. The lowest energy forms of the protein occur when water molecules become stably associated with the surface of the protein and water molecules form a shell around the folded protein (Fig. 3.14).

## PROTEIN–WATER INTERACTIONS IN ENZYMATIC REACTIONS

Basic information of enzymatic catalysis is presented in the chapter on enzymes. Recent research has produced information on the role of water molecules in enzymatic reactions that revises accepted knowledge of enzyme catalysis. The original research was done with a reaction catalyzed by a metalloprotease (enzyme digesting protein that requires a metal). Long-lasting protein–water motions have been shown to persist beyond a single catalytic cycle that is described by steady-state enzyme kinetics. A "**hydration funnel**" appears to form toward the molecular recognition site on the enzyme (the catalytic center or "binding pocket"). The motions of water molecules appear to adapt to the motions of the substrate that are critical for binding of the substrate molecule to the active site of the enzyme. This new phenomenon depends on the substrate and its effective binding mediated by water. Thus the classical theory of enzymatic catalysis has to be revised in the future by taking into consideration the long-lasting protein–water coupled motions into the accepted models of functional enzymatic catalysis. The concept of the **hydration funnel** as it mediates the formation of the enzyme–substrate complex is shown in Fig. 3.15.

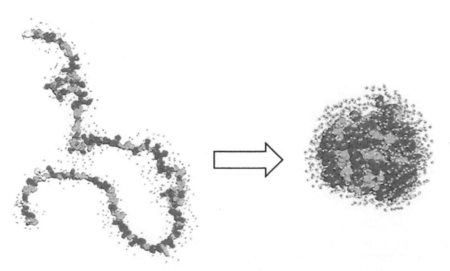

**Unfolded**                    **Folded**

FIGURE 3.14 Folding of a polypeptide chain. The folding allows hydrophobic amino acids (*green spheres*) to huddle together within the protein core, whereas hydrophilic amino acids (*pink spheres*) form the water-accessible surface of the protein. Water molecules engulfing both unfolded and folded proteins are shown. *Reproduced from https://amit1b.files.wordpress.com/2008/03/protein-folding2.png.*

**FIGURE 3.15**   Schematic diagram of the hydration funnel in an enzyme−substrate complex (the enzyme protein is in *gray*; the funnel in *yellow*; substrate in *green*). *Reproduced from http://phys.org/newman/gfx/news/hires/2014/classicalenz.jpg.*

## METABOLIC WATER

The substances derived from ingested food serve as sources of energy. These are the breakdown products of fats, proteins, and carbohydrates. These products, variously, are intermediates in the mitochondrial tricarboxylic acid cycle (TCA, Citric Acid Cycle, Krebs Cycle) or in glycolysis, the central pathways of metabolism (described in later chapters). In the TCA cycle (located in the mitochondrial intermembrane space) certain intermediates, such as succinate and the coenzyme, NADH, are electron donors. Generated protons are accepted by oxygen to form water in the course of oxidative phosphorylation producing ATP from ADP in the terminal reaction (ATP synthase) of the electron transport chain (ETC). The ETC is described in detail in later chapters. The water generated in the turning of the TCA cycle is "metabolic water." In the human, up to 300 mL can be produced in this fashion per day and it satisfies up to 10% the need for water (the rest has to be ingested as liquid water together with the water in foods). Some fuels give rise to more metabolic water than others because they offer more substrate per gram for the TCA cycle. For example, fat, the highest caloric food, will generate slightly more than its mass in water (100 g fat will generate about 110 g water if all the fat is metabolized). Carbohydrate is half as efficient as fat, by weight, in generating metabolic water. Protein will give rise to slightly less than 40% of the water generated by an equal amount of fat. The amount of water produced by the metabolism of the products of protein is sufficient to facilitate the renal excretion of the urea generated from protein degradation (elaborated in future chapters). The breakdown product of fat (via beta oxidation) is the two-carbon fragment, acetyl CoA, which is a direct substrate for the TCA cycle. Protein and carbohydrate generate amino acids and sugars that directly enter either glycolysis or the TCA cycle. The breakdown products of these food sources and how they are substrates for glycolysis and the TCA cycle are described in later chapters. Thus the products of metabolism of fuels through the TCA cycle are ATP, $CO_2$, and "metabolic" water.

## PROTON TRANSFER IN LIQUID WATER

Recent work indicates that the ions within water, the hydronium ion, $H_3O^+$ and the hydroxide ion, $OH^-$, are transported through water in complex ways but these mechanisms are undoubtedly coupled to the properties of the hydrogen bond network in water. These ions are charge defects that can be represented by an excess proton ($H_3O^+$) and a proton hole ($OH^-$) and they exhibit extremely high rates of diffusion relative to other ions, such as $Na^+$ or $Cl^-$. It appears that the ways in which $H_3O^+$ and $OH^-$ move through water must impact mechanisms of enzymatic proton transfer, proton transfer in ion channels, and acid–base chemistry. Although there exist some theories on proton transfer in water, more recent research indicates that *a proton hops in a stepwise fashion from a $H_3O^+$ ion to a water molecule or from a water molecule to $OH^-$* in a time frame of about 1–2 ps (1 ps $= 10^{-12}$ s). An alteration in the local structure of water (solvent reorganization) is required for the successful proton transfer. Thus all protons are believed to diffuse by a similar mechanism as shown in reaction C of Fig. 3.16.

## THE CONCENTRATION OF HYDROGEN IONS (PROTONS) IN SOLUTION (pH)

pH is a measure of the concentration (in molar or moles/liter) of hydrogen ions ($H^+$) in solution where the **concentration of $H^+$** is expressed as $[H^+]$. The relationship between pH and $[H^+]$ is:

$$pH = -\log[H^+]$$

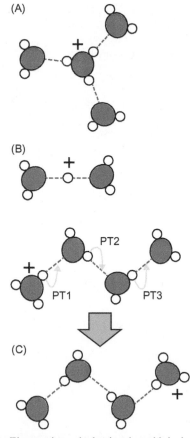

**FIGURE 3.16** Schematic drawings showing (A) the Eigen cation, a hydronium ion with its hydration. (B) The Zundel cation, and (C) a sequence of proton transfers resulting in the displacement of an excess proton along a chain of H-bonded water molecules. In the picture, the proton transfers PT1, PT2, and PT3 are thought to occur as separate steps in the sequence, whereas the work of Hassanali et al. (see "Literature" at end of chapter) reveals that the three transfers can occur in a concerted manner by coupling to the structural fluctuations of the H-bond network. *Reproduced from Colomiu-Hernandez, E., Kusalik, P.G., 2013. Probing the mechanisms of proton transfer in liquid water. Proc. Natl. Acad. Sci. 110, 13697–13698, http://www.pnas.org/content/110/34/13697.full.*

where "p" means taking the negative log of a quantity. In biological systems, pH values usually range about 6 or 7 (5 in the lysosome). However, the actual nonbiological range of pH is from 0 to 14. So, according to this equation, if the pH is (neutrality, 7, where the positive charges equal the negative charges), the $[H^+]$ is $10^{-7}$. Therefore, all pH values are related to the $[H^+]$ as the negative log to the base 10 or 10 to the power of $-pH$ value. Thus, a very acidic pH of 3 (many more $H^+$ species than $OH^-$ species) would equal a $[H^+]$ of $10^{-3}$ and a very basic (alkaline) pH of 13 (many more $OH^-$ species than $H^+$ species) would have a $[H^+]$ of $1 \times 10^{-13}$, and so on, for all pH values in between. Strong acids, such as HCl dissociate completely into $[H^+] + [Cl^-]$ and strong salts (electrolytes), such as NaCl also dissociate completely into $[Na^+] + [Cl^-]$. Strong acids, strong bases, or strong electrolytes have no **buffering capacity** because there is no equilibrium between ionized and unionized species; everything is in the ionized form in water. Thus a 0.1-M aqueous solution of HCl would be represented entirely by $[H^+] + [Cl^-]$ and no species of HCl; therefore, if $[H^+] = 0.1$ M, the pH $= -\log(0.1) = 1$. Water, itself only dissociates to a small, virtually negligible extent:

$$H_2O = H^+ + OH^-$$

so, the dissociation constant can be written as $K = [H^+][OH^-]/1$ and the dissociation constant of water $= K_w = [H^+][OH^-] = 10^{-14}$ (representing the smallest possible dissociation of $H_2O$ into its constituent ions). So, then, if pure water has virtually zero concentrations of $H^+$ ($H_3O^+$) and $OH^-$ and $K_w = 10^{-14}$, the concentration of $[H^+]$ equals $[OH^-]$ and each is equal to $10^{-7}$ ($K_w = 10^{-14} = [H^+][OH^-] = (x)(x)$; and x therefore $= 10^{-7}$) or, stated another way, pure water would have a neutral pH. This would be the case at a temperature of about 25°C; if the temperature were raised significantly, the effect would be to slightly reduce the pH.

## BUFFERS

A buffer is a compound that, in solution, resists changes in pH. This is of critical importance because the human system operates only within a fairly small pH range for most of the internal solutions (except for gastric juice that can be about pH 2 or the lysosome, e.g., that operates at about pH 5 but is separated from the other intracellular environments by a membrane as shown in **Fig. 2.22**). There are three major buffering systems in the body. The primary buffer in **blood** is the **carbonate−bicarbonate buffer**. This system maintains the **pH of blood** between 7.35 and 7.45. **Phosphate buffer** is the main buffer inside the cell and, to a certain extent, proteins in solution can exert a buffering effect. Buffers are salts of a strong acid and weak base or a strong base and a weak acid. All three species (salt and two ions) would appear in solution, in contrast to a strong acid (e.g., HCl) or strong base (e.g., KOH), by itself, that would be ionized completely in solution. Thus in the case of an acid and conjugate base, HA (a conjugate base), the product of removing a proton from a weak acid, such as **lactic acid**: $CH_3CHOHCOOH \rightarrow H^+ + CH_3CHOHCOO^-$ where HA represents $CH_3CHOHCOOH$ and $A^-$ represents $CH_3CHOHCOO^-$:

$$HA \rightleftharpoons H^+ + A^-$$

Then, the dissociation constant of the weak acid, $K_a$, can be written as the ratio of the concentrations of the products over reactant:

$$K_a = \frac{[H^+][A^-]}{[HA]}$$

And, solving for products:

$$[H^+][A^-] = K_a[HA]$$

Solving for $[H^+]$:

$$[H^+] = \frac{K_a[HA]}{[A^-]}$$

Taking the $-\log$ of both sides of the equation:

$$-\log[H^+] = \frac{-\log K_a[HA]}{-\log[A^-]}$$

As $-\log[H^+] = pH$ and $-\log K_a = pK_a$ and converting $-\log([HA]/[A^-])$ to $+\log([A^-]/[HA])$:
The equation can be rewritten:

$$\textbf{pH} = \textbf{pK}_\textbf{a} + \textbf{log}\,([A-]\,/\,[HA])$$

which is the **Henderson–Hasselbalch equation**. A$^-$ can also be referred to as "base" (lacking a proton) and HA can be written as "acid" (with an extra proton). If [A$^-$] and [HA] were in equal concentrations, the log ([A$^-$]/[HA]) would equal log 1 which is zero; thus pH = pK$_a$. Therefore *an unknown pK$_a$ can be determined by titration*.

This equation can be used equally for weak bases:

$$\textbf{pOH} = \textbf{pK}_b + \textbf{log} ([\textbf{B}^+] / [\textbf{BOH}])$$

Some important buffers are listed in Table 3.2.

Acidic molecules are usually carboxylates and basic molecules are usually amines. For optimal pH range control, it is best to select a buffer whose pK$_a$ is less than 2 units distant from the desired pH. The buffer systems operating in the body are listed in Table 3.3.

As was mentioned, the primary buffer in **blood** is the **carbonate–bicarbonate buffer**. The normal pH of blood is 7.4 and this pH is maintained in the face of increases in acids, such as **lactic acid** or bases resulting from ingestion of alkaline substances or ingestion of certain drugs.

In blood plasma, this buffer operates through carbonic acid ($H_2CO_3$) and bicarbonate (hydrogen carbonate; $HCO_3^-$). When protons ($H^+$) are added, they are combined with bicarbonate to form carbonic acid as shown in the reverse of the reaction:

$$H_2CO_3 \rightleftharpoons H^+ + HCO_3^-$$

This equilibrium has a pK$_a$ value of 6.1, the maximal buffering capacity of the carbonic acid–bicarbonate buffering system (Fig. 3.17).

In the titration curve for the carbonic acid–bicarbonate buffer (Fig. 3.17), it can be seen that the curve tends to flatten in the region where the pH approaches the value of the pK, in this case, pH 6.1. As the normal pH of blood is 7.4, this value is out of the range of the maximal buffering capacity of this buffer where the slope of the curve becomes greater. In the region of the pH of blood, a shift in the relative concentrations of bicarbonate and carbon dioxide will produce a sizeable change in the pH. *To maintain the optimal pH range of blood, HCO$_3^-$ and CO$_2$ are supplied from other organs.* Although the concentration of plasma phosphate is too low to have much of a buffering effect on blood pH, **hemoglobin** does play a role by its ability to bond $H^+$ to the globin or to bind $O_2$ to the heme iron. However, when one is bound, the other is released. Consequently, in the case of muscle during exercise when **lactic acid** is being produced, hemoglobin can bind $H^+$ causing the release of $O_2$ that can be used by the muscle for recovery.

**TABLE 3.2** A List of Buffers With Their pK$_a$ Values and the Useful pH Range

| Buffer | pK$_a$ (25°C) | Useful pH Range |
|---|---|---|
| TFA | 0.5 | <1.5 |
| Sulfonate | 1.8 | <1–2.8 |
| Phosphate | 2.1 | 1.1–3.1 |
| Chloroacetate | 29 | 1.9–3.9 |
| Formate | 3.8 | 2.8–4.8 |
| Acetate | 4.8 | 3.8–5.8 |
| Carbonate, pK$_1$ | 6.1 | 5.1–7.1 |
| Sulfonate | 6.9 | 5.9–7.9 |
| Phosphate | 7.2 | 6.2–8.2 |
| Ammonia | 9.2 | 8.2–10.2 |
| Phosphate | 12.3 | 11.3–13.3 |

*TFA*, trifluoroacetate.
Source: Reproduced from http://www.sigmaaldrich.com/technical-documents/articles/reporter-us/reversed-phase-hplc.html.

**TABLE 3.3 The Major Body Buffer Systems**

| Site | Buffer System | Comment |
|------|---------------|---------|
| ISF | Bicarbonate | For metabolic acids |
| | Phosphate | Not important because concentration too low |
| | Protein | Not important because concentration too low |
| Blood | Bicarbonate | Important for metabolic acids |
| | Hemoglobin | Important for carbon dioxide |
| | Plasma protein | Minor buffer |
| | Phosphate | Concentration too low |
| ICF | Proteins | Important buffer |
| | Phosphates | Important buffer |
| Urine | Phosphate | Responsible for most of "Titratable Acidity" |
| | Ammonia | Important—formation of $NH_4^+$ |

*ISF*, interstitial fluid (fluid between cells); *ICF*, intracellular fluid (fluid within cells).
Source: Reproduced from http://www.anaesthesiamcq.com/AcidBaseBook/ab2_2.php.

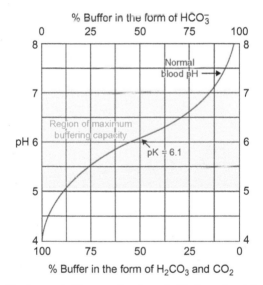

**FIGURE 3.17**  Titration of the carbonate–bicarbonate buffer as a function of pH. *Reproduced from http://www.chemistry.wustl.edu/~edudev/LabTutorials/Buffer/Buffer.html.*

**Carbonic anhydrase** in blood (and in many other tissues) catalyzes a very fast reaction:

$$HCO_3^- + H^+ \rightleftharpoons H_2CO_3 \rightleftharpoons CO_2 + H_2O$$

Water is removed from carbonic acid (derivation of the name of the enzyme). $CO_2$, here, is in the aqueous form but is in the gaseous form in the lungs and is expired by the lungs. *Removal of $CO_2$ by the lungs is part of the bodily control of overall pH* (Fig. 3.17). As $CO_2$ is expired from the body, the pH of blood increases (loss of $CO_2$ is shown by moving from left to right on the bottom *x*-axis of Fig. 3.17). Following the plot of pH, it increases as $CO_2$ is exhaled. If **lactic acid** is being produced by excessive exercise, e.g., lowering the pH of blood, the body increases breathing rate and exhalation (panting) in order to blow off more $CO_2$ and compensates by increasing blood pH. Note that the maximal buffering capacity of the carbonate–bicarbonate buffer is from pH 5.1 to 7.1 with the pK at 6.1. The slightly higher pH of normal blood at pH 7.4 is accomplished by the removal of $CO_2$. As acid is added (Fig. 3.17), the shift (to the left) favors the concentration of carbonic acid and $CO_2$. When a base (alkaline substance) is added to the system, the pH increases (equivalent to removing $CO_2$) the composition shifts toward more bicarbonate ($HCO_3^-$) as shown in the top *x*-axis. In addition to the control of blood pH by the lungs, the kidney is a powerful organ in maintaining pH. When **acidosis** occurs, tubular cells reabsorb more bicarbonate and the collecting duct cells secrete more protons and generate more bicarbonate (Fig. 3.18).

Conversely, when the pH increases in **alkalosis**, the kidney can excrete more bicarbonate by lowering proton secretion from the tubular cells; increased **glutamine metabolism** occurs with the excretion of **ammonia**. The enzyme degrading glutamine is **glutaminase** that catalyzes the reaction:

$$Glutamine + H_2O \rightarrow Glutamate + NH_3^+$$

allowing the excretion of ammonia which is a key mechanism in **renal acid−base regulation**.

In the cell interior, **phosphate buffer** (likely in the mM range) is important in maintaining pH. Dihydrogen phosphate ions ($H_2PO_4^-$) are the acid proton donors ($H^+$) and hydrogen phosphate ions ($HPO_4^{2-}$) are the proton acceptors (bases). These ions are in equilibrium as shown by the reaction in aqueous solution:

$$H_2PO_4^- \rightleftharpoons H^+ + HPO_4^{2-}$$

If there is an acidic reaction in a cell where more protons appear, the reaction moves to the left by **Mass Action (Le Chatelier's Principle)**. If hydroxyl ions ($OH^-$) are generated, they react with $H_2PO_4^-$ to produce $HPO_4^{2-}$ ($+H_2O$) pushing the reaction to the right. As before, the $K_a$ is written:

$$K_a = \frac{[H^+][HPO_4^{2-}]}{[H_2PO_4^-]}$$

And $K_a$ at 25°C is equal to $6.23 \times 10^{-8}$. The value of $[H^+]$ becomes:

$$[H^+] = \frac{K_a[H_2PO_4^-]}{[HPO_4^{2-}]}$$

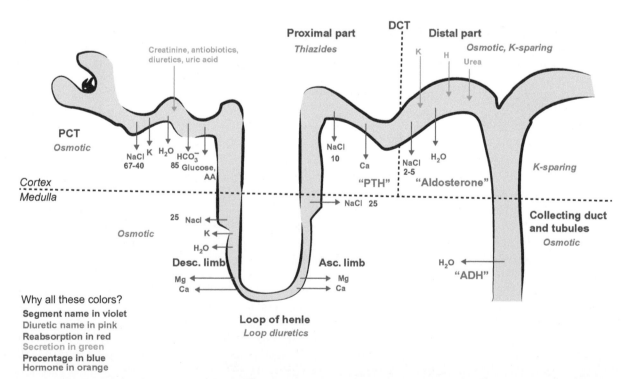

**FIGURE 3.18** Diagram showing the collecting duct and tubules. *Red* arrows indicate reabsorption of the named substances; *green* arrows indicate secretion into the urine of the labeled substances; percentages are given in *blue*; the segment name is in *violet* (*PCT*, proximal collecting tubule; *DCT*, distal collecting tubule); the diuretic name is in *pink*. Note the excretion of protons ($H^+$) in *green* and the reabsorption of bicarbonate ion ($HCO_3^-$) in *red*.

When the values of the two phosphate ions are equivalent, their ratio becomes 1 and the pH is equal to the $pK_a$ which equals 7.21 the greatest buffering capacity of this buffer. This is appropriate for the pH within a cell which ranges from 6.9 to 7.4. Proteins in solution inside a cell have a certain buffering effect in that side chains of amino acid residues contain peptide-$NH_3^+$ and peptide-$COO^-$. These groups can react with protons (H + ):

$$[H^+] + \text{Peptide-COO}^- \rightleftharpoons \text{Peptide-COOH}$$

Likewise, negatively charged ions ($OH^-$) can react with peptide-$NH_3^+$:

$$[OH^-] + \text{Peptide-NH}_3^+ \rightleftharpoons \text{Peptide-NH}_2 + H_2O$$

## RECEPTORS

This is a subject that is related to the effects of AVP on the distal kidney and a subject that occurs in more detail in many of the subsequent chapters. It is introduced broadly because receptors are initial triggers of the mechanisms through which many biologically active substances initiate their subsequent pathways leading to their overall effects on the body.

Interestingly, most receptors are located in the plasma membrane of cells with their ligand-binding domains exposed in the extracellular space to best accommodate a ligand molecule in the blood traveling from the original organ's cell to the extracellular fluid of the target. For the few receptors whose ligand easily can permeate the cell membrane's lipid composition, the receptors for those ligands are located in the aqueous cytoplasm. This is the case for steroidal hormones, e.g., that are able to cross the cell membrane essentially without an energy requirement because of their lipid nature. Otherwise, a membrane receptor for such lipid soluble molecules might have difficulty in concentrating sufficient ligand when most of the steroids would be crossing the membrane randomly in almost any location. Thus, the location of the water environment of the cell in contrast to the membranous lipid environment dictates that a specific ligand that is easily able to cross the cell membrane by virtue of its lipid nature will more economically find its receptor in the aqueous internal region of the cell. Those molecules that are water-soluble cannot diffuse through the cell membrane and will either have receptors in the cell membrane whose binding pockets are exposed to the extracellular space with structures in the cytoplasm or there will be specific channels in the membrane the entrance of which will be exposed to the extracellular space and its exit to the cell interior.

Many types of channels will be encountered later in the text. The channels for the entry of glucose into a cell serve to illustrate this subject. Channels are classified according to the mechanism of transport, the **uniport** being the simplest in that a substance, like glucose is transported in one direction without the movement of ions with it or in opposite directions, although there are channels that accomplish this. There are many glucose uniports named GLUT 1 through GLUT 5, and these will be discussed later in the text. Other transporters are **symports** and **antiports**. Symports transport two different substances in the same direction at the same time, while antiports transport two different substances in opposite directions at the same time.

Receptors are proteins with a binding pocket (active site) that is specific for a given **ligand** (a signal molecule that binds to another molecule, often a target protein, to serve a biological function). Signals to a cell are usually carried from a distant site and are recognized by a specific receptor, generally located in the target cell membrane so that the **binding pocket** of the receptor is available to the extracellular space (so far, there has been no test of the possible role of water in the ligand−receptor interaction compared to enzymatic reactions). The exception to this localization exists in the steroid receptor gene family, whose receptors are located inside the cell either within the cytoplasm or in the cell nucleus (including the thyroid hormone receptor), or in both locations. However, there are small amounts of some steroid-like receptors that appear to be located in the cellular membrane, as well, and these receptors may differ structurally from the cytoplasmic receptors and signal functions different from the major ones that operate in the cytoplasm and the cell nucleus. Examples of ligands that bind to cell membrane receptors are polypeptide hormones, neurotransmitters, viruses, antigens, amino acid-derived hormones, sugars, and others. The distribution of receptors is summarized in Fig. 3.19.

In general, membrane receptors signal other molecules (**second messengers**) inside the cell that trigger signaling pathways culminating in the overall cellular effect of the initial signaling ligand. A typical **seven transmembrane receptor** is the adrenergic receptor that plays a role in sympathetic nervous activity. *The seven transmembrane receptors represent the largest family of cell membrane receptors and they are the G-protein-coupled receptors.* This family is encoded by more than 1000 genes, and the receptors are involved in the regulation of most physiological processes. A cartoon model of the adrenergic receptor is shown in Fig. 3.20.

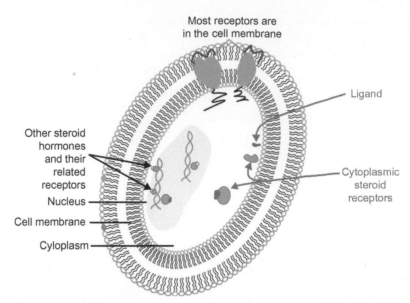

**FIGURE 3.19** Diagram showing that most cellular receptors are located in the plasma membrane with the ligand binding pocket open to the extracellular space. An exception is the steroid receptor gene family that includes the thyroid hormone receptor as well as the receptors for various steroidal hormones that are located inside the cellular cytoplasm, in the nucleus, or in both locations. *Reproduced in part from Figure 1—28 by Litwack, G., 2008. Human Biochemistry and Disease. Academic Press/Elsevier, p. 27.*

Receptors located in the cytoplasm are transported into the nucleus where they bind to specific sites upstream of genes to promote (or repress) their expression. Sometimes, part of the action of receptors is to open (or close) specific **ion channels** indirectly or a receptor may be structurally related to a channel to promote this function. Receptors are the means through which signals outside of the cell are translated to specific pathways within the cell.

## ION CHANNELS

Ion channels are discussed in detail in subsequent chapters but are introduced briefly here. The cell membrane, being primarily lipid in its structure (**Fig. 2.7**), is impermeable to ions which are often small, charged atoms. As with other charged molecules that need to cross the membrane to enter a cell, special structures in the form of specific transmembrane channels have been developed to accomplish this purpose. The same is true for water that does not permeate the lipid cell membrane but water channels, **aquaporins (AQPs)**, fill the requirement for transporting water into cells. The channels are constructed of **glycoproteins** (containing sugars) similar to the **aquaporin channel**, already described (Fig. 3.9). The basic components of an ion channel are shown diagrammatically in Fig. 3.21.

Ion channels regulate the **electrochemical gradient** across a cell membrane. Different ions have different conductance features and the **cell-membrane potential** is a consequence of the flow of these ions and the charges that become separated by the nonconducting cell membrane. Channels are specific for the ion they conduct. For example, calcium channels specifically conduct calcium ion and sodium channels specifically conduct sodium ions, etc. The selectivity filter (3 in Fig. 3.21) selects the ion that is to be transported and the gate at the bottom of the channel (5 in Fig. 3.21) regulates the flow of the ions into the cell and itself may be regulated, e.g., by **phosphorylation—dephosphorylation**. Ion channels often contain one type of subunit protein and several subunits aggregate into a cylindrical conformation that generates a **pore**. There are different mechanisms for opening and closing a channel. A few common models of channels are shown in Fig. 3.22 in open and closed conformations.

More details of ion channels and other types of channels appear later in the text.

## ENZYMATIC PUMPING MECHANISM

There are **energetic pumping mechanisms** in many cell types. The energy of the terminal phosphate of ATP is released by an ATPase and this energy powers the transport of ions. An example of this pump is the sodium—potassium ATPase (**$Na^+$—$K^+$ ATPase**) diagrammed in Fig. 3.23.

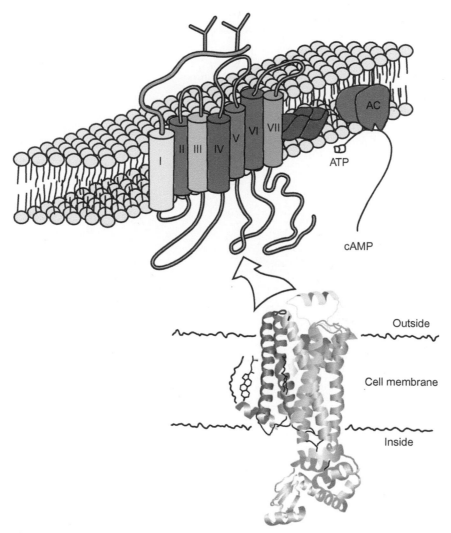

**FIGURE 3.20** Drawing of a membrane adrenergic receptor. The receptor is composed of seven transmembrane subunits and is a G-protein-coupled receptor (the lower part of this figure shows the X-ray crystallographic structure of this receptor and is reproduced from http://www.aps.anl.gov/ Science/Highlights/2007/Images/20071203_Beta2AR.png). Above the lipid membrane is the extracellular space; below the membrane is the aqueous cytoplasmic space. The ligand, epinephrine, binds to the receptor at its upper surface at about the center of the barrel proteins. Also shown are constituent proteins with which the receptor interacts after ligand binding. Alpha- and beta-G -protein subunits are interactive proteins that mediate signaling to adenylate cyclase (AC), which catalyzes the conversion of ATP to cyclic AMP that, subsequently, activates protein kinase A. Activated protein kinase A phosphorylates specific proteins in the cell whose actions potentiate the activities of epinephrine. *Reproduced from http://ardb.bjmu.edu.cn/ image/cover.gif.*

This system would be classified as an **antiporter** (or antiport) because ions are moved in opposite directions. When a channel moves two ions in the same direction, the channel is classified as a **symporter** (or symport). An example of an antiporter is encountered in the cell membrane when potassium ions are pumped into a cell (a cell usually bordering a lumen on the basolateral side) and sodium ions are transported in the opposite direction to the extracellular space and then to the blood. A similar exchange can occur at the lumenal surface (apical) when potassium ions are pumped out of the cell to the lumenal space and sodium ions are pumped into the cell from the lumen. The movement of sodium ions establishes a **sodium-ion gradient** that can be used by other types of transporters to move efficiently molecules like glucose (**$Na^+$-glucose transporter**) and amino acids into the cell. Another function of the $Na^+-K^+$ transporter is to control the **fluid volume** of the cell. Owing to the large number of macromolecules inside the cell, the partial pressure inside tends to be greater than the outside and **water** would be expected to flow into the cell to equalize the partial pressures. This transporter corrects this tendency by moving three sodium ions out of the cell in exchange for two potassium ions and driving water out of the cell. If the cell swells, this transporter is activated in order to move more sodium ions to the outside with water to follow,

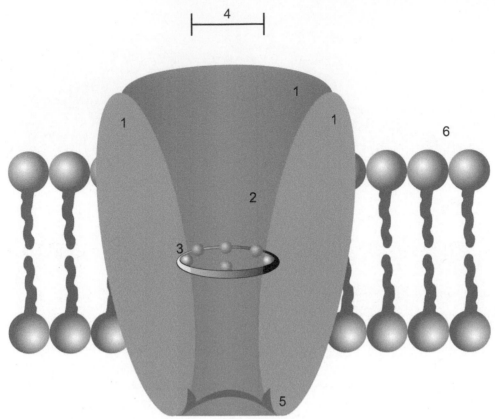

**FIGURE 3.21** Diagram of the basic components of an ion channel. 1 = channel domains, usually 4 per channel; 2 = the outer vestibule; 3 = selectivity filter; 4 = diameter of the selectivity filter; 5 = a phosphorylation site; and 6 = the cell membrane.

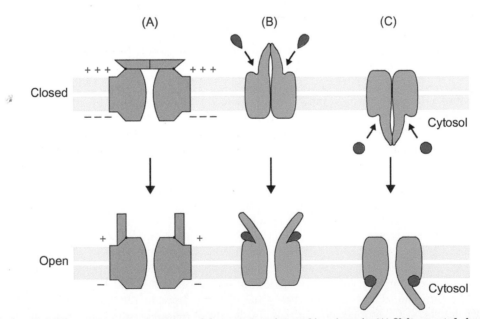

**FIGURE 3.22** Diagrams of closed and open conformations of three common forms of ion channels. (A) **Voltage-gated channel** that may be open depending on the distribution of charges on the membrane. (B) Extracellular **ligand-gated channel** where a specific molecule binds to a specific binding site on an outer subunit. In some cases, this could be a receptor linked to a channel. (C) **Intracellular ligand-gated channel** where the ligand comes from inside the cell and binds to a specific binding site on a channel subunit located on the cytoplasmic domain of the channel. *Reproduced from http://219.221.200.61/ywwy/zbsw(E)/pic/ech5-7.jpg.*

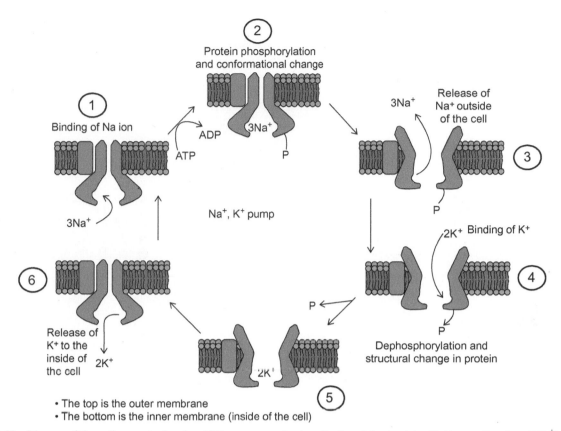

- The top is the outer membrane
- The bottom is the inner membrane (inside of the cell)

**FIGURE 3.23**  Diagram of the sodium–potassium ion ATPase pump mechanism. Starting at the top (step 2), three sodium ions ($3Na^+$) bind to the intracellular side of the pump from the cytosol. This configuration and the affinity of the pump for the sodium ion is made possible by the hydrolysis of ATP in the previous step (step 1). The intracellular channel is phosphorylated as shown in (2) resulting in a conformational change causing the pump to open to the extracellular space (3). Sodium ions are released to the outside (3) because the phosphorylated form of the receptor has a low affinity for $Na^+$. Two extracellular $K^+$ now bind to the opened channel (4) causing dephosphorylation. In this state (5), the pump has a lower affinity for $K^+$ and the $2K^+$ are released into the cytosol and the pump reverts to its initial conformation (closes to the exterior) and again binds $3Na^+$ from the cytosol and the cycle is repeated. *Redrawn in part from http://life.nthu.edu.tw/~b830473/Na-Kpump.gif.*

decreasing the intracellular ionic concentration. Of the many types of pumps inside cells, the **proton pump** in the mitochondrion is important. This pump (Fig. 3.24) is used by **cytochrome oxidase**, shown in three conformations in A, B, and C.

In Fig. 3.24, conformation A illustrates a state in which there is high affinity for a proton ($H^+$) inside the membrane and the proton binds (B). Conformation B is a different conformation and with energy coupled from electron transport (in this particular case), the affinity of the protein for $H^+$ is lowered and the proton is released to the **intermembrane space** in the mitochondrion (conformation C). The protons confined in this space create an **electrochemical potential** that is stored energy for the cell. Transporters will be discussed in more detail in Chapter 14, Human Biochemistry— Metabolism of Fat, Carbohydrate and Nucleic Acids.

## SUMMARY

All of the information in this chapter relates, either directly or indirectly, to the properties, roles and movements of water.

Diabetes *insipidus* is a condition in which there is insufficient arginine vasopressin (AVP) reaching the AVP receptor in the distal kidney or one in which there is insufficient AVP receptor (or, possibly the aquaporins involved in water reabsorption in this kidney location have suffered from a loss of function mutation in the genes from which they derive). The hormone deficiency can be due to a mutation in the gene for **preprovasopressin**. In the presence of sufficient AVP, there can be a mutation in the gene for the AVP receptor that decreases the concentration or function of the AVP receptor in the kidney. Because the AVP-receptor system is responsible for the reabsorption of water from the urine into the bloodstream, there is a loss of large amounts of water from the body through the urine. This condition is

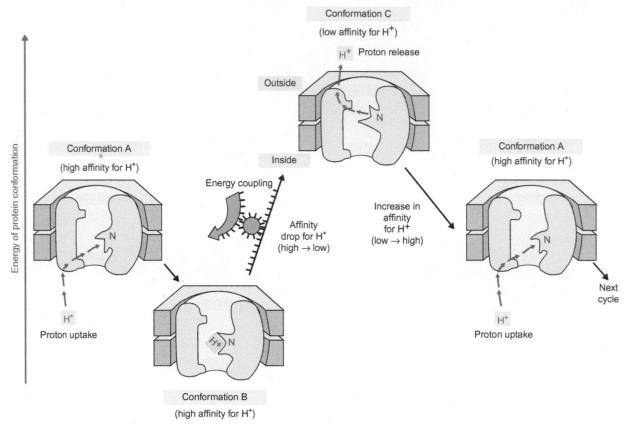

**FIGURE 3.24** Diagram showing a general model for proton (H⁺) pumping from the mitochondrial matrix to the intermembrane space. *Reproduced from http://www.ncbi.nlm.nih.gov/bookshelf/br.fcgi?book=mboc4&part=A2527#A2550.*

different from diabetes *mellitus*, which, when untreated, also results in excessive urination and water loss. In diabetes *mellitus* there is a lack of insulin owing to destruction of the beta cell (type 1) of the pancreas or to reduced insulin secretion or to insulin resistance (type 2). Owing to the deficiency of insulin, there is excessive glucose in the plasma that binds water and is excreted in the urine. Diabetes *insipidus*, however, can be treated with AVP analogs (providing the AVP receptor is suitably expressed) some of which can be taken orally (e.g., desmopressin). Such AVP analogs are somewhat resistant to the action of digestive proteases.

Basic information is presented on the water molecule. Because the plasma membrane of a cell is primarily lipid in nature, water dissolves poorly in the membrane and a special mechanism has evolved to import water in the form of a water channel, named **aquaporins (AQPs)**. AQP structure and function is described and these channels are present in the membranes of different cell types. pH is discussed and defined. The role of buffers in maintaining pH values essential for survival is described. The important buffers in the body are the **carbonate−bicarbonate buffer** in blood, maintaining the pH of blood between 7.35 and 7.45. Inside the cell, **phosphate buffer** maintains the intracellular pH along with dissolved proteins that confer some buffering effect. Quantification of buffer systems involves the **Henderson−Hasselbalch equation** that is derived. The role of the enzyme, **carbonic anhydrase**, catalyzing the removal of water from carbonic acid and liberating carbon dioxide is described. Carbon dioxide liberated by this enzyme can be removed from the overall buffer system by exhalation from the lungs. A graph of changes in carbonic acid, carbon dioxide and bicarbonate describes the functioning of the main buffer in blood. In addition to the lung, the kidneys play a role in the excretion of bicarbonate and protons. The phosphate buffer inside cells has a pKa of 7.21 insuring a buffer capacity in the intracellular medium between pH 6.9 and 7.4.

The general features of receptors include their location within the cell membrane except for the steroid receptor gene family of receptors that reside in the cytoplasm or nucleus or in both compartments (the thyroid hormone receptor is an example of a nuclear receptor).

General aspects of ion channels are reviewed using a model to show their domains including the number of subunits constituting the channel, the outer vestibule, the selectivity filter and its diameter as well as the location of a phosphorylation site. The electrochemical gradient is associated with the functioning of the channel. General types of channels include the voltage-gated channel and the exterior and interior ligand-gated channels. The sodium ion—potassium ion ATPase is an example of an enzymatic pumping channel and its mechanism is described as well as the proton pump in the mitochondrial matrix.

## SUGGESTED READING

### Literature

Behr, R., et al., 2004. Mild nephrogenic diabetes *insipidus* caused by Foxa1 deficiency". J. Biol. Chem. 279, 41936—41941.

Burykin, A., Warshel, A., 2003. What really prevents proton transport through aquaporin? Charge self-energy versus proton wire proposals. Biophys. J. 85, 3696—3706.

Carrillo, D.R., et al., 2014. Crystallization and preliminary crystallographic analysis of human aquaporin 1 at a resolution of 3.28 Ao. Acta Cryst. 70, 1657—1663.

Codomiu-Hernandez, E., Kusalik, P.G., 2013. Probing the mechanisms of proton transfer in liquid water. Proc. Natl. Acad. Sci. 110, 13697—13698.

De Levie, R., 2001. The Henderson—Hasselbalch equation: its history and limitations. J. Chem. Educ. 78, 1499—1503.

Dielmann-Gessner, J., et al., 2014. Enzymatic turnover of macromolecules generates long-lasting protein—water coupled motions beyond reaction steady state. Proc. Natl. Acad. Sci. 111, 17857—17862.

Eriksson, U.K., et al., 2013. Subangstrom resolution X-ray structure details aquaporin-water interactions. Science. 340, 1346—1349.

Eubanks, S., et al., 2001. Effects of diabetes *insipidus* mutations on neurophysin folding and function. J. Biol. Chem. 296, 29671—29680.

Frick, A., et al., 2014. X-ray structure of human aquaporin 2 and its implications for nephrogenic diabetes *insipidus* and trafficking. Proc. Natl. Acad. Sci. 111, 6305—6310.

Gomori, G., 1955. Preparation of buffers for use in enzyme studies. Methods Enzymol. 1, 138—146.

Hassanali, A., Gilberty, F., Parrinello, M., 2013. Proton transfer through the water gossamer. Proc. Natl. Acad. Sci. 110, 13723—13728.

Levin, M.H., Haggie, P.M., Vetrivel, L., Verkman, A.S., 2001. Diffusion in the endoplasmic reticulum of an aquaporin-2 mutant causing human nephrogenic diabetes *insipidus*. J. Biol. Chem. 276, 21331—21336.

Wang, J., Breslow, E., Sykes, B.D., 1996. Differential binding of desmopressin and vasopressin to neurophysin II. J. Biol. Chem. 271, 31354—31359.

### Books

Beynon, R.J., Easterby, J.S., 1996. Buffer Solutions. Oxford University Press.

Hille, B., 2001. Ion Channels of Excitable Membranes. third ed. Sinauer.

Ikeda, M., Matsuzaki, T., 2015. Regulation of aquaporins by vasopressin in the kidney, Hormones and Transport Systems. Vitamins and Hormones, vol. 98. Academic Press/Elsevier.

Litwack, G. (Ed.), 2015. Hormones and Transport Systems. Vitamins and Hormones, vol. 98. Academic Press/Elsevier.

Nicola, J.P., Carrasco, N., Masini-Repiso, A.M., 2015. Expression and regulation of the $Na^+/I^-$ symporter in the intestine, Hormones and Transport Systems. Vitamins and Hormones, vol. 98. Academic Press/Elsevier.

Norman, A.W., Litwack, G., 1997. Hormones. second ed. Academic Press.

Schweitzer, G.K., Pesterfield, L.L., 2010. The Aqueous Chemistry of the Elements. Oxford University Press.

Chapter 4

# Proteins

## PRION DISEASE, A DISEASE OF PROTEIN CONFORMATIONAL CHANGE

The **prion diseases** in man are **Creutzfeldt−Jakob, Gerstmann−Straussler−Scheinker syndrome, fatal familial insomnia, kuru**, and **Alpers syndrome**. Creutzfeldt−Jakob disease is of interest, especially in view of the outbreak in Britain of the **mad cow disease (bovine spongiform encephalopathy, BSE)**. Other important prion diseases of animals are chronic wasting disease in mule deer and elk and scrapie in sheep. These diseases can be contracted by ingestion of **diseased brain** or **lymphoid tissue** even if cooked. The brain is the major site of the pathology of this disease and the major concentration of the **diseased form of the prion protein (scrapie, $PrP^{Sc}$)**. The first description of this disease was in sheep, the scrapie disease. Consequently, the diseased conformation of the normal cellular prion ($PrP^c$) is designated $PrP^{Sc}$. The function of the normal prion protein is still not understood, although there are characteristics of the protein (discussed below) that suggest a normal cellular function. The ingested $PrP^{Sc}$ can be absorbed through the intestine at the **Peyer's patch** (Fig. 4.1A and B).

The associated mucosal lymphoid tissue may be involved in the transport of $PrP^{Sc}$. The scrapie prion can replicate at the spleen or lymphoid nodes, and it can access the nervous tissue and ultimately the spinal cord and brain because there are neural connections to lymphoid sites. Tribes in the Fore Highlands or Papua, New Guinea ingested, ritually, the brains of dead relatives often leading to the prion disease, **kuru**. They were induced to give up this ritual in the 1950s, and the deadly kuru disease disappeared.

**Creutzfeldt−Jakob disease (CJD)** is characterized by loss of motor control leading to paralysis, wasting, dementia, and death sometimes as the result of pneumonia. The clinical and pathologic characteristics of this disease are summarized in Table 4.1.

CJD is of three types: **sporadic form (sCJD), variant form (vCJD)**, or **familial form (fCJD)**. The first two are the most common; vCJD occurs in younger people and has been recently recognized. It seems to be evolving more rapidly than sCJD through a "Ghost prion" gene. The familial form results from inheritance of a mutated gene for $PrP^c$ occurring in about 10%−15% of cases. CJD occurs in about one per million persons each year, affecting individuals between the ages of 45 to 75 years and appearing often at ages 60 to 65 years shown in the first entry in Table 4.1. The prion gene is located on **human chromosome 20** (Fig. 4.2).

The ***PRND* gene** encodes a protein (sometimes called "Doppel") that is biochemically and structurally related to the prion gene, and it is a **glycophosphatidylinositol-anchored** (to the cell membrane) **protein**. In some cells (observed in B lymphocytes, granulocytes, and dendritic cells), the *PRND* mRNA for the protein prnd is expressed with the $PrP^C$. The protein, prnd, can coordinate the binding of $Cu^{2+}$ with high affinity. Mutations and polymorphisms are shown in a cartoon of the **human prion gene** in Fig. 4.3.

**Familial CJD** is inherited from a gene with many possible mutations. It is not clear how these mutations cause CJD. There are two suggestions: either mutations make individuals carrying these mutated genes more susceptible to the prions in the environment or the mutations might produce spontaneous generation of transmissible prions. Some mutations could lead to a structural change or facilitate a structural change in the protein to the scrapie form. The three-dimensional structures of $PrP^c$ and the speculative structure $PrP^{Sc}$ are shown in Fig. 4.4.

Interestingly, the $PrP^c$ is highly conserved among species and is a **copper-binding protein**. Knockout of $PrP^c$ appears to cause **demyelination of Schwann cells**, and the normal cellular protein may have an effect on **hippocampal long-term potentiation** so that the protein could function in long-term memory. The normal protein binds copper ions with multiple low-affinity sites at the N-terminus and a high-affinity site at the center of the protein (Fig. 4.5).

It is likely that $PrP^c$ functions in copper metabolism, possibly as a **copper storage** protein. It also can bind zinc and manganese ions. $PrP^{Sc}$ may lose much of the copper-binding activity and the resulting excess copper in nervous tissue

Human Biochemistry. DOI: http://dx.doi.org/10.1016/B978-0-12-383864-3.00004-1

(A)

Intestinal lumen

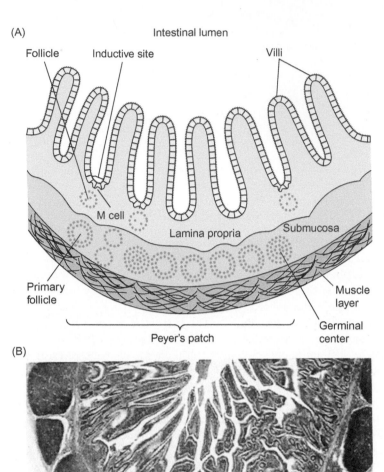

Follicle    Inductive site                      Villi

M cell

Lamina propria          Submucosa

Primary follicle

Muscle layer

Germinal center

Peyer's patch

(B)

**FIGURE 4.1** (A) Diagram of the location of the Peyer's patch in the intestine. (B) Stained intestinal tissue in a microscopic view showing the Peyer's patch in the ileum. Arrows point to germinal centers. *(A) Reproduced from http://www.ugr.es/~oncoterm/csdata/PEYERS-PATCH.html. (B) Reproduced from http://www.ugr.es/~oncoterm/csdata/PEYERS-PATCH.html.*

would contribute to an increase in **reactive oxygen species** that could produce many of the effects of Creutzfeldt-Jakob disease. In addition, the normal prion may have **superoxide dismutase (SOD) activity**. The substrate for SOD is $O_2^{\bullet-}$. SOD catalyzes the reaction: $2O_2^{\bullet-} + 2H^+ \rightarrow O_2 + H_2O_2$. The oxygen radical, $O_2^{\bullet-}$, for example, is derived from the xanthine oxidase reaction: xanthine $+ 2O_2 \rightarrow$ uric acid $+ H_2O_2 + 2O_2^{\bullet-}$. SOD can be coupled to the product to generate $O_2 + H_2O_2$; hydrogen peroxide can be converted to $H_2O + {}_{1/2}O_2$ by the activity of peroxidase or catalase.

CJD is a disease of aggregation of the $PrP^{Sc}$ protein as shown in Fig. 4.6.

The interaction of the normal $PrP^c$ with the disease form, $PrP^{Sc}$ generates a conformational change in the normal partner to produce $PrP^{Sc}-PrP^{Sc}$ that, in turn, incites a chain reaction to form filaments and their aggregates that are presumed to cause the damage in the brain. It is possible that the conformational change of the $PrP^c$ into the $PrP^{Sc}$ form would be catalyzed by an RNA that could perform a function like a chaperone protein. To increase the complexity of this disease, it appears that there are different strains of the $PrP^{Sc}$, so that the CJD can take on different characteristics (incubation period, severity of the disease, patterns of the $PrP^{Sc}$ distribution in nervous tissue, etc.). Thus, the characteristics of the variant strain (vCJD) may be different from the classic CJD because the strain of $PrP^{Sc}$ might be different.

The significance of this disease for this chapter is that a serious, fatal human disease can develop from an alteration in the three-dimensional structure of a normal cellular protein.

**TABLE 4.1** Clinical and Pathological Characteristics of Creutzfeldt–Jakob Disease

| Characteristic | Classic CJD | Variant CJD |
|---|---|---|
| Median age at death | 68 years | 28 years |
| Median duration of illness | 4–5 months | 13–14 months |
| Clinical signs and symptoms | Dementia; early neurologic signs | Prominent psychiatric/behavioral symptoms; painful dysesthesias; delayed neurologic signs |
| Periodic sharp waves on electroencephalogram | Often present | Often absent |
| Signal hyperintensity in the caudate nucleus and putamen on diffusion-weighted and FLAIR MRI[a] | Often present | Often absent |
| "Pulvinar sign" on MRI[b] | Not reported | Present in >75% of cases |
| Immunohistochemical analysis of brain tissue | Variable accumulation | Marked accumulation of protease-resistant prion protein |
| Presence of agent in lymphoid tissue | Not readily detected | Readily detected |
| Increased glycoform ratio on immunoblot analysis of protease-resistant prion protein | Not reported | Marked accumulation of protease-resistant prion protein |
| Presence of amyloid plaques in brain tissue | May be present | May be present |

[a]FLAIR MRI = fluid attenuation inversion recovery MRI to remove effects of fluid from the resulting image.
[b]Pulvinar Sign = hyperintensity of the posterior nuclei of the thalamus.
Reproduced from http://en.wkipedia.org/wiki/Creutzfeldt-Jakob_disease.

Proteins are encoded by messenger RNAs (mRNAs) that are, in turn, encoded in the gene. The progression of information in generating a specific protein is gene (DNA) to messenger RNA to protein. Proteins are composed of amino acids joined together by peptide bonds, a process which takes place in the cellular cytoplasm on ribosomes, the machines of protein synthesis. At the **ribosome**, the protein is assembled by the addition of amino acids directed by mRNA and specific **transfer RNAs** (tRNAs). This discussion begins with the building blocks of proteins, the amino acids and their structures.

## AMINO ACIDS

### Amino-Acid-Related Diseases

There are important **genetic diseases** involving deficiencies of enzymes needed to metabolize amino acids. The more important diseases are **phenylketonuria (PKU)**, **alkaptonuria (AKU)**, **Maple Syrup urine disease (MSUD)**, **tyrosinemia**, and **homocystinuria**. Phenylketonuria occurs in infants (about 1 in 10,000) who are unable to convert phenylalanine to tyrosine owing to a deficiency of the enzyme **phenylalanine hydroxylase**. Consequently, phenylalanine level rises in the blood and is toxic to the brain. These individuals can survive if phenylalanine is restricted from the diet for life. AKU patients excrete urine that turns black when exposed to air. This is a relatively rare disease where 1 person in 100,000 at the most will have it but in Slovakia, 1 person in 19,000 will have Alkaptonuria. These individuals usually develop arthritis of the spine and large joints. This condition is the result of a deficiency of the enzyme **homogentisate 1,2-dioxygenase**, the enzyme that catalyzes the conversion of homogentisate (or homogentisic acid) to 4-maleylacetoacetate (or 4-maleylacetoacetic acid). Maple Syrup urine disease (MSUD) follows from a deficiency of the **branched chain α-keto acid dehydrogenase** needed to metabolize **branched chain amino acids** (**leucine, isoleucine, and valine**). Among Ashkenazi Jews, 1 person in 40,000 will have this disease; worldwide, 1 person in 185,000 will have it but among Older Order Menonites, 1 in 380 will have this disease. This enzyme deficiency leads to the buildup of branched chain amino acids in the blood, and these are toxic to the brain and other organs. Infants are generally screened for this disorder by a blood test for **alloisoleucine** (leucine forms an intermediate and the subsequent transamination generates alloisoleucine). Initial treatment can involve dialysis of the blood. The untreated disease can lead to death. Tyrosinemia is characterized by a lack of enzymes that metabolize tyrosine. There are three types of tyrosinemia:

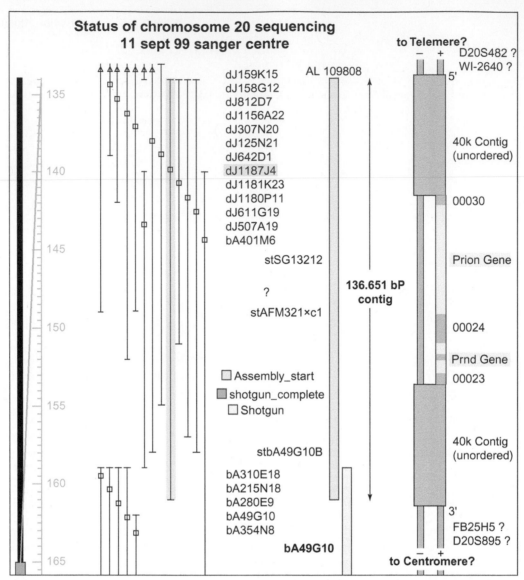

**FIGURE 4.2** Human chromosome 20 showing the prion gene in *green* (*right*) and the *PRND* ghost prion gene about 20 kbp downstream (also in *green*). *Reproduced from http://www.mad-cow.org/chr20_status.gif.*

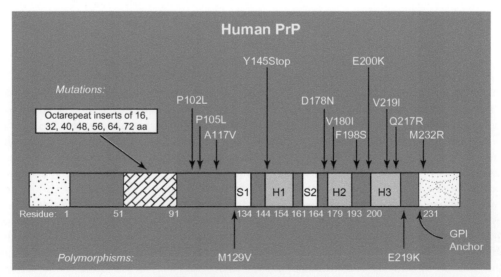

**FIGURE 4.3** Cartoon of the human prion gene showing the locations of eleven or more mutations (single base change) and the locations of two polymorphisms (sequence variations). *Reproduced from http://img.medscape.com/pi/emed/ckb/neurology/1134815-1168941-646.jpg.*

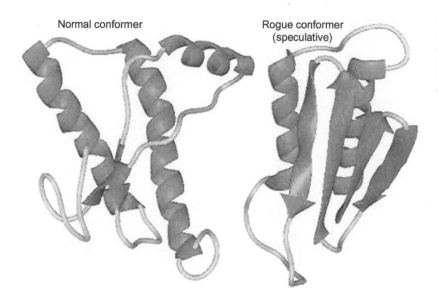

Normal conformer

Rogue conformer
(speculative)

**FIGURE 4.4** The structure of PrP$^c$ (*left*) and the speculative structure of PrP$^{Sc}$ (*right*). PrP$^c$ contains 43% α-helix; PrP$^{Sc}$ has 30% α-helix, and 43% β-sheet structures. As indicated, the structure of PrP$^{Sc}$ is speculative. *Figure reproduced from http://www.stanford.edu/group/virus/prion/normal_rogue.gif.*

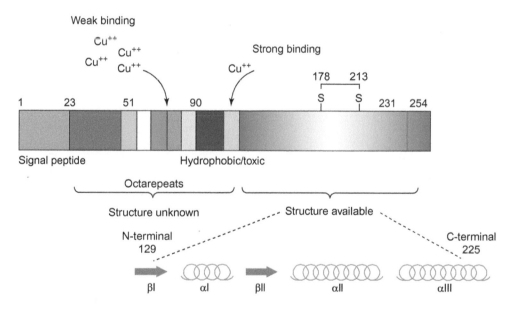

**FIGURE 4.5** Linear cartoon of the normal prion protein, PrP$^c$. There are weak copper-binding sites (octarepeats containing prolines and glycines) in the N-terminus and a strong binding site for copper ion in the center. *Reproduced from G. Litwack,* Human Biochemistry & Disease, *Academic Press/Elsevier, Figure 2–4, page 37, 2008.*

type I, type II, and type III. About 1 child in 100,000 or less has this disease. In Quebec, the incidence can be 1 in 16,000. Type-I tyrosinemia is characterized by a deficiency in **fumarylacetoacetate hydrolase**, the terminal enzyme in the tyrosine catabolic pathway (produces fumarate and acetoacetate from fumarylacetoacetate). Type II is due to a deficiency of **tyrosine aminotransferase**, the initial enzyme in the tyrosine oxidation pathway. Type III, a rare type, is the result of a deficiency of the enzyme **4-hydroxyphenylpyruvate dioxygenase**, the second enzyme in tyrosine catabolism that results in the formation of **homogentisate**. Type-I tyrosinemia affects the liver, kidneys, and nerves and could require liver transplantation. Type I occurs more frequently in children of French Canadian or Scandinavian descent. Type II is less common and can be managed by **restricting tyrosine in the diet** (but this does not solve the symptoms of Type I). Homocystinuria is the result of a deficiency of the enzyme **cystathionine β-synthase**, the first enzyme in the **transsulfuration pathway** catalyzing the conversion of homocysteine (+L-serine) to cystathionine (+water). Although children with this disease are normal at birth, by 3 years of age, the eye lens is dislocated, and they have decreased vision. There follow skeletal deformities, such as curved spine, chest deformities, and elongated limbs and fingers usually with osteoporosis. In addition, there are mental disorders and increased blood clotting. Obviously, just considering these diseases alone, it is of importance to understand the biochemistry of the amino acids.

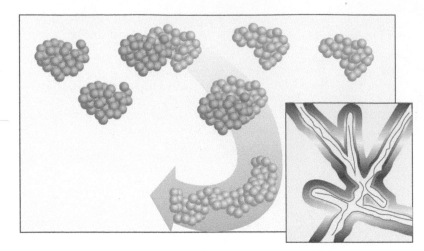

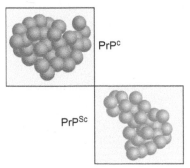

PrPᶜ

PrPˢᶜ

**FIGURE 4.6** Illustration showing how PrPᶜ in the presence of PrPˢᶜ might be converted to a greater number of PrPˢᶜ molecules. The chain reaction results in the formation of fibrils of the PrPˢᶜ molecules that aggregate into fibrils, becomes deposited in the brain, and leads to the symptoms of CJD and death. *Reproduced from G. Litwack*, Human Biochemistry & Disease, *Academic Press/Elsevier, Figure 2–7, page 38, 2008.*

## Biochemistry of the Amino Acids

There are 20 **common amino acids** that comprise the structures of proteins. They can be arranged according to their sizes and net charges in solution as shown in Table 4.2.

The structures of the amino acids are important, and they are shown in Fig. 4.7. For space-filling models (*on right*): nitrogen atom (*blue*); hydrogen atom (*gray*); carbon atom (*green*); oxygen atom (*red*).

The structure of an amino acid can be generalized as Fig. 4.8.

The **essential amino acids** are as follows: valine, leucine, isoleucine, phenylalanine, tryptophan, threonine, methionine, lysine, arginine, and histidine. Arginine and histidine are essential in certain cases. The other amino acids are considered nonessential because they can be synthesized in the body. The **nonessential amino acids** are listed as follows together with their precursors in the body: **pyruvic acid** can give rise to alanine; glutamic acid can form arginine; aspartic acid can form asparagine; **oxaloacetic acid** can form aspartic acid; **alpha-ketoglutarate** (or alpha-ketoglutaric acid) can form glutamic acid which can form glutamine and proline; **serine and threonine** can form glycine; **glucose** can form serine; and phenylalanine can form tyrosine. Some amino acids, once they are in the body, can be converted to other amino acids (e.g., methionine can be converted to homocysteine, and it can be converted to cysteine); phenylalanine can be converted to tyrosine, and arginine is converted to ornithine and citrulline (amino acids, unlisted above, in the urea cycle). These interconversions will be covered in the chapter on metabolism. Selenocysteine, hydroxyproline, lanthionine, 2-aminoisobutyric acid, gamma-aminobutyric acid (GABA), and some others are not common amino acids but occur as metabolites or serve specific functions in the body.

Certain **protein food sources** are **deficient in amino acids**. Wheat and rice are **lysine deficient**; maize is deficient in lysine and tryptophan, and legumes are deficient in tryptophan and methionine or cysteine. Most tissue foods of animal origin are complete in terms of essential amino acid supply. Some plant sources are adequate but most are inadequate in certain amino acids. Balanced protein nutrition is important to obtain a needed supply of

**TABLE 4.2** [a]Names, Groupings by Size and Charge and abbreviations of Amino Acids

| Amino Acid | Molecular Weight | Abbreviation | Letter Name |
|---|---|---|---|
| **Small, Neutral Amino Acids** | | | |
| Lycine[a,c] | 57.05 | Gly | G |
| Alanine[a] | 71.09 | Ala | A |
| Serine[b] | 87.08 | Ser | S |
| Threonine[b] | 101.11 | Thr | T |
| Cysteine[b] | 103.15 | Cys | C |
| **Hydrophobic Amino Acids** | | | |
| **Branched Chain Amino Acids** | | | |
| Valine[a,c] | 99.14 | Val | V |
| Leucine[a,c] | 113.16 | Leu | L |
| Isoleucine[a,c] | 113.16 | Ile | I |
| **Other Hydrophobic Amino Acids** | | | |
| Methionine[b] | 131.19 | Met | M |
| Proline[a,c] | 97.12 | Pro | P |
| **Aromatic Amino Acids** | | | |
| Phenylalanine[c] | 147.18 | Phe | F |
| Tyrosine | 163.18 | Tyr | Y |
| Tryptophan[b] | 186.21 | Trp | W |
| **Carboxylated Acidic Amino Acids** | | | |
| Aspartic acid | 115.09 | Asp | D |
| Glutamic acid | 129.12 | Glu | E |
| **Amino/Amide-Containing Amino Acids** | | | |
| Asparagine[b] | 114.11 | Asn | N |
| Glutamine[b] | 128.14 | Gln | Q |
| **Basic Amino Acids** | | | |
| Histidine[b] | 137.14 | His | H |
| Lysine | 128.17 | Lys | K |
| Arginine | 156.19 | Arg | R |

[a]Amino acids with hydrophobic side chains.
[b]Amino acids that can participate in hydrogen bonding.
[c]Amino acids, having hydrophobic character, found in internal environment of proteins. Table reproduced in part from Litwack, G., Human Biochemistry & Disease, Academic Press/Elsevier, p. 44, 2008

essential amino acids. The requirements of infants and growing children are substantially greater than those for mature adults.

The natural amino acids in the body occur as L-forms. There are also D-forms that occur in lower organisms. These are known chiral forms and the property of bending light either to the left (L-, or *levo*) or to the right (D-, or *dextro*) is **chirality** as applied to amino acids. The L- and D-forms of an amino acid are shown in Fig. 4.9.

An **asymmetric carbon atom** is one that has a different substituent on each of its four bonds. The maximal number of stereoisomers ($n$) relates to the number of asymmetric carbons in a molecule. In amino acids, there is only one asymmetric carbon, so the maximal number of stereoisomers is $2^n$, or $2^1$ or 2 (in a molecule that has 2 asymmetric carbons,

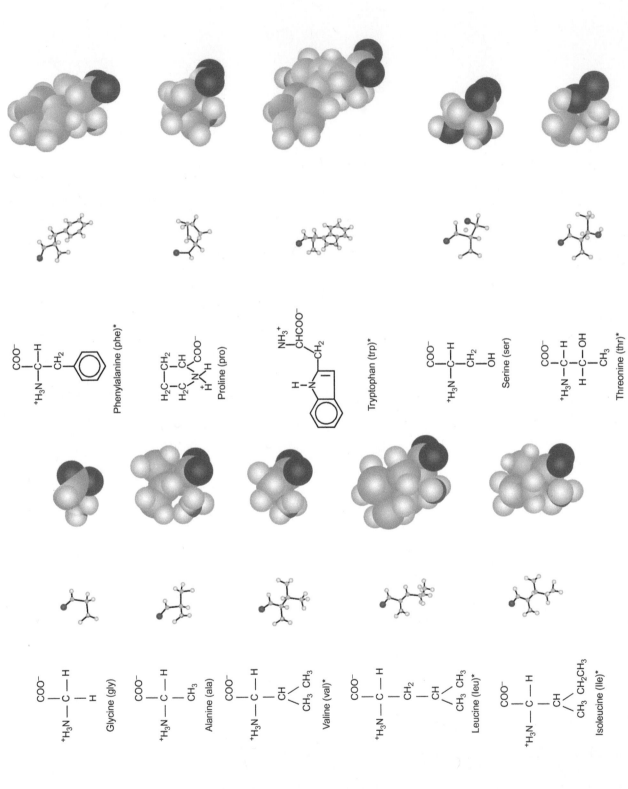

**FIGURE 4.7** The 20 common amino acids. Essential amino acids that are not produced in the body or are only partially produced are denoted by an asterisk after their names. *Reproduced from Litwack, G., Human Biochemistry and Disease, Academic Press/Elsevier. page. 40, 2008.*

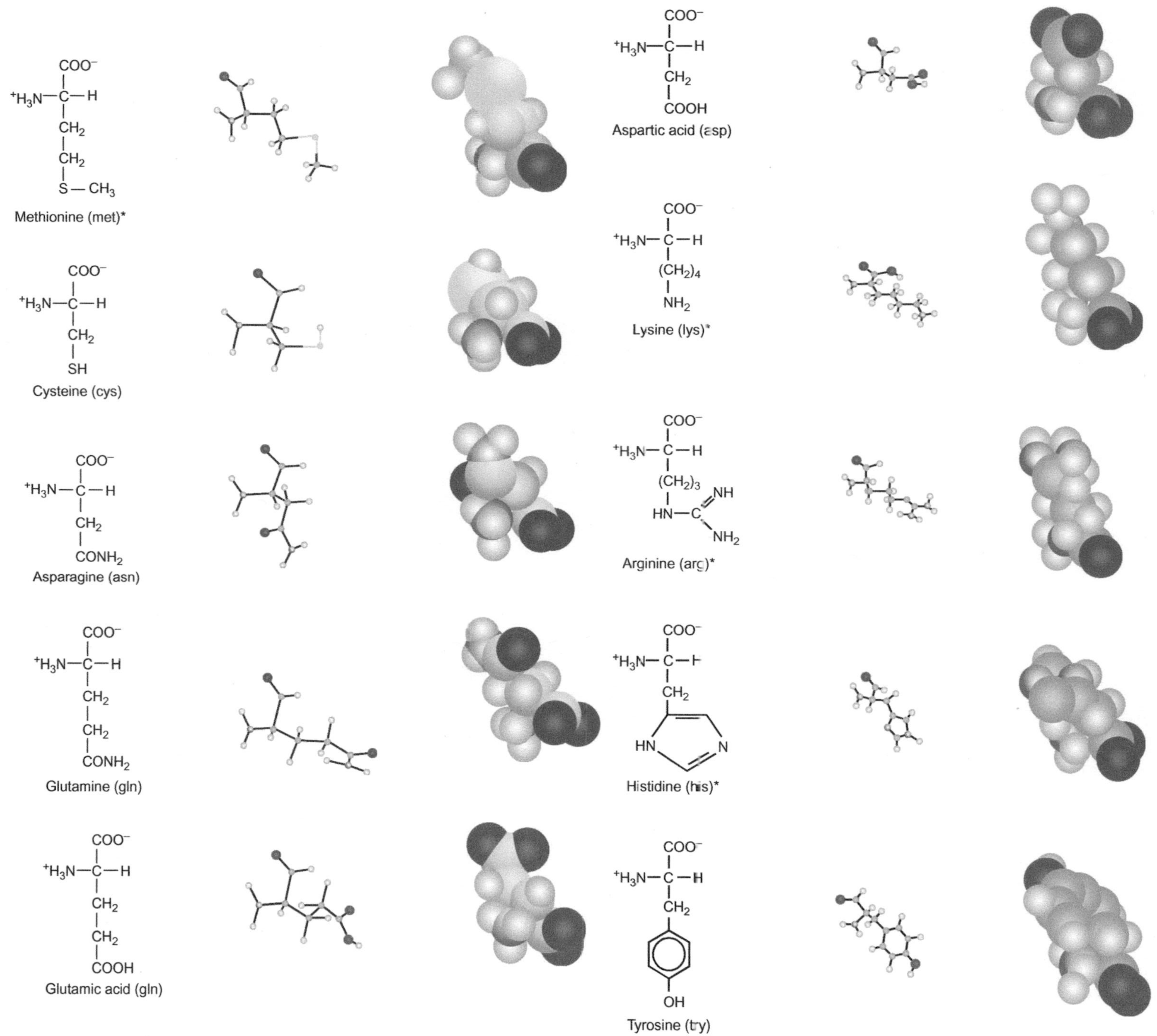

Aspartic acid (asp)

$COO^-$
$^+H_3N-C-H$
$CH_2$
$COOH$

Lysine (lys)*

$COO^-$
$^+H_3N-C-H$
$(CH_2)_4$
$NH_2$

Arginine (arg)*

$COO^-$
$^+H_3N-C-H$
$(CH_2)_3$
$HN-C$
$NH$
$NH_2$

Histidine (his)*

$COO^-$
$^+H_3N-C-H$
$CH_2$

Tyrosine (try)

$COO^-$
$^+H_3N-C-H$
$CH_2$
$OH$

Methionine (met)*

$COO^-$
$^+H_3N-C-H$
$CH_2$
$CH_2$
$S-CH_3$

Cysteine (cys)

$COO^-$
$^+H_3N-C-H$
$CH_2$
$SH$

Asparagine (asn)

$COO^-$
$^+H_3N-C-H$
$CH_2$
$CONH_2$

Glutamine (gln)

$COO^-$
$^+H_3N-C-H$
$CH_2$
$CH_2$
$CONH_2$

Glutamic acid (gln)

$COO^-$
$^+H_3N-C-H$
$CH_2$
$CH_2$
$COOH$

FIGURE 4.7 (Continued)

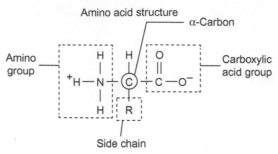

Amino acid structure

α-Carbon

FIGURE 4.8 A model amino acid showing the ionizable groups (−NH$_2$ and −COOH) that become −NH$_3^+$ and −COO$^-$ when ionized, a variable methyl group in the chain (−CH$_2$) and a side chain (R), if it is present.

(A)

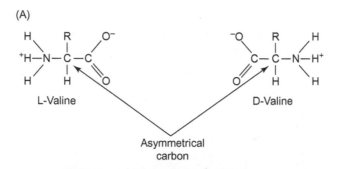

(B)

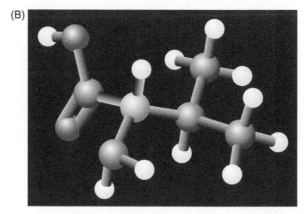

FIGURE 4.9 (A) Pictured are the L- and D-forms of the amino acid **valine**. In the L-form, the alpha-amino group is on the left side of the structure, and in the D-form, the alpha-amino group is on the right side of the structure. The L- form is the **mirror image** of the D-form and vice versa. It is possible to rotate the amino group in this way because the alpha carbon is **asymmetric** (the substituents on each of the four carbon bonds are unique). (B) Model of L-valine. The alpha carbon is highlighted in *green* and emphasizes the tetrahedral character of the alpha carbon. Oxygen is in *red*; hydrogen in *white*; carbon in *gray*, except for alpha carbon; nitrogen in *blue*. *The model is reproduced from http://www.chemeddl.org/resources/models360/files/ 6287/L-valine-jmol.jpeg.*

the maximal number of stereoisomers would be $2^2$ or 4, etc.). Exceptionally, however, isoleucine and threonine have chiral carbon atoms in their side chains. In the case of amino acids, such as valine, the two possibilities from rotation about the alpha carbon are shown in Fig. 4.9, the L- and D-forms. The L-forms are the so-called natural forms; however, D-amino acids are parts of structures (e.g., cell walls) of lower organisms. An exception is the amino acid, glycine:

$$+H_3N \underset{|}{\overset{|}{-}} \overset{H}{\underset{H}{C}} - COO^-$$

Glycine has two hydrogens (identical atoms) as substituents of the alpha-carbon; therefore, only one form of glycine is possible as the alpha-carbon is not asymmetric. L-Forms of amino acids can be converted to D-forms by the catalytic action of the enzyme, **racemase** (from lower organisms); **serine racemase**, however, is present in the human allowing for interconversions between the L- and D-forms. This conversion also can occur spontaneously but at a very slow rate. D-Serine plays a role in the development of the nervous system and may be a cosubstrate agonist for the glycine-binding site and regulator of the NMDA (*N*-methyl-D-aspartate) excitatory glutamate receptor. Aside from the primary role for amino acids as constituents of proteins, certain amino acids give rise to molecules in the body that have specific functions. Examples of these are as follows: the amino acids glutamine, aspartate, and glycine are precursors of **purines** or **pyrimidines**; **nitric oxide**, a biologically important molecule, derives from **arginine**; **heme porphyrins** have a precursor in glycine, and tryptophan can be converted to the neurotransmitter, **serotonin**.

## Amino Acids Have Two or More Potential Charges

Amino acids, typified by glycine in this example, have two potential charges deriving from the alpha-amino and alpha-carboxyl groups (as well as additional charges from ionizable side chains):

$$H_2N-CH_2-COOH + H^+(\text{acid solution}) \rightleftharpoons \, ^+H_3N-CH_2-COOH$$

and

$$H_2N-CH_2-COOH + OH^-(\text{alkaline solution}) \rightleftharpoons H_2N-CH_2-COO^- + H_2O$$

and in approximately neutral solution:

$$^+H_3N-CH_2-COO^-$$

In the first case, the alpha-amino group becomes protonated in acid solution to form a positively charged amino group and in the second case, in the presence of an alkaline solution, the alpha carboxyl becomes deprotonated to form the negatively charged carboxyl group. More complex amino acids, such as lysine or arginine, have side chains with ionizable groups that have an extra amino group. Histidine has an extra secondary amine. Glutamic and aspartic acids have an extra carboxyl group. Thiols and hydroxyls, such as those on cysteine ($-SH$) or serine or threonine ($-OH$), can be deprotonated to produce a negative charge. In **peptide bond** formation (involving the internal amino acid carboxyl group of one amino acid and the internal amino acid amino group of a second amino acid), the groups around the alpha carbons no longer can be charged; charges only will appear in the peptide from the terminal carboxyl and amino groups and from ionizable side chains.

For example, in polyglycine:

$$^+H_3N-CH_2-\overset{\displaystyle O}{\overset{\displaystyle \|}{C}}-O-(NH-CH_2-\overset{\displaystyle O}{\overset{\displaystyle \|}{C}}-O)_n-NH-CH_2-COO^-$$

## Numbering of Carbons in Amino Acids

The alpha carbon has already been described. Depending on the complexity of the amino acid, the numbering starts with the alpha carbon and proceeds: beta, gamma, delta, etc. as shown for lysine in Fig. 4.10.

## Charge State

The charge state of an ionizable group can be determined if the $pK_a$ is known using the **Henderson–Hasselbalch equation** (Chapter 3, Introductory discussion on water, pH, buffers and general features of receptors, channels and pumps): $pH = pK_a + \log[A^-]/[HA]$. If the pH of the solution is two pH units above the $pK_a$, the ionizable group will be almost completely deprotonated. When the pH of the solution is two pH units below the $pK_a$, the ionizable group will be almost completely protonated. *When the pH = pK$_a$, the* **ionizable group** *will be half protonated.* Then, **buffering capacity** of a compound with more than one ionizable group will be effective within two pH values of the $pK_a$ in which

Human Biochemistry

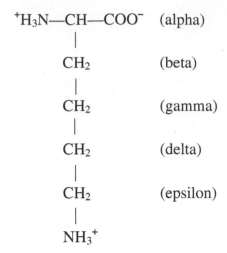

FIGURE 4.10  Structure of L-lysine with the carbons numbered using Greek letters.

TABLE 4.3  p$K_a$ Values for Amino Acid Side Chain Ionizable Groups and Effects on Charge with Protonation or Deprotonation

| Side Chain | p$K_a$ | Amino Acids | Protonated Charge | Deprotonated Charge |
|---|---|---|---|---|
| Carboxyl | −4 | Glu, Asp | 0 | − |
| Amino | 10.5 | Lys | + | 0 |
| Guanido | 12.5 | Arg | + | 0 |
| Thiol | 8.5 | Cys | 0 | − |
| Phenol | 10.5 | Tyr | 0 | − |
| Imidazole | 6 | His | + | 0 |
| Hydroxyl | −13 | Ser, Thr | 0 | |

Reproduced from http://dwb.uml.edu/Teacher/NSF/c10/c10content.html. Reproduced from Litwack, G., *Human Biochemistry and Disease*, Academic Press/Elsevier, page 47, Table 2−2, 2008.

the compound in solution will resist large changes in the pH. For an amino acid with an ionizable side chain or a peptide containing amino acids with ionizable side chains, the charge of the ionizable group of the side chain must be taken into consideration. The charged groups on amino acids in protein structure are often involved in macromolecular interactions. Thus, a positively charged protein (perhaps involving arginine or lysine residues on the surface of the protein) would be appropriate for interacting with the sugar phosphate backbone of DNA. The p$K_a$ values for ionizable groups in amino acid side chains are presented in Table 4.3.

For example, when glycine is protonated ($-NH_2 \rightarrow -NH_3^+$), the amino group has a +1 charge, and the carboxyl group (being uncharged) has a 0 charge. When the carboxyl of glycine is charged ($-COOH \rightarrow -COO^-$), the amino group, being unionized, has a 0 charge, and the carboxyl has a −1 charge. The ionizable groups on amino acid side chains are characterized in the same fashion. The p$K_a$ values for ionizable groups in the side chains of amino acids are summarized in Table 4.4.

Thus, the overall charge on the surface of a protein derives from all of the ionizable groups (terminal amino and carboxyl groups and side chain amino or carboxyl groups). Of the 20 common amino acids, 13 have two ionizable groups, the amino and carboxyl groups substituting the alpha-carbon. In addition, there are seven amino acids that have ionizable groups in their side chains (Table 4.4). The charge of an amino acid is determined by the pH of the solution and the **p$K_a$ values** of the ionizable groups. Titration of an amino acid in solution determines the p$K_a$ values and the value of the **isoelectric point (p$I$)** at which the positive and negative charges of the amino acid are equal so that the net

## TABLE 4.4  p$K_a$ Values of Ionizable Groups in Protein Side Chains

| Group | Acid ⇌ Base + H⁺ | Typical p$K_a$* |
|---|---|---|
| Terminal carboxyl | $-COOH \rightleftharpoons -COO^- + H^+$ | 3.1 |
| Aspartic and glutamic acid | $-COOH \rightleftharpoons -COO + H^+$ | 4.4 |
| Histidine | (see structure) | 6.5 |
| Terminal amino | $-NH_3^+ \rightleftharpoons -NH_2 + H^+$ | 8.0 |
| Cysteine | $-SH \rightleftharpoons -S + H^+$ | 8.5 |
| Tyrosine | (see structure) | 10.0 |
| Lysine | $-NH_3^+ \rightleftharpoons -NH_2 + H^+$ | 10.0 |
| Arginine | (see structure) | 12.0 |

*p$K_a$ values depend on temperature, ionic strength, and the microenvironment of the ionizable group.
Reproduced from Litwack, G., *Human Biochemistry and Disease*, Academic Press/Elsevier, page 47, Figure 2-13, 2008.

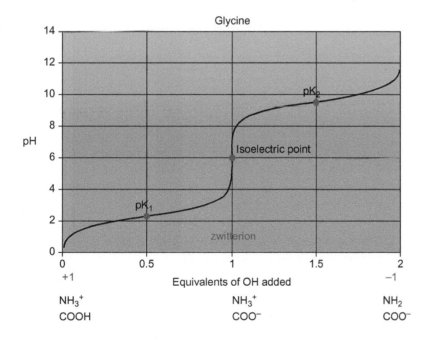

**FIGURE 4.11** Titration of glycine in solution. The titration of glycine with sodium hydroxide proceeds as shown with pH on the ordinate and NaOH (OH⁻) on the abscissa (in equivalents) being added as the titration proceeds. The p$K_a$ values occur at a pH where the amounts of protonated and nonprotonated species are equal. Ionization of the amino acid groups is shown at the bottom. These are the forms predominating at the indicated pH values. The change in charge (+1 or −1) of the amino acid is labeled in *red* (below figure). The **Zwitterion** is the form of the amino acid where the net charge on the molecule is zero. The *blue background* is the region of alkaline pH, and the *red background* is the region of acid pH. *Reproduced from http://molecularsciences.org/files/images/*pI.*gif.*

charge is zero. The same approach determines these values for a protein in solution. When the pH equals the p*I* value, the solute (amino acid or peptide) has the lowest tendency to bind to water (through the dipolar charges of the water molecule; see Chapter 3, Introductory discussion on water, pH, buffers and general features of receptors, channels and pumps), and the p*I* is the best pH condition to precipitate the amino acid or peptide out of solution because of the decreased tendency to interact with water. As an example, the titration of glycine in solution is shown in Fig. 4.11.

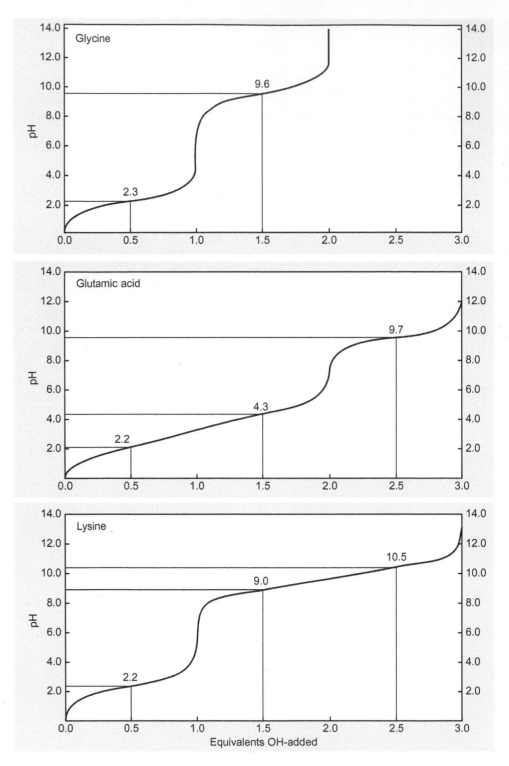

**FIGURE 4.12** Titration of glycine, glutamic acid, and lysine with sodium hydroxide as a function of pH. Titration curves for glycine (*top*), glutamic acid (*middle*), and lysine (*bottom*). Numbers on the curves are the pK values of each ionizable group. Thus, 2.3 is the pKa of the carboxyl group of glycine and 9.6 is the pKa of the amino group of glycine. The pKa falls on the middle part of the titration for each group giving the pH where the charges of the protonated and unprotonated forms are equal. The pI values at the center of the vertical trace where the net charge on the molecule is zero (nothing to titrate). This value can be estimated by connecting the two pKa values for glycine and the intersection on the vertical line will be the pI, about 5.9. pI values for glutamic acid and lysine are about 7 and 5.6. *Reproduced in part from Litwack, G., Human Biochemistry and Disease, Academic Press/Elsevier, page 49, Figure 2−15, 2008.*

In Fig. 4.12, the titrations of glycine, glutamic acid, and lysine are compared.

The p*I* of a protein which would have many ionizable groups in its side chains exposed to the surface could be determined by titration as for amino acids (Fig. 4.7). There are sites on the internet where one can obtain the p*I* of a given protein by supplying the **primary sequence** of that protein. The p*I* value of a protein is useful if one wishes to precipitate the protein from solution or crystallize it.

# Synthesis of Nonessential Amino Acids and Amino Acid Degradation

Nonessential amino acids can be synthesized in the body, whereas essential amino acids must be obtained in the diet. Cells in the body can provide the carbon skeleton of the nonessential amino acids. These carbon skeletons come from intermediates of the glycolytic pathway and from intermediates in the Citric Acid Cycle (Tricarboxylic Acid Cycle or the Kreb's Cycle). The alpha-amino group can be added by enzymatically catalyzed transamination of preexisting amino acids. The enzymes involved are called transaminases or aminotransferases. Thus, nonessential amino acids can be formed from **3-phosphoglycerate** (or 3-phosphoglyceric acid), **pyruvate** (or pyruvic acid), **oxaloacetate** (or oxaloacetic acid), and **α-ketoglutarate** (or α-ketoglutaric acid). These conversions are outlined in Fig. 4.13.

For example, glutamic acid can be formed from α-ketoglutarate by transamination as shown in Fig. 4.14. (The "ate" ending usually refers to the salt of the acid; however, the acid, e.g., glutamic acid, is also referred to as glutamate; thus, α-ketoglutarate is also referred to as α-ketoglutaric acid.)

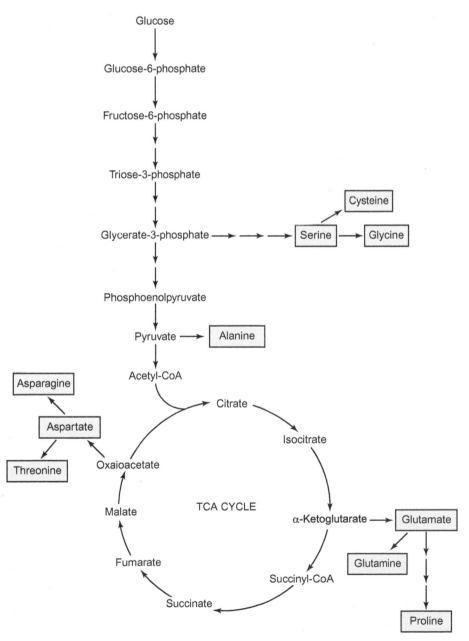

**FIGURE 4.13** Routes of synthesis of nonessential amino acids in the human. Only the starting intermediates in glycolysis and the citric acid cycle are indicated. The specific pathways, enzymes, and cofactors are not shown for simplicity. Clearly the overall metabolism of glucose for energy s, adeno-sine triphosphate (ATP), $CO_2$, and water provides the intermediates for the synthesis of nonessential amino acids. In some cases a suitable ketoacid can be converted to an amino acid with an appropriate enzyme (an aminotransferase). The ketoacid thus provides the carbon skeleton for the amino acid. The requisite amino group is provided by another amino acid or by $NH_4^+$. The metabolic scheme shown here is a somewhat further elaboration of Fig. 1.20. *Reproduced from Litwack, G., Human Biochemistry and Disease, Academic Press/Elsevier, page 51, Figure 2–17, 2008.*

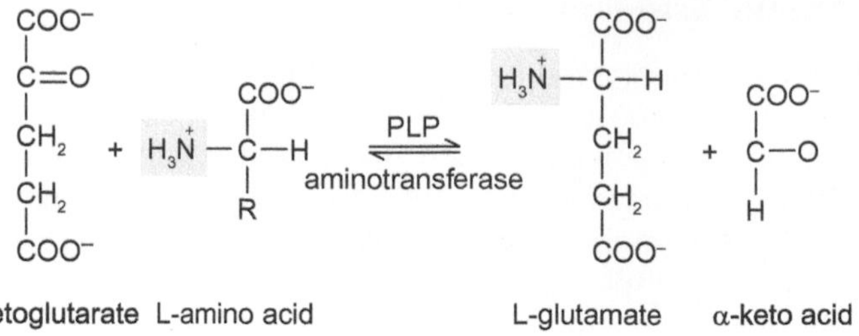

FIGURE 4.14 Reaction of α-ketoglutarate with an amino acid to form glutamate, the amino-derivative of α-ketoglutarate and the keto acid product of the amino-donating amino acid. **Pyridoxal phosphate** (PLP) is the coenzyme for the aminotransferase catalyzing this reaction (see chapter on enzymes).

α-ketoglutarate  L-amino acid          L-glutamate    α-keto acid

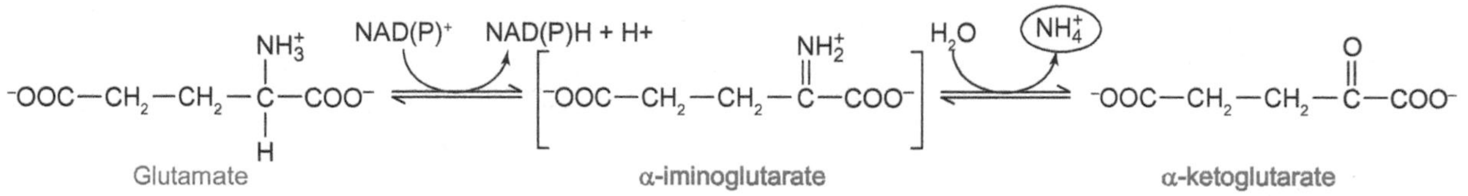

FIGURE 4.15 Oxidative deamination of glutamate, catalyzed by **glutamate dehydrogenase**, to form α-ketoglutarate and ammonium ion.

When there is an excess of amino acids beyond what is required for protein synthesis, amino acids can be degraded to products that can enter the Citric Acid Cycle or be converted to glucose which can be utilized for the production of energy or for storage as fat. **Glucogenic amino acids** are convertible to glucose and **ketogenic amino acids** are those that can enter a metabolic cycle to produce fatty acids and be stored as fat. Also, ketogenic amino acids can be used for **ketone body formation** (acetoacetate, β-hydroxybutyrate, and acetone) in the fasted state. An amino acid also can be degraded by removal of its amino group via **oxidative deamination** to release ammonia in the form of **ammonium ion**, $NH_4^+$. The oxidative deamination of glutamate, for example, is shown in Fig. 4.15.

Ammonium ion generated from this reaction is metabolized to urea through the **urea cycle** ($NH_4^+$ + $HCO_3^-$ + 2ATP → *carbamoyl phosphate* + $P_i$ + 2ADP → + ornithine → *citrulline* + aspartate + ATP → *argininosuccinate* + AMP + $PP_i$ → *arginine* + fumarate + $H_2O$ → ornithine + *urea*). Essential and nonessential amino acids are degraded to products that can be metabolized for energy. All amino acids are able to form glucose (glucogenic) except for leucine and lysine that can form **acetoacetate** and are, thus, uniquely ketogenic. Phenylalanine, tryptophan, tyrosine isoleucine, and threonine can form both glucose and keto acids. The degradative fates of amino acids are shown in Fig. 4.16.

Acetoacetate and **acetyl-Coenzyme A** (**CoA**) can enter the Citric Acid Cycle to produce energy or to be stored as fat.

# PROTEINS

Proteins are composed of amino acids joined through peptide bonds. The **peptide bond** results from an interaction between the alpha-amino group of one amino acid and the alpha-carboxyl group of another amino acid as shown in Fig. 4.17.

The structure of proteins involves four levels of complexity. The first is the **amino acid sequence**. The second is the interactions between amino acid residue side chains based upon their respective location and possible folds; this can be represented in two dimensions. The third level is the three-dimensional structure of the protein often determined by **nuclear magnetic resonance** (**NMR**), if the polypeptide is small enough, or by **X-ray diffraction** analysis of the crystallized protein. The fourth level is the **multimeric form** of the protein if its active structure consists of more than one subunit. This would not obtain for a monomeric protein (single polypeptide chain) with no subunit structure but applies where two (or more) identical subunits are involved in the active structure. The protein would be classified as a **honodimer**, or if the two protein subunits are not identical, the protein is classified as a **heterodimer**. Trimers and tetramers also are important. *The conversion of a monomeric enzyme to a polymeric form is one mode of regulating the enzymatic activity.* An example is the **allosteric regulation** (see chapter on enzymes) of **acetyl CoA carboxylase**. This enzyme catalyzes the committed step in fatty acid synthesis. It is regulated allosterically by high levels of citrate that occur when the levels of acetyl CoA and ATP are high. The elevated [citrate] facilitates the **polymerization** of inactive octamers of acetyl CoA carboxylase into active filaments. Conversely, palmetto; CoA, which is elevated when there is an excess of fatty acids, facilitates the disassembly of filamentous acetyl CoA carboxylase into inactive octamers (constituting a negative feedback). Another level of protein regulation concerns **posttranslational modifications**. These can involve phosphorylation, acetylation, or the addition of carbohydrates or lipids. A vast number of enzymes and other proteins are regulated by phosphorylation—dephosphorylation reactions. Amino acids in the structure of a protein that are potentially subject to phosphorylation are serine, threonine, or tyrosine. Many examples will be discussed in later chapters.

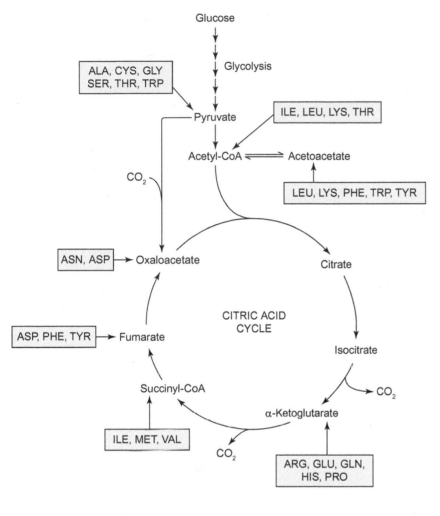

**FIGURE 4.16** Degradation pathways of essential and nonessential amino acids.

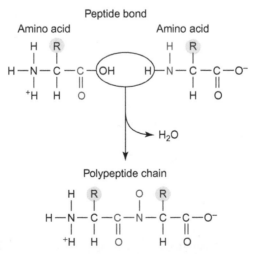

**FIGURE 4.17** Generation of a dipeptide by formation of a peptide bond between two different amino acids. The carboxyl group on the alpha-carbon of one amino acid is linked to the amino group of the other amino acid by exclusion of a molecule of water. The peptide bond eliminates the ionizability of either of those two groups. The ionizable groups are the free amino groups of the left hand (N-terminal) member and the carboxyl group of the right hand (C-terminal) member. The same reaction can be extended to a polypeptide consisting of many amino acid residues. During protein biosynthesis, these reactions occur at the ribosome. R = side chain; the peptide bond is colored in red in the bottom structure. Some peptides are too small to be encoded by an mRNA, although in other cases, smaller polypeptides may be split off from a larger precursor protein that is encoded by an messenger RNA. A case of the former is **glutathione (GSH)**, a tripeptide and its biosynthesis is shown in Fig. 4.18.

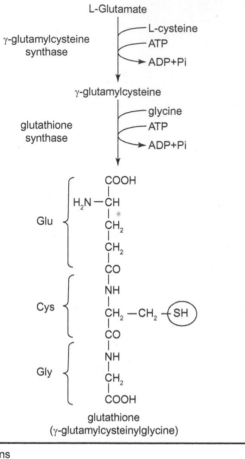

L-Glutamate

γ-glutamylcysteine
synthase

— L-cysteine
— ATP
→ ADP+Pi

γ-glutamylcysteine

glutathione
synthase

— glycine
— ATP
→ ADP+Pi

Glu

Cys

Gly

glutathione
(γ-glutamylcysteinylglycine)

**FIGURE 4.18** The biosynthetic pathway of glutathione. The synthesis of glutathione is dependent on the cellular level of **cysteine**. The enzymes are indicated for each step. The cysteinyl SH group can be oxidized to an S—S bond when two glutathione molecules interact. The S—S bond between the two molecules can be broken by **glutathione reductase**, releasing the two glutathione molecules wherein the cysteine residues now have, again, free SH groups. This represents an important redox system for cells. The redox system is shown at the bottom of the figure. Two molecules of GSH can be oxidized to GS—SG and reduced again to 2 GSH. The enzymes and coenzymes involved in the reactions are indicated.

Redox reactions

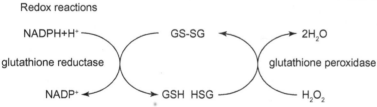

NADPH+H⁺ — GS-SG ← 2H₂O

glutathione reductase        glutathione peroxidase

NADP⁺ ← GSH  HSG — H₂O₂

## Amino Acid Sequence

The amino acid sequence of a protein is specified by the gene encoding it, transcribed to the mRNA and synthesized on the ribosome in the cellular cytoplasm. For the purposes of illustration, a seven-amino-acid peptide is presented in Fig. 4.19 showing the amino acids, their side chains and versions as to how the structure might be abbreviated in the literature.

## Secondary, Tertiary, and Quarternary Structures

Folding of peptide chains or formation of **pleated sheets** or **helices** is determined by the sequence of amino acids. The folding process of proteins is facilitated by specific proteins, **chaperones**, and the process of protein folding occurs as the polypeptide chain exits from the ribosomal machinery to finally achieve its native, active configuration. **Fibrous proteins** typify the predominantly β-**sheet structure,** and **globular proteins** exhibit mostly α-**helical structures.** Usually, proteins contain both types of structures but one type may dominate over the other. The α-helix is stabilized by hydrogen bonding between amide nitrogens and carbonyl carbons of peptide bonds that are spaced apart by four residues as shown in the lower left of Fig. 4.20.

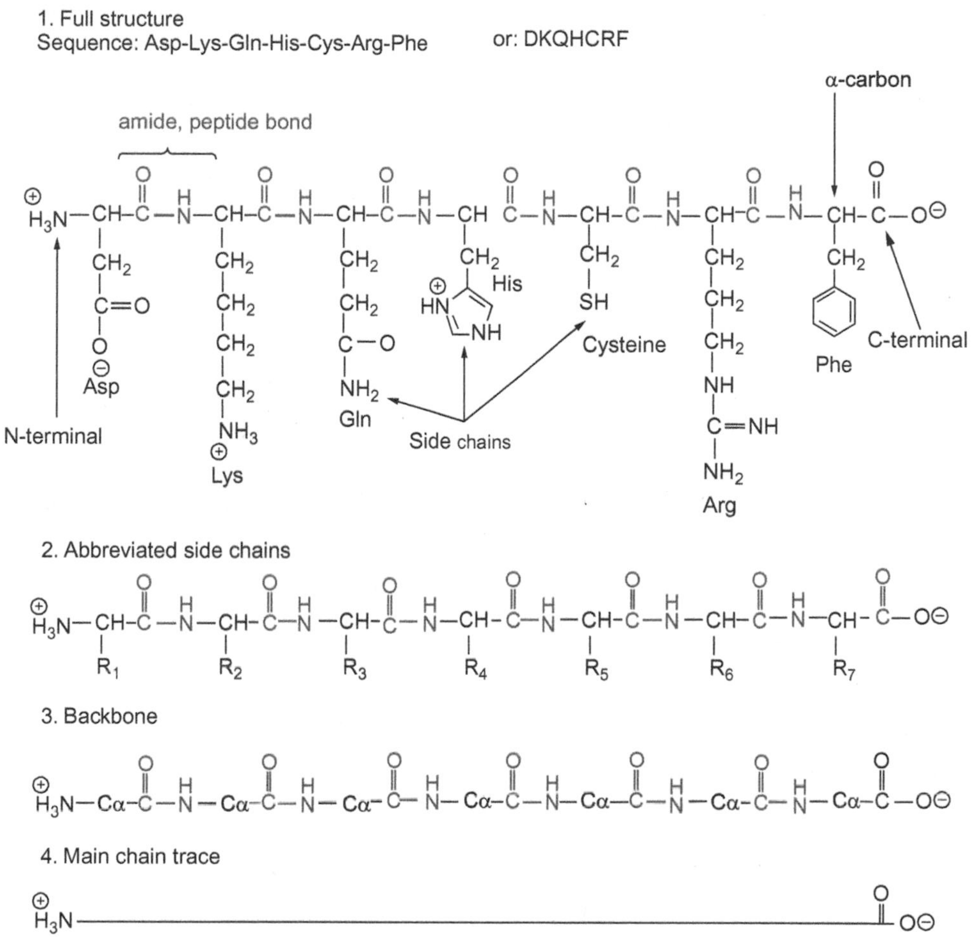

1. Full structure
Sequence: Asp-Lys-Gln-His-Cys-Arg-Phe    or: DKQHCRF

2. Abbreviated side chains

3. Backbone

4. Main chain trace

**FIGURE 4.19** The heptapeptide: aspartyl−lysyl−glutaminyl−histidinyl−cysteinyl−arginyl−phenylalanine is depicted. The N-terminal amino acid, in this case aspartyl, is on the left and the C-terminal phenylalanine is on the right. This representation shows only the sequence (the primary structure) and does not show any potential folding, although for a peptide this short, folding is probably minimal, and the configuration in solution would be random. Ionizable groups would be the N-terminal amino group, the C-terminal carboxyl, the side chain amino group of lysine, the side chain of histidine (secondary amine), and the amino group in the side chain of arginine. Sometimes, a peptide is abbreviated as in 2 where the structures of the side chains are represented by R groups or in 3 where only the backbone peptide is shown or in 4 where only the N-terminal and C-terminal groups are shown.

The **α-helix** is stabilized by hydrogen bonding between amide nitrogens and carbonyl carbons of peptide bonds that are spaced four residues apart, producing a coil where the amino acid side chains lie on the outside of the helix and perpendicular to the axis. Certain amino acids favor **helix formation**; these are alanine, glutamate, aspartate, leucine, isoleucine, and methionine. The coil structure tends to be disrupted by glycine or proline. Proline produces a fold. **Proline** is usually found at a turning point when a chain folds to continue in a new direction. The **pyrrolidine imino group** restricts movement around the peptide bond and interferes with the further extension of the helix. Stretches of five to ten amino acid residues comprise the **pleated sheets** (or β-sheets). These sheets are stabilized by **hydrogen bonding** between amide nitrogens and carbonyl carbons of amino acid residues that are linearly opposite, differing from the α-helix where the hydrogen bonds occur in adjacent regions of the helical backbone (Fig. 4.20). **Electrostatic bonds**, occurring between two oppositely charged groups, also stabilize protein structures. Side chains with ionizable groups interact with dipolar water molecules. Side chains of amino acid residues extend from the surface of proteins, especially globular proteins, and the ionizable substituents in the side chains interact with dipolar water molecules in a charge−charge interaction. When substituent groups of amino acid residue side chains have similar charges and are close together, they repel (van der Waals forces) each other and such repulsions affect protein folding. Weaker van der Waals forces usually outnumber the stronger electrostatic attractions. **Hydrophobic bonding** is a third category of substituent interactions. Such a bond takes place between two amino acid residues that have proximate side chains containing nonpolar groups (such as $-CH_3$) that attract each other and create an environment that excludes water molecules. A strong

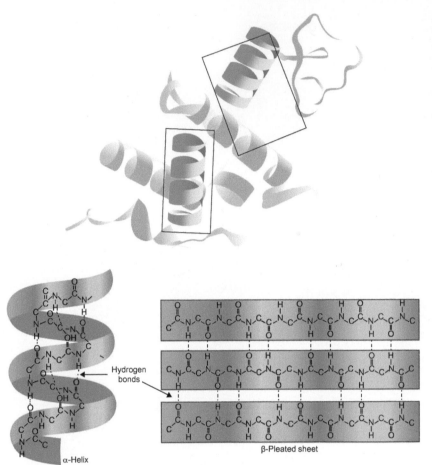

**FIGURE 4.20** Hydrogen bonding between carbonyl carbons and amide nitrogens in the peptide α-helix shown on the bottom *left*. The bonds (*black dotted lines*) form between amino acid residues along the axis of the helix. Hydrogen bonds between amino acid residues in β-sheets form between carboxyl carbons and amide nitrogens of parallel amino acids shown on the lower *right*. The two structures on the bottom represent portions (*boxed*) of the polypeptide shown at the *top*. These are the secondary structure of a protein. *Reproduced from Litwack, G.,* Human Biochemistry and Disease, *Academic Press/Elsevier, page 59, Figure 2−28, 2008.*

disulfide bond can be formed between neighboring amino acids (i.e., cysteines) under oxidizing conditions. These types of bonds are illustrated in Fig. 4.21.

The **tertiary structure** is the assembled and folded polypeptide chain into its natural and active form (unless more than one polypeptide chain is required for it to become active). The tertiary structure of a completed polypeptide chain is visualized in Fig. 4.22.

The **quarternary structure** of a protein obtains when a protein has subunit proteins, more than one polypeptide chain in the native form. Although many proteins are active in the monomeric form (one protein chain), others are made up of subunit polypeptide chains. They can be dimers, trimers, tetramers, and proteins with many more subunits. An example is **hemoglobin** which is a tetramer of two similar subunits ($\alpha_2, \beta_2$) as shown in Fig. 4.23.

## Protein Folding

This process is thought to occur almost automatically as a function of the amino acid sequence and the interaction of amino acid side chains as the polypeptide begins to fold when it is extended from the ribosome. The process of folding is aided by certain proteins, the **chaperones** or **heat shock proteins** (HSPs). Chaperone proteins (HSPs) assist in the folding of the newly forming protein and utilize their inherent ATPase activity to provide energy from ATP for this function (the hydrolysis of the **terminal phosphate group of ATP** to form ADP releases 7.3 kcal of energy). Newly formed proteins are either folded properly with the aid of **HSP40** (heat shock protein of 40,000 molecular weight) and **HSC70 (heat shock cognate 70,000 molecular weight)** or in some cases with **HSP90** (has affinity for nonpolar regions), or, if the folding is imperfect in some way, the protein is polyubiquinated with **ubiquitin (Ub)** and targeted to the **proteasome** where it becomes degraded. A summary of the fates of properly or improperly folded proteins is shown in Fig. 4.24. The proteasome will be described under "protein degradation."

Hydrogen bonding

Hydrophobic interaction

**FIGURE 4.21** Secondary interactions between amino acid residues of adjacent peptide chains of the same molecule that play a role in the tertiary structure of a protein. *Reproduced from Litwack, G.,* Human Biochemistry and Disease, *Academic Press/Elsevier, page 60, Figure 2029, 2008.*

Electrostatic binding

Disulfide bond

van der Waals repulsion

− 2H | Oxidation

Charge-dipole

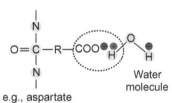

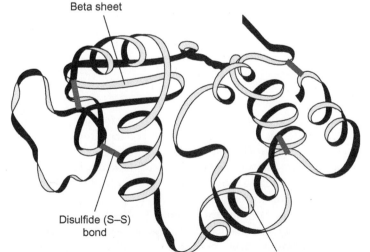

Beta sheet

Disulfide (S–S) bond

Alpha helix

**FIGURE 4.22 Tertiary structure** of a polypeptide chain. After the synthesis of a polypeptide (protein) has been completed, the chain becomes folded from the information dictated by the amino acid sequence. In particular, sulfhydryl amino acids, like cysteine can form an oxidized **disulfide bond** ($-S-S-$) with a cysteine residue in the same chain as the second cysteine moves into the vicinity of the first cysteine residue. This creates a stable bond that helps to shape the three-dimensional structure of the protein. Interactions between neighboring side chains of amino acids also occur, and these involve **hydrogen bonding** and **electrostatic interactions** (e.g., between the side chains of an amino acid amine of arginine or lysine and a carboxyl of a side chain of aspartate or glutamate). Such bonds add to the stability of the three-dimensional conformation. *The figure is reproduced from http://student. ccbcmd.edu/~gkaiser/biotutorials/proteins/images/u4fg1b3.jpg.*

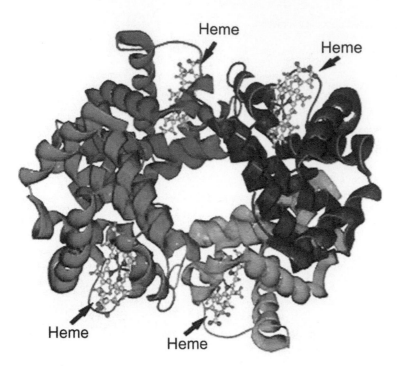

Heme

Heme

Heme

Heme

**FIGURE 4.23** The quarternary structure of hemoglobin consisting of two alpha-chains and two beta-chains. Each subunit is in a different color. There are four hemes, one in each subunit. Each heme has one iron atom at its center. *Reproduced from http://www.chemistry.wustl.edu/~edudev/ LabTutorials/Hemoglobin/images/hhemo_rib1.jpg.*

Many proteins are glycosylated. One postulate is that the terminal residues of a **glycosylated protein** can act as a timing mechanism for protein folding in the endoplasmic reticulum. This involves the activities of the enzymes glycosidases I and II that break the terminal glucose (Glc) residues from newly synthesized *N*-glycosylated proteins. When proteins are misfolded, they undergo a continuous glycosylation−deglycosylation cycle. In this cycle, a Glc is added to the carbohydrate chain by **UDP-glucose glycosyltransferase** and broken off by **glucosidase II**. The chaperones **calnexin** and **calreticulin** recognize the **monoglucosylated state**, and they keep the unfolded protein in the endoplasmic reticulum (ER). The **deglucosylated protein** can enter the **productive folding pathway** and bypass this cycle. If a protein remains in the ER for an extended period of time, **mannosidase I** cleaves a mannose residue and then the protein becomes targeted by the **ubiquitin−proteasome system** for degradation through the **disposal pathway** as shown in Fig. 4.25.

Certain amino acids are found more often within the α-helix structures of a protein; these are alanine, leucine, and glutamic acid. β-Sheet structures often contain isoleucine, valine, and tyrosine, whereas glycine, proline, and asparagine are found at locations where a polypeptide chain bends. Generally, proteins are thought to fold into a conformation that represents the lowest free energy state. For active proteins with more than one polypeptide chain, the subunits associate spontaneously. Protein synthesis (generating an active protein) is a rapid process where it can begin even before transcription of a gene is finished.

## Protein Degradation

Although some proteins are degraded in the **lysosome** (Chapter 2, The Cell), many proteins are degraded by a specific pathway requiring their **ubiquitination**. When proteins are misformed or denatured or synthesized too slowly, they may be directed to a degradation pathway by the addition of several **ubiquitin (Ub)** molecules. Ubiquitin is a small, highly conserved protein of 8564 Da containing 76 amino acids. In this pathway, there are three proteins, a **ubiquitin-activating enzyme**, a **ubiquitin-conjugating protein** and a **ligase**. These act to direct the protein to the **proteasome** where the hydrolysis of the protein takes place. The pathway is summarized in Fig. 4.26.

The ubiquinated protein is moved into the **20S proteasome** that becomes capped at one or both ends with **19S regulatory subunits** to generate the **26S proteasome**. The 19S regulatory proteins recognize and deliver the ubiquitinated protein. Ubiquitin is added to a protein by energy from ATP and when the chain reaches a certain length of four ubiquitins or more, the protein is hydrolyzed down to reusable peptides and free ubiquitin molecules that can be recycled.

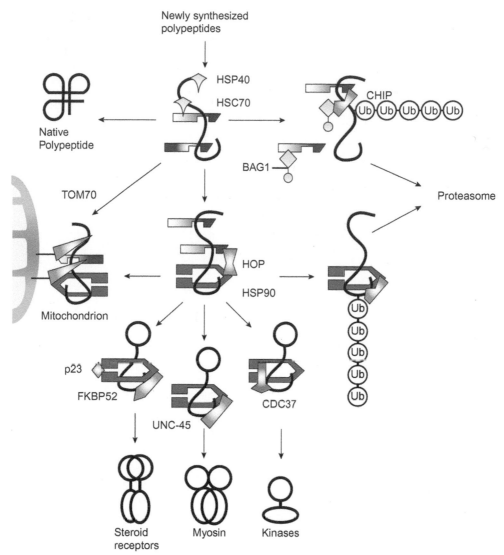

**FIGURE 4.24** Pathways of chaperone-mediated protein folding in the cytosol. Newly synthesized proteins are complexed with heat shock protein 40 (HSP40) and cognate protein 70 (HSC70) (*top*). Some polypeptides fold while interacted with HSC70, whereas some others interact with HSP90. For misfolded polypeptides or for proteins folding too slowly, the U-box ubiquitin (Ub) ligase CHIP (carboxy terminus of HSC70-interacting protein) contacts HSC70 or HSP90 to attach polyubiquitin onto substrate polypeptides that results in targeting to the proteasome for degradation. BAG1 (BCL2-associated athanogene-1) that binds to BCL2, an antiapoptotic factor, can assist in the targeting of the HSC70-bound polyubiquinated polypeptides to the proteasome. The figure shows various protein products associated with other proteins: steroid receptors associate, in addition to HSP90, with p23, HSP90 cochaperone and FKBP52 (immunophilin 52 kDa FK506-binding protein); myosin first associates with UNC-45 (HSP90 cochaperone) in addition to HSP90 and various kinases associate with CDC37 (HSP90 cochaperone) in addition to HSP90. These are generally the forms of the proteins prior to their activation to the functional form. *The figure is reproduced from http://www.nature.com/nrm/journal/v5/n10/images/ nrm1492-f3.jpg from J.C. Young, V.R. Agashe, K. Siegers and F.U. Hartl,* Nat. Rev. Mol. Cell Biol., **5:** 781−791, 2004

Partially degraded proteins expose their inner nonpolar regions that attract them to the inner core of the 20S proteasome, the opening of which measures about 5 to 6 nm (Fig. 4.27).

Some proteins contain a **PEST** sequence (Pro−Glu−Ser−Thr) that accelerates their degradation by this pathway.

## PROTEIN CLASSIFICATION

There are many ways to classify proteins, and there are many different human proteins, although the exact number is not known. If the bacterium ***Escherichia coli*** has 3000 different proteins in its cell, the human cell must have more than the bacterial cell as there are in the range of 25,000 to 30,000 genes in the human cell but a small percentage of

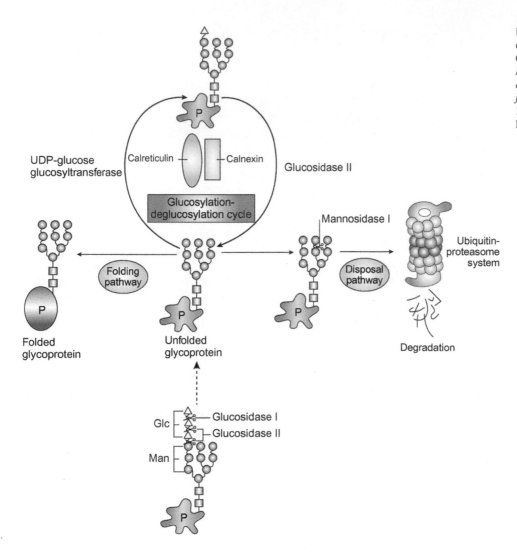

**FIGURE 4.25** Protein-folding cycle of glycosylated proteins. P = protein; Glc = glucose;      Man = mannose. *Reproduced from http://www.nature. com/horizon/proteinfolding/highlights/ figures/s2_spec1-f3.html from B. Tsai, Y. Ye, & T.A. Rapoport,* Nat. Rev. Mol. Cell Biol., *3: 246–255, 2002.*

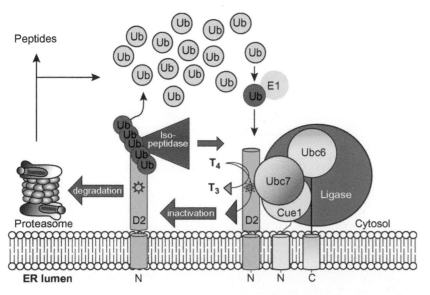

**FIGURE 4.26** The process of ubiquitination of a protein in the disposal pathway. Ub = ubiquitin; E1 = ubiquitin activating enzyme; (E2 = ubiquitin conjugating enzyme); Ubc6 and Ubc7 are E2s involved in D2 ubiquitination (D2 = the protein to be degraded); Cue1 is an ER-docking protein for Ubc7 and the star in the D2 molecule represents the Sec-containing active center. *Reproduced from http:// www.hotthyroidology.com/editorial_74.html.*

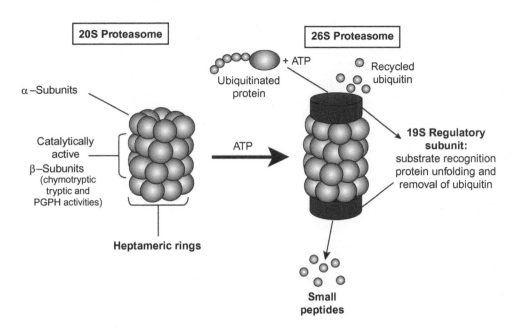

**FIGURE 4.27** Construction of the 26S proteasome from the 20S proteasome. PDPH = peptidyl-glutamyl peptide hydrolyzing (enzyme). *Reproduced from http://www.benbest. com/lifeext/proteasomes.jpg.*

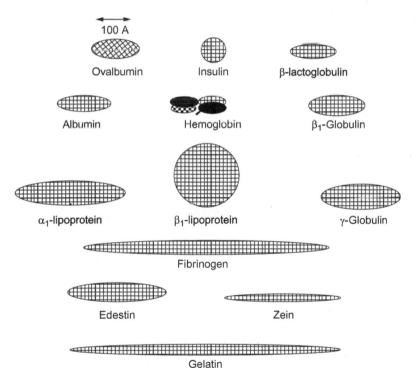

**FIGURE 4.28** Different sizes and shapes of proteins. *Reproduced from G. Litwack,* Human Biochemistry and Disease, *Academic Press/Elsevier, page 86, 2008.*

these genes may be encoding proteins that are expressed. There might be approximately 100,000, or more, different proteins in a human cell.

One way to classify proteins is by their sizes and shapes as shown in Fig. 4.28.

In addition to size and shape, proteins can be classified on the basis of structure and composition (Fig. 4.29).

They also can be classified on the basis of function. One group would be the **enzymes**. Enzymes are classified by the types of reactions they catalyze (see Chapter 5, Enzymes). Many **hormones** are proteins that vary from tripeptides to proteins of relatively high molecular weight (e.g., 44,000 Da). Other categories are **transport proteins, immunoglobulins** or **antibodies, structural proteins, motor proteins, receptors, signaling proteins** and storage **proteins**. Examples of all of these categories will be encountered in later chapters.

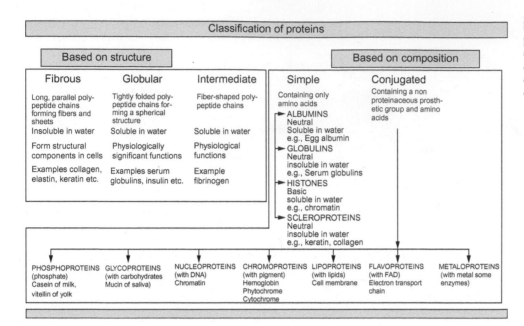

FIGURE 4.29 Classification of proteins on the basis of structure and composition. *Reproduced from http://imagesd.tutorvista.com/content/cellular-macromolecules/proteins-classification.jpeg.*

Some proteins consist of a single polypeptide chain, whereas others are polymeric consisting of two or more polypeptide chains (subunits). Proteins can also be classified on the basis of their modifications that occur within the cell. Examples would be phosphoproteins, glycosylated proteins, etc. (Fig. 4.29).

Proteins can be classified on the basis of their intracellular location. Thus, there are soluble proteins, membrane proteins, secretory proteins, etc. For each type of protein, there will be a further classification within that type.

## Proteomics

The *E. coli* cell has 3000 different proteins. An estimate of the human genes that express proteins suggests that a human cell may have as many as 100,000, or more, different proteins or perhaps many more; the total is not known. In eukaryotes, especially, the proteome is much larger than the genome. There are a greater number of proteins than there are genes that encode the information for proteins. This is due to at least two factors, alternative spicing to create multiple mRNAs from a single gene and posttranslational modifications, such as phosphorylation and glycosylation. *Proteomics is the branch of biology that seeks to identify all of the expressed proteins in a given cell type.* Early approaches to separating and identifying expressed proteins involved two-dimensional electrophoresis (**PAGE; polyacrylamide gel electrophoresis**). Eventually, techniques were devised to pick individual protein spots out from a gel and identify the protein. This could be done by partial hydrolysis and sequencing of the liberated peptides. Another method involves partial sequencing and preparing a cDNA probe that can fish out the RNA or gene whose sequence usually is known. Consequently, there are ways to ascertain the sequence of the peptides in the protein and thus arrive at its primary sequence. The tissue/cell sample of proteins can be first electrophoresed between an anode and a cathode so that proteins are separated in one dimension depending on their isoelectric point (pI) (or their net charge). The same gel is then electrophoresed in a second dimension based (first turning the gel by 90°) on size of the polypeptide or subunit, if run with the surface active agent, SDS (sodium dodecyl sulfate). SDS will generally break down the quaternary structure of proteins into their subunits (single polypeptide chains) unless covalent bonds hold the subunits together. Examples of the two-dimensional polyacrylamide gel technique are shown in Fig. 4.30A and B.

The first dimension of the 2D gel electrophoresis system is an isoelectric focusing gel (IEF). A protein mixture is applied to a circular long gel, and voltage is applied so that proteins distribute at equilibrium at their isoelectric point (pI) between the cathode and anode. At the completion of this separation, the gel is placed atop an SDS square gel so that proteins are separated according to their masses by virtue of the gradient of gel beads with escalating pore sizes; this is referred to as SDS-PAGE, sodium dodecyl sulfate-polyacrylamide gel electrophoresis. This results in a gel where a huge number of proteins are separated. The process is shown in Fig. 4.31.

The apparatuses used for isoelectrofocusing (IEF) and for SDS−PAGE are shown in Fig. 4.32.

(A)    Basic ← Stable pH gradient → Acidic

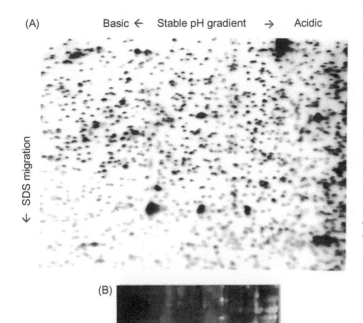

**FIGURE 4.30** (A) A two-dimensional polyacrylamide gel showing a separation of hundreds of proteins that are developed with a protein stain. 2D gel electrophoresis separates molecules in one dimension. In a second dimension in which the gel is shifted by 90° from the first dimension, the components are again separated using a second property. The two properties can be the isoelectric points of proteins in one dimension versus the mass of the proteins in the second dimension. (B) A two-dimensional gel electrophoresis separating mixtures of up to 2000 proteins. This gel has been developed with different stains (different colors) to differentiate between two different protein samples. One stain can distinguish between oxidized and nonoxidized proteins. Individual proteins can be identified using immunoprecipitation and pull-down assays (precipitation of a target protein in a mixture that may involve the use of a matrix-associated antibody; the matrix can be a bead) to distinguish protein–protein interactions. *(A) Reproduced from http://classic.roswellpark.org/files/1_2_1/photos/smiragliaweb2.jpg. (B) Reproduced from http://www.unil.ch/web/dav/site/dbcm/shared/Technologie/Technologie23_BR.jpg.*

**Equilibration**

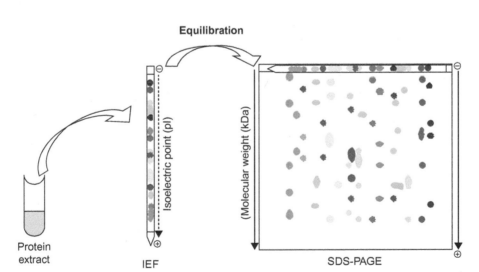

**FIGURE 4.31** The process of SDS-PAGE (sodium dodecyl sulfate-polyacrylamide gel electrophoresis). A mixture of proteins to be separated is applied to a columnar isoelectro focusing gel (IEF) and current is applied so that proteins migrate to the anode or cathode depending on their isoelectric points (pIs). Colors are used to emphasize individual proteins. At equilibrium, the gel is transferred to the top surface edge of a square PAGE gel and current is applied so that the proteins in the IEF gel migrate in the field and become separated in this second dimension based upon their molecular weights. The gel is prepared so that particles of the gel contain sieving pores of increasing cross section (these gels can be obtained commercially) to accommodate separation based on molecular mass. The proteins separated by the IEF gel become arrayed on the SDS–PAGE gel as shown on the *right* and a typical gel developed with a protein stain would appear as shown in Fig. 4.30. *Reproduced from http://www.seed-proteome.com/vars/images/pub_defaut/2d%20electrophoresis.jpg.*

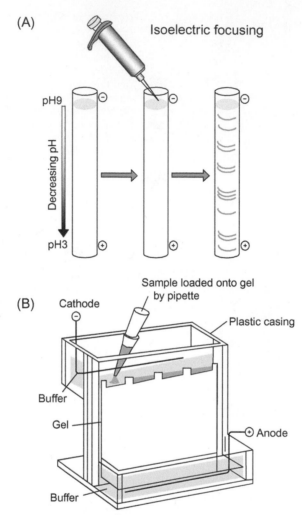

**FIGURE 4.32** Apparatuses for isoelectric focusing (A) and for SDS—PAGE (B). In (A), a stable pH gradient is established in the gel after application of an electric field. A protein solution is added by micropipette and an electric field is reapplied. After staining, proteins are distributed along a pH gradient, according to their isoelectric points. In (B), the SDS—PAGE apparatus is shown with individual slots at the top but when an IEF gel is attached, the SDS—PAGE gel does not contain slots and is smooth edged. The gel from (A) is applied to the top edge of the SDS—PAGE gel. The thin gel is positioned between 2 clear plates of plastic. After an electric current is applied over a predetermined time period, the proteins separate as shown in the figure. An individual protein spot of interest can be identified. The spot can be removed and subjected to trypsin hydrolysis yielding an array of peptides. The mixture of peptides can be separated by mass spectrometry according to their masses and the array of peptide molecular weights can be compared with arrays from other proteins by application of the data into a genomic database that can match the peptide array to a corresponding gene.

The proteome reflects all of the proteins synthesized by a cell. Different cell types express different arrays of proteins. More than one protein can be expressed from a given gene because, of **alternative splicing** of **pre-messenger RNAs**, formation of glycoproteins, phosphoproteins, ubiquinated proteins, and other **posttranslational modifications**, such as methylation, acetylation, oxidation, etc. as well as changes stemming from other alterations in pre-messenger RNAs. The same cell can express different proteins depending on time and conditions. Thus, these analyses are useful for determining the effects of various agents on a specific cell in terms of the array of specific proteins or their levels and extend information gleaned from genomics.

## Protein Microarray

A microarray can accomplish the screening of vast numbers of proteins in terms of their activities instead of experiments with one protein at a time. There are three types of microarrays of proteins: an analytical microarray will screen for binding affinities, specificities, or the expression of protein levels in a complex mixture of proteins. Typically, a library of antibodies or single-stranded RNAs or single-stranded DNAs or engineered binding proteins or protein—phospholipids is arrayed on a glass microscope slide. This array is then probed with the solution mixture of proteins from

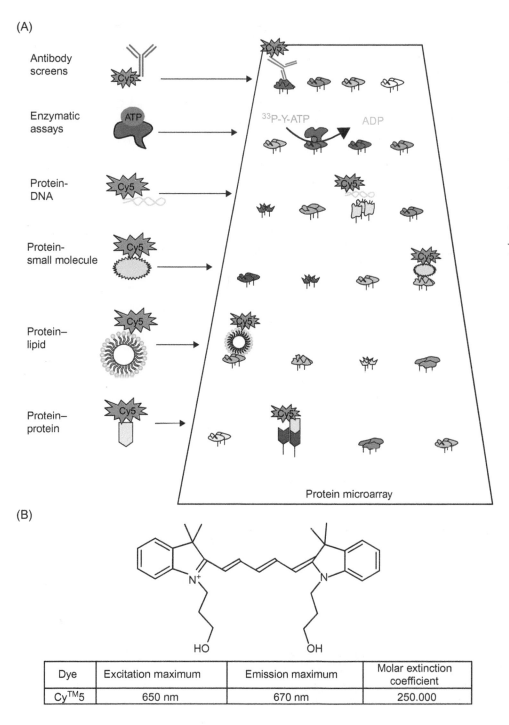

(A)

Antibody screens

Enzymatic assays

Protein-DNA

Protein-small molecule

Protein–lipid

Protein–protein

$^{33}$P-Y-ATP     ADP

Protein microarray

(B)

| Dye | Excitation maximum | Emission maximum | Molar extinction coefficient |
|---|---|---|---|
| Cy$^{TM}$5 | 650 nm | 670 nm | 250.000 |

**FIGURE 4.33** Applications of functional protein microarrays. A representative sample of different assays that have been performed on functional protein microarrays. Proteins are immobilized at high spatial density onto a microscope slide, and the slide can be probed for various interactions. Although Cy5 is the fluorophore shown, many others can be used for detection. *Reproduced from http://www.ncbi.nlm.nih.gov/pmc/articles/PMC1828913/figure/F1/. At the bottom of the figure, the structure of Cy5 is shown together with its optical characteristics. Reproduced in part from http://www.biopeptek.com/images/compounds/flu2.gif.*

which the specific proteins that bind the antibodies (in the first case) or that bind the nucleic acids (in the second case) or that form a complex with the engineered binding proteins (in the third case) will be discovered. This technique is useful in clinical diagnostics. For example, one could profile responses to some specific agent (a drug, stress, or other condition) comparing tissues from a diseased patient versus tissues from a healthy patient. In the case of the administration of a drug, one could test before the drug is given and at various times after administration of the drug. The various applications of protein microarrays are summarized in Fig. 4.33.

Essentially, these microarrays can be used to study specific biochemical activities of the cellular proteome in a single experiment.

In a **reverse phase protein microarray**, specific tissues are used for the source of cells that are lysed, and the lysate is arrayed onto a microscope slide. Such a slide can be probed with specific fluorescently labeled antibodies against the protein of interest. In this way, specific antibody—protein interactions are detected that, in the case of a disease condition, can result from posttranslational modifications affected by the disease state. This can lead to a dysfunctional protein pathway so that a specific therapy can be designed to treat the dysfunctional pathway.

There are many ways in which the slides can be treated to allow for the subsequent attachment of proteins or ligands, etc. For example, proteins can be attached to slides randomly using a glass slide covered by an aldehyde surface as shown in Fig. 4.34A or, in another option, a ligand for the sought after protein(s) can be coated on a slide. This slide can be treated with a mixture of proteins, among which is the protein with a binding site specific for that ligand (Fig. 4.34B).

Proteome slides (chips) can be used to discover previously unknown protein-DNA-binding activity. This has been done with yeast (eukaryotic source), for example, by using both single-stranded and double-stranded Cy3 labeled yeast genomic DNA to probe a yeast proteome array (most of the yeast cell proteins on a single slide). This procedure

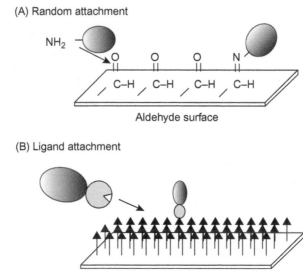

**FIGURE 4.34** (A) Protein attachment methods. Proteins can be attached randomly by way of different chemistries including aldehyde- and epoxy-treated slides that covalently attach protein by their primary amines or by adsorption onto slides coated with nitrocellulose or acrylamide gel pads. (B) Proteins can be uniformly oriented onto slides coated with a ligand. For example, His6X-tagged proteins can be bound to nickel-derivatized slides ($Ni^{2+}$ will bind His at many locations in a protein, whereas $Co^{2+}$, for example, has stricter spatial requirements for binding His in proteins), and biotinylated proteins can be attached to streptavidin-coated slides (biotin and avidin form a specific complex). This leads to attachment through the tag and presumably orients the protein away from the slide surface. *Reproduced from http://www.ncbi.nlm.nih.gov/pmc/articles/PMC1828913/figure/F2/.*

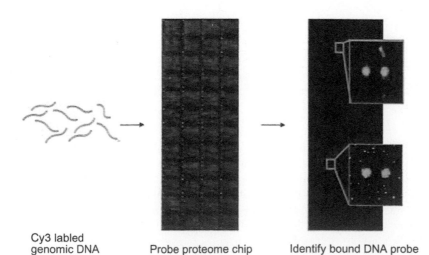

**FIGURE 4.35** Identification of DNA-binding proteins using a functional protein microarray. Genomic DNA (total DNA of a cell) is purified, fragmented, and labeled with Cy3-dCTP (the fluorophore Cy3, similar to Cy5, bound to deoxycytidine triphosphate). A yeast proteome array with the majority of the yeast proteins was incubated with the labeled DNA to identify novel DNA-binding proteins, including Arg5,6 (acetylglutamate kinase), a mitochondrial enzyme. *Reproduced from http://www.ncbi.nlm.nih.gov/pmc/articles/PMC1828913/figure/F3/.*

identified over 200 DNA binding proteins. An unexpected target was the mitochondrial enzyme Arg5,6 (acetylgluta-mate kinase), involved in the biosynthesis of arginine, and this enzyme proved to be associated with specific nuclear and mitochondrial locations, and later experiments indicated that it could have a role in the regulation of gene expression. The original probing experiment is shown in Fig. 4.35.

This technique can be used to discover a wide variety of activities by using different probes.

## SUMMARY

Human prion disease is typified by **Creutzfeldt-Jakob disease**. This disease results from a change in the structure of a normal cellular protein, PrP$^c$, to an abnormal conformation, PrP$^{Sc}$, seemingly without any other infectious agent, such as a virus. The molecular biology of this disease highlights the principles of protein structure developed in this chapter.

Genetic diseases involving amino acids are as follows: phenylketonuria (PKU), alkaptonuria (AKU), Maple Syrup disease, tyrosinuria, and homocystinuria. The enzyme defects in each of these diseases are given.

The 20 common amino acids are summarized in terms of their structures and properties. Essential amino acids are those that are required in the diet because either the body does not synthesize them or they are not synthesized in sufficient quantity. The chiral properties of amino acids are described as well as their ionizable character. The properties of **buffers** are described, and the **Henderson—Hasselbalch equation** is developed which allows predictions of buffer behavior. Titrations of neutral, basic, and acidic amino acids are presented that enable the calculation of **p$K_a$ values** for the ionizable groups on amino acids. This information is extended to peptides and the ionizable groups of the terminal amino and carboxyl groups as well as the ionizable groups in amino acid side chains enabling the understanding of the **pI value** that represents the overall charge of a peptide or protein in solution.

The synthesis of **nonessential amino acids** and the degradation of amino acids in the body are reviewed. These amino acids derive from metabolic intermediates, and amino groups can be added to various metabolites by a process of transamination to produce a given nonessential amino acid. Excess amino acids, not required for protein synthesis, can be degraded into metabolites that enter the glycolytic pathway or tricarboxylic acid cycle for the production of energy in the form of ATP. Certain small peptides, such as **glutathione**, a three-amino acid peptide, are too small to be encoded on a gene, and the route of the biosynthesis of this molecule (gamma-glutamylcysteinylglycine) is shown. The levels of protein structure are discussed including the primary amino acid sequence, secondary ($\alpha$-helix and $\beta$-sheet), tertiary (three-dimensional structure), and quaternary (subunit structure). **Protein folding** is considered and the role of **chaperones** in this process is indicated. **Glycosylation** of proteins may have a function in the timing mechanism of protein folding in the endoplasmic reticulum. Protein degradation in the **proteasome** is signaled by **polyubiquitination** of a protein destined for degradation. There are many ways to classify proteins and several are listed in this chapter. The object of **Proteomics** is to identify all of the proteins in a given cell type. This information, perhaps more complex than the genetic makeup of the genome, can be used to classify diseases and effects of agents on a particular cell type of protein structure developed in this chapter.

A major advance to discover proteins with specific activities is the use of microarray slides (chips). Most of the cellular proteins can be attached to the surface of a microscope slide, and this array can be probed with specific agents, such as fluorescently labeled RNA or DNA to discover previously unknown proteins that bind to nucleic acids. Many other probes can be used to discover specific activities of cellular proteins.

## SUGGESTED READING

### Literature

Bishop, M.T., Will, R.G., Manson, J.C., 2010. Defining sporadic Creutzfeldt—Jakob disease strains and their transmission properties. *Proc. Natl. Acad. Sci.* 107, 12005—12010.

Chiti, F., Dobson, C.F., 2006. Protein misfolding, functional amyloid, and human disease. *Ann. Rev. Biochem.* 75, 333—366.

Chou, P.Y., Fasman, G.D., 1974. Conformational parameters for amino acids in helical, beta sheet and random coil regions calculated from proteins. *Biochemistry.* 13, 211—222.

Doudna, J.A., Cech, T.R., 2002. The chemical repertoire of natural ribozymes. *Nature.* 418, 222—228.

Englander, S.W., Mayne, L., 2014. The nature of protein folding pathways. *Proc. Natl. Acad. Sci.* 111, 15873—15880.

Hall, D.A., Ptacek, J., Snyder, M., 2007. Protein microarray technology. *Mech. Aging Devel.* 128, 161—167.

Hershko, A., Ciechanover, A., 1998. The ubiquitin system. *Ann. Rev. Biochem.* 67, 425—479.

Hidaka, Y., Shimamoto, S., 2013. Folding of peptides and proteins: role of disulfide bonds, recent developments. *Biomol. Concepts.* 4 (6), 597—604.

Jung, T., Hohn, A., Grune, T., 2013. The proteasome and degradation of oxidized proteins: Part II—Protein oxidation and proteasomal degradation. *Redox Biol.* 2, 99—104.

Manuelidis, L., Yu, Z.-X., Barquero, N., Mullins, B., 2007. Cells infected with scrapie and Creutzfeldt–Jakob disease agents produce intracellular 25-nm virus-like particles. *Proc. Natl. Acad. Sci.* 104, 1965–1970.

Martin, L.L., Unrau, P.J., Muller, U.F., 2015. RNA synthesis by *in vitro* selected ribozymes for recreating an RNA world. *Life (Basel).* 5, 247–268.

Morgensen, C.E., 1984. Microalbuminuria predicts clinical proteinuria and early mortality in maturity-onset diabetes. *N. Engl. J. Med.* 310, 356–360.

Notari, S., et al., 2008. Characterization of truncated forms of abnormal prion protein in Creutzfeldt–Jakob disease. *J. Biol. Chem.* 283, 30557–30565.

Pratsch, K., Wellhausen, R., Seitz, H., 2014. Advances in the quantification of protein microarrays. *Curr. Opin. Chem. Biol.* 18, 16–20.

Saa, P., et al., 2014. First demonstration of transmissible spongiform encephalopathy-associated prion protein (PrP$^{TSE}$) in extracellular vesicles from plasma of mice infected with mouse-adapted variant Creutzfeldt–Jakob disease by *in vitro* amplification. *J. Biol. Chem.* 289, 29247–29260.

Tsai, B., et al., 2002. Retro-translocation of proteins from the endoplasmic reticulum into the cytosol. *Nat. Rev. Mol. Cell Biol.* 3, 246–255.

White, J.V., Stulz, C.M., Smith, T.F., 1994. Protein classification by stochastic modeling and optimal filtering of amino-acid sequences. *Math. Biosci.* 119, 35–75.

Wu, M., et al., 2013. Crystal structure of $Ca^{2+}/H^+$ antiporter protein YfkE reveals the mechanisms of $Ca^{2+}$ efflux and its pH regulation. *Proc. Natl. Acad. Sci.* 110, 11367–11372.

Young, J.C., et al., 2004. Pathways of chaperone-mediated protein folding in the cytosol. *Nat. Rev. Mol. Cell Biol.* 5, 781–791.

## Books

Creighton, T.E., 2010. The Biophysical Chemistry of Nucleic Acids and Proteins. Helvetian Press, Faberville, Arkansas.

Crum, A.B., 2016. Glutathione Synthesis: Unraveling the Pleiotropic Paradox and Its Vital Immune Role. Wiley, Hoboken, N.J.

Horwich, A., 2002. *Protein Folding in the Cell*. Academic Press/Elsevier, Academic Press, Cambridge, Mass.

Lesk, A.M., 2001. Introduction to Protein Architecture: The Structural Biology of Proteins. Oxford University Press, Oxford, UK.

Lovric, J., 2011. Introducing Proteomics: From Concepts to Sample Preparation, Mass Spectrometry and Data Analysis. Wiley, Hoboken, N.J.

Nyhan, W.L., Barshop, B.A., Al-Aqeel, A.I., 2012. *Atlas of Inherited Metabolic Diseases*. third ed. Hodder Arnold/Holder & Stoughton/Hachette, London, UK.

Wu, G., 2013. Amino Acids: Biochemistry and Nutrition. CRC Press, Baton Rouge, Florida.

# Chapter 5

# Enzymes

## DIAGNOSTIC ENZYMOLOGY

The principle of diagnostic enzymology is that various disease conditions cause increased **cell membrane permeability** to macromolecules or outright lysis of the cell membrane that allow macromolecules, dissolved in the cytoplasm, to leak into the extracellular space and gain access to the bloodstream. This permits assay of enzyme activity directly (including immune reaction) that would identify abnormally high levels in blood. Some of these enzymes have multiple forms, based on their content of different subunits, where certain forms predominate in specific tissues, making it possible to identify the tissue source of the damage. Invariably, a combination of measurements will identify the tissue source of the elevated enzyme activity. Abnormally released enzymatic activities demonstrate a pattern of release over time after the disease event (tissue damage, cell death, hypoxia, infection, or inflammation) that may be characteristic of the diseased organ in addition to the activity of the released enzyme itself. Moreover, the healing process can be reflected in the course of measurements, and sometimes, prognosis of a disease condition can be reflected by changes in the released activity. The activity of the released enzyme is usually proportionate to the extent of tissue damage and must be sizeable enough to withstand the dilution by the general circulation (about eight quarts). The activities of certain enzymes are of interest in diagnosing certain diseased tissues (Table 5.1).

## Enzymes with Multiple Subunits: Tissue-Specific Isozymes

Enzymes, if they happen to have multiple subunits, can be extremely useful in diagnosis. A case in point is **lactate** (or lactic acid) **dehydrogenase (LDH)**. These isoenzymes can be separated by gel electrophoresis. Fig. 5.1 shows the tetrameric LDH. It consists of four subunits composed to two separate kinds, the heart-type subunit (H) and the muscle-type subunit (M).

A gel electrophoretic separation of these isoenzymes in normal serum is shown in Fig. 5.2.

Specific staining of the electrophoretic spots of LDH isozymes involves the enzymatic assay on the gel by the following method: lactate and $NAD^+$ are added to form pyruvate + NADH + $H^+$. To react with the NADH, phenozine methosulfate is added to form phenozine methosulfate + **tetranitro blue tetrazolium−Formazan** which is the color at the location of the LDH isozyme. LDH1 ($H_4$) is the most negatively charged of the isozymes and migrates closest to the anode. $M_4$ (LDH5) is the most positively charged isoform and migrates most closely to the cathode. The intensity of the color formed is proportional to amount of enzyme present in the spot. It is seen that LDH2 is the predominant isozyme in normal serum. When LDH1 is in a concentration higher than LDH2 in serum, **myocardial infarction** is suspected confirming clinical diagnosis. More recently, **troponin** may be measured in serum as an indicator of myocardial infarction. Elevation of LDH5 suggests the possibility of **liver damage**. LDH1 ($H_4$) derives from cells of the heart; LDH2 ($H_3M_1$) derives from **red blood cells** and the **reticuloendothelial system**; LDH3 ($H_2M_2$) derives from lungs and kidneys; LDH4 ($H_1M_3$) derives from kidneys, and LDH5 ($M_4$) derives from liver and striated muscle.

**Creatine kinase (CPK)** is another enzyme of interest in clinical enzymology. It catalyzes the phosphorylation of **creatine** whose reaction is shown in Fig. 5.3.

**Creatine phosphate** is a major energy source in muscle; hence, creatine kinase in blood can reflect degradative changes in muscle tissue. This enzyme is either a dimer being a homodimer or a mixture of two different subunits: muscle type (M) and brain type (B). Three isozymes exist: MM (CK3), prevalent in skeletal and heart muscles; BB (CK1) prevalent in brain, gastrointestinal tract, and genitourinary tract; and MB (CK2), prevalent in heart muscle. In myocardial infarction, there is an increase of LDH1 over LDH2 in serum, and there are also elevations of MM and MB isozymes of creatine kinase. In skeletal muscular diseases and muscular dystrophy, the MM isozyme is elevated. Fig. 5.4

Human Biochemistry. DOI: http://dx.doi.org/10.1016/B978-0-12-383864-3.00005-3

**TABLE 5.1** Certain Enzymatic Activities are Useful in Determining the Organ from Which They are Derived During a Disease Process. Another Abbreviation Used for Alanine Aminotransferase (AAT) is ALT (Alanine Transaminase)

| Tissue | Useful Enzyme Activity in Serum |
|---|---|
| Heart, liver, muscle | Lactate dehydrogenase (LDH) |
| Muscle, also cardiac muscle | Creatine kinase (CK) |
| Liver | Glutamyl transferase (GT) |
| Heart and liver | Alanine aminotransferase (AAT) |
| | Glutamate-pyruvate transaminase (GPT) |
| | Aspartate aminotransferase (AST) |
| | Glutamate-oxaloacetate transaminase (GOT) |
| Pancreas | $\alpha$-amylase |
| Prostate | Acid phosphatase (AP) (tartrate labile) |
| Bone, intestine (others) | Alkaline phosphatase (ALP) |

Lactate dehydrogenase catalyzed reaction

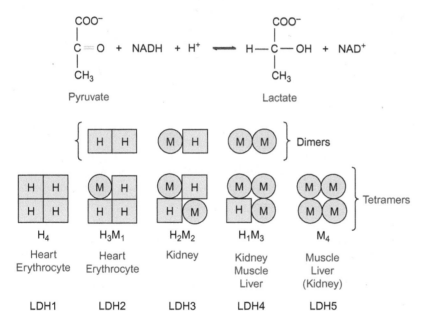

**FIGURE 5.1** Lactate dehydrogenase (LDH) catalyzes the reaction shown at the top of the figure. The models in gray or yellow show the composition of the dimers and tetramers. The $H_4$ (LDH1) enzyme reflects the heart myocardium and erythrocyte, and the $M_4$ (LDH5) reflects liver and muscle primarily.

**Isoenzymes LDH interpretation**

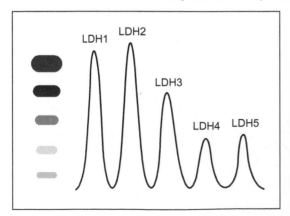

**Normal values range**

| Fraction | Tot. LDH % |
|---|---|
| LDH1 | 20–30 |
| LDH2 | 30–40 |
| LDH3 | 20–25 |
| LDH4 | 7–15 |
| LDH5 | 5–15 |

**FIGURE 5.2** Separation of LDH isozymes by gel electrophoresis. The uppermost spot on the left is LDH1 ($H_4$) and, in descending order: LDH2 ($H_3M_1$); LDH3 ($H_2M_2$); LDH4 ($H_1M_3$) and LDH5 ($M_4$). Note that in normal serum LDH2 is greatest in quantity followed by LDH1 as shown in the table on the right. The tracings on the left, indicating the relative amounts arise from densitometry measurements of the stained enzymes. *Reproduced from http://www.interlab-srl.com/interlabg26/analisi_img/ldh-sotto.gif.*

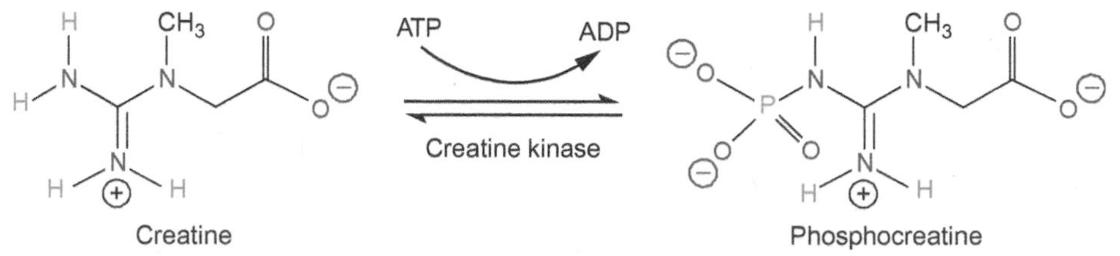

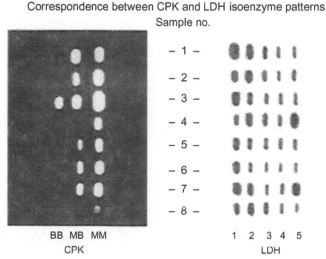

**FIGURE 5.3** Creatine kinase reaction.

Correspondence between CPK and LDH isoenzyme patterns
Sample no.

**FIGURE 5.4** Electrophoretograms of eight different measurements of LDH and CPK isozymes in serum. When LDH1 is low relative to LDH2, as in line 8, the MB and MM isozymes of CPK are also low. Compare these data to those in line 1. *Reproduced from http://pro2services.com/Lectures/Fall/ CardEnz/a6ck&ld.gif.*

shows electrophoretic patterns of both LDH and CPK where the spots of CPK are aligned with the array of serum LDH isozymes. In line 1, for example, LDH1 is higher than LDH2, suggesting myocardial infarction. In confirmation, CPK isozymes MB and MM are elevated confirming results with LDH. *CPK and CK are two abbreviations for the same enzyme (creatine phosphokinase or creatine kinase).*

In myocardial infarction, enzyme activities in serum are followed as a function of time in days after the event. Typical results for activities of lactate dehydrogenase (LDH1), creatine kinase (MB), and aspartate aminotransferase are shown in Fig. 5.5, and times of onset, peak of each enzymatic activity, and duration are summarized in the table below the figure.

Blood draws are made from 18 to 30 hours after the heart attack and 12 hours and 48 hours beyond that. Many diseases can be confirmed or diagnosed by measurement of enzyme activities in serum including hepatitis, jaundice, cirrhosis, muscular dystrophies, and some cancers.

Enzymes also have been used or proposed for the treatment of certain conditions. Some of these are summarized in Table 5.2.

## GENERAL ASPECTS OF CATALYSIS

Virtually, all enzymes in the body are proteins, except for ribozymes. Early life forms were probably based on nucleic acids, rather than proteins and ribozyme must be an ancient vestige and must have been among the first molecular machines. Since the discovery of an RNA that could catalyze an enzymatic reaction is recent, there is much activity into discovering other ribozymes, besides **peptidyl transferase 23 S rRNA** (see Chapter 11: Human Biochemistry: Protein Biosynthesis). There are at least 12 ribozymes known that catalyze different functions, and it is possible to design ribozymes that can cleave any RNA molecule at a specific site. However, the occurrence of ribozymes is rare. The rest of this chapter will concern enzyme proteins that are abundant in the mammalian cell.

Enzymes make possible reactions to occur under bodily conditions that otherwise would require conditions that the body could not tolerate (excessive temperature, pressure, pH, etc.). A catalyst is generally regarded as a substance, in

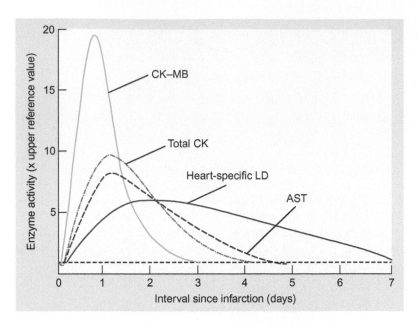

Time course of plasma enzyme activities after
myocardial infarction

| Enzyme | Onset (h) | Peak (h) | Duration (days) |
|---|---|---|---|
| Creatine kinase (MB isoenzyme; CK-MB) | 3–10 | 12–24 | $1\frac{1}{2}$–3 |
| Creatine kinase (total; total CK) | 5–12 | 18–30 | 2–5 |
| Aspartate aminotransferase (AST) | 6–12 | 20–30 | 2–6 |
| Heart-specific lactate dehydrogenase (LD) | 8–16 | 30–48 | 5–14 |

**FIGURE 5.5**   Patterns of serum enzyme activities with time after myocardial infarction. Elevations in enzyme activities are observable as much as 48 h after the attack, and some activities persist for days or even weeks. *AST*, aspartate aminotransferase; *CK-MB*, subunit of creatine kinase; *heart-specific LD*, LDH1, or $H_4$ isozyme of lactate dehydrogenase.

**TABLE 5.2  Enzyme Preparations Used in the Treatment of Some Diseases**

| Enzyme (Preparation) | Disease |
|---|---|
| Pancreatic digestive enzymes (Sometimes as Pancreatin or pancrelipase) | Pancreatic insufficiency Cystic fibrosis |
| [β-Glucocerebrosidase (analog) also called Cerezyme] | Gaucher disease (long-term treatment) |
| Hyaluronidase (human recombinant) and *N*-acetyl-galactosamine-4-sulfatase | Mucopolysaccharidosis VI |
| Myozyme (α-glucosidase) | Pompe disease (α-glucosidase deficiency) |
| Lactade (lactase or β-galactosidase) | Lactose intolerance |
| "Beano." (fungal α-galactosidase) | Prevents gas and bloating after consumption of legumes |
| Enzyme mixture (current research) | Celiac disease (inability to digest gluten) |

small amount compared to the reactants, which modifies and increases the rate of a reaction without itself being consumed. This is generally true for enzymes in that they allow a reaction to occur under bodily conditions, and although they form complexes with substrates and products, they emerge from the reaction in free form just as they started out. The simplest reaction involving an enzyme (E) would be for it to catalyze the conversion of S, the substrate, being converted to P, the product of the reaction:

$$E + S \rightleftharpoons ES \rightleftharpoons E + P$$

Note that many enzymatic reactions are reversible, as indicated by the reversible arrows, so that in the reverse reaction, the enzyme can combine with the product, P to form an enzyme−product complex (EP). In some enzymatic reactions, a relatively large amount of energy is required to initiate the reaction so that the reverse reaction can be quite small compared to the forward reaction. In terms of the first reaction above, the reverse reaction would be:

$$E + P \rightleftharpoons EP \rightleftharpoons ES \rightleftharpoons E + S$$

As will be seen later on, a hydrolytic reaction, where a group is split off from a molecule by addition of water, might be more difficult to reverse than a transfer reaction where one group from one molecule is transferred to another molecule, requiring a smaller amount of energy.

So, a certain amount of energy must be invested to allow the reaction to proceed on its own in the presence of the enzyme. This is called the *energy of activation* (Fig. 5.6).

In order to understand enzymatic reactions and inhibitors of enzymatic reactions, one must resort to a mathematical description of the progress of the reaction. While specifying the conditions of the reaction (pH and buffer, temperature, salt concentration, etc.), the simplest reaction is that of a single substrate to produce a single product (S going to P). When measurements are taken as a function of time, the rate or velocity (ordinate or *y*-axis) can be plotted as a function of substrate concentration ([S], abscissa or *x*-axis) as shown in Fig. 5.7.

The changes in the components of the first-order reaction are shown in Fig. 5.8.

The data of the first-order reaction curve (Fig. 5.7) can be represented in the form of a **straight line** when the reciprocal of the initial velocity ($1/v_i$) is plotted on the ordinate (*y*-axis) as a function of the reciprocal of the substrate concentration ($1/[S]$) plotted on the *x*-axis (abscissa). This representation is known as the **Lineweaver−Burk plot**, and it is shown in Fig. 5.9.

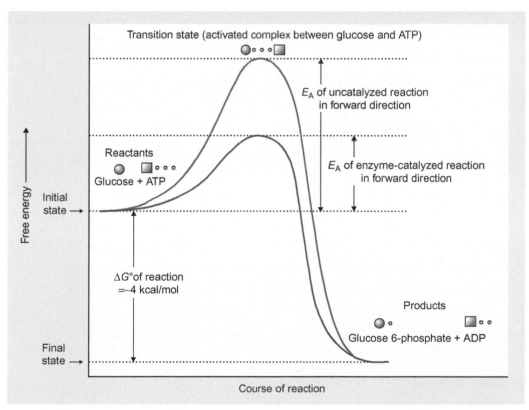

**FIGURE 5.6** A typical energy of activation diagram. In this example, the phosphorylation of glucose is shown: glucose + ATP ⇌ glucose−6−phosphate + ADP. The phosphorylation of glucose by **hexokinase** or by **glucokinase** is the first reaction of glycolysis. Note that the free energy ($\Delta G^0$ = Gibbs free energy) is the same, proceeding from the level of the substrate to the level of the product whether the enzyme is present or not. The superscript $^0$ is for standard reaction conditions, and these can vary dramatically from conditions inside the cell. The difference is that less energy is required to make the reaction proceed in the presence of the enzyme. $E_A$ = energy of activation. *Figure reproduced from http://www.bio.miami.edu/~cmallery/255/255enz/activation_energy%20.jpg.*

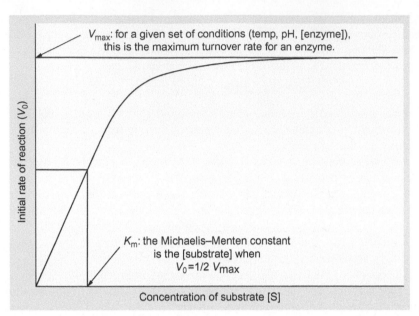

**FIGURE 5.7** Velocity (v) of an enzymatic reaction (y-axis or ordinate) as a function of substrate concentration ([S]) (x-axis or abscissa). Velocity can be quantified as the rate of appearance of the product, P, for which there is usually a direct or indirect measurement. When the various time points are plotted, a first-order curve is obtained as shown in the figure. In this figure, the actual time points are not shown. The straight-line portion of the curve (at lower substrate concentrations) represents a zero-order reaction where the rate of the reaction is proportional to the substrate concentration. The curve reaches saturation (levels off) at the maximal velocity ($V_{max}$). At the **half-maximal velocity**, the value of the molar substrate concentration ([S]) is equal to the **Michaelis−Menten constant** for the given reaction. *Reproduced from G. Litwack,* Human Biochemistry and Disease, *Figure 3−5, page 98, Academic Press/Elsevier, 2008.*

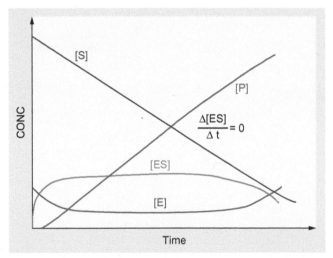

**FIGURE 5.8** Rates of change of the concentrations of substrate (S), product (P), enzyme (E), and enzyme−substrate complex (ES). The substrate concentration, [S], is in great excess over the amount of enzyme, [E]. The concentration of the enzyme−substrate complex, [ES], is low at the start of the reaction, and it is assumed that it remains constant until the end of the reaction when the concentration of substrate is small. *CONC,* concentration.

## Lineweaver−Burk Equation

The **Lineweaver−Burk equation** is derived from the Michaelis−Menten equation (elaborated below) and has the form:

$$1/v = (K_m(1/V_{max}))([S]) + 1/V_{max}$$

or

$$1/v = (K_m/V_{max})(1/[S]) + 1/V_{max}$$

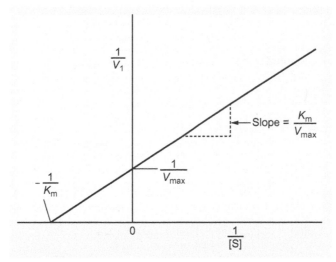

**FIGURE 5.9**  Lineweaver–Burk plot of a first-order enzymatic reaction. $1/v_i$ is plotted on the ordinate as a function of the reciprocal substrate concentration, $1/[S]$, plotted on the abscissa. In this case, the value of the **Michaelis–Menten constant**, $K_m$, is found when the line extrapolates to the second quadrant on the x-axis where the intersection is $-1/K_m$. The $K_m$ is equal to the substrate concentration that promotes 50% of maximal velocity of the reaction. The value of the maximal velocity is determined at the point of intersection on the ordinate that is equal to $1/V_{max}$. Knowing the value of $V_{max}$ from this point, $K_m$ also can be calculated from the slope that is equal to $K_m/V_{max}$. *Reproduced from G. Litwack,* Human Biochemistry and Disease, *Figures 3–6, page 98, Academic Press/Elsevier, 2008.*

that is written in the form of a **straight-line equation**:

$$y = mx + b$$

$y$ is the ordinate value, $m$ is the slope of the straight line, $x$ is the abscissa value, and $b$ is the value at the intersection on the y-axis (when $x = 0$; see Fig. 5.9). The intersection on the x-axis is $-1/K_m$. The straight-line representation facilitates the measurements of $K_m$, and $V_{max}$ compared to the direct plot (Fig. 5.7). Interestingly, the **$K_m$ value** approximates the molar substrate concentration present in the cell.

## Michaelis–Menten Equation

In the simple enzymatic reaction:

$$E + S \underset{k_{-1}}{\overset{k_1}{\rightleftharpoons}} ES \overset{k_2}{\rightarrow} P + E$$

The net effect of the reaction is the conversion of S to P. The term, E, being on both sides of the equation, drops out, reflecting its catalytic action. The enzyme increases the rate of the reaction but is, itself, not altered.

The initial velocity of the reaction, $v_0$, is the rate of appearance of product, P, as a function of time:

$$v_0 = (d[P]/dt)_0$$

and $v_0$ is also proportional to the rate of formation of the enzyme–substrate (ES) complex, ES, and its breakdown to form product:

$$v_0 = d[P]/dt = k_2[ES] = k_2[E]_T[S]/K_m + [S]$$

$[E]_T$ is the total amount of enzyme placed into the reaction, and $K_m$ is the Michaelis constant (as measured in Fig. 5.9). It is the ratio of the rates of the reactions leading to product formation to the rate of the reverse reaction:

$$K_m = k_1 + k_2/k_{-1}$$

When the substrate concentration, [S], is increased significantly so that the rate of reaction is no longer limited by [S], $v_0$ approaches maximal velocity, $V_{max}$ so that, in the equation:

$$v_0 = k_2[E]_T[S]/K_m + [S]$$

all terms in S drop out, including $K_m$ (which is a substrate concentration), then:

$$V_{max} = k_2[E]_T = \text{constant}$$

As $V_{max} = k_2[E]$, $V_{max}$ can substitute for $k_2[E]$ as follows:
In:

$$v_0 = k_2[E]_T[S]/K_m = [S]$$

substituting $V_{max}$ for $k_2[E]$, the Michaelis−Menten equation becomes:

$$v_0 = V_{max}[S]/K_m + [S]$$

This equation describes the plot in Fig. 5.9.

The Lineweaver−Burk equation, describing the straight-line plot shown in Fig. 5.9, is derived from the Michaelis−Menten equation by taking the reciprocals of both sides of the equation. Thus:

$$v_0 = V_{max}[S]/K_m + [S]$$

becomes:

$$1/v_0 = K_m/V_{max}[S] + [S]/V_{max}[S]$$

and

$$1/v_0 = K_m/V_{max}1/[S] + 1/V_{max}$$

is in the form of a straight-line equation:

$$y = mx + b$$

making the determination of $K_m$ and $V_{max}$ direct.

The number of enzyme substrate complexes converted to product per enzyme molecule per unit time is the **turnover number** of the enzyme ($k_{cat}$). The rate of formation of E + P essentially determines the conversion of the enzyme substrate complex to product, this rate, $k_2$, approximates $k_{cat}$:

$$k_{cat} = k_2 = k_2[E]_T/[E]_T = V_{max}/[E]_T = \text{turnover number}$$

The turnover number is expressed as $s^{-1}$, or reciprocal seconds (1/s). The turnover number informs on the rate of conversion of substrate to product that can be useful information for a given enzyme.

## Inhibition of Enzymatic Activity

Within a cell in the body, there may be more than 3000 different enzymes. The rates of many of these enzymes are regulated by other molecules in the cell, and these other molecules participate in the homeostatic milieux. Sometimes, an initial or early enzyme in a metabolic pathway is the rate-limiting step for the function of the entire pathway, and the regulation of this enzyme is of special importance. Sometimes, the activity of an enzyme like this is under **allosteric control** (to be discussed subsequently). Of great interest is the use of medicines/drugs, which function by the inhibition of specific enzymatic activity. Examples of specific drugs will be mentioned later.

There are two types of inhibition of enzymes. One is reversible in which the inhibitor binds to the enzyme noncovalently. In **irreversible inhibition**, the inhibitor binds covalently to the enzyme. A **reversible inhibitor** can bind to the active site of the enzyme where the substrate binds. Consequently, this inhibition is reversible by increasing the amount of the substrate to compete with the inhibitor. Covalent binding of the inhibitor to the enzyme, at any site, is noncompetitive. Also, noncompetitive inhibition can be obtained when a noncovalent inhibitor binds so strongly that the reaction is reversible only to a limited extent.

In addition to competitive and noncompetitive inhibition, a third type, that is less common, is **uncompetitive inhibition**.

### Competitive Inhibition

This type of inhibition is the most important for the use of medicines. A type of drug is desired that can titrate (dosage) the activity of an enzyme competitively while still having the advantage of being able to withdraw the drug and allow

Classical competitive inhibition

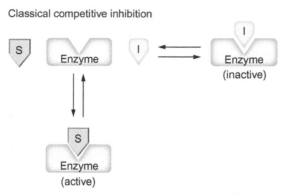

**FIGURE 5.10**   A model showing that a competitive inhibitor binds to the substrate-binding site.

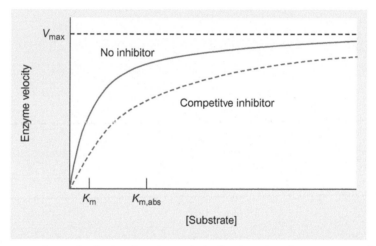

**FIGURE 5.11**   Velocity of the enzymatic reaction is plotted on the $y$-axis (ordinate) vs the substrate concentration on the $x$-axis (abscissa). The values of $K_m$ and v are altered but $V_{max}$ is unaffected (theoretically, if the reaction were to be carried out long enough the rate of the inhibited reaction would reach the same rate as the uninhibited reaction; the inhibited reaction would take much longer to reach the maximal velocity). Or, putting it another way, the same $V_{max}$ can be achieved with the competitive inhibitor present but a higher concentration of substrate would be required. The $K_m$ observed in the presence of the inhibitor is modified by the quantity, $1 + [I]/K_i$, where [I] is the molar concentration of inhibitor, and $K_i$ is the inhibition constant of the inhibitor.

the enzymatic activity to recover. This would be in contrast to a noncompetitive drug inhibitor that would tie up the enzyme in a **dead-end complex**, and withdrawal of the drug would not allow the enzyme to resume its normal activity. Obviously, a drug that is a competitive inhibitor would be the medicine of choice in most cases.

The characteristic of a competitive inhibitor is that it binds in the same site as the substrate and therefore competes with the substrate for the enzyme's binding site or substrate pocket. This is visualized in Fig. 5.10.

A competitive inhibitor resembles the structure of the substrate. Most drugs are competitive inhibitors that bind to the active site of an enzyme more strongly than the substrate. The inhibition by this type of inhibitor is reversible by increasing the amount of available substrate. When the concentration of the inhibitor is increased, the rate of the reaction will decrease accordingly because a greater number of the active sites of the enzyme will be occupied by the inhibitor rather than by the substrate. A plot of the time-course of an enzymatic reaction in the presence or absence of a competitive inhibitor is shown in Fig. 5.11.

If Fig. 5.11 showed a continuing higher amount of substrate on the $x$-axis the curve for the presence of the competitive inhibitor would reach the same value as the control ($V_{max}$) with no inhibitor emphasizing that increased substrate can overcome (replace) the competitive inhibitor. *The inhibition constant, $K_i$, is defined as the dissociation constant in the reversible reaction between enzyme and inhibitor*:

$$E + I \rightleftharpoons EI$$

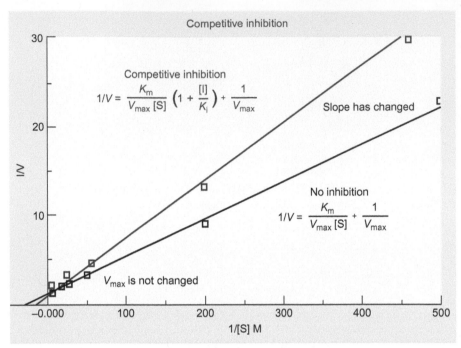

**FIGURE 5.12** Double reciprocal (Lineweaver–Burk) plot of a reaction similar to that shown in Fig. 5.11. The line with the inhibitor is raised above the normal reaction without inhibitor since this is a reciprocal plot. *The straight-line equation is modified by the quantity 1 + [I]/K$_i$ in the presence of the competitive inhibitor, applying to the slope (K$_m$/V$_{max}$) and the value of the x-axis (1/[S]) intercept as shown by the equation in the figure.*

Therefore,

$$K_i = [E][I]/[EI]$$

The double reciprocal plot of the reaction shown in Fig. 5.11 is shown in Fig. 5.12.

As shown in Fig. 5.12, in the presence of a **competitive inhibitor**, the slope and the value of the x-axis intercept are increased by $1 + [I]/K_i$:

$$1/v = K_m/V_{max}(1/[S]) + 1/V_{max}$$

becomes:

$$1/v = K_m/V_{max}(1 + [I]/K_i)(1/[S])(1 + [I]/K_i) + 1/V_{max}$$

or:

$$1/v = K_m/V_{max}[S](1 + [I]/Ki) + 1/V_{max}$$

but the value of $V_{max}$ is unchanged.

## Noncompetitive Inhibition

In the case of a **noncompetitive inhibitor**, the binding of the inhibitor is often to a site distant from the substrate-binding site (the active site). Here, all of the components of the equation are modified by the value, $1 + [I]/K_i$:

$$1/v = K_m/V_{max}[S](1 + [I]/K_i) + 1/V_{max}(1 + [I]/K_i)$$

A comparison of the initial velocity, $v$, as a function of substrate concentration ([S]) between the two types of inhibition is shown in Fig. 5.13.

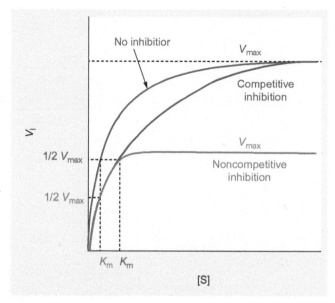

**FIGURE 5.13** Initial velocity, $v_i$, is plotted on the ordinate ($y$-axis) as a function of the molar concentration of substrate ([S]) on the $x$-axis (abscissa). As indicated previously, if the reaction with the competitive inhibitor progresses long enough, it will eventually reach the value of the $V_{max}$. This is not the case in the presence of a noncompetitive inhibitor where some of the enzyme is actually removed from the reaction by forming the dead-end complex, EI, and $V_{max}$ is lower.

The inhibition by a noncompetitive inhibitor will not be reversed by increasing the concentration of substrate because the inhibitor binding site is distant from the substrate-binding site, and the inhibitor's structure may not resemble that of the substrate; thus, the reaction will remove enzyme from the productive pathway:

$$E + S \rightleftharpoons ES \longrightarrow E + P$$
$$+$$
$$I$$
$$\Updownarrow$$
$$ESI$$

This situation is pictured in Fig. 5.14.

When the noncompetitive inhibitor forms a covalent bond with the inhibitor site on the enzyme, the enzyme is essentially irreversibly removed from the reaction (forming a dead-end complex). However, even if the inhibitor binds to the enzyme covalently, there may be a trickle of reversibility (principle of microscopic reversibility). In general, a reversible inhibitor binds to its site noncovalently generating a forward as well as a reverse reaction. If the reversible inhibitor binds to the active site (where the substrate binds), increasing the amount of substrate will displace the reversible (competitive) inhibitor. A covalent inhibitor has a tiny or nonmeasurable reverse reaction so that it essentially removes the enzyme from the reaction by forming a dead-end complex. The double reciprocal plot of noncompetitive inhibition is shown in Fig. 5.15.

## Uncompetitive Inhibition

In another somewhat rare form of noncompetititve inhibition called **uncompetitive inhibition**, the inhibitor binds to a site different from the substrate-binding site, but *interacts only with the enzyme substrate complex*, not with the enzyme alone in the absence of substrate. This could be interpreted to represent the inhibitor binding to a site separate from the active site (where the substrate binds) but also interacting with the enzyme—bound substrate as shown in Fig. 5.16.

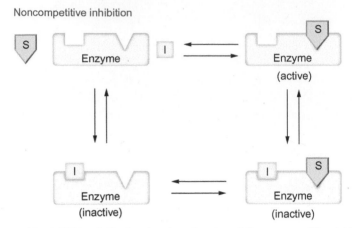

FIGURE 5.14 A model of noncompetitive inhibition indicating that the substrate and the noncompetitive inhibitor bind at different sites on the enzyme.

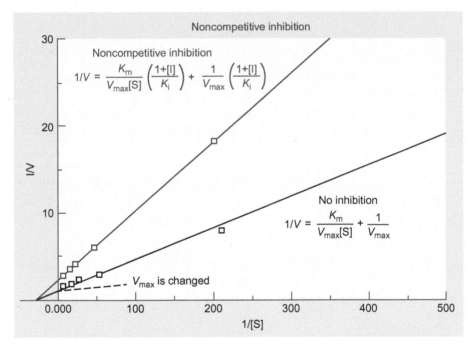

FIGURE 5.15 Double reciprocal plot of noncompetitive inhibition. The upper line represents the system in the presence of the inhibitor (the line is increased because this is a reciprocal plot). The slope and $1/V_{max}$ values are modified by the term, $1 + [I]/K_i$ as shown by the equation in the figure. The x-axis intercept (in the second quadrant) is unchanged by the presence of the inhibitor. This is not true for competitive inhibition (Fig. 5.12) where the value of the x-axis intercept is increased in the presence of the inhibitor (less negative, second quadrant). The y-axis intercept in the presence of the inhibitor is moved upwards, whereas in competitive inhibition, it is unchanged.

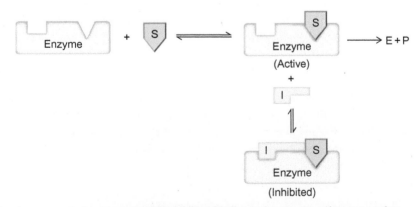

FIGURE 5.16 Scheme showing a scenario for an uncompetitive inhibitor binding to the enzyme–substrate complex.

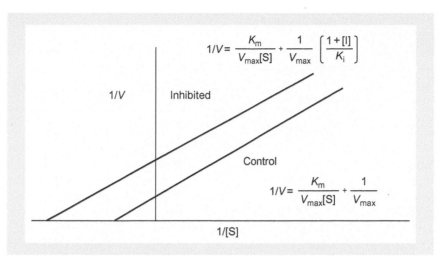

**FIGURE 5.17** Double reciprocal plot of uncompetitive inhibition.

The inhibitor removes enzyme in the form of the ES complex (not the enzyme alone) from the reaction:

$$E + S \rightleftharpoons ES \longrightarrow E + P$$
$$+$$
$$I$$
$$\Updownarrow$$
$$ESI$$

When the reaction is run with increasing levels of the inhibitor, a series of parallel lines emerge in the double reciprocal plots. One such line at one concentration of the **uncompetitive inhibitor** is shown in Fig. 5.17.

In this case only, the "b term" in the equation of a straight line ($y = mx + b$), which is $1/V_{\text{max}}$, is modified by the term: $1 + [\text{I}]/K_{\text{i}}$, and the slope is unchanged, creating the parallel line. Uncompetitive inhibition is a special form of noncompetitive inhibition. Frequently, it occurs when there is more than one substrate.

## Two-substrate Systems

So far, we have been dealing with simple one-substrate enzymatic reactions. In fact, many enzymatic reactions involve more than one substrate. For example,

$$\text{oxidized substrate} + \text{reduced coenzyme} \rightleftharpoons \text{reduced product} + \text{oxidized coenzyme}$$

would pertain to many **oxidoreductases** that use a **dissociable coenzyme** (such as NADH/NAD$^+$). The dissociable coenzyme qualifies as a reactant and is considered a second substrate. Michaelis kinetics are carried out by holding one of the substrates in excess, whereas the other is varied in limiting concentrations. The same is done for the second substrate where the concentration of the first is high, and the concentration of the second substrate is varied. In this way, a set of double reciprocal plots can be produced (a specific case of a multisubstrate system applies to transaminases, and this will be discussed below). The actual chemical reaction takes place, whereas the reactants are in combination with the enzyme (EAB $\rightleftharpoons$ EPQ), so that template mechanism for a random ternary complex (three different molecules bound together) mechanism might be:

In a random mechanism, the substrates can bind in any order. Thus, enzyme could bind to either A or B initially to produce enzyme-A-B complex (EAB).

In an **ordered mechanism**, the substrates would have to bind in a specific order. That is, A would have to bind before B could bind (or vice versa) to form EAB in which case the mechanism would appear as,

$$E + A \rightleftharpoons EA + B \rightleftharpoons EAB \rightleftharpoons EPQ \rightleftharpoons EP + Q \rightleftharpoons E + P$$

In the random mechanism, as per Lineweaver–Burk, when 1/[B] is plotted as a function of 1/$v$ while holding [A] at a high constant concentration and then doing the same analysis when 1/[A] is plotted as a function of 1/$v$ while [B] is held in a high concentration, a set of lines is produced that will intersect.

As applied to some transaminases and other enzymes, a two-substrate mechanism may take the form of a **ping-pong mechanism** rather than a random ternary complex. In this case, the enzyme itself can be modified in its reaction with the first substrate (e.g., formation of a Schiff base between an amino acid substrate and the enzyme) to form an altered enzyme intermediate (E*). The first product would be released before the second substrate binds to E*. The reaction resembles:

$$E + A \rightleftharpoons E^*P \rightarrow P + E^* + B \rightleftharpoons E^*B \rightleftharpoons EQ \rightleftharpoons Q + E$$

The reaction can become more complicated if the coenzyme is dissociable and needs to be treated as a third reactant. Ping-pong mechanisms can be detected by standard methods and by the use of isotopic exchange experiments. For multisubstrate reactions, the rate equations become complex and involve several rate constants describing the formation of the various forms of the enzyme complexes.

These types of enzymatic reactions can be pictured in Cleland representations. W. W. Cleland is an acknowledged pioneer in enzyme kinetics. There are two types of sequential reactions for a two-substrate enzyme system, **random sequential** in which both substrates bind before both products are released, and **ordered sequential** in which there is a specific order of the reactants binding the enzyme and a specific order in which the products are released. These mechanisms are shown in Fig. 5.18A and B.

A two-substrate reaction can follow ping-pong kinetics. These kinetics can represent a transaminase reaction, for example, in a situation where the amino acid being transaminated is one substrate, and the initial keto acid is the second substrate as the other reactant (assuming that the coenzyme is bound tightly enough so that it is not considered a reactant). Here, reactant A binds to the enzyme followed by the release of the product, P. This is followed by the binding of reactant B which is followed by the release of product Q as shown in Fig. 5.19.

Where a transaminase reaction is evaluated in which the coenzyme (pyridoxal phosphate, PLP) is highly dissociable, it may be considered to be a third reactant, generating complex velocity equations. Complicated mechanisms are often approached using matrix algebra.

## Allosterism–Non-Michaelis–Menten Kinetics

This phenomenon occurs in systems usually involving enzymes with multiple subunits. Positive or negative effectors may be involved that bind to sites remote from the active site (catalytic center; substrate-binding site) to produce conformational changes in the enzyme protein that affect the active center where the substrate binds. These changes either reduce (**positive effector**) or increase (**negative effector**) the Michaelis constant ($K_m$). The binding of oxygen to **hemoglobin** (containing four subunits), although not an enzyme, is a good example of this phenomenon. The binding of the first oxygen molecule to hemoglobin begins to promote changes in the conformation of the protein, so that the second oxygen binds with greater affinity and the effect increases with subsequent binding of substrate. Alternatively, the binding of the first oxygen molecule can result in a changed conformation of all the subunits so that oxygen binds subsequently with the highest affinity, in a single step (see Chapter 21: Human Biochemistry: Blood and Lymphatic System). Similar phenomena occur with **allosteric enzymes**. Substrates that produce allosteric effects on an enzyme (similar to hemoglobin binding oxygen) are termed **"homotropic" effectors**. Human enzymes, such as protein kinase, glucokinase and some alkaline phosphatases are allosteric enzymes. Protein kinases, such as protein kinase A, may be a special form of allosteric enzymes because the subunits of this enzyme dissociate as part of the activation process as shown in Fig. 5.20.

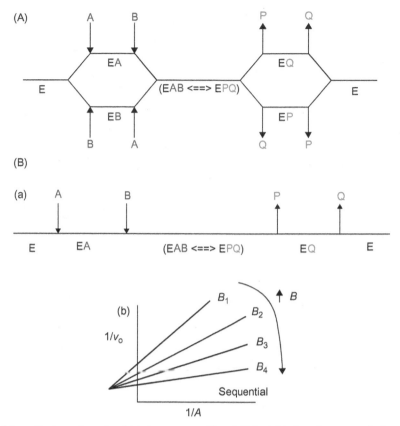

**FIGURE 5.18** (A). Sequential reactions are those in which the reactants (A and B) both bind to the enzyme before both products (P and Q) are released. This figure shows a random sequential mechanism in which there is a random order of reactants binding and a random order in which the products are released. (B). (a) An ordered sequential reaction is one in which there is a specific order of reactants binding and a specific order in which the reactants are released. The appearance of kinetic measurements is shown below in (b). The rates of reactions are measured at increasing levels of reactant A as a function of different concentrations of reactant B that produce the family of lines pictured (*bottom*). The lines intersect in the second quadrant above the *x*-axis and intersect the ordinate at different values. *(A) Reproduced from http://employees.csbsju.edu/hjakuboswki/classes/ ch331/transkinetics/olcomplicatedenzyme.html. (B) Reproduced from http://employees.csbsju.edu/hjakubowski/classes/ch331/transkinetics/olcomplicatedenzyme.html*

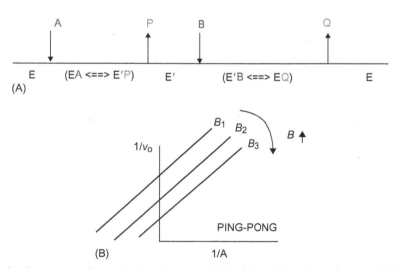

**FIGURE 5.19** (A) Kinetics of a ping-pong reaction mechanism. Reactant A binds, followed by release of product P. This is followed by the binding of reactant B which is followed by the release of product Q. (B) Appearance of the kinetic measurements in which the reactant A is varied in concentrations and the velocity of the reaction is measured as a function of different concentrations of reactant B. A family of parallel lines is observed. *Reproduced from http://employees.csbsju.edu/hjakubowski/classes/ch331/transkinetics/olcomplicatedenzyme.html.*

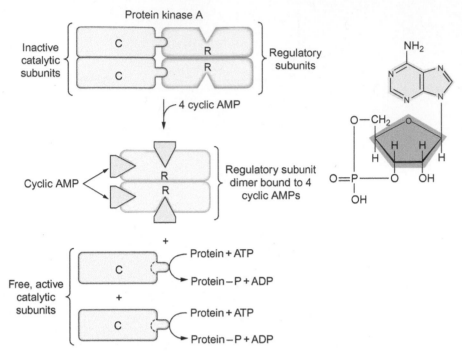

**FIGURE 5.20** Mechanism of activation of protein kinase A showing interaction with the effector, cyclic AMP, to cause the liberation of the active catalytic subunits. On the *right* is the structure of cyclic AMP.

The hallmark of **allostery** is evident from the progress of a reaction when velocity, $v$, is plotted against substrate concentration, [S]. Instead of generating a first-order curve (Fig. 5.7), an **S-shaped curve** showing a lag phase is encountered (Fig. 5.21).

When a **positive effector (heterotropic effector)** is present, it binds to a site distinct from the substrate-binding site and changes the conformation of the enzyme including the active site, so that it is able to bind the substrate more effectively and speed up the reaction. This being the case, less substrate is required to reach the half-maximal rate of reaction, and the rate curve is moved to the left on the graph indicating a reduction in the value of $K_m$. When an allosteric inhibitor is present (**negative effector**), providing the enzyme has a binding site for this effector, the conformational change induced is such that the substrate is less effectively utilized, and the reaction requires more substrate to reach the half-maximal rate. The curve is moved to the right on the graph, increasing the value of $K_m$. These behaviors are shown in Fig. 5.22.

The $K_m$ is a measure of the efficiency with which the enzyme is able to bind and act on the substrate. It is not strictly a measure of **affinity** because affinity is related inversely to the **dissociation constant,** and the production of a product from the enzyme substrate complex adds a complication to the strict definition of affinity. Thus, affinity can be measured for a simple reaction, such as,

$$A + B \rightleftharpoons AB$$

where the affinity is equal to [AB]/[A] [B] or proportional to $k_1/k_{-1}$, where $k_1$ is for the forward reaction and $k_{-1}$ is for the reverse reaction. If AB breaks down further to the right to form a product, P, as in an enzymatic reaction, the reaction is no longer strictly a dissociation of AB into A and B.

A cartoon of what might be a dimeric enzyme having sites for both a positive heterotropic effector and a negative heterotropic effector is shown in Fig. 5.23.

## CLASSIFICATION

There are six major classes of enzymes based on the type of the reaction they catalyze. These classes are (1) oxidoreductases; (2) transferases; (3) hydrolases; (4) lyases; (5) isomerases, and (6) ligases.

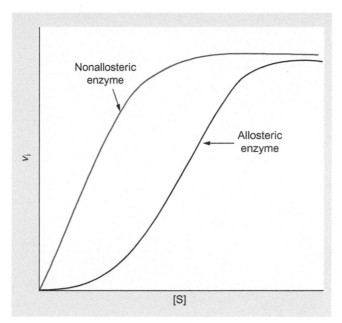

**FIGURE 5.21**  Typical curve of initial velocity, $v_i$, plotted vs [S] for a nonallosteric enzyme and an allosteric enzyme. The allosteric enzyme shows a sigmoidal or S-shaped curve.

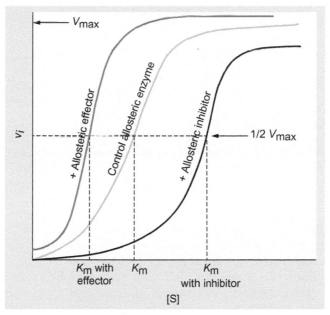

**FIGURE 5.22**  Effects of a positive or negative effector on the S-shaped velocity curve of an allosteric enzyme. A positive effector moves the curve to the left resulting in a lower $K_m$ value of substrate concentration. A negative effector moves the curve to the right resulting in a higher $K_m$ value of substrate concentration.

**Oxidoreductases** either add or remove hydrogen atoms in a given reaction. The direction constituting oxidation refers to the removal of two hydrogen atoms. The reductase activity refers to the direction of the reaction in which there is an addition of hydrogen atoms. **Lactate dehydrogenase** is an oxidoreductase catalyzing the reaction:

$$^-OOC-CO-CH_3 + NADH + H^+ \rightleftharpoons \ ^-OOC-HCOH-CH_3 + NAD^+$$
$$\text{Pyruvate} \qquad\qquad\qquad\qquad \text{Lactate}$$

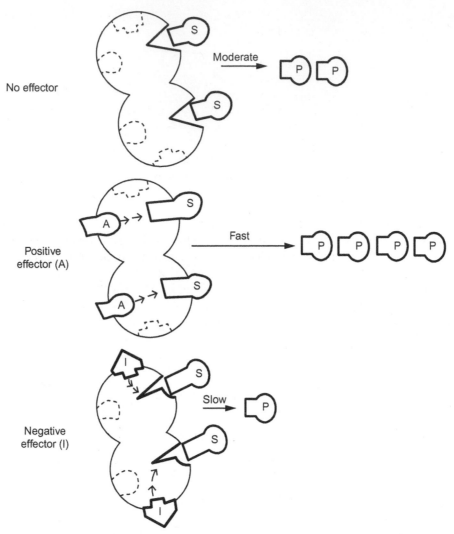

**FIGURE 5.23**   A cartoon showing a hypothetical dimeric allosteric enzyme with positive and negative effector sites. The effects on the conformation of the active site are shown either by a positive effector (A) to increase the efficiency of binding and acting on the substrate or by a negative effector (I) to decrease that efficiency.

The reaction from left to right represents the reductase activity, and the opposite direction of the reaction represents the oxidase activity. The preferred direction of this reaction in the cell would be determined by the availability of pyruvate or lactate and the concentration of $NADH + H^+$ compared to $NAD^+$.

**Transferases** catalyze the transfer of a group from one molecule to another. Typical of this group are the aminotransferases (or transaminases) that catalyze the transfer of an amino group from an amino acid to a keto acid. **Aspartate aminotransferase** is one example, and it catalyzes the reaction:

$$^-OOC - NH_2CH - CH_2 - COO^- \ + \ ^-OOC - CO - CH_2CH_2COO^- \ \leftarrow PLP \rightarrow \ ^-OOC - CO - CH_2COO^- \ + \ -OOC - NH_2CH - CH_2CH_2COO -$$

$L$-aspartate $\qquad\qquad\qquad$ α-ketoglutarate $\qquad\qquad\qquad$ oxaloacetate $\qquad\qquad\qquad$ $L$- glutamate

The amino group of L-aspartate is transferred to the keto group of α-ketoglutarate to form oxalocetate and L-glutamate. The coenzyme is **pyridoxal phosphate (PLP)**.

**Hydrolases** are enzymes that catalyze the hydrolysis of a bond by the addition of water. A protease or peptidase is an example of this group of enzymes. In this case, the peptide bond is hydrolyzed by the addition of water:

$$R - CO - NH - R' \ + \textbf{HOH} \rightleftharpoons R - COOH + \textbf{H}_2N - R' \rightarrow R - COO^- \ + ^+H_3N - R'$$

Peptide $\qquad\qquad\qquad\qquad\qquad\qquad\qquad\qquad\qquad$ peptide 1 $\qquad$ peptide 2

**Lyases** catalyze the addition of water, ammonia, or carbon dioxide to double bonds, or they can catalyze the removal of these substances to create double bonds in the reverse direction. **Citrate lyase** is an example of this group of enzymes. The reaction catalyzed by this reaction is shown in Fig. 5.24.

In this reaction, a double bond (keto group) is created when a hydrogen atom (not by oxidation) is removed from carbon 3 of citrate together with an acetate group (carbons 1 an 2 of citrate). In the reverse reaction, citrate is formed from oxaloacetate (mitochondrial citrate synthase reaction).

**Isomerases** catalyze mutase reactions, such as converting L- to D-forms, shifts of chemical groups and movement of double bonds. Thus, the substrate and the product of an isomerase-catalyzed reaction are a pair of structural isomers, generating the name of the enzyme. **Racemases** (interconverting L- and D-forms of a substrate) and **epimerases** (an isomerase that stereochemically inverts the configuration about an asymmetric carbon in a substrate molecule that has more than one asymmetric center generating epimers). **Triose phosphate isomerase** is an example of an isomerase. It catalyzes the reversible reaction converting dihydroxyacetone phosphate to glyceraldehyde-3-phosphate as shown in Fig. 5.25.

In this reaction, protons are shifted to create a double bond at carbons 1 and 2 of the enediol intermediate. The catalytic center of the enzyme contains glutamate and histidine that extract and donate protons in the reaction.

**Ligases** utilize the energy of the terminal phosphate group of adenosine triphosphate (ATP) to join two chemical groups. **Synthases** are members of this group, and human **aminoacyl-tRNA synthase** is an example. There are twenty such enzymes, each recognizing a specific amino acid. The reaction catalyzed by this enzyme is shown in Fig. 5.26.

These enzymes catalyze the initial step in protein synthesis.

The **International Union of Biochemistry** as part of a Joint Nomenclature Committee adopted a naming system for enzymes, identifying each enzyme with an EC (enzyme classification) number. The **EC number** defines

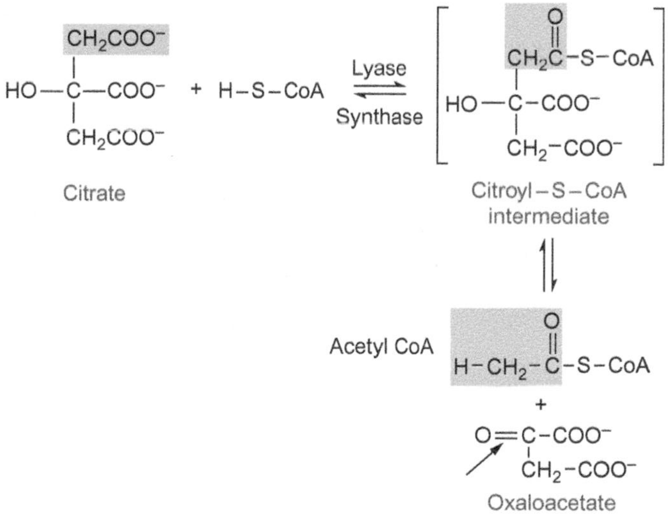

**FIGURE 5.24** Reaction catalyzed by cytosolic ATP-citrate lyase (energized by the hydrolysis of ATP in the fed state). The double bond of oxaloacetate is created by removal of hydrogen from a hydroxyl group and carbons 1 and 2 of citrate (acetyl).

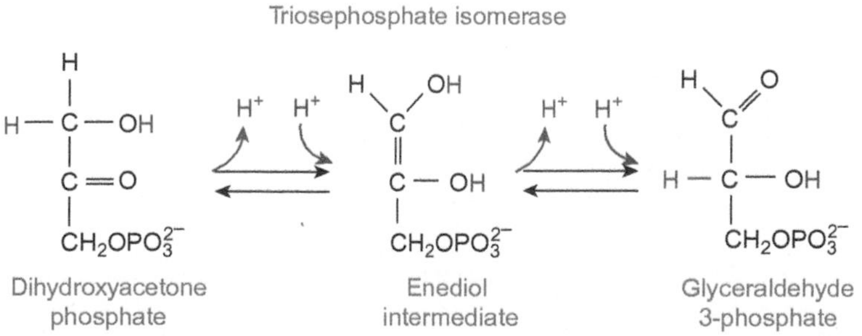

**FIGURE 5.25** The reaction catalyzed by triose phosphate isomerase. Hydrogen atoms moved in the reaction are colored *red*.

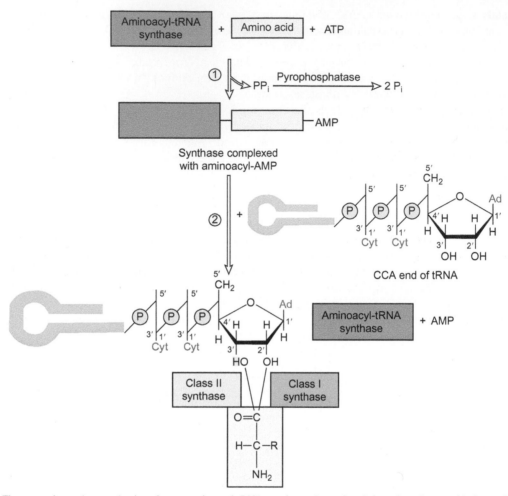

**FIGURE 5.26** The general reaction mechanism for an aminoacyl-tRNA synthase. *Reproduced from http://www.ncbi.nlm.gov/bookshelf/br.fcgi? book = m*

the type of enzyme, its reaction, and its substrate. The EC numbering system for all the classes of enzymes is shown in Table 5.3.

## COENZYMES

Some enzymes are active without coenzymes. However many, require a coenzyme to be active. An enzyme that is inactive in the absence of its coenzyme is called an **apoenzyme**. In the presence of its coenzyme to produce the active form of the enzyme, it is called a **holoenzyme**:

$$\text{apoenzyme} + \text{coenzyme} \rightleftharpoons \text{holoenzyme}$$

Although some enzymes contain a coenzyme that is tightly bound, others may contain a coenzyme that is readily dissociable. In the latter case, the coenzyme can be considered as a reactant or substrate. Thus, in the lactate dehydrogenase reaction, for example, pyruvate and the coenzyme NADH need to be added to the enzyme and, kinetically, this would be considered to be a two-substrate reaction:

$$\text{pyruvate} + \text{NADH} \rightleftharpoons \text{lactate} + \text{NAD}^+$$

In a double reciprocal plot in which 1/velocity (*y*-axis) is plotted against 1/[pyruvate] holding the concentration of NADH high, the plot would give the $K_m$ for pyruvate. A similar experiment in which [pyruvate] was at a saturating level and [NADH] was varied, the reciprocal plot would give the $K_m$ for NADH.

**TABLE 5.3  Enzyme Classification and Nomenclature**

**Class 1. Oxidorectases**

1.1.  Acting on the CH—OH group of donors

1.1.1. With NAD or NADP as acceptor

1.1.2. With a cytochrome as acceptor

1.1.3. With oxygen as acceptor

1.1.4. With a disulfide as acceptor

1.1.5. With a quinone or similar compound as acceptor

1.1.99. With other acceptors

1.2.  Acting on the aldehyde or oxo group of donors

1.2.1. With NAD or NADP as acceptor

1.2.2. With a cytochrome as acceptor

1. 2.3. With oxygen as acceptor

1.2.4. With a disulfide as acceptor

1. 2.7. With an iron—sulfur protein as acceptor

1.2.99. With other acceptors

1.3.  Acting on the CH—CH group of donors

1.3,1. With NAD or NADP as acceptor

1.3.2. With a cytochrome as acceptor

1,3.3. With oxygen as acceptor

1.3.5. With a quinone or related compound as acceptor

1.3.7. With an iron—sulfur protein as acceptor

1.3.99. With other acceptors

1.4.  Acting on the CH—NH$_2$ group of donors

1.4.1. With NAD or NADP as acceptor

1.4.2. With a cytochrome as acceptor

1.4.3. With oxygen as acceptor

1.4.4. With a disulfide as acceptor

1.4.7. With an iron—sulfur protein as acceptor

1.4.99. With other acceptors

1.5.  Acting on the CH—NH group of donors

1.5.1. With NAD or NADP as acceptor

1.5.3. With oxygen as acceptor

1.5.4. With a disulfide as acceptor

1.5.5. With a quinone or similar compound as acceptor

1.5.99. With other acceptors

1.6.  Acting on NADH$_2$ or NADPH$_2$

1.6.1. With NAD or NADP as acceptor

1.6.2. With a cytochrome as acceptor

1.6.4. With a disulfide as acceptor

*(Continued)*

**TABLE 5.3** (Continued)

1.6.5. With a quinone or similar compound as acceptor

1.6.6. With a nitrogenous group as acceptor

1.6.8. With a flavin as acceptor

1.6.99. With other acceptors

1.7.  Acting on other nitrogenous compounds as donors

1.7.2. With a cytochrome as acceptor

1.7.3. With oxygen as acceptor

1.7.7. With an iron–sulfur protein as acceptor

1.7.99. With other acceptors

1.8.  Acting on a sulfur group of donors

1.8.1. With NAD or NADP as acceptor

1.8.2. With a cytochrome as acceptor

1.8.3. With oxygen as acceptor

1.8.4. With a disulfide as acceptor

1.8.5. With a quinone or similar compound as acceptor

1.8.7. With an iron–sulfur protein as acceptor

1.8.99. With other acceptors

1.9.  Acting on a heme group of donors

1.9.3. With oxygen as acceptor

1.9.6. With a nitrogenous group as acceptor

1.9.99. With other acceptors

1.10.  Acting on diphenols and related substances as donors

1.10.1. With NAD or NADP as acceptor

1.10.2. With a cytochrome as acceptor

1.10.3. With oxygen as acceptor

1.10.99. With other acceptors

1.11.  Acting on a peroxide as acceptor (peroxidases)

1.11.1.–A single subclass containing the peroxidases

1.12.  Acting on hydrogen as donor

1.12.1. With NAD or NADP as acceptor

1.12.2. With a cytochrome as acceptor

1.12.99. With other acceptors

1.13.  Acting on single donors with incorporation of molecular oxygen

1.13.11. With incorporation of two atoms of oxygen

1.13.12. With incorporation of one atom of oxygen

1.13.99. Miscellaneous (requires further characterization)

1.14.  Acting on paired donors with incorporation of molecular oxygen

1.14.11. With 2-oxoglutarate as one donor, and incorporation of one atom each of oxygen into both donors

*(Continued)*

**TABLE 5.3** (Continued)

1.14.12. With NADH$_2$ or NADPH$_2$ as one donor, and incorporation of two atoms of oxygen into one donor

1.14.13. With NADH$_2$ or NADPH$_2$ as one donor, and incorporation of one atom of oxygen

1.14.14. With reduced flavin or flavoprotein as one donor, and incorporation of one atom of oxygen

1.14.15. With a reduced iron–sulfur protein as one donor, and incorporation of one atom of oxygen

1.14.16. With reduced pteridine as one donor, and incorporation of one atom of oxygen

1.14.17. With ascorbate as one donor, and incorporation of one atom of oxygen

1.14.18. With another compound as one donor, and incorporation of one atom of oxygen

1.14.99. Miscellaneous (requires further characterization)

1.15.  Acting on superoxide radicals as acceptor

1.16.  Oxidizing metal ions

1.16.1. With NAD or NADP as acceptor

1.16.3. With oxygen as acceptor

1.17.  Acting on –CH$_2$– groups

1.17.1. With NAD or NADP as acceptor

1.17.3. With oxygen as acceptor

1.17.4. With a disulfide as acceptor

1.17.99. With other acceptors

1.18.  Acting on reduced ferredoxin as donor

1.18.1. With NAD or NADP as acceptor

1.18.6. With dinitrogen as acceptor

1.18.99. With H$^+$ as acceptor

1.19.  Acting on reduced flavodoxin as donor

1.19.6. With dinitrogen as acceptor

1.97.–. Other oxidoreductases

**Class 2. Transferases**

2.1.  Transferring one-carbon groups

2.1.1. Methyltransferases

2.1.2. Hydroxymethyl-, formyl-, and related transferases

2.1.3. Carboxyl- and carbamoyltransferases

2.1.4. Amidinotransferases

2.2.  Transferring aldehyde or ketone residues

2.2.1. A single subclass containing the transaldolases and transketolases

2.3.  Acyltransferases

2.3.1. Acyltransferases

2.3.2. Aminoacyltransferases

2.4.  Glycosyltransferases

2.4.1. Hexosyltransferases

2.4.2. Pentosyltransferases

2.4.99. Transferring other glycosyl groups

(Continued)

**TABLE 5.3** (Continued)

2.5. Transferring alkyl or aryl groups, other than methyl groups

2.5.1. A single subclass that includes a rather mixed group of such enzymes

2.6. Transferring nitrogenous groups

2.6.1. Transaminases (aminotransferases)

2.6.3. Oximinotransferases

2.6.99. Transferring other nitrogenous groups

2.7. Transferring phosphorus-containing groups

2.7.1. Phosphotransferases with an alcohol group as acceptor

2.7.2. Phosphotransferases with a carboxyl group as acceptor

2.7.3. Phosphotransferases with a nitrogenous group as acceptor

2.7.4. Phosphotransferases with a phosphate group as acceptor

2.7.6. Diphosphotransferases

2.1.7. Nucleotidyltransferases

2.7.8. Transferases for other substituted phosphate groups

2.7.9. Phosphotransferases with paired acceptors

2.8. Transferring sulfur-containing groups

2.8.1. Sulfurtransferases

2.8.2. Sulfotransferases

2.8.3. CoA-transferases

2.9. Transferring selenium-containing groups

**Class 3. Hydrolases**

3.1. Acting on ester bonds

3.1.1. Carboxylic ester hydrolases

3.1.2. Thiolester hydrolases

3.1.3. Phosphoric monoester hydrolases

3.1.4. Phosphoric diester hydrolases

3.1.5. Triphosphoric monoester hydrolases

3.1.6. Sulfuric ester hydrolases

3.1.7. Diphosphoric monoester hydrolases

3.1.8. Phosphoric triester hydrolases

3.1.11. Exodeoxyribonucleases producing 5′-phosphomonoesters

3.1.13. Exoribonucleases producing 5′-phosphomonoesters

3.1.14. Exoribonucleases producing other than 5′-phosphomonoesters

3.1.15. Exonucleases active with either ribo- or deoxyribonucleic acids and producing 5′-phosphomonoesters

3.1.16. Exonucleases active with either ribo- or deoxyribonucleic acids and producing other than 5′-phosphomonoesters

3.1.21. Endodeoxyribonucleases producing 5′-phosphomonoesters

3.1.22. Endodeoxyribonucleases producing other than 5′-phosphomonoesters

3.1.25. Site-specific endodeoxyribonucleases specific for altered bases

*(Continued)*

**TABLE 5.3** (Continued)

3.1.26. Endoribonucleases producing 5′-phosphomonoesters

3.1.27. Endoribonucleases producing other than 5′-phosphomonoesters

3.1.30. Endonucleases active with either ribo- or deoxyribonucleic acid and producing 5′-phosphomonoesters

3.1.31. Endonucleases active with either ribo- or deoxyribonucleic acid and producing other than 5′-phosphomonoesters

3.2.  Glycosidases

3.2.1. Hydrolyzing *O*-glycosyl compounds

3.2.2. Hydrolyzing *N*-glycosyl compounds

3.2.3. Hydrolyzing *S*-glycosyl compounds

3.3.  Acting on ether bonds

3.3.1. Thioether hydrolases

3.3.2. Ether hydrolases

3.4.  Acting on peptide bonds (peptidase)

3.4.11. Aminopeptidases

3.4.13. Dipeptidases

3.4.14. Dipeptidyl-peptidases and tripeptidyl-peptidases

3.4.15. Peptidyl-dipeptidases

3.4.16. Serine-type carboxypeptidases

3.4.17. Metallocarboxypeptidases

3.4.18. Cysteine-type carboxypeptidases

3.4.19. Omega peptidases

3.4.21. Serine endopeptidases

3.4.22. Cysteine endopeptidases

3.4.23. Aspartic endopeptidases

3.4.24. Metalloendopeptidases

3.4.99. Endopeptidases of unknown catalytic mechanism

3.5.  Acting on carbon–nitrogen bonds, other than peptide bonds

3.5.1. In linear amides

3.5.2. In cyclic amides

3.5.3. In linear amidines

3.5.4. In cyclic amidines

3.5.5. In nitriles

3.5.99. In other compounds

3.6.  Acting on acid anhydrides

3.6.1. In phosphorus-containing anhydrides

3.6.2. In sulfonyl-containing anhydrides

3.7.  Acting on carbon–carbon bonds

3.7.1. In ketonic substances

3.8.  Acting on halide bonds

*(Continued)*

**TABLE 5.3 (Continued)**

3.8.1. In C-halide compounds

3.9. Acting on phosphorus−nitrogen bonds

3.10.−. Acting on sulfur−nitrogen bonds

3.11.−. Acting on carbon−phosphorus bonds

3.12.−. Acting on sulfur−sulfur bonds

**Table 4 Class 4 Lyases**

4.1. Carbon−carbon lyases

4.1.1 − Carboxy-lyases

4.1.2 − Aldehyde-lyases

4.1.3 − Oxo-acid-lyases

4.1.99 − Other carbon−carbon lyases

4.2. Carbon−oxygen lyases

4.2.1 − Hydro-lyases

4.2.2 Acting on polysaccharides

4.2.99 Other carbon−oxygen lyases

4.3. Carbon−nitrogen lyases

4.3.1 Ammonialyases

4.3.2 Amidine-lyases

4.3.3 Amine-lyases

4.3.99 Other carbon−nitrogen lyases

4.4. Carbon−sulfur lyases

4.5. Carbon-halide lyases

4.6. Phosphorus−oxygen lyases

4.99. Other lyases

**Class 5. Isomerases**

5.1. Racemases and epimerases

5.1.1. Acting on amino acids and derivatives

5.1.2. Acting on hydroxy acids and derivatives

5.1.3. Acting on carbohydrates and derivatives

5.1.99. Acting on other compounds

5.2. cis-trans-Isomerases

5.3. Intramolecular oxidoreductases

5.3.1. Interconverting aldoses and ketoses

5.3.2. Interconverting keto- and enol-groups

5.3.3. Transposing C═C bonds

5.3.4. Transposing S−S bonds

5.3.99. Other intramolecular oxidoreductases

5.4. Intramolecular transferases (mutases)

5.4.1. Transferring acyl groups

*(Continued)*

**TABLE 5.3** (Continued)

5.4.2. Phosphotransferases (phosphomutases)

5.4.3. Transferring amino groups

5.4.99. Transferring other groups

5.5. Intramolecular lyases

5.99. Other isomerases

**Class 6. Ligases**

6.1. Forming carbon—oxygen bonds

6.1.1. Ligases forming aminoacyl-tRNA and related compounds

6.2. Forming carbon—sulfur bonds

6.2.1. Acid—thiol ligases

6.3. Forming carbon—nitrogen bonds

6.3.1. Acid—ammonia (or amine) ligases (amide synthases)

6.3.2. Acid—amino-acid ligases (peptide synthases)

6.3.3. Cyclo-ligases

6.3.4. Other carbon—nitrogen ligases

6.3.5. Carbon—nitrogen ligases with glutamine as amido-*N*-donor

6.4. Forming carbon—carbon bonds

6.5. Forming phosphoric ester bonds

Note: The Nomenclature Committee has accepted a recommendation that nicotinamide-adenine dinucleotide and nicotinamide-adenine dinucleotide phosphate should be abbreviated to NAD and NADP, respectively, rather than $NAD^+$ and $NADP^+$, as used previously. In addition to avoiding the erroneous implication that these two compounds will be positively charged at pH values, the reasons for this are given in the Newsletter (1996). The reduced forms of these enzymes are written as $NADH_2$ and $NADPH_2$, rather than NADH and NADPH.
Reproduced from *Encyclopedia of Life Sciences*, Nature Publishing Group, 2001.

In general, *coenzymes are vitamins or derivatives of vitamins*. They are listed in Table 5.4.

# PROSTHETIC GROUPS

As mentioned previously, some coenzymes are dissociable and can act as a reactant in an enzymatic mechanism. By contrast, *a prosthetic group is a nonproteinaceous molecule that is tightly bound to the enzyme and remains associated with it throughout the catalytic reaction* (sometimes, these groups are referred to simply as coenzymes or cofactors). **Metalloenzymes** usually have tightly bound metals, such as iron, zinc, copper, and manganese as prosthetic groups. **Superoxide dismutase** (**SOD**) is an enzyme with a tightly bound metal cofactor, usually bound with iron but also can contain manganese, copper, or zinc. This enzyme metabolizes toxic superoxide radicals:

$$2O_2^- + 2H^+ \rightarrow H_2O_2 + O_2$$

The enzyme is a dimer of 32 kDa per monomer, and the copper—zinc enzyme contains these prosthetic groups as shown in Fig. 5.27.

SOD is present in normal erythrocytes (red blood cells) at a level of about 100 μg/mL. The enzyme can be administered clinically to reduce the presence of superoxide radicals (reactive oxygen species, ROS). ROS are strong oxidants that can cause damage to cellular structures and other molecules. They can arise from cellular respiration, the electron transport chain, ionizing radiation, and other sources. Some of the ROS are: the superoxide anion, $\bullet O_2^-$; hydroxyl radical, $\bullet OH$; hydroxyl ion, $OH^-$; peroxyl radical, $O_2\bullet$, and others. About 5%—10% of ALS (amyotrophic lateral sclerosis)

**TABLE 5.4  The Vitamins, Their Coenzymes, and Their Chemical Functions**

| Vitamin | Coenzyme | Reaction Catalyzed | Human Deficiency Disease |
|---|---|---|---|
| **Water-Soluble Vitamins** | | | |
| Niacin (niacinate) | $NAD^+$, NADP' | Oxidation | Pellagra |
| | NADH, NADPH | Reduction | |
| Riboflavin (vitamin $B_2$) | FAD, FMN | Oxidation | Skin inflammation |
| | $FADH_2$, $FMNH_2$ | Reduction | |
| Thiamine (vitamin $B_1$) | Thiamine pyrophosphate (TPP) | Two-carbon transfer | Beriberi |
| Lipoic acid (lipoate) | Lipoate | Oxidation | – |
| | Dihydrolipoate | Reduction | |
| Pantothenic acid (pantothenate) | Coenzyme A (CoASH) | Acyl transfer | – |
| Biotin (vitamin H) | Biotin | Carboxylation | – |
| Pyridoxine (vitamin $B_6$) | Pyridoxal phosphate (Pl.P) | Decarboxylation | Anemia |
| | | Transamination | |
| | | Racemization | |
| | | $C\alpha$–$C\beta$ bond cleavage | |
| | | $\alpha,\beta$ Elimination | |
| | | $\beta$-Substitution | |
| Vitamin $B_{12}$ | Coenzyme $B_{12}$ | Isomerization | Pernicious anemia |
| Folic acid (folate) | Tetrahydrofolate (THF) | One-carbon transfer | Megaloblastic anemia |
| Ascorbic acid (vitamin C) | | – | Scurvy |
| **Water-Insoluble (Lipid-Soluble) Vitamins** | | | |
| Vitamin A | – | – | – |
| Vitamin D | – | – | Rickets |
| Vitamin E | – | | – |
| Vitamin K | Vitamin $KH_3$ | Carboxylation | |

This table is reproduced from http://wps.purshall.com/wps/media/objects/724/741576/InstructorResources/Chapter_25/Text Images/FC25_TB01.JPG. Reproduced from G. Litwack, *Human Biochemistry and Disease*, Academic Press/Elsevier, Table 3-1, page 119, 2008

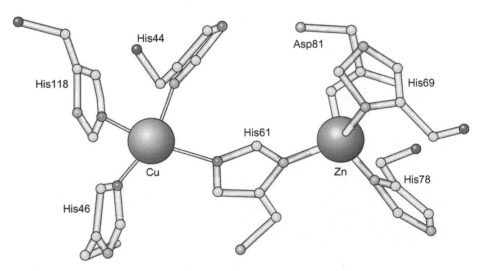

**FIGURE 5.27** Coordination complexes of the superoxide dismutase showing that of copper on the left (*gold central ball*) and that of zinc (*green ball on the right*). Copper ion is coordinated by four histidine residues and zinc by three histidine residues and an aspartate residue. *Reproduced from Litwack, G.,* Human Biochemistry and Disease, *Academic Press/Elsevier, Figures 3–30B, page 124, 2008.*

disease is inherited (involving as many as a dozen genes) and of these, about 20% are known to be caused by a mutation of the gene for $Cu^{2+}-Zn^{2+}-SOD$.

Ribozymes, RNAs that cleave other RNAs, are complexed with magnesium and sometimes with manganese. These metals are coordinated with the phosphate groups of RNA and with nucleoside sugar hydroxyl groups.

## DRUGS THAT OPERATE AS ENZYME INHIBITORS

Drugs that operate by inhibiting a specific enzyme are reversible inhibitors, that is, they are competitive inhibitors. They compete with binding of the natural substrate to the catalytic center (the binding pocket) of the enzyme. The advantage of this type of inhibitor is that the drug can be withdrawn and, as it is cleared from the body, the enzyme can again bind to the natural substrate to regain normal function. Effective drugs are those that have high affinity for the enzyme, have a reasonably long half-life in the body, and exert minimal unwanted side effects.

Some examples of drugs that are enzyme inhibitors are listed in Table 5.5.

## Drug Resistance

Although the resistance to antibiotics and drugs used to kill other invading living organisms come under this heading, drug resistance is known to occur when anticancer drugs are used to kill cancer cells. These drugs invariably are inducers of apoptosis (programmed cell death) that cause the demise of the cancer cell. In the course of these drug treatments, the concentration of a cell membrane transporter in the cancer cell may be increased. The function of this receptor, the **MDR (multidrug resistance) transporter**, is to harness the energy of the hydrolysis of ATP, via an ATPase, to pump the drug out of the cell to diminish the drug's effect. This transporter is known as the MDR (multidrug resistance) transporter, a member of the family of **ABC (ATP-binding cassette) transporter** family. Presently, 49 members of the ABC transporter family have been identified. ABC transporters will be discussed in later chapters.

### Chemotherapeutics Targeting Tyrosine Kinases

Tyrosine kinases, both nonreceptor and receptor tyrosine kinases, are overexpressed in solid tumors and stimulate their growth and spread. One receptor tyrosine kinase that is overexpressed in most solid tumors is **EGFR-TK (epidermal growth factor receptor tyrosine kinase)**. Overexpression results in autophosphorylation (of protein tyrosine residues) that stimulates signal transduction pathways leading to tumor growth and spread. A drug inhibitor of EGFR-TK, Gefitinib (Iressa), has been shown to be beneficial in patients with certain cancers. An inhibitor of Bcr-Abl (a constitutively active nonreceptor tyrosine kinase), Imatinib mesylate (Gleevec), inhibits the growth of leukemia cells by inducing their apoptosis (programmed cell death). Structures of these inhibitors are shown in Fig. 5.28.

Other inhibitors of tyrosine kinases have been shown to be beneficial in cancer patients. Some of these tyrosine kinase inhibitors operate by competing with the kinase for ATP, whereas others prevent tyrosine kinases from interacting with chaperone proteins, an activity resulting in the degradation of the kinase enzyme. Another chemotherapeutic target is **Eph5 (human Ephrin type-A receptor 5)** that is a receptor tyrosine kinase. This receptor is a regulator of cell cycle checkpoints as well as DNA-damage repair induced by ionizing radiation. Eph5 is specifically overexpressed in **lung cancer** and regulates cellular responses to genotoxic insult, such as that induced by ionizing radiation. Thus, in the presence of overexpressed Eph5, radiation treatment of lung cancer is not maximally effective. With the development of inhibitors, such as monoclonal antibodies to Eph5 human lung cancer cells become sensitized to radiotherapy.

## Rational Drug Design

Before the generation of huge libraries of chemical compounds with biological activities, potential drugs were discovered by trial and error methods. With the availability of the structures of thousands of chemical compounds, it becomes possible to screen these libraries with specific filters regarding requirements of shapes and charges. Thus, screening is possible for an enzyme inhibitor knowing the three-dimensional structure of the enzyme in complex with a substrate or an inhibitor. In rational drug design, small molecules will be sought that are complementary in charge and shape to the target. The target could be the active enter of the enzyme to which the substrate binds. In **computer-assisted drug design (CADD)**, the computer is programmed to screen thousands of known molecules for specific three-dimensional structural requirements. Usually, candidates will number only a few potential structures. The candidate chemicals are tested for activity. Design of the candidate drug will involve multiple sites of complementarity interactions with the

**TABLE 5.5 Examples of Drugs that Inhibit Enzymes**

| Drug | Structure | Enzyme Inhibited | Effect |
|---|---|---|---|
| Allopurinol | | Xanthine oxidase | Inhibits formation of uric acid; dissolves urate crystals in joints |
| Captopril | | Angiotensin-converting enzyme | Reduces [Angiotensin II] and some forms of hypertension |
| Celecoxib (or Rofecoxib) | | Cyclooxygenase-2 (inducible) | Reduces inflammation; Ile in cox-1 is replaced by Val in cox-2 allowing entry of drug in cox-2 but not cox-1 |
| Coumadin (Warfarin) | | Vitamin K epoxide reductase | Enzyme recycles oxidized vitamin K to its reduced form after its participation in carboxylating several blood coagulation proteins, mainly Prothrombin and Factor VII. Anticoagulant |
| Methotrexate | | Dihydrofolate reductase | Drug blocks conversion of dihydrofolate (H2 folate) to tetrahydrofolate (H4 folate). Reduces pyrimidine synthesis and compromises growing (cancer) cells |
| Penicillin | | Bacterial DD transpeptidase | Blocks enzyme that links peptidoglycan molecules in bacteria; cytolyzes bacterial cell |

*(Continued)*

**TABLE 5.5** (Continued)

| Drug | Structure | Enzyme Inhibited | Effect |
|---|---|---|---|
| Sildenafil (Viagra) | 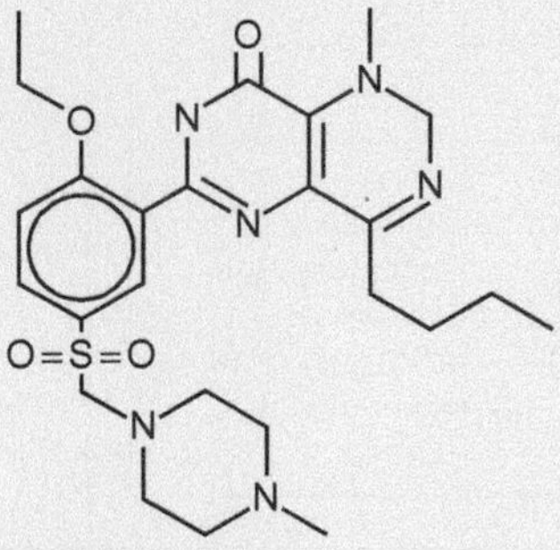 | cGMP-specific phosphodiesterase type 5 | Inhibits degradation of cyclic guanosine monophosphate that triggers smooth muscle relaxation and allows blood flow into *corpus covernosum* to generate erection |
| Statins (e.g., Simvastatin) | 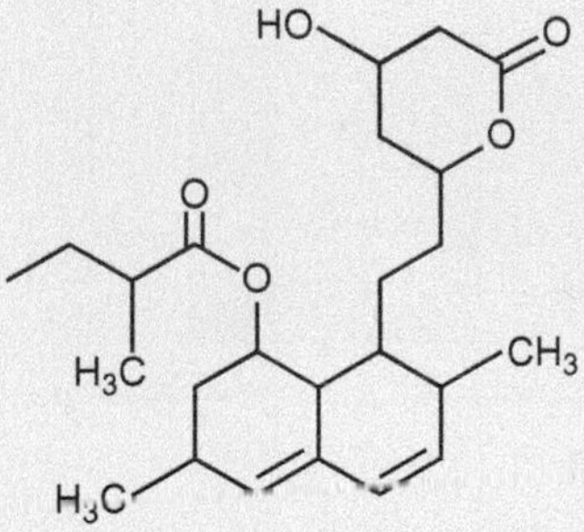 | HMGCoA reductase (3-hydroxy-3-methyl-glutaryl-CoA reductase) | Rate-limiting enzyme in cholesterol synthetic pathway in liver; reduces circulating [cholesterol] |

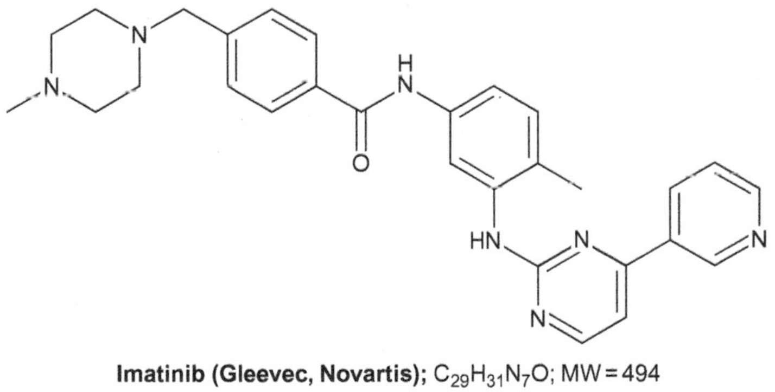

**Imatinib (Gleevec, Novartis);** $C_{29}H_{31}N_7O$; MW = 494

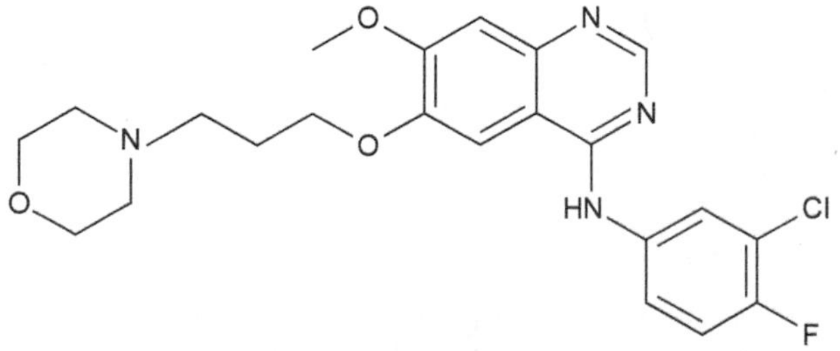

**Gefitinib (Iressa, Astrazeneca):** $C_{22}H_{24}ClFN_4O_3$; MW = 447

**FIGURE 5.28** Chemical structures of Imatinib (Gleevec) and Geftinib (Iressa). *Structure of Imatinib is reproduced from http://www.brimr.org/PKI/Imatinib.jpg. Structure of Geftinib is reproduced from http://www.brimr.org/PKI/Geftinib.jpg.*

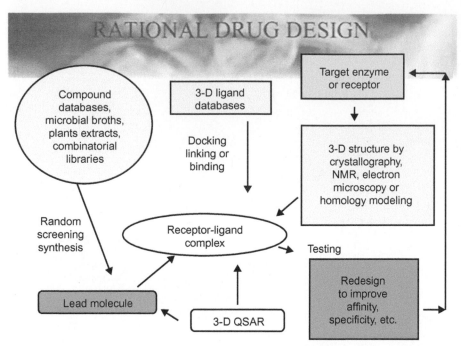

**FIGURE 5.29** Summary of the activities included in rational drug design. *3-D QSAR*, three-dimensional quantitative structure–activity relationships. *Reproduced from http://www.slideshare.net/naresh2231989/rational-drug-design?related = 1.*

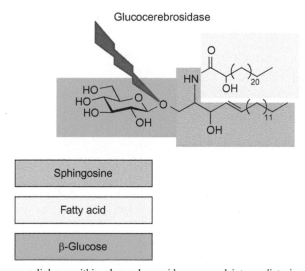

**FIGURE 5.30** Glucocerebrosidase cleaves a linkage within glucosylceramide, a normal intermediate in glycolipid metabolism. *Reproduced from http://www.ncbi.nlm.nih.gov/pmc/articles/PMC529421/figure/pmed-0010021-g001/.*

enzyme [**quantitative structure–activity relationships (QSAR)**]. Information on the active compound may lead to the synthesis of a new or modified small molecule available for further study. A summary of activities involved in rational drug design is presented in Fig. 5.29.

## ENZYME REPLACEMENT THERAPY—GAUCHER DISEASE

Gaucher disease is the most common human **lysosomal storage disorder**. It is caused by mutations in the gene encoding the enzyme, **glucocerebrosidase (GCase)** or **glucosylceramidase**. The action of this enzyme on the substrate, glucosylceramide, is shown in Fig. 5.30.

The loss of GCase activity in Gaucher disease, rather than a dimunition of enzyme function, is due to a decrease in the amount of the enzyme and the GCase that has been expressed is increasingly complexed with ubiquitin ligase,

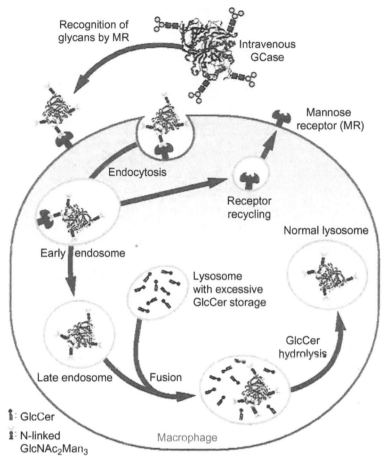

**FIGURE 5.31**  Mannose receptor (MP) uptake of intravenously administered glucocerebrosidase (GCase). The recombinant GCase is modified to ensure that N-linked glycans contain terminal mannose residues, thus conferring high affinity for the mannose receptor (MR) resident in the macrophage and other cell types. Internalization of the enzyme occurs through endocytosis of the MR/GCase complex. Early endosomes containing GCase mature into late endosomes, whereas receptor sorting leads to the recycling of the MR back to the cellular membrane. Late endosomal and lysosomal fusion delivers the therapeutic enzyme to the lysosome that subsequently stores GlcCer (glucosylceramide) to normal levels. *Reproduced from http://www.pnas.org/content/107/24/10842/F2.large.jpg.*

causing its degradation. Recombinant GCase, processed so as to expose mannose, allows the enzyme (Ceradase), injected intravenously, to bind to macrophage cell membrane **mannose receptors**. The action of injected GCase on the macrophage (and other cell types, not in this figure) is shown in Fig. 5.31.

This treatment produces clinical improvement in Gaucher disease. The symptoms of this disease are shown in Fig. 5.32 as a function of the time of appearance of the disease.

## SUMMARY

Except for ribozymes that are scarce in nature, all enzymes are proteins. There may be 3000 to 4000 enzymes in a human cell. Enzymes are catalysts that allow reactions to occur under bodily conditions which otherwise would require intolerable conditions.

Enzymatic kinetics are quantified using **Michaelis−Menten equations**. Manipulation of velocity measurements, such as the double reciprocal plot render quantitation of $K_m$, $K_i$, and $V_{max}$. One-substrate and two-substrate reactions are considered. Three-substrate reactions can be managed with a similar approach. Three types of enzymatic inhibition are discussed: competitive, noncompetitive, and uncompetitive. Most drugs (medicines) operate as competitive inhibitors of enzymes. **Allosteric enzymes** are considered as well as the operation of positive and negative effectors upon them.

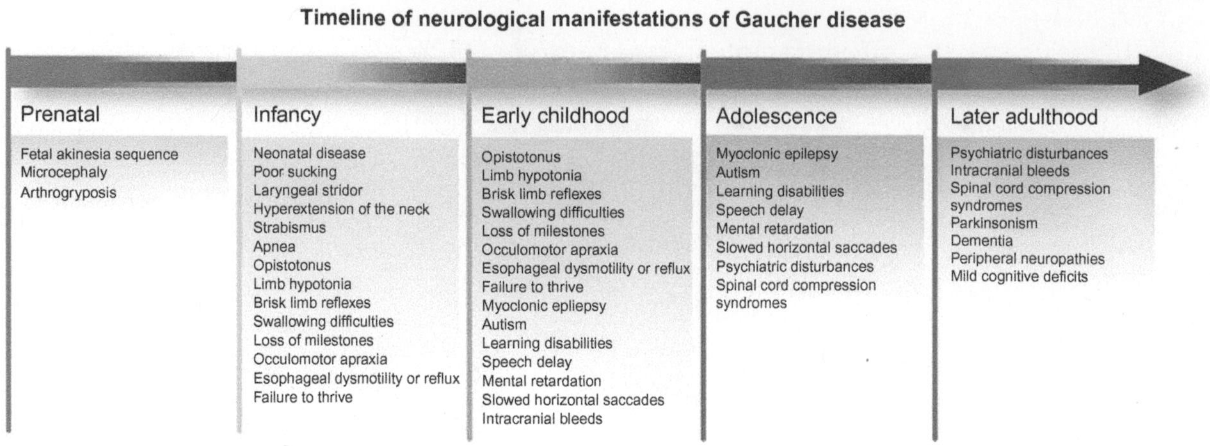

**Timeline of neurological manifestations of Gaucher disease**

| Prenatal | Infancy | Early childhood | Adolescence | Later adulthood |
|---|---|---|---|---|
| Fetal akinesia sequence<br>Microcephaly<br>Arthrogryposis | Neonatal disease<br>Poor sucking<br>Laryngeal stridor<br>Hyperextension of the neck<br>Strabismus<br>Apnea<br>Opistotonus<br>Limb hypotonia<br>Brisk limb reflexes<br>Swallowing difficulties<br>Loss of milestones<br>Occulomotor apraxia<br>Esophageal dysmotility or reflux<br>Failure to thrive | Opistotonus<br>Limb hypotonia<br>Brisk limb reflexes<br>Swallowing difficulties<br>Loss of milestones<br>Occulomotor apraxia<br>Esophageal dysmotility or reflux<br>Failure to thrive<br>Myoclonic epliepsy<br>Autism<br>Learning disabilities<br>Speech delay<br>Mental retardation<br>Slowed horizontal saccades<br>Intracranial bleeds | Myoclonic epilepsy<br>Autism<br>Learning disabilities<br>Speech delay<br>Mental retardation<br>Slowed horizontal saccades<br>Psychiatric disturbances<br>Spinal cord compression<br>syndromes | Psychiatric disturbances<br>Intracranial bleeds<br>Spinal cord compression<br>syndromes<br>Parkinsonism<br>Dementia<br>Peripheral neuropathies<br>Mild cognitive deficits |

FIGURE 5.32 Timeline of neurological manifestations of Gaucher disease. *Reproduced from http://www.discoverymedicine.com/Ellen-Sidransky/ files/2012/10/discovery_medicine_no_77_ellen_sidransky_figure_1.png.*

The basic **classification of enzymes** is presented. Enzymes are classified based on the types of reactions they catalyze: oxidoreductases, transferases, hydrolases, lyases, isomerases, and ligases. E.C. numbers are elaborated for enzymes in each class based on their substrates.

**Coenzymes** are required by some enzymes for activity. An inactive enzyme, in the absence of a coenzyme, is an **apoenzyme**; in the presence of the coenzyme, and therefore active, it is a **holoenzyme**. In general, **vitamins function as coenzymes,** and they may be altered in the body to the appropriate coenzyme form. Some coenzymes are relatively tightly bound to the enzyme, whereas others are dissociable and may be treated, kinetically, as substrates. **Prosthetic groups** are tightly bound molecules or atoms and many of this group are metal ions.

**Diagnostic enzymology** involves blood draws following a disease event, such as myocardial infarction. Enzymes of interest are transaminases, creatine kinase, lactate dehydrogenase, and others. Most useful are enzymes with multiple subunits that form different **isozymes**, determined by their subunit composition, and which derive from different tissues. The premise is that infection, inflammation, or tissue damage involving increased cell membrane permeability or outright membrane lysis will allow leakage of macromolecules into the extracellular space and concentration in the bloodstream. The measurement of enzymes deriving from specific tissues will diagnose or confirm disease in a specific tissue. Similar measurements can be used to follow prognosis. Enzyme preparations can be used to treat certain disease conditions, such as **Gaucher disease**.

**Drug resistance** develops in some diseases. This is observed with drugs used to kill cancer cells. The cancer cell pumps the drug out of the cell using the energy of the terminal phosphate group of ATP. The transporter in this function is the **MDR (multiple drug resistance) transporter**, a member of the **ABC (ATP-binding cassette) transporter** family.

Drugs, such as **Gleevec** and **Iressa,** are used to target **receptor tyrosine kinases** and **nonreceptor tyrosine kinases** in order to fight solid tumors. These enzymes phosphorylate protein tyrosine residues and the pathways responding to these phosphorylation events lead to growth and spread of the solid tumors.

**Rational drug design** can facilitate a search for new enzyme inhibitors. This approach employs computers (**CADD, computer-assisted drug design**) to search among thousands of known chemical structures for lead compounds, often following upon the knowledge of the three-dimensional structure of the protein/enzyme target.

# SUGGESTED READING

## Literature

Arora, A., Scholar, E.M., 2005. Role of tyrosine kinase inhibitors in cancer therapy. J. Pharmacol. Exp. Ther. 315, 971−979.

Baldo, B.A., 2015. Enzymes approved for human therapy: indications, mechanisms and adverse events. BioDrugs. 29, 31−55.

Chen, Y.P., Chen, F., 2008. Identifying targets for drug discovery using bioinformatics. Expert Opin. Ther. Targets. 12, 383−389.

Cleland, W.W., 1967. Enzyme kinetics. Ann. Rev. Biochem. 36, 77−112.

Drury, J.E., Costanzo, L.D., Penning, T.M., Christianson, D.W., 2009. Inhibition of human steroid 5β-reductase (AKR1D1) by finasteride and structure of the enzyme−inhibitor complex. J. Biol. Chem. 284, 19786−19790.

Emery, A.E.H., 1967. The determination of lactate dehydrogenase isoenzymes in normal human muscle and other tissues. Biochem. J. 105, 599−604.

Honarparvar, B., Govender, T., Maguire, G.E.M., Soliman, M.E.S., Kruger, H.G., 2014. Integrated approach to structure-based enzymatic drug design: molecular modeling, spectroscopy and experimental bioactivity. Chem. Rev. 114, 493–537.

Infusino, I., Schumann, G., Ceriotti, F., Panteghini, 2010. Standardization in clinical enzymology: a challenge for the theory of metrological traceability. Clin. Chem. Lab. Med. 48, 301–307.

Johnson, K.A., 2013. A century of enzyme kinetic analysis, 1913 to 2013. FEBS Lett. 587, 2753–2766.

Johnson, K.A., Goody, R.S., 2011. The original Michaelis constant: translation of the 1913 Michaelis–Menten paper. Biochemistry. 50, 8264–8269.

Klebe, G., 2000. Recent developments in structure-based drug design. J. Mol. Med. (Berl.). 78, 269–281.

Kubinyi, H., 1998. Structure-based design of enzyme inhibitors and receptor ligands. Curr. Opin. Drug Discov. Dev. 1, 4–15.

Lee, C.-H., Huang, H.-C., Juan, H.-F., 2011. Reviewing ligand-based rational drug design: the search for an ATP synthase inhibitor. Int. J. Mol. Sci. 12, 5304–5318.

Neal, J.L., Lowe, N.K., Corwin, E.J., 2013. Serum lactate dehydrogenase profile as a retrospective indicator of uterine preparedness for labor: a perspective, observational study. BMC Pregnancy Childbirth. 13, 128–137.

Nipakis, M.J., Cravatt, B.F., 2014. Enzyme inhibitor discovery by activity-based protein profiling. Ann. Rev. Biochem. 83, 341–377.

Sayers, E.W., et al., 2010. Database resources of the national center for biotechnology information. Nucleic Acids Res. 38, D5–D16.

Tso, S.-C., et al., 2014. Benzothiophene carboxylate derivatives as novel allosteric inhibitors of branched chain α-ketoacid dehydrogenase kinase". J. Biol. Chem. 289, 20583–20593.

Wroblewski, F., Gregory, K.F., 1961. Lactic dehydrogenase isoenzymes and their distribution in normal tissues and plasma and in disease states. Ann. N. Y. Acad. Sci. 94, 912–932.

Whittle, P.J., Blundell, T.L., 1994. Protein structure-based drug design. Ann. Rev. Biophys. Biomol. Struct. 23, 349–375.

## Books

Begley, T., Litwack, G. (Eds.), 2001. Cofactor Biosynthesis: A Mechanistic Perspective, Vitamins & Hormones, volume 61. Academic Press, Cambridge, MA.

Dixon, M., Webb, E.C., 1958. Enzymes. Longmans Green, London.

Kazmierczak, S.C., Azzazy, H.M.E., 2014. Diagnostic Enzymology. de Gruyter Textbook, Birmingham, AL.

Lark, S.M., 2013. Enzymes: The Missing Link to Health. Womens Wellness Publishing, Emmaus, PA.

Nature Publishing Group, 2001. Encyclopedia of Life Sciences; Enzyme Classification and Nomenclature. Nature Publishing Group, London, UK.

Segel, I.H., 1975. Enzyme Kinetics. John Wiley and Sons, Inc, Hoboken, NJ.

# Chapter 6

# Insulin and Sugars

## DIABETES

Diabetes was recognized as early as 1500 BC. Diabetes is a disease in which glucose cannot be utilized properly. There are basically two types of diabetes. **Type 1 diabetes** occurs, typically in the young, when the beta cells of the pancreas have been destroyed (by viruses or autoimmunity or both) so that insulin secretion cannot take place when the glucose level in the blood is high. **Type 2 diabetes** occurs when beta cells are at least somewhat compromised (insufficient insulin is produced) or when glucose utilization is impaired, perhaps by a fault in the insulin receptor (IR) (insulin resistance) or in the glucose transporter. Several variant genes contribute to both types of diabetes. Obesity is a major factor in the development of type 2 diabetes. Five to 10% of diabetics are type 1 and the rest are type 2. About 4% of pregnant women (about 135,000 each year in the United States) experience **gestational diabetes**. **Chronic stress** can cause diabetes by the responsive release of **cortisol** from the adrenal cortex so that glucose is increased in the blood by about 10% (cortisol affects peripheral tissues to suppress glucose uptake in those tissues), enough to elicit insulin release from the beta cell. **Continuous demand for insulin release** can exhaust the beta cell over time and cause diabetes. About 20 million persons in the United States have a condition known as "**prediabetes**" where the blood glucose is elevated but not high enough to be classified as type 2 diabetes. An American-Caucasian with a blood glucose value of 200 mg/dL (1 dL is 10% of a liter or 100 mL) has a 20% chance of developing diabetes. Lack of exercise and a high sugar (and fat) diet are contributors. The **glucose tolerance test** is used to indicate diabetes or prediabetes. An oral dose of glucose (usually 75 g for adults ingested within 5 minutes) is followed by blood sampling at zero time (fasting) and at 2 hours (or an intervening measurement could be made at 30 to 90 minutes). The test is performed in the morning after an 8 to 12 hour fast. Typical results are shown in Table 6.1.

The genetic background to type 1 diabetes and type 2 diabetes is complicated. Many genes seem to be involved in both cases. Of interest is the relationship between obesity and type 2 diabetes. Both obesity and diabetes have become epidemic in the United States. For adults as well as for children, 80% of type 2 diabetes is due to obesity. Recent research in Australia indicates that fat cells produce **pigment-epithelium-derived factor** (**PEDF**) that generates events leading to type 2 diabetes. As a result of PEDF action, the response to insulin in liver and muscle is reduced (insulin resistance). The pancreas responds to this situation by producing more insulin, eventually leading to exhaustion of the beta cells so that insulin release is slowed or stopped altogether, generating type 2 diabetes. Thus, the more **fat tissue**, the greater production of this factor and the greater likelihood of developing type 2 diabetes. Inhibition of the action of PEDF, interestingly, reverses its negative effects. Likely, there are other factors (e.g., reduction in circulating **adiponectin**, a protein hormone derived from fat cells; less adiponectin is secreted by fat cells in obesity) issuing from **obesity** (inflammation, chemokines, and others) that contribute to the development of type 2 diabetes.

The cellular utilization of glucose depends on the normal functioning of the insulin receptor and the insulin receptor-signaling pathway. Insulin binds and activates the **insulin receptor** whose actions reduce circulating [glucose] by stimulating glycogen synthesis. In type 1 diabetes, owing to the destruction of the beta cells, insulin is not present and [glucose] increases in the blood. High **blood glucose** can cause all sorts of problems. For example, hemoglobin can become glycosylated, affecting its function and possibly contributing to heart disease (even insulin, itself, can become glycosylated). Ingestion of **fructose** (half of the sucrose molecule and the major component of corn syrup and honey) will be converted quickly to fatty acids and to fat (since its metabolism, in glycolysis, skips two enzymatic steps, compared to glucose, facilitating conversion of fructose to pyruvate, acetyl CoA, and fatty acids). In type 2 diabetes, insulin utilization is subnormal and blood glucose is not converted appropriately to glycogen storage leading to a rise in blood [glucose]. This reduced glucose utilization could be the result of an altered or "masked" insulin receptor that functions poorly or due to some other defect in glucose utilization. Possible overall scenarios are summarized in Fig. 6.1.

Human Biochemistry. DOI: http://dx.doi.org/10.1016/B978-0-12-383864-3.00006-5

**TABLE 6.1** Blood Glucose Levels During Fasting or After Oral Glucose Ingestion That Are Diagnostic of Diabetes or the Prediabetic Condition Compared to Normal Values

| | Blood Glucose (mg/dL) | | |
| --- | --- | --- | --- |
| | Fasting | 30–90′ | 120′ |
| Normal | <115 | <200 | <140 |
| Diabetic | >140 | >200 | >200 |
| Impaired glucose tolerance | <140 | >200 | 140–199 |

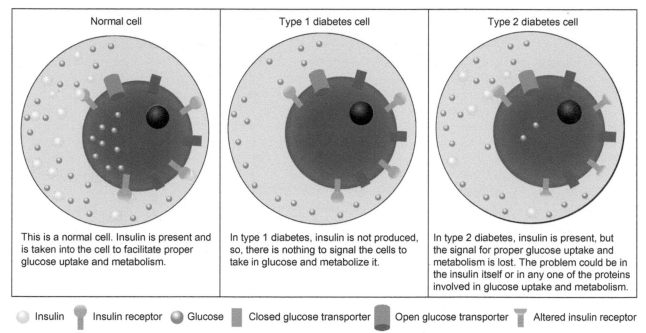

| | | | | | |
| --- | --- | --- | --- | --- | --- |
| ○ Insulin | ⌐ Insulin receptor | ○ Glucose | ▮ Closed glucose transporter | ▮ Open glucose transporter | ⊤ Altered insulin receptor |

**FIGURE 6.1** Characteristics of tissue cells in normal conditions and in type 1 and type 2 diabetes. *Reproduced from G. Litwack,* Human Biochemistry and Disease, *Academic Press/Elsevier, Figure 4-2, page 134, 2008.*

**Beta cells** are located in the pancreas that is located below and behind the stomach. The pancreas is divided between the **exocrine pancreas** and the **endocrine pancreas**. The exocrine pancreas produces **digestive enzymes** that are secreted into the intestinal tract. The endocrine pancreas produces hormones, primarily **insulin** (β-cell), **glucagon** (α-cell) and **somatostatin** (δ-cell). Cells of the endocrine pancreas are located in **Islets of Langerhans**. A drawing representing an Islet of Langerhans is shown in Fig. 6.2.

After insulin is secreted by the β-cell in response to signals (discussed below) insulin circulates in the bloodstream and binds to insulin receptors in the cell membranes of various tissues (e.g., liver and muscle) where, through signaling pathways, various effects can be generated, including those leading to the reduction in circulating [glucose]. Depending on the pathway activated by the insulin–insulin receptor complex, there can be mitogenic effects (DNA synthesis, transcription) and important metabolic effects (protein synthesis, lipid synthesis, and glycogen synthesis). These effects occur through the activation of the insulin receptor and its cytoplasmic tyrosine kinase, followed by the phosphorylation of IRS1 (insulin receptor substrate-1) and IRS-2 (insulin receptor substrate-2) and a phosphorylation cascade. IRS stimulates PI3-kinase that activates glucose transport by GLUT 4 (glucose transporter 4) and the conversion of intracellular glucose to glucose-6-phosphate, a precursor of UDP-glucose (uridine diphosphate glucose). These effects are summarized in Fig. 6.3.

As shown in Fig. 6.3C, **phosphoinositol-3 kinase (PI3K)** is activated through **IRS1** that was phosphorylated by the activated insulin receptor. The phosphorylated IRS1 serves as the docking site for the **SH2 domains** (Src homology 2

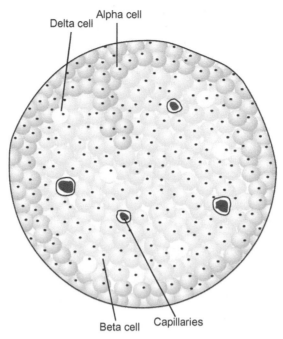

Delta cell   Alpha cell

Beta cell   Capillaries

**FIGURE 6.2** Cell types in an Islet of Langerhans of the endocrine pancreas.

domains) of the regulatory subunit (p85) of PI3K which leads to the formation of **PI(3,4,5)P$_3$** (phosphatidylinositol *tri*-*s*phosphate). In turn, PI3K stimulates glucose transport into the cell (more GLUT4 from the cytoplasm to the cell membrane) and glycogen synthesis. The activation of PI3-kinase by the insulin receptor is shown in Fig. 6.4.

The **SH2 domain** (Src homology 2 domain) is contained in the **Src oncoprotein** and in some 115 human proteins. It has two alpha helices and 7 beta strands and is about 100 amino acids long. It has a high affinity for **phosphorylated tyrosine** in proteins and the optimal motif it recognizes to which it binds is: **Tyr-P-Met-Glu-Pro** (in general, the binding motif is Tyr-hydrophilic residue-hydrophilic residue-hydrophobic residue).

Following the binding of insulin to the IR and IR activation leading to phosphorylation of the component **tyrosine kinase**, multiple signaling pathways can be activated. These include glycogen synthesis, signal transduction, and growth regulation (insulin can act as a **mitogen**) and these effects are summarized in Fig. 6.5.

A mitogen is a substance that stimulates cell division. These signaling pathways are complex and involve other kinases (MEK, MAP kinase, PI3K, PDK, PKC, GSK3, and others). The pathway on the left describes the growth stimulation that can be affected by the activated IR and pertains to cells that are capable of dividing. On the right, the pathway is described in which the activated IR leads to glucose utilization. **GLUT 4**, a glucose transporter is moved from the cytoplasm to the cell membrane (muscle and adipose tissues) where it facilitates the movement of glucose from outside of the cell to its interior. Once inside, glucose can be acted on by **hexokinase** to form glucose-6-phosphate, an intermediate in glycogen formation. If there is an immediate demand for energy, glucose will transit through glycolysis and the citric acid cycle.

## INSULIN

Insulin is synthesized in the beta cells of the pancreas as the preproprotein, **preproinsulin**. At the N-terminal end, there is a **signal peptide** of 24 amino acids that is connected to the B chain at its N-terminus (signal peptide Ala to B chain Phe). The signal peptide is first cleaved at Ala−Phe to produce **proinsulin** (connecting peptide, B chain and A chain). Three disulfide bonds are formed (B7 to A7, B19 to A20, and A6 to A11). This is transported into vesicles budded from the Golgi apparatus. In this location the **connecting peptide** (containing 33 amino acids) which has cleavable amino acids on either end, −R−R− (−Glu−Arg−Arg−Thr−) on the amino end next to amino acid 31 (Glu) and −K−R− (−Gln−Lys−Arg−Gly−) at the N-terminal end next to amino acid 1 (Gln) is removed by **convertases** 1 and 2 at the indicated cleavage sites (Fig. 6.6).

The released, active monomeric form of insulin is shown in Fig. 6.7.

Insulin exists in solution as monomers, dimers or as crystals of hexamers in the presence of zinc ions. Acidity favors the formation of dimers. It is primarily in the monomeric form in the blood. However, in the storage form in the

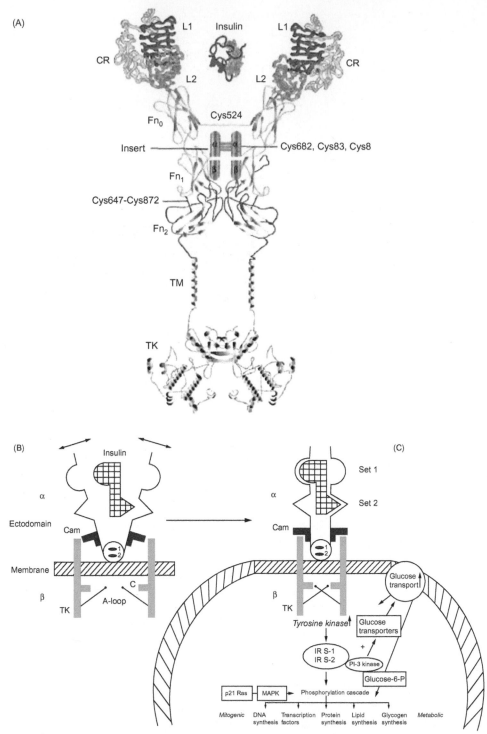

**FIGURE 6.3** Binding of insulin to the insulin receptor and its signaling pathways. Part (A) shows a diagram of the insulin receptor dimer. When a molecule of insulin approaches the active site of the receptor, it binds to the *yellow* surface. The insulin backbone is shown in *blue*. When insulin binds, the two arms of the receptor dimer close about the insulin molecule as shown in the postulated transition indicated in (B) and (C). It may be possible for an additional insulin molecule to bind the receptor, possibly with a different affinity but these possibilities require further investigation. Some amino acid residues in the insulin molecule are thought to represent the binding domain that interacts with the binding domain of the insulin receptor, possibly amino acids 2 (Ile) and 19 (Tyr) of the A chain and 13 (Glu), 23 (Gly), and 24 (Phe) of the B chain. In (A) *TM*, transmembrane; *TK*, tyrosine kinase; *Fn*, fibronectin type repeat; *L1* and *L2*, leucine-rich domains of the alpha subunit of the insulin receptor; *CR*, cysteine-rich domain of the insulin receptor. *(A) is reproduced from Macmillan Publishers Ltd.: "Generation and annotation of the DNA sequences of human chromosomes 2 and 4," Nat. Rev. Drug Discovery, 1: 769, 2002. (B) and (C) are redrawn from F.P. Ottensmeyer, et al., "Mechanism of transmembrane signaling: insulin binding and the insulin receptor," Biochemistry 39: 12103–12112, 2000. Copyright (2000) American Chemical Society.*

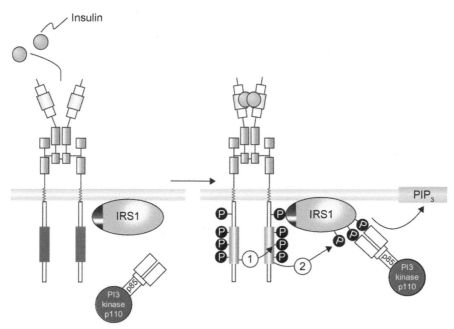

**FIGURE 6.4** Activation of PI3-kinase by the insulin receptor. IRS-1 binds and is phosphorylated by the activated insulin receptor. IRS-1-P then serves as the dock for the SH2 domains of the regulatory subunit (p85) of PI3-kinase, leading to the generation of PI(3,4,5)P$_3$ (phosphatidylinositol *tris*phosphate). *Reproduced from http://bioweb.wku.edu/courses/biol566/Images/pi005.jpg.*

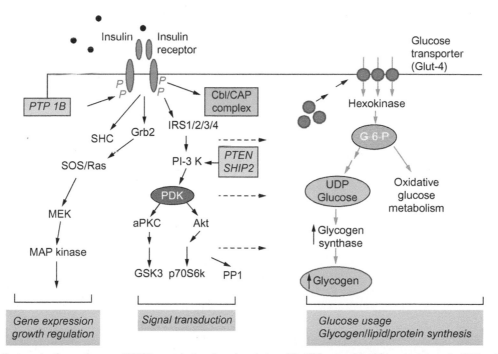

**FIGURE 6.5** Insulin transduction pathways. *PTP1B*, protein tyrosine phosphatase 1B; *Cb1*, cannabinoid receptor type 1; *CAP*, cAMP binding protein; *SHC*, SRC homology 2 containing protein; *Grb2*, growth factor receptor bound protein 2; *SOS*, son of sevenless; *MEK*, mitogen activated protein kinase kinase; *MAP*, mitogen activated protein; *aPKC*, activated protein kinase C; *Akt*, Ser-Thr kinase; *PDK*, phosphoinositide-dependent kinase; *PI3K*, phosphoinositide 3-kinase; *PTEN*, tumor suppressor protein (phosphatase); *SHIP2*, inositol 5-phosphatase; *GSK3*, glycogen synthase kinase-3; *PP1*, protein phosphatase 1.

beta cell of the pancreas, insulin may be in the hexameric form with 2 zinc ions per hexamer and it may be secreted in this form. The two zinc ions are in the center of a doughnut-like shape and are chelated by three His residues in the 10th position of the B chain (see Fig. 6.7A). A molecular three-dimensional model of the insulin hexamer is shown in Fig. 6.8.

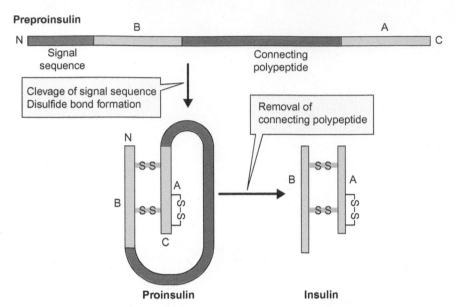

**FIGURE 6.6** Insulin is split out from a preproinsulin precursor. The amino terminal signal sequence is cleaved during transfer of the growing polypeptide chain to the endoplasmic reticulum. This yields the second precursor, proinsulin, which is converted to insulin by further proteolysis that removes the connecting peptide. The disulfide bonds in insulin are formed prior to the removal of the connecting peptide. *Reproduced from G.M. Cooper,* The Cell, a Molecular Approach, *second edition, Sinauer Associates, Sunderland, MA, 2000; http://www.ncbi.nlm.gov/bookshelf/br.fcgi?book = cooper.*

(A)

## Amino acid sequences of human insulin

### A-chain

1            S–S     10                      20

GLY ILE VAL GLU GLN CYS CYS THR SER ILE CYS SER LEU TYR GLN LEU GLU ASN TYR CYS ASN

### B-chain

1                        10                    20                            30

PHE VAL ASN GLN HIS LEU CYS GLY SER HIS LEU VAL GLU ALA LEU TYR LEU VAL CYS GLY GLU ARG GLY PHE PHE TYR THR PRO LYS THR

(B)

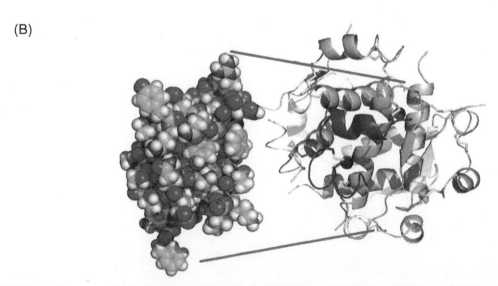

**FIGURE 6.7** The monomeric form of insulin. (A) Active insulin. Shown are the amino acid sequences of the A and B chains. (B) Molecular models of the monomeric form of insulin. *(B) is reproduced from http://itech.dickinson.edu/chemistry/wp-content/uploads/2008/04/insulinmonomer1.jpg.*

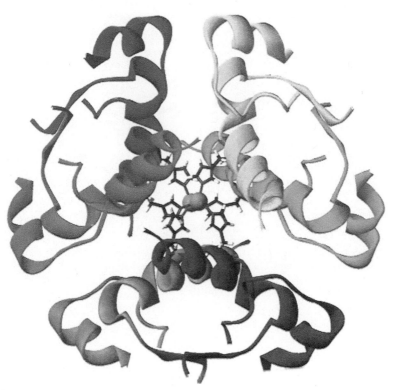

**FIGURE 6.8** A three-dimensional model of the doughnut-like shape of the insulin hexamer. The two zinc ions (in blue-gray) are located in the center of the structure shown each interacting with three His residues. *Reproduced from http://upload.wikimedia.org/wikipedia/commons/5/57/Human-insulin-hexamer-3D-ribbons.png.*

Insulin is in the monomeric form at **physiological concentrations** of about 1 ng/mL. At higher concentrations, it forms dimers and in the presence of zinc, it forms hexamers. Two insulin monomers bind to the α-subunits of the insulin receptor. Pockets in the α-subunits of the insulin receptor interact with two regions of the insulin molecule (**legend,** Fig. 6.3) generating a conformational change in which the extracellular portions of the IR approach each other (Figs. 6.3 and 6.4). This results in the activation of the receptor's **tyrosine kinase** which autophosphorylates the β-subunits. The β-subunit tyrosine kinase, in turn, phosphorylates various IR substrates in the cell, such as **IRS-1** and **IRS-2** and associated binding protein **Grb** that are intermediates in the signal transduction pathway (Fig. 6.5). A motif in the juxtamembrane (inner cell membrane region) is **NPEY** (Asn-Pro-Glu-Tyr) contacts the receptor tyrosine kinase to various signaling molecules. Thus, IRS-1 and Shc both interact with this motif and mutation of this sequence impairs the phosphorylation of these molecules and interferes with the normal movements of the IR and of insulin. Aberrant **NPEY motif** may cause **type 2 diabetes**.

## THE PANCREATIC β-CELL AND INSULIN SECRETION

*Insulin is produced only in the β-cell.* Elevated levels of circulating glucose can trigger the β-cell to secrete insulin. Elucidation of the pathway connecting elevated glucose to the release of insulin by the β-cell has been laborious and difficult. All of the answers may not yet be known; however, there is now a good working model to explain the phenomenon. The signaling pathway leads to an increase in the intracellular concentration of calcium ions that cause the release (**exocytosis**) of insulin from the secretory granule. *An increase in the level of glucose in the blood of 10% or more is sufficient to evoke the release of insulin from the β-cell.* Electron micrographs show the release of insulin from an insulin-containing granule in Fig. 6.9.

The overall mechanism of insulin secretion in response to elevated extracellular glucose is shown in Fig. 6.10. Glucose is transported into the β-cell by **GLUT2** and phosphorylated by **glucokinase** to glucose-6-phosphate that is metabolized through glycolysis (to pyruvate) and the Citric Acid Cycle to produce ATP. The **elevated ATP/ADP ratio** inhibits the **ATP-sensitive $K^+$ channel** which causes **depolarization** of the cell membrane leading

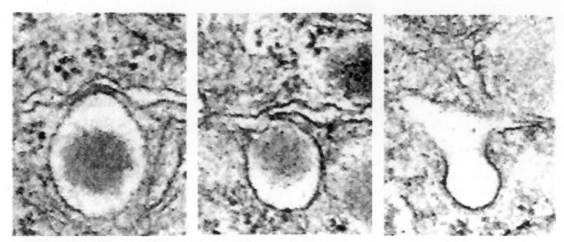

**FIGURE 6.9**  Electron micrographs showing the release of insulin from an insulin-containing granule in the pancreatic β-cell. The granule first docks with the inner cell membrane (*left panel*). Insulin secretory granule membrane fuses with the inner cell membrane and membranes become continuous (*center panel*). The insulin secretory contents are expelled into the extracellular space (*right panel*). *Reproduced from G. Litwack,* Human Biochemistry and Disease, *Figure 4-16, page 144, Academic Press/Elsevier, 2008.*

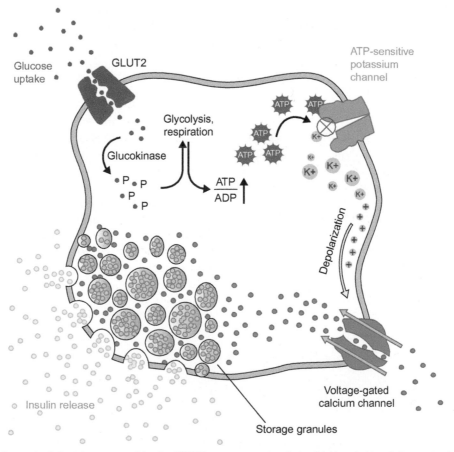

**FIGURE 6.10**  After glucose (*red dots*) is transported by the GLUT2 transporter, glycolytic phosphorylation of glucose (*red dot-P*) and metabolism through glycolysis and the Tricarboxylic cycle leads to an increase of the ATP/ADP ratio which inactivates the potassium channel causing depolarization of the membrane and opening the calcium channel to allow inward flowing of calcium ions (*green dots*) from the extracellular space. The increase in intracellular $Ca^{2+}$ causes granule fusion with the plasma membrane and **exocytotic release of insulin** (*yellow dots*) to the extracellular space. *Reproduced from http://www.betacell.org/content/articles/articlepanel.php?aid = 1&pid = 2.*

to the uptake of $Ca^{2+}$ from the outside. Calcium ions are the trigger for fusion of the insulin-containing granules to the plasma membrane.

An important glycoprotein in the granule membrane is 64 kDa **Phogrin**. The interaction of the insulin-containing granule and the inner cell membrane occurs in three phases, docking, priming and fusion. Other proteins involved in docking are **SNAP-25** and **SNARE**. **Phogrin** is phosphorylated in response to secretory stimuli and this phosphorylation inhibits its activity as a **phosphatidylinositol phosphatase**, dephosphorylating PIP2. **Phosphorylated Phogrin** in the insulin granule membrane attaches to the inner plasma cell membrane. The events of the secretion mechanism involve several components. After docking, $\alpha$-**SNAP** activates **NSF** that hydrolyzes ATP and activates SNARE proteins. A second ATP-dependent step generates PIP2. **Synaptotagmin** is the proposed $Ca^{2+}$ sensor and a fusion clamp that prevents fusion of the granule until there is a $Ca^{2+}$ signal. Thus, after granule priming, $Ca^{2+}$ is the limiting signal for granule fusion with the inner plasma membrane. **Autoantibodies** to Phogrin are apparent in most **prediabetics** and these antibodies are diagnostic of the **prediabetic state**.

A powerful system supporting the release of insulin from the beta cell is the activity of the intestine following a meal. It is known that measuring the increase in plasma [insulin] with time after parenteral injection of glucose compared to oral ingestion of glucose results in far greater levels of plasma insulin after oral ingestion. The agents responsible for the higher plasma insulin levels after ingesting glucose are the **incretins** that consist of the **glucose-dependent insulinotropic polypeptide** (**GIP**) and **glucagon-like peptide-1** (**GLP-1**). In response to ingested nutrients in the gut, the **K cells** of the duodenum and proximal jejunal linings as well as the **L cells** of the ileum and colon linings secrete GLP-1 and GIP. These hormones are absorbed through the intestine, enter the bloodstream and bind to their respective receptors on the membranes of the beta cells. They stimulate the production of **IP3** and **diacylglycerol** that result in an increase in secretion of $Ca^{2+}$ from the internal storage sites of $Ca^{2+}$ in the ER (IP3 activity) to the cytosol and the activation of protein kinase A (PKA). Uptake of $Ca^{2+}$ is increased from the outside. The increase of intracellular $Ca^{2+}$ causes the immediate secretion of insulin from the storage granules. The activation of PKA results in the phosphorylation of MEK-ERK that stimulates the production of **preproinsulin mRNA** in the nucleus, accounting for an increase in the contents of the insulin storage granules. These events are summarized in Fig. 6.11.

In addition to the direct effects of incretins on the beta cells, there may be effects of GLP-1 on neuronal pathways. Thus, there are two pathways, at least, contributing to the secretion of insulin from the beta cells. The direct effects of glucose (Fig. 6.10) are supported by the effects of the **incretins**. The sequences of **GIP** and **GLP-1** are shown in Fig. 6.12.

Cleavage and inactivation of the incretins occurs by the action of **dipeptidyl peptidase IV** (**DPP IV**), cutting between the amino acids, A-E, in the N-terminus of GIP and between the amino acids A-E near the N-terminus of GLP-1. **Inhibitors of DPP IV** become important for the potential treatment of **type 2 diabetes** since inhibiting the inactivating enzyme and prolonging activity of the incretins increases the secretion of insulin from the pancreas. Some of the inhibitors being tested clinically are shown in Fig. 6.13.

These inhibitors could be important because, in addition to stimulating the glucose-dependent insulin secretion from the beta cell, incretins induce **beta cell proliferation** and enhanced **resistance to apoptosis** and GLP-1 is known to inhibit **glucagon secretion**. In addition to inhibitors of this enzyme, direct treatment of type 2 diabetes patients with incretins is being tested.

Glucagon-like peptide 1 and 2 are encoded by the **proglucagon gene** (on the long arm of chromosome 2; Fig. 6.14A) and GIP is encoded by the **ProGIP gene** (on the long arm of chromosome 17; Fig. 6.14B).

## DETRIMENTAL EFFECTS OF DIABETES

Both type 1 and type 2 diabetes may lead to serious medical conditions as shown in Table 6.2.

Although a number of experimental approaches have been tried in animals to restore functioning beta cells, the potential use of adult progenitor/stem cells seems most encouraging. Future clinical research in humans will determine this possibility.

## SYNTHETIC SWEETENERS

Sucrose is cane sugar and the common sweetener. Emphasis on dieting has spawned a number of synthetic sweeteners. Although they contain no calories, experimentally there are downsides to their ingestion. Most of the studies showing pathology have utilized animals, especially rodents. It is not clear what the tolerance level is in terms of amounts

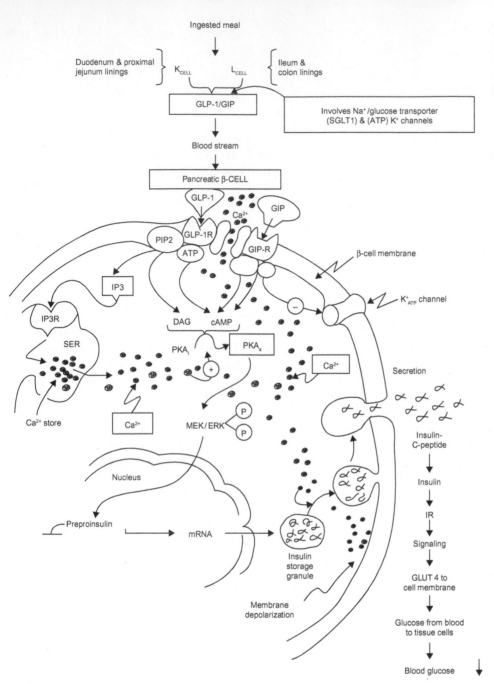

Ingested meal

Duodenum & proximal jejunum linings ⎱ K_CELL   L_CELL ⎰ Ileum & colon linings

GLP-1/GIP

Involves Na⁺/glucose transporter (SGLT1) & (ATP) K⁺ channels

Blood stream

Pancreatic β-CELL

GLP-1        GIP

Ca²⁺

PIP2   GLP-1R
ATP        GIP-R

β-cell membrane

IP3

K⁺_ATP channel

IP3R

DAG   cAMP

SER        PKAᵢ

PKAₐ

Ca²⁺

Ca²⁺ store

Ca²⁺

Secretion

MEK/ERK   P P

Insulin-C-peptide

Nucleus

Insulin

Preproinsulin        mRNA

IR

Signaling

Insulin storage granule

GLUT 4 to cell membrane

Membrane depolarization

Glucose from blood to tissue cells

Blood glucose

FIGURE 6.11 Actions of GLP-1 and GIP on the beta cell of the pancreas. *GLP-1*, glucagon-like peptide-1; *GIP*, glucose-dependent insulinotropic polypeptide; *PIP2*, phosphatidylinositol *bis*phosphate; *IP3*, inositol *tris*phosphate; *SER*, smooth endoplasmic reticulum; *R*, receptor; *DAG*, diacylglycerol; *cAMP*, cyclic AMP; *PKAa*, active protein kinase A; *IR*, insulin receptor; *MEK*, MAP-ERK kinase and *ERK*, extracellular signal-regulated kinase.

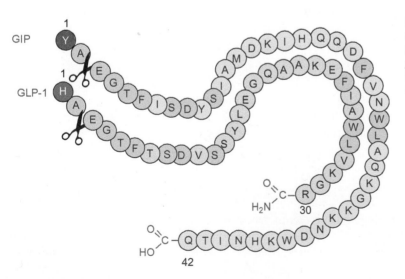

FIGURE 6.12 Amino acid sequences of glucose-dependent insulinotropic polypeptide (GIP) and glucagon-like peptide-1 (GLP-1). Conserved amino acids are in *yellow*. The cleavage sites for dipeptidyl peptidase IV that inactivate the peptides are shown. *Reproduced from http://www.bioscience.org/2008/v13/af/2797/figures.htm.*

## Dipeptidyl peptidase IV inhibitors used in preclinical or clinical studies

### 1. Reversible product analogs

Isoleucine-thiazolidide
P32/98 (probiodrug)
$K_1 = 80$ nM

### 2. Covalently modifying product analogs

NVP DPP728
(Novartis)
$IC_{50} = 22$ nM

LAF237
Vildagliptin
Galvus (Novartis)
$IC_{50} = 3.5$ nM

BMS-477118
Saxagliptin
(BMS/AstraXeneca)
$K_1 = 0.45$ nM

### 3. Reversible nonpeptide heterocyclic compounds

MK-0431
sitagliptin
Januvia (Merck)
$IC_{50} = 18$ nM

R1438
Aminomethylpyridine
(Roche)
$K_1 = 0.1$ nM

### 4. Compounds in clinical trails

ABT-279, ABT-341 (Abbott)
ALS 2-0426 (Alantos/Servier)
BI 1356 (Boehringer Ingelheim)
Denagliptin (GSK)
GRC8200 (Glenmark)
PSN-9301 (OSI)
PHX 1149 (Phenomix)
SSR-162369 (Sanofi-Aventis)
TS-021 (Taisho)
Alogliptin (Takeda)
TA-6666 (Tanabe)

FIGURE 6.13 Dipeptidyl peptidase IV inhibitors used in clinical studies. *Reproduced from http://www.bioscience.org/2008/v13/af/2797/figures.htm.*

(A)

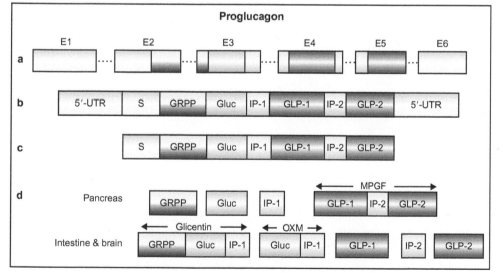

(B)

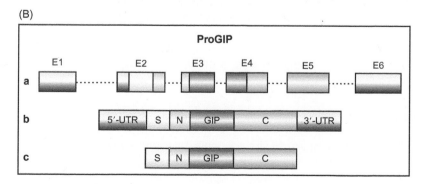

FIGURE 6.14 (A)—Structures of the (a) proglucagon gene, (b) mRNA and (c) protein. (d) Tissue-specific posttranslational processing of proglucagon in the pancreas leads to the generation of Glicentin-related polypeptide (GRPP), glucagon (GLUC), intervening peptide-1 (IP-1) and major proglucagon fragment (MPGF), whereas glicentin, oxyntomodulin (OXM), intervening peptide-2 (IP-2), and GLP-1 and GLP-2 are liberated after proglucagon processing in the intestine and brain. (B)—(a) ProGIP gene, (b) mRNA and (c) protein. Bioactive GIP is generated from its proGIP protein precursor by posttranslational cleavage at single arginine residues that flank GIP. *Reproduced from L. L. Baggio and D.J. Drucker, Gastroenterology,* **132:** *2131–2157, 2007.*

**TABLE 6.2 Medical Problems Associated with Type 1 and Type 2 Diabetes**

| Problem | Diabetes | Indication |
|---|---|---|
| Mortality | Both types 1 and 2 reduce lifespan; type 1 reduces lifespan by 5–8 years; heart disease & stroke are leading causes of death | Control blood glucose & glycosylates Hb |
| Heart & circulation | 60% of deaths due to heart attacks; 25% of deaths due to stroke in types 1 and 2 | Intensive control of blood sugar; reducing blood pressure; reducing [cholesterol] and lipids; lowering LDLs |
| Kidney damage | Glomerular disease in both types | Blood glucose control & antihypertensive drugs; reduce blood clots (drugs) |
| Neuropathy | Especially in legs and feet in both types; blood pressure and bowel and bladder controls; impotence in men; foot ulcers and amputations (types 1 and 2) | Topical drugs for pain antidepressants; antiseizure drugs; antibiotics |
| Retinopathy and eye problems | Leading cause of blindness in types 1 and 2; 40% of type 1 have retinopathy within 10 years of diagnosis | Intensive control of blood glucose; surgery |
| Mental function and dementia | In type 2, higher risk of dementia due to blood vessel damage or Alzheimer's disease; attention and memory loss | |
| Depression | Both types: double risk of depression | Medication and psychotherapy |
| Bone quality | Type 1: slightly reduced bone density; risk of osteoporosis and bone fracture; type 2: increased bone density but also fractures; older patients: risk of falling | Bisphosphonates |
| Other | Both types: higher hearing loss; fatty liver disease; higher risk for uterine cancer and in both sexes, colon and rectal cancer risk higher; both types: higher risk of periodontal disease | |
| Women | Gestational diabetes: existing diabetes can cause birth defects and excessive fetal growth; preeclampsia (dangerously high blood pressure); risk for retinopathy; blunts some effects of estrogen | |
| Adolescents | Type 1 diabetes: self-destructive behaviors; eating disorders in young women (anorexia/bulimia) | |

Source: Data taken from About.com: http://adam.about.com/reports/000009_9.htm?p = 1.

ingested in the human before harmful effects may occur. Clearly, the usage of **artificial sweeteners** is now common. Some of these are derived from sugars and some of their structures are shown in Fig. 6.15.

In experimental models, many using rodents, some of the artificial sweeteners have pathological effects, however, it is difficult to translate the amount of intake required to generate pathologies in humans. Saccharin was discovered in 1879 and is 300 times sweeter than sucrose and does not affect blood insulin level. Saccharin in rodents causes cancer, presumably at levels that would be unrealistically high if translated to the human. Sorbitol is metabolized more slowly than sucrose. In Fig. 6.16A and B are shown: the location of sweet receptors near the tip of the tongue and details of the taste pore in A. In B, a surface model of the taste receptor is shown. It is thought that there is a three-point attachment of the sweet molecule to the receptor. In C, is shown the structure of **lugduname** (Fig. 6.16C) that is reported to be 225,000 times sweeter than sucrose.

## CHEMISTRY OF SIMPLE SUGARS

Simple sugars are single molecules, whereas more complex sugars are more than one molecule joined together by chemical bonds. Sugars are made up of carbon, oxygen and hydrogen; no nitrogen, except for amino sugars. The simplest sugars are **monosaccharides**. Six-carbon monosaccharides are prevalent but monosaccharides can have from 3 to

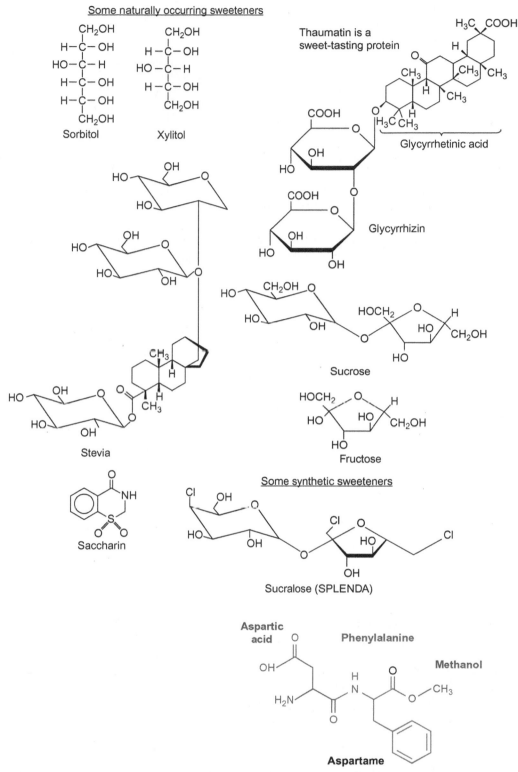

**FIGURE 6.15** Structures of some natural and artificial sweeteners. Sorbitol, xylitol, fructose, glycyrrhizin derivatives, stevia (noncaloric plant source) and sucrose are natural sweeteners. The others shown here are artificial sweeteners. Aspartame, a dipeptide derivative, can release methanol (toxin) in the body.

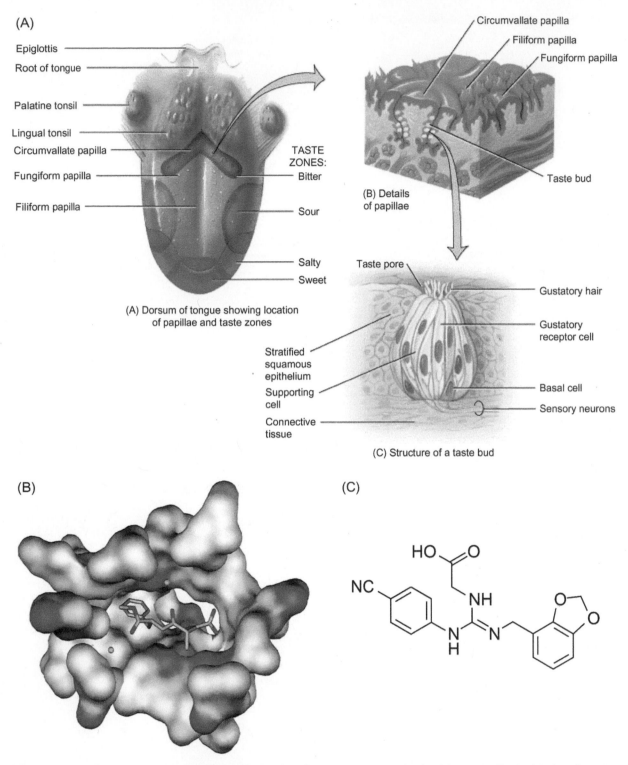

**FIGURE 6.16** (A) Surface of the tongue (left) showing the location of sweet receptors near the tip of the tongue. To the right is a close up section of the tongue showing the location of taste buds and at the bottom right is a close up of the structure of a taste bud. Note that the cells near the pore are supporting cells and gustatory receptor cells that contain the sweet receptor. (B) A surface model of the sweet taste receptor with a molecule of aspartame in the receptor-binding site. (C) Structure of lugduname, a molecule that is reported to be 225,000 times sweeter than sucrose. *(A) Reproduced from http://www.lrn.org/Graphics/Senses/figure%208.8.gif. (B) Reproduced from http://www.physiology.vcu.edu/research/mbrg/images/Aspartame_STR_MCUI.jpg.*

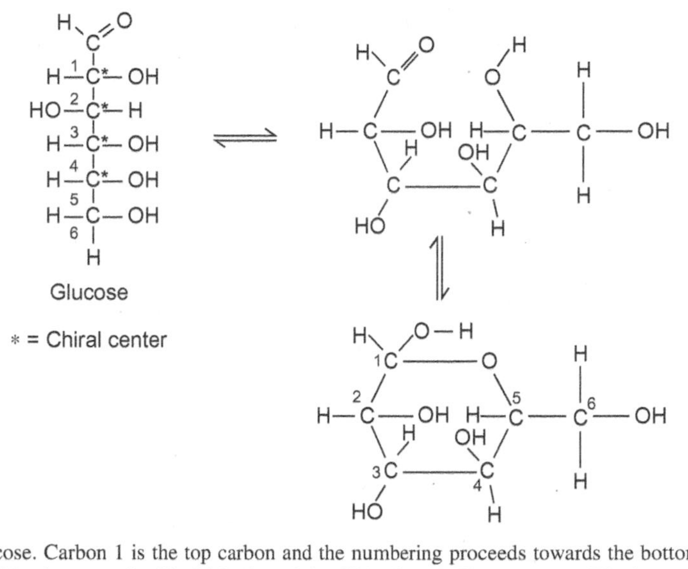

**FIGURE 6.17** Stick model of glucose. Carbon 1 is the top carbon and the numbering proceeds towards the bottom; the lowest is carbon 6. The ring is closed through a nucleophilic addition between the C1 aldehyde and the C5 hydroxyl. The proton on C5 migrates to the oxygen of C1.

7 carbons in their structures (3 carbons, *triose*; 4 carbons, *tetrose*; 5 carbons, *pentose*, 6 carbons, *hexose* and 7 carbons, *heptose*). The hexoses are common and important. **Glucose** is the most representative member of the hexoses. The structure of glucose is shown in Fig. 6.17 in a simple stick model (**Fischer projection**) and how it is closed into a ring structure (**Haworth projection**). Sugars contain *chiral carbons which are those carbons having 4 different substituents.* If 2 of the substituents of a carbon atom are the same (e.g., double bond to an oxygen or single bonds to 2 hydroxyls or to 2 hydrogens) that carbon is achiral. Inspecting the open stick model of glucose, for example, there are 4 chiral centers at C2, C3, C4, and C5. *The number of chiral carbons determines the number of stereoisomers*, thus, for an aldotetrose with 2 chiral centers, the number of stereoisomers is $2^2$ or 4 stereoisomers; for glucose with 4 chiral centers, it has $2^4$ or 16 stereoisomers.

There are a number of simple sugars ranging from three carbons (triose) to seven carbons (heptose), although sugars of six-carbon length (hexose) are considered here. A five-carbon sugar is a pentose and a four-carbon sugar is a tetrose. Sugars are either aldoses or ketoses; an aldose sugar has an aldehyde group (e.g., the C1 of D-glucose seen clearly in the open stick model); a ketose has a ketone group (e.g., the C2 of D-fructose clearly seen in the open stick model). The simple stick structures (Fischer projections) are shown in Fig. 6.18.

The number of chiral carbons determines the number of stereoisomers. The 4 stereoisomers of a tetrose are shown in Fig. 6.19.

In the process of cyclization of glucose, the carbon 1 carbonyl can be attacked from two sides generating the possibility of α- or β-forms, with the rearrangement of a proton in the process as shown in Fig. 6.20.

The α- and β-forms in the ringed form of glucose are named based on the position of the hydroxyl on carbon-1 as shown in Fig. 6.21.

**β-D-glucose** has the C-1 hydroxyl on the left and **α-D-glucose** has the C-1 hydroxyl on the right. "D" refers to *dextro*, or right (as with the amino acids). When the sugar is in the form of a ring, the carbon bonds can bend into either one of two forms: a **"chair" configuration** or a **"boat" configuration**. One form may be favored over the other, depending on hydroxyl substituents in the ring. The general forms of these structures for a six-carbon sugar skeleton are shown in Fig. 6.22.

Sugar rings do not form a flat ring like benzene; the benzene ring contains three double bonds that are shorter (1.34 Å) than carbon-to-carbon single bonds (1.40 Å). Simple sugars can be represented in four ways (not including the boat structure) as shown in Fig. 6.23.

**Disaccharides** are formed using the α- (axial bond down) or the β-(equatorial bond up) hydroxyl on the ring as shown in Fig. 6.24. *The location of a specific substituent (in this case, hydroxyl group) as up or down stems from the stereoisomeric center (anomeric carbon) of the sugar.* In the ring forms (chair, e.g.), *the anomeric carbon can be located as the carbon next to the ring oxygen **not attached** to CH$_2$OH.* In the chair configuration, one hydrogen on each carbon is equatorial and one hydrogen is axial. The equatorial hydrogens radiate from around the ring while the axial hydrogens point along an axis or parallel to an axis; axial bonds are upwards or downwards along an axis through the center of the ring. Also the hydroxyl substituent of the anomeric carbon can be either up or down. If it is down (axial), the sugar is in the α form: if it is up (equatorial), the sugar is in the β form (e.g., α-D-glucopyranose or β-D-glucopyranose).

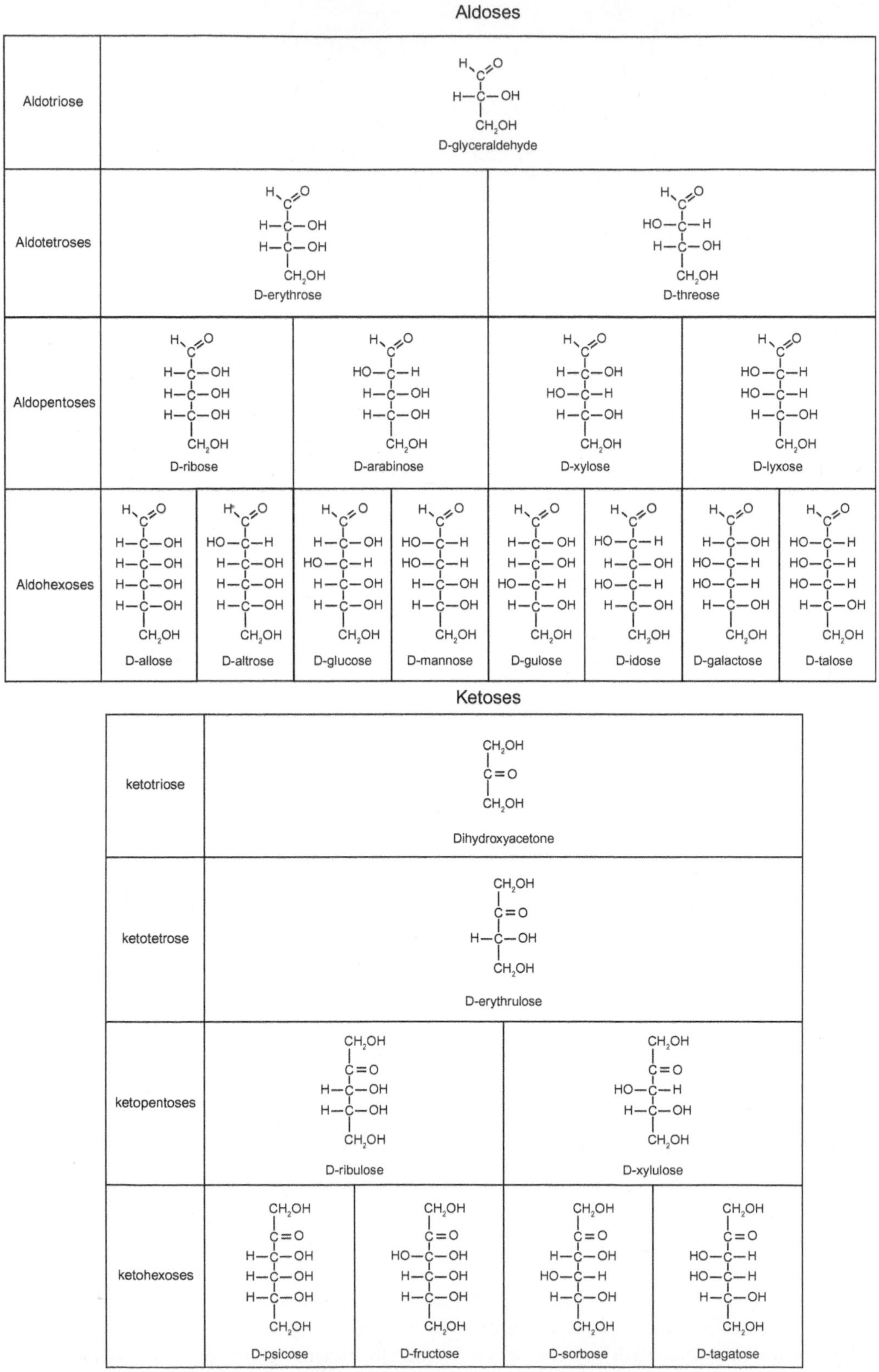

## Aldoses

## Ketoses

**FIGURE 6.18**  Monosaccharides, either natural or synthetic showing aldose sugars (*top*) and ketose sugars (*bottom*).

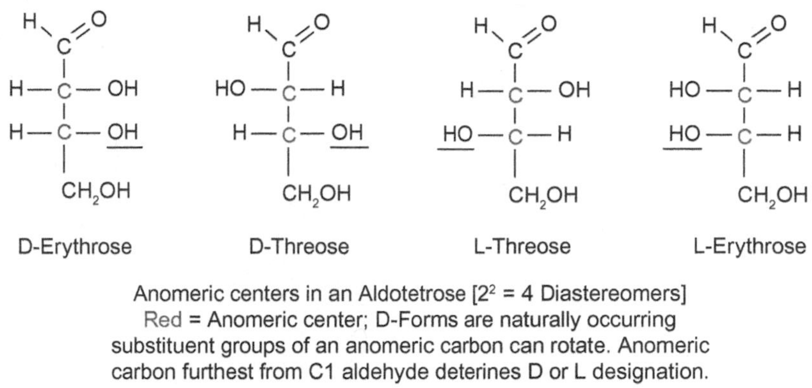

Anomeric centers in an Aldotetrose [2² = 4 Diastereomers]
Red = Anomeric center; D-Forms are naturally occurring
substituent groups of an anomeric carbon can rotate. Anomeric
carbon furthest from C1 aldehyde deterines D or L designation.

**FIGURE 6.19** The four stereoisomers of a tetrose. The anomeric carbons are in *red*.

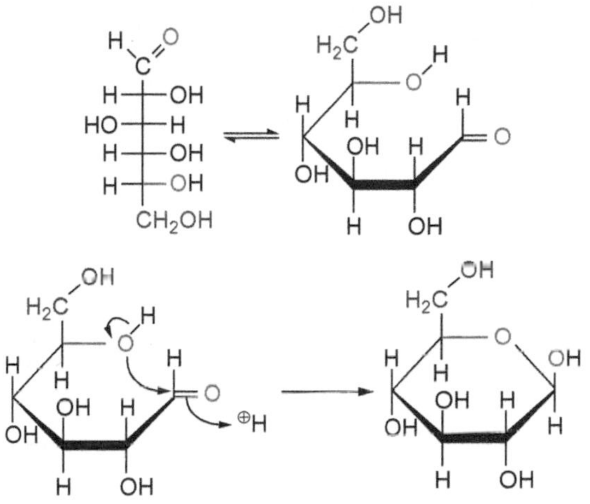

**FIGURE 6.20** D-Glucose, in the Fischer projection, is converted to the Haworth projection with glucose as a five-member ring structure (pyranose). The thick lines at the bottom of the ring structures indicate extension of that part of the structure outward from the page toward the reader.

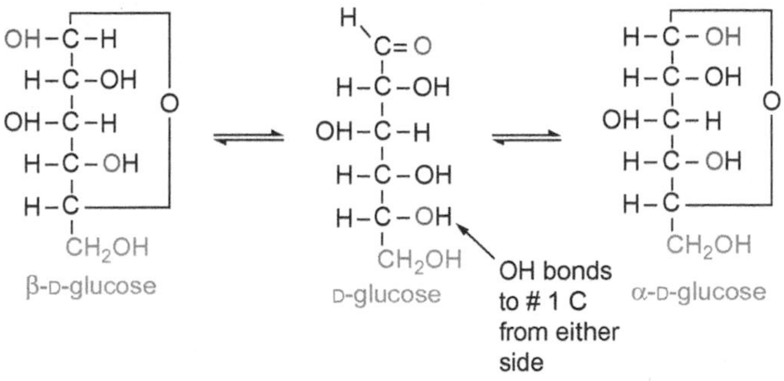

**FIGURE 6.21** β-D-glucose and α-D-glucose ringed forms.

In Fig. 6.25 are shown some common monosaccharides and disaccharides in chair configurations.

## GLUCOSE TRANSPORT

Glucose transporters accomplish the movement of glucose from the extracellular space (deriving from the bloodstream) into cells. The reduction of glucose in the blood results from the action of insulin. Glucose enters the **beta cells of the pancreas** through the glucose transporter **GLUT2** and this is a major signal leading to the release of insulin from these cells that reaches the bloodstream. Insulin in the blood is taken up through a transporter **GLUT4**, in **adipose and muscle cells** resulting in the reduction of blood glucose.

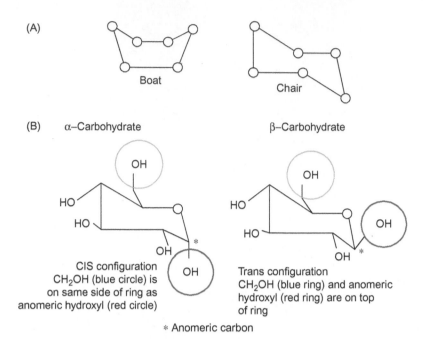

**FIGURE 6.22** (A) Permitted configurations for six-member sugars. (B) Alternate chair conformations for 6-membered sugars: An "up" carbon (e.g., axial) shown on *right* (*circled in red*) and a rotated version showing the axial carbon as "down" on *left* (*circled in red*). α- and β-Forms are determined by the position of the hydroxyl group attached to the anomeric carbon (*asterisk*) and the CH$_2$OH attached to the other carbon next to the ether. α-Carbohydrates have a *cis* configuration between the OH group attached to the anomeric carbon and the CH$_2$OH group (*circled in blue*). The OH group and the CH$_2$OH group are on opposite sides of the ring. β-Carbohydrates have a *trans* configuration between the OH group attached to the anomeric carbon and the CH$_2$OH group (*circled in blue*). The OH group and the CH$_2$OH group are on the same side of the ring.

| Name of molecule | Fischer projection (open-chain form) | Fischer projection (ring form) | Haworth projection | Line form/wireframe form |
|---|---|---|---|---|
| D-glucose | 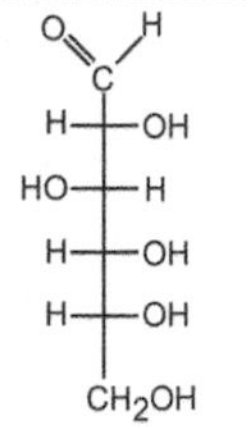 | 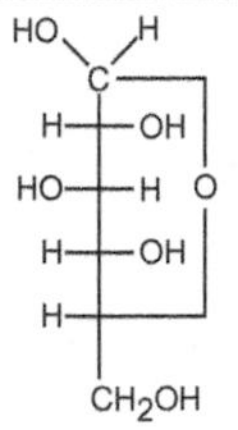 | 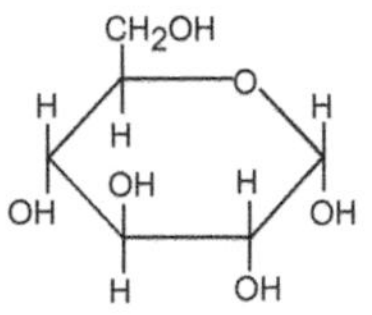 | 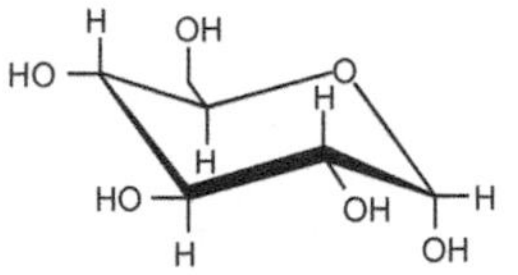 |
| D-ribose | 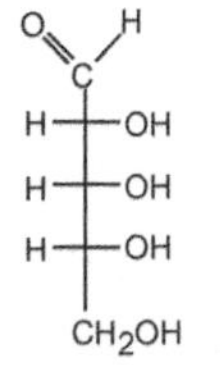 | 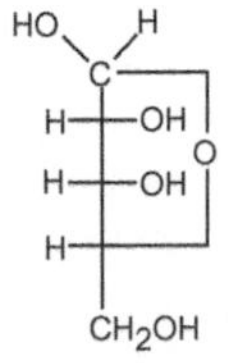 | 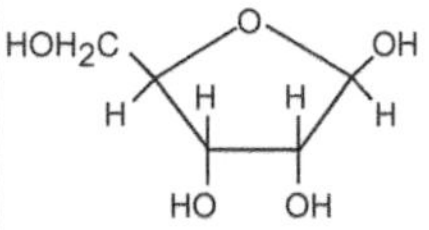 | 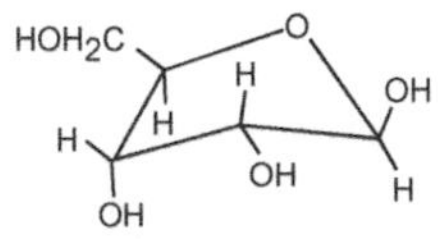 |

**FIGURE 6.23** Four different ways of writing the structures for the hexose, glucose, or for the pentose, ribose, including the chair configuration.

There are two families of glucose transporters. The **sodium coupled glucose transporters** (**SGLT**, **s**odium **gl**ucose **t**ransporter; symporters) consist of three members: SGLT1, SGLT2, and SGLT3. SGLT1 and SGLT2 function as glucose transporters in the intestine, heart, and kidney, whereas SGLT3 functions as a glucose sensor, mainly in intestine, spleen liver, kidney, and muscle. Other members of this family, SGLT4 and SGLT6, serve as inositol and multivitamin transporters and SGLT5 is the thyroid iodide transporter.

The second family consists of **glucose facilitative transporters**, the **GLUT** (**glu**cose **t**ransporter) family. There are 14 members in this family: GLUT 1 through GLUT12, HMIT (**H$^+$** driven **m**yo**i**nositol **t**ransporter, also GLUT13) and GLUT14. The GLUT family and the SGLT family both belong to the larger family of **s**olute **c**arrier gene series (**SLC**) which comprise 43 families.

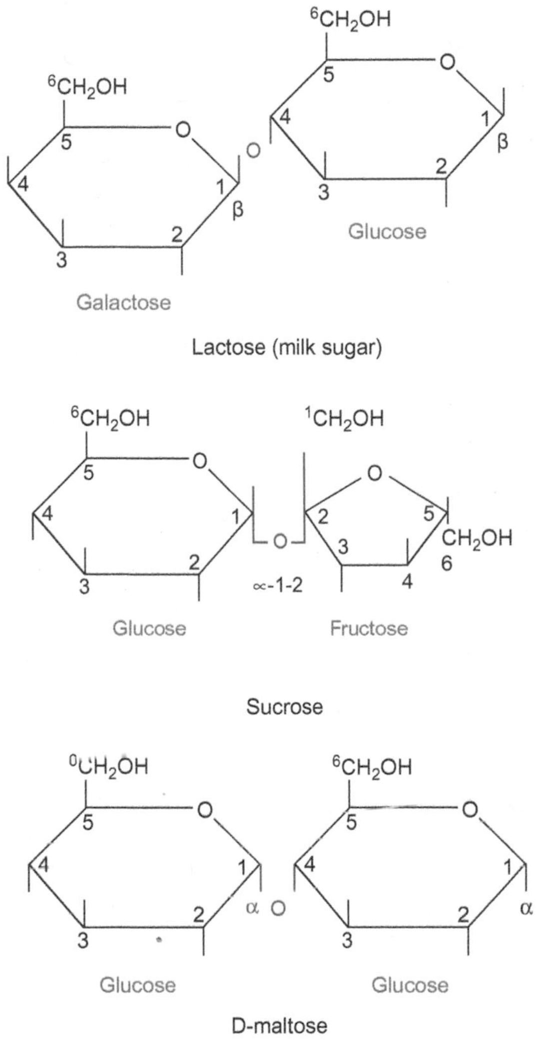

**FIGURE 6.24** Monosaccharides interact using α-hydroxyls or β-hydroxyls on the ring structure to produce some naturally occurring disaccharides. In lactose, the glucose moiety can open and function as a reducing sugar, whereas, in sucrose both rings are locked making it a nonreducing sugar. The glucose moiety on the left of D-maltose cannot open, whereas the glucose moiety on the right is able to open making maltose a reducing sugar. Maltose is the repeating sugar unit in starch.

The properties of the five major **glucose transporters** are summarized in Table 6.3A and the expression of all the other GLUT transporters is shown in Table 6.3B.

The insulin receptor (IR) is activated by the binding of insulin (Fig. 6.3) and the activated receptor phosphorylates itself on tyrosine residues and phosphorylates a number of substrates, such as the IRS (insulin receptor substrate) family of proteins leading to the activation of kinases, such as Akt and PKC as well as the APS adapter protein. APS activates TC10 (a G protein located on lipid rafts) and TC10 causes changes in the actin skeleton. These pathways culminate in the control of the recycling of GLUT4 so that there is an increase in the number of GLUT4 transporters on the membranes of adipose and muscle cells, leading to increased uptake of glucose from the blood. These actions are summarized in Fig. 6.26A and B.

## PENTOSE PHOSPHATE PATHWAY

**Glucose-6-phosphate** is a precursor for the formation of **ribose-5-phosphate**. This involves a **cytosolic pathway** (soluble cytoplasm) called the pentose phosphate pathway that, in abbreviated form, is summarized in Fig. 6.27.

There are two phases to the pentose phosphate pathway, the oxidative phase and the nonoxidative phase. In the oxidative phase, starting with glucose-6-phosphate, the end products are ribose-5-phosphate, $CO_2$, 2NADPH, and $2H^+$.

In the nonoxidative phase, the enzymes, transketolase and transaldolase, are required leading to the interconversion of phosphorylated sugars, including pentoses and hexoses. Specifically, **transketolase** catalyzes the following reaction: ribose (5C) 5-phosphate + xylulose (5C) 5-phosphate $\rightleftharpoons$ sedoheptulose (7C) 7-phosphate + glyceraldehyde (3C)

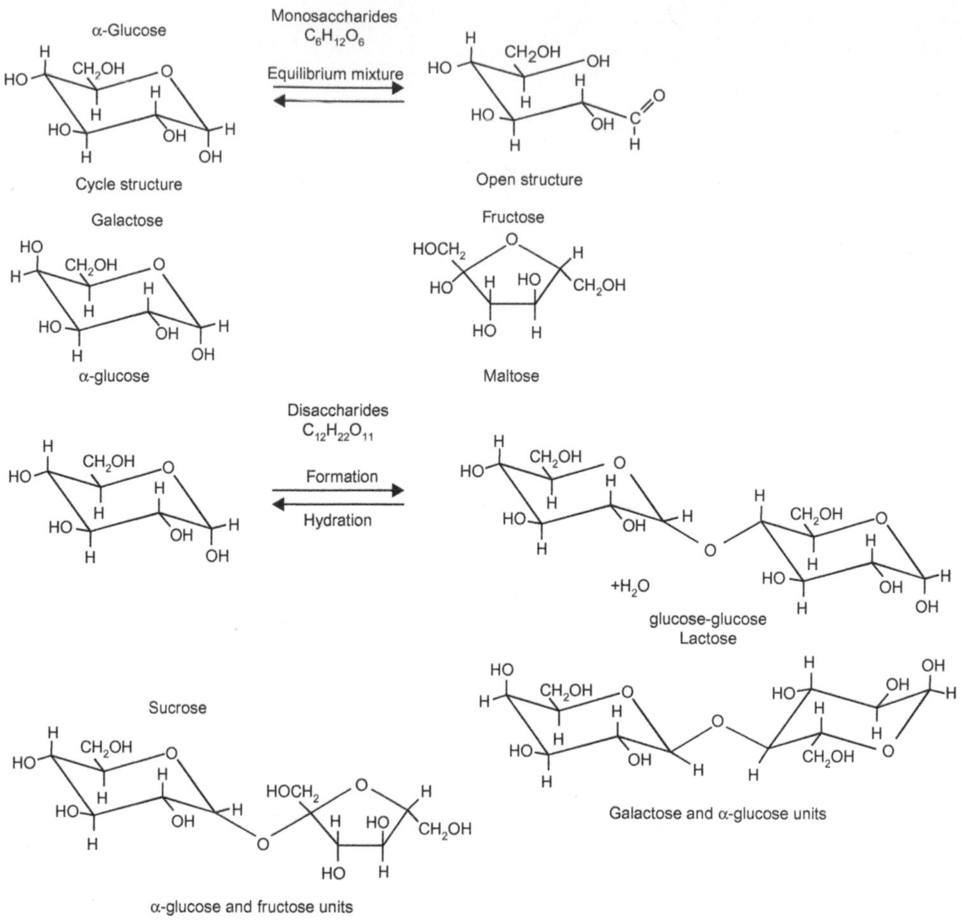

**FIGURE 6.25**   Common monosaccharides and disaccharides in the chair configuration.

3-phosphate. Glyceraldehyde 3-phosphate is an intermediate in glycolysis and is a link to that pathway. **Transaldolase** utilizes these products by catalyzing the reaction: sedoheptulose (7C) 7-phosphate + glyceraldehyde (3) 3-phosphate $\rightleftharpoons$ erythrose (4C) 4-phosphate + fructose (6C) 6-phosphate (an intermediate in the glycolytic pathway). In a third reaction, transketolase catalyzes the conversion of a tetrose and a pentose to intermediates in glycolysis: erythrose (4C) 4-phosphate + xylulose (5) 5-phosphate $\rightleftharpoons$ fructose 6-phosphate (6C) + glyceraldehyde (3C) 3-phosphate (both are intermediates of the glycolytic pathway). The sum of these reactions is:

Ribose-5-phosphate (5C) + 2 xylulose 5-phosphate (5C) $\rightleftharpoons$ glyceraldehyde 3-phosphate (3C) + 2 fructose (6C) 6-phosphate.

When ribose is ingested or when ribose 5-phosphate is in excess, they can be converted by the PPP to glycolytic intermediates. **Phosphoribosylpyrophosphate (PRPP)** in this pathway is used in the formation of nucleotides, as will be discussed later. The pentose phosphate pathway (**phosphogluconate pathway** or **hexose monophosphate shunt**) generates **NADPH** during the oxidative phase of the pathway. One molecule of NADPH is generated from $NADP^{+}$ in the following reactions of the pathway: conversion of glucose-6-phosphate to 6-phosphogluconate by glucose-6-phosphate dehydrogenase; conversion of 6-phosphogluconate to ribulose-5-phosphate by phosphogluconate dehydrogenase and conversion of ribulose-5-phosphate to ribose-5-phosphate (2 NADPH). A **hemolytic anemia** is produced in patients that have a genetic **deficiency of glucose-6-phosphate dehydrogenase (G6PDH)**, resulting in an inadequate supply of NADPH to the **red blood cell**. *This is the most common defect resulting from an enzyme deficiency.* This disorder involves 400 million people worldwide (especially in Africa, Asia, Mediterranean countries and South America), approximately the number of persons afflicted with the malaria parasite. The mutation is X-recessive-linked and is polymorphic with more than 300 variants. Neonatally, the disorder can present as hyperbilirubinemia. Some individuals with this mutation can be asymptomatic while others have episodes of **hemolysis** (destruction of red blood cells), varying in intensity, as there is a range of severity. G6PDH deficiency, however, is protective against malaria. The mutation is dangerous if antimalarial drugs (e.g., Primaquine) are used because of the loss of protection in the red blood cell against hemolyzing reactive oxygen species ($H_2O_2$) that are induced by the drug. Consequently, it is vital to test for G6PDH deficiency before administering the antimalarial drug.

**TABLE 6.3** The Five Major Glucose Transporters and Their Properties

| Transporter | Location | Properties |
|---|---|---|
| **(A)** | | |
| GLUT1 | Almost universal | High capacity, relatively low $K_m$ |
| GLUT2 | Liver, β-cells, hypothalamus, basolateral membrane of small intestine | High capacity but low affinity (high $K_m$ 15–20 mM); part of the glucose sensor in β-cells. Carrier for glucose and fructose in liver and intestine |
| GLUT3 | Neurons, placenta, testes | Low $K_m$ (1 mM) and high capacity |
| GLUT4 | Skeletal and cardiac muscle, fat | Activated by insulin, $K_m$ 5 mM |
| GLUT5 | Mucosal surface in small intestine, sperm | Primarily fructose carrier in intestine |
| **(B) Further Information and Other GLUT Transporters** | | |
| **Class I** | | **Expression** |
| GLUT1 | | Erythrocytes, brain (blood–brain barrier) |
| GLUT4 | | Adipocytes, muscle |
| GLUT3 | | Brain (neuronal), testis |
| GLUT14 | | Testis |
| GLUT2 | | Liver, islet cells, kidney, small intestine |
| **Class II** | | |
| GLUT5 | | Testis, intestine, muscle |
| GLUT7 | | Intestine, testis, prostate |
| GLUT9 | | Liver, kidney |
| GLUT11 | | Pancreas, kidney, placenta, muscle |
| **Class III** | | |
| GLUT6 | | Brain, spleen, peripheral leukocytes |
| GLUT8 | | Testis, brain (neuronal), adipocytes |
| GLUT10 | | Liver, pancreas |
| GLUT12 | | Heart, prostate, breast cancer |
| HMIT | | Brain |

(A) These are facilitative transport proteins (they transport glucose (or fructose) from areas of high concentration to areas of low concentration. The sugar is bound by the protein, the membrane direction of the sugar–protein complex is reversed by a flip-flop mechanism; the sugar is released and the protein flips around to start a new cycle. Extent of transport is determined by sugar concentrations and the number of transport proteins in the outer cell membrane. Since, in most tissues, the internal glucose concentration is low, transport can proceed only from the outside to the cell interior. The GLUT family can transport glucose either into or outside cells. In liver and kidney which are gluconeogenic (can form glucose from other molecules), the intracellular [glucose] can exceed blood [glucose] in postabsorptive or fasting states and glucose can be exported from these tissues through GLUT2.
Source: (A) The information in this table is taken from http://www.medbio.info/Horn/Time%203-4/glucose_transport_proteins.htm. (B) Reproduced from Scheepers, A., Joost, H.G., Schurmann, A., 2004. The glucose transporter families SGLT and GLUT: molecular basis of normal and aberrant function. JPEN J. Parenter. Enteral. Nutr. 28, 364–371.

Also in the red blood cell, the enzyme, **glutathione peroxidase**, functions to degrade **hydroperoxides** that arise because of the oxygen-rich environment. This enzyme reduces the hydroperoxides by the use of two molecules of reduced glutathione (GSH), converting it to oxidized glutathione (GSSG):

$$ROOH + 2\,GSH \rightarrow ROH + GSSG + H_2O$$

or:

$$H_2O_2 + 2GSH \rightleftharpoons H_2O + GSSG$$

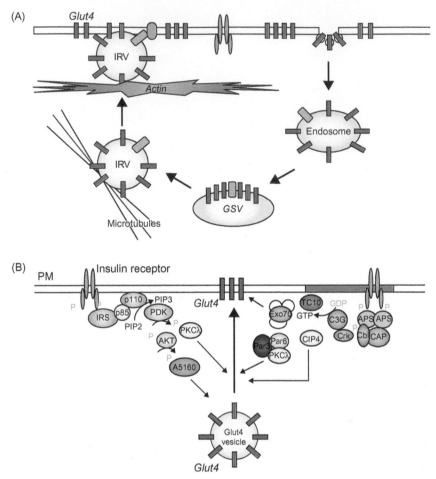

**FIGURE 6.26 (A)** Recycling of GLUT4. The cellular location of GLUT4 is governed by a process of regulated recycling, in which the endocytosis, sorting into specialized vesicles, exocytosis, tethering, docking, and fusion of the protein are tightly regulated. In the absence of insulin, GLUT4 slowly recycles between the plasma membrane and vesicular compartments within the cell, where most of the GLUT4 resides. Insulin stimulates the translocation of a pool of GLUT4 to the plasma membrane, through a process of targeted exocytosis. The microtubular network and actin cytoskeleton play a role in GLUT4 trafficking, either by linking signaling components or by directing movement of vesicles from the perinuclear region to the plasma membrane in response to insulin. Once at the plasma membrane, the GLUT4 vesicles dock and fuse, allowing for extracellular exposure of the transporter. *IRV*, insulin receptor vesicle; *GSV*, GLUT4 storage vesicle. (B) A model for diverse signaling pathways in insulin action. Two signaling pathways are required for the translocation of the glucose transporter GLUT4 by insulin in fat and muscle cells. Tyrosine phosphorylation of the IRS proteins after insulin stimulation leads to an interaction with and subsequent activation of PI3-kinase, producing PIP3, which, in turn, activates and localizes protein kinases, such as PDK1. These kinases then initiate a cascade of phosphorylation events, resulting in the activation of Akt and/or atypical PKC. AS160, a substrate of Akt, plays an as yet undefined role in GLUT4 translocation through its Rab-GTPase activating domain. A separate pool of insulin receptor can also phosphorylate the substrates Cbl and APS (*upper right*). Cbl interacts with CAP, which can bind to the lipid raft protein flotillin. This interaction recruits phosphorylated Cbl into the lipid raft, resulting in the recruitment of CrkII. CrkII binds constitutively to the exchange factor C3G, which can catalyze the exchange of GDP for GTP on the lipid-raft-associated protein TC10. Upon its activation, TC10 interacts with a number of potential effector molecules, including CIP4, Exo70, and in a GTP-dependent manner PM, plasma membrane; IRS, insulin receptor substrate. The figures are reproduced from http://www.ncbni.nlm.nih.gov/pmc/articles/PMC1431367/figure/f1-mol10_2p065/.

In order to regenerate GSH from GSSG, NADPH is required:

$$GSSG + NADPH + H^+ \rightleftharpoons 2\,GSH + NADP^+$$

In the absence of sufficient NADPH, GSSG would accumulate and there would be insufficient GSH to remove harmful **hydroperoxides**.

## CONVERSION OF RIBOSE TO DEOXYRIBOSE

Ribose is the sugar in ribonucleic acids (RNAs), however, deoxyribose is the sugar in DNAs. To achieve this transformation, ribose is not converted directly to deoxyribose but is converted in the **nucleotide form** by the enzyme **ribonucleotide reductase** that converts a **ribose nucleotide** to a **deoxyribose nucleotide** as shown in Fig. 6.28.

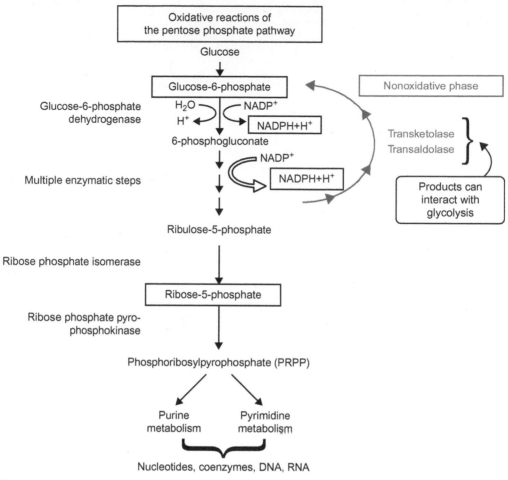

**FIGURE 6.27** Abbreviated pentose phosphate pathway.

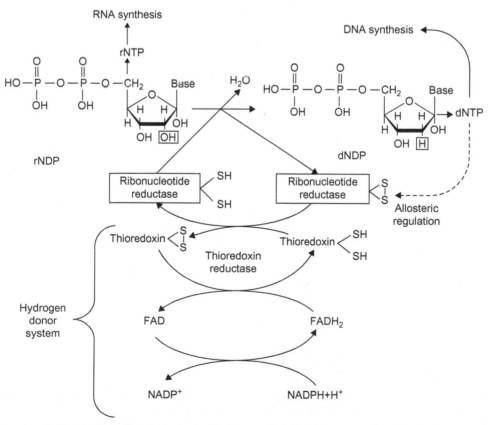

**FIGURE 6.28** Conversion of ribose to deoxyribose in the nucleotide forms catalyzed by the enzyme, ribonucleotide reductase. *rNDP*, ribodinucleotide; *dNDP*, deoxyribodinucleotide; *FAD*, flavin adenine dinucleotide; *NADP*, nicotinamide adenine dinucleotide phosphate.

# CARBOHYDRATE CONSTITUENTS OF PROTEINS—GLYCOPROTEINS

Carbohydrate constituents of proteins are seen frequently protruding from the cell surface (e.g., Fig. 6.29). They are the molecular components of self-recognition and, often, are foreign antigens when encountered in different structures or composition.

Sugar substitutions of completed proteins are made in the **Golgi apparatus** and can serve to signal destinations outside of the cell. Specific **blood groups** are defined by their sugar-containing components as will be seen. Also, substitution with sugars renders a protein more soluble in aqueous media. Sugar substitutions in proteins are through the asparagine amide group (N-linked substitution) or through the hydroxyl group (O-linked substitution) of threonine or serine. These sites are substituted by *N*-acetylglucosamine and *N*-acetylgalactosamine as shown in Fig. 6.30. *N*-acetylglucosamine is linked through the asparagine's amide nitrogen of the evolving protein. The **target sequence** on the protein for *N*-glycosylation is Asn-X-Ser/Thr where X is any amino acid except Pro or Asp.

There are three types of **N-linked carbohydrate substituents** of proteins: a complex type, a hybrid type or a high mannose type (Fig. 6.31).

These oligosaccharides contain at least five sugar residues and there are many variants of these types. In O-linked oligosaccharides, usually there are one to four sugar residues. Examples of O-linked substitutions are shown in Fig. 6.32.

Glycosylation of proteins occurs in the endoplasmic reticulum or in the Golgi apparatus. Glucosylation of proteins offers a means to monitor the folding of proteins or their degradation if they remain in the ER for an extended period of time (Fig. 4.21). The *N*-linked oligosaccharides contain **mannose** and *N*-acetylglucosamine. They often have branches

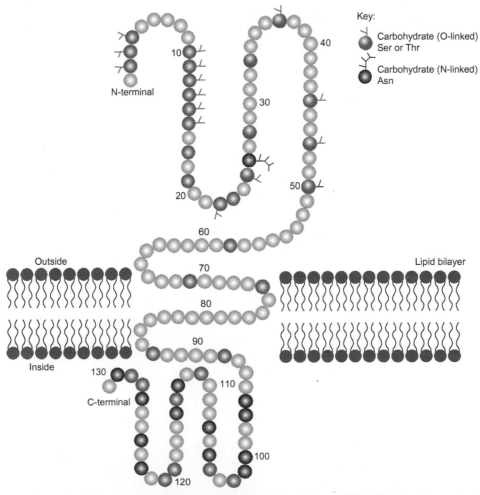

**FIGURE 6.29** Diagram of a transmembrane protein with sugar substitutions in the extracellular domain. Sugars that frequently appear in glycoproteins are glucose, galactose, mannose, fucose and acetylated amino sugars, such as *N*-acetylgalactosamine, *N*-acetylglucosamine, and *N*-acetyl neuraminic acid (NANA or sialic acid). Sugars can be O-linked through Ser or Thr or N-linked through Asn.

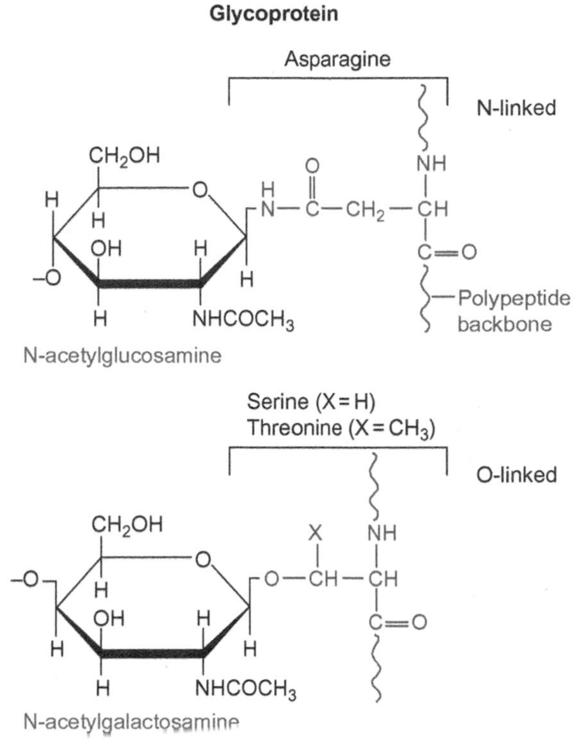

**Glycoprotein**

FIGURE 6.30 N-linked and O-linked sugar substituents of proteins.

and each branch may terminate in a **sialic acid** (**N-acetylneuraminic acid**) residue. The proteins residing in the soluble cytoplasm and in the nucleus are, generally, not glycosylated. Each O-linked sugar is added by a specific glycosyltransferase, one at a time, and each sugar is transferred from a nucleotide precursor, such as GDP-mannose or UDP-galactose (UDP-Gal). This contrasts with N-linked sugars that can be added to the protein with a preformed oligosaccharide.

## TRANSFER OF NUCLEOTIDE SUGARS INTO THE GOLGI *CISTERNAE*

UDP-Gal and UDP-GalNAc (UDP-*N*-acetylgalactosamine) are transferred to the Golgi apparatus (Fig. 2.18) from the soluble cytoplasm by an **antiporter** (transporter that exchanges solutes in opposite directions). These sugar nucleotides are transported into the Golgi by exchange with UMP (uridine monophosphate) as shown in Fig. 6.33.

Oligosaccharide is added to a growing protein chain in the lumen of the endoplasmic reticulum. Carbohydrate moieties are added continuously and trimming takes place until the final mannose has been attached. The final carbohydrate modification occurs in the Golgi apparatus. The folding of the protein would have been facilitated in the ER lumen by molecular chaperones that interact with the carbohydrate portion of the glycoprotein. An improperly folded protein would be degraded by the cytoplasmic proteasome. If improperly folded proteins escape the protective degradative mechanism, they can produce neurodegenerative diseases. The **calnexin-calreticulin cycle** operates to ensure the correct disulfide bonding in proteins and these agents are lectin-containing protein chaperones that contain carbohydrate moieties and they are specific for monoglucosylated proteins. Subsequent to their action, a **ERp57** protein, having **disulfide isomerase** activity, interacts with the protein and ensures the correct disulfide bond formation in the newly synthesized glycoprotein.

## SUGARS IN BLOOD GROUP PROTEINS

There are three blood group antigens in the human, O, A, B, and AB system. Every individual is able to synthesize the O antigen. The O antigen is a carbohydrate attached to a lipid, the carbohydrate being the antigenic region. From the lipid outward, the O antigen is a sequence of glucose, galactose, *N*-acetylglucosamine, galactose and **fucose**. Fucose is a methylated pentose (Fig. 6.31). The blood group antigens are visualized in Fig. 6.34.

N-LINKED OLIGOSACCHARIDES

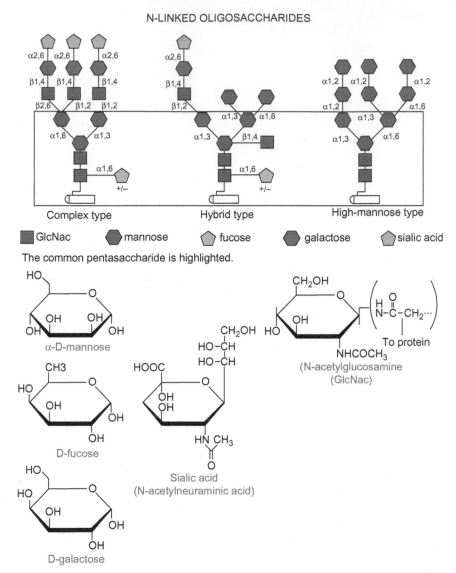

FIGURE 6.31   Types of carbohydrate substituents in glycoproteins. The protein is represented by the *open box* at the bottom of each type. Below are the structures of the sugar substituents. The Greek symbols and numbers in the top figure represent the type of bonds between the sugars.

These antigens are found on glycoproteins and glycolipids. When they occur on the cell surface, they are linked to **sphingolipids**. In the serum, they represent the secreted forms and they are linked to glycoproteins. **Secreters**, constituting about 80% of the people in the United States, secrete their blood antigen into various body fluids, including saliva and mucus. **Nonsecreters** have low levels of glycoproteins and secrete little or no blood antigen into body fluids. This distinguishing feature of a secreter versus a nonsecreter can be used forensically in the courts, especially in cases of rape.

## LACTOSE INTOLERANCE

This condition is marked by an allergic reaction to dairy products that contain lactose. More than half the people in the world are lactose intolerant. Lactose intolerance is marked by the lack of expression of the enzyme **lactase** or **β-galactosidase**. Lactose is the disaccharide, galactosylglucose, where the monosaccharides are joined in a β-1-4 linkage (Fig. 6.24). Even though a small amount of β-galactosidase may be produced in the gastrointestinal tract of lactose intolerants, it is insufficient to handle ingested lactose. The extent of lactose intolerance may be related to the relative amounts of the enzyme produced in these individuals. The lactose that is ingested by these individuals is broken down by the bacteria in the large intestine resulting in diarrhea, dehydration and bloat. Many dairy products are available that are lactose-free. Intolerant individuals also can ingest "**Lactaid**" which contains lactase prior to ingesting lactose-containing foods.

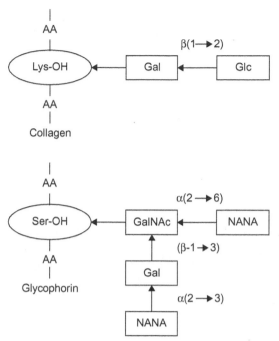

**FIGURE 6.32** Examples of O-linked oligosaccharides. *Lys-OH*, hydroxylysine in **collagen**; *AA*, amino acid in peptide chain; *Gal*, galactose; *Glc*, glucose; *Ser-OH*, hydroxyl group of serine; *GalNAc*, N-acetylgalactosamine; *NANA*, N-acetylneuraminic acid. **Glycophorin** is a sialoglycoprotein in the membrane of a red blood cell, such as shown in Fig. 6.29. This carbohydrate moiety allows the red blood cell to circulate in the bloodstream without sticking to other cells or to the vessel walls.

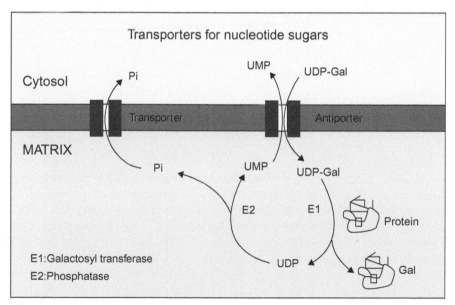

**FIGURE 6.33** Diagram showing the transporter (antiporter) for UMP and UDP-Gal. By this mechanism, UDP-Gal is transported into the Golgi from the soluble cytoplasm (cytosol) in exchange for UMP and the galactose moiety is attached to a protein in the Golgi by **galactosyl transferase**. *Reproduced from http://www.cryst.bbk.ac.uk/pps97/assignments/projects/emilia/trans.gif.*

# GLYCOBIOLOGY

Glycobiology concentrates on the modification of proteins to **glycoproteins** or **proteoglycans** and the modification of lipids. Heavily glycosylated glycoproteins are called proteoglycans and they have one or more chains of **glycosamino-glycans** that are long, linear polymers bearing a negative charge (sulfate and uronic acid groups). They are found in connective tissues and may contain chondroitin sulfate and dermatan sulfate, or heparin and heparin sulfate or keratin sulfate. Glycobiology includes an understanding of the cellular and molecular biology of glycans (Fig. 6.35).

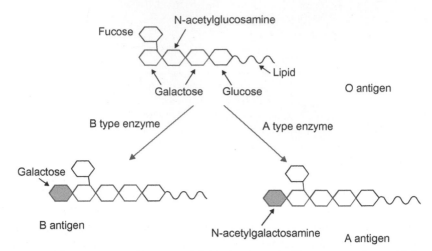

FIGURE 6.34 Carbohydrate constituents attached to proteins and lipids that form the blood group antigens. All humans can synthesize the **O antigen**. Type **A antigen** is characteristic of an individual expressing an enzyme that adds an *N*-acetylgalactosamine to the terminal galactose of the O antigen (*lower right*). The **B antigen** occurs in an individual expressing an enzyme that adds another galactose to the terminal galactose of the O antigen (*lower left*). The **AB type** (mixture of both A and B antigens) expresses both enzymes and the O type lacks expression of these enzymes.

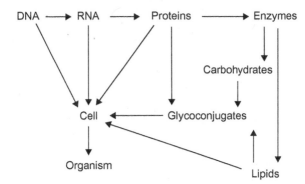

FIGURE 6.35 Glycobiology relates modifications of proteins and lipids and the roles these macromolecules play in cellular functions.

## SUMMARY

**Diabetes** is an ancient disease characterized by the inability of the **beta cells of the pancreas** to produce insulin (type 1 diabetes) or by some defect in the utilization of insulin usually occurring in the adult (type 2 diabetes). A component of type 2 diabetes can be the inadequate secretion of insulin from the beta cell in addition to other causes. Insulin functions by binding to the insulin receptor and activating signaling pathways. These pathways can lead to the utilization of glucose to form **glycogen in liver and muscle** (and lowering blood glucose level) or to growth promotion in tissues where cell growth is possible; *insulin is a mitogen*. Insulin is active in physiological concentrations as a monomer; it can form dimers at higher concentrations and, in the presence of zinc ions, it can form hexamers. In the **beta cell**, insulin is secreted in response to high levels of glucose in the extracellular milieu. Metabolism of glucose produces ATP to a point where the ATP-sensitive $K^+$ channel is blocked and the cell membrane is depolarized, allowing the inward flow of $Ca^{2+}$ and availability of ER-stored $Ca^{2+}$ to the cytosol which triggers the fusion of the insulin-containing granule and its secretion. **Incretins (glucose-dependent insulinotropic polypeptide, GIP**, and **glucagon-like peptide, GLP-1)** from cells lining the intestine facilitate insulin secretion. They operate through receptors on the cell membrane of the beta cell by producing cAMP, diacylglycerol and IP3 (inositol *tris*phosphate). Incretins are inactivated by the enzyme, **dipeptidyl peptidase IV** by cutting between the amino acids, Ala−Glu at or near the N-terminus. Inhibitors of this enzyme promise to be useful drugs in the treatment of type 2 diabetes and incretins, themselves, may be useful in direct treatment of type 2 diabetes. Both types of diabetes cause many severe medical problems that are outlined.

**Structures** of monosaccharides, disaccharides and some complex sugars are presented as well as the various configurations of sugars.

The **transport of glucose across a cell membrane** can occur through two types of transporters: those using sodium ion in tandem with the movement of glucose (sodium/glucose symporter, **SGLT** or **s**odium **gl**ucose **t**ransporter) or the glucose facilitating transporters (**GLUT**s) that transport glucose alone. Both types are members of families. There are five SGLTs of which three transport glucose and 14 types of GLUTs although a smaller number of GLUTs are major factors.

**Glucose-6-phosphate** can form **ribose-5-phosphate** through the **pentose phosphate pathway** and can comprise part of nucleotides through the intermediate, **phosphoribosylpyrophosphate** (**PRPP**). In the case of deoxyribonucleotides, the conversion of the ribose moiety to deoxyribose occurs in the nucleotide form by the enzyme **ribonucleotide reductase**. Under anaerobic conditions, the pentose phosphate pathway generates intermediates of glycolysis so that ATP can be produced under limiting oxygenation.

**Glycoproteins** have their sugar additions in the Golgi apparatus. They are linked to the protein as N-linked or O-linked. N-linked sugars are through asparagine amide groups and *O*-linked sugars are through serine or threonine hydroxyl groups. The target sequence of a protein for *N*-glycosylation is Asn-X-Ser/Thr. Carbohydrate addition to proteins can play a role in the timing of protein folding or of protein degradation. Sugar substituents are key to the differentiation of blood group proteins. **Blood group types** are distinguished by expression of enzymes that determine the addition of sugar derivatives to the structure by adding *N*-acetylgalactosamine to galactose of the O antigen to produce the A antigen or by adding another galactose to the terminal galactose of the O antigen to determine the B antigen. Both enzymes are present in the AB individual and both enzymes are lacking in the O-type individual.

**Lactose intolerance** results from the failure to express β-galactosidase (lactase) in the intestinal tract. Half of the population in the world has this condition.

The modification of proteins to glycoproteins or proteoglycans constitutes the field of **glycobiology**.

## SUGGESTED READING

### Literature

Baggio, L.L., Drucker, D.J., 2007. Biology of incretins: GLP-1 and GIP. Gastroenterology. 132, 2131−2157.

Brammark, C., et al., 2013. "Insulin signaling in type 2 diabetes: experimental and modeling analyses reveal mechanisms of insulin resistance in human adipocytes". J. Biol. Chem. 288, 9867−9880.

Buchner, D.A., 2015. Zinc finger protein 407 (ZFP407) regulates insulin-stimulated glucose uptake and glucose transporter 4 (Glut 4) mRNA. J. Biol. Chem. 290, 6376−6386.

Chang, L., Chiang, S.H., Saltiel, A.R., 2004. Insulin signaling and the regulation of glucose transport. Mol. Med. 10, 65−71.

Guo, H., et al., 2014. Inefficient translocation of preproinsulin contributes to pancreatic β cell failure and late-onset diabetes. J. Biol. Chem. 289, 16290−16302.

Krokowski, D., et al., 2013. A self-defeating anabolic program leads to β-cell apoptosis in endoplasmic reticulum stress-induced diabetes via regulation of amino acid flux. J. Biol. Chem. 288, 17202−17213.

Kuo, T., et al., 2012. Genome-wide analysis of glucocorticoid receptor binding sites in myotubes identifies gene networks modulating insulin signaling. Proc. Natl. Acad. Sci. U.S.A. 109, 11166−11171.

Nyman, E., et al., 2014. A single mechanism can explain network-wide insulin resistance in adipocytes from obese patients with type 2 diabetes. J. Biol. Chem. 289, 33215−33230.

Ottensmeyer, F.P., Beníac, D.R., Luo, R.Z.T., Yip, C., 2000. Mechanism of transmembrane signaling: insulin binding to the insulin receptor. Biochemistry. 39, 12103−12112.

Rui, L., 2014. Energy metabolism in the liver. Compr. Physiol. 4, 117−197.

Sano, R., et al., 2012. Expression of ABO blood group genes is dependent upon an erythroid cell-specific regulatory element that is deleted in persons in the $B_m$ phenotype. Blood. 119, 5301−5310.

Scheepers, A., Joost, H.G., Schurmann, A., 2004. The glucose transporter families SGLT and GLUT: molecular basis of normal and aberrant function. JPEN J. Parenter. Enteral. Nutr. 28, 364−371.

Segurel, L., et al., 2012. The ABO blood group is a trans-species polymorphism in primates. Proc. Natl. Acad. Sci. U.S.A. 109, 18493−18498.

Sims, P.A., 2014. Big Picture" worksheets to help students learn and understand the pentose phosphate pathway and the Calvin cycle. J. Chem. Educ. 9, 541−545.

Stincone, A., et al., 2014. The return of metabolism: biochemistry and physiology of the pentose phosphate pathway. Biol. Rev. Camb. Philos. Soc. 10, 1111/brv.12140.

Wang, M.Y., et al., 2015. Glucagon receptor antibody completely suppresses type 1 diabetes phenotype without insulin by disrupting a novel diabetogenic pathway. Proc. Natl. Acad. Sci. U.S.A. 112, 2503−2508.

Wittaker, J., et al., 2012. α-Helical element at the hormone-binding surface of the insulin receptor functions as a signaling element to activate its tyrosine kinase. Proc. Natl. Acad. Sci. U.S.A. 109, 11166−11171.

Xu, B., et al., 2004. Diabetes-associated mutations in insulin: consecutive residues in the B chain contact distinct domains of the insulin receptor. Biochemistry. 43, 8356–8372.

Xu, X.J., et al., 2012. Insulin sensitive and resistant obesity in humans: AMPK activity, oxidative stress, and depot-specific changes in gene expression in adipose tissue. J. Lipid Res. 53, 792–801.

## Books

Cooper, G.M., 2000. The Cell, A Molecular Approach. 2nd edition Sinauer Associates, Sunderland, MA.

Frayn, K.N., 2010. Metabolic Regulation: A Human Perspective. Wylie-Blackwell, Hoboken, NJ.

Garg, H., Courmant, M., Hales, C. (Eds.), 2008. Carbohydrate Chemistry, Biology and Medical Applications. Elsevier, Amsterdam, NL.

Litwack, G. (Ed.), 2008. Insulin and IGFs, 80. Vitamins & Hormones, Academic Press/Elsevier, Amsterdam, NL.

Litwack, G. (Ed.), 2010. Incretins and Insulin Secretion, 84. Vitamins & Hormones, Academic Press/Elsevier, Amsterdam, NL.

Litwack, G., 2014. The Pancreatic Beta Cell, 95. Vitamins & Hormones, Academic Press/Elsevier, Amsterdam, NL.

Robyt, J.F., 2008. Essentials of Carbohydrate Chemistry. 399. Springer-Verlag, New York.

Stick, R., Williams, S., 2009. Carbohydrates: The Essential Molecules of Life. 496. Elsevier, Amsterdam, NL.

Varki, A., Cummings, R.D., Esko, J.D., Freeze, H.H., Stanley, P., Bertozzi, C.R., Hart, G.W., Etzler, M.E. (Eds.), 2009. Essentials of Glycobiology. 2nd edition Cold Spring Harbor Laboratory Press, Cold Spring Harbor, NY.

Chapter 7

# Glycogen and Glycogenolysis

## GLYCOGEN STORAGE DISEASE (GSD) TYPE I, VON GIERKE DISEASE (AND OTHERS: AT LEAST 11 TYPES OF GSD)

This disease occurs as a result of an inherited deficiency of an enzyme involved in the synthesis or breakdown of glycogen, the storage form of glucose. Glycogen is formed primarily in the liver and muscle and, secondarily, in many other tissues as well. There are a variety of subtypes of this disease based on the deficient enzyme that will dictate the primary organ affected, usually, as either liver or muscle. The frequency of occurrence of glycogen storage disease (GSD) is one baby in 20,000 to 40,000. The main type of GSD ($\sim$90%) is type I, **von Gierke's disease**. It is transmitted from the parents to the offspring by an autosomal recessive mechanism. One example would be two unaffected "carrier" (recessive condition) parents could produce four offspring (in this example); among them are one child with the overt disease, one unaffected child with no carrier gene and two children who are unaffected carriers (recessives). There are two genetic patterns through which GSD or carrier status are inherited (Fig. 7.1).

For a child to be overtly diseased, both parents must carry the mutated gene as shown in (A); in this case, one child in four (25%) will have the overt GSD. Unaffected children who are carriers will be two of four (50%) and a normal child will be one of four (25%). In type VI (and IX) GSD, the inheritance is X-linked (B). For example, a carrier mother with a normal father will produce one normal son in four, one normal daughter in four, one carrier daughter in four, and one affected son in four. *This pattern of inheritance to generate overt GSD only affects the male offspring.* Females carry the mutated gene on one of their two X chromosomes. The overt disease generated by the mutated gene on one X chromosome is masked by the normal gene on the other X chromosome. However, the male has only one X chromosome so that a mutated gene on that chromosome will be expressed as overt disease.

In GSD, there is an accumulation of glycogen that may be specific for one primary organ or another. The types of GSD, the mutated enzyme involved, and the symptoms for each type are displayed in Table 7.1.

For most situations, the **aim of treatment** is to stabilize blood glucose levels as circulating glucose will be low (hypoglycemia) in many of the forms of GSD. This can be accomplished by supplementing glucose or cornflower (starch). Sometimes, a high-protein diet is helpful. Those patients who do not benefit from supplements may require a liver transplant. When the immune system is compromised in some patients, antibiotics are indicated. One type of GSD has been cured in a mouse model by gene therapy where the gene for the mutated enzyme was replaced with the gene for the normal enzyme. To a certain extent, this approach has worked in patients with GSD type II (**Pompe's disease**); however, infantile Pompe's disease is difficult to treat and can limit life expectancy. GSD can affect muscles, including the heart and liver. Consequent breathing problems and heart disease in children sometimes can lead to death.

In von Gierke's disease, the most prevalent form of GSD, the loss of the activity of glucose-6-phosphatase results in many metabolic disruptions that are captured in Fig. 7.2.

## GLYCOGEN—THE STORAGE CARBOHYDRATE

Just as **starch** is the carbohydrate storage form in plants and is composed of linear **amylose** and the branched polysaccharide, **amylopectin**, glycogen is the carbohydrate storage form in eukaryotes including humans. Like amylopectin, glycogen is a branched polysaccharide. Glycogen forms granules in the cytoplasm of different cells, especially liver and skeletal muscle. It can reach levels of over 100 g (ca. 8%) in the liver cells of an adult human but reaches lower levels in muscles, to about 1.5%. Because the muscle mass is much greater, the total amount of glycogen in muscle is greater than the total amount in liver. *Only liver glycogen can be made available to other tissues via the bloodstream in the form of glucose.* However, the level of glycogen depends on many factors including metabolic rate, food intake,

Human Biochemistry. DOI: http://dx.doi.org/10.1016/B978-0-12-383864-3.00007-7

(A)   **Autosomal recessive inheritance pattern**

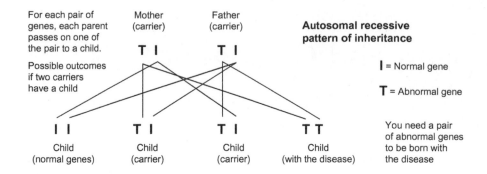

(B)   **x-linked inheritance pattern**

This occurs in some people with glycogen storage disorder type IX. (and type VI)

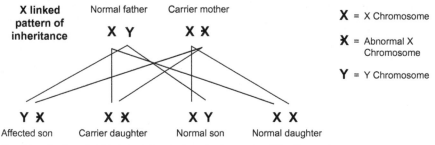

FIGURE 7.1   Most GSDs are inherited by autosomal recessive inheritance (A), whereas GSD type VI (and the equivalent IX) is inherited through an X-linked pattern (B). *Reproduced from http://www.patient.co.uk/health/glycogen-storage-disorders-leaflet.*

**TABLE 7.1** Types of Glycogen Storage Disease, the Inherited Deficient Enzyme, Mutated Gene, Primary Organ Affected and the Symptoms Accompanying Each Type of the Disease

| Type: Name | Enzyme Affected | Gene | Primary Organ | Manifestations |
|---|---|---|---|---|
| GSDGA | Liver isozyme of glycogen synthase | GYS2 | Liver | Hypoglycemia, early death, hyperketonia, low blood lactate and alanine |
| GSD1A von Gierke | Glucose-6-phosphatase | G6PC | Liver | Hepatomegaly, severe fasting hypoglycemia, hyperlipidemia |
| | | | | Hyperuricemia, kidney failure (Fanconi syndrome), thrombocyte dysfunction |
| GSD1b | Microsomal glucose-6-phosphate transporter (G6PT1): This protein is a member of the solute carrier protein family and is identified as SLC37A4 | SLC37A4 | Liver | Like Ia, also neutropenia, bacterial infections |
| GSD2 Pompe | Lysosomal acid α-glucosidase also called acid maltase | GAA | Skeletal and cardiac muscle | Infantile form = death by 2 |
| | | | | Juvenile form = myopathy adult form = muscular dystrophy like |

*(Continued)*

**TABLE 7.1** (Continued)

| Type: Name | Enzyme Affected | Gene | Primary Organ | Manifestations |
|---|---|---|---|---|
| GSD3 Cori or Forbes | Glycogen debranching enzyme | AGL | Liver, skeletal, and cardiac muscle | Infant hepatomegaly, myopathy |
| GSD4 Andersen | Glycogen branching enzyme | GBE1 | Liver, muscle | Infantile hypotonia, hepatosplenomegaly cirrhosis |
| GSD5 McArdle | Muscle phosphorylase | PYGM | Skeletal muscle | Excercise-induced cramps and pain, myoglobinuria |
| GSD6 Hers | Liver phosphorylase | PYGL | Liver | Hepalomegaly, mild fasting hypoglycemia, hyperlipidemia, and ketosis. Improvement with age |
| GSD7: Tarui | Muscle-specific subunit of PFK-1 | PKFM | Muscle, RBC | Like V, also hemolytic anemia |
| GSD9A1/A2 | α Subunit of hepatic phosphorylase kinase | PHKA2 | Liver | Mildest form of GSD, hepatomegaly, growth retardation, elevated plasma AST and ALT, hypercholesterolemia, hypertriglyceridemia, fasting hyperketosis |
| GSD9B | Common β subunit of phosphorylase kinase | PHKB | Liver and muscle | Marked heptomegaly in early childhood, fasting hypoglycemia |
| GSD9C | γ Subunit hepatic phosphorylase kinase | PHKG2 | Liver | Increased glycogen in muscle as well as liver, hepatosplenomegaly, short stature, hypoglycemia, muscle weakness |
| GSD9D | α Subunit muscle phosphorylase kinase | PHKA1 | Muscle | Nighttime muscle cramping in childhood, late-onset exercise-induced muscle fatigue and cramping |
| GSD10 | Phosphoglycerate mutase | PGAM2 | Muscle | Exercise-induced cramps, occasional myoglobinuria, exercise intolerance |
| GSD11 | Muscle-specific subunit of lactate dehydrogenase | LDHA | Muscle | Exercise-induced myoglobinuria, easily fatigued |
| Fanconi−Bickel (hepatorenal glycogenosis with renal Fanconi syndrome); was referred to as GSD11 but term no longer valid for this disease | Glucose transporter-2 (GLUT2) | SLC2A2 | Liver | Is a GSD secondarily related to nonfunctional glucoase transport; failure to thrive, hepatomegaly, rickets, proximal renal tubular dysfunction; also associated with a form of permanent neonatal diabetes mellitus |
| GSD12 | Aldolase A | ALDOA | Liver, RBC | Hepatosplenomegaly, nonspherocytic hemolytic anemia |
| GSD13 | Muscle predominant form of enolase: β enolase | ENO3 | Muscle | Myalgia, exercise intolerance |
| CDG1T (once called GSD14) | Predominant form of phosphoglucomutase | PGM1 | Multiple affected tissues | This disease is a type-1 congenital disorder of glycosylation; associated with cleft lip, bifid uvula, short stature, hepatomegaly, hypoglycemia, exercise intolerance |
| GSD15 | Muscle predominant form of glycogenin | GYG1 | Muscle | Muscle weakness, glycogen accumulation in heart, cardiac arrhythmias |

Reproduced from http://themedicalbiochemistrypage.org/glycogen.php.

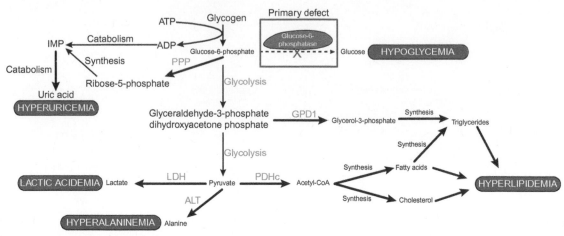

**FIGURE 7.2**   Interrelationships of metabolic pathway disruption in von Gierke's disease. In the absence of glucose-6-phosphatase activity, free glucose cannot be released from the liver contributing to severe fasting hypoglycemia. The increased levels of glucose-6-phosphate lead to increased activity of the pentose phosphate pathway (PPP) and increased glycolysis generating pyruvate. Increased (pyruvate) increases lactate via lactate dehydrogenase (LDH) and alanine via alanine transaminase (ALT). Increased pyruvate is oxidized by the pyruvate dehydrogenase complex (PDHc) generating increased acetylCoA that is used for the syntheses of fatty acids and cholesterol. Excess glycolysis increases the production of glycerol-3-phosphate (G3P) from dihydroxyacetone phosphate (DHAP) by glycerol-3-phosphate dehydrogenase (GPD1). Elevated G3P and fatty acids lead to triglyceride synthesis that, together with increased cholesterol, generates hyperlipidemia as well as fatty infiltration in hepatocytes contributing to hepatomegaly and cirrhosis. *Reproduced from http://themedicalbiochemistrypage.org/glycogen.php#clinical.*

exercise, etc. Some glycogen is found in red blood cells as well as in the heart and smaller amounts in the kidney and brain glia.

The ingestion of a **meal** results in **carbohydrate breakdown** in the gut, absorption of glucose into the bloodstream, and utilization of glucose for energy through **glycolysis** and the **Citric Acid Cycle**. In the fed state, glucose is taken up by the liver cell through the glucose transporter, **GLUT2**, and converted to glycogen. When energy is required some time after a meal, the water-insoluble liver glycogen is broken down to glucose that enters the bloodstream and is utilized by other tissues, such as muscles or brain. The storage and release of glucose from glycogen are processes that are under hormonal control. *Glycogen stored in muscle, in contrast to the liver, is not made available to other tissues because muscle cells do not express **glucose-6-phosphatase** that permits transfer of glucose into the bloodstream.* This enzyme catalyzes the reaction:

$$\text{glucose-6-phosphate} \rightarrow \text{glucose} + P_i$$

*While free glucose is taken up by the liver through the GLUT2 transporter, it is also exported from the liver cell by the same transporter (GLUT2) in the cell membrane.*

## GLUCOSE METABOLISM IN MUSCLE

Glucose metabolism in the muscle cell is summarized in Fig. 7.3.

## GLYCOGENIN AND FORMATION OF GLYCOGEN

Glycogenin is an enzyme, classified as a **glycosyltransferase**. It is a homodimer composed of two 37 kDa subunits. It has the initial role in the formation of glycogen from UDP-glucose in muscle and liver. There are two isoforms of the enzyme: in muscle, it is **glycogenin-1** encoded by the gene, GYG1, and in liver and cardiac muscle, it is **glycogenin-2**, encoded by the gene, GYG2. Linear chains of glucose are attached to glycogenin and up to 10 glucose molecules can be added. At this level, **glycogen synthase** and the **branching enzyme** take over to complete the synthesis of glycogen that may contain as many as 30,000 glucose residues. **Glycogenin** is essential to this process and a patient who has a defective gene (encoding the mRNA for glycogenin) cannot store glycogen and will display muscle weakness and cardiac disease. In the initial reactions, the glycogenin dimer reacts with linear chains of glucose molecules. Then, the **glycogenin-(glucose)$_n$ complex** reacts with **UDP-glucose** to form glycogen through the actions of glycogen synthase

MUSCLE CELL METABOLISM OF GLUCOSE

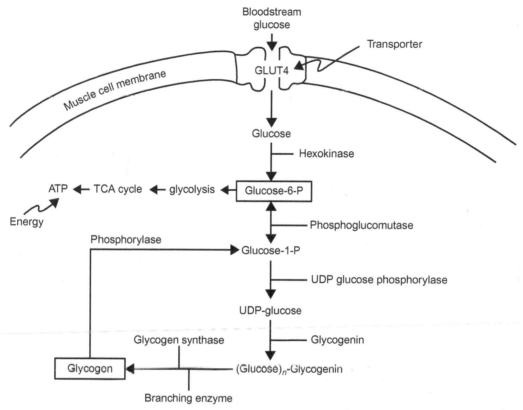

**FIGURE 7.3** Glucose metabolism in the muscle cell. Glucose is transported from the bloodstream into the muscle cell by the transporter, **GLUT4**. If glucose is needed for the formation of energy, it is converted to glucose-6-phosphate by hexokinase and metabolized through the glycolytic pathway to form pyruvate that enters the Citric Acid Cycle through pyruvate dehydrogenase to generate ATP energy (and $CO_2$ and $H_2O$). If there is no immediate need for energy, glucose-6-phosphate is further metabolized to UDP-glucose and then to the storage form, glycogen, through the agency of the enzyme **glycogenin** and the enzymes, **glycogen synthase,** and the **branching enzyme**. When energy is needed in the muscle, glycogen can be broken down to form glucose-1-phosphate (G-1-P) by phosphorylase; this can be converted to glucose-6-phosphate by phosphoglucomutase, and it can proceed through glycolysis to pyruvate and enter the tricarboxylic acid cycle (TCA) to generate ATP energy. *UDP*, uridine diphosphate.

and the branching enzyme. Glucose-1-phosphate is converted to UDP-glucose by the action of UDP-glucose pyrophosphorylase, a reaction that is powered by the hydrolysis of **inorganic pyrophosphates** as shown in Fig. 7.4.

A partial structure of glycogen is shown in Fig. 7.5, showing a straight chain of glucose residues and a branched chain.

In Fig. 7.6, a partial structure of the **glycogen particle** is shown. Five layers are shown here but the completed glycogen particle has 12 layers.

The fully formed glycogen particle resembles the structure shown in Fig. 7.7.

UDP-glucose interacts with each glycogenin monomer through a manganese ion and amino acid residues in the active site of glycogenin (Fig. 7.8).

Glycogenin forms the center of the growing glycogen molecule and attaches UDP-glucose molecules to itself, a step in which glycogenin acts as a primer. Exactly how the enzyme catalyzes the addition of glucose is not completely clear although the UDP-glucose is bound to the hydroxyl group of tyrosine-194 before seven more glucose residues can be added to the chain. When about eight residues are extended in a chain, glycogen synthase takes over to extend the chain further and the branching enzyme creates the side chains. The transfer of the first glucose residues is intermolecular and subsequent glucose molecules are attached by intramolecular reaction within the **glycogenin dimer**. The straight chains of glycogen are made up of $\alpha$-1,4 linkages between the glucose units, except for the branches that are made through $\alpha$-1,6 linkages. **Glycogen synthase** catalyzes the reaction: UDP-glucose + glycogen$_{(n\ \text{glucose units})}$ → UDP + glycogen$_{(n+1\ \text{glucose units})}$. In this way, glucose molecules are added to nonreducing ends of glycogen (Fig. 7.9).

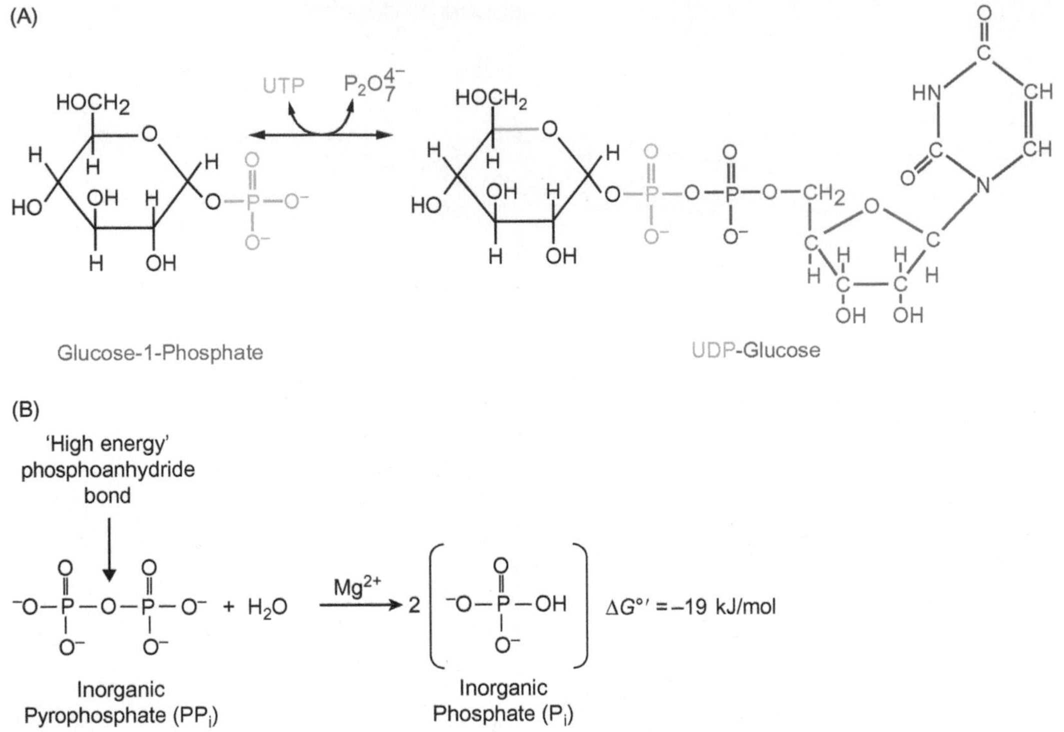

(A)

Glucose-1-Phosphate

UDP-Glucose

(B)

'High energy'
phosphoanhydride
bond

$\Delta G^{\circ\prime} = -19$ kJ/mol

Inorganic
Pyrophosphate (PP$_i$)

Inorganic
Phosphate (P$_i$)

**FIGURE 7.4** Conversion of glucose-1-phosphate to UDP-glucose (UDPG). (A) UDPG (uridine diphosphate glucose) pyrophosphorylase reaction. (B) Inorganic pyrophosphatase reaction; the hydrolysis of inorganic pyrophosphate by inorganic pyrophosphatase provides the energy that drives the synthesis of UDPG.

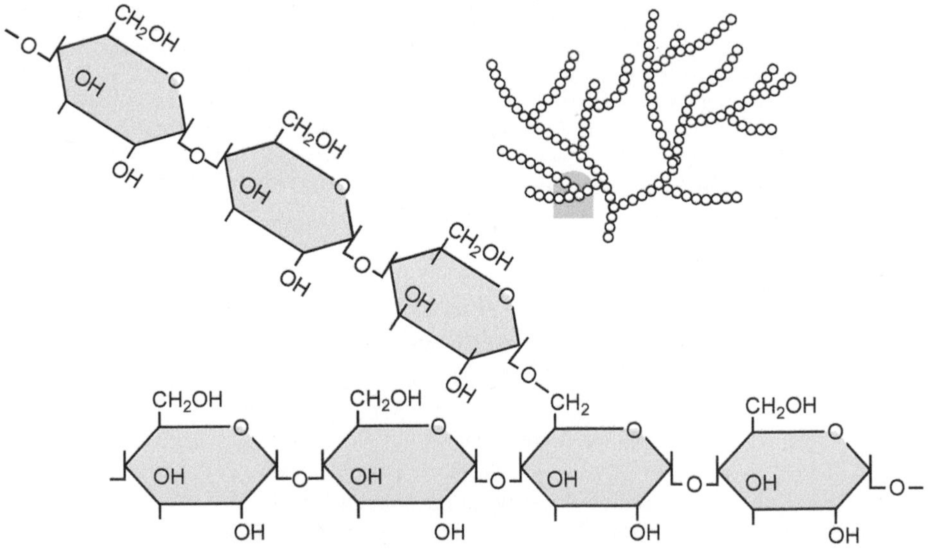

**FIGURE 7.5** The straight chain of glucose residues and a branched chain indicate a partial structure of glycogen. This structure is represented in the *upper right hand figure.*

A nonreducing end of a sugar is one that contains an **acetal group**, whereas a reducing sugar end is either an alde-hyde or a **hemiacetal group** (Fig. 7.10).

## GLYCOGENOLYSIS (RELEASING GLUCOSE FROM GLYCOGEN)

The enzyme, **phosphorylase**, catalyzes the following **phosphorolysis** reaction:

$$\text{glycogen} + P_i \rightarrow \text{glycogen}(n-1) + \text{glucose-1-phosphate}$$

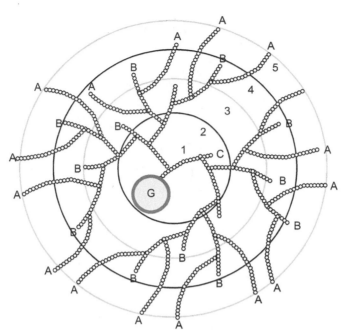

**FIGURE 7.6**  Partial structure of the glycogen particle showing 5 of the 12 layers in the completed particle. Each B chain has two branch points. All chains have the same length. A and B chains have the same distribution. Note that glycogenin (G) is near the center of the particle. *Reproduced from G. Litwack,* Human Biochemistry & Disease, *Academic Press/Elsevier, Figure 4-37, page 158, 2008.*

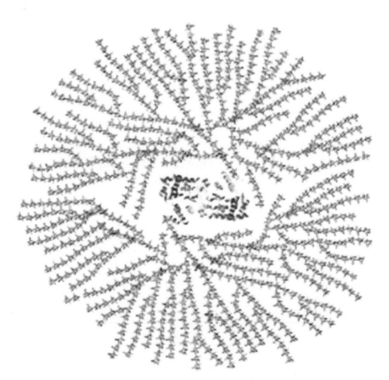

**FIGURE 7.7**  A 2-dimensional crosssectional view of the glycogen particle. The core protein of glycogenin is located in the center and is surrounded by branches of glucose residues. The entire globular granule may contain as many as 30,000 glucose units. *Reproduced from http://www.diapedia.org/ metabolism/glycogenolysis-and-glycogenesis.*

Phosphorolysis occurs to within four residues of a branch point to produce one molecule of glucose-1-phosphate for each glucose unit released. The **debranching enzyme (transglucosylase activity,** $\alpha$-1,4 to $\alpha$-1,4) then catalyzes the transfer of a trimer from a branch to the free end of the glycogen molecule. The $\alpha$**-1,6 glucosidase** activity of the debranching enzyme then cleaves the $\alpha$-1,6 glucose at the branch point of glycogen (Fig. 7.11). 11 to 12 glucose units are

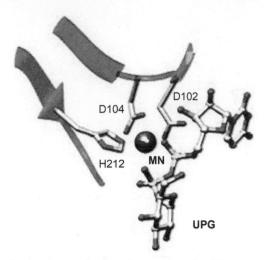

**FIGURE 7.8** The active site of glycogenin showing the coordination of a manganese ion. The manganese ion is coordinated through two aspartate residues and a histidine. *UPG*, UDP-glucose.

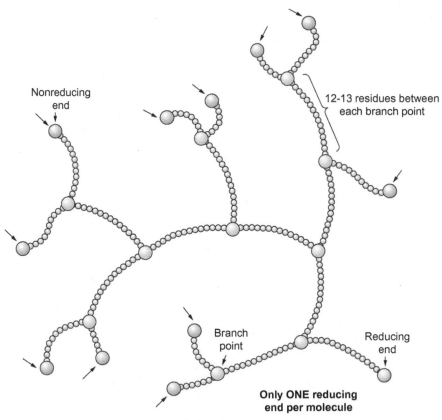

**FIGURE 7.9** Glycogen has several nonreducing ends and one reducing end. Addition of new glucose molecules occurs at the nonreducing ends, and these same ends, in the completed glycogen molecule, are attacked to liberate glucose-1-phosphate during the breakdown process. The single reducing end has the C1 carbon of the glucose residue free from the ring and able to react.

released from glycogen in the form of **glucose-1-phosphate** by phosphorolysis, and these can enter glycolysis through the **phosphoglucomutase reaction** that yields **glucose-6-phosphate**.

Glucose-6-phosphate (G-6-P) is in a key position between glycogen, free glucose, **ribose-5-phosphate** (can be incorporated into ribonucleotides), and pyruvate as shown in Fig. 7.12.

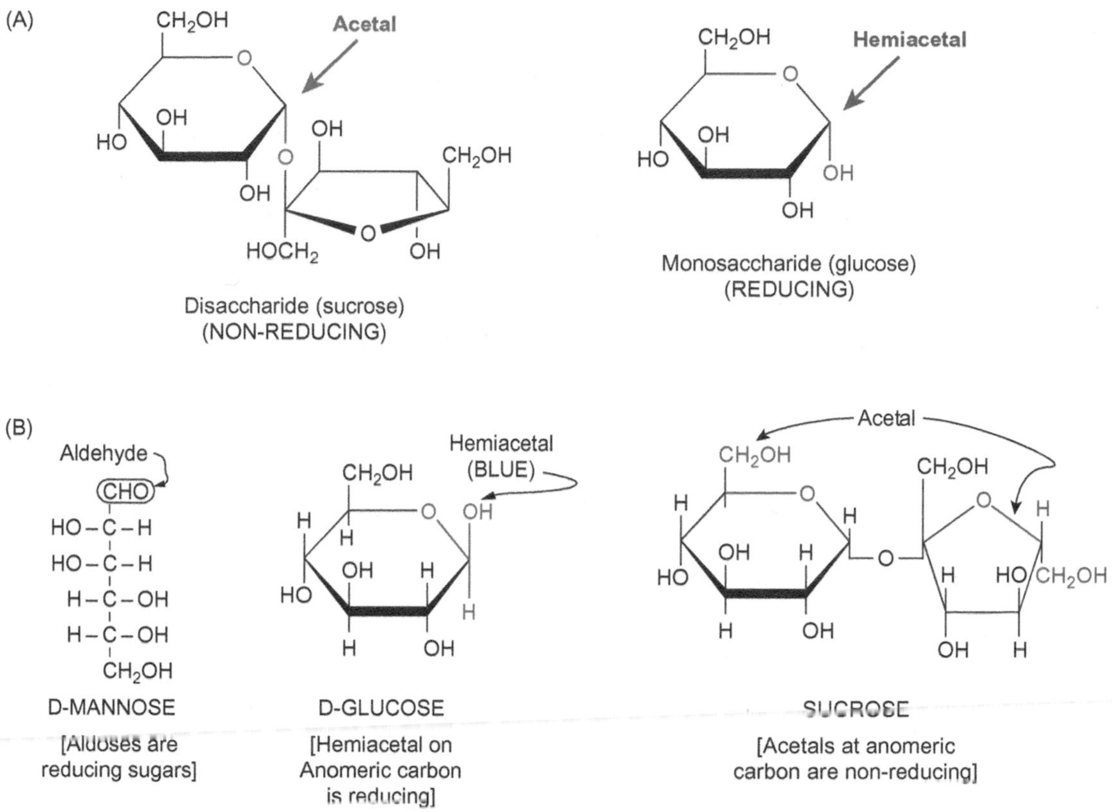

**FIGURE 7.10** (A) A nonreducing sugar (e.g., sucrose) compared to a reducing sugar (glucose). The acetal group of a nonreducing sugar is shown in *blue*. The reducing sugar with a hemiacetal end is shown in *red* on the *right*. (B) Examples of reducing sugars (*left*) and a nonreducing sugar (*right*). On the *left* is shown two reducing sugars: D-mannose with an open chain structure having an aldehyde group at C1 (*circled*) and D-glucose, in a ring structure, having a free hemiacetal group (*blue*). The nonreducing sugar, sucrose, on the *right* has acetal groups at the ends (*red*). *Reproduced from http://chemistry2.csudh.edu/rpendarvis/2feb23.gif.*

In the liver, **glucose-6-phosphatase** converts glucose-6-phosphate back into free glucose. A mutation in this enzyme can lead to a **glycogen storage disease**. For export of free glucose into the bloodstream for use by other tissues, glucose moves out of the liver through the transporter, **GLUT2**. This occurs in fasting or in the postabsorptive state (after a meal has been absorbed) when the concentration of glucose is higher in the liver than it is in the blood.

**Phosphoribosylpyrophosphate (PRPP)** in this pathway is used in the formation of nucleotides, as will be discussed later. The pentose phosphate pathway (**phosphogluconate pathway** or **hexose monophosphate shunt**) generates **NADPH** during the oxidative phase of the pathway. One molecule of NADPH is generated from $NADP^+$ in the following reactions of the pathway: conversion of glucose-6-phosphate to 6-phosphogluconate by **glucose-6-phosphate dehydrogenase**; conversion of 6-phosphogluconate to ribulose-5-phosphate by **phosphogluconate dehydrogenase,** and conversion of ribulose-5-phosphate to ribose-5-phosphate by **ribose-5-phosphate isomerase** (2 NADPH). A **hemolytic anemia** is produced in patients that have a genetic **deficiency of glucose-6-phosphate dehydrogenase (G6PDH)**, resulting in an inadequate supply of NADPH to the **red blood cell**. *This is the most common defect resulting from an enzyme deficiency.* This disorder involves 400 million people worldwide (especially in Africa, Asia, Mediterranean countries, and South America), approximately the number of persons affected with the malaria parasite. The mutation is X-recessive-linked and is polymorphic with more than 300 variants. Neonatally, the disorder can present as hyperbilirubinemia. Some individuals with this mutation can be asymptomatic, whereas others have episodes of hemolysis (destruction of red blood cells), varying in intensity, as there is a range of severity. *G6PDH deficiency, however, is protective against malaria.* Thus, the mutation is dangerous if antimalarial drugs (e.g., Primaquine) are used because of the loss of protection in the red blood cell against hemolyzing reactive oxygen species ($H_2O_2$) that are induced by the drug. Consequently, it is vital to test for G6PDH deficiency before administering the antimalarial drug. Also in the red blood cell, the enzyme, **glutathione peroxidase**, functions to degrade **hydroperoxides** that arise because of the oxygen-rich environment. This enzyme reduces the hydroperoxides by the use of two molecules of reduced glutathione (GSH), converting it to oxidized glutathione (GSSG):

$$ROOH + 2GSH \rightarrow ROH + GSSG$$

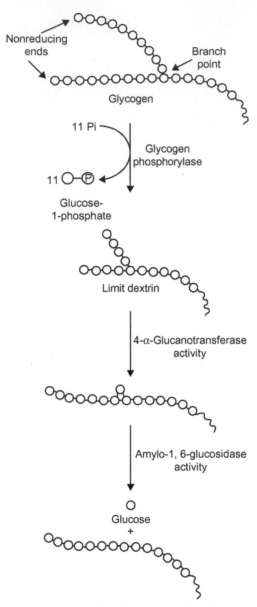

**FIGURE 7.11**   Breakdown of glycogen. The first two steps are catalyzed by glycogen phosphorylase and the last step by the debranching enzyme.

In order to regenerate GSH from GSSG, NADPH is required:

$$GSSG + NADPH + H^+ \rightleftharpoons 2GSH + NADP^+$$

In the absence of sufficient NADPH, GSSG would accumulate, and there would be insufficient GSH to remove harmful hydroperoxides.

## HORMONAL CONTROL OF GLYCOGEN METABOLISM AND BLOOD GLUCOSE LEVEL

The hormones, **glucagon** and **epinephrine**, stimulate the **breakdown of glycogen** (glycogenolysis). Glucagon is secreted by the α-cells of the **pancreas** (Islets of Langerhans). It is a single helix protein of twenty-nine amino acids ($H_2N$-HSQGTFTSDYSKYLDSRRANDFVQWLMNT-COOH). The cells of the **Islet of Langerhans** in the pancreas are diagrammed in Fig. 7.13.

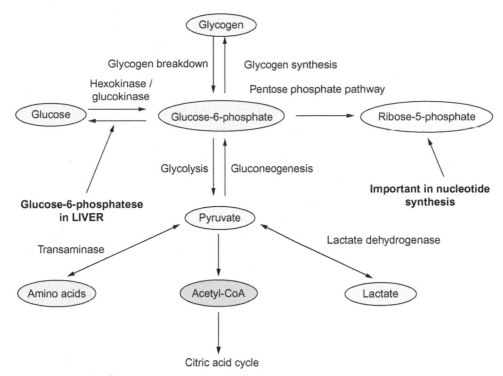

**FIGURE 7.12**  Glucose-6-phosphate is a key intermediate in the metabolism of glucose and glycogen.

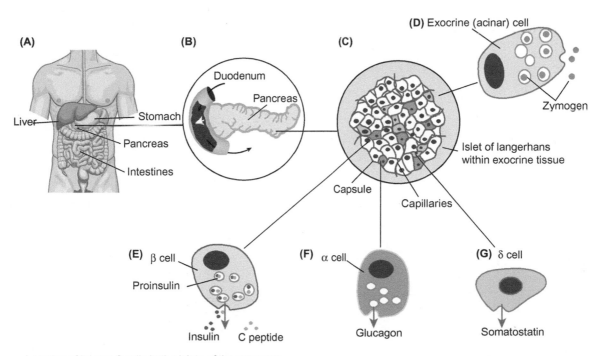

Location of human β cells in the islets of the pancreas

**FIGURE 7.13**  The pancreas is located in the upper abdomen behind the stomach and close to the liver (A). An islet is a collection of endocrine cells (hormone secreting cells) in (C). (B) Location of the pancreas with respect to the duodenum. The hormone secreting cells are shown in (E), (F) and (G). Glucagon is secreted from the **α-cell**; insulin from the **β-cell** and somatostatin from the **δ-cell**. *Reproduced from http://journals.cambridge.org/full-text_content/ERM/ERM2_06/S1462399400001861sup019.gif*

# GLUCAGON

The action of glucagon is to raise the level of blood glucose, in contrast to the action of **insulin** that is to lower blood glucose. Glucagon acts on the liver cell to cause the breakdown of glycogen to release free glucose into the bloodstream. It accomplishes this by binding to and activating a G protein membrane receptor (**glucagon receptor**) that stimulates adenylyl (or adenylate) cyclase to produce cyclic adenosine monophosphate (**cAMP**). cAMP activates **protein kinase A** whose action, through the phosphorylation of **CREB** (cAMP response element binding protein), induces enzymes in the **gluconeogenesis pathway** (formation of glucose from pyruvate). Also activated through a G protein linked to the receptor is the stimulation of **phospholipase C** that converts **phospatidylinositol *bis*phosphate, PIP2**, (in the membrane) to **inositol *tris*phosphate, IP3**, in the cytoplasm that, through binding of IP3 to the IP3 receptor, activates the release of **calcium ions** from the **endoplasmic reticulum** (ER)**store**. Calcium enhances the action of protein kinase A. Finally, **glycogen phosphorylase** is activated to release glucose-1-phosphate. Glucose-1-phosphate is converted to glucose-6-phosphate by the action of **phosphoglucomutase**. Glucose-6-phosphate is then converted to free glucose by the action of elevated levels of **glucose-6-phosphatase**. Free glucose is exported from the hepatocyte by the glucose transporter **GLUT2** to the extracellular space and ultimately to the bloodstream (Fig. 7.14).

Glucose is normally transported into the α-cell from the bloodstream and the extracellular space by the transporter **GLUT1**. When the level of glucose impinging on the α-cell falls, the level of the activity of $K_{ATP}$ **channels** affects the cell membrane potential to allow the opening of voltage-dependent **T- and N-type calcium ion channels** and the opening of voltage-dependent sodium ion channels and voltage-dependent calcium ion channels. These effects generate action potentials and calcium ion influx causing the exocytosis of **glucagon-containing granules**.

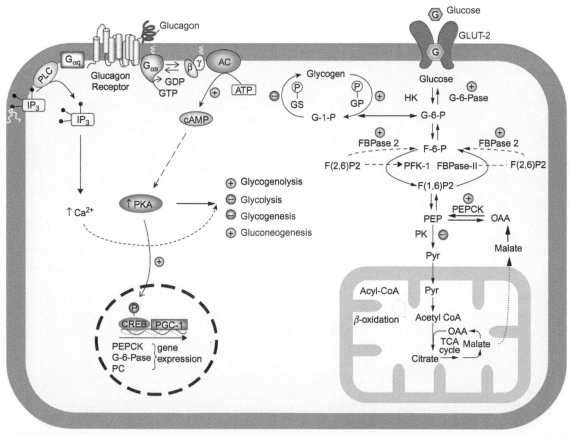

**FIGURE 7.14**   The mechanism of action of glucagon on the hepatocyte to cause the breakdown of glycogen to release free glucose into the bloodstream. *AC*, adenylate cyclase; *HK*, hexokinase; *FBPase*, fructose *bis*phosphatase; *PFK-1*, phosphofructokinase-1; *PEP*, phosphoenolpyruvate; *OAA*, oxaloacetate; *PEPCK*, phosphoenolpyruvate carboxykinase; *G-6-Pase*, glucose-6-phosphatase; *PC*, pyruvate carboxylase; *PK*, pyruvate kinase; *CREB*, cyclic AMP response element binding (protein); *PGC-1*, PPARγ coactivator-1 (PPAR, peroxisome proliferator-activated receptor). *Reproduced from I. Quesada, E. Tuduri, C. Ripoll and A. Nadal, J. Endocrinol., **199**: 5−19, 2008.*

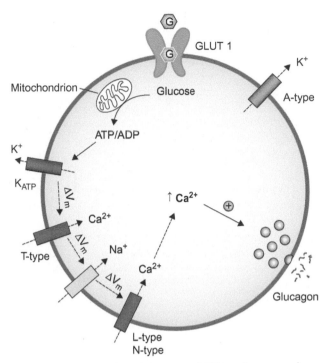

**FIGURE 7.15** Shematic model for glucose dependent regulation of glucagon secretion in the pancreatic α-cell by the GLUT1 transporter. At **low glucose** levels, $K_{ATP}$ channels generate a membrane potential allowing for the opening of voltage-dependent T- and N-type $Ca^{2+}$ and voltage-depndent $Na^+$ channels. Activation of these channels triggers action potentials, influx of $Ca^{2+}$ and exocytosis of glucagon granules. The opening of A-type $K^+$ channels is necessary for action potential repolarization. At **elevated levels of glucose**, the ATP/ADP ratio increases blocking $K_{ATP}$ channels and depolarizing membrane potential to a range where the inactivation of voltage-dependent channels takes place leading to a decrease in electrical activity, the influx of $Ca^{2+}$ and the secretion of glucagon. The function of L-type channels predominates when [cAMP] is elevated. *Reproduced from I. Quesada, E. Tuduri, C. Ripoll and A. Nadal, J. Endocrinol., 199: 5—19, 2008.*

When glucose concentrations are high, on the other hand, the **ATP/ADP ratio** in the cell is elevated causing a block of the $K_{ATP}$ channels and depolarization of the membrane potential where the voltage-dependent channels become inactivated. Electrical activity is inhibited as are the influx of calcium ions and the secretion of glucagon. **L-type channels** predominate when cAMP is elevated. A summary of the mechanism of glucagon release from the α-cell is shown in Fig. 7.15.

Glucagon is split out from a precursor, **preproglucagon** (glucagon is within amino acids 30—69 of the precursor) that encodes three other proteins besides glucagon. These are **GRPP** (amino acids 1—69 of preproglucagon; **glicentin-related pancreatic peptide**, or "**enteroglucagon**" that is thought to increase release of insulin from beta cells of the pancreas); **GLP-1** (amino acids 69—108 of preproglucagon; it is **glucagon-like peptide-1**, an "**incretin**" that increases release of insulin from beta cells); and **GLP-2** (amino acids 126—158 of preproglucagon; **glucagon-like peptide-2** increases the secretion of glucagon and increases the proliferation of **astrocytes** in the CNS). In its action on the liver cell, the interaction of glucagon with its G protein receptor results in an increase of cAMP which, in turn, activates protein kinase A. The mechanism of this activation is shown in Fig. 5.20. Glucagon activates protein kinase A in the liver cell. This results in the activation of phosphorylase kinase that, in turn, activates **glycogen phosphorylase** and the breakdown of glycogen to form G-1-P. G-1-P is converted to G-6-P and then to free glucose that is exported from the cell by the **GLUT2 transporter** to elevate [glucose] in the bloodstream. In synchrony with this effect, activated **protein kinase A** is able to convert **glycogen synthase** to the phosphorylated, inactive form (**glycogen synthase *b***), although the principal enzyme for this function is glycogen synthase kinase 3 (GSK3). Glucagon stimulates casein kinase 2 (CK2), possibly by causing the autophosphorylation of CK2, although the mechanism of its stimulation by glucagon is unclear. These activities in the liver are summarized in Fig. 7.16.

## EPINEPHRINE

Epinephrine also acts on the liver cell by binding to the α-**adrenergic receptor** (G protein receptor) in the cell membrane that results in the activation of phospholipase C causing conversion of phosphatidylinositol *bis*phosphate (PIP$_2$)

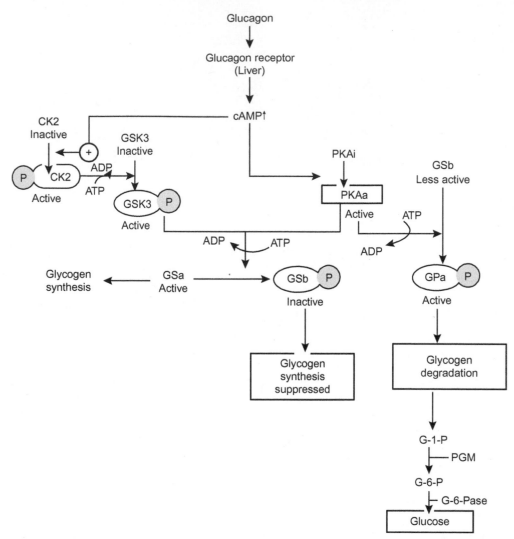

**FIGURE 7.16** Action of glucagon on the synthesis and breakdown of glycogen in liver cells. Although PKA is able to phosphorylate glycogen synthase, glycogen synthase kinase 3 is the major enzyme governing the activity of glycogen synthase. Glycogen synthase kinase 3 is activated by casein kinase 2 that is thought to be activated indirectly by glucagon. *cAMP*, cyclic adenosine monophosphate; *GSK3*, glycogen synthase 3; *CK2*, casein kinase 2; *GSa*, glycogen synthase a (active); *GSb*, glycogen synthase b (less active or inactive); *PKAi*, protein kinase i (inactive); *PKAa*, protein kinase a (active); *GPb*, glycogen phosphorylase b, (less active); *GPa*, glycogen phosphorylase a (active); *G-1-P*, glucose-1-phosphate; *G-6-P*, glucose-6-phosphate; *PGM*, phosphoglucomutase; *G-6-Pase*, glucose-6-phosphatase.

to inositol *tris*phosphate (IP$_3$) and diacylglycerol. IP$_3$ binds to its receptor on the endoplasmic reticulum (ER) evoking the store of calcium ions in the ER to be released into the cytoplasm. These calcium ions enhance calmodulin-mediated phosphorylation of phosphorylase kinase that phosphorylates the enzyme, **phosphorylase**, to its active form (glycogen phosphorylase a). Active phosphorylase converts nonreducing ends of glycogen to glucose-1-phosphate. G-1-P is converted to G-6-P by **phosphoglucomutase,** and G-6-P is converted to free glucose by **G-6-P phosphatase**. These actions are shown in Fig. 7.17.

Overall, **epinephrine** acts on alpha- and beta-receptors. These are categorized as $\alpha_1$, $\alpha_2$, and $\beta_1$, $\beta_2$, and $\beta_3$. $\alpha$-**Receptors** increase glycogenolysis in liver and muscle and increase glycolysis in muscle while decreasing the secretion of insulin from the beta cell of the pancreas. $\beta$-**receptors** increase lipolysis in adipose cells, glucagon secretion from the pancreas and ACTH secretion by the pituitary. They also increase (glucose) in the bloodstream [in addition to the activation of phosphorylase through the action of the alpha receptor (Fig. 7.17)], activated by epinephrine. The beta-receptor, activated by epinephrine, produces **cAMP** (Fig. 7.18) that activates protein kinase A resulting in the activation of phosphorylase kinase and glycogen degradation. Overall, $\beta$-adrenergic receptors ($\beta_1$, $\beta_2$, $\beta_3$) are located primarily in the heart, lung, and adipose tissues. They interact with **heterotrimeric G proteins** ($\alpha$, $\beta$, $\gamma$ subunits), favoring direct interaction with G proteins containing $G_\alpha$ (also known as $G_s$) and $G_{\alpha i}$ (also known as $G_i$) subunits. $G_\alpha$ activates

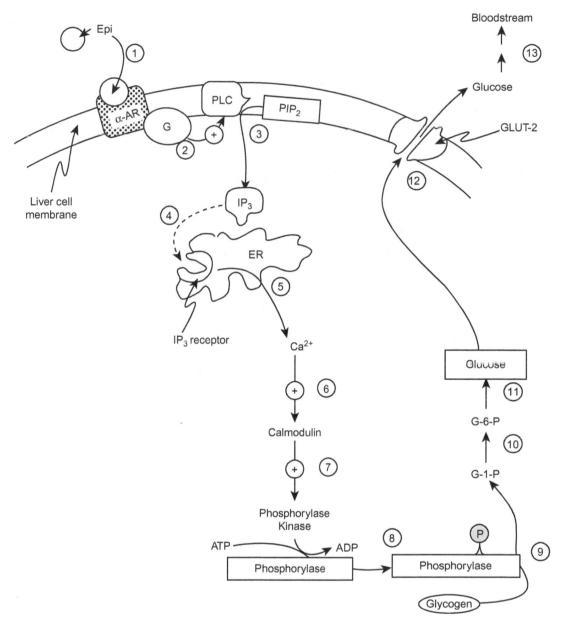

**FIGURE 7.17**  Actions of epinephrine on the α-adrenergic receptor of the liver cell. *Epi*, epinephrine; *G*, G protein; *PLC*, phospholipase C; *PIP₂*, phosphoinositol-4,5-*bis*phosphate; *IP₃*, inositol *tris*phosphate; *ER*, endoplasmic reticulum; *GLUT2*, glucose transporter-2.

adenylate kinase and increases [cAMP]. $G_\alpha$ and $G_i$ stimulate the src family of tyrosine kinases. $G_\alpha$ and $G_{\beta\gamma}$ regulate L-type calcium ion channels. Summarizing the α-receptors, there are two types: $\alpha_1$ and $\alpha_2$. There are three types of $\alpha_1$ receptors, $\alpha_{1A}$, $\alpha_{1B}$, and $\alpha_{1D}$. There are three types of $\alpha_2$ receptors: $\alpha_{2A}$, $\alpha_{2B}$, and $\alpha_{2B}$-like. They interact with $G_{\alpha q}$ subunits, the primary activator of phospholipase Cβ (PLC) that catalyzes the reaction: $PIP_2 \rightarrow DAG + IP_3$. DAG (diacylglycerol) and $IP_3$ (inositol *tris*phosphate) lead to the activation of protein kinase C (PKC). $\alpha_1$ receptors also can stimulate adenylate cyclase to produce cAMP through the subunits $G_\alpha$ or $G_i$.

## INSULIN

The action of insulin is opposite that of glucagon and epinephrine. Because insulin stimulates **protein phosphatase-1** that removes a phosphate group from inactive glycogen synthase *b* to the active dephosphorylated form, glycogen synthesis is activated. Insulin's stimulation of protein phosphatase-1's activity converts active phosphorylated glycogen phosphorylase *a* to the less active form, glycogen **phosphorylase *b*** (Fig. 7.19).

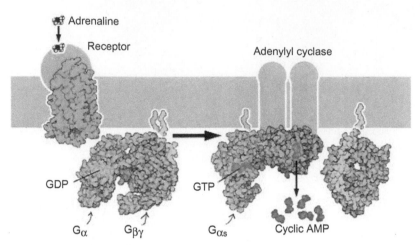

**FIGURE 7.18**   Action of epinephrine (adrenaline) on a β-receptor to produce an increase in cAMP. *Reproduced from http://upload.wikimedia.org/wikipedia/commons/2/2d/G_protein_signal_transduction_(epinephrin_pathway).png.*

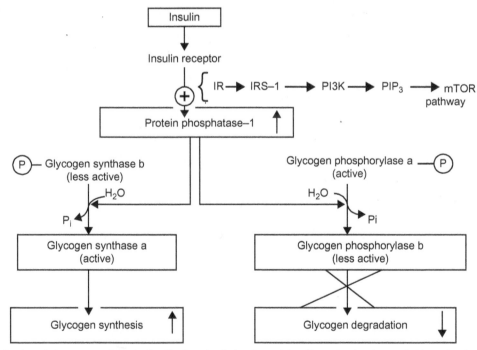

**FIGURE 7.19**   Activation of glycogen synthesis by insulin. By stimulating the activity of protein phosphatase-1 (PP-1), insulin promotes the synthesis of glycogen and inhibits glycogen breakdown. Insulin acts through the insulin receptor (IR); *IRS-1*, insulin receptor substrate-1; *PI3K*, phosphoinositide 3-kinase; *PIP₃*, phosphatidylinositol-3,4,5-*tris*phosphate; *mTOR*, mammalian target of rapamycin or FK506 binding protein pathway (FK506 is an immunosuppressant).

Epinephrine binds to the $\alpha_1$-adrenergic receptor in liver and interferes with the functioning of the **insulin receptor**, thus causing inhibition of glycogen synthesis. Therefore, epinephrine can produce the same effect in liver by acting through two different receptors. Obviously, the resulting regulation of glycogen breakdown and synthesis depends upon the levels of hormones available to the liver cell. These effects are summarized in Fig. 7.20.

## GLYCOGEN PHOSPHORYLASE

When the glucose level is high and a cell needs to store energy in the form of glycogen, the incoming glucose binds directly (somewhat similarly to the product of the enzymatic reaction, glucose-1-phosphate) to glycogen phosphorylase

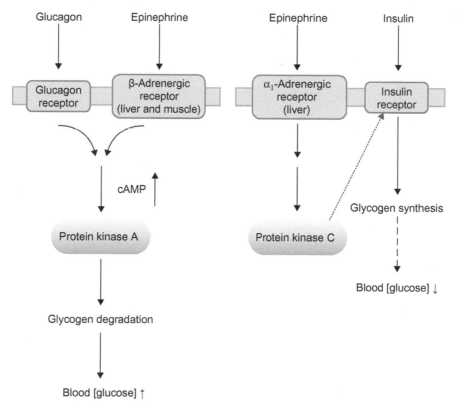

**FIGURE 7.20**  Summary of the actions of glucagon, epinephrine, and insulin on the breakdown and synthesis of glycogen. *Green background* of a label indicates the active form of the enzyme. The *red line* with the *minus sign* indicates inhibition.

*a* causing dissociation of protein phosphatase-1 from its complex with **glycogen phosphorylase *a***. In the complex with glycogen phosphorylase *a*, **protein phosphatase-1** is not active but release from the complex causes its activation. Then, activated protein phosphatase-1 removes the phosphate from glycogen phosphorylase *a* converting it to less active **glycogen phosphorylase *b***, thus reducing glycogen degradation. In addition to its state of phosphorylation, glycogen phosphorylase is also regulated by allosteric effectors. There are two conformations of the enzyme, the **T (taut) form** that is relatively inactive and the **R (relaxed) form** that is active. The allosteric effector for the enzyme in the T form is AMP; AMP converts the enzyme conformation to the active R form. The enzyme may be phosphorylated to a covalently inactive (M form) by phosphorylase kinase and when (glucose) is high, the internally phosphorylated R (active) form is converted to the inactive M form. Thus, glucose can act as a negative allosteric regulator of the internally phosphorylated R form. The active, nonphosphorylated R form can be converted to the inactive conformation (O form) by the negative effectors ATP and/or G-6-P. These effects are summarized in Fig. 7.21.

Activated protein phosphatase-1 also removes the phosphate group from inactive glycogen synthase *b* converting it to the active form, **glycogen synthase *a***, to stimulate glycogen synthesis. These activities are summarized in Fig. 7.22A. In addition to the glucose molecule, **glucose-6-phosphate** is also an inhibitor of glycogen phosphorylase (in some figures, it is referred to only as phosphorylase) and an activator of glycogen synthase. Glucose binding to the catalytic site of phosphorylase *a* promotes a conformational change in the enzyme stabilizing it in the inactive T (taut) state so that the action of protein phosphatase-1 is promoted and glycogen synthase is stimulated (Fig. 7.22B).

Thus, the overall effects on the liver cell when (glucose) is low or when (glucose) is high are summarized in Fig. 7.23.

## DIFFERENT GLUCOSE TRANSPORTERS (GLUTs) IN DIFFERENT TISSUES

In this discussion, it was seen that different glucose transporters (GLUTs) occur in different organs. GLUT2 is used in the liver cell; GLUT4 is used in the muscle cell and GLUT1 is used in the pancreatic alpha-cell. GLUT2 is a low-affinity (for glucose) transporter with a high capacity able to move large amounts of glucose in two directions in or out of the cell. It would be moving glucose into the cell under the action of insulin and out of the cell under the action of

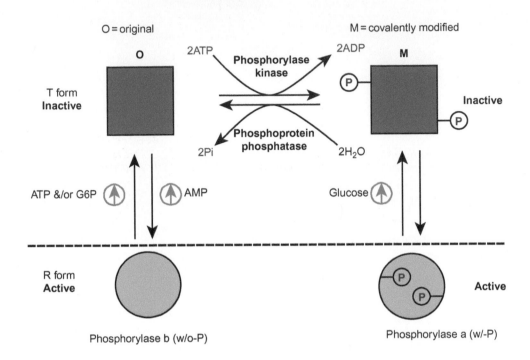

FIGURE 7.21 Regulation of glycogen phosphorylase by phosphorylation and by allosteric effectors. Both phosphorylation and allosteric effectors change the activity of phosphorylase by changing the shape of the enzyme. *Reproduced from http://employees.csbsju.edu/hjakabowski/classes/ch112/pathways-charts/phosphorylasereg2.gif.*

glucagon or epinephrine. GLUT4 in the muscle cell has a high affinity for glucose and operates in the inward direction. It is the insulin-responsive transporter for glucose storage. GLUT1 is a high-affinity transporter of glucose responsible for taking up a low level of glucose that is sufficient for sustaining respiration in the alpha-cell and in other cells.

## SUMMARY

**Glycogen storage disease** (GSD) is a relatively rare disorder appearing in one baby in 20,000 to 40,000. 90% of these disorders are diagnosed as von Gierke's disease in which the deficient enzyme is glucose-6-phosphate. Here, G-6-P accumulates and cannot be metabolized to free glucose that can enter the bloodstream from the liver through the GLUT2 transporter. Von Gierke's disease is inherited from both parents through an autosomal recessive inheritance. There are several other forms of GSD, and in each disorder, there is a missing enzyme activity either in the synthesis or degradation of glycogen. In all these cases, there is an accumulation of glycogen affecting many different organs, the liver and muscles, in particular.

*Glycogen stored in the liver is available to the other organs in the form of free glucose. This is not the case with muscle glycogen because the muscle lacks glucose-6-phosphatase and does not form free glucose, thus, in muscle, glucose-6-phosphate becomes utilized in glycolysis.*

Glycogen is a branched chain polysaccharide forming in granules in many cell types, especially liver and muscle. In liver, glucose is taken up from the blood (e.g., after a meal) by the GLUT2 bidirectional transporter and is converted to glycogen. In muscle, glucose is taken up from the blood (after release from the liver through GLUT2) through the insulin-sensitive GLUT4 transporter. It is converted to UDP-glucose if there is no immediate need for energy, and UDP-glucose is converted to glycogen by the enzymes, **glycogenin, glycogen synthase,** and the **branching enzyme**.

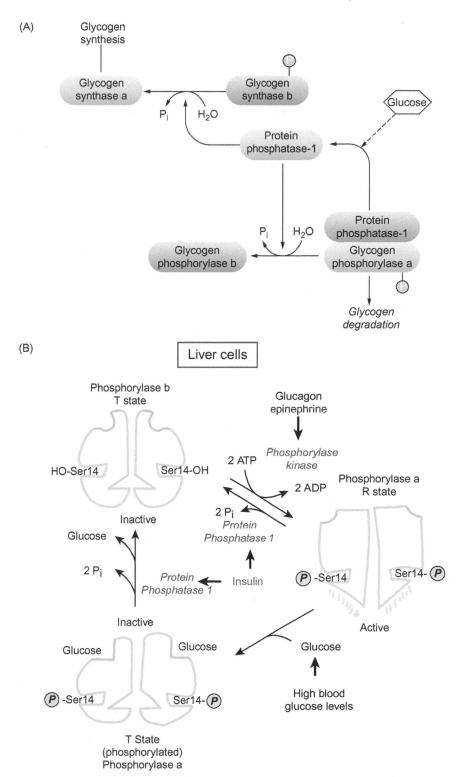

**FIGURE 7.22**   (A). Effects of incoming glucose on glycogen synthesis through the direct binding of glucose to glycogen phosphorylase *a*. (B). High levels of blood glucose (*bottom right*) bind to phosphorylase *a* in liver cells, converting the R (relaxed, active) state of phosphorylase *a* to the T (taut, less active) state of phosphorylase *a*. Protein phosphatase-1 then dephosphorylates phosphorylase *a* in the T state, releasing the bound glucose, through a conformational change, to generate phosphorylase *b* in the inactive T state. Thus, the effect of glucose binding is an inhibitory effector and produces a less active phosphorylase so that glycogen is not degraded to its product and glucose is not released from glycogen into the bloodstream. This represents a negative feedback by glucose on glycogen breakdown. This figure is redrawn from https://classconnection.s3.amazonaws.com/679/flashcards/817679/png/liver1319368796037.png.

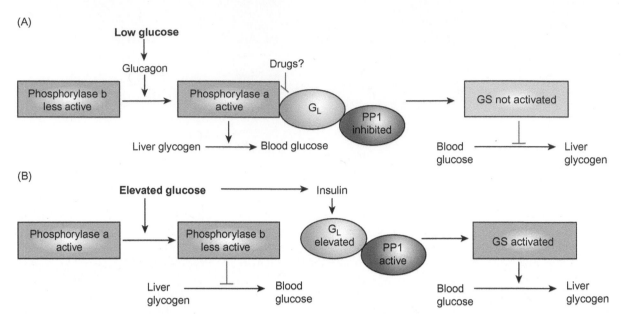

**FIGURE 7.23**    Regulation of the synthesis of liver glycogen and blood glucose level when (glucose) is either high or low. *Reproduced from http:// www.nature.com/nrm/journal/v7/n11/images/nrm2043-f3.jpg.*

When there is a need for energy, glycogen is degraded by phosphorylase to glucose-1-phosphate and then converted to glucose-6-phosphate by phosphoglucomutase. Glucose-6-phosphate is metabolized for energy (ATP) through glycolysis and the tricarboxylic acid cycle. During glycogen phosphorolysis, the debranching enzyme catalyzes a transfer of a trimer from a branch to the free end of the glycogen molecule.

The metabolism of glycogen, involving both its synthesis and degradation, is under the control of insulin, glucagon, and epinephrine. Insulin causes free glucose to generate glycogen. Glucagon and epinephrine stimulate glycogen breakdown.

The release of **glucagon** from the pancreatic $\alpha$-cells occurs **when glucose levels are low** in the blood. Lowered glucose suppresses metabolism and the ATP/ADP ratio in the $\alpha$-cell and the $K_{ATP}$ channels have a membrane potential of -60 mV, causing the opening of T-type channels and the depolarization of the cell membrane potential. This allows the $Na^+$ and $Ca^{2+}$ channels to become activated and stimulate regenerative action potentials. $Ca^{2+}$ entry occurs through the N-type channels and glucagon secretion is induced.

When **glucose is elevated in the blood**, it is transported into the pancreatic $\alpha$-cell through the GLUT1 transporter. The internal glucose level generates an increase in the ATP/ADP ratio in the cell. The $K_{ATP}$ channel becomes blocked leading to depolarization of the membrane potential and inactivation of voltage-dependent channels. Consequently, the influx of calcium ions is suppressed and glucagon cannot be secreted.

The precursor of glucagon (39 amino acids) is **preproglucagon** (158 amino acids). Besides glucagon, other peptides split out of this precursor are: **enteroglucagon**, the incretin **GLP-1** and **GLP-2**, glucagon-like peptide 2.

**Epinephrine** acts on the hepatocyte membrane $\alpha$-adrenergic receptor. Activation of this receptor elevates **phospholipase C$\beta$** that converts PIP2 to diacylglycerol and $IP_3$ (phosphatidylinositol *tris*phosphate). IP3 binds to and activates the $IP_3$ receptor in the endoplasmic reticulum (ER) causing the release of $Ca^{2+}$ from the ER store. The elevated cytoplasmic $Ca^{2+}$ concentration stimulates **calmodulin-mediated phosphorylation** of phosphorylase kinase that phosphorylates phosphorylase to its active form. Phosphorylase acts on the reducing ends of glycogen to release glucose-1-phosphate (G-1-P) that is converted by **phosphoglucomutase** to glucose-6-phosphate that, in turn, is acted on by **glucose-6-phosphatase** to generate free glucose for export by GLUT2 to the extracellular space and ultimately to the bloodstream so that glucose becomes available to the other organs.

Glucagon binds to the **hepatocyte membrane glucagon receptor**, a G protein-coupled receptor, generating elevated levels of cAMP. cAMP activates protein kinase A (PKA). PKA phosphorylates glycogen phosphorylase (GP) to its active form (GP*a*P). GP*a*P degrades glycogen to glucose-1-phosphate that is metabolized further to free glucose. **Casein kinase 2** is activated by glucagon (and inactivated by insulin). **Glycogen synthase kinase 3** (GSK3) is activated by casein kinase 2 to its phosphorylated active form (GSK3P). GSK3P phosphorylates active glycogen synthase a (GS*a*) to its inactive, or less active, phosphorylated form, GS*b*P so that glycogen synthesis is suppressed. Although PKA is able to phosphorylate glycogen synthase, *glycogen synthase kinase 3 is the major regulator of glycogen synthase.*

**Insulin** decreases free glucose in the liver by inhibiting glycogen degradation and stimulating glycogen synthesis. Insulin binds to and activates the hepatocyte membrane insulin receptor. The activated receptor phosphorylates the insulin receptor substrate (IRS-1) activating a pathway through PI3kinase and mTOR to protein phosphatase-1 (PP-1). PP-1 removes the phosphate group from active glycogen phosphorylase *a* to form the less active glycogen phosphorylase *b*, reducing glycogen degradation. On the other hand, PP-1 dephosphorylates inactive glycogen synthase *b*-P to the active glycogen synthase *a*. Glycogen synthase *a* catalyzes the synthesis of glycogen from free glucose. These activities lower the level of circulating glucose.

In addition to regulation by hormones affecting the phosphorylated state of the enzyme, **glycogen phosphorylase** is regulated by **allosteric effectors**. Glycogen phosphorylase is converted from its active R (relaxed) form into its inactive T (taut) form by the negative allosteric effectors **ATP** and **glucose-6-phosphate** (G-6-P), whereas it is converted back into its active relaxed (R) form by the allosteric effector **AMP**. Phosphorylase kinase converts the inactive T form to a modified M form by phosphorylation. The M form can be converted to an active form with internalized phosphates, and this form can be converted back to the inactive M form by **glucose** acting as a negative allosteric effector. The inactive M form can be converted to the inactive T form by phosphoprotein phosphatase. These effectors act by altering the shape of the enzyme.

With elevated glucose levels, glucose can act as an inhibitor by binding directly to glycogen phosphorylase *a* (active R state) to form glucose-bound inactive phosphorylated T state. Protein phosphatase-1 can remove the phosphate groups on phosphorylase *a* (phosphorylated T state) converting it to the inactive phosphorylase *b* T state. Thus, elevated glucose, itself, can act as a negative feedback inhibitor of phosphorylase preventing the breakdown of glycogen.

## SUGGESTED READING

### Literature

Grande, J., Perez, M., Plana, M., Itarte, E., 1989. Acute effects of insulin and glucagon on hepatic casein kinase 2 in adult fed rats: correlation of the effects on casein kinase 2 with the changes in glycogen synthase activity. Arch. Biochem. Biophys. 275, 478−485.

Hiraiwa, H., et al., 2001. The molecular link between the common phenotypes of type I glycogen storage disease and HNF1α-null mice. J. Biol. Chem. 276, 7963−7967.

Patient.co.uk, Glycogen Storage Disorders. http://www.patient.co.uk/health/glycogen-storage-disorders-leaflet, 2014.

Hiraiwa, H., et al., 1999. Inactivation of glucose 6-phosphate transporter causes glycogen storage disease type Ia. J. Biol. Chem. 274, 5532−5536.

Lin, B., et al., 1998. Cloning and characterization of cDNAs encoding a candidate glycogen storage disease type Ib protein in rodents. J. Biol. Chem. 273, 31656−31660.

Quesada, I., Tuduri, E., Ripoll, C., Nadal, A., 2008. Physiology of the pancreatic α-cell and glucagon secretion: role in glucose homeostasis and diabetes. J. Endocrinol. 199, 5−19.

Rayasam, G.V., et al., 2009. Glycogen synthase kinase 3: more than a namesake. Br. J. Pharmacol. 156, 885−898.

Shieh, J.-J., Pan, C.-J., Mansfield, B.C., Chou, J.Y., 2003. A glucose-6-phosphate hydrolase, widely expressed outside the liver, can explain age-dependent resolution of hypoglycemia in glycogen storage disease type Ia. J. Biol. Chem. 278, 47098−47103.

Shieh, J.-J., et al., 2002. The molecular basis of glycogen storage disease type Ia: structure and function analysis of mutations in glucose-6-phosphatase. J. Biol. Chem. 277, 5047−5053.

Watson, K.A., et al., 1995. Glucose analogue inhibitors of glycogen phosphorylase: from crystallographic analysis to drug prediction using GRID force-field and GOLPE variable selection. Acta Crystallogr. D: Biol. Ctrystallogr. 51, 458−471.

Zingone, A., et al., 2000. Correction of glycogen storage disease type Ia in a mouse model by gene therapy. J. Biol. Chem. 275, 828−832.

### Books

Acharya, K., 1991. Glycogen Phosphorylase b. World Scientific Publishing Co. Pte, Ltd, Singapore, Singapore.

Hecht, S.M., 1998. Bioorganic Chemistry: Carbohydrates. Oxford University Press, Oxford, UK.

Weiss, P.L., Faulkner, B.D., 2013. Glycogen. Nova Science Publishers, Inc, Hauppauge, NY.

# Chapter 8

# Glycolysis and Gluconeogenesis

## HEMOLYTIC ANEMIA: GLYCERALDEHYDE-3-PHOSPHATE DEHYDROGENASE (G3PDH) DEFICIENCY (A RARE DISEASE)

### Hemolytic Anemia

*In the process of its maturation, the* **red blood cell** *(RBC) eliminates many of its subcellular structures, including the cell nucleus, mitochondria, endoplasmic reticulum and Golgi apparatus in order to maximize storage capacity for* **hemoglobin** *(Hb).* Because of the lack of mitochondria for oxidative metabolism, *the RBC is dependent on* **glycolysis** *for its energy* (in the form of ATP [adenosine triphosphate]) and its metabolism needs to be intact for the function of the cell membrane as well as for the structure and functions of Hb. *Most of the energy for the RBC is generated by glycolysis (up to 90%) where 2ATP are generated from the metabolism of a glucose molecule.* Other pathways in the RBC are the **pentose phosphate pathway** **(PPP)** (or hexose monophosphate shunt) that metabolizes 5%−10% of glucose available and generates **NADPH**, the **methemoglobin (mHb) reductase** pathway and the **Rapoport-Luebering Shunt**.

Virtually any deficiency of an enzyme in the glycolytic pathway would seriously imperil the lifetime (normally 120 days) of a RBC owing to the fact that the red blood cell depends upon this pathway almost entirely for energy. Glyceraldehyde-3-phosphate dehydrogenase (G3PDH, 37 kDa) catalyzes the sixth reaction in the glycolytic pathway, the conversion of glyceraldehyde-3-phosphate to 1,3-*bis*phosphoglycerate and the generation of $NADH + H^+$ from $NAD^+$. In addition to catalyzing the glycolytic step, G3PDH may function as an activator of transcription in that it has been shown to be part of the *OCA-S* (complex is a gene(s) encoding histones) coactivator complex (in addition to lactate dehydrogenase, also a part of the complex). G3PDH has additional functions: it can move between the cytosol and the nucleus; it is able to initiate apoptosis and it is also able to bind to a nuclear ubiquitin ligase directing a target protein for degradation.

The absence of G3PDH means that the RBC dies before its natural lifetime. Impairment of the function of the RBC causes many symptoms, such as shortness of breath, dizziness, headache, cold extremities, pale skin, chest pain, fatigue, arrhythmias, enlarged heart, and heart failure. Also, because RBCs are dying early, hemoglobin is released into the bloodstream (whereas it is usually collected by the spleen at the end of a normal lifetime) and it is metabolized to bilirubin (yellow color) that can represent jaundice. Excessive bilirubin, especially when combined with high levels of cholesterol, can generate stones in the gallbladder, spleen enlargement, and abdominal pain. Treatment of hemolytic anemia may require blood transfusions. Severe hemolytic anemia can be fatal.

### Methemoglobin Reductase Pathway

Methemoglobin (mHb) contains oxidized iron ($Fe^{3+}$), a form that is unable to bind oxygen. It is only a trace in blood representing about one percent. The NADPH (nicotinamide adenine dinucleotide phosphate, reduced form) methemoglobin reductase pathway maintains the oxygen-binding form of hemoglobin that contains iron in the $Fe^{2+}$ ferrous form (Fig. 8.1).

### Rapoport-Luebering Shunt

**2,3-*bis*phosphoglycerate** (**2,3-BPG**) is an intermediate between 1,3-*bis*phosphoglycerate and 3-phosphoglycerate. This intermediate, **2,3-*bis*phosphoglycerate**, is an allosteric effector of hemoglobin that regulates the affinity of hemoglobin for oxygen and facilitates the release of oxygen to the tissues (e.g., lungs). 2,3-BPG binds to the beta subunit of the T (taut) state of hemoglobin, deoxyhemoglobin, the less active form. The pocket in which 2,3-BPG binds measures 11

**Human Biochemistry. DOI: http://dx.doi.org/10.1016/B978-0-12-383864-3.00008-9**

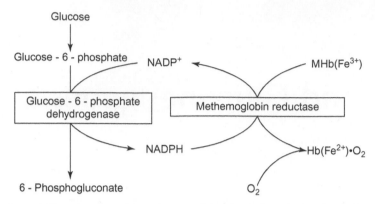

**FIGURE 8.1** The NADPH methemoglobin reductase pathway. As part of the glycolysis pathway, glucose is converted to glucose-6-phosphate (G6P). G6P can continue through the glycolysis pathway, first forming fructose-6-phosphate and eventually forming pyruvate. G6P also can be converted to 6-phosphogluconate by the action of glucose-6-phosphate dehydrogenase, a step that generates NADPH from $NADP^+$. NADPH is utilized by methemoglobin reductase to catalyze the conversion of methemoglobin (mHb) ($Fe^{3+}$), the nonoxygen-binding form, to the oxygen-binding form of hemoglobin (Hb) ($Fe^{2+}$).

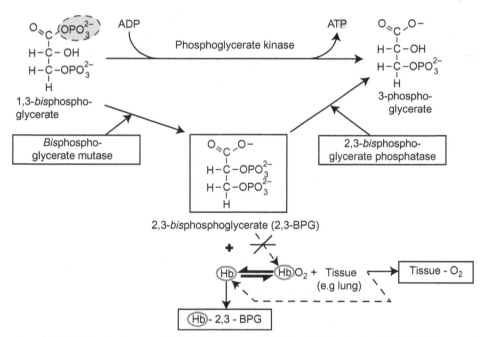

**FIGURE 8.2** Formation of 2,3-*bis*phosphoglycerate (2,3-BPG) in the red blood cell. The level of 2,3-BPG is controlled by the activities of *bis*phosphoglycerate mutase and 2,3-*bis*phosphoglycerate phosphatase. The greater affinity of 2,3-BPG for hemoglobin compared to oxyhemoglobin allows oxygenated hemoglobin to release its oxygen to needy tissue, such as the lungs.

Angstroms for deoxyhemoglobin (T state), whereas the same pocket in oxyhemoglobin (R state, relaxed) measures 5 Angstroms; 2,3-BPG itself measures about 9 Angstroms, so it can fit into the T state pocket but not in the R state pocket of the hemoglobin beta subunit. The complex enhances the ability of oxyhemoglobin to release oxygen to the needy tissues (e.g., lungs). The concentration in the red blood cell of 2,3-*bis*phosphoglycerate is determined by the activities of *bis*phosphoglycerate mutase and 2,3-*bis*phosphoglycerate phosphatase (Fig. 8.2).

A lowered pH in the red blood cell is inhibitory to 2,3-*bis*phosphoglycerate mutase and stimulatory to 2,3-*bis*phosphoglycerate phosphatase. Consequently, under this condition, there would be more 1,3-*bis*phosphoglycerate available to form 3-phosphoglycerate through the phosphoglycerate kinase reaction and generate ATP.

## THE PENTOSE PHOSPHATE PATHWAY (PPP)

The pentose phosphate pathway is outlined in **Fig. 6.26**. It is summarized in its connections to other major pathways: glycolysis, glycogen metabolism and the tricarboxylic acid (TCA) cycle in Fig. 8.3.

The various functions of the PPP can provide NADPH from $NAD^+$ and ribose-5-phosphate for the ultimate synthesis of nucleic acids. NADH is needed to reduce GSSG (2 glutathione molecules joined by a disulfide bridge; the oxidized form of glutathione) to GSH (glutathione), particularly in cells, such as the red blood cell that are subject to oxidative stresses and the production of $H_2O_2$ and free peroxy-radicals. The PPP is, in a sense, elastic in that it can adapt to the needs of a particular cell at a point in time when the metabolism of a cell is requiring reducing equivalents in the form of NADPH, or needing to divide that requires DNA and RNA and the production by the PPP of ribose-5-phosphate, or needing to synthesize lipid from the same 3-carbon intermediates of glycolysis, or needing energy in the form of ATP. Thus, the four modes of PPP function are shown diagrammatically in Fig. 8.4.

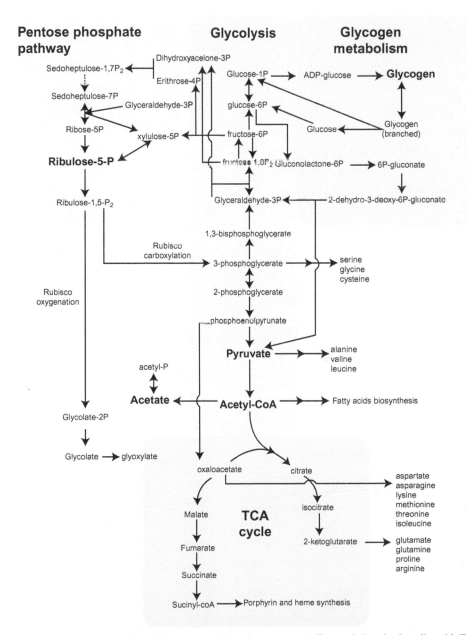

**FIGURE 8.3**    The pentose phosphate pathway in relation to glycolysis, glycogen metabolism and the tricarboxylic acid (TCA) cycle. *Reproduced in part from http://www.biomedcentral.com/content/figures/1471-2164-9-597-6-1.jpg.*

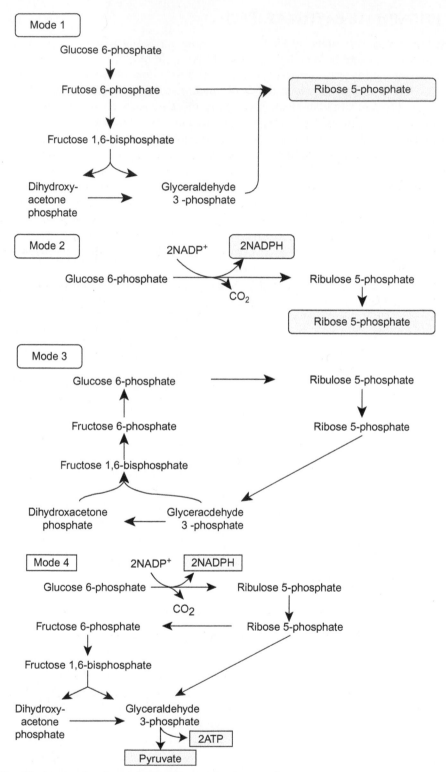

**FIGURE 8.4** Four modes of the pentose phosphate pathway, each geared to the generation of a product needed by a particular cell. In mode 1, 1 molecule of glucose-6-phosphate (G6P) makes 5 molecules of ribose-5-phosphate (5R5P). In mode 2, 1 molecule of G6P makes 1 molecule of R5P and 2 NADPH. In mode 3, 1 G6P makes 12 NADPH (starting with 6 molecules of G6P oxidized to 6 ribulose 5-phosphates, the ribulose 5-phosphates can be "rearranged by the pathway to form 5 G6Ps." The overall stoichiometry is: 6 G6P + 12 NADP$^+$ → 5 G6P + 12 NADPH + 6CO$_2$ + Pi). In mode 4, 3 molecules of G6P make 6 NADPH, 8 ATP and 5 pyruvate and NADH that can be used to create more ATP. *Redrawn from http://synergyhw.blogspot. com/2015/02/the-pentose-phosphate-pathway-missing.html.*

## GLYCOLYSIS, THE EMDEN—MEYERHOF PATHWAY

Glycolysis is a metabolic process in which glucose (or other sugars that can funnel into the pathway) is converted to a series of intermediates leading to the formation of **pyruvic acid**. In muscle, for example, as oxygen becomes used up through activity, pyruvic acid, so formed, can be converted to **lactic acid** under anaerobic conditions by the **lactate dehydrogenase** reaction. In glycolysis, two molecules of pyruvic acid are formed per glucose molecule. The reverse direction of the pathway, leading to the formation of glucose is gluconeogenesis. *Gluconeogenesis is not strictly a reversal of glycolysis because there are some different unique enzymatic steps involved at the points of irreversible reactions.* The pathways of glycolysis and gluconeogenesis are shown in Fig. 8.5(A) and (B).

*In glycolysis, two ATPs are used and four ATPs are generated providing a net of 2 ATP in the conversion of one molecule of glucose to 2 molecules of pyruvate.* One ATP is used in the hexokinase/glucokinase reaction converting glucose to fructose-6-phosphate. **Hexokinase** is primarily used in muscle and **glucokinase** (an isozyme of hexokinase, **hexokinase IV**) is used preferentially in liver. Glucokinase has a larger Km (10 mM) for glucose than does hexokinase (0.2 mM) allowing glucokinase to handle larger amounts of glucose. *Glucokinase in liver shuttles between the nucleus and the cytoplasm.* When the cellular level of glucose is low or when fructose-6-phosphate level is high, glucokinase is transported into the nucleus. Conversely, when the level of glucose is elevated in the cell, glucokinase is transported to the cytoplasm to initiate glycolysis.

Another ATP is used in the phosphofructokinase reaction converting fructose-6-phosphate to fructose-1,6-*bis*phosphate (F-1,6,BP). Two ATPs are produced in the phosphoglycerate kinase reaction converting 2 molecules of 1,3-*bis*phosphoglycerate to 2 molecules of 3-phosphoglycerate. Two molecules of ATP are produced in the pyruvate kinase reaction converting 2 molecules of phosphoenolpyruvate to 2 molecules of enolpyruvate. Thus, there are produced 2 molecules of ATP from the conversion of glucose (6C) to two pyruvate molecules (3C each),

Two molecules of NADH + H$^+$ are produced in the glyceraldehyde-3-phosphate dehydrogenase reaction converting glyceraldehyde 3-phosphate to 1,3 *bis*phosphoglycerate. The overall reaction of **aerobic glycolysis** from glucose to 2 pyruvate molecules is:

$$\text{Glucose} + 2\text{NAD}^+ + 2\text{Pi} + 2\text{ADP} \rightarrow 2\text{ pyruvate} + 2\text{NADH} + 4\text{H}^+ + 2\text{ATP} + 2\text{H}_2\text{O}$$

The overall reaction of **anaerobic glycolysis** (absence of oxygen) converting glucose to 2 molecules of lactic acid is:

$$\text{Glucose} + 2\text{Pi} + 2\text{ADP} \rightarrow 2\text{ lactate} + 2\text{H}^+ + 2\text{ATP} + 2\text{H}_2\text{O}$$

Like aerobic glycolysis, anaerobic glycolysis yields 2 molecules of ATP but does not generate 2NADH from 2NAD$^+$. For recycling lactate through glycolysis (in muscle), it can be converted back to pyruvate by lactate dehydrogenase with the conversion of NAD$^+$ to NADH + H$^+$ and another cycle of glycolysis to yield two more ATP molecules and these reactions can be continued.

## PHOSPHOFRUCTOKINASE ENZYMES INVOLVED IN THE CONVERSION OF FRUCTOSE-6-PHOSPHATE TO FRUCTOSE-1,6-*BIS*PHOSPHATE

*In glycolysis, phosphofructokinase (PFK) is a key regulator of the overall reactions.* It exists as a tetramer and each subunit has two binding sites for ATP. This enzyme catalyzes the first unique step in glycolysis, converting fructose-6-phosphate to **fructose-1,6-*bis*phosphate**. This step is catalyzed by **phosphofructokinase 1 (PFK1)**. The second isoform, **phosphofructokinase 2 (PFK2)** catalyzes the conversion of fructose-6-phosphate to **fructose-2,6-*bis*phosphate**. Fructose-2,6-bisphosphate is a stimulator of PFK1 by its ability to increase the affinity of PFK1 for fructose-6-phosphate and to decrease the ability of ATP to inhibit the reaction. *PFK2 is a bifunctional enzyme in that it has both kinase and phosphatase activities.* The kinase activity is inhibited by phosphorylation and the phosphatase activity is stimulated by phosphorylation. Thus, *the same enzyme can convert fructose-6-phosphate to fructose-2,6-bisphosphate in the nonphosphorylated state and convert fructose-2,6-bisphosphate to fructose-6-phosphate in the phosphorylated state* (Fig. 8.6).

PFK2 is regulated by the hormones **glucagon** in the liver, **epinephrine** in muscle and by **insulin**. Both glucagon and epinephrine stimulate adenylate cyclase and cAMP-dependent protein kinase (PKA) in liver. PKA phosphorylates phosphofructokinase2 (PFK2) in liver, activating its phosphatase activity which decreases the concentration of fructose-2,6-*bis*phosphate (converting it back to fructose-6-phosphate) resulting in an inhibition of glycolysis (and stimulation of gluconeogenesis). This effect is opposite in muscle in that epinephrine elevates cAMP and stimulates glycolysis. The

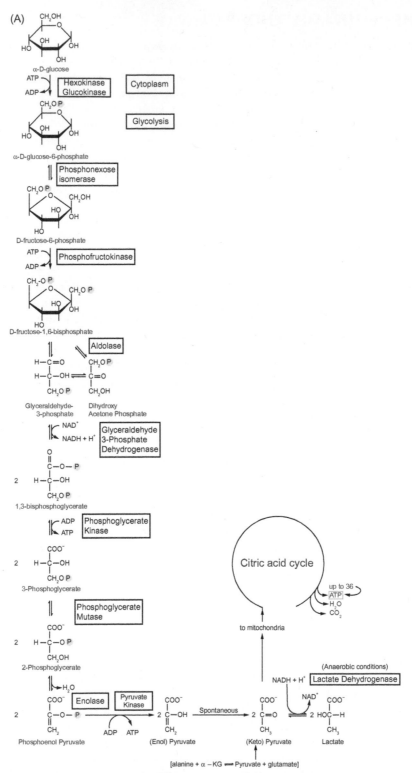

**FIGURE 8.5** (A) Pathway of glycolysis with the enzymes listed for each step. (B) Pathways of glycolysis and gluconeogenesis. The pathways are not strictly reversible because there are some specific enzymes that are used in each direction at irreversible steps. The *red arrows* indicate three **irreversible reactions** in glycolysis involving **hexokinase, phosphofructokinase-1 (PFK-1)** and **pyruvate kinase**. In the reverse of the pathway (gluconeogenesis), the *blue arrows* indicate enzymatic reactions that are **unique to gluconeogenesis: pyruvate carboxylase, phosphoenolpyruvate carboxykinase, fructose-1,6-bisphosphatase,** and **glucose-6-phosphatase**.

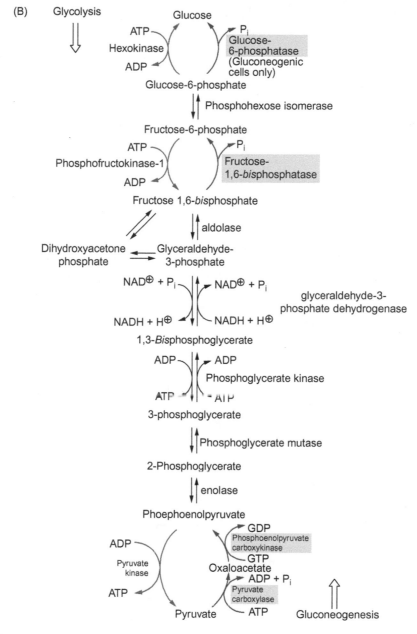

**FIGURE 8.5**  (Continued).

effect of insulin is to decrease the concentration of cAMP, resulting in an increase in the level of fructose-2,6-*bis*phosphate generating an increase in the level of fructose-1,6-*bis*phosphate and subsequent stimulation of glycolysis.

The controls on this step in glycolysis are complicated by a subcycle of enzymes as shown in Fig. 8.7.

Glycolysis is closely regulated by a variety of small molecules. These regulators are shown in Fig. 8.8 and they are listed for their effects on key enzymes in the glycolytic pathway.

In the process of gluconeogenesis, four ATPs and two GTPs (total of 6 ATP equivalents) are used starting with pyruvate and two NADH + H$^+$ are used to generate two NAD$^+$ (two three-carbon molecules are needed to form one six-carbon glucose molecule and vice versa). Glycolysis leads to the formation of pyruvic acid that can enter the mitochondrial **tricarboxylic acid cycle (TCA cycle)** directly, where it is metabolized by mitochondrial **pyruvate dehydrogenase** to **acetyl CoA** (acetyl coenzyme A). Acetyl CoA then combines with oxaloacetate to form citrate and CoA-SH in the **citrate synthase** reaction and the TCA cycle proceeds.

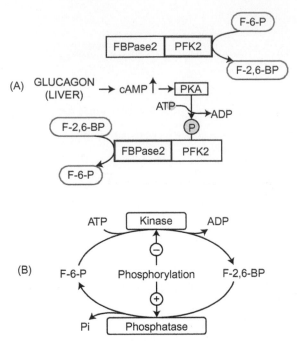

FIGURE 8.6 (A) Phosphofructokinase 2 (PFK2)/fructose bisphosphatase 2 (FBPase2) is a **bifunctional enzyme**. PFK2 is inactive in the phosphorylated form. When PFK2 is in the phosphorylated inactive form, FBPase2 is active. (B) Activities of the bifunctional enzyme showing the effects of phosphorylation.

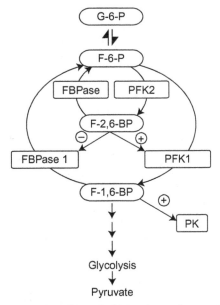

FIGURE 8.7 An enzyme subcycle involved in the conversion of fructose-6-phosphate to fructose-1,6-*bis*phosphate. *FBPase*, fructose *bis*phosphatase; *PFK2*, phosphofructokinase 2; *PFK1*, phosphofructokinase 1; *G-6-P*, glucose-6-phosphate; F-6-P, fructose-6-phosphate; *F-2,6-BP*, fructose-2,6-*bis*phosphate; *PK*, protein kinase.

## CELL PROLIFERATION AND TUMOR GROWTH—THE WARBURG EFFECT

In the early 1920s, Otto Warburg discovered that cancer cells, in contrast to normal differentiated tissue cells, metabolize glucose differently. Normal tissue cells metabolize glucose to pyruvate that enters the Tricarboxylic Acid cycle and oxidative phosphorylation permitting efficient generation of ATP (up to 36 ATP per glucose molecule), $CO_2$ and water. Tumor cells, Warburg found, on the other hand, prefer to convert ("ferment") glucose into lactate even when there is sufficient oxygen (**aerobic glycolysis**) and perfectly functional mitochondria to support oxidative phosphorylation. The

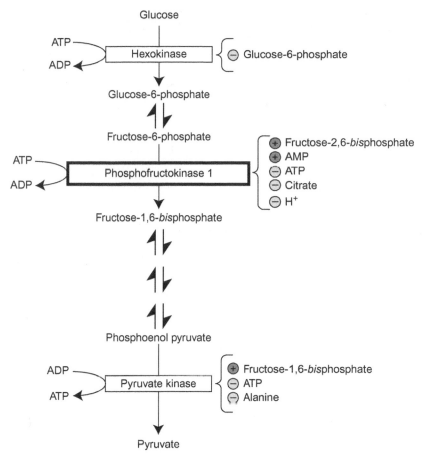

**FIGURE 8.8**  Regulators of key enzymes and their products in the glycolytic pathway. The small molecule regulators are listed next to the enzyme that they control. Phosphofructokinase 1 is a critical regulatory enzyme in the glycolytic pathway as indicated by the heavy outline.

differences between normal differentiated cells and tumor cells/normal proliferating cells in the utilization of glucose are emphasized in Fig. 8.9.

Tumor cells choose to rely on aerobic glycolysis in spite of the fact that their mitochondrial function is not impaired. Such a selection must rest on the requirements of cell proliferation (of normal proliferating cells and tumor cells) in contrast to the needs of differentiated tissue cells. Apparently, proliferating cells have critical metabolic requirements that extend beyond the production of ATP so that glucose is used to generate components (macromolecules) of the new cell mass, such as fatty acids for triglycerides (via acetyl CoA), nonessential amino acids for proteins (via glycolytic intermediates) and nucleotides for nucleic acids (via ribose). Some recent evidence shows that certain cultured cancer cells convert nearly 90% of glucose and 60% of glutamine into lactate or alanine (one of the more common amino acids in proteins). The controls of the various metabolic pathways by the factors involved in cell proliferation are indicated in Fig. 8.10.

## GLUCONEOGENESIS

Gluconeogenesis is the opposite direction of the glycolysis pathway generating glucose from 2 pyruvate molecules as shown in Fig. 8.5B.

### Alanine Cycle

Alanine is a **glucogenic amino acid** that constitutes a high percentage of the amino acids in most proteins. Also, other amino acids, in particular, branched chain amino acids (BCAA), such as valine, leucine, and isoleucine can be converted to alanine. When muscle proteins break down in fasting, for example, alanine is released in relatively large quantity into the bloodstream and is taken up by the liver. Since alanine is a glucogenic amino acid it is readily converted in

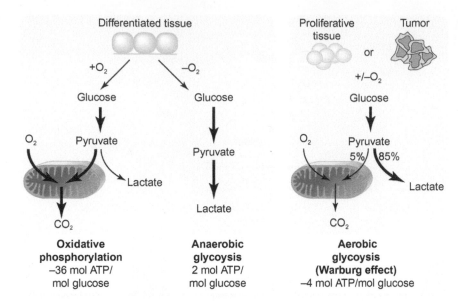

**FIGURE 8.9** Schematic representations of the differences between oxidative phosphorylation, anaerobic glycolysis and aerobic glycolysis (Warburg effect). In the presence of oxygen, nonproliferating (differentiated) tissues first metabolize glucose to pyruvate via glycolysis, followed by complete oxidation of most of that pyruvate in the mitochondria to $CO_2$ during the process of oxidative phosphorylation. Because oxygen is required as the final acceptor to completely oxidize glucose, oxygen is essential for this process. When oxygen is limiting, cells can redirect the pyruvate generated by glycolysis away from mitochondrial oxidative phosphorylation by generating lactate (anaerobic glycolysis). This generation of lactate during anaerobic glycolysis allows glycolysis to continue (by cycling NADH back to $NAD^+$), but results in minimal ATP production (2 molecules of ATP/glucose molecule) when compared with oxidative phosphorylation (up to 36 ATPs/glucose molecule). Warburg observed that cancer cells tend to convert most glucose to lactate regardless of whether oxygen is present (aerobic glycolysis), a property shared by normal proliferative tissue cells. Mitochondria remain functional and some oxidative phosphorylation continues in both cancer cells and normal proliferating cells. Nevertheless, aerobic glycolysis is less efficient than oxidative phosphorylation for generating ATP. In proliferating cells, $\sim 10\%$ of the glucose is diverted into biosynthetic pathways upstream of pyruvate production. *Reproduced from http://www.ncbi.nlm.nih.gov/core/lw/2.0/html/tileshop_pmc_inline.html?title = Click%20on% 20image%20to%20zoom&p = PMC3&id = 2849637_nihms165713f2.jpg.*

the liver by the catalytic action of **glutamate-pyruvate transaminase (GPT)** also known as **alanine transaminase, ALT** with $\alpha$-ketoglutarate to form glutamate and pyruvate. Pyruvate is converted to glucose by the **gluconeogenic pathway** (Fig. 8.5B). Liver glucose, so formed can access (via glucose transporter 2 (GLUT2)) the bloodstream and be taken up by the muscle and used for energy. The amino group of alanine, initially transported into the liver, is converted to **urea** in the urea cycle and excreted. These reactions are shown in Fig. 8.11.

## Glucose Can Be Formed From Glycerol

When fat (triglyceride) is degraded, glycerol is produced. Glycerol can be converted to **dihydroxyacetone phosphate** that can be converted to glucose through the gluconeogenic pathway. **Glycerol kinase** converts glycerol to **glycerol-3-phosphate** that, in turn, can be converted to dihydroxyacetone phosphate by **cytosolic** (and/or mitochondrial) **glycerol 3-phosphate dehydrogenase**. This product can then be converted directly to fructose-1,6-bisphosphate (or through glyceraldehyde-3-phosphate) and then to glucose via the gluconeogenic pathway (Fig. 8.5B). Glycerol derived from adipose tissue provides about 20% of glucose synthesized by this route. This conversion is shown in Fig. 8.12.

Adipose tissue provides glucose from other noncarbohydrate sources through the gluconeogenic pathway from pyruvate plus lactate (60%), from amino acids (20%) and from glycerol (20%) as mentioned above.

Gluconeogenesis takes place mainly in the liver (90%) and in the kidney (10%). About 160 grams of glucose are used daily in the body, with the brain using most of it (120 g, 75%). Red blood cells also use glucose ($\sim 30$ g/day). In muscle, glucose from the liver is metabolized through the glycolytic pathway to pyruvate and then to lactate. The lactate is released from the muscle cell into the bloodstream and is utilized by the liver to generate more glucose through gluconeogenesis. Liver glucose is then released into the bloodstream and is taken up by muscle for its work and the cycle is repeated. The conversion of 2 lactate molecules to glucose in the liver requires six high energy phosphates, four from ATP [pyruvate carboxylase (PC)], and two from GTP [phosphoenolpyruvate carboxykinase (PEPCK)

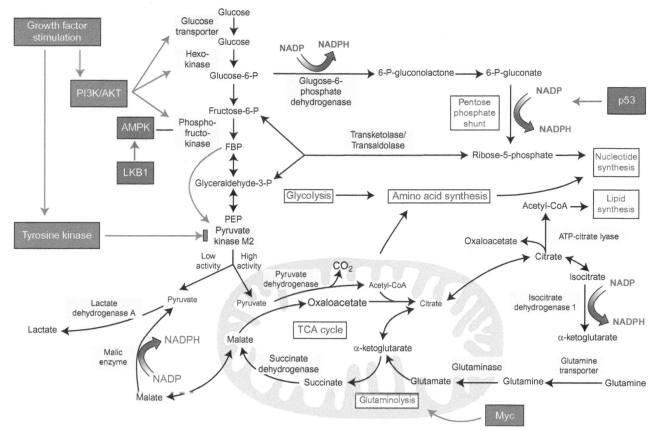

**FIGURE 8.10**  The metabolic pathways active in proliferating cells are directly controlled by signaling pathways involving known oncogenes and tumor suppressor genes. This schematic shows our current understanding of how glycolysis, oxidative phosphorylation, the pentose pathway and gluta-mine metabolism are interconnected in proliferating cells. This metabolic wiring allows for both NADPH production and acetyl CoA flux to the cyto-sol for lipid synthesis. Key steps in these metabolic pathways can be influenced by signaling pathways known to be important for cell proliferation. Activation of growth factor receptors leads to both tyrosine kinase signaling and PI3K (phosphatidylinositol-3-kinase) activation. Via Akt (protein kinase B) PI3K activation stimulates glucose uptake and flux through the early part of glycolysis. Tyrosine kinase signaling negatively regulates flux through the late steps of glycolysis, making glycolytic intermediates available for macromolecular synthesis as well as supporting NADPH production. Myc (multifunctional nuclear phosphoprotein, oncogene product) drives glutamine metabolism that also supports NADPH production. LKB1/AMPK (liver kinase B1/5′AMP-activated protein kinase) signaling and p53 (tumor suppressor, transcriptional activator) decrease metabolic flux through gly-colysis in response to cell stress. Decreased glycolytic flux in response to LKB/AMPK or p53 may be an adaptive response to shut off proliferative metabolism during periods of low energy availability or oxidative stress. Tumor suppressors are shown in *red*, and oncogenes are in *green*. Key meta-bolic pathways are labeled *purple* with *white boxes*, and the enzymes controlling critical steps in these pathways are shown in *blue*. Some of these enzymes are candidates as novel therapeutic targets in cancer. Malic enzyme refers to NADP$^+$-specific malate dehydrogenase [systematic name (S)-malate: NADP$^+$ oxidoreductase (oxaloacetate-decarboxylation)]. *Reproduced from http://www.ncbi.nlm.nih.gov/core/lw/2.0/html/tileshop_pmc_inline.html?title + Click%20on%20image%20to%20zoom&p = PMC3&id = 2849637_nihms165713f2.jpg.*

reaction] as shown in Fig. 8.5B. The utilization of glucose through glycolysis in muscle and the shuttling of lactate from muscle to the liver with the return of glucose synthesized through gluconeogenesis is depicted in Fig. 8.13.

## Small Molecule Regulation of Gluconeogenesis

As was shown for the enzymes involved in glycolysis (Fig. 8.8), the enzymes of gluconeogenesis, are regulated by small molecules. Fig. 8.14 compares the small molecule regulation of both pathways.

The enzymatic reactions that are unique to **gluconeogenesis** are **irreversible reactions** involving the following enzymes: **pyruvate carboxylase (PC), phosphoenolpyruvate carboxykinase (PEPCK), fructose-1,6,bisphosphatase** (F-1,6,BPase) and **glucose-6-phosphatase** (G-6-Pase). In synthesizing glucose from pyruvate, 6 ATP equivalents are consumed (PEPCK reaction uses GTP). The pyruvate carboxylase reaction, converting pyruvate to oxaloacetate, the ini-tial step in gluconeogenesis, occurs in the mitochondrion. Oxaloacetate is reduced to malate that is carried through the malate transporter to the cytosol where it becomes decarboxylated and phosphorylated by PEPCK in the second unique

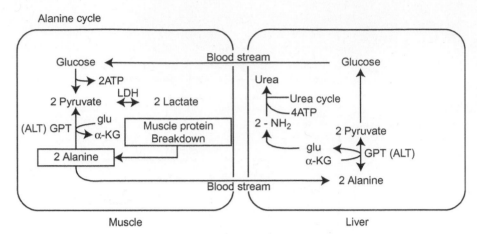

**FIGURE 8.11** The alanine cycle. On the left are the reactions that take place in skeletal muscle. Alanine is a major product of the breakdown of skeletal muscle during fasting or other stresses. Also some other amino acid products from this breakdown can be converted to alanine. Alanine then accesses the bloodstream and is carried to the liver where, through the action of glutamate-pyruvate aminotransferase (ALT), it is converted to pyruvate. Pyruvate can then be converted to glucose through the gluconeogenic pathway. The amino group of alanine is converted to urea, by the urea cycle, and excreted. The glucose formed in the liver from alanine may then enter the skeletal muscle again through the bloodstream and serve as an energy supply. *Glu*, glutamate; *α-KG*, alpha-ketoglutarate; *GPT*, glutamate-pyruvate aminotransferase (or ALT).

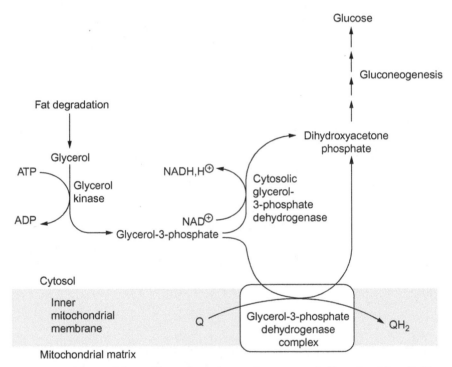

**FIGURE 8.12** Glycerol from the breakdown of fatty acids can form glucose. *Q*, coenzyme Q. *Reproduced from G. Litwack,* Human Biochemistry and Disease, *Figures 4—65, page 182, Academic Press/Elsevier, 2008.*

enzymatic reaction of gluconeogenesis. The product of the PEPCK reaction is phosphoenolpyruvate that is further metabolized by the enzymes of glycolysis, in reverse reactions, to the point where the next irreversible reaction occurs, the fructose-1,6-*bis*phosphatase reaction converting fructose-1,6-*bis*phosphate ($+H_2O$) to fructose-6-phosphate ($+Pi$). The initial reactions taking place in the mitochondrion are shown in Fig. 8.15.

The glucose-6-phosphatase reaction produces free glucose that is transported through the GLUT2, out of the liver cell, ultimately to the bloodstream where it is made available to the other tissues. The glucose-6-phosphatase reaction is localized in the cytoplasm to the lumen of the endoplasmic reticulum (Fig. 8.16).

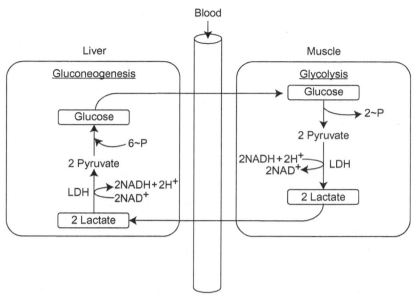

**FIGURE 8.13**  The cycle of glycolysis in muscle ending in lactate and the shuttling of lactate to the liver for gluconeogenesis with the glucose product transferred to muscle.

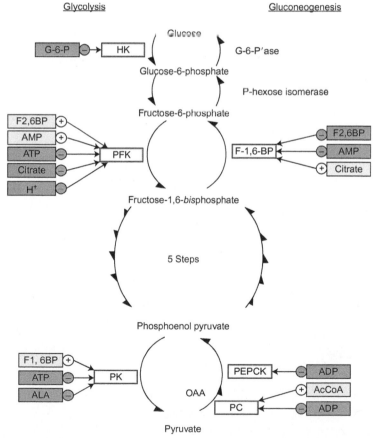

**FIGURE 8.14**  Regulated enzymes catalyzing individual reactions of glycolysis and gluconeogenesis. The regulators are small molecules. Some of the regulators are substrates and products while others are metabolic intermediates and nucleotides. *HK*, hexokinase; *G-6-Pase*, glucose-6-phosphate phosphatase; *P-hexose isomerase*, phosphohexose isomerase; *PEPCK*, phosphoenolpyruvate carboxykinase; *PC*, pyruvate carboxylase; *PK*, pyruvate kinase; *PFK*, phosphofructokinase; *G-6-P*, glucose-6-phosphate; *F-1,6-BP*, fructose-1,6-*bis*phosphate; *F-2,6-BP*, fructose-2,6-*bis*phosphate.

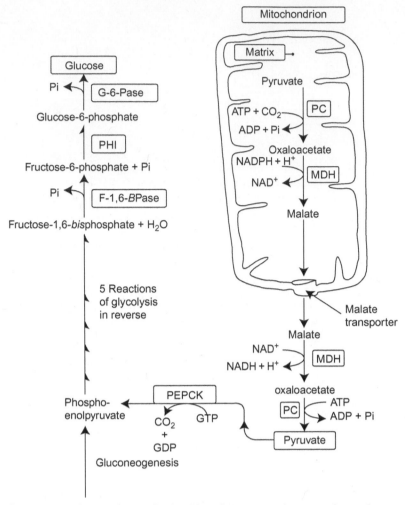

**FIGURE 8.15** Initial steps of gluconeogenesis occur in the mitochondrion where pyruvate is converted to oxaloacetate in the pyruvate carboxylase (PC) reaction and oxaloacetate is reduced to malate by malate dehydrogenase (mitochondrial MDH) utilizing $NADH + H^+$. Malate thus formed travels from the mitochondrion into the cytosol through the malate transporter where it is again oxidized to oxaloacetate by cytosolic malate dehydrogenase reaction (MDH) in reverse. Next, oxaloacetate is converted to phosphoenolpyruvate by PEPCK and gluconeogenesis proceeds.

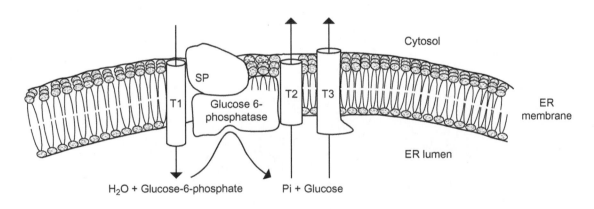

T1 = G-6-P transporter

T2 = Pi transporter

T3 = Glucose transporter (GLUT2 in liver)

SP = $Ca^{2+}$– Binding stabilizing protein

**FIGURE 8.16** The superstructure of the glucose-6-phosphatase in the lumen of the endoplasmic reticulum (ER). *Reproduced from a set of slides by Elizabeth Neufeld.*

## GLUCOSE TRANSPORTERS

GLUTs 1, 3, and 4 are transporters that have high affinity for glucose ranging in Km of 2−5 mM glucose. Consequently, the functions of these transporters align with the physiological concentration of glucose of about 5 mM. On the other hand, GLUT2 has a low affinity for glucose with its Km of about 15−20 mM glucose. GLUT2, therefore, is able to move glucose into the liver cell and the pancreatic beta cell in proportion to the plasma level of glucose. This low affinity provides a variable rate of transport (proportional to the increasing glucose concentration) so that the high level of glucose following a carbohydrate-rich meal can be accommodated.

GLUT3 has a low Km for glucose of 1.6 mM. It transports glucose into brain cells at a rate that is independent of the plasma level of glucose when it exceeds the physiological range of 4−10 mM.

GLUT4 functions for the insulin-dependent translocation of glucose. Thus, insulin stimulates the uptake of glucose by GLUT4 in the muscle cell where hexokinase converts it to glucose-6-phosphate so that the cell may utilize it for either glycolysis for energy or for the formation of glycogen when glucose is abundant. In the adipocyte, GLUT4 also responds to insulin for the uptake of glucose where it is again converted to glucose-6-phosphate by hexokinase and then converted to glycerol-3-phosphate for conversion to trigylcerides. Also under the control of insulin, lipoproteins in the plasma form fatty acids within the cell under the action of lipoprotein lipase and these fatty acids also are used for triglyceride synthesis. In addition, insulin inhibits the breakdown of triglycerides to fatty acids in these cells.

GLUT5 is located in the plasma membranes of small intestinal epithelial cells where it is responsible for the uptake of fructose. Some cells like the small intestinal epithelial cell contain more than one hexose transporter. This cell absorbs glucose, galactose, and fructose from the intestinal lumen. The apical membrane contains a sodium−glucose symporter (SGLUT1) that functions to transport glucose and galactose into the cell together with sodium ion. GLUT2 is located on the basolateral side of the cell where it moves the three hexoses, glucose, galactose, and fructose, out of the cell into the extracellular space and ultimately move further to access the bloodstream. All of the known GLUTs are reviewed in **Table 6.3A** and **B**.

## SUMMARY

The **red blood cell** depends mainly on glycolysis for ATP energy (using 90% of available glucose) that is required for the functions of the cell membrane and the structure and functions of hemoglobin. Maximal storage of hemoglobin in the RBC requires the loss of subcellular structures; the nucleus, mitochondria, endoplasmic reticulum and Golgi apparatus, during maturation. In addition to glycolysis, other metabolic functions in the red blood cell are the **pentose phosphate pathway** requiring 10% of the glucose used by the cell and generating NADPH from NADP + and the enzyme system methemoglobin **reductase** that maintains hemoglobin in the ferrous iron form, the one form able to bind oxygen. NADPH is used to convert oxidized glutathione (GSSG) to the reduced form (GSH) so that it can protect the cell from injury by reactive oxygen species arising from a highly oxygenated environment.

The loss of any enzyme in the glycolytic pathway jeopardizes red blood cell function and increases mortality of this cell before its lifespan is completed. This is the case with **glyceraldehyde 3-phosphate deficiency**. The consequent early demise of the red blood cell dumps degraded cells into the bloodstream (rather than being completely collected by the spleen when the lifespan has been completed) and the hemoglobin in the bloodstream is converted to the yellow pigment bilirubin (jaundice). Hemolytic anemia leads to respiratory problems and in severe cases can result in death.

The **pentose phosphate pathway** of metabolism utilizes glucose-6-phosphate for the development of seven carbon sugars and the five carbon sugar, ribose-5-phosphate, is essential for the synthesis of nucleic acids, especially when there is a demand for cell division. The pathway produces NADPH for cellular metabolism and protection from reactive oxygen species. It also produces three-carbon compounds that are intermediates in glycolysis and under certain circumstances can give rise to fatty acids.

2,3-*Bis*phosphoglycerate is an allosteric effector of hemoglobin that binds to hemoglobin subunits producing conformational changes and functions to allow the transfer of oxygen from Hb-$O_2$ to tissues.

The pathways for the metabolism of glucose to pyruvate and lactate (glycolysis) and the reverse process (different enzymes are involved at irreversible steps) to form glucose (gluconeogenesis) are described with enzymes that are common to each path. In the liver, gluconeogenesis is stimulated when circulating glucose levels are low. Glucose, so formed, is transported from the liver cell through the GLUT2 transporter to the extracellular space and eventually to the bloodstream. In glycolysis, the pyruvate formed, when the cell needs energy, is metabolized through the TCA cycle with the possibility of generating as many as 36 ATP molecules plus $CO_2$ and water (by combination of protons with oxygen). Both glycolysis and gluconeogenesis are regulated by small molecules (metabolites, nucleotides, and others).

For a small molecule that regulates glycolysis either positively or negatively, that same molecule may regulate gluconeogenesis in the opposite direction. The alanine cycle is a means to convert amino acids secreted from muscle, during muscle breakdown, to glucose in the liver. When fat is being degraded, the resulting glycerol can be converted to glucose.

Tumor cells and normal proliferating cells rely mainly on glycolysis for the production of ATP, a system that is far less efficient in ATP production than reliance on mitochondrial oxidation. These cells may require the use of glucose for other demands in addition to energy, such as for cell division and the development of biomass. The recognition that aerobic glycolysis is the main metabolic pathway for ATP production, as limiting as it is, while the mitochondria (and the TCA cycle) are perfectly functional, is known as the "Warburg effect."

Glucose is transported from the blood into the **muscle cell** by way of the GLUT4 transporter. If it is needed for energy, glucose flows through the glycolytic pathway to form pyruvate that is further metabolized in the TCA cycle. If there is no immediate need for energy, glucose is metabolized to UDP-glucose and forms glycogen.

The GLUT transporters of hexoses have either high or low affinity for glucose. The low affinity forms, such as GLUT2 in the liver, are able to handle large amounts of glucose, a situation that occurs normally after a meal.

## SUGGESTED READING

### Literature

Demetrius, L.A., Coy, J.F., Tuszymski, J.A., 2010. Cancer proliferation and therapy and the Warburg effect and quantum metabolism. Theor. Biol. Med. Model. 7, 2.

Eunsook, S.J., Sherry, A.D., Malloy, C.R., 2004. Interaction between the pentose phosphate pathway and gluconeogenesis from glycerol in the liver. J. Biol. Chem. 289, 32593−32603.

Fang, R., et al., 2012. MicroRNA-143 (miR-143) regulates cancer glycolysis via targeting hexokinase 2 gene. J. Biol. Chem. 287, 23227−23235.

Jelen, S., et al., 2011. Aquaporin-9 protein is the primary route of hepatocyte glycerol uptake for glycerol gluconeogenesis in mice. J. Biol. Chem. 286, 44319−44325.

Holyoak, T., Sullivan, S.M., Nowak, T., 2006. Structural insights into the mechanism of PEPCK catalysis. Biochemistry. 45, 8254−8263.

Horecker, B.L., 2002. The pentose phosphate pathway. J. Biol. Chem. 277, 47965−47971.

Kim, D.-K., et al., 2012. Orphan nuclear receptor estrogen-related receptor $\gamma$ (ERR$\gamma$) is key regulator of hepatic gluconeogenesis. J. Biol. Chem. 287, 21628−21639.

Kwan, B.-K., Emr, S.D., 2013. The phosphatidylinositol 3,5-bisphosphate (PI(3,5)P2)-dependent Tup1 conversion (PIPTC) regulates metabolic re-programming from glycolysis to gluconeogenesis. J. Biol. Chem. 288, 20633−20645.

Landis, J., Shaw, L.M., 2014. Insulin receptor substrate 2-mediated phosphatidylinositol 3-kinase signaling selectively inhibits glycogen synthase kinase 3$\beta$ to regulate aerobic glycolysis. J. Biol. Chem. 289, 18603−18613.

Lenzen, L., 2014. A fresh view of glycolysis and glucokinase regulation: history and current status. J. Biol. Chem. 289, 12189−12194.

Lin, R., et al., 2012. Profound hypoglycemia in starved ghrelin-deficient mice is caused by decreased gluconeogenesis and reversed by lactate or fatty acids. J. Biol. Chem. 287, 17942−17950.

Pankrantz, S.L., et al., 2009. Insulin receptor substrate-2 regulates aerobic glycolysis in mouse mammary tumor cells via glucose transporter 1. J. Biol. Chem. 284, 2031−2037.

Patra, K.C., Hay, N., 2014. The pentose phosphate pathway and cancer. Trends Biochem. Sci. 39, 347−354.

Shi, D.-Y., et al., 2009. The role of cellular oxidative stress in regulating glycolysis energy metabolism in hepatoma cells. Mol. Cancer. 8, 32.

Shistov, A.A., et al., 2014. Quantitative determinants of aerobic glycolysis identify flux through the enzyme GAPDH [glyceraldehyde-3-phosphate dehydrogenase] as a limiting step. eLIFE. 3, e03342.

Silva, J.E., 2006. Thermogenic mechanisms and their hormonal regulation. Physiol. Rev. 86, 435−464.

Stark, R., et al., 2014. A role for mitochondrial phosphoenolpyruvate carboxykinase (PEPCK-M) in the regulation of hepatic gluconeogenesis. J. Biol. Chem. 289, 7257−7263.

Vander Heiden, M.G., Cantley, L.C., Thompson, C.B., 2009. Understanding the Warburg effect: the metabolic requirements of cell proliferation. Science. 324, 1029−1033.

Warburg, O., 1956. On the origin of cancer cells. Science. 123, 309−314.

Warburg, O., Wind, F., Negelein, E., 1927. The metabolism of tumors in the body. J. Gen. Physiol. 8, 519−530.

Zhou, Y., et al., 2011. Metabolic alterations in highly tumorigenic glioblastoma cells: preference for hypoxia and high dependency on glycolysis. J. Biol. Chem. 286, 32843−32853.

### Books

Hanson, R.W., Mehlman, M.A., 1976. Gluconeogenesis: Its Regulation in Mammalian Species. John Wiley & Sons, New York, NY.

Jacobs, J., 2012. Metabolism Basics: A Walkthrough Guide to Fermentation and Cellular Respiration. Kindle.

Lithaw, P.N., 2009. Glycolysis. Nova Biomedical Books, New York, NY.

McKenzie, S.B., Williams, J.L., 2010. Clinical Laboratory Hematology. Pearson Education, Inc, Upper Saddle River.

Chapter 9

# Lipids

## HYPERCHOLESTEROLEMIA

**Cholesterol**, a critical structural element, comprises about 25% of the cell membrane and occurs in an equal proportion to phospholipid constituents (Fig. 2.7). It provides some rigidity to an otherwise completely flexible membranous structure. Its presence in the membrane facilitates the diffusion of nonpolar molecules. Cholesterol is essential to life. However, disease situations occur when blood cholesterol cannot be handled properly, that is, when the transporters of cholesterol, the **low-density lipoproteins** (**LDL**s), cannot effectively move cholesterol into the peripheral tissues. Movement of cholesterol from LDLs into peripheral tissues occurs through the binding of LDL to the **LDL receptor site** on the tissue cell membrane. A model of the LDL receptor structure is shown in Fig. 9.1.

The **ligand-binding domain** of the **LDL receptor** recognizes a lipoprotein on the surface of the LDL. A drawing of an LDL particle encircled by the **ApoB-100 protein** is shown in Fig. 9.2A and B. The figure shows the positively charged region of the molecule that constitutes the receptor-binding domain.

There is a **negative cholesterol feedback** mechanism which, when cholesterol is high in liver cell membranes, on the one hand, leads to the reduction of the number of LDL receptors in cell membranes and, on the other, leads to the repression of the synthesis of cholesterol in the liver. This results from the failure of a fragment of the **SREBP (sterol regulatory element binding protein)** to enter the nucleus, activate the **LDLR gene** and induce the enzymes of cholesterol synthesis. This system will be discussed in connection with the biosynthesis of cholesterol.

**Familial hypercholesterolemia** (**FH**) is the inherited form of hypercholesterolemia and is semidominant. There are two alleles for the **LDL receptor (LDLR)** in FH. In the heterozygous condition (1 in 500 individuals), there is one defective gene and one normal gene. This condition would result in a lower number of LDLRs on the membranes of tissue cells resulting in higher levels of circulating cholesterol. Table 9.1 lists the levels of **circulating cholesterol** and **LDL cholesterol** that are either normal or characteristic of disease.

The worst case is the homozygous disease that can occur when both parents are heterozygous and the offspring is homozygous (two defective genes). In this case (1 in 1 million individuals) virtually no (or few) LDLRs are produced in the membranes of tissue cells. These patients rarely live beyond 20 years and can experience heart attacks as early as age 2 years. There are many sites on the LDLR gene that can be mutated (Fig. 9.3) leading to the possibility for many forms of the disease with varying degrees of severity. There can be more than 300 gene defects in the **LDLR gene** and an allele can carry more than one mutation. Not every mutation is pathogenic. In FH, patients can have cholesterol levels that are six times the normal.

These elevated levels can lead to the formation of **plaques** in arterial walls that, in turn, can lead to myocardial infarcts.

In terms of function of the **LDL receptor (LDLR)**, there are five classes of mutations (Fig. 9.3): class 1: mutation in this region affects the synthesis of the LDL receptor in the endoplasmic reticulum; class 2: prevention of proper transport of the LDLR to the **Golgi apparatus** for modification of the receptor [truncation, missing domains of the **EGF precursor domain** (Fig. 9.1); domains 3, 4, and 5]; class 3: mutation in repeat 6 in the ligand-binding domain (N-terminal in the extracellular fluid) is deleted; class 4: inhibition of internalization of the **LDLR complex** (residue C, 807). This domain is responsible for recruitment of **clathrin** and other proteins involved in endocytosis of LDL and therefore the mutation lowers LDL internalization into the tissue cells; class 5: this mutation causes improper receptor recycling. A **deletion mutation** indicates that part of the DNA or chromosome is missing; a **missense mutation** is a change of a single nucleotide; an **insertion mutation** is the addition of one or more nucleotide base pairs into a DNA sequence and a **nonsense mutation** is a change in a base to a stop codon so that the product translated from a mRNA is shorter than the normal product. Other genetic factors in hypercholesterolemia include a deficiency in **lecithin-**

Human Biochemistry. DOI: http://dx.doi.org/10.1016/B978-0-12-383864-3.00009-0

The human LDL receptor

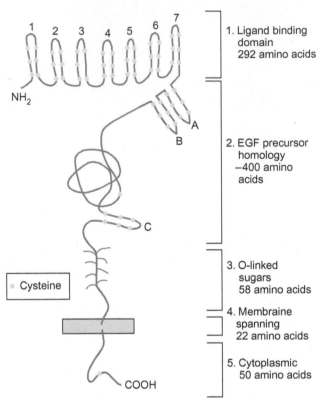

1. Ligand binding
   domain
   292 amino acids

2. EGF precursor
   homology
   ~400 amino
   acids

3. O-linked
   sugars
   58 amino acids

4. Membraine
   spanning
   22 amino acids

5. Cytoplasmic
   50 amino acids

Cysteine

NH₂

COOH

**FIGURE 9.1**   A model of the LDL receptor showing the N-terminus at the *top* and the C-terminus inside the cell at the *bottom*. The structural and functional domains are labeled in the figure. *Redrawn from http://www.web.archive.org/web/20040702012607/http://www.hhmi.org/biointeractive/museum/exhibit98/images/ldl.gif.*

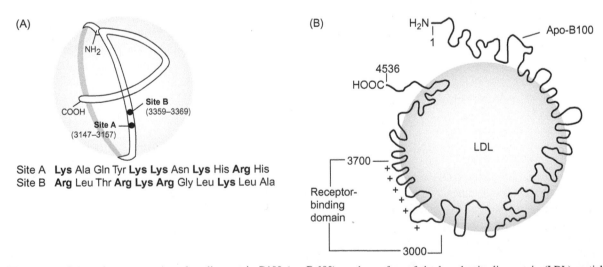

(A)

NH₂

COOH

Site B
(3359–3369)

Site A
(3147–3157)

Site A   **Lys** Ala Gln Tyr **Lys Lys** Asn **Lys** His **Arg** His
Site B   **Arg** Leu Thr **Arg Lys Arg** Gly Leu **Lys** Leu Ala

(B)

H₂N
1

Apo-B100

4536
HOOC

LDL

3700

Receptor-
binding
domain

3000

**FIGURE 9.2**   (A) Schematic representation of apolipoprotein B100 (**apoB-100**) on the surface of the low-density lipoprotein (LDL) particle. The receptor-binding domain is composed of a cluster of positively charged lysine (site A) and arginine (site B) residues that can interact with the acidic residues (glutamate and aspartate) in the ligand-binding domain of the LDL receptor. (B) This schematic shows the ApoB-100 protein. *(A) This image is reproduced from http://openi.nlm.nih.gov/imgs/512/12/22911326/2291326_vhrm0304-491.png.*

**TABLE 9.1** Levels of Circulating Cholesterol and Circulating LDL Cholesterol Characteristic of Normal or Diseased Individuals

| Circulating Cholesterol Levels (mg/dL) | Diagnosis |
| --- | --- |
| <200 | Normal |
| 200–239 | Borderline high |
| >240 | High |
| **Circulating LDL cholesterol** | |
| <100 | Optimal |
| 101–129 | Above optimal |
| 130–159 | Borderline high |
| 160–189 | High |
| >190 | Very high |

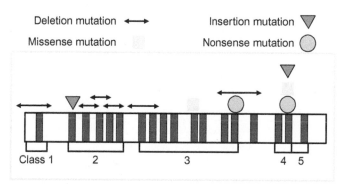

**FIGURE 9.3**   Possible mutations in the LDL receptor gene. Vertical blue lines represent the 18 exons of the gene and various combinations of known mutations are shown (representing 12 mutations). The mutations fall into functional classes; there are 5 and they are shown in the brackets at the bottom of the figure. *Reproduced from http://www.sloppynoodle.com/integen.html that was adapted from JL Goldstein and MS Brown, "Familial Hypercholesterolemia" in CR Scriver, AL Beaudet, WS Sly, D Valle, Editors,* The Metabolic Basis of Inherited Disease, *sixth ed. (New York, McGraw Hill, 1989) page 1232.*

cholesterol acyltransferase (**LCAT**), apolipoprotein that has been altered in some way so as to decrease its ability to bind to the LDL receptor as well as the point mutations in the LDLR mentioned above. As a consequence of mutations of the LDLR, LDLs in the blood increase and they are not taken up in the liver (or other tissues) as shown in Fig. 9.4.

The **LDL receptor gene** is located on human chromosome 19 and the inheritance is autosomal dominant. If one parent has FH and the other does not, half of the offspring will be affected. Death from heart failure in the 40s is not unusual and death can occur in the 50s, 60s, and 70s depending on the severity of FH. Today, regulation of diet coupled with exercise and the use of new drugs (**statins**) to inhibit cholesterol synthesis in the liver form a treatment for this disease. Also, surgical intervention using the bypass technique is needed in some cases.

## Autosomal Recessive Hypercholesterolemia (ARH) Protein

This protein interfaces with the clathrin coat system (soluble clathrin trimers, to clathrin adapter proteins and to phosphoinositides regulating clathrin bud assembly) and is required for the proper functioning of LDLR internalization into a tissue cell. ARH contains a phosphotyrosine-binding domain at its N-terminus that binds the sequence, FXNPXY (Phe-any amino acid-Asn-Pro-any amino acid-Tyr) in the cytoplasmic tail of LDLR. Mutations in the phosphotyrosine-binding domain of ARH show defective internalization of LDLR and are recessive hypercholesterolemics. There are also mutations in the FXNPXY motif that abolishes binding activity. Patients that have ARH mutations exhibit

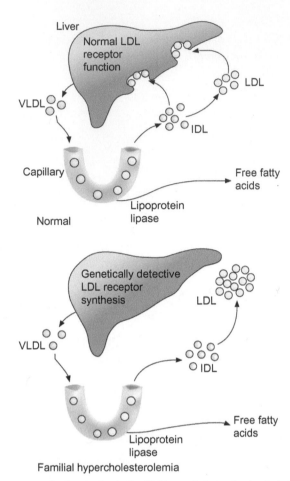

**FIGURE 9.4**    Effects of mutated LDL receptor on its functioning in familial hypercholesterolemia. As LDLs are unable to bind to the LDL receptor (or the receptor is otherwise nonfunctional) in the liver cell membrane, LDLs remain in the bloodstream and become highly elevated. *IDL*, intermediary density lipoprotein; *VLDL*, very low-density lipoprotein. *Reproduced from G Litwack,* Human Biochemistry and Disease, *Figure 5-3, page 192, Academic Press, 2008.*

defective adapter function in hepatocytes generating defective LDLR trafficking and hypercholesterolemia. Thus, mutations in the ARH protein, an adapter protein, cause defects in the binding of LDLR to the endocytic machinery.

## ApoB-100 Protein Mutations

ApoB-100 is embedded in the phospholipid outer layer of LDL particles (Fig. 9.2) and is recognized by the LDLR. LDLR also recognizes the **apoE protein**, as a ligand, in **chylomicron remnants** and in **VLDL**s. The normal LDLR binding to apoB-100 involves an interaction between arginine 3500 (R3500) and tryptophan 4369 (W4369) spanning the region between R3500 to W4369 as shown in Fig. 9.5.

LDLR binds to the apoB-100 protein (ApoB-100) in the phospholipid outer layer of LDL particles. There are two forms of apoB protein, a 48-kDa form synthesized exclusively in the intestinal tract and the 100-kDa form made in the liver. Both ApoB-48 and ApoB-100 are encoded within a single gene that transcribes a long mRNA and both proteins have an identical N-terminal sequence. ApoB-100 is a component of very low-density lipoproteins (VLDLs), intermediate density lipoproteins (IDLs), and LDLs all of which are transporters of fats and cholesterol in the blood. ApoB-100 facilitates the attachment of these particles to specific liver cell membrane receptors. These receptors transport LDLs into the cell where they are degraded to release free cholesterol that is either stored in a lipid droplet or removed. There are more than 90 mutations possible in the APOB gene and these mutations cause **hypobetalipoproteinemia** that impedes the bodily absorption and transport of fat. This inherited disease is **familial hypobetalipoproteinemia (FHBL)**. Many of the mutations of the APOB gene that cause FHBL generate an abnormally shortened form of the ApoB protein. **Inherited hypercholesterolemia** can result from any of about 5 **point mutations** (changes in an

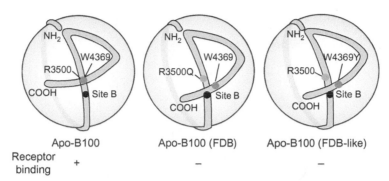

| | Apo-B100 | Apo-B100 (FDB) | Apo-B100 (FDB-like) |
| --- | --- | --- | --- |
| Receptor binding | + | − | − |

**FIGURE 9.5**   ApoB-100 protein of LDL binding to LDLR. Normal binding of the ligand, apoB-100, to the LDLR requires an interaction between arginine 3500 (R3500) and tryptophan 4369 (W4369) as shown in the *left figure*. Mutation of R3500 (the FDB mutation) or the W4369 (FDB-like mutation) disrupts binding to the receptor. The interaction (R3500−W4369) is needed for the correct folding of the carboxyl terminus of apoB-100. When R3500 in mutated to glutamine (R3500Q) there is diminished binding to the LDLR (*middle figure*). Also, when tryptophan 4369 is mutated to tyrosine (W4369Y), there is diminished binding of the LDL to the LDLR (*right figure*). FDB, familial defective apolipoprotein B-100. *Reproduced in part from J. Boren et al., J. Biol. Chem., 276: 9214−9218, 2001.*

individual amino acid in the resulting protein) in the APOB gene and it is called **familial defective apolipoprotein B-100 (FDB)**. This produces severely elevated blood cholesterol and greatly increases the risk of heart disease. In FDB, LDLs are unable to bind effectively to cell membrane receptors, resulting in the excess of circulating cholesterol that is abnormally deposited in the skin, the walls of arteries (especially those supplying blood to the heart) and tendons.

## Normal LDL Function at Tissue Cells

LDLs transfer their cholesterol content as **cholesteryl esters** (**CE**s) to the interior of tissue cells. First, the LDL binds to the **LDL receptors** on the surface of the tissue cell. This is accomplished by the recognition of the LDL through the ligand-binding domain of the apoprotein on the surface of the LDL particle (Fig. 9.2). The process of **endocytosis** begins when the LDL is encapsulated by the invaginated membrane covered on the interior with **clathrin** proteins and other proteins involved in the inward transport. This forms an **endosome** inside the cell and the LDL dissociates from the receptor in the endosome caused by its acidic environment. Eventually the LDL is moved into the **lysosome** where it is degraded and releases cholesterol or CE. The residual LDL receptor is returned to the cell surface by the **recycling vesicle** in preparation for the next event of LDL binding. The forward and return passage of the endosome between the cell membrane at the beginning and its return takes about 10 minutes and it lasts for about 20 hours before it is degraded. The CE in the lysosome can be released as unesterified cholesterol (eventually located mainly in the membrane) by the action of **cholesteryl ester hydrolase** after it has become activated (phosphorylation by PKA). The unesterified free cholesterol in membranes in part can be returned as the CE to the liver by way of **HDL** (Fig. 9.27) (with a protein on its surface) and **acylCoA cholesterol acyltransferase** (**ACAT**) attached in the **reverse cholesterol transport** system that will be shown later. Cholesterol synthesis in the liver will contribute to the unesterified cholesterol pool. As the cholesterol pool in the cell increases there will be a negative feedback on cholesterol synthesis and a positive effect on ACAT that will catalyze the conversion of the free form of cholesterol to the CE. In specific types of cells, CE can be stored in a **fat droplet** (Fig. 9.40A and B) and CE can be made available as free cholesterol to the mitochondria in cells that synthesize steroid hormones, for example, by the action of cholesteryl ester hydrolase. The normal functioning of LDLs is capitulated in Fig. 9.6.

The process of moving cholesterol from the bloodstream into the tissues reduces the circulating LDL cholesterol.

## BIOSYNTHESIS OF CHOLESTEROL

Although most tissue cells have the capacity to synthesize cholesterol, the main organs that do so are the liver, the adrenal cortex and the ovaries and testes. The latter three involve cholesterol as the starting molecule for the synthesis of **adrenal cortical hormones** and the **sex hormones**. *The 4-ring system of cholesterol cannot be broken down in the body* but the side chains, double bonds and substituted groups all can be removed or altered. About 75% of the body's cholesterol is formed in the liver. The diet is also important and supplies the rest, although absorption into the bloodstream involves about 25% of ingested cholesterol, mostly in animal fats and sterols from plant sources. The cholesterol

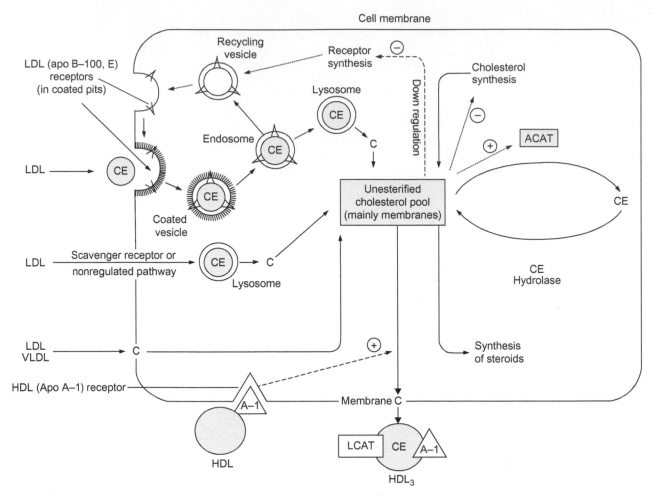

**FIGURE 9.6** The normal operation of LDL on tissue cells to release its cholesteryl ester content to the interior of the cell. Uptake of LDLs from the bloodstream makes cholesterol available to the interior of a tissue cell, such as the liver cell. *ACAT*, acylCoA cholesterol acyltransferase; *apoA-1-receptor*, HDL receptor; *C*, free cholesterol; *CE*, cholesteryl ester; *HDL*, high-density lipoprotein; *LCAT*, lecithin-cholesterol acyltransferase; *LDL*, low-density lipoprotein; *VLDL*, very low-density lipoprotein. *Reproduced from G Litwack, Human Biochemistry and Disease, Figure 5-2, page 191, Academic Press, 2008.*

molecule is derived from the **acetate** of acetyl-CoA. The synthesis can be visualized from the point of view of the number of carbons involved in the intermediates as shown in Fig. 9.7.

Three molecules of acetyl-CoA (three times C2 for acetyl) are converted to **3-hydroxy-3-methylglutarylCoA (HMGCoA)** in a few steps and HMGCoA is then converted to **mevalonate** (C6 compound) in the endoplasmic reticulum. *This step is catalyzed by hydroxyl-methylglutarylCoA reductase (HMGCoAR), the rate-limiting enzyme in the overall pathway for the synthesis of cholesterol.* As indicated, this enzymatic activity is stimulated by **insulin** (stimulates phosphatases, e.g., Fig. 6.30) and **thyroxine** (increases **HMGCoA reductase mRNA** production) and inhibited by the action of **glucagon** (inhibits protein phosphatases and elevates cAMP leading to activation of PKA and the phosphorylation of **inhibitor PPI-1** to its activated form which then inhibits the dephosphorylation of HMGCoAR-P to its active form). Mevalonate is converted to **isopentenyl diphosphate** (C5), the "active isoprene". Six isopentenyl diphosphates are polymerized to form **squalene** (C30). Squalene is converted to **lanosterol** (C30) in two steps. Finally, the 2-methyl groups at C4 of the A ring are removed in multiple steps and cholesterol (C27) is formed. Note the numbering of the carbons and the labeling of the 4-ring system as A, B, C, and D in the cholesterol molecule. The upper limit of the normal level of plasma cholesterol is 200 mg/dL. Values of 240 mg/dL or above are considered to be elevated. When the circulating cholesterol level is above 300 mg/dL the risk of a fatal heart attack is five times that of a person with a level of 200 mg/dL or below.

HMGCoAR catalyzes the rate-limiting step and is the regulated enzyme in this pathway. The enzyme is inactive in the phosphorylated form and active in the nonphosphorylated form. Phosphorylation of the enzyme to the less active form is carried out by **AMP kinase (AMPK)**. In turn, AMPK, itself, is active in a phosphorylated form and less active

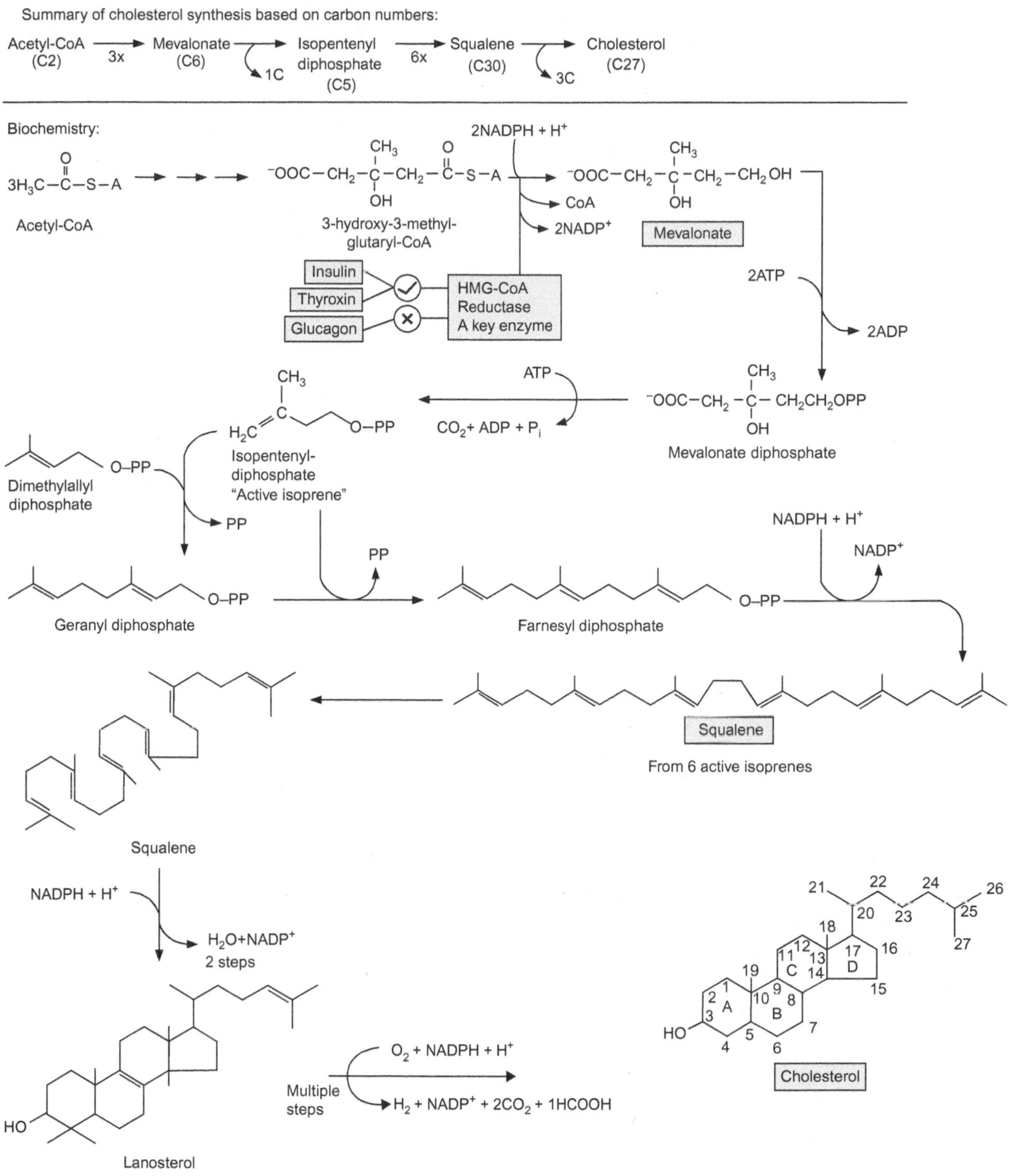

Summary of cholesterol synthesis based on carbon numbers:

FIGURE 9.7  Cholesterol synthesis based upon carbon numbers of the intermediates and the biochemistry of cholesterol synthesis. The numbering system of carbons and the letters of the ring system in cholesterol are shown in the resulting molecule.

in the nonphosphorylated form. Phosphorylation is catalyzed by either of two kinases and dephosphorylation occurs through the action of **protein phosphatase 2C**. The dephosphorylation of HMGCoAR-P occurs through the action of **HMGCoAR phosphatase** that, in turn, is controlled by a protein phosphatase inhibitor (PPI-1). The inhibitor is active in its phosphorylated form. These elaborate reactions involving phosphorylation—dephosphorylation to control the activity of HMGCoAR are outlined in Fig. 9.8. This is another facet in the regulation of the synthesis of cholesterol in the liver and, in the overall sense, may be less determinative for circulating cholesterol levels than the mutations in the LDLR gene.

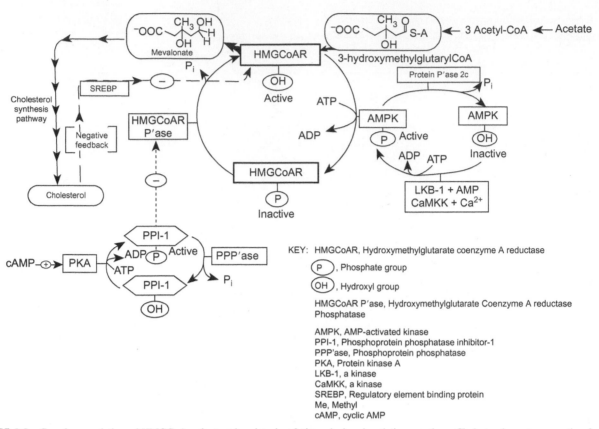

**FIGURE 9.8** Complex regulation of HMGCoA reductase by phosphorylation–dephosphorylation reactions. Cholesterol exerts a negative feedback reaction on its synthesis through a pathway that diminishes the activity of HMGCoAR and involves the **SREBP (steroid regulatory element binding protein)** transcription factor.

## INHIBITION OF LIVER HMGCoA Reductase by Drugs (Statins)

This enzyme, catalyzing the rate-limiting reaction of **cholesterol synthesis** in the liver, is a prime chemotherapeutic target. A group of drugs called **statins** mimic the structure of **3-hydroxy-3-methyl glutarylCoA (HMGCoA)**, the substrate of the enzyme. These drugs are **competitive inhibitors** (Figs. 5.5 and 5.7) of HMGCoAR. Some of the structures of the statin drugs are shown in Fig. 9.9.

The main feature of a competitive inhibitor is that it should resemble the natural substrate enough to compete with it for binding to the active site but not so alike that it will form a product in the reaction. To ascertain that this is the case with a statin, Fig. 9.10 shows the comparison of the structures of lovastatin to that of the natural substrate **HMGCoA** in the active site of **HMGCoAR**. As a result of the action of statin inhibition, the synthesis of cholesterol in the liver is reduced (to as low as 50%). When this occurs, as elaborated by the work of Michael S. Brown and Joseph L. Goldstein, the transcription factor **SREBP (steroid regulatory element binding protein)** is transported to the Golgi apparatus for proteolytic processing that yields an active fragment **bHLH** that enters the nucleus and activates the **LDLR gene** and the genes for enzymes of cholesterol synthesis. This results in the production of more LDLRs that clear the blood of LDL cholesterol and replaces the deficit in cellular cholesterol. The action of statins counteracts the situation occurring when high fat diets are ingested and the influx of cholesterol causes elevated cholesterol levels in liver cell membranes. When an excess of cholesterol develops in cell membranes **SREBP** remains in the endoplasmic reticulum and it is not processed to the fragment that activates the **LDLR gene**. Consequently, the production of LDLR is reduced and **LDL** accumulates in the bloodstream. The mechanism, elaborated by Brown and Goldstein for the action of SREBP is shown in Fig. 9.11.

## THE ARH PROTEIN

The **autosomal recessive hypercholesterolemia (ARH) protein** is involved with the **clathrin coat** apparatus. It is required for the activity of the LDLR in the endocytic process and directly binds soluble **clathrin trimers** and to

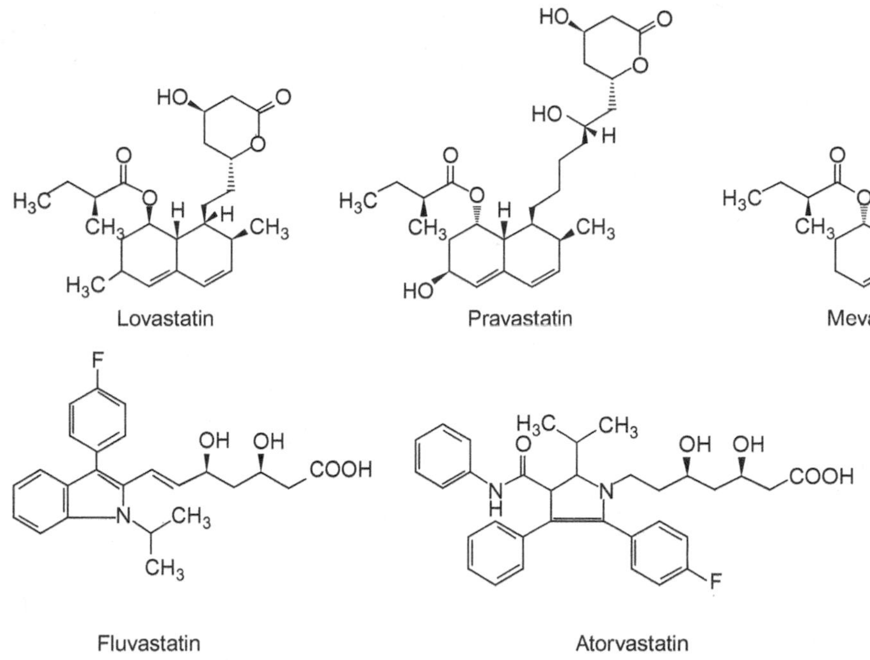

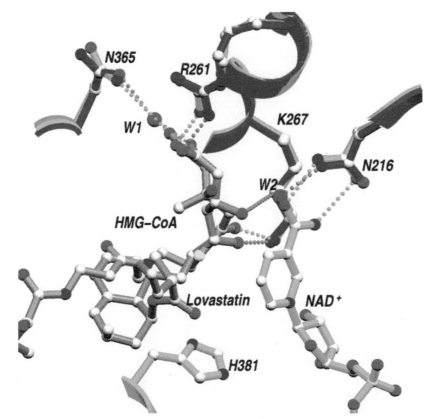

Lovastatin

Pravastatin

Mevastatin

Fluvastatin

Atorvastatin

**FIGURE 9.9**   Some statin competitive inhibitors of HMGCoA reductase used to treat high levels of circulating cholesterol.

N365

R261

W1

K267

HMG–CoA

W2

N216

Lovastatin

NAD⁺

H381

**FIGURE 9.10**   Comparison of the binding of lovastatin to the natural substrate HMGCoA in the active site of HMGCoAR. The amino acids (single letter) in the active site are numbered. *N*, asparagine; *R*, arginine; *K*, lysine; *H*, histidine; *W1* and *W2* are domains of the enzyme; *NAD⁺*, nicotinamide adenine dinucleotide, a coenzyme. The natural substrate, HMGCoA is shown in the *blue* stick model, whereas lovastatin is shown in *orange*. *Reproduced from L. Taberbero, V.W. Rodwell and C.V. Stauffacher,* J. Biol. Chem., **278:** *19933–19938, 2003 and from http://www.jbc.org/content/278/22/19933/F3.large.jpg.*

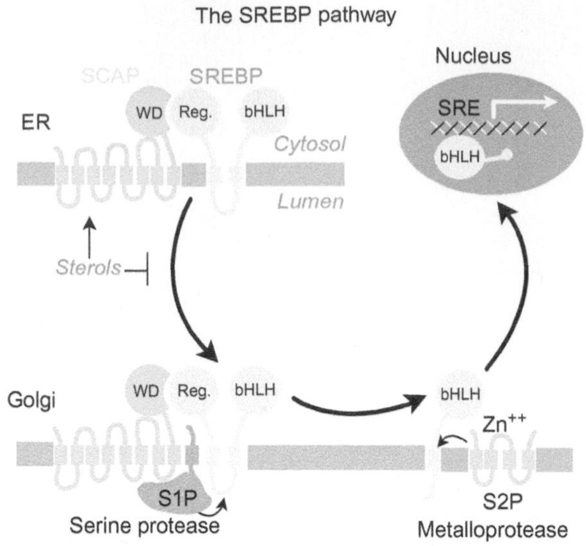

The SREBP pathway

**FIGURE 9.11**    The SREBP (steroid regulatory element binding protein) pathway as elaborated by Brown and Goldstein. SREBP consists of two fragments, **Reg.**, bHLH and **SCAP**, a **sterol-regulated escort protein** that carries SREBP to the **Golgi** apparatus. A protein called **INSIG** (not shown) anchors SCAP and SREBP in the ER when intracellular cholesterol levels are high so that the proteolytic product of SREBP cannot enter the nucleus and activate genes for the LDLR and the enzymes of cholesterol synthesis. This would represent the negative feedback effect of high cellular cholesterol. When cholesterol levels in the liver are very low, SREBP is transported to the Golgi where a serine protease, **S1P**, cleaves the bHLH fragment from the Reg. fragment and bHLH is further cleaved by a metalloprotease, **S2P**, to release it from the Golgi so that it can enter the nucleus and activate the LDLR gene and the genes encoding the enzymes of cholesterol synthesis. In this way, the cellular deficiency of cholesterol is repaired. *Reproduced from http://www.8.utsouthwestern.edu/utsw/cda/dept14857/files/114532.html.*

**clathrin adapter proteins**. The N-terminal region of ARH contains a **phosphotyrosine domain**. This domain interacts with the **internalization sequence**, **NPVY**, in the cytoplasmic tail of **LDLR**. Mutations in the internalization sequence motif of LDLR abolish its binding to ARH. The sequence in ARH that binds to clathrin is **LLDLE**. ARH also binds to phosphoinositides that regulate **clathrin bud assembly** at the cell surface. In the C-terminal domain of ARH, there is a highly conserved sequence of 20 amino acids that binds to the β-2-adaptin subunit of the adapter protein, **AP-2**. Patients with **ARH mutations** have hepatocytes with a defective adapter function leading to aberrant LDLR trafficking in the cell, specifically in the formation of **endosomes**. Mutations in the phosphotyrosine-binding domain cause **recessive hypercholesterolemia**.

## BILE ACIDS

Cholesterol forms **bile acids** in the liver (Fig. 9.12).

They are transported to the intestine where they solubilize ingested fats through **micelle formation** (Fig. 9.22). In the circulation of bile acids, about 10% is lost through the feces. Increasing the loss of bile acids from the body is one measure to reduce cholesterol levels because more liver cholesterol would be diverted to the synthesis of bile acids. The beneficial effects of fiber in the diet, in part, accomplish this in that bile acids combine with **fiber** (they also bind to certain resins, e.g., **cholestyramine**) and are excreted through the intestine. The circulation of bile acids is shown in Fig. 9.13.

**Bile acids** are secreted from the liver into the gallbladder. Following a meal, bile acids enter the intestine through the **bile duct**. When bile acid formation is high, there is a negative feedback on further bile acid formation (elevating the level of cholesterol in the liver). The negative feedback occurs by suppression of **cholesterol 7α-hydroxylase** activity (Fig. 9.12). This may occur by **bile acid repression** of the transcription of human **cholesterol 7α-hydroxylase gene**, *CYP7A1*, resulting from effects on certain transcription factors regulating the expression of this gene. There are two lipid-activated nuclear receptors involved in the regulation of bile acid synthesis in the liver: the **liver X receptor** (**LXR**) and the **farnesoid X receptor** (**FXR**). The LXR binds **oxysterols** to activate it and the FXR binds bile acids. Both receptors form dimers with the **retinoid X receptor** (**RXR**) and these activated receptors bind to gene promoters to generate mRNAs that are translated into proteins that prevent bile salt toxicity (the function of FXR) and the **overproduction of cholesterol** (LXR). When bile acid formation is low, more cholesterol would be diverted to formation of bile acids. The mechanisms involved in the synthesis and removal of cholesterol and its products theoretically should

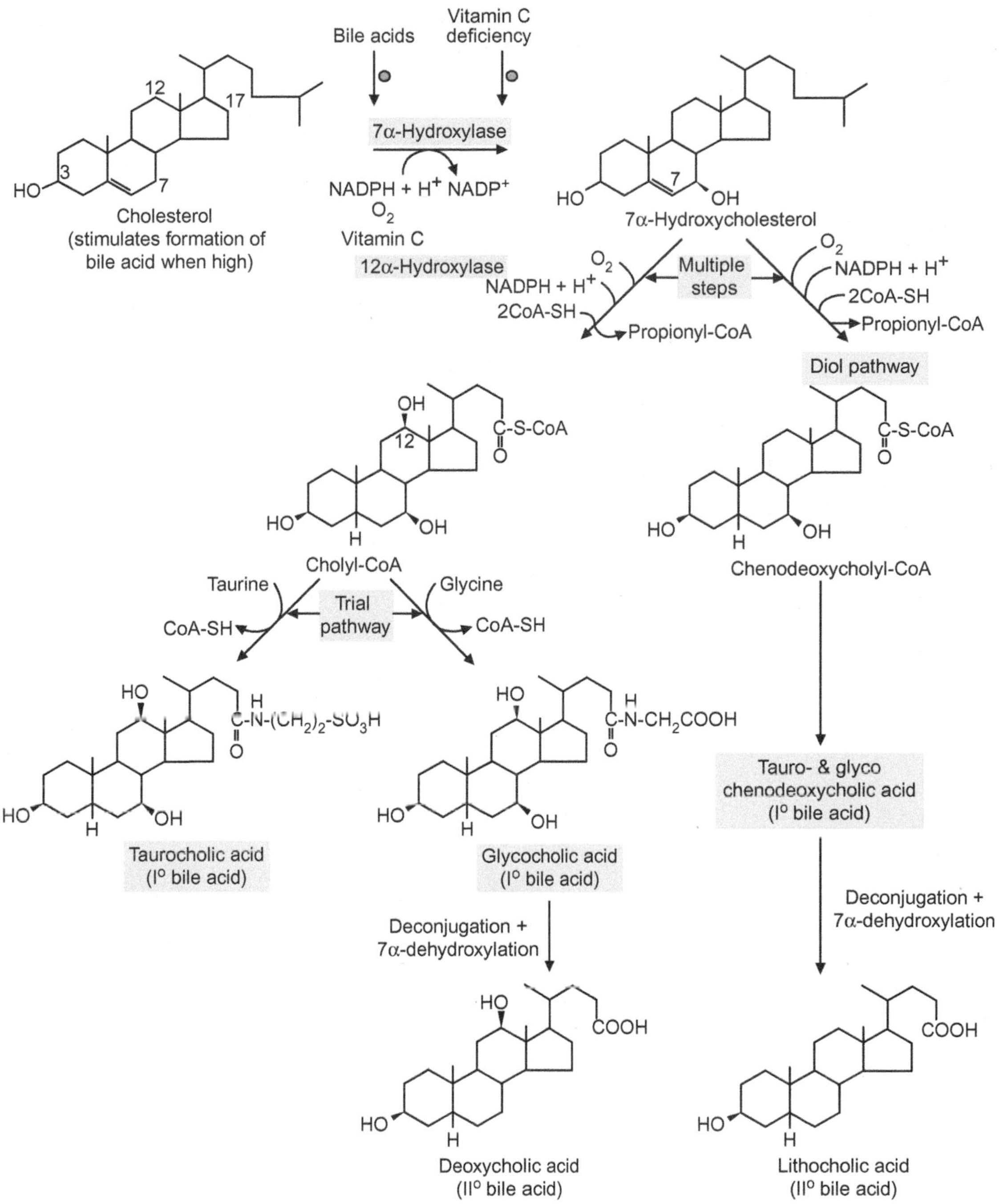

**FIGURE 9.12** Conversion of cholesterol to bile acids in the liver. The three highlighted structures near the bottom are the primary bile acids. Two secondary bile acids are labeled at the bottom.

maintain cholesterol at a homeostatic level. Possibly the maneuvers to cause higher excretion of cholesterol from the large intestine might result in enhanced cholesterol synthesis in the liver, the limiting factor in the synthesis still being HMGCoA reductase.

## FATTY ACIDS AND FAT

Fatty acids are hydrocarbon chains with a carboxyl group at the end of the molecule. The hydrocarbon chain is a saturated fatty acid unless it contains one or more double bonds, in which case, it is an unsaturated fatty acid. Fig. 9.14 shows two fatty acids, **stearic acid** and **oleic acid**. Both contain eighteen carbons but oleic acid has one **double bond** in the 9−10 position. The double bond creates a bend in the chain of about 120°.

Circulation of bile acids

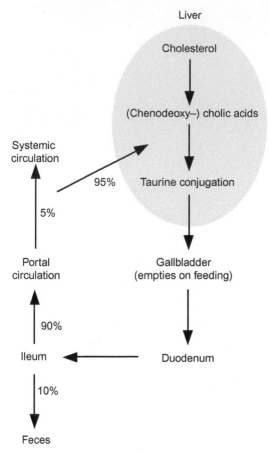

**FIGURE 9.13** Enterohepatic circulation of bile acids. Bile acids are synthesized in the liver and stored in the gallbladder. Upon eating, the gallbladder empties bile acids into the small intestine and then they move into the large intestine. Ten percent are removed through the feces and the remainder is recirculated to the liver although some remain in the bloodstream. *Reproduced from G. Litwack,* Human Biochemistry & Disease, *Academic Press, Figure 5-9, page 197, 2008.*

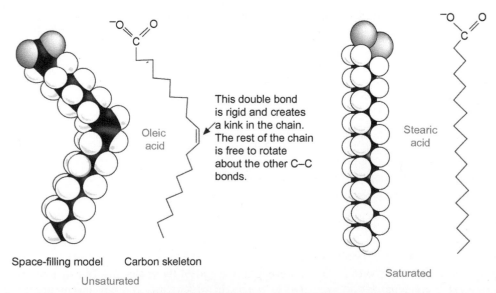

**FIGURE 9.14** Structures of a saturated fatty acid (stearic) on the right and an unsaturated fatty acid (oleic) on the left. Both contain 18 carbons. The double bond creates an angle in the chain of 120°. *Reproduced from http://www.bio.miami.edu/~mallery/255/255chem/p2x4x1a.jpg.*

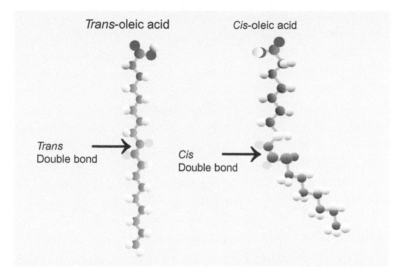

**FIGURE 9.15** Oleic acid structures when the double bond is in *cis* or *trans* configurations. The H atoms in yellow highlight their positions around the double bond. *Reproduced from http://course1.winona.edu/sberg/IMAGES/Oleic AcidCis-Trans.jpg.*

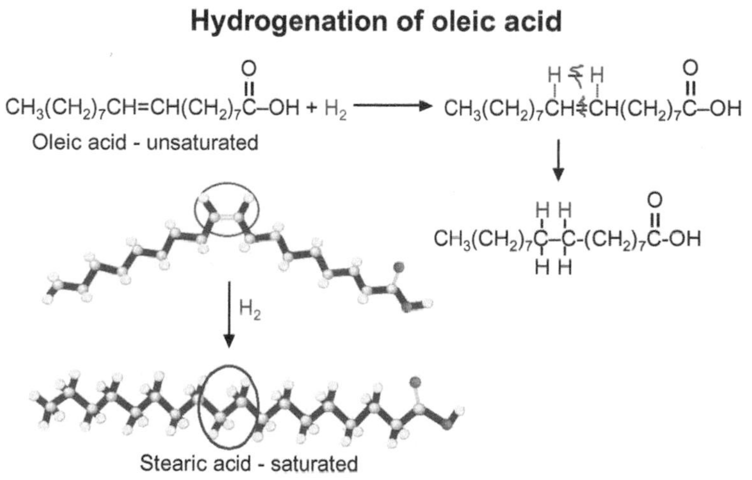

## Hydrogenation of oleic acid

$$CH_3(CH_2)_7CH=CH(CH_2)_7\overset{\overset{O}{\parallel}}{C}-OH + H_2 \longrightarrow CH_3(CH_2)_7\overset{\overset{H}{\vert}}{C}H\overset{\overset{H}{\vert}}{C}H(CH_2)_7\overset{\overset{O}{\parallel}}{C}-OH$$

Oleic acid - unsaturated

$$CH_3(CH_2)_7\overset{\overset{H}{\vert}}{\underset{\underset{H}{\vert}}{C}}-\overset{\overset{H}{\vert}}{\underset{\underset{H}{\vert}}{C}}-(CH_2)_7\overset{\overset{O}{\parallel}}{C}-OH$$

$H_2$

Stearic acid - saturated

**FIGURE 9.16** Hydrogenation of *cis* oleic acid to form stearic acid. *Reproduced from http://www.elmhurst.edu/~chm/vchembook/images/558hydrogenation.gif.*

The bend in the chain around a double bond occurs when the double bond is in *cis*, that is, when the hydrogen atoms on either side of the double bond are on the same side. When the hydrogen atoms on either side of the double bond are in *trans*, that is, on opposite sides of the double bond, there is no bend and the chain appears straight (Fig. 9.15).

Oleic acid can be converted to stearic acid by hydrogenation (saturating the double bond) as shown in Fig. 9.16.

Thus, in **linoleic acid**, which has eighteen carbons and two double bonds, the chain appears straight when both double bonds are in *trans* (bottom of Fig. 9.17), with a single bend when one double bond is in *cis* and the other is in *trans* (middle) or with two bends in the chain when both double bonds are in *cis* (top).

Consequently, the more *cis* double bonds in a hydrocarbon chain, the more bends appear in the structure. If we examine fatty acids with no double bonds (e.g., arachidic, stearic, and palmitic) compared with chains that have one *cis* double bond (e.g., erucic or oleic) compared to a fatty acid having two *cis* double bonds (e.g., linoleic) or a fatty acid having three **cis** double bonds (e.g., linolenic) or a fatty acid having four *cis* double bonds (e.g., arachidonic), we see that there is comparable bending, as shown in Fig. 9.18.

Multiple double bonds in fatty acids often occur at three-carbon intervals ($-C=C-C-C=C-$). The double bond is usually in the *cis* configuration but **partial hydrogenation** can yield double bonds in the *trans* configuration. The latter is harmful in that ingestion of fats that have *trans* fatty acids (the so-called **transfats**) increase LDLs (low density lipoproteins) and decrease HDLs (high density lipoproteins). When fatty acids are left at room temperature, they can become **rancid** by oxidative breakdown into hydrocarbons, aldehydes and ketones. Shorter chain fatty acids, like **formic acid** or **acetic acid** are soluble in water and can produce acidity:

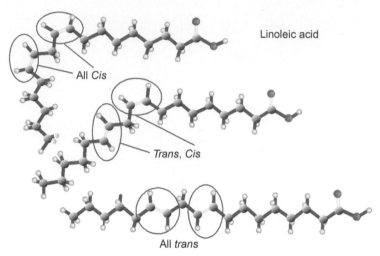

**FIGURE 9.17** Appearance of the hydrocarbon chain or linoleic acid, depending on the configuration of the hydrogen atoms on either side of the two double bonds. *Reproduced from http://www.elmhurst.edu/~chm/vchembook/images/558cistranslino.gif.*

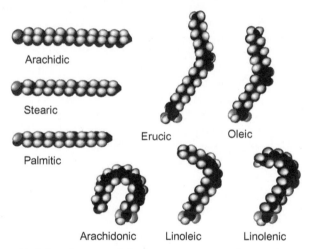

**FIGURE 9.18** Appearance of the fatty acid chains of compounds having none, one, two, three, or four *cis* double bonds. *Reproduced from http://www.chemistrydaily.com/chemistry/upload/1/19/Rasyslami.jpg.*

$$CH_3COOH \rightleftharpoons CH_3COO^- + H^+ \quad \text{(acetic acid; } pK = 4.76\text{)}$$

As the chain length of fatty acids increases, they become insoluble in water and exert little or no effect on the pH of a solution. Chemically, fatty acids behave like other carboxylated compounds; they can undergo esterification and can be reduced to fatty alcohols. When fatty acids are esterified with glycerol they form either fats or oils depending on the types of fatty acids. Substitution with saturated fatty acids produces solid fats. Substitution with unsaturated fatty acids produces oils. Liquid fats are obtained mostly from plants. Unsaturated fatty acids, such as oleic or linoleic acids are **essential fatty acids** and must be obtained in the diet.

Fatty acids are referred to by name and information following the name in parentheses. For example arachidonic acid is referred to as arachidonic acid (20:4) where the number 20 refers to its number of carbon atoms and the number 4 refers to its number of double bonds. Sometimes the double bonds are numbered from the carboxyl (as number 1) and indicated by "$\Delta$" followed by the numbers of the double bonds starting at the carboxyl side of the double bond. Thus arachidonic acid is 20:4 ($\Delta$5,8,11,14). Some of the common fatty acids are shown in Fig. 9.19.

Arachidonic acid is an important fatty acid located in the plasma membrane. It is a precursor of prostaglandins, leukotrienes, lipoxins and anandamide as well as many other important active biological compounds.

## Glycerol and Its Substituents; Triglycerides

Glycerol is a trialcohol whose structure is shown in Fig. 9.20.

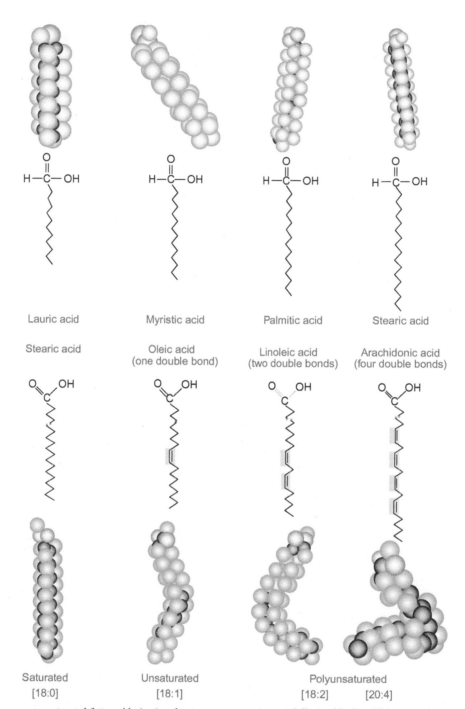

FIGURE 9.19 Some common saturated fatty acids (*top*) and some common unsaturated fatty acids (in addition to saturated stearic acid on the left) at the *bottom*.

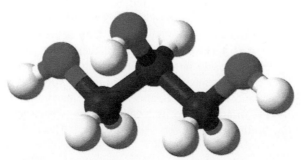

FIGURE 9.20 A ball and stick structure of glycerol.

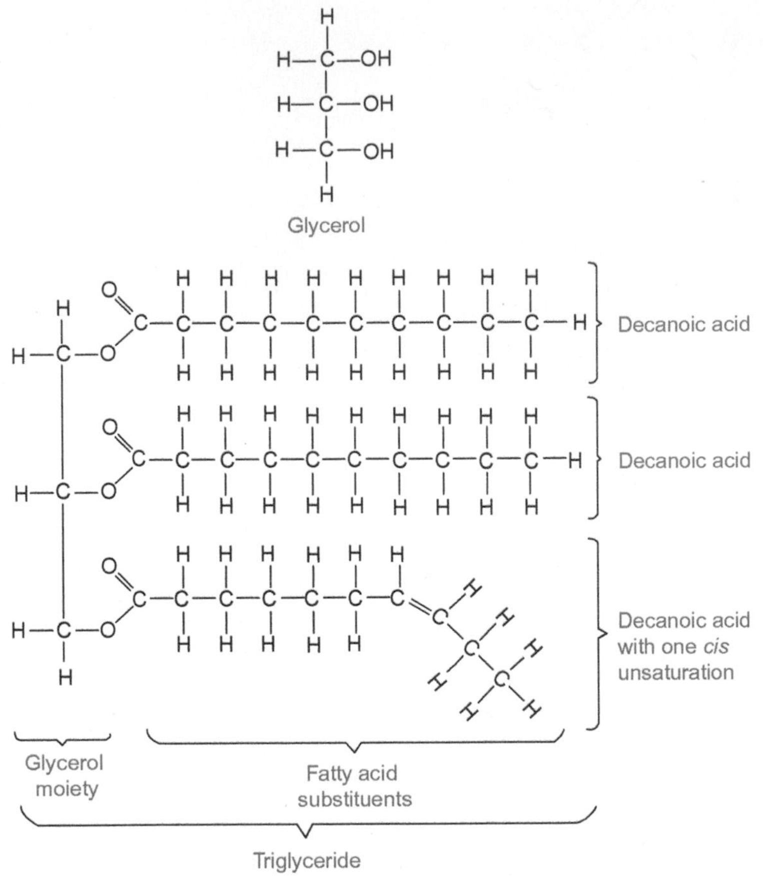

**FIGURE 9.21**    Structures of glycerol (top) and of a triglyceride of decanoic acid. Triglycerides can have different fatty acid components.

Glycerol is 1,2,3-propanetriol, also referred to as **glycerin**. As a 70% solution of glycerol, it freezes at $-37.8°C$ and therefore can act as an **antifreeze compound**, like ethylene glycol. When glycerol is substituted with one fatty acid, it is a **monoglyceride**, with two fatty acids, a **diglyceride** and with three fatty acids, a triglyceride. A triglyceride is shown in Fig. 9.21.

## Fat (Triglyceride) Digestion and Uptake

Ingested lipids are emulsified by **bile acids**. Bile acids are synthesized in the hepatocyte from cholesterol and transported through an ATP requiring transport system to the **biliary canaliculus**. They flow down through the biliary tract and about half of the bile acids reach the **gallbladder**. The bile acids, released from the gallbladder (bile) and the liver, emulsify the fat in the form of **micelles** that are units containing triglycerides (triacylglycerols) in the center surrounded by bile acids. Fats can be broken down by **pancreatic lipase** that gains access to the triglyceride through gaps between the bile salts. The formation of a micelle is shown in Fig. 9.22.

The fats are degraded by **pancreatic lipase** and **phospholipase A₂** (secreted by the pancreas and activated by trypsin) in the intestine to a mixture of monoglycerides and diglycerides. The action of pancreatic lipase is shown in Fig. 9.23. The action of pancreatic lipase releases two free fatty acids and one 2-monacyl-*sn*-glycerol that are rapidly absorbed through the intestinal wall. The transporter for long-chain fatty acids in the apical side of the mature enterocyte is **fatty acid transporter protein 4 (FATP4)**. In the human, there is a family of six homologous transporters (FATP1 through FATP6) and they occur variously in all the fatty acid utilizing tissues in the body. FATP4, the only fatty acid transporter in the small intestine, is located in the apical brush border and is encoded on the human chromosome 9q34. It is a protein of 71,000 Da and contains 641 amino acids. Unsaturated fatty acids are more readily transported than saturated fatty acids. FATP4, in addition to its location in the plasma and internal membranes of the small intestine, is also found in adipocytes, brain, kidney, liver, skin, and heart. The isolated FATP4 also has the enzymatic

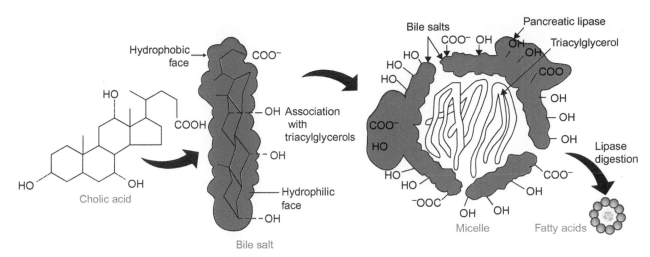

**FIGURE 9.22**  Formation of a micelle, containing triacylglycerol, bile acids, such as cholic acid and pancreatic lipase.

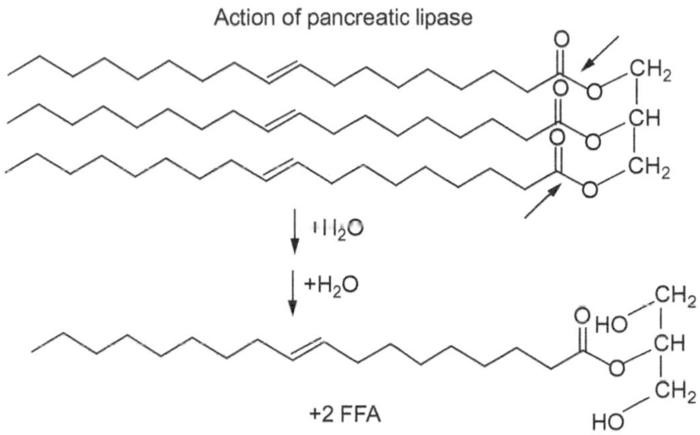

Action of pancreatic lipase

+2 FFA

**FIGURE 9.23**  Action of pancreatic lipase on a triglyceride (triacylglycerol); FFA, free fatty acid.

activity of (long chain) **AcylCoA synthase** and the two activities of transporter and enzyme may work in concert to facilitate the influx of fatty acids across biomembranes. Although the brain contains FATP4, it does not use fatty acids as a source of energy and this requirement, when necessary, can be fulfilled by the action of **glucagon** on **hormone-sensitive lipase** (discussed later) to release free fatty acids and glycerol from stored triglycerides and glycerol can be metabolized to glucose via gluconeogenesis to feed the brain (see Fig. 6.14).

As seen from Fig. 9.23, lipase hydrolyzes at the 1 and 3 positions of glycerol. **Phospholipids** are degraded by **phospholipase A$_2$** at position 2 of glycerol to release free fatty acid and lysophospholipid as shown in Fig. 9.24. Complete digestion of a phospholipid is accomplished by phospholipase A$_1$, phospholipase D, and phospholipase C in addition to phospholipase A$_2$. These phospholipases all originate from the pancreas. The digestion products are absorbed by the intestinal mucosal cells (apical side of mature enterocytes) where resynthesis of triacylglycerides occurs.

## Chylomicrons

In order to become soluble, these resynthesized triacylglycerides are combined with proteins to form soluble **lipoprotein complexes** and these are incorporated into chylomicrons that are comprised by lipid droplets, polar lipids, and a layer of proteins as shown in Fig. 9.25.

In the endoplasmic reticulum, triacylglycerols, phospholipids (to be discussed later), and **apolipoproteins** are combined into chylomicrons and the chylomicrons are moved by exocytosis into the **lymphatic circulation** (Fig. 1.15). This circulation enters the liver and then on to the bloodstream. In the capillaries, the chylomicrons are exposed to

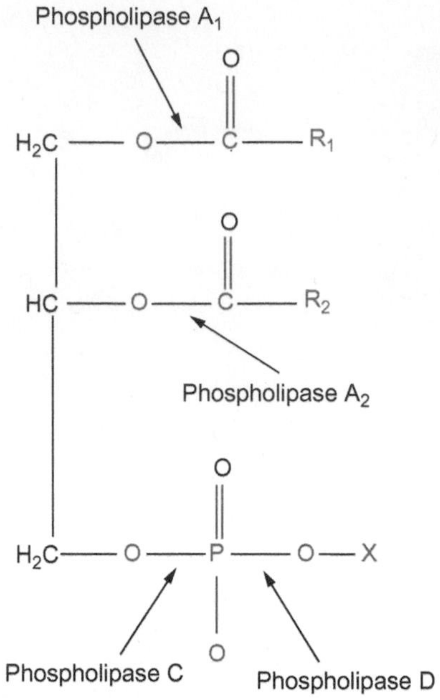

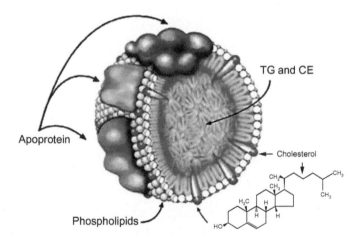

**FIGURE 9.24**   Digestion of a phosphoglyceride by pancreatic phospholipase.

**FIGURE 9.25**   Illustration of a chylomicron. Solubilizing proteins are complexed on the surface of the chylomicron over a layer of phospholipids. Internally are located triacylglycerols, cholesteryl esters, and free cholesterol.

**lipoprotein lipase** (on the vascular epithelium) that digests some of the triglycerides into fatty acids which are taken up by adipose and muscle tissues. The chylomicrons so reduced are now called **chylomicron remnants** that enter the liver through the **remnant receptor** on the surface of the liver cells. The liver cell adds cholesterol to the remnant to form **very low density lipoproteins (VLDLs)**. VLDLs are rich in triglycerides and have lower amounts of cholesterol and phospholipids. VLDLs enter capillaries where lipoprotein lipase again acts upon them to form **intermediary density lipoproteins (IDLs)** that, in turn, are transported back to the liver, entering the liver cell through the **LDL receptor**. IDLs that are not immediately taken up by the liver remain in the blood where more of their triglycerides are hydrolyzed to form **low density lipoproteins (LDLs)**. IDLs have lost most of their triacylglycerides but retain cholesteryl esters. Further hydrolysis of triacylglycerides by hepatic lipase and shedding of lipoproteins E and C generate LDLs. Again, the fatty acids (usually complexed with albumin) released are taken up by adipose and muscle tissues. LDLs (consisting of one apoB-100 molecule, large amounts of cholesteryl esters, smaller amounts of free cholesterol and some triglycerides and phospholipids) are taken up by peripheral tissues by way of plasma membrane LDL receptors that bind to apoB on the surface of LDL, removing LDL from the bloodstream (see Figs. 9.4 and 9.27). The peripheral

**TABLE 9.2 Approximate Densities of Circulating Lipoproteins**

| Lipoprotein | Density (g/mL) |
| --- | --- |
| Chylomicron | Less than 0.94 |
| Very low-density lipoprotein (VLDL) | 0.94−1.006 |
| Intermediate density lipoprotein (IDL) | 1.006−1.019 |
| Low-density lipoprotein (LDL) | 1.019−1.063 |
| High-density lipoprotein (HDL) | 1.063−1.210 |

**Source**: Data taken from http://ethesis.helsinki.fi/julkaisut/laa/kliin/vk/lindbohm/review.html.

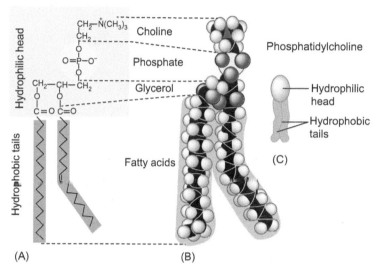

**FIGURE 9.26** The structure of phosphatidylcholine. (A) Structural formula. (B) Space-filling model. (C) Phospholipid symbol. *Reproduced from http://www.bio.miami.edu/~cmallery/150/chemistry/c8.5x13plipid.jpg.*

tissues consume much of the cholesterol some of which may be stored as **cholesteryl esters** in fat droplets in the cytoplasm (especially in cells that utilize cholesterol for synthesis of steroid hormones). **High density lipoprotein (HDL)** precursor (preβ-HDL) is formed in liver and intestinal cells and HDL precursors are able to bind cholesterol from peripheral cells. The cholesterol forms cholesterol esters that are transported from the HDLs to VLDLs and taken up by liver LDL receptors. HDL contents are mainly protein (apoA-I, apoA-II, and apoA-IV) and phospholipid. Once in the bloodstream, it accumulates more phospholipid and free cholesterol. When cholesterol is mobilized from the peripheral plasma membrane to the HDL, the reaction is catalyzed by the enzyme **phosphatidylcholine-sterol *O*-acyltransferase** or **lecithin-cholesterol acyltransferase (LCAT)**. The reaction is: cholesterol + phosphatidylcholine ⇌ cholesteryl ester + lysophosphatidylcholine. These become HDLs that transport cholesterol from the peripheral tissues back to the liver (**reverse cholesterol transport**). The density of the lipoproteins is related to the protein:lipid ratio. Thus, the more protein relative to the amount of lipid raises the density. The **density of protein** is in the range of 1 g/mL, whereas the **density of lipid** is about 0.88 g/mL. As the lipids are broken down or removed from the circulating lipoprotein, the density of the residual lipoprotein increases. Table 9.2 gives the approximate densities of the various circulating lipoproteins.

The structure of phosphatidylcholine is shown in Fig. 9.26.

The overall pathway for lipids from the intestine (where fat comes from the diet primarily; bile acids and cholesterol come from the liver) to the peripheral tissues and recycled back to the liver is shown in Fig. 9.27.

These alterations and transportations are shown in more detail in Fig. 9.28.

A discussion of lipoproteins will occur later in this chapter and they will be emphasized in connection with the discussion on hypercholesterolemia.

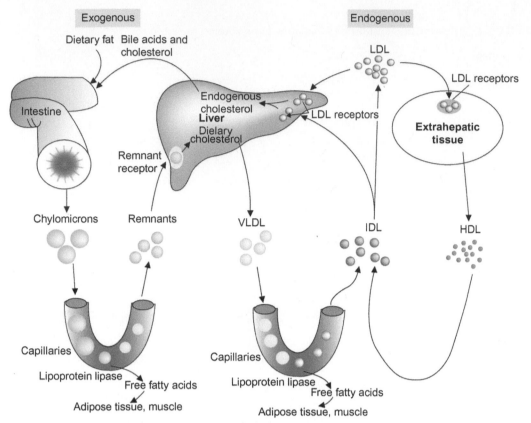

**FIGURE 9.27** Exogenous and endogenous pathways for lipids including cholesterol. Fat is ingested and processed into **chylomicrons** in the intestine; the chylomicrons are absorbed into the circulation where lipoprotein lipase releases some fatty acids from the chylomicrons and the free fatty acids form a complex with albumin and are carried to adipose and muscle tissues. The partially degraded chylomicrons (**chylomicron remnants**) are taken up into the liver by a membrane **remnant receptor** and the cholesterol is packaged into VLDLs. These enter the circulation where **lipoprotein lipase** releases more fatty acids that are carried to peripheral tissues as before and the **IDLs** (partially degraded VLDLs) are carried to the liver through the LDL receptor on the hepatocyte. Some IDLs are partially degraded by the **phosphoprotein lipase** to form **LDLs** and these are carried to peripheral tissues and enter cells by way of the **LDL receptor**. **HDLs** from the peripheral tissues bind cholesterol in the circulation forming **IDLs**, and return to the liver by way of **LDL receptors**. Some of the cholesterol in the liver is converted to bile acids and discharged, along with some cholesterol into the intestine. A small amount of the bile acids are excreted in the feces but most are recirculated to the liver (**enterohepatic circulation**). *VLDL*, very low-density lipoprotein; *IDL*, intermediary density lipoprotein; *LDL*, low-density lipoprotein; *HDL*, high-density lipoprotein. *Reproduced form G. Litwack,* Human Biochemistry and Disease, *Figure 5-17, page 204, Academic Press/Elsevier, 2008.*

# β-Oxidation

β-Oxidation is the process by which the fatty acids released from dietary triglycerides are broken down into two-carbon fragments (as **acetyl-CoA**) that can enter the TCA cycle (and combine with oxaloacetate to form citrate in the **citrate synthase** reaction) for the production of ATP (adenosine triphosphate) energy (Fig. 9.29).

When the fatty acids consist of an even number of carbons, the entire chain can be broken down as increments of acetyl-CoA two-carbon fragments. If the fatty acid chain has an odd number of carbons the chain can be broken down as acetyl-CoA fragments until the terminal three-carbon fragment which results in the three-carbon fragment, propionyl-CoA. Propionyl-CoA is converted to succinyl-CoA which can enter the TCA (Tricarboxylic Acid) Cycle to form **malate** which can either proceed through the Cycle for energy production or can be converted to oxaloacetic acid and proceed through gluconeogenesis to glucose through **phosphoenolpyruvate** (Fig. 6.24) as shown in Fig. 9.30A and B.

The fatty acid to be oxidized is generated in the cytoplasm (of the liver cell, for example). The fatty acid and CoA from the cytoplasm are joined on the surface of the outer mitochondrial membrane by **acylCoA synthase** located within the **outer mitochondrial membrane**. This reaction produces acylCoA + AMP + PPᵢ (from ATP used in the reaction). The pyrophosphate (containing a high energy bond) produced in the reaction is hydrolyzed (with water) to 2 molecules of inorganic phosphate in the cytoplasm. The energy of the gamma phosphate group of ATP and that of the

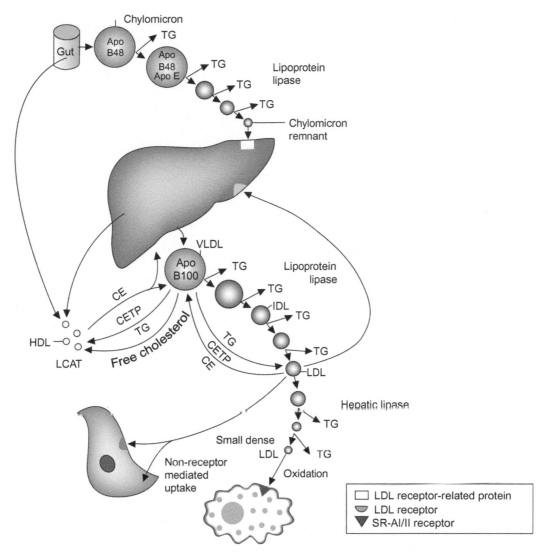

**FIGURE 9.28** The major metabolic pathways by which chylomicron remnants and LDLs are formed from chylomicrons and VLDLs respectively and subsequently catabolized. HDLs are secreted by the gut and liver and receive additional components during the metabolism of triglyceride-rich lipoproteins. *IDL*, intermediate density lipoprotein; *apo*, apolipoprotein; *TG*, triglyceride; *CE*, cholesteryl ester; *CETP*, cholesteryl ester transfer protein; *SR*, scavenger receptor; *LCAT*, lecithin-cholesterol acyltransferase. *Reproduced from http://www.cmglinks.com/asa/lectures/Part_2/lecture/figs_full/03.htm.*

pyrophosphate bond released in the reactions drives the acylCoA synthase reaction (making it virtually irreversible). AcylCoA crosses the outer mitochondrial membrane into the **intermitochondrial membrane space** (Fig. 9.31A and B).

The acylCoA synthase is, apparently, identical to the **fatty acid transport protein 4 (FATP4)**, so that this protein has both activities of the acylCoA synthase and the transporting activity for moving the acylCoA across the outer mitochondrial membrane to the intermitochondrial membrane space.

## Activation and Transport of Fatty Acids Into Mitochondria

The transfer of acylCoA into the mitochondrial matrix occurs through the agency of **carnitine**. Carnitine acyltransferase is another component in the outer mitochondrial membrane. This enzyme catalyzes the formation of a complex between acylCoA and carnitine to produce **acylcarnitine** in the intermembrane space. Through the activity of **carnitine/acylcarnitine translocase**, a component of the inner mitochondrial membrane, **acylcarnitine** is transferred into the mitochondrial matrix, whereas a molecule of carnitine is transported back to the intermitochondrial space. Acylcarnitine is then converted back to carnitine as catalyzed by inner mitochondrial membrane **carnitine acyltransferase II** so that, on the

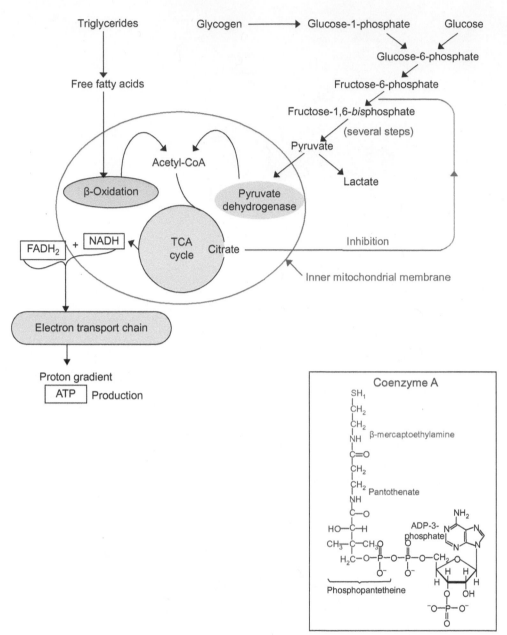

**FIGURE 9.29**  β-Oxidation produces acetyl-CoA from fatty acids derived from ingested triglycerides in addition to the metabolism of glucose that provides acetyl-CoA from pyruvate through the pyruvate dehydrogenase reaction. *Reproduced in part from http://www.bmb.leeds.ac.uk/illingworth/ bioc1110/pathways.gif.*

one hand, carnitine can recombine with another molecule of acylCoA in the membrane space and the acylCoA produced by **carnitine acyltransferase II** can now enter the beta-oxidation pathway of fatty acids inside the matrix. The system is known as the **carnitine cycle** (Fig. 9.32).

In the **β-oxidation pathway**, two carbons are cleaved from the carboxyl end of acylCoA by breakage of the bond between the α-carbon and the β-carbon. This location of the break informs the reaction as β-oxidation. The **products of β-oxidation** are acetyl-CoA, NADH, and FADH$_2$. The **acetyl-CoA** product(s) can enter the TCA cycle for complete oxidation. The reduced coenzymes take part in the electron transport chain for the production of ATP. The resulting acylCoA, now reduced by a two-carbon fragment, enters the cycle again for further reaction until all pairs of carbons in the fatty acid chain have been converted to acetyl-CoA. If the original fatty acid has an odd number of carbons, **propionylCoA** is the final product instead of acetyl-CoA and becomes converted to **succinylCoA** in order to enter the TCA cycle. Through this path, propionylCoA can give rise to glucose (Fig. 9.30A). There is a concensus that glucose

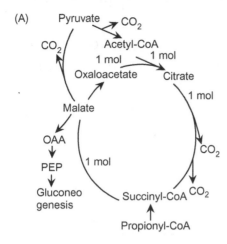

(A)

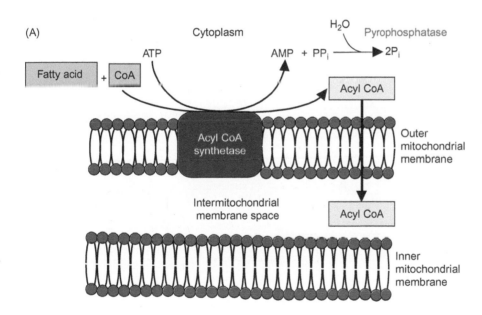

(B)

FIGURE 9.30 (A) Conversion of propionyl-CoA (*bottom*) to succinyl-CoA which can form malate and then course through the TCA cycle or be converted to oxaloacetic acid ultimately to glucose through phosphoenolpyruvate and the gluconeogenesis pathway. *OAA*, oxaloacetate; *PEP*, phosphoenolpyruvate. (B) Three enzymes are involved in the conversion of propionyl-CoA to succinyl-CoA: a carboxylase that converts propionyl-CoA to D-methylmalonyl-CoA, a racemase that converts D-methylmalonyl-CoA to L-methylmalonyl-CoA and finally a mutase that converts L-methylmalonyl-CoA to succinyl-CoA. Atoms in *red* show the fate of the hydrogens on the methyl group. (B) *Reproduced from http://www.rpi.edu/dept/bcbp/molbiochem/MBWeb/mb2/part1/aacarbon.htm.*

FIGURE 9.31 (A) Activation of fatty acid by attachment of CoA on the outer mitochondrial membrane. Part (B) shows the acyl CoA synthase reaction where the structures of the fatty acid (palmitate) and CoA are written out. *Reproduced from http://www.dentistry.leeds.ac.uk/biochem/lecture/faox/faox.htm.*

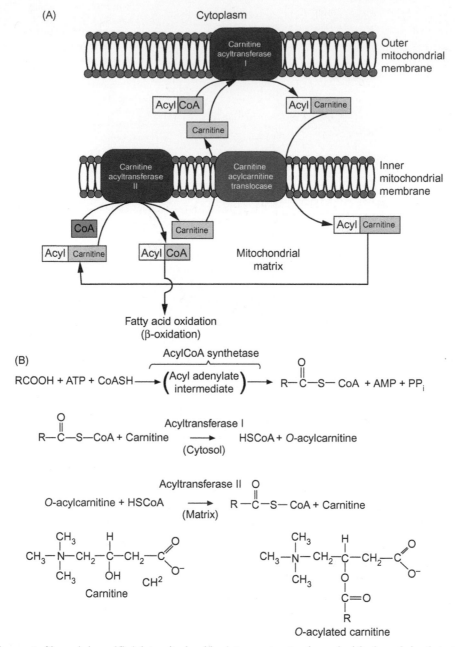

**FIGURE 9.32** (A) Transport of long-chain acylCoA into mitochondria. A transport system is required for long chains that, otherwise, cannot permeate the inner mitochondrial membrane. The net result is that acylCoA is transported as acylcarnitine into the mitochondrial matrix where the enzymes of the β-oxidation pathway are located. In the mitochondrial matrix the acylcarnitine is converted to acylCoA again and free carnitine. The regenerated acylCoA can enter the β-oxidation path. (B) Shown are the chemical reactions to produce acylCoA by acylCoA synthase, where RCOOH represents the long-chain fatty acid. The reaction catalyzed by acyltransferase I, in the cytosol, produces CoA and *O*-acylcarnitine (*O*-acylcarnitine is the condensation product of the fatty acid and carnitine in the form that crosses the inner mitochondrial membrane). *O*-acylcarnitine recombines with CoA as catalyzed by acyltransferase II in the mitochondrial matrix to liberate carnitine and acylCoA as shown in (A). *(A) Reproduced from http://www.dentistry. leeds.ac.uk/biochem/lecture/faox/faox.htm.*

cannot be formed from even-numbered fatty acids in the human or other animals because they lack the glyoxylate cycle (that some plants and bacteria have) that makes such a conversion possible. The way this works is that even-numbered fatty acids are reduced to acetyl-CoAs that enter the (incomplete) TCA cycle. At the level of isocitrate, isocitritase converts isocitrate to glyoxalate and, with CoA, malate synthetase converts glyoxalate to malate (α-ketoglutarate is not formed in many lower organisms or α-ketoglutarate dehydrogenase may be absent). Then, malate can be converted to oxaloacetate that can enter gluconeogenesis and form glucose. So, odd-numbered fatty acids yielding a final propionylCoA are the only candidates for a precursor for glucose derived from fatty acids in the human. The β-oxidation pathway, starting with a 16-carbon fatty acid, palmitic acid as palmitoylCoA, is shown as an example in Fig. 9.33.

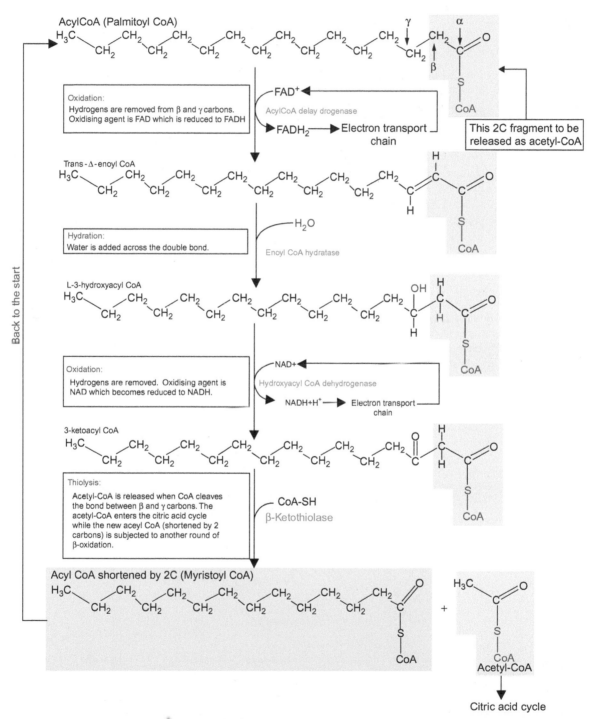

FIGURE 9.33  β-Oxidation of acylCoA. In this case the acylCoA is palmitoylCoA derived from the 16-carbon fatty acid, palmitic acid. As a result of the first cycle, a 2-carbon fragment is removed from the carboxyl end of the chain resulting in a molecule of acetyl-CoA and the shortened chain of myristoylCoA (myristic acid is a 14-carbon fatty acid). MyristoylCoA enters the cycle again to be shortened to a 12-carbon chain with the release of a second molecule of acetyl-CoA and, so on, until all of the carbons from the original fatty acid are accounted for as molecules of acetyl-CoA. *Reproduced from http://www.dentistry.leeds.ac.uk/biochem/lecture/faox/faox.htm.*

Thus, the **degradation of palmitate** (16 carbons) by β-oxidation requires 7 rounds after which there have been 8-acetyl-CoA molecules accumulated. The summary for the complete breakdown of palmitate is:

$$palmitylCoA + 7FAD + 7NAD + 7CoA + 7H_2O \rightarrow 8 \text{ acetyl} - CoA + 7FADH_2 + 7NADH + 7H^+$$

After the 8-acetyl-CoA molecules have entered the TCA cycle, the products will be as follows:

24 NADH through oxidative phosphorylation yield 72 ATP molecules
8 FADH$_2$ through oxidative phosphorylation yield 16 ATP molecules
8 GTP that will yield 8 ATP

7 NADH are generated by β-oxidation that will yield 21 ATP through oxidative phosphorylation.
7 FADH$_2$ are generated by β-oxidation that will yield 14 ATP through oxidative phosphorylation.

Thus, the total ATPs generated from one molecule of **palmitate** equal 131 (adding up the above). As the formation of **palmitylCoA** from palmitate + CoA requires the equivalent of 2 ATP, the actual net ATP yield is 129.

Virtually every tissue in the body can utilize fatty acids for energy except brain, the adrenal medulla and red blood cells. Fatty acids do not cross the blood—brain barrier. Red blood cells have no mitochondria to oxidize fatty acids.

**Peroxisomes** (Fig. 2.21) can carry out β-oxidation of fatty acids longer than 22 carbons as these chains are too long for mitochondria. Peroxisomal oxidation stops when **octanoylCoA** is reached. The residual oxidation can be carried out by mitochondria. Peroxisomal oxidation of fatty acids is not coupled to ATP but, instead, electrons, rather than flowing through an electron transport chain as is the case in mitochondria, are transferred to oxygen to produce **hydrogen peroxide (H$_2$O$_2$)**. **Catalase**, which is a plentiful enzyme in peroxisomes converts H$_2$O$_2$ to H$_2$O and O (Fig. 9.34).

## Unsaturated Fatty Acids

*Cis* double bonds in unsaturated fatty acids (Fig. 9.15) require that they be converted to *trans* double bonds in order to proceed through β-oxidation. There exists in mitochondria a **cis-trans isomerase** that can convert *cis* fatty acidCoA molecules to *trans* fatty acid CoA molecules. This conversion would allow β-oxidation to continue beyond the original *cis* double bond. Odd-numbered double bonds in an acylCoA are acted on by an isomerase (e.g., a 3,2-enoylCoA isomerase) creating a suitable substrate. The conversion of linoleoyl CoA containing two *cis* double bonds to

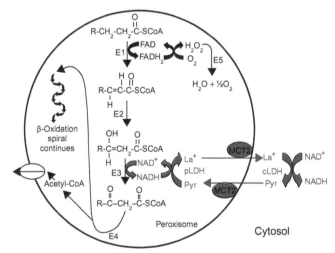

**FIGURE 9.34** β-Oxidation of very long-chain fatty acids in the peroxisome including a "lactate shuttle". When fatty acyl CoAs are oxidized down to eight carbons, or so, they can be released to the cytosol and transported to the mitochondria for complete oxidation. Key elements of the lactate shuttle are shown in red indicating the NADH is reoxidized to NAD$^+$ inside the peroxisome by the conversion of pyruvate (Pyr −) to lactate (La −). La − then leaves the peroxisome via the monocarboxylate carrier, MCT2, the primary pyruvate transporter in man. In the cytosol, La − is converted back to pyruvate with concomitant conversion of NAD + to NADH. Thus, reducing equivalents are delivered from the peroxisome to the cytosol. The resulting pyruvate returns to the peroxisome via the MCT2 transporter to continue the shuttle. The shuttle provides an avenue for NADH reoxidation in the peroxisome, a necessary process for the continuation of peroxisomal β-oxidation of fatty acids. *E1*: acyl-CoA oxidase; *E2*: enoyl-CoA hydratase; *E3*: L-3-hydroxyacyl-CoA dehydrogenase; *E4*: thiolase; *E5*: catalase; *pLDH*: lactate dehydrogenase located inside the peroxisome; *cLDH*, LDH located in the cytosol (outside the peroxisome); *Pyr −*: pyruvate; *La −*, lactate. *Reproduced from http://jp.physoc.org/content/558/1/5/F5. expansion.*

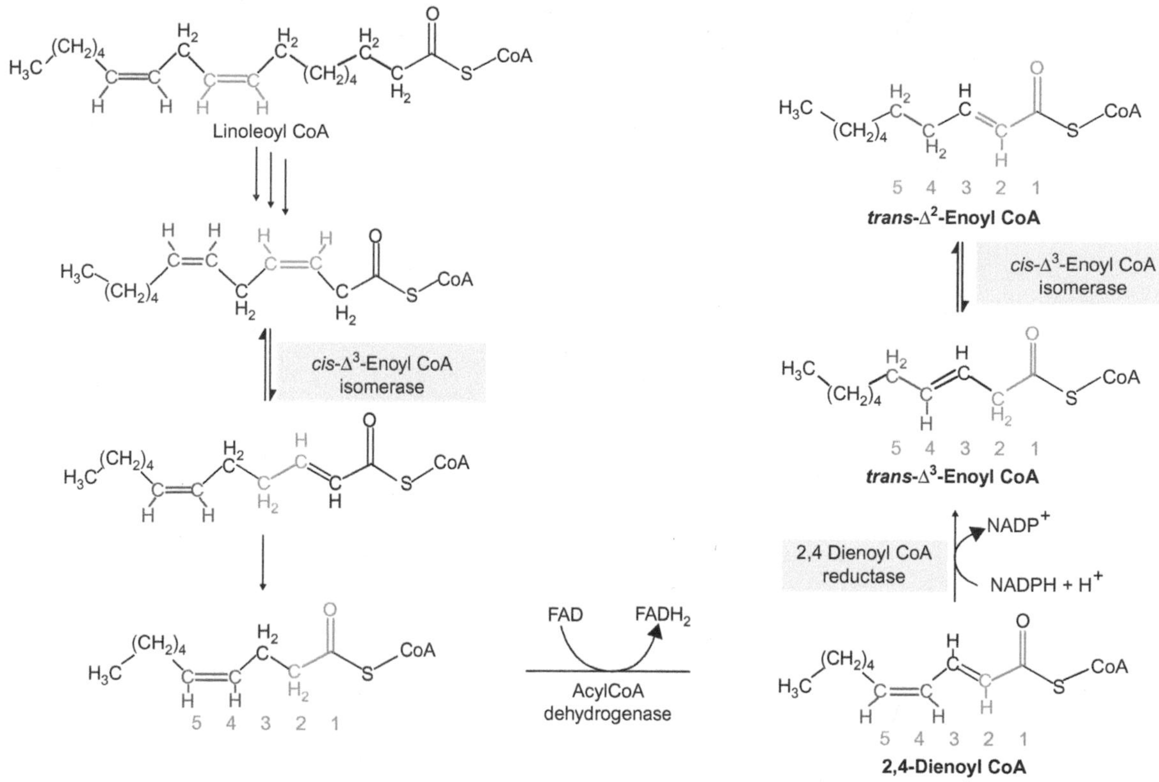

**FIGURE 9.35** Oxidation of linoleoyl CoA, a diunsaturated fatty acid containing two *cis* double bonds is converted to compounds containing *trans* double bonds, enabling subsequent β-oxidation. This is accomplished by enoyl CoA isomerase and 2,4-dienoylCoA reductase. *Reproduced from http:// www.ncbi.nlm.nih.gov/bookshelf/br.fcgi?book = stryer&part = A3061.*

*trans*-Δ3-enoylCoA and *trans*-Δ2-enoylCoA is shown in Fig. 9.35. Even-numbered double bonds are acted on by a reductase (using NADPH) to create an odd number of double bonds and then the isomerase acts on it.

An overview of **fatty acid metabolism** is shown in Fig. 9.36.

## Ketone Bodies

There are situations where **acetyl-CoA** accumulates. For example, in diabetes or in fasting, **oxaloacetate** is used up in the gluconeogenic pathway. There is reduced incorporation of acetyl-CoA into citrate and acetyl-CoA is converted into ketone bodies: **acetone**, **acetoacetate**, and β**-hydroxybutyrate** (Fig. 9.37).

The enzymatic reactions forming acetone, acetoacetate, and β-hydroxybutyrate are shown in Fig. 9.38.

**Ketone bodies**, produced in the liver and kidneys, can be used as energy sources by the heart and the brain. The **Atkins diet** is a low-carbohydrate, high-protein diet. This causes a shift in the major energy source from glucose to fat and fatty acids. As a result, there are elevated levels of acetyl-CoA and the formation of ketone bodies that can be used for energy by the heart and brain. Consequently, the fat storage sites become depleted and the dieter loses weight dramatically over a relatively short period of time. Thus, ketone bodies are sources of acetyl-CoA. Acetoacetate produces two molecules of acetyl-CoA as shown by the reactions in Fig. 9.39.

β-Hydroxybutyrate, although labeled a ketone body, is not a ketone like acetone and acetoacetate; it is a carboxylated acid.

## Lipid Metabolism, Lipid Droplets, and Hormonal Control

Triacylglycerides are ingested in the diet and are broken down in the intestine by lipases derived from the pancreas as already described (Fig. 9.24). The fatty acids and glycerol are converted to substrates for energy or stored as the case may be. In liver, glycerol is converted to **glycerol-3-phosphate** and can enter glycolysis or gluconeogenesis depending on the sufficiency of glucose in the cell. When glucose is limiting, **fatty acids** will be converted to **acetyl-CoA** through

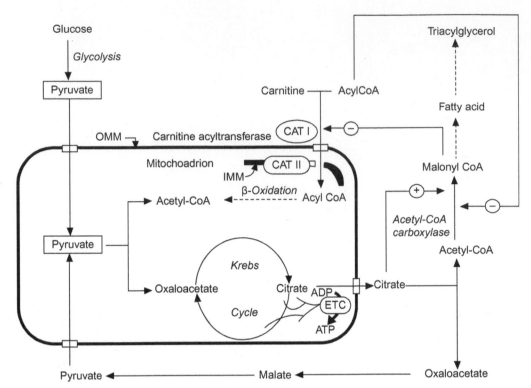

**FIGURE 9.36** An overview of fatty acid metabolism. *CAT I*, carnitine acyltransferase I; *CAT II*, carnitine acyltransferase II; *OMM*, outer mitochondrial membrane; *IMM*, inner mitochondrial membrane; *ETC*, electron transport chain.

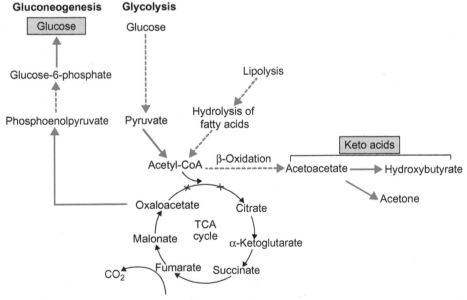

**FIGURE 9.37** Formation of ketone bodies: acetone, acetoacetate and β-hydroxybutyrate. Xs indicate loss of oxaloacetate to the gluconeogenic pathway (to form glucose) under conditions of fasting or diabetes. As a consequence, less citrate is formed from oxaloacetate in the TCA cycle and acetyl-CoA accumulates. The accumulated acetyl-CoA forms ketone bodies: acetoacetate, acetone, and β-hydroxybutyrate. *Reproduced from http://eutils.wip. ncbi.nlm.nih.gov/bookshelf/br.fcgi?book = endocrin&part = A43&rendertype = box&id = A66.*

degradation in the **β-oxidation pathway** in mitochondria. The acetyl-CoA can be used for fuel by entrance into the TCA cycle with **oxaloacetate** to generate ATP ($+CO_2$ and $H_2O$, metabolic water). During **starvation**, oxaloacetate becomes limiting and this prevents the entrance of acetyl-CoA into the TCA cycle. The limiting oxaloacetate will be used to form glucose. For the generation of energy, fatty acid breakdown will occur and ketone bodies will be formed. **Acetoacetate** and **β-hydroxybutyrate** from the liver can be used as energy sources for the heart, skeletal muscle and kidney

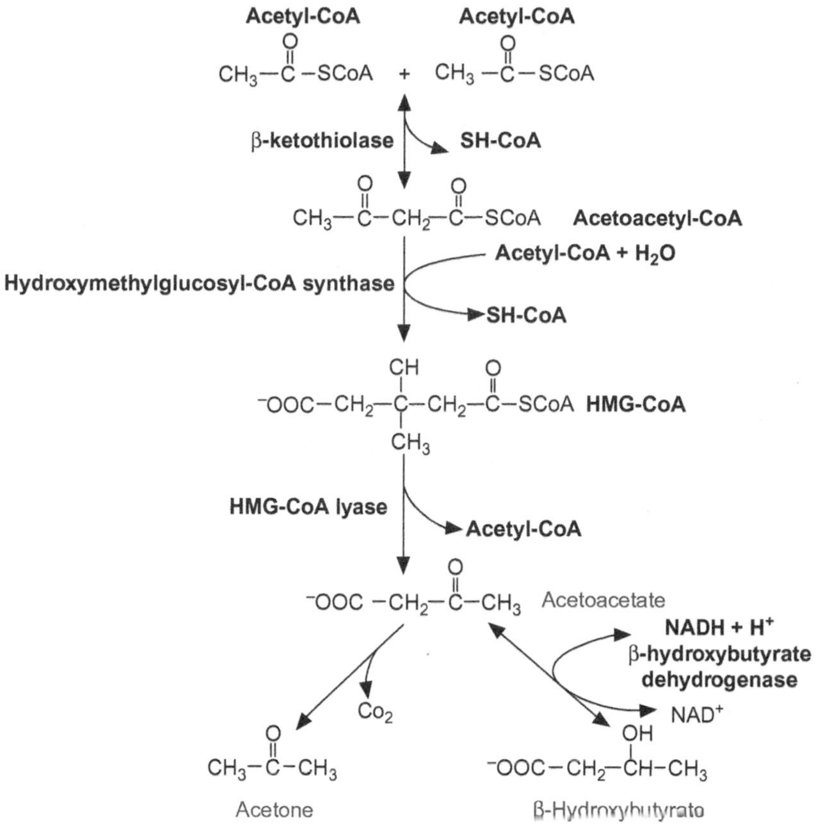

FIGURE 9.38 Enzymatic reactions in the formation of ketone bodies (acetone, acetoacetate and β-hydroxybutyrate from acetyl-CoA). *Reproduced from http://themedicalbiochemistrypage.org/mobilei-mages/m-ketonesynthesis.jpg.*

and provide good energy sources for the brain, in place of glucose, during starvation. **Acetone** is transported to the lungs providing a characteristic odor in ketosis (this occurs in Atkins dieters who have become ketotic). If ketogenesis is carried to an extreme, generating high levels of acidic ketone bodies, **acidosis** can result. Ketone bodies are formed in the liver but the liver cannot use them, only tissues peripheral to the liver are able to do so. In untreated insulin-dependent diabetes, ketoacidosis results from a reduced supply of glucose to the tissues (low insulin). Ketone bodies are excreted in the urine.

When glucose is plentiful and fatty acids are not needed for energy, they are transported as a complex with **albumin** in the bloodstream and converted to **triglyceride** in the **adipocyte** and stored there in **lipid droplets**. Lipid droplets in adipose cells are shown in Fig. 9.40A and B.

Lipid droplets have multiple functions in storing energy, regulation of **cholesterol homeostasis**, biosynthesis of **membrane lipids**, the formation of **steroid hormones** and **eicosanoids** (precursors of prostaglandins and leukotrienes) and a transient storage site of certain proteins. *Many, if not all cell types, can produce lipid droplets if cells are confronted with high levels of fatty acids.* Proteins associated with the lipid droplet are the **PAT family of proteins** consisting of **Perilipin (PLIN)**, **Adipophilin (ADRP)** and **TIP47** (TATA binding protein-Interacting Protein, also referred to as Cargo Selection Protein and IGF receptor binding protein). Although the mechanism of the formation of lipid droplets is not completely understood, there is evidence that it forms from the interior of the surface leaflets of the **endoplasmic reticulum** as shown in Fig. 9.41.

Perilipin is the major protein on the surface of the droplet that prevents other proteins, such as the **hormone-sensitive lipase (HSL)** from entering the droplet. As we shall see, altering the conformation of PLIN on the droplet surface is the regulatory step that will allow the HSL to enter the droplet and break down triglycerides into free fatty acids. More than likely, there are differing populations of LDs with different arrays of proteins on their surfaces. However, here we consider LDs covered mainly by PLINs and the action of **hormone-sensitive lipase (HSL)**.

The main hormones regulating the breakdown of triglyceride in the fat droplet of the adipose cell are **epinephrine** and **insulin**. In particular, epinephrine (catecholamine) is the important direct positive regulator and its effect is countered by the effects of insulin, as will be shown. The effects of **glucagon**, more akin to those of epinephrine, occur in liver relating to its antiinsulin effects but have little effect on adipose tissue. The overall pathway, starting with the binding of epinephrine to the cell membrane adrenergic receptor is shown in Fig. 9.42.

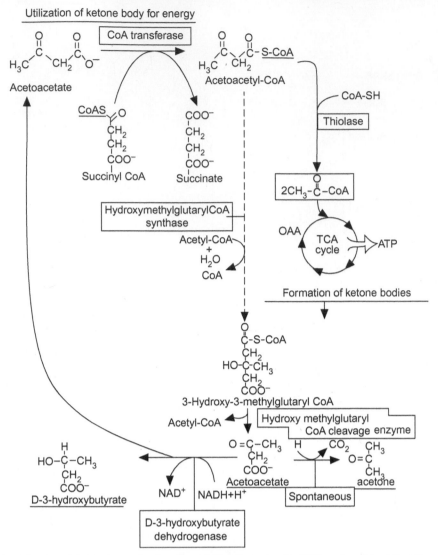

**FIGURE 9.39** Conversion of acetoacetate into two molecules of acetyl-CoA that enters the TCA cycle and is utilized for energy. Upper part of diagram shows the utilization of acetoacetate for energy. The first reaction is catalyzed by CoA transferase to form acetoacetyl-CoA. This is followed by the thiolase reaction converting acetoacetyl-CoA to two molecules of acetyl-CoA by the addition of one molecule of CoA. Acetyl-CoA enters the TCA cycle with oxaloacetate to form citrate and courses through the cycle to yield energy in the form of ATP. The bottom part of the diagram shows the formation of ketone bodies from acetoacetyl-CoA. Through the action of hydroxymethylglutarylCoA synthase and the addition of acetyl-CoA and water, 3-hydroxy-3-methylglutarylCoA is formed. This is converted by hydroxymethylglutarylCoA junctions enzyme to form acetoacetate which converts to acetone through spontaneous decarboxylation or to D-3-hydroxybutyrate catalyzed by D-3-hydroxybutyrate dehydrogenase. Subsequently, the ketone bodies can be used for the production of energy when glucose is limiting.

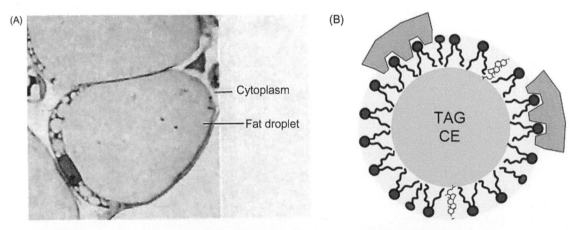

**FIGURE 9.40** (A) Microscopic view of lipid droplets in an adipose cell. Reproduced from http://www.tutorvista.com/topic/fat-stored-in-adipose-tissue. (B) Cartoon of a lipid droplet. The core (in light *blue*) consists mainly of fat (triacylglycerol, TAG) and **cholesteryl esters** (also written as cholesterol esters, **CE**). The droplet is surrounded by a monolayer consisting of **phospholipid-cholesterol** and associated proteins (**perilipins, PLINS**) in *red*. Lipid droplets are composed mainly of neutral lipids (compared to the cell membrane). *Reproduced from http://www.bio.uu.nl/enmeta/burger/Lipid% 20Droplet%20Background.htm.*

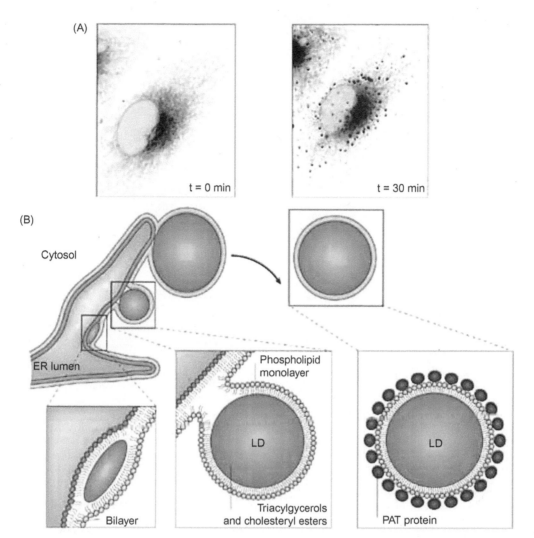

**FIGURE 9.41** Formation of **lipid droplets** (**LDs**) as monitored by the use of caveolin-truncation-mutant–green fluorescent protein (Cav3$^{DGV}$–GFP). (A) Time zero is shown before the addition of fatty acids. The GFP localizes to the endoplasmic reticulum (ER) and the Golgi region. Following addition of fatty acids, LDs appear in minutes, although the picture after 30 minutes is shown here. (B) In this view of LD formation, neutral lipids are synthesized between the leaflets of the ER membrane. The mature LD is thought to bud from the ER membrane to form an independent organelle bounded by a limiting monolayer of phospholipids and LD-associated proteins that are members of the PAT family. Some other evidence suggests that **caveolin** proteins are found on the surface of LDs, as well, implicating the mechanism just described. *Reproduced from S. Martin and R.G. Parton,* Nature Reviews Molecular Cell Biology, *7: 373–378, 2006 and directly from http://www.nature.com/nrm/journal/v7/n5/ fig_tab/nrm1912_F1.html.*

Fig. 9.43 shows several phosphorylatable sites in the C-terminus of the hormone-stimulated lipase. There are four sites that can be phosphorylated by PKA: Ser-563, Ser-565, Ser-659, and Ser-660. There are other sites that can be phosphorylated by **AMPK** (5'-adenosine monophosphate-activated protein kinase) or by **ERK** (extracellular signal-regulated kinase).

**Insulin action** counters the direct activity of epinephrine because insulin signaling results in the expression of **protein phosphatase-1** (Fig. 6.30) that can dephosphorylate the phosphorylated HSL and inactivate it. Glucagon has effects in the liver similar to those of epinephrine on the adipocyte in that its action through its receptor results in the activation of adenylate cyclase to generate cAMP and activates protein kinase A (Fig. 6.28). However, glucagon has little effect on the adipocyte in the human.

Interestingly, chronic high levels of circulating **free fatty acids (FFAs)** have a negative effect on the β-cell causing the **suppression of insulin secretion**. In this situation, the levels of FFAs within the β-cell itself accumulate and the number of insulin-containing granules and the content of insulin decreases. The chronic exposure of the β-cell to high levels of FFAs causes the $K_{ATP}$ **channels** to close (Fig. 6.52), first by increasing sensitivity of these channels to ATP that leads to continuous excitation of the cell, **exhaustion of insulin storage** and suppression of **glucose-stimulated insulin release**.

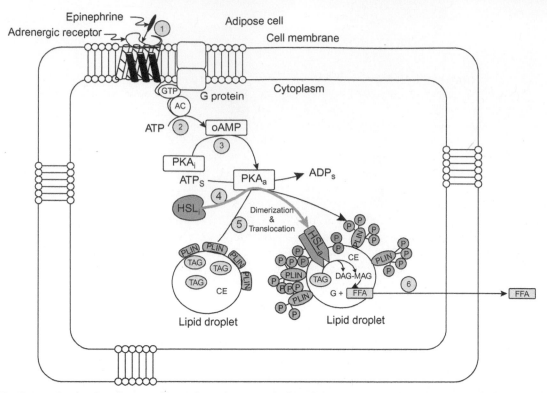

**FIGURE 9.42** Cartoon showing the adipocyte with an adrenergic receptor in the cell membrane to which epinephrine binds with high affinity (*step 1*). This is a G protein-coupled receptor and epinephrine binding activates the receptor and its associated G protein that then activates adenylate cyclase (*AC*). AC catalyzes the conversion of ATP to cyclicAMP (*cAMP*) (*step 2*). cAMP activates protein kinase A (*PKA*) (*step 3*) as shown in **Fig. 6.26**. Active PKA phosphorylates inactive hormone-stimulated lipase (*HSL*) at four sites on the protein (*step 4*). As a result, the phosphorylated HSL dimerizes and translocates to the lipid droplet (*LD*). Simultaneously PKA multiply phosphorylates PLINs on the surface of the LD (*step 5*). The phosphorylated PLINs undergo conformational changes and open gaps on the surface of the LD wide enough to allow the entry of hormone-stimulated lipase, *HSL*. HSL then proceeds to hydrolyze the resident triglycerides (*TAGs*) to diacylglycerol (*DAG*), monoacylglycerol (*MAG*) and eventually to glycerol and free fatty acid. Most believe that the enzyme can carry out hydrolysis to the monoglyceride stage and some believe that the HSL can also hydrolyze the monoglyceride to free glycerol and fatty acid. Presumably, if HSL is unable to accomplish the last hydrolysis step, other resident lipases (MAG lipase) are able to do so. Free fatty acids then translocate across the cell membrane into the extracellular space (*step 6*) and gain access to the bloodstream. *GTP*, guanosine triphosphate; *PKA_i*, inactive protein kinase A; *PKA_a*, active protein kinase A; *HSL_i*, inactive hormone-stimulated lipase; *HSL_a*, activated hormone-stimulated lipase that dimerizes; *PLIN*, perilipin; *P*, phosphate; *TAG*, triglyceride; *DAG*, diacylglycerol; *MAG*, monoacylglycerol; *CE*, cholesteryl ester; *FFA*, free fatty acid. Not shown in missing from this drawing are some details: perilipin is originally complexed to CGI-58 (an enzyme that facilitates hydrolysis of the fat in LDs but is not directly involved in the hydrolytic process. CGI-58 has the activity of 1-acylglycerol-3-phosphate O-acyltransferase). Also, the HSL dimer that catalyzes the hydrolysis of triacylglycerol is complexed with FABP (fatty acid binding protein).

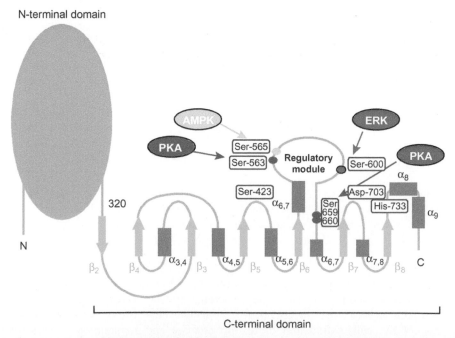

**FIGURE 9.43** Cartoon of hormone-stimulated lipase showing the phosphorylation sites on the C-terminal domain. The amino acids involved in the catalytic domain are Ser-423, Asp-703, and His-733. Helices are in *purple* and β-sheets are in *cyan*. The regulatory module is made up of 150 amino acids and contains important serine residues that can be phosphorylated. *Reproduced from http://www.jlr.org/content/vol143/issue10/images/large/R200009f1.jpeg.*

Insulin has a direct effect on the adipose cell because the insulin receptor resides in the membrane of the adipose cell and insulin binding to its receptor (IR) stimulates glucose uptake in **adipose cells** (as does insulin binding in **muscle cells**). Glucose is transported into the cell from the bloodstream by the transporter in the cell membrane, **GLUT4**. Insulin suppresses glucose production in the liver and stimulates the conversion of glucose to glycogen (Figs. 6.30 and 6.31). In addition to the uptake of free fatty acids from the bloodstream by adipose cells, glucose provides the energy for fat stores. *In the adipose tissue of obese animals (also in starvation) GLUT4 expression is decreased generating a suppression of glucose uptake. Concomitantly, there is an increased expression of fat-derived **RBP4** (**retinol-binding protein 4**).* Under normal conditions, there is a low level of RBP4 secretion by adipose cells. RBP4 in the bloodstream suppresses **insulin signaling** in skeletal muscle. In the liver, RBP4 induces the expression of **PEPCK (phosphoenolpyruvate carboxykinase)**, causing an increase in the output of glucose (PEPCK is a key enzyme in gluconeogenesis converting oxaloacetate to phosphoenolpyruvate which is metabolized on up to glucose a shown in Fig. 6.12). These actions affected by RBP4 (in combination with **retinol**) result in an increase in the level of **blood glucose**. Other factors altered in obesity (and in type 2 diabetes), such as adiponectin, TNFα (tumor necrosis factor α) and resistin may play a role in blood glucose regulation. These concepts are diagrammed in Fig. 9.44.

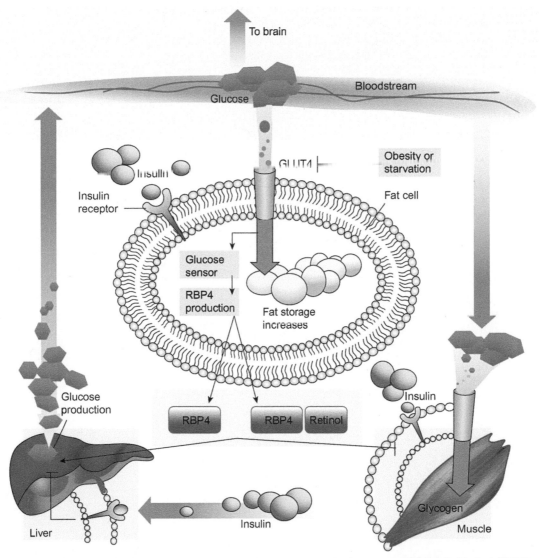

**FIGURE 9.44** Diagram indicating the changes in the adipocyte in obesity or starvation. Under these conditions the level of GLUT4 expression is reduced, suppressing the uptake of glucose. As GLUT4 expression declines, the expression of RPB4 (retinol-binding protein) increases and is secreted into the bloodstream. This protein suppresses the action of insulin on the hepatocyte and raises the level of expression of liver PEPCK (phosphoenolpyruvate carboxykinase) that results in stimulation of gluconeogenesis and the enhanced formation of glucose that is secreted via GLUT2 into the bloodstream. As in liver, the effect of RBP4 suppresses the action of insulin and the formation of glycogen. These effects cause an elevated level of glucose in the blood. *Reproduced from http://www.nature.com/nature/journal/v436/n7049/images/436337a-f1.2.jpg.*

## Fatty Acid Synthesis

Fatty acid synthesis occurs in the liver and in adipose cells. The rate-limiting reaction in fatty acid biosynthesis is that of **acetyl-CoA carboxylase** that catalyzes the reaction of acetyl-CoA to **malonylCoA** in two steps shown in Fig. 9.45.

The coenzyme, **biotin**, is attached to the epsilon ($\varepsilon$) amino group of a lysine residue in the enzyme. Lysine has six carbons (Fig. 4.9) and the **epsilon amino group** is the terminal amino group farthest from the $\alpha$-carbon, the $\alpha$-carbon being the atom that is substituted by the carboxyl group and the amino group of the amino acid. The terminal carbon containing the $\varepsilon$-amino group is four carbons from the $\alpha$-carbon, therefore, $\alpha$, $\beta$, $\delta$, $\gamma$, and $\varepsilon$. The structure of biotin showing its attachment to the enzyme acetyl-CoA carboxylase is shown in Fig. 9.46. Two subunits of the enzyme are located opposite each other with a **biotin carboxyl carrier protein** connecting both at their apex. One subunit is the **carboxylase subunit** and the other is the **transcarboxylase subunit**. In the course of the enzymatic reaction, biotin, extending from the carrier protein, swings first to the active site of the carboxylase subunit where biotin is carboxylated and then swings across to the transcarboxylase subunit's active site where it reacts with acetyl-CoA to form malonylCoA.

The regulation of **acetyl-CoA carboxylase (ACC)** is complex. There exist two isozymes, ACC1 and ACC2 and these derive from two different genes. Each enzyme in the active form consists of two subunits, $\alpha$ and $\beta$. There are 2 atoms of magnesium bound per subunit. The regulation of both isozymes appears to be the same. They produce two pools of **malonylCoA**. One of these (ACC2) inhibits $\beta$-**oxidation**, whereas the other (ACC1) stimulates **lipid biosynthesis**. ACC is inactivated by phosphorylation. The phosphorylation is catalyzed by **AMP protein kinase (AMPPK)**. Phosphorylation causes the enzyme to dissociate into two inactive monomers. ACC1 is a cytosolic enzyme, whereas

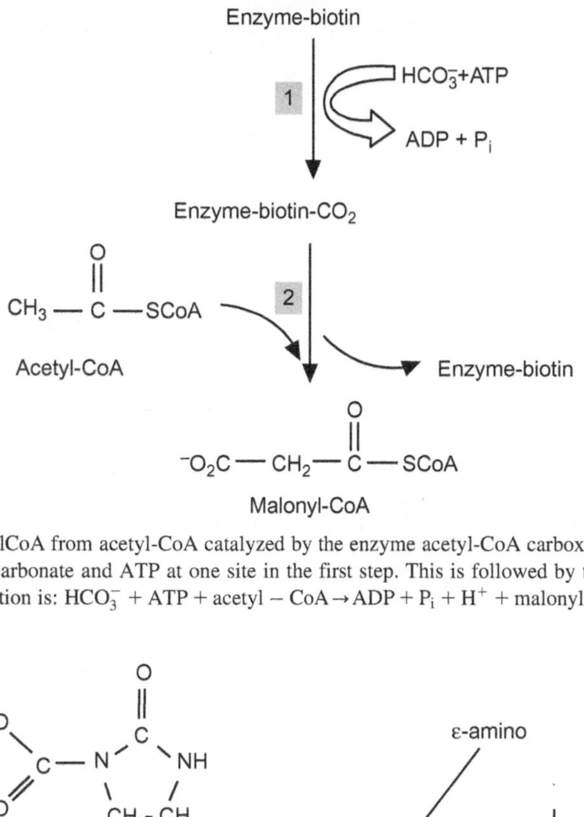

**FIGURE 9.45**    The formation of malonylCoA from acetyl-CoA catalyzed by the enzyme acetyl-CoA carboxylase. The enzyme-biotin complex is carboxylated biotin (the coenzyme) from bicarbonate and ATP at one site in the first step. This is followed by the transfer of the carboxyl group to acetyl-CoA at a second site. The overall reaction is: $HCO_3^- + ATP + acetyl - CoA \rightarrow ADP + P_i + H^+ + malonylCoA$.

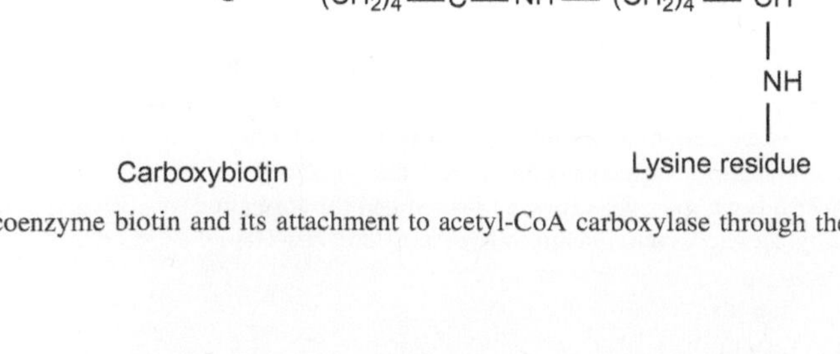

**FIGURE 9.46**    The structure of the coenzyme biotin and its attachment to acetyl-CoA carboxylase through the $\varepsilon$-amino group of a lysine residue in the enzyme.

ACC2 has 20 N-terminal hydrophobic amino acids that direct this isozyme to the **mitochondria**. ACC1 stimulates lipid biosynthesis by supplying malonylCoA and ACC2 inhibits β-oxidation by inhibition through malonylCoA of mitochondrial **carnitine palmitoyltransferase 1**. The inactive form can be reversed by dephosphorylation catalyzed by **protein phosphatase 2 (PP'ase 2)**. Notably, the inactive phosphorylated form of ACC (ACC-P) can bind **citrate** that acts as an **allosteric activator** binding to a site remote from the catalytic center. This produces a phosphorylated enzyme that has partial activity. This system is also controlled by hormones. When the blood glucose level is low, **epinephrine** and **glucagon** stimulate the phosphorylation of ACC but when blood glucose is high **insulin** stimulates the dephosphorylation of ACC-P. Insulin enhances the activity of protein phosphatase 2 (Fig. 6.30). These overall activities are summarized in Fig. 9.47.

It should be noted that the linkage between biotin and the ε-lysine of ACC involves the terminal carboxyl group of **biotin**. The length of this biotin side chain is sufficient to allow for the transit of biotin between the two active sites on the enzyme.

In the **heart**, which does not synthesize fatty acids, **malonylCoA** inhibits fatty acid oxidation. When the concentration of ATP is low and AMP is high, the production of malonylCoA is reduced allowing for fatty acid oxidation, the production of acetyl-CoA and energy from the TCA cycle.

MalonylCoA is produced as shown in Figs. 9.44−9.46. **MalonylCoA synthesis** is summarized:

$$HCO_3^- + ATP + acetyl - CoA \rightarrow ADP + P_i + malonylCoA$$

As AAC generates malonylCoA representing the committed step in fatty acid synthesis, it is a highly regulated enzyme and the regulation is complex as described in Fig. 9.46. The remaining pathway of fatty acid synthesis is carried out by cytoplasmic **fatty acid synthase (FAS)**. FAS is a dimer of two multifunctional polypeptides. It is a single enzyme that has resulted from gene fusion of several individual enzymes. The substrates are acetyl-CoA and malonylCoA, linked as thioester derivatives of CoA through the **β-mercaptoethylamine** of the coenzyme. There are three catalytic domains in the N-terminus and four catalytic domains in the C-terminus (one of which is a carrier protein) as shown in Fig. 9.48 which is a diagram of one of two popular models.

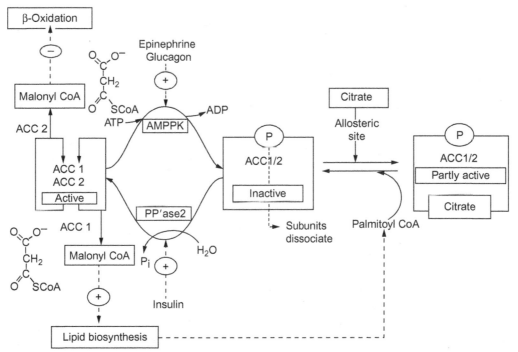

**FIGURE 9.47** Acetyl-CoA carboxylase (ACC) consists of two isozymes, ACC1 and ACC2 (*left, middle*). There are two pools of the product of the enzymatic reaction, malonylCoA. One pool suppresses β-oxidation of fatty acids and the other pool stimulates lipid biosynthesis. ACCs are inhibited by phosphorylation (at multiple sites) catalyzed by AMP protein kinase (AMPPK). The inactive, phosphorylated (P) form can be reversed to generate the active form by protein phosphatase 2 (PP'ase2). The phosphorylated inactive enzyme (AAC-P) is able to bind the citrate that acts as an allosteric activator so that the citrate bound phosphorylated enzyme (AAC-P-citrate) is partially active. Citrate binds at a site on the enzyme remote from the catalytic center and its binding may generate a favorable conformational change. PalmitoylCoA, one of the end products of lipid biosynthesis, can convert a partially active form of the enzyme to the inactivated form. Hormones also regulate the activity of ACC. Epinephrine and glucagon stimulate the phosphorylation of ACC, whereas insulin stimulates the dephosphorylation of ACC-P by enhancing the activity of **protein phosphatase 2 (PP'ase2)**.

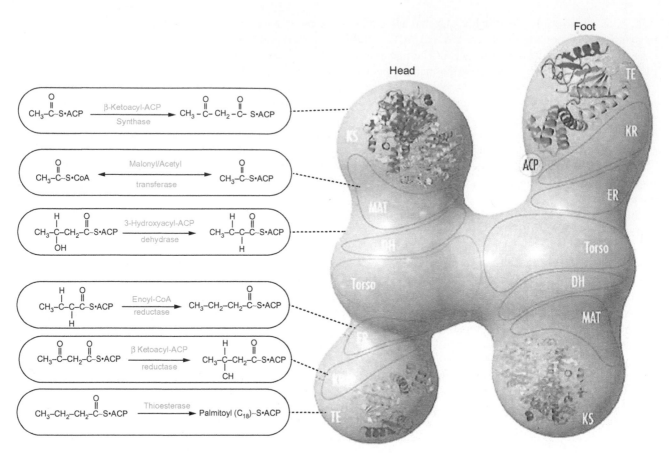

**FIGURE 9.48**    One of two models of the fatty acid synthase enzyme (FAS). FAS is a multienzyme with two identical multifunctional polypeptides of 272 kDa (some reports indicate 250 kDa) connected in antiparallel fashion. The two component proteins are arranged from head (*top*) to tail (or foot) on the *left* and from tail (*foot*) to head on the *right*. Substrates are moved from one functional domain to the next. In the N-terminus (*head*), proceeding downwards the activities are: -ketoacyl synthase (*KS*), malonyl/acetyl transferase (*MAT*) and dehydrase (*DH*). Interposed is a core region (*Torso*) of 600 amino acid residues followed by four functions in the C-terminal domain: enoyl reductase (*ER*), -ketoacyl reductase (*KR*), acyl carrier protein (ACP and thioesterase (*TE*)). The acyl carrier protein (*ACP*) is shown on the *right* hand partner to the *left* of KR. The reactions catalyzed by each component activity are shown on the *left* and are recapitulated below in the text. *Reproduced from http://en.wikipedia.org/wiki/File:FASmodel2.jpg.*

**Pantothenic acid**, a vitamin, is the cofactor in the fatty acid synthase enzyme system. It is also a component of Coenzyme A. **Phosphopantetheine (Pant)** (Fig. 9.29) is attached covalently through a phosphate ester to a serine hydroxyl on the acyl carrier protein (Fig. 9.49). Pant has a long flexible arm that allows its thiol to move from one active site to another within FAS.

The initial substrates of FAS are acetyl-CoA and malonylCoA. The individual steps of the FAS sequential reactions are shown in Fig. 9.50.

In this set of reactions, after all the cycles have been completed, the final product is **palmitate**, a 16-carbon fatty acid. If the fatty acid to be synthesized is larger than 16 carbons, the lengthening is carried out in the mitochondria and the endoplasmic reticulum. To accomplish elongation, the fatty acid oxidation system runs in reverse and malonylCoA is the two-carbon fragment donor. The final electrons are donated by NADPH. In the synthesis of palmitate, the overall reaction (accounting for the ATP synthesis of malonate) is summarized as follows:

$$8\text{ acetyl} - \text{CoA} + 14\text{ NADPH} + 14\text{ H}^+ + 7\text{ ATP} = \text{palmitate} + 14\text{ NADP}^+ + 8\text{ CoA} + 7\text{ ADP} + 7\text{ P}_i$$

FAS set of reactions can be represented as a cycle in which each step is delineated (Fig. 9.51).

Proceeding from the N-terminus of fatty acid synthase toward the C-terminus, the active centers are: condensing enzyme, malonyl/acetyl-CoA transacylase, dehydrase, enoyl reductase, β-ketoyl reductase, acyl carrier protein, and thioesterase. The complex enzyme is regulated at the level of DNA by **upstream stimulatory factor (USF)** and the **sterol regulatory element binding protein (SREBP)**. Transcription of fatty acid synthase is blunted by polyunsaturated fatty acids through **suppression of SREBP-1**. In adipose cells in white adipose tissue, **leptin** (130 amino acid containing

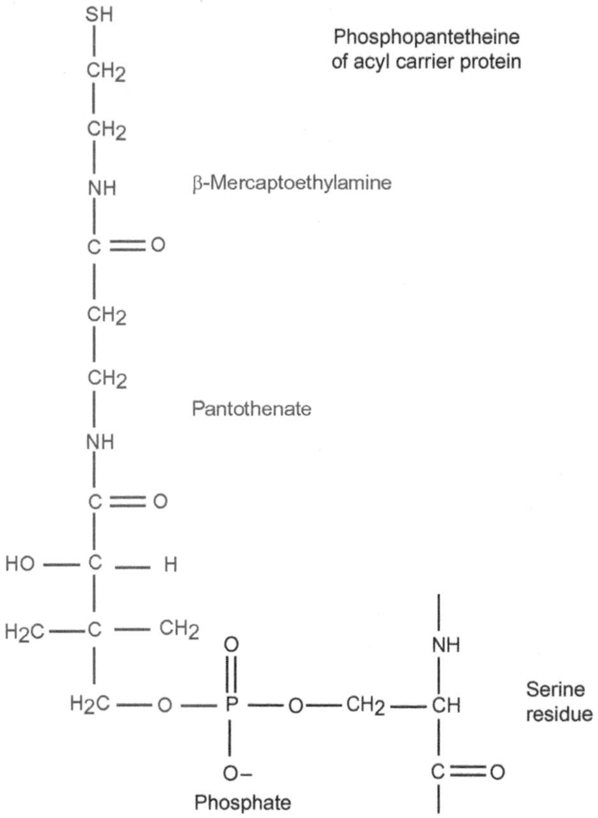

**FIGURE 9.49** Phosphopantetheine is covalently linked through a phosphate ester to a serine hydroxyl of the acyl carrier protein component of FAS.

Steps 1-3 of the Fatty Acid Synthase reaction pathway are catalyzed by the catalytic domains listed in the diagram at right.

Shown are the cysteine of one protein subunit and the acyl carrier protein phosphopantetheine (Pant) of the other subunit of the dimeric complex

In steps 4-6:

The β-Ketone is reduced to an alcohol, by electron transfer from NADPH.

Dehydration yields a trans double bond.

Reduction at the double bond by NADPH yields a saturated chain.

Following intersubunit transfer of the fatty acid from phosphopantetheine to cysteine sulfhydryl, the cycle begins again, with reaction of another malonyl CoA.

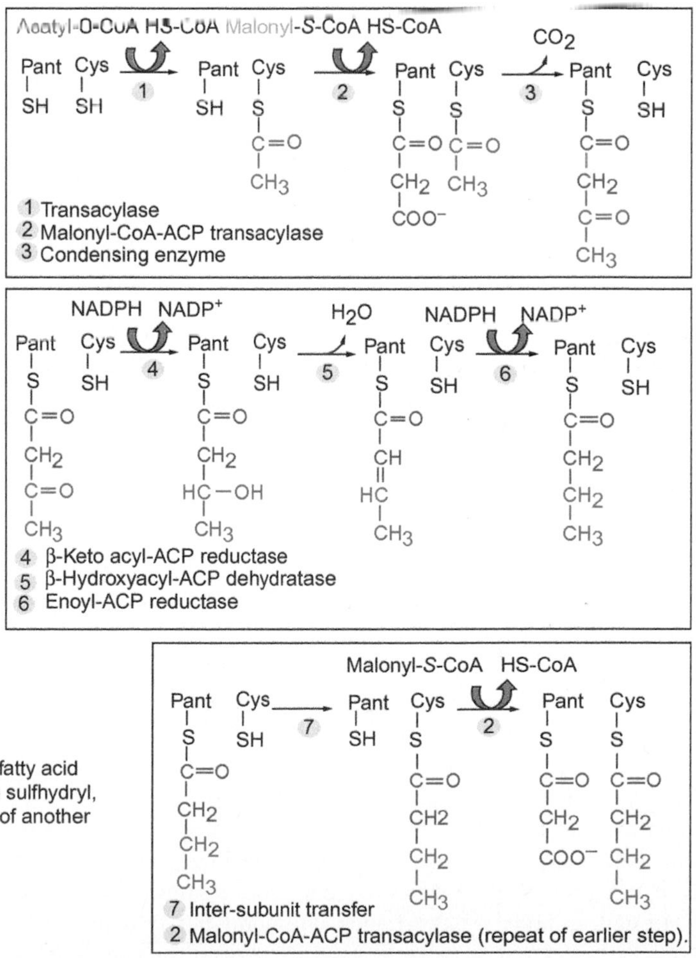

**FIGURE 9.50** Individual steps in fatty acid synthesis. *Pant*, **phosphopantethiene**. Colored structures are self-described. Each step is detailed to the *left of each set of reactions. Reproduced from http://rpi.edu/dept/bcbp/molbiochem/MBWeb/mb2/part1/fasynthesis.htm.*

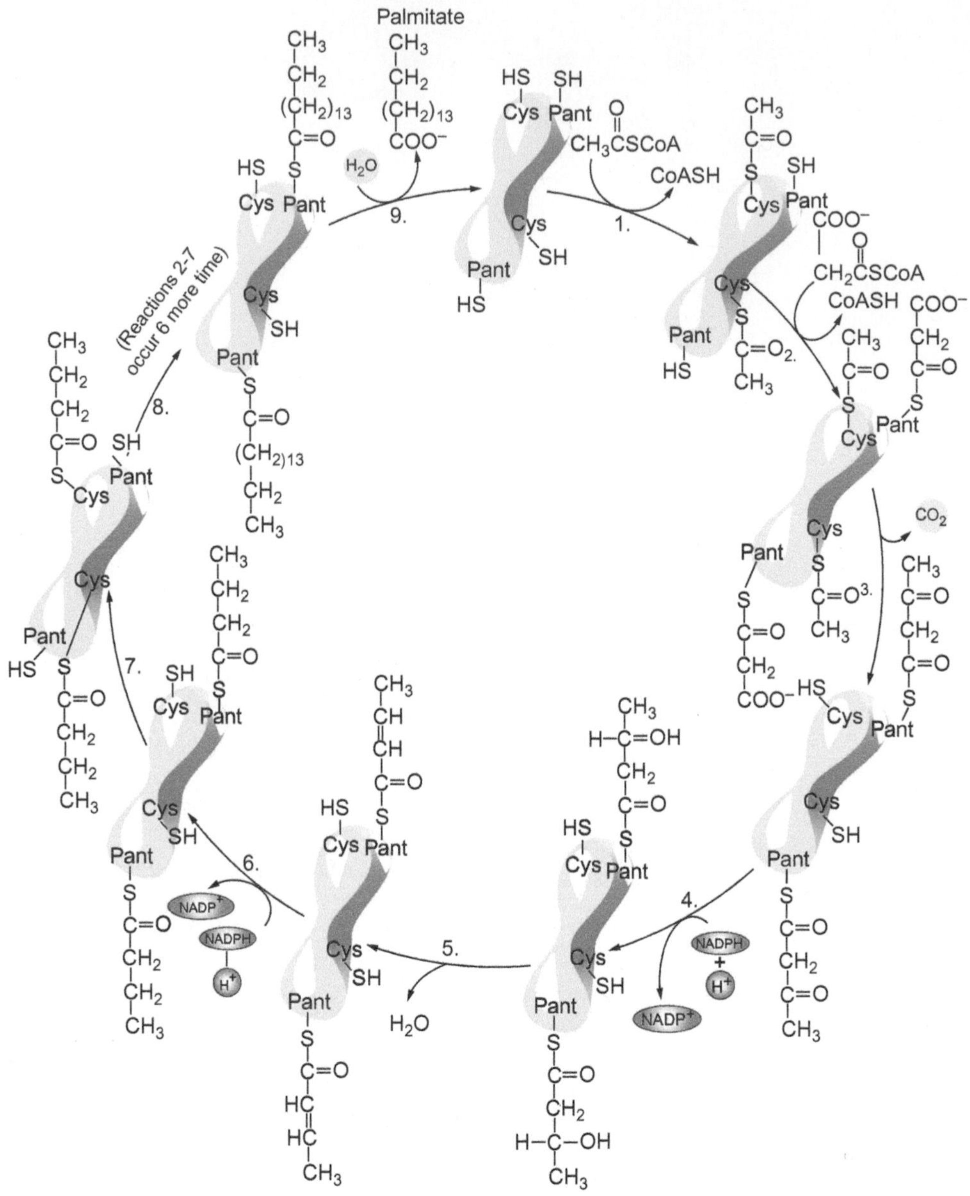

**FIGURE 9.51** Fatty acid synthase reactions shown, stepwise, leading to the synthesis of palmitate. The reactions are numbered starting in the *upper right center*. For convenience, only one of the two subunits of fatty acid synthase is represented. *Redrawn with permission from McGraw-Hill, McKee, Biochemistry, 1996, DuBuque, Iowa.*

protein) is produced and amount of leptin in blood is proportional to the total body fat. In liver and muscle mitochondria, leptin stimulates the oxidation of fatty acids, decreasing fat storage in those tissues. Leptin enters the central nervous system proportionally to its plasma concentration and interacts with leptin receptors on neurons in the **mediobasal hypothalamus** that are involved in energy intake and expenditure. Activation of certain leptin receptors leads to production of α-**melanocyte stimulating hormone** (α-**MSH**) that causes **appetite suppression**. In the **arcuate nucleus** leptin binds to **neuropeptide Y** (**NPY**)-producing neurons and decreases their activity, resulting in **satiety**. Leptin can prevent the secretion of **anandamide**, an endogenous cannabinoid, which binds to its receptors and stimulates **feeding**. Leptin and NPY are polypeptides and anandamide (cannabinoid) is derived from arachidonic acid (Fig. 9.52).

Homozygous mutations in the leptin gene result in the constant desire for food and resultant **obesity**. Mutations in the leptin gene can lead to obesity, however, it is uncertain that all obesity is the result of leptin gene mutation. Consumption of large amounts of **fructose** (e.g., corn syrup and honey) can lead to **leptin resistance**, the elevation of triglycerides and gain in body weight. Leptin will undoubtedly play an important role in the regulation of body weight.

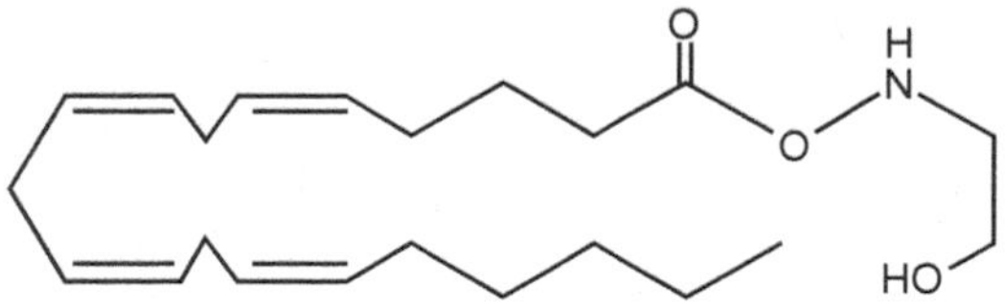

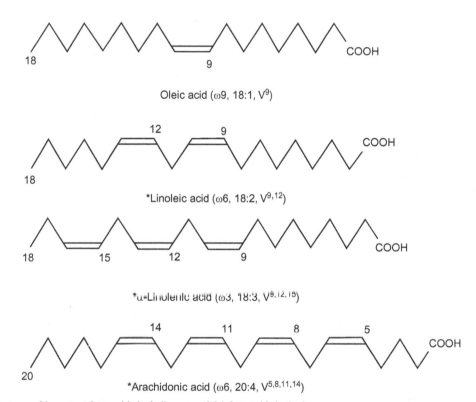

**FIGURE 9.52** Anandamide is arachidonylethanolamide, an endogenous cannabinoid derived from arachidonic acid.

Oleic acid ($\omega$9, 18:1, $V^9$)

*Linoleic acid ($\omega$6, 18:2, $V^{9,12}$)

*$\alpha$-Linolenic acid ($\omega$3, 18:3, $V^{9,12,15}$)

*Arachidonic acid ($\omega$6, 20:4, $V^{5,8,11,14}$)

**FIGURE 9.53** Structures of important fatty acids including essential * fatty acids in the human.

There is evidence in the human for a cytokine called **resistin** ($\sim$ 12.5 kDa polypeptide). It is secreted from adipose tissue and also from immune and epithelial cells. It has been proposed that resistin is linked to **insulin resistance** in obesity and in type 2 diabetes. Its concentration in serum increases with obesity. Resistin also is thought to play a role in inflammation and energy balance.

**Double bonds** are introduced at specific positions in the chain by desaturases. This requires enzymes of the endoplasmic reticulum: NADH cytochrome $b_5$ reductase, cytochrome $b_5$ and **desaturase**. The desaturases are mixed function oxidases that catalyze 4-electron reduction of oxygen to form 2 water molecules as a double bond is introduced into the fatty acid. For introduction of a double bond in **stearate** (18:0; indicating 18 carbons in length with no double bonds) to form **oleate** (18:1 *cis* $\Delta$-9), the overall reaction is as follows:

$$\text{stearate} + \text{NADH} + \text{H}^+ + \text{O}_2 = \text{oleate} + \text{NAD}^+ + 2\text{H}_2\text{O}$$

Oleic acid has a double bond at C9$-$10 and other fatty acids have double bonds in different positions. *Certain fatty acids need to be ingested through the diet as humans have a limited capacity for synthesizing long-chain fatty acids. Therefore the ingestion of linoleic and $\alpha$-linolenic acids is essential.* Linoleic acid is an $\omega$-6 (omega-6) fatty acid, containing a double bond located at the 6th carbon from the methyl end of the molecule. The first carbon ($\alpha$-carbon) is numbered from carbon bearing the terminal carboxyl group so the omega carbon would be the end carbon at the terminal methyl group. Linolenic acid is a component of fish oils and is an $\omega$-3 fatty acid, having a double bond at the third carbon from the terminal methyl group. Arachidonic acid (20:4) can be formed from linoleic acid (18:2), however little arachidonic acid is formed in this way. Arachidonic acid has to be ingested and is therefore essential. The structures of essential fatty acids and other important fatty acids are shown in Fig. 9.53.

## Phospholipids and Membranes

Phospholipids, like triglycerides are usually substituted glycerols called **glycerophospholipids**. One of the three carbons in glycerol is substituted by a phosphate group and, in turn, the phosphate group can be substituted by another group. In the cell membrane the predominant glycerophospholipid is **phosphatidylcholine** shown in Fig. 9.54.

There are a variety of phospholipids in addition to phosphatidylcholine, where instead of choline the phosphate group can be linked to serine, ethanolamine, inositol, amide or glycerol (Fig. 9.55). In **diphosphatidylglycerol** the 1- and 3-carbons of glycerol are substituted by phosphate groups that, in turn, are linked to other groups.

Phospholipids also can act as emulsifiers for solubilizing hydrophobic molecules.

## Lipins

Lipins are a family of three related proteins, lipin-1, lipin-2, and lipin-3. Lipin-1, itself, has 2 isoforms of 98 and 102 kDa formed by alternative mRNA splicing. Lipin-1 is the most studied as it is expressed in many tissues with major expression in adipose tissue, skeletal muscle and testis. These proteins play a role in glycerolipid biosynthesis and in gene regulation as well. Mutations in the lipin genes can result in lipodystrophy, myoglobinuria, inflammatory disorders, insulin resistance and enhanced susceptibility to atherosclerosis. Lipin-1 acts as a **phosphatidic acid phosphatase** (**PAP**) (Fig. 9.55) to produce **diacylglycerol** (**DAG**) and Pi. The enzyme requires a **DxDxT motif** in one of its domains. This phosphatase activity is common to all 3 lipins. Lipin-1 is known to be able to translocate from the cytoplasm to the cell nucleus, requiring a nuclear translocation motif in its sequence. When the level of fatty acids is high in the cell, lipins translocate from the soluble cytoplasm (cytosol) to the endoplasmic reticulum membrane where they exert their

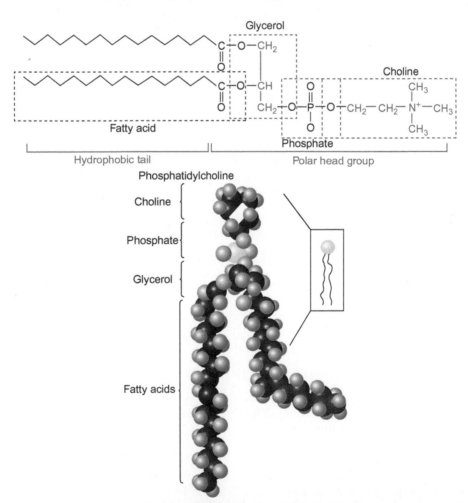

**FIGURE 9.54** Chemical structure of phosphatidylcholine (*top*). The polar head group is formed by glycerol, phosphate and choline. This part of the molecule is hydrophilic. The two fatty acid substituents are lipophilic. An atomic model is shown at the bottom.

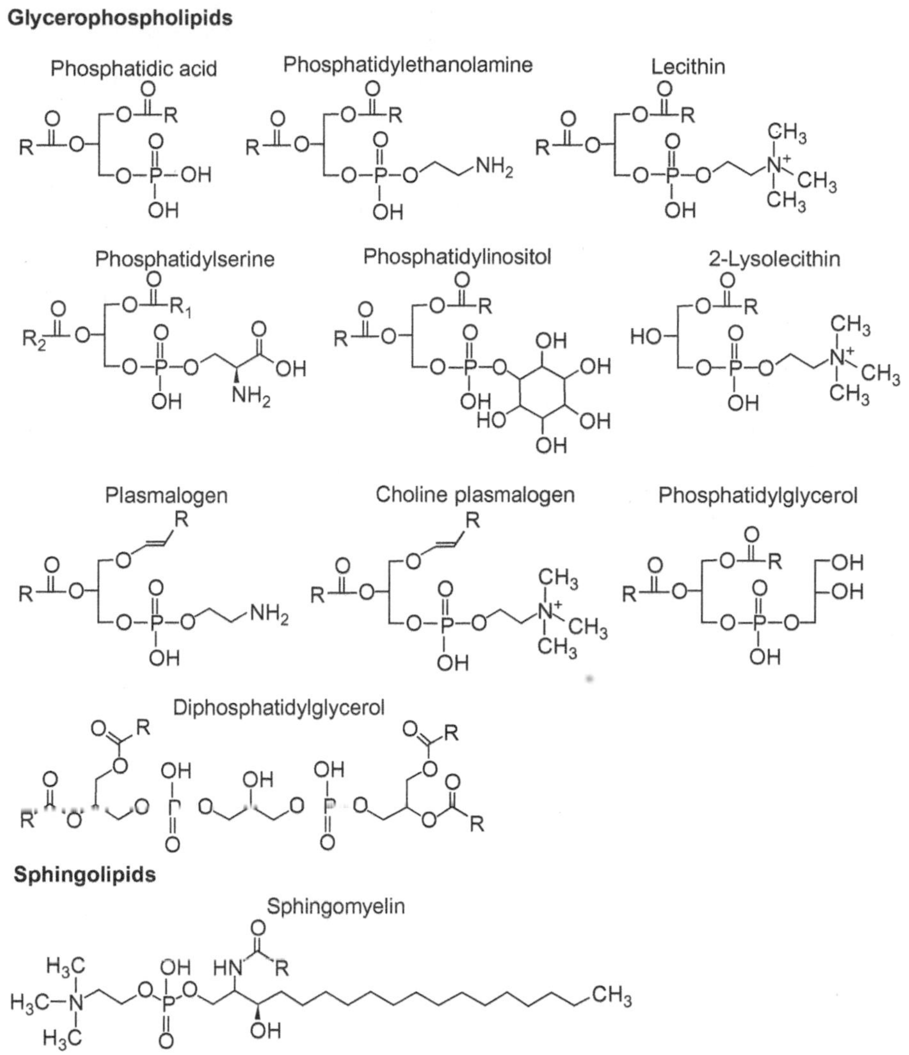

**Glycerophospholipids**

Phosphatidic acid   Phosphatidylethanolamine   Lecithin

Phosphatidylserine   Phosphatidylinositol   2-Lysolecithin

Plasmalogen   Choline plasmalogen   Phosphatidylglycerol

Diphosphatidylglycerol

**Sphingolipids**

Sphingomyelin

**FIGURE 9.55** Structures of a variety of phospholipids. They are generally found in membranes and phosphatidylcholine is a critical component of the plasma cell membrane and forms a part of the lipid bilayer as shown in Fig. 9.56.

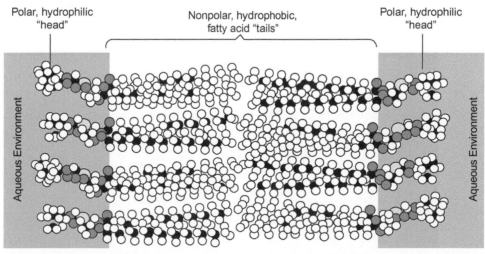

**FIGURE 9.56** Glycerophospholipids in membranes. The polar hydrophilic heads (glycerol, phosphate, and choline) interact with the aqueous environment on either side of the membrane.

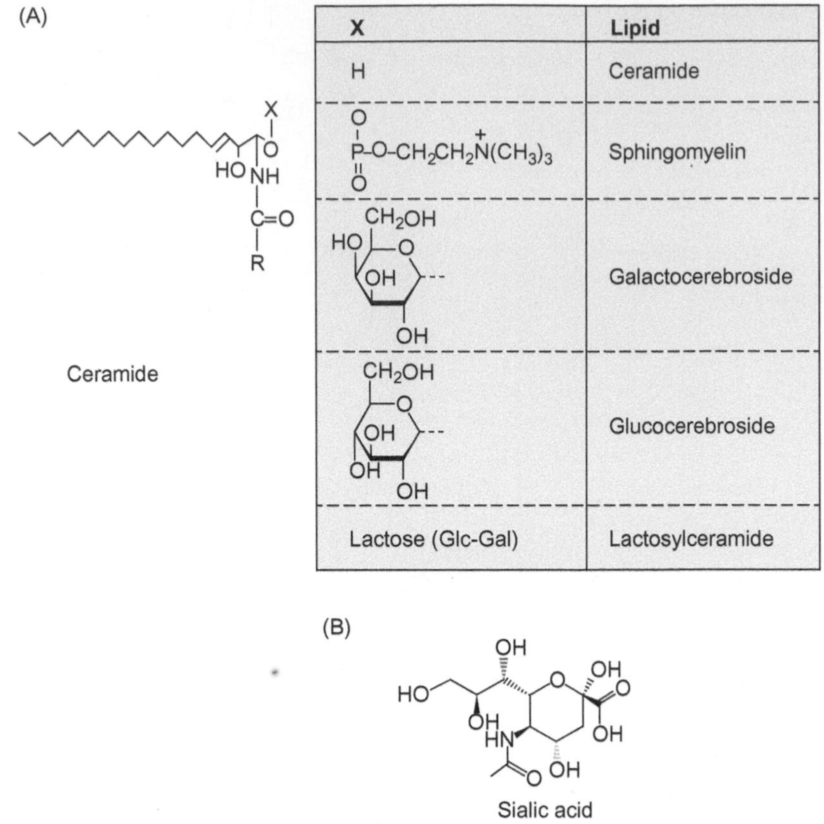

(A)

Ceramide

(B)

Sialic acid

**FIGURE 9.57**   (A) Structures of some ceramide derivatives. The long-chain aminodiol sphingosine (ceramide) serves as the lipid backbone. R on the ceramide structure (*upper left*) is a 15-carbon long-chain hydrocarbon of the upper chain of the ceramide structure. X indicates the substituent location and the various substituents for X are shown in the column so labeled. The compound without a substituent is ceramide. Substituting **sialic acid (B)** in X is the structure of a **ganglioside**. The relevant structures substituting X to generate the named compounds are shown in the figure. (B) The structure of sialic acid.

phosphatidic acid phosphatase activity. Lipin-1 is active by itself in adipose and skeletal muscle but is partly active in the liver, kidney, heart and brain. Lipins-2 and -3 also may be active in addition to lipin-1 in these latter tissues. DAG produced in the enzymatic reaction may be a substrate leading to many other derivatives (Fig. 9.55). DAG can activate PKC and enzymes like PKA as well.

**Sphingomyelin** is an important phospholipid (Fig. 9.55) in the brain. It is a **sphingolipid** and has a polar head group and two lipophilic tails like the phospholipids. The sphingomyelins (part of the myelin sheath; Fig. 9.55) are the only phospholipids of the sphingolipids. Among the sphingolipids are sphingomyelin, the glycosphingolipids and **sphingolipid sulfatides**. In the brain, the glycophospholipids are phosphatidylcholine, **phosphatidylethanolamine**, **phosphatidylinositol**, and **phosphatidylserine**. **Sphingosine** is a long-chain amino alcohol and is a major substance among the sphingolipids. In addition, **glycosphingolipids** are comprised by cerebrosides, sulfatides, globosides and gangliosides. Some of structures of sphingolipids are shown in Fig. 9.57.

As the name implies the **cerebrosides** are located mainly in neuronal cell membranes. They often have **galactose** as a single sugar linked to ceramide (**galactocerebroside**; Fig. 9.57). Glucose also appears in **glucocerebroside** and this appears in the synthesis or degradation of complex glycosphingolipids. Sulfatides are sulfuric esters of galactocerebrosides. These structures are shown in Fig. 9.58.

Thus, **ceramide** is the key structural unit found in all sphingolipids. Sphingolipids, consisting of cerebrosides, sulfatides, globosides and gangliosides, are a class of lipids derived from sphingosine (Fig. 9.59). They are important in cell recognition and signal transmission.

Properties of the sphingolipids are listed in Table 9.3.

## Synthesis and Degradation of Ceramide

Ceramide is synthesized from serine and palmitoylCoA (Fig. 9.60).

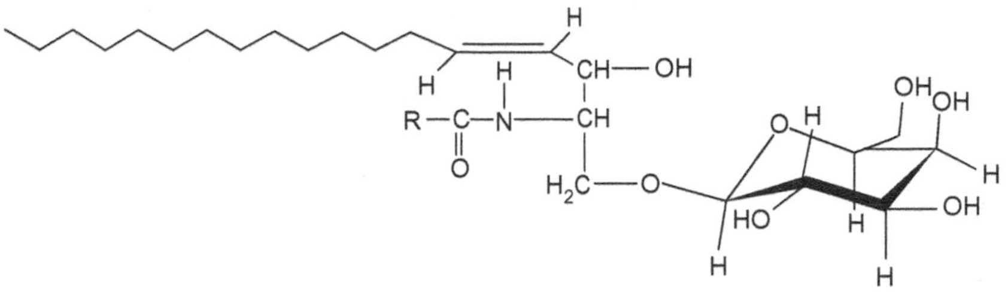

Structure of a glucocerebroside

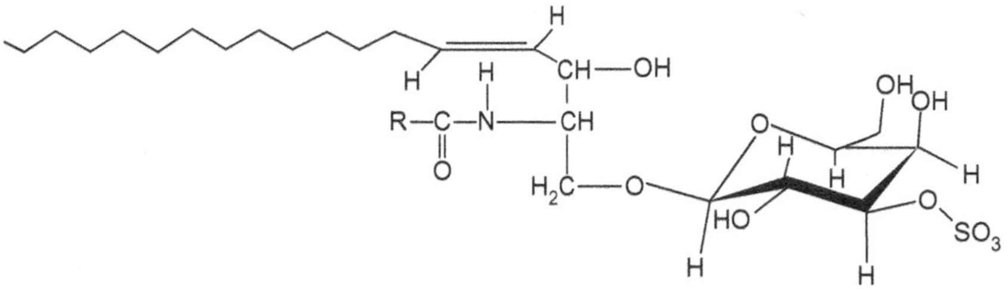

Structure of a sulfatide

**FIGURE 9.58** Structures of a glucocerebroside (*top*) and of a sulfatide (*bottom*).

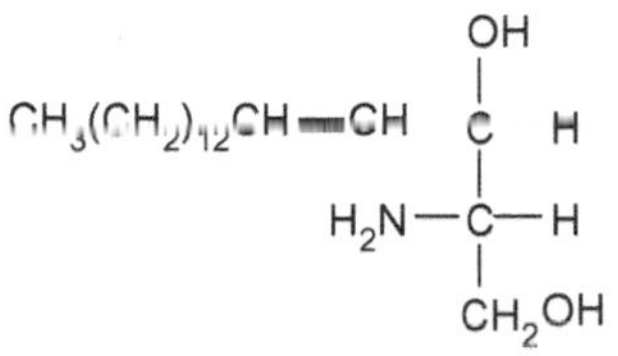

**FIGURE 9.59** Structure of sphingosine.

**TABLE 9.3 Properties of Sphingolipids**

| Sphingolipid | Properties |
|---|---|
| Ceramide | Structural unit common to all sphingolipids (a fatty acid chain substituting an amide linkage to sphingosine); there can be 3 different types of head groups |
| Sphingomyelins | Phosphorylcholine or phosphoroethanolamine in an ester linkage to the 1-hydroxyl group of ceramide |
| Glycophospholipids | Head groups can have different substituents; they are ceramides with one or more sugar residues in a β-glycosidic linkage to the 1-hydroxyl position |
| Cerebrosides | Contain a single glucose or galactose residue in the 1-hydroxyl position |
| Sulfatides | Sulfated cerebrosides |
| Gangliosides | Contain at least 3 sugar residues, one of which is sialic acid |
| Globoside | Glycosphingolipid with N-acetylgalactosamine side chain; cleavable by β-hexosaminidase |

Sphingomyelins contain mostly palmitic acid or stearic acid substituted at C2 of sphingosine by N-acylation. The formation of sphingomyelins from ceramide occurs by the donation of phosphorylcholine from phosphatidylcholine. This is catalyzed by **sphingomyelin synthase** (Fig. 9.61).

**Sphingomyelinase** is the key enzyme in the breakdown of sphingomyelin. There is an **acid sphingomyelinase** located in the lysosome. A deficiency of this enzyme leads to **Niemann−Pick disease (NPD)** in which sphingomyelin and cholesterol accumulate in the lysosome of the brain, leading to severe neurological disease. A **neutral**

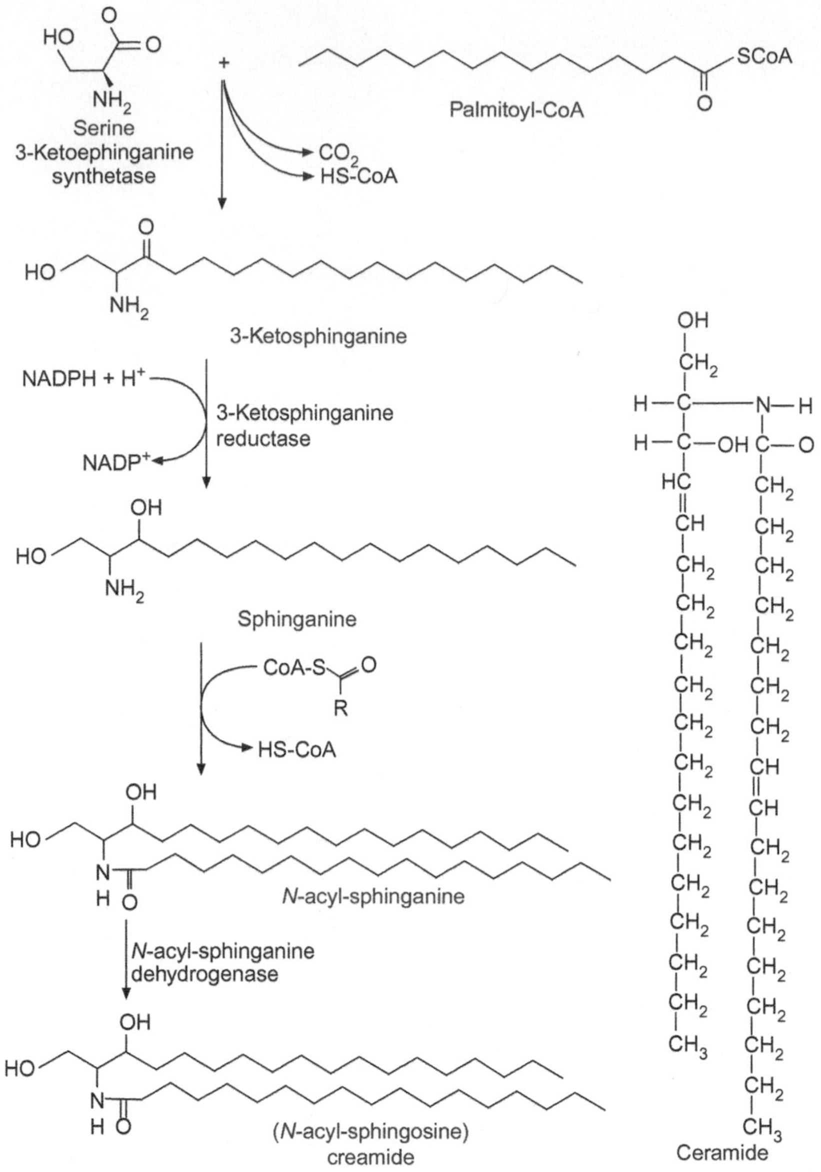

**FIGURE 9.60**   Biosynthesis of ceramide (*N*-acylsphingosine) from serine and palmitoylCoA.

**sphingomyelinase** occurs in liver and spleen (and possibly other tissues). The degradation products are essentially the same for both enzymes (Fig. 9.62).

The resulting ceramide can be further broken down to **sphingosine** by **ceramidase** and sphingosine can be phosphorylated by **sphingosine kinase** to form **sphingosine-1-phosphate (S1P)**. The human neutral **ceramidase** is located in the plasma membrane. There are as many as five ceramidases reported: one neutral, one acidic and three isoforms of alkaline ceramidase. Ceramide is proapoptotic (leads to programmed cell death), whereas sphingosine-1-phosphate has the opposite effect on the cell (involved in cell proliferation, survival, cell migration, angiogenesis and allergic responses). Ceramide binds to the **ceramide receptor**, a calcium ion sensing receptor that is G protein (G$_{\alpha i}$)-coupled. This leads to stimulation of the **intrinsic pathway of apoptosis** (programmed cell death). The signal transduction pathway of ceramide-ceramide receptor complex is: stress-activated protein kinase (SAPK)/cJun, N-terminal kinase (JNK), phosphorylation of cJun (Bax may be involved), and caspase-3 activation leading to cleavage of DNA and cell death. The cell can resynthesize ceramide through a recycling pathway. This involves the sequential phosphorylation of sphingosine-1-phosphate by **sphingosine-1-phosphate kinase**, dephosphorylation by **sphingosine-1-phosphate phosphatase** to form sphingosine which is then converted to ceramide. Alternatively, sphingosine can be converted to

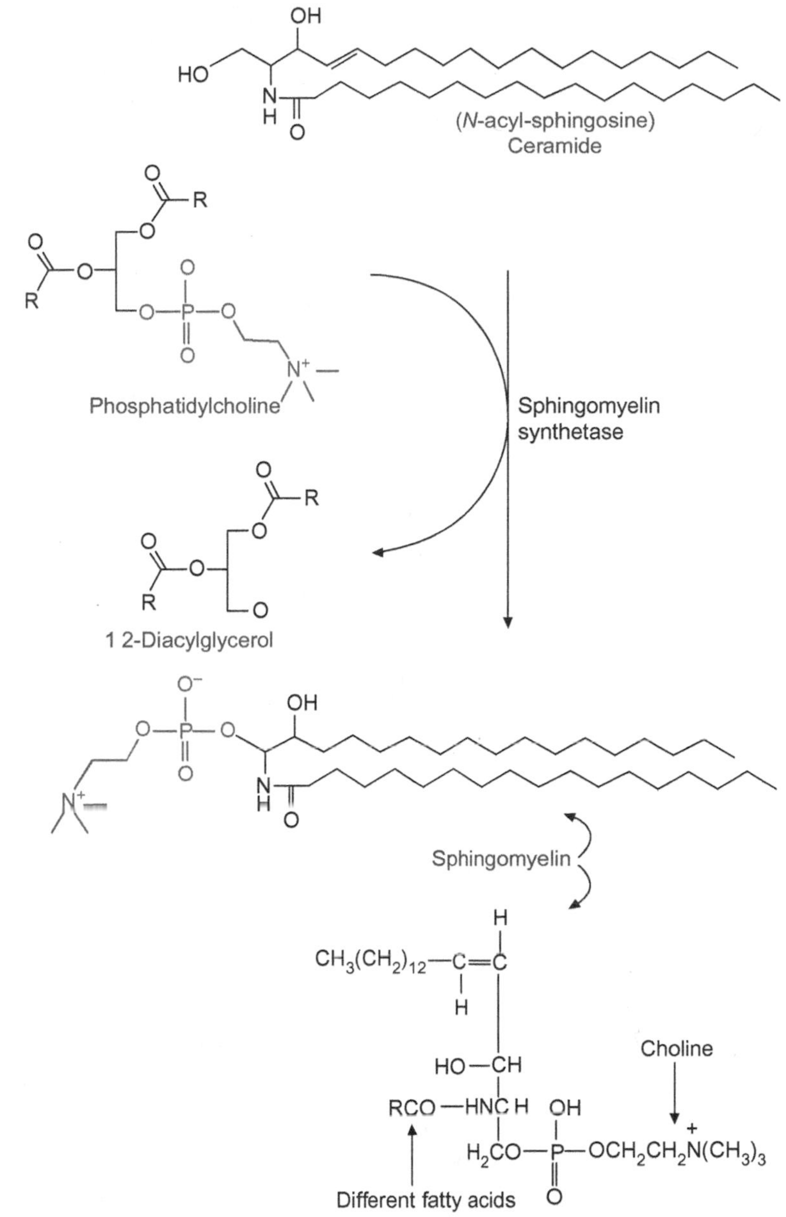

**FIGURE 9.61** Biosynthesis of sphingomyelin from ceramide.

ceramide (C16-ceramide is the active form as shown in Fig. 9.63) in the mitochondrion by **ceramide synthase**. *The balance between the concentrations of ceramide and sphingosine-1-phosphate may determine the life or death of a cell.*

In addition to its ability to reform ceramide through the recycling pathway, **sphingosine-1-phosphate (S1P)** can bind to **S1P receptors** whose actions affect a number of important cellular functions as already suggested (Fig. 9.64A and B). Furthermore, S1P regulates **histone deacetylases (HDACs)** and through this activity regulates gene expression. S1P also regulates **tumor necrosis factor (TNF)** and **nuclear factor kappa B (NFκB)** signaling. Receptors for S1P are endothelial ($S1P_1R$), smooth muscle ($S1P_2R$), vascular ($S1P_3R$), hematopoietic ($S1P_4R$), and neuronal ($S1P_5R$).

**Acid sphingomyelinase** is found in the **lysosome** (where the pH of the internal lysosomal environment is about 5) and it degrades sphingomyelin when it is transported to this organelle. Mutation that inactivates or weakens the activity of this enzyme can lead to the lysosomal storage disease, **Niemann-Pick disease (NPD)**. Of the five types of Niemann-Pick disease, two are caused by defects in acid sphingomyelinase. NPD is fatal and death usually occurs by 18 months of age. Some symptoms of NPD caused by sphingomyelin accumulation are: unsteady gait (ataxia), slurring of speech (dysarthria), uncoordinated swallowing (dysphagia) and in more advanced disease involving the cerebral cortex,

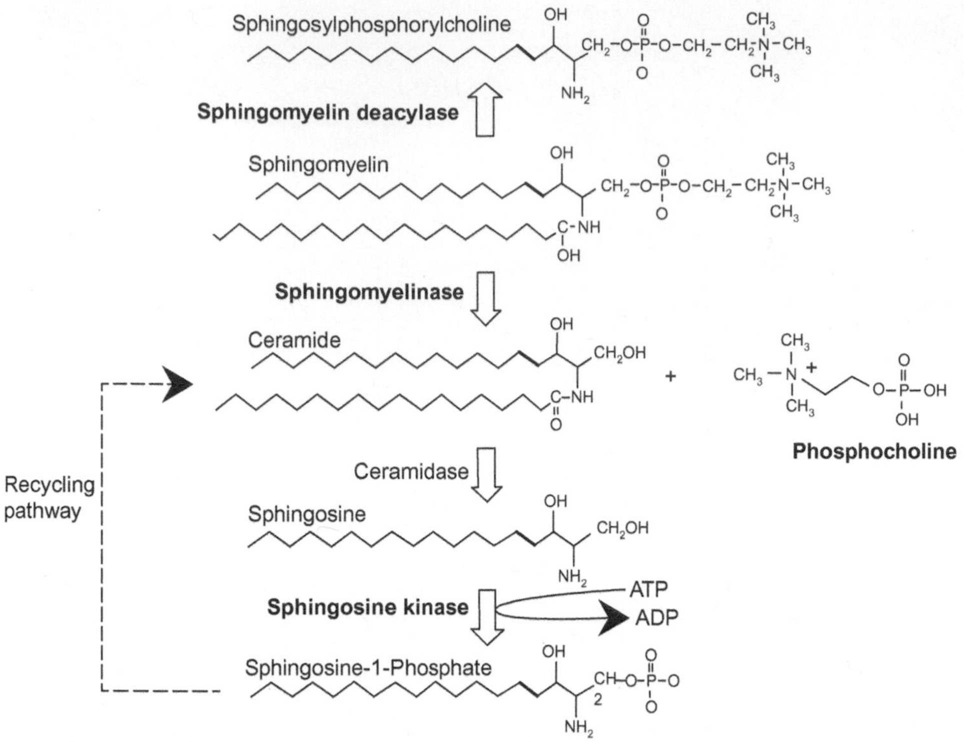

**FIGURE 9.62** Degradation of sphingomyelin. The action of sphingomyelinase converts sphingomyelin to ceramide and phosphocholine. Ceramide can be further degraded by ceramidase to sphingosine which, in turn, can be converted to sphingosine-1-phosphate through the action of sphingosine kinase and ATP. The cell can synthesize ceramide from sphingosine-1-phosphate through a recycling pathway.

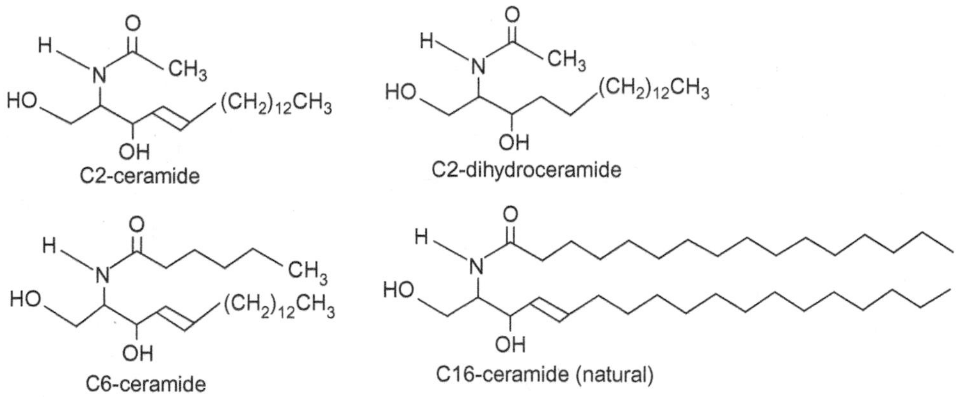

**FIGURE 9.63** The structures of ceramide (C2-ceramide, C6-ceramide, biologically inactive C2-dihydroceramide, and the natural active C16-ceramide). *Reproduced from http://www.molvis.org/molvis/v13/a180/samadi-fig1.html.*

involuntary rapid eye movements, gradual loss of intellectual ability leading to dementia and seizures. Table 9.4 lists several diseases of abnormal sphingolipid metabolism.

## Glycosphingolipids

Glycosphingolipids (**GSL**s) are formed from ceramide with the transfer of a sugar (e.g., **UDP-glucose**) to form glucosylceramide. **Glucosylceramide (GlcCer)** can be further transformed to **lactosylceramide (LacCer)** that is a precursor for several types of glycosphingolipids (Fig. 9.65).

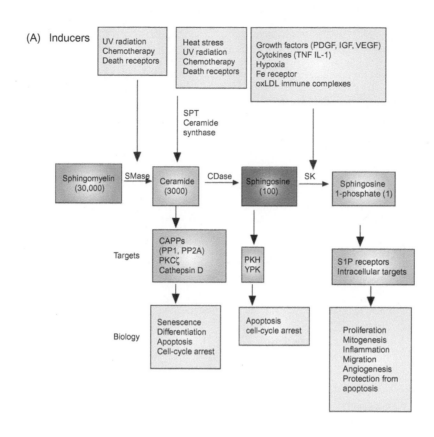

(A) Inducers

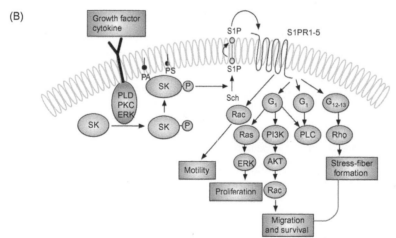

**FIGURE 9.64** (A) Pathway of conversion of sphingomyelin to sphingosine-1-phosphate (S1P) shown in the *center*. Above are inducers of enzymes in this pathway (refer to Fig. 9.62). *SMase*, sphingomyelinase; *CDase*, ceramidase; *SK*, sphingosine kinase. These substances act on various targets (shown in third horizontal items in the figure) to produce biological effects (shown on the bottom line). S1P acts on its receptors to produce a wide variety of effects (right hand column). (B) A growth factor or cytokine can bind to its cell membrane receptor (*top, left side*) leading to the stimulation (via phosphorylation) of sphingosine kinase (*SK*) that catalyzes the conversion of sphingosine (Sch) to sphingosine-1-phosphate (*S1P*). S1P can cross the cell membrane and bind to its receptor (*S1PR_{1-5}*) whose signaling actions (depending on the cell type and the specific cell type receptor) lead to important biological effects (motility, proliferation, migration and survival and stress fiber formation which inhibits migration). *Redrawn from http:// salamano-giovanni.blogspot.com/2009/06/endometriosis-pathophysiological-model.html.*

Glucosylceramide is converted to LacCer by the enzyme **glucosylceramide β4-galactosyltransferase** that is the **lactosylceramide synthase**. LacCer is involved in many cellular activities including cell proliferation, adhesion, migration and angiogenesis. LacCer synthase is located in the **Golgi** apparatus and is regulated by **NFκB** and other factors. As shown in the figure, LacCer can be converted to other glycosphingolipids. The **ganglio-series** of glycosphingolipids are complex oligosaccharides found in the human **blood group B** and they contain galactose, fucose, N-acetylgalactose

**TABLE 9.4 Several Diseases Associated With Abnormal Sphingolipid Metabolism**

| Disorder | Enzyme Deficiency | Accumulating Substance | Symptoms |
|---|---|---|---|
| Tay–Sachs disease | Hexosaminidase A | GM$_2$ ganglioside | Mental retardation, blindness, early mortality |
| Gaucher's disease | Glucocerebrosidase | Glucocerebroside | Hepatosplenomegaly, mental retardation in infantile form, long bone degeneration |
| Fabry's disease | α Galactosidase A | Globtriaosylceramide; also called ceramide trihexoside (CTH) | kidney failure, skin rashes |
| Niemann–Pick disease, more info | Sphingumyelinase | Sphingomyelin | All types lead to mental retardation, hepatosplenomegaly, early fatality potential |
| | | LDL-derived cholesterol | |
| Types A and B | | LDL-derived cholesterol | |
| Type C1 | | | |
| Type C2 | | | |
| Type D | | | |
| Krabbe's disease: globoid leukodystrophy | Galactocerebrosidase | Galactocerebroside | Mental retardation, myelin deficiency |
| Sandhoff–Jatzkewitz disease | Hexosaminidase A and B | Globoside, GM2 ganglioside | Same symptoms as Tay–Sachs, progresses more rapidly |
| GM$_1$ gangliosidosis . | GM$_1$, ganglioside-β-galactosidase | GM$_1$ ganglioside | Mental retardation, skeletal abnormalities, and hepatomegaly |
| Sulfatide lipodosis; metachromatic leukodystrophy | Arylsulfatase A | Sulfatide | Mental retardation, metachromasia of nerves |
| Fucosidosis | α-L-Fucosidase | Pentahexosylfucoglycolipid | Cerebral degeneration, thickened skin, muscle spasticity |
| Farber's lipogranulomatosis | Acid ceramidase | Ceramide | Hepatosplenomegaly, painful swollen joints |

**Source**: Reproduced with permission from http://www.med.unibs.in/-marchesi/lipsynth3.html.

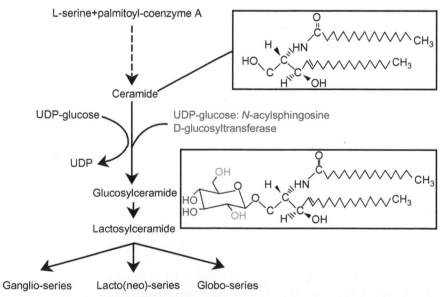

**FIGURE 9.65** Biosynthesis of glycosphingolipids in humans. *Reproduced from F.M. Platt and T.D. Butters,* Expert Reviews in Molecular Medicine, *Cambridge University Press, OSSN 1462-3994, 1999. And from http://journals.cambridge.org/fulltext_content/ERM/ERM2_01/ S1462399400001484sup002.htm.*

and glucose linked to ceramide. The **lacto(neo)-series** can be found in **red blood cells** and in some **leukocytes** and contain lactose (e.g., lactotriosyl-ceramide), polylactosamine long-chain structures and tri-*O*-methylgalactose. The **globo-series** are associated with the monocytic lineage of human myeloid cells. They are neutral glycosphingolipids and can be found in monocytes, granulocytes and monoblastic leukemia cells. Some contain globotetraosylceramide and globotriaosylceramide. The latter has one glucose residue and two galactose residues and accumulates in **Fabry's disease** (Table 9.4). Gangliosides resemble globosides except that they have an additional *N*-acetyl neuraminic acid residue. The structures of a **ganglioside** and of single and **complex glycosphingolipids** are shown in Fig. 9.66.

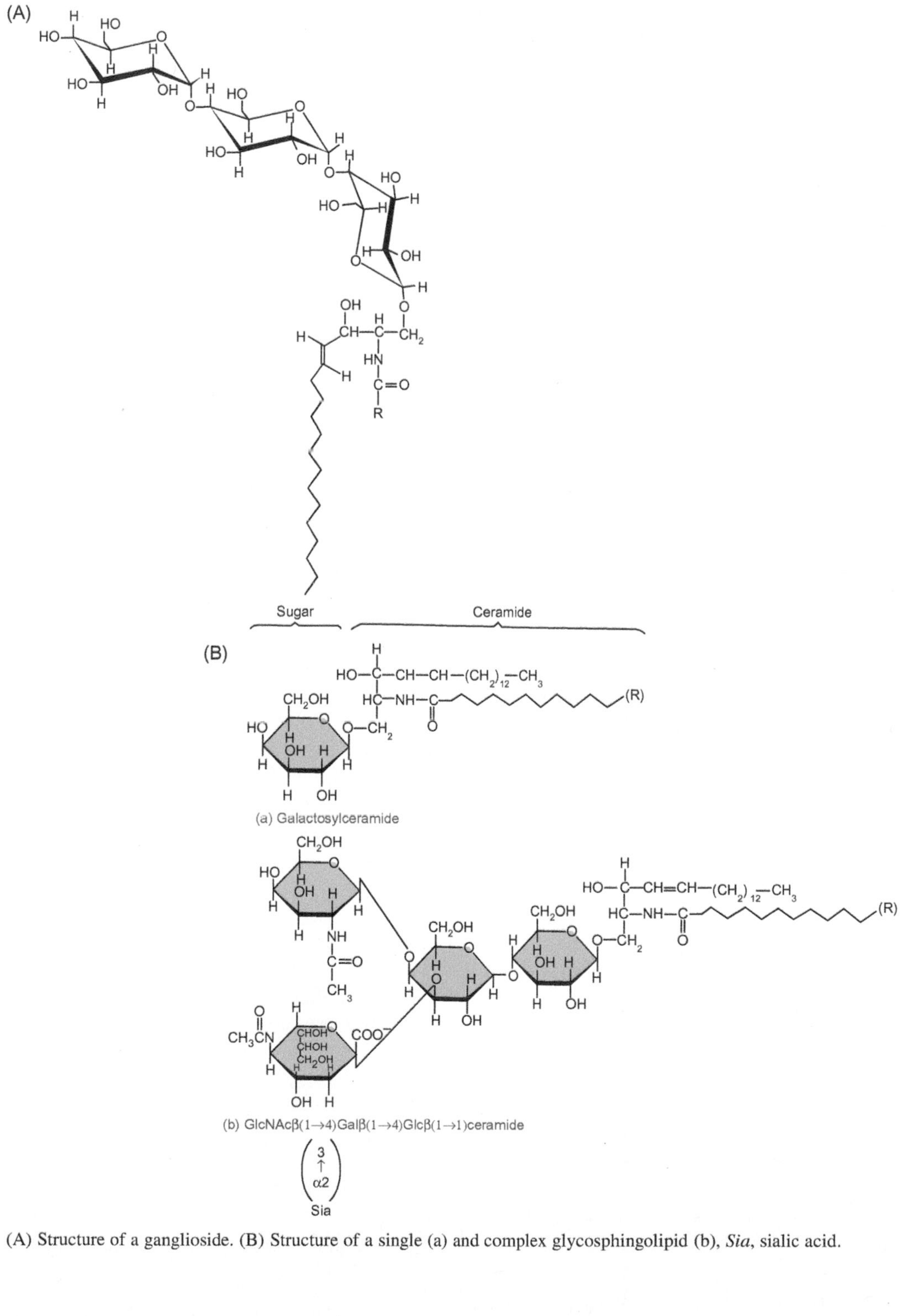

**FIGURE 9.66** (A) Structure of a ganglioside. (B) Structure of a single (a) and complex glycosphingolipid (b), *Sia*, sialic acid.

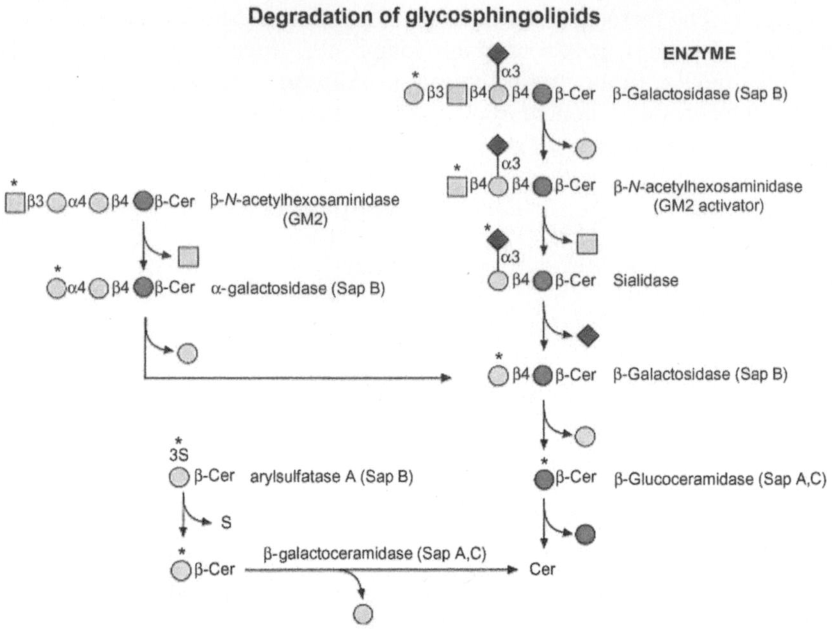

**Degradation of glycosphingolipids**

**FIGURE 9.67** Degradation of glycosphingolipids. *Yellow-filled square*, N-acetylgalactosamine (GalNAc); *blue-filled circle*, glucose; *yellow-filled circle*, galactose; *purple-filled diamond*, N-acetylneuraminic acid (Neu5Ac). Required activator proteins are shown in parentheses: *Sap*, **saposin** (small lysosomal protein, of four types, that isolates the lipid substrate from membranous surrounding so as to facilitate access to the soluble degradative enzyme); *Cer*, ceramide; *GM2*, a glycosphingolipid marker in cell membranes relates to cell—cell interactions. The enzymes involved in each degradative step are named for the reaction. *Reproduced from* Essentials of Glycobiology, *second ed., Figure 41.6, Edited by A. Varki, R.D. Cummings, J.D. Esko, et al., CSH Press, 2009.*

The degradation of glycosphingolipids is shown in Fig. 9.67.

The biosynthesis of gangliosides is shown in Fig. 9.68. This outlines the formation of **GM$_3$-ganglioside** starting with **ceramide** and **UDP-glucose**.

The catabolism of gangliosides is shown in Fig. 9.69.

Lipid storage diseases also can occur as the result of a deficiency of **lysosomal enzymes** that degrade the carbohydrate portions of gangliosides (Table 9.4).

## PROPERTIES OF LIPOPROTEINS

Because of the insolubility of lipids, such as cholesterol and phospholipids they are transported in the bloodstream in lipoprotein particles. Fusing proteins to lipids renders lipids more water-soluble. Whereas lipids have a density (g/mL of solute) of about 0.88 (they float in water), proteins have a density of about one g/mL or slightly higher (have solubility in water). The polar ends of phospholipids provide the interface with the aqueous medium, whereas the nonpolar portion of the lipids is located inside the particle and does not contact water molecules. The drawing of a typical lipoprotein particle can be seen in Fig. 9.25. **Cholesteryl esters** are cholesterol molecules with the fatty acid esterified through the 3-hydroxyl position of cholesterol. There are five classes of lipoproteins from the largest to the smallest: chylomicrons, very low density lipoprotein (VLDL), intermediate density lipoprotein (IDL), low density lipoprotein (LDL), and high density lipoprotein (HDL). The higher the density of the particle, the higher the ratio of protein to lipid and the density of the particle becomes lower as the lipid content increases relative to the protein content (Table 9.2, Fig. 9.70).

The larger particles, such as the **chylomicron remnants**, transport cholesterol from the bloodstream to the liver and the **LDL**s transport cholesterol to the extrahepatic tissues (heart, muscle, kidney and others). **HDL**s (converted to **IDL**s) transport cholesterol from the tissues back to the liver (Fig. 9.27). The lipoprotein particles can be classified by size and composition as shown in Table 9.5.

In the dietary lipid transport system, **chylomicrons** are formed from fats ingested in food processed by the intestine. They contain mostly triglycerides (Table 9.5) and they transport digested fat to the extrahepatic tissues, whereas **chylomicron remnants** transport somewhat diminished contents to the liver. In the triacylglycerol secretion system, **very low density lipoproteins (VLDLs)**, formed in the liver, transport triglycerides to fat cells from the liver and VLDL

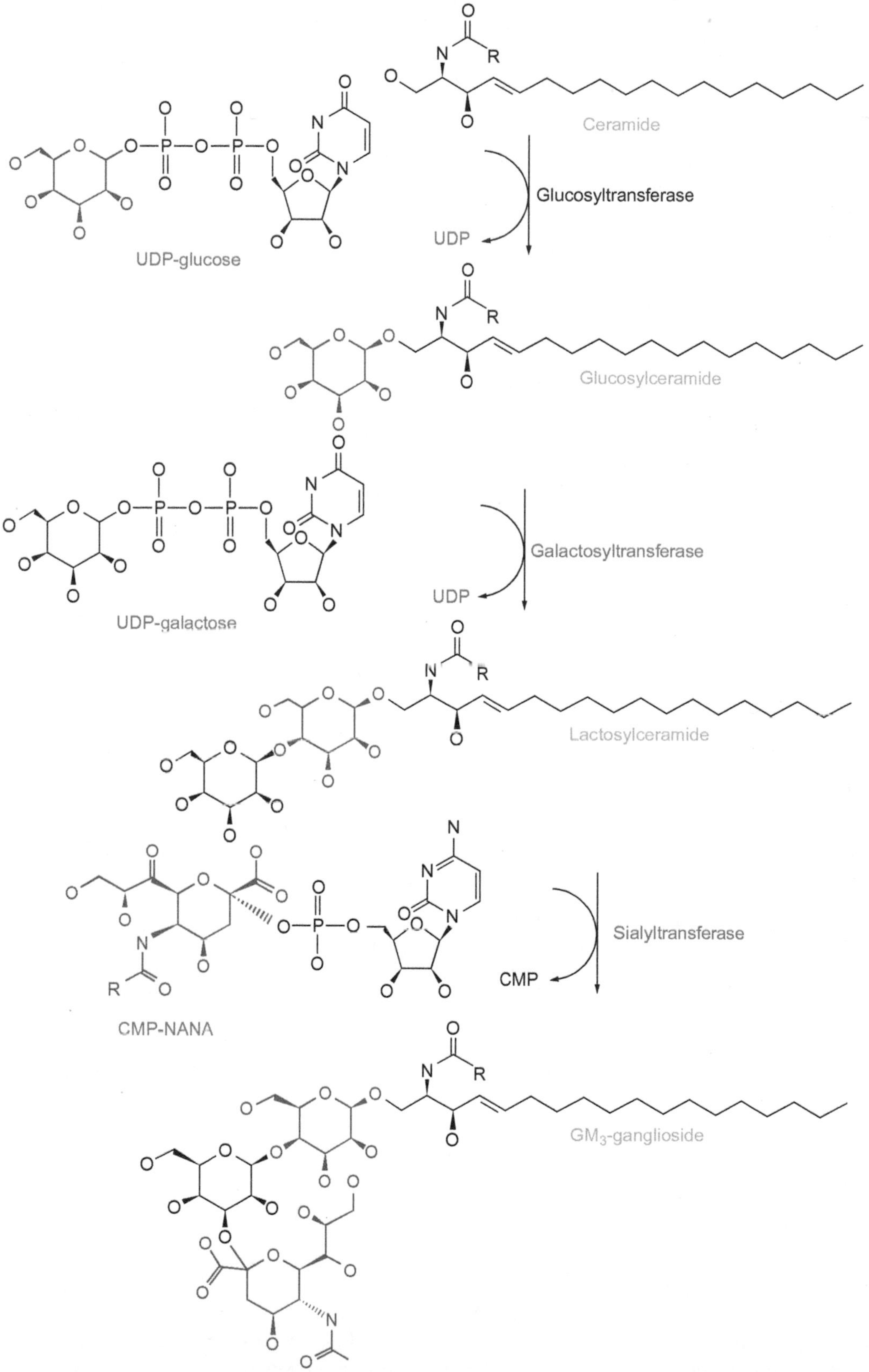

**FIGURE 9.68** Biosynthesis of the ganglioside, GM₃-ganglioside. G, ganglioside; M, monosialic acid (Fig. 9.57). Subscript 3 indicates three **sialic acid** residues or trisialic acid; **NANA**, *N*-acetylneuraminic acid.

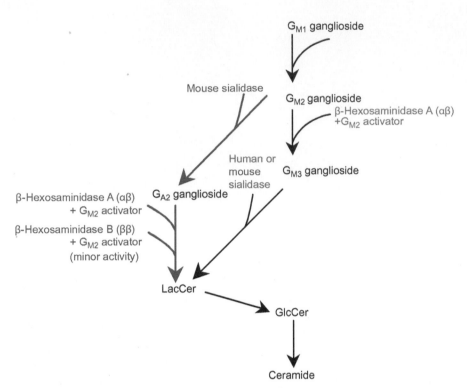

**FIGURE 9.69**  The catabolism of **gangliosides**. G, ganglioside; M, monosialic acid; M with subscript number indicates the number of sialic acid residues. The **$G_{M2}$ activator**, a small human protein encoded by the **GM2A gene** that binds molecules of $G_{M2}$, moves them from cell membranes to facilitate the interaction with **β-hexosaminidase A** to cleave **$N$-acetyl-D-galactosamine** and conversion to $G_{M3}$. It can transport molecules with an **$N$-acetylhexosamine**. *Redrawn from FM Platt & TD Butters,* Expert Reviews in Molecular Medicine, *Cambridge University Press, 1999 or from http://journals.cambridge.org/fulltext_content/ERM/ERM2_01/S1462399400001484sup021.gif.*

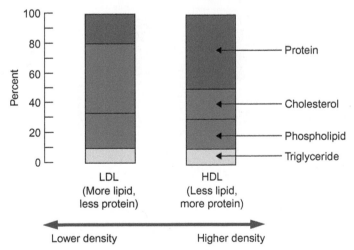

**FIGURE 9.70**  Distribution of the components of lipoproteins that determine high and low densities.

contents can be transported back to the liver via **IDLs** or **LDLs**. In the **reverse cholesterol transport system**, the extrahepatic tissues return cholesterol to the liver via **HDLs** and LDLs. LDLs are the main transporters of cholesterol to various cells and *HDLs are the only transporters of cholesterol from the extrahepatic tissues back to the liver*. High levels of HDLs are protective and forecast low risk of myocardial infarction. On the other hand, high levels of LDLs that transport cholesterol to nonhepatic tissues are considered cautionary. Formation of fatty plaques in arteries occurs through the action of **foam cells** that are derived from **macrophages** with a membrane **scavenger receptor**. The scavenger receptor binds LDL particles, imports the lipid to facilitate the transformation of the macrophage to a foam cell.

TABLE 9.5  Characteristics of Lipoprotein Particles

| Characteristics | Chylomicrons | Very Low-Density Lipoproteins | Low-Density Lipoproteins | High-Density Lipoproteins |
|---|---|---|---|---|
| Density (g/mL) | <0.95 | 0.95–1.006 | 1.019–1.063 | 1.063–1.210 |
| Particle diameter (nm) | >75 | 30–80 | 18–25 | 5–12 |
| Protein composition (% dry weight) | 1–2 | 8–10 | 20–25 | 52–60 |
| Triacylglycerol composition (% dry weight) | 80–88 | 45–53 | 5–9 | 2–3 |
| Cholesterol composition (% dry weight) | 2–4 | 17–27 | 43–50 | 12–25 |
| Phospholipid composition (% dry weight) | 7–9 | 17–19 | 19–21 | 17–24 |
| Function | Transport of exogenous triacylglycerol and cholesterol | Transport of endogenous triacylglycerol | Cholesterol transport to all tissues | Reverse cholesterol transport |

Chylomicrons have a particle diameter in the range of 100 to 1000 nm.
**Source**: Reproduced with permission from *Nature Reviews Drug Discovery*, **7**: 84–99 (January 2008), 2010 Nature Publishing Group, Macmillan Publishers, Ltd. and at http://www.nature.com/nrd/journal/v7/n1/fig_tab/nrd2353_T2.html.

TABLE 9.6  Susceptibility to Coronary Artery Disease (CAD) Based on Disorders of Circulating Lipoproteins

| Disorder | Biochemical Disorder | Susceptibility to CAD |
|---|---|---|
| Familial combined hyperlipoproteinemia | ↑VLDL and LDL particle number, ↑apoB | Increased |
| Familial hypertrig, with low HDL | ↑ Triglycerides, ↓ HDL | Increased |
| Familial hypoalphalipoproteinemia | ↓ HDL | Probably not increased |
| Familial dyslipidermic hypertension | ↑ Triglycerides, ↓ HDL, High blood pressure | Increased |
| Atherogenic lipoprotein profile | ↑VLDL TG; ↓ HDL, small-dense LDL | Increased |
| Familial hyperchylomicronemia[a] | ↑ chylomicrons | Not increased or slightly increased |
| Familial dysbetalipoproteinemia[b] | ↑β-VLDL | Increased |

[a]Caused by lipoprotein lipase deficiency or deficiency of apoCII.
[b]Associated with the apoE 2/2 genotype.
**Source**: Reproduced originally from http://www.acclakelouise.com/acc99/htm/gentbl3.jpg which is no longer available; from G. Litwack, *Human Biochemistry and Disease*, Academic Press/Elsevier, Table 5–2, page 235, 2008.

The foam cell is essentially a macrophage cell loaded with cholesterol. Susceptibility of coronary artery disease related to circulating lipoproteins is briefly outlined in Table 9.6.

## LIPID ANCHORING OF PROTEINS TO MEMBRANES

There are many ways in which a protein is attached or associated with a membrane. Some involve proteins that have transmembrane domains that are internal. There are **glycolipid anchors** in which the lipid portion at the end of the carbohydrate chain inserts (dissolves) into the membrane. Protein-protein interactions can occur when an external protein interacts with the extracellular domain of a membrane protein. Here, we discuss the attachment of nonpolar lipids to proteins to anchor the protein to the cell membrane. A specific amino acid residue (**cysteine** or **serine/threonine**) in a protein to be anchored is esterified or amidated to a fatty acid, such as **myristate** or **palmitate** (Fig. 9.71) or it can be

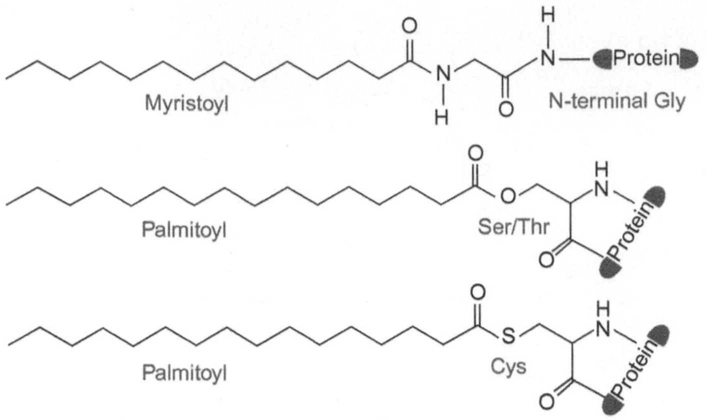

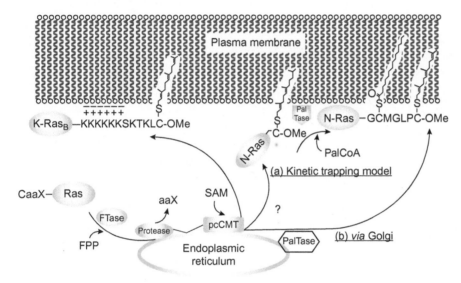

**FIGURE 9.71**  Protein anchors through esterification or amidation with fatty acids, such as palmitate or myristate.

**FIGURE 9.72** Anchoring of the ras protein to the cell membrane through lipidation of a cysteine residue. Caax, motif in *ras* that interacts directly with farnesyl in the active site of farnesyl transferase and the cysteine residue of this motif coordinates a zinc ion in the active site of the enzyme; *FTase*, **farnesyl transferase**; *FPP*, **farnesyl pyrophosphate**; *SAM*, *S*-**adenosylmethionine**; *PalCoA*, **palmitoylCoA**; *PalTase*, **palmitoyl transferase**; *pcCMT*, **prenylcysteine carboxyl methyltransferase**. *Redrawn with permission from the Internet article: "**Farnesylation** of Proteins and Peptides": http://www.mpi-dortmund.mpg.de/deutsch/abteilungen/abt4/praktikum/v3ss2004.pdf (no longer retrievable) and from G. Litwack, Human Biochemistry and Disease, Figure 5–62, page 237, Academic Press/ Elsevier, 2008.*

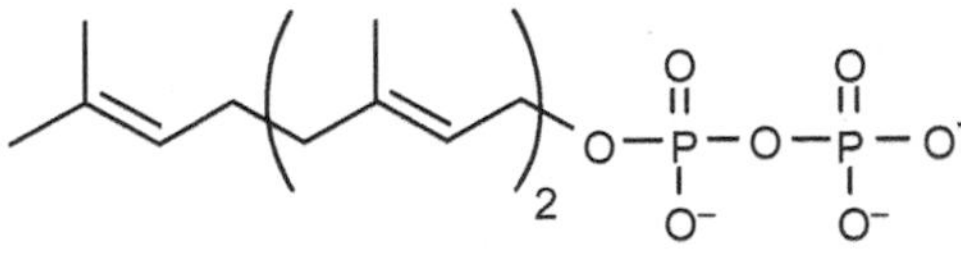

**FIGURE 9.73**  The structure of the 15-carbon farnesyl pyrophosphate. The parentheses indicate the isoprene unit.

an **isoprene residue**, such as farnesyl or geranylgeranyl which are intermediates in the biosynthesis of cholesterol. These are posttranslational modifications of proteins. Where proteins are destined to be anchored to a membrane, modification by **myristoylation** can occur through the N-terminal of a glycine residue, **S-palmitoylation** through a cysteine residue and **S-prenylation** of near C-terminal cysteine residues with farnesyl or geranylgeranyl units. Prenylated proteins comprise about 2% of the cellular proteins. *Ras* **proteins** can be prenylated and proteins involved in intracellular transport are prenylated to ensure localization in the membrane. Membrane insertion of *ras* protein is reversible as long as the protein is farnesylated but believed to be irreversible if the protein is palmitoylated (Fig. 9.72).

**Farnesyl**, usually in the form of the pyrophosphate, was discussed along with geranylgeranyl in the context of cholesterol synthesis in the discussion on hypercholesterolemia. The structure of **farnesyl pyrophosphate** is shown in Fig. 9.73.

*Ras* is an oncogene that plays a role in the generation of certain cancers. On that basis, inhibition of farnesyl transferase becomes a potential target for chemotherapeutic intervention. A sizeable number of proteins belonging to the

Ras, Rab and Rho families together with lamins and other proteins, especially the gamma subunit of G proteins are modified posttranslationally by farnesyl transferase and geranylgeranyl transferase. Inhibitors of these enzymes are used as treatments for cancer, viral infection and serious parasitic diseases.

## SUMMARY

The disease related to high cholesterol in the circulation is hypercholesterolemia. Cholesterol in LDLs that cannot be successfully transported to the tissue cells can be the result of several kinds of biochemical problems. Mutations of the LDL receptor account for many such problems but there are also possibilities for mutations in the protein ligand (e.g., apoB-100 mutations) of the LDL receptor. The rate-limiting enzyme of cholesterol synthesis is HMGCoA reductase. It is controlled by an elaborate complex of regulating proteins subject to phosphorylation—dephosphorylation. There also can be mutations affecting proteins involved in forming an endosome that permits the LDL—LDL receptor complex to enter the tissue cell. All of these mechanisms can lead to hypercholesterolemia. The diet is important because a fat-rich diet leads to high circulating cholesterol and high levels of cholesterol in cell membranes that affect a lowering of the LDL receptor in cells. There is an inherited form of hypercholesterolemia (familial hypercholesterolemia) that is semi-dominant. As there are two alleles for the LDL receptor, there can be one defective gene and one normal gene (one in 500 individuals). This condition results in a lower number of LDL receptors in tissue cells, resulting in higher levels of circulating cholesterol. When both genes are mutated (one in one million persons) few or no LDL receptors are present in tissue cell membranes and heart attacks and death can occur early in life.

Treatment of high levels of circulating cholesterol lately involves the use of statins, many of their structures are presented. These are competitive inhibitors of the rate-limiting enzyme of cholesterol synthesis, HMGCoA reductase. The actions of statins are related to the control of cholesterol synthesis by the sterol regulatory element binding (SREBP) protein and this system elaborated by Brown and Goldstein is reviewed.

Fatty acids and fats are described. The structure of fatty acids regarding the orientation of double bonds and length of the carbon chain are defined. The ingestion of fats, the process of solubilizing them and breaking them down in the intestine and their transport into the circulation are discussed. Fatty acid transporter 4, in particular, and other fatty acid transporters are discussed. The enzymes that digest triacylglycerols and phospholipids are described. The structure and properties of chylomicrons explain the process by which fats are taken up and the partial enzymatic digestion of the chylomicron that leads to its remnant that is taken up by the liver by a membrane receptor. There are major metabolic pathways by which chylomicron remnants and LDLs are formed from chylomicrons and VLDLs and subsequently metabolized.

β-Oxidation of fatty acids produce two-carbon fragments as acetyl-CoA until the total fatty acid is degraded. There is a special mechanism for handling odd-chain fatty acids where the fatty acid is broken down through 2-carbon fragments until the final 3-carbon that results in propionylCoA. This product is converted to succinylCoA which enters the TCA cycle to form malate which can either proceed through the cycle to form ATP or form glucose (gluconeogenesis) through phosphoenolpyruvate.

The carnitine cycle is responsible for moving acylCoA into the mitochondria to complete the beta-oxidation process. The pathway includes acylCoA, FAD, NAD⁺ and CoA. The entire fatty acid is degraded to 2-carbon fragments with as many acetyl-CoAs as there are pairs of carbons in the fatty acid. If all of the resulting carbons are burned in the TCA cycle, the yield of ATP can be calculated. For fatty acids having more than 22 carbons, beta-oxidation is carried out in the peroxisome to the product, octanoylCoA, which is transported to the mitochondria to complete the degradation. Fatty acids containing *cis* double bonds are converted to *trans* double bonds before they can proceed through β-oxidation.

When acetyl-CoA accumulates, ketone bodies (acetone, acetoacetate and β-hydroxybutyrate) are formed. These can be used as energy sources for the heart and brain. Acetoacetate produces two molecules of acetyl-CoA that can enter the TCA cycle and produce energy.

Cholesterol and cholesteryl esters can be stored in lipid droplets in adipose cells and in other kinds of cells when the levels of fatty acids are high, especially those cells that are committed to the synthesis of cholesterol derivatives (bile acids and steroid hormones). The structure of the droplet, evolving from the ER membrane, involves proteins, especially those of the PAT family consisting of perilipin, adipophilin and a cargo selection protein. These are located on the surface of the droplet. In order to break down the triglycerides in the droplet the entry of hormone-sensitive lipase occurs when perilipins are altered in their conformation, a protein kinase A regulated step. Breakdown of triglycerides in the droplet is regulated by epinephrine and insulin that control phosphorylation—dephosphorylation events. The regulation of HSL is defined in terms of its phosphorylatable sites.

Fatty acid synthesis starts with acetyl-CoA and is followed by the rate-limiting reaction, acetyl-CoA carboxylase to form malonylCoA. This is a complex reaction that involves biotin as coenzyme that is attached to the enzyme through a ε-amino group of lysine. The enzyme exists in 2 isozymes to produce 2 pools of malonylCoA. One isozyme inhibits beta-oxidation, whereas the other stimulates lipid synthesis. After the formation of malonylCoA, the remaining synthesis of fatty acid is carried out by cytoplasmic fatty acid synthase that has pantothenic acid as a coenzyme. Double bonds are introduced at specific positions in the carbon chain by desaturases.

Glycerophospholipids are defined by the presence of a phosphate group substituting one of the three carbons of glycerol in the triglyceride structure. The predominant glycerophospholipid in the cell membrane is phosphatidylcholine. Instead of choline, the phosphate group can be linked to serine, ethanolamine, inositol, amide or glycerol.

Lipins are multifunctional proteins involved in lipid metabolism. There are 3 lipins and lipin-1 has two isoforms generated through mRNA splicing. Under certain conditions lipin-1 can translocate the cell nucleus where it may act in gene regulation. When fatty acid levels are high in the cell, lipins translocate from the cytosol to the endoplasmic reticulum membrane where they act as phosphatidic acid phosphatases. Mutations of the genes for lipins cause many disorders.

Sphingomyelin is an important phospholipid in the brain. This sphingolipid has a polar head group and two lipophilic tails like the phospholipids. They are part of the myelin sheath. Ceramide is a key structural unit found in all sphingolipids. Sphingolipids consist of cerebrosides, sulfatides, globosides, and gangliosides and they are important in cell recognition and signal transmission. The synthesis (from serine and palmitylCoA) and degradation (through sphingomyelinase) of ceramide are presented. Deficiency of sphingomyelinase leads to cholesterol accumulation in the lysosome of the brain resulting in severe neurological disease.

Glycosphingolipids are formed from ceramide and UDP-sugar (often UDP-glucose). The degradation of glycosphingolipids is outlined.

The properties of lipoproteins (chylomicrons, chylomicron remnants, LDL, IDL, VLDL, and HDL) are surveyed. The relationship of lipoproteins in the bloodstream to the susceptibility to coronary artery disease is presented.

Lipid anchoring of proteins to membranes involves either serine or cysteine in the protein and a fatty acid, such as myristate or palmitate or an isoprene residue (farnesyl or geranylgeranyl). The protein usually is anchored through esterification or amidation with fatty acids. Myristoylation can occur through the N-terminal of a glycine residue. S-Palmitoylation occurs through a cysteine residue and S-prenylation of a near C-terminal cysteine residue is connected with farnesyl or geranylgeranyl units. An example is the *ras* protein that is prenylated to ensure localization in the membrane.

Cholesterol is a precursor of bile acids in the liver. The synthesis of the major and minor bile acids is reviewed. Bile acids are transported from the liver to the intestine where they solubilize ingested fats through the formation of micelles. The enterohepatic circulation of bile acids is presented. The regulation of bile acid synthesis occurs through the control of the cholesterol 7α-hydroxylase gene and involves the liver X receptor and the farnesoid X receptor.

# SUGGESTED READING

### Literature

Bielska, A.A., Schlesinger, P., Covey, D.F., Ory, D.S., 2012. Oxysterols as non-genomic regulators of cholesterol homeostasis. Trends Endocrinol. Metab. 23, 99−106.

Boren, J., Ekstrom, U., Agren, B., Nilsson-Ehle, P., Innerarity, T.L., 2001. The molecular mechanism for the genetic disorder familial defective apolipoprotein B100. J. Biol. Chem. 276, 9214−9218.

Cai, J., et al., 2009. Human liver fatty acid-binding protein: solution structure and ligand binding. Biophys. J. 96, p600a.

Csaki, L.S., et al., 2014. Lipin-1 and lipin-3 together determine adiposity *in vivo*". Mol. Metab. 3, 145−154.

Dvir, H., et al., 2012. Atomic structure of the autosomal recessive hypercholesterolemia phosphotyrosine−binding domain in complex with the LDL receptor tail. Proc. Nat. Acad. Sci. 109, 6915−6921.

Goldstein, J.L., Brown, M.S., 2015. A century of cholesterol and coronaries: from plaques to genes to statins. Cell. 161, 161−172.

Goldstein, J.L., Brown, M.S., 1989. Familial hypercholesterolemia. In: Scriver, C.R., Beaudet, A.L., Sly, W.S., Valle, D. (Eds.), The Metabolic Basis of Inherited Disease, sixth ed McGraw Hill, New York, NY, p. 1232.

Hertzel, A.V., Bernlohr, D.A., 2000. The mammalian fatty acid-binding protein multigene family: molecular and genetic insights into function. Trends Endocrinol. Metab. 11, 175−180.

Lodhi, I.J., Semenkovich, C.F., 2014. Peroxisomes: a nexus for lipid metabolism and cellular signaling. Cell Metab. 19, 380−392.

Martin, S., Parton, R.G., 2006. Lipid droplets: a unified view of a dynamic organelle. Nature Revs. Molec. Cell Biol. 7, 373−378.

Martin, S.S., Jones, S.R., Toth, P.P., 2014. High-density lipoprotein subfractions: current views and clinical practice. Trends Endocrinol. Metab. 25, 329−336.

Phillips, M.C., 2014. Molecular mechanisms of cellular cholesterol efflux. J. Biol. Chem. 289, 24020−24029.

Platt, F.M., Butters, T.D., 2000. Substrate deprivation: a new therapeutic approach for the glycosphingolipid lysosomal storage diseases. Expert Revs. Molec. Med. 2, 1—17.

Sene, A., Apte, S., 2013. Eyeballing cholesterol efflux and macrophage function in disease pathogenesis. Trends Endocrinol. Metab. 25, 107—114.

Sharpe, L.J., Brown, A.J., 2013. Controlling cholesterol synthesis beyond 3-hydroxy-3-methylglutarylCoA reductase (HMGCR). J. Biol. Chem. 288, 18707—18715.

Sharpe, L.J., Cook, E.C.L., Zelcer, N., Brown, A.J., 2014. The UPS and downs of cholesterol homeostasis. Trends Biochem. Sci. 39, 527—535.

Tabernerol, L., Rodwell, V.W., Stauffacher, C.V., 2003. Crystal structure of a statin bound to a class II hydroxymethylglutaryl-CoA reductase. J. Biol. Chem. 278, 19933—19938.

Tarling, E.J., Aguiar, T.Q., Edwards, P.A., 2013. Role of ABC transporters in lipid transport and human disease. Trends Endocrinol. Metab. 24, 342—350.

Ye, Z.-J., et al., 2012. LRP6 protein regulates low density lipoprotein (LDL) receptor-mediated LDL uptake. J. Biol. Chem. 287, 1335—1344.

Zhou, L.C., Yang, H., Okoro, E.U., Guo, Z., 2014. Up-regulation of cholesterol; absorption is a mechanism for cholesystokinin induced hypercholesterolemia. J. Biol. Chem. 289, 12989—12999.

## Books

Ehnholm, C., 2009. Cellular Lipid Metabolism. Springer, New York, NY.

Gurr, M.I., Frayn, K.N., Harwood, J., 2002. Lipid Biochemistry. Blackwell Publishing, Oxford.

Quinn, P.J., Wang, X. (Eds.), 2008. Lipids in Health and Disease (Subcellular Biochemistry, vol. 49. Springer, New York, NY.

Reckless, J., Morrell, J., 2006. Lipid Disorders. Churchill Livingstone, London, UK.

Vance, D.E., Vance, J.E. (Eds.), 2008. Biochemistry of Lipids, Lipoproteins and Membranes. fifth ed. Elsevier, Amsterdam, NL.

Varki, A., Cummings, R.D., Esko, J.D. (Eds.), 2009. Essentials of Glycobiology. second ed. CSH Press, Cold Spring Harbor, NY.

# Chapter 10

# Nucleic Acids and Molecular Genetics

## HUNTINGTON'S DISEASE, A SINGLE GENE MUTATION

There are 4000 or more human diseases caused by a single gene mutation. Huntington's disease is an example of an **autosomal dominant disease**. There are many others like this: **Marfan syndrome, neurofibromatosis, retinoblastoma, myotonic dystrophy, familial hypercholesterolemia, adult polycystic kidney disease, familial adenomatous polyposis, hypertrophic obstructive cardiomyopathy,** and **osteogenesis imperfecta**, to name a few. In **autosomal recessive diseases**, one in four offspring will be overt bearers of the disease, two of four children will be silent carriers, and one child will be perfectly normal, whereas in **autosomal *dominant* diseases**, each child has a 50% chance of inheriting the disease with no skipping of generations. Only one parent needs to have the dominant allele. Only rarely is the (sporadic) disease manifest in a person whose parents do not have the mutation. The **pedigree** in two successive generations from a husband and wife, where the male has the dominant gene, is shown in Fig. 10.1.

Huntington's is a rare disease affecting one person of European descent in 10,000. Death occurs in one person in 600,000. The disease is less common amongst Chinese, Japanese, and African descendents. There are childhood onset and adult onset forms of the disease. The early onset of the disease progresses more rapidly than the adult onset. Early onset occurs in childhood or adolescence and these individuals live 10 or 15 years after disease symptoms (movement problems, mental and emotional changes, clumsiness, falling down, slurred speech, and drooling) occur. Up to 50% of diseased children have **seizures**. **Suicides** can occur when the disease strikes in the second decile. The most common form of Huntington's is the adult onset form, which has many of the symptoms of the early onset. Involuntary jerking or twitching in the adult form is referred to as **Huntington's chorea** (from the Greek "khoreia" meaning choral dance). Adults live for 15 to 20 years after symptoms begin.

The human gene involved is the **HTT gene** that is transcribed to messenger RNA (mRNA) that translates to a protein called **Huntingtin**, which, in mice, has been shown to have a role in the normal functioning of the **basal ganglia**. Huntingtin is expressed in all types of brain neurons, is a soluble protein of about 350 kD, having 3144 amino acids and containing about 36 alpha-helical **HEAT** (His-Glu-Ala-Thr) **repeats**. Although the exact function of this protein is unknown, experiments in which the equivalent gene to human Htt in mice (Hdh) is deleted, there is a suppression of the calcium ion mobilization response to **inositol *tris*phosphate**. This suggests that Huntingtin plays a direct role in **neuronal Ca$^{2+}$ signaling** by reducing the sensitivity of the **inositol *tris*phosphate receptor** in the endoplasmic reticulum to its ligand, inositol *tris*phosphate. Deletion of this gene in mice is embryonic lethal.

The HTT gene contains a **CAG trinucleotide** repeat segment. This trinucleotide is repeated 10 to 35 times within the gene, coding for **polyglutamine**. In Huntington's disease, the CAG segment is repeated 36 to an excess of 120 times. While individuals with 27 to 35 CAG repeats in the HTT gene do not develop the disease, their offspring may develop the disease. An intermediate form exists in which the CAG segment is repeated 36 to 40 times and these individuals may or may not develop the symptoms. Those who have an excess of 40 repeats in the HTT gene always develop the disease. When the number of CAG repeats exceeds 60, juvenile onset is predicted, whereas mutations showing 40 to 55 repeats result in adult onset. The **CAG triplet repeat** is defined as a **short tandem repeat polymorphism (STRP)**. In experimental animals, the normal huntingtin (Htt) plays a role in the development of the **basal ganglia** in the brain. In the abnormally elongated version of huntingtin (Htt*) in neurons, the pathologic proteins form and are broken down enzymatically into smaller toxic fragments that aggregate and disrupt cellular functioning and eventually cause the death of the cell (Fig. 10.2).

Another analysis indicates that there is reduced **brain-derived neurotrophic factor** (BDNF) in Huntington's disease. Htt* disrupts BDNF induction of the **mitogen-activated protein kinase** (MAPK) by reduction in the expression of adapter proteins (**p52/46Shc**) and the interference with **Ras activation**. This leads to a reduction in the growth

Human Biochemistry. DOI: http://dx.doi.org/10.1016/B978-0-12-383864-3.00010-7

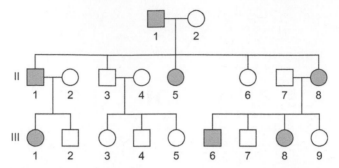

**FIGURE 10.1** A typical pedigree for an autosomal dominant disease. *Squares*, males; *circles*, females; *filled square*, male having the autosomal dominant gene; *filled circle*, female having the autosomal dominant gene. Those individuals with the dominant gene have the overt disease. In the first set of offspring (II), there are five children, three of whom have the overt disease one male (1) and two females (5 and 8) in II. The other two children (3 and 6) are normal. A normal son (3) in II marries a normal woman (4) and they bear three normal children in III (3, 4, and 5). A diseased daughter (8) marries a normal male (7) in II and they have four children (III), two of whom are normal (7 and 9), a diseased son (6), and a diseased daughter (8). Also, a diseased son (1) marries a normal woman (2) in II and they have two children, a diseased daughter (1) and a normal son (2) in III.

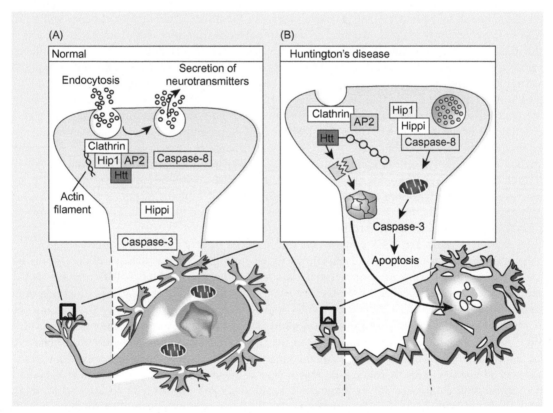

**FIGURE 10.2**   In nerve cells, normal huntingtin (Htt) complexes with proteins **Hip1**, clathrin and **AP₂**. Clathrin, AP₂ (with Htt) are involved in endocytosis as shown in (A). The abnormally long tract of glutamines (Qs) in diseased Htt (Htt\*, elongated version of Huntingtin) cause abnormal endocytosis and secretion in neurons. **Striatal neurons** die by **apoptosis** as Htt\* cannot interact well with the protein Hip1 enabling the free Hip1 to interact with the protein **Hippi** (**Hip1−Hippi**). Hip1−Hippi complex activates **caspase-8** which causes the activation of **caspase-3** and the degradation of Htt\* to produce fragments. The Htt\*fragments aggregate and form inclusions in the neuron and in its nucleus. *Reproduced from http://www.nature.com/nature/journal/v415/n6870/fig_tab/415377a_F1.html.*

pathway through **ERK 1/2**, or MAPK, and tips the balance between cell survival and cell death in favor of cell death. These relationships are shown in Fig. 10.3.

Other evidence involves a direct effect of Htt in the **pretranscriptional complex**. The transcriptional factor **Sp1** (specificity protein 1) binds to DNA elements (GC boxes) in cellular promoters. A protein−protein interaction between the glutamines of Htt and the diglutamines of Sp1 and **TAF II** 130 subunit (Tata-associated factor 130, cofactor for NFAT) is needed to recruit the transcriptional machinery (other proteins including **RNA polymerase II**). In Huntington's disease,

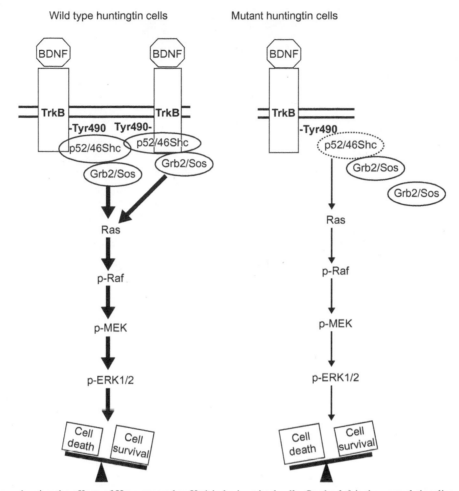

**FIGURE 10.3**   Diagram showing the effects of Htt compared to Htt* in brain striatal cells. On the *left* is the normal signaling pathway leading to cell survival. On the *right* is the pathway resulting from the reduced expression of adapter proteins (p52/46Shc) apparently caused by pathogenic huntingtin (Htt*). Reduced thickness of the arrows indicates lower activity compared to the normal cell with normal Htt. *BDNF*, bone-derived neurotrophic factor; *TrkB*, a tyrosine kinase receptor; *Grb*, adapter protein; *Sos*, Son of Sevenless enzyme; *Ras*, a GTPase; *Raf* and *MEK*, **mitogen-activated kinases**; *ERK1/2*, also known as *MAPK*, mitogen-activated protein kinase. *Reproduced from http://www.jbc.org/content/285/28/21537/F9.large.jpg.*

with the extension of polyglutamine in Htt*, RNA polymerase is not able to bind well and transcription does not take place. This work has been done in the context of the transcription of the **dopamine receptor**. (Reproduced from http://www.bio.davidson.edu/Courses/Molbio/MolStudents/spring2003/McDonald/Huntington.htm.) A **transgene** (a gene placed experimentally in the germline that functions as a normal gene) in mice containing expanded **polyglutamine tracts** generates the central nervous system disorder resembling Huntington's disease, whereas a transgene with a normal number of polyglutamines does not cause the disease. In addition to Huntington's disease, **spinobulbar muscular atrophy** shows expansion of the CAG triplet expansion in the gene for the **androgen receptor**. The **fragile X syndrome** is marked by an expansion in a **polyarginine tract** (**CGG** triplet expansion) and there are other diseases in this category.

An abnormal protein followed by an aggregation phenomenon could classify Huntington's disease as a familial **prion disease**. The alteration of the gene (*HTT*) to the disease form (*HTT**) occurs during the development of the **sperm**. **Neuronal loss** in the neostriatum and the cortex of the brain occurs at age 35 to 44 years, the variation depending on the state of methylation of the *HTT* locus. Neuronal loss leads to a decrease in metabolic activity of the brain (Fig. 10.4), especially where the uptake and metabolism of glucose is concerned.

Detection of the disease can be made with a blood sample by sequencing the CAG repeat region in the DNA.

## PURINES AND PYRIMIDINES

Purines and pyrimidines make up the bases in both RNA and DNA. The two common purines are **adenine** and **guanine**. Purine structure consists of two adjoined rings with five carbons and four nitrogens. The common pryimidines are **cytosine**, **thymine**, and **uracil**. Pyrimidines have one ring with four carbons and two nitrogens. The structures are shown in Fig. 10.5.

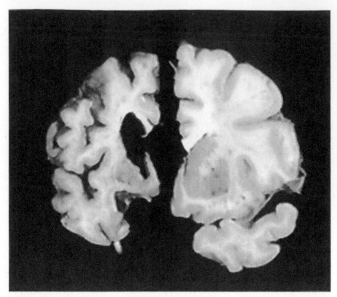

**FIGURE 10.4**    **Section of a human brain** from Huntington's disease (left) compared to a normal section (right). The brain represented by the section on the left has lowered metabolic activity. *Reproduced from http://www-tc.pbs.org/wgbh/nova/assets/img/rnai-cure/image-04-large.jpg.*

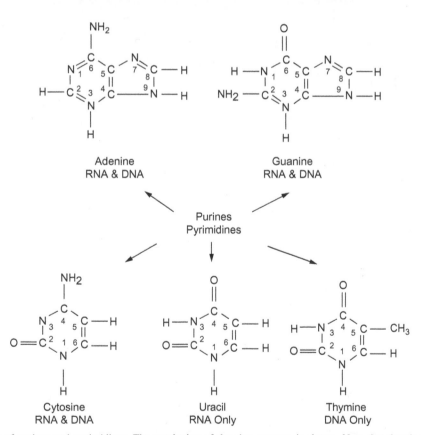

**FIGURE 10.5**    Structures of purines and pyrimidines. The numbering of the ring systems is shown. Note that thymine is a pyrimidine in DNA whereas uracil replaces thymine in RNA.

In the nucleotide form in DNA, adenine, guanine, cytosine, and thymine attach to **deoxyribose**, whereas in RNA adenine, guanine, cytosine and **uracil** attach to ribose. In RNA, uracil replaces thymine because if uracil appeared in DNA, it would be recognized by the base excision repair mechanism and it would be removed. The combination of a purine or a pyrimidine with a sugar is a **nucleoside**. When the sugar is phosphorylated, it is a **nucleotide**. One phosphate group renders it a **mononucleotide** (this form is contained in RNA and DNA); two phosphate groups, a dinucleotide and three phosphate groups, a trinucleotide. In nucleic acids, these bases are bound with ribose (RNA) or deoxyribose (DNA) and they are phosphorylated (mononucleotides). The nucleosides (nonphosphorylated) appear as shown in Fig. 10.6. Nucleotides are more prevalent in the cell than nucleosides.

**Deamination** of cytosine to form uracil can occur through accident or by a mutation as shown in Fig. 10.7.

If this occurs in DNA, uracil will be base-paired with guanine and uracil could be removed directly and replaced by cytosine for the correct C-G base-pairing. In rare cases, uracil could be paired with adenine after replication and a repair mechanism could replace the uracil with a thymine to give a T-A base pair resulting in a mutated DNA. Base pairing will be discussed. Deamination of other bases can occur: guanine can be deaminated to form xanthine and adenine can be deaminated to form hypoxanthine. **Xanthine** and **hypoxanthine** are metabolic products and do not generally appear in nucleic acids. **Orotic acid**, a pyrimidine carboxylic acid, is a metabolite. In nucleosides and nucleotides, the base can exist in two possible orientations about the **N-glycosidic bond** (Fig. 10.8). The *anti*form is the one that predominates in nucleic acids.

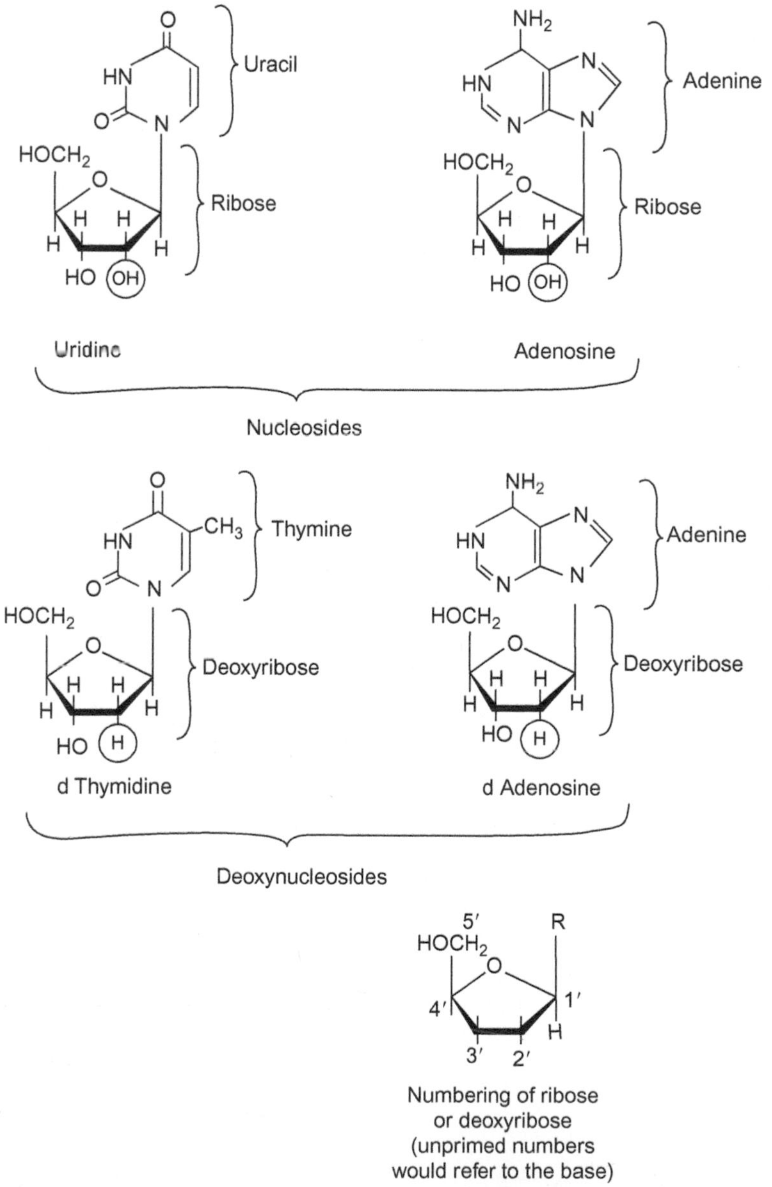

**FIGURE 10.6** Structures of nucleosides of uracil, as uridine, and of adenine, as adenosine. Also shown are thymine, as thymidine and adenine as deoxyadenosine. The other purines and pyrimidines, not shown here, have similar structures as nucleosides. *d*, deoxy.

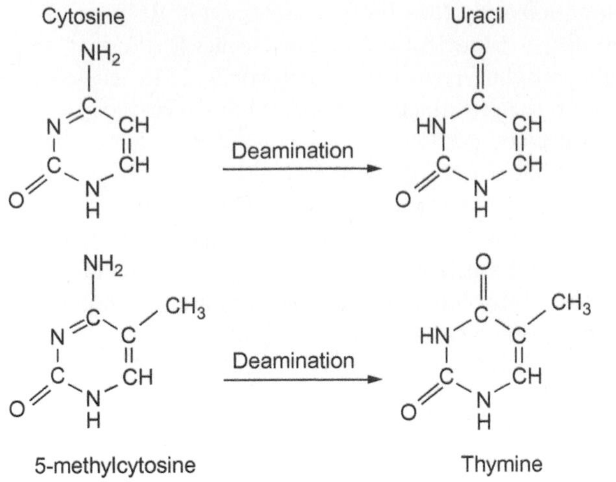

**FIGURE 10.7**  Deamination of cytosine to produce uracil. As uracil is not part of DNA, it can be detected and repaired by a DNA repair mechanism if cytosine is accidentally deaminated. Also shown is the deamination of 5-methylcytosine to produce thymine. Methylcytosine can sometimes be found in nucleic acids.

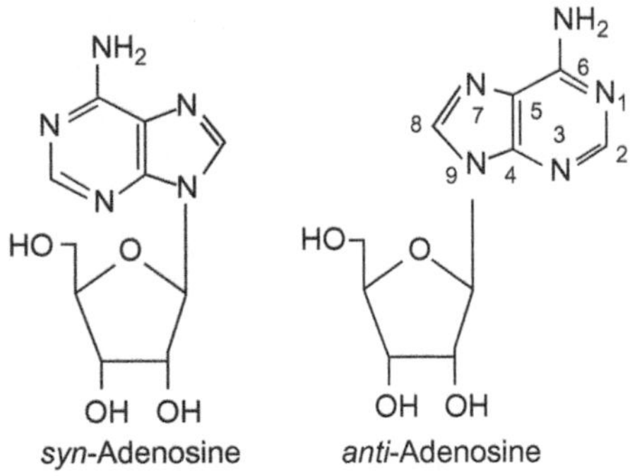

**FIGURE 10.8**  Conformation of *syn-* or *anti*adenosine.

The structures and naming of the bases and their derivatives appear in Fig. 10.9A shows the naming of the base alone, the monosaccharide derivative and the monosaccharide monophosphate derivative. Fig. 10.9B shows the potential structures of the derivatives of adenosine or deoxyadenosine. Fig. 10.9C shows the structures of a common pyrimidine (orotic acid) and of hypoxanthine and xanthine (purines), all of which are metabolites rather than members of RNA or DNA. Orotic acid is an important intermediate in pyrimidine synthesis. It has been labeled as **vitamin B13** because it can partially compensate for vitamin B12 deficiency and is active in folic acid metabolism.

If a nucleoside or nucleotide contains deoxyribose as compared to ribose, its designation is preceded by a *d*, as in dAMP.

There are cyclic forms of adenosine and guanosine: these are cyclic AMP and cyclic guanosine monophosphate (cGMP). Cyclic AMP is formed from ATP by the action of **adenylate cyclase** and cyclic GMP is formed from GTP by the action of **guanylate cyclase** (Fig. 10.10). These are separate enzymes. Adenylate cyclase is a 12 transmembrane protein while guanylate cyclase is a single transmembrane protein. Guanylate cyclase is part of a receptor for the **natriuretic peptide**, a hormone (Chapter 11: **Protein Biosynthesis**), and there are four related receptors. The enzymatic activity of a cyclase is classified as a **lyase** (Chapter 5: **Enzymes**).

## BASE PAIRING

Adenine forms a **base pair** with thymine in DNA and with uracil in RNA. Guanine pairs with cytosine. Base pairing is shown in abbreviated fashion in Fig. 10.11.

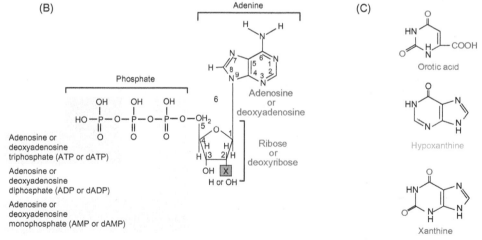

**FIGURE 10.9** (A) Table showing the naming of pyrimidine and purine compounds (base, nucleoside, and nucleotide). (B) The structure of the phosphate derivatives of adenosine or deoxyadenosine. (C) The structures of orotic acid, hypoxanthine, and xanthine that are metabolites and not components of RNA or DNA.

The pairs of bases are held together by **hydrogen bonding**. Other forces contribute to the stability of the double-stranded DNA. There are ionic charges (van der Waals forces) between the tightly stacked pairs and hydrophobic attractions occur between base pairs in the interior of the helix as the nonpolar nitrogenous bases are packed so tightly to exclude water molecules and form a stable nonpolar interior. The bases are located in the interior of the double-stranded

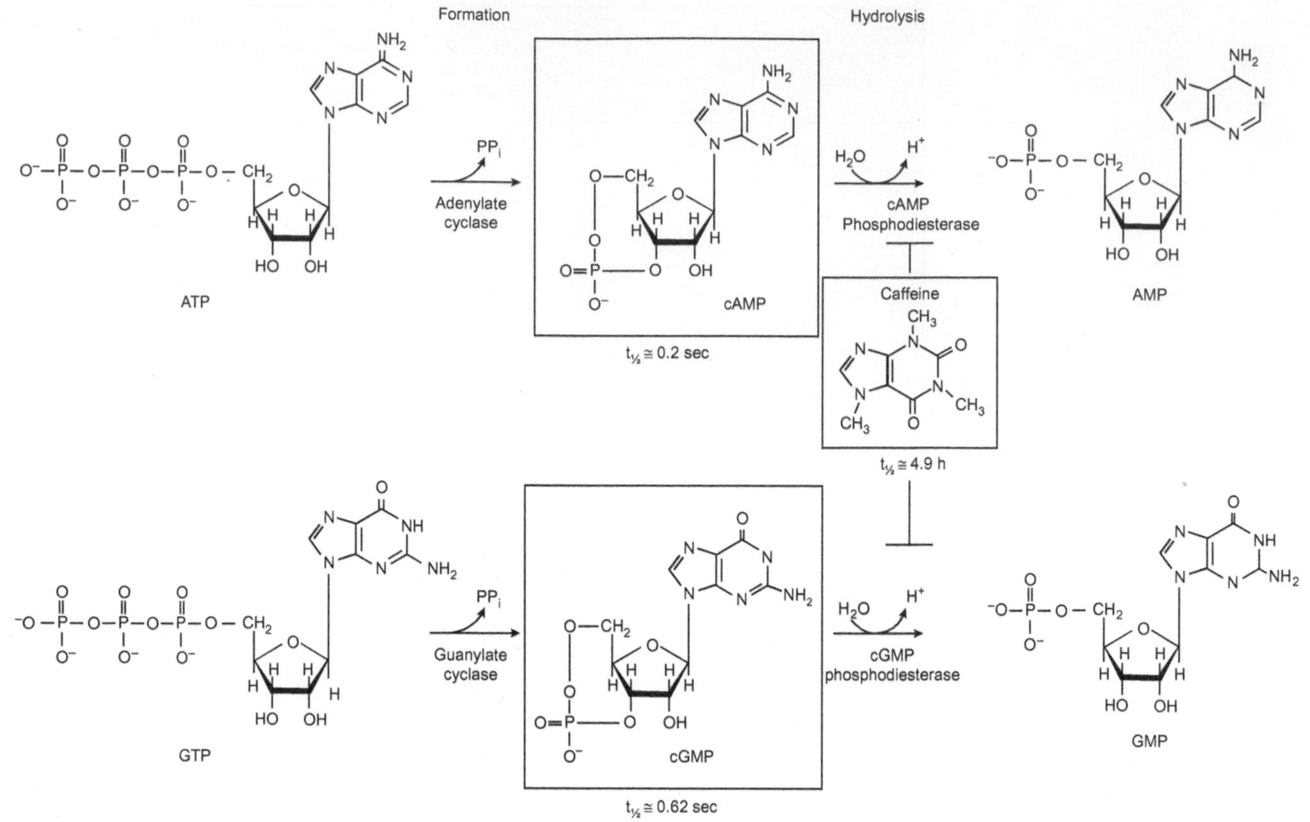

FIGURE 10.10   Conversions of ATP to cyclic AMP and of GTP to cyclic GMP. Adenylate cyclase (also adenylyl cyclase) converts ATP to cAMP by the removal of pyrophosphate and the cyclization as shown. Guanylate cyclase (also guanylyl cyclase) removes pyrophosphate from GTP and cyclizes the methyl oxygen to the phosphate group and the 3'-hydroxyl of ribose to form cGMP. The approximate normal half-lives ($t_{1/2}$) of these compounds are shown. The cyclic compounds are inactivated by hydrolysis catalyzed by phosphodiesterase. This enzyme is inhibited by caffeine ($t_{1/2} \sim 4.9$ h) and related products (paraxanthine, theobromine, and theophylline) that have similar effects on phosphodiesterase. These compounds are dimethyl variants compared to the trimethyl of caffeine. Theobromine and theophylline are present in tea in addition to caffeine.

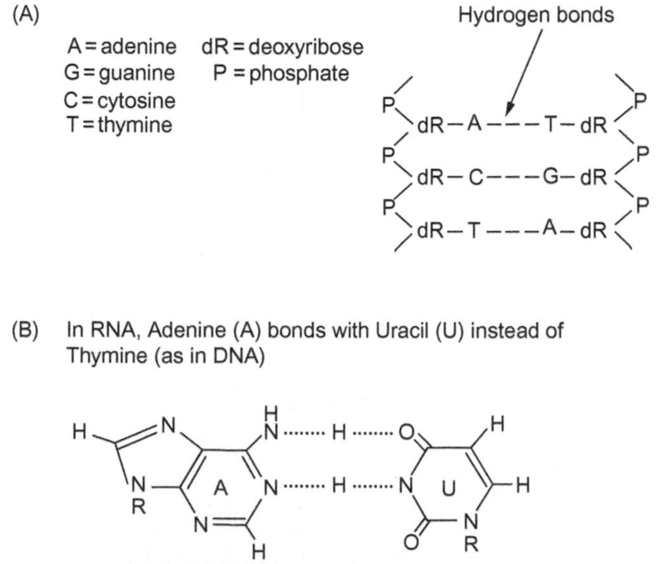

FIGURE 10.11   (A) Base pairing in DNA. In the double strand, adenine pairs with thymine and guanine pairs with cytosine. (B) In RNA, adenine bonds with uracil. dR, deoxyribose.

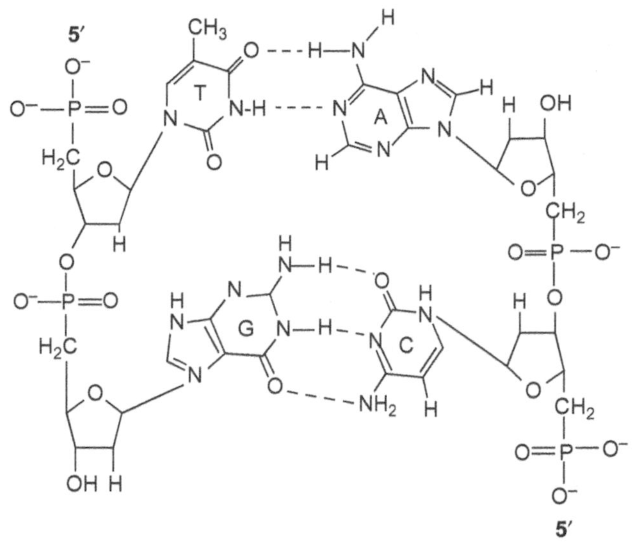

FIGURE 10.12 Base pairing as it occurs in DNA. The bases are joined by hydrogen bonding (dashed lines). *T*, thymine; *A*, adenine; *G*, guanine and *C*, cytosine.

DNA and the pentose sugars and phosphate groups are on the outside as shown in the chemical structures in Fig. 10.12. Van der Waals forces usually involve dipolar molecules where the concentration of positive charges is separated from a concentration of negative charges so that there are exerted attractive (or repulsive) forces between molecules (that do not arise from covalent or ionic bonds).

The 3′ carbon of one sugar is linked to the 5′ carbon of the next sugar through a phosphodiester bond. A single turn of the **DNA helix** (of the naturally occurring abundant B DNA) requires 10 base pairs and the height of the turn measures 3.4 nanometers (nm); the diameter of the helix measures 1.9 nm. The double helix is formed by two antiparallel DNA strands. The 5′ end of the left strand is at the top of the figure and the 3′ end is on the bottom. *Replication of DNA as well as transcription to form mRNA from the sense strand of DNA both proceed in the direction from the 5′ end to the 3′ end.*

## THE STRUCTURE OF DNA

A small portion of a single strand of DNA (AGACC) is shown in Fig. 10.13. Deoxyribose, a nucleoside component and the phosphate backbone are indicated.

Transcription of DNA to form the corresponding mRNA proceeds from the 5′-end to the 3′-end of the RNA molecules. Normally, DNA resides in the cellular nucleus with two strands in the form of a helix. One strand of DNA would appear as shown in Fig. 10.13 and the second strand would be antiparallel where the bases in both strands are joined together by base pairing, each pair of bases being held together by hydrogen bonding. **Double-stranded DNA** appears as in Fig. 10.14.

Double-stranded DNAs of any length appear as helices that is, the two strands together are twisted into a helix as shown in Fig. 10.15.

Again, the base pairing in DNA is dR-A-T-dR (or R-A-U-R in RNA), and dR-G-C-dR (or R-G-C-R in RNA). *dR* is deoxyribose and *R* is ribose. Although not connoted here, phosphate groups are attached to the sugar residues. DNA double strands can be melted apart into single strands *in vitro* by heating to 50 to 60°C. Hydrogen bonds as well as the other forces holding the two strands together are loosened as the temperature is increased. The individual chains reform the double helix when the temperature is returned to normal at about 37°C. The **melting temperature** is an important characteristic of each double-stranded DNA molecule. *The preponderant form of DNA in the cell is the B form* that was originally described by Watson and Crick [Fig. 10.15(B)]. Three forms are known in addition to the B form. Forms A, B, and C each has a right-handed helix (chains proceed from the bottom toward the top in a series of right-hand turns) while **Z-DNA** has a left-handed helix (from the bottom to the top the chains proceed in a series of left-hand turns). The Z-form is found *in vitro* for the most part under nonphysiological high salt conditions, however, Z-forms can occur in the cell along with other strands of the B-form when there are stretches of poly GC or poly AT or nearby

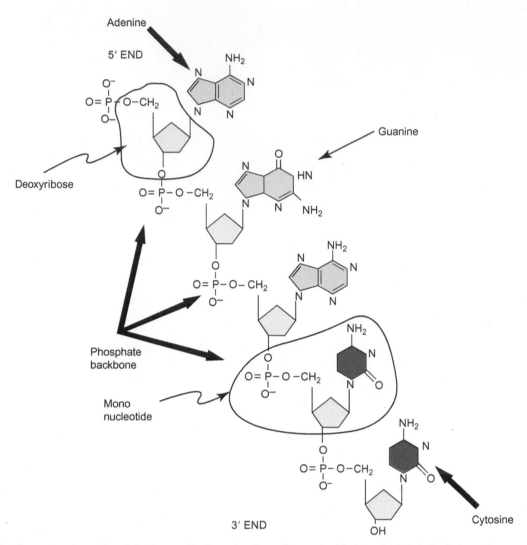

**FIGURE 10.13** Shown here is a series of five bases with alternating deoxyribose units (circled and labeled). Each deoxyribose sugar is esterified with phosphate (phosphodiester bond) at its 3'- and 5'-hydroxyl groups. In the figure, *red* color denotes adenine; *green*, guanine; *dark blue*, cytosine; *light green* deoxyribose. The mononucleotide structure is encircled and the phosphate backbone is indicated. This DNA sequence can be denoted: *p*A*p*G*p*A*p*C*p*C where *p* represents the phosphate group. The sequence is commonly denoted as AGACC.

methylcytosine. These conditions apparently can stabilize runs of Z-DNA. For comparison, the structures of **A-DNA,** **B-DNA,** and Z-DNA are shown in Fig. 10.16.

The **haploid human genome** consists of 24 pairs of **chromosomes** (22 autosomal chromosomes plus X,X sex chromosomes in the female and X,Y sex chromosomes in the male). All cells of the body contain the complete genome, except for the **red blood cell** that does not have a nucleus and, therefore, contains no chromosomes. The microscopic spread of chromosomes is shown in Fig. 2.14. Each chromosome varies in its number of base pairs from about 50 to 250 million. The human genome consists of about 3 billion base pairs. Genes account for about 2% of the human genome and encode the information for about 30,000 genes. Usually there are long stretches of **noncoding DNA** that separate the genes. Aside from that DNA, much of the DNA outside of the genes has been ascribed a few specific functions.

## BIOSYNTHESIS OF PURINES AND PYRIMIDINES

Purines are derived from **aspartate, carbon dioxide, glycine,** $N^{10}$-**formyl-tetrahydrofolate** ($N^{10}$-formylTHF), and **glutamine** as diagrammed in Fig. 10.17.

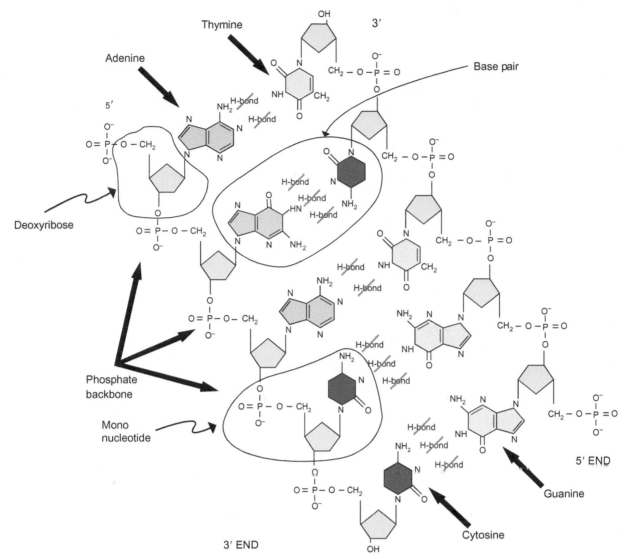

**FIGURE 10.14** Double-stranded DNA. The strand on the left is the same one shown in Fig. 10.13. The strand on the right is antiparallel to the left-hand strand. Base pairs are held together by (weak) hydrogen bonds (shown in *pink* lines). The left strand begins at the top at its 5′-end (5′ on deoxyribose) and ends with its 3′-hydroxyl group. This is only a small component of DNA as cellular DNAs are huge. *Green* is used to highlight guanine; *yellow*, thymine; *red*, adenine; *dark blue*, cytosine and *light blue*, deoxyribose. Circled are deoxyribose, a base pair consisting of cytosine and guanine, a mononucleotide (of cytosine) and the 5′ and 3′ ends of both strands.

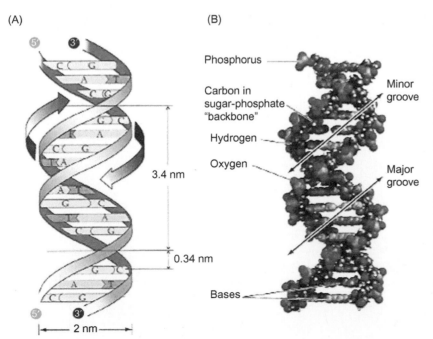

**FIGURE 10.15** (A) A right-handed DNA double helix. Down the center of the structure is the axis of symmetry. The diameter of the helix is about 2 nm. One 360 degree turn of the helix takes about 8−10 base pairs and occupies a space of 3.4 nm. The antiparallel character of the two strands is shown by the 5′-end of one strand opposing the 3′-end of the antiparallel strand at the top and the bottom. In (B) the various groupings are indicated. There is a major groove and a minor groove. Many of the **transactivating proteins** (DNA-binding proteins with a specific function) bind in one of these grooves of DNA to the **promoter** of a specific gene. As mentioned previously, the deoxyribose-phosphate backbone occurs on the outside of the helix and the hydrophobic bases are located in the interior. The negative charges on the phosphate groups in the backbone attract positively charged motifs on **DNA-binding proteins**. *Reproduced from http://bioserv.fiu.edu/~walterm/GenBio2004/chapter11_DNA/figure%2011-06.jpg.*

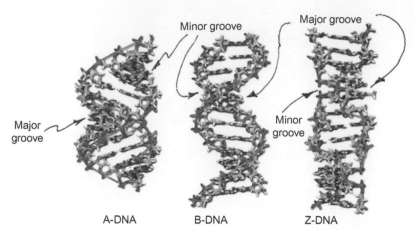

**FIGURE 10.16** The structures of A-DNA (right-handed helix), B-DNA (right-handed helix), and Z-DNA (**left-handed helix**) are shown. There are other differences between different forms of DNA besides the direction of the helical turn. In the A-form, there are 11 bases per 360° turn of the helix and the diameter of the helix is 23 Å (**Angstrom**s; 1 Å = 0.1 nm). In the B-DNA, there are 10 bases per turn with a helical diameter of 19 Å. In the Z-form, there are 12 bases per turn with a helical diameter of 18 Å. In the C-form, another **right-handed helix**, not shown in this figure, there are slightly more than nine bases per turn with a helical diameter of 19 Å. *This figure is reproduced from http://www.richardwheeler.net/contentpages/ image.php?gallery=Scientific_Illustration&img=DNA_Types_Assembled&type=jpg.*

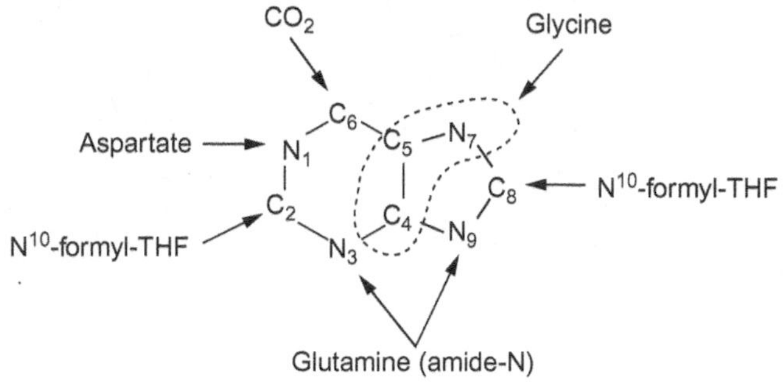

**FIGURE 10.17** The sources of the carbons and nitrogens of the purine ring. THF, tetrahydrofolate.

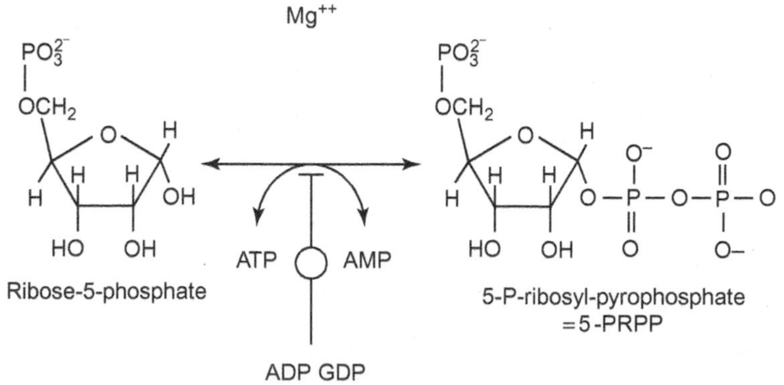

**FIGURE 10.18** The conversion of ribose-5-phosphate to 5-phospho-D-ribosyl-1-pyrophosphate (PRPP) catalyzed by **PRPP synthase**.

The biosynthetic route begins with **ribose-5-phosphate** that is converted to **5-phospho-D-ribosyl-1-pyrophosphate** (**PRPP**) in a reaction catalyzed by **PRPP synthase** (Fig. 10.18). PRPP is the starting substrate for **purine biosynthesis**.

In all, there are 11 enzymatic steps leading to the formation of **inosine 5′-monophosphate** (**IMP**). The 11 catalytic steps are described briefly in Fig. 10.19 where the structures of only the initial substrate and final product are given and the name of the enzyme involved in each step is given in the figure legend.

The synthesis of the purine nucleotide, **IMP**, makes possible the conversion to two major purine nucleotides, **AMP** and **GMP**, as shown in Fig. 10.20.

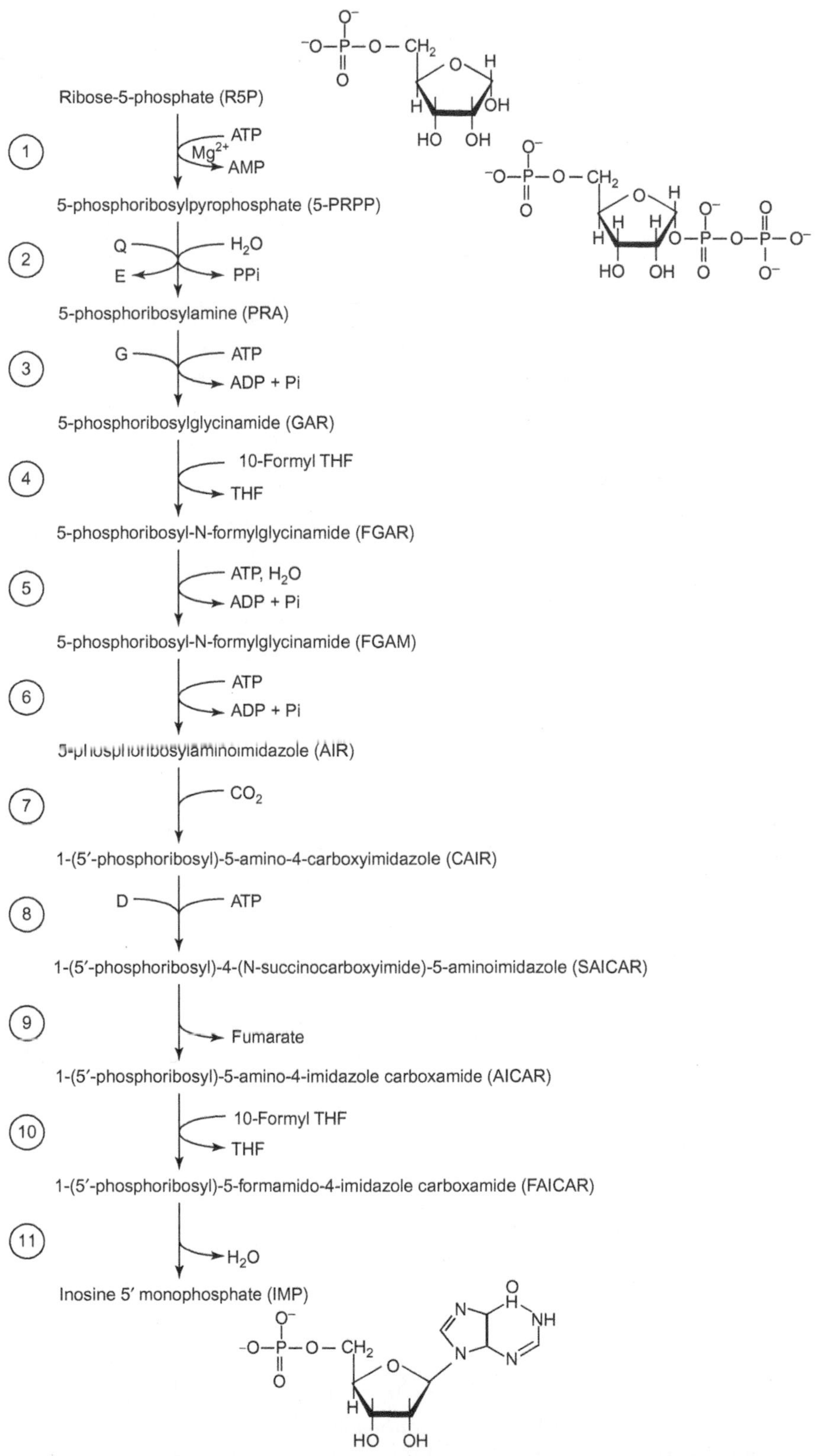

**FIGURE 10.19** The 11 catalyzed steps in the synthesis of inosine 5′-monophosphate (IMP) starting from ribose-5-phosphate. The enzymes involved in each step are as follows: (1) PRPP synthase; (2) glutamine phosphoribosylpyrophosphate amidotransferase; (3) glycinamide ribonucleotide synthase (GAR synthase); (4) phosphoribosylglycinamide formyltransferase (GAR transformylase); (5) phosphoribosylformylglycinamidine synthase (FGAM synthase); (6) phosphoribosylaminoimidazole synthase (FGAM synthase); (7) phosphoribosylaminoimidazole carboxylase (AIR carboxylase); (8) phosphoribosylaminoimidazole succinocarboxamide synthase (SAICAR synthase); (9) adenylsuccinate lyase; (10) phosphoribosylaminoimidazole carboxamide formyltransferase (AICAR transformylase); (11) IMP cyclohydrolase (IMP synthase) to yield IMP. *PPi*, inorganic pyrophosphate; *THF*, tetrahydrofolate. *Q*, glutamine; *E*, glutamate; *G*, glycine; *D*, aspartate.

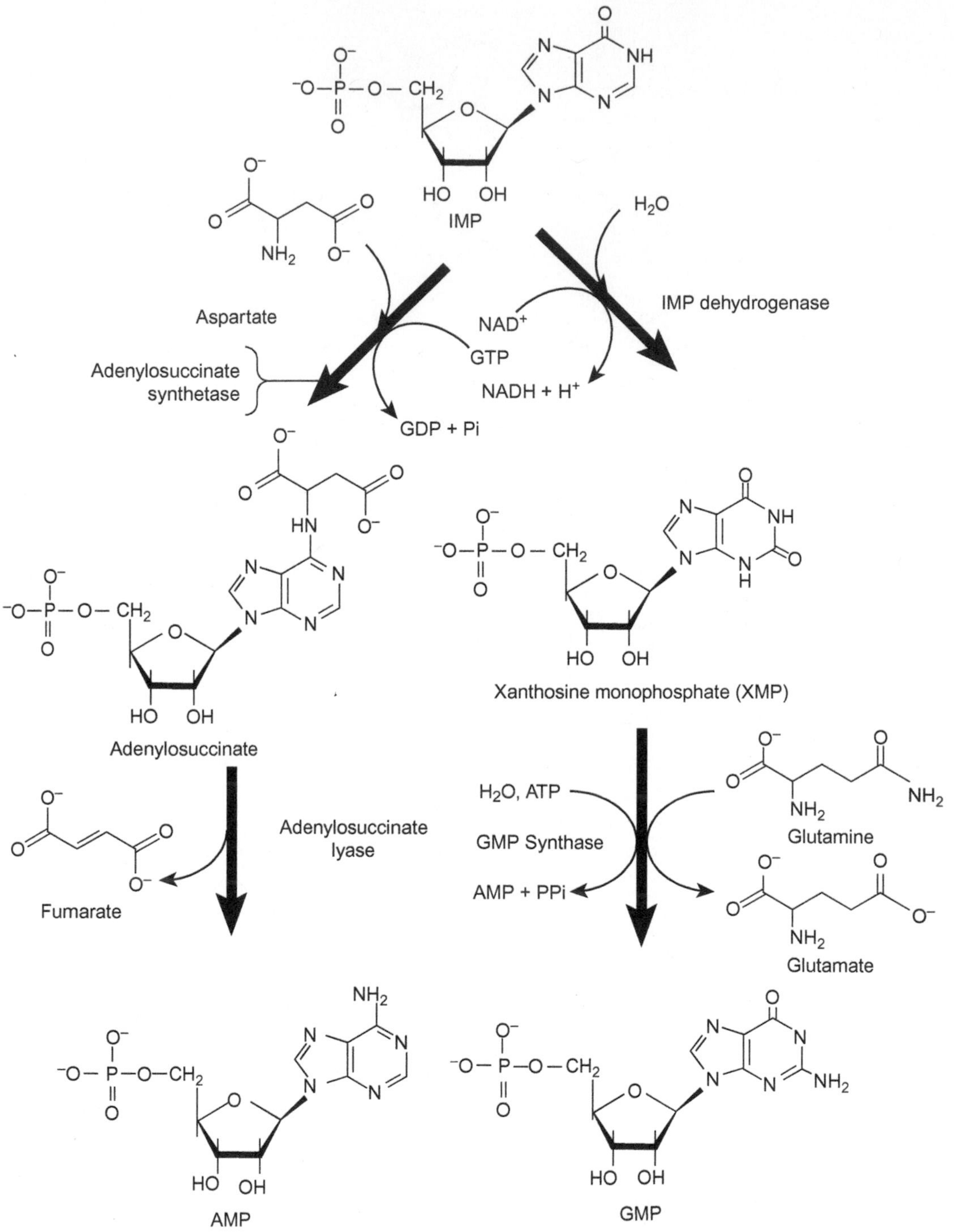

**FIGURE 10.20** Conversion of IMP to AMP and GMP.

**Adenylosuccinate** is formed from IMP as an intermediate by the addition of aspartate and donation of a phosphate group from GTP and then fumarate is split out to form AMP. GMP is formed from IMP with the addition of water and participation of oxidized NAD to form **xanthosine monophosphate**. Then glutamine, water, and ATP participate to form nucleotides (ADP and GDP) and trinucleotides (ATP and GTP). In Fig. 10.21, the feedback and feed-forward effects of the various components of the system are seen.

The **pyrimidine structure** is derived from the amide group of glutamine, aspartate and $HCO_3^-$. **Carbamoylphosphate synthase II** catalyzes the synthesis of **carbamoylphosphate** from 2ATP, $HCO_3^-$, glutamine, and water. In a series of reactions involving the enzymes: carbamoylphosphate synthase II, aspartate **transcarbamoylase (ATCase)**, **dihydroorotase**, **dihydroorotate dehydrogenase**, **orotate phosphoribosyltransferase**, and **OMP decarboxylase** the pyrimidine mononucleotide, uridine monophosphate (UMP) is derived (Fig. 10.22).

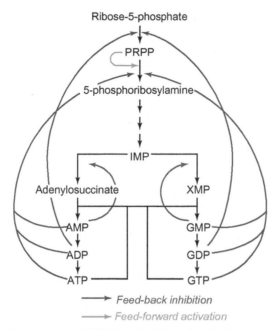

**FIGURE 10.21** Regulation of purine synthetic pathways. PRPP is a feed-forward early intermediate, specifically activating its conversion to **5-phosphoribosylamine**. Excepting IMP, all of the successive products are negative feedback inhibitors: **ADP** and **GDP** feed back negatively on the first step, the conversion of ribose-5-phosphate to PRPP. AMP, ADP, and ATP and GMP, GDP, and GTP all feed-back negatively on the conversion of PRPP to 5-phosphoribosylamine. In addition, ADP and GDP feed back negatively on the conversion of PRPP to 5-phosphoribosylamine.

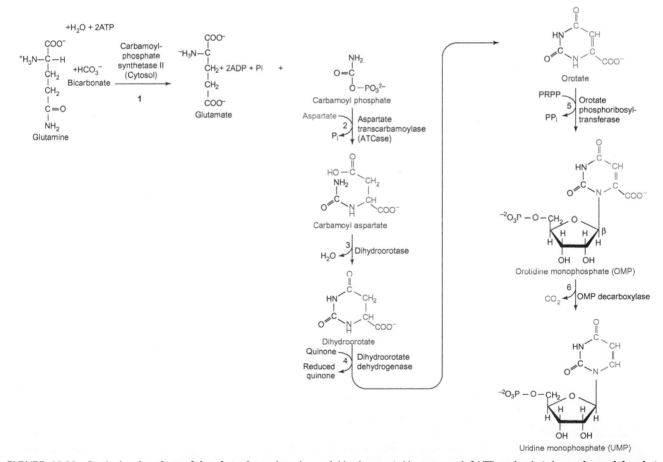

**FIGURE 10.22** Synthesis of **carbamoylphosphate** from glutamine and bicarbonate (with water and 2ATP molecules) by **carbamoylphosphate synthase II (CPSII)**, a cytosolic enzyme. At the *top* is the carbamoylphosphate synthase II reaction generating carbamoylphosphate, the starting compound for the synthesis of the pyrimidine nucleotide, **UMP, uridine monophosphate**. The intermediate steps and the enzymes involved in each step are shown.

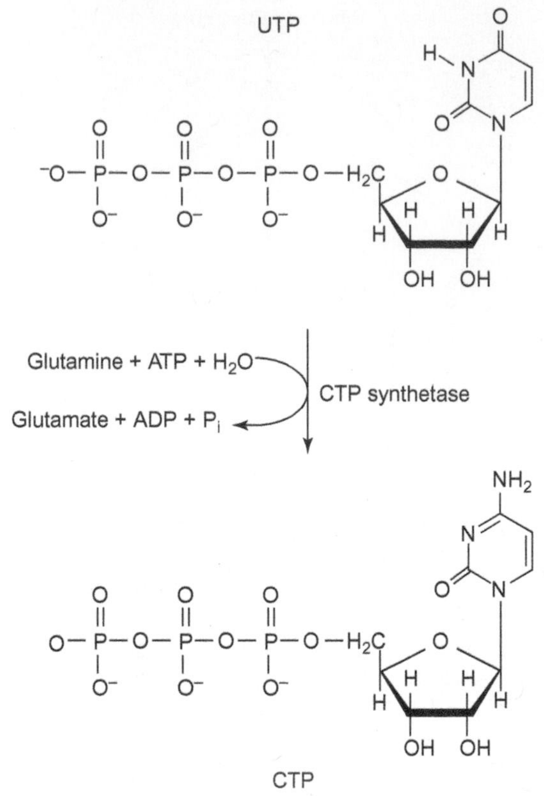

**FIGURE 10.23**   Conversion of UTP to CTP catalyzed by CTP synthase.

CPSII is present in most tissues outside of the liver. There is at least one other enzyme, **carbamoylphosphate synthase I** that is involved in the urea cycle. It catalyzes the formation of carbamoylphosphate from ammonia and bicarbonate in the mitochondrion.

Once UMP is formed, it can be converted to **UDP** by **nucleoside monophosphate kinase**:

$$UMP \, + \, ATP \rightleftharpoons ADP \, + \, UDP$$

UDP can be phosphorylated further to form the trinucleotide:

$$UDP \, + \, ATP \rightleftharpoons UTP + \, ADP$$

UTP can be converted to the **cytidine triphosphate** by **CTP synthase** as shown in Fig. 10.23.

The amine group of cytosine is derived from the amide group of **glutamine**. The purine trinucleotide, **GTP** activates CTP synthase so as to affect a balance between pyrimidine nucleotides and purine nucleotides in the cell. CTP synthase is reversibly inhibited by the product of the reaction, CTP, which also inhibits by **product feedback**, the formation of carbamoylphosphate as shown in the overall pyrimidine biosynthetic pathway (Fig. 10.24).

**Orotic aciduria** is a disorder associated with pyrimidine metabolism. There are variations of this disorder but the cause is a defective **OMP decarboxylase** whose normal function would yield UMP from OMP (Fig. 10.24). This condition can result in growth retardation, megaloblastic anemia and leukopenia. Drugs, such as **allopurinol** (used for gout to block **xanthine oxidase** activity) and **6-azauridine** produce products that inhibit OMP decarboxylase.

## PURINE INTERCONVERSIONS

The formation of **PRPP** from **ribose-5-phosphate** initiates the **purine synthetic pathway** to generate **phosphoribosyl-pyrophosphate (PRPP)**. Nine enzymatic steps follow to finally produce **inosine-5′-monophosphate (IMP)** (Fig. 10.19). IMP, the primary product of purine biosynthesis, can be converted to adenine and guanine nucleoside monophosphates (Fig. 10.25).

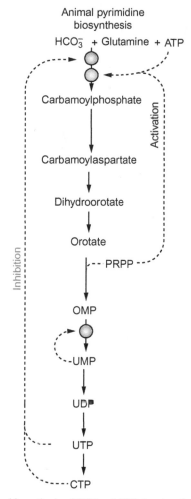

**FIGURE 10.24** The overall synthetic pathway of the biosynthesis of UTP and CTP showing the feedback inhibition by UTP, CTP, and UMP.

**IMP** is first converted to **adenylosuccinate** catalyzed by **adenylosuccinate synthase**. This is converted to **AMP** by **adenylosuccinate lyase**. For conversion to **GMP**, IMP is converted to **xanthosine monophosphate** by the action of **IMP dehydrogenase**. **Xanthosine monophosphate** is converted to GMP by **GMP synthase**. These mononucleotides can be converted to the dinucleotides by the appropriate kinase:

$$GMP + ATP \leftrightharpoons GDP + ADP$$

by guanylate kinase, and:

$$AMP + ATP \leftrightharpoons 2\,ADP$$

by **adenylate kinase**, and nucleoside diphosphates can be converted to the triphosphates by **nucleoside diphosphate kinase**, for example (also Fig. 10.21)

$$GDP + ATP \leftrightharpoons GTP + ADP$$

## CATABOLISM OF PURINE AND PYRIMIDINE NUCLEOTIDES

The purine derivatives AMP and IMP are degraded by **nucleotidase** that removes the phosphate group to yield the corresponding nucleoside. The sugar is then removed by the action of **nucleotide phosphorylase** to generate the free bases, adenine or **hypoxanthine**. Adenine can be deaminated to form hypoxanthine as well. The resulting hypoxanthine is oxidized by **xanthine oxidase** to form **xanthine** that is further oxidized by xanthine oxidase to form **uric acid**, the ultimate product of purine nucleotide catabolism. Likewise, guanine can also be deaminated and the product

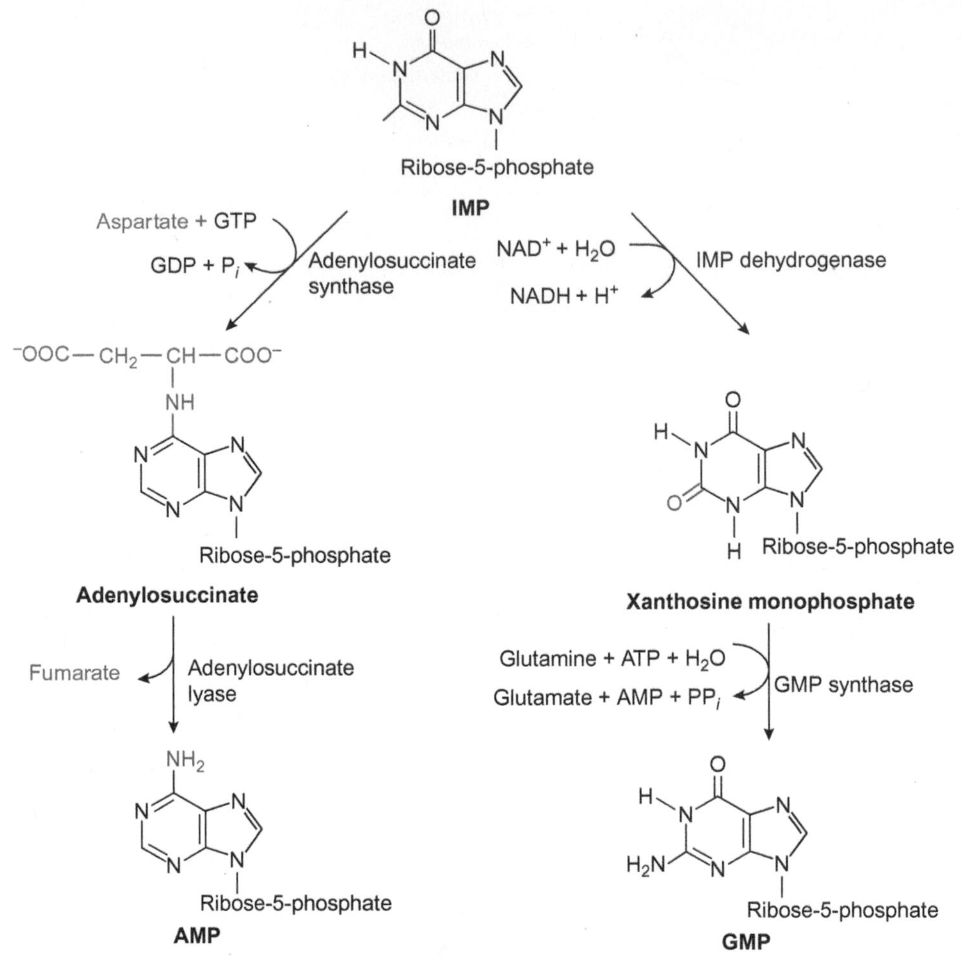

**FIGURE 10.25**   Conversion of IMP to AMP and GMP.

subsequently oxidized to form uric acid. Thus, xanthine oxidase is the terminal enzyme in the catabolic pathway and uric acid is the final circulating product. As has been mentioned, too high a level of circulating uric acid can result in its crystallization in toe and other joints producing the disease of **gout**. The catabolic pathway for purine nucleotides is shown in Fig. 10.26.

## SALVAGE PATHWAY

Catabolism of purines generates purine bases and hypoxanthine as the penultimate product. Adenine, guanine and hypoxanthine can be recovered through a salvage pathway by **phosphoribosylation**. The enzymes involved in the recovery process are adenine **phosphoribosyltransferase (APRT)** and **hypoxanthine-guanine phosphoribosyltransferase (HPRT)**. The reaction catalyzed by APRT is as follows:

$$\text{adenine } + \text{ PRPP} \rightleftharpoons \text{AMP } + \text{ PPi}$$

The reactions catalyzed by HPRT are:

$$\text{hypoxanthine } + \text{ PRPP} \rightleftharpoons \text{IMP } + \text{ PPi}$$

and:

$$\text{guanine } + \text{ PRPP} \rightleftharpoons \text{GMP } + \text{ PPi}$$

The conversion of IMP to AMP and the salvage of IMP from the catabolism of AMP involve the deamination of aspartate to fumarate through the **purine nucleotide cycle** shown in Fig. 10.27.

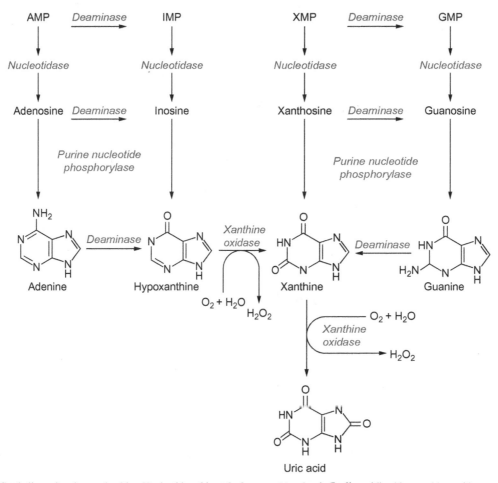

**FIGURE 10.26**  Catabolism of purine nucleotides. Nucleotides ultimately form xanthine that is finally oxidized by xanthine oxidase to generate uric acid.

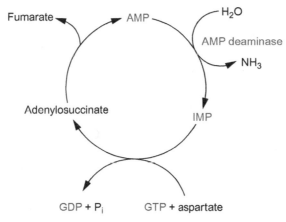

**FIGURE 10.27**  The purine nucleotide cycle. The conversion of AMP to IMP and the conversion of IMP to GDP result in the deamination of aspartate to form fumarate.

For thymine nucleotides (thymine is exclusive to DNA while uracil is exclusive to RNA), the sugar is deoxyribose and it appears in DNA as **dTMP**. dTMP can be formed from **dUMP** by thymidine kinase that can catalyze the following reactions:

$$\text{thymidine} + \text{ATP} \rightleftharpoons \text{dTMP} + \text{ADP}$$

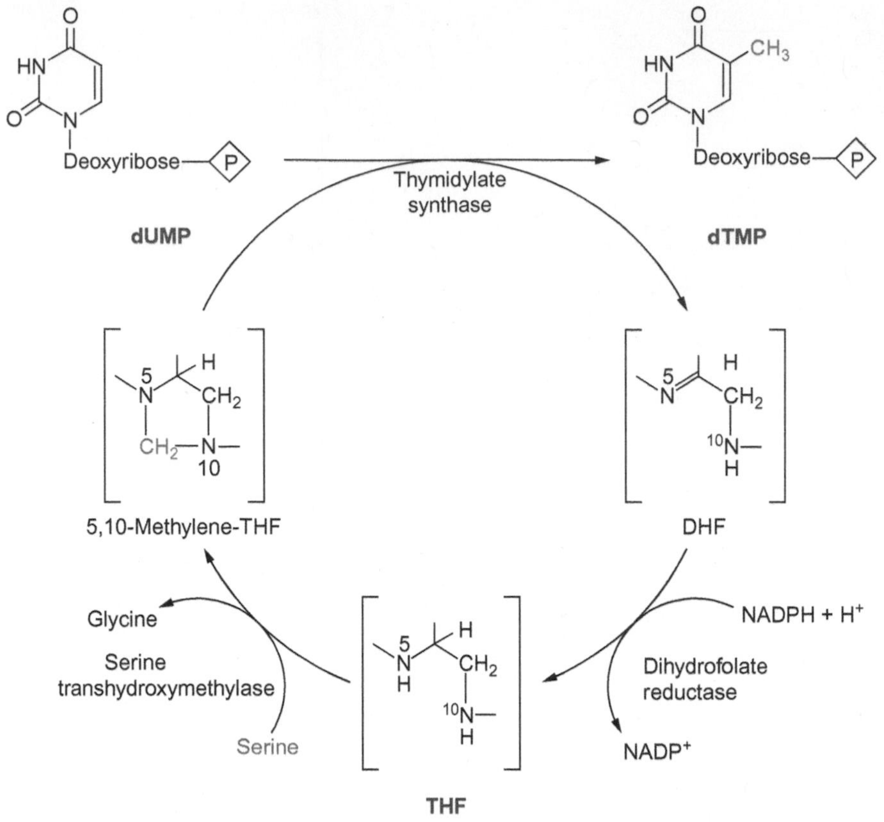

**FIGURE 10.28** To form dTTP, dUMP, from the metabolism of UDP or CDP, is converted to dTMP by thymidine synthase. As dTMP is methylated dUMP, the methyl group addition is derived from $N^5$, $N^{10}$-methylene THF. THF is converted to DHF after donation of methyl. Then THF is regenerated from DHF by DHFR. *THF*, tetrahydrofolate; *DHF*, dihydrofolate; *DHFR*, dihydrofolate reductase. **Methotrexate** is a cancer drug used to interfere with the action of DHFR.

or

$$deoxyuridine + ATP \leftrightharpoons dUMP + ADP$$

dUMP can be converted to dTMP by the action of **thymidylate synthase** as shown in Fig. 10.28.

## Pyrimidine Catabolism

A product of pyrimidine catabolism is β-**alanine**, derived from uracil. Uracil is a product of the breakdown of CMP and UMP (CMP can be converted to UMP). CMP and UMP derive from **RNA catabolism**. β-**aminoisobutyrate** derives from the breakdown of DNA through dCMP and dTMP to evolve thymine. These reactions are shown in Fig. 10.29.

During the catabolism of RNA, **uracil** is released (Fig. 10.29) and it can be salvaged back to UMP by **uridine phosphorylase** and **uridine kinase**:

$$uracil + ribose\text{-}1\text{-}phosphate \xrightleftharpoons{\text{uridine phosphorylase}} uridine + Pi$$

and

$$uridine + ATP \xrightleftharpoons{\text{uridine kinase}} UMP + ADP$$

Thymine can also be salvaged in similar fashion:

$$thymine + deoxyribose\text{-}1\text{-}phosphate \xrightleftharpoons{\text{thymine phosphorylase}} thymidine + Pi$$

and

$$thymidine + ATP \xrightleftharpoons{\text{thymidine kinase}} dTMP + ADP$$

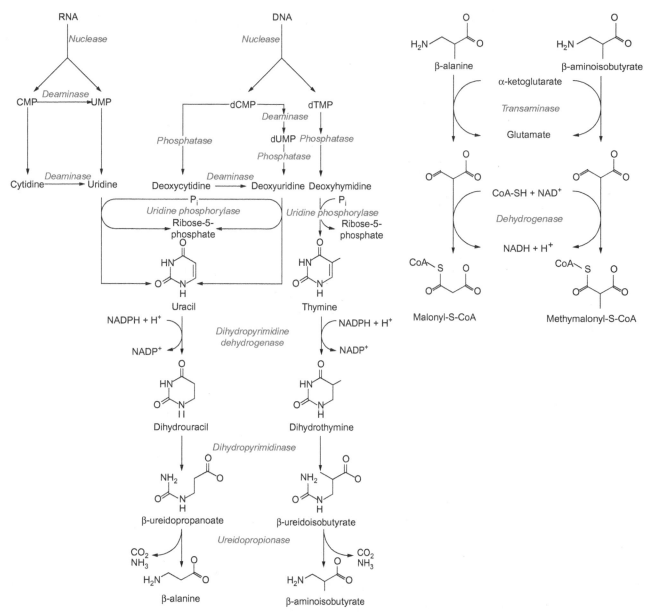

**FIGURE 10.29** Pyrimidine catabolism. The final products, **malonyl-*S*-CoA** and **methylmalonyl-*S*-CoA** can enter the TCA cycle.

**Thymidine kinase** is an important enzyme; its concentration changes during the cell cycle and is at its peak during DNA synthesis.

**Deoxycytidine kinase** allows for the **salvage of deoxycytidine**, its preferred substrate. However, deoxyadenosine and deoxyguanosine are also substrates for this enzyme. The reaction with deoxycytidine is as follows:

$$\text{deoxycytidine} + \text{ATP} \underset{}{\overset{\text{deoxycytidine kinase}}{\rightleftharpoons}} \text{dCMP} + \text{ADP}$$

Because of the low concentrations of both ribose-1-phosphate and the pyrimidine nucleosides, this pathway is a minor one for the salvage of pyrimidines.

dTMP can be formed from dUMP as shown in Fig. 10.28. The reaction is catalyzed by **thymidylate synthase**. 5,10-methylene THF (**5,10-methylene tetrahydrofolate**) coenzyme donates a methyl group to dUMP to form dTMP.

## Deoxyribose-Containing Nucleotides

The syntheses of **deoxyribose trinucleotides** needed for the synthesis of DNA are summarized in Fig. 10.30.

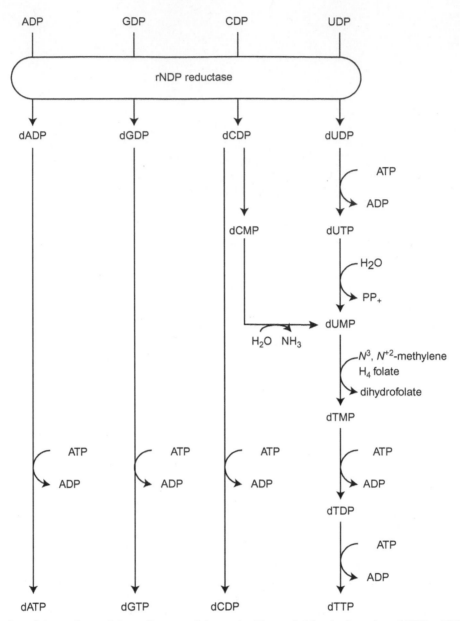

**FIGURE 10.30** Overview of the syntheses of deoxyribose-containing nucleotides needed for the formation of DNA. *rNDP reductase* (or *RNR*), ribonucleotide reductase; *PPi*, pyrophosphate *THF*, tetrahydrofolate; *DHF*, dihydrofolate.

## Formation of Deoxyribose From Ribose Only in Its Nucleotide Form

As indicated in Fig. 10.30 **ribonucleotide reductase (RNR)** converts the ribose-containing nucleotides: ADP, GDP, CDP, and UDP to the corresponding deoxynucleotides. In the case of thymine nucleotides, dTMP is not synthesized by RNR but is the product of the action of thymidine kinase as shown in the reaction above. Deoxyribonucleotide precursors are in much lower concentrations than ribose-containing nucleotides: 0.013 mM dATP, 0.005 mM dGTP, 0.022 nM dCTP, 0.023 mM dTTP. The corresponding ribose-containing nucleotides are at the following concentrations: 2.8 mM ATP, 0.48 mM GTP, 0.21 mM CTP, and 0.48 mM UTP in the cell. Consequently, *RNR becomes a regulatory factor in the synthesis of DNA.*

## Disorders of Purine and Pyrimidine Metabolism

The end product of purine metabolism is **uric acid**. Frequently, the level of uric acid in plasma is high and this condition can lead to **gout** (normal **uric acid concentration**, 3.6−8.3 mg/dL; levels as high as 9.6 mg/dL can occur without the generation of gout). Levels of plasma uric acid can be high enough (**hyperuricemia**) to cause crystallization in

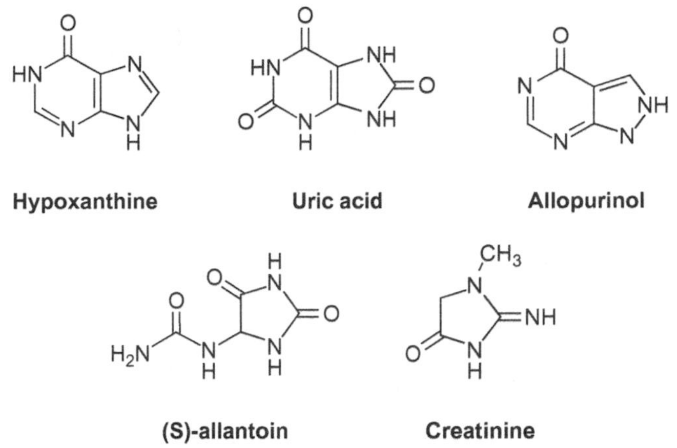

**Hypoxanthine**        **Uric acid**        **Allopurinol**

**(S)-allantoin**        **Creatinine**

**FIGURE 10.31**   Xanthine and **hypoxanthine** (*top left*) are substrates for xanthine oxidase that converts them to uric acid (*top middle*). Allopurinol (*top right*) is a competitive inhibitor of xanthine oxidase and is a favored drug for the treatment of gout. Uric acid can be converted by peroxisomal urate oxidase to (S)-allantoin (*bottom left*). Plasma **creatinine** (*bottom right*) is used to measure kidney function.

various joints that is common in the ball joint of the large toe. Crystals can form in the kidney (sometimes leading to **kidney stones**) and in capillaries. High blood **creatinine** is associated with high uric acid and may reflect **decreased glomerular filtration**. Gout is considered to be a form of arthritis and some believe that high plasma uric acid is a predictor of cardiovascular disease. Curiously, uric acid is an antioxidant, the highest level of an antioxidant in blood. It may be an indicator of oxidative stress. Uric acid is quite insoluble in water, whereas its metabolite, **(S)-allantoin**, is 10 times more water-soluble than uric acid. It is unclear whether uric acid is actually functioning as an antioxidant in blood. Uric acid metabolism occurs in the **peroxisome** by **urate oxidase** (or by catalase) and through two intermediates, produced through two other enzymes, leads to **(S)-allantoin**. There can be genetic alterations in the genes for these enzymes that can account for high circulating **uric acid** and such studies are underway. High **uric acid** can be treated with the drug, **allopurinol** which is a competitive inhibitor of **xanthine oxidase** (Fig. 10.31), however, in some cases of gout, the enzyme, **urate oxidase** has been used effectively as a treatment which seems superior to **allopurinol** unless an allergic reaction to the enzyme protein develops.

Diseases associated with disorders of purine or pyrimidine metabolism are listed in Table 10.1.

**Hypoxanthine-guanine phosphoribosyltransferase** (**HGPRT**) is an important enzyme in the **purine salvage pathway**. It catalyzes the conversion of hypoxanthine to **inosine monophosphate** (**IMP**) and the conversion of guanine to **guanosine monophosphate** (**GMP**). Thus, it plays a major role in generating purine nucleotides through the purine salvage pathway (see "Salvage pathway"). It is possible to lose the function of this enzyme, located on the X chromosome. This trait translates into patients, primarily males, with a relatively rare recessive disease characterized by severe **gout** and a central nervous system disorder (**Lesch−Nyhan syndrome**; see Table 10.1). The disease is independent of geography and race and occurs in 1 of 380,000 births. The severe form of this disease is characterized by **self-mutilation**. There is no direct treatment for this condition except to use devices that will limit self-mutilation and therapy for gout, primarily the use of allopurinol.

Diseases associated with **pyrimidine metabolic disorders** are not as problematical as those associated with dysfunction of purine metabolism because the products are more water-soluble than uric acid. Some of these diseases are described in Table 10.1.

## Biosynthesis of Deoxyribonucleic Acid in the Nucleus

The structure of DNA and the nature of the individual strands of DNA have been discussed (Figs. 10.12−10.16). DNA is a large polynucleotide. In the haploid number (23) of human chromosomes, their length is about 3 billion base pairs with a content of about 25,000 different genes. The earliest form of genetic information may have been in the form of RNA that could have developed before the existence of DNA.

Some evidence suggests that the beginning of DNA synthesis occurs in a small number of perinucleolar sites (within the nucleus) that are regulated coordinately and which are selected in the G1-phase of the cell cycle. In forming the DNA strand, the phosphate group of a mononucleotide reacts with the free 3′-hydroxyl group of a second nucleotide to form a **dinucleotide** joined through a **phosphoric acid ester** and the DNA strand is built by a continuation of this process (Fig. 10.32).

Given that there is an existing duplex of template and primer, DNA polymerase acts. The first nucleotide in the strand will have a free 5′-phosphate group and the last nucleotide added in the completed strand will have a free 3′-hydroxyl

**TABLE 10.1** Diseases Associated With Disorders of Purine or Pyrimidine Metabolism

| Disorder | Defect | Nature of Defect | Comments |
|---|---|---|---|
| Gout | PRPP[f] synthetase | Increased enzyme activity due to elevated $V_{max}$ | Hyperuricemia |
| Gout | PRPP synthetase | Enzyme is resistant to feedback inhibition | Hyperuricemia |
| Gout | PRPP synthetase | Enzyme has increased affinity for ribose 5 phosphate (lowered $K_m$) | Hyperuricemia |
| Gout | PRPP amidotransferase | Loss of feedback inhibition of enzyme | Hyperuricemia |
| Gout | HGPRT[a] | Partially defective enzyme | Hyperuricemia |
| Lesch–Nyhan syndrome | HGPRT | Lack of enzyme | |
| SCID[b] | ADA[c] | Lack of enzyme | |
| Immunodeficiency | PNP[d] | Lack of enzyme | |
| Renal lithiasis | APRT[e] | Lack of enzyme | 2,8-Dihydroxyadenine renal lithiasis |
| Xanthinuria | Xanthine oxidase | Lack of enzyme | Hypouricemia and xanthine renal lithiasis |
| von Gierke's disease | Glucose-6-phosphatase | Enzyme deficiency | |

**Pyrimidine Metabolism**

| Disorder | Defective Enzyme | Comments |
|---|---|---|
| Orotic aciduria, type I | Orotate phosphoribosyltransferase and OMP decarboxylase | |
| Orotic aciduria, type II | OMP decarboxylase | |
| Orotic aciduria (mild, no hematological component) | The urea cycle enzyme, ornithine transcarbamoylase, is deficient | Increased mitochondrial carbamoyl phosphate exits and augments pyrimidine biosynthesis; hepatic encephalopathy |
| β-Aminoisobutyric aciduria | Transaminase, affects urea cycle function during deamination of α-amino acids to α-keto acids | Benign, frequent in Asians |
| Drug-induced orotic aciduria | OMP decarboxylase | Allopurinol and 6-azauridine treatments cause orotic acidurias without a hematological component; their catabolic by-products inhibit OMP decarboxylase |

OMP Orotidine monophosphate.
[a]HGPRT—hypoxanthine-guanine phosphoribosyl transferase.
[b]Severe combined immunodeficiency.
[c]Adenosine deaminase.
[d]Purine nuclentide phosphorylase.
[e]Adenosine phosphoribosyl-transferase.
[f]5-Phospho-d-ribosyl-l-pyrophosphate.

(Fig. 10.13), thus, information in the DNA strand will proceed from the 5′-end to the 3′-end. A hypothetical strand of DNA could be written (starting from the 5′-end):

$$5'\quad ATGCTACGC\quad 3'$$

and the antiparallel strand together with the strand above would generate the double strand:

$$5'\quad ATGCTACGC\quad 3'$$
$$3'\quad TACGATGCG\quad 5'$$

remembering that each letter represents a deoxyribose mononucleotide.

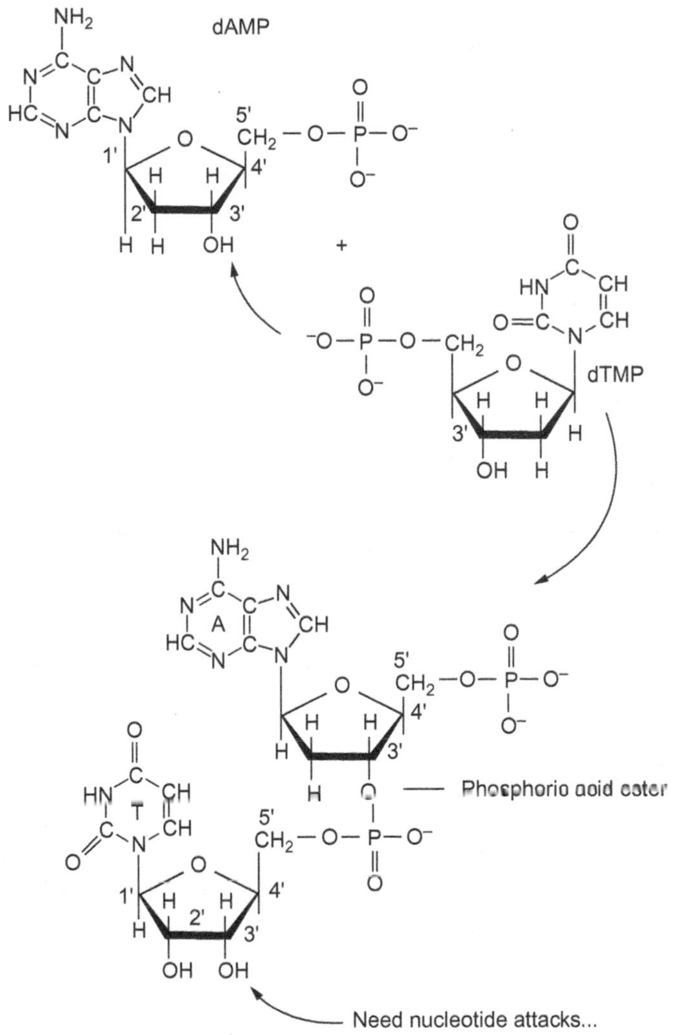

**FIGURE 10.32** Formation of the first dinucleotide in DNA.

In the case of an mRNA formed from the DNA sequence above which is:

5′ ATGCTACGC 3′ (DNA sequence)

the mRNA formed from that sequence would be:

5′ AUGCUACGC 3′ (RNA sequence)

remembering that U replaces T in RNA and that each letter represents a ribose mononucleotide.

The corresponding amino acids derived from the mRNA sequence would be:

*N*-Met-Leu-Arg-*C*

When A forms a base pair with T, there are two sets of hydrogen bonds involved, whereas the interaction between G and C involves three sets of hydrogen bonds (Fig. 10.12), making the G−C interaction stronger than the A−T, or the A−U (Fig. 10.11B) interactions.

During **cell division**, the genetic information has to be copied so that the daughter cell contains the same DNA as the parental cell. The two strands of DNA separate and **DNA polymerase** synthesizes a complementary strand (see the antiparallel strand above) so that for every A there is a T and for every G there is a C and vice versa following the rules of **base pairing** in the newly synthesized strand (Fig. 10.33).

The synthesis of new strands of DNA (**DNA replication**) is the function of **DNA polymerase III (DNA pol III)**. In human cells, the rate of addition of nucleotides to the growing chain is about 100 per second. In order to attach a dNTP,

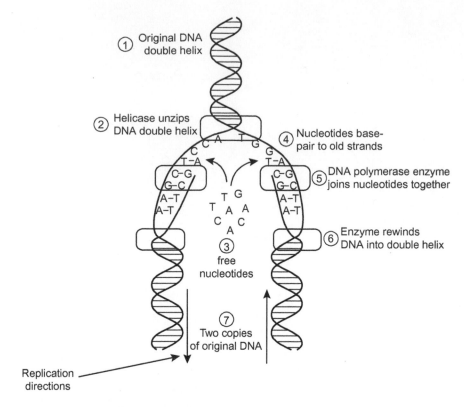

1. Replication starts at a specific sequence on the DNA molecule.
2. An enzyme unwinds and unzips DNA, breaking the hydrogen bonds that join the base pairs, and forming two separate strands.
3. The new DNA is built up from the four nucleotides (A, C, G and T) that are abundant in the nucleoplasm.
4. These nucleotides attach themselves to the bases on the old strands by complementary base pairing. Where there is a T base, only an A nucleotide will bind, and so on.
5. The enzyme DNA polymerase joins the new nucleotides to each other by strong covalent bonds, forming the sugar-phosphate backbone.
6. A winding enzyme winds the new strands up to form double helices.
7. The two new molecules are identical to the old molecule.

**FIGURE 10.33**   Figure showing the opening of the parental double helix and synthesis of the complements of each opened strand by DNA polymerase and **ligase**. In the human there are 13 or more DNA polymerases. Several enzymes are involved in the replication of DNA. The start of the unwinding process of double-stranded DNA is accomplished by **topoisomerase I**. This enzyme nicks the supercoiled DNA in a single strand releasing the coiled tension of the double helix and resulting in the untwisting of the strands. **Topoisomerase II** nicks both strands and the enzyme remains bound to DNA. After the relaxation of the double strand, the enzyme, **helicase**, unwinds the double strands. Helicase utilizes ATP to provide the energy to break the strands apart and overcome the hydrogen bonding forces holding the base pairs together. DNA polymerase travels in the $5'$- to $3'$-direction along the single strand and uses the free **deoxynucleotide triphosphates (dNTPs)** to form a base pair (through hydrogen bonding) with the next dNTP on the single strand. In this way, A binds to T and G to C forming a **phosphodiester bond** with the previous nucleotide of the same strand. The triphosphate provides the energy source for the binding of the next nucleotide, as shown in Fig. 10.34.

a **RNA primer** with a free $3'$ hydroxyl is needed for the polymerase. The DNA polymerase has a **proofreading** capability that ensures the addition of the correct base. Through its inherent $3',5'$-**exonuclease activity**, the polymerase can remove a mistake. One of the subunits of the enzyme acts like a **clamp** to fasten the DNA polymerase to the DNA template.

There is a group of proteins, known as the **primeosome**, that contains a **primase** whose function is to attach a small RNA primer to the single-stranded DNA. For DNA polymerase to start its synthetic activity, this primer provides a substitute $3'$ hydroxyl group. At some point later on, the RNA primer is deleted by **RNaseH**, leaving a gap that becomes filled by DNA pol I. When the RNA primer is removed, the unattached gap is filled in by **DNA ligase** that generates the formation of a phosphodiester bond at the $3'$-end. **Single-stranded binding proteins (SSBPs)** maintain the stability of the **replication fork** by helping **helicase** to make available the single-stranded template for DNA pol III. While the DNA polymerase moves easily on the leading strand in the direction of $5'$ to $3'$, the movement from the $3'$-end toward the $5'$-end of the **lagging strand** is more difficult. To continue in the $5'$ to $3'$-direction, the enzyme synthesizes stretches

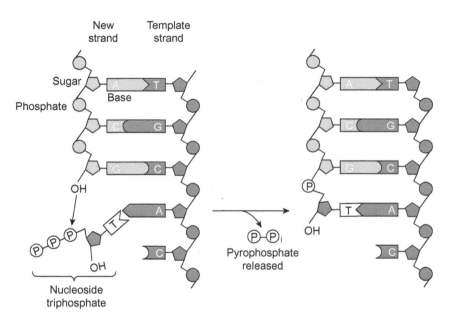

New strand   Template strand

Sugar

Base

Phosphate

OH

OH

P P P

OH

Nucleoside triphosphate

P — P<sub>i</sub>

Pyrophosphate released

OH

**FIGURE 10.34** The terminal phosphate of the next nucleotide added to the growing chain of DNA by DNA polymerase provides the energy for the formation of a new bond.

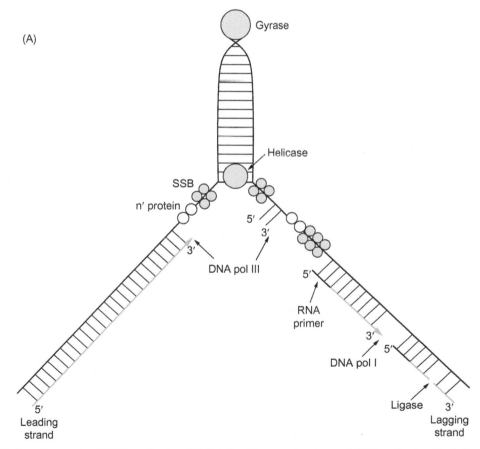

(A)

Gyrase

Helicase

SSB

n′ protein

5′

3′

DNA pol III

5′

RNA primer

3′

5′

DNA pol I

Ligase   3′

3′

5′

Leading strand

Lagging strand

**FIGURE 10.35** (A) Overall summary of DNA replication. (B) Details of the overall process of DNA replication (*top*). Synthesis of the lagging strand on the DNA template (*bottom*).

of the lagging strand in the 5′ to 3′-direction called **Okazaki fragments**. To complete the lagging strand, these stretches are filled by DNA pol I and the ligase.

At the beginning of the process, a topoisomerase called **gyrase** induces the unwinding of the double stranded DNA. The processes of DNA replication are explained by Fig. 10.35A and B.

(B)

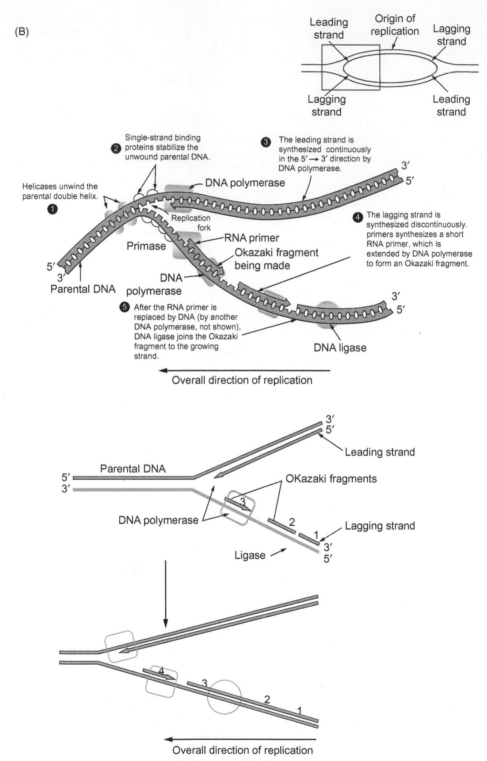

**FIGURE 10.35**   (Continued).

A DNA proofreading and **repair mechanism** is part of the process that can recognize an improper base pair and replace it with the correct base, otherwise, there might be one incorrect base pair per thousand. **DNA pol II** is more involved in proofreading than the other polymerases. Because of this protective mechanism, the actual error rate in DNA synthesis is between one in one million and one in one billion. A repair enzyme can remove a lesion in one

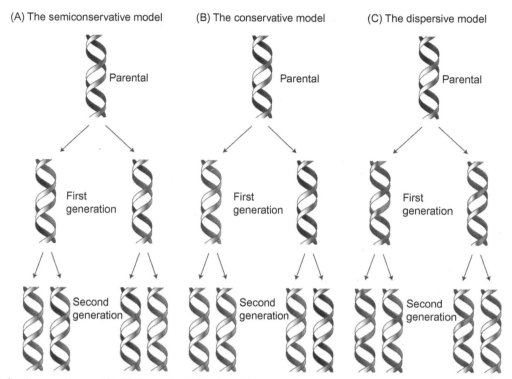

(A) The semiconservative model

Parental

First generation

Second generation

(B) The conservative model

Parental

First generation

Second generation

(C) The dispersive model

Parental

First generation

Second generation

**FIGURE 10.36**  Semiconservative model of DNA synthesis. Starting with a parental hybrid strand (one in *red* and the other in *blue*), the first genera-tion of DNA molecules results in two double strands, each having one parental strand and one new strand. In the succeeding generation, there are four DNA molecules, two hybrids of a parental strand and the new strand and two DNA molecules in which both strands are derived from the new strand of the first generation.

of the two DNA strands and the correct base pair can be added by reference to the undamaged strand. Corrections can occur as follows: **DNA glyoxylase** removes the altered base after the removal of the deoxypentose phosphate (base excision repair) or, in an alternate case, a small oligonucleotide stretch surrounding the damage is removed (**nucleotide excision**). In either case, the gap is filled by DNA polymerase and DNA ligase. The adventitious conversion (by deamination) of cytosine to uracil is one case of a mismatched base pair. In this case, uracil could be removed and replaced with cytosine or, in another case, uracil might form a base pair with adenine to give rise to an **unacceptable base pair**. If the repair mechanism might, in error, excise the uracil and replace it with an adenine, a mutation would result. The experimentally proven model of DNA replication is called the **semicon-servative model**. In this model, two strands of the parental DNA separate with each strand functioning as a tem-plate for the synthesis of a new complementary strand. The resulting first generation would be hybrid molecules, one strand from the parent and one new strand. In the second generation, there would result four double-stranded DNAs, two of which are hybrid molecules and two of which derive from the new DNA in the first generation as shown in Fig. 10.36.

## MITOCHONDRIAL DNA SYNTHESIS

There are 37 genes within a circular DNA in human mitochondria. *The inheritance of these genes is exclusively maternal (the lesser concentration of paternal mitochondria are possibly diluted out during the process of embryogen-esis) and the mitochondrial genetic material is not replicated coordinately with nuclear DNA (although 2 proteins involved in mitochondrial DNA replication are encoded by nuclear genes).* When a mitochondrion has grown to a point when it begins to divide by fission, replication of the DNA will begin, a process that is similar to bacterial DNA replication (fortifying the idea that mammalian mitochondria are derived from bacteria. It is theorized ("**endo-symbiotic hypothesis**") that during evolution some form of bacteria, possibly purple nonsulfur bacteria, survived the endocytotic process of some early cell type and eventually became incorporated into the cytoplasm as mitochondria, complementing the cytoplasmic metabolism with a more efficient oxidative metabolism. Thus, the replication of

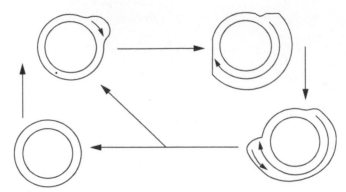

**FIGURE 10.37** Mechanism of mitochondrial DNA replication. Mitochondrial DNA, containing 37 genes is a double-stranded circle (*lower left*). One strand is replicated as the leading strand, pauses, and waits for a signal to complete the synthesis, generating a D loop (*upper left*). When the replication fork passes through the origin of the second strand that is also a leading strand, replication of the second strand begins. The result is the generation of two double strands available for each daughter mitochondrion.

mitochondrial DNA is similar to bacterial replication of its DNA. Mitochondrial genes do not code for all the enzymes and other proteins of the mitochondrion. This may have been the case long before the evolution of mammalian cells progressed and many of the genes in the early genome of the mitochondrion may have been lost while some of the essential genes became relegated to the nuclear compartment. Of the mitochondrial DNA, 80% codes for functional mitochondrial proteins.

In mitochondrial DNA replication, both DNA strands are continuously synthesized as leading strands in the 5' to 3'-direction. After the first strand begins to be synthesized, there is a pause until a signal is generated for completion of the strand. This pause causes the formation of loops (called **D loops**) as shown in Fig. 10.37 that outlines the mechanism of mitochondrial DNA replication.

The human mitochondrion has 37 genes [encoding: 2 ribosomal RNA (rRNAs), 22 transfer RNAs (tRNAs) and 13 polypeptides that are subunits of enzymes in the oxidative phosphorylation system]. Nuclear genes control mitochondrial replication. Two proteins in the mitochondrial DNA replication system are encoded by nuclear genes: these are a **D-loop endoribonuclease** and **DNA primase**. These proteins are presumably synthesized in the cytoplasm and transported into the mitochondria. DNA primase initiates mitochondrial replication at an origin.

When the heavy strand of mitochondrial DNA is transcribed, a transcript results that contains information for **transfer RNA (tRNA)**, **ribosomal RNA (rRNA)** and **messenger RNA (mRNA)** on a polycistronic molecule and the full-length transcript is cut into the functional RNA units. The **initiation of transcription** requires three proteins: mitochondrial RNA polymerase, and four **mitochondrial transcription factors**, A, B1, and B2 (consisting to two factors). These proteins are assembled on the mitochondrial promoters to initiate transcription.

The synthesis of the second strand of DNA begins after the fork of the first strand synthesis passes through the origin of the second strand. The origins for left- and right-strand replication are separate for mitochondrial DNA but coincident for nuclear DNA. When synthesis has been completed, the products are two daughter circular double-stranded mitochondrial DNA molecules to be included in two daughter mitochondria. Some will be smooth circles and others will have **D-loops** (Fig. 10.37).

## DNA MUTATIONS AND DAMAGE

Mutations in human DNA can occur from a number of different sources including radiation and exposure to toxic chemicals. Many **carcinogens** (some are DNA intercalating agents) are incorporated into the structure of DNA and cause mutations. **Cosmic radiation** (gamma radiation) rarely strikes the DNA molecule directly to cause a mutation. Mutations also can occur by viruses. Base deamination and depurination also can result in mutation. Adaptation from a very warm climate to a cold climate may require mutations in mitochondrial DNA affecting the usage of ATP, either primarily toward energy or the production of bodily warmth. These mutations can be passed to future generations, in the case of climatic change, adapting the human from the warmth of Africa to the colder climbs of Europe and North America, for example.

Some alterations in DNA can be corrected by repair mechanisms, although others are irreversible. As already discussed, DNA polymerase has a **proofreading system** that reduces mistakes in replication. Exposure to **ultraviolet light**

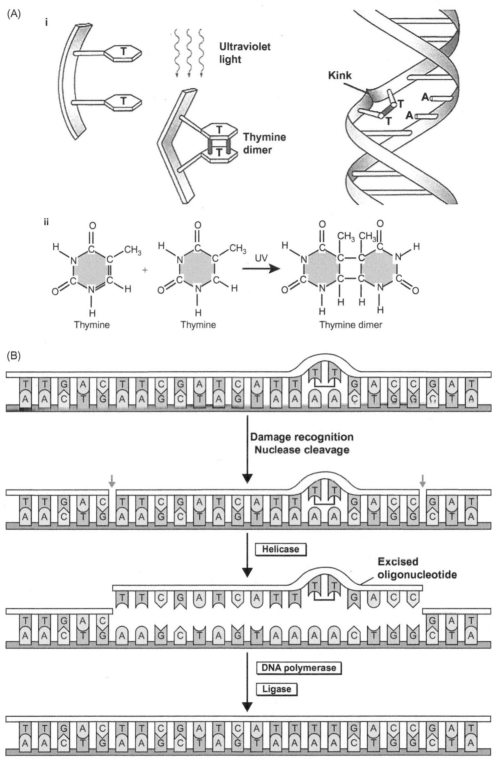

**FIGURE 10.38** (A) Generation of a thymine dimer caused by ultraviolet light. (B) Repair of a thymine dimer in one strand of DNA by nuclease activity, DNA polymerase and ligase.

can cause the appearance of a **thymine dimer** formed between two adjacent thymine bases in the same strand of DNA. This lesion can be repaired by actions of a nuclease, DNA polymerase and ligase as shown in Fig. 10.38.

*Xeroderma pigmentosum* is a rare autosomal recessive genetic disease. Such individuals are extremely sensitive to UV radiation and often develop **basal cell carcinomas** at an early age. Although there are several forms of *xeroderma*

**TABLE 10.2 Types of Damage to DNA**

I. Single base alteration
   A. Depurination
   B. Deamination of cytosine to uracil
   C. Deamination of adenine to hypoxanthine
   D. Alkylation of base
   E. Insertion or deletion of nucleotide
   F. Base analog-incorporation
II. Two base alteration
   A. UV light-induced thymine-thymine (pyrimidine) dimer
   B. Bifunctional alkylating agent crosslinkage
III. Chain breaks
   A. Ionizing radiation
   B. Radioactive disintegration of backbone element
   C. Oxidative free radical formation
IV. Cross linkage
   A. Between bases in same or opposite strands
   B. Between DNA and protein molecules (e.g., histones)

Source: Data from *Harper's Illustrated Biochemistry*, R. K. Murray et al., 26th ed. (2000), p. 335.

*pigmentosum*, most of the variants of this disease involve deficiencies in the **nucleotide excision repair mechanism**. Resulting DNA lesions are mainly formation of thymine dimers and other photoproducts. This disease can result in continuing changes in the eyes and skin to progressive neurological degeneration. *Xeroderma pigmentosum* is six times more prevalent in Japanese than in other groups. **Carcinogens** are chemicals that are either direct acting or indirect acting. Examples of the direct acting group (chemicals that interact directly with DNA) are: β-**propiolactone**, **ethylmethane sulfonate**, **nitrogen mustard,** and **methyl nitrosourea**. Some indirect carcinogens (chemicals that must be metabolized in the body to forms that interact directly with DNA) are: **benzo(*a*)pyrene**, **dibenzanthracene**, **2-napthylamine**, **dimethylnitrosamine**, **vinyl chloride**, **acetylaminofluorene,** and **aflatoxin B$_1$**. The several types of DNA damage are listed in Table 10.2.

## RESTRICTION ENZYMES

**Nucleases** that can cut DNA at specific sites have been discovered in bacteria and the enzymes are commercially available. These enzymes are called **restriction endonucleases** or simply, **restriction enzymes**. Endonucleases cut DNA in the interior of the molecule while **exonucleases** cut DNA at the ends of the molecule. Restriction enzymes bind to the double-stranded DNA and then travel along the molecule until the signal sequence (substrate) is encountered where the enzyme stops traveling and performs its nuclease function on both strands. The target site, recognized by the enzyme, usually consists of four, six, or eight bases in sequence from the 5′ to 3′-direction. The restriction enzymes are named for the organisms from which they are derived using an abbreviated form (e.g., *Eco* **R1** from *E. coli*). As an example, *Eco* R1 catalyzes the cleavage of DNA between the G and the A of the sequence 5′-GAATTC-3′ to generate the products: 5′...G and AATTC...3′, as shown in Fig. 10.39 along with the descriptions of three other restriction enzymes.

In Fig. 10.39, *Hae* **III** and *Sma* **I** produce blunt-ended products, that is, products whose ends align. *Eco* R1 and *Hind* III produce "sticky ends" from the cleavage in that one set of the bases overlaps with the other. Fig. 10.40 shows atomic models of *Eco* R1 bound to double-stranded DNA and the release of the cleaved DNA strands.

Sticky ends are available for the insertion of foreign or specifically designed DNA; these can be engineered to have the complementary sticky end sequences. TTAA is the overhanging end in the case of *Eco* R1 and this can be glued to another DNA with an overhang of AATT, because TT would form base pairs with AA and AA would form base pairs with TT to appear like,

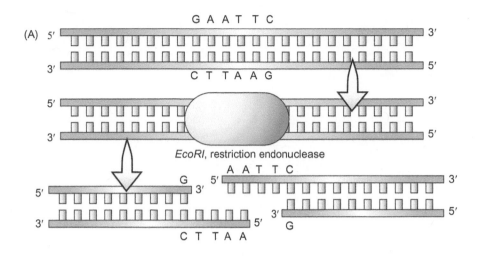

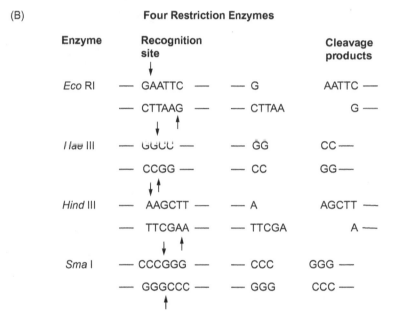

**FIGURE 10.39** (A) Action of *E. coli* restriction endonuclease on double-stranded DNA. (B) The recognition sites and cleavage products of four restriction enzymes. The cleavage products are either blunt-ended or overhanging "**sticky ends**."

$$\text{DNA1 \quad TTAA} + \text{AATT \quad DNA2} \rightleftharpoons$$
$$\text{TTAA \quad DNA1}$$
$$\text{AATT \quad DNA2}$$

where hydrogen bonds would form between T:A, T:A, A:T, and A:T. In a given span of a DNA there may be a number of **restriction sites** for a specific **restriction enzyme**. For example, in a 50-kilobase DNA, there might be a half a dozen *Eco* **R1 target sites**. Hydrolysis by *Eco* R1 would create a signature of the products.

There are hundreds of restriction enzymes with widely varying specificities, each one having its own target sequence as shown in Table 10.3.

Restriction enzymes are useful to isolate a specific fragment from linear or circular DNA. A DNA molecule can be characterized by its array of restriction sites. The products of digestion of DNA with restriction endonucleases can be separated by **sizing gel electrophoresis**. After electrophoresis, the gel can be stained with **ethidium bromide** that binds to the fragments and the bound complexes become visible (yellow-green fluorescence) after exposure to ultraviolet light. In this way, **restriction maps** are developed; this is especially useful for

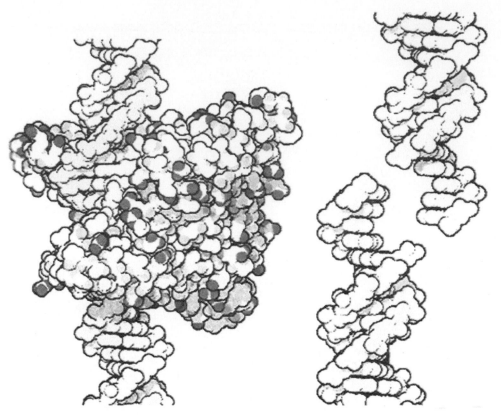

FIGURE 10.40  *Left*: an atomic model of *Eco* R1 restriction endonuclease bound to double-stranded DNA. The intact *Eco* R1 target sequence is shown in the *upper left* of the left model. On *right* is shown the products of the specifically cleaved DNA strands. *Reproduced originally from the Protein Data Bank: http://www.rcsb.org/pdb/molecules/pdb8_2.html.*

vectors that are used to carry a specific DNA of interest. A restriction map of the vector **pBR322** is shown in Fig. 10.41.

The insertion of a foreign DNA into a plasmid by use of an ***Eco* R1 restriction site** is shown in Fig. 10.42.

This is known as **recombinant DNA**. This can also be generated using restriction enzyme cleavage that results in **blunt ends** rather than overhanging **sticky ends**. When blunt ends are the products, synthetic sticky ends can be added. Using **terminal transferase**, poly d(G) can be added to $3'$ blunt ends of the vector while poly d(C) can be added to the $3'$-ends of the foreign DNA to be inserted. The new chimeric (hybrid) DNA cloning vector can be amplified in host cells (Fig. 10.43).

Specific human proteins can be expressed in relatively large quantities for therapeutic use. As the sequences expressed are human, little, if any, antigenic response should be encountered after administration. By the usage of restriction enzymes and different cloning vectors, the complete genome of an organism can be incorporated into a vector. As the therapeutic use of substances from animal tissues (e.g., hormones isolated from animal brains, such as growth hormone), in some cases, cause an antigenic reaction or can carry the danger of disease (e.g., **prion disease** via the scrapie protein transmitted from brains of diseased animals) the use of synthetically derived human sequences would be desirable. As an example of the preparation of a human protein for clinical use, cloning and reproducing the **human insulin gene** is shown in Fig. 10.44.

A collection of recombinant clones constitutes a **genomic library** that contains the total DNA from a specific cell. Genomic DNA can be cut by restriction endonucleases so that large pieces of DNA fragments are obtained increasing the chances that the full sequences of individual genes are conserved. Certain vectors are useful in this respect because they can incorporate large fragments of DNA. Among the vectors that accept large DNAs are **YAC (yeast artificial chromosome)**, **BAC (bacterial artificial chromosome)** and **P1 (bacteriophage-derived vector)**. YAC can carry DNAs from 100 to 3000 kilobase pairs and can be propagated in *E. coli*. BAC can incorporate DNAs from 100 to 300 kilobase pairs and can be propagated in bacteria. P1 can incorporate DNAs from 130 to 150 kilobase pairs. A vector that contains a gene for a protein that can be expressed in active form is an **expression**

**TABLE 10.3 A List of Many Restriction Endonucleases With Their Recognition Sites in DNA**

| Enzyme | Recognition Site | Enzyme | Recognition Site | Enzyme | Recognition Site |
|---|---|---|---|---|---|
| Aat II | GACGI▼C | Cla I | AT▼ CGAT | Nde I | CA▼ TATG |
| AccI | GT▼ |A/T||T/C|AC | Csp I | CG▼G(A/T)CCG | NgoM I | G▼ CCGGC |
| AccIII | T▼CCGGA | Csp 45 I | TT▼ CGAA | Nhe I | G▼ CTAGC |
| Acc65 I | G▼GTACC | Dde I | C▼ TNAG | Not I | GC▼ GGCCGC |
| AccB7 I | CCANNNN▼NTGG | Dpn I | G$^{me}$A▼ TC | Nru I | TCG▼ CGA |
| AcyI | C[A/G]▼ CG[T/C]C | Dra I | TTT▼ AAA | Nsi I | ATGCA▼ T |
| Age I | A▼ CCGGT | EclHK I | GACNNN▼ NNGTC | Pst I | CTGCA▼ G |
| Alu I | AG▼ CT | Eco47 III | ACG▼ GCT | Pvu I | CGAT▼ CG |
| A/w26 I | G▼ TCTC(1/5) | Eco52 I | C▼ GGCCG | Rvu II | CAG▼ CTG |
| A/w44I | G▼TGCAC | Eco72 I | CAC▼ GTG | Rsa I | GT▼ AC |
| Apa I | GGGCC▼ | Eco I CR I | GAG▼ CJC | Sac I | GAGGCT▼ C |
| Ava I | C▼(T/C)CG|A/G|G | Eco RI | G▼ AATTC | Sac II | CCGC▼ GG |
| Ava II | G▼ G(A/T)CC | Eco RV | GAT▼ ATC | Sal I | G▼TCGAC |
| Ba/I | TGG▼ CCA | Fok I | GGATG|9/13| | Sau3A I | ▼ GATC |
| BamH I | G▼ GATCC | Hae II | (A/G)GCGC▼(TC) | Sau96 I | G▼ GNCC |
| Ban I | G▼ G(T/C)|A/G)CC | Hae III | GG▼ CC | Sca I | ACT▼ ACT |
| Ban II | G(A/G)GC(T/C) ▼ | Hha I | GCG▼C | Sfi I | GGCCNNNN ▼ NGCCC |
| Bbu I | GCATG▼ C | Hinc II | GT|T/C|▼|A/G|AC | Sgf I | GCGAT▼ CGC |
| Bcl I | T▼ GATCA | Hind III | A▼ AGCTT | Sin I | G▼ G[A/T]CC |
| Bgl I | GCCNNNN▼ NGGC | Hinf I | G▼ ANTC | Sma I | CCC▼ GGG |
| Bgl II | A▼ GATCT | Hpa I | GTT▼ AAC | SnaB I | TAC▼ GTA |
| BsaM I | GATTGCN▼ | Hpa II | C▼ CGG | Spe I | A▼ CTAGT |
| BsaO I | CG(A/G)|T/C|▼ CG | Hsp92 I | G[A/G]▼ CG|T/C|C | Sph I | GCATG▼ C |
| Bsp1286 I | G(G/A/T)GC(C/A/T)▼ C | Hsp92 II | CATG▼ | Ssp I | AAT▼ ATT |
| BsrBR I | GATNN▼ NNATC | I-Ppo I | CTCTCTTAA▼ GGTAGC | Stu I | AGG▼ CCT |
| BsrS I | ACTGGN▼ | Kpn I | GGTAC▼ C | Sty I | C▼ C[A/T]|T/A]GG |
| BssH II | G▼ CGCGC | Mbo I | ▼ GATC | Taq I | T▼ CGA |
| Bst71 I | GCAGC(8/12) | Mbo II | GAAGA[8/7] | Tru9 I | T▼ TAA |
| Bst98 I | C▼ TTAAG | Mlu I | A▼ CGCGT | Tthill I | GACN▼ NNGTC |
| Bst E II | G▼ GTNACC | Msp I | C▼ CGG | Vsp I | A▼ TAAT |
| Bst O I | CC▼ [A/T]GG | MspA I | C(A/C)G▼ (G/T)G | Xba I | T▼ CTAGA |
| Bst XI | CCANNNNN▼ NTGG | Nac I | GCC▼ GGC | Xho I | C▼ TCGAG |
| Bst ZI | C▼ GGCCG | Nar | GG▼ CGCC | Xho II | [A/G]▼ GATC[T/C] |
| Bsu36 I | CC▼ TNAGG | Nci I | CC▼ (G/C)GG | Xma I | C▼ CCGGG |
| Cfo I | GCG▼ C | Nco I | C▼ CATGG | Xmn I | GAANN▼ NNTTC |

Source: Data in this table were taken originally from http://www.bioscience.org.

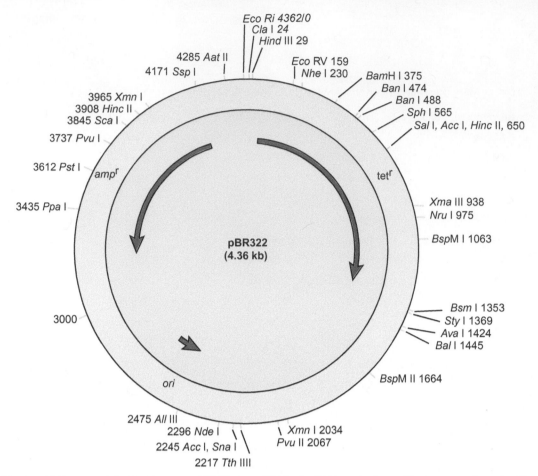

**FIGURE 10.41** A restriction map of the cloning vector pBR322. A specific foreign DNA can be inserted in the vector after a specific restriction site has been cleaved by hybridizing with the complementary sticky ends remaining after cleavage. This figure shows only about 10% of the actual number of restriction sites in this vector. The size of DNA inserts into this vector range from 0.01 to 10 kbases. There are many other cloning vectors and they have capacities for inserting foreign DNA ranging from 10 kbases to 3000 kbases.

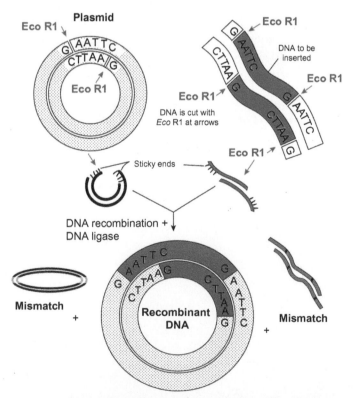

**FIGURE 10.42** Diagram showing how a foreign DNA is inserted into a plasmid using an *Eco* R1 restriction site.

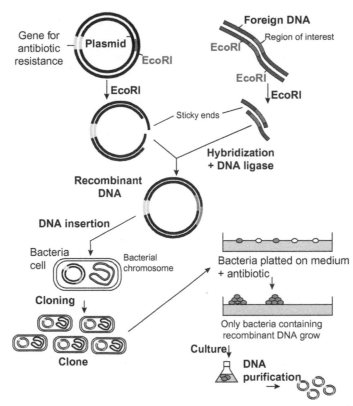

**FIGURE 10.43** Introduction of a foreign DNA into a plasmid, known as gene cloning, using an Eco R1 restriction site and amplification in a bacterial cell.

**vector**. New technologies increase the efficiency and decrease the cost of producing relatively large amounts of human proteins. Although some regions of the human genome are unstable and difficult to clone, a **human artificial episomal chromosomal (HAEC) system** has been developed that appears to be a good candidate for storage and expression.

## PROBING LIBRARIES FOR SPECIFIC GENES

A **cDNA probe** can be generated from a specific mRNA. The mRNA, encoding a specific protein, is a template. By the action of **reverse transcriptase** and **DNA polymerase**, a cDNA is formed that can be used as a probe to hybridize with a specific gene sequence (Fig. 10.45).

Generally, the cDNA probe will be labeled, more recently, with a fluorescent tag that does not interfere with the hybridization reaction. Such a cDNA can be used to probe a library of cDNAs for a complementary sequence, either to find a longer sequence containing more information or to search out a full coding region of the gene.

## HYBRIDIZATION TECHNIQUES

A **Southern blot** (named for its inventor) is used to detect a specific DNA in DNA mixtures. The mixture of DNAs is separated by electrophoresis (negatively charged DNAs migrate to the anode with the smaller fragments moving faster toward the anode than the larger fragments) on a membrane and the separated fragments are hybridized (after transfer to **nitrocellulose**) with the fluorescently labeled cDNA probe. Only those DNAs on the membrane that have complementary sequences will light up and be recorded on film. A similar technique, named the **Northern blot**, is used for the detection of mRNAs. In this case, the sample RNAs are separated by electrophoresis and probed with a cDNA probe. As RNA is less stable than DNA, this method is more demanding. An analogous method is used to identify a specific protein in a mixture of proteins. This technique is called a **Western blot**. Here, the protein mixture is separated by **gel electrophoresis**, and, after transfer to paper is probed with a labeled antibody. Other variations of this general procedure are called the **Eastern blot** for

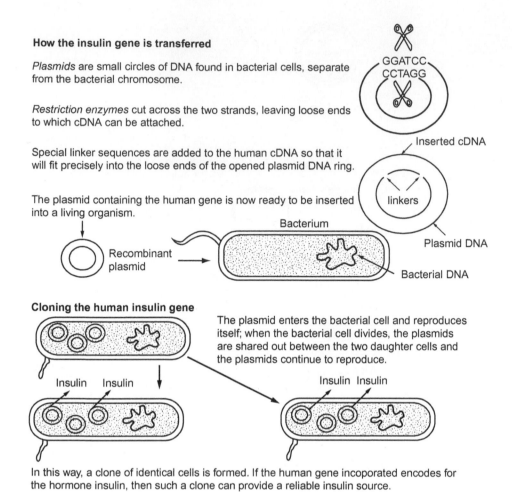

**How the insulin gene is transferred**

*Plasmids* are small circles of DNA found in bacterial cells, separate from the bacterial chromosome.

*Restriction enzymes* cut across the two strands, leaving loose ends to which cDNA can be attached.

Special linker sequences are added to the human cDNA so that it will fit precisely into the loose ends of the opened plasmid DNA ring.

The plasmid containing the human gene is now ready to be inserted into a living organism.

GGATCC
CCTAGG

Inserted cDNA

linkers

Plasmid DNA

Bacterium

Recombinant plasmid

Bacterial DNA

**Cloning the human insulin gene**

The plasmid enters the bacterial cell and reproduces itself; when the bacterial cell divides, the plasmids are shared out between the two daughter cells and the plasmids continue to reproduce.

Insulin    Insulin

Insulin    Insulin

In this way, a clone of identical cells is formed. If the human gene incoporated encodes for the hormone insulin, then such a clone can provide a reliable insulin source.

FIGURE 10.44   Cloning the human insulin gene. *Originally redrawn from http://web.mit.edu/esgbio/www/rdna/cloning.htm.*

posttranslational alterations of proteins (separation of a mixture of proteins, transfer to nitrocellulose membrane and use specific probes for carbohydrates, lipids or phosphate groups). The **Southwestern blot** detects **protein−DNA complexes** (proteins denatured by **sodium dodecylsulfate** are separated by polyacrylamide gel electrophoresis and then blotted onto nitrocellulose). DNA sequences of interest are fragmented by restriction enzymes and fragments are end-labeled [$^{32}$P usually] and bound to the separated proteins. Specific protein−DNA complexes can be analyzed further).

## AMPLIFICATION OF DNA SEQUENCES: POLYMERASE CHAIN REACTION

The **polymerase chain reaction (PCR)** is a technique for the amplification of specific DNA sequences. Two oligonucleotide primers are used that will hybridize to sequences adjacent to the target sequence on both strands of double-stranded DNA after the strands are separated by heat denaturation (melting). Large amounts of the primers are added to drive the annealing process. After annealing, each strand is extended (5′ to 3′) by a DNA polymerase in the presence of the four nucleotide triphosphates (dATP, dGTP, dCTP, dTTP). Two double strands appear in the first cycle. After repeating the cycle again, four double strands will appear and after the third cycle, eight double strands are generated with a doubling after each cycle. Using a heat stable DNA polymerase, the cycles can be run at elevated temperature so that new polymerase does not have to be added after each heat denaturation step. Twenty to thirty such cycles can be run for a given sample providing a huge amplification ($2^{20}$ for 20 cycles or $2^{30}$ for thirty cycles). Thirty cycles represent a billion-fold amplification. There are many applications of this process and it is widely used in scientific research and in forensic science. The PCR reaction through three cycles is shown in Fig. 10.46.

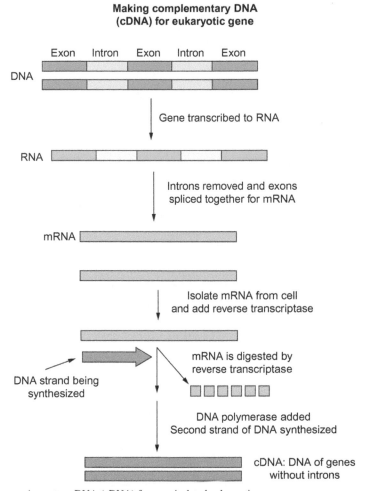

**Making complementary DNA
(cDNA) for eukaryotic gene**

**FIGURE 10.45**   Generation of a complementary DNA (cDNA) from an isolated eukaryotic gene.

## IDENTIFICATION OF A SPECIFIC GENE ON A CHROMOSOME

A **metaphase chromosome spread** is shown in Fig. 2.14. Localization of a specific gene on a specific chromosome is possible using **fluorescence *in situ* hybridization**. To accomplish this, the chromosomal spread on a microscope slide is treated with a specific radioactive ($^{32}$P) DNA probe that hybridizes to the gene containing that sequence. The excess probe that does not hybridize with other chromosomes is washed off and the specific hybridization complex is detected by exposure of the adhered radioactivity (or fluorescent activity) using a photographic emulsion. Upon development of the film, the exact position of the radioactive probe aligns with its **chromosomal location**. In this way, genes are localized to their specific positions in chromosomes as already alluded to in previous chapters.

## DETERMINING DNA SEQUENCE

If one has a single strand of DNA for which the base sequence needs to be determined, a complementary strand will be synthesized in the laboratory by the use of **DNA polymerase**. To initiate synthesis, a primer (similar to that shown in Fig. 10.42) is hybridized to the strand undergoing sequencing. The initial addition of a base in this synthesis is **dideoxy ATP (ddATP)** in place of dATP in addition to the normal 2′-deoxynucleotides of the other three bases. In ddATP both oxygen's on the 2′ and the 3′-positions of the ribose are absent (chemical methods have been developed to generate **dideoxyribose** with the oxygens missing in both the 2′ and 3′-positions). In the absence of a hydroxyl in the 3′-position in the terminal nucleotide, further addition of deoxynucleotides cannot take place so the chain can continue growing until an A is encountered when ddATP is added and the growth stops. The template strand would

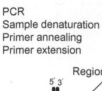

PCR
Sample denaturation
Primer annealing
Primer extension

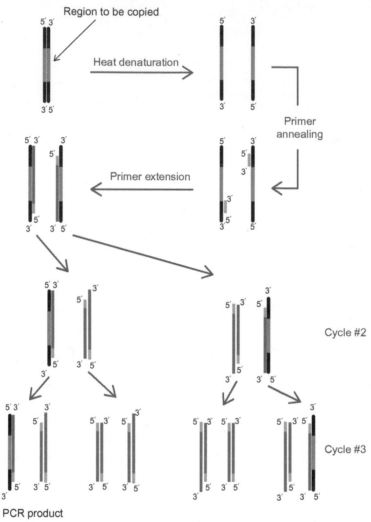

**FIGURE 10.46**   The first three cycles of the polymerase chain (PCR) reaction. A specific region of DNA to be amplified is used with flanking sequences. The double-stranded DNA is subjected to heat denaturation at 95°C. Primers (*green*) are annealed to flanking sequences (*black*). Base pairing occurs as the temperature is gradually cooled to about 60°C. Primers are extended in the 5′ to 3′-direction by addition of a heat-stable DNA polymerase. The first cycle results in two double-stranded DNAs containing the target sequence (*purple*). Each additional cycle produces a doubling of the target DNA sequence. Twenty to thirty cycles can be run to amplify the target DNA one million to one billion times. *Reproduced from http://www. myxxy.com/2010/0822/7887.html or http://www.web-books.com/MoBio/Free/Ch9E.htm.*

have a T opposite the ddATP. The same reaction is carried out individually with **ddGTP** in place of dGTP, **ddCTP** in place of dCTP and **ddTTP** in place of dTTP. The base opposite ddGTP would be a C; the base opposite ddTTP would be an A and the base opposite ddCTP would be a G. Thus a series of DNA fragments will be synthesized until the dideoxy nucleotide is added, representing the last base of each fragment. The newly synthesized molecules will have the same sequence and all will have the same 5′ end (starting position after the primer). The reactions are similar to PCR except for the stopping nucleotides. The fragments are separated on a **sizing electrophoretic gel** and exposed to reflect the addition of the radioactive dideoxy nucleotides (or each of the four dideoxynucleotides is conjugated to a different colored dye and the determination of each individual fluorescence can be used to determine the specific dideoxynucleotide). The size of the fragments indicates their order and the number of bases added until each dideoxynucleotide is reached, allowing for the construction of the sequence of the complementary strand. The rules of base pairing dictate the sequence of the corresponding original template strand. In a single experiment, it is possible to sequence 700 or more bases.

## INHIBITORS OF DNA SYNTHESIS

There are many **inhibitors of DNA synthesis**. Inhibition of the syntheses of nucleotides results in the inhibition of DNA synthesis. These are mainly **purine analogs** and **pyrimidine analogs** and their effects are visualized during the S phase of the cell cycle. Importantly, **folic acid analogs**, such as **methotrexate (MTX)** are competitive inhibitors of **dihydrofolate reductase (DHFR)** and folate-dependent enzymes, such as **thymidylate synthase**. The action of MTX causes depletion of the cellular stores of **tetrahydrofolate (THF)** and blocks both purine and deoxythymidylate syntheses. MTX lowers amino acid metabolism. Together with glutamine, the complex of MTX and **polyglutamine** binds potently to folate-dependent enzymes and the large size of the complex prevents its loss from the cell. Another folic acid analog, **Premetrexed**, is a more potent antagonist of thymidylate synthase than MTX. Other inhibitors of DNA synthesis are: **hydroxyurea, Cytarabine,** and **Gemcitabine**. Hydroxyurea inhibits **ribonucleotide reductase**, thus reducing the concentrations of dNTPs. **Pyrimidine analogs** include **5-fluorouracil (5-FU)** and **Capecytabine** whose actions depress nucleotide synthesis. **Cytarabine** inserts into DNA blocking elongation by reducing the activity of DNA polymerase. Gemcitabine is another inhibitor of DNA synthetic enzymes (DNA polymerase and ribonucleotide reductase) and terminates DNA strands, often leading to **apoptosis**. The inhibition of **purine synthesis** by **MTX** is shown **in** Fig. 10.47.

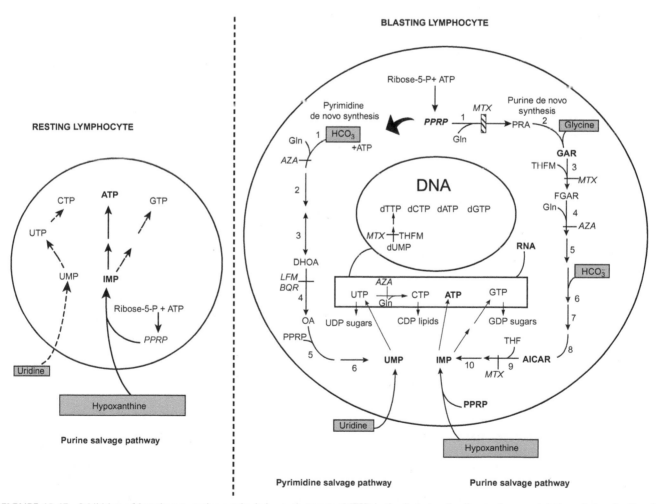

**FIGURE 10.47** Inhibition of lymphocyte purine synthesis by methotrexate (MTX) in the first step (leading to the accumulation of phosphoribosyl pyrophosphate [PRPP]) and in other enzymatic steps and the inhibition at various points in the system by other inhibitors. *THFM*, N$^{10}$-formyltetrahydrofolate; *AZA*, azaserine; *THF*, tetrahydrofolate; *LFM*, leflunomide; *BQR*, brequinar. The intermediates in purine biosynthesis are spelled out in Fig. 10.19. *Reproduced from http://www.biochemj.org/bj/342/0143/bj3420143.htm.*

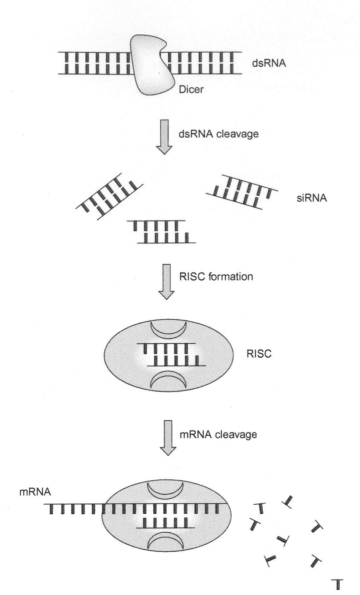

**FIGURE 10.48** Summary of the mechanism of RNA interference (RNAi). dsRNA, double-stranded RNA; siRNA, short inhibitory RNA; RISC, RNA-induced silencing complex. *Reproduced from http://www.translational-medicine.com/content/2/1/39/figure/F1?highres=y.*

## RNA INTERFERENCE

Another external approach to "knocking out" a gene has been the use of **antisense DNA** that has a sequence that is complementary to the sense strand of the expressing DNA. As DNA transcription results in a single-stranded mRNA, the antisense DNA forms a duplex with the mRNA, preventing its translation into protein. A more recent approach has been the use of **double-stranded RNA** to knock out specific genes, a method referred to as **RNA interference** (**RNAi**). The sense strand of double-stranded mRNA encodes the protein to be formed through translation. In mammalian cells, dsRNAs are broken down (by the enzyme "**dicer**") into short (about 20 nucleotide pairs) inhibitory RNAs (**siRNAs**) after delivery of the dsRNA by a virus or other artificial means. These RNA pairs bind to a cellular enzyme called the **RNA-induced silencing complex** (**RISC**) that uses one strand of the siRNA to bind to single-stranded mRNA of complementary sequence. The nuclease activity of the RISC degrades the mRNA to suppress gene expression. The cell's genetic machinery uses this strategy to control the expression of its mRNAs representing another mechanism of post-transcriptional regulation. The mechanism of RNAi is summarized in Fig. 10.48.

A newer technique for developing siRNA is called **esiRNA** which is **endonuclease-prepared siRNA**. In this technique a PCR fragment is transcribed *in vitro* and annealed to form a double-stranded DNA. This is digested with *E. coli*

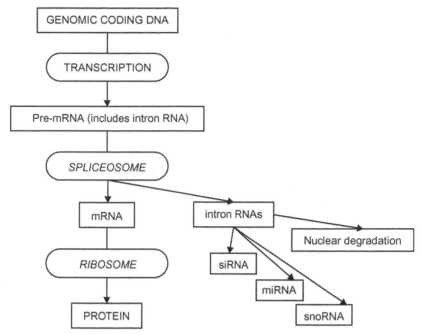

**FIGURE 10.49**  Transfer of information from the gene's DNA through mRNA to the expression of protein. The types of RNA are transcribed from the gene and from intronic RNA.

**RNase III** and the small siRNA products are subsequently purified. The use of RNase III mimics the action of dicer in the mammalian cell.

When dsRNA exceeding 30 nucleotides is introduced, an **antiviral response** by the organism (including an **interferon** response) may occur leading to nonspecific gene suppression. This nonspecific response may be avoided by the use of smaller RNAs of 20 or so nucleotides with two nucleotide 3′ overhangs (sticky ends). Transient transfection of this type of siRNA can lead to 90% or more reduction of the target RNA and the specific protein level.

There are natural components within the mammalian cell that contribute to **gene silencing** as part of the cell's normal regulation. These are: histones, chromatin, **micro RNA**, siRNA, dsRNA, dicer, and **transposons**. The roles of histones are discussed in the chapter on transcription.

## CODING DNA

Only about 1.5% of the total human DNA is considered coding; that is, it carries the information for all the genes through the production of mRNA. The remainder of the DNA has long been considered **"junk" DNA** because functions were not ascribed to it. However, it is becoming apparent that the junk-, or **noncoding-DNA** does have functions, in particular those genes for transfer RNA (tRNA) and ribosomal RNA (rRNA). Other sequences of noncoding DNA are becoming known and continue to be discovered. Coding DNA that carries the information of the genome is expressed as **introns** and **exons**, the exons of which encode **messenger RNA (mRNA)** as summarized in Fig. 10.49.

## NONCODING DNA

Noncoding DNA makes up about 98.5% of the total DNA. While it was previously thought to have no function, newer information is beginning to shed light on the many functions of this mass of DNA. It is involved in the cutting and splicing of large amounts of DNA, is involved in **transposon reassembly**, **genome rearrangements** and the production of small RNAs, some of which may serve as a source for new exons. It is also possible that noncoding DNA was used as a source of new genes needed for adaptation or for functions during evolution. These ideas are summarized in Fig. 10.50.

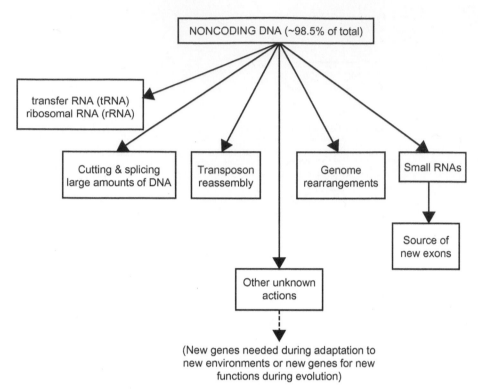

**FIGURE 10.50** Summary of ideas suggesting functions for the noncoding DNA. Both tRNA and rRNA are transcribed from noncoding DNA.

## Introns

Introns also qualify as noncoding DNA. Introns are the sequences between the exons and the intron messages are spliced out by the **spliceosome** in the formation of the coding RNAs (mRNA, rRNA and tRNA). Although they may be considered as noncoding, some introns enhance the expression of the genes in which they are contained and on occasion do code for parts of proteins. There are about 139,418 introns in the human genome and as one proceeds up the evolutionary scale, the number and length of introns increases with the complexity of the organism. In the human genome, there is an average of 8.4 introns per gene. Introns contain short sequences that are useful for efficient splicing. Introns may promote **genetic recombination**. This recombination or **crossing over** occurs in paired chromosomes, for example. These chromosomes can break and exchange a portion of DNA from one chromosome to a matching portion in the other chromosome. After reformation of the chromosomes, each one has acquired new genetic sequences from the other.

**Intron RNAs** that are **spliceosome** products may give rise to small RNAs, such as **siRNA (small interfering RNA)**, **miRNA (microRNA)**, and **snoRNA (small nucleolar RNA)** (Fig. 10.48), or they may be degraded in the nucleus and serve as a reusable pool of nucleotides.

## TRANSPOSONS

One of the functions of noncoding DNA is the reassembly of transposons by transposases leading to genome rearrangement. **Transposases** are enzymes involved in the cutting and splicing of large quantities of DNA. Transposons, themselves, are segments of DNA that are able to move to different positions in the genome of the same cell. Transposons have been referred to as "**jumping genes**" which can create duplicate genes or mutations. There are two classes of transposons based upon mechanism. Class I (**retrotransposon**) uses a copy and paste system in which a specific segment of DNA is transcribed to form a mRNA that, through the action of reverse transcriptase, produces the DNA that is inserted into a new position in DNA. Class II transposons operate without making an RNA intermediate. In this case, a **transposase** cuts the target DNA in such a way as to leave sticky ends, then cuts it out and it is ligated to a target site elsewhere in the DNA. The gaps left from the sticky ends are filled by DNA polymerase and the sugar-phosphate backbone is closed by DNA ligase. By these systems duplicate genes can be produced at specific sites or mutations can occur that may be necessary in the evolutionary process. A cartoon describing a class II transposon is shown in Fig. 10.51.

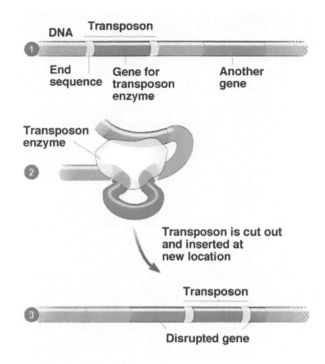

**FIGURE 10.51** A cartoon describing a class II transposon. *Reproduced from http://3.bp.blogspot.com/-wmQB_1b8cpal/AAAAAAAAA94/ l-Z68Kr1lNO/s1600/transposons1.jpg.*

## ALU ELEMENTS

These elements may be part of class I **retrotransposons**. They are small segments of DNA arising from the action of the **Alu restriction endonuclease**. They are derived from a small **7SL RNA** in the cytoplasm that is part of the signal recognition system. The Alu family is represented by about 300 base pair DNA interspersed elements (**short interspersed elements, SINEs**) in the human genome. There are more than one million Alu elements in the human genome representing about 10% of the total. They represent the link between transposable elements and mutated copies of active elements. Alu elements may be involved in many diseases, such as type II diabetes and some types of cancer.

## 5′ CAPPING OF RNA CONTAINING EXONS AND INTRONS

The 5′-cap is added to the end of the precursor mRNA (containing both exons and introns) in the nucleus. The 5′-cap renders the mRNA resistant to the action of **exonucleases**. The 5′-nucleotide of pre-mRNA loses its terminal phosphate group through the action of **RNA terminal phosphatase**. GMP is added to the remaining terminal phosphate by **guanylate transferase** that acts on **GTP** and removes two of its phosphates. **Methyl transferase** adds a methyl group to $N^7$ of guanine creating the **7-methylguanosine ($m^7G$) cap** as shown in Fig. 10.52.

This **capping enzyme complex** is specifically bound to RNA pol II prior to the start of transcription. The capped RNA is then bound by a **cap-binding complex** that specifically binds to the **nucleopore** for export to the cytoplasm. In the cytoplasm the cap-binding complex is replaced by two initiation factors and the binding of these initiation factors competes with **decapping machinery** whose action determines the **half-life of the mRNA**. The 5′-cap on mRNA interacts with the spliceosome (which removes the introns).

## POLYADENYLATION OF PRE-mRNA

The pre-mRNA usually contains the sequence **AAUAAA** that is the **polyadenylation signal**. The cleavage enzyme, **CPSF** (is in contact with RNA pol II and signals the polymerase to stop transcription), binds to this signal sequence and, with several other proteins in a multiprotein complex, cleaves the 3′-end of newly synthesized RNA just after AAUAAA. Polyadenylation is catalyzed by **polyadenylate polymerase** that adds AMP units (derived from ATP and releasing

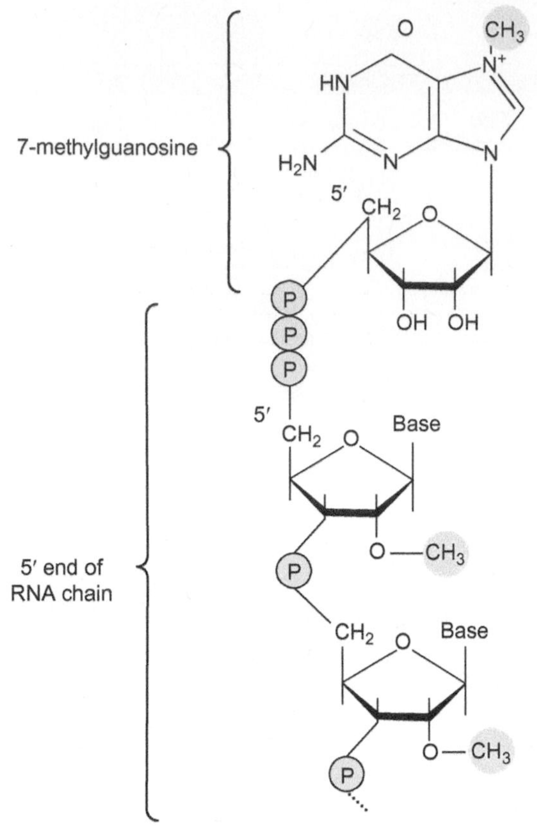

FIGURE 10.52 Structure of the 5′-cap (7-methylguanosine) of mRNA. The linkage between the cap and the 5′-end of RNA is 5′ to 5′ (relatively rare linkage). *Reproduced from http://www.mun.ca/biology/desmid/brian/BIOL2060-21/2118.jpg.*

pyrophosphate in the process). A new short polyA tail is generated that binds the protein **PAB2**. **PAB2** increases the affinity for the polyA polymerase (in contact with the spliceosome) allowing the length of the tail to increase to a maximum of about 250 nucleotides at which point polyadenylation ceases. Most mammalian mRNAs contain a polyA tail that facilitates the transport from the nucleus to the cytoplasm, contributes to the stability of the RNA (by protecting from degradation in the cytoplasm in addition to the 5′-cap) and increases the efficiency of the translation process.

## OVERALL TRANSCRIPTION–TRANSLATION PROCESS

Using the ovalbumin gene, as an example, the overall **transcription–translation process** is summarized in Fig. 10.53.

## INTRON EXCLUSION FROM PRE-mRNA

An intron is removed from the pre-mRNA by a **lariat mechanism**. The requirements of this mechanism are: a 5′-splice site at the 5′-beginning of the intron that is GU and a 3′-splice site at the termination of the intron which is AG. Upstream from the 5′-GU is a **pyrimidine-rich tract** (Cs and Us) and upstream of the polypyrimidine tract is a branch point that includes AMP. The cleavage of the 5′-end of the intron involves a specific 2′ hydroxyl group (branch site). A **phosphodiester bond** is formed with the 5′-phosphate of the intron to generate a **lariat structure**. Exon 1 is now separated from the intron. In the second step the phosphodiester bond between the intron and Exon 2 is converted to an Exon 1-Exon 2 phosphodiester. These reactions are summarized in Fig. 10.54.

## CODING RIBONUCLEIC ACID (RNA)

Coding RNAs are those that carry information that is translated into protein. Messenger RNA (mRNA) carries the coded information of the gene that is translated into protein, therefore mRNA is a coding RNA. In contrast, there are

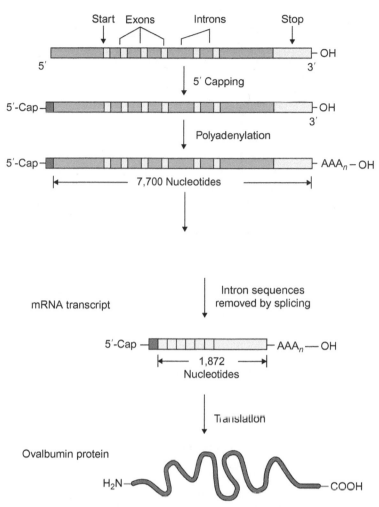

**FIGURE 10.53**  Cartoon of the overall process of transcription–translation of the ovalbumin gene as an example. The gene (*top*) is pictured indicating a start site, exons, introns and a stop signal. 5′ capping takes place (*second from the top*) followed by polyadenylation of the 3′-end at the termination of transcription. The spliceosome removes the intron RNAs and the mature mRNA transcript, containing 1,872 nucleotides, is produced. The capped mRNA with a polyA tail is transported to the cytoplasm through the nucleopore to the ribosome where it is translated into the ovalbumin protein. *Reproduced from http://www2.estrellamountain.edu/faculty/farabee/biobk/BioBookGENCTRL.html.*

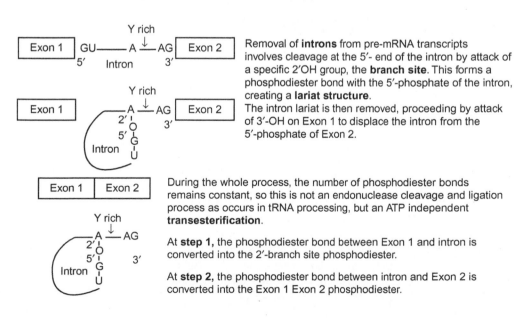

Removal of **introns** from pre-mRNA transcripts involves cleavage at the 5′- end of the intron by attack of a specific 2′OH group, the **branch site**. This forms a phosphodiester bond with the 5′-phosphate of the intron, creating a **lariat structure**.
The intron lariat is then removed, proceeding by attack of 3′-OH on Exon 1 to displace the intron from the 5′-phosphate of Exon 2.

During the whole process, the number of phosphodiester bonds remains constant, so this is not an endonuclease cleavage and ligation process as occurs in tRNA processing, but an ATP independent **transesterification**.

At **step 1,** the phosphodiester bond between Exon 1 and intron is converted into the 2′-branch site phosphodiester.

At **step 2,** the phosphodiester bond between intron and Exon 2 is converted into the Exon 1 Exon 2 phosphodiester.

**FIGURE 10.54**  Removal of an intron existing between hypothetical Exon 1 and Exon 2 by a lariat mechanism. *Reproduced from http://www.hixonparvo.info/mRNA%20processing.html.*

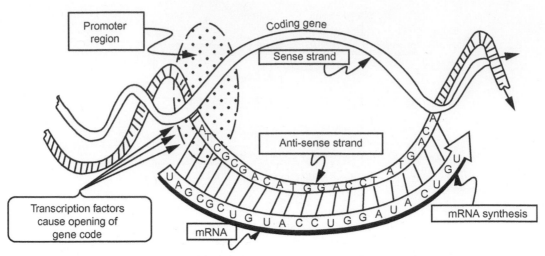

**FIGURE 10.55** Transcription of the antisense strand of DNA forms mRNA.

many RNAs that have functions related to translation as well as other regulatory functions and these are classified as noncoding RNAs (and will be discussed below).

## Messenger RNA

mRNA has been described in terms of cytoplasmic protein synthesis. Maturation from pre-mRNA to mRNA through the action of the **spliceosome** also has been described. The coding region of a gene to be transcribed is opened, causing strand separation, in response to positive controls, such as transcription factors (proteins), separating the sense and antisense strands of DNA. The **antisense strand** is transcribed into mRNA as shown in Fig. 10.55.

Sometimes there occurs an abnormal mRNA called a **pseudo mRNA**. Although the pseudo mRNA resembles a protein-coding mRNA, it cannot code for a full-length protein owing to a disruption in the reading frame; such an mRNA is analogous to a pseudo gene where a protein is made but one or more of its functions is absent. These RNAs do not strictly correspond to the genetic elements of DNA. They exist significantly, up to 10% of the messages, in some organisms. Functional mRNAs are degraded in the cell by protein complexes, called **exosomes** or **Lsm** (Like **Sm** proteins) **complexes**. Exosomes contain **poly(A) ribonuclease**, **decapping factors** and **5′-3′ exonucleases**. **Lsm complex**es refer to a complex of seven proteins that resemble Sm proteins. These seven Sm proteins form a ring-like structure at the core of the spliceosome and the proteins come in contact with the bases of the RNA. At any time, the levels of mRNA in a cell are a balance between transcriptional activation and repression.

## NONCODING RNAs

### Transfer RNA (tRNA)

There exist tRNAs that are specific for each amino acid. Through its anticodon, the function of tRNA is to carry a specific amino acid as encoded by mRNA (codon) to the ribosome where it is incorporated into the growing polypeptide chain. tRNA molecules are clover-leaf-shaped, consist of about 80 nucleotides and each has two functional groupings. At one end of the molecule is an **anticodon** that forms base pairs with the triplet **codon** in mRNA. The codon in mRNA coincides to the anticodon in tRNA. At the opposite end of the molecule is a sequence where the 3′-terminus is the site of attachment by esterification of the amino acid. A specific **aminoacyl tRNA synthase** catalyzes the formation of the ester bond between the 3′-hydroxyl of tRNA and the cognate amino acid. A cartoon of a proline tRNA is shown in Fig. 10.56.

The specific aminoacyl tRNA synthase recognizes the cognate tRNA by: the anticodon, the base adjacent to the CCA sequence at the 3′-end of the tRNA, a single base pair in the acceptor stem and the three dimensional structure of the enzyme, itself. In Fig. 10.57 is shown the general mechanism by which a specific aminoacyl tRNA synthase esterifies the cognate amino acid to its tRNA.

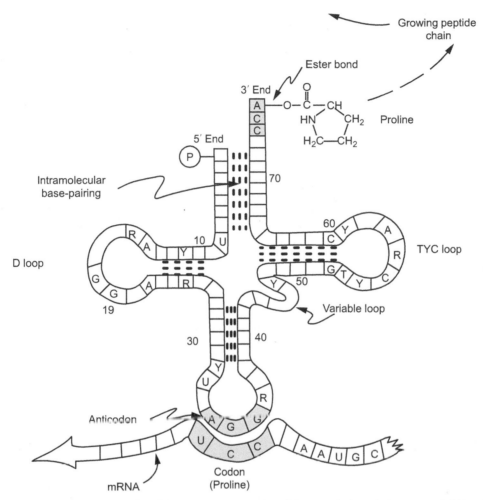

**FIGURE 10.56**   A cartoon of a hypothetical proline tRNA. This is representative of the general clover-leaf structure of tRNAs. The anticodon triplet at the bottom of the tRNA is complementary to the triplet codon in mRNA causing them to be complexed by base-pairing. At the top 3′-end of the acceptor stem, the amino acid is covalently attached by the action of the specific aminoacyl tRNA synthase which recognizes the specific tRNA by the sequence 5′ CCA-OH 3′ that is uniform for all tRNAs.

The enzyme first binds ATP and the specific amino acid and forms an intermediate aminoacyl-adenylate wherein a covalent linkage is formed between the 5′-phosphate group of ATP and the amino acid carboxyl group. The hydrolysis of ATP (releasing pyrophosphate) provides the energy to activate the amino acid and form the enzyme-bound (energy containing) **aminoacyl-AMP** as shown in the first step of Fig. 10.54. In the second step, shown in the figure, the amino acid is transferred to the specific tRNA and it becomes covalently bound to the 2′ or 3′-hydroxyl of the terminal adenosine of the tRNA. The charged tRNA moves through the three sites on the large ribosomal subunit and **transpeptidation** occurs to join the specific amino acid to its predecessor through the formation of a new peptide bond. The formation of the peptide bond with the previous amino acid residue of the growing peptide chain is catalyzed by ribosomal RNA. The protein components of the ribosome are remote from the site of transpeptidation, whereas they are close to the RNA. The catalytic function of RNA is called a **ribozyme** and, in addition to catalyzing the formation of a peptide bond (Fig. 10.58), ribozymes can cleave RNAs.

Presumably, the catalytic functions of RNA are remnants of beginning life in which nucleic acids, particularly RNAs, were present and functional before the advent of DNA.

## RIBOSOMAL RNA (rRNA)

Some of the noncoding RNAs are involved in translation on the **ribosome**. The ribosome (25−30 nm in diameter) is composed of two subunits. The subunits are comprised by rRNAs and proteins. The ribosome can be either free in the

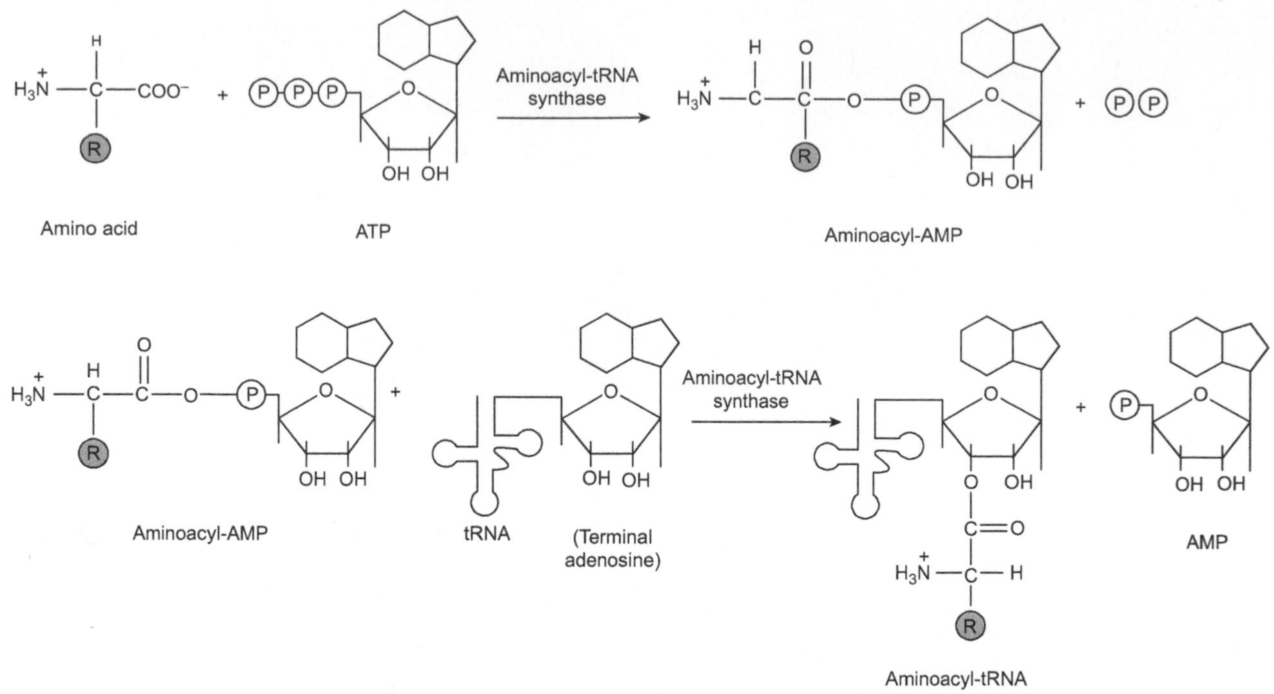

**FIGURE 10.57** Reaction mechanism of aminoacyl tRNA synthase. *From http://www.cs.stewards.edu/chem/Chemistry/CHEM43/CHEM43/tRNA/Function.htm.*

cytoplasm or attached to the ER (**rough ER**). Binding of a ribosome to a membrane is determined by the presence of a signaling sequence targeting the protein being synthesized to the ER. The proteins play a structural role in the ribosome together with rRNA but some rRNA can have enzymatic properties (ribozyme). In the human, the 40S small ribosomal subunit contains about 33 proteins and 18S rRNA (1900 nucleotides). The large 60S ribosomal subunit contains about 49 proteins and three rRNAs: 5S (120 nucleotides), 5.8S (160 nucleotides) and 28S (4700 nucleotides). Both subunits together constitute an **80S ribosome**. The genes for rRNAs occur in the **nucleolus** (**nucleolar organizer regions**). There are five groups of rRNA genes located on human chromosomes 13, 14, 15, 21, and 22. The number of repeated group sequences, encoding rRNA, can vary between individuals. Each person has about 10 groups or "clusters" of genes encoding rRNAs and the frequency of recombination (rearrangements) is such that each human bears a unique fingerprint. A large ribosomal RNA cluster can consist of 12,900 kilobases. **Mitochondrial ribosomes**, contain large and small subunits, as well, complexed into a 70S particle. However, mitochondrial ribosomes resemble those of bacteria, and contain two rRNAs, 12S and 16S that are somewhat smaller than bacterial rRNAs and have some structural differences from bacterial rRNAs.

mRNA binds to the small ribosomal subunit and tRNA, charged with its specific amino acid, binds to the large subunit. When translation is complete, the subunits dissociate. The formation of the peptide bond between the incoming amino acid on tRNA and the previously bonded amino acid of the growing polypeptide chain is catalyzed by an RNA **ribozyme** (**peptidyl transferase**) instead of a proteinaceous enzyme.

## SMALL NUCLEOLAR RNA (snoRNA)

These RNAs function in the formation of rRNA in the nucleolus. They guide the modification of pre-ribosomal RNA near target sites by base pairing, an activity that may have descended from ancient forms of life. Some other snoRNAs may act by cleaving **pre-rRNA** or may help to modify small nuclear RNAs or even mRNA. snoRNAs can regulate the **alternative splicing** of mRNA encoding specific proteins, e.g., the serotonin receptor. Some snoRNAs can be produced by salvage of the **introns** of pre-mRNAs. **Telomerase RNA** contains a snoRNA domain that is essential for its function.

The snoRNAs exist in two classes, the **C/D box snoRNAs** and the **H/ACA box snoRNAs**. Both groups are involved in the maturation of rRNA, the C/D box snoRNAs associated with **methylation** and the H/ACA box snoRNAs associated with **pseudouridylation** (see Fig. 10.59).

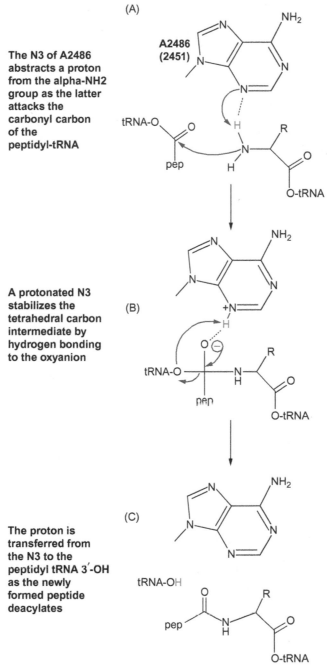

(A)

**A2486
(2451)**

The N3 of A2486
abstracts a proton
from the alpha-NH2
group as the latter
attacks the
carbonyl carbon
of the
peptidyl-tRNA

tRNA-O

pep

R

O-tRNA

(B)

A protonated N3
stabilizes the
tetrahedral carbon
intermediate by
hydrogen bonding
to the oxyanion

tRNA-O

pep

R

O-tRNA

(C)

The proton is
transferred from
the N3 to the
peptidyl tRNA 3′-OH
as the newly
formed peptide
deacylates

tRNA-OH

pep

R

O-tRNA

**FIGURE 10.58** Mechanism for peptide bond formation catalyzed by ribozyme during translation. *From http://bass.bio.uci.edu/~hudel/bs99a/ lecture22/lecture3_4b.html*. Pep, *growing peptide chain*.

Most of the actions of **C/D box snoRNA** lead to methylation of 2′O-ribose. Human rRNA contains about 115 methyl group modifications. The action of **H/ACA box snoRNA** results in the isomerization of uridine to **pseudouri-dine**. There are about 95 pseudouridine modifications in human rRNA. The structures of uridine and pseudouridine in the isomerization reaction are shown in Fig. 10.60.

Each snoRNA is a guide for a modification in the target RNA molecule. At least four proteins (including one or more enzymes) associate with the snoRNA (RNA/protein complex, snoRNP) to facilitate the reaction. Depending on its type, each snoRNA associates with different proteins. The target in the pre-RNA molecule is recognized by a 10−20-nucleotide sequence within the snoRNA that is complementary to the sequence surrounding the targeted base. When the snoRNP binds to the target site, the associated proteins catalyze the chemical modification.

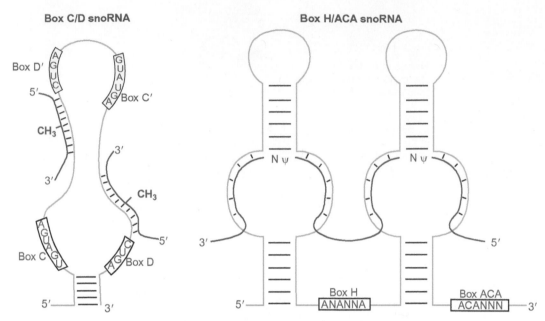

**FIGURE 10.59** Structures of the two classes of snoRNAs. On the *left* is a diagram of the structure of C/D box snoRNA with partial sequences indicated. On the *right* is a diagram of H/ACA snoRNA. $\psi$, pseudouridine; N, some base. *Reproduced from http://biochem.ncsu.edu/faculty/maxwell/ snoRNA.jpg.*

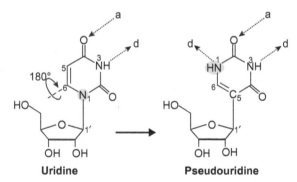

**FIGURE 10.60** Isomerization of uridine to pseudouridine. The glycosidic bond at N1C1' between the base and the sugar ring is first broken. The base is then rotated 180° (*arrow*) along the C6N3 axis leading to the formation of a new bond (C5C1') that links the base to the sugar. *a*, hydrogen bond acceptor; *d*, hydrogen bond donor. *Reproduced from http://wires.wiley.com/WileyCDA/WiresArticle/wisid-WRNA77.html.*

## MicroRNA (miRNA)

As the name implies, miRNAs are small in the range of about 22 nucleotides long. Mature miRNAs, located in the cytoplasm, are posttranscriptional regulators of gene expression through base pairing with target mRNAs. miRNAs are derived from much longer molecules of RNA. In the nucleus, these are processed into 70−100 base hairpin molecules by a double-stranded RNA-specific ribonuclease called **Drosha**. The biogenesis of miRNA is shown in Fig. 10.61.

The mature ∼22mer miRNA is bound to a complex called the **RNA-induced silencing complex** (**RISC**) that is the interfering RNA (RNAi). The miRNAs bind to mRNA causing reduced gene expression of potentially hundreds of genes. miRNAs often act on a mRNA molecule in the 3′-untranscribed region. **RNaseP** is a noncoding RNA that associates with **RNA polymerase III** and is needed for the efficient transcription of various other noncoding RNAs (such as, 5S rRNA, tRNA, and U6 snoRNA genes) that are transcribed by RNA polymerase III. miRNAs bind to proteins, those of the **Argonaute family** and **Gemin3** and **Gemin4** proteins to form mi-ribonucleoprotein complexes. The Argonaute protein family is divided into two subfamilies, the **Ago subfamily** and the **Piwi subfamily**. The Gemin proteins may be involved in neuromuscular functions. miRNAs along with some other noncoding RNAs show abnormal patterns of expression in cancer tissues.

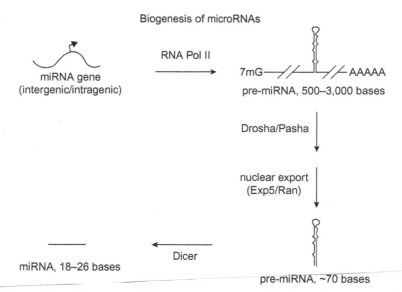

Biogenesis of microRNAs

RNA Pol II

miRNA gene
(intergenic/intragenic)

7mG—//—//—AAAAA
pre-miRNA, 500–3,000 bases

Drosha/Pasha

nuclear export
(Exp5/Ran)

Dicer

miRNA, 18–26 bases

pre-miRNA, ~70 bases

**FIGURE 10.61**   Biogenesis of microRNAs. The gene for miRNA is part of a messenger RNA gene (DNA) (upper left). RNA pol II transcribes the **pre-mRNA** (500–3,000 bases long). In the miRNA portion of the pre-mRNA, miRNA has secondary structures. The enzyme Drosha then cuts the pre-miRNA to the mature pre-miRNA that is exported to the cytoplasm where the enzyme **Dicer** (another double-stranded ribonuclease) matures the pre-miRNA to the 18–26 base miRNA. *Reproduced from http://3.bp.blogspot.com/_OrRF0Q09X-U/ReirA2z1wpl/AAAAAAAAADY/p5xmsoD-9p4/ s1600-h/miRNAbiogenesis.jpg.*

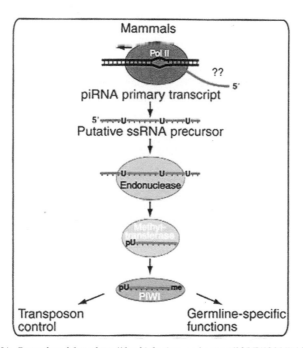

**FIGURE 10.62**   The biogenesis of piRNA. *Reproduced from http://dev.biologists.org/content/135/7/1201/F4.large.jpg.*

## Piwi-interacting RNA (piRNA)

**piRNA** associates with **Piwi proteins**. Piwi proteins are expressed primarily in the germ line and are required for **spermatogenesis**. Piwi proteins stabilize piRNA. piRNA is found only in the **testes**. There can be one million copies in **spermatocyte** or **spermatid** cells and they are located in the nucleus and the cytoplasm. The combination of piRNA and Piwi proteins produces the silencing of mobile genetic elements (transposons). **Transposon elements** are cleaved by Piwi proteins in association with piRNA. If Piwi proteins are reduced in concentration, the number of transposons increases. Fig. 10.62 shows the biogenesis of piRNA.

## Small Interfering RNA or Short Interfering RNA or Silencing RNA (siRNA)

These are double-stranded RNA molecules of 20 to 25 nucleotides in length. siRNA interferes with the expression of a specific gene. It is involved in affecting the structure of chromatin among other effects. siRNAs differ in their effectiveness in different cell types. siRNA is localized to the nucleus or the cytoplasm depending on the location of the target RNA to be silenced. It, like miRNA, functions in the **RNA interference pathway**. In all probability there will be other RNAs discovered that function in the RNA interference pathway.

## RNA Components of the Spliceosome

The spliceosome is the ribonucleoprotein complex responsible for cutting out the introns from pre-mRNA (Fig. 10.49). The spliceosome consists of a major and a minor form. The major spliceosome contains four noncoding RNAs: U1, U2, U4, and U5. The minor form contains U11, U12, U5, U4atac, and U6atac.

## Self-Splicing RNAs

Some introns, separate from the spliceosome, can catalyze self-removal from pre-mRNA and other precursor forms of RNA. There are two groups of self-splicing RNAs called the **group I catalytic intron** and the **group II catalytic intron**. These RNAs can excise themselves from pre-mRNA, pre-tRNA, and pre-rRNA. The structures of these two groups of catalytic introns are complex (Fig. 10.63A and B).

## RNA Secondary Structures

RNAs are usually single stranded. They form double strands when the original strand folds back upon itself to generate regions of base-pairing (e.g., tRNA, Fig. 10.56A). Besides the formation of double strands, RNA can form other secondary structures, such as **hairpins**, **internal loops**, **bulges**, and **junctions** as well as others (Fig. 10.64).

A **hairpin loop** requires a minimum of four bases in length for each loop. A bulge (loop) is generated when one side of the double-stranded structure is unable to form base pairs. Internal loops are formed when bases on both sides of the double strand are unable to form base pairs. **Multiloops**, shown here as three- or four-stems, are formed when two or more regions converge to form a closed structure. Some of these structures are found in tRNA (Fig. 10.56A).

When RNA interacts with DNA, a **pseudoknot** can be formed in the RNA (containing a loop) when the secondary structure pairs with a complementary sequence outside of the loop (Fig. 10.63B, structure A on the *left*). A **pseudoknot** also forms when the loop at the top of the RNA crosses the major groove of DNA and the loop at the bottom of the RNA crosses the minor groove of DNA (*middle* structure B in Fig. 10.63B). Another possibility for pseudoknot formation occurs when one loop of RNA crosses the major groove of DNA and the other group bridges the whole DNA helix (Fig. 10.65, figure on *right*, C).

## GENOMICS

As the name implies, genomics is the study of all the genes in a cell, in fact, the genome of the organism as the genome of the entire organism is repeated in each of its cells. **Functional genomics** stretches the definition to the functions of all the genes ($\sim$25,000 to 30,000 human genes) in the organism. The latest approach to functional genomics has been the **knockout** of each **gene** (or its mRNA) and the effects on the cell or organism so treated. A landmark has been the sequencing of the entire human genome in the **Human Genome Project**. Subsequent work has involved the identification of genes involved in specific human diseases. Another approach has been the mapping of all the human genes to their specific locations on chromosomes. This sequencing of the genome to learn the **mutations** for an individual will allow the prediction of possible disease conditions and the prescription of drugs that would be most effective for an individual's specific disease. In the case of cancer, there may be different classes of the cancer cell and characterization of the specific class based on genetic composition of the tumor cell will allow for the most effective treatment to eradicate it.

One of the most useful tools to understand the effect of some agent upon the genetic expression of a cell is the DNA **microarray**. Changes in genetic expression due to the action of some agent is observed by the changes in the transcribed products of the DNA, the population and amounts of **mRNA**s. A given cell (in tissue culture, for example) can be treated with an agent (e.g., a hormone or carcinogen, etc.). A sample of the cell's total mRNA can be taken at time zero and at subsequent times in order to observe early and late changes in gene expression caused by the agent.

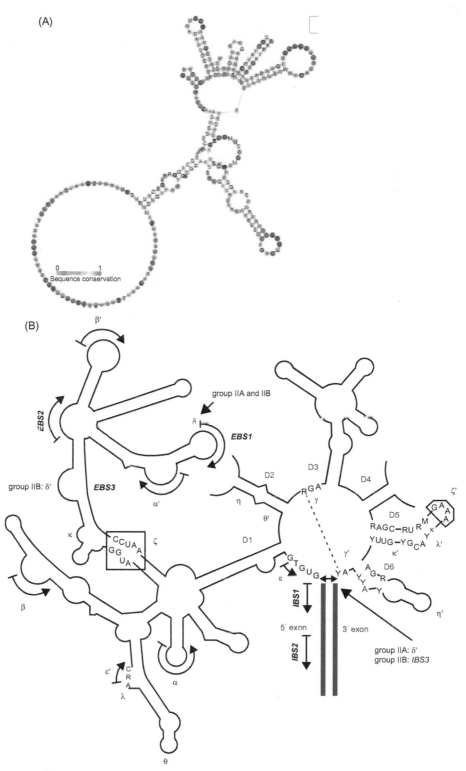

**FIGURE 10.63** (A) Diagram of a structure of a group I catalytic intron. (B) Diagram of a group II catalytic intron. *(A) Reproduced from http://en. wikipedia.org/wiki/File:RF00028.jpg. (B) Reproduced from http://en.wikipedia.org/wiki/File:IntronGroupII.jpg.*

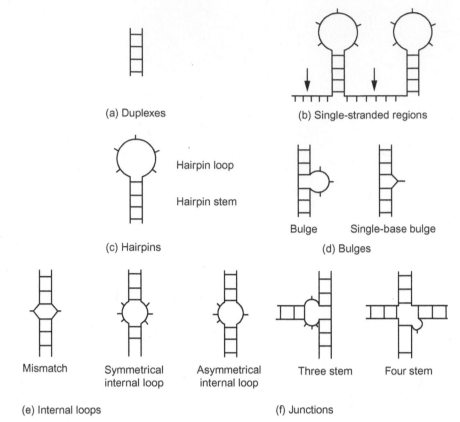

FIGURE 10.64 Secondary structures found in RNAs.

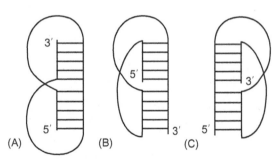

FIGURE 10.65 Formation of three types of pseudoknots in RNA when it interacts with DNA.

Also, the populations of total mRNA can be taken from two different cells (e.g., a normal cell and a cancer cell derived from that cell) and these can be compared to learn how the gene expression is altered in the tumor cell compared to its parent. In the latter case, mRNA is prepared from both cells. Single-stranded cDNAs are made from the total mRNAs. Then, double-stranded cDNAs are formed with DNA ligase, DNA pol I, and RNase H. From these, cRNAs are made. The RNA extracted from one cell is labeled with green fluorescent dye and the RNA extracted from the second cell is labeled with red fluorescent dye. The dye labeled populations are mixed and hybridized to the **microarray**. The microarray has been spotted with cDNA from thousands of genes (nowadays, from genes representing the total genome). Each spot on the microarray represents one gene. The RNA from each sample hybridizes to each spot in proportion to the expression level of each gene in the two samples. The red and green **fluorescence** from each spot is determined after hybridization. The ratio of red to green fluorescence reflects the expression of each gene in the two samples. This can be illustrated by the example shown in Fig. 10.66.

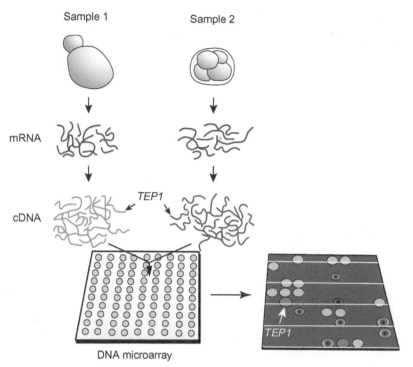

**FIGURE 10.66** Gene expression analysis in two tissue samples. TEP1 is a specific gene. *Reproduced from http://www.bmj.com/content/323/7313/611.full.*

In the case, where the samples are from the same cell (in culture), after treatment with some agent, and two samples are at time 0 and at time 30 minutes of treatment, the differences between the two indicate what genes are expressed or repressed in the 30-minute time period. At later time samples, it becomes possible to identify both early and late gene expression and to identify the genes (and their protein products) involved. As there is some "noise" in the system, it is best to first examine those genes that have expressed messages two- or threefold, or, conversely those that have been repressed two- or threefold. Even with lower levels of induction or repression, the level of expression of a specific gene can be verified using cDNA or Northern measurements in separate experiments.

A more recent development has been devised by **Affymetrix**. This company has developed a gene chip array in which each array has 6.5 million locations in which there are millions of DNA strands (25 base pairs each). This array is treated with a mixture of mRNA fragments containing biotin. The RNA fragments hybridize with appropriate DNA fragments on the gene chip array. Thus, the hybridized DNA–RNA glows (a fluorescent dye binds to the biotin attached to the mRNA fragments) when a laser light is shined on the array; the nonhybridized DNA does not glow. Based on the mating of the fluorescing RNAs with the DNA, the DNA can be analyzed.

## Next-Generation Sequencing

Next generation sequencing (NGS) is used to sequence genomes at high speed and at very low cost. It is used commercially for determining family tree DNA and mitochondrial DNA. This technique generates huge amounts of data by tracking the addition of labeled nucleotides as the DNA chain is copied in parallel fashion. Several types of instruments for NGS are available commercially. NGS involves the transcription of the DNA strand ($5''$ to $3''$) together with the RNA template ($3'$ to $5'$). After degradation of the products and using commercially available adapters to attach to the ends, the first strand of cDNA and mRNA are generated. These strands are then amplified to produce the first strand cDNA with sequencing primer and the template mRNA. The sequenced mRNA can be used to explore the structure of transcribed genes, splice junctions, small RNA and others including the identification of specific transcripts. Small noncoding RNAs are active in gene silencing and posttranscriptional regulation of gene expression. Small RNA sequencing can be applied to miRNAs (micro RNAs). Again, there are commercially available sequencers for this purpose.

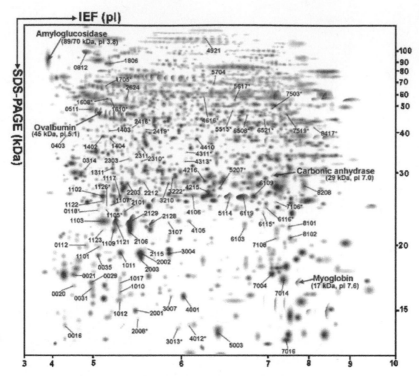

**FIGURE 10.67** **Two-dimensional gel electrophoresis** from a human control pituitary proteome labeled with the 93 differential spots and the 4 protein standard markers. **Isoelectric focusing (IEF)** was done with an 18-cm IPG strip nonlinear pH range 3–10. Vertical SDS–PAGE, that separates proteins based on molecular weight, was done with a 12% polyacrylamide gel. Molecular weight values are shown on the right ordinate. p$I$ values are shown on the abscissa. The four marker proteins (amyloglucosidase, ovalbumin, carbonic anhydrase, and myoglobin) are labeled. p$I$, isoelectric point; *PAGE*, **polyacrylamide gel electrophoresis**. *Reproduced from http://cancerres.aacrjournals.org/content/65/22/10214.full.*

## Proteomics

Proteomics is an extension of genomics, whereas genomics refers to all of the genes in an organism, proteomics refers to the entire protein complement of an organism, including the structures and functions of the proteins as well as their modifications. Posttranslational modifications of proteins include phosphorylation, methylation, acetylation, glycosylation, ubiquination, nitrosylation, and oxidation. Large **two-dimensional gel electrophoresis** of proteins in a cell can be separated and identified. Each gel can usually separate hundreds to thousands of proteins. The most highly expressed proteins in a cell will be prominent and will be most easily identified. The appearance of such a gel is shown in Fig. 10.67.

## Gene Therapy

The goal of gene therapy in humans is to replace defective genes that produce a disease condition, with a normal gene to eradicate the disease. Although there have been some successes in animal experimentation, in general, the results in humans have not been gratifying. Moreover, many inherited diseases are the result, not of a single gene mutation which would be the most desirable situation for gene therapy, but the involvement of several genes. A single gene is responsible for the production of a protein and when that gene bears a mutation, the protein may be poorly functional or not functional at all. The simplest situation involves an inherited single gene defect that could be replaced with a normal gene. Gene therapeutic approaches have mainly involved viruses that carry the replacement gene (Table 10.4).

As shown in the table, a nonviral form of delivery is the **liposome**, an artificial lipid sphere with an aqueous center. Such a construct can carry the repair gene inside a cell. Recently, a very small liposome of 25 nanometers in diameter was shown to carry therapeutic DNA through the nucleopore into the cell nucleus. Other liposomes have been coated with **polyethylene glycol (PEG)** and these have been shown experimentally to cross the **blood–brain barrier**, an achievement impossible with the larger viruses.

**TABLE 10.4** Some Viral and Nonviral Vectors Used or Tested in Gene Therapy Experiments

| Vectors | Advantages | Disadvantages | Notes |
|---|---|---|---|
| **Retroviruses** (e.g., MMLV) | 1. Long-lasting gene expression due to stable integration<br>2. Enters cells efficiently | 1. Only infects dividing cells<br>2. Potential insertional mutagenesis<br>3. Hard to produce (low yield) | Well suited for HSCGT<br>Note: 37% of all GT trials use retroviruses |
| **Lentiviruses** Retroviral subclass (i.e., HIV) | 1. Long-lasting gene expression due to stable integration<br>2. Will infect both dividing and nondividing cells | 1. Potential insertional mutagenesis | 10% of at GT trials |
| **Adenoviruses** (dsDNA virus) | 1. Enters cells efficiently<br>2. High rate of delivery and expression of theraputic gene<br>3. Does not integrate into chromosome<br>4. Infects nondividing cells | 1. Immunogenic—cleared rapidly from body<br>2. Can cause inflammation and tissue damage | Short expression time (~14 days) good for one-time or infrequent treatments 20% of all CT trials |
| **Adenoassociated viruses** (AAV) | 1. Much less immunogenic and toxic than adenovirus<br>2. Long-term expression possible<br>3. Wide host cell range | 1. Difficult to produce in high quantities | Stable in nucleus as a nonreplicating extrachromosomal form 1% of all GT trials |
| **Herpes simplex virus** | 1. Produced at high levels<br>2. Targets nondividing nerve cells<br>3. Can carry a great deal of DNA<br>4. Maintained as a nonreplicating extrachromosomal form in neuronal and bone marrow target tissue | 1. Immunogenic<br>2. Potentially toxic—can cause encephalitis | 6% of all GT trials |
| **Liposomes** | 1. Not immunogenic<br>2. Can deliver large quantities of DNA<br>3. High affinity for uptake into cytoplasm by endocytosis | 1. Low rate of delivery<br>2. Transient expression | Binds to and spontaneously condenses DNA |
| **Plasmid therapy** | 1. No viral component | 1. Transient gene expression<br>2. Difficult to target to specific tissues | 3% of all GT trials |

DNA, deoxyribonucleic acid; GT, gene therapy; HIV, human immunodeficiency virus; HSC, hematopoietic stem cell; MMLV, Moloney murine leukemia virus.
Source: Reproduced from http://www.biology.input.edu/biocourses/Bio154/15genetherapy2kl/html.

There are many genetic diseases that could be approached with this regime. Unfortunately, while some experiments have worked in experimental animals, there have been few successes in humans. Also the positive effects of successful gene therapy, so far, only last for a few to several months, necessitating continual repetitions of the gene therapy. So far, there are no FDA approved commercial gene therapy products. In the 20 years, or so, as the advent of this experimental approach, results in humans have not been hopeful. Although, there are a great many single gene mutations causing human disease, many genetic diseases are not the result of a single malfunctioning gene but usually involve a number of genes. Perhaps **stem cell** replacement of cells that contain mutated genes will be a more fruitful approach in humans.

## SUMMARY

A clinical example of a genetic disease is Huntington's disease. An extended glutamine repeat in the Huntingtin protein appears to be responsible for this disease. The abnormal protein impacts the ability of the inositol *tris*phosphate receptor to evoke the release of calcium from the neuronal calcium ion store. Deletion of the Huntingtin gene is embryonic lethal in mice.

The purines, adenine and guanine and the pyrimidines, cytosine and thymine are the bases in DNA. They are also present in RNA except that thymine is replaced by uracil. In nucleic acids, the bases occur as mononucleotides. Double-helical DNA is formed by base pairing, the bases being held together by hydrogen bonding. The base pairs are adenine-thymine (or uracil in place of thymine in RNA) and guanine—cytosine. Infrequent deamination of bases can occur. When this happens, cytosine forms uracil, deamination of guanine forms xanthine and deamination of adenine forms hypoxanthine.

Purines are derived from aspartate, $CO_2$, glycine, glutamine and $N^{10}$-formyltetrahydrofolate. The starting compound is ribose-5-phosphate that is converted to 5-phospho-D-ribosyl-1-pyrophosphate (PRPP). In 11 enzymatic steps, inosine-5-monophosphate (IMP) is formed. From this monophosphate IMP, AMP and GMP can be formed.

Nucleosides are pentose derivatives of the bases: ribose in RNA and deoxyribose in DNA. Nucleosides are phosphorylated on the sugar residue to form mono-, di-, or tri-nucleotides.

Uridine monophosphate (UMP) is formed from carbamoylphosphate, which, In turn, is formed from glutamine and bicarbonate. UMP can form UDP. UDP can be phosphorylated to form UTP that can be converted to CTP. Inosine monophosphate (IMP) is convertible to AMP or GMP. In the presence of ATP and the appropriate enzymes, the monophosphates can be converted to the trinucleotides.

Cyclized forms of adenosine and guanosine are cyclic AMP and cyclic GMP. They are formed from corresponding trinucleotides by specific cyclase enzymes.

Purines are catabolized to uric acid. Pyrimidines are catabolized to products that can enter the TCA cycle (malonyl-$S$-CoA or methylmalonyl-$S$-CoA). Gout is a disease caused by high circulating levels of uric acid. This disease is usually treated with allopurinol, a competitive inhibitor of xanthine oxidase. Uric acid can be metabolized to (S)-allantoin in the peroxisome by uric acid oxidase, an enzyme sometimes used to treat gout.

Salvage pathways exist that allow the reformation of useable bases.

Deoxyribose is formed, not directly from ribose primarily, but from the mononucleotide form by the enzyme ribonucleotide reductase.

The preponderant form of DNA in the cell is the right-handed helical B form. There are two other right-handed helical forms that occur infrequently. Z-DNA is a left-handed helical form occurring rarely in the cell.

In the human genome, there are 23 pairs of chromosomes (for a total of 46 chromosomes), 22 of which are autosomal. The remaining chromosomes are the XX sex chromosomes in the female and the XY sex chromosomes in the male. There are about 25,000—30,000 genes in the human genome representing about 2% of the total DNA.

The nuclear synthesis of double-stranded DNA is described. The complementary strand is synthesized by DNA polymerase and completed by ligase. Replication of DNA involves unwinding of double-stranded DNA by topoisomerase I. The enzymes topoisomerase II and helicase are also involved. DeoxyNTPs are substrates in the synthesis. Gyrase and DNA polymerase III affect the synthesis of a new strand of DNA. Other enzymes are involved in proofreading. A small RNA primer is added to single-stranded DNA so that DNA polymerase can act. Subsequently, the small RNA primer is removed by RNaseH. Still other proteins and enzymes are needed to complete the new strand of DNA. A repair mechanism fixes improper base pairs if they occur.

In the mitochondrion, there are 37 genes for RNAs and for proteins active in oxidative phosphorylation. Both strands of double-stranded DNA are continuously synthesized as leading strands in the 5′ to 3′-direction. A pause in the process generates the formation of D loops. Nuclear genes control mitochondrial replication. DNA replication begins when the mitochondria enlarge and start to divide, like bacteria.

Initiation of transcription requires: mitochondrial RNA polymerase plus 4 mitochondrial transcription factors assembled on mitochondrial promoters. The products are 2 circular double-stranded mitochondrial DNA molecules.

Hundreds of restriction enzymes cleave specific sites in DNA either internally (endonucleases) or at the ends of the molecule (exonucleases).

Gene cloning is the introduction of a foreign DNA into a plasmid. This allows for human proteins to be produced in relatively large quantities.

RNA interference is an external approach to knocking out a gene. By this method, one may learn about the function of the gene that is incapacitated by the loss.

Coding DNA provides the information in the genome. Noncoding DNA (about 98% of the total DNA) is being found to have many functions in addition to the information for transfer RNA and ribosomal RNA. Several noncoding RNAs are described.

## SUGGESTED READING

### Literature

Bogenhagen, D.F., Rousseau, D., Burke, S., 2008. The layered structure of human mitochondrial DNA nucleoids. J. Biol. Chem. 283, 3665–3675.

Byszewska, M., Smietanski, M., Purta, E., Bujnicki, J.M., 2014. RNA methyltransferases involved in 5′ cap biosynthesis. RNA Biol. 11, 1597–1607.

Chimploy, K., Song, S., Wheeler, L.J., Mathews, C.K., 2013. Ribonucleotide reductase association with mammalian liver mitochondria. J. Biol. Chem. 288, 13145–13155.

DePamphilis, M.L., 2000. Review: nuclear structure and DNA replication. J. Struct. Biol. 129, 186–197.

Dumitrescu, L., Popescu, B.O., 2015. MicroRNAs in CAG trinucleotide repeat expansion disorders: an integrated review of the literature. CNC Neurol. Disord. Drug Targets. 14, 176–193.

Dyawanapelly, S., Ghodke, S.B., Vishwanathan, R., Dandekar, P., Jain, R., 2014. RNA interference-based therapeutics: molecular platforms for infectious diseases. J. Biomed. Nanotechnol. 10, 1998–2037.

Fischer, S., Handrick, R., Aschrafi, A., Otte, K., 2015. Unveiling the principle of microRNA-mediated redundancy in cellular pathway regulation. RNA Biol. 12, 238–247.

Foulkes, W.D., Priest, J.R., Duchaine, T.F., 2014. Dicer 1: mutations, microRNAs and mechanisms. Nat. Rev. Cancer. 14, 662–672.

Gil-Mohapel, J., Simpson, J.M., Ghilan, M., Christie, B.R., 2011. Neurogenesis in Huntington's disease: can studying adult neurogenesis lead to the development of new therapeutic strategies? Brain Res. 1406, 84–105.

Hudson, W.H., Ortlund, E.A., 2014. The structure, function and evolution of proteins that bind DNA and RNA. Nat. Rev. Mol. Cell Biol. 15, 749–760.

Ipsaro, J.J., Joshua-Tor, L., 2015. From guide to target: molecular insights into eukaryotic RNA-interference machinery. Nat. Struct. Mol. Biol. 22, 20–28.

Kalovicova, J., Patel, A., Searle, M., Vorechovsky, J., 2015. The role of short RNA loops in recognition of a single-hairpin exon derived from a mammalian-wide interspersed repeat. RNA Biol. 12, 54–69.

Mattson, M.P., 2002. Huntington's disease: accomplices to neuronal death. Nature. 415, 377–379.

Nolte-Hoen, E.N., et al., 2015. The role of microRNA in nutritional control. J. Intern. Med. Available from: http://dx.doi.org/10.1111/joim.12372.

Rofougaran, R., Vodnala, M., Hofer, A., 2006. Enzymatically active mammalian ribonucleotide reductase exists primarily as an $\alpha_6\beta_2$ octomer. J. Biol. Chem. 281, 27705–27711.

Schraivogel, D., Meister, G., 2014. Import routes and nuclear functions of Argonaute and other small RNA-silencing proteins. Trends Biochem. Sci. 39, 420–431.

Shefer, K., Sperling, J., Sperling, R., 2014. The supraspliceosome—a multi-task machine for regulated pre-mRNA processing in the cell nucleus. Comput. Struct. Biotechnol. J. 11, 113–122.

Shortridge, M.D., Varani, G., 2015. Structure based approaches for targeting non-coding RNAs with small molecules. Curr. Opin. Struct. Biol. 30, 79–88.

Shoulson, I., Young, A.B., 2011. Milestones in Huntington disease. Mov. Disord. 26, 1127–1133.

Skilandat, M., Sigel, R.K.O., 2014. The role of Mg(II) in DNA cleavage site recognition in group II intron ribozymes solution structure and metal ion binding sites of the RNA–DNA complex. J. Biol. Chem. 289, 20650–20663.

Song, M., Kim, Y.H., Kim, J.S., Kim, H., 2014. Genome engineering in human cells. Methods Enzymol. 546, 93–118.

Swinehart, W.E., Jackman, J.E., 2015. Diversity in mechanism and function of tRNA methyltransferases. RNA Biol. 12, 398–411.

Vankatesh, S., Workman, J.L., 2015. Histone exchange, chromatin structure and the regulation of transcription. Nat. Rev. Mol. Cell Biol. 16, 178–189.

Zhu, H., Bilgin, M., Snyder, M., 2003. Proteomics. Annu. Rev. Biochem. 72, 783–812.

Zimmerman, S.B., 1982. The three-dimensional structure of DNA. Annu. Rev. Biochem. 51, 395–427.

### Books

Bloomfield, V.A., Crothers, D.A., Tinoco, I., 2000. Nucleic Acids: Structures, Properties, and Functions. University Science Books, Sausalito, CA.

Fitzgerald-Hayes, M., Reichsman, F., 2009. DNA and Biotechnology. Academic Press, New York.

Hampton, G., Sikora, K. (Eds.), 2006. Genomics in Cancer Drug Discovery. Academic Press, New York.

Mayer, G. (Ed.), 2010. The Chemical Biology of Nucleic Acids. John Wiley & Sons, Hoboken, NJ.

Soll, D., Nishimura, S., Moore, P., 2001. RNA. Pergamon Press, Oxford.

# Chapter 11

# Protein Biosynthesis

## DEFECTS IN MITOCHONDRIAL OXIDATIVE PHOSPHORYLATION AND DISEASE; DEFICIENCY IN MITOCHONDRIAL TRANSLATION

There are 13 proteins encoded by the mitochondrial genome (mitochondrial DNA) that comprise the respiratory chain (Fig. 14.19) located in the inner mitochondrial membrane. About 150 different proteins are required for the translation of these 13 respiratory chain proteins. Translation deficiencies caused by nuclear gene mutations generate mutations in transfer RNA (tRNA), rRNA, and the proteins involved in translation. Early on, some of these were identified as mutations in tRNA modifying enzymes, proteins of the ribosome, aminoacyl-tRNA synthases as well as elongation and termination factors, and these mutations lead to a variety of diseases. The nuclear-DNA-derived proteins are imported from the cytoplasm into the mitochondrion (see Fig. 11.19). The diseases caused by mutations in nuclear genes that are involved in translation or in the regulation of translation are shown in Table 11.1.

Nuclear genes encode most of the mitochondrial proteins that number as many as 1400 proteins. Of these, about 150 different nuclear-derived proteins take part in the translation of the 13 proteins encoded by the mitochondrial genome. These mitochondrial genes encode proteins that are involved in the assembly of the components of oxidative phosphorylation. The nuclear-derived proteins also are involved in many other mitochondrial functions including involvement in DNA replication, metabolism, maintenance, transcription, repair, and dNTP (deoxy trinucleotide) syntheses.

### Protein Synthesis in the Mitochondrion

There are two differences in the mitochondrial genetic code from the nuclear genetic code: in the nuclear genetic code, **UGA** is a **stop codon**, whereas in the mitochondrial genetic code UGA codes for **tryptophan**; AUA codes for **isoleucine** in the nuclear genetic code, whereas AUA codes for **methionine** in the mitochondrial genetic code. **UAG** codes for a stop codon in both systems and the other codons are similar (see **Appendix**). *Although nuclear DNA contains many tRNA genes, there are only 22 tRNA genes in the mitochondrion that are needed to read all the codons using a unique mitochondrial mechanism.* In addition, the mRNAs (messenger RNAs) contain few nucleotides in the 5'-end untranslated regions and cap structure.

Mitochondrial DNA encodes proteins (through mitochondrial mRNA) all of which are hydrophobic and located in the inner mitochondrial membrane. The translation of mitochondrial mRNA occurs in a complex bound to the inner membrane through electrostatic forces and protein interactions. One such protein is LETM1, an inner membrane protein, may serve as an anchor protein in the formation of a complex with the mitochondrial ribosome.

Mutations can occur in any component of the translation machinery to produce deficiencies in protein synthesis presenting in any mechanism of inheritance although *the mode of inheritance of mitochondrial genes is maternal.* These mutations generate a deficiency in oxidative phosphorylation affecting any of the complexes in the respiratory chain containing the proteins of mitochondrial derivation; apparently, only complex II is spared. Mitochondrial-derived diseases are now a rapidly growing area of genetic diseases in medicine.

### Mitochondrial Encephalomyopathy With Lactic Acidosis and Stroke-Like Episodes (MELAS)

Mitochondrial encephalomyopathy with lactic acid and stroke-like episodes (MELAS) is one of the prominent diseases in this category. It is caused by a mutation in any of 5 mitochondrial genes (MTND1, MTND5, MTTH, MTTV, and especially MTTL1). Mutation in the MTTL1 gene causes more than 80% of all the cases of MELAS. The MTTL1 mutation involves a tRNA, the mutation being A3243G (adenine 3243 to guanine) in the tRNA$^{leu\ (UUR)}$ gene. MELAS

Human Biochemistry. DOI: http://dx.doi.org/10.1016/B978-0-12-383864-3.00011-9

**TABLE 11.1** Heritable Defects Caused by Mutations in Nuclear Genes Involved in Translation or Its Regulation

| Gene | Protein | Disease or Phenotype |
|---|---|---|
| **Cytosolic Proteins** | | |
| EIF2B1—5 | eIF2B subunits α—ε | Vanishing white matter, childhood ataxia with central nervous system hypomyelination (chronic progressive, an episodic encephalopathy) |
| EIF2AK3 | eIF2α kinase PERK | Wolcott—Rallison syndrome (neonatal or early childhood diabetes mellitus, epiphyseal dysplasia, kidney and liver dysfunction, mental retardation, central hypothyroidism and dysfunction of the exocrine pancreas) |
| GARS | Glycyl-tRNA synthetase | Charcot—Marie—Tooth type 2D (slowly progressive axonal polyneuropathy) |
| YARS | Tyrosyl-tRNA synthetase | Dominant intermediate Charcot—Marie—Tooth type C (slowly progressive polyneuropathy with a mixed demyelinating-axonal phenotype) |
| RPS19, RPS24 | Ribosomal protein S19, Ribosomal protein S24 | Diamond—Blackfan anemia (abnormalities of the thumb, short stature, ventricular septal defects, kidney hypoplasia, and congenital glaucoma) |
| GSPT1 | Eukaryotic release factor 3 | Gastric cancer |
| PUS1 | Pseudouridine synthase 1 | Mitochondrial myopathy, sideroblastic anemia, mental retardation, microcephaly and dysmorphic features |
| DKCl | Dyskerin | X-linked dyskeratosis congenita (ectodermal abnormalities, bone marrow failure and increased susceptibility to cancer) |
| SBDS | Shwachman—Bodian—Diamond syndrome | Shwachman—Diamond syndrome (exocrine pancreatic insufficiency, bone marrow dysfunction, skeletal abnormalities, and short stature) |
| RMRP | RNA component of ribonuclease mitochondrial RNA processing (RNase MRP) | Cartilage-hair hypoplasia |
| **Mitochondrial Proteins** | | |
| MRPS19 | MRPS19 | Agenesis of corpus callosum and dysmorphism and fatal neonatal lactic acidosis |
| TSFM | Elongation factor Ts | Encephalomyopathy, hypertrophic cardiomyopathy |
| GFMl | Elongation factor G1 | Liver dysfunction, hypoplasia of corpus callosum and delayed growth |
| TUFM | Elongation factor Tu | Lactic acidosis, diffuse cystic leukoencephalopathy, polymicrogyria, liver involvement and early death |
| SPG7 | Paraplegin | Hereditary spastic paraplegia |
| DARS2 | Mitochondrial aspartyl-tRNA synthetase | Leukoencephalopathy with brain stem and spinal cord involvement and elevated lactate |
| LARS2 | Mitochondrial leucyl-tRNA synthetase | Susceptibility to diabetes mellitus |
| PUS1 | Pseudouridine synthase 1 | Mitochondrial myopathy, sideroblastic anemia, mental retardation, microcephaly and dysmorphic features |
| **Mouse Models** | | |
| $AARS^{stl/stl}$ | Alanyl-tRNA synthetase | "Sticky" phenotype (sticky appearance of fur), cerebellar Purkinje cell loss and ataxia |
| $EIF2AK3^{-/-}$ | eIF2α kinase PERK | Diabetes mellitus and exocrine pancreatic dysfunction |
| $GCN2^{-/-}$ | eIF2α kinase GCN2 | Liver steatosis |
| $HRI^{-/-}$ | Hem-regulated inhibitor | Iron-deficiency-induced anemia with erythroid hyperplasia |

*(Continued)*

**TABLE 11.1** (Continued)

| Gene | Protein | Disease or Phenotype |
|---|---|---|
| $4E\text{-}BP1^{-/-}$ | elf 4E-binding protein 1/2 | Sensitivity to diet-induced obesity |
| $4E\text{-}BP2^{-/-}$ | | |
| $EIF2^{+/Ser51Ala}$ | elF2$\alpha$ | Diet-induced obesity, type-2 diabetes mellitus |
| $EEF1A2^{-/-}$ | Eukaryotic elongation factor 1A2 | "Wasted" phenotype: mice are characterized by wasting and neurological and immunological abnormalities |
| $SBDS^{-/-}$ | Shwachman—Bodian—Diamond syndrome | Embryonic lethality |
| $RPS19^{-/-}$ | Ribosomal protein 19 | Lethal prior to implantation |
| $DKC1^{-/-}$ | Dyskerin | Embryonic lethality |

elF, eukaryotic initiation factor; MRP, mitochondrial ribosomal protein; PERK, pancreatic endoplasmic reticulum-resident kinase.
Reproduced from Table 11.1 of Scheper, G.C., van der Knapp, M.S., Proud, C.G., 2007. Translation matters: protein synthesis defects in inherited disease. Nat. Rev. Genet. 8, 711—723.

is relatively rare, occurring in the range of 1 person in 4000. In Fig. 11.1A, the mitochondrial genome emphasizing the gene mutations causing various mitochondrial genetic diseases (MELAS is located on the upper right where the mutation in leu tRNA gene is adenine to guanine at nucleotide 3243) is shown. Fig. 11.1B is a listing of aspects of the diagnosis of MELAS.

The structural changes in mitochondrial tRNAs that lead to disease are shown in two examples in Fig. 11.2.

In Table 11.2 are listed mitochondrial tRNA genes that are mutated in disease.

In early childhood (and also in older patients), MELAS patients have seizures, recurrent headaches, vomiting, anorexia (low body weight and fear of gaining weight), intolerance of exercise, and weakness in arms and legs. Stroke-like episodes are common and they progressively damage motor functions, vision (sometimes resulting in blindness), hearing, and mental ability. MELAS is caused primarily by defects in oxidative phosphorylation through damage to complex I and to complex IV. Eighty percent of MELAS patients harbor the A3243G mutation in the Mt-tRNA$^{Leu}$ gene wobble position (Fig. 11.2). This is an active subject of new research and vital to the diagnosis of new diseases spawned by mutations affecting translation in the mitochondria.

## PROTEIN SYNTHESIS DIRECTED BY THE NUCLEUS

The sequence of events leading to the formation of a protein takes place first in the nucleus and then in the cytoplasm. The major steps are: **transcription, translation**, and **posttranslational modifications**. The process of transcription of a specific gene takes place in the nucleus (see Chapter 12: Transcription). Genetic information is first encoded in a messenger RNA molecule that is exported to the cytoplasm at the ribosome where the new protein molecule is synthesized. mRNA contains three bases coding for a specific amino acid. The amino acid is ferried to the codon site when the anticodon contained in a **transfer RNA (tRNA)** interacts with the codon of the message. The first amino acid to attach is signaled by an **initiation codon**. Each amino acid is annealed to the previous one in a peptide bond. The process (**elongation**) continues until the protein has been completed (signaled by a **stop codon**). The fully formed polypeptide chain is released and is folded into the tertiary structure. Most of these events are summarized in simplified form in Fig. 11.3.

The manner in which the amino acids are recognized through the codons on the mRNA and the interaction with the anticodons on the tRNAs is shown in Fig. 11.4.

mRNA is transcribed from DNA where the base pairing partners are G-C and A-T(U). Guanine (G) pairs with cytosine (C) and adenine (A) pairs with thymine (T) in DNA or with uracil (U) in RNA. Consequently, during transcription, a G in DNA will transcribe for a C in mRNA, and a C in DNA will transcribe for a G in mRNA. An A in DNA will transcribe for a U in mRNA, and a T in DNA will transcribe for an A in mRNA. In Table 11.3, the triplet bases are listed that code for the amino acids.

*The code is degenerate in many cases such that four different triplets in mRNA can code for the same amino acid (e.g., valine). In many of these cases, the first two bases are identical but the third base varies.*

(A)

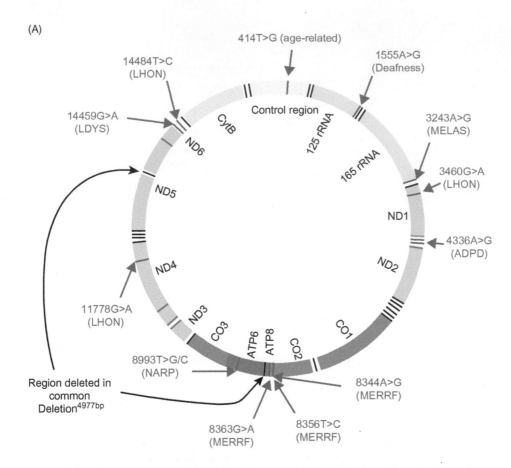

(B)

| MELAS: Diagnosis | |
|---|---|
| Imaging | ▪ Gray and white matter involvement<br>▪ Not conforming to vascular territories<br>▪ Parieto-occiptital predominance<br>▪ Calcification: especially in globus pallidus, but can occur in striaturn, thalamus, internal capsule |
| CSF lactate | ▪ Invariably elevated in encephalopathy |
| Muscle biopsy | ▪ Ragged red fibres<br>▪ SDH positive fibres<br>▪ COX fibres are usually strongly positive |
| Cerebral vessel histology show SDH reactivity indicating angiopathy as a significant component in the pathogenesis of stroke. | |
| Genetics | ▪ 80% have A3243G mutation in the tRNA for leucine.<br>▪ 4 different mutations in the tRNA leucine (UUR) gene |

**FIGURE 11.1** (A) Mitochondrial DNA showing the sites of mutations that lead to diseases in *red*. MELAS (on the *right*) is the mutation of adenine to guanine located at nucleotide 3243. (B) Diagnosis of MELAS. *SDH*, succinate dehydrogenase; *COX*, cyclooxygenase. *(A). Reproduced from Glowacki, S., Synowiac, E., Blasiak, J., 2013. The role of mitochondrial DNA damage and repair in the resistance of BCR/ABL-expressing cells to tyrosine kinase inhibitors. Int. J. Mol. Sci. 14, 16348–16364. (B) Reproduced from www.slideshare.net/drpramodkrishnan/genetics-of-mitochondrial-disorders.*

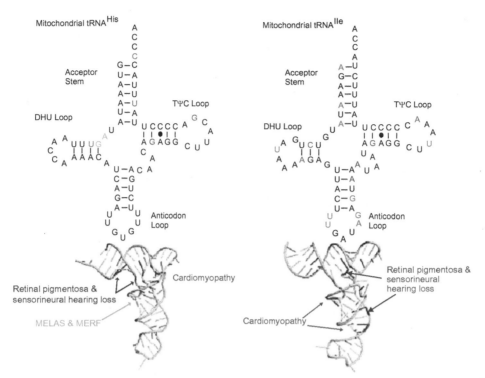

**FIGURE 11.2** Structural analysis of mitochondrial tRNA and disease causing mutations. Cloverleaf structure of Mt-tRNA$^{His}$ and Mt-tRNA$^{Ile}$. Mutations cause neurosensory (*blue*), cardiomyopathy (*red*), MELAS, and MERRF (*green*). *Reproduced from http://www.journal.frontiersin.org/article/10.3389/fgene.2014.00158/full.*

**TABLE 11.2 Mitochondrial tRNA Genes, Their Mutations, and the Diseases Generated**

| Mt-tRNA gene | Mutation | Disease | Structural Location | Aberrant tRNA Biology |
|---|---|---|---|---|
| *MTTL1* | A3243G | MELAS | Anticodon (wobble position WP) | Defect in taurine modification |
| *MTTK* | A8344G | MERRF | Anticodon (WP) | Defect in taurine modification |
| *MTTI* | C4277T | CMH1 | DUH stem | Reduced expression in cardiac tissue |
|  | A4300G |  | Anticodon stem |  |
|  | G4308A | CPEO | T Ψ C stem | Misfolding leads to improper 3′ end processing |
|  | A4302G |  | Variable loop | Disrupt conserved base pairing |
| *MTTH* | G12192 | CM | T Ψ C stem | Disrupt conserved base pairing |
|  | G12183A | NSHL RP | T Ψ C stem | Disrupt conserved base pairing |
|  | T12201C | NSHL | Acceptor stem | Reduced expression of functional tRNA |
|  | A12146G | MELAS | DHU stem | Misfolding |
|  | G12147A | MELAS MERRF | DHU stem | Misfolding and low abundance |
| *MTTE* | G14685A | C,SP,A | T Ψ C stem | Disrupt conserved base pairing |

A listing of the affected mitochondrial tRNA genes with the mutations, diseases names, structural locations, and defects.
Disease abbreviations: MELAS, mitochondrial encephalomyopathy, lactic acidosis, and stroke-like episodes; MERRF, myoclonic epilepsy with ragged red fibers; CMHI, cardiomyopathy familial hypertrophic; CPEO, Chrome progressive external ophthalmoplegia; CM, cardiomyopathy; NSHL, non-sensory hearing loss; RP, Retinitis pigmentosa: C,SP,A, cataracts, spastic paraparesis, and ataxia.
Partial data reproduced from http://www.frontiersin.org/files/articles/93032/fgene-05-00158-HTML/image_m/fgene-05-00158-t001.jpg.

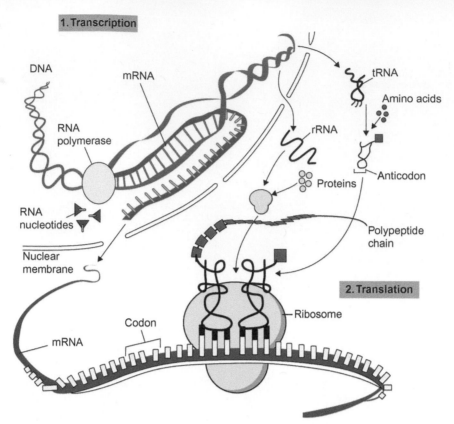

**FIGURE 11.3** A simplified view of protein synthesis. *Reproduced from http://www.rpdp.net/sciencetips_v3/images/L8A1_clip_image002.gif.*

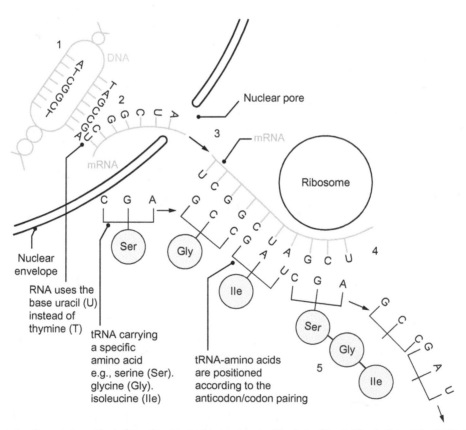

**FIGURE 11.4** Information for protein synthesis from the gene to the growing peptide chain. Step 1: Unwinding of the DNA double helix to expose the transcribable sequence (this may occur after stimulation by a **transcriptional activator** and the assembly of a number of proteins in the **pretranscriptional complex** (Chapter 12: Transcription)). Step 2: mRNA is synthesized from the sense strand of DNA. The completed mRNA is transported through the **nucleopore** into the cytoplasm to the rough endoplasmic reticulum (RER) where it attaches to the **ribosome** (Step 3). Step 4: **Transfer RNAs**, each attaching a specific amino acid interact with the mRNA through the tRNA's anticodon. Step 5: translation takes place and the peptide chain grows (elongation).

**TABLE 11.3** To the Left of Each Amino Acid Are the Sequences That Code for That Amino Acid

**Second Letter**

| First Letter | | U | C | A | G | Third Letter |
|---|---|---|---|---|---|---|
| U | | UUU UUC } Phe<br>UUA UUG } Leu | UCU UCC UCA UCG } Ser | UAU UAC } Tyr<br>UAA Stop<br>UAG Stop | UGU UGC } Cys<br>UGA Stop<br>UGG Trp | U<br>C<br>A<br>G |
| C | | CUU CUC CUA CUG } Leu | CCU CCC CCA CCG } Pro | CAU CAC } His<br>CAA CAG } Gln | CGU CGC CGA CGG } Arg | U<br>C<br>A<br>G |
| A | | AUU AUC AUA } Ile<br>AUG } Met | ACU ACC ACA ACG } Thr | AAU AAC } Asn<br>AAA AAG } Lys | AGU AGC } Ser<br>AGA AGG } Arg | U<br>C<br>A<br>G |
| G | | GUU GUC GUA GUG } Val | GCU GCC GCA GCG } Ala | GAU GAC } Asp<br>GAA GAG } Glu | GGU GGC GGA GGG } Gly | U<br>C<br>A<br>G |

Thus, UUA, UUG, CUU, CUC, CUA, and CUG all code for leucine. Starting with the name for the amino acid, the code can be found from this table with the first letter on the *left*, the second letter on the *top*, and the third letter on the *right*. An anticodon in tRNA to a specific amino acid codon in mRNA will have the three complementary bases. In the case of the mRNA, a codon for leucine might be CUA in which case the anticodon in the leucine tRNA would be GAU. By definition, these opposing bases would hydrogen bond together (Fig. 11.4).

## THE RIBOSOME

The mammalian ribosome consists of two subunits. The large subunit has a sedimentation value of 60S (determined in an analytical ultracentrifuge or by sucrose density gradient centrifugation) equivalent to about 2.8 million Daltons and contains as many as 49 proteins and 4880 nucleotides in the form of 28S, 5.8S, and 5S **ribosomal RNAs** (rRNAs). The small subunit is 40S, weighs 1.4 million Daltons, and contains 33 proteins and 1874 nucleotides in the form of **18S rRNA**. The RNAs can account for half the weight of the ribosome. The complex of both subunits has a sedimentation constant of 80S and weighs 14.22 million Daltons. The **nucleolus** is the site of rRNA production (Fig. 11.5). The **ribosomal proteins**, synthesized in the cytoplasm, are transported into the nucleolus and the nucleoplasm. Four of the rRNAs are synthesized in the nucleolus (Fig. 11.5).

The ribosomal subunits are assembled in the nucleoplasm and transported individually out of the nucleus (through the nucleopore) to the cytoplasm where the ribosomal dimer is found either freely in the cytoplasm or on the rough endoplasmic reticulum (RER). Free ribosomes may be in the form of **polyribosomes**. The 80S ribosome is pictured in Fig. 11.6.

The ribosomal subunits combine in the cytoplasm. **Free ribosomes** in the cytoplasm make proteins that remain in the cytoplasm. The ribosomes that are located within the mitochondria resemble bacterial (prokaryotic, without a defined nucleus) ribosomes at 70S that are smaller than eukaryotic (with a defined nucleus) ribosomes (80S). A human cell may contain a few million ribosomes and cells that are active in protein synthesis would be expected to have more ribosomes than those less active in protein synthesis.

Ribosomes read mRNA from the 5′ end to the 3′ end. Translation can occur either on free polysomes or polysomes associated with a number of cytoplasmic structures, including the endoplasmic reticulum (ER) where proteins destined for export of posttranslationsl modification are synthesized. The polypeptide elongates, forms its tertiary structure, and is released into the endoplasmic reticulum and then to the **Golgi apparatus** where it can be designed for destination to a **secretory granule** (to be transported to the extracellular space), to the **lysosome**, where it can be broken down or to the plasma membrane. The anatomy of a functional ribosome is shown in Fig. 11.7. There are three important sites on the ribosome, the A site, the P site, and the E site. The A site is the aminoacyl acceptor site, the P site is the peptidyl site, and the E site is the location of the exit. The reading of mRNA begins with a **start codon**, **AUG**, and terminates with a **stop codon, UAG** (Table 11.3). The first amino acid added is methionine for the AUG codon. It is transferred to the P site on the large subunit being brought to that site by the **methionine transfer RNA** which has the anticodon, UAC, that forms a complex with mRNA at the AUG site (Fig. 11.8). This is the start of the growing peptide.

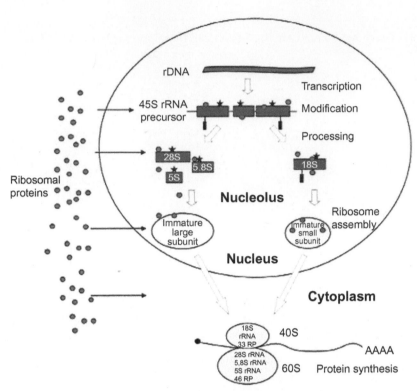

**FIGURE 11.5** This diagram shows the formation of the 80S mammalian ribosome. Ribosomal RNAs are synthesized from genes in the nucleolus and together with numerous proteins from the cytoplasm the preribosomal subunits are formed. These are transported out into the nucleoplasm where the subunits (60S and 40S) are formed and then exported to the RER of the cytoplasm into the functional ribosome. The strand with polyA at the end bound to the ribosome represents mRNA. *Reproduced from http://www.medri.hr/~vsinisa/slike/rib.jpg.*

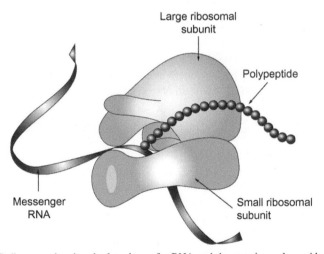

**FIGURE 11.6** A drawing of the 80S ribosome showing the locations of mRNA and the growing polypeptide chain. *Reproduced from Litwack, G., 2008. Human Biochemistry & Disease. Academic Press/Elsevier, Amsterdam/Boston, MA, p. 65.*

## Structure of Transfer RNA (tRNA)

There are at least 20 tRNAs. Each one is specific for one of the amino acids. A tRNA functions to bring a specific amino acid to the ribosome. It attaches to the P site and to the codon specified on mRNA via the tRNA's anticodon and start or add an amino acid, to the growing polypeptide chain. The tRNA molecule consists of about 74 to 95 nucleotides. The amino acid attaches to a $3'$ terminus, catalyzed by an aminoacyl tRNA synthase whose reaction will be shown below. At one end of the molecule resides an attachment site for the amino acid and at the other end is the anticodon that binds to the codon on mRNA. Each tRNA is specific for one amino acid (Table 11.3). The tRNAs have a specific structure (Fig. 11.9) in the form of a clover leaf (when flattened out) that is L-shaped in the three-dimensional structure. This structure permits binding to the ribosome at the A and P sites (Figs. 11.7 and 118).

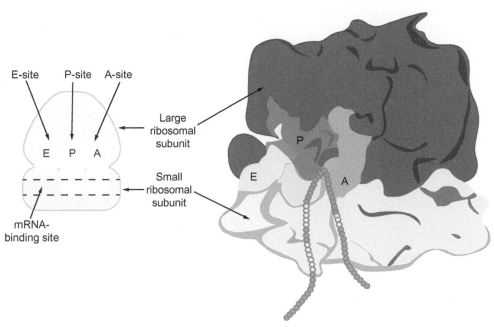

**FIGURE 11.7**   Anatomy of the 80S ribosome. The specific sites used in the process of translation are shown. The **A site** is the **aminoacyl acceptor site**, the **P site** is the **peptidyl site**, and the **E site** is the exit location.

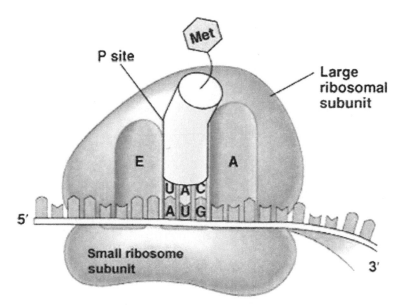

**FIGURE 11.8**   A ribosome in the initial act of starting polypeptide synthesis where methionine tRNA bearing the **anticodon**, UAC, is forming a complex with the mRNA's **start codon**, AUG, at the P site. The mRNA strand is shown at the intersection of the large and small subunit (on the mRNA binding domain) and is read from the 5′ end to the 3′ end. *Reproduced from http://library.thinkquest.org/04apr/00217/images/content/ribosome.jpg.*

More details of the tRNA structure have appeared in Chapter 10, Nucleic Acids and Molecular Genetics on nucleic acids.

### Initiation: Amino Acid tRNA Synthase

The reaction catalyzed by the **aminoacyl tRNA synthase** results in the binding of the cognate amino acid to its specific tRNA (Fig. 11.10). In this sequence, the amino acid first reacts with ATP to form aminoacyl-AMP and pyrophosphate. The aminoacyl-AMP binds to tRNA to produce the charged tRNA plus AMP. The summation of these steps is: amino acid + tRNA + ATP $\rightarrow$ aminoacyl-tRNA + AMP + PP$_i$.

Although there are two classes of this enzyme, differing in the number of subunits and initial point of attachment on AMP, the product ends up to be the same (substitution in the sugar 3′ position of AMP) in either case (Fig. 11.10). Fig. 11.11 shows the charged tRNA.

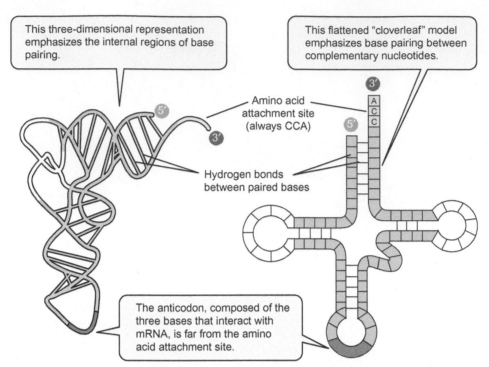

**FIGURE 11.9**    Structure of a typical tRNA molecule in the three-dimensional form (*left*) and in the flattened cloverleaf form (*right*). At the 3′ end is the attachment site for the amino acid (in the "acceptor stem"), is always CCA (specifying proline; Table 11.3). At the bottom is the trinucleotide anticodon that forms a complex with the trinucleotide codon of mRNA (Fig. 11.8). To specify proline, the mRNA would have a codon for this amino acid that could be CCA (acceptable codons would be CCU, CCC, CCA, or CCG), in which case the anticodon in tRNA would be GGU (remember that C pairs with G and A pairs with U in RNA). Note that the tRNA structure is stabilized by hydrogen bonds between paired bases, indicated by lines in the right hand figure. *Reproduced from http://universe-review.ca/l11-21-tRNA2.jpg.*

There are a number of **eukaryotic initiation factors** that have various specific activities in the initiation of the translation process. They are as follows: eIF4 or eIF4F, and this complex consists of eIF4A, eIF4B, eIF4E, and eIF4G; eIF1 and eIF3; eIF2; eIF5, and eIF5B. They are responsible for forming a complex between Met-tRNA and the 40S ribosomal subunit, known as the **43S preinitiation complex**. Other activities of the initiation factors include: recognition of the 5′ cap structure of mRNA; movement of the 43S preinitiation complex to mRNA; recognition of the **AUG initiation codon,** and formation of a complex with the 60S ribosomal subunit to form the 80S ribosome.

### Elongation and Peptidyltransferase Ribozyme

Elongation is the growth of the polypeptide chain as the mRNA progresses through the ribosome from the 5′ to the 3′ ends. Peptide bonds are formed from the preceding amino acid (to be added to the chain) to the incoming amino to be added to the chain. The **peptidyltransferase** is an activity of the 28S RNA component of the 60S ribosome and is called the **ribozyme**. The structure of ribozyme is shown in Fig. 11.12. The reaction catalyzed by the **hammerhead ribozyme** is shown in Fig. 11.13.

So far, it is the only enzyme that is not a protein, although other ribozymes have other enzymatic activities. Because it is believed that the earliest genetic information was encoded in RNA, the enzymatic activity of ribozyme must be an ancient function related to early life forms. A highly conserved ribozyme in the human and in other species is the **hammerhead ribozyme.** Ribozymes are active in the cleavage or in ligation of RNA and DNA as well as participating in various RNA-processing reactions, such as RNA splicing, the biosynthesis of tRNA and in viral replication. In addition to the hammerhead ribozyme, there is a *neurospora* **VS** (Varkud satellite) **ribozyme** and a **hairpin ribozyme** (in plant viruses). The VS ribozyme is the largest ribozyme that cleaves RNA. Because it can cleave an RNA containing a GC-rich stem loop, it performs a base-flipping function previously observed only for DNA-modifying protein enzymes. An **RNA polymerase ribozyme** is known that can catalyze ribozyme (its own) synthesis. Among other known ribozymes are **Leadzyme**, **twister ribozyme,** and many others occurring in nature.

The addition of two amino acids to a growing polypeptide chain is shown in Fig. 11.14.

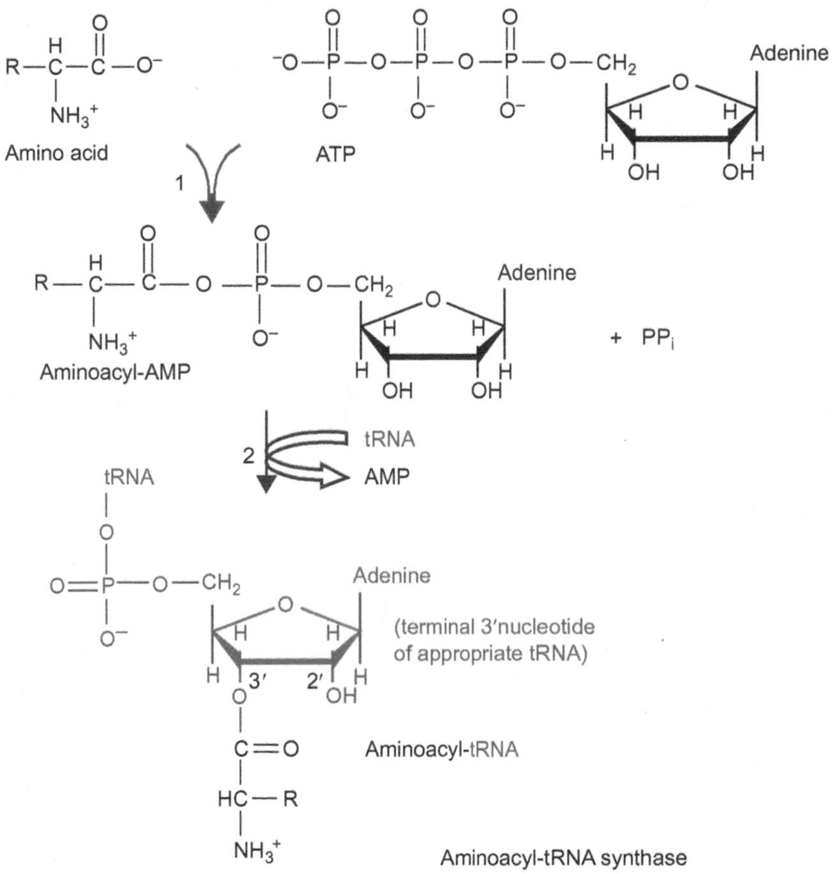

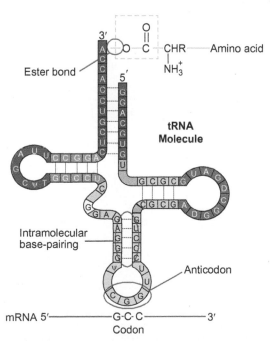

**FIGURE 11.10** Steps in the binding of an amino acid to its specific tRNA. *Reproduced from http://www.rpi.edu/dept/bcbp/molbiochem/MBWeb/mb2/part1/images/aatrna.gif.*

**FIGURE 11.11** Drawing of alanine tRNA (codon GCC) showing the bonding of the amino acid. *Reproduced from http://www.wiley.com/legacy/college/boyer/0470003790/structure/tRNA/trna_diagram.gif.*

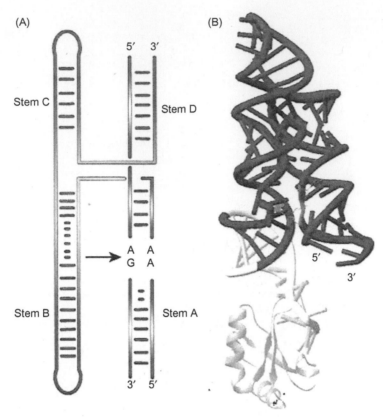

(A) Stem C  5′ 3′  Stem D

A A
G A
A

Stem B  Stem A

3′ 5′

(B) 5′ 3′

**FIGURE 11.12** The three-dimensional structure of a ribozyme, the enzyme catalyzing the formation of the peptide bond, in the process of translation. On the *left* (A) is the secondary structure of the **hairpin ribozyme** showing functionally important nucleotides. Dots indicate noncanonical base pairings. On the *right* (B) is the crystal structure of a precursor form of the hairpin ribozyme. Nucleotides flanking the scissile (breakable) bond are shown in *gold*, whereas the *gray* structure is an RNA-binding domain protein and its cognate (associated, so that the site is specific for the RNA) RNA-binding site, engineered into the construct to assist crystallization (a noncognate site would not recognize the specific ligand). *Reproduced from http://www.nature.com/nature/journal/v418/n6894/fig_tab/418222a_F3.html, originally from Doudna, J.A., Cech, T.R., 2002. The chemical repertoire of natural ribozymes. Nature 418, 222—228.*

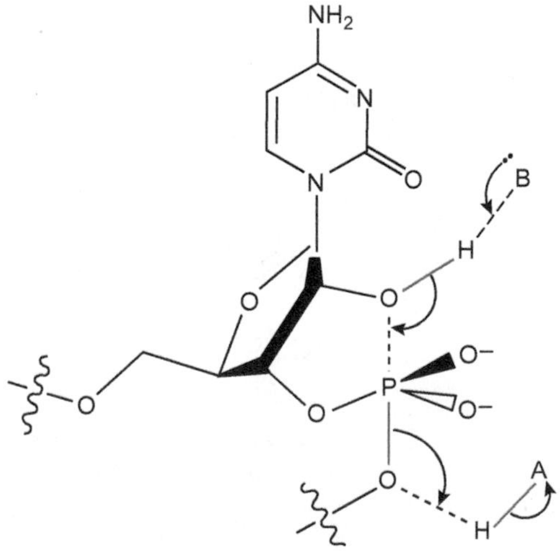

NH₂

B

H

O

O⁻
P
O⁻

O

A

H

**FIGURE 11.13** In-line transition state for the hammerhead ribozyme reaction. A general base (B) abstracts a proton from the 2′-O and a general acid (A), supplies a proton to the 5′-O leaving group as a negative charge accumulates. The reaction product is a 2′,3′-cyclic phosphate. The bonds breaking (shown in *red*) and forming (*dashed lines*) must be in the axial positions and reside approximately 180° apart, as shown. Electron movements are represented by *curly arrows. Reproduced from http://up;oad.wikimedia.org/wikipedia/commons/a/a1/Hammerhead_ribozyme_electron_transfer_ mechanism.svg.*

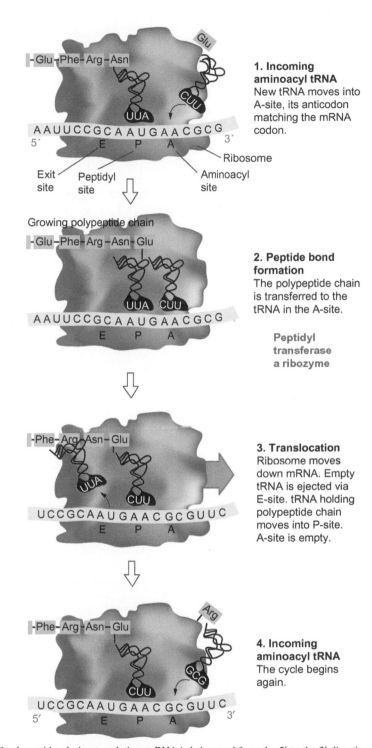

**1. Incoming aminoacyl tRNA**
New tRNA moves into A-site, its anticodon matching the mRNA codon.

Ribosome

Exit site

Peptidyl site

Aminoacyl site

Growing polypeptide chain

**2. Peptide bond formation**
The polypeptide chain is transferred to the tRNA in the A-site.

**Peptidyl transferase a ribozyme**

**3. Translocation**
Ribosome moves down mRNA. Empty tRNA is ejected via E-site. tRNA holding polypeptide chain moves into P-site. A-site is empty.

**4. Incoming aminoacyl tRNA**
The cycle begins again.

**FIGURE 11.14** Elongation of polypeptides during translation. mRNA is being read from the 5′ to the 3′ direction. The peptidyltransferase ribozyme center is in the 60S ribosomal subunit (the structure above the mRNA; the 30S subunit is the structure below the mRNA). *Reproduced from http://www.bio.miami.edu/~cmallery/150/gene/sf13x19.jpg.*

The process of elongation is facilitated by **eukaryotic elongation factors**, **EF-1** and **EF-2**. The locations of their actions in the cycle and the recycling of EF-1 are shown in Fig. 11.15.

The elongation factors monitor the accuracy of the elongation process. The rate of GTP hydrolysis of EF-1 is faster with cognate tRNAs than with noncognate tRNAs, and the tRNA can be rejected if the correct codon−anticodon complex is not formed. This monitoring appears to be detected through a conformational change in the three-dimensional structure of the complex that occurs with the correct interaction of codon and anticodon that is necessary for the GTPase activity of EF-1 to be induced.

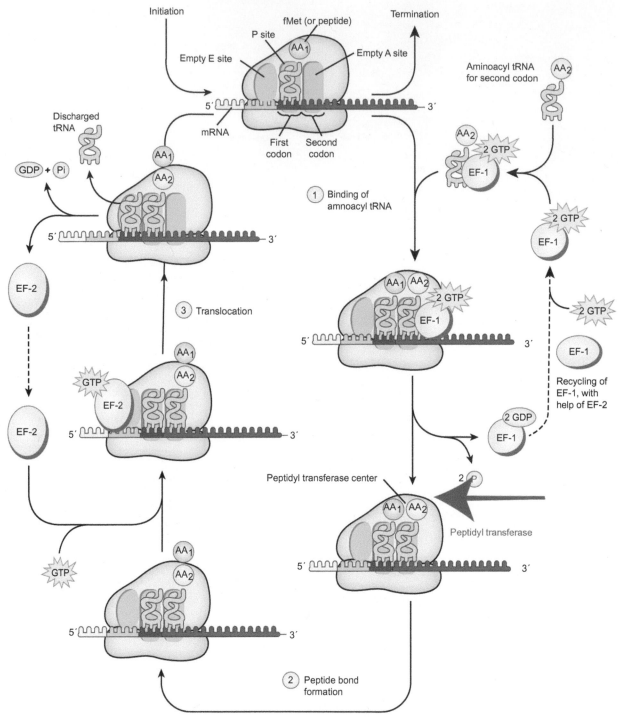

**FIGURE 11.15** Elongation reactions in translation showing the participation of eukaryotic elongation factors, EF-1 and EF-2. *Reproduced from http://course1.winona.edu/sberg/ILLUST/fig20-10B.gif.*

## Termination

The termination of translation is signaled by a **stop codon** (UGA UAA or UAG). When the **UGA** codon moves into the site where a specific tRNA would bind, instead, a specific **release factor** binds, leading to the completion and release of the fully synthesized polypeptide, mRNA, the release factor and the small ribosomal subunit from the large one as shown in Fig. 11.16.

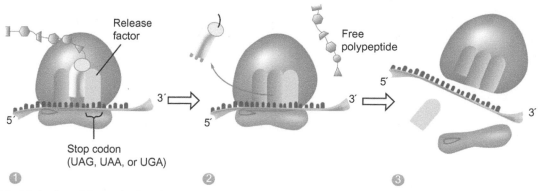

**FIGURE 11.16**   Mechanism of the termination of translation. *Reproduced from http://kvhs.nbed.nb.ca/gallant/biology/translation_termination.html.*

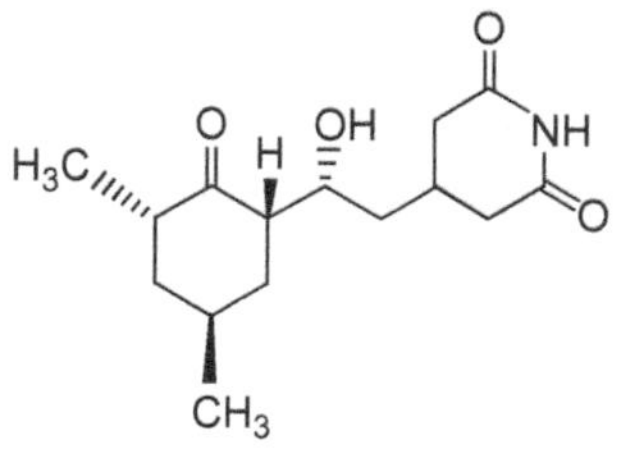

**FIGURE 11.17**   Structure of cycloheximide.

## Inhibitors of Protein Synthesis

There are a large number of compounds that inhibit prokaryotes without affecting protein synthesis in eukaryotes. These compounds are useful in treating bacterial infections in the human and mostly comprise commercial antibiotics (Streptomycin, Puromycin, Tetracycline, Chloramphenicol, and Erythromycin A, for example). There are some agents that inhibit eukaryotic protein synthesis which are especially useful in the laboratory but cannot be used *in vivo* (in the human). **Cycloheximide** (Fig. 11.17), for example, inhibits the eukaryotic **peptidyltransferase** but not the prokaryotic enzyme. It is synthesized by the bacterium *Streptomyces griseus*. Specifically, cycloheximide interferes with the movement of two tRNA molecules and mRNA in relation to the ribosome resulting in the blockade of the elongation process. Cycloheximide can be used to determine the half-life of a protein by measuring the amount of the protein in terms of its RNA (Northern blot) or the incorporation of radioactive amino acids into the specific protein (antibody precipitation) as a function of time with and without cycloheximide.

## Proteins Synthesized in the Cytoplasm but Destined for the Mitochondria

As indicated in Chapter 2 (The Cell), in Cellular Trafficking in Alzheimer's Disease (AD) with reference to the ER, newly formed proteins may be destined for the interior of the cell, including the mitochondria, the cell membrane or for secretion outside the cell. Destination signals are encoded in specific amino acid sequences (**motifs**) located specifically within the primary sequence and on the surface of the protein. For transport into the mitochondrion, the protein is in the form of a preprotein that may be disaggregated so that it can bind to a surface mitochondrial receptor and be transported through a (general) **import pore (GIP)** by a **translocase of the outer membrane (TOM)** using energy from the hydrolysis of ATP. It then crosses the intermembrane space and is translocated by the **translocase of the inner membrane (TIM)** into the mitochondrial matrix (Fig. 2.19). Then, with the assistance of **HSP70**, ATP and **MPP (mitochondrial processing peptidase)**, and **PEP** (mitochondrial **processing enhancing protein**), the proportion of the protein is clipped off, and the protein is folded properly with the assistance of **HSP60 (heat shock protein, 60,000 molecular weight)**. This sequence of events is capitulated in Fig. 11.18. In addition, it has been shown that cytosolic chaperones can protect growing protein chains. **HSP90** and **HSP70** can interact with the **mitochondrial import receptor (TOM70)** at the outer membrane.

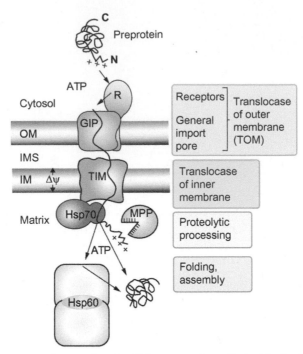

**FIGURE 11.18** Scheme showing the sequence of events through which a preprotein synthesized in the ER and released into the cytosol is transported into the mitochondrial matrix and folded into its native form. *Reproduced from http://www.biochemie.uni-freiburg.de/pfanner/tomtim2001.jpg.*

These HSPs (heat shock proteins) are required for the transport of precursor proteins into the mitochondrion as shown in Fig. 11.19.

## Proteins Destined for the Nucleus

Biosynthesized proteins either have a signaling sequence at the N-terminus or lack such a sequence. Proteins with a signaling sequence are translated in the rough endoplasmic reticulum (rER) containing bound ribosomes, whereas those proteins that have no N-terminal signaling sequence are translated on unattached ribosomes in the cytosol. Proteins with a signal peptide either remain in the endoplasmic reticulum or the Golgi vacuoles, or they are secreted to the plasma membrane or to the extracellular matrix. Cytoplasmic proteins, synthesized on unattached ribosomes that are targeted to the nucleus, contain a **nuclear localization signal (NLS)**. This signal is in the form of a 6-20 amino acid sequence that contains many basic amino acids (Arg, Lys). It can be located at any position in the protein sequence. The NLS is recognized by transporting proteins (**nucleoporins**) that ferry the newly synthesized protein to the **nucleopore** (Fig. 2.15). The newly synthesized protein is then transported into the nucleoplasm. Proteins that have been posttranslationally modified will be directed to elements within the nucleus. For example, certain proteins with basic amino acids in a binding site would be expected to enhance binding to DNA by attraction to its sugar-phosphate backbone. Specific amino acids at strategic locations in the protein-binding site will direct the protein to specific regions of the DNA.

## Proteins Destined for Other Sites Including the Plasma Membrane and Secretion from the Cell

Proteins with destinations to structures within the cell or to the plasma membrane for secretion are those that contain a **signal sequence** in the N-terminus. The signal peptide is made up of a stretch of 16—30 amino acids containing at least one positively charged amino acid (Arg or Lys), followed by a stretch of 6—12 hydrophobic amino acids (e.g., Leu, Ile, Val). The signal sequence is removed from the mature protein, in the ER lumen, and the mature protein undergoes posttranslational modifications, including N-glycosylation, disulfide bond formation between cysteine residues, hydroxylation, or oligomerization. Such proteins may contain targeting regions, such as spanning membrane domains, GPI

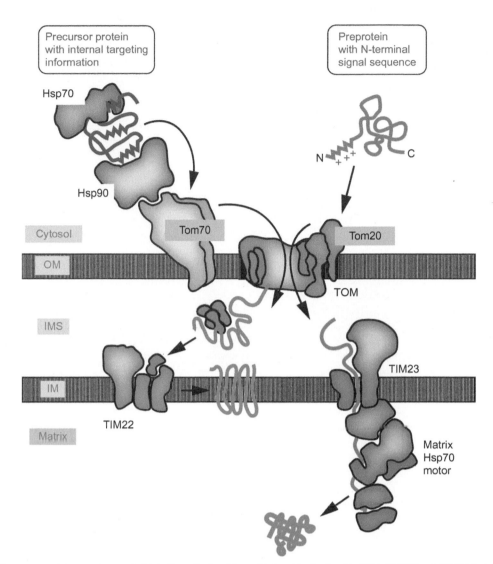

**FIGURE 11.19**  Precursor proteins are transported into the mitochondrion with the aid of cytosolic HSPs HSP90 and HSP70 to interact with TOM70 (*top left*) in a different initial reaction than a preprotein with an N-terminal signal sequence (*top right*). A protein in this pathway may be incorporated into the inner mitochondrial membrane (IM). *Figure reproduced from Voos, W., 2003. A new connection: chaperones meet a mitochondrial receptor. Mol. Cell 11, 1–3.*

(glycosylphosphatidylinositol) anchors, or other sequences allowing attachment to the plasma membrane or to the endoplasmic reticulum membrane. Proteins destined for secretion from the cell may have a secretory signal in the form of one or more hydrophobic sequences. Such proteins are directed to the ER, to the Golgi apparatus or to the plasma membrane where they are secreted. Some proteins destined for secretion will be housed in secretory vacuoles (secretory vesicles). Secretion often occurs following a specific signal such as one that may increase cytosolic concentrations of $Ca^{2+}$.

## SUMMARY

Serious diseases can result from genetic mutations affecting translation in the mitochondria. These aberrations result in deficiency of oxidative phosphorylation. Mutations can occur in nuclear genes that encode enzymes and other factors involved in mitochondrial translation. The nuclear encoded proteins are synthesized in the cytoplasm and transported into the mitochondrial matrix. There are 13 proteins encoded in mitochondrial DNA, and these proteins are components of the respiratory chain located in the inner mitochondrial membrane. The mitochondrial-derived proteins of the respiratory chain require some 150 other proteins for their translation.

The **biosynthesis of proteins** is described detailing events occurring on the **ribosome** including the mechanism of addition of specific amino acids, signaled by mRNA, to the growing peptide chain via each specific tRNA. **Elongation** of the polypeptide and **termination** and release from the ribosome of the completed polypeptide are described. The process of the formation of the **peptide bond** connecting adjacent amino acids as catalyzed by the **ribozyme** is explained. Certain proteins are synthesized in the cytoplasm but destined for the mitochondria and the mechanism of their entry into the mitochondrial matrix is presented.

Cycloheximide inhibits human protein synthesis. Other antibiotics inhibit prokaryotic protein synthesis and are useful in treating human infections.

## SUGGESTED READING

### Literature

Antonicka, H., et al., 2010. Mutations in C12orf65 in patients with encephalomyopathy and a mitochondrial translation defect. Am. J. Hum. Genet. 87, 115–122.

Antonicka, H., Sasaman, F., Kennaway, N.G., Shoubridge, E.A., 2006. The molecular basis for tissue specificity of the oxidative phosphorylation deficiencies in patients with mutations in the mitochondrial translation factor EGF1. Hum. Mol. Genet. 15, 1835–1846.

Borner, G.V., et al., 2000. Decreased aminoacylation of mutant tRNAs in MELAS but not in MERRF patients. Hum. Mol. Genet. 9, 467–475.

Calvo, S.F., Mootha, V.K., 2010. The mitochondrial proteome and human disease. Ann. Rev. Genomics Hum. Genet. 11, 25–44.

Coenen, M.J., et al., 2004. Mutant mitochondrial elongation factor G1 and combined oxidative phosphorylation deficiency. N. Engl. J. Med. 351, 2080–2086.

Costa, I.R., Thompson, J.D., Ortega, J.M., Prosdocimi, F., 2014. Metazoan remaining genes for essential amino acid biosynthesis: sequence conservation and evolutionary analyses. Nutrients. 7, 1–16.

Doudna, J.A., Cech, T.R., 2002. The chemical repertoire of natural ribozymes. Nature. 418, 222–228.

Glowacki, S., Snowiac, E., Blasiak, J., 2013. The role of mitochondrial DNA damage and repair in the resistance of BCR/ABL-expressing cells to tyrosine kinase inhibitors. Int. J. Mol. Sci. 14, 16348–16364.

Hatefi, Y., 1985. The mitochondrial electron transport and oxidative phosphorylation system. Ann. Rev. Biochem. 54, 1015–1069.

Lykke-Andersen, J., Bennett, E.J., 2014. Protecting the proteome: eukaryotic cotranslational quality control pathways. J. Cell Biol. 204, 467–476.

Merrick, W.C., 2010. Eukaryotic protein synthesis: still a mystery. J. Biol. Chem. 285, 21197–21201.

Ning, W., Pei, J., Gonzalez Jr., R.L., 2014. The ribosome uses cooperative conformational changes to maximize and regulate the efficiency of translation. Proc. Nat. Acad. Sci. 111, 12073–12078.

Nissen, P., Hansen, J., Ban, N., Moore, P.B., Steitz, T.A., 2000. The structural basis of ribosome activity in peptide bond synthesis. Science. 289, 920–930.

Petrov, A.S., et al., 2014. Evolution of the ribosome at atomic resolution. Proc. Nat. Acad. Sci. 111, 10251–10256.

Rotig, A., 2011. Human diseases with impaired mitochondrial protein sysnthesis. Biochim. Biophys. Acta—Bioenerg. 1807, 1198–1205.

Scheper, G.C., van der Knapp, M.S., Proud, C.G., 2007. Translation matters: protein synthesis defects in inherited disease. Nat. Rev. Genet. 8, 711–723.

Sebach, M., et al., 2008. Widespread changes in protein synthesis induced by microRNAs. Nature. 455, 58–63.

Voos, W., 2003. A new connection: chaperones meet a mitochondrial receptor. Mol. Cell. 11, 1–3.

### Books

Liljas, A., Ehrenberg, M., 2013. Structural Aspects of Protein Synthesis (Series in Structural Biology). second ed. World Scientific, Hackensack, NJ.

Nierhaus, K.H., Wilson, D.N. (Eds.), 2004. Protein Synthesis and Ribosome Structure. Wiley-VCH, Weinheim, Germany.

Chapter 12

# Transcription

## CONGENITAL HEART DISEASE; MUTATIONS OF TRANSCRIPTION FACTORS

### Normal Heart Development

The progenitor cells for mammalian heart development are located in the anterior lateral plate mesoderm. At about day 15 of human embryonic development, the progenitor cells condense into two lateral heart primordia. These include lineage precursors for the myocardial and endocardial lineages. At 3 weeks of human development, the cardiac precursors move toward the center, forming a primitive linear heart tube. This loops to the right when cells from the second heart field are added to the inflow and outflow locations. Next, within the outflow tract (OFT), endocardial cushions (subset of cells involved in septation) begin to be formed and they are also formed at the common atrioventricular canal (AVC) during the sixth and seventh week of development. These cushions aid in the separation of the heart into four chambers and divide the outflow tract into the aorta and pulmonary artery. At this time, the early conduction system begins developing and contributions from the neural crest and proepicardium (progenitor cells near the venous pole) occur. Later on, extensive remodeling of the heart occurs prior to assuming the mature four-chambered structure with divided inflow and outflow. The developing heart also forms valve loaflets and a functional conduction network. Development of the human heart is summarized in Fig. 12.1.

### Transcription Factors Control Heart Development

Normal development of the heart is dependent on the functions of several transcription factors. The central group of factors involves the GATA family of zinc finger proteins (GATA 4, 5, and 6). Others are the MADS box proteins (the term, MADS, is derived from the first letter of four original members of this group: MCM1, AG, DEFA, and SRF). These proteins recruit other transcription factors into regulatory complexes. Other core factors in heart development are T-box factors (Tbx1, Tbx2, Tbx3, Tbx5, Tbx18, and Tbx20). Isl1 [lim-homeodomain (HD) protein] also is essential. These factors interact with themselves and other transcription factors to regulate the maturation of the cardiac chambers. They are also critical for the development of the conduction system and for remodeling of the endocardial cushion. Of these factors, the most studied are Nkx2-5, GATA4, and Tbx5; these factors are critical for the development of the heart. Congenital heart disease is associated with mutations in the genes for these factors. Fig. 12.2A and B shows transcription factors and their interactions that are involved in myocardial development and heart morphogenesis.

### GATA4

GATA4 is a zinc finger DNA-binding protein of 442 amino acids and a molecular weight of 44,580 Da. Two types of GATA4 are differentiated based on the position of the zinc finger: either at position 217−241 or at 271−295. It is a transactivation factor that binds to the sequence 5′-AGATAG-3′. GATA4 plays a key role in the development of the heart. It is involved in the induction of cardiac specific gene expression mediated by bone morphogenetic protein (BMP) through its binding to the BMP response element DNA sequences within cardiac activating domains. In cooperation with another transcription factor, NKX2.5 (or Nkx2−5), it promotes cardiac myocyte enlargement. GATA4 has several biological functions but a major one is the morphogenesis of the atrial septum including the development of the arterioventricular canal and the associated valve formation.

Through gene mutation, there are nine positions in the GATA4 protein molecule where point mutations occur involving the change in a single amino acid. These cause ventricular septal defects resulting in abnormal communication between the lower two chambers of the heart. Ventricular septal defect may occur alone or in combination with

Human Biochemistry. DOI: http://dx.doi.org/10.1016/B978-0-12-383864-3.00012-0

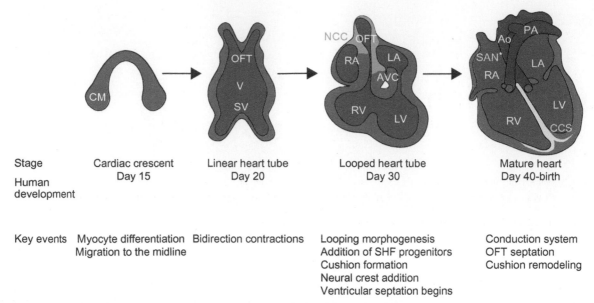

FIGURE 12.1    Schematic representation of several stages of human heart development. *Ao*, aorta; *AVC*, atrioventricular canal; *CCS*, cardiac conduction system; *CM*, cardiac mesoderm; *LA*, left atrium; *NCC*, neural crest cells; *OFT*, outflow tract; *PA*, pulmonary artery; *RV*, right ventricle; *RA*, right atrium; *SAN*, sinoatrial node; *SV*, sinus venosus; *V*, ventricle. *Reproduced from http://www.ncbi.nlm.nih.gov/pmc/articles/PMC3684448/figure/F1/.*

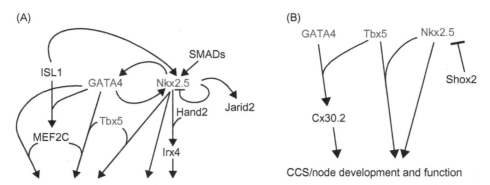

FIGURE 12.2    (A) Transcription factor pathways involved in myocardial development and heart morphogenesis. *Hand2*, member of basic helix–loop–helix family of transcription factors; *Irx4*, homeobox gene product; *ISL1*, required for motor neuron generation; *Jarid2*, interacts with AT-rich domains; *MEF2C*, myocyte specific enhancer factor 2C; *SMADs*, cell signaling proteins. Part (B) shows the transcription factor pathways involved in the development, maturation and function of the cardiac conduction system (CCS). *(A) Reproduced from http://www.ncbi.nlm.nih.gov/pmc/articles/PMC3684448.figure/F2/. (B) Reproduced from http://www.ncbi.nlm.nih.gov/pmc/articles/PMC3684448/figure/F3/.*

other cardiac malfunctions. If these defects go unrepaired, they can result in enlargement of the heart, congestive heart failure, pulmonary hypertension, arrhythmias, and possibly sudden cardiac death. A schematic of the GATA4 protein showing its functional domains and the locations of mutations and phenotypes of congenital heart defects is in Fig. 12.3.

GATA4 is also phosphorylated at position 105, a serine residue, in the sequence PPVSPRFSF by MAPK (also known as ERK). The phosphorylated form is apparently more active in binding to DNA.

## Nkx2.5

NKX2.5 is a homeobox protein of 324 amino acids and a molecular weight of 34,918 Da that is involved in the differentiation of the myocardial lineage. It is a transcriptional activator of the atrial natriuretic factor (ANF) in cooperation with GATA4. It is under transcriptional control by PBX1 (**P**re **B**-cell leukemia homeobo**x** 1). The domains of the NKX2.5 protein are shown in Fig. 12.4.

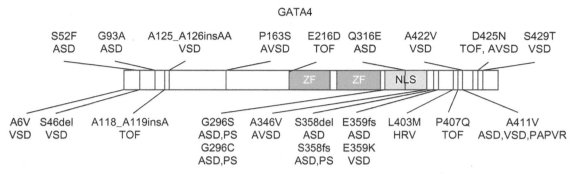

**FIGURE 12.3**  Schematic of GATA4 protein indicating the location of mutations and phenotypes of congenital heart defects. *ASD*, atrial septal defect; *AVSD*, atrioventricular septal defect; *HRV*, hypoplastic right ventricle; *NLS*, nuclear localization signal (containing basic amino acids); *PAPVR*, partial anomalous pulmonary venous return; *PDA*, patent *ductus arteriosus*; *PS*, pulmonary valve stenosis; *PTA*, persistent *truncus arteriosus*; *TOF*, tetralogy of Fallot; *VSD*, ventrivular septal defect; *ZF*, zinc finger domain. The specific point mutations are indicated (e.g., S52F is amino acid 52 serine mutated (in the gene) to form a phenylalanine). *Reproduced from http://www.nature.com/jhg/journal/v55/n10/fig_tab/jhg2010105f1.html.*

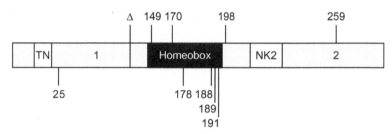

**FIGURE 12.4**  The figure shows mutations causing various cardiac anomalies. NKX2.5 contains two exons encoding a 324 amino acid protein including a tinman domain (TN) [the name derives from *Drosophila*], a homeodomain (*black*) and an NK2 domain. Truncation mutations are shown above, missense mutations below. *NK2* domain [amino acids 212–234]; *TN*, tinman domain [amino acids 10–21]; homeobox domain [amino acids 138–197]. Note the clustering of mutations within the homeobox itself. *Reproduced from http://atlasgeneticsoncology.org/Genes/Images/NKX2-5Fig2.jpg.*

A point mutation within the homeobox domain occurs at amino acid 145 where phenylalanine is converted to a serine (F145S). The homeobox is a DNA-binding domain (DBD) first discovered in *Drosophila*. It is a sequence of about 180 base pairs giving rise to about 60 amino acids. The homeobox domain occurs in proteins that are transcription factors which are involved in the patterns of anatomical development. The homeodomains (HDs) of transcription factors have a characteristic DNA-binding fold. The crystal structure of NKX2.5 is known and it binds to two DNA sequence motifs TGAAGTG/TCAAGAG, straddling them both at the same time (Fig. 12.5).

The NKX2.5 is apparently controlled by phosphorylation—dephosphorylation. The protein occurs both in the cytoplasm and in the nucleus and its subcellular location is probably driven by its phosphorylation status.

## Tbx5

Tbx5, (T-box transcription factor 5) is a protein of 518 amino acids and a molecular weight of 57,711 Da. It is found in the cell nucleus but there are also reports of its location in the cytosol, cytoskeleton and Golgi apparatus. It is involved in the transcriptional regulation of genes specifying mesoderm differentiation, especially in heart development and in the differentiation of cardiac progenitors. Mutations in the gene for this transcription factor are associated with the **Holt—Oram syndrome** which is a developmental disorder affecting the heart and upper limbs. Pictured in Fig. 12.6 is the gene for Tbx5 (A) and the mutations in the Tbx5 protein that cause the Holt—Oram syndrome (B).

Aside from the errors in the formation of the upper limbs, about 75% of persons with the Holt—Oram syndrome have potentially life-threatening cardiac problems. Usually, there is a defect (hole) in the septum separating the right and left sides of the heart. If the hole occurs in the septum between the upper chambers of the heart (atria) it is an atrial septal defect. If it occurs in the septum between the lower chambers of the heart (ventricles), it is a ventricular septal defect (VSD). In addition, some patients have conduction disease involving abnormalities in the heart electrical system that can lead to bradycardia (slow heart rate) or fibrillation (uncoordinated heart rate). The Holt—Oram syndrome is

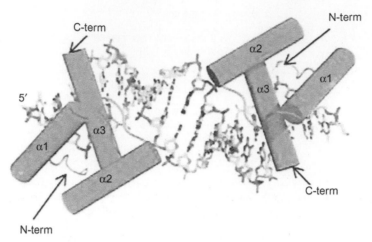

**FIGURE 12.5** NKX2.5 is a homeodomain containing transactivation factor regulating cardiac formation and function. Its mutations are linked to congenital heart disease. In this first report of a crystal structure of NKX2.5 homeodomain in complex with double-stranded DNA of its endogenous target, locating within the proximal promoter −242 site of the atrionatriuretic factor (ANF) gene. The crystal structure was determined at 1.8 Å resolution, demonstrates that NKX2.5 homeodomain occupies both DNA binding sites separated by five nucleotides without physical interaction between themselves. The two homeodomains show identical conformation despite differences in the DNA sequences they bind and no significant binding of the DNA was observed. Tyr54, absolutely conserved in NK2 family proteins, mediates sequence-specific interaction with the TAAG motif. This high-resolution crystal structure of NKX2.5 protein provides a detailed picture of protein and DNA interaction, which allows the prediction of DNA binding of mutants identified in human patients. *Reproduced from http://www.ncbi.nlm.nih.gov/pmc/articles/PMC3448007/.*

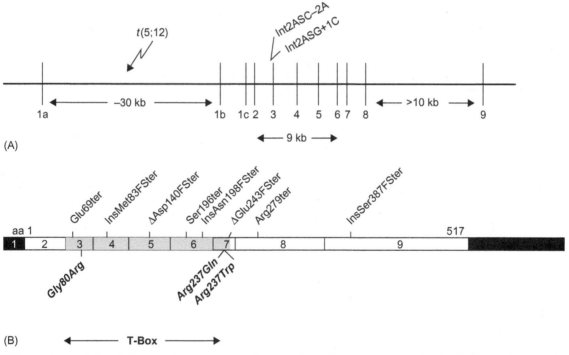

**FIGURE 12.6** Fourteen mutations in TBX5 that cause the Holt−Oram syndrome. (A) *TBX5* genomic structure is shown with approximate intron sizes. Exons 1−9 are shown with vertical bars. Exons 1a, 1b, or 1c are alternatively spliced as the first exon of *TBX5* cDNA. Alternative splicing of the 3′ region of the gene accounts for the variable addition of exon 9. Arrows indicate the translocation t(5;122)(q15;q24) found in the family IIa proband [designated t(5;12)], which disrupts TBX5 in intron 1a and the location of interior acceptor site (AS) mutations Int2ASC-$_2$A and Int2ASG$_{+1}$C. Acceptor site residues were numbered from the splice site with the conserved G residue designated as +1. (B) Schematic representation of *TBX5* cDNA illustrating the target of the alternatively spliced transcripts. Untranslated sequence (*dark shading*), exons 1−9 (*numbered boxes*), and locations of amino acids 1 and 517 are indicated. Codons (*gray shading*) that encode the T-box DNA binding domain [residues 56−238 (exons 3−7)] were defined by homology to other T-box gene family members. *TBX5* mutations that are predicted to truncate are shown above; missense mutations are indicated below (*bold, italics*). Mutations are designated by the name and number of the first substituted amino acid residue: Δ deletion; *FSter*, indicates frameshift mutations with resultant premature stop codons; *Ins*, Insertion; *ter*, indicates nonsense mutation. *Reproduced from http://www.pnas.org/content/96/6/2919.figures-only.*

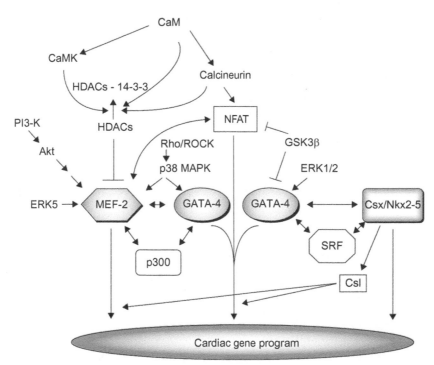

**FIGURE 12.7**   Role of cardiac transcription factors in regulation of cardiac gene program during cardiac hypertrophy. GATA4 transcriptional activity is stimulated through phosphorylation by (ERK1/2) p38 MAPK, although phosphorylation by GSK3β (glycogen synthase kinase 3β) negatively regulates GATA4 activity. In addition, the transcriptional activity of GATA4 is regulated through physical interaction with NFAT (nuclear factor of activated T cells), MEF2 (myocyte enhancer factor 2), SRF (serum response factor) or a cofactor, p300. Most important, is response to hypertrophic stimulation, NFAT is dephosphorylated by calcineurin and translocates into the nucleus, where it activates gene expression partly through forming a complex with GATA4. MEF2 transcriptional activity is enhanced through phosphorylation by p38 MAPK and ERK5 and physical interaction with GATA4, NFAT, and coactivator p300. In addition, MEF2 might be involved in the PI3K/Akt-mediated hypertrophic signal. Most important, MEF2 factors function as important effectors of Ca²⁺ signaling. MEF2 activity is stimulated by constitutively active calcineurin or CAMK (calcium/calmodulin-dependent kinase) in vivo. Activation of MEF2 is dependent of dissociation from class II HDACs. Signal-mediated phosphorylation of HDACs recruits chaperones 14-3-3 to dissociate the HDAC-MEF2 formation, although the endogenous HDAC kinase has not been determined. Csx/Nkx2−5 might regulate cardiac gene expression (1) directly, (2) via association with GATA4 or SRF, and (3) via upregulation of Csl, which activates transcriptional activities of GATA4 and MEF2. Contribution of Csx/Nkx2−5 transcriptional activity to pathophysiological hypertrophic responses remains undefined. *CAM*, calcium/calmodulin; *Csl*, transcription factor; *HDAC*, histone deacetylase; *Rho/ROCK*, small GTPase (Rho) kinase; *SRF*, serum response factor. *Reproduced from http://circres.ahajournals.org/content/92/10/1079.figures-only.*

autosomal dominant, meaning that only one copy of the altered gene is enough to cause the disease. The syndrome occurs in one of 100,000 individuals.

## Transcription Factors Involved in Cardiac Hypertrophy

Hypertrophy of the heart is the abnormal enlargement of the heart muscle that results from increases of the size of the cardiac myocytes as well as changes in other components, such as the extracellular matrix. The condition results from biomechanical stress (including hypertension) that can progress to heart failure or even sudden death. It is a maladaptive process resulting from the hypertrophic signaling cascade of which transcriptional factors are prominent as capitulated in Fig. 12.7.

## TRANSCRIPTION FACTORS AND THE TRANSCRIPTION COMPLEX

Transcription is the synthesis of RNA (mRNA) from DNA (gene) template. Initiation, elongation and termination are the **phases of transcription**. Double-stranded DNA must be opened to enable RNA polymerase to bind to the **gene promoter** (the regulatory region usually upstream from the gene). The **transcriptional apparatus** is complex and requires that transcription factor proteins bind to DNA or associate with other proteins, including **RNA polymerase** at the transcription site. A cartoon of the transcriptional complex is shown in Fig. 12.8.

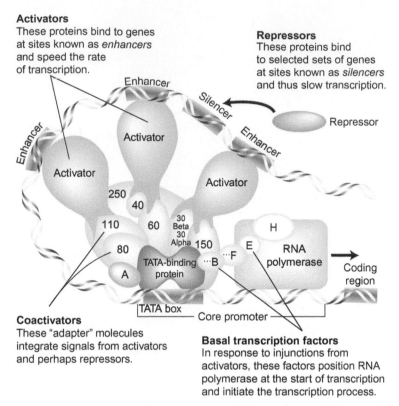

**Activators**
These proteins bind to genes at sites known as *enhancers* and speed the rate of transcription.

**Repressors**
These proteins bind to selected sets of genes at sites known as *silencers* and thus slow transcription.

**Coactivators**
These "adapter" molecules integrate signals from activators and perhaps repressors.

**Basal transcription factors**
In response to injunctions from activators, these factors position RNA polymerase at the start of transcription and initiate the transcription process.

**FIGURE 12.8**    The transcriptional complex in human cells. The **core promoter** DNA contains the **TATA box** (**TATAAAA**) that is often about 50 bases upstream from the start site. It is bound by **transcription factor IID** (**TFIID**) (shown in *red*). Only the TFIID binds directly to the TATA box DNA while the other proteins, shown in *blue*, bind to each other and some directly to the **TATA binding protein**. The TATA binding proteins are the first occupants to be situated on the core promoter. **Basal transcription factors** (A, B, F, E, H) are essential for transcription to occur. These proteins, in their binding to each other, may be responsible for the formation of the loop of the double-stranded DNA and position RNA polymerase at the **transcriptional start site**. There are regulatory molecules, **activators** or **repressors**. Activators (*yellow*) communicate with the basal transcription factors through **coactivators** (in *blue*) that are proteins tightly complexed to the TATA binding protein. *Reproduced from http://scienceblogs.com/pharyngula/2007/01/basics-what-is-a-gene.php.*

While **repressors** (*purple oblong* in Fig. 12.8) can bind to silencer sequences in DNA that can be located very far upstream from the core promoter, **enhancer DNA sequences** that also may be located very far upstream (thousands of bases) from the core promoter form complexes with sequence specific transcription factors, called **enhancer-binding proteins**, such as TFIIB and TFIIA that help to form the transcription complex, bringing the sites into direct contact and increase the rate of transcription. Another example would be the proteins that bind to the CCAT box.

A molecule of **RNA polymerase II** (**pol II**) binds to the transcriptional start site. Pol II is a complex of 10 different proteins (Fig. 12.9).

A simplified model of RNA pol II in a transcriptional complex is shown in Fig. 12.10.

The start site is at the beginning of the information encoding mRNA. The **template strand** of DNA is the **noncoding strand** that is used for the formation of mRNA that becomes a copy of the coding strand of DNA, except that thymine (T) is replaced by uracil (U). Pol II (about 35 base pairs upstream from the start site) lies across the template strand of DNA in the 3′ to 5′ direction and the mRNA produced is an exact copy of the coding DNA strand in the 5′ to 3′ direction. To transcribe a specific gene requires a specific array of regulatory factors that allows transcription of one gene without turning on the expression of other genes that require a different array of transcription factors. In many cases, a group of genes will be expressed by the same **transcription factors**. As mentioned previously, the process of transcription involves 3 phases: **initiation, elongation,** and **termination**. Initiation is the binding of **RNA polymerase** to the double-stranded DNA molecule. The **core transcriptional elements** are the **TFIIB recognition element** (**BRE**), the **TATA box**, the **initiator element** (**INR**), and the **downstream promoter element** (**DPE**) as schematically shown in Fig. 12.11.

In the figure, the sequences defining each motif are shown. The TFIIB recognition element (**BRE**) has the sequence G/C,G/C,G/A,CGCC. Where the 3′ conclusion of BRE is reached, the 5′ of the TATA box starts. Sometimes, at a

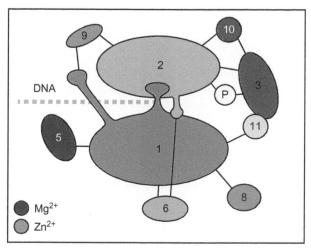

**FIGURE 12.9** Schematic of RNA polymerase II showing its 10 subunits in different colors. The dashed line in *blue* is double-stranded DNA. DNA enters the enzyme complex in a deep cleft formed by two main subunits, **Rpb1** and **Rpb2**. A pair of "jaws" clamps the DNA strands as they enter the complex. Where the cleft ends, the double strand is unwound for a short distance (the "**transcription bubble**") where the template (noncoding strand of DNA) forms a hybrid with the transcribed mRNA. The mRNA transcript leaves the complex through two grooves leading away from the active site. Below the active site is an opening that allows the entry of substrate nucleotides (for continuing mRNA formation) and for the entry of regulatory transcription factors. Proofreading may also take place at this site. *Reproduced from http://www.als.lbl.gov/ala/science/sci_archive/polymerase2.html.*

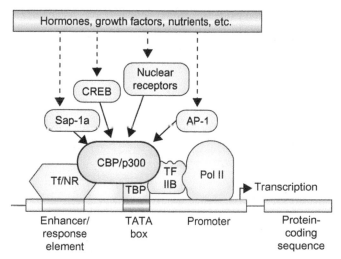

**FIGURE 12.10** A simplified model of a complex for RNA polymerase II (Pol II)-catalyzed transcription. A bridging protein, such as CREB-binding protein (CBP)/p300 would closely contact sequence-specific transcription factors (Tfs) and nuclear hormone receptors, TATA box-binding protein (TBP) and transcription factor IIB (TFIIB). The latter would not contact DNA but would complex with Pol II. A factor, such as CBP/p300 would form complexes with several other transcription factors without the involvement of DNA, such as nuclear receptors, CREB (cyclic AMP response element-binding protein), AP-1 (adapter protein complex 1) and Sap-1a (serum-response-factor 1a). The latter can bind DNA and their activities are modulated by phosphorylation by mitogen-activated protein kinases and protein kinase A, so this allows the networking and integration of plasma membrane and nuclear signaling pathways. *Reproduced from http://www.nature.com/nrm/journal/v3/n9/fig_tab/nrm914_F5.html.*

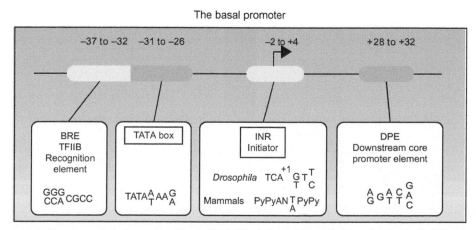

**FIGURE 12.11** Elements of the basal eukaryotic promoter. *Redrawn from Smale, S.T., Kadonaga, J.Y., 2003. The RNA polymerase II core promoter. Ann. Rev. Biochem. 72, 449–479.*

position upstream, a **CCAAT box** can be found close to the initiator, INR. The CCAAT motif is often found when a TATA box is absent (the TATA box occurs in about 25% of human gene promoters); the CCAAT box specifies the binding of **nuclear factor-1** (**NF-1**) and has the sequence: 5'-GGCCAACTC-3'. A CCAAT box is found in about half of the vertebrate promoters.

The TATA-binding protein is required by all RNA polymerases (pol I, pol II, and pol III). The TATA box is about 25–30 base pairs upstream from the start site near the INR. However, transcription can occur when the TATA box is absent from the promoter. The **transcriptional preinitiation complex** is made up of transcription factors, including TFIID during initiation, the TATA binding protein and RNA pol II bound to the promoter. The TATA-binding protein **transcriptional activator factors** (**TAF**s) include the TATA-binding protein, TFIIB, TFIIE, TFIIF, and TFIIH (Fig. 12.12).

To permit the binding of RNA pol II and the process of initiation, other factors play a role as shown in Fig. 12.13.

**RNA pol II** is a complex enzyme consisting of 12 subunits (**Rpb** 1 through 12). The subunits 1, 2, 3, 5, 6, 8, 10, 11, and 12 are conserved in RNA pols I, II, and III. The total mass of all 12 subunits is 513.6 kDa.

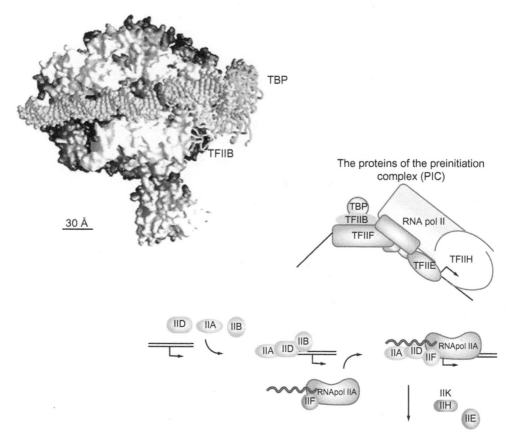

**FIGURE 12.12** Constituents of the transcriptional preinitiation complex. On *left* is a molecular model of the preinitiation complex. RNA pol II is in *white*; TATA binding protein in *green*; TFIIB in *yellow*. Highlighted base pair in *purple* and *red* is the presumed initiation site for DNA strand separation. *Reproduced from http://dev.biologists.org/content/133/22/4393/F1.expansion.html. On* right *is a schematic of the complex. The product of the last reaction is shown in Fig. 12.13.*

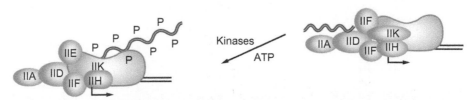

**FIGURE 12.13** RNA pol II in the process of initiation.

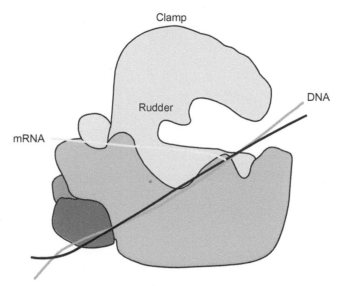

**FIGURE 12.14** Diagram of RNA pol II showing the clamp and the positions of DNA template strand and mRNA emerging. *Reproduced from http://upload.wikimedia.org/wikipedia/commons/f/f0/RNA_polymeraseclamp.jpg.*

The bending of DNA occurs when the **TATA-binding protein** binds to the **TATA box**. This provides a saddle structure for other transcription factors to form a complex (Fig. 12.12). TAFs involve **histone acetyltransferases (HATs)**, **protein kinases**, **coactivators** and other activities. The **INR** interacts with the TATA-binding protein and with the coactivator, SP1 (spacing relative to TATA-binding protein) and the largest subunit of RNA pol II. The sequence of the **INR** is $YYA_{+1}NT/AYY$ where Y is a pyrimidine and A is the transcriptional start site. The transcriptional start site begins with the start codon (usually **AUG**; specifies methionine) one position after the 5′-untranslated region, the 3′-end of the **leader sequence**. The leader sequence often includes information for the destiny of the translated protein but is not part of the coding region.

RNA pol II binds directly to the INR box and initiation occurs when the TATA-binding protein and the TAFs dissociate from RNA pol II and RNA pol II begins a forward movement over the opened strand of DNA (strand opening by **helicase** and **ATPase**). Some of the transcription factors are dissociated but **TFIIB** remains at the initiation site with other transcription factors. During elongation RNA pol II uses a clamp mechanism to move forward. It also has a positively charged saddle structure for the binding of DNA and RNA. Three zinc ions stabilize the fold of the clamp that closes on DNA and RNA in order to trap the DNA template and the RNA transcript (Fig. 12.14).

The active site that includes the hybrid of DNA and the transcribed RNA being formed has a preference for ribose nucleotide triphosphates over deoxyribonucleotide triphosphates as would be expected for the growing RNA. In Fig. 12.15 is shown the addition of a ribonucleotide triphosphate to the growing mRNA chain in the **RNA pol II** overall reaction.

## ENHANCERS

An enhancer is a short sequence of DNA that is recognized by certain transactivating factor proteins. These proteins act over long distances (they can be several thousand base pairs away from the core promoter site as seen from their DNA binding sites pictured in Fig. 12.8 and even may exist on a different chromosome than the gene being affected). Enhancers may be located far upstream or downstream from the transcription start site of the gene being stimulated. The actions of enhancer proteins work through protein—protein contacts, alterations of proteins, changes in chromatin structure, phosphorylation, and others and, by these actions, facilitate transcription.

## COACTIVATORS AND COREPRESSORS

Coactivators and corepressors are proteins that regulate transcription, especially of genes under the control of the **steroid receptor gene family** (as well as other systems) for which there is a great deal of information. It is useful to mention the **nuclear receptor family** here as the glucocorticoid receptor (GR) (member of the nuclear receptor family)

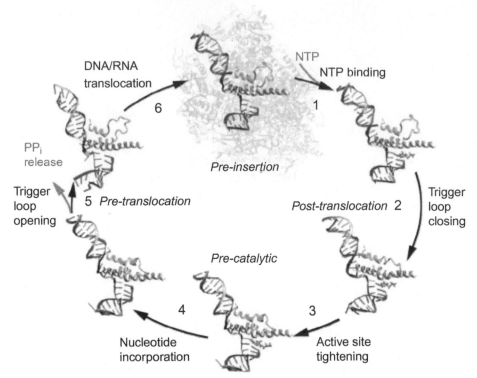

**FIGURE 12.15** Transcription by RNA polymerase II. The opened template strand of DNA is in *pink*. The mRNA chain is in *green*. The molecule incorporated into the growing mRNA chain is a ribonucleoside phosphate produced by the cleavage and release of pyrophosphate (*blue*, step 5). *Reproduced from http://feig.bch.msu.edu/main-research-protein+dna.html.*

mechanism will be used as an illustration of the mechanism of transcription. Generally, corepressor proteins interact with receptors whose ligand-binding domains (LBDs) are empty. In the presence of ligand that binds to the receptor's ligand-binding domain, the corepressor dissociates and is replaced by a coactivator protein that facilitates transcription. Corepressors can be enzymes that methylate or deacetylate histones causing inhibition of transcription. Coactivators are proteins that acetylate histones and phosphorylate **histone H1b** so that it dissociates from the receptor-binding region of DNA. When corepressors are present (and receptor ligands absent), RNA pol II is not active in transcribing the DNA but when ligands fill the ligand binding domains of the receptor dimer (in this case) and coactivators are bound, transcription by RNA pol II proceeds. A general summary of the actions of corepressors and coactivators is illustrated in Fig. 12.16.

It is evident from the figure that many ligands in addition to the ligands of the nuclear receptor family participate in this signaling process. In the **nuclear receptor superfamily**, there are three groups of receptors: the endocrine receptors, the **adopted orphan receptors** and the **orphan receptors**. The endocrine group consists of the **steroid receptors** (receptors for: estrogen, ER, progesterone, PR, androgen, AR, glucocorticoid, GR and mineralocorticoid, MR and for retinoic acid (RA, RX), RARα,β, thyroid hormone, TRα,β,γ and vitamin D, VDR). These represent receptors with high affinity for hormonal lipids. The adopted orphan receptors are receptors for low affinity dietary lipids and include: RXRα,β,γ for 9-*cis* RA, DHA; PPARα,β,γ for prostanoids and fatty acids; LXRα,β (liver X receptor) for oxysterols; FXR (farnesoid X receptor) for bile acids; PXR/SXR (pregnane X receptor) for **xenobiotics**, and CAR (constitutive androstane receptor) for xenobiotics. The orphan receptors have been identified but their ligands have not. There are at least 13 receptors that fall into this classification.

There are three classes of coactivators that interact with nuclear receptors. Class I representatives are: GRIP1, SRC-1, and AIB-1. Class II can be represented by TRAP220 and RIP140. Class III can be represented by RIP140, PGC-1, DAX-1, and SHP. These classes contain an amino acid sequence motif, **LXXLL**, which interacts with nuclear receptors. Class I have a basic amino acid preceding the binding motif and can be represented by SRLXXLL (R is arginine or the amino acid could be K, lysine). Class II motif is **PphiLXXLL** (where phi is an hydrophobic amino acid, such as isoleucine,). Class III motif is S/TphiLXXLL. Major corepressors are **N-CoR** and **SMRT**. These proteins have a nuclear receptor interaction motif that is: LXXIXXXL (where L is leucine, X is any amino acid, and I is isoleucine). Some repressors have a **helix−turn−helix** motif (two helices) where one of the helices interacts with DNA. The nuclear receptor interaction motif in coactivators is located in the central portion of the molecule (another domain in the

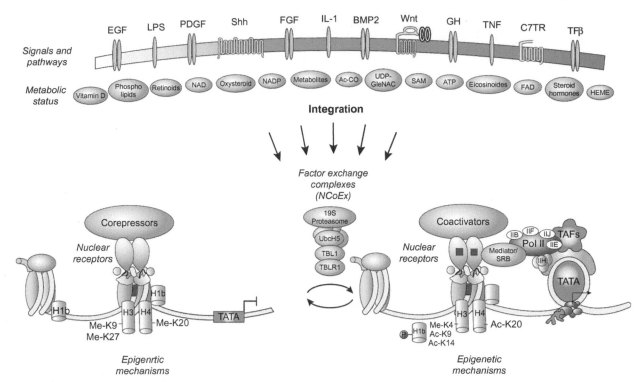

**FIGURE 12.16** Corepressor—coactivator exchange complexes are targets of many extracellular and intracellular signaling pathways. Transcriptional activation, for many systems, requires the removal of **corepressors** and the recruitment of **coactivators**. In the case of nuclear receptors, **TblR1** (nuclear receptor corepressor/HDAC3 complex subunit TBLR1 is one of several names for this protein) is used as a sensor of ligand binding, which activates its E3 **ubiquitin ligase** activity leading to ubiquination, clearance and likely proteosome-dependent degradation of repressor complexes. The clearance of corepressor results in gene derepression and is a prerequisite to the recruitment of coactivator complexes. The exchange of corepressor for coactivator is linked to changes in histones, as shown here, the loss of H (histidine) 3-K (lysine) 9 and K 27 methylation and the gain in H3K9 and K14 acetylation (presumably causing chromatin remodeling). *Reproduced from http://genesdev.cship.org/content/20/11/1405/F5.expansion.*

C-terminal region is for interaction with CBP/p300 (the **CREB binding protein**)), whereas the nuclear receptor interaction motif in corepressors is located at the C-terminal ends of N-CoR and SMRT.

The CREB (cAMP-response element binding protein) binding protein (CBP) interacts with at least 72 other proteins acting as a coactivator of transcription. It has intrinsic histone **acetyltransferase** activity that acetylates both histone and nonhistone proteins. In addition to this activity, it acts as a scaffold to stabilize other proteins of the **transcription complex**. Consequently, it is involved in the regulation of gene expression, differentiation, and cell growth.

In the nuclear receptor gene family, the receptors are either located in the cytoplasm and become activated by ligand binding in order to translocate to the nucleus, typical of many of the steroid hormone receptors, or they reside in the nucleus. An example of the latter is the thyroid hormone receptor. In the absence of the thyroid hormone ligand (triiodothyronine), the receptor is associated with DNA at the **hormone responsive element (HRE)** but it cannot sponsor transcription because it is complexed with a **corepressor**. When thyroid hormone enters the nucleus and binds to the receptor, the corepressor is dissociated and replaced by a coactivator facilitating the binding of RNA polymerase and the commencement of transcription, as summarized in Fig. 12.17.

In the case of the thyroid hormone receptor heterodimer with RXR, RXR may be silent, not binding its own ligand, while the TR binds the thyroid hormone. In this form, the complex is activated. It is also possible that the RXR binds its ligand (**9-retinoic acid**) as well as the TR binding thyroid hormone and the "activated" RXR in the heterodimer may assist in the dissociation of the corepressor.

## THE GLUCOCORTICOID RECEPTOR AS A MODEL TRANSCRIPTION FACTOR

The **nuclear receptor superfamily** consists of four classes: steroid receptors (the glucocorticoid receptor is one of these), RXR heterodimers, dimeric orphan receptors, and monomeric/tethered orphan receptors. As mentioned before, the orphan receptors are known proteins but the *in vivo* ligands are either unknown or uncertain. These groups of receptors are summarized in Fig. 12.18.

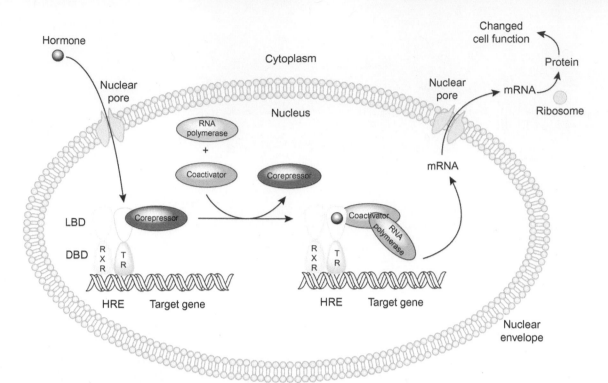

**FIGURE 12.17** The thyroid hormone receptor, as a heterodimer with **RXR**, is located at the thyroid **hormone-responsive element** (**HRE**). In the absence of the thyroid hormone ligand, the receptor is complexed with a corepressor and is inactive. When the thyroid hormone (**triiodothyronine**) enters the nucleus and binds to its receptor, the corepressor is dissociated, allowing a **coactivator** to be recruited, rendering the receptor in the active form and facilitating the positioning of RNA polymerase and the start of transcription. *Reproduced from http://en.wikipedia.org/wiki/File: Type_ii_nuclear_receptor_action.png.*

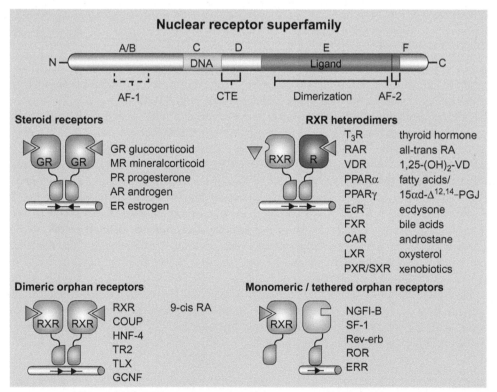

**FIGURE 12.18** Structure/function organization of nuclear receptors. There are 6 domains: (A–F) of nuclear receptors comprise regions of conserved function and sequence. All receptors contain a DNA binding domain (DBD) in region C that is the most highly conserved domain and includes 2 **zinc finger** modules. The **ligand-binding domain (LBD)** is in region E in the C-terminal half of the receptor. Between these two domains is a variable length hinge domain (D). The **N-terminal domain** (A/B) is variable in sequence and contains an activation function as does domain F. The N-terminal domain also contains the major **epitope** (antigenic site). The function of F is not well understood. Nuclear receptors are functional as homodimers or heterodimers. Amino acid sequences involved in **dimer formation** are located in the **DBD** and in the LBD. *Reproduced from http://www.jbc.org/content/276/40/36863/F1.large.jpg.*

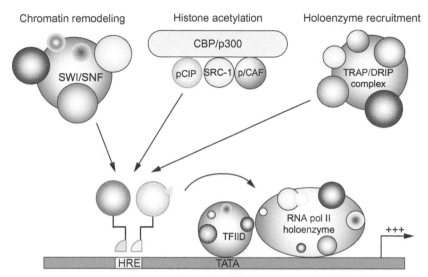

**FIGURE 12.19** Ligand-dependent recruitment of multiple **coactivator complexes**. Upon ligand binding, the receptors recruit different coactivator complexes. The complex CBP/p160/pCAF possesses **histone acetyltransferase** activity; the SWI/SNF complex possesses **ATP-dependent chromatin remodeling activity** and the TRAP/DRIP complex may recruit RNA polymerase II holoenzyme. Recruitment of the complexes may be sequential or combinatorial. Chromatin remodeling complexes may be initially recruited to the promoter. These factors relieve repression imposed by high-order chromatin structure and allow a second acetylation-dependent step on gene activation. Activation would require the combinatorial of subsequent action of additional complexes including the TRAP/DRIP complex. *Redrawn from Auwerx, J., Drouin, J., Laudet, V., 2003. Recepteurs a la Provencale. EMBO workshop on the biology of nuclear receptors. EMBO Rep. 4, 1122−1126.*

In the glucocorticoid receptor , the *in vivo* ligand is cortisol secreted by the adrenal cortex. The active receptor in the nucleus has a ligand attached to each of the two LBD sites in the dimer and the monomers form the dimer in a head-to-head orientation as shown by the model in the *upper left* of Fig. 12.18. The general model of a nuclear receptor is shown at the *top* of the figure. Sections of the structure are labeled A through F. **Nuclear activation function**s (AFs) bind coactivators (such as the p160 coactivators including SRC-1, the steroid receptor coactivator-1) and the **glucocorticoid receptor-interacting protein 1 (GRIP1)**, **p300/CBP-associated protein (pCAF)**, the **switching/sucrose nonferment-ing (SWI/SNF)** complex and the **vitamin D receptor (VDR)-interacting protein/thyroid-hormone-associated protein (DRIP/TRAP)** complex in regions A/B and F. GRIP1 contains three **LXXLL motif**s in the N-terminus for interaction with the liganded receptor at **AF-1** and another site in the C-terminus that interacts with **AF-2**. P160, CBP/p300, and p/CAF all have **acetyltransferase** activity and acetylate promoter histones, enhancing transcription (Fig. 12.19).

In Fig. 12.18, region C houses the **DNA binding domain (DBD)**; D contains the "hinge" region and the C-terminal extension (CTE), and E, the ligand-binding domain (LBD). The **hinge region** contains a **lysine cluster motif (KXKK)** that can be acetylated by **histone acetyltransferase** and deacetylated by **histone deacetylase**. When acetylated, the GR does not promote transcription and when the acetyl groups are removed from the lysine residues, the GR is again able to induce transcription. *This appears to be part of a **clock mechanism** regulating the nuclear activity of GR under various conditions of stress and environmental lighting.* The LBD also contains binding sites for the **heat shock proteins** that block the **nuclear localization motif** (there are two such motifs in the GR and they contain arginines and lysines) prior to activation by the binding of ligand. There are two nuclear localization (NL) signals that contain key basic amino acids. The NL1 signal appears as: KI**RRK**NCPAC**RYRK**CLQAGMNLEA**RKTNNNIK**GIQQ where the underlined letters (N refers to any amino acid) are essential to the **nuclear localization motif**. During activation, the heat shock proteins are dissociated from the monomeric receptor. Thus, there are many points of regulation in the structure of this steroid receptor.

In the DBD, there are two **zinc finger domains** and these bind in the major groove of DNA at the glucocorticoid hormone responsive element. The zinc finger structures are shown in Fig. 12.20.

**Zinc fingers** are between 23 and 28 amino acids in length and the cysteine (sometimes a combination of cysteines and histidines in other steroid receptors) residues stabilize the zinc ion. As mentioned, the zinc fingers interact with the hormone responsive element of DNA. The cytoplasmic **glucocorticoid receptor** monomer becomes activated by binding ligand (**cortisol** or, experimentally, a more active glucocorticoid, such as **dexameth-asone**). The activated receptor dimerizes in the cytoplasm and then is transported through the **nuclear pore** (after docking onto the nuclear pore complex) into the nucleus where the dimer binds to DNA (hormone responsive element) as shown in Fig. 12.21.

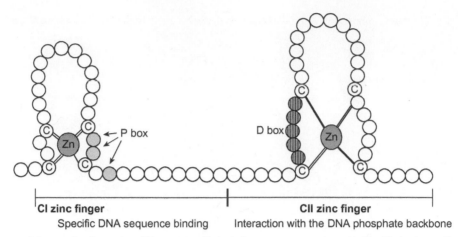

**FIGURE 12.20** Diagram of the two zinc fingers in a GR monomer. Each circle represents an amino acid. The CI zinc is coordinated by four cysteine (C) residues as is the zinc in the CII finger. The CI interacts with five base pairs of the HRE. The three shaded amino acids in CI (arrows) are within the **P** (proximal) **box** and these amino acids discriminate from the regions of DNA occupied by other steroid receptors. The vertically striped amino acids in the CII zinc finger make up the **D** (distal) **box** that is downstream from the CI. The D box is involved in **dimerization** and contact with the phosphate backbone of the DNA. *Reproduced from http://www.glowm.com/resources/glowm/cd/pages/v5/ch004/tops/003f.html?SESSID=1ovgb318n7t0gcprsbiqdq9eu6.*

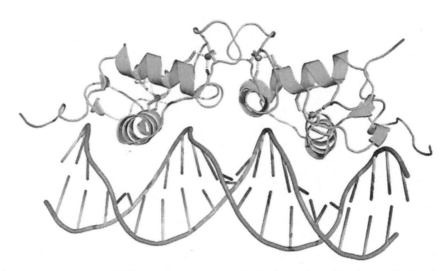

**FIGURE 12.21** Partial crystal structure model of GR (in *green* above) showing the interaction of the dimeric GR zinc fingers in the DBD with the major groove of DNA (in *orange* below). The zinc atoms are represented in *gray* spheres. The nucleosides are in *green* (sticks) and the phosphates of the backbone are in *blue*. *Reproduced from http://en.wikipedia.org/wiki/File:1r4o.png.*

## The Inactive Glucocorticoid Receptor (GR) in the Cytoplasm and its Activation

The GR is located in the cytoplasm of the cortisol-responsive target. Virtually, all cells of the body have either large or small amounts of this receptor except the cells in the space between the anterior and posterior lobes (*pars intermedia* of lower forms) of the pituitary and the **hepatobiliary cells**. The effect of the hormone is more or less proportional to the number of receptors in a target cell although when large amounts of the receptor are present, not all of the receptors are needed to obtain a response. The receptor is inactivated by its binding to the heat shock protein **hsp90** dimer. Hsp90 covers the **nuclear localization signal** so that the receptor is inactive. It may also cover the **ligand-binding site** with a binding constant that is much less powerful than that of cortisol so that the ligand can easily displace the heat shock protein. Other proteins are attached to the inactive receptor besides hsp90. Cortisol circulates in the blood stream bound to proteins, mainly to **transcortin** (corticosteroid binding globulin). About 10% of the blood steroid is in the unbound free form and the free steroid can enter any cell by free diffusion. Because of its hydrophobic character, it is miscible with the hydrophobic cell membrane. It enters the cytoplasm and is retained by binding to the receptor (binding constant for cortisol, about 50 nM; binding constant for dexamethasone or triamcinolone, about 1−10 nM); if little or no receptor is present in the cytoplasm, the hormone can diffuse back out of the cell into the bloodstream. The receptor has five serine residues that become phosphorylated and in this form the receptor is translocated through the nuclear pore. **Cortisol** is secreted by the **adrenal cortex** in response to the **adrenocortical hormone (ACTH)**. In turn, ACTH is secreted by the **corticotropic releasing hormone (CRH)** from the **hypothalamus**. The entire process is started internally by a **serotonergic neuron** or by **environmental stress**es.

Cortisol or other more potent synthetic glucocorticoids, such as dexamethasone (Fig. 12.22) binds to the **ligand-binding site** of GR causing the release of the **hsp90** dimer and other proteins from the complex.

In this form, the GR is the activated monomeric receptor. Two of these form a **homodimer** that is translocated into the nucleus. In the nucleus, it binds to the **glucocorticoid** HRE in the major groove of double-stranded DNA (Fig. 12.21) and **coactivator** binds to the receptor, allowing the formation of the pretranscriptional complex (Fig. 12.16). **RNA polymerase II** is positioned at the start site and transcription commences. The overall process of the activation of the GR, its entry into the nucleus, and gene expression is shown in Fig. 12.23.

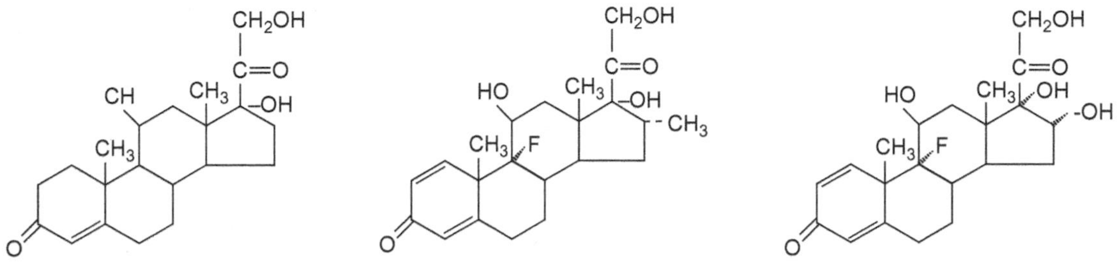

**FIGURE 12.22** Structures of cortisol and the potent synthetic glucocorticoids dexamethasone and triamcinolone.

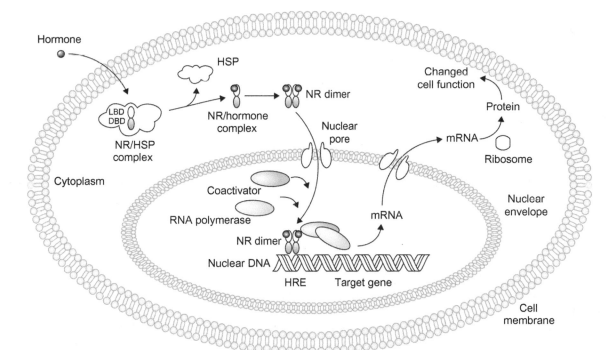

**FIGURE 12.23** Overall mechanism of a class 1 nuclear receptor that includes the glucocorticoid receptor. The free hormone (cortisol, sometimes called hydrocortisone) enters the extracellular space from the bloodstream and crosses the cell membrane by free diffusion. It binds to the inactive receptor (NR/HSP) in the soluble cytoplasm. Binding of the hormone to the receptor causes the dissociation of the heat shock protein 90 dimer (HSP) as well as other associated proteins to generate the activated receptor complex (NR/hormone complex). The activated receptor monomers dimerize in the cytoplasm and are translocated through the nuclear pore into the nucleus. In the nucleus the activated receptor dimer binds a coactivator and binds to DNA at the hormone responsive element (HRE). RNA polymerase is positioned to start the transcription process. This process produces a mRNA that leaves the nucleus through a nuclear pore and is translated in the cytoplasm on a ribosome into a specific protein whose action changes the metabolism or other function in the cell; the summation of these effects of the newly translated protein molecules is the cellular hormonal response. *Reproduced from http://upload.wikimedia.org/wikipedia/commons/3/3f/Nuclear_receptor_action.png.*

## CLASSES OF NUCLEAR RECEPTORS

There are four **classes of nuclear receptors**. Class 1, like the **glucocorticoid receptor**, dissociate heat shock proteins upon binding of the hormonal ligand. The monomeric receptor complex, thus activated, forms a dimer in the cytoplasm that translocates to the nucleus and binds to a HRE of DNA. The hormone responsive element (HRE) consists of two half-sites separated by a variable stretch of DNA, the second half-site of which has an inverted sequence from the first half-site. For the glucocorticoid responsive element the sequences are: 5′ AGAACAnnnTGTTCT 3′ and in the other strand of DNA, at the same location is the following sequence: 3′ TCTTGTnnnACAAGA (where n is any nucleotide). In this class are: **androgen receptor**, **estrogen receptor**, **glucocorticoid receptor**, and **progesterone receptor**. **Class 2 receptors** are located in the nucleus whether the ligand is present or not. These receptors bind DNA as heterodimers where the other receptor is usually **RXR**. When ligand is absent, these receptors may be bound to **corepressors**. In the presence of ligand, the corepressor dissociates and **coactivators** bind to the receptor resulting in the recruitment of additional proteins and RNA polymerase for the start of transcription. **Class 3 receptors** bind DNA as homodimers (like class 1) but their HREs are direct repeats instead of inverted repeats (see class 1 above). Finally, there is a class 4 in which the receptors bind to DNA as monomers or dimers but the DBD of the monomeric receptor binds to a single half site.

## CELL MEMBRANE RECEPTORS

There are reports of steroids acting too quickly to be regulating transcription, a process that usually requires hours. Some very small percentage of "steroid receptors" are associated with the **cell membrane** and they apparently produce these rapid actions that do not require transactivation in the nucleus. These molecules are different from the classical glucocorticoid receptor and it is possible that some of these systems are involved in transport of molecules/ions across the cell membrane. Binding preference of various steroids may differ from the classical cytoplasmic receptor.

## RECEPTOR ISOFORMS

The gene encoding the glucocorticoid receptor can give rise to two transcripts (the second by **alternative splicing**), an mRNA for the full-length classical receptor (**GRα**), and an mRNA for the shorter isoform (GRβ). **GRβ** is truncated by 35 amino acids in the C-terminal region. The two forms are identical to amino acid 727. After amino acid 727, the alpha form adds 50 amino acids to the C-terminal and the beta form adds 15 nonhomologous amino acids in the same direction (Fig. 12.24A).

In the presence of cortisol, the unliganded GRβ inhibits the effects of GRα (with bound hormone) and translocates into the nucleus where it can act as a dominant negative inhibitor of the GRα homodimer.

Many of the receptors in the nuclear receptor family also have isoforms. The thyroid hormone receptor has two isoforms, alpha and beta; both forms use the thyroid hormone as ligand. The retinoic acid receptor (RAR) has three isoforms, alpha, beta, and gamma all of which use vitamin A and related compounds as ligand. The peroxisome proliferator-activated receptor has three isoforms, alpha, beta/delta, and gamma and they bind fatty acids or **prostaglandins** as ligands. The **RAR-related orphan receptor** has three isoforms, alpha, beta, and gamma and these forms bind **cholesterol** or ATRA (**all trans-retinoic acid**). The **liver X receptor-like** (LXR has two isoforms, LXRbeta and FXR Farnesoid X **receptor**). These bind **oxysterols**. The **vitamin D receptor-like** has three isoforms: the vitamin D receptor (VDR) that binds vitamin D; PXR is the **pregnane X receptor** and it binds xenobiotics, and CAR (**constitutive androstane receptor**) that binds **androstane**. The **retinoid X receptor** (RXR) has three isoforms, alpha, beta, and gamma and all forms bind retinoids. In addition to RXR, there are four **retinoid X receptor-like receptors**: hepatocyte nuclear factor-4 (**HNF alpha and gamma**) and these bind fatty acids; **testicular receptor** NR2C1 (TR2) and NR2C2 (TR4) with uncertain ligands; TLX/PNR (TLX is the homolog of *Drosophila* **tailless gene** expressed protein and PNR which is a photoreceptor). COUP/EAR: there are two COUP receptors (**COUP-TFI** and **COUP-TFII**) that are **chicken ovalbumin upstream promoter-transcription factors**. **EAR-2** is **V-erbA** (Viral ERythroblastoma A)-related protein. The **estrogen receptor** (ER) has two isoforms (**ERalpha** and **ERbeta**) that bind estrogen as their ligand. There are two other **estrogen receptor-like proteins**: the estrogen-related receptor (ERRbeta and ERRgamma). There are also other receptors in addition to those mentioned here. Not only do multiple forms of receptors create fine-tuning in regulation but the receptors, themselves, can also bind to other regulatory factors, including transactivation factors, such as **AP-1** and **NFκB** to create another level of regulation of gene expression.

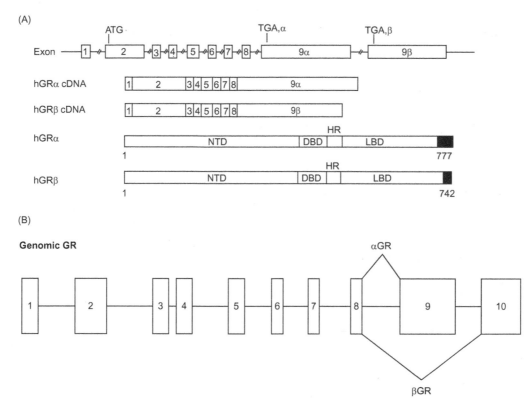

**FIGURE 12.24** (A) Schematic representation of the structure of the human **glucocorticoid receptor gene**. The two GR isoforms are the full transcript (alpha) and alternative splicing (beta). The functional domains and subdomains are indicated beneath the linearized protein structures. *DBD*, DNA-binding domain; *HR*, hinge region; *LBD*, ligand-binding domain; *NTD*, N-terminal domain. *GRβ occurs in lower concentrations than the alpha form in many cell types, is mainly nuclear (and cytoplasmic) and does not bind ligand.* It is transcriptionally inactive but can inhibit the transcriptionally active alpha form. Inhibition of the alpha form by the beta form increases with the concentration of the beta form and may be involved in glucocorticoid resistance. *Figure redrawn from Charmandari, E., Kino, T., Chrousos, G.P., February 2004. Molecular mechanisms of glucocorticoid action. In: Filetti, S.B. (Ed.), Familial/Sporadic Glucocorticoid Resistance. Alternative splicing pattern to produce the beta form of GR. (B) shows the arrangement of exons in the gene for the glucocorticoid receptor.*

## CHROMATIN

Because some of the transactivation proteins affect chromatin structure and function (e.g., methylation, histone acetylation, and histone deacetylation) and these changes are important to gene expression, some of the aspects of chromatin will be discussed here (although there has already been a discussion on chromatin structure in Chapter 1: Organ Systems and Tissues). Chromatin consists of double-stranded DNA, tightly packaged with **histones** contained in loops of DNA and with uniform stretches of DNA between the histone units. The condensed form of a chromatin fiber and how it appears when stretched out to show the repeating histone units is shown in Fig. 12.25.

Octamers of histone repeating groups consist of pairs of histones **H2A, H2B, H3,** and **H4.** The nucleosome is the minimal unit of chromatin structure and it consists of 146 base pairs of DNA and 8 histones with DNA (almost 2 turns) wrapped around the histone core. **H1** and **H5** are **linker histones** critical to the solenoid structure. The "beads-on-a-string" stretched-out chromatin is turned into a **solenoid**. To be transcriptionally active, chromatin must be opened. **Coactivators** and other related molecules are involved in the opening process. While **corepressors** deacetylate histones causing chromatin to tighten, coactivators through their acetylation activity, acetylate histones, and cause chromatin to relax. **Acetylation of histones** takes place on a histone lysine residue:

$$\text{Histone-lys} + \text{acetylCoA} \rightleftharpoons \text{Histone-lys-acetyl} + \text{CoA-SH}$$

Many coactivator proteins have **histone acetyltransferase (HAT)** activity. Acetylated histones **H3** and **H4** become attached to newly formed DNA and soon become deacetylated. The acetylation involves lysine residues of histone tails eliminating the positive charge of the lysine terminus. Histone tails lie outside the core of the **nucleosome** so that they interact with nearby nucleosomes. Acetylation opens this highly ordered structure so that chromatin becomes accessible to large complexes of proteins, thus facilitating **transcription**. ATP energy allows remodeling of the nucleosome so that individual

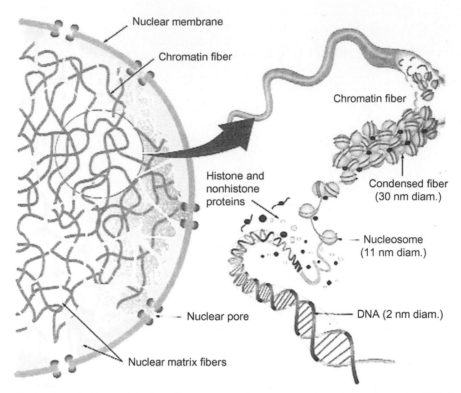

**FIGURE 12.25** Chromatin structure stretched out to show the component histone groups that occur at intervals of 200 base pairs. Between the histone groupings are individual proteins bound to DNA that include histone H1 and nonhistone proteins. The DNA connecting histone groups (**octamers**) is called linker DNA.

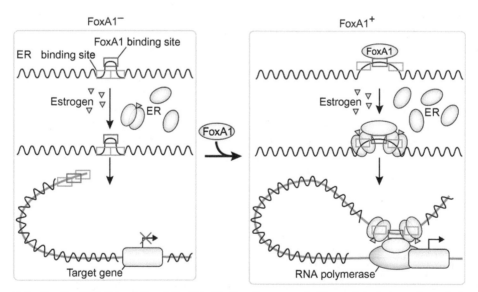

**FIGURE 12.26** In cells lacking FoxA1 (left-hand figure) the ERα binding sites are packed in condensed chromatin excluding the binding of ERα, even in the presence of estrogen, (ligand) and cannot activate target gene expression. When the cell expresses FoxA1, it creates an open conformation at the ERα binding sites. In the presence of estrogen, ERα binds and activates target gene expression (right-hand figure). *Reproduced from http://www.nature.com/ng/journal/v43/n1/full/ng0111-11.html#f1.*

strands of DNA become available to RNA polymerase. Transactivation factor III (**TFIII**) is held in place through the actions of a specific domain (**bromodomain**); these domains are spaced to bind consecutive **acetyllysines** in the tail of H4 that favors propagation of acetylation to immediate neighbors. Other HATs also are recruited by coactivators. **Histone deacetylase** of corepressors return chromatin to its original tightly packaged state that excludes the transcription machinery.

Transcription factor binding can control the actions of receptors on chromatin that lead to transcription. An example is the transcriptional function of the **estrogen receptor** (**ERα**) as demonstrated in human breast cancer cells. Here, the

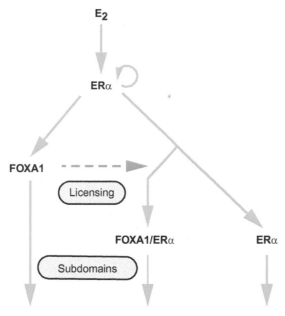

**Biological responses (e.g., cell cycle, signaling)**

FIGURE 12.27   Model illustrating how **FoxA1 licensing** defines subdomains of estrogen action in **breast cancer cells**. *Green* arrows, direct transcriptional activity of ERα and FoxA1; dashed *blue* arrow, action of FoxA1 as a modulator of ERα binding to a subset of promoters. The presence of FoxA1 grants permission to ERα to regulate a subset of the hormonal response that can be further amplified by regulation of FoxA1 expression by estrogen-bound ERα. *Reproduced from Laganière, J., Deblois, G., Lefebvre, C., Bataille, A.R., Robert, F., Giguère, V., 2005. From the Cover: location analysis of estrogen receptor alpha target promoters reveals that FOXA1 defines a domain of the estrogen response, Proc. Natl. Acad. Sci 102, 11651–11656.*

action of **FOrkhead boX A1** (**FoxA1**) protein determines the ability of ERα to activate gene expression. FoxA1 binds to DNA at a concensus sequence (5′-ACAATTAGTTGTAGCTTCT-3′) whose action is permissive for the subsequent binding of ERα to its hormone responsive element (HRE). This is illustrated by Fig. 12.26.

The transcriptional function of ERα in breast cancer cells and in other cells is dictated by FoxA1. It is colocalized to the ERα binding sites and the repression or knockdown of FoxA1 generates a decrease in ERα binding sites by about 50%. Even ERα binding sites that do not have FoxA1 nearby are reduced in availability to ERα by repression of FoxA1. This is possible if FoxA1 also regulates another factor involved in the binding of ERα to its HRE. Transcription of most of the estrogen-regulated genes are reduced when FoxA1 is down-regulated. **FoxA1 licensing** defines the subdomains of **estrogen action** as summarized in Fig. 12.27.

Because estrogen (through the action of the estrogen receptor) is a growth factor for **mammary cancer cells**, FoxA1 may become a useful therapeutic target for breast cancer.

## SUMMARY

Mutations in the genes involved in the development of the heart lead to congenital heart diseases. Although several genes may be involved in processes that damage the development of the heart, the transcription factors GATA4, Nkx2.5, and TBX5 are centrally important. Damage to the resulting heart can vary from hypertrophy to holes in the septa (involving either upper or lower chambers) as well as modification of the electrical conducting system. Some of these conditions are severe enough to lead to heart attack or sudden death.

The **transcriptional complex** consists of a **core promoter** containing the **TATA box** located about 50 base pairs upstream from the transcriptional start site. The transcription factor IID (**TFIID**) binds directly to the TATA box. Other factors bind to each other and some also bind directly to the TATA box. For transcription to take place, there are basal transcription factors required; these are A, B, F, E, and H and they are responsible for the formation of a loop of double-stranded DNA and enable the positioning of RNA polymerase at the transcriptional start site. Activator proteins communicate with basal transcription factors through **coactivators**, proteins that are complexed to the **TATA binding protein**.

**RNA polymerase** is a complex enzyme consisting of 10 subunits. It contains a deep cleft for the entry of DNA and has a pair of jaws that clamp the DNA as it enters the complex. A transcription bubble denotes the unwinding of double-stranded DNA where the cleft ends and where the noncoding strand of DNA forms a hybrid with the transcribed

**mRNA**. The mRNA leaves the complex through two grooves that lead away from the active site. Substrate nucleotides enter the enzyme complex below the active site. The substrate entry site is also used for the entry of regulatory transcription factors and proof-reading takes place at this site, as well.

An **enhancer** is a short sequence of DNA that is recognized by certain transacting factor proteins that act over long distances from the core promoter. They can even exist on a different chromosome than the gene promoter being activated for transcription. Enhancers can be located far upstream or downstream from the transcription start site. Enhancer proteins act through protein–protein contacts, alterations of proteins, changes in chromatin structure, phosphorylation or other actions, by which they facilitate transcription.

**Coactivators** and **corepressors** are proteins that regulate transcription. These are well known for the steroid receptor gene family. Corepressor proteins interact with nuclear receptors whose ligand-binding domains are empty. In this condition, RNA polymerase II is not active in transcribing DNA. In the presence of ligand bound to the ligand-binding domains of the receptor, the corepressor dissociates and is replaced by a coactivator protein that facilitates transcription. Corepressors may be enzymes that methylate or deacetylate histones causing tight packaging of DNA and inhibition of transcription. Coactivators are proteins that acetylate histones and phosphorylate histone H1b causing it to dissociate from the receptor-binding region of DNA and relaxing DNA.

**Thyroid hormone receptor** (TR), a representative of the nuclear receptor family, is an example of a heterodimeric receptor (together with RXR) located in the nucleus at its HRE. In the absence of thyroid hormone (usually **triiodothyronine**), it is complexed with a corepressor and is inactive. When thyroid hormone enters the nucleus and binds to TR, the corepressor is dissociated and a coactivator is recruited causing the activation of the receptor and facilitating the positioning of RNA polymerase II and the start of transcription.

**Nuclear receptors** have 6 domains, A through F that comprise regions of conserved functions and sequences. All contain a **DNA-binding domain (DBD)** in region C that is the most highly conserved domain and includes 2 **zinc finger** modules that contact the major groove of DNA defining the HRE. The **ligand-binding domain (LBD)** is in region E in the C-terminal half of the receptor. There is a variable **hinge domain** between the DBD and the LBD. The N-terminal domain (A/B) and the C-terminus (F) is variable in sequence and contains an activation function for the binding of activator proteins. The N-terminal domain contains a major epitope (antigenic site), and, in the case of the glucocorticoid receptor, this region has a variable structure. The function of F is not well understood. Nuclear receptors are in the form of **homodimers** or **heterodimers**. Cytoplasmic receptors without bound ligands may be monomeric but form homodimers in the cytoplasm after binding ligand. Inactive receptors residing in the nucleus in the absence of ligand may be in the form of heterodimers attached to corepressors but upon ligand-binding are activated by the loss of corepressors and the addition of coactivators. This can happen whether the nonsteroidal ligand (often **retinoic acid**) is complexed to the heterodimeric partner or not. Sequences governing **dimer formation** are located in the DBD and the LBD. **Zinc fingers** in the DBD contain zinc coordinated by 4 cysteine residues or a combination of 2 cysteine and 2 histidine residues. A **proximal box** (**P box**) contains 3 amino acids that discriminate from the regions of DNA occupied by other steroid receptors. Other amino acids in the **distal box** (**D box**; downstream from the P box) are involved in **dimerization** and contact with the phosphate backbone of DNA.

The unliganded **glucocorticoid receptor (GR)** is located in the cytoplasm of a cortisol-responsive target cell. Most cells of the body contain GR in small or large amounts; the cells containing the largest amounts of GR are the main cortisol targets (the human liver cell might contain 25–30 thousand molecules). **Cortisol effects** are more or less proportional to the amount of receptor in the target cell although all of the receptor molecules in a major target may not be required for a full hormonal response. The unliganded receptor in the cytoplasm is bound to two molecules of **heat shock protein 90** (**Hsp90**) that cover the nuclear localization signal motif and may cover some of the ligand-binding domain (in which case the affinity of receptor for cortisol would be significantly larger than the affinity of the receptor for Hsp90 so that Hsp90 would be displaced by the binding of cortisol). There are other proteins bound to the unliganded GR that are dissociated upon ligand binding.

**Cortisol** circulates in the bloodstream after being released from the adrenal cortex as a result of the action of ACTH and circulates mainly in bound form with the protein, **transcortin (corticosteroid binding globulin)**. About 10% of the circulating cortisol is in free (unbound) form and the free form can enter cells by free diffusion owing to the lipophilic property of cortisol and the lipid nature of the cell membrane. In the cytoplasm cortisol binds to the GR (binding constant is about 50 nM). The hormone can diffuse back out of the cell if the GR is absent or in very low concentration (in the latter case, the small amount of GR is rapidly filled and the excess cortisol diffuses back out of the cell). GR contains 5 serine residues that become phosphorylated for translocation into the nucleus through the **nuclear pore**.

There are four **classes of nuclear receptors**: Class 1 receptors dissociate heat shock proteins upon binding of the ligand. The liganded monomeric receptor forms a dimer in the cytoplasm (some figures suggest that dimerization takes place in the nucleus; there is some disagreement on this point) that translocates into the nucleus through the nuclear pore and binds to the hormone responsive element (HRE). The HRE for GR consists of two half-sites separated by a variable stretch of DNA, the second half-site of which has an inverted sequence from the first half-site. The concensus **GR responsive element** has the sequence: 5′-AGAACAnnnTGTTCT-3′ and in the opposing DNA strand the sequence is: 3′-TCTTGTnnnACAAGA = 5′, n being any nucleotide. In this class are: androgen receptor, estrogen receptor, glucocorticoid receptor and progesterone receptor. Class 2 receptors are localized to the nucleus whether the ligand is present or not. These receptors bind to DNA as heterodimers and the second monomeric receptor is often RXR. These receptors are bound to **corepressors** in the absence of ligand; in the presence of ligand, the corepressor dissociates and a **coactivator** binds to the receptor. This activated receptor allows for the recruitment of other proteins and RNA polymerase. Class 3 receptors bind to DNA as homodimers (like Class 1) but their HREs are direct repeats compared to the inverted repeats seen in Class 1. In Class 4, receptors bind to DNA as monomers or dimers, however, the DBD of the monomeric receptor binds to a single half-site.

A very small percentage of some of the steroid hormone receptors occurs in the **cell membrane**. These receptors mediate responses that are too rapid to be accounted for by nuclear transactivation. Apparently, they are different from the GRα.

Some of the steroid receptors have **isoforms** generated by alternative splicing. This is certainly the case for the **estrogen receptor** and the **glucocorticoid receptor** as well as most of the other receptors in the steroid receptor superfamily. The isoform, usually named the β form is smaller by reduction of the LBD. This form is inhibitory and the balance between the α-form and the β-form may account for part of the modulation of receptor action.

**Chromatin structure** is affected by the action of activating proteins and repressor proteins. Some repressors have **deacetylase** activity causing acetyl groups to be removed from **histones**. This results in tighter packaging of chromatin and generally inhibits transcription. Activating proteins possess **histone acetyltransferase** activity which causes histones to be acetylated and allowing chromatin relaxation conducive to transcription. This takes place on **histone lysine residues** in the histone tail, eliminating the positive charge of the lysine terminus. Histone tails lie outside of the core of the **nucleosome** so that they interact with nearby nucleosomes. Acetylation opens this highly ordered structure enabling chromatin to become accessible to large complexes of proteins that stimulate transcription.

In the stimulation of transcription of **estrogen-responsive genes** by the liganded estrogen receptor (ERα), the action of **forkhead box A1 (FoxA1)** protein is permissive for the binding of the receptor to its HRE. In the absence of the expression of FoxA1 by the estrogen target cell, the estrogen response is reduced. FoxA1 binds to its recognition site on DNA and affects the binding of ERα to its HRE.


# SUGGESTED READING

### Literature

Akazawa, H., Komuro, I., 2003. Roles of cardiac transcription factors in cardiac hypertrophy. Circ. Res. 92, 1079—1088.

Aoyagi, S., Trotter, K.W., Archer, T.K., 2005. ATP-dependent chromatin remodeling complexes and their role in nuclear receptor-dependent transcription. Vitam. Horm. 70, 281—307.

Auwerx, J., Drouin, J., Laudet, V., 2003. *Recepteurs a la Provencale*. EMBO workshop on the biology of nuclear receptors. EMBO Rep. 4, 1122—1126.

Basson, C.T., et al., 1999. Different *TBX5* interactions in heart and limb by Holt—Oram syndrome mutations. Proc. Nat. Acad. Sci. 96, 2919—2924.

Charmandari, E., Kino, T., Chrousos, G.P., 2004. Molecular mechanisms of glucocorticoid action. In: Filetti, S.B. (Ed.), Familial/Sproatic Glucocorticoid Resistance, Orphanet encyclopedia.

Chen, J.D., 2000. Nuclear receptor coactivators. Vitam. Horm. 58, 391—448.

Chen, X., Kao, H.-Y., 2009. G protein pathway suppressor 2 (GPS2) is a transcriptional corepressor important for ERα-mediated transcriptional regulation. J. Biol. Chem. 284, 36395—36404.

Huang, R.-T., et al., 2013. A novel NKX2.5 loss-of-function mutation responsible for familial atrial fibrillation. Int. J. Mol. Med. 31, 1119—1126.

Kassel, O., Herrlich, P., 2007. Crosstalk between the glucocorticoid receptor and other transcription factors: molecular aspects. Mol. Cell. Endocrinol. 175, 13—29.

Lageniere, J., et al., 2005. Location analysis of estrogen receptor α target promoters reveals that FOXA1 defines a domain of the estrogen response. Proc. Natl. Acad. Sci. 102, 11651—11656.

Lai, I.-L., et al., 2010. Histone deacetylase 10 relieves repression on the melanogenic program by maintaining the deacetylation status of repressors. J. Biol. Chem. 285, 7187—7196.

McCulley, D.J., Black, B.L., 2012. Transcription factor pathways and congenital heart disease. Curr. Top. Dev. Biol. 100, 253—277.

Nolis, I.K., et al., 2009. Transcription factors mediate long-range enhancer-promoter interactions. Proc. Natl. Acad. Sci. 106, 20222—20227.

Olefsky, J.M., 2001. Nuclear receptor minireview series. J. Biol. Chem. 276, 36863—36864.

Pradhan, L., Genis, C., Scone, P., Weinberg, E.O., Kasahara, H., Nam, H.-J., 2012. Crystal structure of the human NKX2.5 homeodomain in complex with DNA target. Biochemistry 51, 6312−6319.

Pratt, W.B., Mishima, Y., Osawa, Y., 2008. The Hsp90 chaperone machinery regulates signaling by mediating ligand binding clefts. J. Biol. Chem. 283, 22885−22889.

Schott, J.-J., et al., 1998. Congenital heart disease caused by mutations in the transcription factor NKX2-5. Science 281, 108−111.

Schupp, M., Lazar, M.A., 2010. Endogenous ligands for nuclear receptors: digging deeper. J. Biol. Chem. 285, 40409−40415.

Smale, S.T., Kadonaga, J.Y., 2003. The RNA polymerase II core promoter. Ann. Rev. Biochem. 72, 449−479.

Tata, J.R., 2002. A simplified model of a complex for RNA polymerase II (Pol II)-catalyzed transcription. Nat. Rev. Mol. Cell Biol. 3, 702−710.

Vandevyver, S., Dejager, L., Libert, C., 2014. Comprehensive overview of the structure and regulation of the glucocorticoid receptor. Endocr. Rev. 35, 671−693.

Wada, Y., et al., 2009. A wave of nascent transcription on activated human genes. Proc. Natl. Acad. Sci. 106, 18357−18361.

Watt, A.J., Battle, M.A., Li, J., Duncan, S.A., 2004. Gata4 is essential for formation of the proepicardium and regulates cardiogenesis. Proc. Natl. Acad. Sci. 101, 12573−12578.

Witchell, S.F., DeFranco, D.B., 2006. Mechanisms of disease: regulation of glucocorticoid and receptor levels—impact on the metabolic syndrome. Nat. Clin. Pract. Endocrinol. Metab. 2, 621−631.

## Books

Hughes, T.R., 2011. A Handbook of Transcription Factors (Subcellular Biochemistry). Springer, New York, NY.

Litwack, G. (Ed.). 1994. Steroids. Vitamins & Hormones, vol. 49. Academic Press/Elsevier, Amsterdam, NL.

Litwack, G. (Ed.). 2004. Nuclear Receptor Coregulators. Vitamins & Hormones, vol. 68. Academic Press/Elsevier, Amsterdam, NL.

Ma, J. (Ed.). 2006. Gene Expression and Regulation. Springer, New York, NY.

Zhang, C. (Ed.). 2009. Interactions of Steroid Hormone Receptors with Coactivators and DNA, UMI Microform 3347568. Proquest LLC.

# Chapter 13

# Metabolism of Amino Acids

## UREA-CYCLE-RELATED DISEASE: HYPERAMMONEMIA

The urea cycle is one of the metabolic pathways that will be discussed in this chapter. The urea cycle eliminates unneeded nitrogen, derived from nitrogen-containing compounds, from the body in the form of urea. Many amino acids can be converted to glutamate that can, in turn, be converted to **aspartate**. Aspartate can enter the urea cycle to produce **urea** for excretion in the urine. If there is a defect in the urea cycle, excess ammonia (hyperammonemia) accumulates in the blood and results in devastating disease, including death. Defects in the cycle are genetic (develop in infants) and involve deficiency of **carbamoyl phosphate synthase** and/or **ornithine transcarbamylase**. Other enzyme deficiencies also contribute by accumulation of the substrate for the deficient enzyme. Thus, **argininosuccinic acid synthase** (citrulline + Asp + ATP → argininosuccinate + AMP + PP$_i$) **deficiency** leads to **citrullinuria** and **argininosuccinate lyase** (argininosuccinate → Arg + fumarate) **deficiency** causes **argininosuccinic aciduria**.

Normally, the conversion of amino acid-derived ammonia to urea is highly efficient and protects the central nervous system from toxicity. Elevations in the blood level of ammonia occur in an infant when an enzyme activity is low or missing altogether from a genetic defect. In an adult, this can occur from a diseased liver, the organ site of the urea cycle but there appears to be increasing incidence of underlying genetic disease in previously normal adults. Increased blood levels of ammonia produce a variety of symptoms as shown in Fig. 13.1.

The normal **blood level of ammonia** is roughly 500 nmol/L. The normal level of **blood urea nitrogen** is in the range of 300 μmol/L. There is a factor of about 600-fold or more urea than ammonia. **Urea cycle disorders** have a frequency of about 1 in 30,000 in the newborn.

In the cerebellum and striatum of the brain, ammonia hyperactivates the excitatory **N-methyl-D-apartate receptor (NMDAR)** at its glycine site, resulting in the accumulation of **cyclic guanosine monophosphate (cGMP)** [when Glu from a presynaptic nerve ending crosses the synapse and activates the NMDAR on the postsynaptic membrane, it leads to an influx of $Ca^{2+}$ which binds to calmodulin which, in turn, activates neuronal nitric oxide synthase (nNOS) to produce nitric oxide (NO). NO then crosses the synapse back to the presynaptic membrane to activate guanylate cyclase and produce cGMP from GTP].

The ammonia ($NH_3$ or ammonium ion, $NH_4^+$) in the brain is utilized by **glutamate dehydrogenase** wherein $\alpha$-**ketoglutarate** + $NH_4^+$ is converted to glutamate. This lowering of α-ketoglutarate (and subsequently oxaloacetate) depresses the **Tricarboxylic Acid Cycle** and aerobic oxidation and leads to damage and death of the cell (Fig. 13.2).

There are several urea cycle-related disorders. Most derive from genetic abnormalities involving genes for the enzymes in the urea cycle (Table 13.1).

The plasma level of ammonia is determined if **hyperammonemia** is suspected. The condition of the liver is also tested by measuring cellular components that may leak into the bloodstream when liver cells are damaged. These component activities include **serum transaminases** and **alkaline phosphatase**. The level of **serum albumin** can be measured as the level can fall below 3.5 mg/dL in advanced liver disease. **Prothrombin time** is also measured as the proteins of the coagulation system are made in the liver. There is a good correlation between prothrombin time and liver function. Increased plasma levels of **citrulline** or **argininosuccinic acid** can be a signal for primary genetic disease. If there is a general increase in plasma amino acid levels, **liver disease** may be the problem. Partial inhibition of the urea cycle can be generated by dysfunctions of amino acid catabolism (to be discussed later), and this can lead to increased levels of blood ammonia. In this case, intermediates of amino acid catabolism may be increased in the blood, and these would include **propionic acid**, **methylmalonic acid**, and **isovaleric acid**. Amino acid profiles in the urine can characterize **argininosuccinic aciduria**, **hyperornithinemia**, **hyperammonemia**, **homocitrullinuria**, or intolerance to dibasic

Human Biochemistry. DOI: http://dx.doi.org/10.1016/B978-0-12-383864-3.00013-2

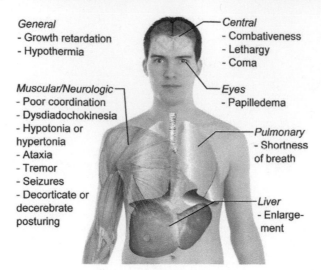

General
- Growth retardation
- Hypothermia

Central
- Combativeness
- Lethargy
- Coma

Muscular/Neurologic
- Poor coordination
- Dysdiadochokinesia
- Hypotonia or hypertonia
- Ataxia
- Tremor
- Seizures
- Decorticate or decerebrate posturing

Eyes
- Papilledema

Pulmonary
- Shortness of breath

Liver
- Enlargement

**FIGURE 13.1** Symptoms of hyperammonemia. *Reproduced from http://upload.wikimedia.org/wikipedia/commons/7/76/ Symptoms_of_hyperammonemia.svg.*

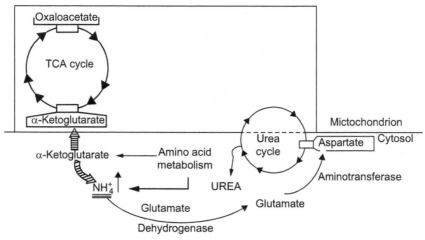

**FIGURE 13.2** High levels of ammonia (as ammonium ion, $NH_4^+$) in the brain drain α-ketoglutarate from entering the Tricarboxylic Acid (TCA) Cycle to combine with ammonia to form glutamate. This lowers the activity of the cycle and consequently reduces the level of **oxaloacetate** accounting for the loss of aerobic metabolism that damages the cell and results in cell death.

**TABLE 13.1** Urea Cycle Disorders

| | Abnormalities of Metabolism | | |
| --- | --- | --- | --- |
| | Excess Metabolites | Reduced Metabolites | Specific Clinical Features |
| CPSI deficiency | Ammonium, glutamate | Citrulline, arginine | – |
| OTC deficiency | Ammonium, glutamate | Citrulline, arginine | – |
| Citrullinemia (classical) | Ammonium, citrulline | Arginine | – |
| Argininosuccinic aciduria | Ammonium, argininosuccinic acid, citrulline | Arginine, arginosuccinic acid, citrulline | Hepatomegaly, twisted hair |
| Argininemia | – | – | Spastic paraplegia |
| Gyrate atrophy of retina (OAT deficiency) | Ammonium (transient), ornithine | – | Retinal degeneration |
| Adult onset citrullinemia type II (citrin deficiency) | Ammonium, citrulline | Arginine | Liver damage |
| Hyperammonemia– hyperornithinemia– homocitrullinemia syndrome (mutations in ORT1) | Ammonium, ornithine homocitrulline | – | – |
| Lysinuric protein intolerance (mutations in SLC25A13) | Ammonium | Lysine, arginine | Hepatosplenomegaly, osteoporosis |

CPSI, carbamyl phosphate synthetase 1; OTC, ornithine transcarbamylase; OAT, ornithine aminotransferase; ORT1, ornithine transporter.
Source: Reproduced from http://jn.nutrition.org/content/134/6/1605S/T1.expansion.html.

amino acids in dietary protein (lysinuric protein intolerance). The **blood lactic acid** concentration can rule out mitochondrial diseases because the failure to utilize pyruvate efficiently by the mitochondrial TCA cycle will cause an increase of lactate in the blood. The pH of blood may be elevated when **blood ammonia level** is increased because it stimulates the respiratory system. The normal range of **blood urea nitrogen** is 8−20 mg/dL, but in disorders of the **urea cycle**, it can be less than 3 mg/dL. The treatment of hyperammonemia requires reduced protein intake and replacing the protein with nonprotein sources. **Sodium phenylacetate** or **Ucephan** (sodium benzoate) is administered intravenously as it stimulates the excretion by the kidneys of nitrogen as **phenylacetylglutamine** and **hippuric acid**. Sodium benzoate conjugates with glycine to form hippuric acid that is also excreted by the kidneys. The combination of the two reagents (**Ammonul**) is also used. Molecular genetics (gene sequencing) confirms the diagnosis of a genetic disease of the urea cycle. If elevated levels of ammonia are not treated, there is damage to the central nervous system in terms of cell death; cerebral edema will follow with increased intracranial pressure, and death will result.

## THE UREA CYCLE

Eighty percent of urea is synthesized from ammonia in the liver in the urea cycle. When **glutamine** is produced in excess in the liver, it is converted to ammonia by a **glutaminase** enzyme found in the periportal hepatocytes and renal epithelial cells (the enzyme also is located in the intestine). Ammonium ion is important in acid−base regulation in the kidney, and during **acidosis**, this enzyme is induced in the kidney to increase the excretion of ammonium ($NH_4^+$). The glutaminase reaction is:

$$\text{glutamine} + H_2O \rightleftharpoons \text{glutamate} + NH_4^+$$

The liver urea cycle functions by converting ammonia (as ammonium ion) to urea that is excreted in the urine. The cycle consists of five enzymes of which the initial two enzymes are located in the mitochondrial matrix, and the other three are located in the liver soluble cytoplasm. The urea cycle is also referred to as the **Krebs−Henseleit cycle** named after the discoverers.

Ammonia is derived from dietary amino acids and proteins (plants fix nitrogen from the atmosphere to form ammonia). Proteins are broken down in the intestinal tract and absorbed as peptides and free amino acids. These become precursors of human proteins and the excess amino acids, not needed for protein synthesis, are deaminated to produce ammonia. The deaminated products enter the Tricarboxylic Acid Cycle. Ammonia (ammonium ion) from amino acid metabolism enters the **mitochondrial matrix** and is combined with bicarbonate and ATP to form **carbamoyl phosphate**, the first step in the urea cycle (Fig. 13.3).

The first step is catalyzed by **carbamoyl phosphate synthase-I**. **Ornithine transcarbamylase** converts carbamoyl phosphate to **citrulline**. These two steps take place in the **mitochondrial matrix**. Citrulline is transported out of the mitochondrial matrix, by a transporter, to the soluble cytoplasm. In the cytosol, citrulline is combined with **aspartate** and is converted to **argininosuccinate** by **argininosuccinate synthase**. Argininosuccinate is converted to **fumarate** and **arginine** by **argininosuccinate lyase**. Fumarate can enter the mitochondrial TCA cycle and arginine is converted by **arginase**, the terminal enzyme of the urea cycle, to one molecule of **urea** and one molecule of ornithine. Ornithine can be transported to the mitochondrial matrix by an ornithine transporter and take part in the **urea cycle**. The interactions between the urea cycle and the mitochondrial matrix are summarized in Fig. 13.4.

The overall reactions of the TCA cycle and the urea cycle can be summarized:

$$2NH_4^+ + HCO_3^- + 3ATP^{4-} \rightarrow \text{urea} + 2ADP^{3-} + 4Pi + AMP^{2-} + 5H^+$$

## AMINO ACID METABOLISM: AMINO AND AMIDE GROUP TRANSFERS

As mentioned, **glutamate dehydrogenase** forms glutamate from α-ketoglutarate and ammonia ($NH_4^+$, ammonium). **Glutamate** is converted to **glutamine** by **glutamine synthase**. The glutamine synthase catalyzed reaction is:

$$\text{glutamate} + NH_4^+ + ATP \rightarrow \text{glutamine} + ADP + P_i + H^+$$

Glutamine can be broken down to glutamate and ammonium ion by **glutaminase**:

$$\text{glutamine} + H_2O \rightarrow NH_4^+ + \text{glutamate}$$

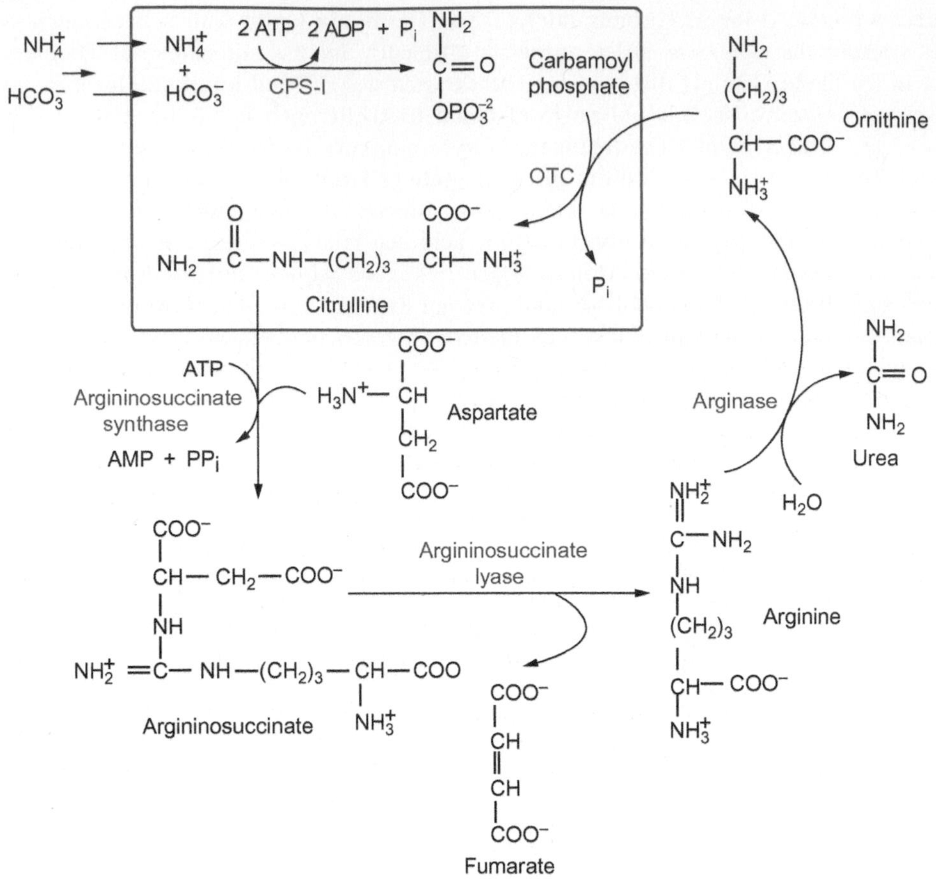

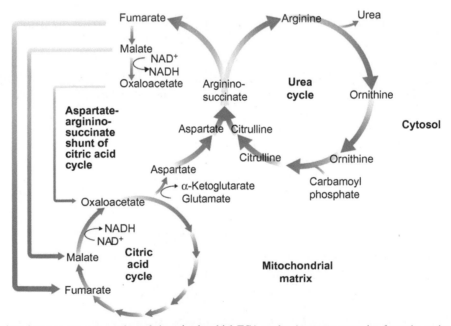

**FIGURE 13.3** Diagram of the urea cycle. Reactions in the rectangle (outlined in *red*) are in the mitochondrial matrix. Participating enzymes are labeled in *red. CPS-I,* carbamoyl phosphate synthetase-I; *OTC,* ornithine transcarbamoylase.

**FIGURE 13.4** Interactions between the urea cycle and the mitochondrial TCA cycle. Aspartate can arise from the amino acid pool or from the conversion of fumarate or malate to oxaloacetate and then, by transamination, to aspartate. Mitochondrial aspartate can, via a transporter, emigrate to the soluble cytoplasm to combine with citrulline to form arginine and then the final product, urea. *Reproduced from http://www.tamu.edu/faculty/bmiles/lectures/urea.pdf.*

**Transamination** accounts for the addition of amino groups from these amino acids to carbon skeletons to form other amino acids. The amide group from glutamine can be transferred to carbon skeletons by **transamidation**. Glutamate dehydrogenase, in its reverse reaction, can convert glutamate into ammonium and $\alpha$-ketoglutarate that can enter the TCA cycle for the production of energy and the reduced coenzymes **NADH + H$^+$** or **NADPH + H$^+$** (glutamate dehydrogenase can use either coenzyme). **Glutamate dehydrogenase** is localized to the **mitochondrial matrix** (although some evidence suggests that the enzyme may not be entirely mitochondrial) and is a branch point that links amino acids to energy metabolism. Glutamate dehydrogenase is a hexameric enzyme consisting of two stacks of trimers. It is in high concentration in liver, brain, kidney, and pancreas but is not in high concentration in muscle. It is an **allosteric enzyme** regulated by the positive effectors ATP and GTP directing the reaction to the formation of glutamate. The positive effectors, ADP and GDP, regulate the enzyme in the reverse direction for the formation of ammonium ion and $\alpha$-ketoglutarate. The ATP concentration in the cell (the cellular level of ATP varies between 1 and 10 mM) determines the direction of the glutamate dehydrogenase reaction. Thus, when [ATP] is high in the cell, the conversion of glutamate to $\alpha$-ketoglutarate is depressed; the keto-acid is not needed as a source of energy. However, when the concentration of ATP is low, glutamate is converted to $\alpha$-ketoglutarate so that it can enter the TCA cycle for the production of ATP.

There are two forms of human glutamate dehydrogenase, **GDH1** and **GDH2**. The two isozymes differ in their endogenous activity, allosteric regulation and stability to heat. The overall reaction catalyzed by glutamate dehydrogenase is:

$$\text{glutamate} \, + \, \text{NADP}^+ + \text{H}_2\text{O} \rightleftharpoons \alpha\text{-ketoglutarate} \, + \, \text{NADPH} \, + \, \text{H}^+ + \, \text{NH}_4^+$$

GDH1 is markedly inhibited by GTP, but GDH2 is not. GDH2 has low basal activity and can be fully activated by ADP or L-leucine. GDH2 is concentrated in **testis** and **brain**. In the testis, the enzyme is localized to the **Sertoli cells**, while in the brain, it is localized to the **astrocytes** and in low concentration in neurons. Astrocytes support neurons and Sertoli cells support germ cells. Because GDH2 is not controlled by GTP, the selective expression of this isoform, allows GDH2 to metabolize glutamate even when the TCA cycle is generating GTP in amounts sufficient to inactivate the GDH1 isozyme thus allowing the supporting cells, astrocytes, and Sertoli cells, to continue functioning even when oxidative metabolism is high.

**Glutamine** is a major amino acid in blood that carries ammonia from various tissues; it is important in the transport of ammonia from peripheral tissues to the kidney. In the kidney, glutamine amide nitrogen is cleaved by **glutaminase** to produce glutamate plus ammonium ion that is excreted in the urine. In tissues that contain the **urea cycle**, **glutaminase** is also present allowing for ammonia to be incorporated either into urea or glutamine. *Both urea and NH$_4$$^+$ are excreted into the urine by the kidney.*

In the case of **acidosis** (when the blood pH becomes acidic), more glutamine is transferred from the liver to the kidney so that HCO$_3$$^-$ is conserved (in the urea cycle, two bicarbonate ions are used per ammonium ion to form urea). Kidney glutaminase releases glutamate and ammonium ion from glutamine, and **glutamate dehydrogenase** can release another mole of NH$_4$$^+$ (plus $\alpha$-ketoglutarate) so that ammonium ion (carrying a proton) can be excreted. The net effect is a reduction of protons (H$^+$) causing an increase in pH to overcome acidosis.

The generation of glutamate dehydrogenase in the **developing brain** is essential for the control of **ammonia detoxification**. When its activity is low or lacking during development, **mental retardation** can occur because the toxic effects of ammonia cannot be dealt with effectively.

**Glutamine synthase** in the liver is localized to the **perivenous hepatocytes** (surrounding a vein other than the portal vein), whereas glutaminase is periportal (surrounding the portal vein). While **glutamine** is the most important transporter of nitrogen between tissues, the conversion of glutamate to glutamine occurs intracellularly. In periportal hepatocytes, glutamine is used for the synthesis of **glucose**: glutaminase converts glutamine to glutamate plus ammonium ion, and glutamate is converted to $\alpha$-**ketoglutarate**, by transamination, which enters the TCA cycle and then through gluconeogenesis to produce glucose. Ammonium ion enters the urea cycle. $\alpha$-Ketoglutarate is converted to **malate** in the TCA cycle and, when cellular concentrations of glucose are low, malate is transported to the soluble cytoplasm via the **malate/pyruvate shuttle**, and malate is converted to **pyruvate** by cytoplasmic **malic enzyme-1** (malic enzyme-2 operates *within the mitochondria* to convert malate to pyruvate), and two molecules of pyruvate are converted to glucose in the **gluconeogenic pathway** (reverse direction of glycolysis).

Glutamine is formed from **lactate** and **arginine** in perivenous hepatocytes. This occurs by the conversion of lactate to $\alpha$-ketoglutarate that is converted to glutamate and then to glutamine via glutamine synthase. Arginine is converted to **ornithine** by **arginase** and ornithine is converted to **pyrroline-5-carboxylate** by **ornithine aminotransferase** and pyrroline-5-carboxylate is converted to glutamate by **1-pyrroline-5-carboxylate dehydrogenase** and then to glutamine by glutamine synthase. A summary of the events in the periportal hepatocyte compared to the perivenous hepatocyte is shown in Fig. 13.5.

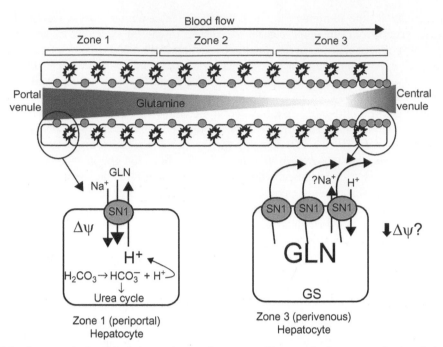

**FIGURE 13.5**  A model for the events in the periportal vs perivenous hepatocytes. **Urea synthesis** occurs predominantly in the **periportal hepatocytes** (zones 1 and 2) where bicarbonate is consumed to create a more acidic intracellular environment. The incoming portal blood contains higher concentrations of glutamine and the outwardly directed proton gradient helps to drive the inward uptake of glutamine through the **SN1** (**system N transporter** that transfers glutamate, histidine, and asparagine only) through its **Na$^+$/H$^+$ exchange mechanism**. Zone 3 contains a small population (5%−7%) of perivenous **glutamine synthase** (GS)-positive hepatocytes with enriched plasma membrane SN1. Glutamine exits these cells aided by diminished plasma glutamine content in zone 3 due to consumption in zones 1 and 2. The perivenous hepatocyte has a less acidic cytoplasm and a high cytoplasmic glutamine level from GS activity. The net glutamine movement is influenced by **transmembrane electrical potential ($\Delta\psi$)**, but it is unclear if there is a difference in this respect in periportal vs perivenous hepatocytes. *Reproduced from http://jn.nutrition.org/content/131/9/2475S/ F1.expansion.*

# TRANSAMINATION

Transamination is the process by which amino groups are removed from amino acids and transferred to acceptor keto-acids to generate the amino acid version of the keto-acid and the keto-acid version of the original amino acid. The reactions are highly reversible, and the forward or reverse direction depends upon the concentrations of substrates or products. This class of enzymes contains **pyridoxal phosphate** as coenzyme, although certain transaminases (also termed aminotransferases) can use **pyruvate** as in glutamate-pyruvate transaminase:

$$\text{glutamate } + \text{ pyruvate} \rightleftharpoons \text{alanine } + \text{ }\alpha\text{-ketoglutarate}$$

Alpha-ketoglutarate is sometimes written as *2-oxoglutarate.*

**Aspartate aminotransferase** is an important enzyme across many species and catalyzes the reaction:

$$\text{L-aspartate } + \text{ }\alpha\text{-ketoglutarate} \rightleftharpoons \text{oxaloacetate } + \text{ L-glutamate}$$

There are two different forms of this enzyme (different primary amino acid sequence), one residing in the mitochondrion and one in the cytosol (soluble cytoplasm). The enzyme also has been named **glutamate-oxaloacetate transaminase**, and the two forms have been referred to as **s-GOT** and **m-GOT** (**see** Chapter 3: Water, pH, Buffers and Introduction to the General Features of Receptors, Channels, and Pumps). In Fig. 13.6 is shown the general reaction mechanism for a transaminase.

Muscle cells rely on **glutamate-pyruvate transaminase** to produce **alanine** from **pyruvate** and an amino acid so that the keto-acid produced (like $\alpha$-ketoglutarate) can be used as fuel for the **TCA cycle** for the production of energy as ATP. The **alanine** is carried to the liver in the bloodstream so that the amino groups from amino acids can be converted to urea in the **urea cycle**. In this way, **muscle cells** can use amino acids as energy sources while relying on the liver to deal with the amino groups (as ammonium ions). Alanine, a predominant amino acid in proteins, is also transported in the bloodstream to the liver where it can be converted to glucose. Transamination of alanine to pyruvate allows pyruvate to form glucose through the **gluconeogenic pathway**. The amino group of alanine is attached to $\alpha$-ketoglutarate

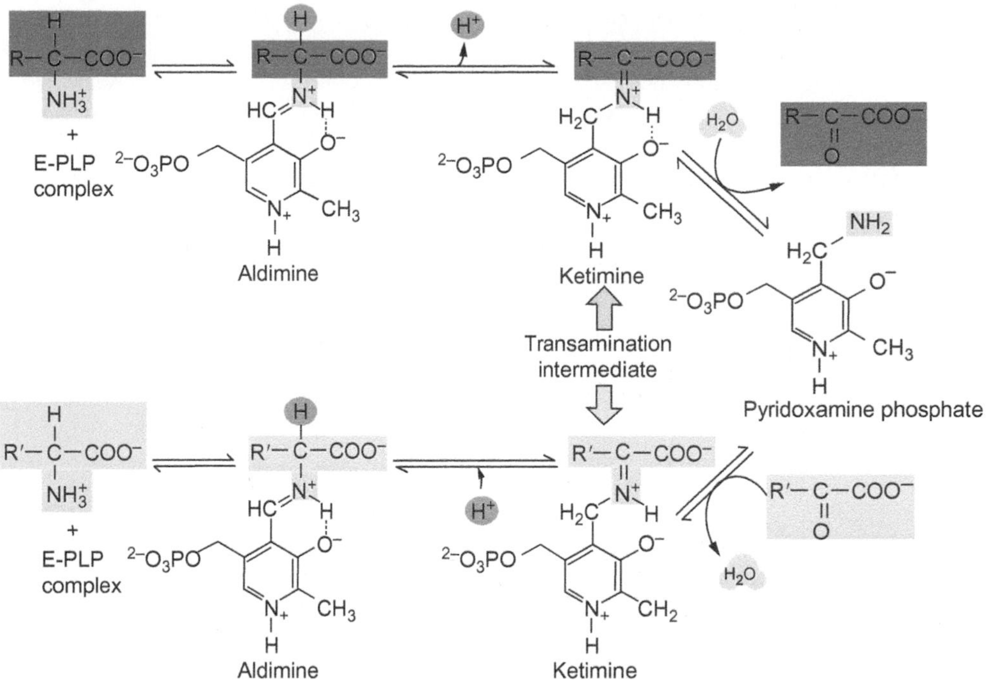

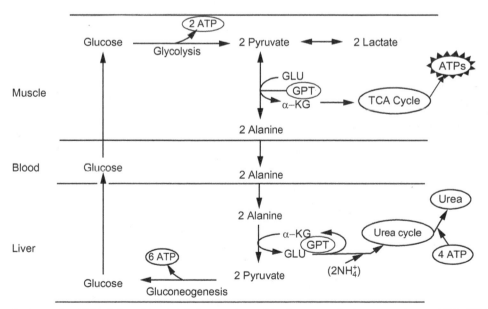

**FIGURE 13.6** A general **transamination mechanism**. The coenzyme, **pyridoxal phosphate** (**PLP**), attaches to the apoenzyme (enzyme lacking coenzyme or cofactor) through an ε-**amino group** (ε = epsilon) of a lysine residue in the active site, as shown in the second *top left* structure; this linkage is known as a **Schiff base (aldimine)**. The *orange color* represents the first amino acid added. The first **ketimine** intermediate is formed (third structure, *top*). The ketimine intermediate is formed followed by the release of the first keto-acid derived from the first amino acid (*orange structure on far right*) and the formation of **pyridoxamine phosphate** (structure on *right, middle*). The second keto-acid (*green*) is added to form the second ketimine derivative (aldimine, *green*) and, in the final step, the amino acid derived from the second keto-acid is released (*bottom, left in green*). *Reproduced from http://www.nd.edu/~aserriann/transam.html.*

**FIGURE 13.7** The alanine cycle. *GPT*, **glutamate-pyruvate transaminase** (also known as alanine transaminase, ALT); *TCA*, tricarboxylic acid cycle; α-*KG*, α-ketoglutarate.

through transamination into glutamate. The amino group of glutamate is removed as $NH_4^+$ by **glutamate dehydrogenase** for incorporation into urea that is cleared through the kidney. These reactions are known as the **alanine cycle**, summarized in Fig. 13.7.

During **stress**, especially prolonged stress, the adrenal cortex secretes cortisol. Even in short events of stress, there is enough **cortisol** secreted to increase the level of circulating glucose (~10% increase) that evokes a release of insulin.

Prolonged stress results in the breakdown of the musculature releasing amino acids into the bloodstream, of which alanine is predominant because of its plentiful occurrence in many proteins. Increased [alanine] is taken up by the liver where it can be converted to glucose (alanine cycle) and, under the influence of **insulin**, can be converted to **glycogen**.

Another important enzyme is **γ-aminobutyric acid** (**GABA**, also 4-aminobutanoic acid) transaminase. GABA is a key amino acid in the central nervous system, being the main inhibitory neurotransmitter. *Although it is technically an amino acid, it is not incorporated into protein*. It is an excitability factor, operating through GABA receptors, in the nervous system, and GABA can be acted upon by **GABA transaminase**:

$$\alpha\text{-ketoglutarate} + 4\text{-aminobutanoic acid} \rightleftharpoons \text{glutamate} + \text{succinic semialdehyde}$$
$$\text{succinic semialdehyde} \rightarrow \text{succinic acid (oxidation reaction)} \rightarrow \text{TCA cycle} \rightarrow \text{energy}$$

This enzyme is one control on the GABA concentration, and GABA can be converted to a form that serves as an energy source. Moreover, the glutamate produced in the reaction also can be converted back to α-ketoglutarate (e.g., **glutamate-oxaloacetate transaminase**) that can enter another round of GABA transaminase activity, or it can enter the TCA cycle to serve as an energy source.

## TRANSAMIDATION

Transamidation is an enzyme-catalyzed reaction. Transamidases catalyze the formation of a covalent bond between a free amine group (e.g., lysine bound to a protein or peptide) and the gamma-carboxamide group (e.g., glutamine bound to a protein or peptide) In this case, the **transamidase** is a **transglutaminase**. A simple transamidase reaction is:

$$
\begin{array}{cc}
 & \mathbf{H} \\
 & | \\
\mathbf{NH_2} & \mathbf{N-R} \\
| & | \\
\text{Glu} + R-\mathbf{NH_2} \rightleftharpoons & \text{Glu} + \mathbf{NH_3} \\
(1) & (2)
\end{array}
$$

In this reaction, the substrate (1) is glutamine and the product (2) is the glutamine cross-reacted with the initial amine containing an R group. The same reaction could apply to lysine as the substrate (1). These enzymes require $Ca^{2+}$ which complexes substrate (1) to the enzyme. Transamidinases catalyze the formation of **γ-glutamyl-ε-lysine bonds** that may be involved in tissue healing (Fig. 13.8).

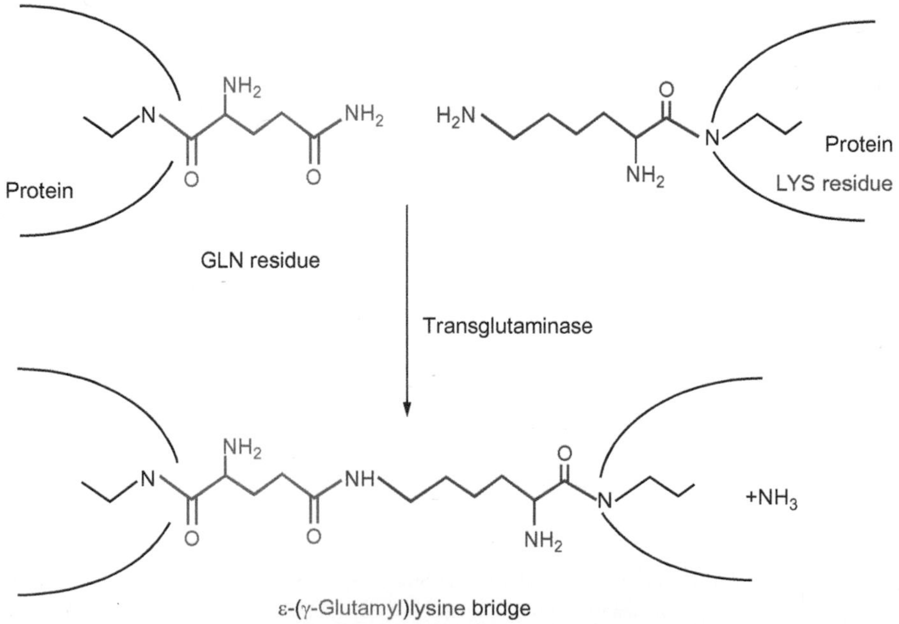

ε-(γ-Glutamyl)lysine bridge

**FIGURE 13.8** The transglutaminase reaction as an example of **transamidation**. At the top (*red*) glutamine, attached to a protein or peptide, in proximity to a lysine residue (*blue*) attached to a protein or peptide, forms an ε-(γ-glutamyl) **lysine bridge** catalyzed by **transglutaminase**. This is an example of the formation of a biological barrier. The bridge can be hydrolyzed to form glutamate and $NH_3$.

There are eight transglutaminases (they are calcium-dependent mammalian enzymes): **Factor VIII** which is the fibrin-stabilizing factor in blood coagulation; three enzymes in skin: **keratinocyte transglutaminase, epidermal transglutaminase**, and **transglutaminase MX**; a widely distributed transglutaminase, **tissue transglutaminase; transglutaminase MZ** in testis and lung; **prostate transglutaminase** and **transglutaminase MY**.

In pulmonary (arterial) smooth muscle cells, serotonin enters the cell through the **serotonin transporter (SERT)** and becomes transamidated to small GTPases, such as **RhoA** that activates **ROCK (rho-associated kinase)**.

Transamidation is important in the synthesis of **aminosugars**. In particular, the C5 ($\varepsilon$) amide of glutamine is contributed to fructose-6-phosphate to form glutamate and **glucosamine-6-phosphate**. Aminosugars are found in certain **proteoglycans**, probably on human chondroitin, but most aminosugars are found in plants and bacteria. Interestingly, there is a human testis **glucosamine-6-phosphate deaminase** ($\sim 33$ kDa) found in human **sperm** that seems to show an oscillation-inducing ($Ca^{2+}$ oscillations) activity in human eggs.

Most important is the transamidase reaction involved in the **glycosylphosphatidylinositol (GPI)** anchoring of proteins to a cellular membrane. The C-terminal amino acid of a protein is attached to the GPI, and the protein is stably anchored to the outer leaflet of the lipid bilayer membrane. Although many proteins are anchored to a membrane by transmembrane hydrophobic polypeptides, the GPI system is the mechanism for many attachments of proteins (including the **Prion protein**) to a membrane. A schematic of the structure of a **GPI-anchored protein** and the transamidase reaction involved is shown in Fig. 13.9.

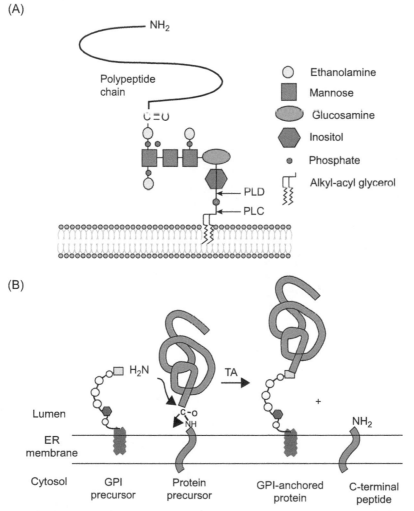

**FIGURE 13.9** (A) Drawing of a GPI-anchored protein to a membrane. The polypeptide chain is linked covalently through its C-terminus to the GPI structure (ethanolamine phosphate-mannose₃—glucosamine-phosphatidyl inositol). The first and third mannose residues contain additional ethanolamine phosphate groups. Sites of potential cleavage by **phospholipase D (PLD)** and **phospholipase C (PLC)** are indicated. (B) Addition of the **GPI anchor** to a protein. The C-terminal hydrophobic peptide transiently tethers the polypeptide chain in the **rough ER membrane** following its translation and translocation into the lumen. Sequential addition of sugars builds the GPI anchor along with **phosphoethanolamine** and **phosphatidylinositol**. The **transamidase** complex (TA) cleaves the polypeptide chain between the first and second mannose residues and at the same time adds the preformed GPI anchor. The released C-terminal peptide is subsequently degraded. *Reproduced from http://what-when-how.com/proteomics/gpi-anchors-proteomics/.*

Human GPI transamidase is a complex enzyme consisting of five subunits with portions of each transiting the ER membrane and portions extending into the lumen of the ER and other portions extending into the cellular cytosol (Fig. 13.10).

It has been suggested that there is a conserved proline residue in the last transmembrane segment of GAA1 creating a **hinge region** that is the structural basis for the interaction of GPI precursor and protein precursor (Fig. 13.9B).

## DEAMINATION

Deamination of amino acids, mainly serine and threonine, is catalyzed by either **serine dehydratase** or **threonine dehydratase** (these enzymes may also be referred to as Ser or Thr deaminase, Ser or Thr dehydratase, or Ser or Thr ammonia lyase). These reactions are nonoxidative. The coenzyme for these enzymes is **pyridoxal phosphate (PLP)** and the reactions catalyzed are:

$$\text{L-serine} \rightarrow \text{pyruvate} \ + \ NH_3(\text{or } NH_4^+)$$

and

$$\text{L-threonine} \rightarrow \alpha\text{-ketobutyrate (or 2-oxobutanoate)} \ + \ NH_3(\text{or } NH_4^+)$$

Both enzymes are classified as **ammonia lyases**.
A more detailed reaction is shown in Fig. 13.11.

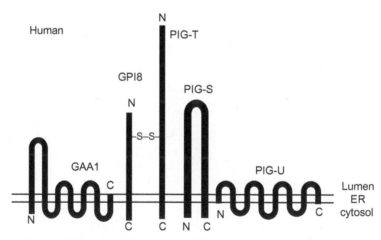

**FIGURE 13.10** Representation of **human GPI transamidase**. GAA1, GP18, and PIG-T/GPI16 are common components across species, whereas PIG-S and PIG-U are specific to the human. The two horizontal lines represent the membrane; the N- and C-termini are indicated. *Reproduced from http://www.pnas.org/content/100/19/10682/F6.expansion.html.*

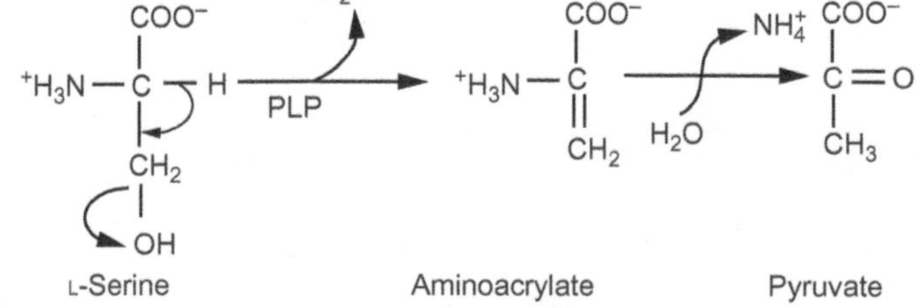

**FIGURE 13.11** The **serine dehydratase** reaction. Serine is deaminated with the involvement of PLP. Water is removed from the amino acid (*arrow*), followed by removal of the amine to produce an **ammonium ion ($NH_4^+$)**, and **pyruvate** is formed by the hydration of the intermediate **aminoacrylate**. The analogous reaction occurs with L-threonine and **threonine dehydratase** generating the product 2-ketobutyrate (2-oxobutanoate; α-ketobutyrate). Serine dehydratase is located in hepatocytes and the enzyme functions mainly to provide a substrate, pyruvate, for gluconeogenesis to form glucose from amino acids.

The enzyme, **histidase** (histidine ammonia lyase), removes the amino group from histidine to form ***trans*urocanic acid**. The enzyme is located in the liver and skin. In skin, UV light causes the isomerization of *trans*urocanic acid into ***cis*-urocanic acid**. The reactions are shown in Fig. 13.12.

Although the above reactions are significant, **glutamic acid** is the major amino acid for deamination reactions. Other amino acids can be deaminated but these reactions occur mainly by transamination and deamination of glutamate where glutamate recycles. Thus, any amino acid can react with α-ketoglutarate to form glutamate plus the keto acid analog of the original amino acid. Then, glutamate can be deaminated to produce ammonia (from the original amino acid) that is cleared through the **urea cycle** and excreted. Since glutamate is the product of many transamination reactions, it is the main substrate for **oxidative deamination** (by glutamate dehydrogenase in liver and other tissues). The **glutamate dehydrogenase (GDH)** reaction is:

$$\text{glutamate} + H_2O + NAD^+ \rightarrow \alpha\text{-ketoglutarate} + NH_3 + NADH + H^+$$

$NH_3$ is converted to urea through the urea cycle and excreted by the kidney. As mentioned previously, GDH is an allosteric enzyme where ADP is an activator and GTP, an inhibitor (Fig. 13.13).

In the chapter on nucleic acids, deamination reactions for **cytosine** (to uracil), **5-methylcytosine** (to thymine), **guanine** (to xanthine), and **adenine** (to hypoxanthine) have been discussed.

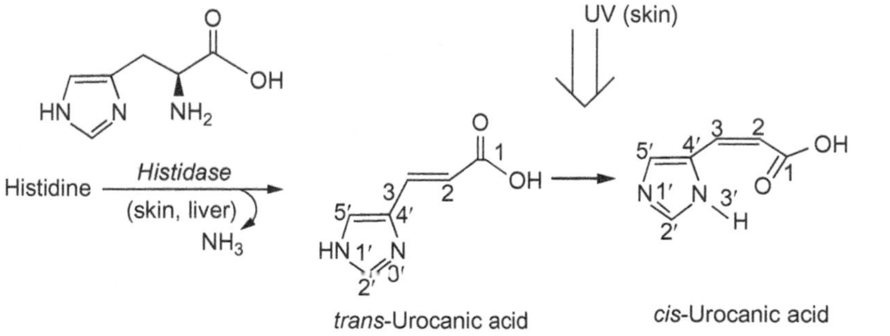

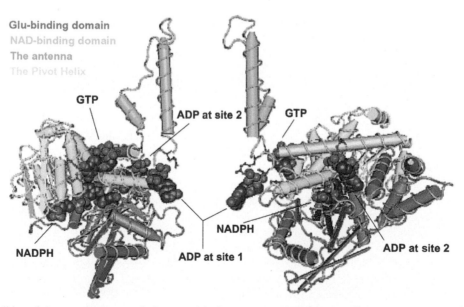

**FIGURE 13.12** The histidase reaction in liver and skin produces *trans*urocanic acid and ammonia. In the skin, the action of UV light is to cause the isomerization of *trans*urocanic acid (the product in the absence of UV light) into the *cis*-form.

**FIGURE 13.13** Rendition of the crystal structure of glutamate dehydrogenase generated from two X-ray structures, one with ADP bound to the enzyme and one with **GTP** bound to the enzyme. Individual domains are color-coded (see key at *upper left* of figure). The cofactor and regulatory molecules are shown in *spheres*; the remainder of the enzyme structure is shown in *ribbons*, *lines*, and *barrels*. **Glutamate** binds to the structures in *purple*. Note that there are two binding sites for allosteric regulator ADP. *Reproduced from http://en.wikipedia.org/wiki/File:GLUD1_f1.png.*

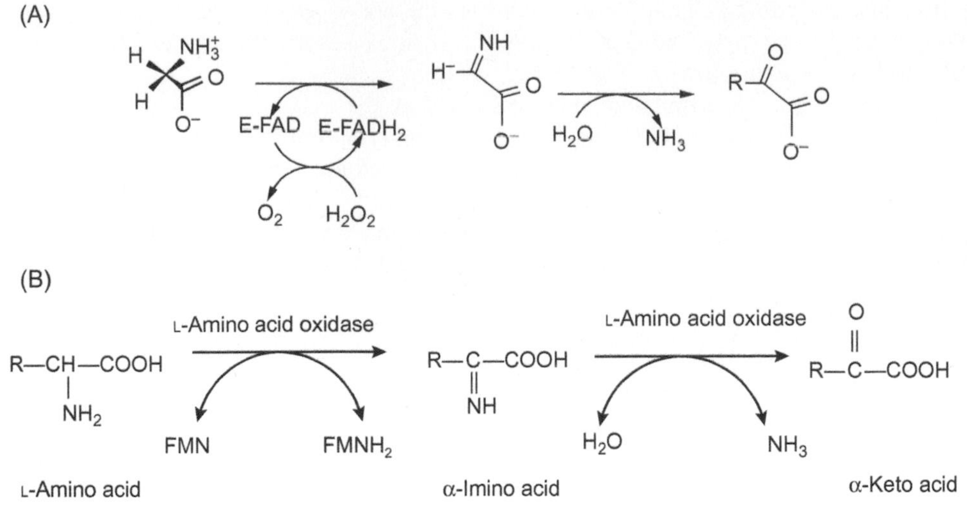

**FIGURE 13.14** (A) Reaction mechanism for D-amino acid oxidase. *E-FAD*, enzyme bound flavin adenine nucleotide. (B) A reaction mechanism for L-amino acid oxidase. The mechanism of L-amino acid oxidase is the same as for D-amino acid oxidase.

## OXIDATION OF AMINO ACIDS

Liver and kidney **peroxisomes** contain **D-amino acid oxidase** (as well as L-amino acid oxidase) that directly oxidizes D-amino acids (D-amino acids are not incorporated into proteins). The coenzyme is **flavin-adenine dinucleotide (FAD)**, and the enzyme exhibits wide specificities. An **imino acid** is produced with $H_2O_2$, and ammonia is removed from the enzyme intermediate as shown in Fig. 13.14. Plants and bacteria do not have this enzyme, so many D-amino acids may occur in the diet. The human also has **amino acid racemase** that converts L-amino acids to D-amino acids.

There are small amounts of L-amino acid oxidase in the liver and kidney peroxisomes. This would come into play when there is an excess of amino acids for all other pathways.

An extensive loss of **motor neurons** occurs in **amyotrophic lateral sclerosis (ALS)**. This may be due to aberrant excitability of motor neurons that contributes to their cell death. There is a mutation in the D-amino acid oxidase gene that causes the downregulation of serine degradation resulting in the accumulation of D-serine, and these effects are associated with familial ALS. D-Serine is a coagonist (besides glycine) of the **N-methyl-D-aspartate (NMDA) receptor** that causes excitability in the brain. **Serine racemase** is also present in the human brain, and it generates D-serine from L-serine. The expression of serine racemase is altered in some mental diseases, such as **schizophrenia**.

## AMINO ACID RACEMIZATION

High levels of **D-serine** in the human brain need to be regulated due to its coagonist action on NMDA receptors resulting in damaging excitability that can lead to cell death. Although **L-amino acid racemase**, the enzyme converting an L-amino acid to a D-amino acid, contributes to the metabolism of glycine, serine, threonine, and cysteine, it is important in the brain for its contribution to the level of D-serine (large amounts of D-serine occur in the hippocampus and *corpus callosum*). Inside the cell, L-serine can be converted to D-serine by the racemase and D-serine can either be transported out of the cell, or it can be converted to pyruvate and ammonia by an **α,β-elimination**. The extracellular D-serine can form a complex with the membrane **NMDA receptor** or can be taken up again by the original cell (or transported out of the brain). This metabolism is shown in Fig. 13.15 and includes the **L-serine racemase reaction mechanism** as well as the mechanism of α,β-elimination to form pyruvate and ammonia ($NH_3$ or $NH_4^+$).

L-Serine racemase would seem to be a good therapeutic target for some of these diseases based on high brain D-serine levels. There would have to be a site different from the coenzyme-binding site (because there are many PLP-requiring enzymes in the body, and there would be the added problem of crossing the blood—brain barrier).

## L-AMINO ACID DECARBOXYLATION

As for other amino acid metabolizing enzymes, the L-amino acid decarboxylases use **pyridoxal phosphate (PLP)** as coenzyme. Most of the amino acids have a decarboxylase that will remove $CO_2$ from their structures. An

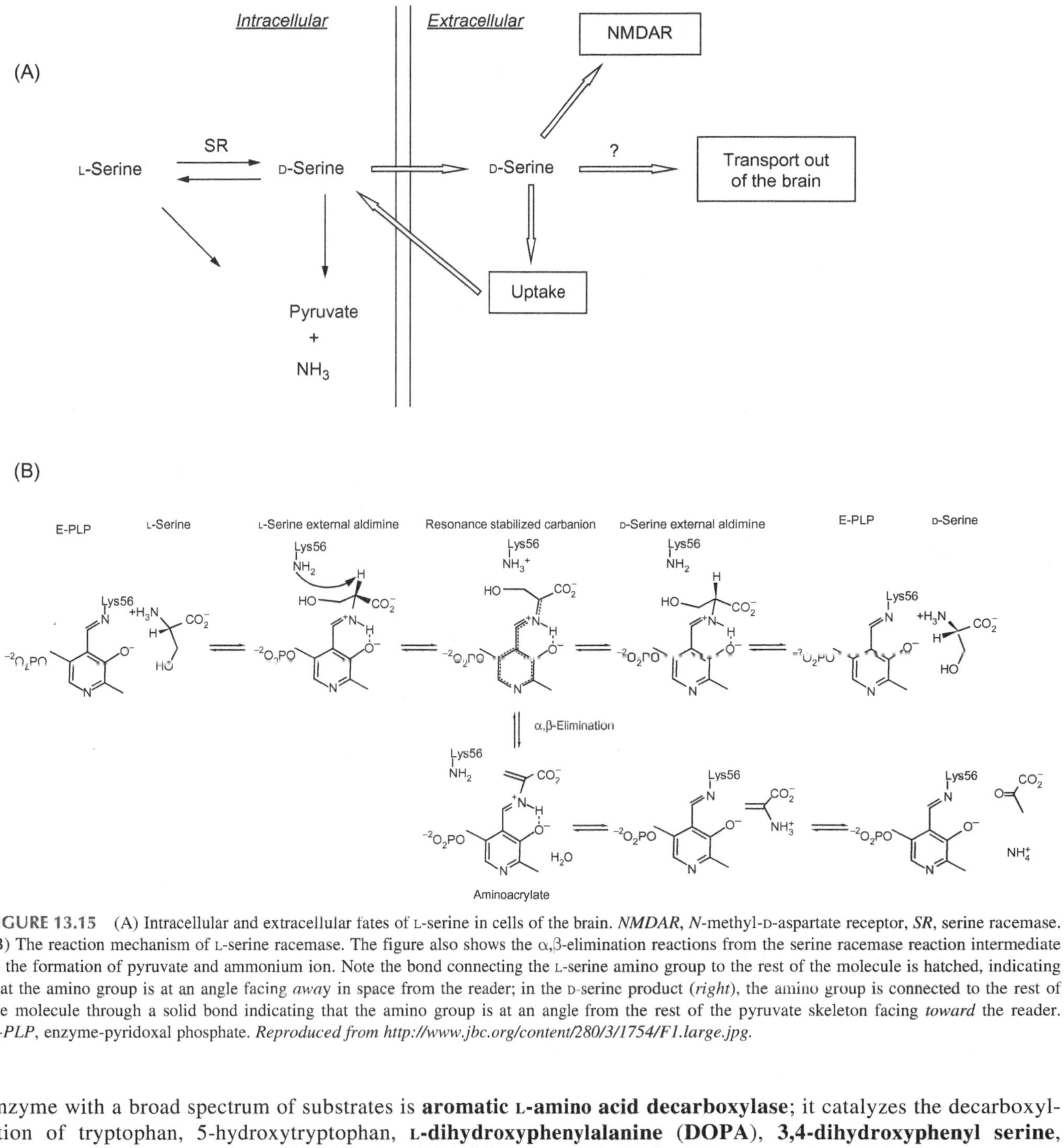

(A)

(B)

**FIGURE 13.15**   (A) Intracellular and extracellular fates of L-serine in cells of the brain. *NMDAR*, *N*-methyl-D-aspartate receptor, *SR*, serine racemase. (B) The reaction mechanism of L-serine racemase. The figure also shows the α,β-elimination reactions from the serine racemase reaction intermediate to the formation of pyruvate and ammonium ion. Note the bond connecting the L-serine amino group to the rest of the molecule is hatched, indicating that the amino group is at an angle facing *away* in space from the reader; in the D-serine product (*right*), the amino group is connected to the rest of the molecule through a solid bond indicating that the amino group is at an angle from the rest of the pyruvate skeleton facing *toward* the reader. *E-PLP*, enzyme-pyridoxal phosphate. *Reproduced from http://www.jbc.org/content/280/3/1754/F1.large.jpg.*

enzyme with a broad spectrum of substrates is **aromatic L-amino acid decarboxylase**; it catalyzes the decarboxylation of tryptophan, 5-hydroxytryptophan, **L-dihydroxyphenylalanine (DOPA)**, **3,4-dihydroxyphenyl serine**, **tyrosine** (ortho-tyr, meta-tyr, and para-tyr), **phenylalanine**, and **histidine**. This enzyme would be important in the kidney and brain and, likely, other organs. The general reaction is:

$$\text{L-aromatic amino acid} \rightleftharpoons \text{aromatic amine} + CO_2$$

Many important amino acid derived amines are generated from the decarboxylase reactions as will be seen later on. There is an associated oxygen-consuming reaction that occurs with **DOPAmine** but which does not occur with most of the other amino-acid-derived amines:

$$\text{aromatic amine} + \tfrac{1}{2} O_2 \rightleftharpoons \text{aromatic aldehyde} + NH_3$$

A list of some **amino acid decarboxylases** is presented in Table 13.2.

**TABLE 13.2** A List of Some L-Amino Acid Decarboxylases With Pyridoxal Phosphate as Coenzyme

| Enzyme | Substrate(s) | Products |
|---|---|---|
| Aspartate α-decarboxylase | L-Aspartate | β-Alanine + $CO_2$ |
| Valine decarboxylase | L-Valine (or L-leucine) | 2-Methylpropanamine + $CO_2$ |
| Glutamic acid decarboxylase | L-Glutamate (in brain: L-cysteate*, 3-sulfino-L-alanine, L-aspartate) | 4-Aminobutanoate + $CO_2$ |
| Lysine decarboxylase | L-Lysine (hydroxyl-L-lysine) | Cadavarine + $CO_2$ |
| Arginine decarboxylase | L-Arginine | Agmatine + $CO_2$ |
| Histidine decarboxylase | L-Histidine (PLP or pyruvate as coenzyme) | Histamine + $CO_2$ |
| Aromatic L-amino acid decarboxylase | Tryptophan (DOPA, hydroxy-tryptophan) | Tryptamine + $CO_2$ |
| Phenylalanine decarboxylase | L-Phenylalanine (tyrosine + other aromatic armino acids) | Phenylethylamine + $CO_2$ |
| Methionine decarboxylase | L-Methionine | 3-Methylthiopropanamine + $CO_2$ |

*2-amino-3-sulfoproprionate.

**Pyridoxal-phosphate-associated enzymes** have a wide variety of half-lives (of the apoenzyme). Apparently, the apoenzyme is more likely to be degraded than when the apoenzyme is in the holoenzyme form (associated with the coenzyme). Thus, one consideration is the rate of dissociation of PLP from the enzyme and whether this coincides with the **half-life** of the enzyme protein (the more rapid PLP dissociation ∼ the shorter the half-life of the apoenzyme).

## METABOLISM OF AMINO ACIDS TO ACTIVE SUBSTANCES

### Methionine

There are eight **essential amino acids** (must be obtained through the diet; not synthesized in the body). One of these is methionine. The others are as follows: tryptophan, lysine, phenylalanine, threonine, valine, leucine, and isoleucine. Histidine and arginine are considered to be essential in children. The **nonessential amino acids** are as follows: glutamate, glutamine, aspartate, asparagine, alanine, cysteine, tyrosine, proline serine, glycine (and ornithine). The genetic code specifies methionine (AUG) and cysteine (UGU, UGC) but not homocysteine or cystine (two oxidized cysteines joined by an −S−S− bond). Methionine and tryptophan (UGG) have single codons but the other amino acids have two or three codons. **Homocysteine** arises from the methyl donor, **S-adenosyl methionine (SAM)**. **Hyperhomocysteinemia (HHCE)** occurs in 5%−7% of the population. Patients with mild HHCE often have premature coronary artery disease accompanied by venous and arterial **thrombosis** in the third and fourth decile. Defects in the enzyme **cystathionine β-synthase** and B vitamin deficiency (especially folate, vitamin B12 or vitamin B6) can lead to HHCE. Moderate HHCE can be caused by chronic disease states and by certain drugs.

**Arterial endothelial cells** can be damaged by high levels of **blood homocysteine**, and smooth muscle growth also can be a response. This can generate the formation of **plaque**, and the blood-clotting mechanism can be disrupted as well, increasing the risk of heart attack or stroke. Apparently, normalizing the blood level of homocysteine does not reverse the potential risks. Polymorphisms in the gene for **methyltetrahydrofolate reductase** produce moderate HHCE.

Because most proteins start their translation with the start codon, AUG, methionine is the N-terminal amino acid of all eukaryotic proteins. It can occur in other positions in proteins, as well. Also, it is possible to remove the **N-terminal methionine** of the completed protein by posttranslational modification.

**Methionine** and **homocysteine** are interconvertible, and **cysteine** can be derived from homocysteine as shown in Fig. 13.16.

These are the **sulfur-containing amino acids**. Also, methionine is involved in the syntheses of cysteine (Fig. 13.16), **carnitine**, **taurine**, **glutathione (GSH)**, and cysteine as a source of sulfide (e.g., iron-sulfide clusters) by the **transsulfuration pathway** (this pathway is summarized in Fig. 13.16 in the reactions from homocysteine to cystathionine to cysteine). The synthesis of taurine proceeds according to the reactions: methionine → **cysteine** → cysteine sulfinate (catalyzed by cysteine dioxygenase) → hypotaurine → taurine (Fig. 13.17).

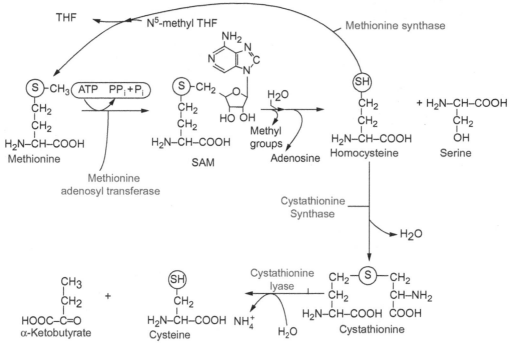

FIGURE 13.16 Metabolism of methionine for the synthesis of homocysteine and cysteine. *SAM*, *S*-adenosylmethionine; *THF*, tetrahydrofolate.

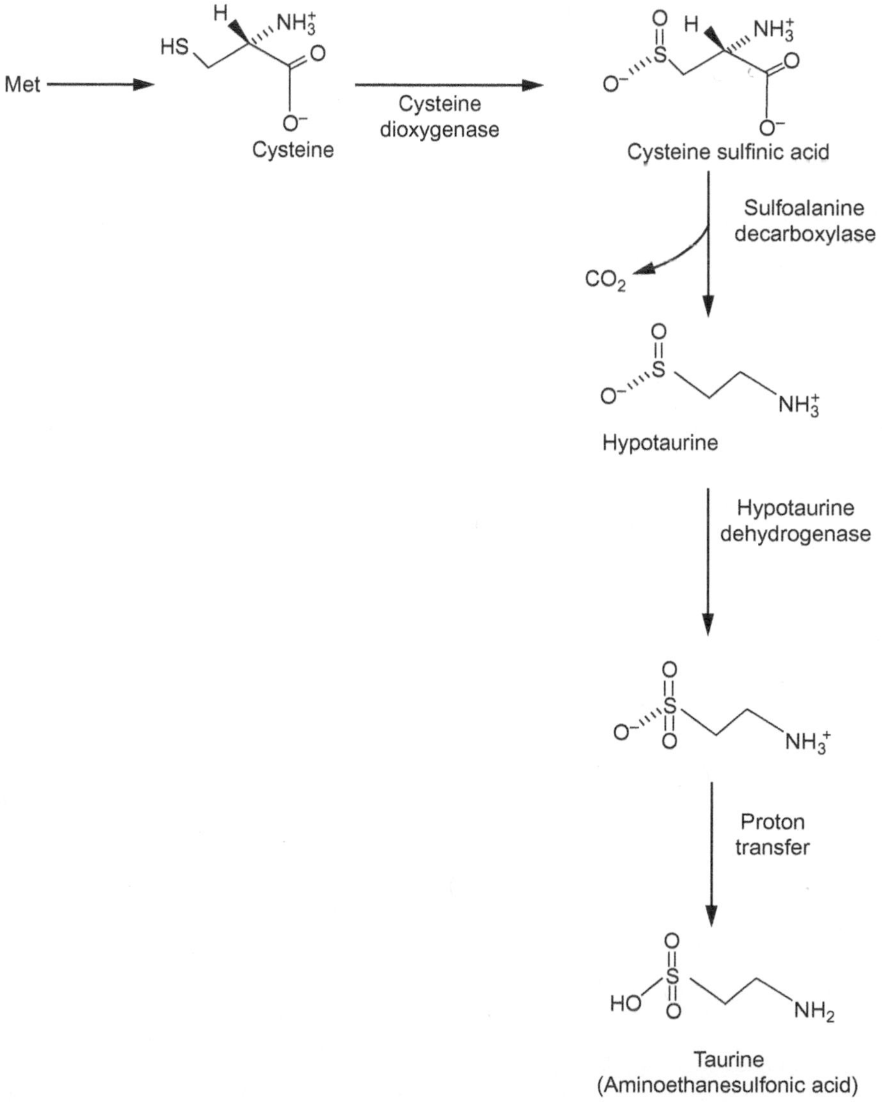

Met ⟶ [Cysteine] → (Cysteine dioxygenase) → [Cysteine sulfinic acid]

Cysteine sulfinic acid → (Sulfoalanine decarboxylase) → $CO_2$ → [Hypotaurine]

Hypotaurine → (Hypotaurine dehydrogenase) → [structure]

[structure] → (Proton transfer) → Taurine (Aminoethanesulfonic acid)

FIGURE 13.17 The synthesis of taurine in man.

Taurine is an amino-group-containing acid but is not considered as an amino acid and is not incorporated into protein. It is a major constituent of **bile** and occurs mainly in the large intestine. It is a conjugate with **bile acids** and has various other activities, such as antioxidant, modulator of calcium signaling, stabilizer of membranes, functioning in the retina, the central nervous system (CNS), and skeletal muscle. It is important in the development and functioning of **skeletal muscle** and is an inhibitor of **apoptosis**.

There is a **transporter for taurine**, a neurotransmitter called **TauT** whose highest affinity is for taurine and β-**alanine**. The transporter will transport 2−3 sodium ions inward in exchange for chloride ions moving outwardly. Taurine also acts as an **osmolyte** (a compound that affects osmosis) for cellular volume control. Normally, taurine is excreted in the urine, and the variable excretion rises when taurine is consumed in the diet.

Methionine, as *S*-adenosylmethionine (Fig. 13.16), is involved in the step-by-step methylation of **phosphatidylethanolamine** to form **lecithin (phosphatidylcholine)** and other phospholipids principally in brain, lung, and spleen. Phosphatidylcholine can be degraded to release **choline** that can be converted to **betaine** $[(CH_3)_3-N^+-CH_2-COO^-]$, and betaine together with homocysteine can reform methionine.

*S*-adenosylmethionine (SAM) is an important **methyl group donor**. **Cystathionine** is formed in the synthesis of cysteine from methionine (Fig. 13.16) where cystathionine is coverted to cysteine by **cystathionase** which cleaves cystathionine to cysteine and α-ketobuyrate. After decarboxylation, α-ketobutyrate is converted to **proprionylCoA**. Cystathionase is regulated by cysteine in a negative allosteric manner and cysteine also inhibits the expression of the gene for **cystathionine synthase**. Both cystathionine synthase and cystathionase contain **pyridoxal phosphate** as coenzyme.

The synthesis of glutathione (GSH) is shown in Fig. 13.18.

Glutathione (glutamylcysteinylglycine, GSH) is formed directly from the amino acid components with energy from two ATP molecules (Fig. 13.18) rather than by protein synthesis. The disulfide bond in oxidized GSH (GSSG) is created by the oxidation of two cysteine residues (from two molecules of GSH) to form a **cystine** (containing the −S−S− bridge). GSH is an important cellular redox agent. Also, it is conjugated to a variety of drugs and **xenobiotics** (components that do not occur naturally in the body) by the family of GSH *S*-transferase enzymes. The GSH conjugate usually

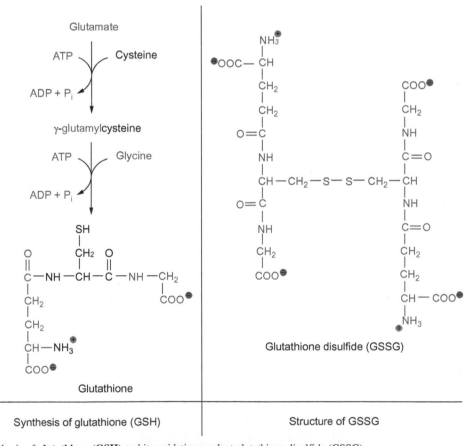

FIGURE 13.18   Synthesis of **glutathione (GSH)** and its oxidation product glutathione disulfide (GSSG).

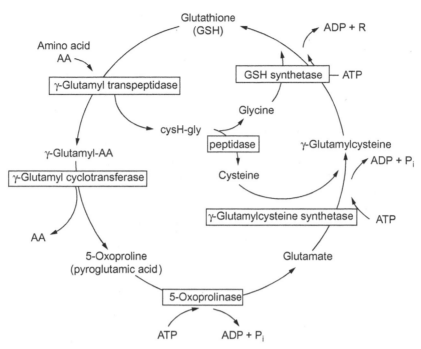

**FIGURE 13.19**   The γ-glutamyl cycle.

solubilizes drugs and xenobiotics so that they can be excreted and, in this way, acts naturally to prevent cancer formation by some xenobiotics that could bind to and damage DNA.

GSH is involved in the **γ-glutamyl cycle** (Fig. 13.19) that can aid the transport of an amino acid from outside the cell to the cell interior.

The first step of the cycle is catalyzed by **γ-glutamyl transpeptidase** that transfers the γ-glutamyl group of **GSH** to an amino acid, a peptide or water (to form glutamate) whose reaction is:

$$\gamma\text{-L-glutamyl-peptide} + \text{amino acid} \rightleftharpoons \text{peptide} + \gamma\text{-L-glutamyl-amino acid}$$

This enzyme occurs in the cell membrane of many tissues (liver, kidney, pancreas, bile duct, heart, brain, spleen, and seminal vesicles). The **γ-glutamyl cycle** is the pathway for the synthesis and degradation of GSH functioning in the detoxification of drugs and **xenobiotics**.

One of the leukotrienes $C_4$ ($LTC_4$) contains GSH. Its synthesis starts with **arachidonic acid** (20 carbons) that is a substrate of **lipoxygenase** to generate **5-hydroperoxyeicosatetraenoic acid** (**5-HPETE**). The same enzyme converts 5-HPETE to **leukotriene $A_4$** (**$LTA_4$**) and **glutathione $S$-transferase** converts, with GSH as substrate, $LTA_4$ to leukotriene C4 (**$LTC_4$**) that contains GSH (Fig. 13.20).

$LTC_4$ has contractile activity on the tissues of the airway and, experimentally, this leukotriene is 20,000 times more potent than acetylcholine on bronchial and tracheal smooth muscle contraction indicating its importance in **airway smooth muscle**.

## Phenylalanine and Tyrosine

Phenylalanine is an essential amino acid meaning that it cannot be made in the body and must be ingested in the diet. Tyrosine is a nonessential amino acid and can be formed by the hydroxylation of phenylalanine in the liver when the intake of tyrosine in the diet is low. In a diet low in tyrosine, as much as half the ingested phenylalanine may be converted to tyrosine in the body. Conversely, if the diet is rich in tyrosine, the requirement for phenylalanine could be reduced by 50%. Tyrosine is metabolized to important substances in the body, such as **catecholamines** (epinephrine and norepinephrine) and the tissue pigment, **melanin**, as well as other metabolites. **Phenylalanine hydroxylase (PAH)** is expressed in the liver and kidney. Mutations in the gene that expresses PAH can lead to **phenylketonuria**, a serious metabolic disease. PAH is the enzyme that metabolizes excess phenylalanine. PAH is characterized as a **mixed function oxidase** because it incorporates one atom of oxygen into the hydroxyl of the product, tyrosine, and another atom of

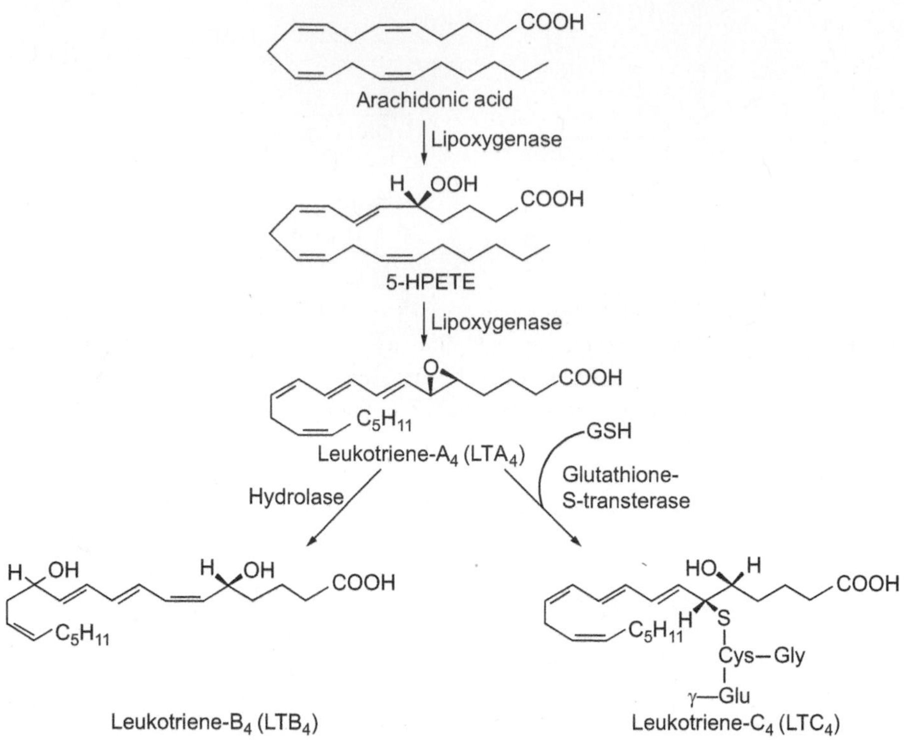

**FIGURE 13.20**    Synthesis of leukotriene $C_4$ ($LTC_4$). Arachidonic acid is released from the cell membrane by phospholipase $A_2$. *HPETE*, hyroperox-yeicosatetraenoic acid. The amino acids contained in the GSH component are released upon degradation of $LTC_4$.

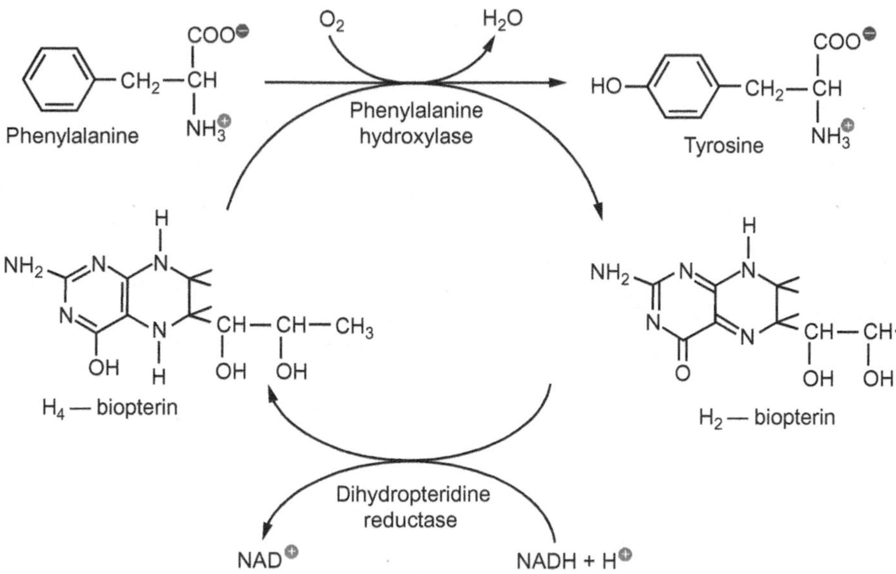

**FIGURE 13.21**    Conversion of L-phenylalanine to L-tyrosine catalyzed by phenylalanine hydroxylase, a mixed function oxidase. Hydrogens from the coenzyme are donated to an oxygen atom for the formation of the hydroxyl group on tyrosine and for the formation of water.

oxygen into water. Its coenzyme is **tetrahydrobiopterin** (**BH4**) which donates two hydrogen atoms in the PAH reaction while BH4, after donating two hydrogens, is in the less reduced state called **dihydrobiopterin** (**BH2**). BH2 is converted back to BH4 by **dihydrobiopterin reductase** with NADH + H$^+$ as coenzyme (Fig. 13.21).

The enzymatic mechanism is substantially more complex than shown in this figure because the coenzyme contains an atom of nonheme iron that forms a superoxide. Fe(III) is reduced to Fe(II) followed by a decrease in affinity for two water molecules near the iron atom. BH4 binds to the enzyme followed by the binding of phenylalanine; then dioxygen (O:O) binds where one of the two water molecules was previously located forming **peroxy-BH4**. The oxygen−oxygen bond breaks down followed by the breakdown of peroxyhydrobiopterin to hydroxybiopterin plus an activated **peroxypterin iron intermediate**. In the final step, L-tyrosine and hydroxytetrabiopterin (BH4OH) are released.

There are more than 400 known mutations in children of the gene for PAH leading to a partially active or inactive PAH and generating phenylketonuria. In this disease, phenylalanine cannot be converted to tyrosine and blood levels of phenylalanine rise. The severity of the disease is reflected by the **serum values for phenylalanine**: normal $=1$ mg/dL (0.061 mM) which is benign hyperphenylalaninemia (HPA); variant HPA $=4-10$ mg/dL (0.24−0.605 mM); classic phenylketonuria $=10-20$ mg/dL ($>20$ mg/dL or $>1.21$ mM). When there is a defect in generating appropriate amounts of the cofactor, BH4 (defects in **GTP cyclohydrolase**, the first enzyme in the pathway and **pyruvoyl tetrahydropterin synthase**, the second enzyme in the pathway), HPA (hyperphenylalaninemia) occurs only in a small percentage of the cases but these are not controlled by limiting the phenylalanine in the diet, a method that is in practice for HPA generated by a deficiency of PAH. In HPA, phenylalanine is transaminated to **phenylpyruvate** (phenylalanine + α-ketoglutarate = phenylpyruvate + glutamate) instead of tyrosine. Accumulation of phenylpyruvate can lead to **mental retardation** (phenylketonuria, PKU) in infants. When phenylalanine is present in excess over the need for tyrosine, it is transaminated with α-ketoglutarate to phenylpyruvate and glutamate, as in PAH deficiency but, in this case, the formation of **phenylpyruvate** would be insufficient to cause problems. Also, when phenylalanine is in excess over the need to form tyrosine, it can be converted to **phenylethylamine** and other metabolites, such as **phenylacetic acid**, **phenylacetylglutamine**, and **phenylactic acid**.

## Formation of Catecholamines DOPA (Dopamine, Norepinephrine, and Epinephrine)

Catecholamines are formed from tyrosine. Both **norepinephrine** (noradrenalin) and **epinephrine** (adrenalin) are secreted from nerve endings. The main mediator at postganglionic sympathetic endings is norepinephrine. Catecholamine synthesis occurs in catecholamine-secreting neurons and in the **adrenal medulla** (the adrenal medulla consists of modified neurons essentially lacking axonal structures) where the tyrosine precursor is imported from the extracellular space (probably through a **tyrosine transporter (LAT1)**). The first step in tyrosine conversion is catalyzed by **tyrosine hydroxylase** in a mechanism resembling that of PAH including having **BH4** as coenzyme (Fig. 13.22).

The overall conversion of phenylalanine to tyrosine and then to the catecholamines is shown in Fig. 13.23.

In addition to the adrenal medulla, some areas of the brain have these reactions ending at the level of **dopamine**. In this sequence, the product of the tyrosine hydoxylase reaction is DOPA (3,4-dihydroxyphenylalanine). DOPA is converted to dopamine by the enzyme DOPA decarboxylase (aromatic amino acid decarboxylase, a **pyridoxal phosphate**-containing enzyme). DOPAmine is converted to norepinephrine by **DOPAmine β-hydroxylase**. This enzyme is an oxidase that has **ascorbate** as a cofactor and, unlike the other cytoplasmic enzymes in this pathway, it is membrane-bound requiring that the synthesis of norepinephrine takes place inside a vesicle in noradrenergic nerve terminals or within the **chromaffin cell** of the adrenal medulla. The subsequent conversion of norepinephrine to epinephrine is catalyzed by **phenylethanolamine N-methyl transferase** (**PNMT**) with **S-adenosylmethionine** (**SAM**) as the methyl donor. PNMT

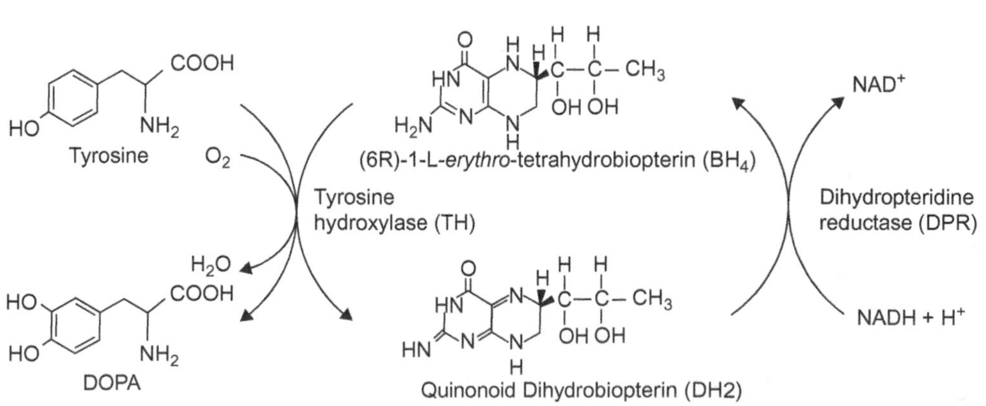

**FIGURE 13.22**  The tyrosine hydroxylase reaction. In the formation of **DOPA** from tyrosine, the second hydroxyl on DOPA derives from molecular oxygen ($O_2$) to produce the hydroxyl and a molecule of water; accordingly, tyrosine hydroxylase is a **mixed function oxidase**. Hydrogens donated from the coenzyme, BH4, and its subsequent reduction from BH2 (designated DH2 here) is similar to the phenylalanine hydroxylase mechanism.

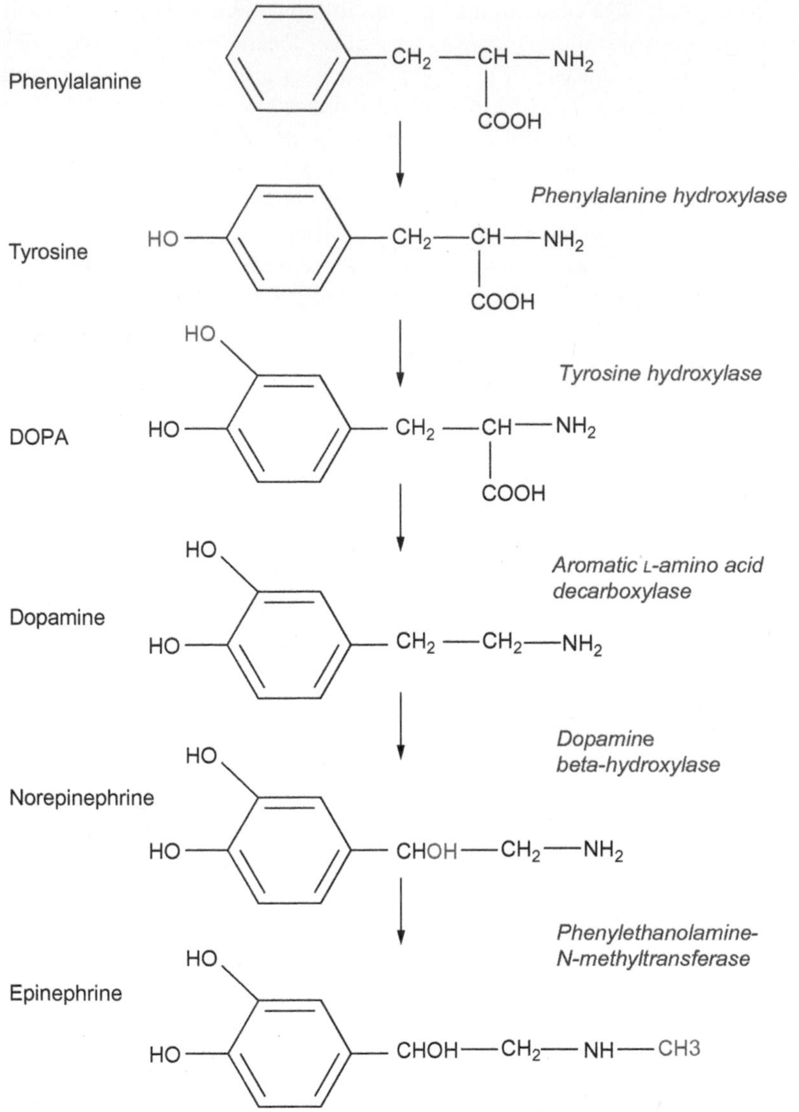

**FIGURE 13.23** Conversion of L-tyrosine to epinephrine and other catecholamines in the adrenal medulla. *Reproduced from http://www.pharmacorama. com/en/Sections/Catecholamines_3.php.*

is induced by **cortisol** secreted by the adrenal cortex. Cortisol, through vascular connections, flows through the adrenal medulla on its way to the general circulation, thus a stress event causes the release of cortisol and the synthesis and release of epinephrine from chromaffin cells.

The degradation products of the catecholamines: dopamine, epinephrine, and norepinephrine are shown in Fig. 13.24.

In addition to the degradation system shown in Fig. 13.24, an enzyme named **renalase** (with a flavin adenine dinucleotide, FAD, coenzyme) has been discovered in blood and oxidizes catecholamines. The enzyme also is located in heart and skeletal muscles and in liver cells. Importantly, catecholamines have short half-lives in the bloodstream on the order of a few minutes.

## Melanin

Melanin is derived from tyrosine, and more directly from **DOPA**. Melanin is a family of pigments having different colors. In this case, DOPA is the product of the enzyme, **tyrosinase (diphenol oxidase)**. Differently from tyrosine hydroxylase, tyrosinase, a copper enzyme, uses molecular oxygen directly [without tetrahydrobiopterin (BH4)] as is the

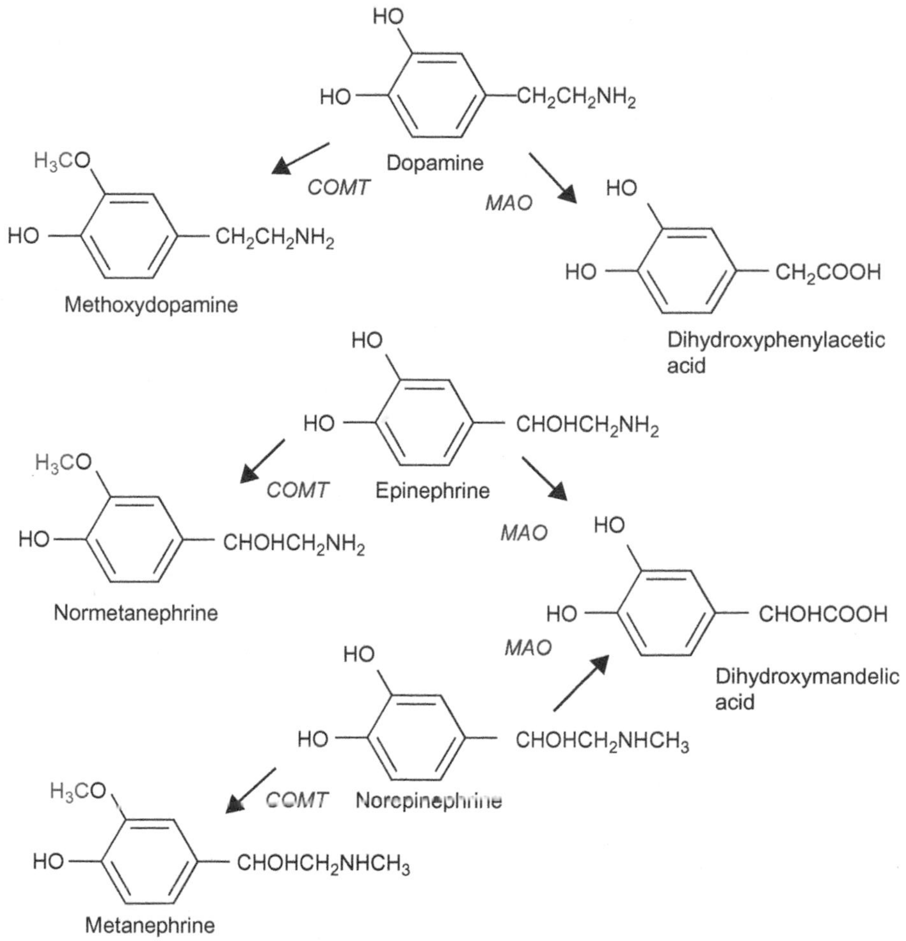

**FIGURE 13.24** Degradation products of catecholamines. *COMT*, catecholamine-*O*-methyl transferase; *MAO*, monoamine oxidase.

case with tyrosine hydroxylase) to form DOPA from tyrosine. The synthesis of melanin occurs in the melanocyte, and the reactions starting with tyrosine are shown in Fig. 13.25.

After the formation of **DOPA** from tyrosine, the further conversion of DOPA to **DOPAquinone** follows. Then, a number of intermediates are formed ending in **indolequinone** that polymerizes to form **melanin**. The more common product is **eumelanin** (*brown*) but in the presence of cysteine, **pheomelanin** can be formed (*red* to *yellow*). Melanin is formed primarily in the **melanocyte**, located in the inner layers of the skin where melanin and **carotene** blend to produce the skin color as well as the color in the eyes and hair. Red hair is produced by pheomelanin in spherical **melanosomes** (melanin granules). Black-colored melanin is formed in oblong melanosomes. Melanin granules are distributed uniformly in the skin cell in order to absorb UV rays from the sun and protect, at least partially, from injurious rays.

## Blockages in the Metabolism of Phenylalanine and Tyrosine and Disease

When the conversions of phenylalanine to tyrosine and subsequent metabolites of tyrosine are blocked, various disease conditions occur. When **phenylalanine hydroxylase** (PAH) is nonfunctional, phenylalanine is converted to **phenylpyruvate** resulting in **phenylketonuria**. Blockage of the conversion of tyrosine to melanin leads to the melanin-deficient condition of **albinism**. When the conversion of tyrosine to iodinated derivatives is blocked, there is insufficient thyroid hormone during development, and **cretinism** can occur. **Tyrosinosis** occurs when the conversion of hydroxyphenylpyruvate to homogentisic acid is blocked. **Alkaptonuria** is manifested when the conversion of homogentisic acid to maleylacetoacetate is blocked. The blockages are usually created when the genes for the enzymes involved have suffered mutations so that the encoded proteins are poorly functional or nonfunctional. These conditions are summarized in Fig. 13.26.

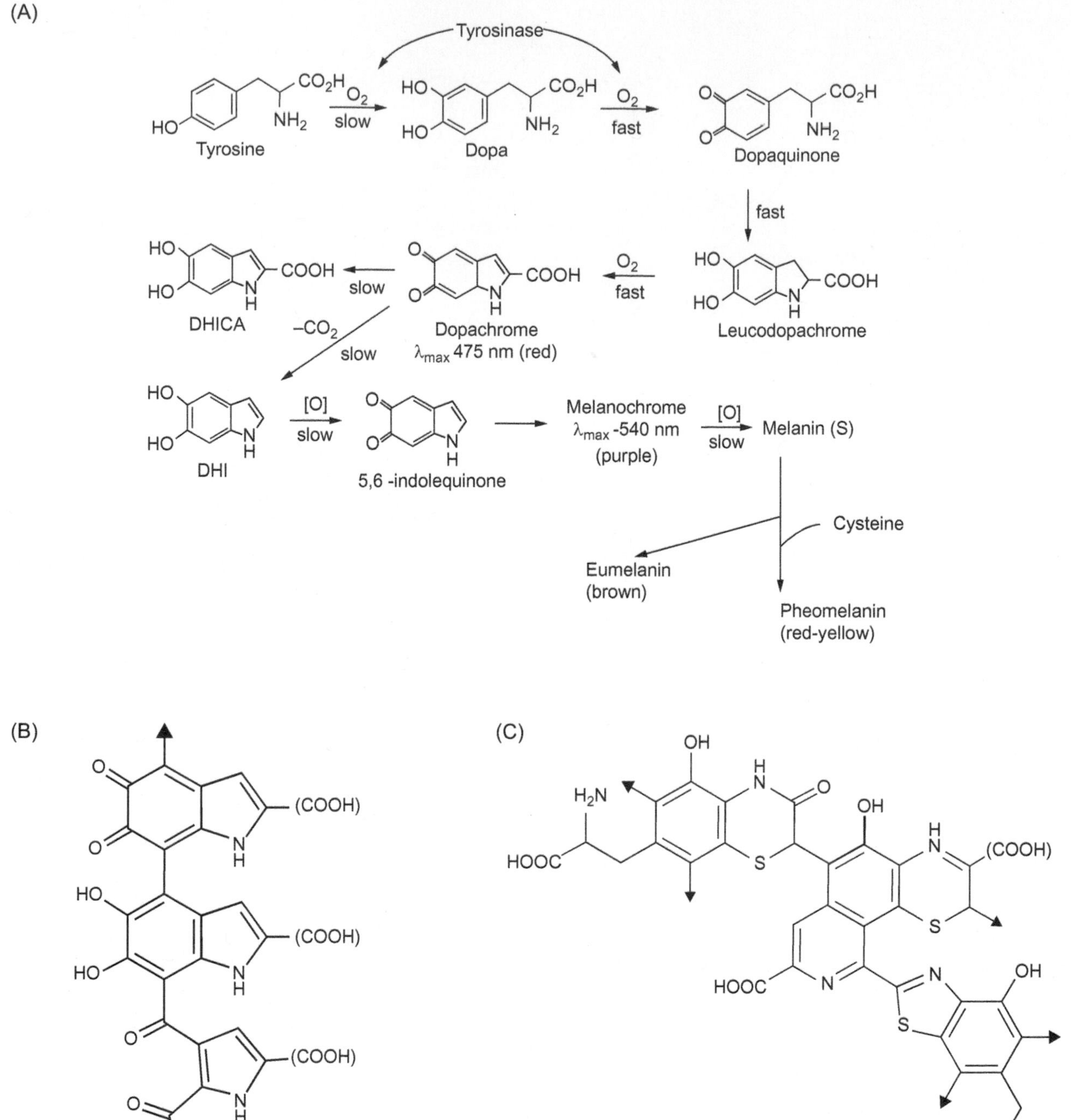

FIGURE 13.25 Synthesis of melanin from tyrosine. *DHI*, dihydroxyindole; *DHICA*, dihydroxyindole catecholamine. In the presence of cysteine another pigment called **pheomelanin** can be formed that has a *red-yellow* color compared to **eumelanin** that has a *brown* color. Melanins are the pigments that produce the color of the eye. The arrows at the top of the structures of eumelanin and pheomelanin indicate the point at which polymerization can occur.

# Tryptophan

Tryptophan is an essential amino acid. In addition to its incorporation into proteins, this amino acid is converted into the neurotransmitters, **serotonin**, and **melatonin**, importantly, in the **pineal gland**. Reactions in the pineal gland are conditioned by light and darkness. Serotonin accumulates in the pineal gland in daylight, whereas in the dark, it is

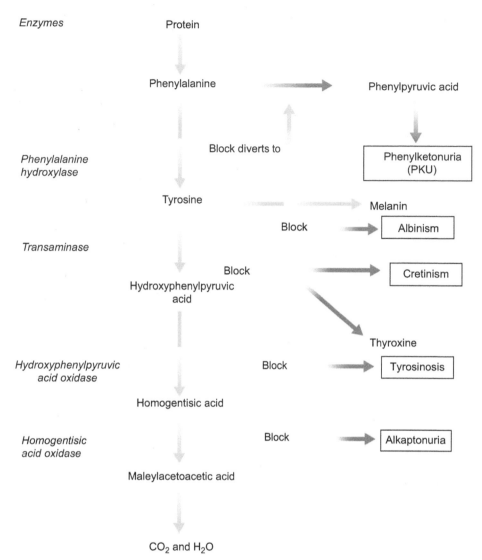

**FIGURE 13.26**  Conversion of phenylalanine to tyrosine and the continued metabolism of tyrosine. Deficits in the functioning of enzymes catalyzing individual steps lead to disease conditions indicated.

converted to **N-acetylserotonin** that is further converted to melatonin. Melatonin is secreted from the gland in the dark. Light is conveyed to the pineal gland by neural connections from the eyes. **Norepinephrine** is released in darkness, and it binds to two types of receptors in the pineal cell (**pinealocyte**). The receptors are alpha- and beta-receptors in the pinealocyte. This action induces the synthesis of **N-acetyltransferase** that converts serotonin (the product during light) to N-acetylserotonin. **Hydroxyindole-O-methyltransferase** converts N-acetylserotonin to melatonin. These reactions are summarized in Fig. 13.27.

The **pineal gland** plays a role in human reproductive physiology and is involved in sexual maturation. Melatonin receptors occur in the gonads. **Melatonin** also is synthesized in the skin where it plays an important role as an antioxidant. People ingest melatonin to counteract jet lag. Melatonin increases **prolactin** concentrations in the blood, and there may be a role for melatonin in **menstrual cyclicity**. The pattern of sleep affects the amount of melatonin produced by the pineal gland, and the amount of melatonin probably plays a role in setting the **biological clock** regarding the sleep circadian rhythm. Melatonin apparently scavenges hydroxy radicals and superoxide anions to protect brain (and probably other tissue) nuclear DNA, proteins and membrane lipids especially in the nighttime (when the hormone is at concentrations 10 times that during daylight).

After **serotonin** is secreted from a serotonergic neuron, most of it is reabsorbed by a reuptake mechanism limiting the amount of effective serotonin. *Certain **antidepressant drugs** (e.g., Prozac) inhibit the reuptake process so that prolonged serotonin effects are produced in the synaptic cleft.*

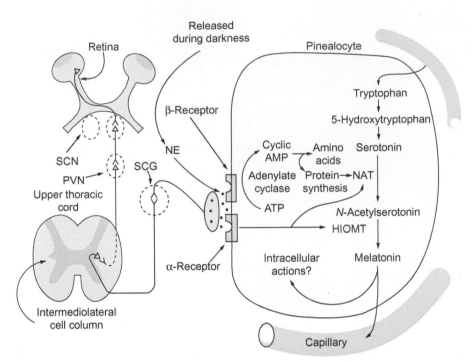

**FIGURE 13.27**   On *right*, tryptophan serves as the precursor for serotonin and melatonin in the pinealocyte. On *left*, connections to the eyes and spinal column. *HIOMT*, hydroxyindole-*O*-methyltransferase; *NAT*, *N*-acetyltransferase; *NE*, norepinephrine; *PVN*, paraventricular nucleus; *SCG*, superior cervical ganglia; *SCN*, suprachiasmatic nuclei.

In addition to the pathway converting tryptophan to serotonin and then to melatonin, other metabolites of serotonin can arise through **monoamine oxidase** that leads to the formation of **5-hydroxyindole acetaldehyde**. This compound can be converted through **aldehyde reductase** to generate **5-hydroxytryptophol** or through **aldehyde dehydrogenase** to generate **5-hydroxyindole acetic acid**. 5-Hydroxytryptophol can be converted to **5-methoxytryptophol** by **Hydroxyindole-*O*-methyltransferase (HIOMT)**. 5-Hydroxyindole acetic acid can be converted to **5-methoxyindole acetic acid** by the action of HIOMT.

The pineal gland has many regulatory roles, and its actions are thought to impact sleeping, locomotor activity, the hypothalamus, pituitary, the parathyroid, and the pancreas.

**Tryptophan** is metabolized to **quinolinic acid** through the **kynurenine pathway**. Quinolinic acid is also converted to **niacin** in small amounts that serves as a required component of the nicotinamide nucleotide coenzymes. Additional niacin is needed in the diet because rather small amounts are produced in the liver from tryptophan, giving niacin a "vitamin context." The major route of metabolism of tryptophan through kynurenine (in addition to the pathway leading to melatonin, already discussed) is shown in Fig. 13.28.

## Arginine

Both arginine and histidine are essential amino acids in childhood. Arginine is incorporated into proteins and is the precursor of creatine and creatinine. In the **urea cycle**, **arginine** forms citrulline (see Fig. 13.3). The metabolic fates of arginine in the human are shown in Fig. 13.29.

Creatinine is produced from creatine as in Fig. 13.30.

Ninety-five percent of the bodily content of creatine is contained in skeletal muscle (the remainder is located in heart, brain, and testes). Creatine is converted to **creatine phosphate**, an energy form, in the muscle. When energy demands are high, as in muscular contraction, the reversible reaction of **creatine phosphokinase** runs in favor of the breakdown of creatine phosphate + ADP to creatine + ATP. When energy demands in muscle are low, the opposite occurs to build up the supply of creatine phosphate. When muscle creatine levels are increased by about 20%, there is improvement in **muscular performance**, and oral supplementation of creatine is used for this purpose.

Creatine phosphate cyclizes to **creatinine** spontaneously (Fig. 13.30). The amount of creatinine in muscle is quite constant. Creatinine is excreted by the kidney and is not effectively reabsorbed by the kidney so that its clearance into the urine provides a measure of **kidney glomerular filtration** that reflects kidney function. If the kidney is not

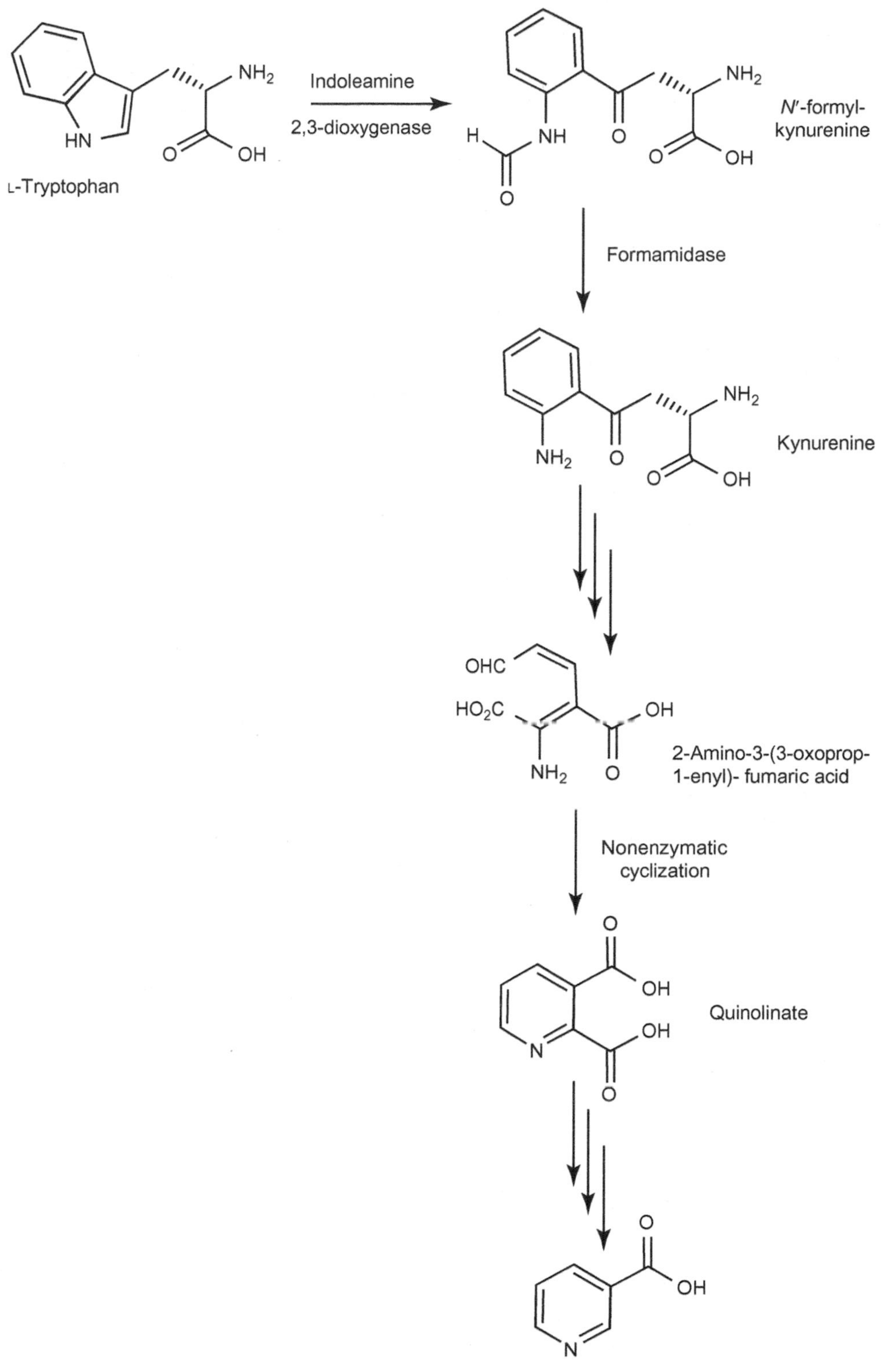

**FIGURE 13.28**  Metabolism of tryptophan to kynurenine and its products leading to niacin (and eventually to the NAD coenzymes).

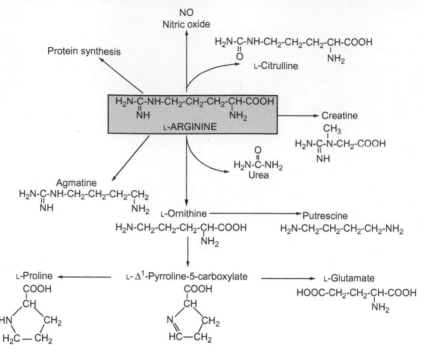

**FIGURE 13.29** Metabolic fates of arginine in the human. The five enzymes on which the central limbs of the pathways are based include (clockwise from the top): **nitric oxide synthase**, **arginine: glycine amidinotransferase**, **arginase**, **arginine decarboxylase**, and **arginyl-tRNA synthetase**. *Reproduced from Wu, G., Morris Jr., M., 1998. Arginine metabolism: nitric oxide and beyond. Biochem. J. 336, 1–17.*

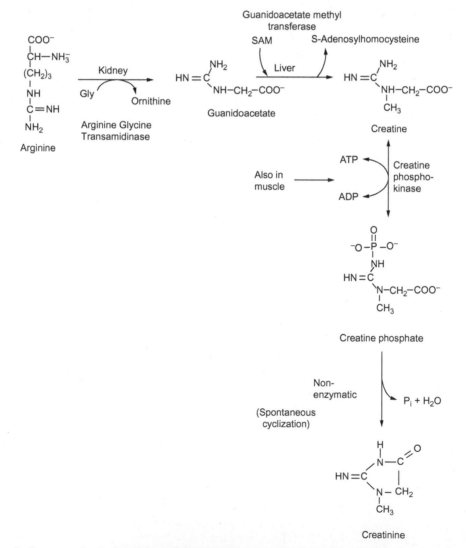

**FIGURE 13.30** Steps leading to creatine from arginine and the conversion of creatine to creatinine. Arginine is converted to guanidoacetate in the kidney, and guanidoacetate is converted to **creatine** in the liver. *SAM*, S-adenosylmethionine.

**TABLE 13.3** Properties of Mammalian Nitric Oxide Synthase Isozymes

| Enzyme | Gene | No. of Exons | No. of Residues | Subcellular Location | Regulation |
|--------|------|--------------|-----------------|----------------------|------------|
| nNos | NOS1 | 29 | 1429–1433 | Mainly soluble (brain); mainly particulate (skeletal muscle) | $Ca^{2+}$/CaM |
| iNos | NOS2 | 27 | 1144–1153 | Mainly soluble | Cytokine-inducible; $Ca^{2+}$-independent |
| eNOS | NOS3 | 26 | 1203–1205 | Mainly particulate | $Ca^{2+}$/CaM |

Source: Reproduced from http://metallo.scripps.edu/PROMISE/NOS.html.

functioning properly, the blood level of creatinine rises. A normal level of creatinine in the blood of adult males is 0.6−1.2 mg/dL and for adult females, 0.5−1.1 mg/dL. Ingestion of meat often is followed by a transient rise in blood creatinine. Kidney malfunctions will cause the blood (and urinary) creatinine to rise above the normal levels cited until a point is reached when the kidney is no longer effectively filtering the blood (values in blood of 10 mg/dL and above); then **kidney dialysis** is required to remove toxins.

Arginine is also the precursor of **nitric oxide (NO)**. NO is formed from arginine through the activity of **nitric oxide synthase** (**NOS**). This enzyme catalyzes a five-electron oxidation of the guanidino nitrogen of L-arginine. The reaction employs 2 mol of $O_2$ and 1.5 mol of NADPH per mole of NO generated. This is a complex enzyme, requiring five-bound cofactors: FAD, FMN, heme, BH4, and $Ca^{2+}$—calmodulin. Three discrete genes generate NOS isozymes: **neuronal NOS (nNOS)**, **cytokine-inducible NOS (iNOS)**, and **endothelial NOS (eNOS)**. Characteristics of these isozymes are listed in Table 13.3.

NO is a biological signal that functions at low concentrations to control blood pressure, neurotransmission, learning, and memory. It also causes vasodilation through the activation of **guanylate cyclase** and the generation of **cyclic GMP**. The half-life of cyclic GMP is very short, on the order of seconds and a **phosphodiesterase** opens the cyclic structure to form GMP. Drugs that block the phosphodiesterase will prolong the vasodilation produced by cyclic GMP. iNOS and nNOS are cytosolic, whereas eNOS is associated with the endothelial membrane through N-terminal myristoylation and posttranslational palmitoylation.

Some arginine can be synthesized from citrulline by the **argininosuccinate synthase** reaction; this enzyme produces L-argininosuccinate and subsequently arginine through the action of **argininosuccinate lyase**. The growing child probably does not produce arginine by this pathway.

Part ($\sim$10%) of the metabolism of arginine and its product, ornithine, gives rise to **polyamines** as shown in Fig. 13.31.

## Histidine

Although histidine is not an essential amino acid in the diet of the adult, it is required in the diet of the growing child. In addition to its incorporation into proteins, histidine is converted to **histamine** by the action of **histidine decarboxylase**. The secretion of **gastric acid** is stimulated by histamine through **histamine $H_2$ receptors**. Drugs, such as **cimetidine** (Tagamet), **ranitidine** (Zantac), **femotidine** (Pepcid), and **nizatidine** (Axid) are used to block the action of histamine at the $H_2$ receptors for the treatment of **gastroesophageal reflux** disease or **esophagitis**.

**Mast cells** synthesize and release histamine to mediate the allergic response by generating vasodilation and bronchoconstriction through **histamine $H_1$ receptors**. There are a large number of drugs that inhibit histamine H1 receptors, including **diphenylhydramine (Benadryl)** and **loratadine (Claritin)**.

The degradation of **histidine** is shown in Fig. 13.32.

Histidine is part of the **carnosine** dipeptide (L-histidine-β-alanine). It is formed in skeletal muscle and brain cells by **carnosine synthase**:

$$ATP + \text{L-histidine} + \beta\text{-alanine} \rightleftharpoons AMP + PP_i + \text{carnosine (L-His-β-Ala)}$$

The structures of carnosine, histidine, **β-alanine**, and histamine are shown in Fig. 13.33.

The transport of β-alanine in skeletal muscle cells requires 2 $Na^+$ and 1 $Cl^-$ per β-alanine transported into the cell. Carnosine has several activities: a chelator of divalent ions, an antioxidant, an inhibitor of glycation of LDLs that cause

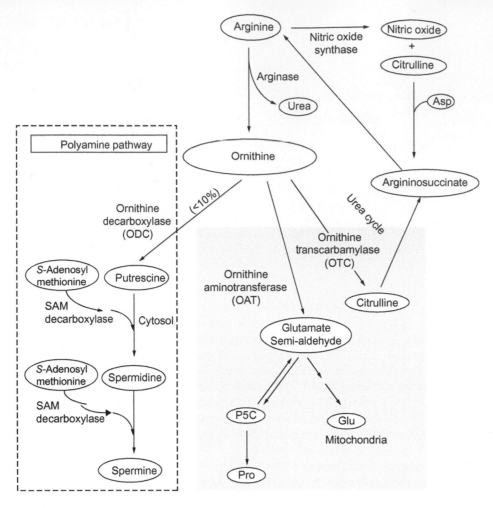

FIGURE 13.31 The metabolism of arginine and ornithine. *P5C*, ʟ-Δ¹-Pyrroline-5-carboxylate.

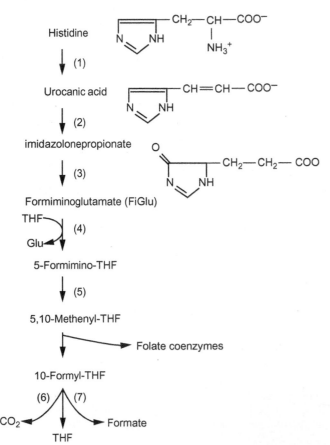

FIGURE 13.32 The catabolism of histidine. Histidine is degraded sequentially to formiminoglutamate (FiGlu) by histidase (1), urocanase (2), and imidazoleproprionate amino hydrolase (3). The formimino group of FiGlu originates from the ring two-carbon of histidine and it enters the one-carbon pool as 5-formiminotetra-hydrofolate (5-formimino-THF) in a reaction catalyzed by FiGlu formiminotransferase (4). The formimino group of 5-formimino-THF is deaminated by 5-formimino-THF deaminase (5) to yield 5,10-methylenyl-THF. The one-carbon moiety of 5,10-methylene-THF can proceed reductively through the one-carbon pool (to 5,10-methylene-THF, 5-methyl-THF and methionine), or oxidatively to the folate coenzyme 10-formyl-THF. The formyl group of 10-formyl-THF can be released as $CO_2$ or formate in reactions catalyzed by 10-formyl-THF dehydrogenase (6) and 10-formyl-THF hydrolase (7), respectively, thereby regenerating the THF molecule used in reaction (4).

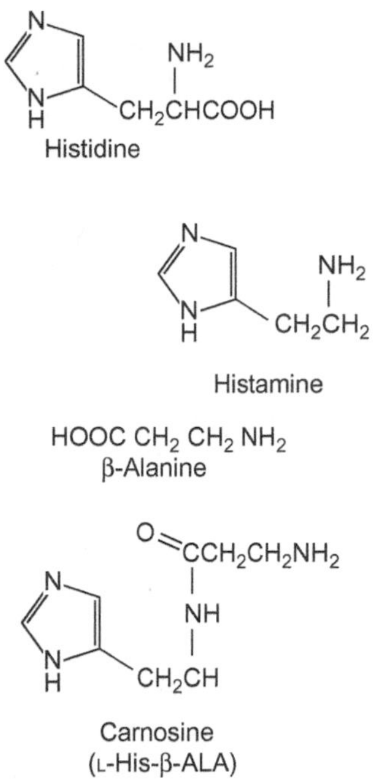

FIGURE 13.33   Structures of histidine, histamine, β-alanine, and carnosine.

the formation of foam cells *in vitro* and an inhibitor of diabetic nephropathy. Carnosine has been formulated as a supplement that is proposed to be an antiaging factor.

In the rare disease (approximately 1 in 12,000 in the United States but in Japan it is the foremost inborn error of metabolism), **histidinemia**, there is a deficiency of **histidine ammonia lyase** (**histidase**) that converts histidine to ammonia and urocanic acid [reaction (1) in Fig. 13.32]. Histidinemia derives from mutations of the *HAL* gene (autosomal recessive disorder) that codes for histidase (mRNA). Symptoms that include speech impediment, developmental delay, hyperactivity, learning problems (and sometimes mental retardation) develop in early childhood. There are elevated levels of histidine, histamine, and imidazole in the blood, urine and cerebrospinal fluid and decreased levels of urocanic acid in the blood, urine, and in skin cells.

The **imidazole group** of histidine is in the active center of several enzymatic reactions in that this group can be in protonated or unprotonated forms.

## Glutamate

As has been discussed, glutamate takes part in generating **glutamine**, in **glutamate dehydrogenase** reactions, in being incorporated into **glutathione**, and in transamination reactions. Another role for glutamate is in the formation of **γ-aminobutyric acid (GABA)**, a major inhibitory neurotransmitter in the brain. GABA is formed in the **glutamate decarboxylase** reaction:

$$\text{glutamate} \rightarrow CO_2 + \text{ γ-aminobutyric acid } (^-OOCC(\alpha)H_2CH_2C(\gamma)H_2NH_3^+)$$

GABA inhibits presynaptic action on excitatory (but not on inhibitory) innervation of certain muscles. GABA interacts with $GABA_A$ receptors and the β subunit of the **$GABA_A$ receptor** is essential for functioning of the receptor. Part of these receptors consists of ion channels transporting chloride ions (Fig. 13.34).

Both glutamate dehydrogenase and glutamate transaminase form **α-ketoglutarate** that can be degraded in four steps to form **proline** and in five steps to form **ornithine** which is convertible to **arginine** in the urea cycle.

## Serine

Because glucose can be converted to serine (Fig. 13.35), this amino acid is nonessential.

**Serine hydroxymethyltransferase** converts serine to glycine:

$$\text{serine } + \textbf{tetrahydrofolate}(FH_4) \rightleftharpoons \text{glycine} + N^5, N^{10}\text{-methylene } FH_4$$

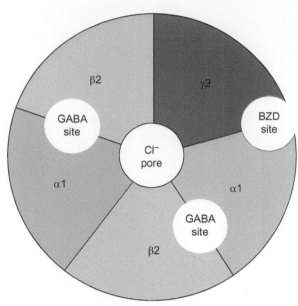

**FIGURE 13.34** Diagram showing the arrangement of subunits and the central ion channel transporting $Cl^-$ ions of the GABAA receptor. *BZD*, benzodiazepine. *Reproduced from http://en.wikipedia.org/wiki/File:GABAA-receptor-protein-example.png.*

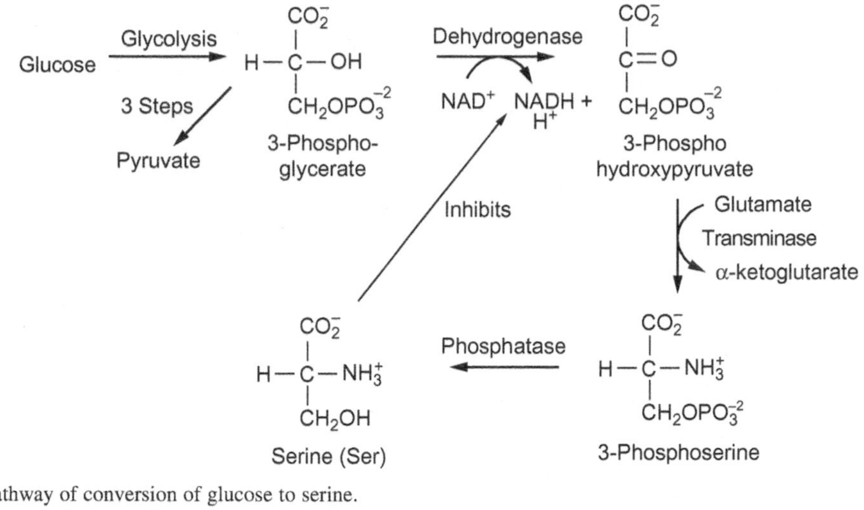

**FIGURE 13.35** The pathway of conversion of glucose to serine.

## CATABOLISM OF AMINO ACIDS

When the pool of amino acids is plentiful enough for the demands of protein synthesis, the excess amino acids are metabolized to compounds that can enter the TCA cycle for the production of energy in the form of ATP, accounting for about 10% of the body's energy supply. Those amino acids that do not enter the TCA cycle are either ketogenic (form fatty acids) or glucogenic (form glucose). **Ketogenic amino acids** form acetoacetate or acetyl CoA. **Glucogenic amino acids** form pyruvate, α-ketoglutarate, succinyl CoA, fumarate, or oxaloacetate. Amino acids that have both properties (ketogenic and glucogenic) are as follows: tryptophan, phenylalanine, tyrosine, isoleucine, and threonine. Lysine and leucine are ketogenic only and the remaining amino acids are solely glucogenic: arginine, glutamate, gluamine, histidine, proline, valine, methionine, aspartate, asparagine, alanine, serine, cysteine, and glycine. Amino acids that are metabolized to **pyruvate** are alanine, cysteine, and serine. Serine dehydratase converts serine to pyruvate as shown in Fig. 13.36.

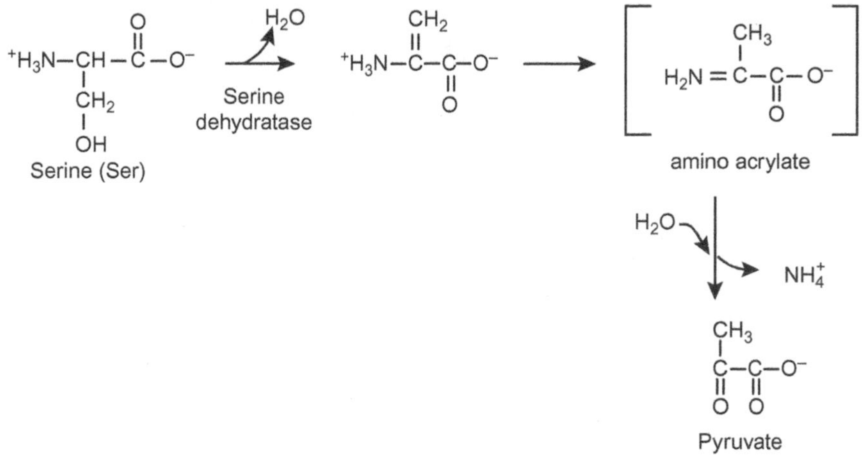

**FIGURE 13.36** The reaction of L-serine with **serine dehydratase** forms an amino acrylate intermediate that tautomerizes to the imine that is hydrolyzed to produce water and ammonia. The hydroxyl of serine is removed by β-**elimination**. Serine dehydratase has pyridoxal phosphate as its coenzyme.

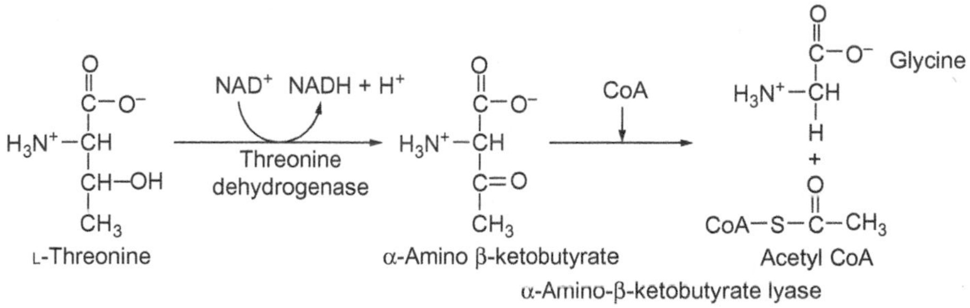

**FIGURE 13.37** Conversion of L-threonine to acetyl CoA and L-glycine by the sequential action of **threonine dehydrogenase** and α-**amino-β-keto-butyrate lyase**.

Serine hydroxymethyltransferase is able to convert glycine to pyruvate by first converting glycine to serine (shown above). Serine is then converted to pyruvate by serine dehydratase. Cysteine also can be converted to pyruvate, and the three alkyl carbons of tryptophan are converted to **alanine**, and then **alanine aminotransferase** converts the alanine to pyruvate.

**Threonine dehydrogenase** ($NAD^+$ coenzyme) converts threonine to α-amino-β-ketobutyrate, and α-**amino-β-keto-butyrate lyase** converts it to acetyl CoA plus L-glycine (Fig. 13.37).

Either aspartate or asparagine can be converted to **oxaloacetate** by **aspartate aminotransferase** (in the case of aspartate); in the case of asparagine, it is converted to aspartate and ammonia by **asparaginase** and then to oxaloacetate as described.

L-Asp + α-ketoglutarate (*aspartate aminotransferase*) → oxaloacetate + L-Glu

L-Asn + $H_2O$ (asparaginase) → L-Asp + $NH_3$

Several amino acids are convertible to α-**ketoglutarate**. **Glutaminase** converts glutamine to glutamate. Glutamate is converted to α-ketoglutarate by aspartate aminotransferase (see above, reverse reaction). Proline can be converted to α-ketoglutarate by the following pathway:

L-Pro (***proline oxidase***) → pyrroline 5-carboxylate (*spontaneous*) → Glu γ-semi-aldehyde (***5-semi-aldehyde dehydrogenase***) → L-Glu + oxaloacetate (aspartate amoinotransferase) → α-ketoglutarate + L-Asp

**Arginase** converts arginine to ornithine that also can form Glu 5-semi-aldehyde. Histidine can form glutamate through the pathway: L-Histidine (deamination by ***histidine ammonia lyase***) → urocanate (addition of $H_2O$ by ***urocanate hydratase***) → 4-imidazolone 5-propionate (hydrolysis by ***imidazole propionase***) → N-formininoglutamate (+ tetrahydrofolate) → glutamate + $N^5$-formimino-tetrahydrofolate (by ***glutamate formiminotransferase***) and glutamate can be converted to α-ketoglutarate by aspartate aminotransferase (above, reverse reaction).

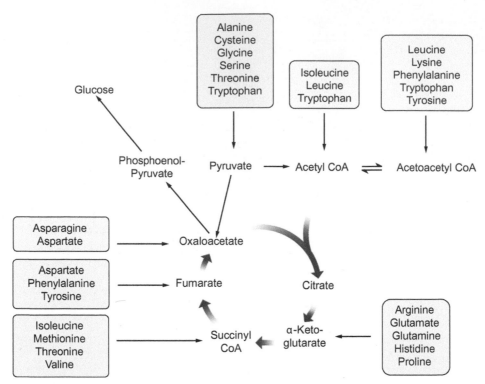

**FIGURE 13.38** Catabolism of amino acids through the **citric acid cycle**. *Yellow* boxes indicate **ketogenic amino acids**; *blue* boxes indicate **gluco-genic amino acids**. Isoleucine, tryptophan, phenylalanine, and tyrosine can form both glucose and fatty acids.

Certain amino acids are convertible to **succinyl CoA**. These are methionine, valine, and isoleucine. In the case of methionine, it is first converted to **S-adenosylmethionine (SAM)** by **methionine adenosyltransferase**. **SAM methyl-ase** converts SAM to S-adenosylhomocysteine and then to homocysteine. **Cystathionine β-synthase** converts homocys-teine to cystathionine. In the next step, cystathionine is converted to α-ketobutyrate and cysteine by the action of **cystathionine-γ-lyase**. α-Ketobutyrate (plus CoASH plus NAD$^+$) is converted to propionyl CoA (plus $CO_2$ plus NADH) by **α-ketobutyrate dehydrogenase**. This is followed in several steps by the conversion to succinyl CoA.

A summary of the **catabolism of amino acids** through the TCA cycle is shown in Fig. 13.38.

**Branched chain amino acids** are catabolized by muscle, adipose, kidney, and brain but not by the liver. The reac-tions involving the other amino acids, as discussed above, take place mainly in the liver. The branched chain amino acids are converted to the corresponding keto acids by **branched chain amino acid aminotransferase**. Together with CoASH and NAD$^+$, **branched chain α-keto acid dehydrogenase (BCKAD)** converts the keto acids to the correspond-ing CoA derivatives. This enzyme is a large multienzyme complex homologous to pyruvate dehydrogenase. The BCKAD is generally in the inactivated phosphorylated form and activated by a protein phosphatase after ingestion of the branched chain amino acids. If BCKAD is mutated (poorly or not expressed), branched chain amino acids accumu-late in the blood and urine. Consequently, the urine smells of maple syrup and the disease is called "**maple syrup disease**." Unless the diet is restricted in early life by the omission of valine, leucine, and isoleucine, **mental retarda-tion** will develop. The catabolic pathways of branched chain amino acids are shown in Fig. 13.39.

Lysine and leucine are solely ketogenic, and leucine is converted to α-ketoisocaproate by transamination, and this is converted to isovaleryl CoA by BCKAD in an oxidative decarboxylation reaction. Isovaleryl CoA is converted to β-methylcrotonyl CoA by **isovaleryl CoA dehydrogenase**. This product is converted to β-methylglutaconyl CoA by the action of **methylcrotonyl CoA carboxylase** (biotin-containing). Methylcrotonyl CoA is then converted to β-hydroxy-β-methylglutaryl CoA by a hydratase and β-hydroxy-β-methylglutaryl CoA is hydrolyzed to acetyl CoA and acetoace-tate by **hydroxymethylglutarate-CoA lyase (HMG-CoA lyase)**.

The aromatic amino acids, phenylalanine and tyrosine, are degraded to homogentisic acid, then to maleylacetoacetic acid and finally to $CO_2$ and $H_2O$ as shown in Fig. 13.26. The third aromatic amino acid, tryptophan, is converted to acetoacetate as shown in Fig. 13.40.

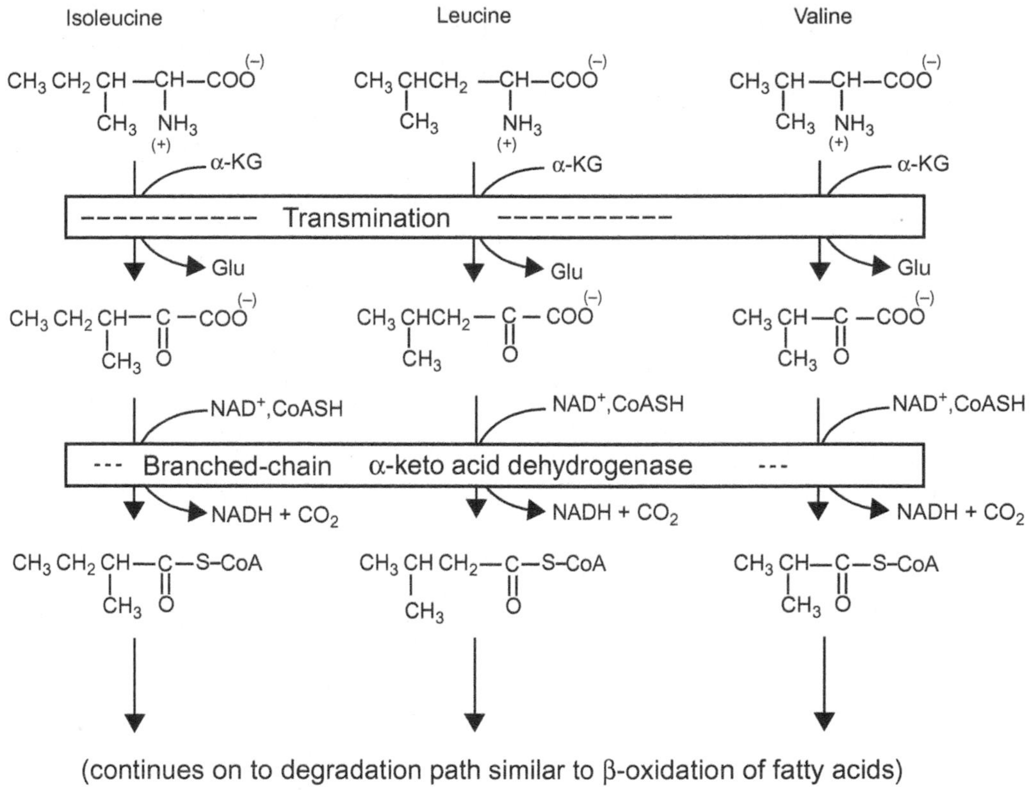

**FIGURE 13.39**  Catabolic pathways of branched chain amino acids.

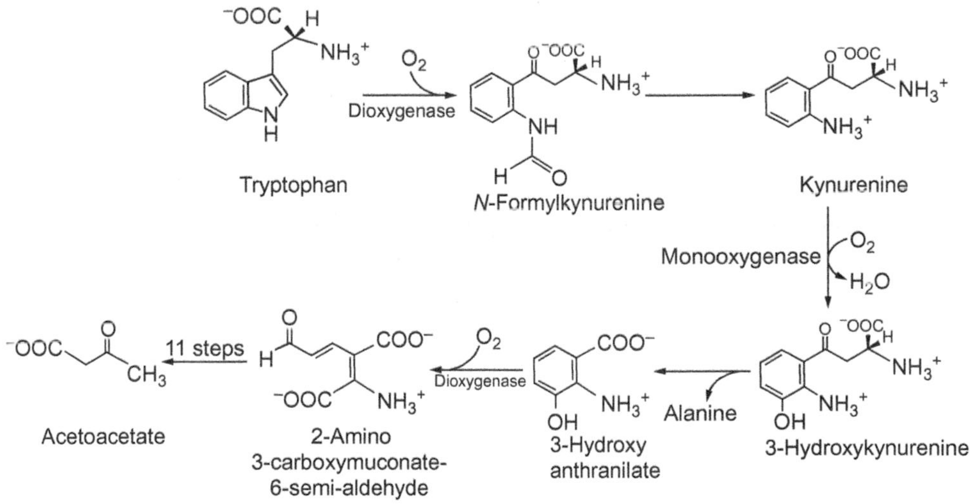

**FIGURE 13.40**  Catabolism of tryptophan.

## SUMMARY

**Urea-cycle**-related disease in the form of **hyperammonemia** is used to illustrate a disease of metabolism, specifically related to a defect in the urea cycle. A deficiency of one of the enzymes of the urea cycle leads to a specific disease condition where **ammonia**, which is a toxin, can accumulate in the blood (hyperammonemia). When glutamine is produced in excess in the liver, it is converted to ammonia by **glutaminase**. The liver urea cycle converts ammonia to urea that is excreted in the urine. The overall reactions of the TCA cycle and the urea cycle can be summarized:

$$2NH_4^+ + HCO^{3-} + 3ATP^{4-} \rightarrow urea + 2ADP^{3-} + 4Pi + AMP^{2-} + 5H^+$$

**Glutamine synthase** and **glutaminase** utilize ammonium ion and produce ammonium ion in their reactions, respectively:

Glutamine synthase: glutamate $+ NH_4^+ + ATP \rightarrow$ glutamine $+ ADP + Pi + H^+$
Glutaminase: glutamine $+ H_2O \rightarrow NH_4^+ +$ glutamate

**Transamination** is a process in which amino groups are removed from amino acids and transferred to acceptor keto acids to generate the keto acid version of the original amino acid.

**Muscle cells** can use amino acids as energy sources, and the liver can detoxify the amino groups (as ammonium ions) via the **urea cycle**. Alanine is a predominant amino acid in most proteins. It can be transported in the bloodstream from peripheral tissues to the liver where it can be converted to glucose. Alanine is transaminated to form pyruvate, and glucose can be formed from pyruvate through **gluconeogenesis**. Glucose can then be shipped to the muscle (for energy utilization) through the bloodstream. This system relating muscle and liver metabolism is known as the **alanine cycle**.

**Transamidation** is the catalytic formation of a covalent bond between a free amine group and a gamma carboxamide group. Transamidinases catalyze the formation of **γ-glutamyl-ε-lysine bonds** involved in tissue healing. Transamidinases are involved in the synthesis of amino sugars (e.g., glucosamine-6-phosphate). These enzymes are also involved in the **glycosylphosphatidylinositol anchoring of proteins** to cellular membranes.

**Deamination** of amino acids is catalyzed by **ammonia lyase enzymes**. Examples are **serine dehydratase** and **threonine dehydratase** both of which have **pyridoxal phosphate** as coenzyme. Serine dehydratase converts serine to pyruvate and ammonia, and threonine dehydratase converts threonine to α-ketobutyrate and ammonia. Another member of this group of enzymes is **histidase** (**histidine ammonia lyase**) that removes the amino group from histidine to form *trans*urocanic acid in liver and skin. **Oxidative deamination** occurs in the liver by **glutamate dehydrogenase** in which glutamate + water + $NAD^+$ forms α-ketglutarate + ammonia + NADH + $H^+$.

Amino acids can be oxidized by **D-amino acid oxidase** located in liver and kidney **peroxisomes**. D-Amino acids occur in the diet especially in plant foods, since plants do not contain D-amino acid oxidase. The catalytic products are an imino acid and $H_2O_2$. The human also has the enzyme **amino acid racemase** that interconverts D- and L-amino acids. Liver and kidney peroxisomes contain small amounts of **L-amino acid oxidase** that would be useful when there is an excess of L-amino acids for protein biosyntheses and for other pathways.

L-Amino acid racemase converts L-amino acids (the natural forms in proteins) to D-amino acids. This enzyme is especially important in the brain since **D-serine** is present in large amounts in the *corpus callosum* and hippocampus.

**L-Amino acid decarboxylase** with pyridoxal phosphate as coenzyme removes $CO_2$ from amino acids to yield the corresponding amines. **Aromatic L-amino acid decarboxylase** catalyzes the decarboxylation of tryptophan, 5-hydroxytryptophan, L-dihydroxyphenylalanine, 3,4-dihydroxyphenylserine, tyrosine, phenylalanine, and histidine.

**Essential amino acids** are those that are absent in the body, or those that are synthesized to an extent that is insufficient for growth and maintenance. They are as follows: methionine, tryptophan, lysine, phenylalanine, threonine, valine, leucine, and isoleucine. In addition, histidine and arginine are essential for children. The **nonessential amino acids** are as follows: glutamate, glutamine, aspartate, asparagine, alanine, cysteine, tyrosine, proline, serine, and glycine (and ornithine). Methionine and tryptophan have single codons but the other amino acids have two or three codons. Most proteins begin their translation with the start codon, **AUG**. AUG codes for **methionine** that is the N-terminal amino acid of all eukaryotic proteins; however, it is possible to remove the N-terminal methionine by posttranslational modification. **Homocysteine**, which derives from *S*-adenosylmethionine, can generate cysteine. **Taurine** is synthesized as follows: methionine → cysteine → cysteine sulfinate → hypotaurine → taurine. Although taurine is an amino acid, it is not incorporated into protein. However, it is a conjugate with bile acids and is active as an antioxidant, modulator of calcium signaling, stabilizer of membranes, and an apoptosis inhibitor.

**Glutathione (GSH)** is a tripeptide (glutamylcysteinylglycine) but it is synthesized without mRNA. With two GSH molecules, the cysteines can be oxidized to form a disulfide ($-S-S-$), and this interconversion ($2GSH \leftarrow \rightarrow$ **GSSG**) represents a critical redox agent in the cell. The **glutathione *S*-transferase** family of enzymes protects cells from damage by **xenobiotics** and certain drugs by forming **GSH** that solubilize them, including certain carcinogens, and allow their excretion. GSH is involved in the **γ-glutamyl cycle** that enhances the transport of amino acids from outside the cell to the cell interior. The first step in this cycle is the transfer of the γ-glutamyl group of GSH to an amino acid, a peptide or to water by cell membrane **γ-glutamyltranspeptidase** (γ-L-glutamylpeptide + amino acid ← → peptide + γ-L-glutamyl-amino acid). This pathway is part of the synthesis and degradation of GSH functioning in the detoxification of drugs and xenobiotics. **Leukotriene C4 (LTC4)** contains GSH, and this leukotriene has contractile activity on airway tissues.

**Tyrosine** is the precursor of **catecholamines** (**epinephrine** and **norepinephrine**) as well as the main body pigment, **melanin**. Tyrosine can be formed from phenylalanine by **phenylalanine hydroxylase (PAH)** in liver and kidney. This enzyme removes any excess phenylalanine. Mutations (there are more than 400 mutations of the gene expressing PAH known in children) in the gene for this enzyme lead to **phenylketonuria**. Catecholamines are synthesized in the **adrenal medulla**: phenylalanine + PAH → tyrosine + **tyrosine hydroxylase** → DOPA + **aromatic L-amino acid decarboxylase** → dopamine + **dopamine β-hydroxylase** → norepinephrine + **S-adenosylmethionine (SAM,** as the methyl donor) + **phenylethanolamine-N-methyltransferase (PNMT)** → epinephrine. In a reaction to **stress**, **cortisol** is produced in the **adrenal cortex**. On its way to the general circulation, cortisol flows through the adrenal medulla and there induces PNMT so as to increase the output of catecholamines that are also elevated in stress.

**Tryptophan**, in the **pineal gland**, is the precursor of the neurotransmitters **serotonin** (in daylight) and **melatonin** (in darkness). **N-Acetyltransferase** converts serotonin to N-acetylserotonin, and **hydroxyindole-O-methyltransferase** converts N-acetylserotonin to melatonin. The pineal gland has a role in sleep, locomotor activity and impacts the hypothalamus, parathyroid, and pancreas. Tryptophan is converted to **quinolinic acid** through the **kyurenine pathway**, and **quinolinic acid** can be converted in small amounts to **niacin**, a precursor of nicotinamide nucleotide coenzymes.

**Arginine** can be converted to **creatine**. The kidney converts arginine to guanidoacetic acid that is converted to creatine in the liver. Most of the creatine in the body is in skeletal muscle where **creatine phosphate** is an energy reserve. When energy demands are high (muscular contraction) **creatine phosphokinase** with ADP converts creatine phosphate to creatine plus ATP. Creatine phosphate spontaneously cyclizes to **creatinine** in muscle where it is maintained at a constant level. *Creatinine is excreted by the kidney without being reabsorbed so that its clearance in the urine provides a measure of **kidney glomerular filtration**.* Arginine is a precursor of **nitric oxide (NO)** in a reaction catalyzed by **nitric oxide synthase (NOS)**. There are three forms of NOS, iNOS, Enos, and nNOS. **NO** is a biological signal controlling blood pressure, neurotransmission, learning, and memory.

**Histidine**, through the action of **histidine decarboxylase**, is converted to **histamine**. Gastric acid secretion is stimulated by histamine through **histamine $H_2$ receptors**. **Mast cells** release histamine to mediate the **allergic response** to generate vasodilation and bronchoconstriction through **histamine $H_1$ receptors**. The **imidazole group** of histidine that can exist in protonated or unprotonated forms is a component of the active sites of many enzymes.

Glutamate generates glutamine through the **glutamate dehydrogenase** reaction. It also takes part in many transaminase reactions and is a component of **glutathione**. Glutamate also forms **γ-aminobutyric acid (GABA)**, an important inhibitory neurotransmitter in the brain. Through the actions of glutamate dehydrogenase and **glutamate transaminase**, glutamate is transformed to α-ketoglutarate that can be a precursor of proline as well as ornithine or can enter the TCA cycle for the production of energy.

Serine can give rise to glycine through the **serine hydroxymethyltransferase** reaction that utilizes tetrahydrofolate coenzyme.

Amino acids, when their concentrations exceed the requirements of protein synthesis, can be metabolized to compounds that can enter the TCA cycle for the production of ATP. Amino acids that do enter the TCA cycle are ketogenic or glucogenic in that the former can be converted to fatty acids, and the latter can be converted to glucose. Branched chain amino acids are not catabolized by the liver but rather by muscle, adipose, kidney, and brain. They are converted to the corresponding keto acids by **branched chain amino acid aminotransferase**. **Branched chain α-keto acid dehydrogenase** converts the keto acids to CoA derivatives. If this enzyme is nonfunctional through mutation of its gene, branched chain amino acids accumulate in the blood and urine and produce "**maple syrup disease.**"

## SUGGESTED READING

### Literature

Auron, A., Brophy, P.D., 2012. Hyperammonemia in review: pathophysiology, diagnosis and treatment. Pediatr. Nephrol. 27, 207–222.

Batshaw, M.L., Tuchman, M., Summar, M., Seminara, J., 2014. A longitudinal study of urea cycle disorders. Mol. Genet. Metab. 113, 127–130.

Baumgart, F., Rodriguez-Crespo, I., 2008. D-Amino acids in the brain: the biochemistry of brain serine racemase. FEBS J. 275, 3538–3545.

Bischoff, R., Schluter, H., 2012. Amino acids: chemistry, functionality and selected non-enzymatic post-translational modifications. Proteomics. 75, 2275−2296.

Blau, N., Shen, N., Carducci, C., 2014. Molecular genetics and diagnosis of phenylketonuria: state of the art. Expert Rev. Mol. Diagn. 14, 655−671.

Braissant, O., 2010. Current concepts in the pathogenesis of urea cycle disorders. Mol. Genet. Metab. 100, S3−S12.

D'Andrea, G., Ceroli, S., Colavito, D., Leon, A., 2015. Biochemistry of primary headaches: role of tyrosine and tryptophan metabolism. Neurol. Sci. 36, 17−22.

Foltyn, V.N., et al., 2005. Serine racemase modulates intracellular D-serine levels through α,β-elimination activity. J. Biol. Chem. 280, 1754−1763.

Hosta-Rigau, L., York-Duran, M.J., Kang, T.S., Stadler, B., 2015. Extracellular microreactor for the depletion of phenylalanine toward phenylketonuria treatment. Adv. Funct. Mater. 25, 3860−3869.

Maniscalco, S.J., Tally, J.F., Harris, S.W., Fisher, H.F., 2003. The direct measurement of thermodynamic parameters of reactive transient intermediates of the L-glutamate dehydrogenase reaction. J. Biol. Chem. 278, 16129−16134.

Muniz, M., Zurzolo, C., 2014. Sorting of GPI-anchored proteins from yeast to mammals − common pathways at different sites?. J. Cell Sci. 127, 2793−2801.

Nagamune, K., et al., 2003. GPI transamidase of *Tripanosoma brucei* has two previously uncharacterized (trypanosomatid transaminase 1 and 2) and three common subunits. Proc. Nat. Acad. Sci. 100, 10682−10687.

Oakley, A.J., 2005. Glutathione transferases: new functions. Curr. Opin. Struct. Biol. 15, 16−23.

Pegg, A.E., 2009. S-Adenosylmethionine decarboxylase. Essays Biochem. 46, 25−45.

Phillips, R.S., 2015. Chemistry and diversity of pyridoxal-5′-phosphate dependent enzymes. Biochim. Biophys. Acta. 1854, 1167−1174.

Sarup, A., Larsson, O.M., Shousboe, A., 2003. GABA transporters and GABA transaminase as drug targets. Curr. Drug Targets CNS Neurol. Disord. 2, 269−277.

Toney, M.D., 2011. Controlling reaction specificity in pyridoxal phosphate enzymes. Biochim. Biophys. Acta. 1814, 1407−1418.

Tsutsui, M., et al., 2015. Significance of nitric oxide synthases: lessons from triple nitric oxide synthases null mice. J. Pharmacol. Sci. 127, 42−52.

Wang, Y.X., Gong, N., Xin, Y.F., Hao, B., Zhou, X.J., Pang, C.C., 2012. Biological implications of oxidation and unidirectional chiral inversion of D-amino acids. Curr. Drug Metab. 13, 321−331.

Woloskler, H., 2011. Serine racemase and the serine shuttle between neurons and astrocytes. Biochim. Biophys. Acta. 1814, 1558−1566.

Wu, G., Morris Jr., M., 1998. Arginine metabolism: nitric oxide and beyond. Biochem. J. 336, 1−17.

Yamaguchi, Y., Hearing, V.J., 2009. Physiological factors that regulate skin pigmentation. BioFactors. 35, 193−199.

Zhang, Y., et al., 2015. Coordinated regulation of protein synthesis and degradation by mTORC1. Nature. 513, 440−443.

## Books

Begley, T. (Ed.), 2001. Cofactor Biosynthesis: A Mechanistic Perspective. Vitamins & Hormones, vol. 61. Academic Press/Elsevier, San Diego, CA.

Litwack, G. (Ed.), 2015. Nitric Oxide. Vitamins & Hormones, vol. 96. Academic Press/Elsevier, Amsterdam.

Michal, G., Schomburg, D., 2012. Biochemical Pathways: An Atlas of Biochemistry and Molecular Biology. second ed. John Wiley & Sons, Inc., Hoboken, NJ.

Salway, J.G., 2004. Metabolism at a Glance. third ed. Blackwell Publishing, Malden, MA.

# Chapter 14

# Metabolism of Fat, Carbohydrate, and Nucleic Acids

## GAUCHER DISEASE: MOST COMMON LIPID STORAGE DISEASE

This is an autosomal recessive disease resulting from a mutation in the gene for **glucocerebrosidase**. The genetic mutation is received from both parents as shown in Fig. 14.1. As shown in the fourth panel of the figure (*lower left*), when both parents are carriers but do not exhibit the overt disease, one in four offsprings will have the overt disease, one in four will be normal, and two in four will be carriers but not have the overt disease.

The symptomology is complex in that there are three types of the disease: Type 1−3, as shown in Fig. 14.2.

The variation in intensity of the disease may be a function of the location of the mutation(s) within the gene. There are more than 250 possible mutations of the glucocerebrosidase gene including 203 missense mutations and 18 nonsense mutations. Certain mutations are more common than others and Fig. 14.3 shows the 15 common mutations.

There are 34 mutations known to cause Gaucher's disease, 4 of which account for 95% of the disease in the Ashkenazi (Germany, France, Poland, Lithuania, Russia but not Spain [Sephardic Jews]) Jewish population and 50% in the general population (among Ashkenazi Jews, 1 in 10 is a carrier; among the general population, 1 in 200 is a carrier). Gaucher disease is the most common genetic defect in the Ashkenazi Jewish population. Both parents must contribute when a child has the overt disease. The genetic mutation occurs equally in males and females (Fig. 14.1).

The **glucocerebrosidase** enzyme catalyzes the hydrolysis of glucocerebroside, yielding ceramide and glucose as shown in Fig. 14.4.

Within 3 months after birth, there are severe symptoms: sucking and swallowing are impaired, the liver and spleen are enlarged (fat accumulates in the spleen, liver, kidneys, lungs, brain, and bone marrow); there can be extensive brain damage with spasticity, seizures, abnormal eye movements, and rigidity of limbs with death occurring before age 2. There also is a chronic milder neuropathological form of Gaucher's disease that can start anytime in childhood or adulthood. This can exhibit symptoms described above, including enlarged spleen and liver together with seizures, poor coordination, skeletal irregularities (bone disease), abnormal eye movements, anemia, and respiratory problems. Persons so affected can survive into the early teen years or, in some cases, into adulthood.

Enzyme (glucocerebrosidase) replacement therapy has been developed and is administered every 2 weeks that alleviates many symptoms. A vector (Fig. 14.5) has been developed that can be injected intravenously. It is able to cross the blood−brain barrier.

In the bloodstream, the recombinant protein binds to the LDLR and transcytoses to the central nervous system. By injection into the bloodstream, the genes also are delivered to the liver and spleen allowing these organs to be the sites of expression and secretion of the therapeutic enzyme.

Prior to gene therapy, the following treatments could have been administered: bone marrow transplant, spleen removal, blood transfusions, and/or joint replacement.

## LIPID METABOLISM

Many aspects of lipid metabolism have been covered in Chapter 9 (Lipids) on lipids. Some of the topics reported in Chapter 9 (Lipids) are the action of **pancreatic lipase** on a triglyceride and on a phosphoglyceride; β-**oxidation** of acyl CoA and the β-oxidation of fatty acids to acetyl CoA as well as the β-oxidation of very long-chain fatty acids in the **peroxisome**; activation of a fatty acid by attachment of CoA on the outer mitochondrial membrane; the transport of **long-chain acyl CoA** into mitochondria; and an overview of fatty acid metabolism (Fig. 9.36).

Human Biochemistry. DOI: http://dx.doi.org/10.1016/B978-0-12-383864-3.00014-4

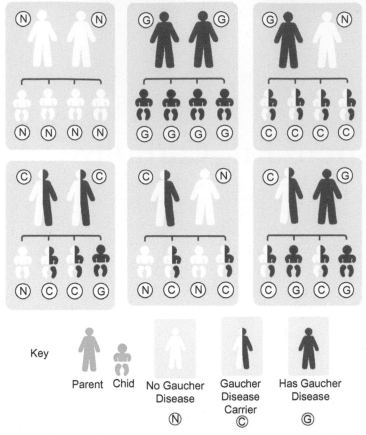

**FIGURE 14.1** Outcomes for the children of parents who are either carriers or who have overt Gaucher's disease. All of the possibilities are shown. Upper *left*: both parents are normal and are not carriers in which case, if there are four children, all are normal. Upper *center* panel shows both parents with the overt disease: all four children would have the overt disease. Upper *right* panel: one parent has the overt disease, the other is normal in which case all children would be carriers but not have the overt disease. Lower *left* panel: both parents are carriers: one in four is normal, one in four has the overt disease, and two in four are carriers. Lower *center* panel: one parent is a carrier, while the other is normal: two children in four will be normal and two will be carriers. Lower *right* panel: one parent is a carrier and the other has the overt disease: two of four will be carriers and two of four will have the overt disease. *Reproduced from http://www.childrensgaucher.org/wp-content/uploads/gaucher_genetics_7.gif.*

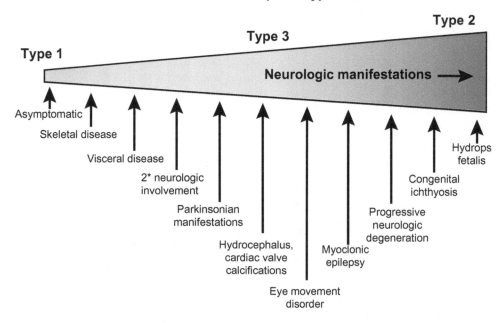

**FIGURE 14.2** A continuum of the symptoms of Gaucher's disease. *Reproduced from http://www.childrensgaucher.org/wp-content/uploads/gaucher_types002.gif.*

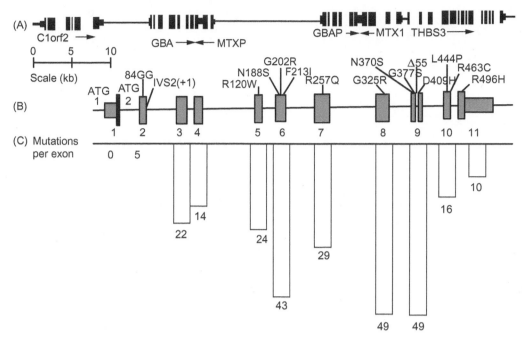

**FIGURE 14.3** Structure and distribution of mutations in the glucocerebrosidase gene. (A) The 62-kb region surrounding the cerebrosidase (GBA) gene (*GBA*) along chromosome 1q showing the known genes and pseudogenes and their transcriptional direction. *C1orf2*, chromosome 1 open reading frame 2 (cote1); *GBA*, glucocerebrosidase; *MTXP*, metaxin 1 pseudogene; *GBAP*, glucocerebrosidase pseudogene; *MTX1*, metaxin 1; *THBS3*, thrombospondin 3. (B) The exonic structure of GBA, with the two ATGs and positions of 15 common mutations indicated. (C) Number of reported substitution, deletion, insertion, and splice-site mutations per exon. *Reproduced from K.S. Hruska, M.E. LaMarca, C.R. Scott and E. Sidransky, Hum. Mutat. 29: 567−583, 2008.*

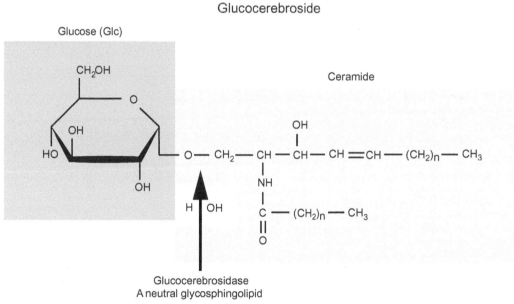

**FIGURE 14.4** Action of glucocerebrosidase. *Reproduced in part from http://fog.ccst.edu/-cpogge/disease/Gaucher/glucocerebroside.gif.*

There are four **phospholipases** capable of degrading a triglyceride. Their actions on the triglyceride structure are shown in Fig. 14.6.

One of the fates of fatty acids is the formation of double bonds in their chains. There are four major human **fatty acid desaturases** [$\Delta$9-desaturase, $\Delta$5-desaturase, $\Delta$4-desaturase and $\Delta$6-desaturase]. Their names reflect the position in the fatty acid chain being desaturated and forming a double bond. The desaturase associates with **cytochrome b5**

(A)

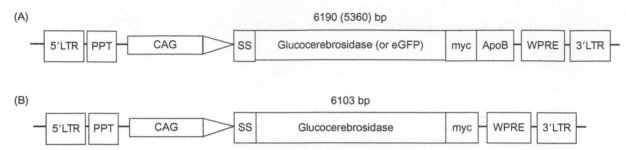

FIGURE 14.5   A schematic drawing of lentiviral vectors. The lentivirus is long-lasting (months−years). The enhanced glucocerebrosidase gene (eGFP) was cloned with the preprotrypsin secretory signal (SS) at the N terminus and the myc epitope tag was cloned at the C terminus. The ApoB LDL receptor (LDLR)-binding domain was cloned at the C terminus of these genes, generating the constructs LV-sGCmApoB and LV-sGFPmApoB. (A) These constructs were cloned into the self-inactivating LV vector under the control of the chicken β-actin/globin (CAG) promoter and containing the central purine tract [polypurine tract (PPT)] as well as the woodchuck hepatitis virus posttranscriptional regulatory element (WPRE) to aid in transduction and transcription of the LV vector. (B) As a control, the sGCm gene without the addition of the ApoB LDLR-binding domain was generated. The LV-sGCmApoB vector is 6190 base pairs, whereas the control LV-sGCm vector is 6103 bp. The LV-sGFPmApoB vector is 5360 bp. *LTR*, long terminal repeat. *Reproduced from http://www.pnas.org/content/104/18/7594/F1.large.jpg.*

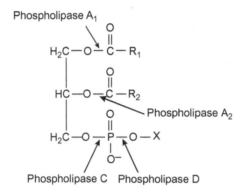

FIGURE 14.6   Sites of actions of various phospholipases (A1, A2, C, and D) in hydrolyzing portions of a triglyceride.

and **cytochrome b5 reductase**; the latter employs NADH and $O_2$ for the introduction of the double bond. An overall reaction would be

$$R_1 - CH_2 - CH_2 - R_2 + O_2 + 2e^- + 2H^+ \rightarrow R_1 - CH = CH - R_2 + 2H_2O$$

A general reaction catalyzed by fatty acid desaturases is represented schematically in Fig. 14.7.

**Oleic acid** and **linoleic acid** are **essential fatty acids** in the diet. The common omega fatty acids with their designations are summarized in Table 14.1.

## Fatty Acid Degradation

The mitochondrion is the site of the **β-oxidation pathway** that generates **acetyl CoA** from fatty acids. Fatty acids, as the CoA derivatives, are transported into the mitochondrion by **carnitine acyltransferase** located in the outer mitochondrial membrane. The product of the reaction is acyl carnitine from acyl CoA plus carnitine derived from the mitochondrial matrix (carnitine is passed from the mitochondrial matrix through the inner mitochondrial membrane to the outer membrane where the carnitine acyltransferase reaction occurs). The system is summarized in Fig. 9.32A. As indicated, the first step in fatty acid degradation is the conversion of the fatty acid to the fatty acyl CoA derivative. A FAD dehydrogenase catalyzes the subsequent oxidative step and this step is followed by hydration, oxidation by $NAD^+$-dehydrogenase, and cleavage of the chain to release acetyl CoA and a fatty acyl CoA reduced in length by two carbons. This set of reactions is repeated until the fatty acid is fully degraded to acetyl CoA (the system is referred to as the fatty acid spiral). The degradation system is reported in Fig. 14.8.

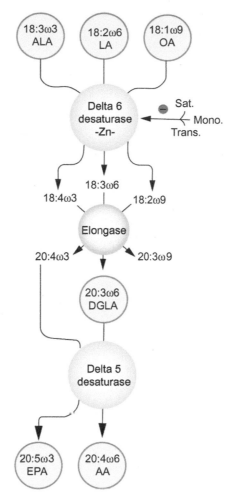

**FIGURE 14.7** General reaction catalyzed by fatty acid desaturases. The first number in the designation of a fatty acid is the total number of carbons; the second number is the number of double bonds in the molecule, and the omega ($\omega$) number is the position of the double bond counting from the last carbon in the chain. *AA*, arachidonic acid; *ALA*, $\alpha$-linolenic acid; *DGLA*, dihomo-$\gamma$-linolenic acid; *EPA*, eicosapentaenoic acid; *LA*, linolenic acid; *OA*, oleic acid. For example, linoleic acid = 9,12-octadecadienoic acid = $18:2\Delta9,12$ = $18:2\omega6$. Originally redrawn from http://bioinfo.pbl.nrc.ca/covello/.

Fatty acid degradation sometimes can occur at a rate faster than glycolysis; in this case, an excess of acetyl CoA would be produced (there would be less pyruvate formed from glycolysis). With a low pyruvate concentration, little oxaloacetate would be produced so that the utilization of acetyl CoA in the TCA cycle would be limited. In this case, the excess acetyl CoA would be converted to **ketone bodies**: acetone, acetoacetate and $\beta$-hydroxybutyrate (Fig. 9.38).

## Fat as Storage Energy

As indicated above, **fatty acid degradation** to acetyl CoA allows this product to enter the TCA cycle for the production of energy (ATP). Fatty acids can be released from chylomicrons and circulate in the bloodstream bound to **albumin**. However, the major source of fatty acids is found in the **adipocyte** in the form of triglycerides. In the adipocyte, **hormone-sensitive lipase** (HSL) converts stored fat into free fatty acids in response to the **lipolytic hormones norepinephrine** and **glucagon**. The action of the opposing hormone, **insulin** leads to the depression of HSL. Norepinephrine and glucagon lead to an increase of the phosphorylation in the regulatory domain of HSL, since this enzyme is regulated by a phosphorylation—dephosphorylation mechanism. Consequently, elevated levels of insulin lead to an inhibition of adipocyte HSL by decreasing the phosphorylation in the regulatory domain of the enzyme. Phosphorylation facilitates a translocation of HSL with the accessory protein, **perilipin**. Perilipin has two forms, 57 and 46 kDa derived from RNA splicing of a single gene, where the 46-kDa product forms from the skipping of exon 6. Perilipin forms a

TABLE 14.1 Structures and Designations of the Common Omega Fatty Acids

| Numerical Symbol | Common Name and Structure | Comments |
|---|---|---|
| $18{:}1^{\Delta 9}$ | Oleic acid | An omega-9 monounsaturated fatty acid |
| $18{:}2^{\Delta 9,12}$ | Linoleic acid | An omega-6 polyunsaturated fatty acid |
| $18{:}3^{\Delta 9,12,15}$ | $\alpha$-Linolenic acid (ALA) | An omega-3 polyunsaturated fatty acid |
| $20{:}4^{\Delta 5,8,11,14}$ | Arachidonic acid | An omega-6 polyunsaturated fatty acid |
| $20{:}5^{\Delta 5,8,11,14,17}$ | Eicosapentaenoic acid (EPA) | An omega-3 polyunsaturated fatty add enriched in fish oils |
| $22{:}6^{\Delta 4,7,10,13,16,19}$ | Docosahexaenoic acid (EPA) | An omega-3 polyunsaturated fatty acid enriched in fish oils |

Source: Reproduced from http://supplementscience.org/pufas.html.

coat on the **lipid storage droplets** in adipocytes and perilipin protects the adipocyte until HSL is transported from the cytoplasm and breaks down the stored fat.

Perilipin is activated through phosphorylation by adipocyte **protein kinase A (PKA)** in response to the stimulating hormones (glucagon and/or norepinephrine). Perilipin, which is confined to adipocytes, is localized to the surface of intracellular neutral lipid droplets. Perilipin is a member of the **"PAT" protein family (P**erilipin, **A**dipophilin, and **T**IP47) and the other members of this family adipophilin and TIP47, in contrast to perilipin, have broad tissue distributions. When HSL is underphosphorylated it resides in the cytosol. Under the influence of glucagon or norepinephrine, cAMP is formed in the adipocyte and PKA is activated and phosphorylates both perilipin and HSL. Phosphorylated perilipin translocates HSL to the surface of the lipid droplet facilitating enzymatic action on the stored triglycerides resulting in a stimulation of lipolysis 30-fold over unstimulated cells (a similar mechanism occurs in steroidogenic cells where, in the case of cortisol production, ACTH acts on the cell membrane, increasing cytoplasmic cAMP, activating a protein kinase that phosphorylates and activates cholesterol esterase to release free cholesterol from cholesterol esters in the lipid droplet). An overview of the **perilipin-HSL system** is shown in Fig. 9.42. The free fatty acids thus liberated from the adipocyte cross the plasma membrane and enter the bloodstream. Various tissues take up the fatty acids from the blood and the fatty acids cross the cell membrane. The uptake of the fatty acids is mediated by three distinct proteins: **fatty acid translocase (FAT), plasma membrane (pm) fatty acid binding protein (FABP-pm)**, and **fatty acid transport protein (FATP**; FATPs may also exist in the mitochondrion to assist in the mitochondrial uptake of fatty acids). Catalysis of the activation of fatty acids to acyl CoA esters by **acyl CoA synthase (ACS)** enhances the cellular uptake process by moving the acyl CoA derivatives into anabolic or catabolic pathways, creating a mass action effect (by removing product) on the plasma membrane uptake process.

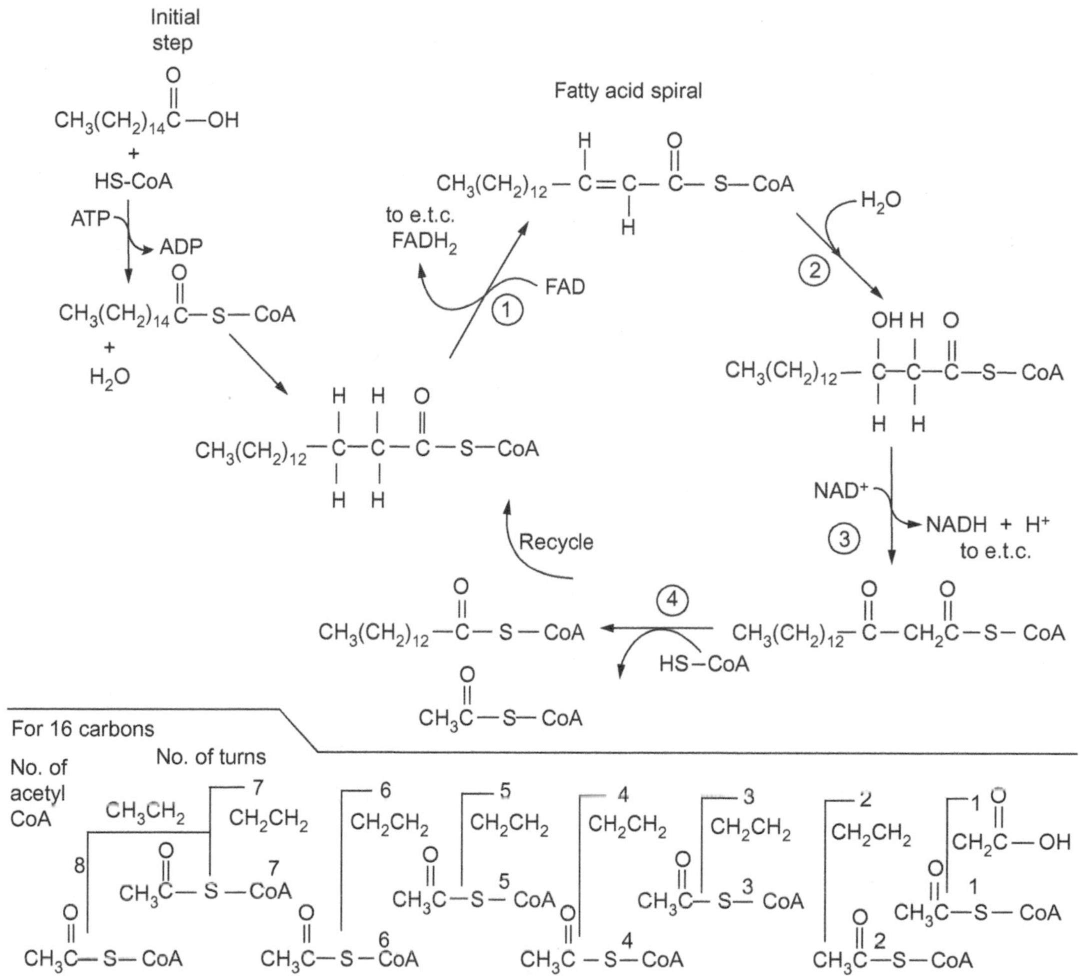

Initial
step

$$CH_3(CH_2)_{14}\overset{\overset{O}{\|}}{C}-OH$$
$+$
HS-CoA

ATP
ADP

$$CH_3(CH_2)_{14}\overset{\overset{O}{\|}}{C}-S-CoA$$
$+$
$H_2O$

Fatty acid spiral

$$CH_3(CH_2)_{12}-\overset{\overset{H}{|}}{C}=\overset{\overset{}{\underset{H}{|}}}{C}-\overset{\overset{O}{\|}}{C}-S-CoA$$

to e.t.c.
FADH$_2$

① FAD

$H_2O$

②

$$CH_3(CH_2)_{12}-\overset{\overset{OH}{|}}{\underset{H}{|}}{C}-\overset{\overset{H}{|}}{\underset{H}{|}}{C}-\overset{\overset{O}{\|}}{C}-S-CoA$$

$$CH_3(CH_2)_{12}-\overset{\overset{H}{|}}{\underset{H}{|}}{C}-\overset{\overset{H}{|}}{\underset{H}{|}}{C}-\overset{\overset{O}{\|}}{C}-S-CoA$$

NAD$^+$

③ NADH + H$^+$
to e.t.c.

Recycle

$$CH_3(CH_2)_{12}-\overset{\overset{O}{\|}}{C}-S-CoA$$

④

HS—CoA

$$CH_3(CH_2)_{12}-\overset{\overset{O}{\|}}{C}-CH_2\overset{\overset{O}{\|}}{C}-S-CoA$$

$$CH_3\overset{\overset{O}{\|}}{C}-S-CoA$$

For 16 carbons

No. of
acetyl
CoA

No. of turns

7    6    5    4    3    2    1

$CH_3CH_2$    $CH_2CH_2$    $CH_2CH_2$    $CH_2CH_2$    $CH_2CH_2$    $CH_2CH_2$    $CH_2CH_2$    $CH_2\overset{\overset{O}{\|}}{C}-OH$

8    $CH_3\overset{\overset{O}{\|}}{C}-S-CoA$    7    5    3    1

$CH_3\overset{\overset{O}{\|}}{C}-S-CoA$    $CH_3\overset{\overset{O}{\|}}{C}-S-CoA$    $CH_3\overset{\overset{O}{\|}}{C}-S-CoA$    $CH_3\overset{\overset{O}{\|}}{C}-S-CoA$    $CH_3\overset{\overset{O}{\|}}{C}-S-CoA$    $CH_3\overset{\overset{O}{\|}}{C}-S-CoA$    $CH_3\overset{\overset{O}{\|}}{C}-S-CoA$

6    4    2

**FIGURE 14.8** The β-oxidation of fatty acids occurring in the mitochondrion. The fatty acid is degraded by two carbons at a time to form acetyl CoA, and the steps are repeated until the entire fatty acid is converted to the acetyl CoA two-carbon fragments. The acetyl CoA produced can enter the citric acid cycle (**Fig. 9.29**) for conversion to ATP.

# Joint Regulation of Lipid and Carbohydrate Metabolism

The metabolism of lipids and carbohydrates varies from one organ to the next. The rates of entry of metabolic products into the brain, muscle and adipose tissues are under the control of the blood level of glucose by the liver and also controlled by the levels of **insulin** and **glucagon** from the pancreas. Fatty acid metabolism in adipose tissue is coordinated with glycolysis in the liver determining the blood levels of glucose that is converted in the adipose cell to glycerol for esterification of fatty acids to triglycerides. Glucose, as well as fatty acids and ketone bodies, can be oxidized in muscle that also forms lactate that can be transported to the liver for gluconeogenesis to generate glucose. Glucose is the main source of energy for the brain unless the body is fasting or starving in which case the brain can utilize **ketone bodies** effectively. The **Atkins diet** uses the condition of low or absent carbohydrate intake (and high fat and protein intake) that leads to **ketosis**, allowing heart and skeletal muscles to burn ketone bodies derived from fat, instead of using glucose. This is quite effective for weight loss but the high protein intake can put a strain on the kidney.

The **energy charge** (ADP/ATP) of the cell is a determinant of β-**oxidation** of fatty acids because this process is coupled to oxidative phosphorylation. With a low cellular energy charge ([ADP] is relatively high compared to [ATP]) the degradation of fatty acids (β-oxidation) to **acetyl CoA** is stimulated. Acetyl CoA enters the TCA cycle and regenerates ATP. Conversely, when the energy charge of a cell is high ([ATP] is relatively high compared to [ADP]) there is a stimulation of the synthesis of fatty acids and phosphatidic acid.

Another point of regulation is the transport of fatty acids across the mitochondrial inner membranes mediated by **carnitine acyltransferase**. When [glucose] is high, **malonyl CoA** levels are high. Malonyl CoA is an inhibitor of mitochondrial carnitine acyltransferase I in the mitochondrial outer membrane. This results in an inhibition of **fatty acid transport** into the mitochondrion and acetyl CoA is used for the synthesis of fatty acids.

## Glucagon and Glucagon-Like Peptides

Glucagon is a 29-amino acid peptide synthesized and secreted from the α-**cells** of the pancreas. It is secreted from these cells when the blood levels of glucose fall (also when blood levels of amino acids are high and when exercise depletes circulating glucose) and its actions are to increase the level of blood glucose. Stimulation of the formation of glucose occurs through the binding of glucagon to its receptor (seven-membrane spanning receptor) to increase the level of cAMP that activates PKA. As a result, **glycogen synthase** *a* is phosphorylated and converted to inactive glycogen synthase *b*, inhibiting the formation of glycogen from glucose. Second, the protein kinase phosphorylates inactive **phosphorylase kinase** converting it to the active phosphorylase kinase that subsequently phosphorylates **glycogen phosphorylase** *b* to the active *a* form that causes the degradation of glycogen ultimately to free glucose. When the level of blood glucose is elevated, **insulin** is released from the β-**cell** of the pancreas and exerts actions that are opposite to those of **glucagon** in terms of the reduction of blood glucose through the stimulation of the formation of glycogen plus its other actions (see Fig. 6.28).

In the intestine, the reaction to a meal is to release **glucagon-like peptides**: **glucagon-like peptide (GLP-1)** and the **gastric inhibitory peptide**. GLP-1 binds to an intestinal receptor and to a receptor in the pancreatic β-cell and stimulates the glucose-dependent release of insulin. A summary of these effects is shown in Fig. 6.53 and the interplay between these hormones at the level of the pancreas is shown in Fig. 14.9.

## Metabolism of Cholesterol

Cholesterol is subjected to oxidation when cholesterol is in relatively large excess. The oxidation products usually are present in very low concentrations in the human compared to the level of cholesterol (1/1000 to 1/1,000,000). The oxidation products of cholesterol are called **oxysterols** that have very short half-lives. They are intermediates in catabolic pathways in the liver and in other organs and lead to the formation of **bile acids** in the liver. **Oxysterol receptors** are part of the nuclear receptor family and regulate sterol-sensitive genes. It has been postulated that **27-hydroxycholesterol** may suppress the accumulation of cholesterol in the presence of **7-hydroperoxycholesterol** or other lipids deemed to be toxic. A summary of the human oxysterols produced from cholesterol is shown in Fig. 14.10. The most important oxysterols appear at the bottom of the figure and are underlined.

The levels of oxysterols regulate the expression of **cholesterologenic enzymes** and thus act as a negative product feedback on the synthesis of cholesterol. The **sterol regulatory element-binding protein-2 (SREBP-2)** is responsive both to sterols and oxysterols and controls the response of **cholesterol-forming enzymes** (lanosterol **14α-demethylase**

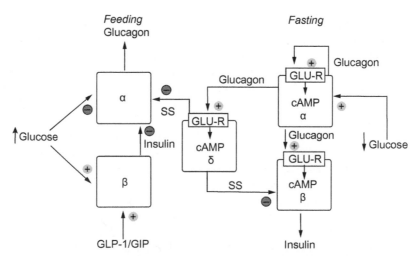

**FIGURE 14.9**   A model depicting the roles of glucagon in insulin-secreting islet β-cells, glucagon secreting α-cells and somatostatin (*SS*) secreting δ-cells. Major actions are either stimulatory (*green circles with + sign*) or inhibitory (*red circles with − sign*). *GIP*, gastric acid inhibitory peptide; *GLP*, glucagon-like peptide; *GLU-R*, glucagon receptor.

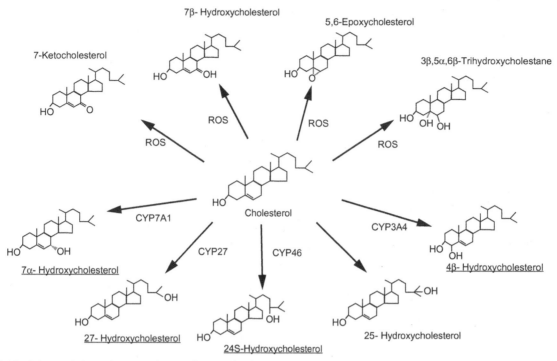

**FIGURE 14.10**   Primary cholesterol oxygenation reactions mediated by different **cytochrome P-450** (CYP) species or occurring nonenzymatically in the presence of **reactive oxygen species** (ROS). The oxygenation of cholesterol into **25-hydroxycholesterol** is catalyzed by **cholesterol 25-hydroxylase**, a nonheme iron protein. The quantitatively most important **oxysterols** present in the human circulation are underlined. *Reproduced from Figure 1 of http://www.jci.org/articles/view/16388.*

**(CYP51A1)** and **squalene synthase (farnesyl diphosphate farnesyl transferase 1)** by way of the negative LXR DNA response elements (**nLXREs**) which are components of each of the genes regulating these enzymes. Both the **sterol regulatory element (SRE)** and nLXRE are involved in the oxysterol-dependent repression of the CYP51A1 gene. This expands information on the negative feedback control of the synthesis of cholesterol discussed in the chapter on lipids. The summary actions of the two liver receptors (LXRα, LXRβ) that form heterodimers with **RXR** (whose ligand can be 9-*cis*-retinoic acid) activate genes that regulate both cholesterol and fatty acid homeostasis.

## Sunlight and Vitamin D

In the bowel wall epithelium, cholesterol is oxidized to 7-dehydrocholesterol that is then transported to the skin. In the skin, sufficient sunlight allows the conversion of **7-dehydrocholesterol** to a precursor (**cholecalciferol**) of the activated form of the vitamin. 7-Dehydrocholesterol, with UV radiation, is converted to cholecalciferol (**vitamin D₃**) in the skin and then to **25-hydroxycholecalciferol** in the liver by **25-hydroxylase**. This product is transported in the blood to the kidney where it is further hydroxylated to **1,25-dihydroxycholecalciferol** (1,25-dihydroxy vitamin D₃, the hormone), the active form that serves for the ligand of the **vitamin D receptor**. The **vitamin D-binding protein** (induced by the vitamin D receptor) carries the hormone in the blood to various cells where dissociation occurs outside the cell, the vitamin D₃ enters the target cell and is carried to the nucleus where the vitamin D receptor resides. The complex (1,25-dihyroxycholecalciferol-vitamin D receptor) then activates specific genes to bring about the vitamin D cellular response. One response is the induction of **calcium-transporting proteins** that transport calcium from the intestinal lumen across the intestinal epithelial cell to the basolateral side of the cell and eventually into the bloodstream. The role of this vitamin is to ensure the proper balance between calcium and phosphorous for the **mineralization of bone**.

Ten minutes of sunlight per day supplies the human requirement for vitamin D (as activated vitamin D is considered in the class of steroids, rather than vitamins, owing to the nature of its receptor; see the chapter on steroids for further elaboration of the vitamin D mechanism). Natural diets do not contain sufficient vitamin D, explaining the need for sunlight to activate the vitamin D precursor in skin. In climates with limited numbers of days of sunlight, children often grow up vitamin D-deficient and remain deficient in adulthood. This condition can lead to an increased predisposition to **colorectal cancer** as this vitamin is known to enhance the function of the immune system among other effects.

# Metabolism of Steroid Hormones Formed From Cholesterol

The steroidal sex hormones and the steroidal adrenal hormones are synthesized from cholesterol as will be discussed in the later chapter on steroid hormones. The major male **sex hormones** (androgens) are testosterone and dihydrotestosterone. The major female sex hormones (estrogens and progestins) are 17β-estradiol and progesterone. The major **adrenal steroidal hormones** are cortisol (glucocorticoid) and aldosterone (mineralocorticoid).

**Aromatase** is an enzyme in female sexual tissues (e.g., mammary gland) that can act on **Δ4-androstenedione** or **testosterone** to form 17β-estradiol or estrone and subsequently to estriol:

cholesterol ⇌ pregnenolone ⇌ 17α-hydroxypregnenolone ⇌ dehydroepiandrosterone ⇌ androstenedione ⇌ *testosterone → (aromatase) ← 17β-estradiol*

pregnenolone ⇌ progesterone ⇌ 17α-hydroxyprogesterone ⇌ *androstenedione → (aromatase) → estrone* ⇌ estriol

Although 17β-estradiol is the preferred ligand for the estrogen receptor, estrone can bind and inhibit receptor action (competitive inhibitor). The conversions of androstenedione to estrone and testosterone to 17β-estradiol are both catalyzed by **aromatase**. The aromatase reaction converting androstenedione to estrone is shown in Fig. 14.11.

Many **mammary cancers** are growth stimulated by **17β-estradiol** and because aromatase can convert **dehydroepiandrosterone** to estradiol, aromatase inhibitors are used to decrease the production of the female sex hormone. Some **inhibitors of aromatase** are: formestane (Lentaron) and exemestane (Aromasin) both of which are steroids and would act as competitive inhibitors as well as nonsteroidal compounds (competitive inhibitors), such as aminoglutethimide, letrozole (Femara) or anastrozole (Arimidex). 17β-Estradiol can be inactivated by oxidation of the 17β-hydroxyl group to the corresponding ketone; hydroxylation at C2 followed by methylation; further hydroxylation or ketone formation at C6, C7, C14, C15, C16, or C18, and by **glucuronidation** of the 3-hydroxyl. The glucuronide of C17 as well as C3 of 17β-estradiol also can occur. These structures are shown in Fig. 14.12. **Estrone** can be derivatized in much the same way.

Testosterone is inactivated by reduction of the 4-ene-3-one and the oxidation of the C17 hydroxyl to the corresponding ketone. Some urinary excretion products of testosterone are androsterone and etiocholanolone (Fig. 14.13).

The **inactivation of progesterone** involves the reduction of the C20 ketone to a hydroxyl and reduction of the 4-ene-3-one resulting in the saturation of the double bond and converting the C3 ketone to a hydroxyl. The excreted products of progesterone are **pregnanediol** and the C3 glucuronide of pregnanediol.

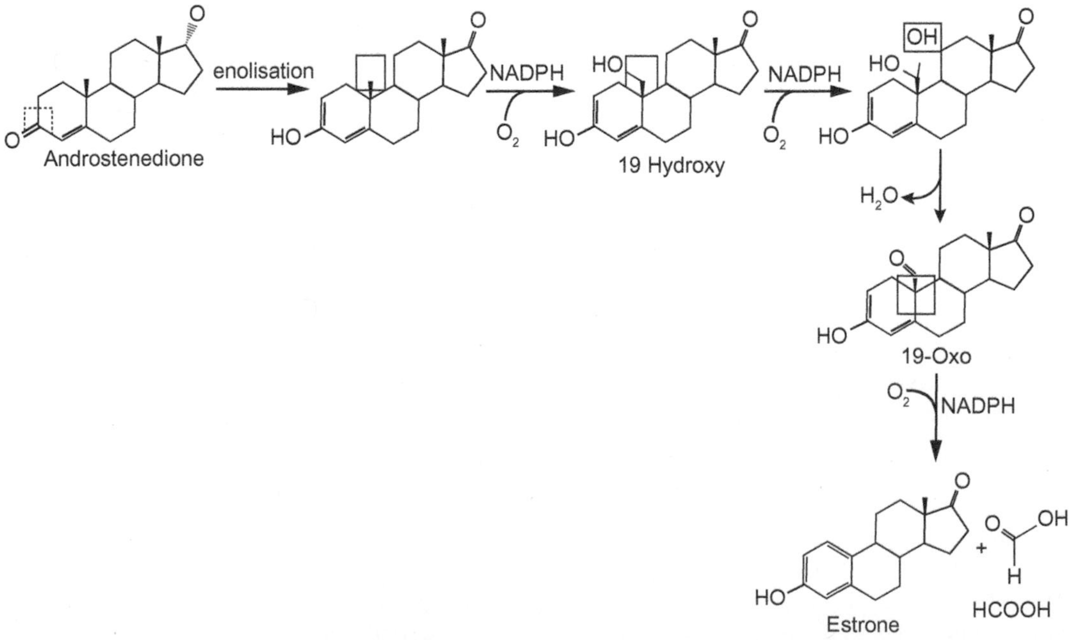

**FIGURE 14.11** Aromatase reaction converting androstenedione to estrone. The reaction is mediated by cytochrome P-450.

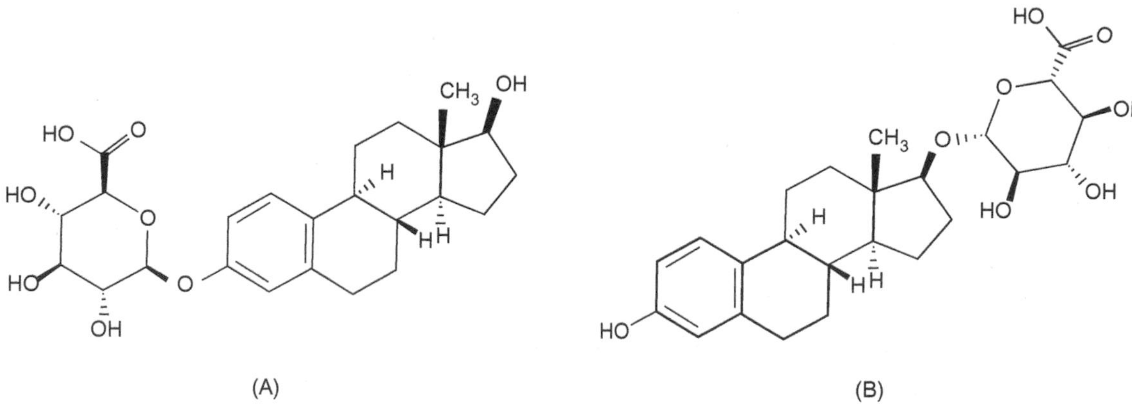

**FIGURE 14.12**  Structures of 17β-estradiol 3-D-glucuronide (A) and of estradiol 17β-D-glucuronide (B).

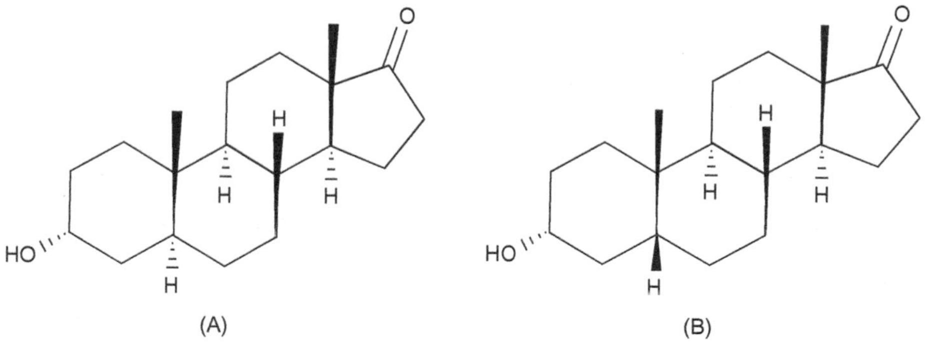

**FIGURE. 14.13**  Structures of **androsterone** and **etiocholanolone**. Both are excreted either as glucuronide- or sulfate-derivatives of the C3 hydroxyl group. (A) Structure of androsterone. (B) Structure of etiocholanolone. The difference between the two structures is in the orientation of the hydrogen atom at C5: in androsterone, the bond to the hydrogen atom is alpha from the plane of the steroid ring system (away from the viewer); in etiocholanolone, the bond to the C5 hydrogen atom is beta from the plane of the steroid ring system (toward the viewer).

Cortisol is the main glucocorticoid secreted by the adrenal cortex. It is inactivated by reduction of the 4-ene-3-one resulting in saturation of the double bond at C4 and conversion of the C3 ketone to a hydroxyl. The C20 ketone is also reduced to a hydroxyl and the side chain can be cleaved. The excretion products of cortisol are **11β-hydroxyandrosterone** and **allotetrahydrocortisone** as well as their C3 glucuronide derivatives.

The major mineralocorticoid aldosterone is inactivated by reduction of the 4-ene-3-one leading to saturation of the double bond at C4 and conversion of the C3 ketone to a hydroxyl. The major excreted form derived from aldosterone is **3α,11β,21-trihydroxy-20-oxo-5β-pregnane-18-al** and the glucuronide derivative of the C3 hydroxyl.

Vitamin D is inactivated by cleavage of the side chain between C23 and C24 to form **calcitroic acid**, the main excretion product.

These various excretion products of the steroid hormones are shown in Fig. 14.14.

Normal levels of excretion products: *pregnanediol*: female urine, 0.2−6 mg/day; in follicular phase, 0.1−1.3 mg/day; luteal phase, 1.2−9.5 mg/day; male urine: 0.2−1.2 mg/day; *17-hydroxycorticosteroids*: female urine, 2.0−6.0 mg/day; male urine, 3.0−10.0 mg/day; *aldosterone metabolites*: 6−25 μg/day on normal salt diet.

## Bile Acid Metabolism

Bile acids are formed from cholesterol in the liver (Fig. 9.12) and undergo considerable circulation through the intestine with 10% being lost in the feces and 90% recovered for further cycling (Fig. 9.13). The cycling process starts after the bile acids are conjugated and leave the liver. For example, **chenodeoxycholic acid** is conjugated with **taurine** in the liver and is then circulated to the gallbladder. It is emptied upon food ingestion and then conjugated cholic acid moves to the duodenum and then to the ileum where partitioning occurs: 10% to the feces and 90% returned to the portal circulation. Of that, 5% enters the systemic circulation and 95% circulates back to the liver and the cycle can begin

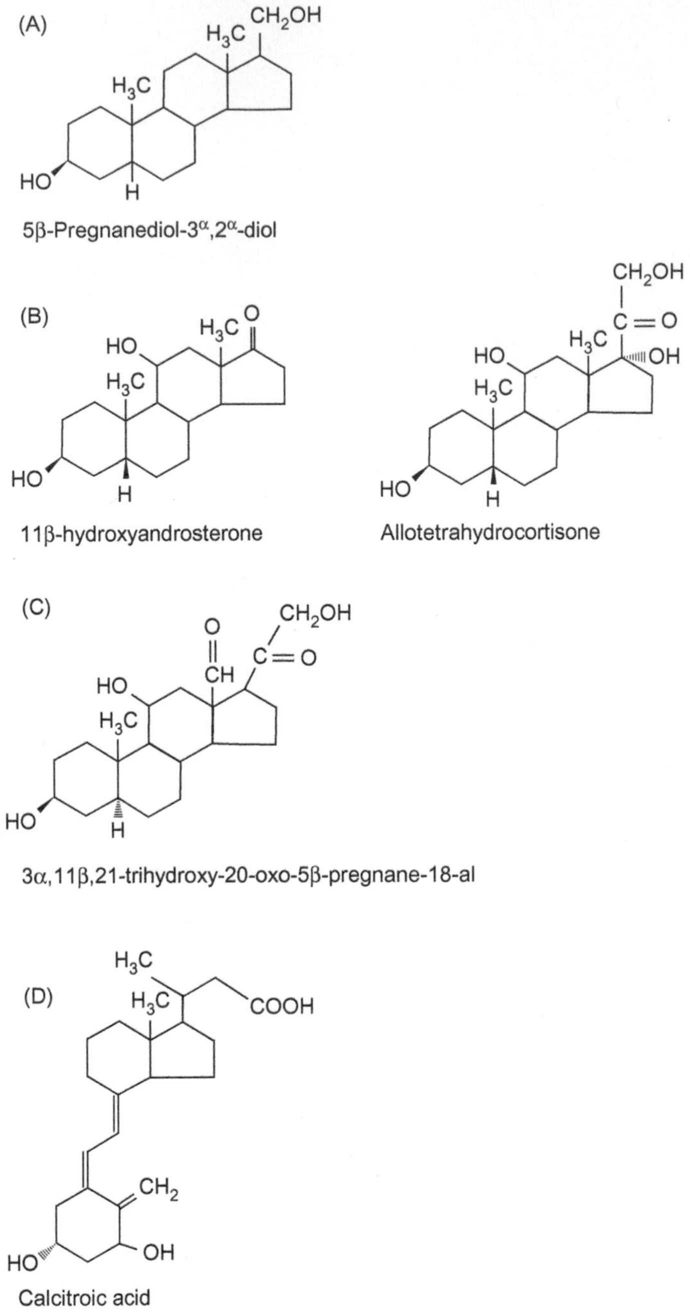

(A)

5β-Pregnanediol-3$^{\alpha}$,2$^{\alpha}$-diol

(B)

11β-hydroxyandrosterone

Allotetrahydrocortisone

(C)

3$\alpha$,11β,21-trihydroxy-20-oxo-5β-pregnane-18-al

(D)

Calcitroic acid

FIGURE 14.14 **Excretion products** of various **steroid hormones**. (A) Excretion product of progesterone as pregnanediol (and its glucuronide). (B) Excretion products of cortisol that also occur as the glucuronide derivatives. (C) Major aldosterone excretion product, also occurring as the glucuronide derivative. (D) Excretion product of vitamin D3.

again. Conjugated cholic acid is metabolized (deconjugated) in the cecum/proximal colon by bacterial **cholylglycine hydrolase** and **7$\alpha$-dehydroxylase**.

## CARBOHYDRATE METABOLISM

Much of the metabolism of glucose in the liver has been presented in Chapter 6, Insulin and Sugars and Chapter 8, Glycolysis and Gluconeogenesis. An overview of **glucose metabolism** in the liver is presented in Fig. 14.15.

Ingested carbohydrate is degraded in the intestine into monosaccharides which cross the intestinal wall, enter the bloodstream, and are taken up by various tissues. The liver is the major organ for the metabolism of glucose, either for

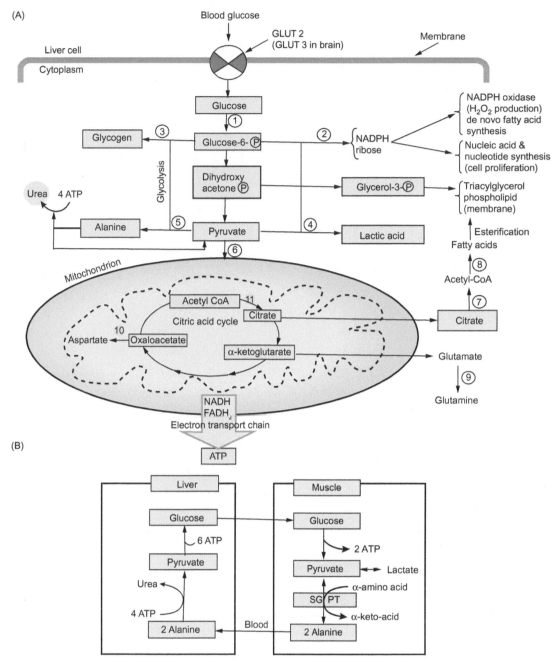

**FIGURE 14.15** (A) Metabolism of glucose in the liver. (1) hexokinase/glucokinase; (2) pentose phosphate pathway; (3) glycogen synthesis; (4) lactate dehydrogenase; (5) alanine aminotransferase = soluble glutamate-pyruvate transaminase (SGPT); (6) pyruvate dehydrogenase; (7) ATP-citrate lyase; (8) fatty acid synthesis; (9) glutamine synthase; (10) aspartate aminotransferase; and (11) citrate synthase. (B) Glucose–alanine cycle by means of which muscle can eliminate nitrogen in the form of alanine, especially during muscle breakdown. Alanine is transported to the liver where it is converted back to pyruvate and then to glucose by way of gluconeogenesis. Glucose can then circulate to muscle for energy formation. There are multiple steps between glucose-6-phosphate and dihydroxyacetone phosphate and pyruvate that are not shown here.

the formation of glycogen (if free, glucose is not needed immediately in the bloodstream) or it enters the bloodstream. Glucose enters the liver cell through the **glucose transporter**, **GLUT2** (GLUT3 is used by the brain to transport glucose). The **normal level of glucose** in the blood varies between 70 and 126 mg/dL; following a meal, the level usually rises but should not exceed 140 mg/dL. There are a dozen or so monosaccharide transporters, but those of interest here are the five glucose transporters (GLUT1, GLUT2, GLUT3, GLUT4, and GLUT5). GLUT1 is present in all human cells. The various tissues and their respective glucose transporters are shown in Table 14.2.

**TABLE 14.2** The GLUT Transporters and Their Specific Distribution in Various Tissues

| Protein | Major Isoform (aa) | $K_m$[a] (mM) | Major Sites of Expression | Proposed Function |
|---|---|---|---|---|
| **Facilitative Glucose Transporters (GLUT)** | | | | |
| $GLUT_1$ | 492 | 3–7 | Ubiquitous distribution in tissues and culture cells | Basal glucose uptake; transport across blood tissue barriers |
| $GLUT_2$ | 524 | 17 | Liver, islets, kidney, small intestine | High capacity low-affinity transport |
| $GLUT_3$ | 496 | 1.4 | Brain and nerves cells | Neuronal transport |
| $GLUT_4$ | 509 | 6.6 | Muscle, fat, heart | Insulin-regulated transport in muscle and fat |
| $GLUT_5$ | 501 | | Intestine, kidney, testis | Transport of fructose |
| $GLUT_6$ | 507 | ? | Spleen, leukocytes, brain | |
| $GLUT_7$ | 524 | 0.3 | Small intestine, colon, testis | Transport of fructose |
| $GLUT_8$ | 477 | 2 | Testis, blastocyst, brain, muscle, adipocytes | Fuel supply of mature spermatozoa; insulin-responsive transport in blastocyst |
| $GLUT_9$ | 511/540 | ? | Liver, kidney | |
| $GLUT_{10}$ | 541 | 0.3 | Liver, pancreas | |
| $GLUT_{11}$ | 496 | ? | Heart, muscle | Muscle-specific; fructose transporter |
| $GLUT_{12}$ | 617 | ? | Heart, prostate, mammary gland | |
| HMIT | 618/629 | ? | Brain | $H^+$/Myo-inositol cotransporter |
| **$Na^+$/Glucose Cotransporters (SGLT)** | | | | |
| $SGLT_1$ | 664 | 0.2 | Kidney, intestine | Glucose reabsorption in intestine and kidney |
| $SGLT_2$ | 672 | 10 | Kidney | Low affinity and high selectivity for glucose |
| $SGLT_3$ | 660 | 2 | Small intestine, skeletal muscle | Glucose activated $Na^+$ channel |

aa, amino acids; ? = unknown.
[a]Net influx for 2-deoxyglucose or glucose.
**Source:** Reproduced from http://www.ncbi.nlm.nih.gov/PMC/articles/PMC2435356/table/T1/.

# Glycolysis

Inside the liver cell, glucose enters the cytoplasmic glycolytic pathway culminating in the formation of pyruvate (Fig. 6.12A). Pyruvate then enters the mitochondrion and together with oxaloacetate forms citrate and continues in the Tricarboxylic Acid Cycle (TCA) (or citric acid cycle or Kreb's cycle) to form ATP, $H_2O$, and $CO_2$. As part of the **glycolytic pathway**, dihydroxyacetone phosphate is formed as an intermediate, and it can also (in addition to the rest of the pathway to form pyruvate) form glycerol 3-phosphate, a precursor of membrane phospholipids. The glycolytic pathway from glucose to pyruvate has been discussed in Chapter 8, Glycolysis and Gluconeogenesis. Hexokinase is the first enzyme in the glycolytic pathway. **Hexokinase** is a family of four enzymes including **glucokinase**, also called **hexokinase IV** in liver and pancreatic islet cells. It catalyzes the conversion of glucose to glucose 6-phosphate. Glucokinase acts as a sensor coupling glucose metabolism to release of **insulin** from the pancreatic β-cell. When glucose binds to the active site of hexokinase, the enzyme undergoes a conformational change so that the 6-hydroxyl of the bound glucose molecule is positioned close to the bound ATP.

In the case of fructose ingested as the primary sugar from sucrose, honey, corn syrup, fruits, and vegetables, it enters the glycolytic pathway by way of hexokinase or fructokinase. **Fructokinase** converts fructose to fructose 1-phosphate by the addition of ATP (producing ADP as product). Fructose 1-phosphate is converted to dihydroxyacetone phosphate plus glyceraldehyde and dihydroxyacetone phosphate is converted to glyceraldehyde 3-phosphate by **triose phosphate isomerase**. Glyceraldehyde is converted to glyceraldehyde-3-phosphate by **triose kinase**. In the starting hexokinase reaction, fructose (plus ATP) yields fructose 6-phosphate (plus ADP) and fructose-6-phosphate (plus ATP) is converted to fructose 1,6-*bis*phosphate by **phosphofructokinase**. The irreversible reactions catalyzed by hexokinase, phosphofructokinase, and pyruvate kinase are all regulated allosterically (see Chapter 5: Enzymes). **Acetyl CoA** inhibits pyruvate kinase, as well as pyruvate dehydrogenase, and activates pyruvate carboxylase so that acetyl CoA is a regulator linking glycolysis to the TCA cycle.

In glycolysis, from one molecule of glucose (6C), the following summarizes the use and production of ATP: in the first step, hexokinase uses one ATP to produce glucose 6-phosphate; following the next step by phosphohexose isomerase to form fructose 6-phosphate, phosphofructokinase produces dihydroxyacetone phosphate (3C) and glyceraldehyde 3-phosphate (3C) with the use of one ATP; then glyceraldehyde 3-phosphate dehydrogenase converts two glyceraldehyde 3-phosphate molecules (one dihydroxyacetone phosphate is converted to one glyceraldehyde 3-phosphate) to 1,3-*bis*phosphoglycerate with the reduction of two $NAD^+$ and generation of six ATP (three ATP/$NAD^+$ reduced) through the electron transport chain (ETC). In the next phosphoglycerokinase reaction forming two molecules of 3-phosphoglycerate from *bis*phosphoglycerate, two ATP are generated from ADP. In the following step, two molecules of 3-phosphoglycerate are converted to two molecules of 2-phosphoglycerate. Then, enolase converts two molecules of 2-phosphoglycerate to two molecules of phosphoenol pyruvate and pyruvate kinase converts two molecules of phosphoenol pyruvate to two enol pyruvates with the generation of two ATP from ADP. Spontaneously, enol pyruvate is converted to keto pyruvate that enters the TCA cycle to form citrate with oxaloacetate. Thus, in all, six molecules of ATP are produced: one ATP lost each in the hexokinase and phosphofructokinase reactions ($-2$ ATP) plus six ATP ($+6$ ATP) produced in the electron transport chain from two $NAD^+$ in the glyceraldehyde 3-phosphate dehydrogenase reaction and the generation of two ATP ($+2$ ATP) in the pyruvate kinase reaction producing enol pyruvate from 3-phosphoenol pyruvate.

In the fructokinase reaction (mentioned above) producing fructose-1-phosphate, the phosphofructose kinase reaction is bypassed; *the phosphofructokinase reaction is the rate-limiting step in glycolysis.* Excessive dietary fructose ($>50$ g/day) (corn syrup-rich drinks, etc.) leads to the accumulation of fructose-1-phosphate that can raise the levels of **acetyl CoA** and **lipogenesis** and may even lead to **gout** because excess dietary fructose can increase the levels of circulating **uric acid** (ingestion of high levels of fructose causes cells to burn ATP rapidly promoting cell death; components of cellular breakdown lead to elevated uric acid in the blood). *Fructose, unlike glucose, maybe does not stimulate efficiently the release of insulin* (fructose is cleared rapidly by the liver and the blood level of glucose is not raised). Lack of insulin response fails to suppress **ghrelin** (the hunger hormone) and fails to stimulate **leptin** (the satiety hormone); consequently, there is a tendency to eat more and develop **insulin resistance**. Thus, some believe that the ingestion of high levels of fructose parallels the incidence of **Type II diabetes**.

**Mannose** and **galactose** are other ingested sugars. Mannose is converted to **mannose 6-phosphate** by **hexokinase**. **Phosphomannose isomerase** converts mannose 6-phosphate to fructose 6-phosphate which is converted to fructose 1,6-*bis*phosphate, thus entering **glycolysis**. Galactose is converted in two steps to **UDP-galactose**, then to UDP-glucose, then to glucose 1-phosphate, then to glucose 6-phosphate then to fructose 6-phosphate and then to fructose 1,6-*bis*phosphate, thus entering glycolysis. Galactose is derived primarily from **lactose** a disaccharide of glucose and galactose. The disaccharide is hydrolyzed in the intestine by **lactase** (**β-galactosidase**). Lactase is poorly expressed or missing altogether in persons with **lactose intolerance**. **Lactaid** is a commercial product (containing lactase) that can be ingested prior to consuming a lactose-containing food (dairy products).

## Gluconeogenesis

Gluconeogenesis is a process converting pyruvate to glucose in the direction opposite to that of glycolysis although the mechanism is not simply the reversal of glycolysis. Having the means to produce glucose from noncarbohydrate sources is important for the **brain** and **red blood cells** since glucose is the primary source of energy for the brain and red blood cells derive all of their energy from glycolysis because they do not possess mitochondria. Liver is the primary site for gluconeogenesis although the kidney can carry out the same steps but contributes only 10% of gluconeogenic activity (the liver produces 90% of gluconeogenic activity). *In glycolysis, the steps catalyzed by* **hexokinase, phosphofructokinase**, *and* **pyruvate kinase** *are essentially irreversible.* These steps need to be bypassed in gluconeogenesis. In gluconeogenesis, six high-energy phosphate compounds are used as the overall reaction shows:

$$2 \text{ pyruvate} + 4ATP + 2GTP + 2NADH + 4H_2O \rightarrow \text{glucose} + 4ADP + 2GDP + 6Pi + 2NAD^+ + 2H^+$$

with a free energy change of $-37$ kJ/mol (J = **Joule**; 1 J = $\sim 0.24$ cal; 1 cal = $\sim 4.2$ J; therefore, $-37$ kJ/mol = $-\sim 2220$ cal or $-\sim 2.22$ Cal).

Pyruvate is first carboxylated to form oxaloacetate by the biotin cofactor-containing **pyruvate carboxylase** in the mitochondrial matrix. **Acyl CoA** activates this enzyme and is bound to an allosteric binding site on the enzyme. As stated previously, the activation by acyl CoA is a factor in physiologic regulation. When cellular [ATP] is low

(2ATP are used in this reaction) and [acyl CoA] is low, pyruvate carboxylase is not much active and pyruvate enters the TCA cycle to generate ATP. When [ATP] and [acyl CoA] are high, the energy level of the cell is high and metabolites are processed to glucose via gluconeogenesis through the activation of pyruvate carboxylase as the first step. In the next step, **Phosphoenol pyruvate (PEP) carboxykinase** (in mitochondria and cytosol) decarboxylates and phosphorylates oxaloacetate to form phosphoenolpyruvate. Guanosine triphosphate (GTP) is used in this reaction as the phosphoryl donor and guanosine diphosphate (GDP) and $CO_2$ are the other products. The same enzymes that are used in glycolysis between PEP and fructose-1,6-bisphosphate are used in **gluconeogenesis** because these reactions are virtually reversible (**phosphoglycerate mutase, phosphoglycerokinase, triose phosphate isomerase,** and **aldolase**). The reaction of fructose-1,6-bisphosphate to fructose-6-phosphate is catalyzed irreversibly by **fructose-1,6-*bis*phosphatase** (to get around the irreversibility of phosphofructokinase in glycolysis). This reaction generates an ATP from ADP and Pi from hydrolysis by $H_2O$. Fructose-6-phosphate can be isomerized by the reversible reaction catalyzed by the same isomerase as used in glycolysis to form glucose-6-phosphate. The last reaction to form free glucose cannot use hexokinase since the phosphorylation reaction is irreversible. Instead, **glucose-6-phosphatase**, an endoplasmic reticulum (ER) membrane-bound enzyme, is used for the final reaction to form free glucose. Glucose-6-phosphate is transported from the cytosol to the lumen of the ER by a carrier protein (**glucose-6-phosphate transporter, G6PT**) for the final irreversible reaction by glucose-6-phosphatase to occur. *Both brain and muscle do not express this carrier protein accounting for the absence of the ability of these organs to produce free glucose, so these organs must import it from the bloodstream.* There is also a reversible glucose transporter in the ER that must account for the return of free glucose into the liver cell cytosol. Gluconeogenesis is summarized by Fig. 14.16.

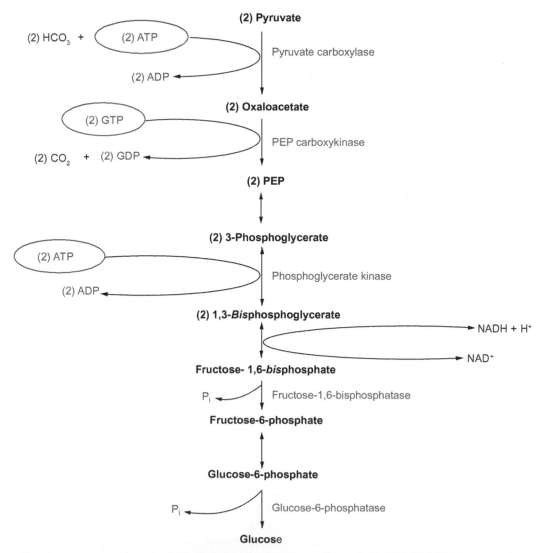

**FIGURE 14.16** Reactions involved in gluconeogenesis. *From http://www.namrata.co/wp-content/uploads/2012/02/pep.jpg?w = 300.*

There are about 190 g of glucose stored as **glycogen** in the liver. The total blood volume of a normal human is about 5 L (5000 mL) and an approximately normal level of blood sugar is about 120 mg/100 mL (on the higher end), so the total amount of glucose in the bloodstream would be in the region of 6 g. Glucose is synthesized from noncarbohydrate precursors to maintain the blood glucose level if glycogen is used up by **strenuous exercise** or by **fasting** for more than a day. The noncarbohydrate precursors of glucose are pyruvate, lactate, oxaloacetate, glycerol, and amino acids, and these substances enter gluconeogenesis through pyruvate, oxaloacetate, or dihydroxyacetone phosphate.

With the exceptions of the enzymes located in the mitochondrion as indicated above, most of the gluconeogenic reactions occur in the cytosol and this is also true for glycolysis. The regulatory processes are such that when glycolysis is active, gluconeogenesis is subdued and vice versa, permitting the synthesis and storage of glucose as **glycogen** when the energy level of the cell is high (high ATP and acetyl CoA). Conversely, when the energy pools of the cell are low, glucose is metabolized through the glycolytic pathway to pyruvate that enters the TCA cycle for the production of ATP energy.

In the liver and kidney TCA cycles, the enzyme, **succinyl-CoA synthase** (also known as **succinate thiokinase**), produces GTP rather than ATP in its reaction, whereas in the TCA cycles of other tissues, ATP, not GTP, is produced by this similar enzyme. The GTP, produced by succinyl-CoA synthase, is used in the **phosphoenolpyruvate carboxykinase (PEPCK)** reaction providing a direct link between the **TCA cycle** activity and **gluconeogenesis**. By this means, removal of **oxaloacetic acid** for gluconeogenesis is prevented and the activity of the TCA cycle is thus ensured. In starvation, the level of circulating glucose decreases. Consequently, glucagon is released from the pancreas when circulating glucose levels are low and **glucagon** stimulates the production of PEPCK that, in turn, stimulates gluconeogenesis.

## Vigorous Exercise and Gluconeogenesis

**Muscle cells** use up their oxygen and become anaerobic during vigorous exercise. Under these conditions pyruvate is converted to **lactate** by **lactate dehydrogenase (LDH)** and in this reaction, NADH is converted to $NAD^+$:

$$\text{pyruvate} + \text{NADH} \rightleftharpoons (\textbf{LDH}) \rightleftharpoons \text{lactate} + \text{NAD}^+ + \text{H}^+$$

The $NAD^+$ is needed by glycolysis specifically to drive the **glyceraldehyde phosphate dehydrogenase** reaction that converts glyceraldehyde 3-phosphate to 1,3-*bis*phosphoglycerate in the pathway converting glucose to pyruvate (see Fig. 6.12A). The lactate resulting from muscle metabolism is carried by the bloodstream to the liver where the $NAD^+$/NADH is high, and it is oxidized back to pyruvate by liver LDH and subsequently converted to glucose by gluconeogenesis using up six high-energy phosphate compounds (ATP and GTP). The glucose, thus formed in the liver, is returned to the muscle by way of the bloodstream where it can enter muscle glycolysis and produce two ATP in each conversion of glucose to pyruvate (and pyruvate is converted to lactate, if vigorous exercise is still going on). This cycle, where metabolism is shared by muscle and liver is known as the **Cori cycle**.

## THE TRICARBOXYLIC ACID CYCLE (TCA CYCLE), CITRIC ACID CYCLE, OR KREB'S CYCLE

The main power generator where fuels are metabolized for the formation of ATP (and $CO_2$ and **metabolic water** [amounting to about 300 mL/day in the adult]) is in the mitochondrial TCA cycle that utilizes pyruvate from glycolysis (and other compounds that can segue into the cycle by conversion to the various TCA cycle intermediates). One complete turn of the cycle, starting from two pyruvates, can generate as many as 30 ATP molecules; from two acetyl CoA molecules, 24 ATPs can be generated. The overall cycle is shown in Fig. 14.17.

The other metabolites (from lipids, carbohydrates, proteins, and amino acids) flow into the TCA cycle that is the ultimate metabolic machine producing most of the ATP for energy-requiring reactions in the cell (Fig. 14.18).

Pyruvate (3C), as the product of glycolysis starting with a **glucose** molecule (6C) generates *four ATP* molecules (six ATP are generated and two ATP are used up in glycolysis). **Acetyl CoA** (2C) can be formed from pyruvate (3C) by **pyruvate dehydrogenase** (reducing one $NAD^+$ to produce *three ATP* through the electron transport chain and generating one $CO_2$) or acetyl CoA can be derived from mitochondrial fatty acid β-oxidation. The **TCA cycle** begins with the condensation of acetyl CoA (2C) and oxaloacetic acid (4C) to form citric acid (6C) catalyzed by **citrate synthase** (using one molecule of $H_2O$). **Aconitase** converts citrate (6C) to *cis* aconitate, generating one molecule of $H_2O$. Aconitase also converts *cis* aconitate to isocitrate, using one molecule of $H_2O$. Next, **isocitrate dehydrogenase** converts isocitrate (6C) to oxalosuccinate with the reduction of one $NAD^+$ (generating *three ATP* through the ETC). Again, isocitrate dehydrogenase converts oxalosuccinate to α-ketoglutarate with the generation of 1 $CO_2$. α-Ketoglutarate is converted to succinyl CoA by the **α-ketoglutarate dehydrogenase** complex with the entry of 1 CoA-SH and the reduction

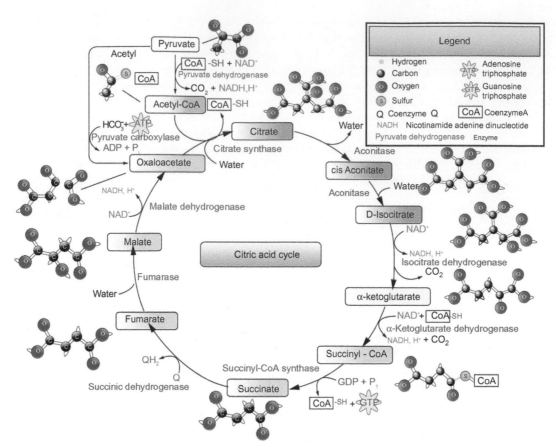

**FIGURE 14.17**   Overview of the TCA cycle. Pyruvate is converted to **acetyl CoA** and the oxidation of the acetyl group generates two molecules of $CO_2$. The reductions of coenzymes ($NAD^+$ to $NADH + H^+$), throughout the cycle become reoxidized through the electron transport chain linked to the formation of ATP. *Reproduced from http://upload.wikimedia.org/wikipedia/commons/0/0b/Citric_acid_cycle_with_aconitate_2.svg.*

of 1 $NAD^+$ (to yield *three ATP* through the ETC) plus one molecule of $CO_2$. **Succinyl CoA synthase** (**succinate thiokinase**) next converts succinyl-CoA to succinate, producing *one GTP* (equivalent to one ATP) from GDP and releasing CoA-SH. **Succinate dehydrogenase** converts succinate to fumarate with the reduction of one FAD to $FADH_2$ (*+2 ATP* through the ETC). Fumarate is converted to L-malate by **fumarase**, using one $H_2O$. Malate is finally converted to oxaloacetate by **malate dehydrogenase** to complete one turn of the cycle with the reduction of one $NAD^+$ to $NADH + H^+$ (to generate *three ATP* in the ETC). Thus, 15 ATPs are generated from 1 pyruvate molecule through the TCA cycle and 12 ATP are produced starting with acetyl CoA. There would be 24 ATP molecules produced from 2 acetyl CoA molecules (from glucose), or 30 ATP molecules produced from 2 pyruvates and 6 ATP are generated while 2 ATP are used up from one glucose molecule, producing 4 ATP through glycolysis; the **total production of ATP** from 1 molecule of glucose through glycolysis and the TCA cycle is 34 ATP.

## The Electron Transport Chain

For each molecule of $NADH + H^+$ oxidized, the following products derive:

$$NADH + H^+ \rightarrow NAD^+ 2H^+ 2e^-$$

The electrons flow through the electron transport chain (ETC) to culminate in the production of ATP from ADP by **ATP synthase** at the terminal end of the chain. The overall process of the oxidation of coenzymes to produce ATP is called **oxidative phosphorylation**. NADH is an electron donor (see reaction above); these electrons flow through the **electron transport chain (ETC)** and end up by combining with molecular oxygen, an electron acceptor, to form water ("metabolic water"). The transport of electrons across the ETC generates free energy and this energy is used to produce ATP from ADP and inorganic phosphate by **ATP synthase** at the end of the chain. **Protons** are pumped from the mitochondrial matrix across the inner mitochondrial membrane to the intermembrane space, creating an electrical gradient

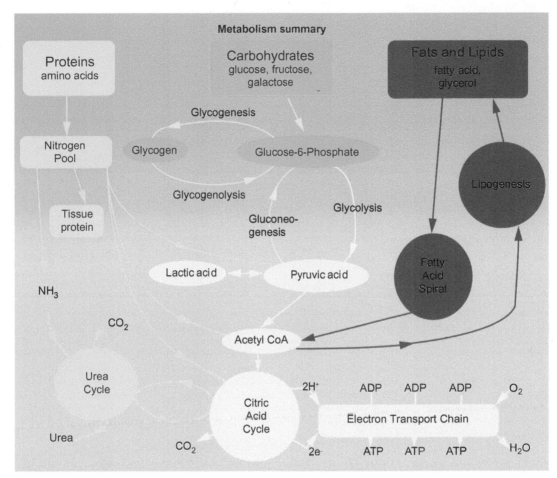

**FIGURE 14.18**  Overview of the general metabolism of proteins, carbohydrates, and lipids that feed into the TCA cycle (citric acid cycle) for the production of ATP energy via the electron transport chain. The production of electrons from the oxidation of reduced coenzymes (**NADH** and **FADH**), after their reduction, are processed through the electron transport chain to generate ATP from ADP. *Reproduced from http://www.elmhurst. edu/~chm/vchembook/images/590 metabolism.gif.*

and a pH gradient (**chemiosmosis**). There are more positive charges on the outer membrane than inside and, correspondingly, there is a lower pH on the outer membrane than inside. Oxidation is coupled to phosphorylation by the proton gradient and electron transport provides the energy to produce ATP.

Because the conversion of succinate to fumarate, catalyzed by **succinic dehydrogenase**, utilizes $FADH_2$ and the oxidation of $FADH_2$ to FAD occurs at complex II, one step beyond complex I where protons from NADH enter, the oxidation of $FADH_2$ produces one less ATP (two ATP) than NADH + $H^+$ (three ATP) because the overall free energy change from $FADH_2$ is less than that from NADH. The electron transport chain is shown schematically in Fig. 14.19.

Converting all of the energy derived from the oxidation of NADH + $H^+$ to $NAD^+$, would lead to the synthesis of about seven molecules of ATP from ADP + Pi. The components of the ETC are the four complexes plus the terminal ATP synthase.

Complex I is **NADH dehydrogenase** (or, **NADH-coenzyme Q reductase**) that accepts electrons from NADH and serves as the link to the ETC, a link to glycolysis, fatty acid oxidation, and the TCA cycle. Complex I consists of about 30 protein subunits and has a molecular weight of about 850,000 Da; its cofactors are **FMN (flavin mononucleotide**) and about 7 **Fe—S clusters** (**iron—sulfur clusters**; up to 26 iron atoms are bound). It contains a substrate-binding site for NADH in the matrix portion and a binding site for CoQ in the lipid core. This complex transfers two electrons to coenzyme Q from NADH. After binding NADH, two electrons are transferred (as a hydride, $H^-$) to FMN to generate $NAD^+$ and $FMNH_2$. The electrons are transferred to a series of iron—sulfur clusters. One electron at a time is transferred to CoQ that can diffuse in the bilipid layer of the inner mitochondrial membrane because it contains an isoprenoid tail structure that has both hydrophobic and hydrophilic properties. The action of complex I results in the transport of protons from the matrix side of the inner mitochondrial membrane to the inter membrane space. Here, protons

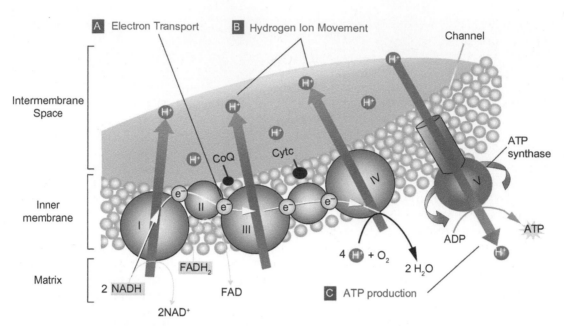

**FIGURE 14.19** Drawing of a hypothetical electron transport chain. There are four protein complexes labeled I, II, III, and IV. A portion of complex I is located in the mitochondrial matrix and the other portion is embedded in the inner mitochondrial membrane. The last member of the chain is the enzyme, ATP synthase. Hydrogen ions from NADH + H$^+$ (at complex I) or from FADH$_2$ (at complex II) move into the intermembrane space along the outside of the inner membrane and are imported through a proton channel that is part of ATP synthase. The electrons from NADH or from FADH$_2$ flow along the complexes and finally interact with molecular oxygen to form water (along with protons). ATP synthase forms ATP from ADP + Pi with energy provided by electron transport. *FADH$_2$*, flavin adenine dinucleotide (reduced); *CoQ*, coenzyme Q or **ubiquinone**; *Cyt c*, cytochrome *c*; e$^-$, electron. *Redrawn from http://www.teachersdomain.org/asset/tdc02_img_electronchai/. From Biology, Kenneth R Miller and Joseph Levine ©2002 by Pearson Education, Inc. Reproduced by permission of the publisher.*

accumulate and generate a **proton motive force** (energy generated by transfer of protons and electrons across an energy-transducing membrane for use in chemical or other kinds of work). There are two H$^+$ transported per electron.

Complex II is **succinate dehydrogenase** (or **succinate-coenzyme Q reductase**), the only enzyme of the TCA cycle that is membrane-bound and is the link in the ETC to the TCA cycle. This enzyme has four subunits with a molecular weight of 140,000 Da. Its cofactors are **FAD** (two subunits are FAD-binding proteins) and two of the subunits are Fe−S proteins. The **Fe−S clusters** are in various stoichiometries: of the three Fe−S units, one is 4Fe−4S, a second is 3Fe−4S, and the third is 2Fe−2S. Complex II has substrate-binding sites for succinate in the matrix portion and for CoQ in the lipid core. Complexes I and II both produce reduced **Coenzyme Q** (CoQH$_2$). Two electrons that are transported from NADH to CoQ are coupled to the transport of four protons across the membrane. First, succinate is bound; then a hydride (H$^-$) is transferred to FAD, producing FADH$_2$ and **fumarate**. Electrons, one at a time, are transferred to the Fe−S units. Then, two electrons are transferred, one at a time, to CoQ to generate CoQH$_2$. The total free energy change is −72.4 kJ/mol; this is insufficient to drive the transport of protons across the inner mitochondrial membrane and accounts for nearly two ATP from FADH$_2$ compared to nearly three ATP from NADH.

Complex III is **coenzyme Q reductase** (also **coenzyme Q-cytochrome *c* reductase**) and transfers electrons from CoQH$_2$ for the reduction of **cytochrome *c***. The complex contains one cytochrome *c* and two **cytochrome *b*** types. A larger portion of the structure of this protein dimer extends into the mitochondrial matrix and the rest extends outward into the mitochondrial inter membrane space. A **Fe−S protein** is contained in the enzyme, and it appears to have mobility within the structure that facilitates the transfer of electrons. The transport of electrons occurs through a complicated set of reactions called the "**Q cycle.**" The result of the steps in the Q cycle is that two electrons are transported to cytochrome *c1* and four protons, in total, are released into the inter membrane space. The electrons on cytochrome *c1* are transferred to cytochrome *c* (the only soluble cytochrome) that carries electrons by diffusing into the inter membrane space carrying electrons from the cytochrome *c1* heme to the Cu$_A$ site of complex IV.

Complex IV is **cytochrome *c* reductase** (also known as **cytochrome *c* oxidase**) that transfers electrons from cytochrome *c* for the reduction of molecular oxygen to form water:

$$4\text{cyt } c \ (\text{Fe}^{2+}) + 4\text{H}^+ + \text{O}_2 \rightarrow 4\text{cyt } c \ (\text{Fe}^{3+}) + 2\text{H}_2\text{O}$$

It has up to 10 subunits and has a molecular weight of 162,000 Da. Complex IV has four prosthetic groups: heme $a$, heme $a_3$, $Cu_A$ (copper A) and $Cu_B$ (copper B). $Cu_A$ is associated with **cytochrome $a$** and $Cu_B$ is associated with **cytochrome $a_3$**. The enzyme contains two substrate-binding sites, one for **cytochrome $c$** (a single polypeptide chain of 13,000 Da whose prosthetic group is **heme $c$**) in the inter membrane side and the other for $O_2$ in the matrix portion. Complex IV has two binding-sites, one for **cytochrome $c_1$** and the other for cytochrome $a$. The copper sites transfer electrons one at a time. Cytochrome c binds to Complex IV facing the side of the inter membrane space and transfers an electron to $Cu_A$ that also is located on the portion of the protein in the inter membrane space. Cytochrome $c$, now oxidized, moves off of the enzyme into the inter membrane space. $Cu_A$ transfers the electron to cytochrome $a$ whose iron component is close to $Cu_A$ on the enzyme. The electron is then passed on to the cytochrome $a_3$, in close range toward the matrix side and then finally, the electron is moved to the "electron center," housing two electrons, and the binding of $O_2$ to the center occurs. Two protons are then bound and the entry of another electron leads to the dissociation of the $O-O$ bond to generate $Fe^{4+}$ in the metal (Fe) center. A fourth electron now forms a hydroxide which is protonated and leaves the center as $H_2O$.

**ATP synthase** is considered as the fifth complex (**Complex V**). It consists of two parts, $F_1$, the enzyme, and $F_0$, the **proton channel** (Fig. 14.19). $F_1$ is located outside of the membrane in the matrix, whereas $F_0$, the proton channel, is located in the inner membrane. These two structures are linked together by central and peripheral vertical structures (stalks). The catalytic center of $F_1$ is joined to a rotary mechanism of the subunits of the central stalk and to proton translocation. The enzyme, $F_1$, has five subunits ($\alpha$, $\beta$, $\gamma$, $\delta$, and $\varepsilon$) and the mitochondrial membranous proton channel has three main subunits (A, B, and C) and six additional subunits (d, e, f, g, F6, and AGL). The overall action of Complex V is to produce water from protons and electron energy and to produce ATP from ADP plus Pi.

Fig. 14.20 shows the **TCA cycle** in terms of the production of NADH from NAD, the sources of high-energy phosphate and respiratory chain (ETC).

## NUCLEIC ACID METABOLISM

The syntheses of purines and pyrimidines, purine interconversions, the salvage pathway (purine conversion of hypoxanthine to IMP; guanine to GMP; adenine to AMP), the **catabolism of purines** (to uric acid), and pyrimidines (to malonyl CoA and methylmalonyl CoA; thymine to thymidine and TMP; uracil to UMP) already have been reviewed in Chapter 10, Nucleic Acids and Molecular Genetics.

Deamination of purine and pyrimidine bases sometimes occurs. **Cytosine** can be deaminated to form **uracil**. **5-Methylcytosine** can be deaminated to form **thymine**. **Guanine** can be deaminated to form **xanthine** and **adenine** can be deaminated to form **hypoxanthine**.

## RNA Degradation

There appears to be an active **RNA degradation** system in cells because they transcribe more RNA than accumulates. The useful lives of RNA species vary: ribosomal RNA would be expected to have a longer lifetime than excised **introns** (nuclear degradation) or **spacer fragments**, for example. The duration of lifetimes of mRNAs must be closely regulated, although the story of how this occurs may not be complete. Obviously, those RNA molecules that exhibit defects from improper processing, folding, or assembly with proteins are rapidly identified by a surveillance system and degraded. **mRNA degradation** is controlled, at least in part, by **microRNAs (miRNAs)** and **small-interfering RNAs (siRNAs)**. These small RNAs ($\sim 21-26$ nucleotides in length) can trigger **endonuclease cleavage** or possibly they can stimulate the rate of mRNA decapping. In humans, the only endonuclease capable of mRNA cleavage is Ago2, an **Argonaute protein**. Base-pairing at the $5'$ end of the miRNA (residues $2-7$) and the mRNA being attacked establishes the interaction. Base-pairing between bases 10 and 11 of miRNA, and the mRNA target establishes the cleavage site. Fragments of mRNA generated by **RISC** (Fig. 10.45) cleavage become degraded by the general mRNA degradation machinery. *Most mRNA degradation occurs using the CCR4/Not complex, targeted to the 3'UTR of the mRNA and directed by elements and proteins binding to this part of the mRNA.* A miRNA can function with RISC and the **TTP protein** (sequence-specific RNA-binding protein) to target a mRNA containing an ARE (AU-rich element) for degradation. These are complex mechanisms bearing on RNA degradation and the repression of translation.

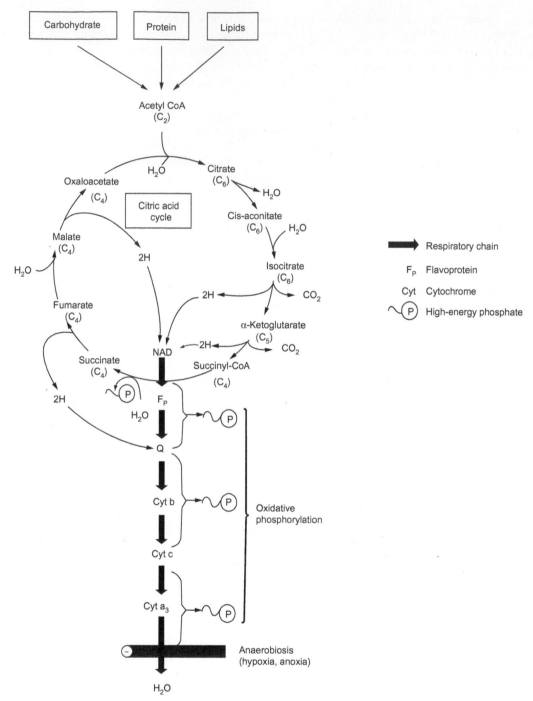

**FIGURE 14.20** The TCA cycle starting with acetyl CoA. The release of reducing equivalents is shown for each relevant step in the cycle, and these are used to reduce NAD$^+$ to NADH + H$^+$. The conversion of succinate to fumarate also produces 2 H but these are shuttled to coenzyme Q10 (Q; complex II) and give rise to two ATPs, whereas the other reducing equivalents each give rise to three ATPs (2 H is shuttled into complex I in each case). The respiratory chain (ETC) is shown with the production of ATP equivalent for each site.

## DNA Degradation

In the adult, cellular DNA is relatively stable, although individual bases may be exchanged without degrading the DNA molecule. However, in the process of **programed cell death** (**apoptosis**), DNase activity is activated to produce a laddering (discrete characteristic sizes of digestion products) of the DNA digestion products. In chromatin DNA, the linker sites are the only part of the DNA that is not tightly wrapped around histones and, therefore they are exposed to potential cleavage. These sites can be cleaved by **caspase-activated DNase** (**CAD**). Normally, this enzyme is complexed

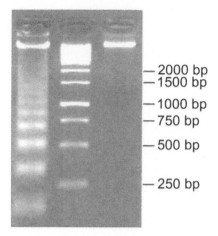

— 2000 bp
— 1500 bp

— 1000 bp
— 750 bp

— 500 bp

— 250 bp

**FIGURE 14.21**  Photograph of a ladder of DNA degradation products from apoptosis. There are three lanes in this gel. On the right is a control, undegraded DNA. In the middle lane are DNA markers showing various fragment sizes as indicated by the numbers on the *right*. On the *left* is a lane showing DNA fragmentation from apoptosis producing a laddering effect of multiples of inter nucleosomal stretches. *Reproduced from http://en.wikipedia.org/wiki/Apoptotic_DNA_fragmentation.*

**TABLE 14.3 Comparison Between Apoptosis and Necrosis**

| Apoptosis | Necrosis |
|---|---|
| **Morphologic Criteria** | |
| Membrane blebbing | Loss of membrane integrity |
| Cell shrinkage and formation of apoptotic bodies | Cell swelling and lysis |
| Lack of inflammatory response | Significant inflammatory response |
| Lysosomal preservation | Lysosomal leakage |
| **Biochemical Criteria** | |
| Induction by physiological stimuli disturbances | Induction by nonphysiological disturbances |
| Energy requirement | Lack of energy requirement |
| Macromolecular synthesis requirement | Lack of macromolecular synthesis requirement |
| *De novo* gene transcription | Lack of *de novo* gene transcription |
| Nonrandom fragmentation of DNA | Random digestion of DNA |

**Source:** Reproduced from http://www.scielo.org.ar/scielo.php?script = sci_arttext&pid = S0327-95452005000200001#tab1.

with an inhibitor (**ICAD**) and, in the process of apoptosis, **caspase 3** cleaves ICAD so that CAD becomes activated. As a result of the cleavage of the linker DNA by CAD, fragments of internucleosomal length and multiples are generated; these are about 180 base pairs. Because of this, the DNA fragments can be separated on a sizing gel displaying the characteristic size and multiples of the minimal degradation product as shown in Fig. 14.21.

In the process of **cell necrosis**, that is, less organized than the events of apoptosis, DNA is more randomly broken down and the digestion products do not produce a laddering effect but rather a smearing of the DNA breakdown products. Generally, apoptosis is a genetically driven program triggered by specific physiological molecules, whereas necrosis is nonphysiological and generated by toxins or other damaging environmental agents (Table 14.3).

An overview of the **process of apoptosis** is shown in Fig. 14.22.

DNA can be damaged and then repaired by various DNA repair enzymes.

## DNA Damage and Repair

The human genome has over 130 repair genes encoding mRNAs for proteins that can screen the genome to eliminate damage. Damage to the genome can occur through radiation, such as by **UV irradiation**, **X-rays**, and by a large variety

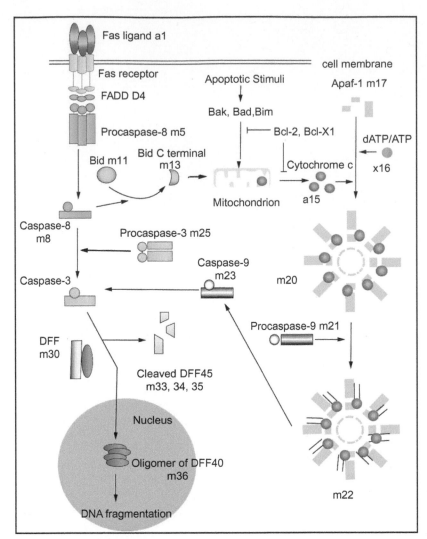

FIGURE 14.22 The proposed steps of apoptosis induced, in this case, by **Fas ligand** [type-II transmembrane protein, a member of the tumor-necrosis family (**TNF**)]: Fas ligand exists as a trimer that binds and activates its receptor by induction of receptor trimerization. Activated receptors recruit adapter molecules, such as **Fas-associating protein with the death domain (FADD)**; FADD recruits **procaspase 8** to the receptor complex where it undergoes autocatalytic activation. Activated **caspase 8** activates **caspase 3** through 2 pathways: caspase 8 cleaves **Bcl-2 interacting protein Bid** and its C-terminal translocates to mitochondria where it triggers **cytochrome c** release. Released cytochrome c binds to **apoptotic protease activating factor-1 (Apaf-1)** together with dATP and **procaspase 9** and activates it. **Caspase 9** cleaves procaspase 3 to its activated form. In an alternate pathway, **caspase 8** cleaves procaspase 3 directly and activates it. Caspase 3 cleaves DNA to generate inter nucleosomal length fragments and multiples thereof (**DNA fragmentation ladder**). **DFF**, generated from caspase 3 cleavage of DFFA; DFF is a heteromeric protein functioning downstream of caspase 3 to trigger DNA fragmentation. *Reproduced from http://genomicobject.net/member3/GONET/apoptosis.html.*

of chemical agents. There are three major mechanisms for repairing DNA damage: **base excision repair**, **nucleotide excision repair**, and **mismatch repair**. Cytosine can be altered to uracil and 5-methylcytosine to thymine by deamination. A common deaminating agent is **nitrous acid** ($HNO_2$) that can convert cytosine to uracil, adenine to hypoxanthine, and guanine to xanthine. Hydrogen bonding of the modified base thus disrupted results in mispairing. **Alkylating agents**, such as ethylmethane sulfonate ($CH_3CH_2OSO_2CH_3$), sulfur mustard (di-(2-chloroethyl)sulfide; $ClCH_2CH_2SCH_2CH_2Cl$)), and nitrogen mustard (di-(2-chloroethyl)methylamine; $ClCH_2CH_2N(CH_3)CH_2CH_2Cl$)), can alter guanine, normally paired with cytosine, to **7-ethylguanine**, which, instead, would pair with thymine and potentially lead to a **mutation**. A base modified by deamination or alkylation is an "**abasic site**" (**AP site**), and this site can be recognized by **DNA glycosylase**, and it can remove the altered base. **AP endonuclease** removes the AP site plus neighboring nucleotides and the gap, so generated, is filled by **DNA polymerase I** and **DNA ligase** (Fig. 14.23).

DNA damage by **ultraviolet light** can result in the formation of a dimer. The dimer is removed enzymatically similar to the removal of a single base (Fig. 14.23). The resulting gap is filled by **DNA polymerase I** and **DNA ligase**.

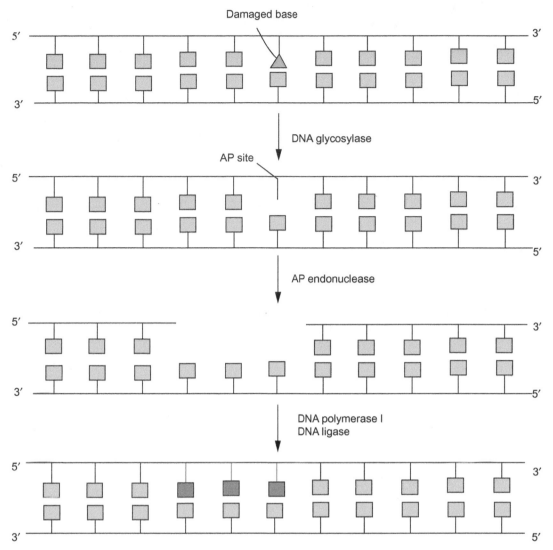

**FIGURE 14.23**   DNA repair by base excision. *Reproduced from Litwack, G.* Human Biochemistry and Disease, *Academic Press/Elsevier, page 574, 2008.*

**Mismatch repair** is often needed for newly synthesized daughter single-stranded DNA in which abnormal pairs, such as G/T or A/C can be formed. This can occur when, during DNA synthesis (S phase of the cell cycle), a base can be tautomerized. Normally the nucleotide base is in the keto form, considering the equilibrium between the tautomers, ketone versus aldehyde: $RCH_2C = O(R') \rightleftharpoons R = COH(R')$. The aldehyde forms could make the abnormal base pairing possible. When this occurs the signal-directed **mismatch proofreading system** seeks out the damaged strand. The signal consists of nicks in the newly synthesized DNA strand that exist prior to the sealing event by DNA ligase. A mismatch would cause a deformity and a few too many bases in the newly formed strand would be removed. The proteins involved are the **Mut proteins** of which there are three. **MutS** first forms a dimer, $MutS_2$, which detects the mismatch on the daughter strand and binds to it. Then $MutL_2$ binds to the $MutS_2$-DNA and mediates the action of $MutS_2$ and MutH causing the activation of MutH. Activated MutH nicks the daughter DNA strand near the methylated site and recruits **DNA helicase II**. This enzyme separates the two strands of DNA in the 3′ to 5′ polarity. The entire complex of MutS—MutH—MutL slides along the DNA toward the mismatch liberating the candidate strand to be cut off. The single-stranded DNA tail is removed by an exonuclease. The process is completed beyond the mismatch site. The site is resealed by DNA ligase and methylated by **Dam methylase (DNA adenine methylase)**. One version of this scenario is illustrated by Fig. 14.24.

The names of the proteins mentioned here derive from early studies with *Escherichia coli*. These names are used because the mammalian process is not as clearly understood as is the bacterial process. In eukaryotes, there are homologs of the *E. coli* proteins: **MSH1** and **MSH5** for MutS and **MLH1, PMS1** and **PMS2 (DNA mismatch repair**

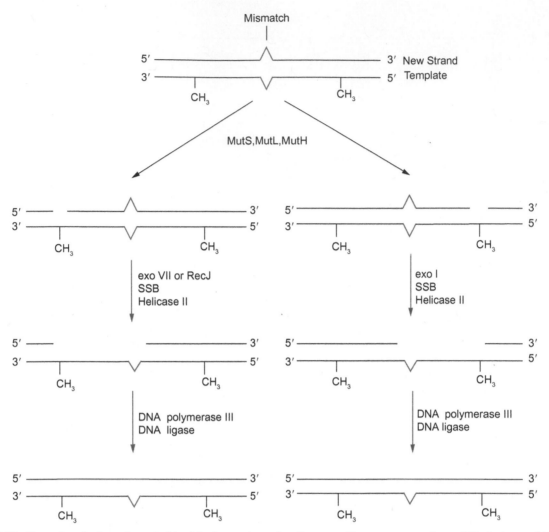

**FIGURE 14.24** Illustration of mismatch repair. *Mut*, Mut (mutator) proteins; *Rec*, recombinase; *exo*, exonuclease; *SSB*, single-strand binding (protein). *Reproduced from Litwack, G.* Human Biochemistry & Disease, *Academic Press/Elsevier, page 576 (2008).*

**protein derived form the PMS1 gene**; *PMS*, **post meiotic segregation**) for MutL. In one case, for example, of mismatched repair in *E. coli*, MutS binds to mismatched base pairs. Then MutL binds to the complex and activates MutH. MutH cleaves the unmethylated strand at the GATC (guanine—adenine—thymine—cytosine) site. Then exonuclease removes the segment from the cleavage site extending to the mismatch with the cooperation of other proteins. When the cleavage occurs on the 3′ side of the mismatch, exonuclease I degrades a single strand in the direction of 3′ to 5′. When cleavage occurs on the 5′ side of the mismatch, **exonuclease VII** or **RecJ** degrades the single-stranded DNA. The gap is filled by **DNA polymerase III** and **DNA ligase** as shown in Fig. 14.24.

A list of **genetic diseases** related to **defects in DNA repair systems** is given in Table 14.4.

## OVERVIEW OF METABOLISM

Ingestion of food provides external proteins, nucleic acids, polysaccharides, and fats and encompasses utilization, catabolism, and resynthesis of the endogenous products. Proteins are digested in the gut to amino acids and small peptides. These are absorbed and transported into the circulation that provides precursors to the tissues for the synthesis of cellular proteins, nucleic acids, polysaccharides, and lipids. Amino acids derived from protein breakdown are reused for protein synthesis or are converted to keto acids and ammonia that can be excreted as urea. Dietary nucleic acids are degraded to pentoses and purines and pyrimidines. These can be reused for the cellular synthesis of nucleotides and nucleic acids. Polysaccharides, such as plant starches are degraded in the gut to monosaccharides, including glucose,

**TABLE 14.4 Genetic Diseases Associated with Defects in DNA Repair Systems**

| Disease | Symptoms | Genetic Defect |
|---|---|---|
| Xeroderma pigmentosum | Frecklelike spots on skin, sensitivity to sunlight, predisposition to skin cancer | Defects in nucleotide-excision repair |
| Cockayne syndrome | Dwarfism, sensitivity to sunlight, premature aging, deafness, mental retardation | Defects in nucleotide-excision repair |
| Trichothiodystrophy | Brittle hair, skin abnormalities, short stature, immature sexual development, characteristic facial features | Defects in nucleotide-excision repair |
| Hereditary nonpolyposis colon cancer | Predisposition to colon cancer | Defects in mismatch repair |
| Fanconi anemia | Increased skin pigmentation, abnormalities of skeleton, heart, and kidneys, predisposition to leukemia | Possibly defects in the repair of interstrand cross-links |
| Ataxia telangiectasia | Defective muscle coordination. dilation of blood vessels in skin and eyes, immune deficiencies, sensitivity to ionizing radiation, predisposition to cancer | Defects in DNA damage detection and response |
| Li–Fraumeni syndrome | Predisposition to cancer in many different tissues | Defects in DNA damage response |

**Source:** Reproduced from http://www.nature.com/scitable/resource?action = showFullimageForTopic&imgSrc = 19346/pierce_table_17_6_FULL.jpg.

which can enter glycolysis, especially in the liver for the formation of pyruvate and ATP. Pyruvate is converted to acetyl CoA, especially when the [ATP] is on the low side, which enters the TCA cycle for the production of NADH and CoQ10H and their subsequent oxidation through the **electron transport chain** for the production of ATP energy. Using the hexose monophosphate pathway, glucose can form pentose sugars. This process also generates NADH. The products of glycolysis can be reused to form glucose through **gluconeogenesis** when glucose is in need. Fats are hydrolyzed to free fatty acids and glycerol. Glycerol can enter glycolysis and fatty acids can be degraded to **acetyl CoA** through the β-oxidation process in mitochondria. Acetyl CoA is also utilized for the synthesis of cholesterol, steroid hormones, bile acids, and isoprenoids. Fig. 14.25 diagrams an overview of metabolism.

## METABOLISM IN STEM CELLS

Currently, there is great interest in stem cells especially for their potential use in regenerative medicine. **Mesenchymal stem cells** (multipotent stromal cells that can differentiate into bone cells, cartilage cells, muscle cells, and fat cells) are proving to be effective in cartilage replacement in the knee. Recently, it has been observed that the status of metabolism in stem cells involves a shift in the balance between glycolysis, oxidative phosphorylation in the mitochondrion and oxidative stress during stem cell maturation, and in the process of reprograming somatic cells into pluripotency. **Totipotent stem cells** are cells in the developing **blastocyst** (embryonic stem cells) that are capable of developing into any tissue. Adult stem cells have been engineered to resemble early developing cells and are at least pluripotent cells, capable of developing into a number of tissues. Totipotent stem cells (undifferentiated cells) have a metabolism different from adult cells (differentiated cells) as shown in Fig. 14.26.

In **pluripotent stem cells**, that can develop into a number of tissues but not all tissues, there are considerable variations from the balance of metabolism observed in totipotent stem cells as shown in Fig. 14.27.

In differentiating stem cells a more normal picture develops where there is abundant ATP synthesis shifts from glycolysis to mitochondrial oxidative phosphorylation. The changes from pluripotent stem cells to developing stem cells are shown in Fig. 14.28.

Thus, the totipotent stem cell is the most dedifferentiated and produces little in the way of energy sources. The pluripotent stem cell, depending on glycolysis for its energy supply, resembles the metabolism of the cancer cell as observed by Warburg except that mitochondrial oxidation is intact in the cancer cell. Finally, the differentiating stem cell resembles the normal mature cell in that energy supplies shift to mitochondrial oxidation and oxidative phosphorylation with enlarged production of ATP.

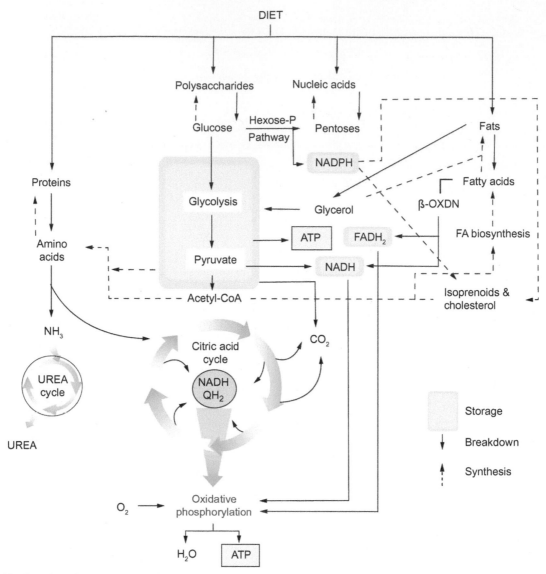

**FIGURE 14.25**    Overview of metabolism. *FA, fatty acid.*

Another type of stem cell is the **long-term hemopoietic stem cell** (**LTHSC**). These cells reside in a pool, called a "**niche**" (most adult tissue cells have niches of progenitor cells) in a relatively quiescent state so that they can be called upon at some later time to replace tissue cells that have been degraded. The niche provides a microenvironment in which cellular factors of the host (in this case, bone marrow) are made available to the LTHSCs for survival in the quiescent state. Among these factors are cytokines, adhesion molecules, and extracellular matrix. The metabolism within these cells is characterized by promotion of glycolysis. The transcription factor, MEIS1 combines with low levels of $O_2$ to activate **hypoxia-inducible factor 1$\alpha$** (**HIF1$\alpha$**) in quiescent LTHSCs. Pyruvate oxidation is suppressed by HIF1$\alpha$-dependent pyruvate dehydrogenase kinases (PDK1-4) that suppress the pyruvate dehydrogenase complex. The peroxisome proliferator receptor $\delta$ drives mitochondrial fatty acid oxidation that is needed for the self-renewal and quiescence of LTHSCs. When fatty acid oxidation is inhibited, LTHSCs proliferate and differentiate.

## SUMMARY

The most common lipid storage disease is **Gaucher disease**. It is caused by the mutation of the gene for the enzyme, **glucocerebrosidase** that cleaves glucocerebroside into glucose and ceramide. This results in the deposition of fats into spleen, liver, kidneys, lungs, brain and bone marrow. Death can occur before age 2. However, there is a spectrum of

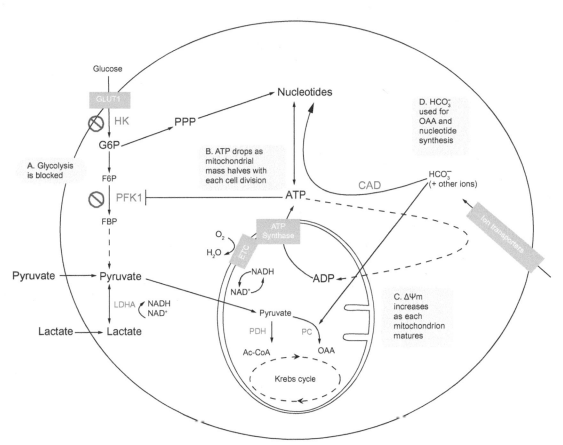

**FIGURE 14.26** Metabolism in totipotent stem cells. (A) Glycolysis is impaired due to the low activities of the rate-limiting enzymes hexokinase (HK) and phosphofructokinase 1 (PFK1). Totipotent stem cells use pyruvate as their major energy and carbon source instead, via pyruvate dehydrogenase (PDH) to generate acetyl CoA (Ac-CoA) and via pyruvate carboxylase (PC) to generate oxaloacetate (OAA) for anaplerosis (forming intermediates in a metabolic pathway) or gluconeogenesis. (B) ATP synthesis is dependent on mitochondrial oxidative phosphorylation driven by the electron transport chain (ETC) and ATP synthase. However, as mitochondrial replication has not yet initiated, the halving of mitochondrial mass leads to a drop in ATP levels during embryo cleavage. (C) Simultaneously, each mitochondrion matures and the inner mitochondrial membrane potential ($\Delta\Psi$m) increases steadily, thus turning the exponential drop of ATP into a linear drop. (D) Bicarbonate ($HCO_3^-$) is needed to buffer the pH and also provides a carbon source to OAA in the Krebs cycle for anaplerosis via PC or to nucleotide synthesis for DNA and RNA via carbamoyl phosphate synthase (CAD). *F6P*, fructose-6-phosphate; *FBP*, fructose-1, 6-*bis*phosphate; *G6P*, glucose-6-phosphate; *LDHA*, lactate dehydrogenase A; *PPP*, pentose phosphate pathway. *Reproduced from http://dev.biologists.org/content/140/12/2535/F1.expansion.html.*

severity that may arise from the many sites of mutations of the gene. Glucocerebrosidase is required for the degradation of the carbohydrate portions of gangliosides and in the absence of this enzymatic activity, gangliosides accumulate in the lysosomal compartment.

Lipids are broken down by **lipases** to glycerol and fatty acids. Fatty acids undergo β-**oxidation** in mitochondria to form acetyl CoA. Fatty acids have double bonds introduced by **fatty acid desaturases** of which there are four major such enzymes in the human. Desaturases associate with cytochrome *b*5 and **cytochrome *b*5 reductase**. **Oleic** and **linoleic** acids are essential. Fatty acids and glycerol are taken up by adipose cells and stored there as triglycerides. When needed, fat is broken down in the adipose cell and released into the bloodstream. This process is controlled by a **hormone-sensitive lipase (HSL)**.

**Cholesterol**, when in excess, is oxidized to various products that are present in low concentrations. These products are **oxysterols** that have a short half-life, and they (**27-hydroxycholesterol**) may bond to **oxysterol receptors** whose action may suppress the accumulation of cholesterol. Cholesterol is also the precursor of the steroid hormones that are either sex hormones or adrenal cortical hormones. These hormones are catabolized to various products that are excreted. Some of these products are solubilized as glucuronides for excretion. Bile acids are also formed from cholesterol. Some are conjugated with taurine in the liver and then circulated to the gallbladder. They mix with ingested food and circulate through the intestine of which 10% is excreted in the feces and the rest are recirculated back to the liver. Conjugated cholic acid is deconjugated in the colon by bacterial cholylglycine hydrolase and 7α-dehydroxylase.

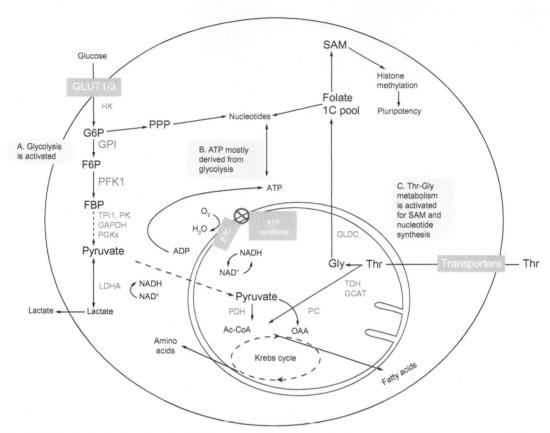

**FIGURE 14.27** Metabolism in pluripotent stem cells. (A) Glucose flux increases with the increase in GLUT 1/3 expression and the hexokinase (HK) and phosphofructokinase 1 (PFK1) enzymes become activated to sharply increase glycolytic flux. As a result, flux into the pentose phosphate pathway (PPP) for nucleotide synthesis increases. (B) ATP synthesis is more dependent on the reactions carried out by glycolytic phosphoglycerate kinases (PGKs) and pyruvate kinases (PKs) and is decoupled from oxygen consumption by the mitochondrial electron transport chain (ETC). (C) Activation of threonine dehydrogenase (TGH), glycine C-acetyltransferase (GCAT) and glycine decarboxylase (GLDC) promotes threonine-glycine catabolism to feed the folate one carbon (1C) pool, which in turn fuels S-adenosylmethionine (SAM) and nucleotide synthesis to maintain pluripotency and proliferation. *Ac-CoA*, acetyl coenzyme A; *F6P*, fructose-6-phosphate; *FBP*, fructose-1,6-*bis*phosphate; *G6P*, glucose-6-phosphate; *GAPDH*, glyceraldehyde-3-phosphate dehydrogenase; *GPI*, glucose-6-phosphate isomerase; *LDHA*, lactate dehydrogenase A; *OAA*, oxaloacetate; *PC*, pyruvate carboxylase; *PDH*, pyruvate dehydrogenase complex; *PK*, pyruvate kinase; *TPI1*, triosephosphate isomerase. *Reproduced from http://dev.biologists.org/content/140/12/2535/F2.expansion.html.*

The metabolism of carbohydrates, especially glucose, has been discussed in the chapter on glycolysis and gluconeogenesis. There are five major glucose transporters (GLUT1 through GLUT5) that move glucose into tissue cells. GLUT1 is present in all human cells. GLUT2 moves glucose into the liver cell. GLUT3 moves glucose into brain cells. Table 13.2 reviews the glucose transporters and their tissue distribution. Glucose metabolism involves glycolysis and entry of the products of glucose metabolism (pyruvate) into the TCA cycle. These topics are discussed in some detail in this chapter. The TCA cycle is also discussed in some detail. One molecule of glucose can give rise to as many as 34 molecules of ATP after moving its products through glycolysis and the TCA cycle. ATP and water ("metabolic water") are produced by the transport of electrons and protons through the electron transport chain and the oxidation of NADH or CoQ10H generated by various TCA intermediates.

Nucleic acid metabolism involves the degradation of mRNAs by microRNAs (miRNAs) and small interfering RNAs (siRNAs) that can trigger endonuclease (**argonaute protein**; **Ago2**) cleavage or decapping. Fragments of mRNA generated by RISC (RNA-induced silencing complex) cleavage are degraded by the general mRNA degradation machinery. DNA is relatively stable in the adult. However, during growth and development, many cells die by programed cell death (apoptosis). Various death signals activate endonucleases to produce internucleosomal DNA breakdown (by caspase 3) and generate a characteristic DNA degradation laddering consisting of multiples of the internucleosomal stretch of DNA. DNA is also broken down by the random process of cell necrosis when cells die from toxins or other random events. In this case, the breakdown of DNA is not ordered, as in apoptosis, but generates a broad spectrum of degradation products from the actions of several nucleases. In the process of DNA replication, many mistakes can occur and

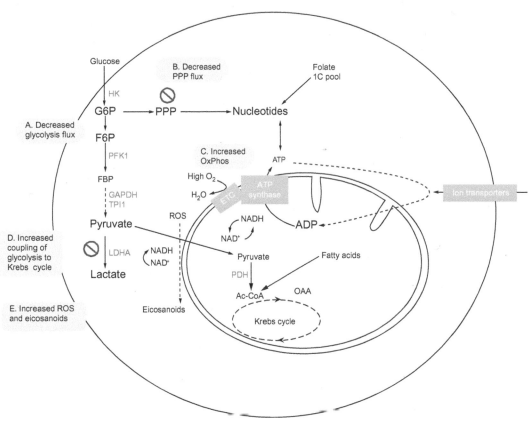

**FIGURE 14.28**  Metabolism in differentiating embryonic stem cells. (A) Glycolytic flux and lactate production drop rapidly upon embryonic stem cell differentiation. (B) Flux into the pentose phosphate pathway (PPP) decreases as a result. (C) $O_2$ consumption increases sharply as the electron transport chain (ETC) again becomes coupled to ATP synthase to fulfill the need of cell differentiation. (D) Glycolysis also becomes more coupled to the Krebs cycle, as pyruvate is transported more efficiently into mitochondria. (E) Increased ETC activity leads to increased reactive oxygen species (ROS) and eicosanoid signaling, which promote cell differentiation. *Ac-CoA*, acetylcoenzyme A; *HK*, hexokinase; *F6P*, fructose-6-phosphate; *FBP*, fructose-1,6,*bis*phosphate; *G6P*, glucose-6-phosphate; *GAPDH*, glyceraldehyde-3-phosphate dehydrogenase; *LDHA*, lactate dehydrogenase A; *OAA*, oxaloacetate; *Ox Phos*, oxidative phosphorylation; *PDH*, pyruvate dehydrogenase complex; *PFK1*, phosphofructokinase 1; *TPI1*, triosephosphate isomerase. *Reproduced from http://dev.biologists.org/content/140/12/2535/F3.expansion.html.*

these are detected by a surveillance system consisting of repair enzymes. Involving different types of repair: base excision repair, nucleotide excision repair, and mismatch repair. Gaps in the modified strand of DNA are repaired by DNA polymerase I and DNA ligase.

Stem cell biology represents a new frontier is which the replacement of tissue cells to repair degraded tissue conditions is foreseen. There are differences in the balance of metabolic activities within totipotent, pluripotent, and differentiating stem cells. Totipotent stem cells, representing an undifferentiated state, have low energy production attributed to minimal activities of glycolysis and of mitochondrial oxidative phosphorylation. Pluripotent stem cells, having some differentiation, essentially rely on glycolysis for the production of ATP, similar to the conclusion of Otto Warburg that tumor cells select glycolysis as the main source of energy (in spite of the fact that the mitochondria are fully active). In differentiating embryonic stem cells, both glycolysis and mitochondrial oxidative phosphorylation are active reflecting the development of a mature functioning tissue cell.

## SUGGESTED READING

### Literature

Bar-Even, A., Flamholz, A., Noor, E., Milo, R., 2012. Rethinking glycolysis: on the biochemical logic of metabolic pathways. Nat. Chem. Biol. 8, 509–517.

Fabian, M.R., Sonenberg, N., 2012. The mechanics of miRNA-mediated gene-silencing: a look under the hood of miRISC. Nat. Struct. Mol. Biol. 19, 586–593.

Feng, L., Sheppard, K., Nangoong, S., Ambrogelly, A., Polycarpo, C., Randau, L., et al., 2004. Aminoacyl tRNA synthesis by pre-translational amino acid modification. RNA Biol. 1, 16–20.

Germann, M.N., Johnson, C.N., Spring, A.M., 2012. Recognition of damaged DNA: structure and dynamic markers. Med. Res. Rev. 32, 659−683.

Goldstein, J.L., Brown, M.S., 2015. A century of cholesterol and coronaries: from plaques to genes to statins. Cell. 161, 161−172.

Hruska, K.S., LaMarca, M.E., Scott, C.R., Sidransky, E., 2008. Gaucher disease: mutation and polymorphism spectrum in the glucocerebrosidase gene. Hum. Mutat. 29, 567−583.

Motamed, M., et al., 2011. Identification of luminal loop 1 of Scap protein as the sterol sensor that maintains cholesterol homeostasis. J. Biol. Chem. 286, 18002−18012.

Musumeci, G., et al., 2014. New perspectives for articular cartilage repair treatment through tissue engineering: a contemporary review. World J. Orthop. 5, 80−88.

Papa, S., Martino, P.L., Capitanio, G., Gaballo, A., DeRasmo, D., Signorile, A., et al., 2012. The oxidative phosphorylation system in mammalian mitochondria. Adv. Exp. Med. Biol. 942, 3−37.

Paulick, M.G., Bertozzi, C.R., 2008. The glycosylphosphatidylinositol anchor: a complex membrane-anchoring structure for proteins. Biochemistry. 47, 6991−7000.

Planey, S.L., Litwack, G., 2000. Glucocorticoid-induced apoptosis in lymphocytes (Review). Biochem. Biophys. Res. Commun. 279, 307−312.

Raymundo, N., Baysal, B.E., Shadel, G.S., 2011. Revisiting the TCA cycle: signaling to tumor formation. Trends Mol. Med. 17, 641−649.

Shuck, S.C., Short, E.A., Turchi, J.J., 2008. Eukaryotic nucleotide excision repair: from understanding mechanisms to influencing biology. Cell Res. 18, 64−72.

Shyh-Chang, N., Daley, G.Q., Cantley, L.C., 2013. Stem cell metabolism in tissue development and aging. Development. 140, 2536−2547.

Spencer, B.J., Verma, I.M., 2007. Targeted delivery of proteins across the blood-brain barrier. Proc. Nat. Acad. Sci. 104, 7594−7599.

Vanderperre, B., Bender, T., Kunji, E.R.S., Martinou, J.C., 2015. Mitochondrial pyruvate import and its effects on homeostasis. Curr. Opin. Cell Biol. 33, 35−41.

Walther, T.C., Farese Jr, R.V., 2012. Lipid droplets and cellular lipid metabolism. Annu. Rev. Biochem. 81, 687−714.

Wurstle, M.L., Laussmann, M.A., Rehm, M., 2012. The central role of initiator caspase-9 in apoptosis signal transduction and the regulation of its action and activity on the apoptosome. Exp. Cell Res. 18, 1213−1220.

## Books

Ehnholm, C. (Ed.), 2009. Cellular Lipid Metabolism. Springer, NY, New York.

Knoepfler, P., 2013. Stem Cells: An Insider's Guide. World Scientific, Singapore, Singapore.

Kohlmeier, M.H., 2015. Nutrient Metabolism. second ed. Academic Press, San Diego, London, Salt Lake City, UT.

Litwack, G., 2008. Human Biochemistry and Disease. Academic Press, San Diego, London, Salt Lake City, UT.

Litwack, G. (Ed.), 2012. Stem Cell Regulators, Vitamins & Hormones, Volume 87. Academic Press, San Diego, London, Salt Lake City, UT.

Stipanuk, M.H., Caudill, M.A. (Eds.), 2013. Biochemical, Physiological, and Molecular Aspects of Human Nutrition. Elsevier Saunders, Philadelphia, PA.

# Chapter 15

# Polypeptide Hormones

## PANHYPOPITUITARISM: MALFUNCTION OF THE HYPOTHALAMUS—PITUITARY-END ORGAN AXIS

The pituitary is the source of several hormones that control many essential terminal hormonal functions. These include signals form the **anterior pituitary** that govern the stress hormone, **cortisol** from the adrenal cortex, the **thyroid hormone** from the thyroid gland, **growth hormone** from the anterior pituitary, **sex hormones** from the gonads of both male and female and from the posterior pituitary, the reabsorption of water from the kidney as well as milk secretion from the mammary gland and other actions. A loss or dimunition of all of these functions is called "**panhypopituitarism.**" A loss of one of these functions is called "**hypopituitarism.**" Thus, panhypopituitarism is an appropriate clinical example to introduce the subject of polypeptide hormones.

The secretion of the anterior pituitary hormones (the **adrenocorticotropic hormone, ACTH**; **thyroid-stimulating hormone** (or **thyrotropin**), **TSH**; **growth hormone, GH**; and the **gonadotrophic hormones**: luteinizing hormone, **LH**; and **follicle-stimulating hormone, FSH**). The hormones of the **posterior pituitary** are the **antidiuretic hormone, ADH** (or **vasopressin, VP**) for water reabsorption in the kidney and **oxytocin, OT** (for muscular contraction in the secretion of milk in the mammary gland and other functions). The stimulus for the secretion of each of these hormones from the anterior pituitary comes from **releasing hormones** in the hypothalamus that are transported to the appropriate secretory cells of the anterior pituitary (**corticotroph** for ACTH; **thyrotroph** for TSH; **sommatotroph** for GH; and **gonadotroph** for LH and FSH). The "troph" ending translates to a growth stimulus for the end organ secretory cell and the ending, "trope," is also used that translates to causing a change in the end organ cell (e.g., thyrotroph and thyrotrope; the terms can be used interchangeably).

Incoming electrical or chemical signals to a specific region of the **hypothalamus** will cause a target neuron to secrete a releasing hormone specific to that neuron. If the stimulation is to a neuron secreting TRH (the thyrotropin-releasing hormone), it would be labeled a thyrotropinergic neuron (*ergic* referring to the ability of that specific neuron to synthesize and secrete TRH). The releasing hormones are: **corticotropic-releasing hormone, CRH** (for the release of ACTH from the corticotropic cell of the anterior pituitary); **thyrotropin-releasing hormone, TRH** (for the release of TSH from the thyrotrope of the anterior pituitary); **growth hormone-releasing hormone, GHRH** (for the release of GH from the somatotrope of the anterior pituitary); **gonadotropin-releasing hormone, GnRH** (for the release of LH and FSH from the gonadotrope of the anterior pituitary). *LH and FSH are synthesized in the same cell (gonadotrope) and are released together although they are packaged separately inside the cell.*

**ADH/AVP** is formed in a hypothalamic neuron (**vasopressinergic neuron**) and is secreted in response to a depletion of blood volume (or by certain drugs) along the axon of the secreting cell all the way to the posterior pituitary where it is released. **Oxytocin (OT)** is formed in an **oxytocinergic** (magnocellular) **neuron** of the paraventricular and supraoptic regions of the hypothalalmus and is transported through long axons to the posterior pituitary where it is secreted. One signal to the oxytocinergic neuron is stimulation of the pregnant mother's nipple by the feeding infant. Oxytocin release also can be stimulated by estrogen and by drugs.

The **releasing hormones** only can reach the pituitary through a fragile **stalk** connecting the hypothalamus to the pituitary in a closed portal system. The narrow closed portal system carrying the releasing hormones from the hypothalamus to the anterior or posterior pituitary is enveloped by the delicate (narrow) stalk. The releasing hormones do not enter the general circulation but are confined to the portal circulation, and these hormones have a very short half-life on the order of a few minutes. The releasing hormones bind to cognate (bearing a structural relationship; in this case, the structure of the releasing hormone to the binding site of the receptor) receptors in the cell membrane of the target anterior pituitary cell (**CRH** binds to its receptor on the **corticotropic cell** of the anterior pituitary; the other releasing hormones bind to their receptors on their respective target cells in the pituitary). **Panhypopituitarism** can occur when the fragile stalk is severed

Human Biochemistry. DOI: http://dx.doi.org/10.1016/B978-0-12-383864-3.00015-6

through head trauma (an automobile accident or impacting the skull with a heavy object, for example). When the stalk is damaged, the blood supply connecting the hypothalamus to the pituitary is disrupted, leading to an infarction of the anterior pituitary. In this case, the releasing hormones do not reach the pituitary and a deficiency of the pituitary hormones begins to take place in a sequence that depends on the amounts of the native pituitary hormones that remain in cellular storage.

A picture of the hypothalamus, the stalk, and the pituitary is shown in Fig. 15.1.

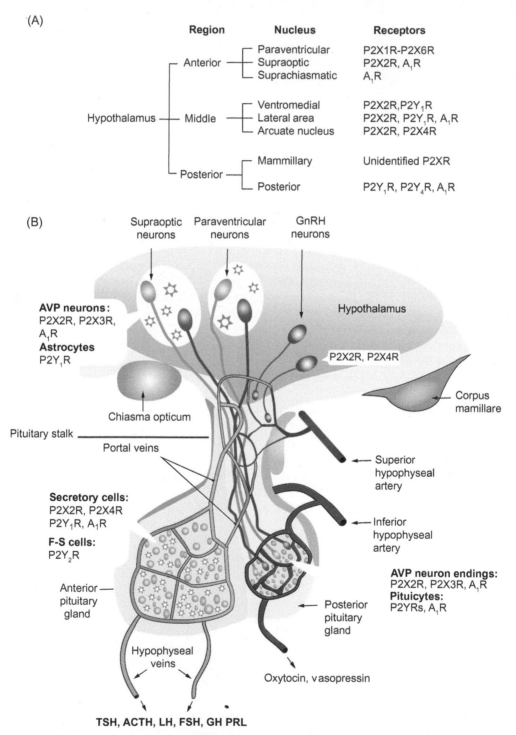

FIGURE 15.1 The **hypothalamic–pituitary system** connected by the pituitary stalk. Releasing hormones from neurons in the hypothalamus course down the neuronal axons in the pituitary stalk to the anterior or posterior pituitary, where they encounter their receptors on the membranes of target cells. *P2X1R*, ATP-gated channel; *P2X2R*, purigenic receptor (hypothalamic orexin neurons); *A1R*, adenosinergic receptor; *P2X2R*, purinergic receptor; *P2X4R*, ATP receptor (regulator of anxiety); *P2Y1R*, purinergic receptor (modulates stimulatory actions of ATP & ADP); *P2Y4R*, pyrimidinergic receptor. *Reproduced from http://www.cell.com/cms/attachment/603935/4780364/gr1.jpg.*

The target pituitary cell responds to the activation of the releasing hormone receptor through binding the releasing hormone and transducing the signal to affect biochemical changes in the cell leading to changes in transcription (activation of the gene for the respective pituitary hormone leading to resynthesis of the hormone) and the release from the cell of the target pituitary hormone (e.g., **ACTH** from receptor binding of **CRH**; **TSH** from receptor binding of **TRH**, etc.). The biochemical changes in the different target cells usually consist of an increase of **cAMP** or a stimulation of **inositol *tris*phosphate**, or both. The pituitary hormone is released into the general circulation through thin areas (**fenestrations**) in the vessels at the bottom of the pituitary gland.

**Traumatic brain injury** occurs at a frequency of about two million people per year in the United States. About 80,000 of these have **neuroendocrine disorders**. Additionally, 50% of people with traumatic brain injury had neuroendocrine disorder that was discovered postmortem (problem undiagnosed while they were alive). There may be as many as 6.5 million people alive with consequences of this kind of injury. Symptoms of **panhypopituitarism** do not manifest for weeks or months following brain injury, until the effective supply of hormones has been used up. This condition is signaled by decreased vital signs and malaise with lethargy, slow heart rate, hypothermia, and sometimes with hypotension with low blood sodium. The decreased blood levels of the end organ hormones: **cortisol** (this hormone is essential for life), **testosterone**, **triiodothyronine (T3)**, **tetraiodothyronine (thyroxine, T4)**, and the pituitary hormone, **thyroid-stimulating hormone (TSH)**, are confirmatory.

When congenital at birth, panhypopituitarism can occur in childhood as a result of poor development in the midline brain structure (*septum pellucidum*). Sometimes, tumors of the hypothalamic−pituitary region arise, of which **childhood craniopharyngioma** is an example. **Growth hormone**, the driver of stature and growth rate, is often most affected in childhood panhypopituitarism. A deficiency of TSH from the anterior pituitary leads to central hypothyroidism. The loss of the gonadotropic hormones, **luteinizing hormone (LH)** and **follicle-stimulating hormone (FSH)** is not visualized until puberty when the development of breasts and menstrual cycles in the female are affected. Ultrasound can detect the abnormality in females. In males, the enlargement of the penis and testicles at puberty is interfered. Decreased availability of the **antidiuretic hormone (ADH** or **VP)** from the posterior pituitary leads to *diabetes insipidus* with excessive urination.

During pregnancy, **hypopituitarism** can occur. Often as a result of profound blood loss during and following childbirth, known as **Sheehan's syndrome**. Blood loss can precipitate the death of anterior pituitary cells. There is a chronic form of this syndrome that can become diagnosed months or years following childbirth, and a more rare form of the syndrome appears shortly after delivery. **Autoimmune antibodies** to the pituitary have been reported in some cases of Sheehan's syndrome. *In most cases, panhypopituitarism is treated by replacement of the end organ hormones.* In some cases, surgery is required to remove a tumor.

A male adult, for example, has **panhypopituitarism** as a result of a catastrophic automobile accident. The hormones of his **anterior pituitary** are low due to a severed stalk, so that the **releasing hormones** of the **hypothalamus** are not reaching their target receptors on the cells of the pituitary. **TSH** is low so that the end organ hormones of the thyroid, **T3** and **T4**, are also low because there is lacking stimulation by **TSH**. As a result of the low thyroid hormones, there is little or no negative feedback by T3 and T4 on the hypothalamus (as we shall see later) so that the level of TRH is increased (without stimulation of the anterior pituitary thyrotrope). Because of the lack of effective **GHRH**, there is a paucity of **somatotropin** and a consequent lack of **GH** secretion from the somatotrope of the anterior pituitary. This deficiency causes **low blood glucose** (as we shall see later on) and if the problem started in boyhood, he might be a victim of **dwarfism**. Because the availability of **CRH** is low or lacking altogether, circulating **ACTH** is low and, therefore, the secretion of **cortisol** from the adrenal cortex is also low or nearly absent. This causes low blood glucose (to be explained later) and a lack of negative feedback to the hypothalamus resulting in increased CRH. Because cortisol is a key hormone for stress adaptation, it is essential that it be replaced (usually orally) in order to prevent death resulting in failure to adapt to extreme stress or shock as in a catastrophic accident of some sort. The gonadotropic hormones of the anterior pituitary are also low because **GnRH** is not reaching the gonadotropes of the anterior pituitary. Low **FSH** results in decreased production of sperm, and low **LH** results in decreased testosterone levels (as will be seen later on). Decreased circulating **testosterone** lowers the negative feedback to the hypothalamus and increased secretion of GnRH (although it would not be reaching the gonadotropes). Decreased **ADH (VP)** could result in *diabetes insipidus* with resulting increased urinary output. Decreased **prolactin** would have little consequence for the male. It is a challenging situation for the endocrinologist to control because the end organ hormones have to be correctly titrated and, in the case of cortisol, the problem for the patient is that he/she must be able to predict stressful situations in order to adjust the dose of cortisol (or cortisol-like drug). Accidents could still be a dangerous and a life-threatening situation for the patient with panhypopituitarism.

One clinical test is to inject a specific releasing hormone. If there is a response of the appropriate anterior pituitary hormone, then one can conclude that the problem resides in the hypothalamus. If the anterior pituitary does not respond to the releasing hormone, then the problem (such as a tumor) may reside at the pituitary.

A tumor of the pituitary can also secrete (ectopically) an excess of a pituitary hormone. Thus, if the production of GH by the tumor, for example, is very high, it would cause a pronounced increase in blood glucose and would increase the growth of soft connective tissues, causing enlargement of the nose, fingers, and possibly compressing nerves in the adult (symptoms of **giantism**).

## HORMONAL SIGNALING PATHWAYS

The **releasing hormones** from the hypothalamus travel down the long axons of the particular releasing hormonergic neuron to the vicinity of the portal circulation of the stalk in which the hormone is transported to the specific cell of the anterior pituitary. In the case of hormone of the posterior pituitary, the neuron in the hypothalamus that becomes excited moves the hormone (e.g., ADH) down a long axon of the same cell to the vicinity of the end of the small vessels in the posterior pituitary. The releasing hormones are confined to the **portal circulation** of the **stalk** that is composed of a very small volume of blood. Accordingly, the concentration of a releasing hormone, in any event where the hormone is released from the hypothalamic neuron, is very small on the order of nanograms (ng; $10^{-9}$ g) or less. In the closed portal circulation of the stalk, the releasing hormone must be transported to its receptor on the cell membrane of the tropic cell of the anterior pituitary. The half-life of the releasing hormone is of the order of a few minutes so that a single event is short-lived. The releasing hormone is rather quickly inactivated in the closed portal circulation by resident enzymes that either cleave the releasing hormone or remove its **C-terminal amide** group (in those releasing hormones that contain a C-terminal amide, this substituent is essential for activity) accounting for the short half-life. The latter enzyme is an **amidase**. A list of the releasing hormones of the hypothalamus is presented in Table 15.1.

In general, the secretion of the releasing hormones is episodic; they enter the closed portal circulation in extremely small amounts that are large enough to activate their receptors (the amount reaches the binding constant of the cognate receptor, or higher), and these activated receptors set off a signaling chain of events (elevated cAMP or phosphatidyl inositol, or both) that culminate in the release of the pituitary hormone and usually affect the activation of the gene that

**TABLE 15.1  Releasing Hormones of the Hypothalamus**

| Hormone | Number of Amino Acid (AA) Residues | Action | AA Sequence (Left, N-Terminal) |
|---|---|---|---|
| (CRH) Corticotropin-releasing hormone | 40 | Binds to receptor on corticotrope to release ACTH, β-endorphin & lipotropin | SQEPPISLDTFHLLREVL EMTCADQLAQQAHSNR KLLDI-Ala-NH$_2$ |
| (GHRH) Growth hormone releasing hormone | ~44 | Binds to receptor on somatotrope for GH secretion | YADAIFTNSYRKVLGQL SARKLLQDIMSRQQGES NQERGARAR-Leu-NH$_2$ |
| Somatostatin | 14 | Binds to receptor on somatotrope to inhibit release of GH | AGCKNFFWKTFTSC —S—S— between Cys residues |
| (GnRH) Gonadotropin-releasing hormone | 10 | Binds to receptor on gonadotrope to release LH & FSH | pyro-EHWSYGLRPG-NH$_2$ |
| (TRH) Thyrotropin-releasing hormone | 3 | Binds to receptor on thyrotrope to release TSH | pyro-EHP-NH$_2$ |
| (PRF) Prolactin-releasing factor (unknown, but TRH, OT, and estrogen can release prolactin) | ? | Binds to receptor on lactotrope to release prolactin | ? |
| (PIF) Prolactin release inhibiting factor; probably dopamine; leukemia inhibitory factor also inhibits | ? | | |

*H*, hormone (releasing hormones are known structures); *F*, factor; these are postulated but structures, if a peptide, are not well known. The first amino acid of a peptide sequence is the N-terminal amino acid. *ACTH*, adrenocorticotropic hormone; *GH*, growth hormone; *LH*, luteinizing hormone; *FSH*, follicle stimulating hormone; *TSH*, thyroid stimulating hormone.

encodes the mRNA of the pituitary hormone so that amounts of the pituitary hormone are synthesized for storage in the cell. The pituitary hormones circulate in the blood and bind to their receptors in the tissue cells of their targets generating a response at the tissue level. The combination of all the tissue responses produces the bodily response to the hormone. The signaling system for hormones starting with the hypothalamus is summarized in Fig. 15.2.

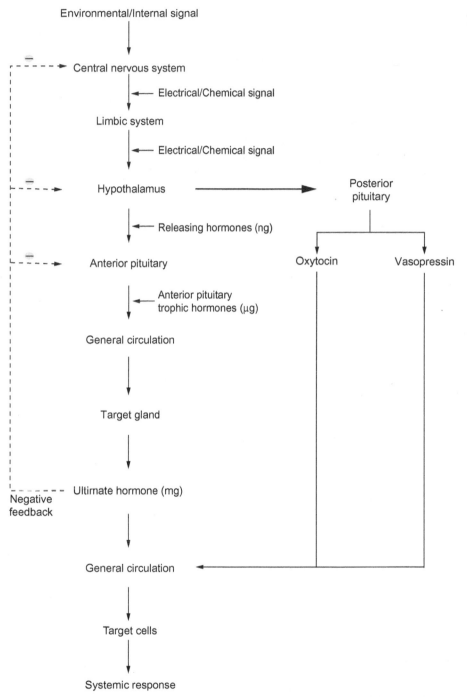

**FIGURE 15.2** The humoral mechanism connecting hormonal secretions from the **hypothalamus**, the **pituitary** and end organs that are targets of the pituitary hormones. The target gland is the last organ in the pathway. For the **corticotropin-releasing hormone (CRH)**, for example, it is released from the CRH-ergic neuron in the hypothalamus, causing the release of **ACTH** from the anterior pituitary which finds its receptor on cell membranes of the middle layer of cells of the **adrenal cortex** (target gland) from which **cortisol** is released into the bloodstream. In turn, cortisol circulates in the blood, crosses many cell membranes, and is tethered to a target cell by **steroid receptors** *in the cytosol*. Almost every cell of the body has some cortisol receptors (**glucocorticoid receptors**), except the *pars intermedia-like* **cells** of the pituitary and the **hepatobiliary cells**; the liver has the highest concentration of these receptors, followed by the kidney; thus, the **stress response** affects nearly every cell in the body to some extent. Glucocorticoid receptors become activated and are transported to the cell nucleus where they activate target genes. Within parentheses are the relative amounts of each hormone released.

## SIGNALING FROM HYPOTHALAMUS TO POSTERIOR PITUITARY

**Oxytocin (OT)** is released from nerve endings in the posterior pituitary by stimulation of the nipples and in lactating women will aid in release of milk from the breast. There are hypothalamic **interneurons** (neurons impinging on the cell body of the **oxytocinergic** (OTergic) **neuron**) that electrically stimulate the OTergic neuron to release OT from the posterior pituitary nerve endings (site of storage) by the process of **exocytosis** (discharge of components stored in vesicles) when the nerve endings are depolarized (alteration of charges on the surface of the nerve ending, usually a loss of outside negative charges). OT is initially released as a large polypeptide that includes the sequence of the OT carrier protein **neurophysin I**. The polypeptide precursor is broken down by a number of proteases, the final one being **peptidylglycine α-amidating monooxygenase (PAM)** releasing the nine-amino acid containing peptide, OT, and the intact neurophysin I (Fig. 15.3).

**OT** and **neurophysin I** form a complex for the stabilization of OT in the bloodstream. The complex dissociates before OT binds to its receptor on the surface of the plasma membrane of the target cell. OT also can be released from the hypothalamic OTergic neuron by estrogen. There are extrahypothalamic organs that contain OT, however, the functions of OT in these tissues are not well known. Tissues, besides the brain, that contain OT and/or its receptor are: *corpus luteum*, Leydig cells, retina, placenta, pancreas, adrenal medulla, heart, and thymus.

**Vasopressin** (usually **arginine vasopressin, AVP**) is stored in vasopressinergic (VPergic) nerve endings in the **posterior pituitary**. The release of AVP is stimulated by a fall in blood pressure, a decrease in blood volume and, in general, any condition where the blood volume is decreased. Because the action of AVP on the kidney is to cause the reabsorption of water into the bloodstream, the initial signals are neutralized by this activity. AVP, but not OT, can be released by **nicotine** (smoking cigarettes can have this activity). The gene encoding the mRNA for AVP contains responsive elements at its N-terminus: these are **osmolar response element (OsRE), glucocorticoid response element (GRE), estrogen response element (ERE),** *fos/jun* **response element (API-RE)**. The polypeptide precursor containing vasopressin and **neurophysin II** (its carrier protein) and another polypeptide (**copeptin**) are hydrolyzed by enzymes, like the case of OT precursor maturation, to release the mature products, vasopressin and neurophysin II. While AVP has a short half-life (ca. 5−15 minutes) and is somewhat unreliable as a measure of osmolality for that reason, copeptin is much more reliable because of its greater stability. The level of copeptin in blood is a direct indicator of the level of AVP and also mirrors the levels of stress-induced **cortisol**. Copeptin level appears to be a predictor of the severity of **hemorrhagic and septic shock** and high levels of copeptin predict an elevated mortality risk for heart failure patients.

AVP and neurophysin II form a complex and remain in complex form for the stability of AVP until AVP binds to its receptor on target cells of the kidney to increase reabsorption of water, ultimately into the bloodstream. The gene for vasopressin-neurophysin II and copeptin and the maturation products are shown in Fig. 15.4.

The structures of **OT** and **AVP** are similar with differences in only two amino acids out of nine (both hormones are nanopeptides) as shown in Fig. 15.5.

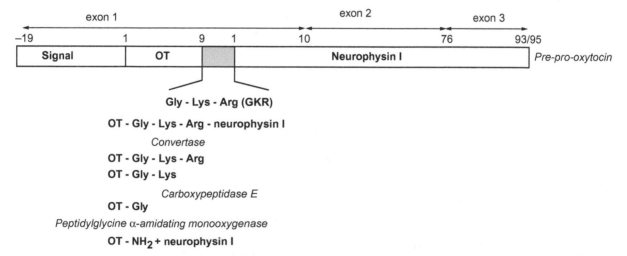

**FIGURE 15.3**  Gene product of the mRNA encoding the polypeptide precursor of oxytocin and neurophysin I. The tripeptide Gly−Lys−Arg separates the nonapeptide OT from neurophysin I. *Reproduced from Gutkowska, J., Jankowski, M., 2008. Oxytocin revisited: it is also a cardiovascular hormone. J. Am. Soc. Hypertens. 2, 318−325; also available at http://www.scincedirect.com/science/article/pii/S1933171108000491.*

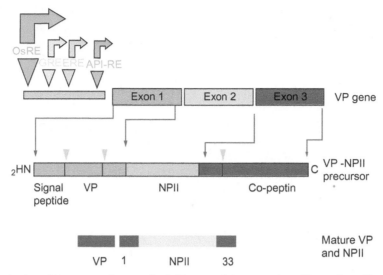

**FIGURE 15.4** Structural organization of the vasopressin-neurophysin II gene and the processing of its products. The gene has three exons. The transcribed mRNA yields a large preprohormone precursor that is modified subsequently through posttranslational modifications. The VP gene is similar to the OT gene (Fig. 15.3). *VP*, vasopressin; *NPII*, neurophysin II; *OsRE*, osmolarity response element; *GRE*, glucocorticoid response element; *ERE*, estrogen response element; *AP1-RE*, *jun/fos* response element. *Reproduced from http://www.endotext.org/neuroendo/neuroendo2/neuroendo2.htm.*

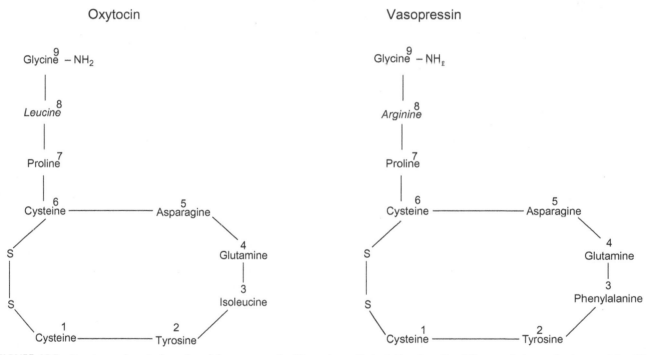

**FIGURE 15.5** Structures of **oxytocin** and **arginine vasopressin**. The amino acids in italics show the differences between the two peptides. The amino acids are numbered as shown: cysteine (1) to tyrosine (2), etc. The positions of the variant amino acids are 3 and 8. Note that the C-terminal amino acid is **glycine amide**. The cysteine residues automatically form a disulfide bridge.

## Hormonal Signaling From the Hypothalamus to the Anterior Pituitary

The **releasing hormones** from the **hypothalamus** are secreted into the closed portal system contained in the stalk and reach their cognate receptors on the cell membrane of the target cell as outlined in Table 15.1.

Stimulation for release of releasing hormones in the hypothalamus derives from either aminergic or peptidergic hormones that are released from **interneurons** that impinge upon the cell bodies of neurons containing the releasing hormones. These earlier signals would generate in response to internal or external signals, and some of the effects elicited by them would be inhibitory as well as stimulatory. Some specific stimuli are known for the secretion of certain releasing hormones. In the case of the secretion of the **growth hormone releasing hormone (GHRH)** and **somatostatin (SS)** of the hypothalamus, **cAMP** stimulates the release of both hormones. Another stimulant is a calcium ionophore indicating that the **calcium messenger** system could be involved. **Thyrotropic-releasing hormone (TRH)** released from neurons in the **paraventricular nucleus** of the hypothalamus is stimulated directly by the hormone, **leptin**. Leptin increases **melanocortin (α-MSH)** that is required for TRH expression. The melanocortin system activates the TRH promoter on DNA through the phosphorylation of the **signal transducer and activator of transcription 3 (Stat3)**. The Stat response element in the TRH promoter is required for the effects of leptin. Undoubtedly, there are many other signals that cause the release of the releasing hormones into the closed portal system leading to the pituitary.

Releasing hormones bring about the release of the relevant hormones from the **anterior pituitary** into the general circulation and, in some cases, the resynthesis of the anterior pituitary hormone is activated. The anterior pituitary hormones and their general actions are listed in Table 15.2.

**ACTH** and the **melanocyte-stimulating hormone**s (**MSH**s) arise from the same protein precursor as shown in Fig. 15.6.

The *pars intermedia*-like cells are scattered between the anterior and posterior pituitary in the human, whereas in lower forms the *pars intermedia* forms a discrete organ. Note that in the *pars intermedia*-like cells, there is an absence of ACTH since it is broken down into α-MSH and the hormone CLIP. Absence of ACTH coincides with the fact that these cells also do not contain the **glucocorticoid receptor** that would be needed for **negative feedback** by **cortisol** on the production of ACTH (which *does* occur in the corticotrope of the anterior pituitary). In the negative feedback mechanism, the terminal hormone (e.g., **cortisol**) feeding back binds to its receptor (**glucocorticoid receptor**) in the cells producing the signal (ACTH or **CRH** in the **hypothalamus**) for the release of the terminal hormone. The receptor in those cells sets off a signal pathway leading to an inhibition of further release of the hormone normally secreted by that cell. CLIP appears to act on the endocrine pancreas and causes the release of **insulin** from the β-cell. In the **exocrine pancreas**, CLIP acts like the hormone **secretin** and stimulates enzyme release (this has been shown for the secretion of the enzyme, **amylase**).

Hormones are split out of the precursor peptide by proteases that are specific for the Arg−Lys or Lys−Arg linkages, and the location of these sites is shown at the top of Fig. 15.6. Some hormones are generated primarily in the anterior pituitary cells and some others arise in the *pars intermedia*-like cells. The release of γ-MSH, ACTH, and β-LPH is mainly from anterior pituitary cells. In the *pars intermedia*-like cells, ACTH is broken down further to release α-MSH and CLIP. β-**LPH** is broken down to γ-**LPH** and β-**endorphin**. The latter two serve as precursors for β-**MSH** and **Met-enkephalin**.

**TABLE 15.2 Hormones of the Anterior Pituitary**

| Hormone | Number of amino Acid Residues (MW) | Action |
|---|---|---|
| Adrenocorticotropic hormone (ACTH) | 39 (4540 Da) | Signals adrenal cortex to release/synthesize cortisol (also 2° stimulus for aldosterone) |
| Growth hormone (GH, somatotropin) | 191 (22,124 Da) | Stimulates growth, lipid, & carbohydrate metabolism especially in liver & adipose; releases IGF-1: stimulates bone sulfation |
| Follicle-stimulating hormone (FSH) | 82 amino acids in α-subunit; 118 amino acids in β-subunit; (35,500 Da for holoFSH) | Stimulates growth & maturation of ovarian follicles; with estrogen, stimulates formation of LH receptors on granulosa cells in late follicular phase; with testosterone, supports spermatogenesis (FSH, LH, TSH all share same α-subunit: β-subunit differs in each) |
| Luteinizing hormone (LH) | α-subunit, 96: β-subunit, 121 amino acids (~28,000 Da for holoLH) | Stimulates secretion of sex hormones from male & female gonads; in males, LH acts on Leydig cells to stimulate synthesis & secretion of testosterone; in females, LH stimulates theca cells to secrete testosterone that is converted to estrogen in nearby *granulosa* cells |
| Prolactin (luteo-otropin, LH) | 199 amino acids (23,000 Da) | Growth/development mammary gland, synthesis of milk, & maintenance of milk secretion; stimulates progesterone secretion; role in immune response; acts on PRL receptor & certain cytokine receptors; suppresses gonadotropins (its secretion inhibited by dopamine) |
| Thyroid-stimulating hormone (TSH) | 211 amino acids 2 subunits (α & β) (28,500 Da for holoTSH) | Stimulates secretion & synthesis of thyroid hormone (T4 & T3) from thyroid gland (α subunit identical to that for FSH & LH) |
| Melanocyte-stimulating hormone (MSH) | α-MSH = 13 (1665 Da) β-MSH = 18 (2661 Da) γ-MSH = 12 (1571 Da) | Skin darkening; central nervous system actions |

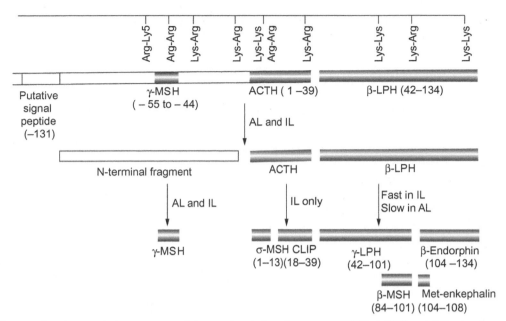

**FIGURE 15.6 Preproopiomelanocortin** is a precursor of the anterior pituitary hormones γ-**MSH, ACTH,** and β-**lipotropin** as well as hormones (α-**MSH** and **CLIP**) specific to *pars intermedia*-**like cells**. Preproopiomelanocortin polypeptide occurs in pituitary cells, neurons, and other tissues. Numbers in parentheses identify the location of the hormone according to amino acid numbers in the polypeptide sequence (amino acid 1 starts here at the N-terminus of ACTH and continues toward the C-terminus of the parent molecule; sequences to the left of ACTH have negative numbers). The locations of Lys−Arg and other pairs of basic amino acid residues are also indicated; these are the sites of proteolytic cleavage for the formation of smaller fragments from the parent molecule. *AL*, anterior lobe; *IL*, intermediate lobe, *LPH*, lipotropin, *CLIP*, corticotropin-like intermediate peptide.

The **anterior pituitary** (also, adenohypophysis) is composed of five cell types. The **somatotrophs (somatotropes)** represent 50% of the cell population and produce **growth hormone (GH)** (its release is signaled by the hypothalamic **growth hormone releasing hormone, GHRH**). The **lactotrophs (lactotropes)** produce **prolactin (PRL)** and represent 20% of the anterior pituitary cells. Another 20% of the cell population is represented by the **corticotrophs (corticotropes)** that synthesize the **adrenocorticotropic hormone** (corticotropin, **ACTH**). **Thyrotropic hormone (thyroid stimulating hormone, thyrotropin, TSH)** is produced by the **thyrotrophs (thyrotropes)** that occupy 5% of the cells in the anterior pituitary. The **gonadotrophs (gonadotropes)** occupy another 5% of the cells and produce both **follicle-stimulating hormone (FSH)** and **luteinizing hormone (LH)** stored in separate compartments in the same cell. The release of both FSH and LH is signaled by the **gonadotropin-releasing hormone, GnRH**. The "troph" ending on the name of the anterior pituitary cell type indicates that its secretory product induces a growth response in the target cell, whereas the "trope" ending on the name of the cell type indicates that its secretory product induces changes in the target cell; the two terms are often used interchangeably. A hormone like **ACTH** causes the synthesis and release of **cortisol** (tropic effect) from the **adrenal cortex cell** target and is also a growth factor for the cells producing cortisol (trophic effect). In the absence of ACTH, the adrenal cortical cells producing cortisol will die.

Virtually, *all polypeptide hormones and neurotransmitters, including those derived from amino acids (e.g., norepinephrine), are ligands for receptors located in the cell membrane or on the surface of the target cell, whereas steroid hormones pass through the cell membrane in both directions but are tethered within the cell by binding to cytosolic receptors or receptors within the cell nucleus* (a very small portion of some steroid hormone receptors are located in the cell membrane and bring about rapid, nontranscriptional effects, such as stimulating ion transport), however, these receptors have been shown to be different from the classical cytosolic receptors. The **thyroid hormones, thyroxine ($T_4$)** and **triiodothyronine ($T_3$)**, bind to the **thyroid hormone receptor** that is localized within the nucleus, forming an exception since it is not located in the cell membrane and is grouped as a member of the **steroid receptor gene family**. In Fig. 15.7, a generalized scheme is shown of the signal pathways from the hypothalamic hormones through the anterior pituitary hormones to the end target cells and the actions produced by the terminal hormones.

## MODELS OF HORMONE ACTION OF ANTERIOR PITUITARY HORMONES

### ACTH

The adrenocorticotropic hormone (ACTH) is released from the anterior pituitary **corticotrope** by the **hypothalamic-releasing hormone, CRH** (the **corticotropic releasing hormone**). ACTH activates its receptor on the cell membrane

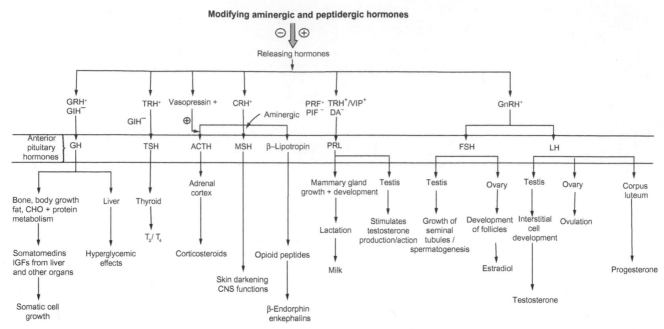

**FIGURE 15.7** Overview of the **anterior pituitary hormones** showing the connections between the aminergic hormones and neurotransmitters of the CNS, the releasing hormones from the hypothalamus, and the anterior pituitary hormones together with the organs upon which they act and the general actions produced by the terminal hormones. *ACTH*, adrenocorticotropic hormone; *CHO*, carbohydrate; *CRH*, corticotropic-releasing hormone; *FSH*, follicle-stimulating hormone; *GH*, growth hormone; *GIH* (**somatostatin**), growth hormone release inhibiting hormone; *GnRH*, gonadotropic-releasing hormone; *GRH*, growth hormone releasing hormone (also GHRH or **somatocrinin**); *IGF*, insulin-like growth factor; *LH*, luteinizing hormone; *MSH*, melanocyte-stimulating hormone; *PIF*, prolactin release inhibiting factor; *PRF*, prolactin releasing factor; *PRL*, prolactin; $T_3$, triiodothyronine; $T_4$, thyroxine; *TRH*, thyroid-stimulating hormone releasing hormone; *TSH*, thyrotropic-stimulating hormone (**thyrotropin**); *DA*, dopamine; *VIP*, vasoactive intestinal peptide. Plus or minus signs as superscripts of plus or minus signs within circles refer to positive or negative actions.

of the *zona fasciculata* cell of the adrenal cortex (the intermediate layer of the cortex) and the signaling pathway that follows generates the synthesis of **cortisol** and its release into the bloodstream (through the adrenal medulla) as shown in Fig. 15.8.

## α-MSH (Melanocyte-Stimulating Hormone); Melanotropin, Intermedin

There are three MSH molecules: α-**MSH** (13 amino acids); β-**MSH** (18 amino acids); and γ-**MSH** (11 amino acids). α-MSH is secreted from cells located between the anterior and posterior pituitary lobes. Whereas lower forms have discrete structures resembling an **intermediary lobe**, the human has only scattered cells in the same region, having the same properties as if they were located in a discrete lobe. The precursor polypeptide generating these MSH molecules as well as other molecules, including **ACTH**, β- and γ-lipotropins (LPH), **CLIP** and β-**endorphin** (β-**END**), is **proopiomelanocortin** (**POMC**). Whereas ACTH remains intact in the corticotropic cell of the anterior pituitary, in the intermediary cells ACTH is degraded completely to α-MSH and CLIP; consequently, in these cells, ACTH acts as an intermediate.

There are five **melanocortin receptors** (**MCRs**). **MC1R** binds α-MSH primarily but can also bind ACTH, β-MSH, and γ-MSH with much lower affinity. The function of MC1R is in the regulation of **melanin** production. **MC2R** binds ACTH almost exclusively for the production of cortisol (and to a small extent aldosterone) in cells (primarily *zona fasciculata* cells) of the adrenal cortex. **MC3R** binds γ-MSH with a lower affinity toward α-MSH and β-MSH for the action upon energy homeostasis and energy partitioning. **MC4R** primarily binds β-MSH, to a slightly lesser extent α-MSH, and has a reduced affinity for γ-MSH. **MC5R** prefers to bind α-MSH, to a lesser extent ACTH, β-MSH, and γ-MSH for the production of **sebum** (oily secretion of sebaceous glands). The generation of important peptides from proopiomelanocortin in the pituitary and the overall roles of the five melanocortin receptors with their binding preferences are summarized in Fig. 15.9.

The actions of MSH in the production of **melanin pigment** and in the neural regulation of **feeding** and **antiinflammation** are of special importance. With regard to the action of α-MSH, in the production of melanin, Fig. 15.10 shows that α-MSH binds to the MC1R (receptor) on the cell membrane of the **melanocyte** to increase the production of the skin-darkening pigment, melanin, in its various forms.

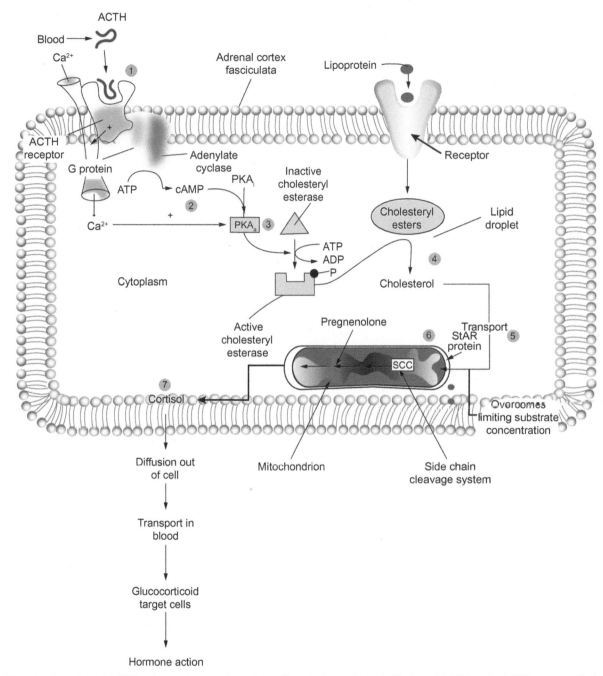

**FIGURE 15.8** Overview of ACTH action on the *zona fasciculata* cell producing cortisol. (1) Binding of ACTH to the ACTH receptor; (2) Activated receptor activates **adenylate cyclase** through G protein; (3) **cAMP** from adenylate cyclase reaction activates **protein kinase A** (**PKA**) and PKA phosphorylates inactive **cholesteryl esterase** to the active form; (4) Cholesteryl esterase hydrolyzes cholesteryl esters in the **lipid droplet** (derived from lipoproteins in the blood) to form free **cholesterol**; (5) Cholesterol is transported into the mitochondrion and activates substrate-limited **cortisol synthesis** in a side chain cleavage step catalyzed by the **StAR protein** (**steroid acute response protein**); (6) Cortisol synthesis proceeds and finally in (7) cortisol is released into the blood circulation.

In the melanocytes, α-MSH regulates the synthesis of melanin forms in the pigment granules. The other forms of MSH (β and γ) can also stimulate melanin synthesis but to a much lesser extent than the α-form as witnessed by the binding preferences for MC1R shown in Fig. 15.9. Melanocytes are located in the basal layer of the skin epidermis where they comprise 5%−10% of the cells in this layer. **Melanosomes** (the granules in the melanocyte containing melanin) are transferred from melanocytes to neighboring **keratinocyte**s (Fig. 15.11). Keratinocytes are located in the surface of the skin epidermis and constitute 95% of the cells in this layer. Keratinocytes, consequently, can secrete α-MSH.

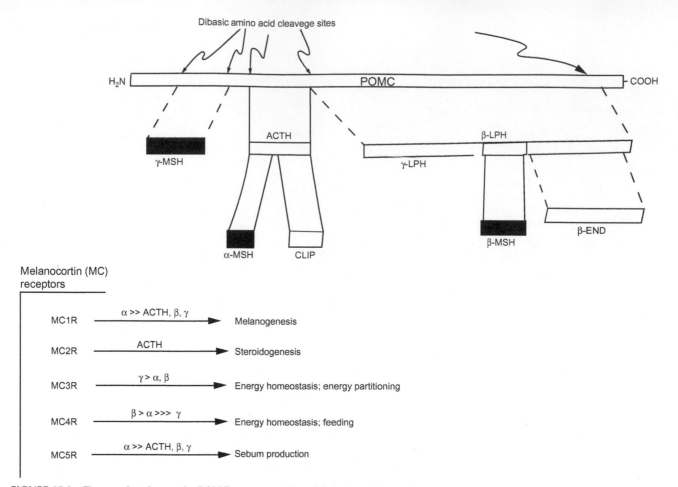

**FIGURE 15.9** The proopiomelanocortin (POMC) precursor polypeptide is cleaved at dibasic amino acid (combinations of arginine and lysine) cleavage sites into three MSH molecules (α-MSH, β-MSH, and γ-MSH), ACTH (adrenocorticotropin hormone), CLIP (corticotropin-like intermediary peptide), two lipotropin (LPH) molecules (γ-LPH and β-LPH), and β-endorphin (β-END). This is shown in the upper diagram. The lower part of the figure shows the information about the five MSH receptors, their ligand preferences, and their general actions. Greek symbols over arrows at bottom left represent the specific isoforms of MSH.

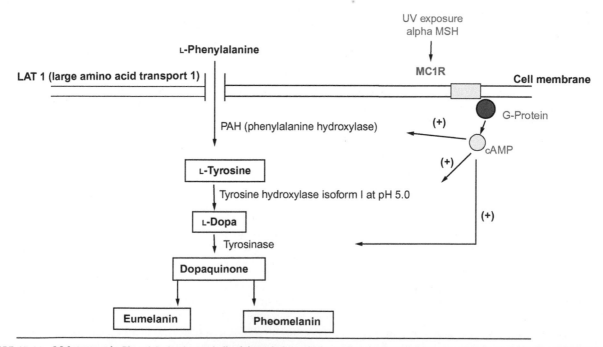

**FIGURE 15.10** **Melanogenesis.** Phenylalanine is metabolized through the melanin synthetic pathway catalyzed by individual steps: **phenylalanine hydroxylase** converts phenylalanine to tyrosine. A **tyrosine hydroxylase** isoform (isoform I) converts tyrosine to **L-DOPA** (dihydroxyphenylalanine), and **tyrosinase** converts DOPA to **DOPAquinone** that is, converted into the major forms of the pigment melanin, **eumelanin** (brownish-black), and **pheomelanin** (reddish-yellow). Signaling of the activated MC1R (by the binding of α-MSH or the presence of UV light) proceeds through a G protein and adenylate cyclase to produce cAMP from ATP which activates **protein kinase A** and stimulates the activities of all of the enzymes of the melanogenesis pathway, thereby increasing skin pigmentation. *MC1R*, melanocortin 1 receptor; *cAMP*, cyclic adenosine monophophate. *Reproduced from http://www.medscape.com/viewarticle/728367_2.*

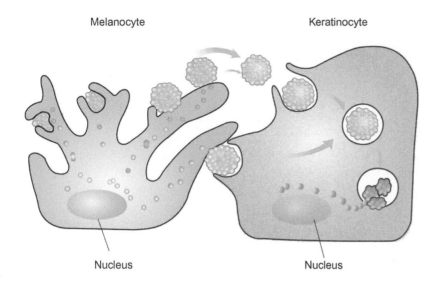

Melanocyte     Keratinocyte

Nucleus     Nucleus

FIGURE 15.11 Melanosomes are transferred from melanocytes to keratinocytes through packaging, release, and dispersion. Melanosomes are packed in globules enclosed by the melanocyte plasma membrane and are released into the extracellular space from areas of the melanocyte dendrites (thread-like projections of the cellular cytoplasm). The globules are phagocytozed by keratinocytes and are dispersed around the perinuclear area. *Reproduced from http://www.nature.com/jid/journal/v132/n4/fig_tab/jid2011513f4.html#figure-title.*

**Melanin** absorbs all of the **UV-B light** (315−280 nm) to prevent the damaging effects of those rays on DNA. **Melanocytes** are also found in the **uvea** (middle layer) of the eye, in the **inner ear** and elsewhere and melanin provides the coloration of skin, eyes, and hair. In the case of **blue eyes**, it is thought that the uvea contains less melanin, and the light passes to deeper layers of the eye to be scattered by resident proteins. The scattered light reflects back through the iris and appears blue to the onlooker. It is possible that in the case of **green** or **hazel eye** colors, there is slightly more melanin than in blue eyes to account for the difference in color.

In **Cushing's disease**, excess ACTH is produced providing more precursor for α-MSH. Accordingly, there is usually a **hyperpigmentation** of the skin. In **Addison's disease** where insufficient cortisol is produced by the adrenal cortex, there is a reduced negative feedback on the production of ACTH providing more α-MSH and a darkening of the skin even in areas not exposed to sunlight.

α-MSH has potent antiinflammatory effects. These are mediated by antagonizing the effects of **proinflammatory cytokines** as well as by decreasing the concentrations of key inflammatory mediators. The antiinflammatory actions of α-MSH operate through the **MC4 receptor (MC4R)** to produce cAMP and the subsequent activation of **protein kinase A (PKA)**. PKA phosphorylates **CREB (cAMP-responsive element binding protein)** and the activated CREB binds to cAMP responsive element sequences in target genes. Signals for inflammation lead to transfer of **NFκB** to the nucleus and subsequent production of inflammatory cytokines. The action of PKA (initiated by activated MC4R) prevents the breakdown of the inhibitor of NFκB nuclear transport and reduces the production of proinflammatory cytokines, such as **iNOS (induced nitric oxide synthase)** and **cox-2 (cyclooxygenase-2)**. Both of these enzymes lead to the production of inflammatory cytokines (although the role of iNOS in inflammation is incompletely understood). Cox-2 increases the production of inflammatory **prostaglandins**.

The neural effects of α-MSH are mediated by the activation of MC4R. Normally, **NPY (neuropeptide Y)** stimulates feeding, whereas α-MSH inhibits NPY-stimulated feeding by activating the MC4R in the regulatory hypothalamic neurons. α-MSH, operating through this receptor, inhibits the tonic release of NPY. Theoretically, the balance between NPY and α-MSH regulates **appetite** and feeding behavior.

## Growth Hormone

The pulsatile release of GH from the somatotropic cell of the anterior pituitary is signaled by the hypothalamic **GH-releasing hormone (GHRH)**. The release of GH is negatively controlled by **somatostatin** that is called into play when GH circulates and feeds back positively on the **somatostatinergic neuron** (neuron secreting somatostatin). A third regulator of GH release is **ghrelin** from the stomach that stimulates the release of GH from the anterior pituitary. *The two central activities of GH are to increase the production of growth factors from the liver and other organs, principally IGF-1 and to promote hyperglycemia through effects in the liver and other effects on adipose tissue and skeletal muscle.* To stimulate the production of IGF-1 and related factors, circulating GH (released from the **somatotropic cell** of the anterior pituitary) binds to the GH receptor on the plasma membrane of a liver cell and the GH receptor so activated sets off a chain of events culminating in the increased transcription and translation of **IGF-1 (insulin-like growth factor-1)**, **IGFBP-3 (insulin-like growth factor binding protein-3)** and other growth-promoting proteins (Fig. 15.12).

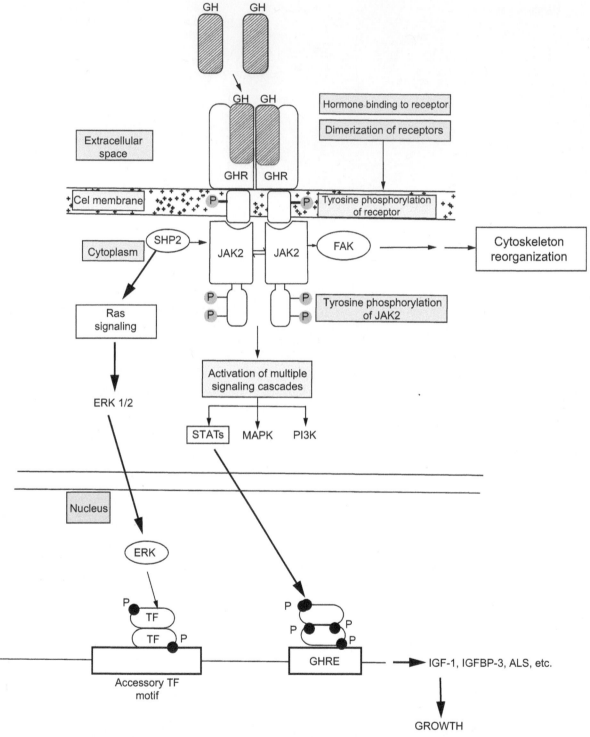

**FIGURE 15.12** Signaling pathway for GH following activation of the GH receptor (GHR). Two molecules of GH bind to the dimeric receptor which autophosphorylates on tyrosine residues. This causes the tyrosine phosphorylation of associated **JAK2** resulting in the activation of multiple signaling cascades, of which **STAT**s are a major pathway. In one scenario, JAK2 activates **FAK** (**focal adhesion kinase**) whose subsequent actions result in the reorganization of the **cytoskeleton**. STAT is translocated from the cytoplasm to the cell nucleus where a dimer is phosphorylated and binds to the **growth hormone response element** (**GHRE**). Simultaneously, **Ras** signaling can be activated to cause **ERK**s to translocate from the cytoplasm to the nucleus where they become phosphorylated (P) and bind to an accessory transactivation factor (TF) motif. These actions increase the transcription of the genes for IGF-1, IGFBP, ALS, and other growth-promoting factors, resulting in the growth of tissues. *ALS*, acid labile subunit; *ERK*, extracellular signal-regulated kinase; *GHRE*, GH response element; *IGFBP*, IGF-binding protein; *JAK*, Janus family tyrosine kinase; *TF*, transcription factor; *FAK*, focal adhesion kinase; *SHP2*, SH2-containing protein tyrosine phosphatase-2; *Ras*, monomeric G protein oncogene; *STAT*, signal transducer and activator of transcription.

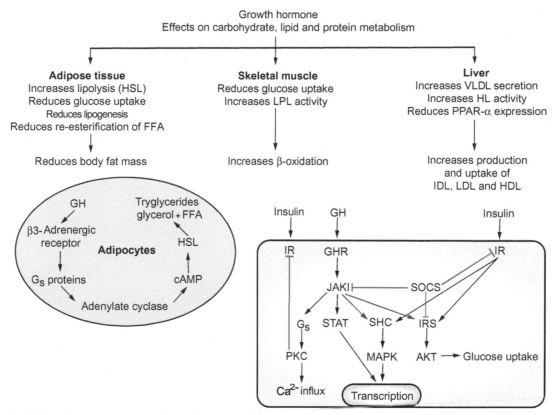

**FIGURE 15.13** Pleiotropic effects of GH on adipose tissue, skeletal muscle, and liver involving carbohydrate, lipid, and protein metabolism. Effects of GH in adipocytes are mediated by the β3-**adrenergic receptor,** whereas in skeletal muscle and liver, GH effects operate through the **GH receptor** (**GHR**) and subsequent activation of the **JAK-STAT** (Janus family tyrosine kinase-signal transducer and activator of transcription) pathways. In these tissues, GH signaling interacts with insulin signaling (*blue* arrows from JAK and *red* T-bar from **PKC; SOCS, suppressor of cytokine signaling**) interferes with GH and insulin signaling. GH antagonizes insulin action on carbohydrate metabolism both directly (through cell signaling blockade mechanisms) and indirectly (by enhancing lipolysis in adipocytes). Free fatty acids (FFA) released from adipocytes, secondary to GH interference, induce **insulin resistance** at other sites, such as muscle. Effects on liver might also involve fat metabolism. *AKT*, protein kinase B; *cAMP*, cyclic AMP; *G_s*, stimulatory G protein; *HL*, hepatic lipase; *HSL*, hormone-sensitive lipase; *IDL*, intermediate density lipoprotein; *IR*, insulin receptor; *IRS*, insulin receptor substrate; *LPL*, lipoprotein lipase; *MAPK*, mitogen-activated protein kinase; *PKC*, protein kinase C; *PPARα*, peroxisome-proliferator-activated receptorα; *SHC*, Src homology domain 2-pleckstrin homology domain; *SOCS*, suppressor of cytokine signaling; *STAT*, signal transducer and activator of transcription. *Reproduced from http://www.nature.com/nrendo/journal/v3/n3/fig_tab/ncpendmet0427_F1.html.*

Other important effects of GH involve the liver, adipose tissue, and skeletal muscle. These are pleiotropic effects (multiple effects of a single substance) of GH as shown in Fig. 15.13.

## Prolactin

Prolactin is the hormone from the **lactotrope** of the anterior pituitary that stimulates and maintains **lactation** in the mother. Principally, it is responsible for the formation of human milk proteins: **casein**, β-**lactalbumin,** and **whey acidic protein**. However, there are many other actions of PRL, such as the regulation of **sexual activity** and **behavior**. After **orgasm** in either the male or the female, there is a transitory increase of PRL in the blood that is thought to diminish **sexual arousal**. In experimental animals, there have been observed over 300 separate functions of PRL including effects on water and electrolyte balance, growth, development, behavior, reproduction, and immunoregulation to mention a few. In pregnant mice, PRL increases **pancreatic β-cells**, availing increased insulin and increased brain neurogenesis.

The monomeric form of PRL consists of 198 amino acids and has a molecular weight of 23 kDa. Interestingly, there are other forms of PRL each of which may have different physiological functions. There is a fragment of 16 kDa, a dimer of 40−50 kDa and a trimer of 70−80 kDa.

Regulation of the secretion of PRL from the lactotrope of the anterior pituitary is summarized in Fig. 15.14.

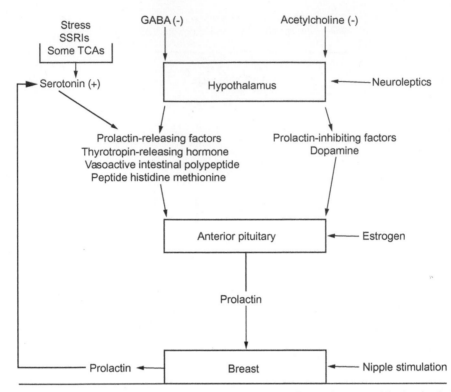

**FIGURE 15.14**   Factors involved in controlling the secretion of PRL from the anterior pituitary lactotrope. **Nipple stimulation** by the feeding infant causes the release of **serotonin** that causes the release of **prolactin-releasing factors** from the hypothalamus; the principal releasing factor is the **thyrotropin-releasing hormone**. The prolactin-inhibiting factor is **dopamine** from the hypothalamus that decreases the secretion of PRL. PRL binds to a membrane receptor on the mammary gland cell, and the signaling mechanism of PRL increases the transcription of genes for milk proteins as shown in Fig. 15.15. *SSRI*, selective serotonin reuptake inhibitor; *TCA*, tricyclic antidepressant; *GABA*, γ-aminobutyric acid. *Reproduced from http:// www.medscape.com/viewarticle/468929_2.*

## Thyroid Stimulating Hormone (TSH), Thyrotropin

TSH is released from the anterior pituitary **thyrotropic cell** by the hypothalamic hormone, **thyrotropin-releasing hormone (TRH)**. TSH binds and activates its receptor on the **thyroid follicle cell,** and the following signaling system results in the release of thyroid hormones (**triiodothyronine, $T_3$** and **thyroxine, $T_4$**) into the general circulation. The thyroid hormone, in the form of $T_3$, binds to its receptor (considered a member of the steroid receptor gene family even though the ligand is not a steroid) in the nucleus of target cells to result in increased metabolic activity, increased levels of ATP and oxidations. These activities and substances that either stimulate or inhibit various steps in the signaling process are summarized in Fig. 15.16.

The **TSH receptor** on the plasma membrane of the **thyroid follicle cell** is complex. The ligand, TSH, is a dimer of an α- and a β-subunit. The α-subunit is common to TSH, FSH (follicle-stimulating hormone), LH (luteinizing hormone), and HCG (human chorionic gonadotropin), all dimers of α- and β-subunits. The β-subunit of TSH determines the specificity for its receptor as shown in Fig. 15.17. The TSH receptor is comprised of two subunits, one a glycoprotein receptor and the other a ganglioside receptor. After binding TSH, the receptor undergoes a conformational change of TSH causing its α-subunit to intercalate into the bilayer (bottom figure). Two associated G proteins signal a stimulation of the phosphatidylinositol cycle to produce **diacylglycerol, DAG**, and **inositol *tris*phosphate, IP₃**, and the production of **cAMP** (resulting in the activation of protein kinase A) and stimulating the production and ultimate release of thyroid hormones into the bloodstream.

It is possible that PKA phosphorylates and activates the **NIS cotransporter (sodium ion and iodide transporter)** that permits the entry into the cell of **iodide ($I^-$)** along with two atoms of **sodium ion.** Iodide moves across the cytoplasm to the basolateral membrane transporter, **Pendrin**, which moves iodide to the colloid of the thyroid follicle. **Thyroglobulin** is being synthesized in the rough ER (endoplasmic reticulum) and enters the luminal colloid by exocytosis from the cell. The iodide ion ($I^-$), once in the colloid, is oxidized by **thyroid peroxidase** to form **iodine ($I^0$)**. Many of the tyrosine residues ($\sim 120$) in thyroglobulin are iodinated by $I^0$. $T_4$ and $T_3$ are cut out of the iodinated thyroglobulin by proteases, and these thyroid hormones ($[T_4] \gg [T_3]$) enter the bloodstream and are taken up by cellular targets of the hormones. $T_4$ is converted to $T_3$ before entering the (liver) target cell nucleus. This system is diagramed in Fig. 15.18.

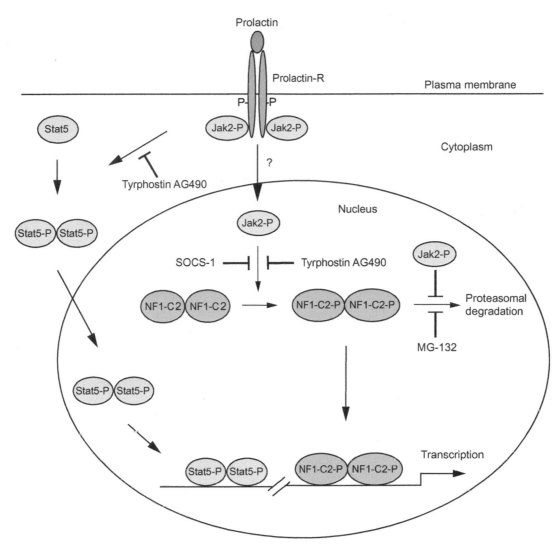

**FIGURE 15.15** PRL binds to its dimeric receptor on the plasma membrane of a mammary gland cell. The receptor, thus activated, autophosphorylates on tyrosine residues and then phosphorylates **Stat5** and **Jak2**. Stat5 dimerizes and translocates to the nucleus and Jak2-P phosphorylates the dimer of nuclear factor **NF1-C2**. The phosphorylated dimeric NF1-C2-P binds to genes for milk proteins as does dimeric Stat5-P to increase the mRNAs and translate the milk proteins: **casein**, β-**lactalbumin**, and **whey acidic protein**. *Prolactin-R*, prolactin receptor; *Tyrphostin*, tyrosine kinase inhibitor; *SOCS*, suppressor of cytokine signaling; *NF*, nuclear factor; *Stat*, signal transducer and activator of transcription; *Jak*, Janus family of protein kinases; *MG-132*, proteasome inhibitor. *Reproduced from http://mcb.asm.org/content/26/15/5663/F9.large.jpg.*

## The Gonadotropins: Luteinizing Hormone (LH) and Follicle Stimulating Hormone (FSH)

Both of these hormones are synthesized and secreted from the same cell, the **gonadotrope** of the anterior pituitary. Both hormones are dimers and share the same α-subunit. The β-subunit determines the specificity for receptor binding. There are 118 amino acids in the β-subunit of FSH and 121 amino acids in the β-subunit of LH. The α-subunit, common to FSH, LH, human chorionic gonadotropin (HCG), and TSH contains 92 amino acids. In the female, LH and FSH drive the ovarian cycle and the release of estrogens from the follicle. In the male, LH and FSH are responsible for the formation of testosterone and spermatogenesis. The structures of LH and HCG, secreted from accessory tissues during pregnancy, are similar enough that they bind to the same receptor and generate the same signal cascade. LH, FSH, and TSH receptors are members of a superfamily of seven membrane G protein-coupled receptors. These receptors have long extracellular domains with leucine rich repeats forming the ligand-binding domain. In the male, LH binds to and activates its receptor on the plasma membrane of the **Leydig cell**; in turn, the receptor activates adenylate cyclase through a G protein to produce cAMP from ATP. cAMP activates protein kinase A which phosphorylates inactive cholesteryl ester hydrolase and activates it so that it can convert cholesteryl esters in the cytoplasmic lipid droplet to free cholesterol. Free cholesterol enters the mitochondrion, through the agency of the StAR (steroid-activated response) protein, for its conversion to testosterone. Cholesterol cannot enter the outer mitochondrial membrane by itself, and the

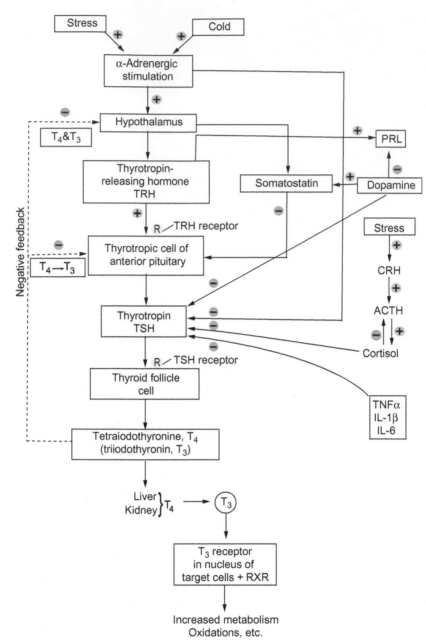

**FIGURE 15.16** Pathway of thyroid hormone release from TRH, TSH, and other regulators. *Green circles*, activators or stimulators; *red circles*, inhibitors or negative regulators; *ACTH*, adrenocorticotropic hormone; *CRH*, corticotropin-releasing hormone; *IL-1β*, interleukin-1β; *IL-6*, interleukin-6; *PRL*, prolactin; *SST*, somatostatin; *T₃*, triiodothyronine; *T₄*, tetraiodothyronine; *TNFα*; tumor necrosis factorα. Negative feedback by thyroid hormones on TSH and TRH causes a decrease in their syntheses.

StAR protein is needed so that cholesterol can penetrate the mitochondrial membrane. The roles of the StAR protein in steroid hormone synthesizing cells are shown in Fig. 15.19.

In the Leydig cell, under the initial stimulation by LH, cholesterol in the mitochondrion is converted to pregnenolone by the **side chain cleavage system (SCC)** that will be discussed in the next chapter. Pregnenolone is converted to progesterone by **3β-hyroxysteroid dehydrogenase**. Progesterone is converted to androstenedione by microsomal P450c17, and androstenedione is converted to testosterone by **17β-hydroxysteroid dehydrogenase**. Both hydroxysteroid dehydrogenases are located in mitochondria and microsomes. Testosterone generated in the Leydig cell is used by the Sertoli cell to amplify spermatogenesis in the seminiferous tubules in concert with FSH as shown in Fig. 15.20.

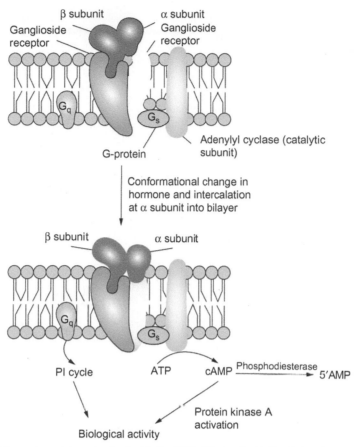

**FIGURE 15.17** Model of the TSH receptor and its activation by binding TSH. *PI*, phosphatidylinositol cycle.

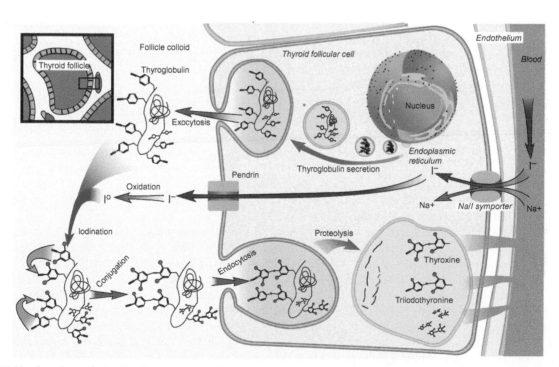

**FIGURE 15.18** Overview of thyroid hormone synthesis in the thyroid follicle cell. *Reproduced from http://en.wikipedia.org/wiki/File: Thyroid_hormone_synthesis.png.*

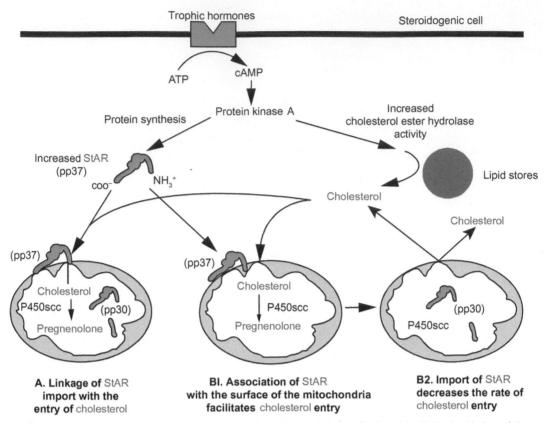

**A. Linkage of** StAR
**import with the**
**entry of** cholesterol

**BI. Association of** StAR
**with the surface of the mitochondria**
**facilitates** cholesterol **entry**

**B2. Import of** StAR
**decreases the rate of**
cholesterol **entry**

**FIGURE 15.19** Two proposed mechanisms for the acute regulation of steroidogenesis by the StAR protein. Following binding of the trophic hormone (LH, ACTH, etc.) to a cell surface receptor, cAMP is increased and protein kinase A is activated. PKA increases the synthesis of StAR and its phosphorylation giving rise to pp37. In proposal A, StAR is imported into the mitochondria and processed into the mature pp30. During the import process, it is proposed (B) that cholesterol is transported to the inner mitochondrial membranes. Proposal B1 takes into account data suggesting that the interaction of StAR protein with the mitochondrial surface leads to cholesterol movement. In B2, it is proposed that importation of StAR and cleavage into pp30 is a mechanism of inactivating cholesterol flux into the mitochondria. *Reproduced from http://www.pnas.org/conten/93/24/13552/F1.expansion.html.*

The androgen receptor can bind testosterone but has a higher affinity for 5α-dihydrotestosterone (where the 4−5 double bond of testosterone is saturated). The structures of these two steroids are shown in Fig. 15.21. Testosterone produced in the Leydig cell has a negative feedback on the release of LH and interferes with the action of GnRH in the anterior pituitary. These feedback mechanisms are complex.

FSH binds and activates its receptor on the plasma membrane of its target cells. As mentioned, it is a seven-membrane receptor with a large extracellular domain that forms the ligand-binding site, similar to the LH receptor. As with the LH receptor, the FSH receptor is coupled to a G protein that activates adenylate cyclase to produce cAMP that activates PKA. The activity of PKA opens calcium channels admitting calcium ions into the cell and also phosphorylates transcription factors that stimulate transcription of specific genes.

In the female, the level of FSH increases at the beginning of the follicular phase through ovulation and then diminishes in the luteal phase as shown in Fig. 15.22.

On the first day of the 28-day cycle, the blood levels of LH and progesterone are low. The blood level of estrogen also is low but begins to rise to stimulate the developing follicle; this constitutes the follicular phase. FSH levels rise slowly and the level of estrogen peaks around day 12, and LH increases sharply (known as the **LH spike**) to drive ovulation. After ovulation, the residual follicle (the Graafian follicle) secretes progesterone until day 28 where the level is low, and at this point **menstruation** occurs. If fertilization occurs at ovulation, the placenta elaborates human **chorionic gonadotropin (HCG)** in order to maintain a luteal phase, and the placenta produces progesterone until pregnancy terminates when the level of progesterone is very high. This high level of progesterone competes with endogenous cortisol for binding to the glucocorticoid receptor (progesterone is a glucocorticoid antagonist); the resulting effect is the inhibition of lactation plus other functions as these depend on the activated glucocorticoid receptor. The overall actions of gonadotropins in the female are illustrated in Fig. 15.23, and the progression of the 28-day ovarian cycle is pictured in Fig. 15.24.

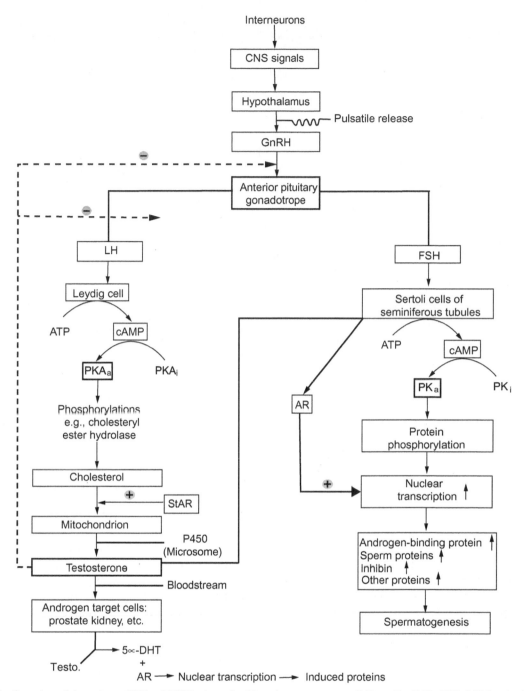

**FIGURE 15.20** Overview of the actions of LH and FSH in the male. *AR*, androgen receptor; *cAMP*, cyclic AMP; *FSH*, follicle-stimulating hormone; *GnRH*, gonadotropin-releasing hormone; *LH*, luteinizing hormone; *PKAₐ*, activated protein kinase A; *PKAᵢ*, inactive protein kinase A; *StAR*, steroid acute response protein; *testo*, testosterone; *5α-DHT*, 5α-dihydrotestosterone.

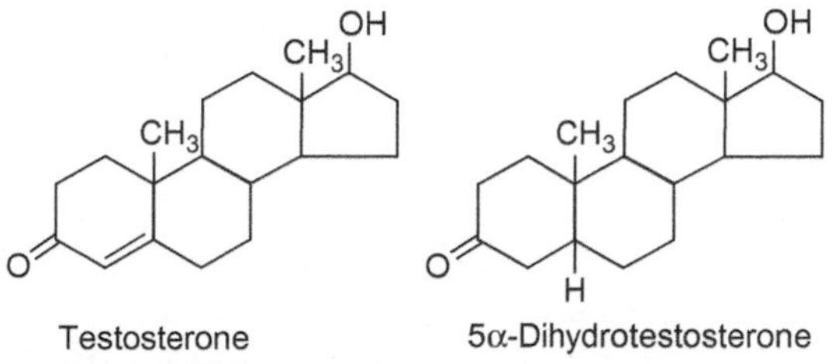

**FIGURE 15.21** Structures of testosterone and 5α-dihydrotestosterone.

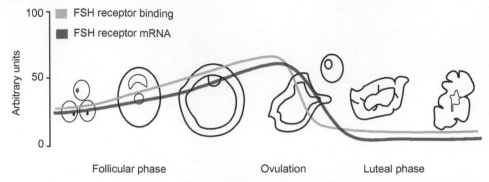

**FIGURE 15.22** Expression of the FSH receptor during estrus. Follicle maturation is indicated by the development of primary follicles into a Graafian follicle. After ovulation, the *corpus luteum* and the *corpus albicans* are shown. Expression levels of the FSH receptor are given in an arbitrary scale ranging from 0 to 100 units. *Reproduced from Simoni, M., Gromoll, J., Nieschlag, E., 1997. The follicle-stimulating hormone receptor: biochemistry, molecular biology, physiology and pathophysiology. Endocr. Rev. 18, 739—773.*

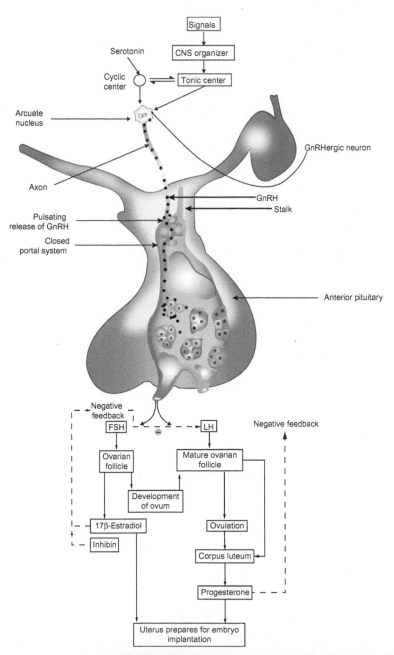

**FIGURE 15.23** Ovarian cycle showing the generation of GnRH, LH, FSH, estrogen, and progesterone. The hormone **inhibin** inhibits FSH secretion. 17β-estradiol inhibits LH release. The *red* dot indicates negative feedback.

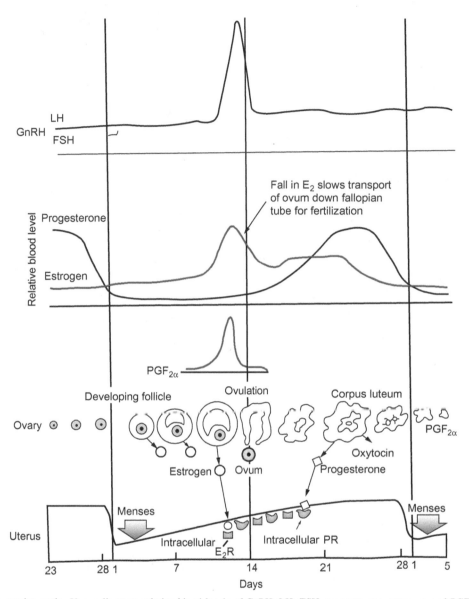

**FIGURE 15.24**   The ovarian cycle. *Upper* diagram: relative blood levels of GnRH, LH, FSH, progesterone, estrogen, and $PGF_{2\alpha}$ are shown. In the *lower* diagram, events in the ovarian follicle, *corpus luteum,* and uterine endometrium are diagramed. $E_2$, 17β-estradiol; $E_2R$, intracellular estrogen receptor; *FSH*, follicle-stimulating hormone; *GnRH*, gonadotropin-releasing hormone; *LH*, luteinizing hormone; $PGF_{2\alpha}$, prostaglandin $F_{2\alpha}$, *PR*, intracellular progesterone receptor.

## Overview of the Release and Functioning of the Vasopressin (AVP) and Oxytocin (OT) Hormones from the Posterior Pituitary

Vasopressin can serve as a model for the posterior pituitary hormones. The biological controls for the release of arginine vasopressin (AVP) are shown in Fig. 15.25.

The release of the **posterior pituitary hormones** can be mediated by the *N*-methyl-D-aspartate (NMDA) receptor or by the α-amino-3-hydroxy-5-methyl-4-isoxazole proprionic acid (AMPA) receptor (both are **glutamate receptors**) when the NMDA receptor is not involved. There are differences between the signals that result in release of the two hormones as would be expected.

**Oxytocin release** from the posterior pituitary is stimulated by **estrogen** or by the **suckling response** as shown in Fig. 15.26.

Both AVP and OT have similarly constructed receptors. They contain seven membrane-spanning subunits. The amino acid sequences of both receptors are similar; about 16% of the amino acids in the receptors are identical, sharing

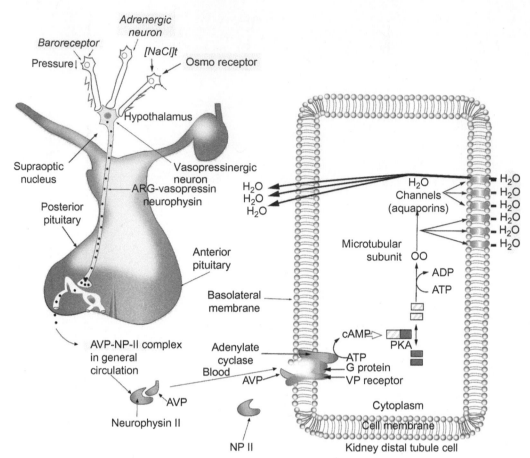

**FIGURE 15.25**    The release and functioning of the posterior pituitary vasopressin hormone. Signals impinging on the cell body of the neuron synthesizing arginine vasopressin (**AVP**) and **neurophysin II** (**NPII**). The signals that affect the disposition of AVP synthesized in the cell body are the **baroreceptor** that can sense a fall in blood pressure, an **adrenergic interneuron** and an **osmoreceptor** sensing an increase in the salt content of blood. Upon responding to any of these signals, the complex of **AVP-NPII** is transported down the long axon to the nerve ending in the posterior pituitary. AVP-NPII is released into the general circulation and the complex dissociates later on. The free AVP binds to its receptor in the cell membrane of the distal kidney tubule cell. The activated receptor is coupled to a G protein and to **adenylate cyclase** that converts ATP to cyclic AMP (**cAMP**). cAMP activates protein kinase A (PKA) that phosphorylates microtubular subunits that form **aquaporins** (water channels) in the apical membrane. Water is taken up and transported across the cell into the basolateral space outside the cell; it ultimately dilutes the blood and increases its partial pressure causing a rise in blood pressure. *These actions take place in response to the initial signals that are a fall in blood pressure, or a decrease in the blood salt content or stimulation of the AVP cell body by an adrenergic interneuron.*

the same positions in the sequence. In both cases, amino acid residues of receptors that are involved in binding the hormones are located on the surface of the plasma membrane although the cyclic portion of OT interacts with receptor residues just below the target cell membrane (the same may be true for AVP since the structures in that part of each receptor are similar in both cases).

## OREXINS (HYPOCRETINS): HYPOTHALAMIC HORMONES CONTROLLING SLEEP AND FEEDING

There are two hypothalamic hormones, **Orexin-A** and **Orexin-B** (also known as **hypocretin 1** and **hypocretin 2**). Orexin-A is made up of 33 linear amino acid residues with a molecular weight of 3561 Da, and Orexin-B is a 28-amino acid linear peptide with a molecular weight of 2937 Da. The production of the mature hormones from **preproorexin** is shown in Fig. 15.27.

These hormones act through G coupled receptors (**OX1 receptor** and **OX2 receptor**). **Orexins** are expressed (probably exclusively) in the hypothalamic **paraventricular nucleus** and **median eminence,** and their receptors occur in anterior pituitary **corticotropes**, the **adrenal cortex** and **adrenal medulla,** and probably in the alpha-cells and

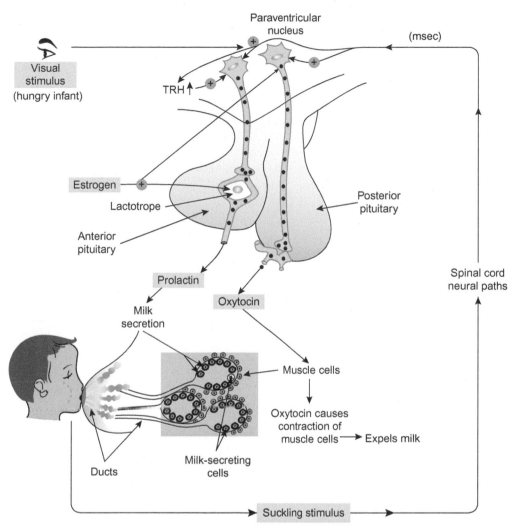

**FIGURE 15.26** The release of oxytocin (OT) and **prolactin (PRL)** in the nursing mother in response to **suckling** or **visual stimuli** (sight of the child desiring feeding). The stimulation of neuronal cells in the hypothalamus affects the release of OT from the posterior pituitary and PRL from the anterior pituitary. PRL binds to its receptor in plasma membrane of the **milk-secreting cells** in the **mammary gland** to stimulate the release of **milk**. OT binds to its receptor in muscle cells surrounding the milk-producing cells causing contraction and expulsion of milk into the mammary ductal system and out through the nipple.

beta-cells of the endocrine pancreas. Although orexins may not enhance excretion of aldosterone from the adrenal cortex to any great extent, they do stimulate the production of **cortisol** in the adrenal cortex by stimulating the **corticotropin-releasing hormone (CRH)** release of **adrenocorticotropic hormone (ACTH)** from the anterior pituitary corticotrope that further stimulates the production and release of cortisol from the adrenal cortex. OX1 receptors in the membrane of adrenal cortical cells also are coupled to the **adenylate cyclase** pathway that increases **cAMP** which activates **protein kinase A (PKA)** whose phosphorylating activity results in increasing the hydrolysis of **cholesteryl ester** stored in the **lipid droplet**, and the resulting free cholesterol enters the mitochondria where most of the reactions producing cortisol are located, an overall system that is substrate (cholesterol)-limited. There is experimental evidence that orexins also can enhance the growth of adrenocortical cells. It is further believed that orexins play a role in the **hypothalamic−pituitary−adrenal (HPA)** axis in response to **stress** and also a role in the pathophysiology of **cortisol-secreting adrenal adenomas**.

Fasting and levels of **plasma glucose** at 2.8 mmol/L or below cause the stimulation of Orexin-A-ergic neurons, and this can lead to **acetylcholine release**, a decreased release of **LHRH**, an increased release of **glucagon**, a decrease in plasma **insulin**, an increase in plasma glucose, and the release of **adiponectin** from adipose cells with a decrease in **lipolysis**.

The control of **sleep cycles** may be related to the effects of orexins on the acetylcholine (cholinergic), dopamine, norepinephrine, and histamine systems that function together to stabilize sleep cycles. The absence of orexin-A in

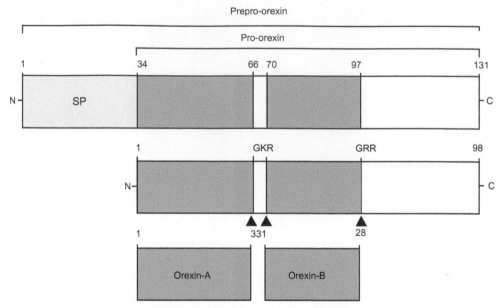

**FIGURE 15.27**    Diagram illustrating the posttranslational processing of preproorexin. Arrowheads indicate the sites where, after the removal of the signal peptide (SP), proorexin is cleaved by **prohormone convertase** at sites with basic amino acid residues (GKR and GRR) to produce mature Orexin-A and Orexin-B. *Reproduced from Spinazzi, R., Andreis, P.G., Rossi, G.P., Nussdorfer, G.G., 2006. Orexins in the regulation of the hypothalamic—pituitary—adrenal axis. Pharmacol. Revs. 58, 46—57; also http://pharmrev.aspetjournals.org/content/58/1/46/F1.large.jpg.*

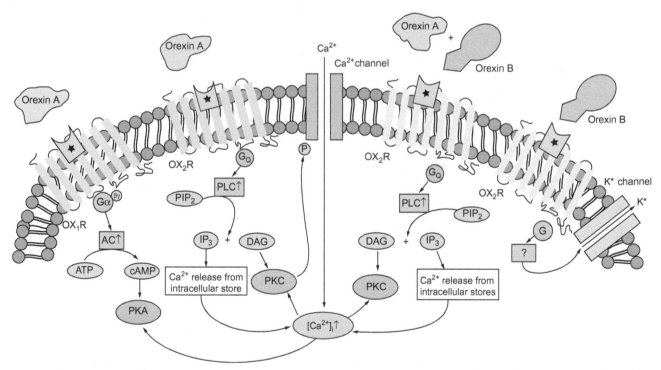

**FIGURE 15.28**    Overall scheme of signaling pathways of orexin receptors OX$_1$ and OX$_2$ upon the activation by orexin-A or orexin-B. *AC*, adenylate cyclase; *cAMP*, cyclic AMP; *DAG*, diacylglycerol; *G*, G protein; *IP$_3$*, inositol *tris*phosphate; *P*, phosphorylation site; *PIP$_2$*, phosphatidylinositol *bis*phosphate; *PKA*, protein kinase A; *PKC*, protein kinase C; *PLC*, phospholipase C. Asterisks denote orexin recognition receptor sites of OX$_1$R and OX$_2$R (1 = A and 2 = B). *Reproduced from Shahid, I.Z., Rahman, A.A., Pilowsky, P.M., 2012. In: Litwack, G. (Ed.), Sleep Hormones. Vitamins and Hormones, vol. 89. Academic Press/Elsevier, pp. 163—182.*

the human causes **narcolepsy** (sleep deprivation; extreme drowsiness during the day) and treatment with orexins is useful in counteracting this condition. Narcolepsy is thought to be genetic and affects individuals between the ages of 15 and 30.

The main signaling pathways involving the orexin-A and orexin-B receptors are shown in Fig. 15.28.

## ADIPONECTIN FROM ADIPOSE (FAT) TISSUE

It has been realized that fat tissue, consisting of fat (adipose) cells is more than just a storage depot for fat (triglycerides) but, in fact, is an endocrine tissue and secretes the hormones (**adipokines**), **adiponectin**, **resistin**, **leptin**, and others, adiponectin being the major secreted hormone. Adiponectin has important antiinflammatory and insulin-sensitizing activities, and its circulating level is downregulated in conditions of **obesity**, **hyperlipidemia**, **insulin resistance**, and related diseases. Dietary **long chain n-3 polyunsaturated fatty acid** (**n-3PUFA**) consumption and weight loss increase the synthesis and secretion of adiponectin. Adiponectin is a 244 amino acid long polypeptide with a molecular weight of 30,000 Da. There are four domains in the adiponectin molecule: a short signal sequence at the N-terminus that targets the hormone for secretion to the outside of the cell, followed by a second species specific region, a third region of 65 amino acids that has similarities to collagen and a fourth globular domain at the C-terminus. The three-dimensional structure of the globular domain is similar to **TNFα** even though the amino acid sequence is different. Adiponectin circulates in trimeric, hexameric, or higher molecular weight forms. The high molecular weight (hMW) forms contain as many as 12 to 18 subunits. The amounts of the hMW forms correlate with whole bodily **sensitivity to insulin**. Adiponectin is protective against **obesity**, **type2 diabetes**, and **cardiovascular disease**. It counteracts **atherosclerosis** by inhibiting both the activation of **adhesion molecules** and the formation of **foam cells** while increasing the production of **nitric oxide** (a beneficial "hormone").

Proteolytic cleavage of **full-length adiponectin** (**fAd**) produces a **globular form of adiponectin** (**gAd**; about 1% of total adiponectin) that is particularly effective in **skeletal muscle**. It promotes **glucose uptake** in skeletal muscle and decreases **gluconeogenesis** in **liver**, leading to its insulin-sensitizing effects. Circulating adiponectin concentrations, especially the hMW forms, are inversely correlated with adiposity (fatness), the fatter or more obese the individual, the lower the levels of circulating adiponectin. Apparently, the circulating trimers of adiponectin are positively correlated with **diabetes**.

There are two receptors for adiponectin, **AdipoR1** and **AdipoR2**. Both are seven transmembrane receptors. AdipoR1 is expressed in most tissues and is highest in skeletal muscle. This receptor has the highest affinity for the globular form of **adiponectin** (gAd) and a lower affinity for **full-length adiponectin** (**fAd**). AdipoR2 is expressed primarily in **liver** and has moderate affinity for both gAd and fAd; this indicates that fAd has its effects predominantly in liver. In both **muscle** and liver cells, adiponectin stimulates the activation of **AMP-activated protein kinase** (**AMPK**) and fatty acid oxidation. The latter effect lowers triglyceride concentrations in tissues. **T-cadherin** acts as third receptor for adiponectin exclusively in endothelial and smooth muscle cells. It is a receptor for the hexameric and high molecular weight forms (hMW) of adiponectin. Hyroxylation of lysine residues, the formation of disulfide bonds and glycosylations play a role in the formation of the **hMW** forms of **adiponectin**. Since T-cadherin is a **glycosylphosphatidylinositol-anchored extracellular protein**, it may be a coreceptor for another, more membranous or internal, receptor (**APPL1**) that transmits the metabolic signals from adiponectin.

Fig. 15.29 shows how **Adiponectin receptor 1** (**AdipoR1**) functions when activated by binding adiponectin.

AdipoR1 is downregulated by its **endocytosis** with **clathrin**, and this function depends on Rab5. Adiponectin colocalizes with its receptor and becomes internalized as well. These proteins may be degraded in the **lysosome**. Inhibiting the endocytotic cellular uptake increases the basal- and adiponectin-mediated phosphorylation of **AMPK**, thus *endocytosis is the regulator of adiponectin activity.*

Fat intake and obesity are known risk factors for the development of **breast cancer**. The lowered levels of adiponectin in **obesity** and in patients with breast cancer are suggestive of *adiponectin's role in the prevention of the development of breast cancer.*

## HORMONES OF THE GASTROINTESTINAL (GI) TRACT

The hormones of the GI tract are shown in Table 15.3 that also summarizes their actions.

Some **gastrointestinal hormones** are also synthesized in the brain (referred to as **"brain-gut" hormones**). One such hormone is **galanin**. Galanin is a peptide of 30 amino acids ($\sim$3211 Da) that is split out of its precursor, **preprogalanin** (123 amino acids long). It connects the sympathetic nervous system with **vagal stimulation** by adrenergic neurotransmitters, such as **neuropeptide Y** (**NPY**). A diagram hypothesizing the action of galanin is shown in Fig. 15.30.

**Galanin**, expressed in the brain, spinal cord and gut, signals through three G-protein-coupled receptors. In addition to the above, galanin is involved in the regulation of sleep and waking and in eating disorders. Galanin colocalizes with the following neurotransmitters: **acetylcholine** and **norepinephrine**, **neuropeptide Y** (e.g., Fig. 15.30), and also **serotonin** as well as **substance P** and **vasoactive intestinal peptide** (**VIP**).

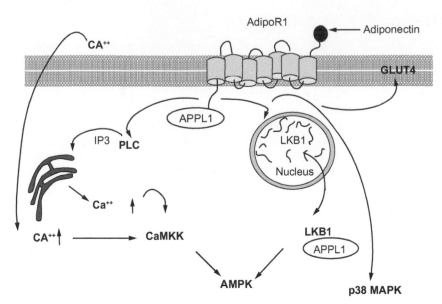

**FIGURE 15.29** The function of **APPL1** in AdipoR1. APPL1 (and **APPL2**, the inhibitor of APPL1 activity by complexing AdipoR1 into an inactive form) is a homologous **Rab5** effector protein. APPL1 forms a complex with AdipoR1. Knockdown of APPL1 inhibits adiponectin-mediated phosphorylation of AMPK and **p38 MAPK**. GLUT4 translocation to the plasma membrane is induced. Adiponectin increases AMPK by promoting the nuclear export of **LKB1** that subsequently binds to APPL1 to stabilize the cytosolic localization. Adiponectin also increases **calcium-calmodulin-dependent protein kinase (CaMKK)** by **phospholipase C (PLC)**-mediated increase in $Ca^{2+}$. LKB is the major AMPK kinase, whereas CaMKK plays a minor role in AMPK phosphorylation. Furthermore, adiponectin increases extracellular $Ca^{2+}$ influx and subsequent activation of AMPK (it is unclear whether APPL1 is involved here). *APPL1*, an adapter protein with phosphotyrosine interaction, PH domain and leucine zipper-containing 1; *Rab5*, a GTPase that can regulate **endosome fusion**; *AdipoR1*, adiponectin receptor 1; *MAPK*, mitogen-activated protein kinase; *GLUT4*, glucose transporter4; *LKB1*, liver kinase B1; *AMPK*, AMP-activated protein kinase. *Reproduced from Buechler, C., Wanninger, J., Neumeier, M., 2010. Adiponectin receptor binding proteins − recent advances in elucidating adiponectin signaling pathways. FEBS Lett. 534, 4280−4286; also http://www.sciencedirect. com/miamilmageURL/1-s2.0-S0015579310007726-f15-02-9780123838643_1rg.jpg/0?wchp=dGLbVlt-z5kWb.*

## Gastrin

**Gastric acid (HCl)** secretion is stimulated by the hormone, gastrin (linear peptide of 17 amino acid residues ($P_1$GPWLEEEEE$_{10}$AYGWMDF$_{17}$); molecular weight 2216 Da). It is split out of a prepropeptide precursor that also gives rise to a family of proteins having the same carboxyl terminus. The major circulating form of gastrin is called **"big gastrin,"** a peptide of 34 amino acids (possibly a dimer of the 17 amino acid form of gastrin). The greatest activity of gastrin is inherent in **"mini gastrin"** containing 14 amino acids. The five C-terminal amino acids of gastrin are identical to the five C-terminal amino acids of **cholesystokinin (CCK)** that explains their overlapping biological effects. Gastrin binds to the **gastrin/CCK$_B$ receptor**, a G protein-coupled receptor that activates **protein kinase C, phosphatidylinositol *bis*phosphate (PIP$_2$)**, hydrolysis (to form **inositol *tris*phosphate (IP$_3$)**), and **diacylglycerol (DAG)** and increases the cytosolic concentration of **calcium ion**.

Gastrin also stimulates the proliferation of the gastric epithelium. It is released from the **G cells** in the stomach antrum in response to distention. Its release is also stimulated in response to partially digested proteins (peptides and amino acids) and other food digestion products in the gastric lumen. The secretion of gastrin is inhibited by **somatostatin** and also if the pH of the gastric lumen falls below 4.

Gastrin affects the **enterochromaffin-like (ECL)** cells in the acid-producing region of the stomach to induce **histidine decarboxylase** that produces **histamine** and the secreted histamine stimulates the **parietal cells** to secrete acid. Histamine activates the **histamine H$_2$ receptor** in the gastric parietal cells resulting in the secretion of acid. There are two additional histamine receptors, histamine H$_1$ and histamine H$_3$ receptors. While the **H$_1$ receptor** depresses **synaptic transmission** (e.g., in the circular muscle in the oral compartment), at low concentrations, it activates the **enteric excitatory ascending pathways** at high concentrations. The activated H$_2$ receptor stimulates the enteric excitatory ascending pathways, and activation of the histamine H$_3$ receptor decreases synaptic transmission. In addition to elevating the release of HCl from the acid-producing part of the stomach, gastrin increases the secretion of **pepsinogen** from the **Chief cells** of the stomach and also increases motility in the stomach. The **gastrin receptor (CCK2 receptor)** in complex with gastrin stimulates calcium ion entry through calcium N- and L-type channels. The **L-type channel** is activated by **adenylate cyclase-activating peptide**. Gastrin also is a growth factor for the **enterochromaffin-like (ECL)** cells and for the stomach stem cells that produce acid. **Somatostatin, misoprostol (prostaglandin E$_1$)** and **galanin** bind to receptors whose actions, through inhibitory G proteins, block the L-type calcium channels.

**TABLE 15.3 Gastrointestinal Hormones Overview**

| Hormone (MW) | Source | Action | | | | | Other | Regulation | | | | | |
|---|---|---|---|---|---|---|---|---|---|---|---|---|---|
| | | Stomach | | Intestine | Pancreas | | | Stomach Distention | Acidity | Glucose | Fatty Acids | Amino Acids | Vagal Stimulation |
| | | MOTILITY | ACID | MOTILITY | ENZYMES | $HCO_3$ | | | | | | | |
| Gastrin (2098 Da) | G cells stomach antrum | ++ | ++ | | | | ++ Growth of gastric mucosa | ++ | − | | | ++ | ++ |
| CCK (cholecystokinin) (3945 Da) | I cells duodenum | − | | | ++ | | Contraction of gall-bladder | | | | ++ | ++ | |
| Secretin (3055 Da) | S cells duodenum | | − | | | ++ | −Bile secretion | | ++ | | ++ | | |
| GIP (glucose dependent insulinotropic polypeptide) (4984 Da) | K cells duodenum | | − | | | | ++ insulin | | | ++ | ++ | ++ | |
| Somatostatin (1638 Da) | D cells: pancreatic islets; GI mucosa | | − | | − | − | − Pepsinogen − Insulin & glucagon | | ++ | | | | − |
| Nitric oxide (30 Da) | | − | | − | | | Relaxes Les | | | | | | |
| VIP (vasoactive intestinal polypeptide) (3326 Da) | Para sympathetic ganglia in sphincters, gallbladder, small intestine | | | ++ | | | ++ Intestinal fluid secretion | ++ | | | | | ++ |
| Motilin (2698 Da) | Small intestine | | | ++ | | | Increases MMCs | Increases during fasting | | | | | |

Eight of the major hormones of the GI tract, including nitric oxide, are presented in this table. The column on the left names the relevant hormone and its approximate molecular weight in Daltons (Da) in parentheses. The molecular weight of CCK (cholesystokinin) is for the sulfated form. Detailed information on GIP (glucose-dependent insulinotropic polypeptide or gastric inhibitory polypeptide) is presented in **Fig. 6.53** and **Fig. 6.54** and is discussed in the **Carbohydrate chapter**. Note that the molecular weights of these polypeptide hormones vary from about 2100 to 4900 with the exception of nitric oxide, a compound (gas) that produces hormonal effects.

*Les*, lower esophageal sphincter; *MMC*, migrating myoelectric complex.

++, Stimulates; −, Inhibits.

Source: Reproduced from http://imgusmlestep1.blogspot.com/2010/05/ultimate-study-guide-for.html.

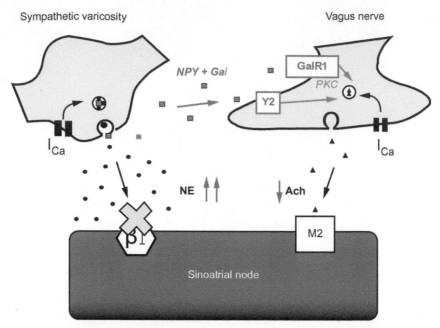

**FIGURE 15.30** A diagrammatic hypothesis of the mechanism by which the adrenergic cotransmitters, neuropeptide Y (NPY, *red* square) and galanin (Gal, *red* square), are released during high frequency sympathetic stimulation in the presence of beta blockade (**norepinephrine, NE**, *black* circle) and **metoprolol (Toprol,** used to decrease heart rate in treatment of certain cardiovascular conditions; *yellow cross*). NPY and Gal act via the $Y_2$ and $GalR_1$ **receptors** (both in *red* boxes), respectively, to reduce **acetylcholine (ACh,** *black* triangle) release and subsequent **vagal bradycardia** (a too slow heart rate) at the **sinoatrial node** (a group of cells in the heart that produce electrical impulses to make the heart beat). Both $GalR_1$ and $Y_2$ receptor signaling directly or indirectly involve **protein kinase C (PKC)**. *Reproduced from http://www.sciencedirect.com/science/article/pii/S0022282811004822.*

## Food Intake and Hormones

There are four phases in response to **eating**: cephalic, gastric, intestinal, and interdigestive. The **interdigestive phase** is comprised of events between meals. The **cephalic phase** is the sensing phase (smell, taste, swallowing, etc.). Cholinergic reflexes are activated by **parasympathetic nerves** causing **salivation**. Reflexes of the **vagal nerve** relax the proximal stomach, stimulate secretion of acid from the **parietal cells** of the stomach, and stimulate the release of gastrin from the **G cells. Hydrochloric acid** and **histamine** (also stimulating secretion of HCl) are secreted by the ECL cells. The cellular composition of the **gastric gland** is shown in Fig. 15.31.

The sequence of events leading to HCl secretion in the stomach, therefore, is first the smell, taste (brain activity), and the mechanical action of swallowing, the **vagal nerve stimulation** coupling the **enteric nerve plexus** which stimulates the G cells to release gastrin that, in turn, evokes histamine from the ECL cells. The enteric nerve plexus also stimulates the release of **acetylcholine** in the stomach which stimulates the secretion of histamine from the ECL cells and the release of HCl from the parietal cells. **Gastrin** from the G cells also directly affects the parietal cells to stimulate the release of HCl.

The cellular events in the parietal cell leading to the secretion of HCl are shown in Fig. 15.32.

The **gastric phase** involves stomach distention that is a signal for some further events. When **HCl** is secreted, the pH of the lumen is reduced and **somatostatin** is secreted from the stomach **D cells.** Somatostatin inhibits **gastrin secretion** from the **G cells** and inhibits **histamine secretion** from the **ECL** cells thereby preventing excessive acid production. Release of **pepsinogen** from the **chief cells** is stimulated by cholinergic reflexes. Although the major degradation of ingested proteins occurs in the intestine, there is some protein breakdown in the stomach by HCl and **pepsin**. The precursor, **pepsinogen**, is activated by the acid cleavage of the inhibitory peptide in pepsinogen to produce active pepsin. Partially digested food (**chyme**) from the stomach enters the small intestine to start the **intestinal phase,** and this is marked by the secretion of digestive enzymes (**trypsin** from trypsinogen, **chymotrypsin** from chymotrypsinogen, **lipase, amylase,** and other enzymes, such as **ribonuclease, deoxyribonuclease, gelatinase,** and **elastase**) through the **pancreatic duct** (to the small intestine) of the exocrine pancreas. **Enterokinase** from the intestinal mucosa activates trypsinogen to produce trypsin that activates chymotrysinogen to produce active chymotrypsin. Both trypsin and chymotrypsin cleave proteins into smaller peptides. Peptidases from the intestinal epithelial cells and enzymes like carboxypeptidase (activated from procarboxypeptidase from the pancreas by trypsin) further digest the smaller peptides into free amino acids. Free **amino acids** (together with sodium ions) are moved through transporters into the cytoplasm of

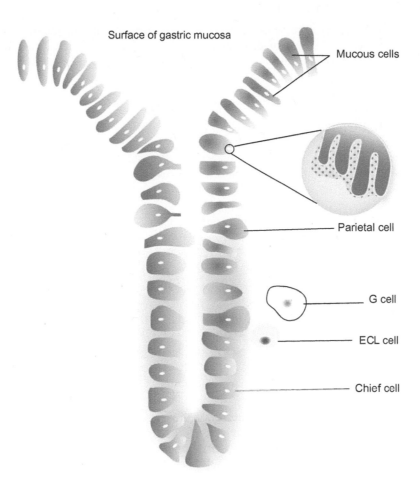

**FIGURE 15.31** Cellular composition of the gastric gland.

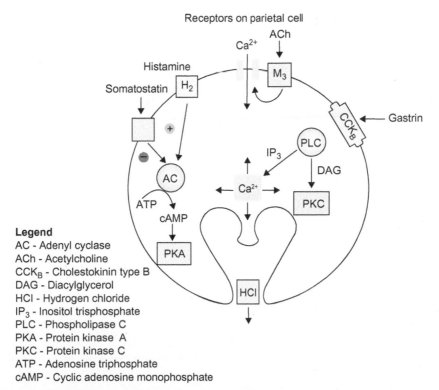

**Legend**
AC - Adenyl cyclase
ACh - Acetylcholine
$CCK_B$ - Cholestokinin type B
DAG - Diacylglycerol
HCl - Hydrogen chloride
$IP_3$ - Inositol trisphosphate
PLC - Phospholipase C
PKA - Protein kinase A
PKC - Protein kinase C
ATP - Adenosine triphosphate
cAMP - Cyclic adenosine monophosphate

**FIGURE 15.32** Cellular events in the parietal cell leading to the secretion of HCl.

lumenal cells of the intestine. There are four types of these transporters (**sodium-dependent amino acid transporters (symports)**), one type for each type of amino acid: neutral, acidic, basic, and branched chain. **Dipeptides** and **tripeptides** are absorbed independently of sodium ions and are digested within the cytoplasm of the **enterocyte** by peptidases. When amino acids are absorbed at the apical side (closest to the lumen) of the enterocyte, an **osmotic gradient** is generated requiring the absorption of water. Within the cell, amino acids are transported to the basolateral side (adjacent to the extracellular space) of the enterocyte and subsequently to the extracellular space on route to the bloodstream through sodium ion-independent transporters.

The hormone, **secretin**, is secreted from the duodenal mucosal **S cells** in response to the entrance of chyme into the small intestine. Secretin causes pancreatic ductal cells to secrete water and **bicarbonate**. Bicarbonate secretion is similar to the secretion of HCl from the **parietal cells** of the stomach but depends on **carbonic anhydrase** which catalyzes the reaction: $CO_2 + H_2O \rightleftharpoons HCO_3^- + H^+$. Carbonic anhydrase has a molecular weight of 29 kDa and contains an atom of zinc. Secretin also stimulates the **duodenal I cells** and the upper jejunum to secrete **cholesystokinin (CCK)**. Enzyme secretion from the pancreatic acinar cells is stimulated by CCK and gastrin.

There are a number of other gastrointestinal hormones. Examples are **ghrelin**, **motilin**, and **gastric inhibitory peptide (GIP)**.

The **P/D1 cells** of the stomach fundus secrete **ghrelin** that induces organized motor activity in the stomach and small intestine during the fasting state. Ghrelin is released by the **epsilon cells** of the pancreas. Ghrelin is a 28 amino acid peptide with the sequence: GSSFLSPEHQRVQQRKESKKPPAKLQPR. The third amino acid from the N-terminus, serine 3, is octanoylated (acylated) posttranslationally. The gene for ghrelin, *ghrl*, encodes an mRNA for **preproghrelin** (117 amino acids long) whose signal peptide becomes cleaved away to produce **proghrelin** that, in turn, is cleaved into two peptides: ghrelin (28 amino acids long) and **C ghrelin** which is further cleaved to produce **obestatin** (23 amino acids long). In contrast to ghrelin that increases food intake, obestatin suppresses food intake and gastrointestinal activity. Obestatin reduces gastric emptying, jejunal contractions, and body weight gain and could be a factor in **anorexia**. Ghrelin binds to its receptors on vagal afferent nerve terminals that activate brain neurons secreting **neuropeptide Y (NPY)**. NPY, however, is not involved in the ability of ghrelin to induce fasted motor activity. Ghrelin levels are elevated before meals and reduced after meals in contrast to **leptin**, secreted from adipose cells, which, at elevated levels after meals, increases **satiety**.

The ghrelin receptor is named the **growth hormone (GH secretagogue receptor) (GHSR1)** owing to its action in stimulating the release of growth hormone from the anterior pituitary. GHSR1 is a G protein-coupled receptor in the hypothalamus, pituitary, vagal afferent cell bodies, and nerve endings along the gastrointestinal tract. The activity of ghrelin to release GH from the anterior pituitary **somatotrope** is shown in Fig. 15.33.

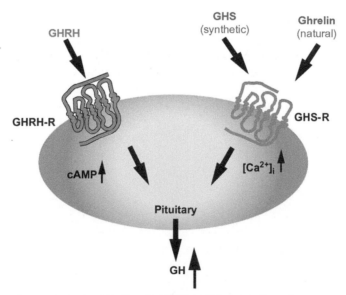

**FIGURE 15.33** Action of ghrelin at its receptor to increase the release of growth hormone from the somatotrope of the anterior pituitary. *GHRH*, growth hormone releasing hormone; *GHRH-R*, growth hormone releasing hormone receptor; *GHS*, growth hormone secretagogue; *GHS-R*, growth hormone secretagogue receptor; *cAMP*, cyclic AMP; *GH*, growth hormone. *Reproduced from http://physrev.physiology.org/content/85/2/495/F2.large.jpg.*

In addition to these activities, **ghrelin** plays a role in **cognitive adaptation** and the process of **learning** both of which are functions of the **hippocampus**. It also can induce **endothelial nitric oxide synthase** (**eNOS**) apparently mediated by GH-dependent mechanisms and increases the production of **nitric oxide** in aortic epithelial cells. The signaling pathway of ghrelin may involve activation of AMP-activated protein kinase (AMPK) and Akt (serine-threonine protein kinase) in epithelial cells and blood vessels as well as calmodulin-dependent kinase. Activation of eNOS (probably through phosphorylation of eNOS) is important for the reduction of **vascular inflammation** by **ghrelin**.

**Motilin** is a 22 amino acid long peptide having a molecular weight of 2698 Da with the following amino acid sequence: FVPIFTYGELQRMQEKERNKGQ. Motilin is secreted by the **M cells** of the small intestine (M cells are specialized epithelial cells lying above lymphoid follicles in the small and large intestine. They are involved in the initiation of the mucosal immune response to antigens and microorganisms that they transport to underlying lymphoid tissue) and binds and activates the **motilin receptor**, a G protein-coupled receptor. Motilin reduces **gastric motor activity** at low pH, and at high pH, it stimulates gastric motor activity (stomach phase III contractions) by way of neural pathways involving a cholinergic pathway, **5-hydroxytryptamine**, the **hydroxytryptamine3 receptor** (**HT3R**) and alpha receptors. Thus, motilin is considered a brain-gut peptide. It regulates the **interdigestive migrating contractions** (**IMC**) that comprise the contraction pattern of the fasting gut. The IMC has four phases: I, the resting state; II, irregular contractions; III, intense **peristalsis** (a series of organized contractions occurring throughout the GI tract of which motilin plays its major role); and IV, decreasing activity until the resting state is achieved. The **motilin receptor** in smooth muscle has different properties from the neuronal motilin receptor. Patients with **ulcerative colitis** or **Crohn's disease** have elevated levels of blood motilin.

The **incretins**, **gastric inhibitory peptide** (**GIP** also known as the **glucose-dependent insulinotropic peptide**), and **glucagon-like peptide-1** (**GLP-1**) stimulate the release of **insulin** from the $\beta$-cells of the pancreas (Fig. 6.53). GIP is secreted by the **K cells** of the mucosa of the jejunum and duodenum, and GLP-1 is secreted from the **L cells** (enteroendocrine cells), lower down in the small intestine, within minutes of ingesting a meal even before blood glucose is elevated. The primary inducer of the secretion of GIP is **ingested fat**. GIP does not induce the secretion of GLP-1 but both hormones are secreted in response to the same stimuli. These hormones also reduce the rate of gastric emptying, reduce the intake of food, and inhibit the release of glucagon from pancreatic $\alpha$-cells.

The **incretins** are inactivated rapidly by the enzyme **dipeptidyl peptidase-4** (**DPP-4**). The net effect of incretins is to lower **blood glucose** levels by increasing the release of **insulin** and decreasing the release of **glucagon**. By inhibiting the enzyme (DPP-4) that inactivates incretins, their activities can be extended and such inhibitors (Fig. 6.55) may be useful in treating type-2 diabetes.

On the other hand, it has been shown experimentally that mice lacking the **GIP receptor** (**GIPR**), inactivating the action of **GIP**, and fed a **high fat diet** were protected from **obesity** and **insulin resistance** (possibly through the suppression of hyperinsulinemia and subsequent insulin resistance). Thus, GIP is a potential target for development of anti-obesity drugs and may be related to the development of **type 2 diabetes**.

GIP is a 42 amino acid peptide with a molecular weight of 5105 Da that is structurally related to **glucagon** and **secretin** and has the sequence: GIPEGTFISDY$_{10}$SIAMDKIHQQ$_{20}$DFNNWLLAQK$_{30}$GKKNDWKHNI$_{40}$TQ. GIP is matured from a precursor peptide, **prepro-GIP** of 153 amino acids, having a signal peptide of 21 amino acids and an N-terminal propeptide of 30 amino acids. GIP is released from the propeptide by hydrolysis at arginine residues. The release of GIP and its actions in generating **hyperinsulinemia** and, secondarily, insulin resistance are summarized in Fig. 15.34.

**GLP-1** has 31 amino acids and has a molecular weight of 4112 Da having the sequence: HAEGTFTSDV$_{10}$SSYLEGQAAK$_{20}$EFIAWLVKGR$_{30}$G. It is secreted from the intestinal **L cell**, slightly lower down in the small intestine from the K cells, and is matured from **proglucagon** by enzymatic hydrolysis by **proglucagon convertase-2**. Proglucagon is located both in the L cells of the small intestine and in the $\alpha$-cells of the endocrine pancreas (Fig. 15.35).

The matured hydrolytic products from proglucagon in L cells are, in addition to GLP-1, glicentin, IP-2, and GLP-2. Glicentin (hormone with uncertain function) is further broken down to yield GRPP (glicentin-like polypeptide) and oxyntomodulin. GRPP is a 30 amino acid peptide located at the N-terminal of glicentin; oxyntomodulin is a 37 amino acid peptide containing 29 amino acids in the sequence of glucagon and an 8 amino acid extension at its C-terminus. As such, it binds weakly to the glucagon receptor in liver and pancreas. Experimentally, it elevates cAMP in the stomach and inhibits meal-stimulated gastric acid secretion. The amino acid sequence of glicentin is: RSLQDTQQKS$_{10}$RSFSASQADP$_{20}$LSDPDQMNQD$_{30}$KRHSQGTFTS$_{40}$DYSKYLDSRR$_{50}$AQDFVQWLMN$_{60}$ TKRNRNNIA$_{69}$. The amino acid sequences of GRPP and other products of the proglucagon polypeptide are shown in Fig. 15.35(B).

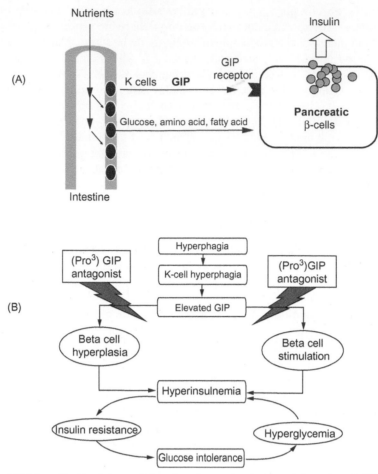

**FIGURE 15.34** (A) Release of GIP from **K cells** of the small intestine after nutrient intake and the action of GIP on the β-cell to release insulin. (B) The means by which GIP induces hyperinsulinemia and insulin resistance. *Hyperphagia* = increased appetite. *Reproduced from http://www.phoenixpeptide.com/catalog/pnxfoget.php?id=pnxnews_000000479&title=Compound&sum=Function.*

GLP-1 acts to release **insulin** from the pancreatic β-**cell** through its G protein-coupled receptor, which, when activated elevates cellular **cAMP** which activates **protein kinase A (PKA)** operating through **CREB** to initiate the transcription of the gene for **proinsulin** (Fig. 15.36A). A general scheme showing all of the major actions of GLP-1 is illustrated in Fig. 15.36B.

## Hormones Affecting Food Intake

There are several hormones that affect food intake, and many of these actions occur in the central nervous system. These gastrointestinal hormones and the locations of their effects are shown in Table 15.4.

## SUMMARY

**Panhypopituitarism** is a clinical model in which the releasing hormones become ineffective, often through an accident or trauma to the head when the **pituitary stalk** is severed and the releasing hormones can no longer reach the pituitary to release the pituitary hormones. Thus, panhypopituitarism results in the suppression of the hypothalamic–pituitary-end organ axis. The pituitary provides hormones that activate functions of many end organs: ACTH, liberated from the corticotropic cell of the anterior pituitary by hypothalamic CRH, acts on its plasma membrane receptor in the *zona fasciculata* cell of the adrenal cortex to stimulate the synthesis of cortisol and its release into the general circulation to be absorbed by its many tissue targets (nearly all tissues); thyrotropin (TSH) is released from the thyrotropic cell of the anterior pituitary and binds to the thyroid follicle cell membrane receptor that, with sodium ions, allows iodide ion to enter the cell and iodinate, after conversion to iodine, thyroglobulin to form the thyroid hormones, $T_3$ and $T_4$; growth

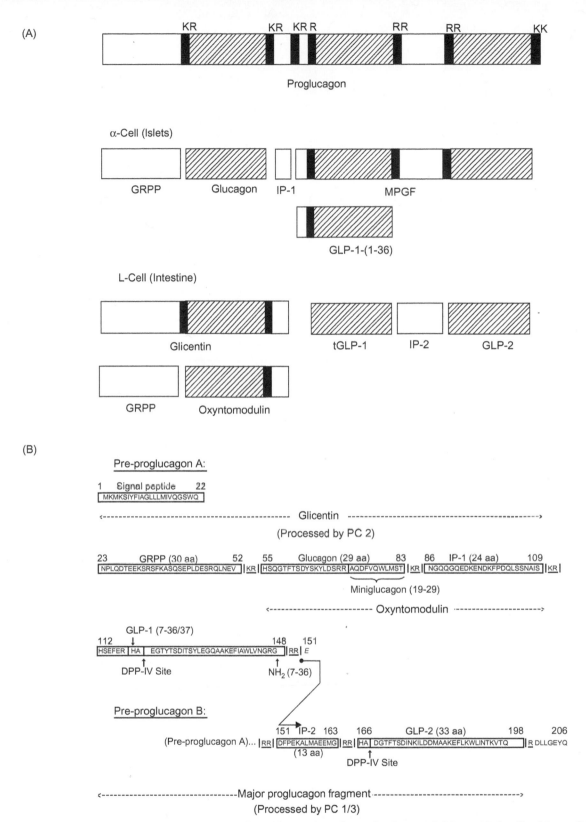

**FIGURE 15.35** (A) The protein domains in the proglucagon polypeptide located both in α-cells of pancreatic islets and in L-cells of the small intestine. In L-cells, proglucagon is broken down by **proglucagon convertase-2** to form **glicentin, tGLP-1, IP-2 (intervening peptide-2)**, and **GLP-2**. Glicentin is further broken down to form **GRPP** and **oxyntomodulin**. In α-cells, proglucagon is broken down to form **GRPP, glucagon, IP-1 (intervening peptide-1)**, and **MPGF (major proglucagon fragment)**. Formation of an N-terminal fragment of MPGF is GLP-1. (B) The amino acid sequence and proposed post-translational processing mechanisms for the **preproglucagon** precursor proteins, and the peptide fragments derived from preproglucagon precursor proteins are shown including: signal peptide, **glicentin, oxyntmodulin**, glucagon, **miniglucagon** (glucagon 19–29), intervening peptide-1 (IP-1), **glucagon-like peptide (GLP-1), IP-2,** and GLP-2 (cosecreted with GLP-1 from L-cell and also from neurons in the CNS; it is a growth factor for the intestine, an inhibitor of bone breakdown and a neuroprotective agent). The basic amino acid recognition sites for cleavage by prohormone convertase enzymes (PC 1/3 and PC2) are highlighted by bold underlined text. Two sites for proteolytic cleavage by **dipeptidyl peptidase IV (DPP-IV)** that is known to deactivate GLP-1 and GLP-2 are shown. *(A) Reproduced from http://www.jbc.org/content/276rf/29/27197/F1.large.jpg and (B) reproduced from http://www.sciencedirect. com/science?_ob=MiamiCaptionURL&_method...urlVersion=0&_userid=260508&md5=fcf82cc1c5d5da0aa8394855ea37ee74.*

(A)

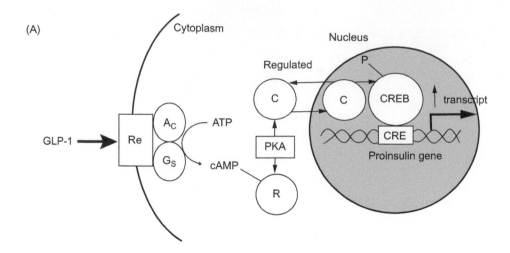

(B)

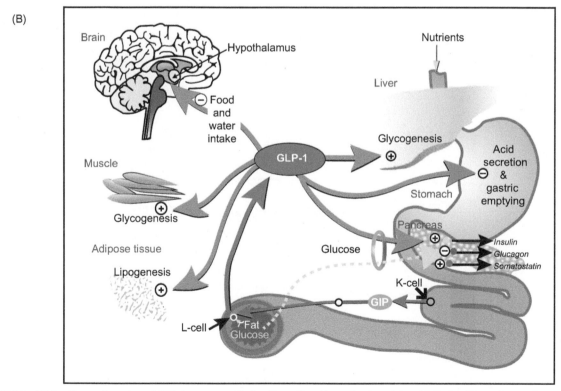

**FIGURE 15.36** (A) Insulinotropic action of GLP-1 on pancreatic β-cells mediated by activation of the cAMP-signaling pathway. GLP-1 binds to its receptor (Re) and activates **adenylate cyclase** ($A_c$) resulting in the formation of cAMP that binds to the regulatory subunit of (R) PKA to release the catalytic subunit (C). C translocates to the nucleus and activates the nuclear transcription factor CREB (**c**yclic AMP **r**esponse **e**lement **b**inding protein) by phosphorylation bound to the CREB response element (CRE) in the promoter of the proinsulin gene. Subsequently, proinsulin gene transcription is stimulated resulting in increased synthesis of insulin to replete the insulin previously released in response to nutrients, including glucose and incretins. (B) The general actions of GLP-1. GLP-1 is released from the K-cell of the small intestine in response to nutrients. It elevates glycogenesis in liver and muscle and inhibits stomach acid secretion and gastric emptying. It releases insulin and somatostatin and inhibits the secretion of glucagon all from the endocrine pancreas. GLP-1 increases lipogenesis in adipose cells, stimulates glycogenesis in muscle, and inhibits the intake of food and water (under glutamate receptor blockade). *(A) Reproduced from http://edrv.endojournals.org/content/20/6/876/F14.expansion.html and (B) reproduced from http://edrv.endojournals.org/content/20/6/876/F15.large.jpg.*

**TABLE 15.4** Overview of Gastrointestinal Hormones Known to Affect Food Intake

| Cut Hormone | Principal site(s) of Release | Examples of Factors Stimulating or Suppressing Secretion | Hormone Receptor(s) | Sites of Action Thought to Mediate Effects on Food Intake | Role of Hormone in Body Weight Regulation |
|---|---|---|---|---|---|
| Peptide YY | Distal gut (L-cells) | ↑Macronutrients | Y2 | Vagus | ↑Satiation |
| | | ↑Exercise | | Brainstem | ↑Satiety |
| | | ↑Bariatric surgery | | Hypothalamus | ↑Energy expenditure |
| | | | | Reward circuits (e.g., OFC/VTA/insula) | Fuel partitioning |
| | | | | | Long-term regulation of body weight |
| | | | | | Possible pathogenic role in obesity |
| Pancreatic polypeptide | Pancreatic islets (F-cells) | ↑Macronutrients | Y4 > Y1, Y5 | Vagus | ↑Satiation |
| | | ↑Exercise | | Brainstem | ↑Satiety |
| | | ↑Gastric distension | | | ↑Energy expenditure |
| | | ↑CCK, gastrin | | | |
| | | ↓Somatostatin | | | Possible pathogenic role in obesity |
| Glucagon-like peptide 1 | Distal gut (L-cells) | ↑Macronutrients | GLP-1R | Vagus | ↑Satiation |
| | | ↓Bariatric surgery | | Brainstem | ↑Satiety |
| | | ↓Somatostatin | | Hypothalamus | Potentiates insulin release |
| | | ↓Calorie restriction | | | Possible pathogenic role in obesity |
| Oxyntomodulin | Distal gut (L-cells) | ↑Macronutrients | GLP-1R | Hypothalamus | ↑Satiation |
| | | | | | ↑Satiety |
| | | | | | ↑Energy expenditure |
| Ghrelin | Gastric mucosa (X/A-like cells) | ↑↓Circadian rhythms | GHS-R1a | Vagus | Meal initiator |
| | | ↑Lack of sleep | | Brainstem | Long-term regulation of body weight |
| | | ↑Calorie restriction | | Hypothalamus | Fuel partitioning |
| | | ↓Macronutrients | | Reward circuits (e.g., OFC/amygdala/VTA/insula) | Possible pathogenic role in obesity |
| | | ↓Bariatric surgery | | | |
| Amylin | Pancreatic islets (β-cells) | Coreleased in a molar ratio with insulin | AMY1(a) | Brainstem | ↑Satiation |
| | | | AMY2(a) | Hypothalamus | ↑Satiety |
| | | | AMY3(a) | | |
| Cholecystokinin | Proximal small intestine (I-cells) | ↑Fat and protein-rich cyme | CCK-1 | Vagus nerve | ↑Satiation |
| | | ↑Coffee | | Brainstem | Possible pathogenic role in obesity |
| | | ↓Bile acids | | Hypothalamus | |

CCK, cholecystokinin; GHS-R1a, growth hormone secretagogue-receptor; GLP-IR, glucagon-like peptide 1 receptor, OFC, orbital frontal cortex; PYY, peptide YY; VTA, ventral tegmental area. "Satiation" refers to meal termination, whereas "satiety" refers to postponing the initiation of the next meal.
Source: Reproduced in part from http://www.precisionnutrition.com/wordpress/wp-content/uploads/2010/04/Appetite-hormone-table-1.jpg.

hormone is secreted from the somatotropic cell of the anterior pituitary in response to the action of hypothalamic GHRH and GH acts on its target organs, particularly the liver, to produce growth factors, such as IGF-1 and IGF-1 binding protein. IGF-1 activates mitogenic pathways to stimulate cell growth and proliferation; the gonadotropic hormones, LH and FSH, are released from the gonadotrope of the anterior pituitary by the action of the hypothalamic gonadotropin releasing hormone (GnRH) and LH, in the male, binds to its receptor on the Leydig cell membrane and sets off a chain of events leading to the formation of testosterone and the more active male sex hormone 5α-dihydrotestosterone; FSH in the male binds to its receptor on Sertoli cells of the seminiferous tubules and signals the nuclear transcription of several proteins required by sperm during spermatogenesis; in the female, LH binds to its receptor on the membrane of the mature ovarian follicle to stimulate ovulation and generate progesterone in the corpus luteum; FSH binds to a more immature ovarian follicle to promote the synthesis of 17β-estradiol. Prolactin (PRL) is released from the lactotropic cell of the anterior pituitary by the releasing hormone, hypothalamic TRH, or by stimulation of the anterior pituitary by estrogen. Its action is to generate the transcript of genes for milk proteins, such as casein and whey protein and to stimulate the release of milk.

**Melanocyte stimulating hormone** (α-MSH) derives from the cleavage of ACTH in cells of the intermediate space between the anterior and posterior pituitary. This and other hormones derive from the precursor protein, **proopiomelanocortin** that is broken down by hydrolases acting at the sites of basic amino acid couples. MSH acts on its receptors in melanocytes to stimulate the formation of the two major forms of melanin that account for skin color, eye color, and hair color.

The hormones of the posterior pituitary are the antidiuretic hormone, **arginine vasopressin** (AVP), and **oxytocin** (OT). The release of AVP from the hypothalamus vasopressinergic neuronal cell body and subsequent travel down the cell's axon to posterior pituitary nerve endings where it is released in complex with neurophysin II is triggered by signals from hypothalamic baroreceptors, adrenergic neurons or an osmoreceptor. AVP is released from its complex with neurophysin II near the kidney distal tubule cell where it binds to its membrane receptor and, through activated PKA, creates channels in its apical membrane for the transport of water molecules across the cell to ultimately enter the general circulation and potentially increase the partial pressure of the bloodstream. The stimulation for the release of oxytocin can be the sight of a hungry child, through suckling or through estrogen stimulation. The hypothalamic paraventricular nucleus contains oxytocinergic neurons that produce OT and move it down the axons, complexed to neurophysin I, and releases OT-neurophysin I stored at the nerve endings. Free OT, through OT membrane receptors, contracts muscle cells of the milk-secreting system ejecting milk through mammary ducts.

**Orexins** (hypocretins) are peptide hormones that control sleep and feeding. Two hormones, orexin A (hypocretin-1) and orexin B (hypocretin-2), are released from proorexin by cleavage at basic amino acid sites. Orexins stimulate the release of cortisol from the adrenal cortex by increasing the release of ACTH through the stimulation of hypothalamic CRH release. Orexins are responders to **stress**. **Sleep cycles** are affected by orexins by their actions on acetylcholine, dopamine, norepinephrine, and histamine systems functioning together to stabilize sleep cycles.

Adipocytes secrete **adiponectin**, resistin, and leptin. Adiponectin is the prominent hormone from this source, and it circulates in different forms: monomer, trimer, hexamer and higher molecular weight forms; the last correlate with sensitivity to insulin. Adiponectin protects from obesity, type-2 diabetes, and cardiovascular disease. It increases the production of nitric oxide and inhibits activation of adhesion molecules and the formation of foam cells as the means to counteract atheroslerosis. Circulating trimers of adiponectin correlate with diabetes.

Some gastrointestinal hormones are produced in the brain as well as in the gut. **Galanin** is one such hormone, and it connects the central nervous system with vagal stimulation by adrenergic neurotransmitters, such as neuropeptide Y (NPY).

**Gastrin** from the G cells of the stomach stimulates HCl secretion from the stomach and proliferation of gastric epithelium. A large polypeptide precursor gives rise to a family of proteins having the same carboxyl terminus. The major circulating form is called "big gastrin" containing 34 amino acids, while "mini gastrin" contains 14 amino acids. Because the 5 C-terminal amino acids in gastrin are identical to the 5 C-terminal amino acids of **cholesystokinin,** their biological activities overlap.

Eating is followed by four phases: cephalic, gastric, intestinal, and interdigestive. The cephalic phase involves smell, taste, and swallowing; salivation is a cholinergic reflex activated by parasympathetic nerves; the gastric phase involves the release of HCl from the stomach, and the interdigestive phase is comprised of events occurring between meals.

**Ghrelin**, acting through its receptor, causes the release of growth hormone from the somatotrope of the anterior pituitary. Ghrelin also plays a role in cognitive adaptation and learning functioning in the hippocampus. Through effects on GH, ghrelin can induce **endothelial nitric oxide synthase** to increase nitric oxide in aortic epithelial cells and the reduction of vascular inflammation.

**Motilin**, a brain-gut hormone, is secreted from the M cells of the small intestine, and it reduces gastric motor activity at low pH and stimulates gastric motor activity at high pH. It regulates contraction patterns of the fasting gut.

The **incretins** consist of gastric inhibitory peptide (GIP) and glucagon-like peptide (GLP-1). GIP is secreted from the K cells of the mucosa of the jejunum and duodenum and GLP-1 is secreted by the L cells of the small intestine. Both hormones are secreted in response to ingestion of a meal. These hormones reduce the rate of gastric emptying, reduce food intake, and inhibit the release of glucagon from pancreatic alpha-cells while enhancing the release of insulin from pancreatic beta-cells.

## SUGGESTED READING

### Literature

Ando, H., Niki, Y., Ito, M., Akiyama, K., Matsui, M.S., Yarosh, D.B., Ichihashi, M., 2012. Melanosomes are transferred from melanocytes to keratinocytes through the processes of packaging, release, uptake and dispersion. J. Invest. Derm. 132, 1222–1229.

Buechler, C., Warninger, J., Neumeier, M., 2010. Adiponectin receptor binding proteins – recent advances in elucidating adiponectin signaling pathways. FEBS Lett. 584, 4280–4286.

Busnelli, M., et al., 2012. Functional selective oxytocin-derived agonists discriminate between individual G protein family subtypes. J. Biol. Chem. 287, 3617–3629.

Camina, J.P., et al., 2003. Regulation of ghrelin secretion and action. Endocrine 1, 5–12.

Caraty, A., Lormet, D., Me, S., Guillaume, D., Beltramo, M., Evans, N., 2013. GnRH release into the hypophyseal portal blood of the ewe mirrors both pulsatile and continuous intravenous infusion of kisspeptin: an insight into kisspeptin's mechanism of action. J. Neuroendocrinol. 10, 12030.

Chia, D.J., Varco-Merth, B., Rotwein, P., 2010. Dispersed chromosomal Stat5b-binding elements mediate growth hormone-activated insulin-like growth factor-1 gene transcription. J. Biol. Chem. 285, 17636–17647.

Estabrook, R.W., Rainey, W.E., 1996. Twinkle, twinkle little StAR, how we wonder what you are. Proc. Natl. Acad. Sci. 93, 13552–13554.

Fan, Y., et al., 2009. Liver-specific deletion of the growth hormone receptor reveals essential role of growth hormone signaling in hepatic lipid metabolism. J. Biol. Chem. 284, 19937–19944.

Funkelstein, L., et al., 2008. Major role of cathepsin L for producing the peptide hormones ACTH, β-endorphin, and α-MSH, illustrated by protease gene knockout and expression. J. Biol. Chem. 283, 35652–35659.

Furuta, M., et al., 2001. Severe defect in proglucagon processing in islet A-cells of prohormone convertase 2 null mice. J. Biol. Chem. 276, 27197–27202.

Gutkowska, J., Jankowski, M., 2008. Oxytocin revisited: it is also a cardiovascular hormone. J. Am. Soc. Hypertens. 2, 318–325.

Hoffert, J.D., et al., 2008. Vasopressin-stimulated increase in phosphorylation at Ser[269] potentiates plasma membrane retention of aquaporin-2. J. Biol. Chem. 283, 24617–24627.

Kleinau, G., et al., 2011. Defining structural and functional dimensions of the extracellular thyrotropin receptor region. J. Biol. Chem. 286, 22622–22631.

Kojima, M., Kangawa, K., 2005. Ghrelin: structure and function. Physiol. Rev. 85, 495–522.

Le Roith, D., Yakar, S., 2007. Mechanisms of disease: metabolic effects of growth hormone and insulin-like growth factor 1. Nat. Clin. Prac. Endocrinol. Metab. 3, 302–310.

Nilsson, J., Bjursell, G., Kannius-Janson, M., 2006. Nuclear Jak2 and transcription factor NF1-C2: a novel mechanism of prolactin signaling in mammary epithelial cells. Mol. Cell Biol. 26, 5663–5674.

Romero, A., et al., 2008. Role of a pro-sequence in the secretory pathway of prothyrotropin-releasing hormone. J. Biol. Chem. 283, 31438–31448.

Shahid, I.Z., Rahman, A.A., Pilowsky, P.M., 2012. Orexin and central regulation of cardiorespiratory system. Vitam. Horm. 89, 163–182.

Simoni, M., Gromoll, J., Nieschlag, E., 1997. The follicle-stimulating hormone receptor: biochemistry, molecular biology, physiology and pathophysiology. Endocr. Rev. 18, 739–773.

Spinazzi, R., Andreis, P.G., Rossi, G.P., Nussdorfer, G.G., 2006. Orexins in the regulation of the hypothalamic–pituitary–adrenal axis. Pharmacol. Rev. 58, 46–57.

Sviridonov, L., et al., 2013. Differential signaling of the GnRH receptor in pituitary gonadotrope cell lines and prostate cancer cell lines. Mol. Cell. Endocrinol. 10, 1016.

Wen, S., et al., 2010. Embryonic gonadotropic releasing hormone signaling is necessary for maturation of the male reproductive axis. Proc. Natl. Acad. Sci. 107, 16372–16377.

Zhang, S., et al., 2012. Sertoli cell-specific expression of metastasis-associated protein 2 (MTA2) is required for transcriptional regulation of the follicle-stimulating hormone receptor (FSHR) gene during spermatogenesis. J. Biol. Chem. 287, 40471–40483.

### Books

Kastin, A., 2013. Handbook of Biologically Active Peptides. Academic Press, San Diego, London, Salt Lake City.

Litwack, G., 2008. Human Biochemistry and Disease. Academic Press/Elsevier, San Diego, London, Salt Lake City.

Litwack, G. (Ed.), 2012. Sleep Hormones. Vitamins and Hormones, vol. 89. Academic Press/Elsevier, San Diego, London, Salt Lake City.

Litwack, G. (Ed.), 2012. Adiponectin. Vitamins and Hormones, vol. 90. Academic Press/Elsevier, San Diego, London, Salt Lake City.

Litwack, G. (Ed.), 2015. Hormones and Transpport Systems. Vitamins and Hormones, vol. 98. Academic Press/Elsevier, San Diego, London, Salt Lake City.

# Chapter 16

# Steroid Hormones

Steroid hormones play a major role in bodily functioning because they are required for many critical physiological processes including survival of stress, injury (and illness), metabolism, inflammation, salt and water balance, immune functions, and development of sexual characteristics. Notably, the steroid hormone, **cortisol**, is required for adaptation (in particular, to **stress** and to changing environments) and human life is not possible in its absence. In the previous chapter, it was pointed out that the release of these hormones from the cells in the end organ, synthesizing the particular steroid hormone, is under the control of the anterior pituitary hormones (except in the case of activated vitamin D, functioning as a steroid hormone, which will be discussed in this chapter). The pituitary hormones are released from specific anterior pituitary cells by releasing hormones from the hypothalamus. The releasing hormones are signaled by other neurons in the **hypothalamus**.

## STRESS

In most situations, stress is harmful. Although there is so-called healthy stress, the emphasis here is on stress events that can lead to damage. There are acute and chronic stresses. **Acute stress** is short-lived and usually invokes a "flight or fight" response. For example, walk through the woods and suddenly see a rattlesnake in your path. **Catecholamines** would pour out and you would turn and run as fast as possible. After the stressor has disappeared, the internal systems would normalize again; **epinephrine, norepinephrine**, and cortisol would have entered the bloodstream during the stress and ultimately, these secretions would normalize again after the stress event is over.

**Chronic stress**, on the other hand, lasts over a long period with suppression of the "flight or fight" response. Long recoveries from surgery, long-term relationship problems, financial worries, and pressured work environment are examples. Long-term stresses can lead to elevated blood pressure, heart disease, suppression of the immune system, and even to diabetes (stress-induced diabetes) as well as other pathologies. Over long periods, stress causes an increase in the level of blood **cortisol** (also seen during **aging**; so, aging itself may constitute a type of stress). Cortisol feeds back negatively on the **hippocampus** and **amygdala** of the **limbic system** that can lead to depression. As this chronic stress continues, the increase in circulating cortisol can cause the hippocampus, via the glucocorticoid receptor (GR), to shrink (cortisol induction of programmed cell death). Since the hippocampus is the center of **memory**, this also can account for memory losses. Even in acute stress, there can be short-term memory losses that are reversed during recovery. Another factor in the development of depression relates to the fact that part of serotonin's action is to create a feeling of well-being and cortisol acts to suppress **serotonin release**. This form of depression is typical in aging.

## Nociceptin

**Nociceptin** (also **orphanin FQ**) is a 17-amino acid peptide (FGGFTGARKSARKLAMQ) anxiolytic that experimentally modulates anxiety produced by acute stress. While it is classified as an **opioid**, it binds to its receptor that is separate from the classical plasma membrane **opioid receptors** (kappa, mu, and delta); it binds to the **nociceptin receptor** called **NOP1**. The **kappa receptor** (κ), KOR (another name for the kappa receptor) binds dynorphins (there is "big dynorphin," dynorphin A and dynorphin B; big dynorphin and dynorphin A are more potent and selective than dynorphin B); the **mu receptor** (μ), MOR (another name for the mu receptor) binds β-endorphin and endomorphins (endomorphin 1 is YPYF); and the delta receptor (δ), DOR, binds enkephalins (Met enkephalin is YGGFM and Leu-enkephalin is YGGFL). Remember that in the anterior pituitary, preproopiomelanocortin is the precursor of β-**endorphin** and **Met-enkephalin** (Fig. 15.6). The precursor of nociceptin in nociceptinergic neurons is **prepronociceptin**. It is a polypeptide of 176 amino acids that is degraded by enzymes that cleave at basic amino acid sites into a signal peptide and active peptides: amino acids 1−19 constitute a signal

Human Biochemistry. DOI: http://dx.doi.org/10.1016/B978-0-12-383864-3.00016-8

peptide, amino acids 20—95 constitute a spacer peptide, amino acids 98—127 constitute **neuropeptide 1** ("**nocistatin**"), amino acids 130—146 constitute nociceptin, amino acids 149—165 constitute **neuropeptide 2** (**Nocil**), and amino acids 169—176 constitute the C-terminal peptide fragment. The nociceptin receptor (NOP) is distributed throughout the brain, including the limbic system (e.g., hippocampus and amygdala), and spinal cord and the nociceptin-NOP system are present both in the central nervous system and in the peripheral nervous system where it modulates **nociception**, the transmission of **pain**. Nociceptin binds to NOP in neurons where it reduces the activation of **adenylate cyclase**, reduces the activity of **calcium channels**, and also opens **potassium channels** (similarly to the action of opioids).

## Responses to Stress

The release of **catecholamines** in short-term stress and the release of glucocorticoid in long-term stress is shown in Fig. 16.1.

A **stress event**, evoked from the external or internal environment, is filtered in the brain by some mechanism, probably **sensory gating** (involving the thalamus) so that certain stimuli are rejected from further reaction. However, many forms of stress, especially **psychological stressors** (job insecurity, abuse, marital problems, etc.) or acute stressors (loud noises, sudden fright, etc.), set into motion the short-term and long-term effects of stress shown in Fig. 16.1. Some sort of signal (possibly an electrochemical signal), following a stress event, impacts the **hypothalamus** and it sends impulses through the spinal cord to preganglionic fibers that innervate the **adrenal medulla** resulting in the secretion of **epinephrine** and **norepinephrine** into the bloodstream. The elevation of these catecholamines in the circulation causes rapid increases in heart rate, blood pressure, and metabolic rate and stimulates the dilation of the bronchioles as well as causing the conversion of glycogen to free glucose for energy use.

The hypothalamus releases **CRH** into the portal blood system, enclosed by the stalk, and generates the release of **ACTH** from the **corticotropic cells** of the anterior pituitary. ACTH circulates to the **adrenal cortex** where it binds and activates its receptor on the plasma membranes of adrenal cortex cells resulting in the synthesis and release of the glucocorticoid **cortisol** and the mineralocorticoid **aldosterone**. Aldosterone is also released by **acetylcholine** during stress and by **angiotensin II**.

Cortisol circulates in the bloodstream and crosses the plasma membranes (presumably by free diffusion owing to its lipid solubility) and freely passes back out to the extracellular space if there is little receptor available in the cytoplasm

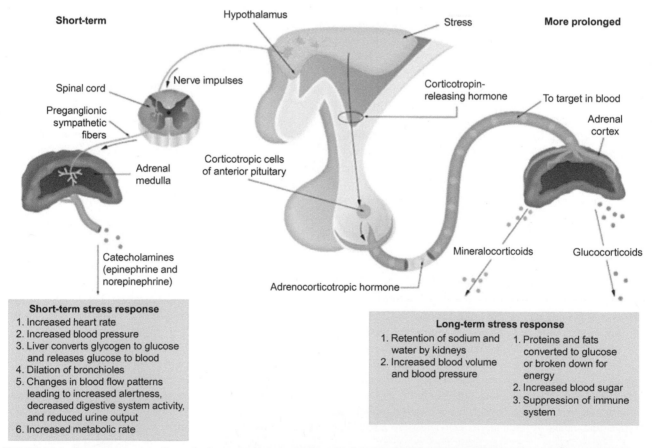

**FIGURE 16.1**  Responses to stress. Sympathetic nervous system responses are shown on the *left* and hormonal responses are shown on the *right*.

of the cell. Target cells, like the hepatocyte, have large amounts of the cytoplasmic receptor and cortisol is sequestered there by binding to the receptor molecules. Kidney cells also have large amounts of the receptor. Virtually, all the cells in the body have some receptor molecules, except **hepatobiliary cells** and cells in the space between the anterior and posterior pituitary, so it can be said that some cortisol is taken up by all cells except the two forms indicated. As a result, there is protein and lipid breakdown whose products are converted to blood glucose. Also glucose uptake by peripheral tissues is inhibited resulting, in the short term, of about a 10% increase in the blood level of glucose, enough to cause a release of **insulin** from the **pancreatic β-cell**. In itself, it is not a harmful response but if continued over a long period, the pressure to release **insulin** from unrelenting stress events could lead to **stress-induced diabetes** through exhaustion of the β-cells' ability to produce and secrete insulin. Also, in prolonged stress, suppression of the **immune system** and breakdown of muscle protein together with decreased protein synthesis can occur.

   **Aldosterone**, a true stress hormone, causes the retention of sodium by its actions on the kidney and increases blood volume (by increasing water reabsorption) and blood pressure. Whereas **cortisol** secretion (primarily from *zona fasciculata* cells of the adrenal cortex) occurs endogenously under the control of **serotonin**, in stress, its secretion is added to the endogenous level. This accounts for the nearly 1000 times higher circulating level of cortisol (up to 25 μg/dL) compared to aldosterone (up to about 30 ng/dL). Aldosterone is secreted primarily as the result of stress. Abnormally **high salt ingestion** will increase the blood level of aldosterone. In addition to the action of ACTH on the *zona glomerulosa* cell (outer cells of the adrenal cortex), aldosterone is also released by the action of **angiotensin II** and **hypokalemia** as shown in Fig. 16.2.

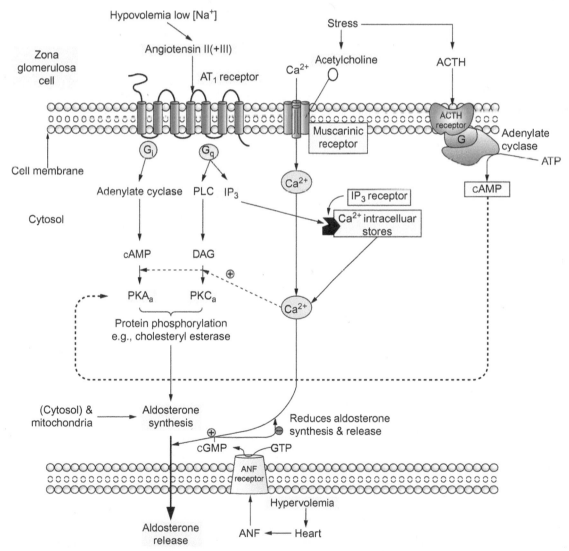

**FIGURE 16.2** Activities of stress, hypokalemia, and angiotensin II on the synthesis and release of aldosterone from the cells of the outer layer of the adrenal cortex, the *zona glomerulosa*. The actions of ACTH and ANF also are shown. Many effects result in the mobilization of calcium ions that stimulate the release of aldosterone. *AT*, angiotensin; *G*, G protein; *IP₃*, inositol *tris*phosphate; *DAG*, diacylglycerol; *PKA*, protein kinase A; *PKC*, protein kinase C; *ANF*, atrionatriuretic factor; *GMP*, guanosine monophosphate; *GTP*, guanosine triphosphate; *c*, cyclic.

Besides causing the release of aldosterone from the *zona glomerulosa*, angiotensin II has many other activities. It constricts blood vessels, especially the efferent arterioles of the kidney causing increased **sodium ion** and water retention. **Sodium retention** also is promoted by the direct action of angiotensin II on proximal tubules. It increases **thirst** by acting on the hypothalamus and stimulates the secretion of **antidiuretic hormone, ADH** (AVP) from the posterior pituitary, thus increasing **water reabsorption**, among other actions. Angiotensin II stimulates **tyrosine kinases** (pp60$^{\text{c-src}}$) **focal adhesion kinases** and **JAK2, TYK2** kinases that are involved in vasoconstriction of vascular smooth muscle, proto-oncogene expression, and protein synthesis. The **angiotensin precursor** is formed in the liver and converted to angiotensin II in the lungs.

**Aldosterone** is secreted from the *zona glomerulosa* cell into the bloodstream and circulates to the tissues. The cellular targets, especially the **distal kidney tubular cell**, contain cytoplasmic **mineralocorticoid receptors** to which aldosterone binds (there is a very small percentage of mineralocorticoid-like receptors in the plasma membrane which, when activated, probably play a small role in ion transport; plasma membrane "mineralocorticoid receptors" may be proteins different from the classical receptor). The binding of the steroid hormone activates the receptor complex, causing the release of bound **heat shock proteins** and the **dimerization** of the monomeric receptor protein. The dimer translocates to the cellular nucleus and binds to the appropriate steroid response element in various promoter elements to activate genes encoding the **serum and glucocorticoid-induced kinase (SgK)**, the **Kirsten-ras A oncogene protein**, and **CHIF**, the **corticosteroid hormone-induced factor** that stimulates the formation of potassium uptake channels. The actions of aldosterone on the **principal cells** of the kidney are shown in Fig. 16.3.

The activation of the **ACTH receptor** by binding ACTH occurring on the *zona fasciculata* cell (the intermediate layer of cells in the adrenal cortex) membrane of the adrenal cortex to generate the synthesis of **cortisol** and its secretion is shown in Fig. 16.4.

**Cortisol** is synthesized and released from the *fasciculata* cell, enters the bloodstream, and is carried through the adrenal medulla to the general circulation. In the medulla, the hormone induces the formation of **catecholamines** by inducing a key enzyme, **phenylethanolamine *N*-methyltransferase**. Cortisol circulates to the tissues entering most of the cells; in the target cells, there are many molecules of the cytoplasmic **glucocorticoid receptor** that bind cortisol. The retention of

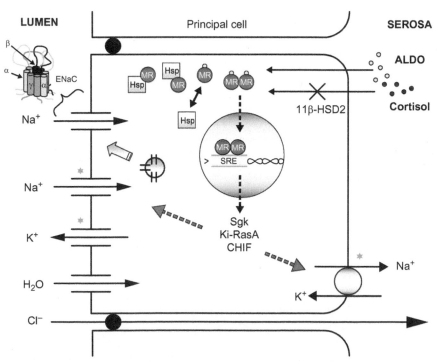

**FIGURE 16.3**    Mineralocorticoid receptor-mediated effects of aldosterone in the principal cells of the kidney. The kidney collecting duct cells are of two types: the principal cell and the intercalated cell. The result is the increase of sodium reabsorption and potassium excretion. *ALDO*, aldosterone; *Hsp*, heat shock protein; *SRE*, steroid response element; *CHIF*, corticosteroid hormone-induced factor; it is a small epithelial specific transmembrane protein interacting with a subunit of **Na$^+$–K$^+$-ATPase** that increases the pump's affinity for cellular Na$^+$ and participates in the regulation of **ion transport**. *MR*, mineralocorticoid receptor; *ENaC*, epithelial sodium channel; *11β-HSD2*, 11β-hydroxysteroid dehydrogenase 2. *Reproduced from http://www.clinsci.org/cs/113/0267/cs1130267f01.htm?resolution = HIGH.*

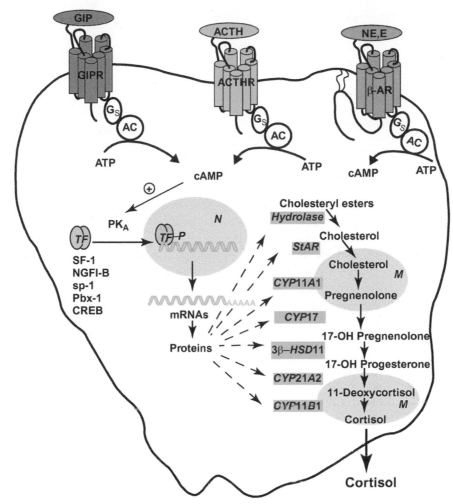

**FIGURE 16.4** Regulation of steroidogenesis by plasma membrane hormone receptors in *zona fasciculata* cells of the adrenal cortex. ACTH is the physiological modulator of cortisol production in the adrenal cortex. Binding to its receptor (ACTHR; also known as the melanocortin 2 receptor, MC2R) activates **adenylate cyclase (AC)** leading to the production of cAMP from ATP and the activation of **protein kinase A (PKA)** that phosphorylates specific protein transcription factors (TF) listed on the *left* of the figure. mRNAs produced are translated into the proteins involved in the synthesis of cortisol. PKA phosphorylates and activates **cholesteryl ester hydrolase** that converts **cholesteryl ester** in the droplet to free cholesterol that can enter the mitochondrion. The other ectopic receptors for **GIP (gastric inhibitory peptide)** and the catecholamines, **epinephrine (E)**, and **norepinephrine (NE)** on the cell membrane may contribute to the synthesis of cortisol by generating cAMP to increase the activation of **PKA**. *N*, nucleus; *M*, mitochondrion. *Reproduced from http://edrv.endojournals.org/content/22/1/75/F1.expansion.*

cortisol within a cell depends on the concentration of its receptor, the liver and kidney cells having the largest amounts. The receptor complex becomes activated, enters the nucleus as a dimer (some models show the receptor as forming a dimer in the nucleoplasm) and binds to the glucocorticoid response elements of various genes to produce many different mRNAs and subsequently their protein counterparts. These events are pictured in Figs. 16.5 and 16.6.

It is uncertain as to in what compartment, the receptor complex dimerizes since some mechanistic schemes show the receptor entering the nucleus as a dimer, whereas in the illustrations here, the GR enters the nucleus as a monomer. In Fig. 16.6, examples are shown of the response elements in gene promoters in which the GR dimer is involved: the **GRE (glucocorticoid response element)** where the receptor dimer positively transcribes the message for genes like **tyrosine aminotransferase**; the **STAT** (signal transducer and activator of transcription) **response element** positively regulating the protein, casein and the negative regulator, **nGRE**, that downregulates the osteocalcin gene; the AP-1 response element where *Fos/Jun* bind and then the GR binds to *c-Jun* to downregulate collagenase and the **NFκB response element** where the receptor binds to the subunits of NFκB to downregulate IL-1β.

The endogenous secretion of cortisol under the control of a **serotonergic neuron** is highest at about 8 a.m. and falls to lower levels during the day; it will rise again to its peak at 8 a.m. on the next day. During individual **stress events**, the released cortisol in the blood is added to the circulating cortisol derived from the serotonin-induced endogenous

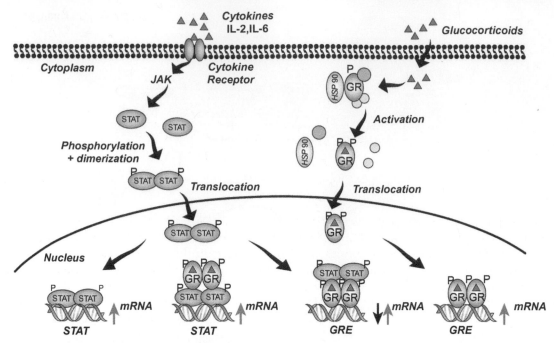

**FIGURE 16.5** The glucocorticoid receptor (GR) mechanism is shown in the *upper right*. Unbound cortisol in the blood enters the target cell by free diffusion and binds to the inactive receptor complex (one molecule of receptor is bound to a dimer of **HSP90 (heat shock protein 90)** and other proteins). Upon binding the steroid, a conformational alteration in the receptor occurs along with the dissociation of HSP90 and other proteins. The receptor becomes phosphorylated by **mitogen-activated kinase (MAPK)**, or by **cyclin-dependent kinase (CDK)**, or by both on threonine and serine residues. MAPK and CDK phosphorylate different amino acid residues. The receptor, thus activated, is (dimerized and then) translocated to the cellular nucleus where it can affect the steroid response elements in various ways as shown in the *left bottom* of the Figure. On the *left* side of the figure is shown the JAK-STAT pathway that, in some cases, interacts with GR on DNA. **Cytokines,** such as IL-2 and IL-6, induce **Janus kinase (JAK)** and activate **STAT** proteins. JAK phosphorylates monomeric STAT proteins in the cytoplasm, and the phosphorylated STAT proteins dimerize and translocate to the nucleus where they bind to response elements and regulate gene expression for growth, survival, apoptosis, host defense, stress, and differentiation. In some cases, the glucocorticoid receptor interacts with members of the STAT family to enhance STAT-mediated gene expression. STAT proteins can either enhance or repress GR-mediated gene activation depending on the stimulus and the specific STAT protein involved. At the *bottom* of the figure, *STAT* refers to the STAT response element and *GRE* refers to the **glucocorticoid response element**. *Reproduced from http://pats/atsjournals.org/content/1/3/239/F5.large.jpg.*

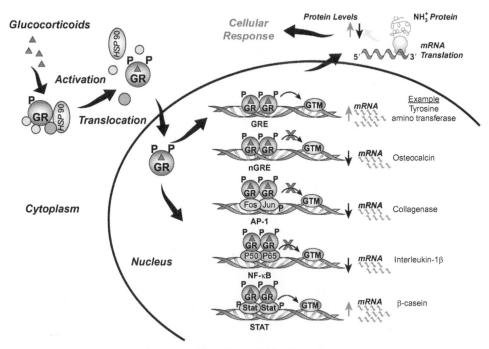

**FIGURE 16.6** Basic mechanisms of glucocorticoid receptor (GR) action. *Left*: the GR resides in the cytoplasm complexed with several **chaperones** including **Hsp90** and **immunophilin p59**. When **cortisol** binds to the receptor complex, the accessory proteins dissociate, and the receptor dimerizes and translocates as a dimer to the nucleus (although dimerization is not shown to occur in the cytoplasm in this figure). *Middle*: the GR regulates the expressions of genes by several basic modes of action. From *top to bottom*: the dimeric GR binds to glucocorticoid response elements in target genes to activate gene transcription; the GR binds to negative GREs (nGREs) and inhibits target gene transcription; the GR physically interacts with the *c-Jun* subunit of the AP-1 complex to inhibit AP-1-mediated gene transcription; the GR physically interacts with the p65 (RelA) subunit of NFκB and represses NFκB-regulated gene expression; the GR physically interacts with members of the STAT family (STAT1, STAT3, and STAT5) and synergistically enhances **STAT-regulated gene transcription**. *GTM*, general transcription machinery; *P*, phosphate. *Right*: examples of genes regulated by the GR by the various mechanisms are shown. *Reproduced from http://pats.atsjournals.org/content/1/3/239/F2.expansion.*

release. The incidence of heart attacks may be associated with the highest endogenous level of cortisol in the blood and many occur in the early morning.

Although these responses to stressors reoccur after each stress event, eventually, an adaptive process must take place allowing the survival of the individual. Such adaptation may occur in the brain in the form of **epigenetic modifications** in the brain cells' DNA to allow the expression of genes that are required for the stressed individual to adapt.

## Production of High Levels of Cortisol (Cushing's Disease) and Subnormal Levels of Cortisol (Addison's Disease)

Normally, in response to **serotonin** for the endogenous production of cortisol and in response to a stress event, **CRH** (corticotropin releasing hormone) is released from the hypothalamus causing the output of **ACTH** (adrenocorticotropin hormone) from the **corticotrope** of the anterior pituitary. ACTH enters the bloodstream and binds to ACTH receptors on the cells of the *zona fasciculata* of the adrenal cortex resulting in the synthesis and release of **cortisol** into the bloodstream. In **Cushing's disease**, there is an overproduction of cortisol. The cause is often a tumor in the anterior pituitary that is producing abnormally high amounts of ACTH. This results in overstimulation of the adrenal *zona fasciculata* cells that synthesize and secrete abnormally high levels of cortisol. The usual negative feedback occurs by cortisol on the CRH neurons and the corticotrope but the tumor itself may not respond to this negative feedback (it may express little or no **glucocorticoid receptor**). A high level of cortisol in the blood is the result. **Hypercortisolism**, typical of Cushing's disease, also can be caused by **ectopic tumors** that produce ACTH and occasionally a tumor can produce CRH resulting in the abnormally elevated production of ACTH. Overproduction of ACTH can lead to **hyperpigmentation** because α-**MSH** is produced from ACTH by cells in the space between the anterior and posterior pituitary. Hypercortisolism generates symptoms of rapid weight gain, especially of the trunk and face and growth of fat pads along the collar bone, excessive sweating, dilation of capillaries often generating purple or red *striae* (stripes or lines) due to hemorrhage, elevation of blood glucose level that can lead to **diabetes**, **hypertension**, and others.

Underproduction of cortisol (**Addison's disease**) is due to chronic adrenal insufficiency in which there is abnormally low production of adrenal steroids (cortisol and **aldosterone**). In this somewhat rare disease, cells of the adrenal cortex may have been destroyed by autoimmunity or by infectious agents or the adrenal cortex can be poorly formed from underdevelopment. There are many symptoms: weight loss, anxiety, nausea, headache, diarrhea, mood swings, sweating, low blood pressure, and hyperpigmentation of the skin (this time caused by insufficient negative feedback by cortisol on the corticotrope so that ACTH is overproduced and its breakdown by cells in the *pars intermedia* area generates α-MSH). There are symptoms of muscle weakness, light headedness, fatigue, fever, muscle and joint pains, vermillion border of the lips, and a low concentration of blood sodium with elevated calcium ions and potassium ions. Addison's disease is treated by oral administration of adrenal steroids so that a near normal existence can be obtained.

## Adrenal Cortex

The adrenal gland is located within masses of fat above the kidneys. The adrenal has two structures, the outer adrenal cortex and the inner **adrenal medulla**. The adrenal cortex consists of three layers of cells internal to the outer capsule. The layer directly beneath the capsule is the *zona glomerulosa* and this layer of cells secretes mainly **aldosterone** in response to stress signals. The thick middle layer of cells is the *zona fasciculata* and its main secretion product is **cortisol** in response to **ACTH** and indirectly to **serotonin** (through **CRH** and to ACTH) forming the endogenous production of cortisol. The innermost layer of the cortex is the *zona reticularis* and its main secretion products are weak androgens, in particular, **dehydroepiandrosterone** (**DHEA**) plus some **androstenedione**. Reactions leading to all of the individual steroid hormones are shown in Fig. 16.7.

The **fetal adrenal gland** consists mainly of *zona reticularis* type cells. The other layers of cells develop near birth.

Aldosterone levels, similar to cortisol, fluctuate during the day because ACTH also releases some aldosterone in addition to cortisol. The stimulation of aldosterone by ACTH is short term, whereas the stimulation of cortisol by ACTH is longer term. Consequently, the blood concentrations of aldosterone are very much smaller than those of cortisol. Also, cortisol levels respond both to the endogenous pathway elicited by serotonin and to stress events, whereas the secretion of **aldosterone** is mainly the result of stress events. In the absence of **ACTH**, **sodium depletion** activates the **renin−angiotensin system** that elevates the synthesis of aldosterone as shown in Fig. 16.8.

Details of the action of the hormones **angiotensin II** and ACTH and **acetylcholine** in response to stress are shown in Fig. 16.2. The hormone angiotensin is derived from the precursor, **angiotensinogen** (480 amino acid polypeptide), synthesized in the liver. Angiotensinogen is cleaved by the kidney proteolytic enzyme, **renin**, to produce **angiotensin I** (10 amino acids). Angiotensin I is further cleaved by the lung enzyme, **angiotensin-converting enzyme** (**ACE**) to yield

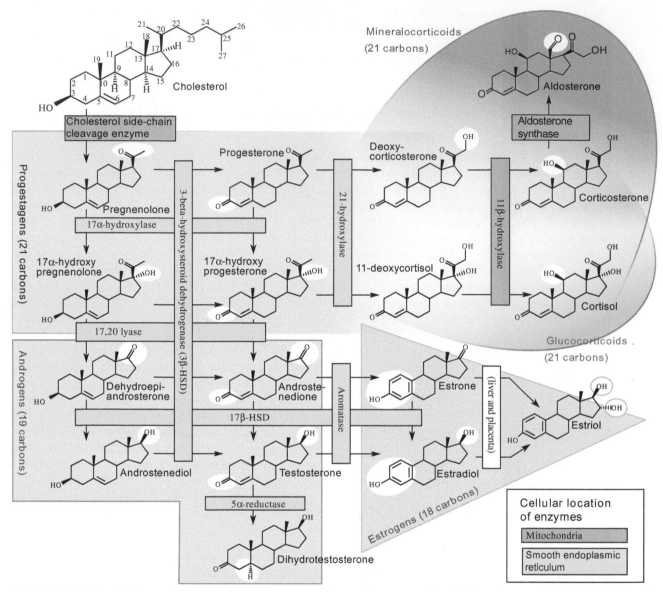

**FIGURE 16.7**    Pathways of the syntheses of **steroid hormones** showing intermediates, enzymes involved in each step and their locations: *pink*, mitochondria; *green*, smooth endoplasmic reticulum. *Yellow background* shows pathways for **progestogens** (21 carbons); *light blue background* shows pathways for **androgens** (19 carbons); *purple oval background*, **mineralocorticoids** (21 carbons); *green oval background*, **glucocorticoids** (21 carbons). *All steroid hormones are formed from* **cholesterol** *and cholesterol forms* $\Delta^5$**-pregnenolone** *which is the mandatory intermediate.* Progestogens are formed in the ***corpus luteum*** (see discussion of the ovarian cycle) and in the accessory tissue of pregnancy; dehydroepiandrosterone is formed in the ***zona reticularis*** of the adrenal cortex and the stronger androgens, **testosterone** and **dihydrotestosterone**, are formed in the **Leydig cells** of the testes; **estradiol** and related products are formed in the ovary and in the mammary gland; aldosterone is formed mainly in the *zona glomerulosa* of the adrenal cortex and cortisol is formed in the *zona fasciculata* cells of the adrenal cortex. Cholesterol **side chain cleavage** (*upper left*) occurs in the mitochondria. *Reproduced from http://upload.wikimedia.org/wikipedia/commons/1/13/Steroidogenesis.svg.*

an octapeptide containing amino acids 1−8. This (*right side*) form can be further converted to **angiotensin IV** (amino acids 3−8) and **angiotensin III** (amino acids 2−8). *Angiotensin I is converted to* ***angiotensin II****, the major form of the hormone, by ACE* and angiotensin II binds to its receptor (AT1R) to produce **hypertension**, **vasoconstriction**, and **cardiovascular** (CDV) **disorders**. Alternatively, angiotensin I can be cleaved by ACE2 to produce angiotensin 1−9 or it can be cleaved by **neprilysin** (NEP) to produce angiotensin 1−7. If angiotensin 1−7 binds to the **Mas receptor** (**MasR**) or if angiotensin II binds to the renal AT2 receptor (AT2R), opposite reactions are obtained (antihypertension, vasodilation, and cardiovascular protection). These reactions are depicted in Fig. 16.9.

There are abundant physiological effects of angiotensin II (**Ang II**) directly and through its stimulation of the secretion of **aldosterone**. Ang II causes increased blood pressure through its constriction of blood vessels. Ang II has important effects on the **kidney microcirculation** and acts directly on the **proximal tubules** to cause **sodium retention.**

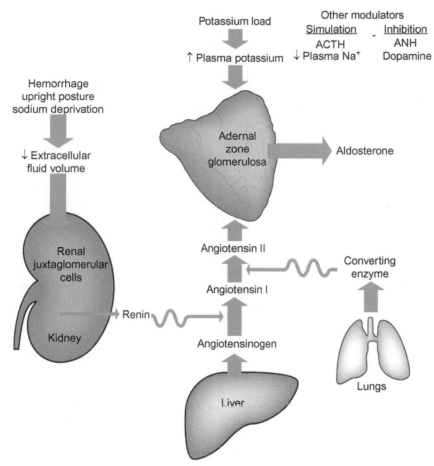

Regulation of aldosterone secretion: Activation of renin angiotensin system in response to hypovolemia is predominant stimulus for aldosterone synthesis.

**FIGURE 16.8** The renin−angiotensin system is stimulated by a decrease in blood volume or a decrease in sodium ion concentration. This causes an increase in the synthesis and release of aldosterone from the *zona glomerulosa* of the adrenal cortex. *ANH*, atrionatriuretic hormone.

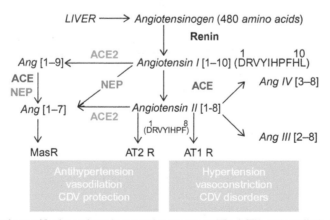

**FIGURE 16.9** Processing of **angiotensin peptides** by angiotensin-converting enzyme (ACE), ACE2, and neprilysin (NEP) as part of the renin−angiotensin system. Kidney renin cleaves angiotensinogen from liver to produce angiotensin I (DRVYIHPFHL). ACE, in lungs, converts angiotensin I to angiotensin II (DRVYIHPF). In a second processing axis, angiotensin I is cleaved by ACE2 resulting in **angiotensin 1−9** (DRVYIHPFH) which is cleaved by either **ACE** or **neprilysin** to produce **angiotensin 1−7** (DRVYIHP). This product also can result from the action of **NEP** on **angiotensin I** or the action of **ACE2** on **angiotensin II**. Binding of angiotensin II to its receptor **AT1R** activates **vasoconstriction**. In contrast, binding of angiotensin II to the **AT2R** mediates **vasodilation** that also can be initiated by the binding of angiotensin 1−7 to the **Mas receptor**. *Reproduced from http://circgenetics.ahajournals.org/content/5/2/265//F2.expansion.html.*

Through the release of aldosterone and its effects on the collecting tubules, there are increases in plasma sodium ion concentration and reductions in potassium ions and hydrogen ions. Aldosterone acts on the **hypothalamus** to cause **thirst** and the secretion of ADH (AVP). Ang II potently constricts both afferent and efferent arterioles in the **kidney microvasculature** [its responses can be regulated by both paracrine (from neighboring cells) and autocrine (from its own cell) factors derived from the endothelium and from the *macula densa*]. The concentration of Ang II determines the extent of **tubuloglomerular feedback** activity, especially when the concentration of the protease **renin** (from the *macula densa*) is elevated. The vascular effects of Ang II are primarily mediated by the AT1 receptor (AT1R). In the fetus, AT2R is expressed in the kidney and is critical for **kidney development** but seems to be expressed less strongly in the adult. As shown in Fig. 16.9, the AT2R−Ang II complex can mediate vasodilation.

Regarding the effects of Ang II on the afferent arteriole (blood vessel carrying blood to the glomerulus), it constricts this vessel by stimulating the entry of calcium ions through **voltage-sensitive L-type channels** while the constriction of efferent arterioles (blood vessel carrying blood away from the glomerulus) results from the release of calcium ions from intracellular stores and the uptake of calcium ions through **voltage-independent channels** to cause back pressure at the glomerular capillaries.

Significantly, angiotensin receptors are expressed in the brain, kidney, adrenal, vascular wall, and in the heart. Angiotensin-converting enzyme 2 (ACE2), related to ACE, is expressed in the endothelium of the coronary, intrarenal vessels, and in the epithelium of the renal tubules. Ang II can stimulate **tyrosine kinases** [pp60$^{c\text{-}src}$, focal adhesion, and Janus kinases (JAK2, TYK2)]. These tyrosine kinases may be involved in the **vasoconstriction** of vascular smooth muscle, protein synthesis, and protooncogene expression.

**Aldosterone**, released from the *zona glomerulosa* cell, circulates in the bloodstream and binds to its receptors in the cytoplasm of its cellular targets, principally the **distal kidney tubular cell** and exerts the effects shown in Fig. 16.3. The action of aldosterone causes **sodium ion reabsorption** from the tubular urine while **potassium ion** is secreted. Aldosterone signal transduction increases **epithelial sodium ion channels (ENaC)** in the apical membrane while increasing **Na$^+$/K$^+$-ATPase** in the basolateral membrane and these represent slow responses (taking place about 6−24 hours after aldosterone). The rapid and genomic responses to aldosterone (Fig. 16.3) are the elevation of **serum and glucocorticoid-inducible kinase (Sgk),** the **corticosteroid hormone-induced factor (CHIF, and Kirsten *ras* (Ki-*ras*)** and these actions occur in the first 6 hours after aldosterone. Very rapid ion transport responses to aldosterone that occur before gene regulation are possible and could be due to a direct effect on a small fraction of "aldosterone receptors" located in the cell membrane. Fig. 16.10 explains recent information on the genomic and nongenomic effects

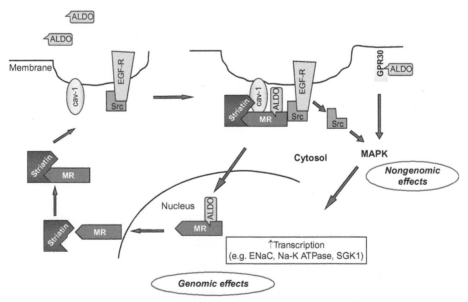

**FIGURE 16.10** An hypothetical model describing membrane-initiated rapid aldosterone signaling. The classical mineralocorticoid receptor (Mr) is translocated to and is associated with the membrane in a signaling complex including **striatin** (calmodulin-binding protein), **caveolin 1 (Cav 1)**, src, and the **epidermal growth factor receptor (EGF-R)**. Similar associations have been described for the estrogen receptor, androgen receptor, and progesterone receptor. Upon aldosterone stimulation, Mr is released from the membrane-signaling complex and EGF-R is translocated by aldosterone initiating **c-src-mediated mitogen-activated protein kinase (MAPK)** activation. In addition, G protein-coupled receptor (GPR) 30 may function as an aldosterone receptor mediating rapid mineralocorticoid receptor-independent signaling. *ALDO,* aldosterone. *Reproduced from http://hyper.ahajournals. org/content/57/6/1019/F1.expansion.html.*

of aldosterone indicating that the "**membrane aldosterone receptor**" may be due to the action of a protein (possibly G protein-coupled receptor 30) that is different from the **classical aldosterone (mineralocorticoid) receptor (Mr)** that operates genomically.

Rapid responses after 6 hours elevate the number of transport proteins allowing the reabsorption of sodium ion from the tubular urine. The uptake of sodium ions is followed by the passive reabsorption of water to maintain a constant concentration of sodium causing the extracellular volume to expand and increasing blood pressure.

If the plasma volume is decreased (**hypovolemia**), the protease, **renin**, is secreted from the granular cells of the **juxtaglomerular apparatus**. Renin degrades liver-expressed **angiotensinogen** to yield **angiotensin I** and pulmonary **ACE (angiotensin converting enzyme)** cleaves a dipeptide from the C-terminus of decameric angiotensin I to produce octomeric **angiotensin II** (Fig. 16.9). The action of angiotensin II on the *zona glomerulosa* cell of the adrenal cortex generates the synthesis and release of **aldosterone** whose effects also increase blood volume and pressure. One of the effects of aldosterone is to cause $K^+$ to move into the extracellular space (Fig. 16.3) and ultimately into the urine. *Potassium ion depolarizes the zona glomerulosa cell and because its entry into the cell stimulates the synthesis of aldosterone, the removal of $K^+$ into the urine acts as a negative feedback [there is also a negative effect of the **atrionatriuretic hormone** (ANH or ANF) on the synthesis of aldosterone].*

## STRUCTURES OF STEROID HORMONE RECEPTORS

The structures of the **glucocorticoid** (cortisol) **receptor** and the **mineralocorticoid** (aldosterone) **receptor** are similar; their N-terminal domains, however, are quite different. Fig. 16.11 shows a cartoon depicting the domains of several of the hormone receptors in the **steroid hormone receptor gene family**.

There are many other receptors assigned to the **steroid hormone receptor gene superfamily**. In addition to those listed in Fig. 16.11, the following also are members: thyroid hormone receptors (in addition to TRβ): TRα1 and TRα2;

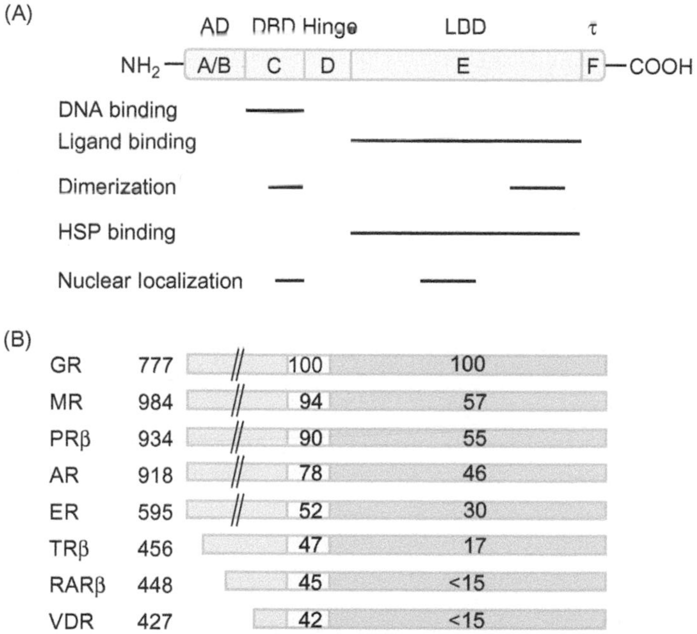

FIGURE 16.11    (A) The hormone receptor is pictured as a horizontal bar divided into the functional domains A through F. Regions of the various activities of the receptors are shown below. (B) The genomic organization of some of the receptors in the steroid receptor hormone gene family is shown (*GR*, glucocorticoid receptor; *Mr*, mineralocorticoid receptor; *PRβ*, the beta form of the **progesterone receptor**; *AR*, **androgen receptor**; *ER*, **estrogen receptor**; *TRβ*, the beta form of the **thyroid hormone receptor**; *RARβ*, the beta form of the **retinoic acid receptor** and *VDR*, the **vitamin D receptor**). There are other receptors that are members of this gene family and some will be addressed in separate discussions. Genes for these receptors encompass about 60 kb and are interrupted by numerous introns. Promoters of these genes resemble housekeeping genes and may be embedded in **GC-rich islands**. There are multiple sites of transcriptional initiation. Numbers to the *left* of each entry indicate the numbers of amino acid residues in the protein. The numbers in *yellow* indicate the percentage homology in the **DNA-binding domain** referring to the GR arbitrarily taken as the standard. The numbers on the right over the long *blue* region of E also indicate the extent of homologies of the **ligand-binding domains** (**LBDs**) referring to the GR arbitrarily taken as standard. Note that the DNA-binding regions (**DBDs**) of the GR (100) and MR (94) are similar and, in many cases, both receptors bind to the same promoters in target genes. The ligand-binding domains somewhat overlap for GR, MR, PRβ, and AR and all four bind the same ligands to different degrees, although a ligand activating one receptor (e.g., PRβ) may be an antagonist for another (e.g., GR). *AD*, antigenic domain; *τ*, tau (transactivation function). *Redrawn from Tsai, MJ and O'Malley, BW, "Steroid Hormone Receptors," in Ann. Rev. Biochem., **61**: 451−486, 1994.*

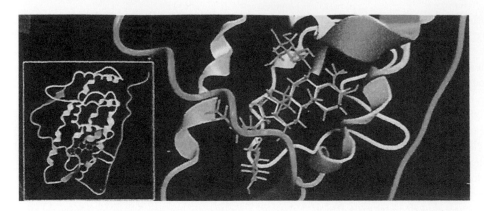

**FIGURE 16.12** The LBD of the glucocorticoid receptor. The ligand, dexamethasone, a synthetic steroid for which the receptor has much higher affinity ($\sim$50-fold) than for cortisol, is shown within the LBD in *gray*. A similar structure is superimposed upon **dexamethasone** (in *purple*). The inset shows dexamethasone (structure) bound in the LBD relative to the entire glucocorticoid receptor. *Reproduced from http://www. ugc.edu.hk/rgc/rgcnews13/west/09.htm.*

c-erbA1 and Rev-Erb; RAR$\alpha$ and RAR$\gamma$ in addition to RAR$\beta$ listed in the Figure; RXR$\alpha$, RXR$\beta$, and RXR$\gamma$ (retinoic acid receptors activated by 9-*cis*-retinoic acid, whereas RARs are activated by **all *trans*-retinoic acid** and **9-*cis*-retinoic acid**; and PPAR$\alpha$, PPAR$\beta$, and PPAR$\gamma$ (peroxisome proliferator-activated receptor) that bind **fatty acids**, including $\omega$6 **atherogenic lipids**, and fatty acid derivatives; there are numerous synthetic compounds (e.g., thiazolidine diones) that bind with high affinity but all of the physiological ligands are not known.

There are a whole series of receptors that fall into this class that are labeled as **orphan receptors**; some of these now have ligands that have been identified (the benzoate X receptor; the pregnane receptor; the liver X receptor; CAR$\beta$, a constitutive androstane receptor; FXR, a farnesoid receptor and hSXR, a steroid and xenobiotic receptor). Additionally, there are a number of orphan receptors whose human physiological ligands are unknown (original definition of an "orphan" receptor). Among these are the COUP receptors (COUP-TFOI/EAR and COUP-TFII/ARP-1), DAX-1, SHP, nur77/NGF1-B, ERR1 and ERR2, HNF-3 and HNF-4, and SF-1.

The **ligand-binding domain** (**LBD**) of the **glucocorticoid receptor** is shown in Fig. 16.12. This receptor will be used as a model, although the LBDs of other receptors in this family also have been reported.

The binding of the ligand induces conformational changes in the receptor causing the associated nonreceptor proteins to dissociate (Fig. 16.5). The relative affinities of the receptor for various steroid ligands are presented in Table 16.1. Remembering that the LBDs of some of the receptors in this class are homologous to a certain extent, some of the ligands listed are bound, not only to the GR, but also to the MR and to other receptors.

Structures of the **ligands** for many of the receptors in this class are shown in Fig. 16.13.

As Table 16.1 shows, there are cross-overs in ligand binding between the **GR** (antiinflammatory) and the **MR** (salt-retaining). *Both receptors bind cortisol equally well.* They both bind **prednisolone** to differing degrees. While there is cross-over in certain binding of ligands, the overall physiological actions of these receptors appear distinct, although, in some cases, the activated receptors may bind to the same gene promoters. Presumably, in addition to common promoters, there exist specific promoters for MR and for GR. When there is a **gene knockout** of either of these receptors, different physiological consequences are encountered but death occurred in both cases. When the glucocorticoid receptor is knocked out in mice, most of the experimental animals die shortly after birth, primarily due to pathologies in the lungs but also from pathologies in the liver, adrenal glands, brain, thymus, and bone marrow (all of these tissues normally express the GR), and there is interference in the feedback regulation of the hypothalamic—pituitary axis because the negative feedback of cortisol requires the activation of GRs in the responding tissues. In the case where a similar **knockout of the MR**, by targeted gene disruption, was performed, the newborns die of **dehydration** as a major effect of renal sodium and water loss with increased potassium ion in plasma, decreased sodium ion concentration, and elevated levels of components of the angiotensin system, namely, **renin**, **angiotensin II** and **aldosterone** and elevated expression of renin, **angiotensinogen**, and angiotensin II receptor (ACE, angiotensin I-converting enzyme, in the kidney was not affected). It should be remembered that although there exist similarities in the **ligand-binding domains** and in the DNA-binding domains, the **N-termini** of these receptors are widely different and must contribute to the differing actions of the two receptors, perhaps by attracting different activation factors to the **Tau1 domain** in the N-terminus. The **hyperphosphorylation**s of GR occur in the N-terminus and regulate the negative charge in this region.

The **DNA-binding domains** (**DBDs**) of receptors in the steroid receptor gene family have some homology as shown in Fig. 16.10. In the case of the glucocorticoid receptor (and the MR and AR), the DBD consists of two zinc fingers. The left zinc finger is closer to the N-terminus and the right zinc finger is downstream from the left zinc finger (Fig. 16.14).

**TABLE 16.1** Compounds Related to Cortisol which have Glucocorticoid Activity (Antiinflammatory) and/or Mineralocorticoid Activity (Salt-retaining)

| | | Mineralocorticoid | Antiinflammatory |
|---|---|---|---|
| | Cortisol | 1 | 1 |
| | Aldosterone | 800 | 1 |
| | Fludrocortisone | 800 | 5–40 |
| | 11-deoxycorticosterone | 40 | 0 |
| | Prednisolone | 0.6 | 4 |

(*Continued*)

**TABLE 16.1** (Continued)

| | | Mineralocorticoid | Antiinflammatory |
|---|---|---|---|
| | 6a-methyl prednisolone | 0 | 5 |
| | Betamethasone | 0 | 5–100 |
| | Dexamethasone | 0 | 10–35 |
| | Flumethasone | 0 | >100 |

The zinc fingers of the **homodimer** of **GR** and other receptors associate with DNA in the major groove as shown in Fig. 16.15.

**Steroid receptors** bind to response elements in the promoters of the genes they affect. In the case of the glucocorticoid receptor, the response element is a sequence of two half-sites: 5′ AGAACAnnnTGTTCT 3′ (or on the complementary strand: 3′ TCTTGTnnnACAAGA 5′) where n is any nucleotide. This is the general response element for MR, AR, and PR as well as for GR. There are differences in the DNA base contact points when the receptors other than the GR bind to this **palindromic sequence** (same sequence whether read 5′ to 3′ on one strand or 5′ to 3′ on the complementary strand). The glucocorticoid response element is located on a large number of different gene promoters. In many cases,

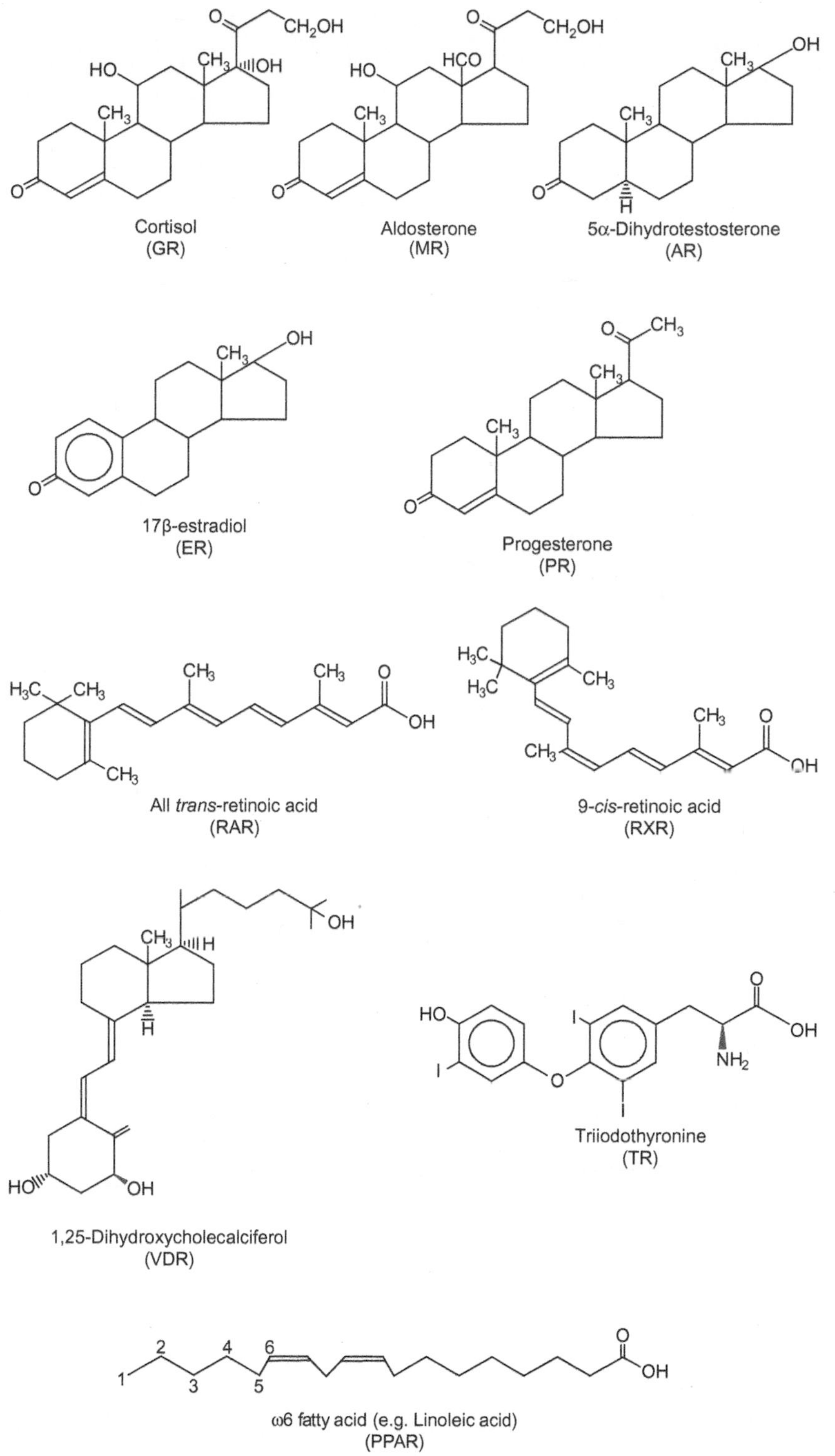

**FIGURE 16.13** The chemical structures of ligands for some of the receptors in the steroid hormone nuclear receptor family. *GR*, glucocorticoid receptor; *MR*, mineralocorticoid receptor; *AR*, androgen receptor; *ER*, estrogen receptor; *PR*, progesterone receptor; *RAR*, retinoic acid receptor binding all-*trans* retinoic acid; *RXR*, retinoic acid receptor also binding 9-*cis* retinoic acid; *VDR*, vitamin D receptor; *TR*, thyroid hormone receptor; *PPAR*, peroxisome proliferator-activated receptor.

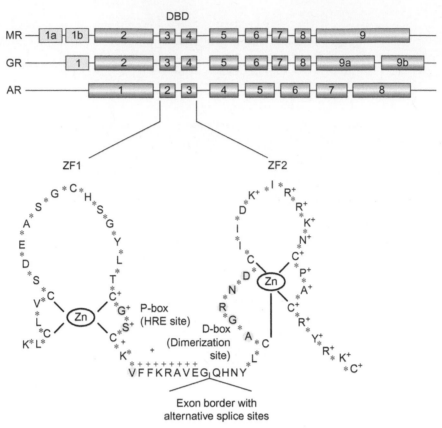

**FIGURE 16.14** Organization and structure of the DNA-binding domain of the MR, GR, and AR genes. Above, *beige boxes*, untranslated exons, *blue boxes*, translated exons. Below: amino acid sequence of the partial MR is shown in one-letter code. Amino acids with a plus sign form α-helices. *DBD*, DNA-binding domain; *D-box*, distal box; *black letters*, DNA identification sequence; *HRE*, hormone responsive element; *P-box*, proximal box; *ZF*, zinc finger. *Redrawn from L. Wickert and J. Selbig,* J. Endocrinol., ***173,*** *429–436, 2002.*

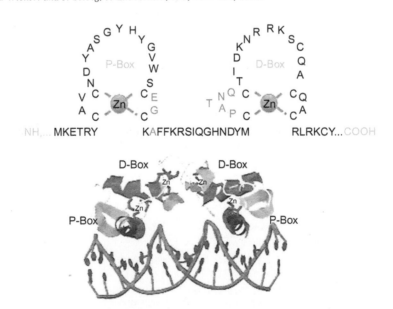

| Receptors | P-box | Half-site | Response element |
|---|---|---|---|
| ER | cEGckA | AGGTCA | AGGTCAnnnTGACCT |
| GR, MR, PR, AR | cGSckV | TGTTCT | AGGACAnnnTGTCCT |
| PPAR, RAR, VDR, PPAR | cEGckG | AGGTCA | AGGTCAnAGGTCA |

**FIGURE 16.15** Sequence-specific recognition of DNA by nuclear receptors. The core half-site recognized by a nuclear receptor is based on amino acid residues within the **P-box**. The sequence shown in this Figure is for the **ER**. The helical structure of the P-box provides contacts with the major groove of the DNA helix. The *shaded* residues in the **D-box** are important for interactions with the phosphate groups of the DNA helix as well as for dimerization. *Reproduced from http://nrresource.org/_Media/dna_binding_domain-3-2.png.*

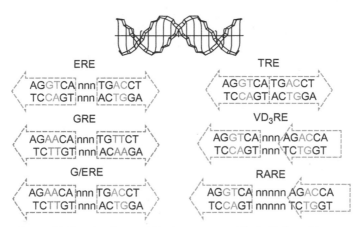

**FIGURE 16.16** Hormone response element-binding sites for steroid/nuclear receptors. The **half-sites** of the response elements are shown for the various receptors (*ERE*, estrogen response element; *GRE*, glucocorticoid response element; *G/ERE*, glucocorticoid or estrogen response element; *TRE*, thyroid hormone response element; *VD₃RE*, vitamin D response element; *RARE*, retinoic acid response element). Note the orientations of the half-sites that are indicated by the directions of the arrows.

the response element is positive and promotes the synthesis of a mRNA that leads to the synthesis of a new protein or, conversely, the response element could be negative and generate repression of specific gene activation. Examples of positive glucocorticoid response elements are those located in genes for tyrosine aminotransferase and enzymes of gluconeogenesis. Examples of negative GREs are those that reside in genes for gonadotropin releasing hormone (GnRH), prolactin, or proopiomelanocortin.

The major receptors of the **steroid hormone receptor gene family** and their **hormone response elements** are shown in Fig. 16.16.

There are four **classes of nuclear receptors**. **Class I** (also referred to as Type 1) receptors are those located in the cytosol and when the ligand binds to the unliganded receptor complex, ancillary proteins (e.g., heat shock proteins, HSP90) dissociate, homodimerization takes place followed by nuclear translocation and binding to hormone responsive elements (see Fig. 16.16). GR, AR, ER, and PR are representatives of class I. **Class II** receptors are retained in the nucleus with or without ligand and bind to DNA as heterodimers, often with RXR, and are in complex with **corepressor** proteins in the absence of the ligand. When the ligand binds, the corepressor proteins dissociate and **coactivator** proteins are recruited. Then, an array of other proteins forms a complex along with RNA polymerase and transcription of DNA follows to form a mRNA. **Class III** receptors resemble class I because they bind DNA as homodimers that bind to direct repeats instead of inverted repeats. **Class IV** receptors bind to DNA in the form of dimers or monomers; however, there is only one DNA binding domain that binds to a single half-site hormone response element. These distinctions are further clarified in Fig. 16.17.

The sequences flanking the palindromic (a **palindrome** is a nucleic acid sequence that is the same whether read 5′ to 3′ on one strand or 5′ to 3′ on the complementary strand of a double helix) responsive element (e.g., for the GR or ER) can play a role in the tightness of binding between the **responsive element** and the receptor.

## Coactivators and Corepressors

Coactivator proteins interact directly with certain steroid hormone receptors and stimulate transcription. There are at least 15 coactivators, so far. A summary of several of the coactivator proteins and the receptors to which they bind is shown in Table 16.2.

Coactivators bind to **nuclear steroid receptors** after the ligand binds to the receptor and a corepressor protein is dissociated from the receptor complex as shown in Fig. 16.18.

The major corepressors are **NCoR** and **SMRT** as shown in Table 16.3 where it describes with which specific receptors each corepressor interacts as well as additional information.

These corepressors have a HDAC- binding site and function by recruiting histone deacetylases (HDACs) with other proteins to the receptor complexes at the promoter of gene target sequences. Unlike coactivators, corepressors themselves do not appear to have enzymatic activity but they are essential for the full activity of HDAC. When the nuclear receptor, like TR (thyroid receptor) or RAR (retinoic acid receptor) complexed with DNA, binds its ligand, the attached

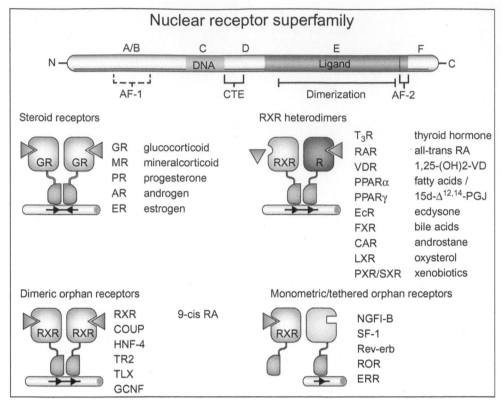

**FIGURE 16.17** The nuclear receptor superfamily. *Top*: Cartoon of the activity domains of the typical steroid receptor. *Below* are the classes of receptors in this superfamily. Class I, *upper left*; class II, *upper right*; class III, *lower left*, and class IV, *lower right*. Ligands are indicated by *green* triangles. Orientations of the half-sites are indicated by *arrows. Reproduced from http://www.jbc.org/content/276/40/36863/F1.large.jpg.*

corepressor, NCoR, is dissociated allowing the attachment of coactivator complexes with HAT activity to open DNA and enhance transcription.

## PHYSIOLOGICAL FUNCTIONS OF STEROID HORMONES FROM SPECIFIC RECEPTOR KNOCKOUTS

Members of the steroid and nuclear receptor family are present in various concentrations in many or most tissues of the body. In the case of the **glucocorticoid receptor**, it has been shown to reside in the cytosols of all tissues, in various concentrations, except that it is absent from **hepatobilary cells** and cells located in the region of the *pars intermedia (in the human, there is no specific organ as in lower forms, only scattered cells with the described activities)* of the anterior pituitary. Experimentally, it has been possible to "knock out" the genes for certain steroid receptors to create a **homozygous knockout mouse**. Following the knockout, the physiological effects of the loss in specific hormonal functions on the experimental mouse and its tissues are indicative of the actual roles of the steroids in the body. Knockout of the **estrogen receptor-α (ERα)** did not result in death of the newborn but did result in **infertility** in both male and female animals. The uteri, ovaries, and mammaries of F-1 females were abnormal. Plasma levels of **estradiol** and **LH** (luteinizing hormone) were substantially higher than in wild-type animals although FSH (follicle-stimulating hormone) was relatively unaffected. There appeared to be heightened aggression towards other females. In males, lacking the estrogen receptor, there was abnormal reproductive tract histology, infertile sperm, and smaller testicles than in the wild types. These males have reduced aggressive behavior over the normal wild types. Loss of the estrogen receptor (and estrogen action), although not lethal, affects both female and male offspring in sexually related physiology and behavior.

There are several reproductive abnormalities in mice lacking the **progesterone receptor**. In females, the **mammary gland** does not develop normally and there is hyperplasia of the **uterus**. This **knockout** also was not lethal and sexual differentiation was normal, although females were infertile, whereas male fertility was unaffected. Circulating levels of

**TABLE 16.2 Coactivator Proteins That Interact with Steroid Hormone Receptors**

| Names | Interacts with SN/NR | Effect of SR Ligand on Direct Interaction | Effect of Coexpression on Transcription | Other Information |
|---|---|---|---|---|
| ACTR/ SRC-3/ | ERα | Requires agonist ligand | Stimulates E$_2$-dependent transcription | AIB1 expression elevated in human breast and ovarian cancers |
| RAC3/p/ CIP/AIB1 | RXRα | | | |
| | TR | | | |
| ARA$_{70}$ (ELElα) | ER | Androgens and antiandrogens promote AR-ARA$_{70}$ interaction—also genistein and RU486 | Stimulates AR transcription with DHT or E$_2$ | No intrinsic HAT activity |
| Truncated | ERα | | | Interacts with p/CAF that has HAT activity |
| variant ELE1 β | GR | | | Interacts with TFIIB |
| | PPAR | | | Highest ELElα and ERE1β expression in testis |
| CBP/ p300/ p270 | AR, ERα, GR, PPARγ, RAR, RXR, TR, HNF4 | Requires agonist ligand, except for AR | Stimulates transcription | Intrinsic HAT activity CBP/p300 is also a cofactor for AP-1, c-myb, STAT1, E1A, p53, and Myo D |
| RIP140 | ERα, TR, RXR, PPARα, PPARγ | Requires agonist ligand; antagonists tamoxifen and ICI 164,384 block interaction with ERα | Stimulates ERα-induced transcription Inhibits PPAR and RXR activities | Identical to ERAP140 |
| SRC-I/ NoA-l | AR, ERα, ERβl, PR, GR, TR, RARβ, RXRα, PPARγ, HNR4 | Requires agonist ligand; antagonist inhibits interaction | Stimulates PR, GR, and ERα induced transcription | Identical to ERAP160 and p160; intrinsic HAT activity…; interacts with p300/CBP, TBP, and TFIIB |
| SWI/SNF | ER, GR RAR, HNF-4 | | Stimulates transcription | Chromatin remodeling complex in yeast with human homologues |
| TIF1α | ERα, ERβ, PR, RXR, VDR | Requires agonist ligand | Stimulates transcription-requires agonist ligand for ERα, but for ERβ 4-OHT acts as an agonist | Interacts with heterochromatin proteins, including hSNF2b of the SWI/SNF complex and TIF1β |
| TIF2/ GRIP1/ NCoA-2 | ERα, GR, AR, PR, RAR, RXR, TR, VDR HNF4 | Requires agonist ligand | Stimulates ERα, AR, PR, TR, RAR, and RXR but not GR, or VDR | 40% sequence homology to SRC-1 |

*AIB1*, "Amplified In Breast" is NCOA3, a steroid receptor coactivator; *AR-ARA70*, a coactivator, especially of the AR; *RU486*, a synthetic antagonist of GR and PR; *P/CAF*, a histone acetylase and a nuclear receptor coactivator; *TFIIB*, transcription factor, reacts with members of the steroid receptor superfamily; *ELE1α*, AR-specific coactivator; *HNF4*, hepatocyte nuclear factor 4 nuclear transcription factor binds DNA as homodimer; *HAT*, histone acetyltransferase; *CBP/p300*, CREB (cAMP response element-binding protein) binding protein; *AP-1*, heterodimeric protein transcription factor; *c-myb*, transcriptional activator and human protooncogene; *STAT-1*, transcriptional factor and signal transducer; *E1A*, transactivating protein; *p53*, tumor suppressor DNA-binding protein; *MyoD*, myogenic-regulating factor involved in muscle differentiation; *ERAP140*, tissue-specific nuclear coactivator; *ICI164*, antiestrogen; *TBP*, TATA-binding protein; *ERB4*, receptor protein tyrosine kinase; *hSNF2b*, mitotic growth and transcription activator; *SWI/SNF*, "SWItch/SucroseNonFermentable" nucleosome-remodeling complex; *SRC-1*, transcriptional activator of nuclear receptors.
http://glowm.com/resources/glown/cd/pages/v5/v5c004.html?
SESSID = 51aj43ff0q4c80haavr5k6r217.

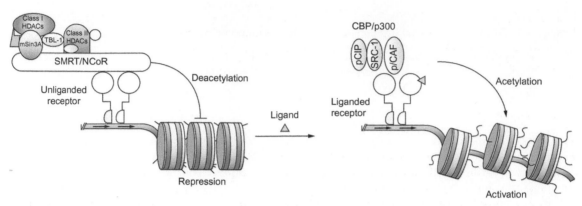

**FIGURE 16.18** Coactivator and corepressor complexes and **histone acetylation**. In the absence of ligand, the nuclear hormone receptor heterodimer (in this case) is bound directly with the corepressor complex (shown on the *upper left*). The corepressors **SMRT/NCoR**, associated with **corepressor complexes**, recruit **histone deacetylases** (**HDACs**) either directly or through their interaction with **Sin3** (a large protein that provides structural support for Sin3/HDAC complex). Many other proteins are involved in this complex. Deacetylation of histone tails leads to chromatin compaction and **transcriptional repression**. Ligand binding causes the release of the **corepressor complex** and allows the AF-2-dependent recruitment of a coactivator complex that contains p160 coactivators (e.g., **p/CIP** or **SRC-1**), **CBP/p300**, and **PCAF**. All of these proteins possess **histone acetyltransferase** (**HAT**) activity that allows chromatin decompaction and gene activation. Multiple protein—protein interactions exist among the different components: CBP/p300 contacts the receptor, the p160 coactivators, and PCAF through independent domains. The receptor itself binds CBP/p300, p160 coactivators, and PCAF. PCAF also can bind directly to CBP/p300, p160 coactivators, and to the receptor.

**TABLE 16.3  Corepressor Proteins That Interact with Steroid/Nuclear Receptors**

| Name | Interacts With | Effect of Coexpression on Transcription | Other Information |
|------|----------------|------------------------------------------|-------------------|
| NCoR | TR RAR RXR PR ERα | Increased transcriptional activity of unliganded PR | NCoR levels were reduced in many breast tumors that had acquired resistance to the antiproliferative effects of tamoxifen |
| | | At high levels, NCoR increased RAR transcription | |
| SMRT | TR RXR RAR ERα | Increased transcriptional activity of unliganded PR | HDAC1 is associated with SMRT |
| | | Overexpression strongly reduced basal and 4-OHT-stimulated gene expression with no effect on E$_2$ activity | |

Partial data from http://glowm.com/resources/cd/pages/v5/v5c004.html?SESSID = 51ej43ffoq4c80haavr5k6r217.

LH were elevated about twofold (lack of negative feedback) but FSH levels were unchanged compared to normals. Thus, the lack of progesterone function underscores the role of progesterone receptors in the regulation of hypothalamic and pituitary functions in the secretion of gonadotropins.

When the **glucocorticoid receptor** (**GR**) is knocked out, the offspring die shortly after birth by damage to the lung where the GR is essential for the development of **surfactant**. If neonatal animals were to survive in **early development** (which is not the case), the lack of ability to adapt to stress and environmental change would be lethal. In addition, there are effects on many other tissues whose functions rely on glucocorticoids, such as liver, brain, adrenals, thymus, bone marrow, and especially in the negative feedback regulation (via the GR) of the hypothalamic—pituitary axis. Thus, **cortisol** and its receptor are essential for survival.

When the **mineralocorticoid receptor** (**Mr**) is knocked out, offspring (mice) die within 2 weeks after birth. The animals die from loss of kidney reabsorption of water and sodium ions and they exhibit dehydration, hyperkalemia, loss of circulating sodium, and elevated levels of angiotensin II, aldosterone, and renin. There were elevations in expression of renin, angiotensinogen, and angiotensin II receptor (ST1) but the kidney level of angiotensin I-converting enzyme (ACE) was unchanged over normal controls.

**Androgen receptor deficiency** is characteristic of the **Tfm mouse** (X-linked inherited disease) and these mice display the **androgen insensitivity syndrome** causing male mice to be infertile.

## STEROID TRANSPORTING PROTEINS IN PLASMA

Steroid hormones are hydrophobic and they are substantially bound to proteins in the blood both to maintain a solubilized form and to stabilize them. The major protein in blood that binds both corticosteroids and progestins is **transcortin** or **corticosteroid-binding globulin** (**CBG**, an $\alpha$-globulin) or **serpin A6**. It has a molecular weight in plasma of 53,800 and contains a signal peptide and carbohydrate (the carbohydrate moiety is not essential for binding cortisol). In the blood, 75% of **cortisol** is bound to transcortin, about 4% is in the free form and the remainder is bound to **serum albumin**. *It is the free form of cortisol that enters cells that crosses the cell membrane by free diffusion owing to its hydrophobic character.* Circulating **corticosterone** is 78% bound to transcortin. Aldosterone binds mainly to serum albumin (about 47%) while 17% is bound to transcortin and 2% is in the free form. Some **aldosterone** may be bound in the blood to $\alpha$1-glycoprotein. In the case of the **sex hormones**, about 4% of either **testosterone** or **estradiol** is bound to transcortin and the remainder is bound to serum albumin except for a small percentage (about 2%) that is in the free form. About half of circulating **progesterone** is bound to serum albumin and the remainder is bound to transcortin. *The steroid hormones enter many different tissue cells in the free form and when there is little or no receptor in a cell to retain the hormone, the hormone crosses the cell membrane outwardly to reenter the bloodstream.* After hormones are retained in target cells, more hormones dissociate from the transporting proteins so that the level of circulating free hormone remains fairly constant.

Some research also suggests that cortisol, bound to CBG, may travel as the complex, directly to sites of inflammation where proteolytic enzymes may attack CBG to release cortisol.

## ENZYMATIC INACTIVATION OF CORTISOL

In addition to its positive effects on **gluconeogenesis** and the elevation of **blood glucose** (cortisol inhibits the uptake of blood glucose by peripheral tissues), cortisol elevates blood pressure. This increase in blood pressure, particularly **systolic blood pressure**, is not explained by mineralocorticoid effects on induced salt and water retention. There seems to be a suppression of sympathetic nervous system activity and an increase in the pressor responsiveness to the catechols, **epinephrine**, and **norepinephrine**. However, *the latest information relates the action of cortisol to the suppression of the synthesis of* **nitric oxide** *that exhibits* **vasodilator activity**. Most of the other mechanisms relating to elevation of blood pressure have been ruled out as direct effects of cortisol.

Under normal conditions, the body balances the levels of cortisol by the enzymatic conversion to its inactive counterpart, **cortisone**. In contrast, the same system is responsible for the conversion of inactive **cortisone** to active **cortisol** and this system is summarized in Fig. 16.19.

## CORTISOL AND ALDOSTERONE

Cortisol binds equally well to the glucocorticoid and mineralocorticoid receptors and cortisol binds to the MR as well as **aldosterone**; however, cortisol is about 1000 times more concentrated in blood than aldosterone. Aldosterone target tissues are protected from excessive cortisol by the enzyme 11$\beta$-hydroxysteroid dehydrogenase 2 (11$\beta$-HSD 2) that catalyzes the conversion of cortisol to inactive cortisone (inactive both for GR and MR) as shown in Fig. 16.17. Aldosterone is not a substrate for 11$\beta$-HSD 2.

The overproduction of aldosterone (**hyperaldosteronism**) is known as **Conn's syndrome**. It can be caused by a tumor or by hyperplasia of the adrenal gland. Both hyperaldosteronism and apparent mineralocorticoid excess can be treated with the MR competitive inhibitors **spironolactone** or **eplerenone** both of which are steroidal derivatives lacking an 11$\beta$-hydroxy function.

Activated GR is important at the termination of pregnancy to start the formation of milk in the mammary gland. During **pregnancy**, there is a high level of **progesterone** until termination when there is a sharp drop in circulating progesterone (Fig. 16.20).

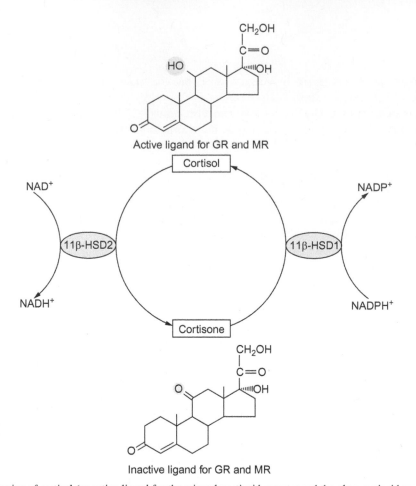

Active ligand for GR and MR

**FIGURE 16.19** The conversion of cortisol (an active ligand for the mineralocorticoid receptor and the glucocorticoid receptor) to inactive cortisone is catalyzed by **11β-hydroxysteroid dehydrogenase 2 (11β-HSD 2)**. 11β-HSD 2 in the kidney prevents **hypertension** by converting cortisol to cortisone which is an inactive ligand for the **mineralocorticoid receptor** [cortisol binds and activates both receptors (GR and MR) but cortisone does not because the 11-hydroxy function is required for receptor binding and activation; in cortisone, there is a keto group in the 11-position in place of the required hydroxyl]. Mutations in 11β-HSD 2 that reduce or obliterate the resulting enzymatic activity can lead to hypertension in a condition known as "apparent mineralocorticoid excess" that can be treated with the MR inhibitor spironolactone or similar compounds. 11β-Hydroxysteroid dehydrogenase 1 (11β-HSD1) catalyzes the reverse reaction in which cortisone is converted to cortisol.

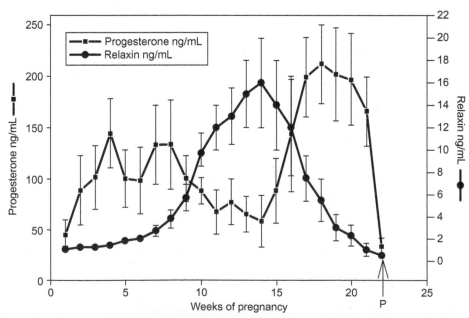

**FIGURE 16.20** Circulating levels of progesterone and **relaxin** ( ± SEM) in the macaque monkey during pregnancy. The data resemble those for the human except that the gestation period in the human is 39 weeks. *Reproduced from http://www.bioreprod.org/content/61/2/512/F2.large.jpg.*

The high level of **progesterone** binds and competitively inhibits the **glucocorticoid receptor** in mammary gland until the termination of pregnancy when the level of progesterone falls dramatically allowing cortisol to bind and activate the glucocorticoid receptor that initiates **milk production**.

The most important cellular targets are defined by the number of receptors, although some cells have a surplus of receptors over the number required to generate a cellular response. The range is somewhere in the neighborhood of 1000 to about 50,000 receptor molecules per cell. Primary targets for cortisol in the case of GR are the liver and kidney that have the highest numbers of receptor molecules per cell. In the **developing lung**, cortisol induces **surfactant protein** that makes it possible for the newborn to breathe on its own. Surfactant is a complex of several substances synthesized in type II alveolar cells of the lung. The major components of surfactant are **dipalmitoylphosphatidylcholine** ($\sim 40\%$), phospholipids ($\sim 40\%$), and surfactant proteins, induced by cortisol, that are **SP-A**, SP-B, SP-C, and SP-D (SP-A is the principal protein) and these constitute approximately 5% plus cholesterol and small amounts of other substances. Uniquely, the **surfactant complex** has regions that are hydrophobic (facing the air) and hydrophilic (head groups in water). The function of surfactant is to reduce surface tension so that the lung can inflate easily and the work of breathing is reduced. **Premature infants** born earlier than 28 weeks of gestation usually do not have cortisol available to induce surfactant and suffer from **respiratory distress syndrome** (**RDS**). Cortisol induces the formation of alveolar SP-A, the major surfactant component, and SP-A generates the release of the prostaglandins $PGE_2$ and $PGF_{2\alpha}$ via the enzyme COX2 (inducible cyclooxygenase) in the chorionic trophoblast. The actions of cortisol and the released **prostaglandins** (together with **oxytocin**) trigger **parturition**. The treatment of RDS is to replace surfactant directly into the lung and use other supports, such as a ventilator and oxygen therapy. If the physician is aware that there might be a premature birth, glucocorticoids are sometimes given to the mother before birth to aid in the formation of SP-A in the infant lung.

As well as its role in adapting to stress and changing environments (as described in the beginning of this chapter), cortisol is a major defense against **inflammation**. Apparently, the activated GR inhibits inflammation by somehow interfering with the nuclear action of **nuclear factor kappaB** (**NF$\kappa$B**). There are several theories as to how the activated glucocorticoid receptor actually inhibits the action of NF$\kappa$B (NF$\kappa$B causes the transcription of many primary inflammatory molecules). GR can induce an inhibitor of NF$\kappa$B, namely **I$\kappa$B** (binds to the translocating component of NF$\kappa$B and prevents translocation into the nucleus) but some investigators report that activated GR can physically interact with cytoplasmic NF$\kappa$B and prevent the translocation of NF$\kappa$B into the nucleus where it normally transcribes mRNAs for many **inflammatory factors** [e.g., TNF$\alpha$, IL-1, **IL-2**, **COX2**, **intercellular adhesion molecule** (**ICAM**), and **cytokine-producing immune cells**, etc.]. Inflammatory mediators cause vascular endothelial cells to increase expression of adhesion molecules (e.g., ICAMs) and **vascular cell adhesion molecules** (**VCAMs**). Vascular endothelial cells produce **IL-8**, a proinflammatory molecule, which can trigger the activation of **integrins** on the leukocyte surface. **Rolling leukocytes** bind to ICAMs and VCAMs on the inner surface of **vascular endothelial cells**, enabling leukocytes to flatten and squeeze between the endothelial cells and pass through the blood vessel (Fig. 16.21).

These freed leukocytes can infiltrate tissues to reach an infected inflammatory site. If prolonged, tissue damage and scarring can occur.

**NF$\kappa$B** induces the inflammation-producing cyclooxygenase 2 (COX2), as mentioned, and this enzyme is critical for the synthesis of inflammatory **prostaglandins**, **leukotrienes**, and **lipoxins** as shown in Fig. 16.22.

Cortisol inhibits the synthesis of COX2 induced by inflammation via the action of NF$\kappa$B in the nucleus. The activated **glucocorticoid receptor** complex binds to the active inflammatory gene (e.g., **GM-CSF** and **COX2**) regulators between CBP [cyclic AMP response element binding protein (**CREB**)-binding protein] with **HAT** (**histone acetyltransferase**) activity that causes acetylation of histones and opens DNA for transcription. The **GR** represses HAT and activates **HDACs** (**histone deacetylase**s), causing the deacetylation of histones and the compaction and repression of **inflammatory genes** (Fig. 16.23).

**COX1** is the **constitutive cyclooxygenase** required for the generation of minimal amounts of prostaglandins, leukotrienes, and lipoxins that are needed for the operation of the cell. In contrast, the induced COX2 generates high levels of inflammatory prostaglandins and their relatives. Consequently, COX2, and not COX1, becomes the drug-therapeutic target. Such a targeted drug design is possible because there is a structural difference between the two enzymes as shown in Fig. 16.24.

In addition to the action of cortisol in inhibiting induced COX2, both COX1 and COX2 are inhibited by **nonsteroidal antiinflammatory drugs** (**NSAIDs**), such as **aspirin**, and these NSAIDs acetylate the active site of the enzymes.

Also, in many tissues subject to inflammation, activated GR induces **lipocortin I** (also named **annexin I**). This protein (now a family of proteins) possibly interacts directly with cytoplasmic **phospholipase A$_2$** (**PLA$_2$**) and inhibits its activity. One domain of lipocortin could interact with the cell membrane while the other might contact PLA$_2$, in its

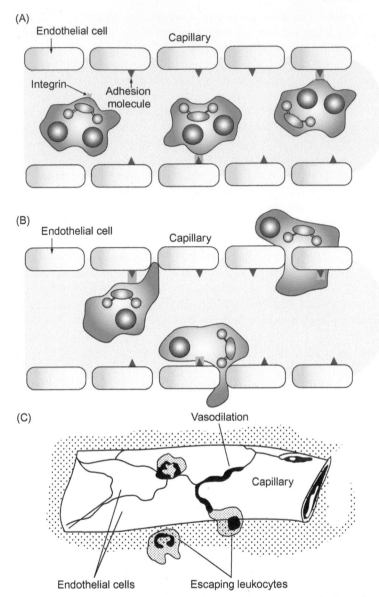

**FIGURE 16.21**   Role of adhesion molecules in the escape of leukocytes, from blood vessels, to invade tissue sites of inflammation. (A) Integrins on the surface of leukocytes bind to adhesion molecules on the inner surface of vascular endothelial cells. (B) Leukocytes flatten and squeeze out between endothelial cells, increasing vascular permeability. (C) External view of leukocytes squeezing out of a blood vessel. *Redrawn with permission from http://www.cat.cc.md.us/courses/bio141/lecguide/unit1/prostrucl/inflam.htm. Doc Kaiser's Microbiology home page. Reproduced directly from G. Litwack,* Human Biochemistry and Disease, *Academic Press/Elsevier, Figure 9−30, page 471, 2008.*

lipid-binding domain, which releases arachidonic acid, hydrolytically, from the cell membrane to the cell cytoplasm where it is a substrate for the formation of **inflammatory prostaglandins**, especially **thromboxane**, leukotrienes, and lipoxins (Fig. 16.22).

Another protein that is induced by glucocorticoids is **vasocortin** whose action is to inhibit **histamine release** from mast cells. The actions of histamine are reviewed in Table 16.4.

The actions of glucocorticoid-induced **lipocortin** and vasocortin lead to potent antiinflammatory effects.

## DEHYDROEPIANDROSTERONE (DHEA)

DHEA circulates in man at levels higher than any other steroid and is found largely as the sulfate derivative (**DHEA-S**). While DHEA or DHEA-S binds to a subunit of **glucose-6-phosphate dehydrogenase (G6PD)** and is a noncompetitive inhibitor of the enzyme, it does not bind significantly to a hormone receptor. *However, DHEA has been found to*

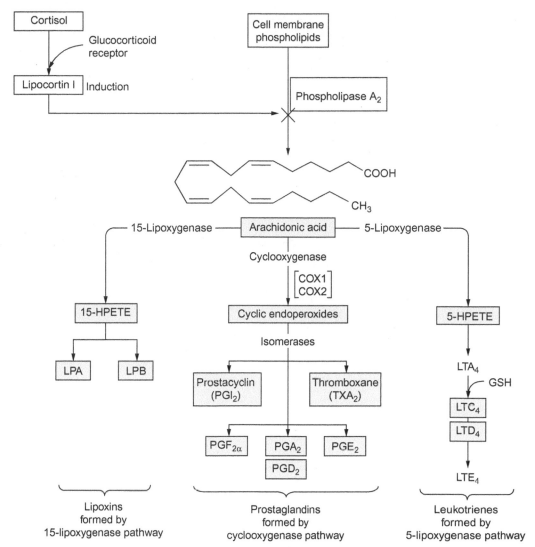

**FIGURE 16.22**   Summary of the antiinflammatory effect of **cortisol** on the release of **arachidonic acid** from cell membrane phospholipids by **phospholipase A₂** (**PLA₂**) and overview of enzymatic pathways giving rise to major prostaglandin relatives. The subscript number refers to the number of double bonds in the structure. **COX1** is the **constitutive cyclooxygenase**; COX2 is the inflammation-induced cyclooxygenase. The conjugated leukotrienes are LTC4 and LTD4. Alpha in **PGF2α** refers to the steric position of the C-9 substituent. *PG*, prostaglandin; *LT*, leukotriene; *LP*, lipoxin; *GSH*, glutathione.

*bind to a **receptor on the membrane of certain endothelial cells** to activate associated G proteins (Gα_{i2} and Gα_{i3}) resulting in the activation of endothelial nitric oxide synthase (eNOS).*

DHEA is synthesized by cells of the *zona reticularis*, the innermost layer of cells of the adrenal cortex. Its structure is shown in Fig. 16.7 (fourth structure down from the top structure on the *left*).

G6PD is the first enzyme in the **pentose phosphate pathway** and an important source of cellular **NADPH**. Besides the important reducing equivalents accorded to NADPH, it is a cofactor for **nitric oxide synthase** (**NOS**) in the reaction:

$$L\text{-Arginine} + NADPH + O_2 \rightleftharpoons \text{citrulline} + \text{nitric oxide} + NADP^+$$

There are three forms of NOS: **neuronal NOS** (**nNOS**), **epithelial NOS** (**eNOS**), and **induced NOS** (**iNOS**). NOS is located in a cell attached, through myristoylation or palmitoylation, to the cytoplasmic side of the endoplasmic reticulum, Golgi, or plasma membrane. *iNOS is induced when cytokine secretion is stimulated by **endotoxins** or **cytotoxins**. nNOS and eNOS are constitutive and are regulated by **calmodulin** interaction with **calcium** reflecting the calcium concentration.* The NOS monomer contains activities of both a reductase and oxidase with **FAD** (**flavin adenine dinucleotide**), NADPH, and **FMN** (**flavin mononucleotide**) or **heme** and **tetrahydrobiopterin** (**BH4**) cofactors. BH4 is

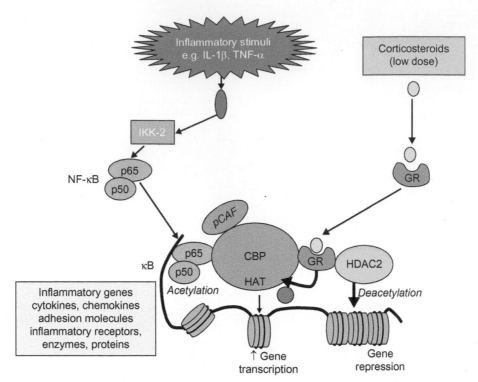

**FIGURE 16.23** Activated GR binds between CBP (CREB-binding protein) with HAT (histone acetyltransferase) activity and HDAC (histone deacetylase) to inhibit HAT activity and facilitate HDAC activity generating the deacetylation of histones connecting inflammatory genes (including COX2) and repressing inflammatory gene transcription. *Reproduced from P.J. Barnes,* Pharmaceuticals, *3: 514–540, 2010.*

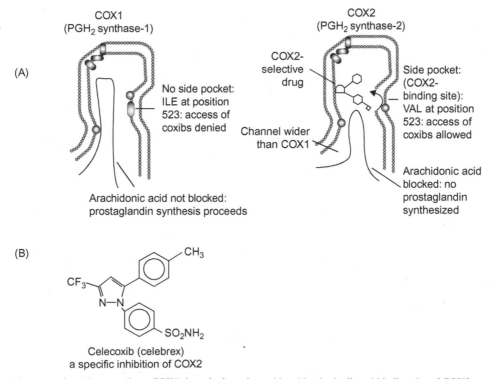

**FIGURE 16.24** (A) Because of a difference (from COX1) in a single amino acid residue in the ligand-binding site of COX2, specific drugs, such as **Celecoxib** (B), can inhibit specifically COX2 but not COX1. *Redrawn from B.N. Cronstein,* Cleveland Clinic J. of Medicine, *1: 13–19, 2002. (Also reproduced from HBD, Figure 9–31, page 472).*

**TABLE 16.4** Summary of Many Effects of Histamine and the Histamine Receptor (H1 or H2) Involved

| Histamine Effect | Type of Histamine Receptor |
|---|---|
| **Vascular Effects** | |
| Capillary dilation | H1, H2 |
| Increased permeability of venules | H1 |
| **Cardiac Effects** | |
| Increased force of contraction | H1, H2 |
| Increased heart rate | H2 |
| Slowed AV conduction | H1 |
| **Visceral Smooth Muscle** | |
| Bronchoconstriction | H1 |
| Intestinal stimulation | H1 |
| **Endocrine Glands** | |
| Gastric acid secretion | H2 |
| Minor stimulatory effect on salivary, lacrimal, pancreatic, intestinal, and bronchial exocrine glands | H1 |
| **Nervous System** | |
| Stimulation or sensitization of peripheral nerve endings | H1 |
| cAMP accumulation in brain | H2 |

Reproduced from http://classconnection.s3.amazonaws.com/835/flashcards/411835/jpg/histamine1355804892515.jpg.

**TABLE 16.5** Adult Human Plasma Concentrations of Adrenal Cortical Steroids and Their Secretion Rates

| | Plasma Concentration (μg/dL) | Secretion Rate (mg/dL) |
|---|---|---|
| Cortisol | 13 | 15 |
| Corticosterone | 1 | 3 |
| 11-Deoxycortisol | 0.16 | 0.40 |
| Deoxycorticosterone | 0.07 | 0.20 |
| Aldosterone | 0.009 | 0.15 |
| 18 OH corticosterone | 0.009 | 0.10 |
| Dehydroepiandrosterone sulfate | 115 | 15 |

Average Plasma Concentrations at 8 a.m.

required for the dimerization of NOS. The inhibition of G6PD (and consequently NADPH) by DHEA promotes **endothelial cell oxidant stress** and can decrease bioavailability of nitric oxide in endothelial cells.

The level of DHEA-S is about 10-fold higher than cortisol as shown in Table 16.5.

In the fetal adrenal gland, **DHEA** is the major secretory product. The secretion of other adrenal cortical steroid hormones begins to be significant following birth.

## STRUCTURAL CONSIDERATIONS OF STEROID HORMONES

Features of the structures of steroid hormones were developed by **X-ray crystallography** shown in Fig. 16.25.

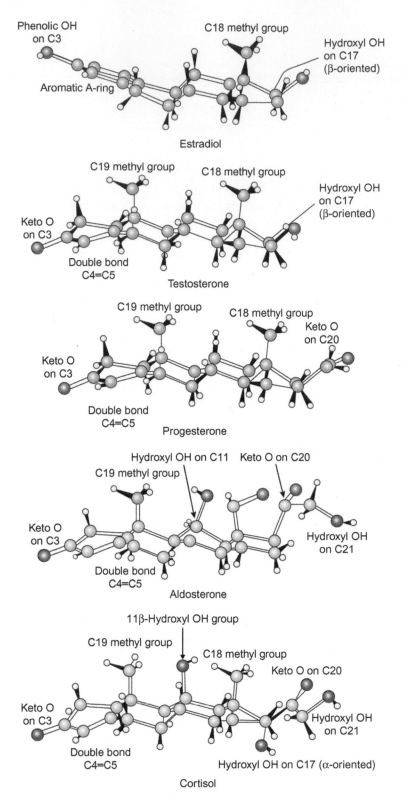

**FIGURE 16.25** Ball and stick representations of steroid hormones determined by X-ray crystallography. Details of each structure are labeled. In **aldosterone**, the **acetal grouping** is $OR_1$ and the **hemiketal grouping** is

$$
\begin{array}{ccc}
& OR_1 & \\
& | & \\
R - & C - H & \\
& | & \\
& OR_2 &
\end{array}
\qquad\qquad
\begin{array}{ccc}
& OR_3 & \\
& | & \\
R_2 - & C - R_1 & \\
& | & \\
& OH &
\end{array}
$$

ACETAL              HEMIKETAL
(2ethers/carbon)      (1ether-1alcohol/carbon)

where $R_1$, $R_2$, and $R_3$ refer to different substituents. *Redrawn from J. Glusker in* Biochemical Actions of Hormones, *G. Litwack, Editor, Academic Press, New York, 1979, pp 121−204 (also reproduced from Human Biochemistry & Disease, by G. Litwack, Figure 9−34, page 477).*

The A, B, C, and D rings for all the structures shown, excepting **estradiol**, lie virtually on a planar surface. The A ring of estradiol, in contrast to the other hormones shown, is unsaturated, having three double bonds. *The double bond between two carbons is shorter than a single bond between two carbons.* Thus, the A ring of estradiol is contracted and this contraction raises the ring above the planar surface compared to the other steroid hormones. All of these steroids enter the **ligand-binding pocket** of their cognate receptors A-ring first. Consequently, there are clear structural differences between the A rings of estradiol and **testosterone** (and dihydrotestosterone) so that there is no overlap in ligand binding between the estradiol receptor and the testosterone receptor (estradiol does not bind to the androgen receptor and testosterone does not bind to the estrogen receptor because of this structural difference in the A rings of these compounds). Also, while testosterone has a **C-19 methyl group**, estradiol does not, another difference affecting the entry into the ligand binding pockets of their cognate receptors. Testosterone, **progesterone**, **aldosterone**, and **cortisol** are all similar in terms of their A rings and C-19 methyl groups. The steroid rings, A−D, all are similarly planar and differ importantly in the substitutions of the D rings. These substitutions will be important in interacting with groupings near the lips of the **ligand-binding pocket**s; some will enhance binding while others will obstruct it depending on the receptor involved. Because of the similarities emphasized, these compounds will bind to each other's receptors to greater or lesser extents. Some of the noncognate ligands will be agonists and others will be antagonists. In the case of the **glucocorticoid receptor**, for example, **progesterone** binds to the active site and is a competitive inhibitor of **cortisol**. The structures of both cortisol and **aldosterone** suggest that they would bind well to the **mineralocorticoid receptor** and act as agonists. In contrast, aldosterone does not bind as well as cortisol to the glucocorticoid receptor and this difference may be explained by the architecture around the C and D rings possibly signifying that the binding pocket is deep enough so that those substituents of the latter rings entering the pocket are interacting with receptor substituents. Physiologically, there is about 1000 times more cortisol in the circulation than aldosterone.

During **pregnancy**, as already pointed out, there is a rather huge excess of **progesterone** present in the circulation from the accessory tissues that competitively inhibits the glucocorticoid receptor in the **mammary gland**, thus preventing the synthesis of milk proteins until termination of pregnancy when the level of progesterone falls rapidly allowing cortisol to enter the binding pocket, activate the receptor, and initiate milk production.

The cellular location of **unliganded steroid hormone receptors** varies. Many of the unactivated receptor complexes reside mainly in the nucleus, requiring that their ligands travel through the cytoplasm into the nucleus to activate them. Among these are receptors for thyroid hormone, retinoic acid, estrogen, and androgen. Some others, like the **vitamin D receptor**, are divided between the cytoplasm and the nucleus. The unliganded glucocorticoid receptor complex resides in the soluble cytoplasm as is the case with the mineralocorticoid receptor. Aldosterone seems to drive the mineralocorticoid receptor into the nucleus more efficiently than cortisol.

## RECEPTOR ACTIVATION

**Class I steroid hormone receptors** reside in the cytoplasm in complexes with other proteins. These proteins block the **zinc fingers** of the DNA-binding domains. Typical of this class is the glucocorticoid receptor and include, besides the GR, the **estrogen receptor**, **androgen receptor**, **progesterone receptor**, and mineralocorticoid receptor. This group requires the formation of **homodimers** for interaction with DNA (Fig. 16.17). Although some report that the **homodimerization process** occurs in the cytoplasm (e.g., Fig. 16.3), recent work suggests that the majority of the hormone receptor dimers are formed in the **nucleoplasm**. Formation of the dimers after nuclear transfer allows for the smaller version of the receptors to pass through the nucleopore, although the pore could probably accommodate a dimer. The models shown in Figs. 16.5 and 16.6 support the view that **homodimerization** of the **glucocorticoid receptor** occurs in the nucleoplasm prior to DNA binding. A generalized view of the **unactivated receptor complexes** for receptors that form homodimers involved in DNA binding and for **heterodimeric receptors** is shown in Fig. 16.26.

The **activation mechanism** involves multiple phosphorylations of the receptors. The specific phosphorylations sites in the receptors have been mapped to serine residues (S203, S211, S226) in the N-terminus. A specific phosphorylation can induce an interaction between the receptor and a protein that can form a bridge between the general transcription factors and the receptor. Although the roles of specific phosphorylations of GR in the N-terminus are incompletely understood, it has been suggested that phosphorylation of S211 and SA226 modifies the binding of cofactor and phosphorylation of S211 is gene specific.

Interestingly, **dopamine**, acting through a **dopamine receptor**, can modulate the transcriptional activity of PR, ER, TR, and COUP-TF (chicken ovalbumin upstream promoter-transcription factor) but not the GR or other steroid hormone receptors. Also, dopamine can activate (followed by translocation to the nucleus) certain steroid receptors in the absence of their cognate ligand and there is evidence for this with regard to the progesterone and estrogen receptors.

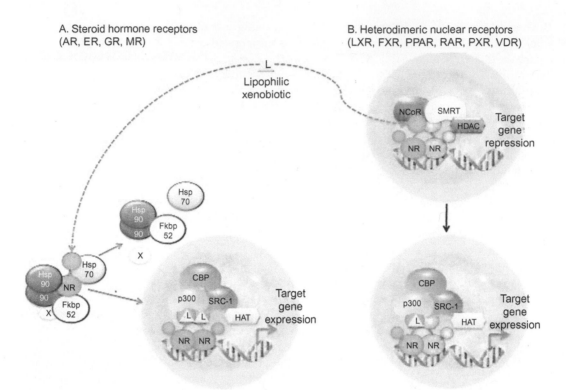

**FIGURE 16.26** Mechanism of action of nuclear receptors. *L*, ligand; *NR*, nuclear receptor; *Hsp*, **heat shock protein**; *HAT*, **histone acetyltransferase**; *HDAC*, **histone deacetylase**; *CBP*, *p300*, coactivators; *FKBP*, **FK506-binding protein**; *SMRT & NCoR*, **corepressors**; *NR:NR in blue*, homodimer; *NR (blue) NR (pink)*, heterodimer. The ligand can be a lipophilic steroid hormone or a xenobiotic compound. For the **glucocorticoid receptor subfamily** (*left*), the unactivated receptor complex contains several proteins: *Hsp90 dimer, Fkbp52, Hsp70,* and another protein (*X*), sometimes p23. Upon ligand binding (**cortisol**, in the case of GR), the receptor becomes activated by the ligand, accompanied by a conformational alteration. Nonreceptor proteins of the unactivated complex are dissociated. The cytoplasmic receptor, as a monomer (e.g., GR), is phosphorylated and translocated to the nucleus through the nucleopore and forms a homodimer in the nucleoplasm. It binds to the hormone response element of DNA and is decorated with coactivator complexes, including histone acetyltransferase (HAT), whose action "opens" DNA so that transcription by **RNA polymerase II** can proceed. The high affinity recognition of the "**core half-site**" of the hormone response element is accomplished by the α-helical structure of the P-box (with the left-hand zinc finger). The second **zinc finger** in the D-box is an α-helix perpendicular to the α-helix of the P-box and is the site mediating **receptor dimerization**. The steroid hormone receptors (GR, PR, ER, AR, and MR) bind to DNA as homodimers and recognize a **palindromic response element**. Thyroid, retinoid, peroxisome proliferator, and vitamin D receptors (TR, RAR, VDR, PPAR, and most orphan receptors), shown in the *upper right*, bind to DNA as heterodimers with RXR. Many of these receptors are located in the nucleus in the absence of ligand where they exist as heterodimers complexed with **corepressor complexes** (*upper right*). When ligand binds, the heterodimers are activated, the corepressor complexes are dissociated and are replaced with coactivator complexes containing histone acetyltransferase (HAT) to "open DNA" and allow transcription to occur. Other details of binding to hormone response elements are shown in Figs. 16.17 and 16.18. *Reproduced from http://nrresource.org/general_information/general_information.html.*

This may be of consequence in the developing mammalian brain whereby dopamine can organize a certain set of behaviors.

When the GR, in the unactivated form, has bound to it a dimer of **HSP90**, a **p23** protein and an **immunophilin** (other than FKBP51), the GR has the greatest affinity for **glucocorticoid ligand**. When the immunophilin **FKBP51** is bound in the complex, the unactivated GR has a somewhat lowered affinity for the glucocorticoid ligand. This general configuration of the unactivated receptor is typical for the steroid receptor class when the receptor is located in the cytoplasm or in the nucleus. The main features of activation of steroid receptors are related in Fig. 16.26.

The trafficking of the glucocorticoid receptor in the cytoplasm, the nucleoplasm, and back out of the nucleus into the cytosol to reform the unactivated receptor complex is shown in Fig. 16.27.

## VITAMIN D HORMONE

Vitamin D is available in several food sources but because the synthesis of the vitamin D precursor occurs in skin as triggered by sunlight (UVB rays), it is not strictly a dietary vitamin. In the skin, **7-dehydrocholesterol** (synthesized from acetate, a precursor for the formation of cholesterol) is the precursor of vitamin D in its active form that is the

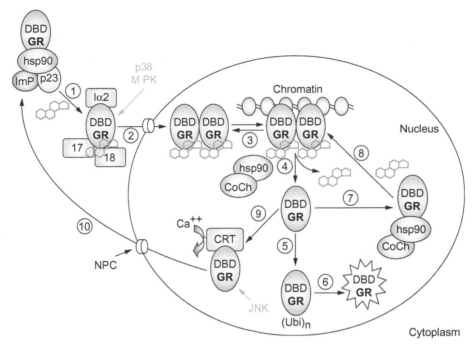

**FIGURE 16.27** Subcellular trafficking of the glucocorticoid receptor. GRs capable of binding hormone exist in the cytoplasm in a heteromeric complex (*upper left*) associated with dimeric chaperone hsp90, CoCh p23, and alternative ImP proteins. Upon ligand binding (step 1), GRs translocate to the nucleus (step 2) by use of nuclear transport factors I$\alpha$2, I7, or I8. In this diagram, GR is depicted as binding chromatin DNA as a dimer, but various transcription factors can cooccupy specific DNA sites with monomeric GRs. GR rapidly exchanges with high affinity DNA-binding sites and might use chaperone−CoCh complexes that contain hsp90 to facilitate dynamic interactions with target sites (step 3). Hormone release from GR and/or GR release from chromatin (step 4) might also require complexes with hsp90. After its ubiquitylation (step 5), GR is degraded by the proteasome (step 6). Unliganded nuclear GR requires a complex with hsp90 (step 7) in order to regain hormone-binding activity (step 8) and reassociate with chromatin. The binding of CRT to GR (step 9) triggers its export from the nucleus and reuse (step 10). *CoCh*, chaperone−cochaperone complex; *CRT*, calreticulin; *DBD*, DNA-binding domain; *GR*, glucocorticoid receptor $\alpha$; *hsp90*, heat-shock protein 90; *I7*, importin; *I8*, importin 8; *I$\alpha$2*, importin $\alpha$2; *ImP*, immunophilin; *JNK*, c-Jun N-terminal kinase; *MAPK*, mitogen-activated protein kinase; *NPC*, nuclear pore complex; *(Ubi)n*, multiple ubiquitin moieties. *Reproduced from http://www.nature.com/nrendo/journal/v2/n11/fig_tab/ncpendmet0323_F1.html.*

ligand for the **vitamin D3** (1,25-dihydroxycholecalciferol or calcitroic acid or calcitriol) **receptor** (**VDR**). UVB rays of sunlight convert 7-dehydrocholesterol to previtamin D3, characterized by the opening of the B ring and methylating the 4 position of the A ring. At the temperature of the skin, previtamin D3 is converted to vitamin D3, in which the components of the B ring are rearranged so that two carbons connect the C and D rings on the upper side and the A ring on the lower side by double-bonded carbons. Vitamin D3 is circulated in the blood to the liver where it is acted on by 25-hydroxylase to form 25-hydroxy D3. This form is circulated to the kidney where it is acted on by 1$\alpha$-hydroxylase to form 1,25-dihydroxyvitamin D3 and in a final step, this precursor is converted to 1,25-dihydroxycholecalciferol (or calcitroic acid or calcitriol) the active form of vitamin D that now acts like a steroid hormone and binds to the vitamin D receptor (VDR) and activates target genes in the cell nucleus. The biosynthesis of calcitriol is shown in Fig. 16.28.

Calcitriol binds to the VDR in many different tissue cells. VDR forms a heterodimer with RXR (Fig. 16.17) and binds to its hormone response element (AGGTCAnnnAGACCA where the antisense strand contains TCCAGTnnnTCTGGT) in the promoter of many different genes. Vitamin D has many physiological effects on mineral metabolism and bone growth. Importantly, the activated receptor facilitates the absorption of calcium and activates the gene that translates the mRNA for the proteins **calbindin** and **TRVP6** that are critical for the intestinal absorption of calcium ions. In addition, VDR facilitates the absorption of phosphate and $Mg^{2+}$.

Reduced blood levels of phosphate or the action of **parathyroid hormone** induce **1$\alpha$-hydroxylase** in the kidney (Fig. 16.29). This figure summarizes calcium regulation in the body.

Growing up in a location where there is little sunlight often precipitates a deficiency of vitamin D and, in adults, this may predispose to **colon cancer**. In contrast, if circulating levels of vitamin D are too high, this condition may predispose to **pancreatic cancer**. The role of the transporter of vitamin D in the blood may be important here since high levels of the vitamin D binding protein (transporter in plasma) could account for elevated levels of the circulating vitamin.

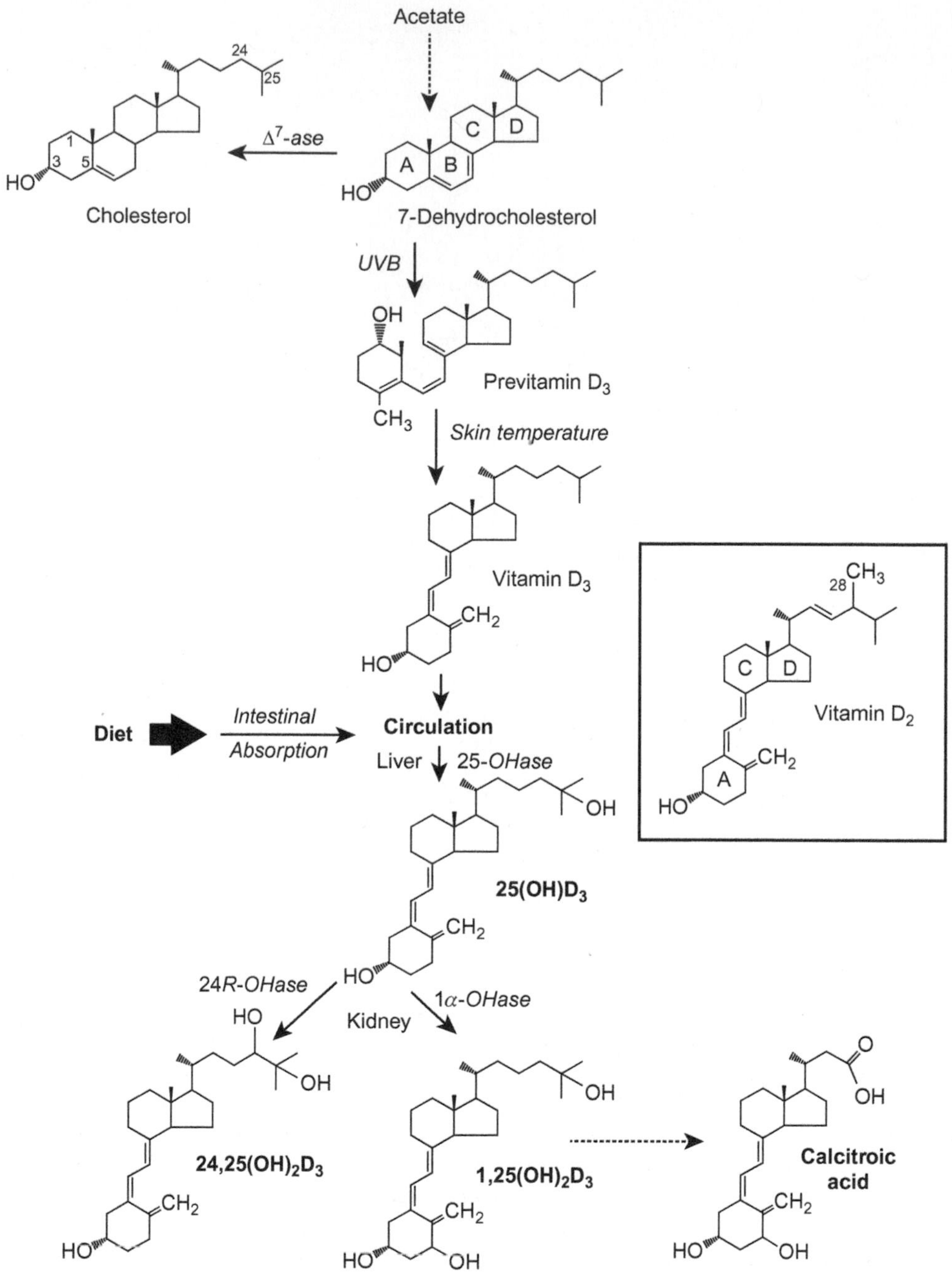

**FIGURE 16.28**  Biosynthesis of the active form of vitamin D in the body. The precursor in the skin is activated by UVB rays in sunlight. The active form that is the ligand for the vitamin D receptor is calcitroic acid (or 1,25-dihydroxycholecalciferol or calcitriol). $\Delta^7$-ase, 7-dehydrocholesterol reductase; *25-OHase*, vitamin D 25-hydroxylase; *1α-OHase*, 25-hydroxyl-D-1α-hydroxylase; *24R-OHase*, 25-hydroxyl D-24R-hydroxylase. Inset, the structure of vitamin D2. Taken from M.F. Holick, 1996.

## THYROID HORMONE

The thyroid hormone receptor is a member of the steroid/thyroid hormone receptor gene family (Fig. 16.17) and it binds to the hormone response element of double-stranded DNA (AGGTCATGACCT:TCCAGTACTGGA) as shown in Fig. 16.16 as a heterodimer with RXR. Note that there are no intervening random nucleotides (n) between half-sites.

The thyroid hormones, $T_3$ and $T_4$, arise in the **thyroid follicle cell**. $T_3$ is the active form that moves into the nucleus of the liver cell, for example, and binds to the thyroid hormone receptor located on its response element and bound to a corepressor complex. When the hormone binds to the receptor, the corepressor complex is dissociated and is replaced

# Calcium regulation

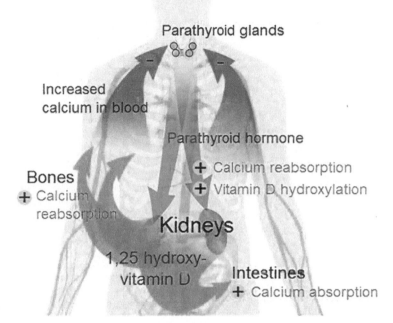

**FIGURE 16.29**   Calcium regulation in the body. *Reproduced from http://upload.wikimedia.org/wikipedia/commons/2/23/Calcium_regulation.png.*

by a coactivator complex and then transcription occurs with RNA polymerase II. The thyroid hormone increases metabolic rates of many systems in most cell types.

## ENVIRONMENTAL XENOBIOTICS THAT AGONIZE OR ANTAGONIZE THE ESTROGEN RECEPTOR

There are several synthetic chemicals in the environment with structures having a resemblance to the **A-ring of estradiol** (Fig. 16.25). These chemicals can bind to the **estrogen receptor**, competitively with **estradiol**, and either agonize or antagonize the estrogen receptor. Some of these compounds are **organochlorine pesticides**, such as DDT, toxaphene, dieldrine, and chlordecone. Polychlorinated biphenyls and polycyclic aromatic hydrocarbons also bind to the receptor. Many of these compounds are **carcinogens** that can produce tumors resulting from overstimulation of the estradiol receptor pathway. Pesticides used in agriculture can antagonize the estrogen receptor to produce deleterious effects on female agricultural workers when they work in field sprayed with pesticides. Antagonism of the estrogen receptor can produce masculinization (e.g., hirsutism). Some of the compounds that can affect the estrogen receptor are shown in Fig. 16.30.

## CROSSTALK BETWEEN STEROID RECEPTORS AND PEPTIDE HORMONES

Various steroid receptors, excluding the glucocorticoid receptor, can be stimulated, in terms of activation of transcription, by the effects of certain **peptide hormones** that lead to **specific phosphorylation** of the **steroid receptor**, often in the AF-1 domain in the N-terminal region. Steroid receptors that can be affected in this way are ER, PR, AR, RAR, RXR, and the VDR.

   **Epidermal growth factor (EGF), insulin-like growth factor (IGF-1)** and **TGFα** can stimulate ER transcriptional activity in an estrogen-dependent manner. These effects may involve **PKA** and **PKC**. Also, some steroid receptors can

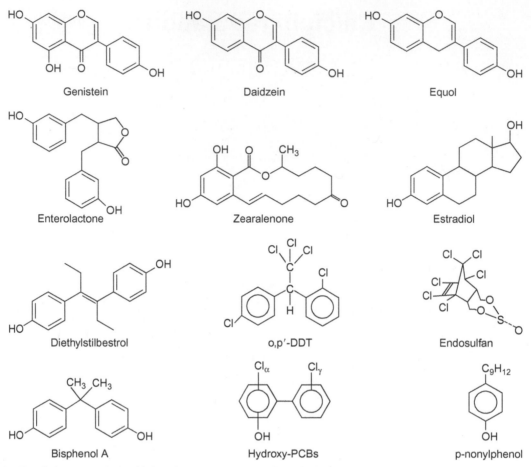

**FIGURE 16.30** Chemical compounds that bind to the estrogen receptor. These include structures with an unsaturated ring and a hydroxyl group (phenolic ring). They also contain a hydrophobic group(s) para to the phenolic hydroxyl (three carbons away from the hydroxyl group).

be activated in the absence of their ligands by growth factors or neurotransmitters. These peptide hormones modulate kinase or phosphatase pathways affecting the receptor and the generation of mRNAs. The **GnRH receptor**, for example, stimulates a phosphorylation cascade that can modulate **progesterone receptor** activity independently of progesterone. *Thus, polypeptide hormones or growth factors, under the right conditions, can generate specific phosphorylation of steroid receptors (the ER is a good example) to stimulate ligand-independent action.*

## SEX HORMONES

The specific roles of **estrogens** and **progestins** have been discussed in terms of the **ovarian cycle**. The roles of the androgens, **testosterone**, and **dihydrotestosterone** were discussed in connection with **spermatogenesis**. The blood level of estradiol in the premenopausal adult is $23-361$ pg/mL (1 pg $=10^{-12}$ g); the postmenopausal adult has a level of $<30$ pg/mL. The prepubertal female has a level of $<20$ pg/mL. After puberty, the androgen levels in the blood increase dramatically during adolescence so that muscle growth and lean body mass can increase. In the male adult, the circulating level of testosterone is $300-1100$ ng/dL (1 ng $=10^{-9}$ g). In the female, the level of circulating testosterone is $20-90$ ng/dL. When males age to $70-80$ years, the androgen levels in the blood fall to a range of $450-500$ ng/dL. This is accompanied by decreased lean body mass and a tendency to falling and fracturing bones. After the age of 40, the stimulatory effects of androgens may increase the incidence of **prostate cancer**. Like the case of the estrogen receptor, there are a number of synthetic compounds that can bind in the ligand-binding site and agonize or antagonize the **androgen receptor**.

The **DNA-binding domain of the androgen receptor** is similar for the glucocorticoid receptor subfamily as shown in Fig. 16.31.

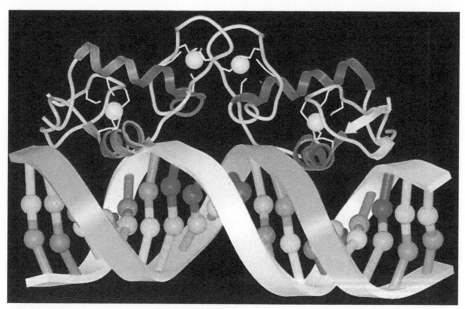

**FIGURE 16.31**  A model of the androgen receptor's DNA-binding domain interacting with the androgen response element of DNA. *Reproduced from http://www.bcgsc.ca/project/discovery-of-new-drug-candidates-for-the-clinical-management-of-prostate-cancer/AR/image_view_fullscreen.*

## PEROXISOME PROLIFERATORS AND THEIR RECEPTORS

Industrial and pharmaceutical chemicals, such as **plastisizers**, **herbicides**, and **hypolipidemic drugs** (drugs that inhibit lipids, lipid derivatives, or functions: **statins**, **fibric acid derivatives**, **bile acid resins**, and **nicotinic acid**), certain **fatty acids**, **rodent carcinogens**, comprise the peroxisome proliferators. The substances, variable in structure, bind to the **peroxisome proliferator-activated receptor (PPAR)**, the central receptor of which is **PPARα** (**PPARδ** and **PPARγ** are the other receptors in the PPAR family). PPARα is located mainly in liver and kidney and binds saturated fatty acids and peroxisome proliferators. PPARβ is in high concentration in kidney, brain, muscle, spleen, lung, adrenal, and immune systems and has highest affinity for prostaglandin J2. PPARγ is most concentrated in white adipose and the immune system and its highest affinity is for PGA1. The PPAR monomer has domains very similar to the glucocorticoid receptor family (Fig. 16.11). All of the PPARs form heterodimers with **RXR** and act in this form upon DNA (Fig. 16.15). The response element for this heterodimer is TGACCTXTGTCCT and the action of the transcriptional complex directs the synthesis of mRNA. *The regions surrounding the response element may determine the extent of transcriptional activity.* PPAR, through its activation by fatty acids, is an important regulator of lipid metabolism and thus it is no longer classified as an **orphan receptor** (a receptor with uncharacterized ligand or function). Orphan receptors are listed in Fig. 16.17. As the name PPAR implies, the action of this receptor is to lead to an increase in the size and number of cellular **peroxisomes** that are important sites of lipid metabolism (β-**oxidation** of fatty acids) and reactive oxygen species (especially the metabolism of $H_2O_2$). In the process of increasing the number of peroxisomes, they progress through three stages: elongation, constriction of the peroxisomal membrane, and fission. These stages are complicated and involve several proteins and the homeostatic mechanisms involve proteins called **peroxins**.

In addition to **linoleic acid, 8S-HETE** (8-hydroxyeicosatetraenoic acid), and **carba-prostacyclin** that act as peroxisome proliferators, there are a number of compounds in this category; some of these are shown in Fig. 16.32. Peroxisome proliferators can cause **hepatocyte hypertrophy**.

Note that the structures of these compounds are somewhat diverse, making it difficult to assign the exact structural requirements for ligands.

Recently, it has been discovered that **peroxisomes** can interact physically with **mitochondria** and the **endoplasmic reticulum**. This interaction can mediate developmental decisions by modulating concentrations of **second messengers** (e.g., $Ca^{2+}$, cAMP, phosphatidylinositol, etc.) in organelles to which peroxisomes attach. The **dysfunction of peroxisomes** may be related to cancer, type 2 diabetes, and neurodegenerative diseases (e.g., Alzheimer's disease and Parkinson's disease) as well as amyotrophic lateral sclerosis.

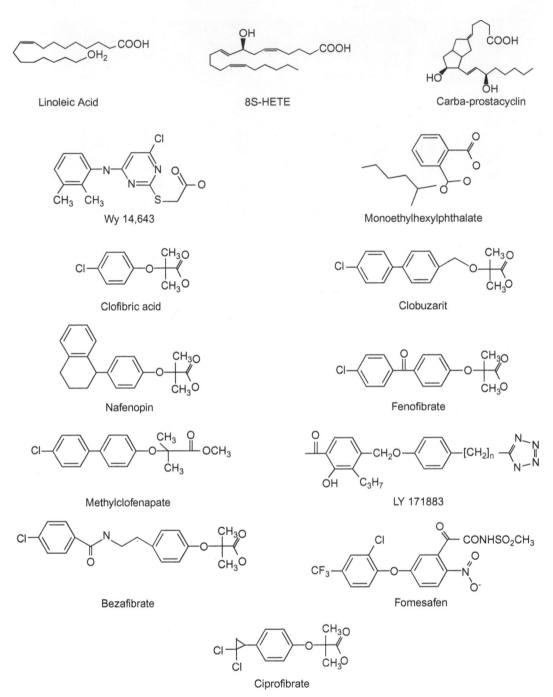

**FIGURE 16.32** Structures of some peroxisome proliferators.

## GLUCOCORTICOID INDUCTION OF PROGRAMMED CELL DEATH (APOPTOSIS)

Programmed cell death or apoptosis constitutes the balance to cell proliferation and is an important mechanism in tissue development when one class of cells needs to be replaced by another. The design of drugs and the use of natural proteins that can attack cancer cells and cause them to undergo this suicide program are important approaches to **chemotherapy of cancer**. Most of the drugs or agents used to kill cancer cells operate by inducing apoptosis. If a physiological process of apoptosis is disrupted, cells that should have died may become cancerous. Agents that induce apoptosis stimulate the transcription of genes whose protein products are involved in programmed cell death. **Hemopoietic cells**, such as T cells, and cancer cells derived from the hemopoietic system can be killed by apoptosis

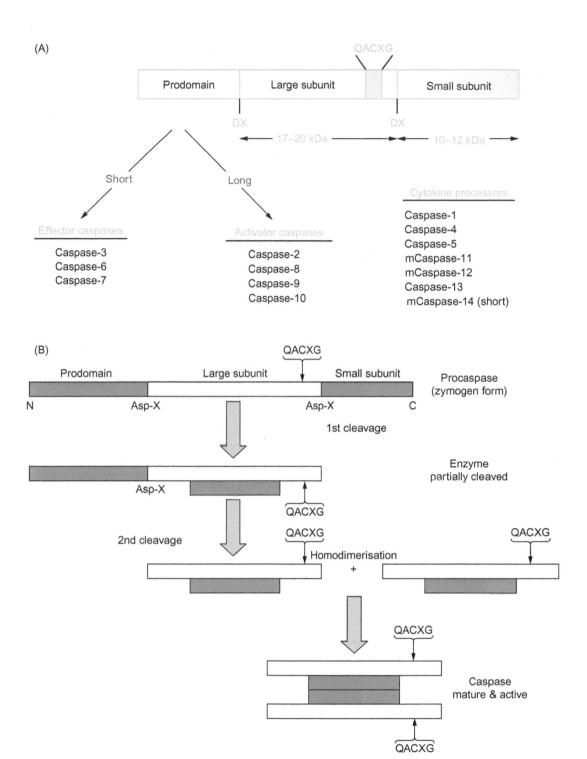

**FIGURE 16.33** (A) Caspases have 3 domains: amino terminal prodomain, a large subunit, and a small subunit. Each caspase contains a conserved pentapeptide motif, QACXG, which includes the active site cysteine residue. Activation of caspases involves proteolytic processing between domains at critical aspartic acid residues (DX) resulting in the removal of the prodomain and self-association of large and small subunit heterodimers to form an active tetramer. Caspases are characterized based on the length of their prodomains. Caspases 2, 8, 9, and 10 are activators or initiators, whereas caspases 3, 6, and 7 are effectors (final steps in apoptosis). The remaining caspases are classified as cytokine processors with the majority having long prodomains. Of these, caspases 11, 12, and 14 have been found only in the mouse. (B) Activation of a caspase by proteolytic cleavage of the **procaspase precursor**. The small subunit is split out from the prodomain in the initial cleavage and it binds to the attached large subunit, forming the precursor of a heterodimer. In the second cleavage, the large subunit (as a heterodimer) is split from the prodomain. Cleavages occur at the caspase-specific site, Asp-X. Two heterodimers dimerize to form the catalytically active enzyme. *Reproduced from S.L. Planey and G. Litwack,* Biochem. Biophys. Res. Commun.*, **279**: 307–312, 2000.*

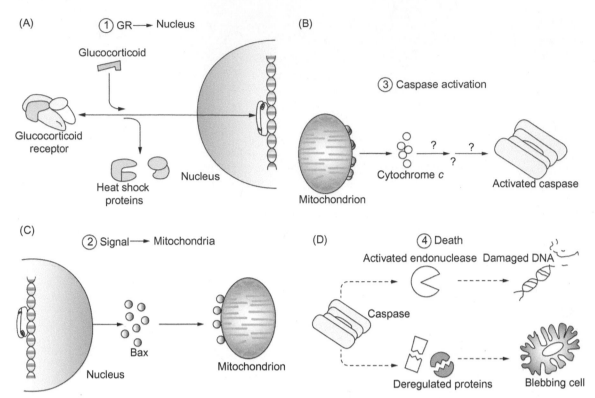

**FIGURE 16.34** Glucocorticoids induce the intrinsic pathway of apoptosis in **hemopoietic cells**. (A) Activation of the glucocorticoid receptor by a glucocorticoid molecule (e.g., **cortisol**). (B) Expression of the **Bax** protein by the glucocorticoid receptor through transcription−translation of the Bax gene. (C) Caspase activation by cytochrome c released from the mitochondrion by the action of Bax. (D) Cell death following proteolysis of key proteins in the cell and the activation of a nuclease that cleaves DNA.

induced by the glucocorticoid, **cortisol**, and more potently by synthetic glucocorticoids (dexamethasone, prednisolone, triamcinolone, and others) that may be 50−100 times more potent than cortisol. In all cases, the **apoptotic program** is mediated by the activation of the proteases called **caspases** that are **aspartate-specific cysteine proteases**. The structural composition of caspases and their activation mechanism are shown in Fig. 16.33A and B.

There are two distinct **apoptotic pathways** that initiate the **activation of caspases** as well as other proapoptotic activities. At the cell membrane, there exists a death receptor that becomes activated by a ligand (e.g., **TNF-related polypeptide**) and recruits adapters and **caspase-8** or **caspase-10** and these activate downstream caspases, such as **caspase-3**, -6, or -7. This would represent the **extrinsic pathway**. In the second, **intrinsic pathway**, cellular stress causes the release of **cytochrome c** from the **mitochondria**. Cytochrome c binds **Apaf-1** and the complex activates **caspase-9**. Caspase-9 then activates downstream caspases that cleave critical proteins and activates an **endonuclease** that cleaves cellular DNA resulting in the death of the cell, as shown in Fig. 16.34.

Several other proteins are involved in the mitochondria as part of the intrinsic pathway: Bax (as already mentioned), **BAK, BCL-xl, BCL-2, BH3** protein, cytochrome c (as already mentioned), and **SMAC**. The simplified reaction mechanism is presented in Fig. 16.34.

Although this process was thought to be exclusively the activity of glucocorticoids, recent reports on **anabolic−androgenic steroids** used extensively by some athletes show, experimentally, that these androgens induce apoptosis in adult rat **ventricular myocytes** and may be a factor in the **sudden cardiac death** associated with **anabolic−androgenic steroid abuse**.

## SUMMARY

Stress is used to demonstrate facets of the hypothalamic−pituitary-end organ axis. Cortisol is the key hormone elaborated in response to ACTH, which, in turn, is released by the hypothalamic hormone CRH in response to environmental or internal stress events. Nociceptin is another hormone involved in the modulation of anxiety produced by acute stress as well as pain. It is an opioid, distinct from the classical opioids that bind to the mu, kappa, and delta opioid receptors.

Nociceptin binds to the NOP1 receptor. Short-term stress causes the release of catecholamines while longer term stress results in the release of cortisol and also aldosterone from the adrenal cortex. Most tissues have receptors for cortisol to a greater or lesser extent, the liver, and kidney having the highest number of receptors, while hepatobiliary cells and cells located in the *pars intermedia* area between the anterior and posterior pituitaries have no receptor.

The mechanism of the glucocorticoid receptor and members of the glucocorticoid subfamily of the nuclear receptor superfamily involves the decoration of the nonliganded receptor by several proteins, including a dimer of HSP90, a molecule of HSP70, an immunophilin, and other proteins like p23. Upon ligand binding, the receptor becomes activated which involves a conformational change in the receptor and the dissociation of the nonreceptor proteins. The monomeric receptor becomes phosphorylated and forms a homodimer, either in the cytoplasm or in the nucleoplasm after translocation to the nucleus. There, the homodimer binds to the hormone response element to elicit transcription of a target gene.

Receptors in the steroid nuclear receptor family are similar in their activity domains. They have N-termini that are important sites for transcriptional factor activity, DNA-binding domains, ligand binding domains, and a C-terminal site for transcriptional factor activity. Receptors have very different N-termini and there are differences in the sequences in the other domains so that specific hormone response element-binding sites are appropriately attributed to the specific receptor and differences in the sequences and structures in the ligand-binding sites to promote ligand specificity. Often specificity is not absolute so that there is cross-over in the binding of ligands and also cross-over in the binding to the hormone response elements. Once activated, receptors that translocate from the cytoplasm to the nucleus form a complex with coactivators in a complex that leads to transcription by RNA polymerase II. Unliganded receptors, already in the nucleus, are complexed with corepressor proteins; when the ligand enters the nucleus and binds and activates the receptor, the corepressor complex dissociates and is replaced by a coactivator complex that facilitates transcription of the target gene.

Disease states follow from the overproduction or underproduction of cortisol. Cushing's disease is the overproduction of cortisol by a tumor of the anterior pituitary overproducing ACTH. There can be a tumor of the adrenal gland overproducing cortisol. These cases usually need surgical intervention. When the body can deal with slight overproduction of cortisol, it can be enzymatically inactivated by conversion to cortisone, in which the 11-hydroxyl is replaced by a 11-keto group that is an inactive ligand for the receptor. Underproduction of cortisol (Addison's disease) can result from an adrenal gland that has experienced disease of some sort. This can be treated by the oral administration of the steroid hormone. The absence of cortisol in the body will result in death because of a failure to adapt to stress and environmental changes.

The renin—angiotensin system responds to a loss of blood volume or a decreased sodium ion content of the blood. The action of angiotensin results in increased blood volume and pressure and its overproduction can be a factor in hypertension. The action of aldosterone on the kidney results in the reabsorption of sodium ions into the blood and also increases blood volume and blood pressure. In addition to the other activities of cortisol, one of its major effects is to reduce inflammation. The glucocorticoid interferes with the nuclear actions of NFκB that causes the transcription of many inflammatory factors (e.g., cytokines). Cortisol, through its receptor, induces lipocortin 1, a protein inhibitor of phospholipase $A_2$. Normally, this enzyme cleaves arachidonic acid out of the cell membrane and arachidonic acid serves as the precursor of many proinflammatory agents, such as prostaglandins, leukotrienes, and lipoxins. The activated glucocorticoid receptor also suppresses the induction of COX2 (responding to inflammation), a cyclooxygenase that is key to the production of proinflammatory prostaglandins.

Dehydroepiandrosterone (DHEA) circulates in the blood, mainly as the sulfate derivative, in concentrations higher than any other steroid. Although it is not known to have a hormone receptor, it does bind to a subunit of glucose-6-phosphate dehydrogenase (G6PD) that gives rise to cellular NADH and this coenzyme is key to the production of nitric oxide which can suppress endothelial cell oxidant stress. More recently, DHEA has been found to bind to epithelial cell membrane receptors that may be involved in the production of nitric oxide.

Structures for ligands of receptors in the glucocorticoid subfamily are somewhat similar and there are situations where several ligands can bind to a receptor as either agonists or antagonists. However, in the case of estrogen, the only hormone in this group to have an unsaturated A-ring, the opposite sex hormone, testosterone or dihydrotestosterone, cannot bind to the estrogen receptor and vice versa because the unsaturated A-ring is constricted and lifts the A ring away from the planar surface characteristic of the other hormones. Certain environmental pollutants and insecticides have rings that allow them to act as ligands for the estrogen receptor. This situation causes deleterious changes in the female through the activation or inhibition of the estrogen receptor.

Peroxisome proliferators bind to receptors in the nuclear receptor family to cause the increase in the number of peroxisomes. Among several kinds of xenobiotic ligands, the PPAR receptors also bind fatty acids physiologically and play a role in lipid metabolism.

Cortisol can activate the intrinsic cell death program (apoptosis) in which hemopoietic cells and tumors resulting from those cells can be destroyed.

## SUGGESTED READING

### Literature

Barnes, P.J., 2010. Inhaled corticosteroids. Pharmaceuticals 3, 514–540.

Biddie, S.C., Conway-Campbell, B.L., Lightman, S.L., 2012. Dynamic regulation of glucocorticoid signaling in health and disease. Rheumatology 51, 403–412.

Berghagen, H., et al., 2002. Corepressor SMRT functions as a coactivator for thyroid hormone receptor T3Rα from a negative hormone response element. J. Biol. Chem. 277, 49517–49522.

Bledsoe, R.K., Stewart, E.L., Pearce, K.H., 2004. Structure and function of the glucocorticoid receptor ligand-binding domain. Vitam. Horm. 68, 49–91.

Chen, T.C., Chimeh, F., Lu, Z., Mathieu, J., Person, K.S., Zhang, A., et al., 2007. Factors that influence the cutaneous synthesis and dietary sources of vitamin D. Arch. Biochem. Biophys. 460, 213–217.

Chen, W., et al., 2008. Glucocorticoid receptor phosphorylation differentially affects target gene expression. Mol. Endocrinol. 22, 2007–2019.

Cronstein, B.N., 2002. Cyclooxygenase-2 selective inhibitors: translating pharmacology into clinical utility. Cleve. Clin. J. Med. 1, 13–19.

Donica, C.L., et al., 2013. Cellular mechanisms of nociceptin/orphanin FQ (N/OFQ) peptide (NOP) receptor regulation and heterologous regulation by N/OFQ. Mol. Pharmacol. 83, 907–918.

Evans, R.M., Mangelsdorf, D.J., 2014. Nuclear receptors, RXR & the big bang. Cell 157, 255–266.

Feelders, R.A., Hofland, L.J., 2013. Medical treatment of Cushing's disease. J. Clin. Endocrinol. Metab. 98, 425–438.

Hawkins, U.A., et al., 2012. The ubiquitous mineralocorticoid receptor: clinical implications. Curr. Hypertens. Rep. 14, 573–580.

Hinds Jr., T.D., et al., 2011. Protein phosphatase 5 mediates lipid metabolism through reciprocal control of glucocorticoid receptor and peroxisome proliferator activated receptor γ (PPARγ). J. Biol. Chem. 286, 42911–42922.

Klieber, M.A., et al., 2007. Corticosteroid-binding globulin, a structural basis for steroid transport and protease-triggered release. J. Biol. Chem. 282, 29594–29603.

Liu, D., Dillon, J.S., 2002. Dehydroepiandrosterone activates endothelial cell nitric-oxide synthase by a specific plasma membrane receptor coupled to $G\alpha_{i2,3}$. J. Biol. Chem. 277, 21379–21388.

Mamenko, M., et al., 2012. Angiotensin II increases activity of the epithelial sodium channel (ENaC) in distal nephron additively to aldosterone. J. Biol. Chem. 287, 660–671.

Nishi, M., 2011. Dynamics of corticosteroid receptors: lessens from live cell imaging. Acta Histochem. Cytochem. 44, 1–7.

Olefsky, J.M., 2001. Nuclear receptor minireview series. J. Biol. Chem. 276, 36863–36864.

Papadimitriou, A., Priftis, K.N., 2009. Regulation of the hypothalamic-pituitary-adrenal axis. Neuroimmunomodulation 16, 265–271.

Planey, S.L., Litwack, G., 2000. Glucocorticoid-induced apoptosis in lymphocytes. Biochem. Biophys. Res. Commun. 279, 307–312.

Tata, J.R., 2013. The road to nuclear receptors of thyroid hormone. Biochim. Biophys. Acta 1830, 3860–3866.

Tsai, M.J., O'Malley, B.W., 1994. Molecular mechanisms of action of steroid/thyroid receptor superfamily members. Annu. Rev. Biochem. 63, 451–486.

Wichert, L., Selbig, J., 2002. Structural analysis of the DNA-binding domain of alternatively spliced steroid receptors. J. Endocrinol. 173, 429–436.

Witchel, S.F., DeFranco, D.B., 2006. Mechanisms of disease: regulation of glucocorticoid and receptor levels − impact on the metabolic syndrome. Nat. Clin. Pract. Endocrinol. Metab. 2, 621–631.

### Books

Glusker, J.P., Trueblood, K.N., 2010. Crystal Structure Analysis (A Primer), Intern. Union of Crystallography. third ed. Oxford Science Publications, Oxford University Press, Oxford, UK.

Larsen, P.R., Kronenberg, H., Melmed, S., Polonsky, K., 2011. Williams Textbook of Endocrinology. twelfth ed. Saunders/Elsevier, Philadelphia, PA.

Litwack, G., 2008. Human Biochemistry and Disease. Academic Press/Elsevier, San Diego, Amsterdam, London, Salt Lake City.

Litwack, G. (Ed.), 2004. Nuclear Receptor Coregulators. Vitamins & Hormones, vol. 68. Academic Press/Elsevier, San Diego, Amsterdam, London, Salt Lake City.

Litwack, G. (Ed.), 1994. Steroids. Vitamins & Hormones, vol. 49. Academic Press, San Diego, Amsterdam, London, Salt Lake City.

Melmed, S., 2011. The Pituitary. Academic Press/Elsevier, San Diego, Amsterdam, London, Salt Lake City.

Roy, A.K., Clark, J.H., 2011. Gene Regulation by Steroid Hormones. Springer-Verlag, NY, New York.

# Chapter 17

# Growth Factors and Cytokines

## PROSPECTS FOR CYTOKINE TRAIL (TNF-RELATED APOPTOSIS INDUCING LIGAND) AND OVARIAN CANCER

Among **gynecological cancers**, ovarian cancer causes more deaths than the others. One of the problems is that ovarian cancer is often diagnosed when it has already progressed to a late stage. Even when initial treatment seems successful, the cancer can reoccur and when this happens the cancer is usually not curable. During a lifetime, 1 female in 72 will be diagnosed with this disease. Most deaths occur when women are diagnosed in the age range of 60 to 70. For the year 2013, it is estimated that there were more than 14,000 deaths from ovarian cancer of more than 22,000 women diagnosed. About 5%−10% of these cancers are hereditary. The cure rate of ovarian cancer can be as high as 95% if the disease is diagnosed early. At present, of the women diagnosed with ovarian cancer, only 46% will survive for 5 years or longer.

Work is proceeding on tests for the **early diagnosis**, but there is little in the way of early diagnosis available in practice. Recently, researchers are using cells shed into the female gynecological tract from ovarian cancers to detect mutaztions in the DNA of those cells using automated systems. It is hoped to combine this approach with the frequently used Pap test. So far, experiments show that 40% of ovarian cancers can be detected, with no false positives and this test can be used to detect early stage disease.

Increased risk of developing ovarian cancer is associated with a family history of this disease or of breast cancer. Other possible conditions are early menopause and no history of pregnancy. In general, the more children a woman has, the lower the risk of developing ovarian cancer. The presence of the **BRCA1 gene** (on chromosome 17) and the **BRCA2 gene** (on chromosome 13), especially hypermethylated BRCA1 (**mutated BRCA1**) is related more to ovarian cancer than mutated BRCA2. When they are mutated, BRCA1 and BRCA2 genes become **oncogenes**, predisposing to cancers. In particular, BRCA1, when it is mutated, increases the risk of developing breast or ovarian cancer. There are many types of mutations that BRCA1 can undergo and some of these, in addition to **hypermethylation**, are **11-base pair deletion**, **1-base pair insertion**, a misplaced **stop codon** or a **missense substitution**. These genes are carried in the germline and are expressed in many tissues in addition to breast and ovary. Normally the BRCA1 gene encodes (through its mRNA) for a zinc-finger containing protein of 1863 amino acids, a fairly large protein. Both BRCA1 and BRCA2 normally function in the **repair of DNA damage** and in the **regulation of transcription**. If there is DNA damage and it cannot be repaired, the BRCA1 gene will aid in the destruction of the cell in which the damage has occurred. Because of high densities of **repetitive elements** in both genes, there is the risk of **gene instability**. Consequently, there can be large **genomic rearrangements** generating inherited and somatic mutations. In the face of the proliferative actions of **estrogen** in both breast and ovarian tissues, the normal BRCA1 gene is needed to control tissue proliferation. **Loss of heterozygosity** (**LOH**) is a further indication and may be a factor in tumor spread. LOH occurs when a somatic cell contains only one copy of an allele. This can be the result of segregation during recombination, deletion of a chromosomal segment or due to nondisjunction during mitosis. This occurrence becomes critical when the surviving allele contains a mutation that results in an **inactive gene**. This frequently is the case when a **tumor suppressor gene** is mutated.

For the formation of ovarian tumors, the oncogenes **HER2/neu**, **K-*ras***, **p53**, BRCA1, and some tumor suppressor genes on chromosome 17 may be involved in a complex pathway. Tumors at grades 2 or 3 have induced expression of genes associated with the cell cycle: **signal transducer and activator of transcription (STAT)** 1 or 3 or **Janus tyrosine kinase (JAK)** 1 or 2.

**Hereditary ovarian cancer** involves the mutation of the BRCA1 tumor suppressor gene (produces a protein through an mRNA that is a tumor suppressor). Mutations in BRCA1 and BRCA2 genes account for 5%−10% of all

Human Biochemistry. DOI: http://dx.doi.org/10.1016/B978-0-12-383864-3.00017-X

507

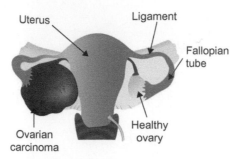

**FIGURE 17.1** Appearance of ovarian carcinoma compared to the normal ovary.

**breast cancers** in females and mutations in either gene result in a fivefold higher risk than normal for developing breast cancer and a 10−30-fold risk for developing **ovarian cancer**. Mutations in these genes can be inherited from either parent and being a **dominant mutation**, can be passed on to both male and female children, so that half the population with BRCA mutations are males meaning that each child has a 50% chance of inheriting the mutated gene from the one parent who carries it. In the male offspring with the mutation, there is only a slightly higher risk of developing breast cancer but there is an increased risk of developing other cancers, such as colon cancer and cancers of the prostate and pancreas.

Symptoms of ovarian cancer are somewhat unspecific: vaginal bleeding, abnormal intestinal gas and nausea. An **enlarged ovary** is probably the first sign but the ovaries are located deep in the pelvic cavity so that it may not be detected for some time and when enlargement is detected, it can also be due to benign fibrosis and not cancer. In advanced ovarian cancer, there is a swollen abdomen with lower abdominal and leg pain and a sudden change in body weight (increase or decrease), change in bowel and bladder function, nausea, and swelling of the legs. An ovarian cancer can shed cells that can grow in other tissues including the uterus, bladder and bowel and these **secondary tumors** can develop even before the primary cancer is diagnosed. Treatment of advanced **ovarian cancer** involves surgery to reduce the tumor bulk and administration of **Taxol** and **carboplatin** for chemotherapy. Unfortunately, **drug resistance** develops in a majority of patients and they may die as a result. Expression of **STAT1** by the tumor leads to resistance to carboplatin (or cisplatin). The location of the ovaries and the appearance of an ovarian carcinoma are shown in Fig. 17.1.

There are four **stages of ovarian cancer**. In **stage 1**, the cancer is limited to one or both ovaries. Ninety percent of women with this stage cancer have a 5-year survival rate. **Stage 2** is characterized by spread of the tumor into the pelvic region (uterus, fallopian tubes, sigmoid colon, or rectum) but not to the abdomen. So far, there has been a poor record of diagnosis at this stage. Of the patients with stage 2 ovarian cancer, 80% will survive for 5 years. In **stage 3**, the cancer will have spread beyond the pelvis to the abdomen, to the abdominal wall, small bowel, lymph nodes or surface of the liver. Of women with this stage cancer, 20%−50% will survive for 5 years. The most advanced condition of the disease is **stage 4** in which the cancer has metastasized to the liver, spleen or lung. Only 10%−20% of these patients will survive for 5 years. The grade of the tumor corresponds to the stage of the cancer and this assignment occurs when surgery takes place and then treatment is adjusted to the stage. The **grade 1** tumor is well differentiated and appears much like normal tissue. **Grade 2** is somewhat differentiated and **grade 3** is poorly differentiated and is clearly abnormal. After completion of therapy for a given grade of the tumor the ovarian cancer can reoccur in its original stage. Many clinical trials have been completed with the administration of TRAIL as a form of treatment.

## TRAIL and Its Mechanism

**The tumor necrosis factor-related apoptosis-inducing ligand** (**TRAIL**) is a protein related to the **TNF** (tumor necrosis factor) **family** of cytokines that can stimulate the growth of certain cells (e.g., **vascular smooth muscle** cells via activation of **NFκB** and the induction of **IGF-1**; **human glioma cells** via cFLIP$_L$-mediated activation of **ERK 1 and 2 kinases** (cFLIP$_L$ is an inhibitor of **caspase-8**)) and kills most types of tumor cells without, seemingly, affecting normal cells. Normal cells may have low levels of the **TRAIL receptors** or high levels of **TRAIL decoy receptors** to buffer the effects of the cytokine. TRAIL is effective in killing ovarian cancer cells in culture but the cancer cells secrete **interleukin-8** (**IL-8**) that inhibits the killing effects of TRAIL. Inhibition by IL-8 is mediated by the p38-MAPK pathway. Patients with ovarian cancer form **ascites** in which the tumor cells float in **peritoneal fluid**. This forms a microenvironment in which the cancer cells may synthesize and secrete factors that affect the killing process and the peritoneal fluid also may contain factors that affect the action of **TRAIL** [this would pertain to clinical trials in which soluble TRAIL is administered (it is possible that insoluble TRAIL might be more effective than the soluble form)]. Treatment of **ovarian cancer** with TRAIL is augmented by combining TRAIL with **paclitaxel** (Taxol relative).

There are 5 **cellular receptors for TRAIL**: R1–R5. R1 is **death receptor** 4 (Dr4) and R2 (Dr5) are TRAIL receptors that have transmembrane domains and cytoplasmic domains. The TRAIL ligand binds to these receptors to generate a signaling mechanism from the membrane to the cytoplasm that results in the programmed death of the cell. R3 (DcR1) and R4 (DcR2) are **decoy receptors** that bind TRAIL. R3 is attached to the cell membrane by a glycophospholipid anchor but lacks a transmembrane domain so that the death signal cannot reach the cytoplasm. Although R4 does have a cytoplasmic domain it lacks the **death domain (DD)**. Thus, R3 and R4 can bind TRAIL but such complexes are unproductive and for that reason these receptors are named decoy receptors (DcRs). **Osteoprotegerin** (**OPG**) is a secreted receptor for TRAIL that lacks a transmembrane domain, a membrane anchor and a cytoplasmic domain, so this soluble receptor also is a decoy. If the levels of the decoy receptors in a TRAIL target cell exceed the levels of Dr4 and Dr5, **TRAIL resistance** will occur. Conversely, if the productive receptors for TRAIL exceed the numbers of decoy receptors, the TRAIL program will be activated. The types of TRAIL receptors are summarized in Fig. 17.2.

TRAIL initiates a signal transduction system leading to **apoptosis** (programmed cell death) when the TRAIL trimer binds to a productive cell membrane receptor. Following this binding, cytoplasmic **death domains (DDs)** form an aggregate, called a **DISC** (death-inducing signaling complex). DISC reacts with another intracellular protein, **fas-associated death domain protein (FADD)** that causes the activation of **procaspase-8** to the activated enzyme (caspase-8). Active **caspase-8** cleaves **proapoptotic Bid** to **tBID**; tBID translocates to the mitochondria and promotes

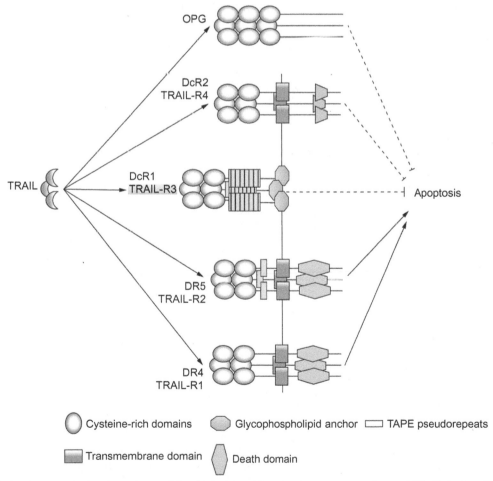

**FIGURE 17.2** TRAIL receptor system. Homotrimers of TRAIL interact with homotrimers or heterotrimers of TRAIL-R1 and TRAIL-R2 to induce apoptosis through their cytoplasmic death domains. Trimers of TRAIL-R3 and TRAIL-R4 also can bind TRAIL but do not trigger the apoptotic signal. Osteoprotegerin (OPG) is a soluble receptor that binds TRAIL and inhibits its activity. *Redrawn from de Almodovar, C.R., et al., 2004. Transcriptional regulation of the TRAIL R3 gene. Vitam. Horm. 67, 51–63.*

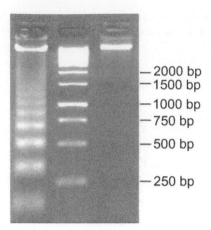

**FIGURE 17.3**   Apoptotic laddering of fragments of cellular DNA visualized on an agarose gel stained with ethidium bromide and visualized under UV light. *Left lane*, apoptotic DNA fragment laddering; *middle lane*, marker DNAs of known molecular weights; *right lane*, unhydrolyzed cellular DNA control. *Reproduced from http://upload.wikimedia.org/commons/7/75/Apoptotic_DNA_Laddering.png.*

the assembly of **Bax-Bak** (*Bax*, **B**cl-2-**a**ssociated **X** protein; *Bak*, **B**cl-2 homologous **a**ntagonist/**k**iller) oligomers (both are proapoptotic members of the Bcl-2 family of proteins) and permeability alterations in the mitochondrial outer membrane. Bax-Bak causes the release of cytochrome *c* from mitochondria and Smac/DIABLO (mitochondrial protein that neutralizes the activity of **IAP**s, **i**nhibitors of **ap**optosis) from the mitochondria into the soluble cytoplasm. With cytochrome *c* and Apaf-1 (**a**poptotic **p**rotease-**a**ctivating **f**actor-1) in the cytoplasm, the **apoptosome** is formed and facilitates the activation of caspase-9. Caspase-9 promotes the activation of the executioner caspases (caspases-3, -6, and -7) that cleave key cellular proteins resulting in DNA cleavage and cell death. DNA fragmentation in apoptosis occurs when caspase-3 activates a nuclear DNase of about 40 kDa. DNA fragmentation in apoptosis can be visualized on an agarose gel that separates nucleic acids based on size (Fig. 17.3).

The overall extrinsic pathway and the intrinsic pathway of apoptosis are summarized in Fig. 17.4.

Active **caspase-9** leads to the activation of **procaspase-3** and other **executioner caspases** and these cleave key proteins in the cell that precipitate **DNA fragmentation** and apoptosis. The **extrinsic pathway** of apoptosis does not involve the mitochondria whereas the **intrinsic pathway** does involve the mitochondria.

## THE TUMOR NECROSIS FACTOR (TNF) SUPERFAMILY

There are an extensive number of ligands related to the **tumor necrosis factor** and these form a superfamily (TNFSF) of which **TRAIL** is a member. Correspondingly, there are a number of receptors for these ligands called the **tumor necrosis factor receptor superfamily** (**TNFRSF**). This combination of ligands and receptors is involved in the regulation of the **immune system**, **inflammation** and, in some cases, **antitumor activity** as the name "tumor necrosis factor" implies. **Spontaneous regression** of tumors sometimes occurs and the activities of these factors may be responsible. Some of the TNF family of ligands and receptors is listed in Table 17.1.

At present, there are 19 soluble and membrane-bound ligands and 32 receptors in the **TNFSF** (Fig. 17.5).

The gene for **TRAIL** is THF-like-2 or *TL2* that is located on chromosome 3. TRAIL is considered as the 10th member of the TNF ligand superfamily (**TNFSF10**). It occurs in the cell in two forms, soluble and insoluble (membrane bound). It can kill a number of cancer cells and transformed lines of cells and, in these cells, TRAIL induces DNA fragmentation and apoptosis. Some investigators posit that the insoluble form is the more active form of the two. TRAIL is expressed in many tissues including the spleen, thymus, prostate, lung, kidney, and intestine. There are **alternative splice variants** that occur in both neoplastic and normal tissues and these are designated **TRAIL**β (lacks exon 3 of the TRAIL gene) and **TRAIL**γ (lacks exons 2 and 3 of the TRAIL gene). Both tumor cells and normal cells express the **TRAIL receptor, Dr4;** tumor cells (transformed cells) suffer apoptosis by TRAIL but most normal cells (untransformed) are resistant to TRAIL. As mentioned before, the situation in normal cells may be an excess of **decoy receptors** over normal receptors. Some tumor cells may become partially resistant to TRAIL by producing **IL-8** (or some other factor) either from the tumor cell or from the medium (in the case of ovarian tumors, they form **ascites** so that the medium is an important factor *in vivo*).

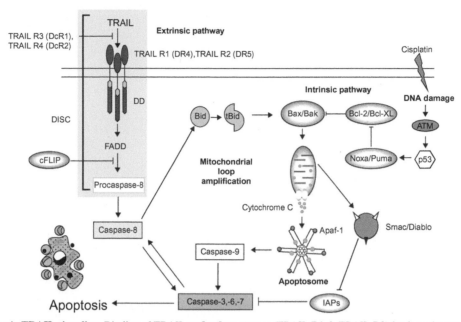

**FIGURE 17.4** Apoptotic **TRAIL signaling**. Binding of TRAIL to **death receptors** (TRAIL R1 & TRAIL R2) leads to the recruitment of the adapter molecule, **FADD (fas-associated death domain protein)**. **Procaspase-8** binds to FADD leading to **DISC (death-inducing signaling complex)** formation and resulting in its activation. Activated caspase-8 directly activates **executioner caspases** (caspase-3, -6, and -7) and cleaves **Bid**. Translocation of the **truncated Bid (tBID)** to the mitochondria promotes the assembly of Bax-Bak oligomers and changes the permeability of the mitochondrial outer membrane. Cytochrome *c* is released into the soluble cytoplasm resulting in **apoptosome** assembly. The apoptosome is a platform-like structure for **procaspase-9 activation** that is formed in the following manner: **Apaf-1** and cytochrome *c* coassemble in the presence of dATP to form the apoptosome. A central ring containing 7 **caspase recruitment domains (CARDs)** is formed within the apoptosome. There are also 7 copies of a **nucleotide-binding and oligomerization domain** (**NOD**) associated laterally to form the hub of the apoptosome. The hub forms a circle around the CARD ring. A helical domain, resembling an arm, links to each NOD to a pair of **propellers** and these propellers bind a single molecule of cytochrome *c*. Active caspase-9 then propagates a proteolytic cascade of **effector caspases** activation that leads to the morphological hallmarks of **apoptosis**. Further cleavage of **procaspase-8** by **effector caspases** generates a mitochondrial amplification loop that further enhances apoptosis. When **FLIP (FLICE-like inhibitory protein; *FLICE*, FADD-like ICE)** levels are elevated in cells, **caspase-8** preferentially recruits FLIP to form **caspase-8-FLIP** heterodimer which does not trigger apoptosis. *DD*, death domain; *ATM*, serine-threonine kinase that phosphorylates proteins that activate DNA damage. *Reproduced from http://www.ncbi.nlm.nih.gov/pmc/articles/PMC3236573/figure/fig01/.*

**TABLE 17.1** The TNF Superfamily (TNFSF) and the TNF Receptor Superfamily (TNFRSF)

| Ligands/Coreceptors[a] | Amino Acid Residues (AAs); Molecular Weight | Comments and Characteristics |
|---|---|---|
| NGF (Nerve Growth Factor) | 120AA; 12.5 kDa | From a propeptide with signal sequence; homodimer 1525 kDa; binds to LNGFR member of the TNFRSF |
| CD40L | 261AA; 39 kDa; 22AA cytoplasmic domain; 215AA extracellular domain | Membranes of B cells, CD4+ & CD8+ T cells, mast cells, basophils, eosinophils, dendritic cells, monocytes, NK cells, and gd cells. Also as proteolytically cleaved, cytoplasmic form of 15–18 kDa with biological activity; forms trimeric structures like TNFα |
| CD 137 L/4-1BBL | 309AA; 50 kDa; 82AA and 34-kDa cytoplasmic region; 21AA transmembrane segment; 206AA extracellular domain | Expressed in B cells, dendritic cells, and macrophages |

*(Continued)*

**TABLE 17.1** (Continued)

| Ligands/Coreceptors[a] | Amino Acid Residues (AAs); Molecular Weight | Comments and Characteristics |
|---|---|---|
| TNFα | 233AA; 26 kDa in membrane with 29AA cytoplasmic domain; 28AA transmembrane domain; 176AA extracellular domain | Either transmembrane or soluble protein is biologically active; expressed in many cell types, including macrophages, CD4$^+$ and CD8$^+$ T cells, adipocytes, keratinocytes, mammary and colon epithelia, osteoblasts, mast cells, dendritic cells, pancreatic β-cells, astrocytes, neurons, monocytes, and steroid-producing cells of *zona reticularis* |
| CD134/OX40L | 183AA; 21AA cytoplasmic domain; 23AA transmembrane segment; 139AA extracellular domain | OX40L exists as a trimer; limited expression: activated CD4$^+$ and CD8$^+$ T cells, B cells, and vascular epithelial cells |
| CD27L/CD70 | 50 kDa; 193AA transmembrane glycoprotein; 20AA cytoplasmic segment; 18AA transmembrane segment; 155AA extracellular domain | Expressed by NK cells, B cells, CD45RO$^+$, CD4$^+$, and CD8$^+$ T cells, gd T cells, and some leukemic cells; may be involved in antibody production in B cells |
| FasL (Fas Ligand) | 40-kDa transmembrane protein 281AA; 80AA cytoplasmic domain; 179AA extracellular domain | Can occur as circulating trimer; can be cleaved by a protease to give an active 70-kDa trimer of 26-kDa monomers; F273L mutation results in gld/gld generalized lymphoproliferative disease; expressed by: type II pneumocytes and bronchial epithelium, monocytes, LAK cells, NK cells, dendritic cells, B cells, macrophages, CD4$^+$ and CD8$^+$ T cells, colon, and lung carcinoma cells |
| CD30L | 40-kDa; 234AA transmembrane glycoprotein: 46AA cytoplasmic domain; 21AA transmembrane segment; 172AA extracellular domain | Expressed by monocytes and macrophages, B cells, activated CD4$^+$ and CD8$^+$ T cells, neutrophils, megakaryocytes, resting CD2$^+$ T cells, erythroid precursors, and eosinophils |
| TNFβ/LT-α | Circulates as 171AA, 25-kDa glycosylated polypeptide; a larger form (205AA) exists, suggesting proteolytic processing; no transmembrane form, but it can be membrane associated as it can bind to membrane anchored LTβ, forming a heterotrimer | Circulating TNFβ is ~150 pg/mL heterotrimer binds to LTβR and TNFRI receptor, but TNFRI activation will not occur |
| LTβ | 33-kDa type II transmembrane glycoprotein; 244AA; 16AA cytoplasmic segment; 31AA transmembrane domain; 197AA extracellular region | LTβ forms a heterotrimer with TNFβ on membrane; LTβ is not secreted |
| TRAIL | 32 kDa, 281AA; 17AA cytoplasmic domain; 21AA transmembrane segment; 243AA extracellular domain | Homotrimer in membrane; many tissues express TRAIL, including lymphocytes; may have anticancer cell activity |
| **RECEPTORS (TNFRSF)[b]** | | |
| (LNGFR/p75 human low-affinity nerve growth factor receptor) | 75 kDa; 427AA with extracellular N-terminus; 25AA signal sequence; 225AA extracellular domain; 23AA transmembrane segment; 154AA cytoplasmic domain | Transmembrane-glycoprotein; can appear as a 200-kDa disulfide-linked homodimer; neurotrophins bind to LNGFR with $K_D \sim$ l–3 mM, no inherent tyrosine kinase activity; death domain in cytoplasmic domain; protease cleavage-35 to 45-kDa LNGFR; cells expressing LNGFR: oligodendocytes, B cells, bone marrow fibroblasts, autonomic and sensory neurons, Schwann cells, follicular, dendritic cells, select astrocytes, and mesenchymal cells |

*(Continued)*

**TABLE 17.1**  (Continued)

| Ligands/Coreceptors[a] | Amino Acid Residues (AAs); Molecular Weight | Comments and Characteristics |
|---|---|---|
| CD40 | 50 kDa; 277AA transmembrane glycoprotein (B cell proliferation and differentiation); 20AA signal sequence; 173AA extracellular domain; 22AA transmembrane segment: 62AA cytoplasmic domain | 4 Cys-rich motifs in extracellular region with juxtamembrane sequence rich in SER and THR: CD40 up-regulates FAS to prime cells for subsequent FAS-mediated apoptosis; CD40 pathway involves NFκβ and protein kinase (LYN) activation: cells expressing CD40: monocytes, basophils (not mast cells), eosinophils, endothelial cells, interdigitating dendritic cells, Langerhans cells, blood dendritic cells, fibroblasts, keratinocytes, and Reed-Sternberg cells of Hodgkin's disease and Kaposi's sarcoma cells |
| CD137/4-IBB/ILA | 30−35 kDa; monomer and dimer on cell surface 255AA; 17AA signal sequence, 169AA extracellular region, 27AA transmembrane segment; 42AA cytoplasmic domain | Cys-rich motif in extracellular domain CD 137 binds its ligand, CD137L, at $K_D$ of ~30 pM; alternative splicing event can give rise to soluble form; CD137 ligation can interrupt cell apoptotic program associated with activation-induced cell death; cell expressing CD 137: fibroblasts, thymocytes, monocytes, and CD4$^+$ and CD8$^+$ T cells |
| TNFR1/p55/CD120a | 55 kDa; 455AA transmembrane glycoprotein; 190AA extracellular domain; 25AA transmembrane segment; 220AA cytoplasmic domain | Expressed in all nucleated mammalian cells; four Cys-rich motifs in extracellular region; first Cys-rich motif required for binding; 80AA death domain in cytoplasmic region; NFκB is activated by TNFR1; TNFR1 binds both TNFα and TNFβ; $K_D$ ~20−60 pM; for soluble TNFα; $K_D$ for TNFβ = 650 pM; TNFR1 most important for circulating TNFα; membrane bound TNFα associates with TNFR2; soluble TNFR1 blocks TNFα activity (decoy) and occurs in blood and urine at 1−3 ng/mL; protease activity gives soluble forms of 32 kDa and 48 kDa; cells expressing TNFR1: hepatocytes, monocytes, and neutrophils; cardiac muscle cells; endothelial cells; and CD34$^+$ hematopoietic progenitors |
| TNFR2/p75/CD120b | 75 kDa; 461AA transmembrane glycoprotein; 240AA extracellular region; 27AA transmembrane: segment; 173AA cytoplasmic domain | TNFR2 binds TNFα and transfers it to TNFR1, which becomes activated; TNFα binding to TNFR2 induces apoptosis in rhabdomyosarcoma cells and cell migration in Langerhans cells; soluble TNFα binds TNFR2 with A $K_D$ of 300 pM; TNFα levels are usually at 100 pM so that it should normally bind to TNFR1; therefore, TNFR2 acts as a decoy; cells expressing TNFR2: monocytes, endothelial cells, Langerhans cells, and macrophages |
| CD134/OX40/ACT35 | 48 kDa; 250AA; 188AA in extracellular region; 26AA transmembrane segment; 36AA cytoplasmic domain | Expressed in CD4$^+$ and CD8$^+$ T cells only |
| CD27 | 50−55 kDa; mature CD27 is 27 kDa, 242AA; 175AA in extracellular domain; 21AA transmembrane segment; 46AA cytoplasmic domain | Expressed as homodimer on cell surface; it has no death domain but induces apoptosis by associating with Siva cytoplasmic protein, which has a death domain; blood and urine contain a soluble 32-kDa CD27 (probably from proteolysis) cells expressing CD27: NK cells, B cells, CD4$^+$ and CD8$^+$ T cells, and thymocytes |

*(Continued)*

**TABLE 17.1 (Continued)**

| Ligands/Coreceptors[a] | Amino Acid Residues (AAs); Molecular Weight | Comments and Characteristics |
|---|---|---|
| FAS/CD95/APO-1 | 43 kDa; 335AA; 156AA extracellular region; 20AA transmembrane segment; 144AA cytoplasmic domain | On fibroblasts FAS ligation can lead to either proliferation or apoptosis depending on number of expressed FAS molecules; 3 Cys-rich motifs; 68AA death domain in cytoplasmic region identical to one found in TNFR1 cytoplasmic domain; death domain associates FADD protein (with FAS) or TRADD protein with TFNR1; both transmit apoptotic signals; alternative gene splicing produces soluble forms of FAS |
| | | Soluble blood FAS circulates as dimer and trimer at low mg/mL concentrations; cells expressing FAS: CD34$^+$ stem cells, fibroblasts, NK cells, keratinocytes, hepatocytes, B cells and B cell precursors, monocytes CD4$^+$ and CD8$^+$ T cells, CD45RO$^+$ gd T cells eosinophils and thymocytes |
| CD30/Ki-1 | 105–120-kDa transmembrane glycoprotein; mature CD30 is 577AA; 18AA signal sequence, 365 extracellular region, 24AA transmembrane segment; 188AA cytoplasmic domain | 6 Cys-rich motifs in extracellular region- patients with CD30$^+$ lymphomas have 85-kDa soluble CD30 in blood; cells expressing CD30 Reed-Sternberg cells, CD8$^+$ T cells and CD4$^+$ T cells |
| LT-βR | 75-kDa transmembrane glycoprotein; 201AA extracellular domain; 26AA transmembrane segment; 187AA cytoplasmic domain | 4 Cys-rich motifs in extracellular domain; LT-βR binds heterotrimers (1 TNFβ + 2 LT-β) over LT-β homotrimers: first 2 Cys-rich motifs resemble TNFR1, and third & fourth Cys-rich motifs resemble TNFR2; LT-βR activates NF−κB and induces cell death via TRAF-3; LT-βR activates genes for IL-8 and Rantes; LTβR expressed by monocytes, fibroblasts, smooth muscle, and skeletal muscle cells |
| DR3/WSL-1/TRAMP/APO-3/LARD (Death Receptor 3) | 54 kDa; 417AA; 24AA signal sequence; 178AA extracellular domain; 23AA transmembrane segment; 192AA cytoplasmic domain | Can activate both NF$_k$B and induce apoptosis like TNFR1; four Cys-rich motifs in extracellular region, many alternate splice forms of DR3, some of which may be soluble, cells expressing DR3: T and B cells and human umbilical vein endothelial cells |
| DR4 (Death Receptor 4) | 468AA; 23AA signal sequence; 226AA extracellular domain; 19AA transmembrane segment, 220AA cytoplasmic domain | One of the three known receptors for TRAIL; two Cys-rich motifs in extracellular domain; expressed by activated T cells |
| DR5 (Death Receptor 5) | 411AA; 51AA signal sequence; 132AA extracellular domain; 22AA transmembrane segment; 206AA cytoplasmic domain | Second receptor for TRAIL; triggers apoptotic program like DR4 without FADD participation; two Cys-rich motifs in extracellular domain |
| DcR1/TRID (decoy receptor 1) (TRAIL receptor without an intracellular domain) | 259AA; 23AA signal sequence; 217AA extracellular domain; 19AA transmembrane domain | Membrane receptor for TRAIL with no intracellular domain two Cys-rich motifs in extracellular region 50%–60% identical to AA sequences in same regions of DR4 and DR5; inhibits responsiveness to TRAIL (decoy receptor) |
| TR2 | 32 kDa; 283AA; 36AA signal sequence; 165AA extracellular region; 23AA transmembrane segment; 59AA cytoplasmic domain | No known ligand as yet; found in T cells, B cells, monocytes, and endothelium; four Cys-rich motifs in extracellular domain |
| GITR (Glucocorticoid-Induces TNFR Family-Related) | 228AA; 19AA signal sequence; 134AA extracellular domain; 23AA transmembrane segment; 52AA cytoplasmic domain | Inducible during T cell activation; three Cys-rich motifs in extracellular region; ligation interrupts TCR-DC3-induced apoptosis in T cells |

*(Continued)*

**TABLE 17.1** (Continued)

| Ligands/Coreceptors[a] | Amino Acid Residues (AAs); Molecular Weight | Comments and Characteristics |
|---|---|---|
| OPG (Osteoprotegerin) | 55 kDa; 380AA | Inhibits osteoclasts and protects bone from breakdown; secreted member of TNFRSF; similar to TNFR2 and CD40; no transmembrane segment circulates as a disulfide-linked homodimer; ligand unknown |
| TL1A (TNF-like ligand 1A) | | TL1A is a ligand for the DR3 receptor. DR3 (TNFRSF25) is a TNFR receptor expressed primarily on lymphocytes and is the receptor for TL1A (TNFSF15). DR3 is a costimulator of T-cell activation (which produces a variety of cytokines) signaling through an intracytoplasmic death domain (DD) and the adapter protein TRADD (TNFR-associated death domain). DR3 stimulates T-cell proliferation and inflammation of tissue sites. |
| TWEAK (TNF-like weak inducer of apoptosis) | 102 amino acids plus 6 cysteine Residues in its extracellular domain | TWEAK is a member of the TNF family of ligands. The TREAKR cytoplasmic domains bind TRAFs-1, -2, and -3 (TRAF, TNFR-associated factor, that is a scaffold or adapter protein linking the IL1R/Toll to TNFR). The TWEAKR inhibits endothelial cell migration *in vitro* and inhibits corneal angiogenesis *in vivo* |

[a]*TNFSF usually forms trimeric structures, ligands and receptors of the TNSF and TNFRSF undergo clustering during signal transduction, monomers of TNFSF are two-sheet structures composed of β-strands.*
[b]*TNFRSF is usually trimeric or multimeric, stabilized by intracysteine disulfide bonds; it exists in both membrane-bound and soluble forms; many forms transduce apoptotic signals in a variety of cells.*
**Source**: Much of the information in this table was taken from http://www.rndsystems.com/asp/g_sitebuilder.asp?bodyId = 227#top (no longer retrievable on the internet).

There is an insertion of 12−16 amino acids in the sequence of TRAIL that creates an elongated loop that other members of the TNF family do not have. It turns out that the extracellular region of Dr4 has extensive interactions with this extended loop of TRAIL accounting for the specificity for TRAIL.

Trimers of TNFα, like TRAIL, bind to the TNF receptor (TNFR1) forming an aggregate on the cell membrane. The form of a trimer may amplify the signal or facilitate the clearance of the aggregate from the membrane after functional completion, or both (Fig. 17.6A and B).

TNF-related **cytokines**, such as **TNFα** and **TRAIL** appear as monomers, dimers, and trimers. In soluble form, the trimer seems to be the most active specie. An aggregate of **TNF trimers** binds to parallel receptor dimers (sTNFR1) as shown in Fig. 17.7.

Many other cytokines and their receptors are summarized briefly in Table 17.1 and in Fig. 17.5.

# GROWTH FACTORS

Growth factors and **cytokines** are somewhat similar. Cytokines can have negative effects, such as the production of **inflammation**. Here, it is emphasized that some cytokines, like TRAIL and TNF are able to kill cells by activating **apoptosis**, although certain cytokines, in the right conditions, can also stimulate the growth of cells. In general, growth factors inherently cause cells to grow and divide. Like cytokines, growth factors interact with cell membrane receptors and activate them but, invariably, they induce a signal induction process that leads to cell division and proliferation. Some of these proteins are like cytokines but produce growth effects, e.g., many of the **interleukins**. However, here the

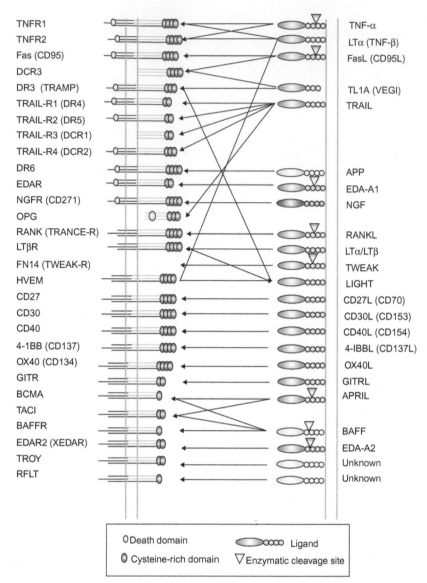

**FIGURE 17.5** The TNF and TNF receptor superfamily. Receptors are illustrated in the *left* column and ligands are depicted in the *right* column. Arrows from each ligand (*on right*) point to receptors (*on left*) that bind the indicated ligand. *Reproduced from http://www.enzolifesciences.com/browse/immunology/tnf-tnf-receptor-superfamily/.*

interleukins are considered along with the growth factors. A great many growth factors have been identified. For convenience, a few of the most common growth factors will be considered first and they are listed in Table 17.2.

## Platelet-Derived Growth Factor

**PDGF** is a stimulator of cell division (**mitogen**) for different cell types, such as cells in **connective tissue** and **developing nervous system**. It has five different polypeptide chains (subunits), labeled AA, AB, BB, CC and DD. The AA and BB chains are somewhat similar (AA chain, 211 amino acids; BB chain, 241 amino acids). The dimerized subunits are stabilized by disulfide bonds. There is some specificity for either chain at the receptor level. There also exist two distinct subunits of the PDGF receptor that also occurs in the cell membrane as a homodimer ($\alpha\alpha$, $\beta\beta$) or a heterodimer ($\alpha\beta$). The $\alpha\alpha$-receptor [PDGFR$\alpha$] recognizes the AA, CC, AB and BB subunits of PDGF. The $\beta\beta$-receptor (molecular weight about 180 kDa) recognizes subunits BB and DD of PDGF. The binding of **PDGF** by the receptor requires that

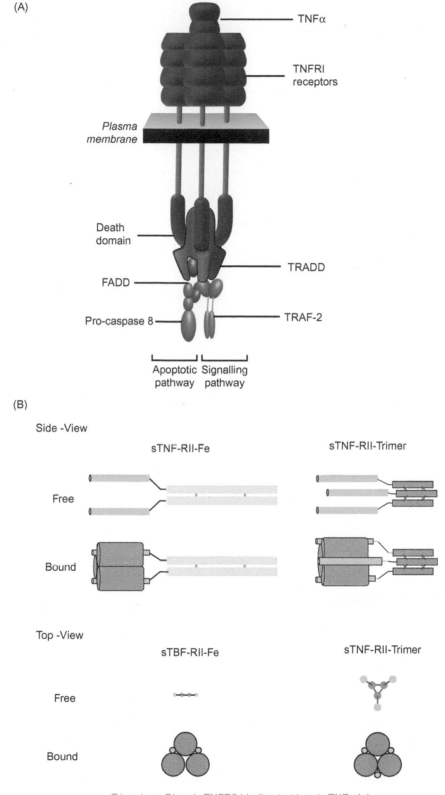

**FIGURE 17.6**   (A) A model of **TNFα receptor**. TNFα is shown bound to the extracellular ligand-binding domain of the transmembrane receptor. (B) Models of trimeric versus dimeric TNFR2 binding to **trimeric TNFα**. *(A) Reproduced from http://php.med.unsw.edu.au/cellbiology/index.php? title = File:TNFalpha_Receptor.JPG. (B) Reproduced from http://www.genhunter.com/cgi-local/news_display.pl?action = article&date = 1189549571.*

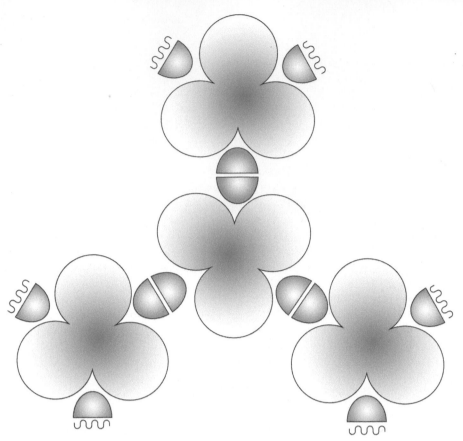

**FIGURE 17.7** Schematic representation of an aggregate of TNF homotrimers bound to parallel dimers of TNF receptor 1 (sTNFR1; *s* = soluble). TNF is shown as triangles and the receptor as semicircles. *Redrawn from Naismith, J.H., et al., 1996. Seeing double crystal structures of type 1 TNF receptor. J. Mol. Recog. 9, 113−117.*

### TABLE 17.2 Some Common Growth Factors and Their Properties

| Factor | Principal Source | Primary Activity | Comments |
|---|---|---|---|
| PDGF | Platelets, endothelial cells, placenta | Promotes proliferation of connective tissue, glial and other cells | Two different protein chains form three distinct dimer forms; AA, AB, and BB |
| EGF | Submaxillary gland, Brunners gland | Promotes proliferation of mesenchymal glial, and epithelial cells | |
| TGF-α | Common in transformed cells | May be important for normal wound healing | Related to EGF |
| FGF | Wide range of cells; protein is associated with the extracellular matrix | Promotes proliferation of many cells; inhibits some stem cells; induces mesoderm to form in early embryos | At least 19 family members, 4 distinct receptors |
| NGF | | Promotes neurite outgrowth and neural cell survival | Several related proteins first identified as protooncogenes, trkA [trackA], trkB, trkC |
| Erythropoietin | Kidney | Promotes proliferation and differentiation of cerythrocytes | |
| TCF-β | Activated TH1, cells\|T-helper\| and natural killer \|NK\| cells | Ami-inflammatory (suppresses cytokine production and class II MHC expression), promotes wound healing, inhibits macrophage and lymphocyte proliferation | At least 100 different family members |
| IGF-I | Primarily liver | Promotes proliferation of many cell types | Related to IGF-II and proinsulin, also called somatomedin C |
| IGF-II | Cells | Promotes proliferation of many cell types primarily of fetal origin | Related to IGF-I and proinsulin |

EGE, epidermal growth factor, FGF, fibroblast growth factor, IGF-I, insulin-like growth factor I, IGF-II, insulin-like growth factor II, MHC, major histo-compatability complex, NGF, nerve growth factor, PDGF, platelet-derived growth factor, IGF-α, insulin growth factor α, TGF-β, transforming growth factor β.
**Source:** Reproduced from http://web.indstate.claytheme/mwking/growth factors.html.

the receptor also exists in the cell membrane as a dimer. **PDGFRαβ** binds the AB and the BB forms of PDGF. **PDGFRββ** binds the BB and DD forms of PDGF (Figs. 17.8 and 17.9).

The three different PDGF receptors and the five different forms of PDGF are shown in Fig. 17.9.

Activation of these receptors by the binding of the PDGF dimer generates cell growth. The effects of PDGF can change the shape of the cell and affect its motility. PDGF-AA, PDGF-AB, PDGF-BB, PDGF-CC and PDGF-DD all act through the PDGF receptor α- and β-tyrosine kinases. **PDGF-AA** and **PDGF-BB** undergo intracellular activation during transport in the exocytotic pathway for subsequent secretion, whereas **PDGF-CC** and **PDGF-DD** are secreted as latent products that can become activated by **extracellular proteases**. PDGF-AA and PDGF-BB are active in many physiological and disease processes in target cells, especially those that originate from mesenchymal or neuroectodermal tissues. PDGF also mediates interactions between **glial cells** and it may play a role in the communication between **neurons** and glial cells and also between neurons. PDGF receptors (PDGFRs) are present on neurons of the **developing nervous system** and therefore PDGF is a mediator of intercellular signaling during the process of neuronal development. PDGFR may be present transiently as in the case of **ganglion cells**, for example, that possess **PDGFR** only when active outgrowth is occurring.

In **hepatic stellate cells** (versatile mesenchymal cells of the liver that are active in fibrosis and repair and contain large amounts of vitamin A in lipid droplets), PDGF activates its receptor that, in addition to stimulating growth through *Ras*-ERK, can induce **cyclooxygenase-2 (COX-2)** giving rise to increased **prostaglandin synthesis** and the elevation of **cAMP** (cyclic adenosine monophosphate) that inhibits the effects of *Ras*-ERK on **cell proliferation**. The activated PDGFR can also act through **PI3-K (phosphatidylinositol-3-kinase)** and **protein kinase B** to stimulate **cell survival** as well as proliferation. These activities are enhanced by intracellular sodium and calcium ions whose cellular transporters are also activated by signaling of the phosphorylated PDGFR.

In addition to its receptors, PDGF binds to other soluble proteins. **α₂-Macroglobulin** binds PDGF-BB but not PDGF-AA. This is a regulatory mechanism for controlling the amount of PDGF available to its receptor (much like the decoy receptors that bind TRAIL). **PDGF-associated protein** in the **neural retinal cell** binds PDGF with low affinity and enhances the activity of **PDGF-AA** but lowers the activity of PDGF-BB.

As PDGF is a dimer (Fig. 17.9), it binds two receptor monomers at once forming a bridge between them. The dimeric receptor complex is further stabilized, in addition to disulfide bonds, by one of the Ig (immunoglobulin) domains (domain 4) of the receptors. **PDGF-BB** activates **protein kinase C** that leads to the activation of the **mitogen-activated protein kinase (MAPK)** pathway.

In some cells, PDGF signaling can lead to the phosphorylation of ERK 1/2 (extracellular signal-regulated kinase 1/2) that can further phosphorylate **cytoplasmic phospholipase A₂ (cPLA₂)**. The activated cPLA₂ releases **arachidonic acid** from the cell membrane phospholipids and stimulated PKC (by PDGF-BB) activates membrane **NADPH oxidase**

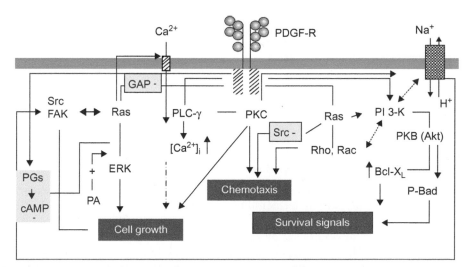

**FIGURE 17.8 PDGF signaling** in hepatic stellate cells. PDGF binds to a receptor with intrinsic **tyrosine kinase** activity. Receptor dimerization leads to autophosphorylation with formation of high affinity binding sites for signaling proteins with **SH-2** (*src* homology 2; phosphotyrosine binding) or **PTB** (phosphotyrosine-binding) domains. The downstream pathways are differentially implicated in the regulation of the biological activities of PDGF. *Reproduced from http://www.bioscience.org/2002/v7/d/pinzani/figures.htm.*

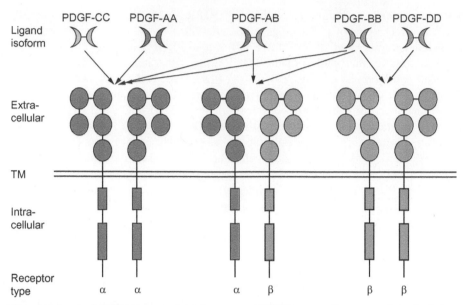

**FIGURE 17.9**   The different **isoforms of PDGF** dimers and the three types of PDGF receptor dimers are schematized. Arrows indicate the forms of PDGF that bind to the particular receptor type. *TM*, transmembrane. *Reproduced from Heldin, C.H., 2012. Platelet-derived growth factor. In: Encyclopedia of Cancer. Springer-Verlag, Berlin, pp. 2908–2910.*

that can affect **reactive oxygen species** (**ROS**) and ROS activates the **p38 MAPK pathway** leading to cell division and proliferation.

Some recent evidence suggests that, at least in certain cancer cells, PDGF regulated gene transcription involves alterations in **micro RNA** (noncoding RNA).

**Termination of the PDGF signal** occurs with the recycling of the PDGF receptors and this seems to be mediated by **protein tyrosine phosphatases** that remove the phosphate groups from the activated receptors thus inactivating them. This may not occur with all forms of the receptors. It is not completely *clear how PDGF itself is removed from the surface of the target cell but, clearly, the action of PDGF is initiated once its receptor has been activated.*

## Epidermal Growth Factor

There exist high affinity cell surface receptors for EGF having **tyrosine kinase** catalytic centers in the cytoplasmic domain. When EGF binds to and activates the receptor, its tyrosine kinase becomes activated and autophosphorylates the receptor and other proteins in the signaling pathway. Cells that are derived from **mesoderm** and **ectoderm** respond to EGF and proliferate. In the case of **PDGF**, it becomes cleared away from the target cell on its surface. EGF, on the other hand, is absorbed into the cell by **endocytosis**. This uptake process permits the down-regulation of the ligand-receptor complex after the **signal for mitosis** has been transmitted through the signal pathway. The cellular uptake process is mediated by the protein **clathrin**. This process involves the formation of a clathrin-coated vesicle. Different proteins interact with the vesicle that interacts with the cell membrane to form a **clathrin-coated pit** that becomes the internalized **endosome** (Figs. 17.10 and 17.11).

The internalized endosome is either degraded by transport into the **lysosome** or it is recycled to the cell surface via the **Golgi network** (Fig. 17.11). The clathrin molecule forms a triskelion as shown in Fig. 17.12. This figure also illustrates the clathrin-coated pit.

At first, the **clathrin network** is developed as an outline structure that is subsequently filled in to form a coat surrounding the complete vesicle. The network contains 36 triskelia in a structure composed of pentagons and hexagons (Fig. 17.12C). Ultimately, the removal of clathrin is accomplished by **uncoating ATPase** (**hsc70**, heat shock cognate 70). Three of these ATPases bind to one triskelion when ATP is absent but not when ATP is present. In the presence of ATP, the ATPase is activated and the clathrin molecules disassemble.

The signal transduction pathway initiated by the activation of EGF receptors upon binding EGF is shown in Fig. 17.13.

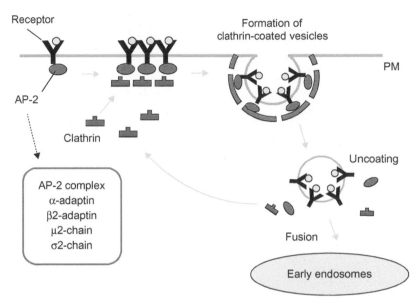

FIGURE 17.10 **Clathrin-dependent endocytosis**. Clathrin and cargo molecules are assembled into clathrin-coated pits on the plasma membrane (PM) together with an adapter complex called **AP-2** that links clathrin with transmembrane receptors concluding in the formation of mature **clathrin-coated vesicles (CCVs)**. CCVs are then actively uncoated and transported to **early sorting endosomes**.

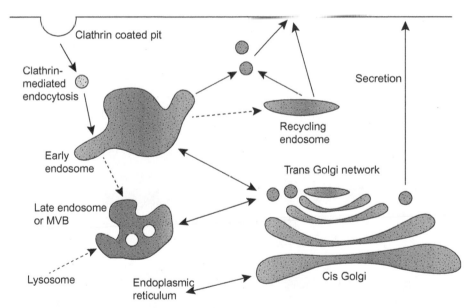

FIGURE 17.11 Diagram of the movement of the clathrin-coated pit, containing a ligand-receptor complex, from the cell surface to interior structures. *Solid arrows* indicate **trafficking events** mediated by vesicle transport, whereas *dashed lines* show events mediated by direct fusion or fission of organelles. *MVB*, multivesicular body. *Reproduced from Figure 1 of Owen, D.J., 2004. Linking endocytic cargo to clathrin: structural and functional insights into coated vesicle formation. Biochem. Soc. Trans. 32, 1—14.*

Two molecules of **EGF** bind to EGF receptors that dimerize on the surface of the target cell. The activated receptor dimer autophosphorylates in the receptor cytoplasmic domains and the activated receptor phosphorylates other proteins as well. A number of pathways become activated including **mTOR (mechanistic target of rapamycin)**, **STAT (signal transducer and activator of transcription)**, and the proliferation pathway through **ERK (extracellular signal-regulated kinase)** or **MAPK (mitogen-activated protein kinase)**. These activated signaling pathways lead to cell proliferation and to other effects as shown in the figure.

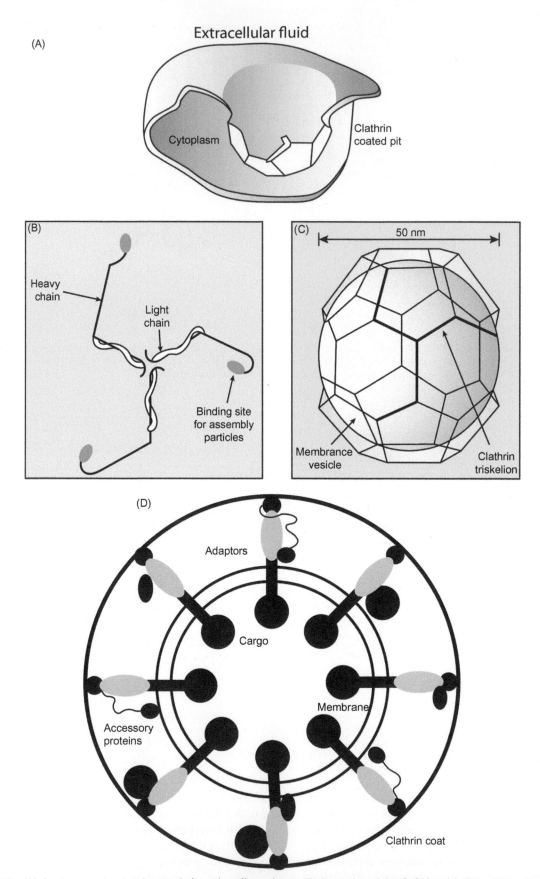

**FIGURE 17.12**   (A) Appearance of a clathrin-coated pit on the cell membrane. (B) Formation of the **clathrin triskelion**. (C) Appearance of a clathrin membrane vesicle formed from clathrin triskelia. (D) An electron micrographic reconstruction of a clathrin-coated vesicle.

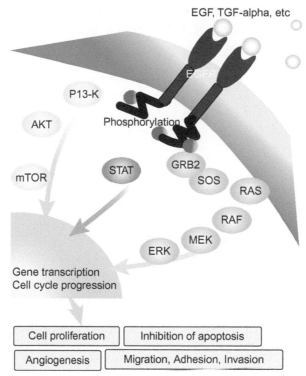

**FIGURE 17.13**    The signaling pathway of certain growth factors, such as EGF.

## Transforming Growth Factor

There are two forms of this growth factor, **TGFα** and **TGFβ**. Several other proteins are related to this growth factor by having extensive similarities in their amino acid sequences. Among these are the **activins** (activins A, B, and AB) and **inhibins** (inhibins A and B), the **Mullerian-inhibiting substance** and **bone morphogenic protein**. There are about 100 other proteins that contain stretches of sequence homology to TGFβ. Receptors for TGFβ have **serine/threonine kinase** activity in their cytoplasmic domains [differing from receptors for EGF, PDGF and FGF (fibroblast growth factor) all of which have protein tyrosine kinase activities]. TGFβ can inhibit growth of endothelial cells, macrophages, and T- and B-lymphocytes.

The **TGFβ gene family** evolved from a common ancestral gene and the family consists of: TGFβ1 (identical to TGFβ4), TGFβ2, TGFβ3, and TGFβ5. TGFβ2 is a homodimer where the two monomers are joined by disulfide bonds. The monomer is made up of 112 amino acid residues having a molecular weight of 12,720. TGFβ2 has 71% identity in amino acid sequence to TGFβ1 and has about 35% identity to the inhibins and activins and to the Mullerian-inhibiting substance. A comparison of the amino acid sequences of **human TGFβ1** and **TGFβ2** is shown in Fig. 17.14.

There are three types of receptors to which TGFs bind. TGFβ binds to the type II receptor, a binding that can be amplified by the presence of the type III receptor. TGFβ binds to TGFβRII dimer and TGFβRII recruits TGFβRI to form a heterodimer and the receptor II phosphorylates the receptor I and activates it. As mentioned, the receptors are serine/threonine kinases and the catalytic domain is located in the cytoplasmic domains of the receptors. The extracellular domains are rich in cysteine residues. There is a region called the GS domain in which about 30 serine-glycine repeats reside.

In the **SMAD signaling pathway** that is another pathway than the ones shown in Fig. 17.13, there are several regulated SMADs. The TGFβ family of growth factors interact sequentially with two membrane receptors, first with type II receptor and then the ligand-receptor complex interacts with the type I receptor. The type II receptor phosphorylates the type I receptor on serine residues mainly. To continue the signaling process downstream, activated TGFβ type I receptor (TGFR I) then phosphorylates Smad 2 and SMAD 3 and these form heterodimeric or heterotrimeric complexes with Smad 4 that translocate to the nucleus and initiate the transition from epithelial to mesenchyme and other effects. This pathway is pictured in Fig. 17.15.

Some genes are upregulated and a number are downregulated; these effects lead to **S-phase progression** in the cell cycle as would be expected for the effects of a growth factor.

```
                1     5        10        15        20
hTGF-β1   A L D T N Y C F S S T E K N C C V R Q L
hTGF-β2   A L D A A Y C F R N V Q D N C C L R P L


                      25        30        35        40
hTGF-β1   Y I D F R K D L G W K W I H E P K G Y H
hTGF-β2   Y I D F K R D L G W K W I H E P K G Y N


                      45        50        55        60
hTGF-β1   A N F C L G P C P Y I W S L D T Q Y S K
hTGF-β2   A N F C A G A C P Y L W S S D T Q H S R


                      65        70        75        80
hTGF-β1   V L A L Y N Q H N P G A S A A P C C V P
hTGF-β2   V L S L Y N T I N P E A S A S P C C V S


                      85        90        95        100
hTGF-β1   Q A L E P L P I V Y Y V G R K P K V E Q
hTGF-β2   Q D L E P L T I L Y Y I G K T P K I E Q


                    105       110
hTGF-β1   L S N M I V R S C K C S
hTGF-β2   L S N M I V K S C K C S
```

**FIGURE 17.14**  Amino acid sequences of human TGFβ1 and TGFβ2. The amino acids that are underlined in the sequence of TGFβ2 are the ones that deviate from the sequence of TGFβ1.

Transforming growth factor α (TGFα) is similar to EGF (30% sequence homology) and contains 50 amino acid residues and three disulfide bridges. TGFα binds to the EGF receptor (EGFR) and is a competitor of EGF at the receptor level. TGFα is secreted by human cancers. TGFα can act in concert with other growth factors, including TGFβ to produce cellular alterations. With EGF and its relatives, TGFα can play a role in wound healing and in tumor formation. The EGF pathway can be dysregulated when there is an excess of TGFα. Binding to the EGF receptor by TGFα either agonizes or dysregulates the EGF pathway and the direction of the effect depends upon the number of TGFα-EGFR complexes formed compared to the number of EGF-EGFR complexes formed. *Dysregulation of the EGF pathway, when an excess of TGFα forms many EGFR-TGFα complexes, may lead to tumor formation.* TGFα is expressed only in a few cell types, such as activated macrophages, keratinocytes (it is a growth factor for the keratinocyte), some other epithelial cells and a variety of carcinomas. The signaling pathway for TGFα is shown in Fig. 17.16.

## Fibroblast Growth Factor

FGFs in the human are numerous and number 23 factors, so far, that are active in proliferation and differentiation of many cells and tissues. All of these factors contain a core region consisting of 120 amino acids with six identical ones distributed within the core of all the FGFs. Stimulation of cell proliferation by these factors occurs in endothelial cells, chondrocytes, smooth muscle cells, melanocytes, and adipocytes. FGF induces specific interkeukins in some cell types and prolongs the survival of neuronal cells. In order for FGF signal transduction to occur, FGF must first associate with heparan sulfate proteoglycans on the target cell surface. The heparan sulfate links to the FGF receptor

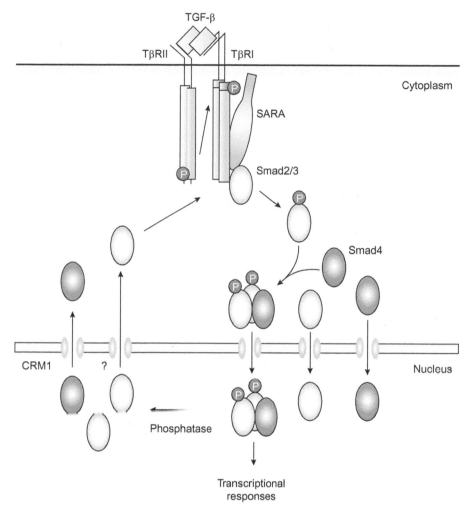

**FIGURE 17.15** The canonical transforming growth factor (TGF)-β/Smad signaling pathway. Members of the TGFβ family of growth factors [TGFβs, **activins**, **nodals** (members of the TGFβ family)] interact sequentially with two membrane receptors. TGF binds first to the constitutively active type II receptor (R) and then the ligand-receptor complex associates with type I TGFR. TGF IIR (TβIIR) phosphorylates TGF-IR (TβIR) on a cluster of serine/threonine residues. Activated TGF-RI propagates the signal downstream by directly phosphorylating **Smad 2** and **Smad 3**. These form heterodimeric or heterotrimeric complexes with **Smad 4** and translocate to the nucleus where, in combination with LEF-1/T cell factor (TCF) family transcription factors, they down-regulate **E-cadherin genes** and initiate **epithelial-mesenchymal transition**. Complexes of **Smad 7** and **Smurf (Smad ubiquitin regulatory factor)**1 or Smurf 2 promote **ubiquination** and degradation of activated receptors limiting the intensity and duration of signaling. *P*, phosphorylation site; *SARA*, small anchor for receptor activity. *Reproduced from http://arthritis-research.com/content/8/3/210/figure/F2.*

(FGFR) through an **Ig3 domain**. There are three Ig-like domains in the extracellular regions of the four EGFRs (EGFR1, EGFR2, EGFR3, and EGFR4). There is a single transmembrane domain and an intracellular split domain containing **protein tyrosine kinase activity**. The **domain structure of the FGFRs** is shown in Fig. 17.17.

FGFs are involved in promoting **cell division** but they also function in development, angiogenesis, hematopoiesis and tumorigenesis. FGFs have a mass of 18,000 Da and also there exist higher molecular weight forms of 21.5 and 22 kDa. The higher molecular weight forms have cell proliferation activity but also have additional activities. Some FGFs have molecular weights on either side of the range just indicated and FGF-1 is 7 kDa and FGF-5 is 389 kDa. The 22 human FGFs each have a fold in their structures consisting of 12 antiparallel β strands called a **β-trefoil motif** that is involved in FGF receptor recognition. In Table 17.3, the human FGFs and their receptors are recorded showing the binding of each human FGF to a specific receptor, although there are a few assignments still missing.

FGF-2 is an 18s-kDa protein with both intracellular and extracellular activities. FGF-2 also has a 24-kDa form that increases the expression of IL-6, whereas the 18-kDa monmeric form decreases IL-6 production. The 18-kDa form is secreted and binds to a cell membrane receptor and the 24-kDa form targets the cell nucleus. The secreted 18 kDa FGF binds to the target cell surface **heparan sulfate** or to matrix **glycosaminoglycans** where it can dimerize. FGF-2 target

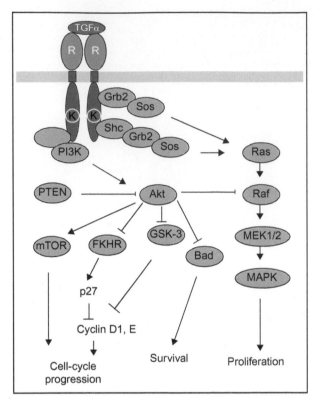

**FIGURE 17.16** Signaling pathways for TGFα. EGF receptors and members of the same receptor family, such as **HER-2 (human epidermal growth factor receptor-2**, a transmembrane **protein tyrosine kinase receptor)**, HER-3, and HER-4 are activated through dimerization. After receptor dimerization and activation of intrinsic protein tyrosine kinase activity, tyrosine autophosphorylation occurs, resulting in recruitment and phosphorylation of several intracellular substrates, leading to **mitogenic signaling** and other cellular activities. *Reproduced from http://imaging.ubmmedica.com/cancer-network/journals/oncology/images/unwinding/200404-480-1.gif.*

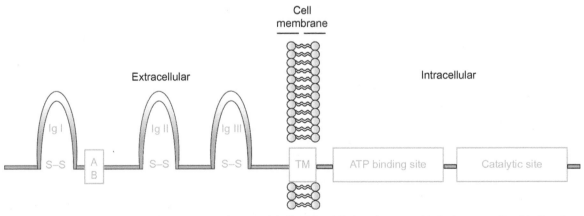

**FIGURE 17.17** Domain structure of the FGF receptor. *Ig*, immunoglobulin, *AB*, acidic box (interacts with the heparan sulfate binding site to prevent receptor activation in the absence of the FGF ligand), *TM*, transmembrane sequence. *Reproduced from http://jorde-lab.genetics.utah.edu/people/reha/Reha.html (no longer available on the Internet).*

cells can block the dimerization if they ribosylate the 18-kDa form in the receptor binding domain. Following cellular internalization, FGF-2 either can be translocated to the nucleus through a nuclear translocation motif in the N-terminus or degraded to active fragments of 4−10 kDa.

FGF-1 is expressed by cells derived from the three germ layers: ectoderm, mesoderm and endoderm. FGF-1 is released from the cell of origin as a dimer stabilized by disulfide linkages. It can be reduced to the monomeric form by reducing agents, such as glutathione. The monomeric form can bind either to a target cell membrane receptor or to

**TABLE 17.3** A List of the 23 Human FGFs and Their Binding Preferences for Four Receptors (FGFR1, FGFR2, FGFR3, FGFR4)

| Ligands | Receptors | | | | | | |
|---|---|---|---|---|---|---|---|
| | FGFR1 | | FGFR2 | | FGFR3 | | FGFR4 |
| | IIIb | IIIc | IIIb | IIIc | IIIb | IIIc | |
| FGF-1 | • | • | • | • | • | • | • |
| FGF-2 | • | • | | • | | • | • |
| FGF-3 | • | | • | | | | |
| FGF-4 | | • | | • | | • | • |
| FGF-5 | | • | | | | | |
| FGF-6 | | • | | • | | | • |
| FGF-7 | | | • | | | | |
| FGF-8a | | | | | | | |
| FGF-8b | | | | • | | • | • |
| FGF-8c | | | | | | • | • |
| FGF-8f | | | | • | | • | • |
| FGF-9 | | | | • | • | • | • |
| FGF-10 | • | | • | | | | |
| FGF 11 | | | | | | | |
| FGF-12 | | | | | | | |
| FGF-13 | | | | | | | |
| FGF-14 | | | | | | | |
| FGF-16 | | | | | | | • |
| FGF-17b | | | | • | | • | • |
| FGF-18 | | | | | | | |
| FGF-19 | | | | | | | • |
| FGF-20 | | | | | | | |
| FGF-21 | | | | | | | |

•, Binding.
FGF-1 Binds to All Four Receptors Including the Subtypes of FGFR1, FGFR2 and FGFR3.
**Source**: Reproduced from Table 3 of "Fibroblast Growth Factors," R & D Systems and from former Internet site http://www.rndsystems.com/asp/g_sitebuilder.asp?bodyid = 308.

heparan sulfate of the extracellular matrix that acts as a storage depot. FGF-1 can be released from this depot as needed and can interact with any of the four **FGF receptors** (Table 17.3). After binding to the cell membrane receptor, FGF-1 is internalized and translocated to the nucleus via the **nuclear localization motif** located between amino acids 22 and 30 of its sequence.

The FGF receptor consists of three immunoglobulin-like domains in its extracellular part, an acidic box at the amino terminus and a tyrosine kinase catalytic center in the cytoplasmic domain (Fig. 17.17). Alternative splicing can give rise to up to 14 isoforms. There are several known mutations that can occur in FGF receptors and these mutations are associated with various diseases some of which are summarized in Table 17.4.

The **signaling pathway of FGFs** is shown in Fig. 17.18.

**TABLE 17.4** Mutations of Human FGF Receptors Associated With Various Disease Syndromes

| Affected Receptor | Syndrome | Phenotypes |
|---|---|---|
| FGFR1 | Pfeiffer | Broad first digits, hypertelorism |
| FGFR2 | Apert | Midface hypoplasia, fusion of digits |
| FGFR2 | Beare Stevenson | Midface hypoplasia, corrugated skin |
| FGFR2 | Crouzon | Midface hypoplasia, ocular proptosis |
| FGFR2 | Jackson-Weiss | Midface hypoplasia. foot anamolies |
| FGFR2 | Pleiffer | Same as for FGFR1, mutations |
| FGFR3 | Crouzon | Midface hypoplasia, *acanthosis nigricans*, ocular proptosis |
| FCFR3 | Nonsyndromatic craniosynostosis | Digit defects, hearing loss |

**Source**: Reproduced from http://web.indstate.edu/theme/mwking/growth-factors.html (file no longer available on the Internet).

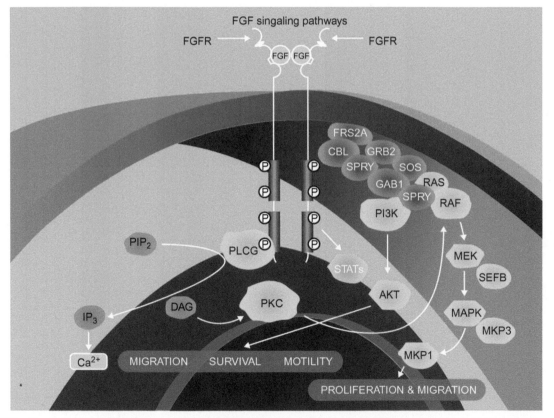

**FIGURE 17.18** The dimerized FGFR is shown at the *top* with two molecules of FGF bound. There are two tyrosine kinase domains in the cytoplasmic domain that become activated following autophosphorylation by the activated receptor. The MAPK sequence (*right* side of Figure) follows the phosphorylation of key nuclear proteins and leads to proliferation and migration of the target cell. Protein kinase C also can become activated by the FGFR complex through the stimulation of phospholipase C$\gamma$ (PLCG) by the activated FGFR. *Reproduced from http://www.lillyoncologypipeline. com/Pages/cancer-signaling.aspx.*

FGFRs are members of families of **tyrosine kinase-containing receptors**. In addition to **FGFR**, there are several other members including EGFR, PDGFR (platelet-derived growth factor receptor), VEGFR (vascular endothelial growth factor receptor), human growth factor receptor, **TrKR (tropomyosin receptor kinase receptor)**, **Elk/ERKR (ETS domain-containing protein/extracellular regulated kinase)** and **Axl (AXL tyrosine kinase receptor)**, all members of this family (Fig. 17.19).

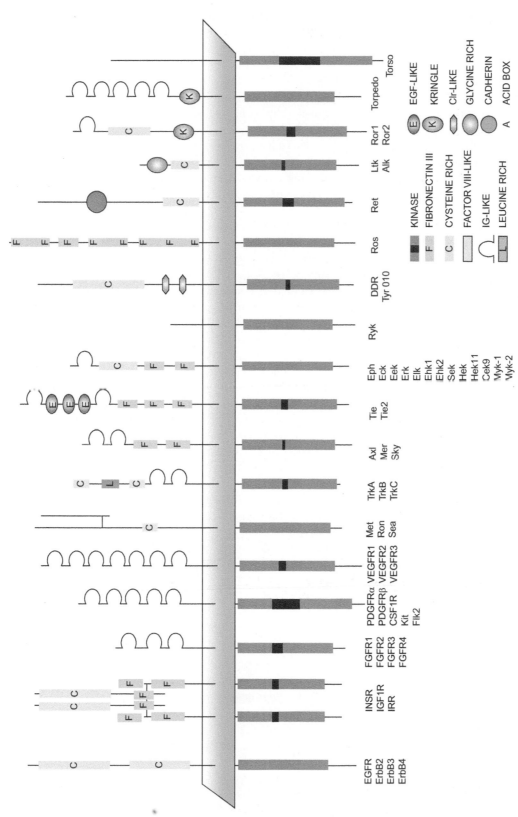

**FIGURE 17.19** Receptor tyrosine kinase families. The *blue* structure in the center represents the cell plasma membrane. The structural features are indicated in the key (*lower right*). *Reproduced from http://www.ludwig.edu.au/angiogenesis/image35.gif (no longer available on the Internet).*

## Neurotrophins and Nerve Growth Factor (NGF)

The neurotrophins (NTs) consist of **nerve growth factor** (**NGF**), **brain-derived neurotrophic factor** (**BDNF**), **neurotrophin-3** (**NT-3**) and **neurotrophin 4/5** (**NT-4/5**). These are involved in some positive activity on nervous tissue. In essence, *they protect neurons from **apoptosis** so that their presence or deficiency determines the number of neurons in a tissue or organ.*

Neurotrophins are released from a **postsynaptic cell** after an action potential has occurred in that cell. The released neurotrophins then diffuse backwards across the synapse and become absorbed by the presynaptic neuronal terminals. The neurotrophins in the **presynaptic cell** terminals evoke the branching of those terminals so that new synaptic connections are made. The neurotrophins thus cause growth in the presynaptic terminals.

During development, the pattern of connectivity is determined by the infusion of neurotrophins from a postsynaptic partner. Those neuronal endings that do not receive the stimulation from a partner postsynaptic cell would evoke cell death (apoptosis) of the presynaptic partner. The surviving terminals, under the influence of the neurotrophins, would sprout new connections and these processes lead to differentiated growth. *These activities occur in both the peripheral and central nervous systems.*

The neurotrophins act by binding to receptors on nerve terminals and also on the neuronal cell body wherein the nucleus is located. If the binding occurs on the surface of the cell body, the signal can be carried through the long axon to the target cell body where it can leave its instructions. There are two types of receptors for neurotrophins, a **low affinity neurotrophin receptor** (**LANR**), named **p75** and three high affinity receptors: TrkA, TrkB and TrkC that contain protein tyrosine kinase catalytic centers. The high affinity receptors have a binding constant in the range of $10^{-11}$ (0.01 nM). **TrkA** is specific for **NGF**; **TrkB** binds both **BDNF** and **NT-4** (NT-4 also binds to TrkA with lower affinity). **TrkC** binds **NT-3** (NT-3 can also bind to TrkA and TrkB with lower affinity than it binds to TrkC). There appear to be **isoforms of both TrkB and TrkC** that lack the activity of protein tyrosine kinase. These isoforms could resemble the decoy activity corresponding to the TRAIL decoy receptors. This buffering effect may be part of the activity of the p75 receptor that also lacks protein tyrosine kinase activity, although p75 liganded with NGF has a signal transduction pathway leading to the activation of **NFkB** and the activation of **genes for survival**. p75 also activates other proteins that stimulate the activity of **sphingomyelinase** leading to the production of **ceramide** and through further signaling of ceramide to the activation of genes promoting **apoptosis**. Also, ligand binding to **p75** results in an increase in the number of TrkA binding sites and an increase in TrkA levels of autophosphorylation. Activated p75 additionally leads to the activation of **JNK** kinase. Conversely, TrkA activation inhibits p75 promoted signaling. The **neurotrophin receptors** and their specificities are summarized in Fig. 17.20.

**Nerve growth factor** (**NGF**) is a survival factor for neurons, especially for pain-transmitting neurons (nociceptive neuron). Moreover, NGF is upregulated in adults responding to pain stimuli and is required for **hyperalgesia** (extreme sensitivity to pain) that is associated with inflammation. Although **inflammatory hyperalgesia** can be reduced by the binding of NGF, a high level of NGF is required for the full hyperalgesic response. Among the neurotrophins, this activity is specific to NGF. Interestingly, **snake venom** is a rich source of NGF.

NGF forms a homodimer and this neurotrophin is a member of the **"cystine knot" family** that has as members NGF, TGFβ, PDGF, and hCG (human chorionic gonadotropin). In the case of the NGF dimer, it is stabilized mainly by hydrophobic interactions, whereas the other members of this family have dimers stabilized chiefly by disulfide bonds. The productive binding of NGF to its high affinity receptor, **TrkA**, leads to a signaling cascade within the target cell as shown in Fig. 17.21.

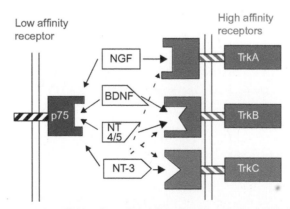

**FIGURE 17.20**   Neurotrophin receptors and their specificities. *Reproduced from http://betarhythm.blogspot.com/2008/09/neurotrophins.html.*

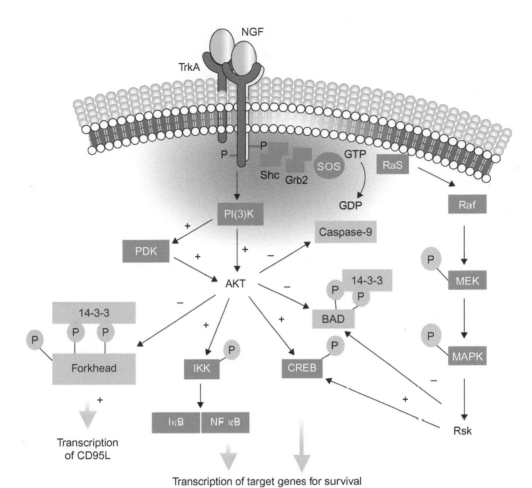

**FIGURE 17.21** Neuronal survival pathways induced by the binding of NGF to its receptor TrkA. NGF induces the autophosphorylation of TrkA which provides the docking sites for signal transduction molecules, such as **phospholipase Cγ (PLCγ)**, **phosphoinositide 3-kinase, (PI(3)K)** and the adapter protein **Shc**. Activated PI(3)K induces the activation of **Akt** through **3'-phosphorylated phosphatidylinositol** as well as **phosphoinositide-dependent kinase (PDK)**, which, in turn, phosphorylates and activates Akt. The phosphorylation of **CREB** and **IKK** stimulates the transcription of **prosurvival factors**, whereas the phosphorylation of **Bad**, **Forkhead** and **caspase-9** inhibits the proapoptotic pathway. In a parallel pathway, the interaction of **Shc-Grb2** and **SOS** activates the **Ras-Raf-MEK-ERK pathway** resulting in the activation of **Rsk (ribosomal s6 kinase)**. Bad and **CREB (cAMP response element binding [protein])** are also the targets of Rsk that might act synergistically with Akt to activate the survival pathway. *Reproduced from http://www.nature.com/nature/journal/v407/n6805/fig_tab/407802a0_F2.html.*

Each of the genes for the **neurotrophins** has been knocked out in order to further understand the physiological functions of a specific neurotrophin (Table 17.5).

Research is being conducted to utilize neurotrophins for the treatment of **macular degeneration** and **Parkinson's disease**.

## Colony Stimulating Factor

CSFs are involved in the differentiation of ancestral cells/parental cells into **hemopoietic cells** and are required for survival and proliferation. Each CSF is named for the cell type that it stimulates. Consequently the CSFs are named as follows: **macrophage CSF (M-CSF or CSF-1)**, **granulocyte CSF (G-CSF)**, **granulocyte-macrophage CSF (GM-CSF, CSF-2)** stimulates both cell types. In addition, there is a multi-CSF that is **interleukin-3 (IL-3)**. CSFs are soluble glycoproteins produced in the cells that line the blood vessels. In general, they stimulate **bone marrow** and they are used clinically for this purpose to mobilize stem cells to form **white blood cells**. CSFs have their receptors (members of the **IL-1 receptor family**) (CSF-1R) or the **cytokine receptor superfamily** (G-CSF-R and GM-CSFRβ) located on the surface of **hemopoietic stem cells**. When activated by ligand binding, signal pathways that ultimately result in

**TABLE 17.5** The Knockout of Each Member of the Neurotrophins to Shed Light on Physiological Functions

| Neurotrophin Knocked Out (Mice) | Consequences |
| --- | --- |
| Nerve growth factor (NGF) | Losses of certain peripheral sympathetic neurons |
| Brain-derived neurotrophic factor (BDNF) | Deficiency of sensory neurons |
| Neurotrophin 3 (NT-3) | Loss of proprioceptive neurons |
| Neurotrophin 4 (NT-4) | Loss of particular sensory neurons |

**Source**: Information in this table taken from http://betarhythm.blogspot.com/2008/09/neurotroohins.html.

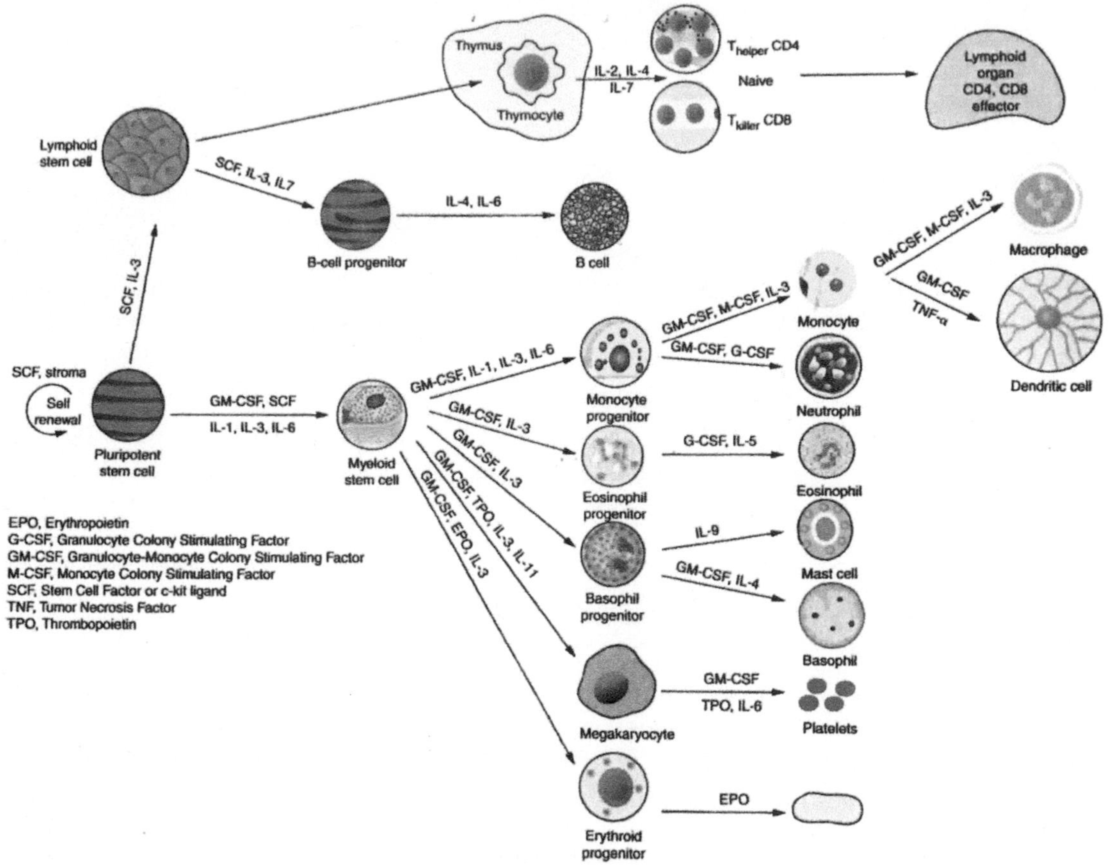

**FIGURE 17.22** **Formation of hemopoietic cells** from stem cell precursors showing the roles of **growth factors** and **cytokines** in the process of differentiation. GM-CSF is involved in many of the early steps in this process: conversion of **pleuripotent stem cell** to **myeloid stem cell**; conversion of myeloid stem cell to: **monocyte progenitor, eosinophil progenitor, basophil progenitor, megakaryocyte,** and **erythroid progenitors**. GM-CSF is further utilized in the transition of: monocyte progenitor to **monocyte**; basophil progenitor to **basophil**; megakaryocyte to **platelet**; monocyte to **macrophage** and monocyte to **dendritic cell**. G-CSF is involved in the conversion of monocyte progenitor to **neutrophil** and eosinophil progenitor to **eosinophil**. M-CSF is involved in the conversion of monocyte to **macrophage**. *This is a redrawn figure taken originally from http://edu.med.image. ch/apprentissage/module4/dif-imm/apprentissage/problemes/hemoporse-csf.jpg (no longer available on the Internet).*

proliferation and differentiation usually lead to a white blood cell. The receptor for M-CSF is CSF-1R is similar to **Flt3** (a tyrosine kinase receptor type III that plays a key role in **hematopoiesis**). **Flt3** is closely related to PDGF receptors. **GM-CSF-R** binds **IL-3** and **IL-5**, in addition to **GM-CSF**.

The formation of hemopoietic cells from precursor stem cells and the involvement of CSFs are shown in Fig. 17.22.

M-CSF (CSF1) participates in the formation and activity of **osteoclasts**, along with **RANKL (receptor activator of NFκB ligand)**. Osteoclasts function to break down bone; they are multinucleate cells that absorb bone matrix and act in bone remodeling in the formation of cavities or canals (Fig. 17.23).

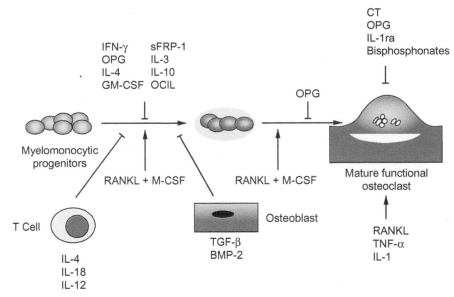

**FIGURE 17.23**  Formation of osteoclasts, their activity and the direct and indirect actions of inhibitors. This depicts **osteoclast differentiation** in response to M-CSF (CSF1) and RANKL. Direct inhibitors are listed above the proposed differentiation pathway and indirect inhibitors are listed below the pathway and they signal through intermediate cell types. *RANKL*, receptor activator of NFκB ligand; *IFN-γ*, interferon gamma; *OPG*, osteoprotegerin; *IL*, interleukin; *sFRP-1*, secreted frizzled-related protein 1; *OCIL*, osteoclast inhibitory lectin; *CT*, calcitonin; *BMP*, bone morphogenic protein. *Redrawn from Figure 1 of Quinn, J.M.W., Gillespie, M.T., 2005. Modulation of osteoclast formation. Vertebr. Skeletal Biol. 328, 739–745.*

As a differentiation factor, **M-CSF (CSF1)** increases **osteoclast motility**, reduces bone resorption by osteoclasts and is responsible for a number of signaling pathways involving cell survival, proliferation, attenuation of signaling by internalization and ubiquitination, differentiation and other activities, such as cell adhesion, spreading, polarization, motility and phagocytosis. Macrophage and **osteoclast development** as regulated by CSF1 (M-CSF) is schematized in Fig. 17.24.

The various signaling pathways under the control of colony-stimulating factor-1 (CSF1) or M-CSF are elaborated in Fig. 17.25.

**Granulocyte-stimulating factor** (**G-CSF**) signals target cells by activating the **JAK/STAT pathway**. In **cardiomyocytes**, for example, *G-CSF prevents cardiac remodeling after a myocardial infarction and thus plays an important role in recovery of the heart*. The action of G-CSF can produce **neuronal survival** by activation of the **ERK** system and, through activation of the **STAT pathway**, can prevent apoptosis through inhibition of caspases and induction of the antiapoptotic factors, such **as Bcl-xL**. These features are illustrated in Figs. 17.26 and 17.27.

**GM-CSF** promotes proliferation of granulocytes and macrophages. It operates through the GM-CSF receptor that recognizes GM-CSF, IL-3 and IL-5 as ligands. The receptor is a dodecamer (a dimer of two hexamers) and the activated receptor activates JAK2 which phosphorylates STAT5 and utilizes other paths including **ERK1/2**, **NFκB**, and **Akt** that bring about cell survival, proliferation and differentiation (Fig. 17.28).

**Cytokines** are produced by cells of the **immune system** and often affect the immune response. In some circumstances, cytokines can elicit cell proliferation, whereas **growth factors** generally induce mitosis, differentiation and proliferation. Because of the occasional overlap in cell proliferation, the distinction between the two may become blurry at times.

CSFs are **pleiotropic** (having multiple phenotypic actions) and elicit a range of biological responses including the following: initiation of cell division in responsive cells; suppression of programmed cell death (apoptosis); initiation of lineage commitment; and maturation in certain hemopoietic subpopulations and, additionally, they promote survival and functional activity of mature target cells.

## Erythropoietin

EPO is a glycoprotein cytokine-like hormone produced mainly in two locations, principally from the peritubular capillary epithelium of the **kidney** and to a lesser extent from liver **hepatocytes**. *It regulates the formation of red blood cells (erythropoiesis) by stimulating the proliferation and differentiation of immature erythrocytes to functional red blood cells*. It also induces the growth of erythroid progenitor cells and causes the differentiation of erythrocyte colony-forming units into **erythroblasts**.

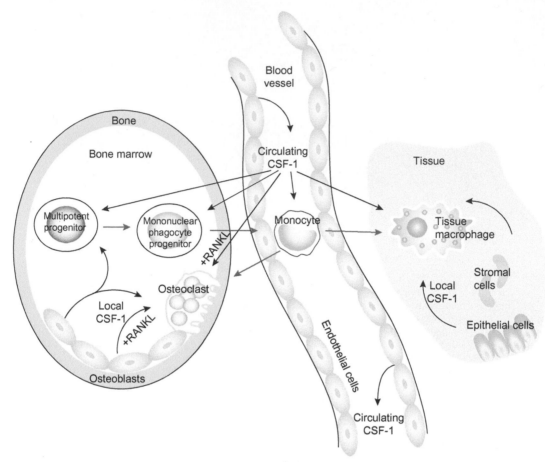

**FIGURE 17.24** Regulation of **macrophage** and osteoclast development by CSF-1. Circulating CSF-1 produced by endothelial cells in blood vessels, together with locally produced CSF-1, regulates survival, proliferation and differentiation of mononuclear phagocytes and osteoclasts. CSF-1 synergizes with **hemopoietic growth factors** to generate mononuclear progenitor cells from multipotent progenitors, and with **RANKL (receptor activator of NFκB ligand)** to generate osteoclasts from **mononuclear phagocytes**. *Red arrows* indicate steps in cell differentiation; *black arrows* indicate cytokine regulation.

The two organs responsible for the production of red blood cells are the **bone marrow** (production of red blood cell precursors from which maturation occurs) and the **kidney** that produces EPO.

EPO, consisting of 165 amino acids, binds to its receptors on the surface of bone marrow erythroid precursor cells signaling the rapid maturation of **erythroid precursors** to form mature red blood cells. This action results in an increase in the red blood cell count and an increase in blood oxygen content. The stimulation of the kidney's production and release of **EPO** is a result of reduced **red blood cell count** (**anemia**) that results in **hypoxia** which is the signal for the release of EPO (Fig. 17.29).

*Low oxygen content in the blood (anemia), where the red blood cell count is low, is a signal for the production and release of EPO from the kidney.* This reaction requires another kidney iron-containing protein that senses the inadequate supply of oxygen. This protein is a transcription factor called the **hypoxia-inducible factor** (**HIF-1**). HIFs are dimers of HIF-1α and HIF-1β subunits. The α-subunit is regulated by oxygen levels but the β-subunit is not. When the oxygen tension is normal the α-subunits are degraded rapidly so that they are transcriptionally inactive. In the face of lowered levels of oxygen, the α-subunits become stabilized and they translocate to the nucleus where they form heterodimers with HIF-βs and bind to **hypoxia response elements** (**HREs**) to activate gene transcription. The **oxygen sensors** that regulate HIFs have been found to be **oxygen-dependent hydroxylases (prolyl hydroxylase (PHD)** and **asparaginyl hydroxylase)**. Under conditions or normal oxygen tension, PHD modifies specific proline residues in HIF proteins that result in the recruitment of another protein, the **Von Hippel Lindau protein (pVHL)** and the HIF-α proteins are ubiquinated and degraded. Under normal oxygen conditions, asparaginyl hydroxylase (**FIH-1**) modifies HIF proteins, suppressing their transcriptional activity by preventing the interaction with transcriptional coactivators, such as **CBP/p300**. Under limiting oxygen conditions, both of the hydroxylases fail to modify the HIFs, so that HIFs are

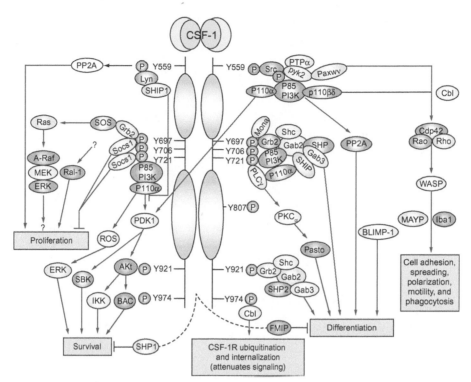

**FIGURE 17.25**   Signaling pathways regulated by colony-stimulating factor 1 receptor (CSF-1R) in **myeloid cells** (from bone marrow). Binding of CSF-1 stabilizes the dimeric form of **CSF-1R** and leads to activation of the **CSF-1R kinase**, its tyrosine autophosphoylation and the direct association of signaling molecules with the receptor through their **phosphotyrosine-binding domains**. Molecules shown touching specific phosphotyrosines are those that associate directly with a particular phosphotyrosyl sequence motif. The plasma membrane associations of several molecules are not shown for the sake of clarity. The precise involvement of the **Ras-MEK-MAPK pathway** in the CSF-1-regulated proliferation and differentiation of myeloid cells is not clear but **Raf-1** apparently signals independently in this pathway. Differences in signaling pathways are also expected to exist between **macrophage progenitor cells** and macrophages 25 and 26. BLIMP, B-lymphocyte-induced maturation protein; Cb1, Casitas B lineage; FMIP, FMS-interacting protein; Iba, ionized $Ca^{2+}$-binding adapter protein; IKK, IKB kinase; MAYP, macrophage actin-associated and tyrosine phosphorylated protein; MAPK, mitogen-activated protein kinase; MEK, MAPK kinase; Mona, monocyte adapter; P, phosphate; PDK1, 3'-phosphoinositide-dependent kinase; PI3K, phosphatidylinositol 3-kinase; PK, protein kinase; Pkare, PKA-related gene; PLC, phospholipase C; PP, protein phosphatase; Pyk, proline-rich and $Ca^{2+}$-activated tyrosine kinase; ROS, reactive oxygen species; SH, Src homology domain; SHIP, SH2-domain-containing polyinositol phosphatase; SHP, SH2-domain-containing phosphatase; SOS, Son of sevenless. *Redrawn from Figure 3 of Pixley, F.J., Stanley, E.R., 2004. CSF-1 regulation of the wandering macrophage: complexity in action. Trends Cell. Biol. 14, 628–638.*

transcriptionally active and the production of EPO and other gene products occurs. These activities are summarized in Fig. 17.30.

**EPO** consists of four helices stabilized by two disulfide bonds and other interactions in a topology that is shared with other cytokines. EPO binds to its receptor EPOR on the surface of cells in **blastocyst forming units** and **colony forming units** in the cell maturation pathway forming adult **red blood cells**. EPO binding results in the dimerization of two homodimers of EPOR (EPOR preexists as homodimers). The activated EPOR signals an intracellular cascade of phosphorylations through JAK2 and STAT5. The binding of EPO to EPOR and the functions of the activated receptor are summarized in Fig. 17.31.

## Interferon

Viral infection causes tissue cells, macrophages and lymphocytes to release interferons (small proteinaceous cytokines). IFN released from the producer cell binds to receptors on the surface of adjacent cells to produce a signal cascade generating a specific protein inhibitor of viral protein synthesis, thus preventing the spread of the virus. There are three types of interferon: **IFN-α, IFN-β** (referred to as type I interferons) and INF-γ (referred to as type II interferon). IFN-α has as many as 20 subtypes but there are only single forms of IFN-β and IFN-γ. Information concerning these types of interferons appears in Table 17.6.

As interferons are not synthesized in normal, uninfected cells (there exists a repressor of the IFN gene promoter that may compete with the binding of coactivator), transcription of the **genes encoding INF (INF-α, INF-β)** is turned on

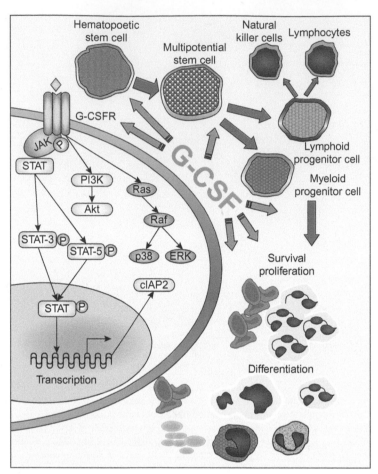

**FIGURE 17.26** Hemopoietic actions of granulocyte-colony-stimulating factor (G-CSF). G-CSF plays an important role in the survival, proliferation and differentiation of the hemopoietic cells. Also illustrated is the signaling pathway generated by the activated G-CSFR which activates **Janus kinase (JAK/STAT)**, the **Ras/mitogen-activated protein kinase (MAPK)** and **phosphatidylinositol 3-kinase (PI3-K)/Akt pathways**. Activation of JAK2/STAT3 leads to increased expression of **cellular inhibitor of apoptosis protein2 (cIAP2)** in human neutrophils resulting in their survival. Activation of STAT3 by G-CSF also has been linked to **myeloid cell differentiation**. STAT5, however, may be involved in G-CSF-dependent cell proliferation. *Reproduced from http://www.bioscience.org/2007/v12/af/2095/figures.htm.*

after a cell has been exposed to an inducer, such as viral infection, the presence of double-stranded RNA, lipopolysaccharide or some bacterial components. **Mitogens** and **antigens** that would activate lymphocytes would induce the production of **INF-γ**. The general events following viral infection causing the induction of INF that binds to its receptor on a neighboring cell to block **viral reproduction** in that cell are shown in Fig. 17.32.

More detail on the signaling pathways of viral infection and the generation of IFNα (in this case) and the action of IFNα on a neighboring cell is given in Fig. 17.33.

Following **viral infection**, the cell under siege can produce INF and the result of INF binding to its receptor on a neighboring cell is to produce a variety of **antiviral proteins** that will prevent the replication of the virus in that cell. The activated INF receptor turns on transcription to achieve the appearance of the antiviral proteins. These proteins remain in the neighboring cell until they are degraded several days later. One of these proteins is **2′5′ OligoA synthase** that converts ATP into a unique polymer (**2′5′ OligoA**) containing **2′−5′ phosphodiester bonds**. This polymer activates **RNAse** that degrades viral mRNA. In addition, a protein kinase is produced which is activated by **autophosphorylation** in the presence of **double-stranded RNA**. The activated kinase phosphorylates elongation factor **eIF-2** to its inactive form. These events lead to the inhibition of protein synthesis that stops the replication of virus. Although the infection is halted the survival of the cell is jeopardized by the inhibition of its own protein synthesis, so there is something of a balancing act going on. While some of the cells may die in the process of self-protection from the viral infection, others, in which the replication of virus has been halted, will survive. IFNs can affect the immune response as the three forms of **IFN** can increase the expression **of class I MHC molecules** on all cells leading to enhanced recognition by Tc cells (**cytotoxic T cells**) that are capable of destroying virus infected cells. In addition, **IFN-γ** increases the

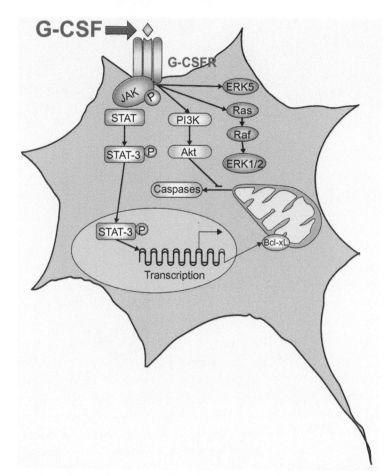

FIGURE 17.27 Signaling pathways in G-CSF mediated **neuronal survival**. Activation of the extracellular signal-regulated kinase (ERK) family enhances neuronal survival. Activated PI3K/Akt and STAT3 signaling pathways prevent apoptotic cell death by inhibiting activation of caspases and increasing antiapoptotic protein members, such as Bcl-xL. *Reproduced from http://www.bioscience.org/2007/v12/af/2095/figures.htm.*

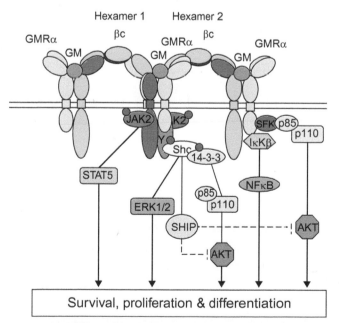

FIGURE 17.28 The high affinity complex of **GM-CSF receptor** (GMR in this figure) is a dodecamer comprising two ligand bound hexamers (the receptors of other CSFs are usually dimers of monomers). The central structure in the dodecamer complex enables **JAK2** transphosphorylation and the activation of **STAT5**- and **Shc**-mediated pathways. The proposed signaling through GMR-α occurs in αβ heterodimers (outer structures in the dodecamer complex), which facilitate activation of Akt and NFκB pathways. Also indicated is a possible negative feedback mechanism whereby activation of **SHIP** (**SH2-containing inositol 5-phosphatase**), downstream of JAK2, may suppress Akt activation. *Reproduced from http://bloodjournal. hematologylibrary.org/content/115/16/3346/F5.expansion.*

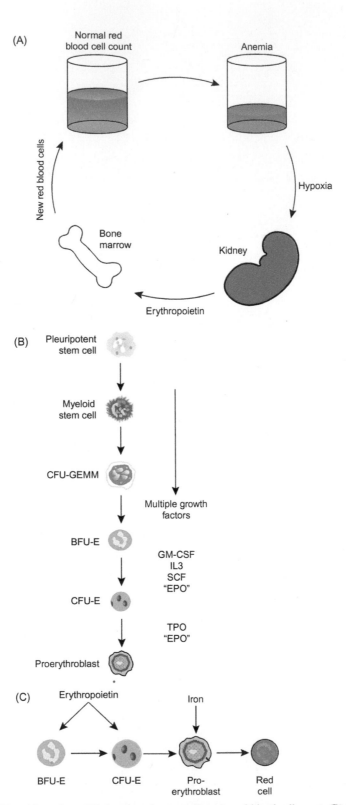

**FIGURE 17.29** The release of EPO and its actions. (**A**) A system that normalizes the red blood cell count. (**B**) Production of proerythroblasts from red cell precursors. EPO is required in the last two steps. *CFU*, colony forming unit; *BFU*, blastocyst forming unit; *GEMM*, granulocyte, erythrocyte, monocyte, megakaryocyte; *TPO*, thrombopoietin; *EPO*, erythropoietin; *SCF*, stem cell factor. (**C**) Maturation of stem cells under the influence of EPO and iron (iron is part of the hemoglobin structure), *EPO*, erythropoietin. *Redrawn from Figures 1, 2 and 3 of "The interaction of iron and erythropoietin", http://sickle.harvard.edu/epovionmodel.gif (no longer available on Internet). Also from Human Biochemistry and Disease by* G. Litwack (2008), *Figures 11−53, page 641.*

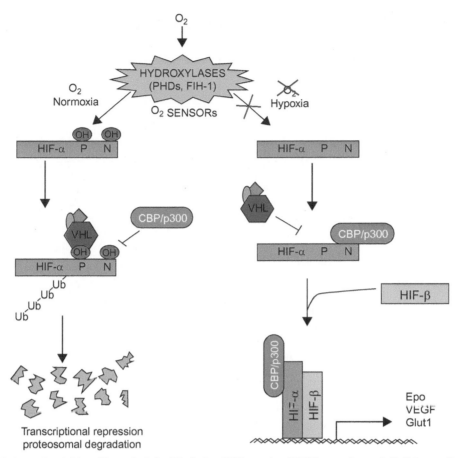

**FIGURE 17.30**  Regulation and activities of **hypoxia-inducible factor** (HIF) proteins. ***VEGF*, vascular endothelial growth factor**; ***Glut1*, glucose transporter** 1. *Reproduced from http://www.adelaide.edu.au/mbs/research/peet/.*

expression of **class II MHC molecules** on **antigen presenting cells** allowing for greater activity of $CD4^+$ **T helper cells** and activates **NK (natural killer) cells** that also kill **virus-infected cells**. As IFN is able to inhibit cellular protein synthesis (as long as its receptor is available on the surface of the cell), it is thought to be useful in inhibiting growth of some cancer cells.

## Insulin-Like Growth Factors (IGF-I, IGF-II)

These factors are structurally related to **insulin** but they function independently and are distinct antigens. IGF-I is 48% related to insulin (in terms of amino acid sequence) and 50% identical to IGF-II. They are induced by **growth hormone** (GH) in the liver (Fig. 17.34) as well as in other tissues (e.g., mammary gland).

GH is secreted from the **somatotrope** of the anterior pituitary following stimulation by hypothalamic **growth hormone releasing hormone (GHRH)**. GH circulates to the liver where it binds to hepatocyte **GH receptor**, causing the release of IGF-I (the mechanism is not clear). This factor is responsible for most of the actions of GH in the adult. IGF-II (important as a fetal growth factor) also is produced and, in addition, there are 6 **IGF binding proteins** (IGFBP-1, IGFBP-2, IGFBP-3, IGFBP-4, IGFBP-5 and IGFBP-6). The binding proteins modulate the actions of IGFs. They may inhibit the action of IGFs by sequestering them away from their receptors (similar to decoy action) or they may transport an IGF to its site of action. Possibly, a specific IGFBP may bind to an IGFBP receptor (by-passing the IGF receptors) to produce a specific action. IGF-I produces growth in a number of tissues, notably bone and muscle and also feeds back positively on the hypothalamic cell producing **somatostatin** that inhibits the release of GHRH, constituting the negative feedback system. When the blood levels of **GH** or **IGF-I** fall significantly, the positive feed forward system operates again causing the release of GHRH and the synthesis and release of GH, as before.

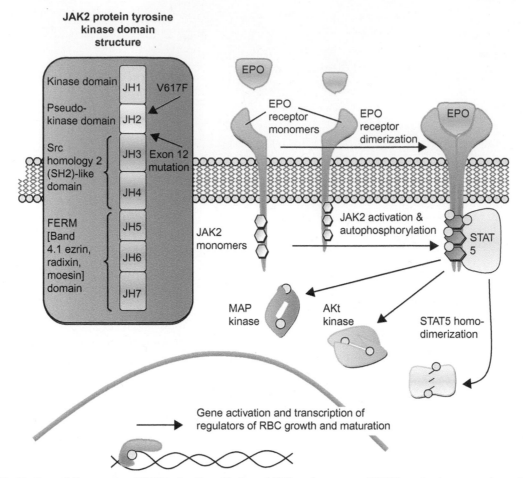

**FIGURE 17.31   Erythropoietin receptor (EPOR)** signaling. Binding of EPO to its receptor (EPOR) results in receptor homodimerization and autophosphorylation of the receptor-associated **Janus kinase 2 (JAK2)**. Activated JAK2 in turn mediates the phosphorylation of key tyrosine residues on the distal cytoplasmic region of EPOR, which then serve as docking sites for downstream effectors, including **signal transducer and activator of transcription protein 5 (STAT5)** and **phosphatidylinositol 3-kinase (PI3 kinase)**. Activated STAT5 homodimerizes and translocates to the nucleus to affect gene transcription. *Reproduced from http://www.nature.com/leu/journal/v23/n5/fig_tab/leu200954f2.html.*

**TABLE 17.6  Types of Interferons, Their Cellular Sources and Properties**

|  | Interferon | | |
| --- | --- | --- | --- |
| Property | Alpha | Beta | Gamma |
| Previous designations | Leukocyte IFN | Fibroblast IFN | Immune IFN |
|  | Type I | Type I | Type II |
| Genes | >20 | 1 | 1 |
| pH2 stability | Stable | Stable | Labile |
| Inducers | Viruses (RNA > DNA) | Viruses(RNA > DNA) | Antigens, Mitogens |
|  | dsRNA | dsRNA |  |
| Principal source | Leukocytes, Epithelium | Fibroblasts | Lymphocytes |

**Source**: Reproduced from http://pathmicro.med.sc.edu/mayer/vir-host2000.htm.

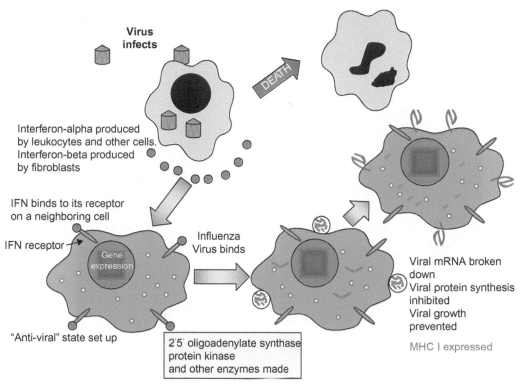

**FIGURE 17.32** Actions of interferon in a virus infected cell and its neighboring cell. *MHC1*, major histocompatability complex class I. *Reproduced from http://pathmicro.med.sc.edu/mayer/v-h3.jpg.*

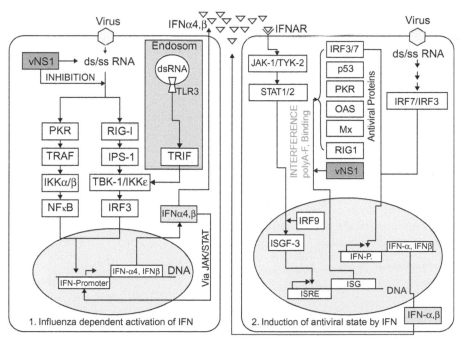

**FIGURE 17.33** Activation of interferon-induced antiviral state by **influenza virus replication** (1). In response to **viral RNA** cells start to produce and secrete type I IFNs. These cytokines activate **IFN receptors (IFNAR)** of other cells, which results in the induction of the **JAK/STAT pathway**, initiating the transcriptional activation of **ISRE (IFN stimulated response element)** controlled genes. These include several genes with antiviral activity. In addition, the IFN-stimulated upregulation of **IRF7 (interferon regulatory factor 7)** further enhances production of IFNs. As a consequence of this positive feedback loop even small stimuli are sufficient for a strong activation of the IFN-signaling. *Reproduced from http://www.mpi-magdeburg. mpg.de/research/projects/bpe/MolBio/IFN/index.en.html?pp = 1.*

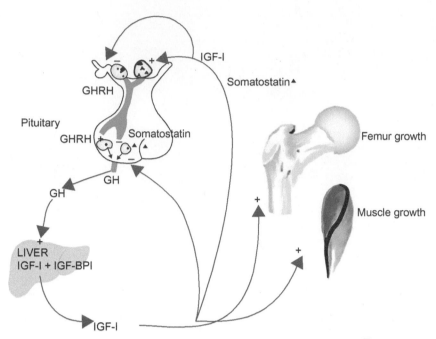

**FIGURE 17.34** Pituitary GH and hepatic IGF-I secretion from the liver in the hypothalamic-pituitary feedback system, IGF-1 feeds back negatively on the hypothalamus by stimulating the somatostatinergic cell that inhibits the release of GHRH. *IGF-BPI*, insulin-like growth factor binding protein 1; *GHRH*, growth hormone releasing hormone. *Reproduced from http://www.zuniv.net/physiiology/book/images/30-1.jpg.*

The six IGF binding proteins, the IGF receptors and a receptor for the binding protein, **IGFBP-3**, are shown in Fig. 17.35.

The **signaling pathway of IGF-I** is of importance in the adult, whereas there would be no signaling from the **IGF-II** molecule expected in the adult. The signaling pathways of IGF-I occur through the activation of the **Akt-mTOR pathway** and through the **ERK-ELK** and **MAPK** pathways for promotion of proliferation and survival that are reviewed in part in Fig. 17.36.

The **IGF-1 receptor** is similar to the insulin receptor. The **IGF-II** is the **mannose-6-phosphate receptor** that operates almost exclusively in embryonic and neonatal tissues as a transport protein. It integrates enzymes containing mannose-6-phosphate residues to the **lysosome**. At birth, the level of the IGF-II receptor falls dramatically. Although in human **keratinocytes**, IGF-II induces the expression of **COX2** through tyrosine kinase-Src-ERK and through tyrosine kinase-PI3K pathways but not through the MAPK pathway suggesting that IGF-II may play a role in the inflammatory process in some cells. In rheumatoid arthritis, IGF-II may spur the growth of **fibroblast-like synoviocytes** to increase inflammation in tissues of low inflammatory synovia.

Fig. 17.37 shows the ligands that bind to cognate receptors in the IGF system.

In addition, the **IGFBP-3** binds to the **IGFBP-3R** whose activation leads to inhibition of cell growth. The half-lives of free IGF-I in the bloodstream, IGF bound to IGF-I bound to IGFBP-1 and IGF bound to IGFBP-2 are about 10 minutes. IGF-I bound to IGFBP-3, however, has a half-life of more than 6 hours. The molecular weight of IGFBP-3 is about 40,000 and it binds an **acid-labile protein** of about 100,000 molecular weight producing a complex so large that it cannot be excreted intact; thus, bound IGF in this complex does not escape from the bloodstream until it is degraded.

## Interleukins

There are some 37 interleukins known, so far, and the list continues to grow. They are cytokines secreted by lymphocytes (also called **lymphokines**). ILs affect cellular processes in cells of hemopoietic origin, especially lymphocytes. Table 17.7 lists 26 of the 37 ILs describing their cellular origins and their primary activities.

IL-1 stimulates the activation of **T cells** leading to the production of T cell IL-2. IL-1 induces INF-γ. IL-1 actually is a family of 11 proteins as shown in Table 17.8.

IL-1α and IL-1β both bind to the same receptor (IL-1RI) along with IL-1Rα that is an **IL-1 receptor antagonist**. This antagonist binds to the receptor, competing with the binding of IL-1α or IL-1β, and its complex with the receptor

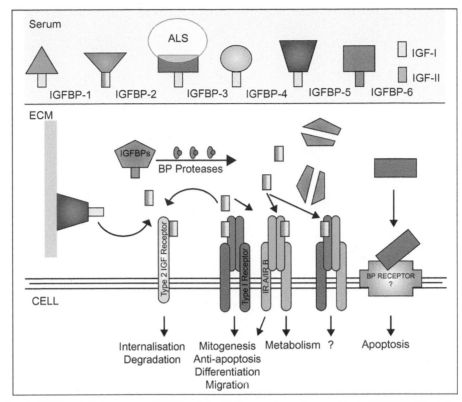

**FIGURE 17.35** The ligands (IGFs) and their receptors (IGFRs) and the binding proteins and a receptor for IGFBP-3. The IGF axis consists of IGF-I and IGF-II, the type I and type II IGF receptors and six known IGF binding proteins (IGFBP-1 through IGFBP-6). The IGFs bind both their receptors and binding proteins with high affinity. IGFs are important mitogenic factors involved in cell proliferation and metabolism. Locally produced IGFs and IGFBPs regulate tissue growth and differentiation. IGFBPs are thought to modulate the action of IGFs in several ways, including an inhibitory model in which the IGFBPs sequester IGFs from their receptors, an enhancing model in which IGFBPs transport IGFs to their site of action, or by a receptor-independent model that may involve direct interaction of IGFBPs with IGFBP receptors. IGFBPs are regulated by various endocrine factors and are expressed in specific ontogenic patterns. The degree of growth inhibition caused by IGFBPs appears to be directly related to their concentrations relative to IGFs. The modulation of IGF levels by IGFBPs is further regulated by IGFBP proteases that cleave the high affinity IGFBPs into fragments with lower affinity for IGFs, thereby increasing free IGF bioavailability. This process lowers the inhibition of cell growth by IGFBPs. In addition, IGF-independent actions of IGFBPs affecting cell survival and apoptosis have been described. *IGF-I*, yellow rectangle; *IGF-II*, *orange* rectangle; *split color rectangle*, IGF-I and IGF-II both bind; *ALS*, acid-labile subunit; *IR-A*, insulin receptor A isoform; *IR-B*, insulin receptor B isoform, *hybrid*, IGF type I receptor coupled with insulin receptor. *Reproduced from http://www.igf-society.org/joomla15/index.php?option = com_content&view = category&layout = blog&id = 1&ltemid = 7 or http:// spider.science.strath.ac.uk/spider/spider/modules/deptWeb/media/40/spider_image.jpg.*

(IL-1RI) does not lead to a signaling path. The ligands for IL-1RI (IL-1α/IL-1β) bind to the extracellular domain of IL-1RI; the receptor complex then recruits a coreceptor (IL-1RAcP) that is required for signal transduction to occur. The signaling pathway is shown in Fig. 17.38.

Other ILs have different signaling pathways. Some activate the **JAK/STAT pathway**; an example is **IL-9**. The sequence of the IL-9 receptor signaling complex is shown in Fig. 17.39A; the distinct T cells that express IL-9 are shown in Fig. 17.39B, and the cellular sources and immune functions of IL-9 are shown in Fig. 17.39C. Table 17.9 lists the sources and targets of IL-9 during immune responses.

The cellular sources and targets of IL-9 and the effects produced are shown in Table 17.9.

## Overview

A general view of the signaling mechanisms for all growth factors, cytokines and related factors is shown in Fig. 17.40.

## SUMMARY

**TRAIL (TNF-related apoptosis inducing ligand)** is a cytokine that can kill certain cancer cells but does not kill normal cells. This selectivity may be due to a higher concentration of **TRAIL decoy receptors** in normal cells or to a low

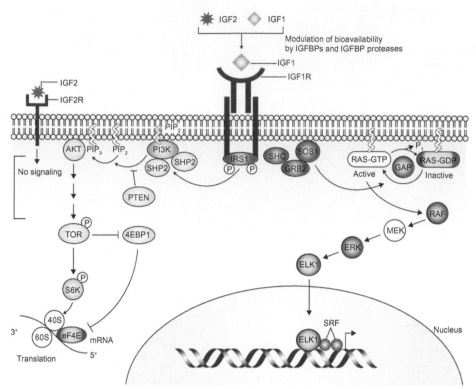

**FIGURE 17.36** Overview of **IGF-I receptor** (a tyrosine kinase cell surface receptor) activation and downstream signaling. The bioavailability of ligands to the receptor site is subject to complex physiological regulation (IGF-I concentration available to its receptor may be high in many cancers). **IGF binding proteins** either enhance or reduce the activity of IGFs as discussed above. Key downstream networks from the activated **IGF-1R** include the **PI3K-AKT-TOR system** and the **RAF-MAPK system** that stimulate proliferation and inhibit apoptosis. Stimulation of AKT relates to **cell survival** and translation of mRNA, whereas stimulation of the **RAS pathway** involves regulation of cell proliferation. *4EBP1*, eukaryotic translation initiation factor 4E binding protein 1; *eIF4E*, eukaryotic translation initiation factor 4E; *ERK*, extracellular signal-regulated kinase; *GRB2*, growth-factor-receptor-bound protein 2; *IRS1*, insulin receptor substrate 1; *MAPK*, mitogen-activated protein kinase; *MEK*, mitogen-activated protein kinase kinase; *PI3K*, phosphatidylinositol 3-kinase; *PIP*, phosphatidylinositol; *PTEN*, phosphatase and tensin homolog; *S6K*, S6 kinase; *SHC*, SRC-homology-2-domain transforming protein; *SHP2*, phosphatidylinositol 3-kinase regulatory subunit; *SRF*, serum response factor; *TOR*, target of rapamycin. *Reproduced from http://www.medscape.com/viewarticle/483288_2.*

content of the productive TRAIL receptors, or both. Consequently, this cytokine may prove to be useful in the clinical treatment of certain cancers. Of particular interest in this connection is **ovarian cancer**.

There are many ligands in the **tumor necrosis factor** (**TNF**) superfamily. TRAIL is a member of this family. These cytokines and their receptors are involved in the regulation of the **immune system**. Some of these factors may be involved in the **spontaneous repression of tumors**. There are 19 soluble and membrane-bound ligands and 32 receptors currently in the TNF superfamily.

Cytokines and growth factors are similar. In some cases, cytokines can stimulate the growth of cells; in other cases they promote **inflammation**, or, in the case of TRAIL and some TNF-related cytokines, they can kill tumor cells.

**Platelet-derived growth factor** (**PDGF**) is a mitogen for certain cell types, especially cells of **connective tissue** and cells of the **developing nervous system**. PDGF has four different subunits. This growth factor membrane receptor is either a homodimer or a heterodimer and has **tyrosine kinase activity**. There are five forms of PDGF and three types of the PDGF receptor. The PDGF signaling system leads to stimulation of cell growth. Termination of the PDGF signaling system is effected by **protein tyrosine phosphatases** that inactivate the autophosphorylated active receptors.

The **epidermal growth factor** (**EGF**) binds to its cell surface receptor that also has tyrosine kinase activity. The receptor autophosphorylates on receptor tyrosine residues and the activated receptor complex activates the signaling pathway. The result is the stimulation of proliferation of cells of mesodermal or ectodermal origin. The mechanism of turning off the activity of **EGF-EGFR complex** involves trafficking in which the receptor complex is moved into the cell interior through **clathrin-coated vesicles**. The resulting **endosome** is transported either to the **lysosome** for degradation or it is recycled to the cell surface by way of the Golgi network.

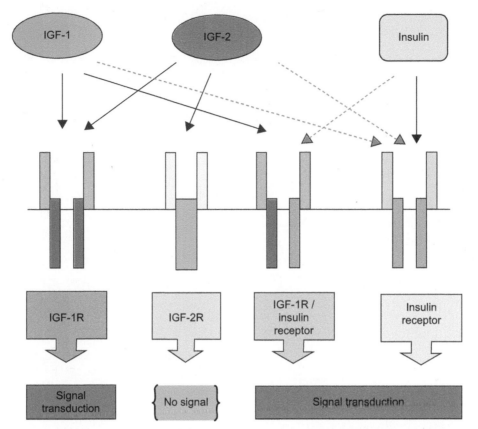

**FIGURE 17.37**   The IGF system. Binding of insulin (I) and IGF ligands to their receptors. Insulin receptor (IR) and IGF-IR are both tyrosine kinases. IGF-IIR functions as a clearance site for IGF-II. IR and IGF-IR are homologous and form hemireceptors. IGF-I binds to IGF-IR and to IGF-IR/IR hemireceptor; it binds to IR only at very high concentrations. IGF-II binds to IGF-IR, IGF-IIR and binds to IR only during early fetal development. I binds to IR and it binds to IGF-IR/IR hemireceptor at high concentration. Signal transduction is activated after the activation of IGF-IR, IGF-IR/IR hemireceptor and IR, however, IGF-IIR activation results in no signal downstream. *Solid lines*, high affinity binding; *dotted lines*, weak binding. *Reproduced from http://www.jhoonline.org/content/4/1/30/figure/F1?highres = y.*

**TABLE 17.7   List of 26 of 37 ILs Currently Known Indicating Their Cellular Sources and Primary Activities**

| Interleukins | Principal Source | Primary Activity |
|---|---|---|
| ILl-α and β | Macrophages and other antigen-presenting cells (APCs) | Costimulation of APCs and T cells, inflammation and fever, acute phase response, hematopoiesis |
| IL-2 | Activated $TH_1$ cells, NK cells | Proliferation of B cells and activated T cells, NK functions |
| IL-3 | Activated T cells | Growth of hematopoietic progenitor cells |
| IL-4 | $TH_2$ and mast cells | B cell proliferation, eosinophil and mast cell growth and function, IgE and class II MHC expression on B cells, inhibition of monokine production |
| IL-5 | $TH_2$ and mast cells | Eosinophil growth and function |
| IL-6 | Activated $TH_2$ cells, APCs, other somatic cells | Acute phase response, B cell proliferation, thrombopoiesis, synergistic with IL-1 and TNF on T cells |
| IL-7 | Thymic and marrow stromal cells | T and B lymphopoiesis |
| IL-8 | Macrophages, other somatic cells | Chemoattractant for neutrophils and T cells |
| IL-9 | T cells | Hematopoietic and thymopoietic effects |
| IL-10 | Activated $TH_2$ cells, $CD8^+$ T and B cells, macrophages | Inhibits cytokine production, promote B cell proliferation and antibody production, suppresses cellular immunity, mast cell growth |

*(Continued)*

**TABLE 17.7** (Continued)

| Interleukins | Principal Source | Primary Activity |
|---|---|---|
| IL-11 | Stromal cells | Synergistic hematopoietic and thrombopoietic effects |
| IL-12 | B cells, macrophages | Proliferation of NK cells, IFN-$\gamma$ production, promotes cell-mediated immune functions |
| IL-13 | TH$_2$ cells | IL-4-like activities |
| IL-15 | Endothelial cells and monocytes | Effects arc similar to IL-2 |
| IL-16 | CD8 T cells | Chemo attracts CD4 T cells |
| IL-17 | Activated memory T cells | Promotes T cell proliferation |
| IL-18 | Macrophages | Induces IFN-$\gamma$ production |
| IL-19 | Belongs to IL-10 family | Produces IL-6 and TNF$\alpha$; TNF$\alpha$ leads to apoptosis |
| IL-20 | Belongs to IL-10 family | Activates STAT pathway; epidermal function |
| IL-21 | Activated CD4$^+$ T cells | Affects NK and T cell responses; proliferation of activated |
|  |  | T cells; maturation of NK cells, with IL-15 or IL-18, enhances INF-$\gamma$ production in NK and T cells |
| IL-22 | Belongs to IL-10 family |  |
| IL-23 |  | Critical cytokine for autoimmune inflammation of the brain; recruits inflammatory cells |
| IL-25 | Related to IL-17 family, produced by mast cells and helper T cells | Induces production of IL-4, IL-5, and IL-13 |
| IL-26 | Member of IL-10 family; transformed T cells | Activates STAT 3 |
| IL-27 | Spleen cells | Proliferation of main CD4$^+$ T cells; enhances IFN-$\gamma$ production by activated T cells and NK cells; induces STAT 1 and STA T 3 phosphorylation |

**Source**: Originally reproduced from http://web.indstatethcme/mwking/growth-factors.html (no longer retrievable from the Internet) and from *Human Biochemistry and Disease* by G. Litwack (2008), Academic Press, Table 11−6, page 655.

**TABLE 17.8** The IL-1 Family. The Names of These Proteins are Given With Their Cognate Receptors, Coreceptors, Properties and Chromosomal Location

| Name | Family Name | Receptor | Coreceptor | Property | Chromosomal Location |
|---|---|---|---|---|---|
| IL-1$\alpha$ | IL-1F1 | IL-1R1 | IL-1RacP | Proinflammatory | 2q14 |
| IL-l$\beta$ | IL-1F2 | IL-1R1 | IL-1RacP | Proinflammatory | 2q14 |
| IL-1Ra | IL-1F3 | IL-1R1 | NA | Antagonist for IL-I$\alpha$, IL-1$\beta$ | 2q14.2 |
| IL-18 | IL-1F4 | IL-18R$\alpha$ | IL-18R$\beta$ | Proinflammatory | 11q22.2-q22.3 |
| IL-36Ra | IL-1F5 | IL-1Rrp2 | NA | Antagonist for IL-36$\alpha$, IL-36$\beta$, IL-36$\gamma$ | 2q14 |
| IL-36$\alpha$ | IL-1F6 | IL-1Rrp2 | IL-1RAcP | Proinflammatory | 2q12-q14.1 |
| IL-37 | IL-1F7 | Unknown | Unknown | Antiinflammatory | 2q12−q14.1 |
| IL-36$\beta$ | IL-1F8 | IL-1Rrp2 | IL-1RAcP | Proinflammatory | 2q14 |
| IL-36$\gamma$ | IL1-F9 | IL-1Rrp2 | IL-1RAcP | Proinflammatory | 2q12−q21 |
| IL-38 | IL-1F10 | Unknown | Unknown | Unknown | 2q13 |
| IL-33 | IL-1F11 | ST2 | IL-1RAcP | Th2 responses, proinflammatory | 9p24.1 |

**Source**: Reproduced from http://en.wikimedia.org/wiki/interleukin_1_family.

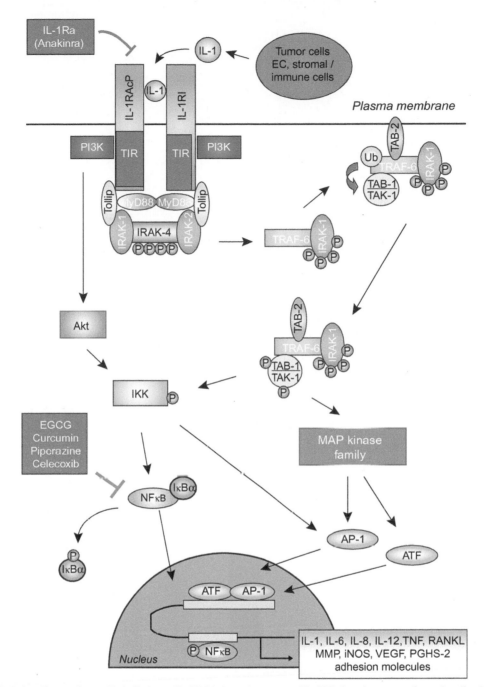

**FIGURE 17.38  IL-1 signaling pathway.** IL-1 (IL-1α or IL-1β) binds to the receptor (IL-1RI) that changes conformationally, followed by **coreceptor (IL-1RAcP)** binding to the complex to initiate signal transduction. Ligand-mediated heterodimerization of the receptor complex leads to recruitment of **dimeric myeloid differentiation protein 88 (MyD88)** via its **TIR (Toll/IL-1R) domain** followed by complex formation between **IRAK-4**, MyD88 and IL-1RAcP and subsequent phosphorylation of IRAK-4. After recruitment of **IRAK-1/Tollip** to the complex, IRAK-1 is initially phosphorylated by IRAK-4. Then IRAK-1 becomes hyperphosphorylated and dissociates into the cytoplasm where it binds **TNF receptor-associated factor 6 (TRAF-6)**. IRAK-1 interacts with membrane-bound **TAK-binding protein 2 (TAB-2)** as well as **TAK-1/TAB-1 complex** followed by translocation of TAB-2 from the plasma membrane to the **signalosome** and subsequent partial activation of TAK-1 by TAB-2. IRAK-1 as dimer or oligomer enables dimerization of TRAF-6 resulting in its **ubiquitination** and activation. TAK-1 is partially activated followed by complete activation through polyubiquitinated TRAF-6 enabling activation of numerous signaling cascades. Polyubiquitination of TRAF-6 occurs through IRAK-2. TAK-1 activates certain members of the **MAP kinase family** leading to activation of **AP-1** and **ATF (activating transcription factor)**, the latter augmenting NFκB-mediated transcription via transactivation. TAK-1 also phosphorylates and activates **IKK** resulting in phosphorylation and inactivation of IκBα. IκBα then dissociates from the complex with **NFκB** and undergoes proteosomal degradation. After phosphorylation, NFκB translocates to the nucleus and activates **NFκB-dependent gene transcription**. IL-1 signaling also involves recruitment of **PI3 kinase (PI3K)** to the IL-1 receptor complex via the p85 regulatory subunit of PI3K and subsequent activation of **AKT/PKB** leading to IKK-dependent activation of NFκB and AP-1. Receptor ligation also can activate numerous G proteins resulting in activation of AP-1 and ATF mediated by several **MAP kinases** and an IκBα-independent transactivation of NFκB. IL-1 signaling finally regulates gene expression of a great variety of tumorigenic factors including proangiogenic factors (**IL-8, VEGF**), growth factors (**IL-6, GM-CSF**), antiapoptotic factors (Bcl-X$_L$, c-FLIP), invasion-promoting factors (**MMP-2, MMP-7, MMP-9, uPA**), prostaglandins, **inflammatory enzymes (PGHS-2, LOX)**, iNOS, chemokines (**CCL2, CCL20, IL-8**) and proinflammatory cytokines (IL-1, IL-6, IL-23, TNF, TGF-β, EGF, RANKL). **Inhibitors of NFκB** are also indicated that suppress the inflammatory network. *Reproduced from http://c431376.r76.cf2.rackcdn.com/17950/fimmu-02-00098-HTML/image_m/fimmu-02-00098-g001.jpg.*

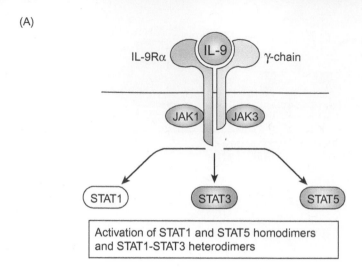

(A)

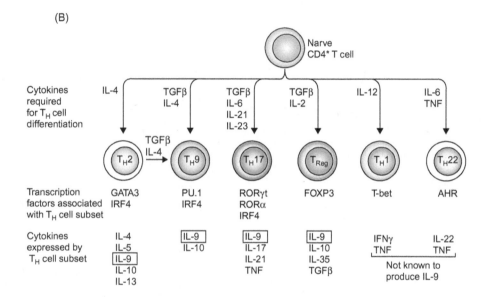

(B)

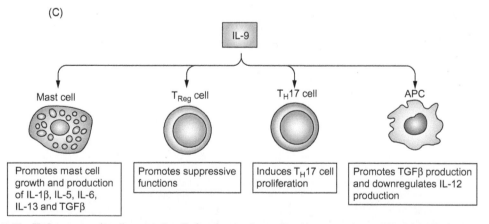

(C)

**FIGURE 17.39** (A) The **IL-9 receptor signaling complex**. IL-9 activates a heterodimeric receptor that consists of the IL-9 receptor α-chain (IL09Rα) and the γ-chain and promotes the crossphosphorylation of Janus kinase (JAK1) and JAK3. This leads to the activation of signal transducer and activator of transcription 1 (**STAT1**), STAT3 and **STAT5** and the up-regulation of IL-9 inducible gene transcription. (B) Differentiation pathways of effector **CD4⁺ T cell subsets** are shown, as well as the transcription factors that are necessary for their differentiation and some of the cytokines that they produce. IL-9 has been reported to be expressed by T helper 2 (T$_H$2), T$_H$9, T$_H$17 and regulatory T (T$_{Reg}$) cell subsets, and all of these subsets, except Th2 cells, require transforming growth factor-β (TGFβ) for IL-9 production. *AHR*, aryl hydrocarbon receptor, *FOXP3*, forkhead box P3, *GATA3*, GATA-binding protein 3, *INF-γ*, interferon-gamma; *IRF4*, interferon-regulatory factor 4; *ROR*, retinoic acid receptor-related orphan receptor, *TNF*, tumor necrosis factor. (C) Cellular sources and **immune functions of IL-9**. IL9 has been shown to have various effects on different cell types. These effects include activating **mast cells** to secrete several products, including **IL-13**, which exerts its effects on epithelial cells of the lung and gut. In addition, IL-9 seems to have a direct effect on regulatory T (T$_{Reg}$) cells, T helper 17 (T$_H$17) cells and antigen-presenting cells (APCs). *TGFβ*, transforming growth factor-β. *(A) Reproduced from http:// www.nature.com/nri/journal/v10/n10/fig_tab/nri2848_F1.html#figure-title. (B) Reproduced from http://www.nature.com/nri/journal/v10/n10/fig.tab/ nri2848_F2.html#figure-title. (C) Reproduced from http://www.nature.com/nri/journal/v10/n10/fig_tab/nri2848_F3.html#figure-title.*

**TABLE 17.9 Sources and Targets of IL-9 During Immune Responses**

| Immune Model | Sources of IL-9 | Targets of IL-9 | Effects of IL-9 |
|---|---|---|---|
| **Allergy** | | | |
| Allergic airway inflammation | NKT cells and $T_H9$ cells | Mast cells | Promotes allergic inflammation |
| Oral antigen-induced anaphylaxis | $T_H2$ cells and $T_H9$ cells | Mast cells | Promotes allergic inflammation |
| **Autoimmunity** | | | |
| EAE | $T_H17$ cells and $T_H9$ cells | Mast cells and $T_H17$ cells | Promotes EAE |
| | $T_H17$ cells and $T_H9$ cells | $T_{Reg}$ cells | Inhibits EAE |
| Type 1 diabetes | $T_H17$ cells | Unknown | Unknown |
| **Parasitic infection** | | | |
| Lung infection with *Schistosoma mansoni* | $T_H2$ cells and $T_H9$ cells | Mast cells | No effect on granuloma formation |
| Intestinal infection with *Trichuris muris* | $TH_2$ cells and TH9 cells | Mast cells | Promotes parasite expulsion |
| **Transplantation** | | | |
| Skin allograft transplantation | $T_{Reg}$ cells | Mast cells and $T_{Reg}$ cells | Promotes allograft tolerance |

EAE, experimental autoimmune encephalomyelitis; NKT, natural killer T; $T_H$, T helper; $T_{Reg}$, regulatory T.
**Source**: Reproduced from Noelle, R.J., Nowak, E.C., 2010. Cellular sources and immune functions of interleukin-9. Nat. Rev. Immunol. 10, 683−687. http://www.nature.com/nri/journal/v10/n10/fig_tab/nri2848_T1.html#figure-title.

The **transforming growth factor** (**TGF**) exists in two forms, **TGFα** and **TGFβ**. There are several other proteins that have similarities in their amino acid sequences to that of TGF. Among these are the **activins, inhibins,** the **Mullerian-inhibiting substance** and the **bone morphogenic protein**. There are a great many other proteins that have some similarity to the amino acid sequence in TGF. The TGFβ receptor has **serine/threonine kinase** activity, differing from the tyrosine kinase activity of many of the other growth factor receptors. TGFβ exists as a gene family of four growth factors: TGFβ1/TGFβ4, TGFβ2, TGFβ3 and TGFβ5. TGFβ is able to inhibit the growth of **endothelial cells, macrophages** and T- and B-lymphocytes. There are three types of receptors for TGFs. TGFβ binds to the type II receptor (dimer) and the presence of the type III receptor can amplify the binding to the type II receptor. The type I receptor-TGFβ complex generates the phosphorylation of **SMAD2** and **SMAD3** that forms a complex with **SMAD4** which translocates to the nucleus. An important resulting effect is the transition of epithelia to mesenchyme.

TGFα is similar to EGF and it binds to the EGF receptor in competition with EGF. TGFα is secreted by human cancer cells. It may be involved in **wound healing** and in **tumor formation**. Tumor formation could be the result of dysregulation of the EGF pathway when EGF is outnumbered by the number of TGFα molecules forming an excess of EGFR-TGFα complexes. EGF dimeric receptors have intrinsic tyrosine kinase activity and, after autophosphorylation activate a signaling pathway leading to cell proliferation.

Fibroblast growth factors (FGFs) currently consist of 23 proteins and they are active in causing cell proliferation and differentiation in many tissues. Cells responding to FGFs to produce proliferation include: **endothelial cells, chondrocytes, smooth muscle cells, melanocytes** and **adipocytes**. In addition, FGFs can induce specific interleukins and prolong the survival of neuronal cells. FGFs bind to cell surface **heparan sulfate proteoglycans**; the heparan sulfate links to the FGF receptor that has protein tyrosine kinase activity. FGFs promote cell division in their target cells and also function in development, angiogenesis, hematopoiesis and tumorigenesis. There are four receptors for FGFs: FGFR1, FGFR2, FGFR3 and FGFR4. FGF-1 binds to all four receptors and to all their subtypes except in the case of FGFR4. FGF-2 exists in two forms, an 18-kDa form and a 24-kDa form. The lower molecular weight form has both intracellular and extracellular activities. The higher molecular weight form increases the expression of **IL-6** whereas, the lower molecular weight form decreases IL-6 production. The 18-kDa form is secreted and binds to a cell membrane

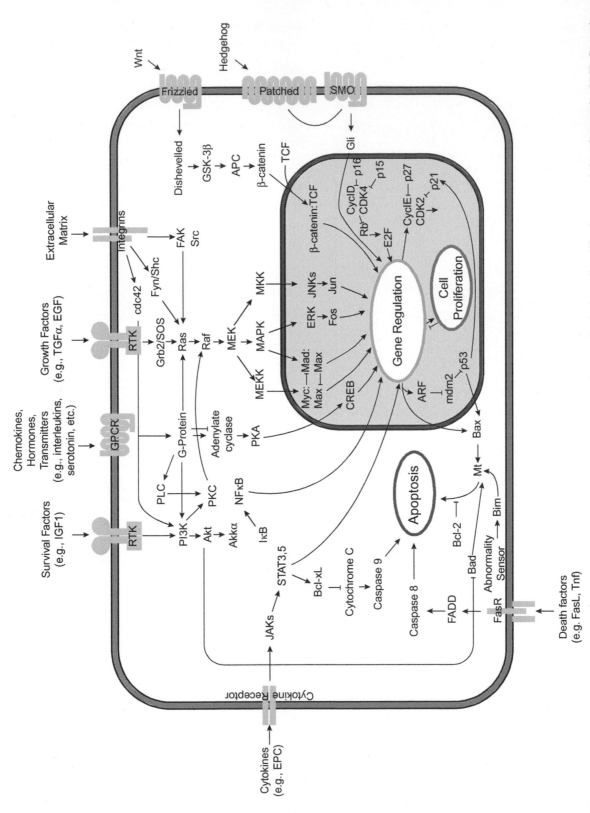

**FIGURE 17.40** Overview of **signal transduction pathways** of growth factors, cytokines and related substances that lead to cell proliferation or regulation of apoptosis. *GPCR*, G protein-coupled receptor; *RTK*, receptor tyrosine kinase; *PLC*, phospholipase C; *PKC*, protein kinase C; *PKA*, protein kinase A; *EPC*, endothelial protein C; *ARF*, tumor suppressor; *mdm*, murine double minute; *Mt*, mitochondria; see glossary for definitions of other terms. *Reproduced from http://upload.wikimedia.org/wikipedia/commons/f/fb/Signal_transduction_pathways.png.*

receptor, whereas the 24-kDa form targets the cell nucleus. Mutations in **FGF receptors** cause various human disfiguring conditions. FGFs signal through **protein kinase C** and the **MAPK pathway** to produce cell proliferation, cell survival and cell migration.

The **neurotrophins** consist of four factors: **nerve growth factor (NGF)**, **brain-derived neurotrophic factor (BDNF)**, **neurotrophin-3 (NT-3)** and **neurotrophin-4/5 (NT-4/5)**. They protect neurons from **apoptosis**. Following an action potential in a **postsynaptic cell**, neurotrophins are released from that cell. They then diffuse backwards across the **synapse** to be absorbed by the **presynatic nerve terminals**. Absence of the neurotrophin results in apoptosis of the presynaptic partner cell. The terminals that survive in the presence of neurotrophins will sprout new connections leading to differentiated growth. Presynaptic nerve terminal surfaces have **neurotrophin receptors** as do the surfaces of cell bodies of these cells. Neurotrophins binding to receptors on the cell body emit a signal that is carried through the **long axon** to the target location. Neurotrophins have high and low affinity receptors. NGF binds to its receptor, **TrkA**, causing autophosphorylation (via **protein tyrosine kinase activity**) of TrkA and signal transduction leading to cell survival. Experimental **knockouts** of the four neurotrophins lead to the loss of a variety of neurons.

**Colony stimulating factors (CSFs)** are involved in the differentiation of ancestral and parental cells into **hemopoietic cells**. A CSF is named for the cell type that it stimulates: **macrophage CSF (M-CSF)**, **granulocyte CSF (G-CSF)**, and **granulocyte-macrophage CSF (GM-CSF)**. **Interleukin-3** acts as a **multi-CSF**. CSFs are soluble glycoproteins produced in cells that line blood vessels. CSFs stimulate **bone marrow** and they can mobilize **stem cells** to form **white blood cells**. CSFs have receptors (members of the **IL-1 receptor family**) on the surface of hemopoietic stem cells that activate signaling leading to proliferation and differentiation, usually forming a white blood cell.

**Erythropoietin (EPO)** is a glycoprotein cytokine-like hormone produced in the **kidney peritubular capillary epoithelium** and, to a smaller extent, in the **hepatocyte**. EPO regulates the formation of **red blood cells** by stimulating proliferation and differentiation of immature erythrocytes to functional red blood cells. Red blood cell precursors derive from the bone marrow. EPO binds to its receptor on the surface of bone marrow **erythroid precursor cells** resulting in their maturation to mature red blood cells. Release of EPO from the **kidney** is triggered by **hypoxia** that is a product of **anemia** (low blood oxygen content). In addition to the release of EPO form the kidney under these conditions, **HIF-1 (hypoxia-inducible factor-1)**, a transcription factor, is produced by the kidney and it acts on **hypoxia response elements (HREs)** to activate gene transcription. HIFs are regulated by **oxygen-dependent hydroxylases** that, under limiting oxygen conditions, fail to be modified by the HIFs so that HIFs are transcriptionally active and produce **EPO** and other gene products. The EPO receptor binds EPO and then dimerizes. The activated receptor autophosphorylates on tyrosine residues and signals an intracellular cascade of phosphorylations through **JAK2** and **STAT5**. Activated STAT5 homodimerizes and translocates to the nucleus to stimulate gene transcription.

**Interferon (IFN)**, a small proteinaceous cytokine, is released from **macrophages** and **lymphocytes** following **viral infection**. The infected cell releases IFN that binds to surface receptors on adjacent cells, producing a signal cascade that results in the production of an **inhibitor of viral protein synthesis** (and also protein synthesis in the viral host) that prevents spread of the virus.

There are three types of IFNs: **IFN-α**, **IFN-β** (Type I) and **IFN-γ** (type II). Uninfected cells have a repressor of the IFN gene promoter that may act to prevent IFN synthesis by competing with **coactivator**. **Viral infection** causes the production of type I IFNs. The IFN-IFNA receptor complex activates the **JAK-STAT pathway** resulting in the activation of the **IFN-stimulated response element** that is located on several genes with antiviral activity.

**Insulin-like growth factors (IGFs-I, IGF-II)** are structurally related to insulin but function independently. IGFs are induced by **growth factor (GH)** mainly in the liver. In the adult, IGF-1 is responsible for most of the actions of GH. IGF-II is primarily a **fetal growth factor**. There are 6 **IGF binding proteins** (IGFBP-1—IGFBP-6) and these modulate the actions of IGF, either by binding and delivering the IGF to a site of action or by preventing the IGF from associating with the IGF receptor (decoy function). IGF can bind directly to an IGFBP and this complex can bind to an IGFBP receptor, thus bypassing the IGF receptor. IGF-I produces growth of **bone** and **muscle** as well as other tissues. IGF-I feeds back positively on the **somatostatin**-producing cell that inhibits the further release of **GHRH** to form a negative feedback on the release of GH from the somatotrope of the anterior pituitary. IGF-I also feeds back negatively on the hypothalamic neurons producing GHRH. The signaling pathways activated by IGF-I-IGF-IAR operate through **Akt-TOR** and through **ERK-ELK** and **MAPK** to produce proliferation and survival.

There are a large number of **interleukins (ILs)**, cytokines secreted by lymphocytes (also called **lymphokines**). They affect cellular processes in lymphocytes and some other cells of hemopoietic origin. Interleukins have a variety of activities and the focus, here, are on two examples: **IL-1** and **IL-9**. The actions of the interleukins and other cytokines, growth factors and related substances are summarized in an overview.

# SUGGESTED READING

## Literature

Andrae, J., Gallini, R., Betsholtz, C., 2008. Role of platelet-derived growth factor in physiology and medicine. Genes Dev. 22, 1276−1312.

Benedict, C.A., Ware, C.F., 2012. TRAIL: not just for tumors anymore? J. Exp. Med., 209. pp. 1903−1906.

de Almodovar, C.R., et al., 2004. Transcriptional regulation of the TRAIL R3 gene. Vitam. Horm. 67, 51−63.

Donovan, J., Abraham, D., Norman, J., 2013. Platelet-derived growth factor signaling in mesenchymal cells. Front. Biosci. (Landmark Ed.). 18, 106−119.

Heldin, C.H., 2012. Platelet-derived growth factor. Encyclopedia of Cancer. Springer-Verlag, Berlin, pp. 2908−2910.

Huang, L.E., et al., 1997. Erythropoietin gene regulation depends on heme-dependent oxygen sensing and assembly of interacting transcription factors. Kidney Int. 51, 548−552.

Johnson, H.M., et al., 2012. Steroid-like signaling by interferons: making sense of specific gene activation by cytokines. Biochem. J. 443, 329−338.

MacFarlane, M., 2003. TRAIL-induced signaling and apoptosis. Toxicol. Lett. 139, 89−97.

Mitra, A., et al., 2012. Signal transducer and activator of transcription 5b (Stat 5b) serine 193 is a novel cytokine-induced phospho-regulatory site that is constitutively activated in primary hemopoietic malignancies. J. Biol. Chem. 287, 16596−16608.

Naismith, J.H., et al., 1996. Seeing double crystal structures of type 1 TNF receptor. J. Mol. Recog. 9, 113−117.

Noelle, R.J., Nowak, E.C., 2010. Cellular sources and immune functions of interleukin-9. Nat. Rev. Immunol. 10, 683−687.

Oshimori, N., Fuchs, E., 2012. The harmonies played by TGF-β in stem cell biology. Cell Stem Cell. 11, 751−764.

Owen, D.J., 2004. Linking endocytic cargo to clathrin: structural and functional insights into coated vesicle formation. Biochem. Soc. Trans. 32, 1−14.

Patnaik, M.M., Tefferi, A., 2009. The complete evaluation of erythrocytosis: congenital and acquired. Leukemia. 23, 834−844.

Perugini, M., et al., 2010. Alternative modes of GM-CSF receptor activation revealed using activated mutants of the common β-subunit. Blood. 115, 3346−3353.

Pixley, F.J., Stanley, E.R., 2004. CSF-1 regulation of the wandering macrophage: complexity in action. Trends Cell. Biol. 14, 628−638.

Quinn, J.M.W., Gillespie, M.T., 2005. Modulation of osteoclast formation. Vertebr. Skeletal Biol. 328, 739−745.

Radaev, S., et al., 2010. Ternary complex of transforming growth factor β1 reveals isoform-specific ligand recognition and receptor recruitment in the superfamily. J. Biol. Chem. 285, 14806−14814.

Zhan, Y., Xu, Y., Lew, A.M., 2012. The regulation of the development and function of dendritic cell subsets by GM-CSF: more than a hemopoietic growth factor. Mol. Immunol. 52, 30−37.

## Books

Fitzgerald, K., et al., 2013. The Cytokine Factsbook and Webfacts. Academic Press, London.

Litwack, G. (Ed.), 2004. TRAIL. Vitamins & Hormones, vol. 67. Academic Press, San Diego, CA.

Litwack, G. (Ed.), 2006. Interleukins. Vitamins & Hormones, vol. 74. Academic Press, San Diego, CA.

Meager, A. (Ed.), 2006. The Interferons. Wiley-VCH Verlag GmBH & Co., Weinheim.

Sarchielli, P., et al., 2011. Nerve Growth Factor and Pain (Pain and its Origins, Diagnosis and Treatments). Nova Sci. Publisher, Inc., New York.

Sherbert, G., 2011. Growth Factors and their Receptors in Cell Differentiation, Cancer and Cancer Therapy. Elsevier, Amsterdam and Boston, MA.

# Chapter 18

# Membrane Transport

## CYSTIC FIBROSIS (MUCOVISCIDOSIS) AND ABERRANT ION TRANSPORT

A mutation in the gene for a **chloride ion transporter** gives rise to this disease. The transporter protein is called the **cystic fibrosis transmembrane conductance regulator** (**CFTR**). CFTR regulates the passage of chloride and sodium ions across epithelial membranes (Fig. 18.1).

CFTR, in addition to its function as a **Cl⁻ channel**, is a regulator of other channels. It regulates the **outwardly rectifying chloride channel** (**ORCC**), the **epithelial Na⁺ channel** (**ENaC**) and at least two **inwardly rectifying K⁺ channels** (**ROMK1, ROMK2**). It also plays a role in ATP transport, affecting exocytosis/endocytosis and the regulation of intracellular organelle pH. These relationships between CFTR and these other channels are shown in Fig. 18.2.

Of its structure in the cell, **CFTR** occupies 4% as extracellular loops, 19% in the membrane-spanning region, and 77% in the cytoplasm. Its preference for anion selectivity is $Br > Cl > I > F$. CFTR, in addition to $Cl^-$, also can conduct $HCO_3^-$, $Na^+$ (follows $Cl^-$ through the tight junctions), and ATP anions. CTFR can form a dual pore for transport of $Cl^-$ or it can heterodimerize with other channels. Fig. 18.3 shows the protein−protein interaction between CTFR and ENaC.

In the presence of ATP, the required phosphorylations take place and the CFTR channel is opened, allowing Cl ions to flow in the direction (either in or out of the cell) of the electrochemical gradient. Thus, in the presence of ATP followed by the required phosphorylations, the channel is open and CFTR inhibits the function of ENaC so that there would be an excess of $Na^+$ on the outside of the cell driving $Cl^-$ movement through the CFTR from the intracellular space to the extracellular space. In the absence of ATP, the channel is closed and $Cl^-$ does not flow in either direction. CFTR is synthesized in red blood cells, microvasculature lining, epithelia of the pancreas and lung, colon, sweat gland, kidney proximal tubules, cortex and medulla, heart myocytes, hypothalamus, and lymphoid cells.

Prior to genetic sequencing, **cystic fibrosis** was indicated by excessive sodium chloride in the sweat, therefore a **sweat test** was the means of identification. In the sweat gland, the reabsorption of salt is abnormal in cystic fibrosis. CFTR is the only anion channel in the duct of the **sweat gland** and it is nonfunctional in cystic fibrosis. Chloride transfer would be absent and because the transport of chloride is followed by water, causing the reduction of water transport, sweat becomes concentrated in sodium chloride ($>60$ mmol/L in sweat is diagnostic), thus, the salty sweat condition of the person with this disease and the deficient secretion of digestive enzymes from the pancreas leads to the development of **thick mucus**. This defect occurs in many tissues and causes great difficulty in the lung where breathing problems and infections develop. In the newborn, stools do not develop within the first 24 to 48 hours of life and the condition known as *meconium ileus* occurs. The low secretions of **digestive enzymes** from the pancreas lead to an obstructive mass (*meconium*) in the intestine (Fig. 18.4).

The *cf* gene occurs on chromosome seven and a gene located on chromosome one encodes for a protein called the **CF antigen** that occurs in high levels in the sera of patients with overt cystic fibrosis as well as in carriers. The CF antigen is part of a complex of proteins and the complex inhibits casein kinase I and II but it does not inhibit protein kinase C or protein kinase A; they are located in the R domain of CFTR. The CF antigen could be associated with myeloid cell functions and either is involved in the causation of cystic fibrosis or is a consequence of the disease. It is possible that CFTR may interact with the CF antigen because the structure of the CF antigen is reminiscent of an ion transport regulator.

Cystic fibrosis is an **autosomal recessive disorder**. The normal individual has two copies of the CFTR gene. If one of these genes is mutated (there may be more than 1000 known mutations) so that the ultimate protein gene product (CFTR protein) is poorly functional or nonfunctional, that condition would be recessive for cystic fibrosis, and the individual would be a healthy carrier and would not have overt signs of the disease. About 70% of the mutations involve a single amino acid residue, phenylalanine 508. One person in 25 of northern European extraction carries a mutation in

Human Biochemistry. DOI: http://dx.doi.org/10.1016/B978-0-12-383864-3.00018-1

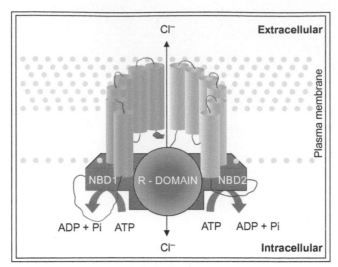

**FIGURE 18.1**   Model of the CFTR (cystic fibrosis transmembrane conductance regulator) protein channel. The CFTR protein consists of 1480 amino acid residues with a molecular weight of 170,000. It is a membrane-bound glycoprotein and a member of the **ATP binding cassette** (ABC) **superfamily** of proteins. In addition to CFTR, this superfamily includes such clinically important proteins as *P*-glycoprotein, **multidrug resistance associated protein,** and the TAP transporters. CFTR contains two six membrane-spanning regions each of which is connected to an ATP-binding domain (NBD1, NBD2). Between NBD1 and NBD2 is a regulatory (R) domain that is unique to CFTR among members of the ABC superfamily. The R domain contains protein kinase A (PKA) and protein kinase C (PKC) activities and is phosphorylated in order to activate the ion channel. There are several sites in the R domain for phosphorylation by cAMP-activated PKA or PKC. *Reproduced from http://www.genemedresearch.ox.ac.uk/cysticfibrosis/protein.html.*

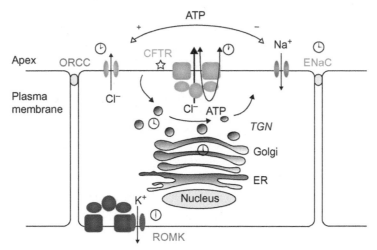

**FIGURE 18.2**   CFTR shown as a multifunctional protein, acting as a chloride ion channel, pumping $Cl^-$ from the intracellular space to the extracellular space in the presence of ATP, and its interactions with other channels that involve the transport of $Na^+$, $Cl^-$, and $K^+$. Thus, when CFTR is pumping $Cl^-$ out of the cell, **ENaC** is inhibited so extracellular $Na^+$ can combine with exported $Cl^-$ to form extracellular NaCl. *TGN*, trans-Golgi network. *Reproduced from http://atlasgeneticsoncology.org/Educ/CisFibID30032ES.html.*

CFTR, whereas in Asian populations only about one in 500 carries the mutation. If both parents had this condition (recessive for cystic fibrosis involving one mutated CFTR gene), 25% of the offspring would be expected to be normal, 50% would be expected to be healthy carriers, and 25% would be expected to have the overt disease. If one parent is normal and the other has overt cystic fibrosis the diseased parent would have both genes mutated and would be homozygous for *cf*, thus *cf/cf*. The combinations of parents and the results for their offspring are shown in Table 18.1.

In a child born with the overt disease, death may follow, except in manageable cases where, with current treatments, a child may live to as long as 30–40 years of age. There are about 30,000 living patients (children and adults) in the United States that have this disease. There is one child in 2000 born with CF and it is the most common lethal hereditary disease of Caucasians. Certain native Americans also are subjected to CF (one Pueblo in 4000; one Zuni in 1500).

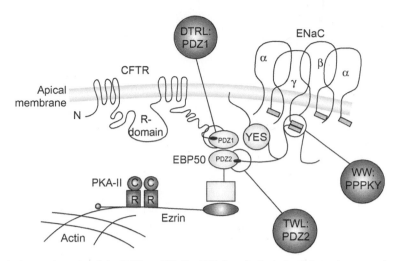

**FIGURE 18.3** Protein–protein interactions that link CTFR to **ENaC. *PDZ* domain** is derived from three proteins: **P**ostsynaptic density protein (PSD95), ***D***rosophila disc large tumor suppressor (Dig1) and ***Z***onula occludens-1 protein (zo-1); PDZ has the amino acid sequence (Gly-Leu-Gly-Phe). The β-sheet in the PDZ domain is lengthened by the addition of a β-strand from the tail of the protein binding to it.

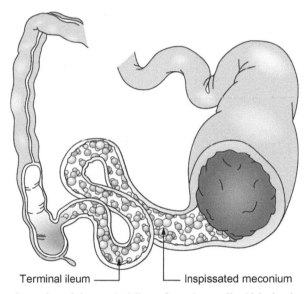

Terminal ileum ——              —— Inspissated meconium

**FIGURE 18.4** *Meconium ileus* causing obstruction of the terminal ileum from abnormally thick, inspissated (congealed) meconium *Reproduced from http://www.surgicalcore.org/popup/55929.*

One Hispanic in 8000; 1 African in 15,000; 1 Asian in 32,000 has this disease. In the United States, 1 in 25 is an unaffected carrier and in Ashkenazi Jews 1 in 29 is a carrier. In all, there are about 10 million carriers.

**Pancreatic digestive enzyme supplements** are fed, therapeutically, chests are percussed (hand-tapping) routinely in attempt to clear buildup of mucus and **mucus-clearing drugs** are given as well as antibiotics. A **lung transplant** is considered when the patient's lungs begin to fail.

The most common mutations in CFTR are summarized in Table 18.2.

The predominant deletion mutation of **phenylalanine 508** is part of the **nucleotide-binding function** in NBD1 (Fig. 18.1). The resulting **mutant CFTR** (CFTR δ-F508) protein is synthesized but fails to be modified in the endoplasmic reticulum (ER) and becomes degraded.

Normal CFTR is also involved in the regulation of the **bicarbonate ion** ($HCO_3^-$) in the **pancreatic duct**. Here, CFTR increases a transporter that exchanges sodium ions and protons. Epithelial cells of the pancreas and lung absorb $Na^+$ through the **epithelial sodium channel (ENaC)**, whereas $Cl^-$ enters through tight junctions between cells and water is absorbed from the lumen to the basolateral side of the cell. Elevated cAMP activates PKA that phosphorylates and activates CFTR allowing $Cl^-$ to pass from the cell interior into the lumen followed by the passage of $Na^+$ through

**TABLE 18.1** Genetic (Autosomal) Inheritance of the Cystic Fibrosis Gene

| Parents | Crosses | Offspring |
|---|---|---|
| **Carrier × normal** | | |
| (Acf) (Ax) | AA Ax | |
| | Acf xcf | |
| | AA Ax Acf xcf | 50% normal + 50% carrier |
| **Carrier × carrier** | | |
| (Acf) (Acf) | AA Acf | |
| | Acf cf/cf | |
| | AA Acf | |
| | Act cf/cf | 25% normal + 50% carrier + 25% |
| | | Overt disease |
| **Overt disease × normal** | | |
| (cf/cf) (Ax) | Acf xcf | |
| | Acf xcf | |
| | Acf xcf | |
| | Acf xcf | 100% carriers |
| **Overt disease × carrier** | | |
| (ct/cf) (Acf) | Acf, Acf, cf/cf, cf/cf | |
| | Acf, Acf, cf/cf, cf/cf | |
| | Acf, Acf, cf/cf, cf/cf | |
| | Acf, Acf, cf/cf, cf/cf | 59% carriers + 50% overt disease |
| **Overt disease × overt disease** | | |
| (cf/cf) (cf/cf) | | 100% overt disease |

**TABLE 18.2** Common *CFTR* Mutations[a]

| Mutation | Type | Frequency (%) |
|---|---|---|
| DeltaF508 | Deletion | 28,948 (66.0) |
| G542ter | Nonsense | 1062 (2.4) |
| G551D | Missence | 717 (l.6) |
| N1303K | Missense | 589 (1.3) |
| W1282ter | Nonsense | 536 (1.2) |
| R553ter | Nonsense | 322 (0.7) |
| 621 + 1G-> T | Splice junction | 315 (0.7) |
| 1717-1G-> A | Splice junction | 284 (0.6) |
| R117H | Missense | 133 (0.3) |
| 3849 + l0kb C-> T | Alternative splice | 104 (0.2) |

[a]Data are from the Cystic Fibrosis Genetic Analysis Consortium based on 43,000 CF chromosomes examined through 1994. Mutations are indicated by amino acid residue or nucleotide position, ter, termination codon.
Source: Reproduced from Table 1 of http://www.bmb.leeds/oc.uk/leaching/icu3/mdcases/ws4/.

the tight junctions between cells. As noted before, the activation of CFTR channels causes inhibition of the ENaC channel reducing the uptake of $Na^+$ into the cell and the uptake of water from the lumen resulting in an osmotic gradient inducing the secretion of water into the lumen. Following activation of CFTR by cAMP, ENaC function is inhibited by an interaction between the two channels (Fig. 18.3). The binding occurs through **ezrin-radixin-moesin binding phosphoprotein** (**EBP50**) that contains **PDZ domains** involved in the linkage of integral membrane proteins to the cytoskeleton.

In the lung, which maintains high levels of **glutathione** (**GSH**), the CFTR is the modulator of **GSH efflux** that occurs as a response to the stress accompanying infections. There may prove to be other functions of the CFTR.

## TYPES OF MEMBRANE TRANSPORT

## Absorption of Large Molecules Binding to Receptors on the Cell Surface

There are two processes by which large molecules are imported from the surface of cells where they may have bound to receptors or other types of molecules attached to the cell surface. One process is **receptor-mediated endocytosis** that has been described for the case of the **epidermal growth factor** (**EGF**) that utilizes **clathrin-dependent endocytosis** for entry into the target cell in order to quench the initial signal of binding to its receptor and activating the signal mechanism. Another more general process is **pinocytosis** whereby macromolecules and fluid are ingested in small vesicles (<150 nm diameter). The vesicles are formed by invagination of the cell membrane and enclose the molecules to be ingested. *Pinocytosis is a process that can be carried out by all cells.*

Where particles (very large structures compared to molecules; >250 nm diameter) are encountered (e.g., bacteria or parts of dead cells), these can be engulfed, usually by a specialized cell (phagocyte) by the process of **phagocytosis**. Phagocytosis involves the extension of the cell membrane to form *pseudopodia* that enclose the particle. *Pseudopodia* are temporary projections of eukaryotic cell membranes. They extend and contract into microfilaments formed by the reversible assembly of **actin subunits**. Contraction of the pseudopodium occurs at its end by interaction with **myosin**.

There are two general types of phagocytes: "professional" phagocytes and "nonprofessional" phagocytes. **Professional phagocytes** have receptors on their cell surface that detect certain **nonself particles**. These professional phagocytes are: neutrophils, monocytes, dendritic cells, macrophages, and mast cells. The **nonprofessional phagocytes** engulf dying cells and foreign organisms and these are: epithelial cells in skin, endothelial cells in blood vessels, fibroblasts in connective tissue, mesenchymal cells, lymphocytes in blood, lymph and lymph nodes, erythrocytes in blood and natural killer (NK) cells and large granular lymphocytes in blood, lymph, and lymph nodes. Pinocytosis is compared to phagocytosis in Table 18.3 and in Fig. 18.5.

The processes of **pinocytosis** and **phagocytosis** are pictured in Fig. 18.5.

## Exocytosis

Exocytosis is a process for moving large molecules out of the cell to the cell exterior. Commonly, these macromolecules originate in **storage vacuoles** inside the cell and are moved to the exterior after an appropriate signal for this action. There is **regulated exocytosis** in which the contents of an intracellular vacuole are secreted in response to a specific signal and there is **constitutive exocytosis** wherein macromolecules are secreted from the cell without having to await a specific signal.

**TABLE 18.3** Comparison of Pinocytosis and Phagocytosis

|  | Pinocytosis | Phagocytosis |
| --- | --- | --- |
| **Substrate** | Fluids, macromolecules | Particles |
| **Mechanism** | Membrane invagination | Pseudopodia formation |
| **Endosome** | Pinocytic vesicle [~150 nm] | Phagosome [>250 nm]—determined by particle |
| **Energy requirement** | Energy independent | Energy dependent |
| **Performed by** | All cells | Phagocytes |

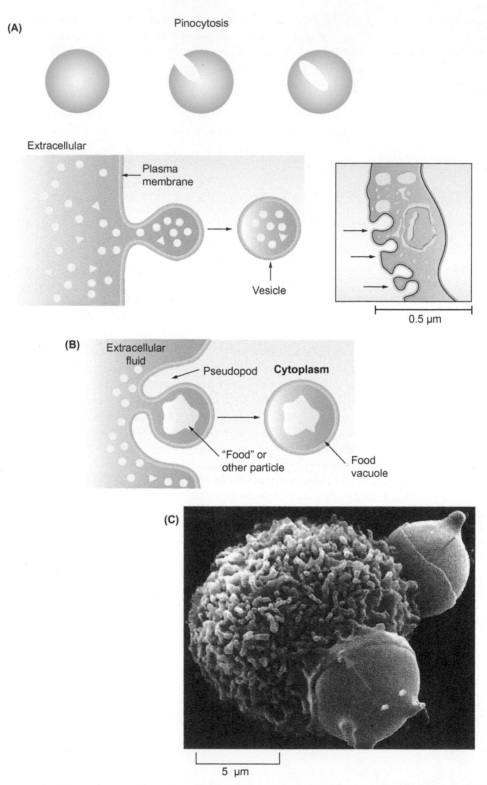

**FIGURE 18.5** The processes of pinocytosis and phagocytosis. (A) *Top*, diagram of pinocytosis. *Lower left*, pinocytosis showing the invagination at the cell membrane to engulf outside particles and form a vesicle inside the cell. *Lower right*, picture of several pinocytic invaginations. (B) Diagram of the process of phagocytosis in which **pseudopodia** are formed from the cell membrane that surround particles to be engulfed. (C) Photograph of a **phagocyte** ingesting two red blood cells. *Part C reproduced with permission from Alberts,* Essentials of Cell Biology, *Garland Publishing Company.*

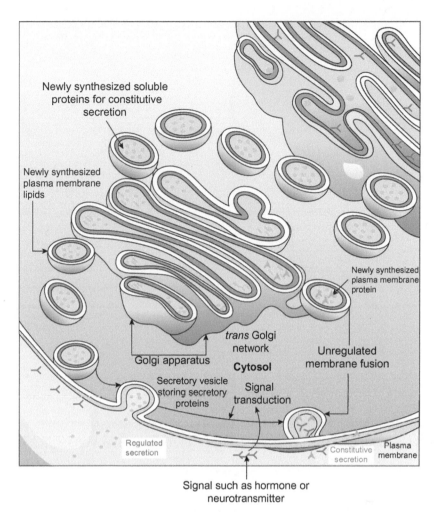

**FIGURE 18.6**   Regulated and constitutive pathways of exocytosis. The two pathways diverge in the **trans-Golgi network**. Many soluble proteins are secreted without interruption from the cell by the **continuous secretory pathway**, which operates in all cells. The pathway also supplies the cell membrane with newly synthesized lipids and proteins. Specialized secretory cells also have a regulated exocytosis pathway by which selected proteins in the trans-Golgi network are diverted into secretory vesicles, where the proteins are concentrated and stored until an extracellular signal stimulates their secretion, often involving calcium ions.

In the case of constitutive exocytosis, a signal for secretion is part of the amino acid sequence of the protein. Amino acid sequences signaling secretion are often located in the N-terminus. Proteins thus destined for secretion transit through various compartments of the Golgi apparatus and are enclosed in a vesicle that will fuse with the cell membrane and be transported to the cell exterior through the agency of cup-shaped structures (**porosomes**) in the cell membrane that will dock the vesicle in the process of fusion with the cell membrane. The porosomes involve **SNARE** (**s**oluble **NSF** [*N*-ethylmaleimide sensitive factor] **a**ttachment protein **re**ceptor) **proteins**. There are many regulatory factors involved in this overall process that regulate transport, the selection of specific proteins, and modification of proteins. The steps in constitutive and regulated exocytosis are shown in Fig. 18.6.

A close-up picture of the **vesicle** fusing with the cell membrane is shown in Fig. 18.7.

## Passive Diffusion (Osmosis)

Some substances, such as water, $CO_2$, can cross a selective semipermeable membrane without a requirement for energy; this defines passive membrane transport or osmosis. The osmosis of water through an artificial semipermeable membrane is determined by the concentration of solutes on either side of the membrane. Thus, if the solute concentration on one side of the membrane is higher than on the other, water tends to move toward the higher solute concentration (water moves down its concentration gradient from where there is more water to where there is less water). The result will be

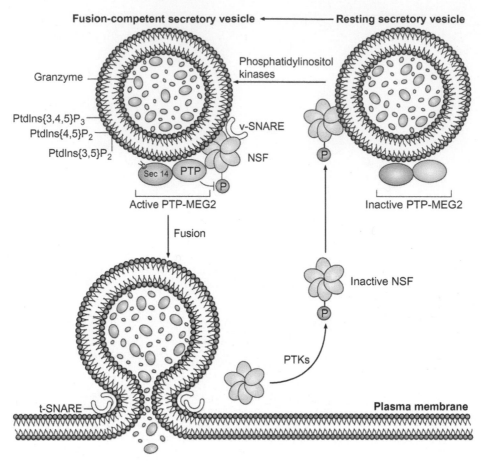

**FIGURE 18.7** Fusion of a vesicle containing proteins for secretion with the cell membrane *Reproduced from http://www.nature.com/nri/journal/v5/n1/images/nri1530-f7.jpg.*

an equal distribution of solutes on either side of the membrane to equalize the **tonicity** of the two solutions (on either side of the membrane). Tonicity is the effective particle concentration, **osmolality**, which exerts an osmotic force across the membrane. Thus, when both compartments are equally concentrated (isotonic), there will be no further movement of water. A **hypotonic solution** has more water and less solute; a cell in a hypotonic solution will swell (take up water). A **hypertonic solution** has a high solute concentration. A cell placed in a hypertonic solution will lose water and shrink. Normally, the intracellular solute concentration is balanced by the extracellular sodium chloride concentration. A 0.9% solution of sodium chloride is considered isotonic.

In cells there are proteinaceous channels for the passive uptake of dissolved molecules, excepting water. Cells contain specific channels (**aquaporins**) for the ATP coupled (phosphorylation of the water channels) uptake of water and its subsequent passage across the cell interior. Fig. 18.8 shows an idealized **aquaporin water channel** through which water molecules pass single file.

This process is different from facilitated diffusion in which specific proteinaceous channels or permeases are involved. In most cases, the movement of charged ions through membranes requires specific transporters.

**Arginine vasopressin** (**AVP**) secreted from the posterior pituitary stimulates the number of aquaporins in the kidney to cause an increase in the reabsorption of water from the **collecting duct** as shown in Fig. 18.9.

## Facilitated Diffusion

This is a modified version of **osmosis** where a **proteinaceous channel** is used for the solute to enter a cell from the extracellular space but energy is not required, similarly to free diffusion. Here, molecules will move into the cell based on the outside concentration of the solute (following the electrochemical gradient). When a substance to be transported to the interior of a cell is impermeable to the **lipid bilayer** of the cell membrane (e.g., glucose and ions), it can be

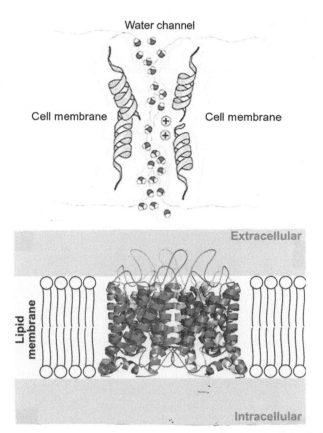

**FIGURE 18.8** *Top*: Passage of **water molecules** single file through the **aquaporin AQP1 channel**. The positive charge at the center of the channel deflects positively charged ions, such as $H_3O^+$, preventing proton leakage through the channel. Water molecules are oriented with the oxygen facing downward toward the cytoplasm (*bottom of figure*). At the middle of the pore, the orientation reverses with the oxygen facing upward toward the extracellular space (*top* of figure). Top: *Reproduced from http://aquaporins.org/peter.htm.* Bottom: *Molecular model of the aquaporin channel. Reproduced from http://www.user.gwdg.de/~aponte/research.html.*

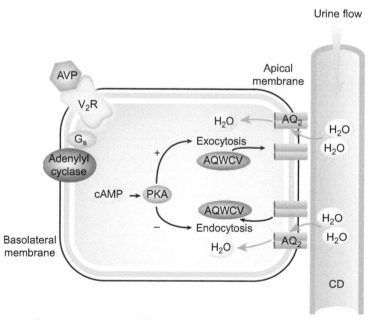

**FIGURE 18.9** Action of arginine vasopressin on the kidney to increase the reabsorption of water. The binding of arginine vasopressin (AVP) to its receptor ($V_2R$) in the kidney (*upper left*) stimulates a $G_s$-coupled protein that activates adenylyl cyclase to produce cAMP (from ATP) that activates **protein kinase A (PKA)**. This pathway (causing **the phosphorylation of aquaporin subunits** to polymerize into channels) increases the exocytosis (moving them from the cell interior to the apical membrane) of **aquaporin water channel-containing vesicles (AQWCV)** and decreases endocytosis of the vesicles (removing them from the membrane to the interior of the cell) both resulting in increases in **aquaporin 2 ($AQ_2$) channel** formation and apical membrane insertion. This allows an increase in the permeability of water from the **collecting duct (CD)** and water is moved from the apical membrane across the cell to the basolateral membrane and ultimately into the extracellular fluid and into the bloodstream. *Reproduced from http://circ.ahajournals.org/content/118/4/410/F2.large.jpg.*

imported through a proteinaceous **membrane-spanning channel** that overcomes the need for the solute to contact the impermeable bilayer. As already mentioned, the importation of the solute follows the concentration gradient, so that a high concentration of the solute on the outside of the cell will drive the solute into the cell which contains less solute than on the outside (Fig. 18.10).

Note in this case, that the channel is closed at one end or the other and not continuously opened between the extracellular fluid and the cytoplasm.

There are three high affinity transporters for glucose: Glut 1, Glut 3, and Glut 4, and one low affinity **glucose transporter**, Glut 2. These operate by facilitated diffusion and in the model presented in Fig. 18.10 a Glut transporter would be equivalent to the carrier protein.

Tissues that exhibit high glycolytic activity contain **high affinity glucose transporters** and these would be found in the liver, intestine, and kidney. Most tissues will contain glucose transporters but in amounts less than liver, intestine, and kidney. Glut 2, the **low affinity transporter**, also will be present in these tissues because these tissues are sites of high glucose fluxes. Glut 5 is primarily a **fructose transporter**. Various Glut transporters, their tissue locations, and functions are listed in Table 18.4.

The concentration of high affinity glucose transporters on the surface of a cell influences the rate of glucose uptake into that cell. Expression of the GLUT transporters is regulated by glucose and by hormones (Table 18.4B). In addition to some substances that can enter the cell by free diffusion, there are molecules that can dissolve readily in the **lipid bilayer** of the cell membrane. These substances, such as neutral steroids, like **cortisol** can enter the cell by passive or free diffusion requiring little or no energy.

Comparing the rates of transport of **passive transport** compared to **facilitated transport**, the passive process has a rate of transport proportional of the extracellular solute concentration while the facilitated process reflects an enhanced rate as the solute concentration increases and this rate is greater than the proportionality to the outside solute concentration (Fig. 18.11).

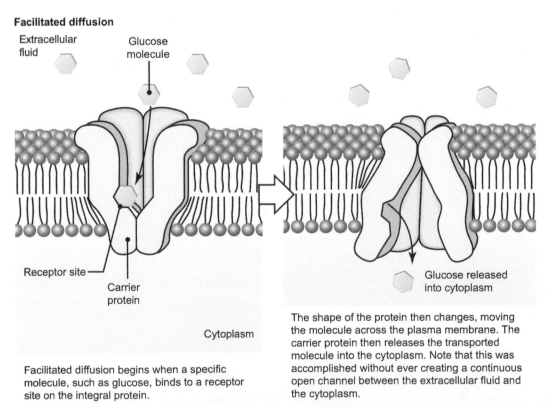

**Facilitated diffusion**

Extracellular fluid

Glucose molecule

Receptor site

Carrier protein

Cytoplasm

Facilitated diffusion begins when a specific molecule, such as glucose, binds to a receptor site on the integral protein.

Glucose released into cytoplasm

The shape of the protein then changes, moving the molecule across the plasma membrane. The carrier protein then releases the transported molecule into the cytoplasm. Note that this was accomplished without ever creating a continuous open channel between the extracellular fluid and the cytoplasm.

**FIGURE 18.10** Facilitated diffusion in the case of glucose transport down its concentration gradient (higher concentration of solute on the outside of the cell than in the inside). *Reproduced from http://www.highlands.edu/academics/divisions/scipe/biology/faculty/harnden/2121/images/facdiff.jpg.*

**TABLE 18.4** A. Tissue-Specific Expression of GLUT Family Members. B. Regulators and Signaling Pathways Involved in Glucose Transport

| Protein | Alias | Expression | Function |
|---|---|---|---|
| GLUT1 | | All tissues (abundant in brain and erythrocytes) | Basal uptake |
| GLUT2 | | Liver, pancreatic islet cells, retina | Glucose sensing |
| GLUT3 | | Brain | Supplements GLUT1 in tissues with high energy demand |
| GLUT4 | | Muscle, fat, heart | Insulin responsive |
| GLUT5 | | Intestine, testis, kidney, erythrocytes | Fructose transport |
| GLUT6 | GLUT9 | Spleen, leukocytes, brain | |
| GLUT7 | | Liver | |
| GLUT8 | GLUTX1 | Testis, brain | |
| GLUT9 | GLUTX | Liver, kidney | |
| GLUT10 | | Liver, pancreas | |
| GLUT11 | GLUT10 | Heart, muscle | |
| GLUT12 | GLUT8 | Heart, prostate | |
| Pseudogene | GLUT6 | | |

| Regulator | Pathway | GLUT Isoform | Cell Type |
|---|---|---|---|
| Insulin | IR, PI3K | GLUT4 | Muscle, fat |
| IGF-I | IGF-IR, PI3K | GLUT4 | Muscle, fat |
| IGF-II | IGF-IR, PI3K | GLUT4 | Muscle, fat |
| Contraction | AMPK | GLUT4, GLUT1? | Skeletal muscle |
| Ischemia | AMPK | GLUT4 | Heart muscle |
| Hypoxia | AMPK? | GLUT4 | Skeletal muscle |
| Nitric oxide | cGMP | Presumed GLUT4 | Skeletal muscle |
| Phorbol ester | PKC | Presumed GLUT4 | Skeletal muscle |
| $\alpha$-Adrenergic agonists | Gs protein | GLUT4 | Brown fat, muscle |
| $\beta$-Adrenergic agonist | Gi protein | Presumed GLUT4 | Heart muscle |
| Bradykinin | Gq protein | GLUT3 | Skeletal muscle |
| Thrombin | Gi protein | GLUT3 | Platelets |
| Adenosine | Gq protein | GLUT4 | White and brown fat |

Source: Reproduced from http://www.scielo.cl/scielo.php?pid = S0716-97602002000100004&script = sci_arttext.

## Active Transport Requiring Energy

This usually takes place when a solute is to be transported into a cell from the extracellular space against a concentration gradient, that is, a condition in which the solute concentration is higher inside the cell than on the outside. Also it can represent the opposite in which the concentration of the solute is greater on the outside than on the inside and the solute is being transported from the inside to the outside. The source of energy for this transport is usually the terminal high-energy phosphate of **ATP**. This requires a proteinaceous channel that is opened by the expenditure of energy (ATP) and closes after the transport is complete. An important example is the **sodium-potassium ATPase (Na$^+$/K$^+$ ATPase)**

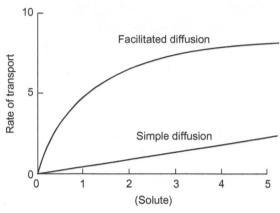

**FIGURE 18.11**   **Kinetics of transport** in simple diffusion and in facilitated diffusion as a function of the concentration of the solute (*x*-axis).

shown in Fig. 18.12A and the 10-ms pumping cycle of $Na^+/K^+$ ATPase shown in Fig. 18.12B. In this example, $Na^+$ is being pumped from the inside of the cell to the outside where the concentration of $Na^+$ is higher than in the cell interior.

*This ATPase system is found in all human cells and is crucial to nerve and muscle cells.* In each cycle, three sodium ions are transported out of the cell and two potassium ions are transported into the cell. The $Na^+/K^+$ ATPase is often located in the **basolateral membrane** and accounts for as much as 30% of the total consumption of ATP by the cell. Thus, the operation of this pump consumes about one third of the body's energy expenditure and it serves to maintain a 10- to 30-fold higher **$Na^+$ concentration** outside the cell compared to inside that creates an inwardly directed **electrochemical gradient** plus a greater $K^+$ concentration inside the cell compared to the outside. The **epithelial sodium channel (ENaC)** together with the sodium-potassium **ATPase** are two key transporters in the distribution of sodium and potassium ions.

## Simple and Coupled Transporters

A simple transporter can be defined as one in which a solute is transported in one direction without the participation of ions or other solutes going in the same or opposite direction. This example would be called a **uniport**. When a solute is transported in one direction and an ion is transported with the solute in the same direction, it is called a **symport**. In the third case, a solute can be transported in one direction and at the same time an ion is transported in the opposite direction; this is called an **antiport**.

When a solute gradient has been established, for example, in the case of the sodium ion whose concentration on the outside of a cell is greater than the concentration on the inside of the cell, the osmotic pressure created by the movement of the sodium ion from the outside to the inside of the cell can be used to drive the transport of a second molecule in the same direction. This is an example of a symport, a coupled transporter. Comparatively, when an ion is more concentrated in the interior of the cell than outside the cell, the transport of this ion from the inside to the outside (down the concentration gradient) can be used to drive another molecule into the cell from the outside by an antiport. These **types of transporters** are diagramed in Fig. 18.13A and an example of coupled transport is shown in Fig. 18.13B.

In the case of **intestinal epithelial cells**, where the concentration of glucose is higher inside the cell, glucose transport into the cell from the lumen can be coupled to the uptake of sodium ions because sodium ion concentration is higher on the outside of the cell than on the inside. This allows the positive electrochemical gradient for sodium ion to be coupled to the transport of glucose (against its concentration gradient) into the cell from the outside. Following the uptake of glucose into the cell, it is further transported to the basolateral side of the cell by a **passive glucose transporter** (uniport) down its concentration gradient as glucose concentration is higher inside the cell than in the extracellular space.

## Ions and Gradients

There are important cellular ions that must be controlled with respect to their intracellular and extracellular concentrations. These are $Na^+$, $K^+$, $Mg^{2+}$, $Ca^{2+}$, $H^+$, and $Cl^-$ and their partitioned concentrations inside and outside cells are listed in Table 18.5.

(A)

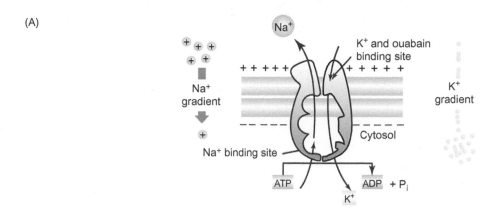

(B)

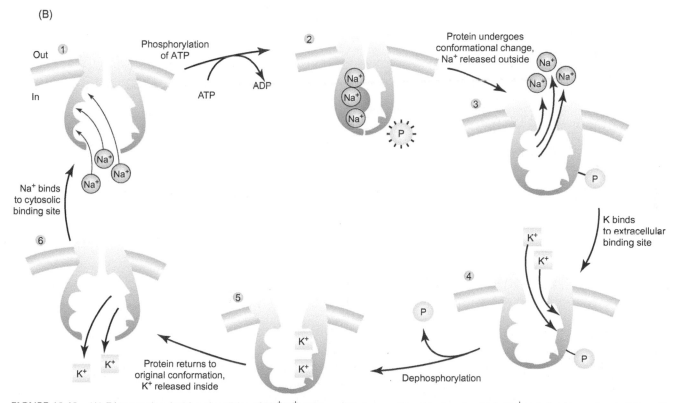

**FIGURE 18.12** (A) Diagram showing functionalities of Na$^+$/K$^+$ ATPase. This is an **antiport** system in which Na$^+$ is being pumped out of the cell against a concentration gradient ([Na$^+$] is higher on the outside of the cell) and K$^+$ is pumped inwardly against a concentration gradient. The enzyme utilizes the energy of ATP terminal phosphate to achieve the transport of these ions. (B) The 10-ms pumping cycle of Na$^+$/K$^+$ ATPase.

The movement of water from low to high solute concentration and the extracellular sodium chloride that balances the intracellular solute concentration are ways in which osmotic balance is maintained. The balance is maintained further by the action of Na$^+$/K$^+$ ATPase (Fig. 18.12). Most cell membranes have a potential derived from a voltage across the membrane and this potential influences the movement of ions. Membranes have different charges on their inner and outer surfaces where the cytoplasmic side of the membrane is usually negatively charged so that negatively charged

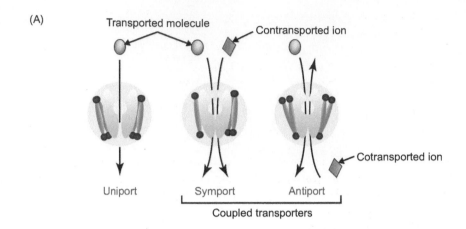

**FIGURE 18.13** (A) Types of transporters. (B) Transporting glucose against its gradient, in intestinal epithelia, by using the positive sodium ion gradient in coupled transport; an example of a symporter.

### TABLE 18.5 Concentrations of Ions Inside and Outside Cells

| Ion | Intracellular | Extracellular |
| --- | --- | --- |
| $Na^+$ | 5–15 mM | 145 mM |
| $K^+$ | 140 mM | 5 mM |
| $Mg^{2+}$ | 0.5 mM | 1–2 mM |
| $Ca^{2+}$ | $10^{-7}$ mM | 1–2 mM |
| $H^+$ | $10^{-7.2}$ M (pH 7.2) | $10^{-7.4}$ M (pH 7.4) |
| $Cl^-$ | 5–15 mM | 110 mM |
| Fixed anions | High | 0 mM |

Fixed anions are negatively charged small and large organic ions (anions) that cannot leave the cell.
Intracellular vs extracellular ion concentrations. [Intracellular] very different from [extracellular] cations [+ve charged species] balanced by anions [−ve].

anions are driven out of the cell while positively charged cations are drawn into the cell. When charged solutes are crossing a membrane the concentration gradient comes into play (Fig. 18.14).

Calcium ion is a **second messenger** and it is involved in many cellular activities. As the concentration of $Ca^{2+}$ is vastly greater outside the cell than inside the cell (Table 18.5), there is a pressure to move $Ca^{2+}$ into the cell. $Ca^{2+}$ can be pumped out of the cell by pumps in the plasma membrane utilizing the energy of ATP and similar pumps are used to pump $Ca^{2+}$ from the intracellular store in the endoplasmic reticulum (ER). $Ca^{2+}$ can be released from the (ER) intracellular store in response to a hormonal signal (releasing $IP_3$ that causes the stored $Ca^{2+}$ to be released into the cytosol). $Ca^{2+}$ may be needed for depolarization to produce exocytosis of another protein to be released from the cell or $Ca^{2+}$ may be required for other functions, such as muscular contraction, cell proliferation, neuronal activity, or cell death.

A hormonal signal acting through a receptor at the cell membrane may operate through the **phospholipase C pathway** to yield **diacylglycerol (DAG)** and **inositol-1,4,5-*tris*phosphate ($IP_3$)** from **phosphoinositol-*bis*phosphate ($PIP_2$)**. $IP_3$ is an intracellular signal for $Ca^{2+}$ release from the ER $Ca^{2+}$ store. $IP_3$ binds to the **$IP_3$ receptor ($IP_3R$)** on the ER membrane to generate the release of $Ca^{2+}$ from the store into the cytosol. In some cell types $Ca^{2+}$ is released from the ER $Ca^{2+}$ store by **cyclic ADP ribose (CADPR)**.

$IP_3$ is not the only evocator of $Ca^{2+}$ from the calcium stores and there are various channels for the entry of $Ca^{2+}$ into the cell as well. There are three types of channels, admitting $Ca^{2+}$ from outside the cell, activated by ligands or voltage-mediated channels. The intracellular pool of $Ca^{2+}$ is regulated by various sensors, like **calmodulin** as well as by the transport of $Ca^{2+}$ into organelles. $Ca^{2+}$ enters the **sarco/endoplasmic reticulum** through the **sarcoplasmic reticulum $Ca^{2+}$ ATPase**, known as the **SERCA pump**. $Ca^{2+}$ is released through $Ca^{2+}$-modulated channels that also require $IP_3$ or other $Ca^{2+}$ releasing agents, such as cyclic ADP ribose (CADPR) or **nicotinic acid $ADP^+$ ($NAADP^+$)**. CADPR acts on **Ryanodine receptors** in the endoplasmic reticulum. Ryanodine receptors are very large ion channels that release $Ca^{2+}$ from the sarco/endoplasmic reticulum (sarcoplasmic reticulum refers to the smooth endoplasmic reticulum in smooth muscle and in striated muscle). $NAADP^+$ mobilizes the **endolysosomes** (lysosomes are important $Ca^{2+}$ storage organelles) through two-pore channels. Both CADPR and $NAADP^+$ are synthesized by a ubiquitous enzyme called CD38. **CD38 (cluster of differentiation 38)** is **cyclic ADP ribose hydrolase,** a novel multifunctional enzyme expressed widely in cells and tissues and is significant in **leukocytes**. The mitochondria also take up $Ca^{2+}$ and release it. $Ca^{2+}$ enters the **mitochondrion** through a low affinity electrophoretic importer and $Ca^{2+}$ can leave the mitochondrion through a **$Na^+/Ca^{2+}$ release exchanger (MNCX)**. The **nuclear envelope** also serves as a $Ca^{2+}$ reservoir and has similar transporters and channels as the ER. The various channels for uptake of $Ca^{2+}$ into the cell, the intracellular stores, and $Ca^{2+}$ movements between cytosol and mitochondria or the nucleus are summarized in Fig. 18.15 that also presents the chemical structures of $IP_3$, CADPR and $NAADP^+$, and the enzymatic reaction of **cyclic ADP ribose hydrolase.**

The classical release of $Ca^{2+}$ from the endoplasmic reticulum store operates through $IP_3$ as discussed. $IP_3$ (Fig. 18.15B) is a product of the **phospholipase C** reaction following a suitable external signal as shown in Fig. 18.16.

A number of **G-protein-linked receptors** give rise to $IP_3$ and lead to an increase in cytoplasmic $Ca^{2+}$. The cellular response to $Ca^{2+}$ and PKC, in many cases, is mitogenesis and this result also can be generated through the agency of

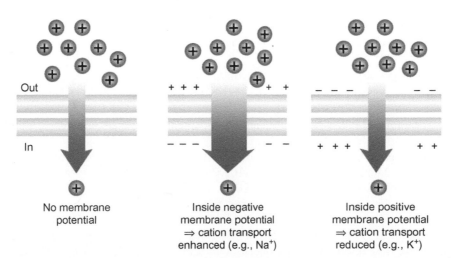

**FIGURE 18.14** Electrochemical gradient established by charges on both sides of the cell membrane. The size of the arrow indicates the relative flux of solute entering the cell.

(A)

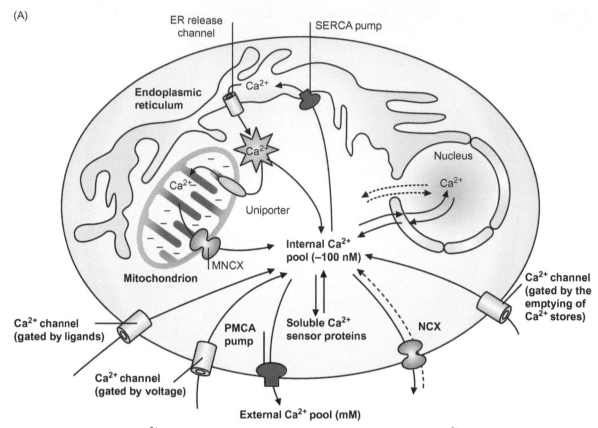

**FIGURE 18.15**   (A) Movements of $Ca^{2+}$ from outside the cell to the cell interior and the movements of $Ca^{2+}$ into intracellular stores and its release from those stores. *Dashed arrows* indicate postulated movements. *Dashed arrows* in the nucleus refer to possible traffick through the nuclear pores. $Ca^{2+}$ is expelled from the cell by a high affinity, **low capacity plasma membrane $Ca^{2+}$ ATPase (PMCA)** pump and a **low affinity/high capacity $Na^+/Ca^{2+}$ exchanger (NCX)**. Under certain conditions NCX can mediate $Ca^{2+}$ influx (*dashed arrow*). The release route from the sarco/endoplasmic reticulum creates $Ca^{2+}$ hotspots (*yellow star*) that activate the low affinity electrophoretic importer of neighboring **mitochondria**. $Ca^{2+}$ can exit through a **$Na^+/Ca^{2+}$ exchanger (MNCX)** and a **$H^+/Ca^{2+}$ release exchanger** has also been described although it is not shown in this figure. *SERCA*, sarco/endoplasmic reticulum $Ca^{2+}$ ATPase. (B) Chemical structures of inositol *tris* phosphate, CADPR, and $NAADP^+$. C. The **cyclic ADP ribose hydrolase (CD38)** reaction. The product split off can be either nicotinic acid or nicotinamide depending on the substrate. *Reproduced from http://www.nature.com/nrm/journal/v4/n4/fig_tab/nrm1073_F2.htm.*

**tyrosine kinase-linked receptors** generating **PIP₃** and activation of **MAP kinase**. These receptors and their pathways are shown in Fig. 18.17.

## Entry of Magnesium and Divalent Ions into Cells

Human receptors for $Mg^{2+}$ are the subject of current studies. There appear to be several transporters for $Mg^{2+}$. Transport of other divalent ions ($Fe^{2+}$, $Co^{2+}$, $Ni^{2+}$, $Cu^{2+}$, $Zn^{2+}$, $Sr^{2+}$, $Cd^{2+}$, $Ni^{2+}$, and $Ba^{2+}$) needs to be studied. Calcium ions represent an important second messenger and have been considered above. $Mg^{2+}$, $Fe^{2+}$, and $Zn^{2+}$ are important components of biological macromolecules and certain other trace metals are cofactors for specific enzymes. There are a number of transporters for $Mg^{2+}$ as summarized in Fig. 18.18.

The **$Mg^{2+}$ transporters** are found in various locations in the cell, including plasma membrane (TRPM7,TRPM6, MagT1,TUsc3, SLC41A1,SLC41A2, SLC41A3,ACDP1,ACDP2, ACDP3, ACDP4, NIPA1, NIPA2, NIPA3, NIPA4), **subplasma membrane vesicles** (HIP14, HIPIL, MagC1), **mitochondria** (Mrs2p), **Golgi apparatus** (MMgT1, MMgT2), and **post-Golgi vesicles** (MMgT1, MMgT2). Some of these channels possess properties of bacterial and yeast proteins, however, more newly discovered transporters (TRPM6/7, MagT, NIPA, and HIP14) do not share counterparts in prokaryotic genomes and these proteins are located in many different tissues and subcellular organelles. These proteins have no obvious similarities in amino acid sequence indicating a diversity of $Mg^{2+}$ transport across cell membranes. These channels also transport a number of different cations across membranes. In addition to $Mg^{2+}$, SLC41A2 transports a range of divalent ions: $Ba^{2+}$, $Ni^{2+}$, $Co^{2+}$, $Fe^{2+}$, and $Mn^{2+}$ but not $Ca^{2+}$, $Zn^{2+}$, or $Cu^{2+}$, however, MagT1

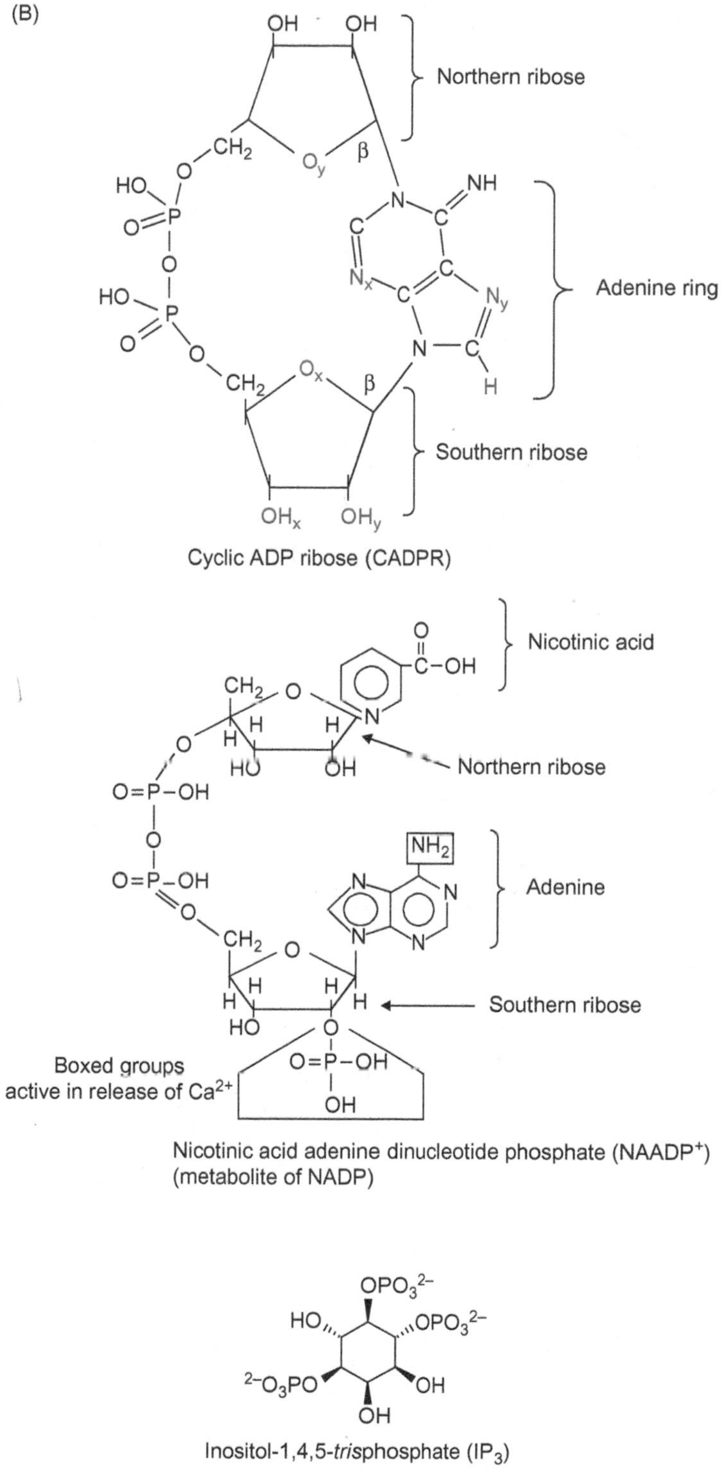
(B)
Northern ribose
Adenine ring
Southern ribose
Cyclic ADP ribose (CADPR)
Nicotinic acid
Northern ribose
Adenine
Southern ribose
Boxed groups
active in release of Ca2+
Nicotinic acid adenine dinucleotide phosphate (NAADP+)
(metabolite of NADP)
Inositol-1,4,5-trisphosphate (IP3)

**FIGURE 18.15** Continued

and NIPA2 are selective only for $Mg^{2+}$. There are congenital disorders (e.g., **ALS**, **Parkinson's**, **dementia**, **Huntington's**) associated with mutated genes for many of these transporters that involve the intestine, kidney, brain, nervous system, and skin.

The **ancient conserved domain protein 2** (**ACDP2**) transports $Mg^{2+}$ with saturable uptake and a Michaelis constant of 0.56 mM in a voltage-dependent manner without being coupled to either $Na^+$ or $Cl^-$ ions. This transporter is

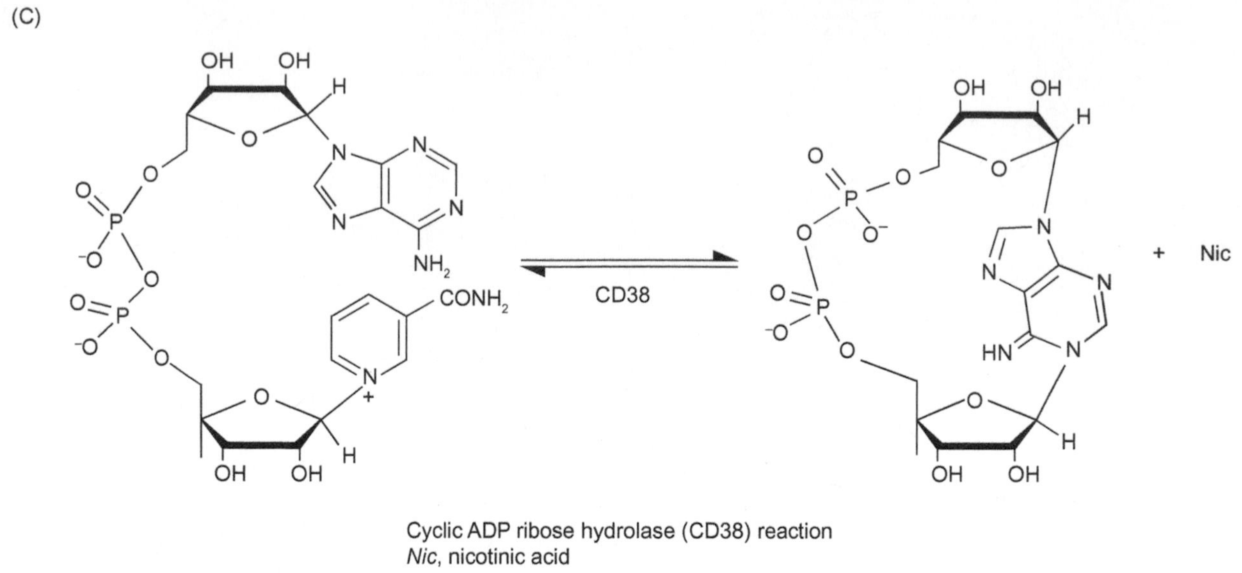

(C)

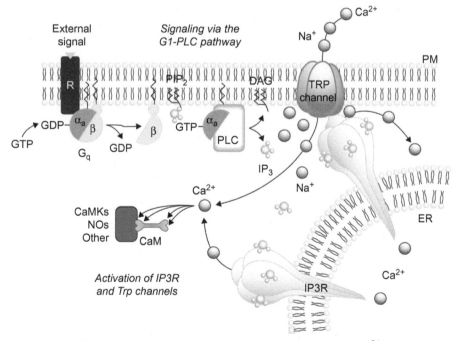

Cyclic ADP ribose hydrolase (CD38) reaction
*Nic*, nicotinic acid

**FIGURE 18.15** Continued

**FIGURE 18.16** External signaling pathway leading to the release of endoplasmic reticulum stores of $Ca^{2+}$ into the cytoplasm as well as uptake of $Ca^{2+}$ from the extracellular space through a **Na$^+$/Ca$^{2+}$ cotransporter** (**TRP channel**). *PM*, plasma membrane; *ER*, endoplasmic reticulum; *PLC*, phospholipase C; *IP3R*. inositol-1,4,5-*tris* phosphate receptor; *CaM*, calmodulin; *CaMKs*, calmodulin kinases; *NO*, nitric oxide *R*, receptor; *DAG*, dia-cylglycerol. *Originally redrawn from http://dir.niehs.nih.gov/dir1st/groups/birnbaumer/images/fig.pathway.jpg (no longer on Internet).*

located in kidney, brain, and heart with lesser quantities in liver, small intestine, and colon. In addition to transporting $Mg^{2+}$, it also can transport $Co^{2+}$, $Mn^{2+}$, $Sr^{2+}$. $Ba^{2+}$, $Cu^{2+}$, and $Fe^{2+}$, but not $Ca^{2+}$, $Cd^{2+}$, $Zn^{2+}$, or $Ni^{2+}$. This trans-porter is up-regulated as a result of **Mg$^{2+}$ deficiency** and this occurs in the distal convoluted tubule cells, kidney, heart, and brain.

The **brush border membrane** of the upper small intestine contains the **divalent ion transporter (DMT1)** that transports $Mn^{2+}$, $Fe^{2+}$, $Co^{2+}$, $Ni^{2+}$, and $Cu^{2+}$. This is a **high affinity copper transport system** powered by ATP that could be the central route for the entry of copper ions in the intestine.

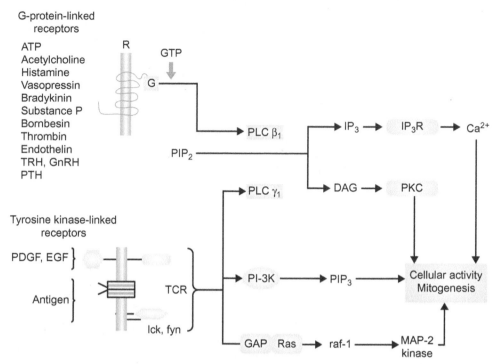

**FIGURE 18.17**   Summary of the two major receptor-mediated pathways for stimulating the formation of inositol *tris*phosphate (IP3) and diacylglycerol (DAG) and for activation of PI3K and MAP kinase, all of which can lead to **mitogenesis**. Many agonists bind to **seven-membrane spanning receptors** that use a **GTP-binding protein** (**G**) to activate phospholipase C-1 (PLC-1) and PLC-1 can be stimulated by tyrosine kinase-linked receptors. The latter activate other effectors, such as phosphatidylinositol-3 kinase (PI3K) that generates the putative lipid messenger **phosphatidylinositol *tris*phosphate** and **GTPase-activating protein** (**GAP**) that regulates *Ras*. *GnRH*, gonadotropic-releasing hormone; *IP3R*, IP3 receptor; *PKC*, protein kinase C; *PTH*, parathyroid hormone; *TRH*, thyrotropin-releasing hormone.

## Proton ($H^+$) Transport

Proton transporters are designed to pump protons from the cellular cytoplasm into a medium where acidification is needed to balance pH. This transporter also regulates intracellular pH. The **sodium-proton antiport** is a transporter in which a sodium ion and a proton move in opposite directions. When these ions move across a lumenal membrane the medium of the lumen becomes acidified. One effect of this transport allows **bicarbonate ion** to move into the cell from the lumen against its concentration gradient. The $Na^+/H^+$ antiport is inhibited by **amiloride**, classifying this transporter as amiloride-sensitive. Consequently, amiloride reduces the uptake of $Na^+$ into cells. As the movement of $Na^+$ is followed by the uptake of water, amiloride, a **potassium-sparing diuretic**, can be used in hypertension and edema. The $Na^+/H^+$ antiporter, a pH regulator, is a component of most cells. A model of acidification of a lumen by the $Na^+/H^+$ transporter and the consequent movement of **bicarbonate ion** are shown in Fig. 18.19 that also presents the chemical structure of amiloride (Fig. 18.19B).

Nine isoforms of **NHE** (**sodium-proton exchanger = $Na^+/H^+$ antiporter**) occur in the human and their locations are in different organelles, such as the **mitochondria** and in different tissues. They all operate to regulate **cytosolic acidification** and **osmotic cell shrinkage** by a similar mechanism (Fig. 18.19C). The C-terminal regulatory domain, containing about 315 amino acid residues, mediates reactions with cytoskeletal components (e.g., **actin**). Transmembrane segments 4, 7, and 9 are important to the **antiporter function**. Although the transport of ions by this antiporter is not directly linked to the hydrolysis of ATP, the effects of ATP depletion in the cell upon osmotic activation and cellular volume regulation affect the functions of at least NHE1, NHE2, and NHE3. Furthermore, there are major phosphorylation sites in the C-terminal segment of the antiporter.

In response to **apoptotic stress**, NHE1 is initially stimulated by **PI(4,5)P$_2$** binding for protection of the cell. The transporter subsequently activates **PI3-kinase** to phosphorylate **PIP$_2$** to form **PIP$_3$** (**PI(3,4,5)P$_2$**). PIP$_3$ phosphorylates and thus stimulates **Akt** to enhance the **cell survival pathway** while depressing NHE1 activity and tends to deplete the pool of PIP$_2$. Thus, phosphoinositide binding can regulate NHE1 and protect the cell in the face of apoptotic stress.

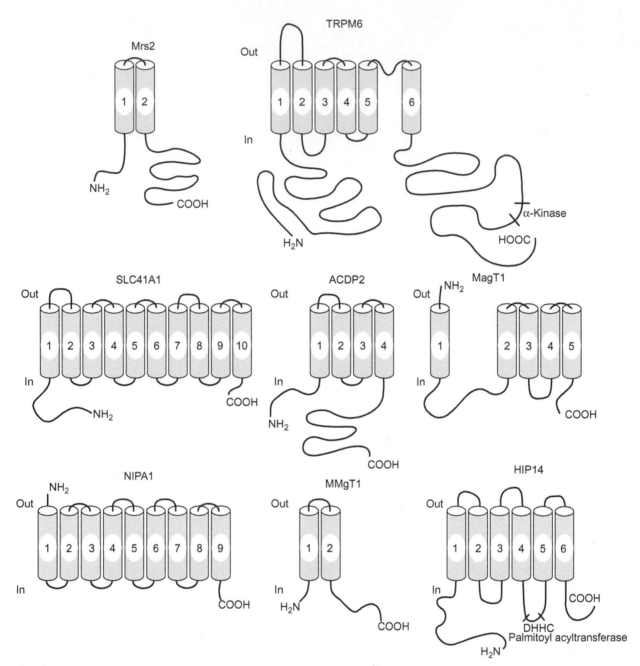

**FIGURE 18.18** Topological models of representative members of mammalian $Mg^{2+}$ transporters showing their structural diversity. Amino acid sequence similarity of transporters within individual families is >85%. *Reproduced from Quamme, G.A., 2010. Molecular identification of ancient and modern mammalian magnesium transporters. Am. J. Cell. Physiol. 298, C407−C429.*

This transporter is involved in cell growth and differentiation, cell migration, and in regulation of **sodium fluxes**. Dysfunction of this transporter is related to hypertension, cardiac hypertrophy, heart failure, and epilepsy.

## Monocarboxylate Transporter (MCT), a Cotransporter (Symporter)

Another proton-linked channel is the **monocarboxylate transporter** (**MCT**). This transporter moves a monocarboxylic acid, such as **lactate** or **pyruvate** together with the transport of a **proton** (or **sodium ion**) in the same direction, defining MCT as a cotransporter or symport (Fig. 18.13A). MCT is essential to cellular metabolism at the plasma membrane and the mitochondrial membrane (Fig. 18.20).

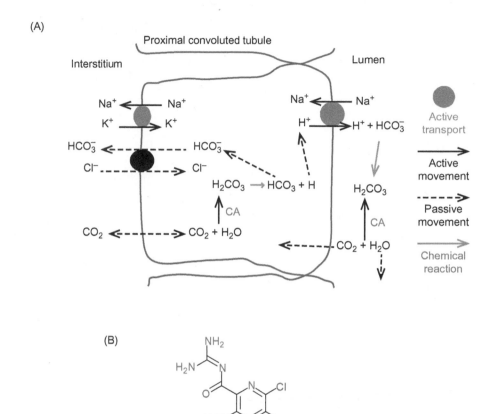

(A)

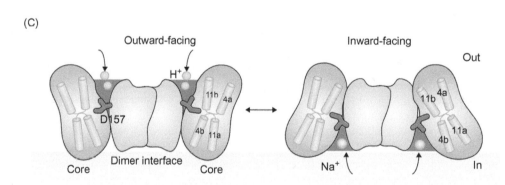

**FIGURE 18.19** (A) Cellular model of acidification involving the $Na^+/H^+$ antiport. (B) Structure of amiloride. (C) Model of a $Na^+/H^+$ antiporter in the plasma membrane of every cell. The transporter consists of two domains, a core domain, and a dimerization domain. A cavity is located between the two domains revealing access to the **ion-binding site**. The transporter is regulated by pH and it becomes active above pH 6.5, producing a conformational change. A negatively charged cavity opens to the outside allowing access to a conserved Asp residue that coordinates ion binding. To alternate access to the ion-binding site requires a large rotation of the core domain ($\sim 20°$) against the dimerization interface. The transporter operates by a 2-domain rocking bundle model and transports about 1500 ions/second. *Left*: NHE1 (sodium-proton exchanger 1) is open to the exterior where $H^+$ is released to the extracellular fluid. *Right*: Conformationally changed transporter now open to the cytoplasm allowing extracellular $Na^+$ to bind to the transporter site for passage to the intracellular cytoplasm. *Reproduced from http://www.nature.com/nature/journal/vaop/ncurrent/full/nature12484. html?WT.ec_id = NATURE-20130905.*

MCTs are important in many tissues, such as basolateral intestinal cells, blood–brain-barrier brain cells, and kidney. Present in human muscles are **MCT1, MCT3,** and **MCT4**. These transporters have different properties and differing functional roles. There is a strong correlation between skeletal muscle MCT1 and **muscle oxidative capacity** as well as capacity to absorb **lactate** from the bloodstream. As slow twitch fibers have a higher capacity than fast twitch fibers, the distribution of the transporter is fiber-dependent. The efflux of lactate and protons resulting from intense muscle

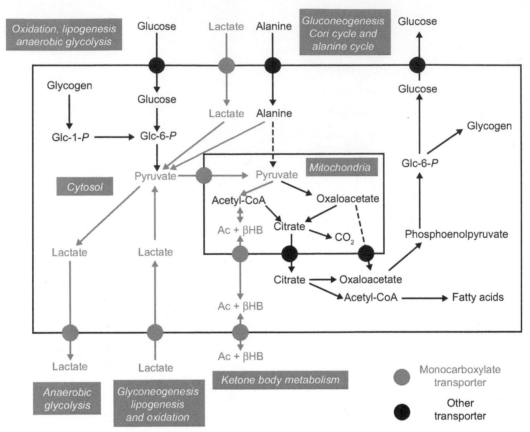

**FIGURE 18.20**    Metabolic pathways involving monocarboxylate transport across the plasma membrane and mitochondrial membrane. *Glc-1-P*, glucose-1-phosphate; *Glc-6-P*, glucose-6-phosphate; *AC + βHB*, acetoacetate + β-hydroxybutyrate. *Redrawn from Halestrap, A.P., Price, N.T., 1999. The proton-linked monocarboxylate transporter (MCT) family: structure, function and regulation. Biochem. J. 343, 281–299.*

activity and recovery are mediated principally by the **lactate-proton transporter** which accounts for the reduction in muscle lactate concentration and the fall in pH involved in **muscle fatigue**. Thus, MCT regulates **muscle cell pH** associated with muscle activity. The uptake of lactate into resting muscle as well as other tissues speaks to the importance of MCT. Intense training and reduced muscle inactivity can direct adaptive changes in this transporter owing to the need to increase the transport of lactate and protons.

The MCT family of transporters has 12 transmembrane domains. The N-terminal region is important for energy coupling, membrane insertion, or correct structural maintenance; the C-terminal domain recognizes the substrate (**pyruvate or lactate**) and binds it. A conserved Asp residue in the N-terminal half of the protein probably binds a proton that is cotransported. The essential Asp resides as the second residue on the cytoplasmic side before the start of the transmembrane domain. There is an Arg residue in position 313 in the C-terminus of human MCT1 that appears to bind the carboxylate anion (of pyruvate or lactate).

## Citrate Transporters in Mitochondria and Plasma Membrane

The **citrate transport protein (CTP)** moves a molecule of citrate from the mitochondrial inner membrane to the soluble cytoplasm. The movement of citrate outwardly from the mitochondrion to the cytoplasm is exchanged for a molecule of **malate** that moves in an **electroneutral exchange** from the cytoplasm to the mitochondrion, a form of antiport action. CTP is one of many transport proteins in the inner membrane of the mitochondrion and these constitute the **mitochondrial carrier family** (MCF) derived from the *SLC25* gene. These transporters probably function as homodimers like many other transporters (e.g., see Figs. 18.19C and 18.20).

Cytosolic citrate becomes a carbon source for **gluconeogenesis** and **fatty acid synthesis**. Citrate also regulates glucose metabolism by virtue of the fact that it exerts a negative allosteric effect on **phosphofructokinase**. The gene for this transporter maps to human chromosome 22 (22q11.21) in a region that is associated with allelic losses that

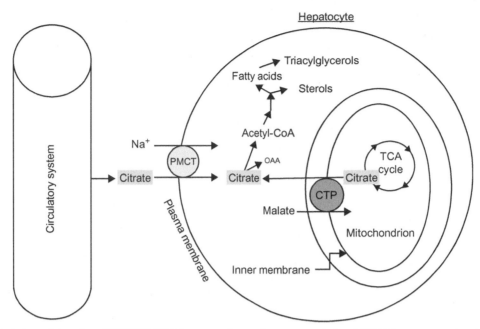

**FIGURE 18.21** Depictions of the roles of PMCT/NaCT (plasma membrane citrate transporter) and CTP (citrate transport protein) in supplying citrate to fuel hepatic fatty acid, triacylglycerol and sterol biosyntheses. Citrate can be transported from the blood across the hepatocyte plasma membrane into the cytoplasm on the PMCT or it can be effluxed from the mitochondrial matrix across the mitochondrial inner membrane on the CTP. Cytoplasmic citrate derived from either source is then broken down to **acetyl-CoA** by **citrate lyase** and the resulting acetyl-CoA provides all of the carbon precursor to fuel the fatty acid, triacylglycerol, and sterol biosynthetic pathways. *Reproduced from Sun, J., et al., 2010. Mitochondrial and plasma membrane citrate transporters: discovery of selective inhibitors and application to structure/function analysis. Mol. Cell. Pharmacol. 2, 101–110.*

give rise to clinical diseases including the **DiGeorge syndrome, velo-cardio-facial syndrome,** and a subtype of **schizophrenia**. The DiGeorge syndrome, producing multiple lesions of the heart, results from a microdeletion of chromosome 22q11. Patients with this syndrome have abnormalities of the palate, including cleft palate and facial irregularities with a small mouth, wide-set and down-slanting eyes, a long face, other structural defects, and sometimes learning disorders. Parents carrying the **22q11 microdeletion** (in the cell nucleus, not a mitochondrial gene) pass the deletion to 50% of their offspring. **Alcohol consumption** during fetal development may be a factor in production of the microdeletion.

Another citrate transporter is coupled to sodium ion transport (NaCT or PMCT) in the same direction (symporter). It functions principally in the liver cell membrane (also found in the brain and testes) for the uptake of citrate (or other dicarboxylates or tricarboxylates with lower affinity) from the bloodstream into the liver cell (Fig. 18.21). This transporter differs in structure and function from the mitochondrial citrate transporter.

The **mitochondrial citrate transporter** moves citrate into the cytoplasm in exchange for **malate** (mainly) functioning as an antiport system. NaCT/PMCT contains 572 amino acid residues and has structural similarities to the $Na^+$-dicarboxylate cotransporter/$Na^+$-sulfate cotransporter gene family. NaCT is important for the uptake of citrate from the blood for fatty acid and cholesterol syntheses in the liver and for the generation of energy in the brain. **Lithium** can replace two of the four $Na^+$ binding sites in the human resulting in enhanced activity of the transporter, a feature that may relate to the use of lithium in **bipolar disease**. Amounts of lithium that stimulate NaCT are equivalent to the concentrations found in humans treated for this disease.

## Intestinal Transport of Di- and Tripeptides

Digested proteins in the stomach and intestine produce tripeptides, dipeptides, and free amino acids. Di- and tripeptides can be absorbed as such in the intestine. Their transport across the human luminal border of the intestinal cell is mediated by the **proton-dependent dipeptide transporter (PEPT1)**. PEPT1 (also known as oligopeptide transporter 1) is located in the apical membrane of intestinal **enterocytes** operating as an **electrogenic proton-peptide** (di- and tripeptides) transporter (**cotransporter**). **PEPT1** is also known as solute carrier family 15 member 1 (SLC15A1). This transporter does not transport free amino acids or peptides containing four or more amino acid residues. In addition to its location in the **intestinal brush border**, it is responsible for the reabsorption of oligopeptides in the kidney. The number of protons cotransported depends on whether the peptide to be transported is neutral, singly charged, or multiply charged. This transporter has a large capacity and, in addition to transport of di- and tripeptides, it can take up **peptidic drugs**, such as the **cephalosporins** as well as **amino acid-conjugated drugs**. Single nucleotide polymorphisms can occur in the **gene for PEPT1** (*SLC15A1* **gene**) that result in reduced cotransport activity.

Another transporter of interest is **PEPT2**, a member of the **proton-dependent oligopeptide transporters** (**POT**) or **peptide transport family** (**PTF**) gene family by virtue of similarities in amino acid sequence. The POT family is a family within the **major facultative superfamily** (**MFS**) of transporters. The POT family has sequences in the range of 600 amino acid residues in length and contains 12 transmembrane $\alpha$-helical spanning regions. There are at least two members of the POT family in the human, **hPHT1** (**human peptide/histidine transporter 1**, or SLC15A4) and **hPHT2**. hPHT1 is located primarily in skeletal muscle and spleen and hPHT2 is in spleen, placenta, lung, leukocytes, and heart.

## Amino Acid Transporters

Amino acid transporters exist in many epithelial cell types, especially in those cells that border a lumen. Functional cooperation occurs between the transporters on the apical membrane and the basolateral membrane of mammalian epithelial cells. In the apical membrane, there are **amino acid-sodium ion symporters** in addition to the **proton motive force** (energy generated by transfer of protons or electrons across an energy transducing membrane) plus the existing gradient of other amino acids that assure absorption of amino acids from the lumen.

**Apical transporters** in the **kidney** and **intestine** supply amino acids to all tissues including blood plasma. In the basolateral membranes of these epithelial cells, there exist **antiporters** for the release of amino acids to the extracellular space and ultimately to the bloodstream without depleting the apical cell of necessary nutrients. Individual amino acids have more than one transporter, providing a reserve capacity when an inactivating mutation occurs in a transport system. However, there are genetic disorders that result from inactivated epithelial transporters and these can lead to diseases, such as **cystinuria, dicarboxylic acid aminoaciduria, iminoglycinuria, lysinuric protein intolerance,** and **Hartnup disorder**. Hartnup disorder is an autosomal recessive disorder affecting the absorption of **nonpolar amino acids**, especially tryptophan, caused by a mutation in the gene *SLC6A19* on chromosome 5. A sodium ion-dependent and chloride ion-independent neutral amino acid transporter, predominantly in **kidney** and **intestine** is affected. The **apical transporters** in the kidney and intestine responsible for moving neutral amino acids are summarized in Fig. 18.22A.

The **amino acid transporter family** has many transporters and they are assembled based on the specificity of the amino acid molecule being transported as well as dependence on sodium ion for transport activity.

The L-**amino acid transporter** (**LAT1**), contains a **12-membrane-spanning domain** (**MSD**) and differs from the other 12 MSD transporters by an additional single membrane spanning protein, **4F2 heavy chain** (**4F2hc:CD98**). **LAT1** transports neutral amino acids (e.g., leucine, isoleucine, and valine) through the plasma membrane, independently of sodium ion. Also, LAT1 transports aromatic (e.g., **DOPA**) and branched chain amino acids. It is a major transporter in the blood–brain barrier (**BBB**) system. The heavy subunit participates in the movement of the heterodimer to the plasma membrane, whereas the light subunit is responsible for the transport function and amino acid specificity. Fifteen similar transporters have been recognized in the human **placenta** functioning to transport amino acids.

The central system for the transport of cationic amino acids is **system y$^+$** that functions independently of Na$^+$. y$^+$ can transport **lysine** from the maternal side to the fetal side.

**hNAT3** is a human amino acid transporter with highest specificity for L-**alanine**. It is a low affinity transporter that is dependent on Na$^+$ and pH. Some of its Na$^+$ binding sites can be occupied by **lithium ion**, a possible occurrence in **bipolar patients** being treated with lithium. Its main location is the liver with lesser amounts in the muscle, kidney, and pancreas. This transporter consists of 547 amino acid residues and its gene is located on human chromosome 12 (12q12-q13).

**ATB$^{o+}$** or **hATB$^{o+}$** is a novel human member of the **Na$^+$/Cl$^-$-dependent neurotransmitter transporter family**. It shares some sequence similarities with **glycine transporters** and **proline transporters**. It is located in trachea, salivary gland, mammary gland, stomach, and pituitary gland. It has its highest affinity for **hydrophobic amino acids** and also transports neutral and cationic amino acids. Its function depends on Na$^+$ and Cl$^-$. Its function is inhibited by inhibitors of the **B$^{o+}$ transporter system**, thus characterizing it as a member of this system of transporters (Fig. 18.22A).

Another system is the **cystine-glutamate antiporter** that contains both heavy (4F2hc) and light (xCT) subunits. xCT has 12 transmembrane domains. It exchanges extracellular anionic L-**cystine** for intracellular L-**glutamate** in a 1:1 exchange (system is also called x(c)($-$) or $x_c^-$). As there is a low intracellular concentration of cystine, this transporter can generate with ease the efflux of a high intracellular concentration of L-glutamate. As intracellular cystine is rapidly reduced to **cysteine** followed by the synthesis of glutathione (**GSH**), xCT provides an **antioxidant defense mechanism**

(A)

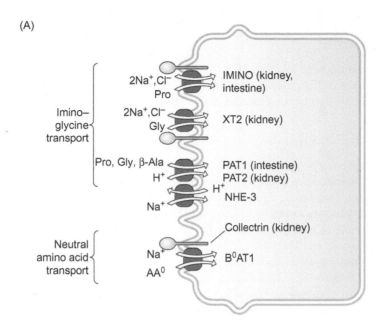

(B)

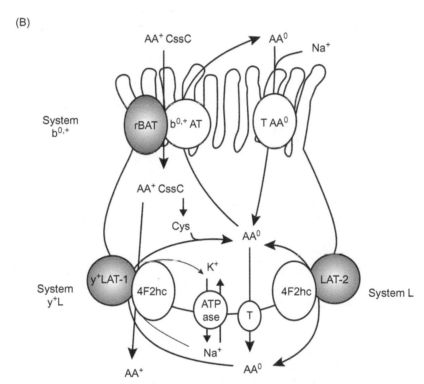

**FIGURE 18.22** (A) **Apical amino acid transporters** of neutral amino acids in intestine and kidney. Their interactions with other proteins are indicated. Tissue specificity is stated in brackets. Substrate specificity and mechanism of each transporter are indicated. *PAT1*, H + /amino acid transporter; *NHE3*, cation proton antiporter; *B⁰AT1*, in the kidney associated with cystinuria, Na⁺/amino acid transporter; *IMINO*, transporter of proline, and hydroxyproline in intestine/kidney; *XT2*, imino amino acid transporter in renal epithelia (an orphan transporter; *Collectin*, C type lectin with collagen regions that binds mannose). Transporters in the **basolateral membrane** are shown in (B). Model for the reabsorption of cystine, dibasic, and neutral amino acids in a **proximal tubule cell line** OK (opposum kidney). Apical membrane is at the *top*; basolateral membrane is at the *bottom*. **rBAT** heterodimerizes with **b⁰·⁺**. **AT** is in the apical membrane of **renal epithelial cells**. The complex mediates the active reabsorption of **dibasic amino acids (AA⁺)** and **cystine (CssC)** coupled to the exchange of intracellular **neutral amino acids (AA⁰)** which enter the cell by **Na⁺-dependent transporters** located at both apical and basolateral plasma membranes (**TAA⁰**) (only those at the apical side are depicted). Membrane potential favors the uptake of dibasic amino acids. **Cystine uptake** is favored by its reduction to **cysteine (Cys)** associated with **glutathione oxidation** (GSH → GssG). Dibasic amino acids are then released through the basolateral membrane by **y⁺LAT-1/4F2hc complex**. This efflux is also coupled to the influx of neutral amino acids plus Na⁺. At the basolateral membrane, **4F2hc/LAT-2** accounts for the **efflux of neutral amino acids** including the cyctine-derived cysteine. When equal concentrations of neutral amino acids are applied on both sides of the epithelium, the direction of the exchange favors the net exit of cysteine and the net influx of alanine, serine, or threonine. Another system (T) at the basolateral membrane would account for the net efflux of neutral amino acids. The heteromeric amino acid transporter cDNAs cloned from American opossum are shaded. *(A) Reproduced from http://physiologyonline.physiology.org/content/23/2/95/F1.expansion. (B) Reproduced from http://jasn.asnjournals.org/content/14/4/837/F7.expansion.*

particularly in regions of **inflammation**. The location of xCT in brain suggests that it maintains the redox state in **cerebrospinal fluid**. It may play a role in certain diseases of the CNS and the eye. As tumors require extracellular cystine/ cysteine, inhibitors of this transport system may arrest **tumor growth**.

## The Glutamate Synapse and Excitatory Amino Acid Transporters

The **glutamate synapse** occurs between the terminus of the axon of a neuron that accumulates L-glutamate from **glutamine** in **secretory vesicles** and the surface of a **dendritic spine** (a protrusion of the membrane from a neuronal projection that receives input from the synapse of an axon). The glutamate synapse contains vesicles that import glutamate from the cytosol in exchange with protons, a **glutamate-proton antiport**. The vesicular antiporter is highly specific for L-glutamate and utilizes a favorable proton gradient to exchange a proton for glutamate against its concentration gradient. ATP drives the protons to the interior of the vesicle and the resulting high proton concentration inside the vesicle is utilized by the **antiporter**. There are three **vesicular glutamate transporters** (VGLUT1, VGLUT2, and VGLUT3). VGLUT1 and VGLUT2 predominate in the adult brain. VGLUT1 is highly concentrated in cerebral and cerebellar cortices and the **hippocampus** while VGLUT2 is highly expressed in the diencephalon, brainstem, and spinal cord. VGLUT3 is less widely expressed than VGLUT1 and VGLUT2 and it is uncharacteristically found in cholinergic, serotonergic, and GABAergic neurons. Thus, VGLUT3 may play a unique role in **unconventional glutaminergic neurotransmission**. VGLUT2 predominates in the first 2 weeks of postnatal development. These transporters mediate glutamate uptake into synaptic vesicles driven by a proton electrochemical gradient generated by $H^+$-**ATPase** in the vesicle. Glutamate transport by all three antiporters is dependent biphasically on the concentration of $Cl^-$. In addition to this antiporter, the high proton level inside the vesicle also is coupled to a **vesicular monoamine transporter** (**VMAT**).

Glutamate is an **excitatory amino acid** and **excitatory amino acid transporters** (**EAATs**) are important in the central nervous system. Glutaminergic neurotransmission at excitatory synapses is terminated by sodium-ion dependent L-glutamate transporters on the plasma membranes of neurons and astroglia. Extracellular L-glutamate is efficiently cleared by glutamate transport and the sodium gradient across the membranes. In addition to glutamate, the other excitatory amino acids, **L-aspartate** and **L-cysteine** are also moved by this transporter. There are five **human excitatory amino acid transporters**: EAAT1-5 and, of these, EAAT1 and EAAT2 are the most abundant. EAAT1 and EAAT2 are located in **astroglia** throughout the central nervous system, excepting the **retina** where EAAT2 and EAAT5 are present. During development, the nervous system contains EAAT2. The **Perkinje cells** of the cerebellum have EAAT4 predominantly. EAAT3 is located in both neurons and astroglia and in other tissues, as well. These transporters remove L-glutamate from the extracellular space, thus preventing glutamate toxicity. **Deficiency of EAAT1** results in visual and movement abnormalities. Dysfunction of these transporters would be expected to be involved in **neurodegenerative diseases**, **central nervous system disorders,** and **psychiatric problems**. The locations and functions of these transporters are shown in Fig. 18.23.

Information about each excitatory amino acid transporter is shown in Table 18.6.

## FATTY ACID TRANSPORT PROTEINS

All tissues that use fatty acids have cell membrane associated proteins that are responsible for the transport of fatty acids from the extracellular space to the cytoplasm; these are **fatty acid transport proteins** (**FATP**) or fatty acid transporters. These transporters provide the link between extracellular fatty acids and **mitochondrial β-oxidation** to the breakdown product of fatty acids, **acetyl-CoA**. These connections are shown in Fig. 18.24.

The **fatty acid transport proteins** (**FATPs**) constitute a six-membered family, designated FATP1, FATP2, FATP3, FATP4, FATP5, and FATP6. Each FATP has a highly **conserved AMP-binding region** involved in the activation of **very long-chain fatty acid** (**VLCFA**) formation of acyl-CoA derivatives. Uptake of VLCFAs is related to **acyl-CoA synthase** activity. CoA esters are necessary for the structural integrity of **lipid rafts** in cellular membranes and these rafts are required for the cellular uptake of LCFAs. **Cholesterol** molecules in the membrane serve as the base for construction of lipid rafts and the formation of rafts is stimulated by **palmitoylation** (Fig. 18.25).

Specific proteins may be anchored to the raft by a **glycosylphosphatidylinositol anchor.**

Rafts can be aggregated in the plasma membrane. In one case, a **CD45 protein** (a receptor-like protein having a glycosylated external domain and a large cytoplasmic domain with two tyrosine protein phosphatases associated) has access to its substrates. CD45 is excluded in the aggregated state, a condition that allows the accumulation of **phosphotyrosines** on raft-associated proteins. Therefore, the **aggregation of rafts** becomes a signaling mechanism. **Fatty acid translocase** (**FAT/CD36**), the enzyme needed for the uptake of LCFAs such as oleic acid in adipocytes, is located

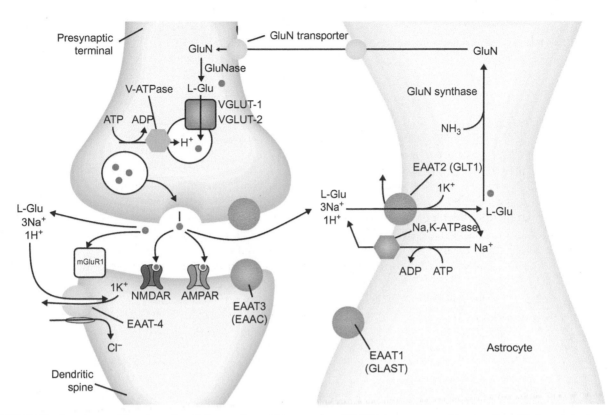

**FIGURE 18.23** Locations and functions of **excitatory amino acid transporters** (**EAATs**) in synapses and in astroglia. L-**Glutamate** is stored within vesicles of the **presynaptic neuronal terminal** by conversion of L-**glutamine** to L-glutamate by the action of **glutaminase**. The glutamate so produced is transported into the **storage vesicle** by the transporters, **vesicular glutamate transporter 1** and **vesicular glutamate transporter 2**. A gradient of protons is generated by **vesicular ATPase** and these protons can be exchanged for glutamate. Following an appropriate signal, glutamate can be released from the vesicle into the synaptic space and subsequently binds to the *N*-**methyl-D-aspartate receptor** (**NMDAR**) and the **AMPA** (α-**amino-3-hydroxy-5-methyl-4-isoxazolepropionic acid**) **receptor** on the membrane of the dendritic spine of the neuron receiving the glutamate signal. *GluN*, glutamine; *GluNase*, glutaminase; *V*, vesicular; *EAAT*, excitatory amino acid transporter. Small orange filled circles represent glutamate. *Reproduced from http://www.google.com/search?q = excitatory + amino + acid + transporter.*

exclusively on lipid rafts. **Transport of LCFAs** from the soluble cytoplasm into mitochondria is a function of the **mitochondrial carnitine system** as shown in Fig. 18.24. The transport system involves **malonyl-CoA-sensitive carnitine palmitoyltransferase** (**CPT-1**) in the outer mitochondrial membrane, **carnitine acylcarnitine translocase** in the inner mitochondrial membrane and **carnitine palmitoyltransferase II** on the matrix side of the inner mitochondrial membrane. The fatty acid is carried to the plasma membrane FAT/CD36 by a **fatty acid binding protein** (**FABP**) that allows entry of the fatty acid into the cytoplasm. The entry of fatty acids into subcellular particles is mediated by **fatty acid translocase** (**FAT/CD36**). In **cardiac myocytes**, the major stimulus for movement of FAT/CD36 from intracellular stores to the plasma membrane covering the outer surface of the muscle fiber (sarcolemma) is contraction. Phosphorylation of the translocase also is a process that can enhance the uptake of fatty acids.

## VOLTAGE-GATED SODIUM CHANNELS

Sodium channels play a role in the initiation and propagation of neuronal **action potentials** and they play important roles in endocrine cells and myocytes. In Fig. 18.26 is shown an action potential and the sequence of events involving the movement of **sodium ions** and **potassium ions**.

**TABLE 18.6**  Excitatory Amino Acid Transporters (EAATs)

| Currently Accepted Name | EAAT1 | EAAT2 | EAAT3 | EAAT4 | EAAT5 |
|---|---|---|---|---|---|
| **Alternate Name** | GLAST | GLT-1 | EAAC1 | None | None |
| **Structural Information** | 542 aa (human) | 574 aa (human) | 525 aa (human) | 564 aa (human) | 561 aa (human) |
| **Uptake Inhibitors** | *trans*-2,4-PDC (P 7575)[a] | *trans*-2,4-PDC (P 7575)[a] Dihydrokainate (D 1064)[b] Kainate (K 0250)[b] TBOA[c] *trans*-2,3-PDC[d] | *trans*-2,4-PDC (P 7575)[a] | *trans*-2,4-PDC (P 7575)[a] | *trans*-2,4-PDC (P 7575)[a] |
| **Radiolabeled Substrates**[e] | L-[³H]-Glutamate | L-[³H]-Glutamate | L-[³H]-Glutamate | L-[³H]-Glutamate | L-[³H]-Glutamate |
|  | D-[³H]-Aspartate | D-[³H]-Aspartate | D-[³H]-Aspartate | D-[³H]-Aspartate | D-[³H]-Aspartate |

***trans*-2,3**-PDC: L-*trans*-Pyrrolidine-2,3-dicarboxylic acid

***trans*-2,4**-PDC: L-*trans*-Pyrrolidine-2,4-dicarboxylic acid

**TBOA:** threo-β-Benzyloxyaspartate

**Abbreviations**

[a]*trans*-2,4-PDC is a nonselective EAAT inhibitor. $K_i$ values for EAATs 2, 4, and 5 are between 5 and 10 $\mu M$ and for EAAT1 and EAAT3 are between 50 and 100 $\mu M$.
[b]Selective inhibitors of the EAAT2 subtype with $K_i$ values between 15 and 60 $\mu M$.
[c]TBOA is a nonsubstrate inhibitor of EAAT2 and has seven-fold selectivity over EAAT1. Not fully characterized with respect to other EAATs. Also inhibits EAAT3. Nonselective.
[d]*trans*-2,3-PDC is a selective nonsubstrate inhibitor of EAAT2 (compared with EAAT1 and EAAT3).
[e]L-Glutamate as the endogenous substrate and D-aspartate, a nonmetabolizable substrate analogue are most frequently employed in uptake studies.
Source: Reproduced from http://www.sigmaaldrich.com/technical-documents/articles/biology/rbi-handbook/transporters/excitatory-amino-acid-transporters.html.

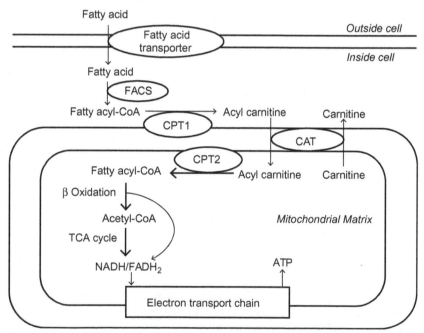

**FIGURE 18.24**  Fatty acid β-oxidation is the process by which fatty acids are broken down to produce energy. Fatty acids enter a cell through fatty acid transporters on the cell surface. Once inside, **FACS (fatty acid CoA synthase)** adds a CoA group to the fatty acid. **CPT1 (carnitine palmitoyltransferase)** then converts the long chain **acyl-CoA** to **long chain acylcarnitine**. The fatty acyl moiety is transported by **CAT (carnitine translocase)** across the inner mitochondrial membrane. **CPT2** then converts the **long chain acylcarnitine** back to **long chain acyl-CoA**. The long chain acyl-CoA can then enter the fatty acid β-oxidation pathway, resulting in the production of one **acetyl-CoA** for each cycle of β-oxidation. This acetyl-CoA then enters the TCA cycle. The NADH and FADH2 produced by both β-oxidation and the TCA cycle are used by the electron transport chain to produce **ATP**. *Reproduced from http://lipidlibrary.aocs.org/animbio/fa-oxid/index.htm.*

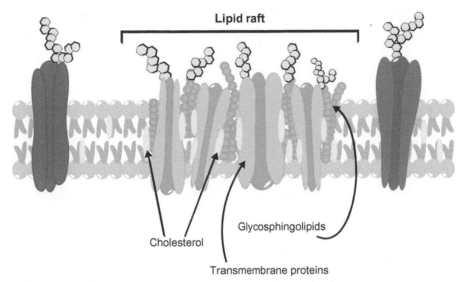

**FIGURE 18.25** Diagram of a lipid raft. The composition of the raft in terms of the types of lipids and proteins varies depending on the location of the membrane and the organelle or cell that it borders. Thus, variations can account for a wide variety of processes that are carried out by cell membranes. The lipids and proteins of the **fluid model of a membrane** indicate that they are capable of lateral movements and are constantly changing positions. The aggregation of patches of lipids and proteins in the membrane form the lipid rafts. They have higher concentrations of cholesterol and **glycosphingolipids** than other regions of the membrane. The reaction between cholesterol and unsaturated acyl-chains may promote the formation of rafts. Acylated proteins are more common in lipid rafts. Some proteins are associated with lipid rafts by way of **glycosylphosphatidylinositol anchors**. The lipid rafts may have a transitory existence in the membrane. *Reproduced from http://en.wikibooks.org/wiki/Structural_Biochemistry/Lipids/Lipid_Rafts.*

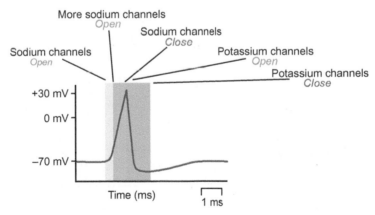

**FIGURE 18.26** An action potential occurs when different ions cross the neuronal membrane. Sodium channels are the first to open and because the extracellular concentration of $Na^+$ is much greater than the intracellular concentration, $Na^+$ enters the cell, causing the neuron to become more positively charged and depolarized. $K^+$ channels open later and $K^+$ leaves the cell and reverses the depolarization. Then the $Na^+$ channels start to close and the action potential reverses toward $-70$ mV (**repolarization**). Actually, the action potential exceeds $-70$ mV (**hyperpolarization**) because the $K^+$ channels remain open a bit longer. Eventually the concentration of ions reaches resting levels, returning the cell to $-70$ mV. *Reproduced from http://faculty.washington.edu/chudler/ap.html.*

The cell membrane can be depolarized by only a few millivolts (mV) resulting in the activation of **voltage-gated sodium ion channels**. These channels are also rapidly inactivated. Sodium ions enter the cell through the activated (opened) channel and depolarize the membrane further, raising the **action potential** (Fig. 18.26). The channel is made up of an α-subunit and one or more β-subunits. The β-subunits (35 kDa each) are responsible for the opening and closing of the channel and the α-subunit (260 kDa) forms the aqueous pore through which the $Na^+$ ions course. The α-subunit folds into four domains (I through IV) containing six α-helical transmembrane segments (S1–S6). The S4 segment contains the **voltage sensor** having a positively charged basic amino acid every third amino acid residue. The structure of the voltage-gated sodium channel is suggested by Fig. 18.27.

Amino and carboxyl termini are part of the face of the sodium channel. The cellular skeletal proteins **ankyrin** and **neurofascin** are associated with the channel. These channels are found in initial segments of axons, postsynaptic folds

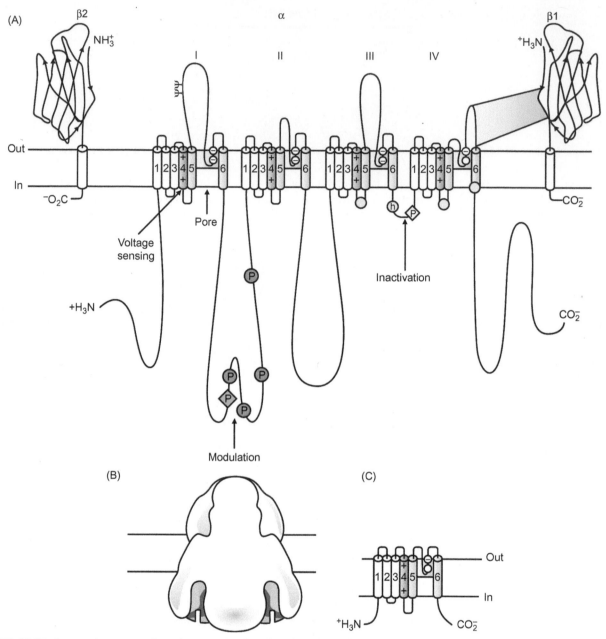

**FIGURE 18.27** Suggested structure of a voltage-gated sodium ion channel. (A) Schematic representation of the sodium channel subunits. The α-subunit of the Na$_v$ 1.2 channel is illustrated together with the β1 and β2 subunits; the extracellular domains of the β-subunits are shown as immunoglobulin-like folds which interact with the loops of the α-subunits as shown. Roman numerals indicate the domains of the α-subunit; segments 5 and 6 (*orange*) are the pore-lining segments and the S4 helices (*yellow*) make up the voltage sensors. *Yellow circles* in the intracellular loops of domains III and IV indicate the inactivation gate IFM (Ile-Phe-Met) motif and its receptor (h, inactivation gate); *P*, phosphorylation sites (in *red circles*, sites for protein kinase A; in *green diamonds*, sites for protein kinase C); ψ, probable N-linked glycosylation site. The circles in the reentrant loops in each domain represent the amino acids that form the ion selectivity filter (the outer rings have the sequence EEDD and the inner rings DEKA). (B) The three-dimensional structure of the sodium channel α-subunit at 20A° resolution compiled from electron micrograph reconstructions. (C) Schematic representation of NaChBac, the bacterial voltage-gated channel. *Redrawn with permission from Mathews, G.G., 1991. Cellular Physiology of Nerve and Muscle, second ed. Blackwell Scientific, Boston, MA. http://genomebiology.com/2003/4/3/207/figure/F1?highres = y.*

and **nodes of Ranvier** (gaps between segments of the myelin sheath representing the junction between adjacent neuroglial cells; the gaps are about 1 mm apart along the length of the axon) where ankyrin is needed for the clustering. High levels of these channels are found in brain, skeletal muscle, heart, uterus, and spinal cord. The **voltage-gated channel superfamily** includes, in addition to the sodium ion voltage-gated channel, voltage-gated potassium ion channels and voltage-gated calcium ion channels.

# EPITHELIAL SODIUM CONDUCTANCE CHANNEL

**ENaC (epithelial sodium channel)** is a sodium conductance channel in the renal tubular cell that is stimulated by the hormone **aldosterone** but this channel is not a voltage-gated sodium channel. Phenomenology of ENaC is shown in Fig. 18.28.

Functional **ENaC** is assembled from closely related subunits: α, β, and γ with a stoichiometry of 1:1:1 in a trimeric structure of dimers as shown in Fig. 18.29.

The epithelial sodium channel, ENaC (also referred to as sodium channel nonneuronal 1 (SCNN1) or amiloride-sensitive sodium channel (ASSC)), is a channel that is sensitive to the inhibitory diuretic, **amiloride** (also **triamterene**) (Fig. 18.19B). There are four related amiloride-sensitive channels: ASSC1-4. ENaC is critical in the transport of sodium ion through hydrophobic pores (high specificity for $Na^+$ but also admits $Li^+$) and in the reabsorption of fluid in the kidney, lung, colon, sweat glands and taste receptor cells (accounting for about 20% of taste in the human). The extent of $Na^+$ **transport** by this channel is proportional to the number of channels on the apical surface of epithelial cells and the proportion of these channels that are open. The number of ENaCs in the apical membrane is determined by the rate

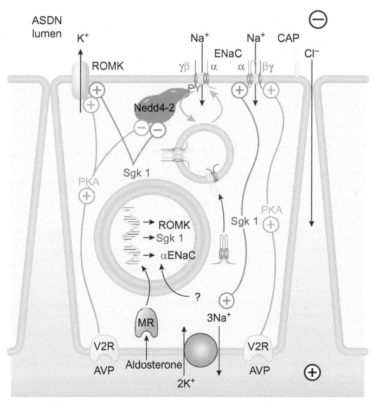

**FIGURE 18.28** A The epithelial sodium channel and its regulatory proteins. This is a model of $Na^+$ **reabsorption** through epithelial sodium channel (ENaC) in the **aldosterone-sensitive distal nephron** (ASDN): insights from genetic mouse models. Aldosterone (Aldo) binds to **mineralocortcoid receptors (MRs)** and stimulates the expression of α-**ENaC** (α-subunit), **serum and glucocorticoid regulated kinase 1 (Sgk1)** and **ROMK (renal outer medullary potassium channel)**. α-ENaC associates with constitutive β and γ subunits to form fully active ENaC, as supported in experiments with **knockout mice**. Normal salt balance in **cortical collecting duct (CCD)**-selective α-ENaC knockout mice underlines the importance of the early ASDN for salt balance. **MR knockout mice** die within days as a result of renal salt loss. Studies in **Sgk1 knockout mice** reveal that the kinase is not required for insertion of ENaC or ROMK into the apical membrane. However, up-regulation of $Na^+$ **reabsorption** to adapt to a reduced dietary NaCl intake or to up-regulate renal $K^+$ excretion involves Sgk1. Sgk1 increases $Na^+$ reabsorption by activating $Na^+/K^+$ ATPase, enhancing the abundance in the cell membrane of ENaC through phosphorylation and inhibition of **ubiquitin ligase Nedd4-2**-mediated internalization of ENaC and probably through direct phosphorylation of α-ENaC. Sgk1 and **arginine vasopressin (AVP)**-activated **protein kinase A (PKA)** phosphorylate different ENaC subunits and the same residues on Nedd4-2 and ROMK and thus can induce inhibition of Nedd4-2 and activation of ROMK independently. This arrangement and the luminal regulation of ENaC through **channel activating proteases (CAP)** may contribute to the mild phenotype of Sgk1 knockout mice. Effects of Sgk1 on ENaC and $Na^+/K^+$-ATPase increase the driving force for $K^+$ secretion through ROMK. In Sgk1 knockout mice, up-regulation of $K^+$ driving force but not the activity and cell membrane expression of ROMK is impaired, consistent with Sgk1-independent regulation of ROMK by Aldo and/or PKA-mediated pathways. In the **Liddle mouse model** which lacks an intact PY motif, internalization of ENaC is impaired and the mice retain salt, causing **salt-sensitive hypertension**. *V2R*, vasopressin V2 receptor. Illustration by Josh Gramling-Gramling Medical Illustration. *Reproduced from http://jasn.asnjournals.org/content/16/11/3160/F1.expansion.*

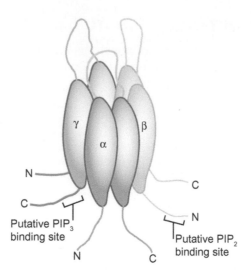

**FIGURE 18.29**   A possible model for the structure of ENaC where there are three closely related subunits: α, β, and γ and these are in a 1:1:1 stoichiometry with each subunit assembled as a dimer, thus ENaC is a trimer of three dimers. There is evidence for the modulation of ENaC by phosphoinositides and the locations of putative binding sites for PIP₂ and PIP₃ are shown. *Reproduced from http://www.nature.com/nrn/journal/v8/n12/box/nrn2257_BX3.html.*

of incorporation of new channels into the membrane vs the removal and degradation of the channels. The C-terminal (and also the N-terminal) of the three subunits of ENaC (Fig. 18.29) is cytoplasmic and contains a proline rich motif (PPxY, prolyl−prolyl-some amino acid-tyrosine). If this motif is inactivated (deletion or mutation) in the β and γ subunits, a specific **ubiquitin ligase** (**Nedd4-2**; see Fig. 18.28) cannot bind to the channel subunits preventing turnover by the **ubiquitin proteasome system**.

In the kidney, ENaC is regulated indirectly by **aldosterone, angiotensin II, vasopressin, insulin, atrial naturetic peptide,** and **glucocorticoids** (Fig. 18.28) so that **sodium balance** between daily intake and urinary excretion of Na⁺ and blood volume and blood pressure can be maintained. In this regard, the extracellular domains of ENaC (containing several cysteine residues) act as receptors for some regulators (e.g., **phosphoinositides**) to control the activity of the channel. In the kidney, aldosterone, operating through the **mineralocorticoid receptor**, induces the formation of **ENaC subunits**, thereby increasing the numbers of ENaC in the apical plasma membrane.

Recent evidence suggests that phosphoinositides modulate ENaC involving channel opening, trafficking and degradation. The activity of ENaC is increased by the products of **phosphatidylinositol 3-kinase** (**PI3K**), **phosphatidylinositol-3,4,-*bis*phosphate** (**PtdIns(3,4)P2**) and **phosphatidylinositol-3,4,5-*tris*phosphate** (**PIP₃**). Specifically, PIP₂ restores lowered ENaC to its normal level of tonic activity. In contrast, PIP₃ increases ENaC activity beyond the tonic level even in the presence of PIP₂ (concentration is ~100 fold higher than that of PIP₃) suggesting that these two factors operate through separate sites (Fig. 18.29). Increased levels of PIP₂ elevate the insertion rate of ENaC into the cell membrane. In the pancreas, testes and ovary there is evidence of a fourth subunit of ENaC called delta (δ) and in these tissues the **δ-subunit** replaces the α-subunit.

## MUTIDRUG RESISTANCE CHANNEL (MDR), A MEMBER OF THE ABC TRANSPORTER SUPERFAMILY

This superfamily has at least seven known subfamilies involving 49 individual transporters. They are divided as ABCA, 1−13; ABCB, 1−11; ABCC, 1−12; ABCD, 1−4; ABCE1; ABCF, 1−3; and ABCG, 1−3. They occur in many tissues and some are located ubiquitously in **mitochondria** and others in **peroxisomes**. Genetic abnormalities with varied types of inheritance in these transporters involve a variety of clinical disorders, such as **cystic fibrosis, neurological disease, retinal degeneration**, defects in the transport of **cholesterol** and **bile acids, anemia** and resistance to drugs. Within this group of ABC transporters are the MDR (multiple drug resistance) channel (also known as **P glycoprotein**, 179 kDa) and the **CFTR channel** (Fig. 18.1) involved in **cystic fibrosis**; the two channels are similar. MDR uses the energy of ATP to pump drugs out of cancer cells, accounting for drug resistance in many cases. CTFR is a cyclic AMP-activated Cl⁻ **channel** and can exchange HCO₃⁻ for Cl⁻. These two channels are discussed as representatives of the ABC transporter superfamily. A comparison of these two transporters is shown in Fig. 18.30.

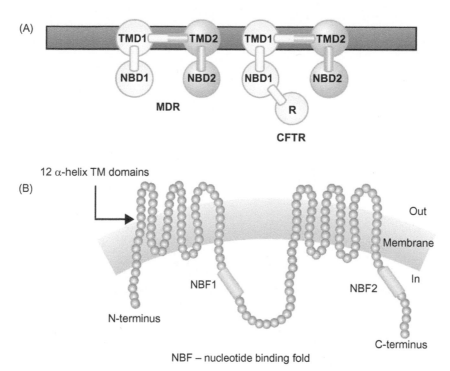

**FIGURE 18.30** (A) Comparison of MDR with CFTR. *TMD*, transmembrane domain; *NBD*, nucleotide-binding domain. (B) Model of MDR1 (P glycoprotein). *Part A redrawn with permission from Figure 1 in part from Luckie, D.B., et al., 2003. CTFR and MDR: ABC transporters with homologous structure but divergent function. Curr. Genomics 4, 109—121.*

MDR transporter expression and function is regulated by **PI3K signaling**. In turn, PI3K activity is regulated by **ERBB2-CD44** interactions. The interactions of ERBB2-CD44 are dependent on **hyaluronan**. Hyaluronan is a high molecular weight polysaccharide that is an anionic, nonsulfated glycosaminoglycan in all tissues and body fluids and widely distributed in connective, epithelial and neural tissues. It is synthesized in the **plasma membrane** of the cell where it exists in a mobile pool. It functions as a lubricant, in **water homeostasis**, as a filter and a regulator of the **distribution of plasma proteins**. It is catabolized by **receptor-mediated endocytosis** and lysosomal degradation with a half-life of about 1 to several days. It is removed from the bloodstream to **epithelial cells** and **liver sinusoids** with a half-life of 2—5 minutes. The relationships of **hyaluronan** to PI3K and the **MDR** are summarized in Fig. 18.31.

## BLOOD—BRAIN-BARRIER

The blood–brain barrier (BBB) limits the substances from the blood that can enter the brain. Epithelial cells lining the blood vessels form a barrier between the blood and the cells of the brain. As lymphocytes, monocytes and neutrophils cannot penetrate the BBB, immune responses in the brain are limited and in this way the **neuronal network** is protected from the damage that could be caused by a full immune response. In the rare case that viral or fungal infections or prions occur in the brain, the capacity for a local immune response exists. The epithelial cells lining blood vessels are encased by a basement membrane composed of **lamin, fibronectin**, and other proteins that function as a mechanical support and a barrier. Fig. 18.32A shows the overall structure of the BBB and Fig. 18.32B illustrates the pathways across the BBB.

## SUMMARY

**Cystic fibrosis** is an **autosomal recessive disease** resulting from mutations in a **chloride ion transporter (CFTR)**. Consequently, the reabsorption of salt is abnormal, the transport of water is reduced and the reduced secretions of pancreatic digestive enzymes lead to obstructions in the intestine (*meconium ileus*) and mucus buildup in other tissues, particularly the lung. Death can occur after birth or a patient may be treatable for 30—40 years of age before death occurs.

Large molecules may be imported into cells by generalized processes of **pinocytosis** and **phagocytosis**. Pinocytosis involves invagination of the cell membrane where particles outside the cell are engulfed forming a vesicle inside the

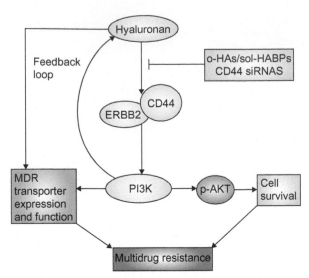

**FIGURE 18.31**  Hyaluronan-dependent **ERBB2-CD44 interactions** regulate **phosphatidylinositol 3-kinase** (**PI3K**) activity. PI3K signaling regulates **multidrug resistance** (MDR) transporter expression and function and also phosphorylates **AKT** (p-AKT) to activate **cell-survival signaling**. The combination of these effects promotes drug resistance. ERBB-PIK3 signaling acts in a positive-feedback loop to stimulate further hyaluronan production which promotes cell survival and drug resistance. These pathways are inhibited by **hyaluronan oligomers** (**o-Has**), soluble **hyaluronan-binding proteins** (sol-HABPs) and **small interfering RNAs** (**siRNAs**) that are directed against CD44. Hyaluronan also interacts directly with MDR transporters during synthesis and secretion. *Reproduced from http://www.nature.com/nrc/journal/v4/n7/fig_tab/nrc1391_F4.html.*

cell. Phagocytosis is a process in which **pseudopodia** are formed from the cell membrane and these surround a particle to be engulfed.

Exocytosis is a process in which large molecules inside the cell (stored in vesicles) are moved to the cell exterior. **Constitutive exocytosis** is a continuous process and occurs without a specific signal, whereas **regulated exocytosis** occurs only in response to a specific signal.

**Osmosis** is a process of **passive diffusion** in which some substances, such as $CO_2$ and certain lipid-soluble substances can cross the cell membrane without a requirement for energy. The extent of diffusion is governed by the concentration of solute on either side of the membrane so that the more concentrated side will have solute movement to the less concentrated side (similar to a mass action effect) tending toward equalization of the solute concentration on either side of the membrane.

The transport of water into a cell occurs by **aquaporins** that are channels inserted into the cellular membrane from inside the cell. In this process, the signal, **vasopressin**, binds to its membrane receptor and **PKA** is activated and it phosphorylates **aquaporin subunits** followed by their polymerization into channels and insertion of channels into the cell membrane. Water is transported by this process across the cell and through the basolateral membrane, reaching the extracellular space and ultimately the blood circulation.

**Facilitated diffusion** is a modification of osmosis involving a proteinaceous channel through which an extracellular solute can enter a cell without an energy requirement. This adaptation is required for water-soluble substances that are not soluble in the lipid membrane. Molecules will move into the cell by this process depending on their outside concentration. The **glucose transporters** move glucose into the cell by facilitated diffusion.

**Energy-requiring transport** is used when a solute is moved either from the outside of a cell against a concentration gradient or from inside a cell to the outside against a concentration gradient. The energy source is usually ATP and the $Na^+/K^+$ **ATPase** is a prime example of this case.

There are three types of transporters: **uniport**, **symport** and **antiport**. As their names imply, the uniport (simple transporter) transports a solute in one direction without involving another substance moving in the same or opposite direction to the solute. The symport involves another substance (such as an ion) moving in the same direction as the solute being transported. The antiport involves another substance moving in the opposite direction from the solute. The symport and antiport are **coupled transporters**.

**Calcium ion** is a key second messenger in the cell and it is stored in a site in the ER. Signals for the release of $Ca^{2+}$ from the ER store occur through membrane receptors operating through **G proteins** to produce **inositol *tris*phosphate** (**IP₃**) and **DAG**). IP₃ binds to its receptor in the ER membrane and signals the release of $Ca^{2+}$ from the ER store. $Ca^{2+}$ is also taken up from the cell exterior through a **TRP channel**, a member of the **TRP channel system**. One

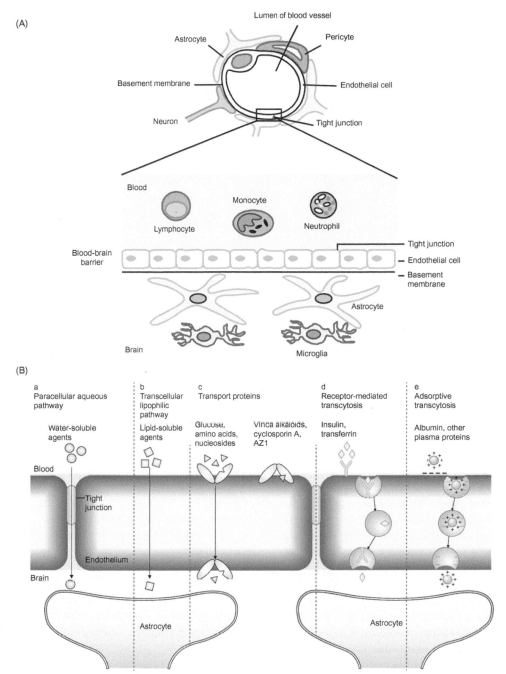

**FIGURE 18.32** (A) The blood—brain barrier (BBB) is created by the tight apposition of endothelial cells lining blood vessels in the brain, forming a barrier between the circulation and the brain parenchyma (**astrocytes**, **microglia**). Blood—borne immune cells, such as lymphocytes, monocytes and neutrophils cannot penetrate this barrier. A thin basement membrane comprising lamin, fibronectin and other proteins, surrounds the endothelial cells and associated pericytes and provides mechanical support and a barrier function. Thus, the BBB is crucial for preventing infiltration of pathogens and restricting antibody-mediated immune responses in the **central nervous system**, as well as for preventing disorganization of the fragile neural network. This, together with a generally muted immune environment within the brain itself, protects the fragile neuronal network from the risk of damage that could ensue from a full-blown immune response. On rare occasions, **pathogens** (e.g., viruses, fungi, or prions) and autoreactive T cells breach the endothelial barrier and enter the brain. A local innate immune response is mounted in order to limit the infection's challenge and pathogens are destroyed and cell debris is removed, a vital process that must precede tissue repair. (B) Pathways across the **blood—brain barrier (BBB)**. Shown is a schematic diagram of the **endothelial cells** that form the BBB and their associations with the perivascular endfeet of **astrocytes**. The main routes for molecular traffick across the BBB are shown. (a) Normally the **tight junctions** severely restrict penetration of water-soluble compounds, including polar drugs. (b) However, the large surface area of the lipid membranes of the endothelium offers an effective diffusive route for **lipid-soluble agents**. (c) The endothelium contains **transport proteins** (carriers) for glucose, amino acids, purine bases, nucleosides, choline and other substances. Some transporters are energy-dependent (for example, **P-glycoprotein**) and act as **efflux transporters**. *AZT*, azidothymidine. (d) Certain proteins, such as insulin and transferrin, are taken up by specific receptor-mediated **endocytosis** and **transcytosis** (membrane-bound carriers selectively transport substances within a cell so that unique environments can be maintained on either side of the cell; endocytosis occurs from the outside to the interior of a cell, whereas transcytosis occurs inside the cell). (e) Native plasma proteins, such as **albumin** are poorly transported but cationization can increase their uptake by **adsorptive-mediated endocytosis** and transcytosis. Drug delivery across the brain endothelium depends on making use of pathways b-c; most CNS drugs enter via route b. *(A) Reproduced from Francis, K., et al., 2003. Innate immunity and brain inflammation: the key role of complement. Exp. Rev. Mol. Med. 5, 1—19, with permission. (B) Reproduced from http://www.nature.com/nrn/journal/v7/n1/fig_tab/nrn1824_F3.html.*

use for the generation of increased levels of $Ca^{2+}$ can be **membrane depolarization** and the release of stored proteins into the extracellular space.

The entry of $Mg^{2+}$ and other divalent ions into cells is carried out by many transporters having different structures.

**Proton transporters** pump protons where **cytosolic acidification** is needed playing a role in the control of pH. One of these cell membrane transporters is the **amiloride-sensitive $Na^+/H^+$ antiport**. Amiloride, by inhibiting this antiport, reduces the uptake of $Na^+$ and also the uptake of water into cells. This antiport also occurs in **mitochondria**.

The **monocarboxylate cotransporter** (symporter) moves **lactate** and **pyruvate** along with either $H^+$ or $Na^+$ in the same direction. It plays an important role in cellular metabolism both at the plasma membrane and the mitochondrial membrane. It is especially important in the metabolism of muscle cells where intense activity promotes the formation of lactate and protons.

The **citrate antiporter (CTP)** moves a molecule of citrate from the mitochondria to the cytoplasm in exchange for a molecule of **malate** from the cytoplasm into mitochondria. CTP is a member of the **mitochondrial carrier family** and functions to move citrate out of the mitochondrion into the cytoplasm where it can be used to form glucose (gluconeogenesis) and fatty acids. There is another **citrate symporter** (together with $Na^+$) found mainly in the liver (to a lesser extent in brain and testes) cell membrane for the uptake of citrate and other di- or tricarboxylates from the bloodstream.

Intestinal transport of di- and tripeptides (**proton-dependent dipeptide cotransporter**) is located in the apical membrane of **intestinal enterocytes** and in the kidney. It transports mainly di- and tripeptides but not free amino acids or peptides containing four or more amino acids.

**L-Amino acid transporter (LAT1)** transports neutral amino acids, aromatic amino acids and branched chain amino acids (without $Na^+$) through the plasma membrane of various tissue cells. It is a component of the **blood–brain barrier** system. Some of the other amino acid transporters transport cationic amino acids and are dependent on $Na^+$.

**Excitatory amino acid transporters (glutamate antiporters)** in the nervous system occur in **vesicles** at the terminus of a neuronal axon where glutamate from the cytoplasm is concentrated in exchange for protons. ATP powers the transport of protons into the vesicle that become used in the exchange for glutamate.

**Fatty acids** are transported into the mitochondrion for β-**oxidation** and are also transported from the extracellular space into the cell. These transporters fall under the heading of **fatty acid transport proteins (FATPs)**. There are six transporters in this family (FATP1-6).

Neuronal **action potentials** are developed by the movements of $Na^+$ and $K^+$. Sodium channels opened first and $Na^+$ enters the cell causing the neuron to become more positively charged and **depolarized**. $K^+$ channels open later causing $K^+$ to leave the cell and reverse the polarization, after which $Na^+$ channels close and **repolarization** occurs. As the $K^+$ channels remain open somewhat longer, **hyperpolarization** occurs and eventually, the concentration of ions returns to resting levels returning the cell to its normal polarization.

The **epithelial sodium conductance channel (ENaC)** of the renal tubular cell of the distal nephron is stimulated by the hormone **aldosterone**. It is not a voltage-gated channel. Aldosterone stimulates the expression of the α-subunit of ENaC and then the α-subunit associates with constitutive β- and γ-subunits (1:1:1) to form the fully active ENaC that is inserted into the apical membrane. There is evidence that ENaC is modulated by **phosphoinositides**. ENaC regulates the sodium ion balance between daily ingestion and urinary excretion so that blood volume and blood pressure can be maintained.

The **multidrug resistance channel (MDR)** is a member of the **ABC transporter family**; members of this family occur in mitochondria and peroxisomes. MDR is of particular interest in cancer as MDR pumps **anticancer drugs** out of the cancer cell to promote drug resistance. MDR has a general structure that is somewhat similar to the **cystic fibrosis chloride channel (CFTR)**. The expression and function of the MDR transporter is regulated by **PI3K signaling**.

The **blood-brain barrier (BBB)** is a unique system that limits the substances in blood that are able to enter the brain. A barrier, consisting of epithelial cells, exists between the blood and brain cells. The **immune responses** in the brain are limited by the exclusion of lymphocytes, monocytes and neutrophils. In rare cases of infection the brain can mount an immune response but it is more subdued than that elicited by a systemic infection involving those immune cells that are excluded from the brain.

# SUGGESTED READING

## Literature

Abbott, N.J., Friedman, A., 2012. Overview and introduction: the blood–brain barrier in health and disease. Epilepsia. 53 (Suppl), 1–6.
Berridge, M.J., 2009. Inositol trisphosphate and calcium signaling mechanisms. Biochim. Biophys. Acta 1793, 933–940.

Bozoky, Z., et al., 2013. Regulatory R region of the CFTR chloride channel is a dynamic integrator of phospho-dependent intra- and intermolecular interactions. Proc. Natl. Acad. Sci. 110 (E44), 27–36.

Catterall, W.A., 1995. Structure and function of voltage-gated ion channels. Ann. Rev. Biochem. 64, 493–531.

Charrier, L., Merlin, D., 2006. The oligopeptide transporter hPepT1: gateway to the innate immune response. Lab. Invest. 86, 538–546.

Cura, A.J., Carruthers, A., 2012. Role of monosaccharide transport proteins in carbohydrate assimilation, distribution, metabolism, and homeostasis. Compr. Physiol. 2, 863–914.

Ehehalt, R., et al., 2006. Translocation of long chain fatty acids across the plasma membrane – lipid rafts and fatty acid transport proteins. Mol. Cell Biochem. 284, 135–140.

Ferre, S., Hoenderop, J.G., Bindels, R.J., 2011. Insight into renal $Mg^{2+}$ transporters. Curr. Opin. Nephrol. Hypertens. 20, 169–176.

Francis, K., et al., 2003. Innate immunity and brain inflammation: the key role of complement. Exp. Rev. Mol. Med. 5, 1–19.

Gees, M., Owslanik, G., Nillus, B., Voets, T., 2012. TRP channels. Compr. Physiol. 2, 563–608.

Germann, U.A., Chambers, T.C., 1998. Molecular analysis of the multidrug transporter, P-glycoprotein. Cytotechnology 27, 31–60.

Halestrap, A.P., Price, N.T., 1999. The proton-linked monocarboxylate transporter (MCT) family: structure, function and regulation. Biochem. J. 343, 281–299.

Halestrap, A.P., Wilson, M.C., 2012. The monocarboxylate transporter family – role and regulation. IUBMB Life 64, 109–119.

Hayashi, K., et al., 2013. LAT1 is a critical transporter of essential amino acids for immune reactions in activated human T cells. J. Immunol. 191, 4080–4085.

Kashian, O.B., Kleyman, T.R., 2011. ENaC structure and function in the wake of a resolved structure of a family member. Am. J. Physiol. Renal Physiol. 301, P684–P696.

Linsdell, P., 2014. Functional architecture of the CFTR chloride channel. Mol. Membr. Biol. 31, 1–16.

Luckie, D.B., et al., 2003. CTFR and MDR: ABC transporters with homologous structure but divergent function. Curr. Genomics. 4, 109–121.

Mercer, J., Helenius, A., 2012. Gulping rather than sipping: macropinocytosis as a way of virus entry. Curr. Opin. Microbiol. 15, 490–499.

Myclelska, M.E., et al., 2009. Citrate transport and metabolism in mammalian cells: prostate epithelial cells and prostate cancer. Bioessays 31, 10–20.

Quamme, G.A., 2010. Molecular identification of ancient and modern mammalian magnesium transporters. Am. J. Cell. Physiol. 298, C407–C429.

Schloth, H.B., Roshanbin, S., Hagglund, M.G., Fredriksson, R., 2013. Evolutionary origin of amino acid transporter families SLC32, SLC36 and SLC 38 and physiological, pathological and therapeutic aspects. Mol. Aspects Med. 34, 571–585.

Sherwood, T.W., Frey, E.N., Askwith, C.C., 2012. Structure and activity of the acid-sensing ion channels. Am. J. Physiol. Cell Physiol. 303, C699–C710.

Strazielle, N., Ghersi-Egea, J.F., 2013. Physiology of blood-brain interfaces in relation to brain disposition of small compounds and macromolecules. Mol. Pharm. 10, 1473–1491.

Sugano, K., et al., 2010. Coexistance of passive and carrier-mediated processes in drug transport. Nat. Rev. Drug. Discov. 9, 597–614.

Sun, J., et al., 2010. Mitochondrial and plasma membrane citrate transporters: discovery of selective inhibitors and application to structure/function analysis. Mol. Cell. Pharmacol. 2, 101–110.

Underhill, D.M., Goodridge, H.S., 2012. Information processing during phagocytosis. Nat. Rev. Immunol. 12, 492–502.

Vandenberg, R.J., Ryan, R.M., 2013. Mechanisms of glutamate transport. Physiol. Rev. 93, 1621–1657.

Xie, Z., 2003. Molecular mechanisms of Na/K-ATPase-mediated signal transduction. Ann. NY Acad. Sci. 986, 497–503.

## Books

Alberts, B., et al., 2016. Essential Cell Biology. fourth ed. Garland Science, Taylor & Francis, Milton Park, Montreal, CA, Abington, UK.

Arguello, J., 2002. Metal Transporters. Academic Press, San Diego.

Friedlander, M., Mueckler, M., Bourne, G.H., Jarvik, J. (Eds.), 1993. Molecular Biology of Receptors and Transporters: Pumps, Transporters and Channels. 137C of International Review of Cytology, Elsevier, Amsterdam, London, Salt lake City.

Hille, B., 2001. Ion Channels and Excitable Membranes. third ed. Sinauer, Sunderland, MA.

Kirk, K.L., Dawson, D.C., 2003. The Cystic Fibrosis Transmembrane Conductance Regulator. Landes Bioscience/ Kluwer Academic/Plenum Publishers, New York, NY.

Linton, K.J., Holland, I.B. (Eds.), 2011. The ABC Transporters of Human Physiology and Disease: Genetics and Biochemistry of ATP Binding Cassette Transporters. WB World Scientific, Hackensack, NJ.

Litwack, G. (Ed.), 2015. *Hormones and Transport Systems. Vitamins & Hormones*, vol. 98. Academic Press/Elsevier, Cambridge, MA.

Mathews, G.G., 1991. Cellular Physiology of Nerve and Muscle second ed. Blackwell Scientific, Hoboken, NJ.

# Chapter 19

# Micronutrients (Metals and Iodine)

## IRON DEFICIENCY ANEMIA

Estimates by the World Health Organization are that 30% of the world's population is anemic, and of these, half is due to the most common form of anemia, iron deficiency anemia (**IDA**). IDA is the result of inadequate consumption of dietary iron. The diminished capacity of persons with IDA results in lowered efficiency of job performance, indirectly affecting the economy. IDA in children inhibits growth and learning.

During **menstruation**, from 4 to 100 mg of iron is lost. A woman loses about 500 mg of iron with a **pregnancy,** and the newborn consumes iron-deficient mother's milk. Infants and young children consuming **cow's milk** have a risk of IDA because the high level of calcium in milk competes with iron for **intestinal absorption**. There may be conditions that result in lowered efficiency in iron absorption due to disease (**Crohn's disease** or **celiac disease**), and there are other situations leading to decreased iron absorption, such as overuse of **antacids** containing calcium, **gastric bypass surgery**, a strict **vegetarian diet,** and poor general nutrition, especially in older adults. With IDA, there is a reduction in the number of red blood cells that require **hemoglobin** (containing heme iron) to carry oxygen to the tissues.

In addition to inadequate dietary consumption of iron, there are other causes of anemia: **internal bleeding** (e.g., from ulcers), **tumors** (e.g., colon cancer), heavy menstruation, and others. Small blood losses from gastrointestinal bleeding can cause anemia that could evade detection through testing hemoglobin in the stool, a test that might not be highly sensitive. When there is a condition of ulcers, **aspirin** or other **nonsteroidal antiinflammatory drugs** (**NSAIDs**) can cause internal bleeding. Dietary IDA is most common in women (20%) and 50% in pregnant women compared to 3% of males (men have larger stores of iron than women).

The uptake of iron from the diet is regulated by absorptive cells of the small intestine. The need for iron is about 20 mg/day. Only 10% of dietary iron is actually absorbed, and about 1 mg is absorbed in a day from a typical diet containing 10−20 mg. Given the 120-day **lifespan of a red blood cell**, about 0.8% of red blood cells are degraded and replaced in a day; 2.5 g of iron are incorporated into hemoglobin with a daily turnover of 20 mg for synthesis of hemoglobin plus 5 mg for other functions. A few tenths of a gram of **iron** are bound to **myoglobin,** and about 20 mg is distributed to proteins involved in electron transfer, especially those in the **electron transport chain** that generate ATP. About 1 g of iron is stored in **ferritin** for future use. Iron is reused through the plasma and amounts of iron in excess of usage are deposited in the iron stores, **ferritin,** or **homosiderin**. Nonheme plant sources of iron in the ferric state are poorly absorbed by the small intestine. Most of the absorbed iron from the intestine is in the form of myoglobin and **hemoglobin**; these forms account for about 60% or more of the iron in the body.

A person who is anemic usually feels cold because iron plays a role in **temperature regulation**. There is a limit to the amount of oxygen that can be delivered to the mitochondria for electron transport and ATP production because in IDA there is a decrease in muscle myoglobin, causing an increase in the production of **lactic acid** and a reduced content of hemoglobin in red blood cells. Enzymes that require nonheme iron as cofactors, such as **NADH dehydrogenase** and **succinate dehydrogenase** of mitochondrial metabolism are affected as is **ribonucleotide reductase**, an enzyme involved in the synthesis of DNA. Other enzymes requiring iron, besides the mitochondrial enzymes, are heme-containing **catalase** and **peroxidase** that protect cells from **reactive oxygen species** derived from hydrogen peroxide.

**Bodily temperature** is a reflection of the ability of tissues to form ATP energy, and iron (and copper) may affect the functioning of the **thyroid gland** that plays a role in the regulation of body temperature.

**Vitamin deficiencies** of **folic acid** and **vitamin $B_{12}$** (cobalamine) play a role in certain anemias. Oxidized iron ($Fe^{3+}$) in the diet is reduced to ferrous iron ($Fe^{2+}$) by **ascorbic acid** (vitamin C), and an absorbable **ascorbic acid-iron complex** is formed that enhances the intestinal absorption of **nonheme iron**. *Oxidized iron is not absorbed as such and must be reduced to ferrous iron to be absorbed.* **Stomach acid** is needed to release the $Fe^{3+}$ in ingested foods, so a

Human Biochemistry. DOI: http://dx.doi.org/10.1016/B978-0-12-383864-3.00019-3

**591**

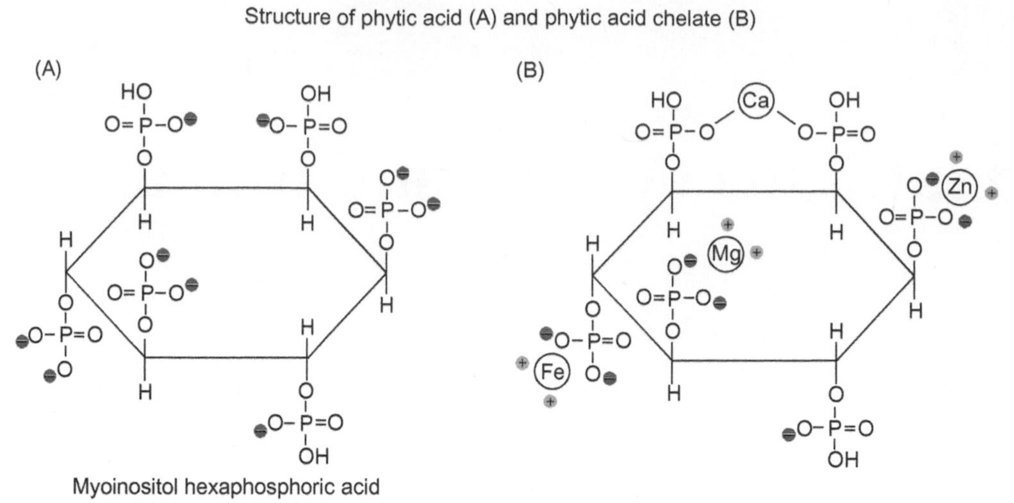

Structure of phytic acid (A) and phytic acid chelate (B)

Myoinositol hexaphosphoric acid

**FIGURE 19.1**    (A) Structure of phytic acid (**myoinositol hexaphosphoric acid**) and its complexes with metals a shown in (B).

**TABLE 19.1  The Content of Iron in Various Foods**

| Food | Serving | Iron Content (mg) |
|---|---|---|
| Beef | 3 ounces, cooked* | 2.31 |
| Chicken, dark meat | 3 ounces, cooked | 1.13 |
| Oysters | 6 medium | 5.04 |
| Shrimp | 8 large, cooked | 1.36 |
| Tuna, light | 3 ounces, canned | 1.30 |
| Blackstrap molasses | 1 tablespoon | 3.50 |
| Raisin bran cereal[a] | 1 cup, dry | 5.00 |
| Raisins, seedless | 1 small box (1.5 ounces) | 0.89 |
| Prune juice | 6 fluid ounces | 2.27 |
| Prunes, dried | ~5 prunes (1.5 ounces) | 1.06 |
| Potato, with skin | 1 medium potato, baked | 2.75 |
| Kidney beans | 1/2 cup, cooked | 2.60 |
| Lentils | 1/2 cup, cooked | 3.30 |
| Tofu, firm[b] | 1/4 block (~1/2 cup) | 6.22 |
| Cashew nuts | 1 ounce | 1.70 |

* – 3 oz of meat—size of a dock$$ of cards.
[a]Cereals contain much nonheme iron, which is not absorbed as well as heme iron.
[b]Tofu also contains phytic acid, which blocks absorption of nonheme iron.
Reproduced from http://ohiohealth.cancersource.com/.

condition in which there is a lack of hydrochloric acid in digestive juices (**achlorhydria**) can lead to poor absorption of iron. Vegetarians who refrain from eating red meat (rich in iron) may also ingest low levels of vitamin $B_{12}$. Cereals, soybeans (tofu and tofu derivatives), and other plant foods contain **phytic acid** that binds iron and other metals rendering them unabsorbable by the intestine. Phytic acid contains six phosphate groups providing a metal sequestering agent (Fig. 19.1).

Table 19.1 lists some common foods that contain iron.

In addition to the foods listed in Table 19.1, broccoli, spinach, and egg yolks are good sources of iron.

The recommended **daily allowances for dietary iron** are listed in Table 19.2.

**Mild anemia** does not produce noticeable symptoms but **intense anemia** will produce many symptoms: pale skin color, shortness of breath, fatigue, palpitations on climbing stairs, leg cramps, dizziness, irritability, sore tongue, weakness, atrophy of taste buds, sores at corners of mouth, brittle nails, frontal headache, difficulty in sleeping, and

**TABLE 19.2 Recommended Daily Dietary Allowances for Iron Intake in Various Individuals**

| Life Stage | Age | Males (mg/day) | Females (mg/day) |
|---|---|---|---|
| Infants | 0−6 months | 0.27 (AI) | 0.27 (AI) |
| Infants | 7−12 months | 11 | 11 |
| Children | 1−3 years | 7 | 7 |
| Children | 4−8 years | 10 | 10 |
| Children | 9−13 years | 8 | 8 |
| Adolescents | 14−18 years | 11 | 15 |
| Adults | 19−50 years | 8 | 18 |
| Adults | 51 years and older | 8 | 8 |
| Pregnancy | All ages | − | 27 |
| Breastfeeding | 18 years and younger | − | 10 |
| Breastfeeding | 19 years and older | − | 9 |

AI, adequate intake, which is established when an RDA cannot be determined.
Reproduced from page 5 of http://lpi.oregonstate.edu/infocenter/minerals/iron.

**TABLE 19.3 Hemoglobin and Hematocrit Values in the Blood in Various Levels of Anemia**

| Normal Values | | Values in Anemia |
|---|---|---|
| Hemoglobin (measured in grams (g) per deciliter (dL)) | | Mild: 9.5−10.9 g/dL (grade 1) |
| | | Moderate: 8−9.4 g/dL (grade 2) |
| | | Severe: 6.5−7.9 g/dL (grade 3) |
| | | Life threatening: less than 6.5 g/dL (grade 4) |
| Men | 14−18 g/dL | Less than 14 g/dL |
| Women | 12−16 g/dL | Less than 12 g/dL |
| | | Mild: 30%−36% |
| Hematocrit | | Moderate: 25%−30% |
| Men | 42%−52% | Severe: less than 25% |
| | | Less than 42% |
| Women | 36%−48% | Less than 36% |

LearnAboutCancer/detail_frame.$$DiseaseID−1&ContentID−22234−1&Page−33&subjected-3&TpeID−2.
Reproduced from Table 1 on page 2 from http://ohiohealth.cancersources.com/.

decreased appetite. In advanced IDA, there can be problems in swallowing due to the formation of webs of tissue in the throat and esophagus (**Plummer−Vinson syndrome**). Iron-deficient persons may consume nonfood items like starch or clay, known as the condition of **Pica**, a behavioral disturbance of iron deficiency. Measurement of the **blood hemoglobin** level generates the diagnosis of iron deficiency. These measurements can be expressed as **hematocrit values** that are measurements of the volume of red blood cells as a percentage of the total blood volume where the normal values for males are 43% to 49% and for females, 37% to 43% (Table 19.3).

Other measurements that can be made to define iron deficiency are **red blood cell size**, **serum level of iron**, and the **iron-binding capacity of blood**. In the differentiation between IDA and anemia caused by chronic disease, the measurements of **serum ferritin** and sustainable iron in tissue stores can be useful. The saturation of ferritin with iron does not become abnormal until the tissue stores of iron have been depleted. When the tissue stores of iron have been depleted, the **hemoglobin** concentration decreases but values of **red blood cells** may not become abnormal for several months (red blood cell half-life = 120 days) after the tissues stores have been depleted. Because conditions of rapid growth require high levels of iron, infants, children, adolescents, and pregnant women are particularly vulnerable to **iron deficiency**. Especially, in children in poor conditions, parasitic worms can cause intestinal bleeding that leads to iron deficiency.

Treatment of IDA involves the oral administration of ferrous iron ($Fe^{2+}$) that corrects the deficiency. Normalization of the iron level is determined by measurement of the blood hemoglobin level. Intravenous transfusion could be required where iron cannot be absorbed, or there are acute attacks of bleeding.

Measurements of serum **ferritin** are the most sensitive tests that can be carried out in the laboratory. Normally, the ferritin complex store houses about 4500 iron atoms in the ferrous form ($Fe^{2+}$) that is stored as the hexahydrate (Fe $[H_2O]_6^{2+}$). When there is less than 5% saturation of iron in ferritin storage, this always indicates iron deficiency. Normal values for females is >12% saturation of iron storage in ferritin and for males the value is >15%. Fig. 19.2 shows various aspects of ferritin and the storage of ferrous iron atoms.

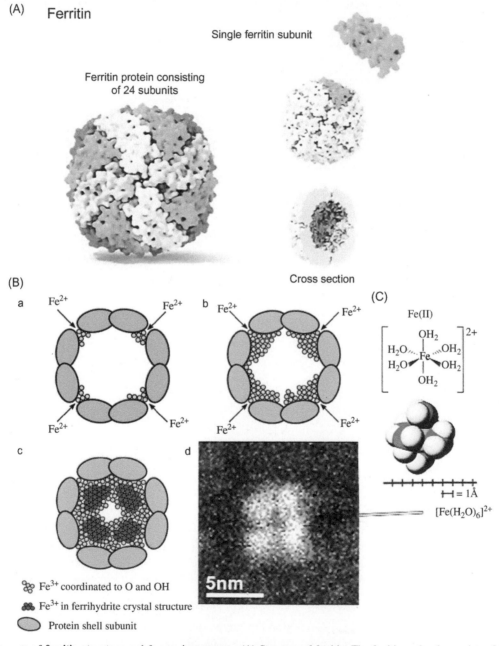

**FIGURE 19.2** Aspects of **ferritin structure** and ferrous iron storage. (A) Structure of ferritin. The ferritin molecule consists of 24 subunits (*left*). A single ferritin subunit is shown on the *right*. (B) A schematic version of the biomineralization process shown through **c**. Shown in **d** is the actual corrected image from X-ray superSTEM (Aberration-Corrected Scanning Transmission Electron Microscopy) of the diagrammed structure in **c**. The scale bar is equal to 5 nm ($5 \times 10^{-9}$ m) and the mineralization core contains about 1500 atoms of ferrous iron hexahydrate. (C) shows the structure of **ferrous iron hexahydrate**, the storage form of iron. In the molecular model structure, the iron atom is in *green* just barely visible in the center and the water molecules are in *light green* and *red* (oxygen). *(A) Reproduced from http://ghr.nim.nih.gov/handbook/illustrations/ferritin, (B) reproduced from http://www.nanofolio.org/research/paper12.php, and (C) reproduced from http://www.pnas.org/content/99/8/5195/F1.expansion.html.*

Disease conditions, such as rheumatoid arthritis, inflammatory bowel disease, HIV, and heart failure are accompanied by anemia.

## UPTAKE OF IRON DURING DIGESTION

In the diet, iron is in the form of both heme and nonheme iron bound to proteins. In the intestinal cells, the **endoplasmic reticulum heme oxygenase** enzyme converts **heme** (heme, by itself, can be toxic) to **biliverdin**, **carbon monoxide** (the only reaction in the body producing CO expired by the lungs), and reduced iron, releasing the iron from heme, and the released iron is transferred into the body as nonheme iron. The ferrous iron can be incorporated into **ferritin** or exported from the cell. Biliverdin is converted to **bilirubin** by **biliverdin reductase**. These reactions are recorded in Fig. 19.3.

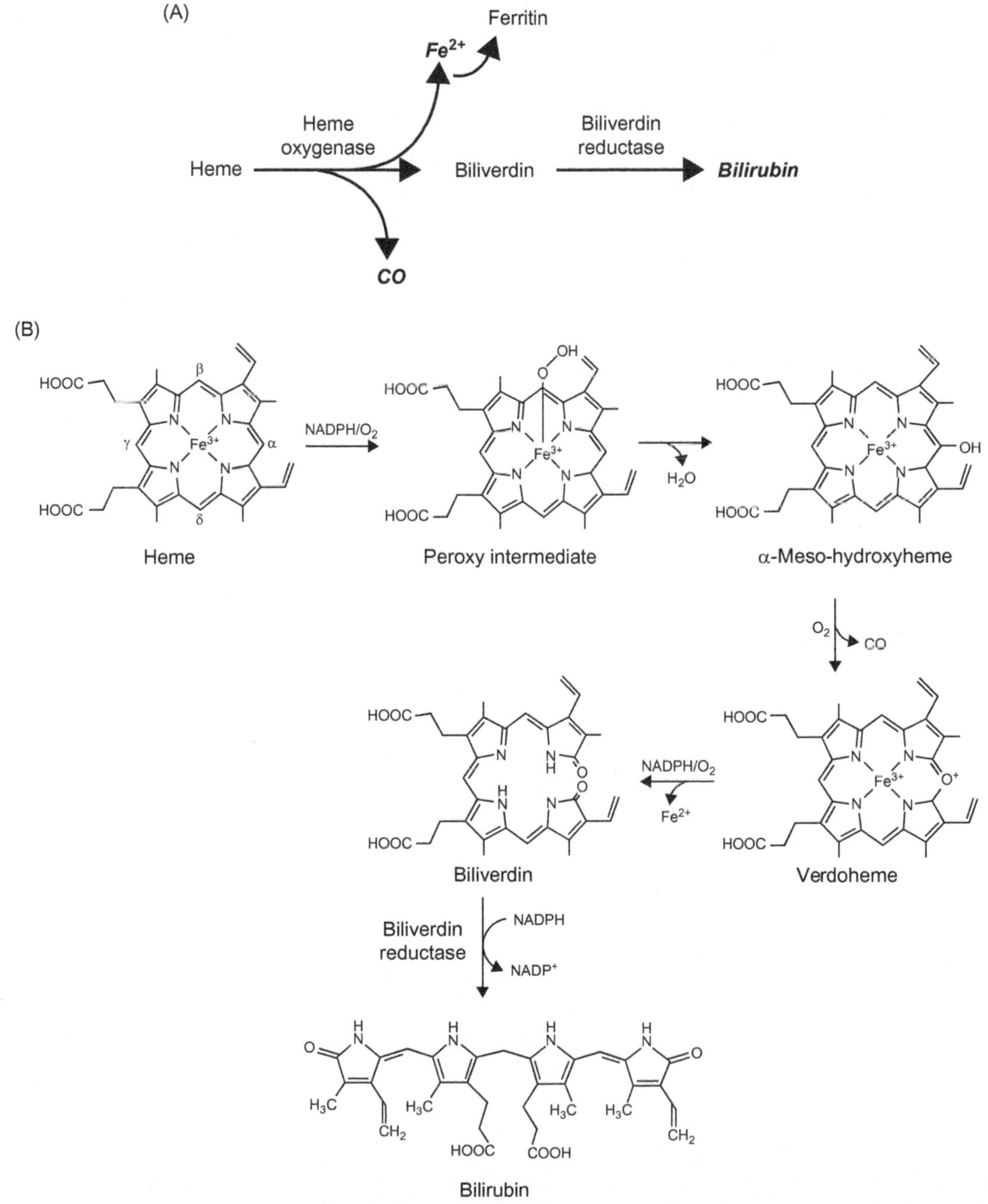

**FIGURE 19.3** (A) Overall reactions converting intracellular heme to biliverdin, CO and reduced iron, and biliverdin to bilirubin. (B) The chemical reactions showing the **heme oxygenase** (anchored to the endoplasmic reticulum membrane) reactions and the action of biliverdin reductase to produce bilirubin.

In quiescent macrophages, the breakdown products of heme are released into the soluble cytoplasm for incorporation into ferritin or for recycling from the cell to **ferroportin** for export. Ferroportin (SLC40A1) is a transporter that plays a role in intestinal absorption of iron and also in the release of iron from the interior of a cell. Major functions of ferroportin occur in intestinal cells for uptake, in hepatocytes and in macrophages for release of cellular iron. A regulatory peptide (**hepcidin**, produced in the liver) binds to ferroportin causing ferroportin to be internalized and degraded. Ferroportin has been found to be essential in early development. Ferroportin-deficient experimental animals accumulate iron in **intestinal enterocytes** (involved in intestinal transport of iron), macrophages (where heme breakdown occurs), and **hepatocytes** (iron storage) that are the chief sites of ferroportin action. **Hephaestin** is highly expressed in the intestine and is required for the removal of iron from the enterocyte to the extracellular space and eventually into the bloodstream. Membrane-bound hephaestin is a ceruloplasmin homolog and is a **multicopper ferrooxidase** converting $Fe^{2+}$, which has been removed from the enterocyte on the basolateral side, to the $Fe^{3+}$ form. The roles of ferroportin and hephaestin in macrophage cells and in enterocytes are pictured in Fig. 19.4.

Thus, **copper** (in **ceruloplasmin**) is important in the oxidation of ferrous iron to ferric iron so that it can be transported in the plasma bound to **transferrin**. **Hephaestrin** is a membrane-bound homolog of ceruloplasmin that is required for iron transport in the intestine. The level of ceruloplasmin is greatly reduced in certain hereditary diseases, such as **Wilson's disease**, the **Melk syndrome,** and **hereditary hemochromatosis**. Mutations in the gene for ceruloplasmin can lead to **aceruloplasminemia**, a rare human genetic disease. In this unusual case, iron is overloaded in brain, liver, pancreas, and retina. **Iron overload** in cells, mostly a result of hereditary hemochromatosis, an autosomal recessive condition, causes an accelerated rate of **iron absorption** in the intestine and progressive tissue iron deposition. This disease affects 0.5% of those of European ancestry mostly in the age range of 30s to 50s, although there is occasional occurrence in children. The pathological effect of iron accumulation in a given organ is in the form of **hemosiderin**. Hemosiderin is a particle representing an iron-storage complex that is formed by the breakdown of hemoglobin or an abnormal metabolic pathway of **ferritin**. Hemosiderin granules are yellowish in color and have diameters in the range of about 10 to 75 Å, and some contain iron hydroxide micelles of ferritin molecules in the form of ferritin and apoferritin and partially hydrated alpha-$Fe_2O_3$. In cases of iron overload, **bloodletting** is used as a treatment until the level of iron reaches a normal range. There are **iron-chelating agents**, such as the drug **Deferoxamine** that binds

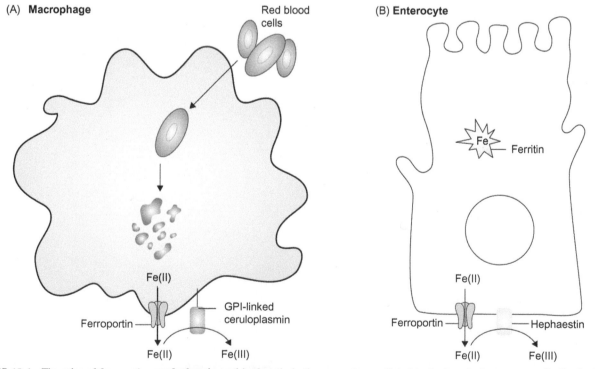

**FIGURE 19.4** The roles of ferroportin, **ceruloplasmin,** and hephaestin in the macrophage cell and in the intestinal enterocyte cell. Ceruloplasmin carries 95% of copper in the plasma. It oxidizes its substrate, $Fe^{2+}$, to the ferric form ($Fe^{3+}$) which can bind to **transferrin** (the plasma transporting protein); transferrin can only bind to the $Fe^{3+}$ form of iron. Prior to the transport of iron in the plasma, the copper in ceruloplasmin plays a key role in the oxidation of iron. Ceruloplasmin is required for the proper localization of ferroportin. Ferroportin, in addition to its localization in macrophages and enterocytes, is also located in astrocytes. *GPI*, glycerophosphatidylinositol. *Reproduced from http://flipper.diff.org/app/pathways/info/2459.*

plasma iron and increases the elimination of iron in the urine and feces. One of the unfortunate effects of iron overload is that it enhances the rate of cancer cell growth and the rate of growth of infectious organisms.

Normally, heme is converted to **bilirubin** (Fig. 19.3), and in hepatocytes, bilirubin is rendered more water soluble by the formation of **bilirubin diglucuronide**, a product of **UDP glucuronyl transferase** (Fig. 19.5).

Bilirubin diglucuronide is conjugated to **cholesterol** and excreted as a bile acid pigment through the intestine.

The **absorption of dietary iron** across intestinal epithelial cells is facilitated by the **divalent metal ion transporter** (**DMT1**). DMT1 is a 12-transmembrane domain protein consisting of about 586 amino acid residues. The fourth transmembrane domain (TM4) is critical for the function of the transporter. Iron in the diet is transported from the intestinal lumen across the **villus cell** and exported from the cell at the basolateral side, mediated by transferrin. Transferrin regulates transport of iron from the villus and crypt cells to body tissues. The **transferrin** molecule (20%−45% saturated with iron) binds two iron atoms. During iron deficiency, crypt cells tend to move toward the villi to increase the efficiency of iron transfer to internal tissues. The transport of iron into the **intestinal cells** is reviewed in Fig. 19.6, and the overall fates of dietary iron are illustrated in Fig. 19.7.

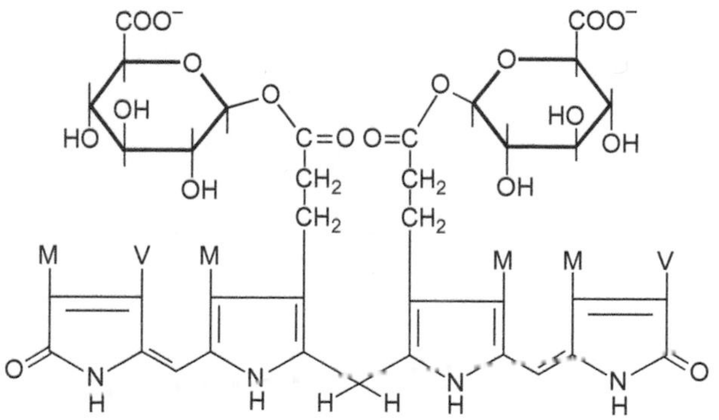

**FIGURE 19.5** Structure of bilirubin diglucuronide. *M*, methyl; *V*, vinyl. *Redrawn from http://themedicalbiochemistrypage.org/heme-porphyrin.php.*

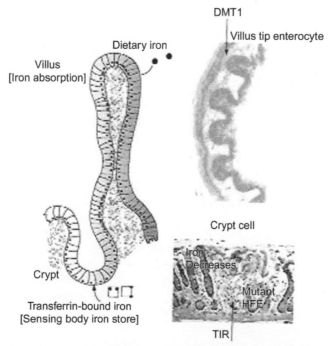

**FIGURE 19.6** Dietary iron is transported into the intestinal **villus cell** by the **divalent metal ion transporter** (**DMT1**). Iron in crypt cells is transported to the blood by the carrier transferrin (Tf) that interacts with a **transferrin receptor** (**TFR**). A **hemochromatosis gene** (**HFE**) encodes a protein (**HFE protein**) that is highly expressed in duodenal crypt cells (and in other cells of the body). When part of the hemochromatosis (HFE) gene is deleted experimentally, depressing the function of HFE protein, **iron overload** occurs. In this situation, DMT1 is increased, suggesting a relationship between DMT1 expression and HFE protein. The HFE protein interacts with and inhibits the transferrin receptor (TFR) on the basolateral membrane of crypt cells. Decreased intracellular iron content leads to increased expression of DMT1 that increases the uptake of dietary iron.

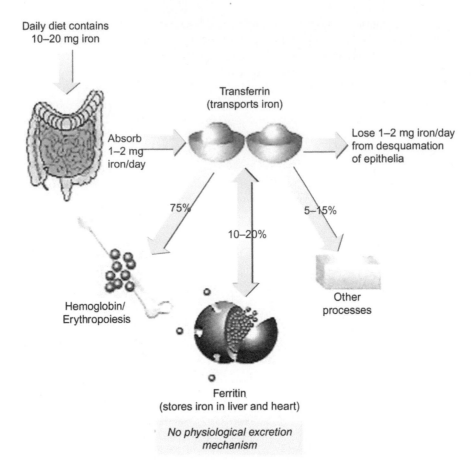

**FIGURE 19.7** Overview of **normal iron absorption**. Iron is bound and transported in the body via transferrin and stored in **ferritin** molecules. Once iron is absorbed, there is no physiological mechanism for excretion of excess iron from the body other than blood loss (i.e., pregnancy, menstruation, or bleeding). *Redrawn from http://www.cdc.gov/hemochromatosis/training/images/iron_cycle.jpg (no longer appearing on the Internet).*

Seventy-five percent of the iron absorbed into the body becomes bound to proteins, such as **hemoglobin** that are involved in the transport of oxygen. The **iron storage pool** takes up 10%−20% of the absorbed iron, and this pool can be used in the formation of red blood cells (erythropoiesis). One molecule of ferritin can store as many as 4000 iron atoms in its mineral core (Fig. 19.2). When excess iron is absorbed in the diet, more ferritin is produced to store the excess. Tissue cells have membrane receptors for transferrin-iron complexes. The transferrin receptors engulf and internalize transferrin-iron complexes.

There exists an **iron-sensing mechanism** for both transferrin receptor and ferritin at the level of mRNA (posttranscriptional mechanism). There are **iron-responsive elements** (**IREs**) in mRNA sequences that code for the translation of ferritin and transferrin receptor. *When there is excess iron over what is in the immediate stores, iron binds to the **IRE-binding protein (IREBP)**, which is an **aconitase**, and this binding event causes a change in the conformation of the **IREBP** so that it no longer binds to **transferrin receptor mRNA**. Consequently, the cell produces more **ferritin** (Fig. 19.8).

ApoIREBP (without iron) binds to the IRE on the mRNA coding for **transferrin receptor**. The binding of IRE without iron stabilizes the mRNA-encoding transferrin receptor and permits its translation to proceed.

When cells become low in iron content, they continue to produce transferrin receptors allowing more **transferrin-iron complexes** to enter the cells. As the level of iron accumulates in the cells, iron binds to IREBPs; as a result, they change their conformation and dissociate from the transferrin receptor mRNA. In the absence of bound IREBP, transferrin receptors are degraded, and cells no longer produce transferrin receptors. In the condition of **low cellular iron**, IREBPs bind to the IRE of **ferritin mRNA** and cause reduced translation of ferritin but the binding of apoIREBPs to transferrin receptor mRNA causes an increase in the translation of **transferrin receptors** increasing the cellular capacity for acquisition of more iron as it becomes available.

There are at least two IREBPs, named **IREBP1** and **IREBP2**. IREBP1, under conditions of elevated cellular content of iron, binds with an iron cluster (4Fe-4S) and adopts the conformation of aconitase that is not permissive for binding

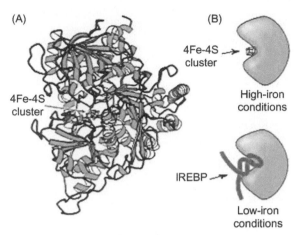

**FIGURE 19.8** (A) **Aconitase** contains a relatively unstable **4FE-4S cluster** at its center. (B) Under conditions of low iron, the 4Fe-4S dissociates and appropriate RNA-binding molecules can bind in its place. *Reproduced from http://www.ncbi.nim.nih.gov/books/NBK22400/figure/A4495/.*

to the IRE. IREBP2, on the other hand, is degraded under conditions of high levels of iron. The affinities of different IREBPs for different IREs vary.

The **iron response element** (**IRE**) is located in untranslated regions of the mRNAs that translate proteins involved in iron metabolism, such as ferritin and transferrin receptor. In ferritin mRNA, a single IRE exists in its 5′ untranslated region. In the case of the transferrin receptor mRNA, several IREs exist in the 3′ untranslated region. The IREs in different mRNAs, that are bound by IREBPs, are conserved short stem-loops, shown in Fig. 19.9.

The posttranscriptional regulation of translation by regulatory proteins is a new area and one that is sure to open up new pharmaceutical approaches.

## Hepcidin, A Peptide Hormone, is the Principal Regulator of Iron Homeostasis

The principal regulator of systemic iron homeostasis is the peptide hormone hepcidin. Hepcidin is produced in hepatocytes through the pathway: **preprohepcidin** (84 amino acids) → **prohepcidin** (60 amino acids) → hepcidin (25 amino acids). The last conversion of prohepcidin to hepcidin is catalyzed by the convertase, **furin**. Secretion of hepcidin produced in the liver is regulated by the proteins: HFE (high iron) protein (encoded by the *HFE* **gene**), **hemojuvelin** (**HJV**), and **transferrin receptor 2** (TFR2). The HFE protein, the protein product of *HFE* gene (chromosome 6.22.2), also referred to as the hemochromatosis protein, appears to regulate the interaction between the transferrin receptor (TfR) and **transferrin**. HFE protein is located mainly on the surface of liver and intestinal cells.

In **iron deficiency,** the amount of hepcidin secreted from the liver is reduced, allowing for the normal functioning of **ferroportin** in releasing iron from the intestinal **enterocytes** and from iron stores in **macrophages** into the bloodstream. Conversely, when iron is in excess, more hepcidin is released from the liver resulting in the blockage of ferroportin so that iron is neither absorbed into the bloodstream from iron in enterocytes nor is it released into the bloodstream from macrophage stores. These events are illustrated in Fig. 19.10.

## Hemojuvelin (HJV), An Anchored Membrane Protein Stimulates Hepcidin Transcription Through Bone Morphogenic (Morphogenetic) Proteins and Smad

Hemojuvelin (HJV; also named RGMc or HFE2) is a regulator of hepcidin production. It is tethered to the cell membrane by a **glycosylphosphatidylinositol (GPI) anchor**. It complexes with **bone morphogenic proteins (BMPs)** that when activated by HJV signals **Smads** to form a complex that enters the nucleus and stimulates the transcription of hepcidin. HJV that is released from the cell to the extracellular fluid lasts there for more than 24 hours, whereas the predominant membrane-associated form (2 chains stabilized by disulfide bonds) is not found in the extracellular fluid and disappears from the cell surface with a half-life of less than 3 hours. HJV consists of bound and soluble forms and three isoforms: **RGMa** (repulsive guidance molecule a) and **RGMb** are found in the nervous system, and **RGMc** is located in skeletal muscle and liver. The **knockout of HJV** in experimental animals produces a depression in the expression of hepcidin. Whereas the membrane-bound form of HJV stimulates the expression of hepcidin, the soluble form may

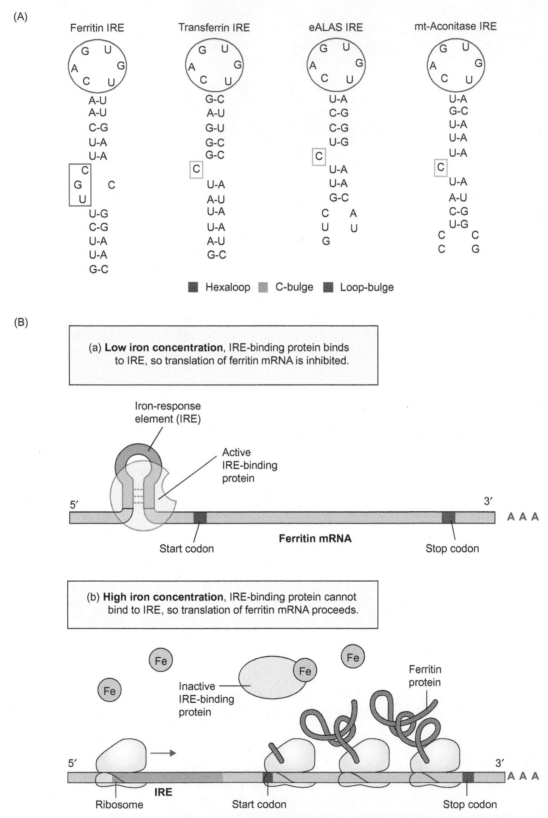

FIGURE 19.9   (A) Sequences of IREs in the untranslated regions of different mRNAs. The proteins encoded by a specific mRNA with the illustrated IRE are labeled at the top of each structure. The hexaloop at the top of each IRE sequence is conserved. In the middle of the structure, there is a bulge where a cytosine or a UGC sequence protrudes from the stem. (B) pictures the inhibition of **translation of ferritin mRNA** when the IREBP is bound to the IRE under low iron concentrations (*left*), and the figure on the *right* shows the promotion of translation of ferritin under conditions of high iron concentration where the IREBP is complexed with iron causing its dissociation from the IRE and relief of the inhibition of translation. *(A) Reproduced from http://bioinformatica,upf.edu/2002/projects/4.2/pages/english/page02.htm and (B) reproduced from http://www.mun.ca/biology/desmid/brian/BIOL2060/BIOL2060-23/23_31.jpg.*

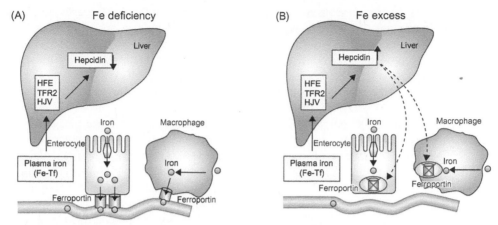

**FIGURE 19.10** The HFE protein, homojuvelin (HJV), and transferrin receptor 2 (TFR2) signaling pathways regulate hepcidin secretion from the liver. (A) In states of iron deficiency, hepcidin levels are low, and the iron transport protein ferroportin allows entry of iron from duodenal enterocytes into the blood and the recirculation of iron from macrophages into the plasma. (B) In states of iron excess, **hepcidin** levels increase, and this promotes the internalization and degradation of **ferroprotein** and results in decreased iron absorption from the gut and decreased release from macrophages. **Mutations in HFE**, HJV, or TFR2 result in low hepcidin levels despite high iron levels and inappropriate continued transport of iron into the plasma. *Fe-Tf*, iron-transferrin complex; *HJV*, hemojuvelin; *TFR2*, transferrin receptor 2. *Reproduced from http://www.rayur.com/hemochromatosis-definition-causes-symptoms-diagnosis-and-treatment.html.*

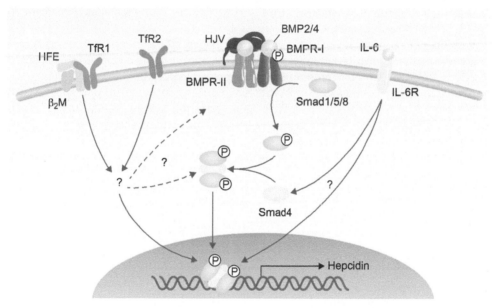

**FIGURE 19.11** Hemojuvelin, a glycosylphophatidylinositol (GPI)-linked membrane protein, acts as a BMP coreceptor and modulates hepcidin expression by stimulating BMP2 and BMP4 signaling. Upon binding by BMPs, BMP receptor type II phosphorylates type I receptors, and this complex, in turn, phosphorylates the receptor Smads 1, 5, and 8. These form heteromeric complexes with Smad4, and this complex translocates to the nucleus and alters the expression of target genes. Hepcidin can also be regulated by HFE which forms a complex with TfR1 and $\beta_2$-microglobulin ($\beta_2$M) and TfR2 but how these proteins exert their effects and whether HJV and/or the BMP signaling pathway are involved remains to be resolved. Inflammatory cytokines, such as **IL-6**, can also stimulate hepcidin expression, and recent data suggest that **Smad4** is involved in regulating this response. *Reproduced from http://www.nature.com/ng/journal/v38/n5/fig_tab/ng0506-503_F1.html.*

suppress hepcidin expression. There appear to be two soluble isoforms and two membrane isoforms of HJV. The action of HJV as a coreceptor of bone morphogenic protein operating through Smads is summarized in Fig. 19.11.

# HEME SYNTHESIS

Heme is a cofactor for several **heme-containing proteins** that function in oxygen binding or metabolism as well as in electron transport. **Hemoglobins** are the primary oxygen-binding proteins, and oxygen metabolism is carried out by oxidases, peroxidases, catalases, hydroxylases and other heme proteins contain the heme prosthetic group, such as **nitric oxide synthase** and **guanylic cyclase**. **Cytochromes**, containing heme, are involved in the electron transport chain and as cofactors in hydroxylation reactions. The biosynthetic pathway for heme is located in the soluble cytoplasm and in mitochondria and this pathway is identical in all mammalian cells; however, the regulation of heme biosynthesis may differ between erythroid and nonerythroid cells. Heme, the final product of the biosynthetic pathway, feeds back negatively on the synthetic process but the specific step at which the negative feedback operates may differ between different cell types.

Cellular levels of **heme** are carefully controlled so that there is a balance between the biosynthesis of heme and its catabolism by **heme oxygenase**. Developing erythroid cells synthesize about 10 times more heme than the liver hepatocytes, the second major producer of heme. **Iron metabolism** in a specific cell type determines the regulation and the rate of heme biosynthesis. The first enzyme in the heme biosynthetic pathway is **5-aminolevulinic acid synthase** (**ALAS**). There are two genes specifying ALAS, one that encodes **ALAS1** that is expressed ubiquitously and one that encodes **ALAS2** that is specific to **erythroid cells**. The mRNA for the translation of ALAS2 contains an **IRE** in its 5′ untranslated region; this controls the induction of translation. The availability of iron controls the level of **protoporphyrin IX** in hemoglobin-synthesizing cells.

In **nonerythroid cells**, the rate-limiting step in heme biosynthesis is catalyzed by ALAS1 whose synthesis is the object of negative feedback by heme (not in the case for erythroid cells). However, the concentration of heme decreases the access of iron from **transferrin** without affecting its use in heme synthesis. In erythroid cells, heme stimulates the transcription of the **globin** gene so that **hemoglobin** can be formed. The enzyme controlling the degradation of heme, heme oxygenase, turns out to be a major enzymatic antioxidant system in that it provides **bilirubin** that is an antioxidant.

The biosynthesis of heme begins with **succinyl-CoA** and **glycine** in the mitochondria to form **γ-aminolevulinic acid** (**ALA**) through the ALAS reaction. ALA is exported to the cytoplasm where **ALA dehydrase** converts it to **porphobilinogen**. Porphobilinogen is then converted to **uroporphyrinogen III** (or uroporphyrinogen I) by **uroporphyrinogen I synthase** and **uroporphyrinogen II cosynthase**. Uroporphyrinogen III is converted to **coporphyrinogen III** by **uroporphyrinogen decarboxylase,** and this product enters the mitochondria where it is converted to **protoporphyrinogen IX** by **coproporphyrinogen III oxidase**. Protoporphyrinogen IX is converted to protoporphyrin IX by **protoporphyrinogen IX oxidase**. In the final step, protoporphyrin IX is converted to **heme** by the action of **ferrochetalase** that incorporates iron into the molecule. The outline of the biosynthetic pathway is shown in Fig. 19.12, and the chemical reactions are shown in Fig. 19.13.

In Fig. 19.14 is shown a **space-filling model of heme,** and in Fig. 19.15 is shown a **space-filling model of hemoglobin**.

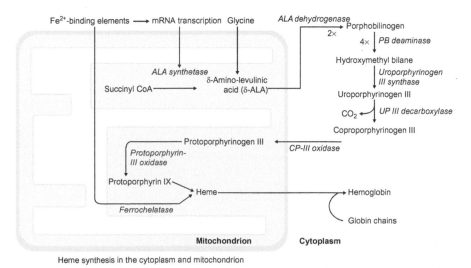

Heme synthesis in the cytoplasm and mitochondrion

**FIGURE 19.12** Overview of the **synthesis of heme** showing the events occurring inside the mitochondrion and in the cytoplasm. The process begins in the mitochondrion with succinyl-CoA from the TCA cycle and glycine from the cytoplasm to form **δ-aminolevulinic acid**. *Redrawn from http://en. wikipedia.org/wiki/heme.*

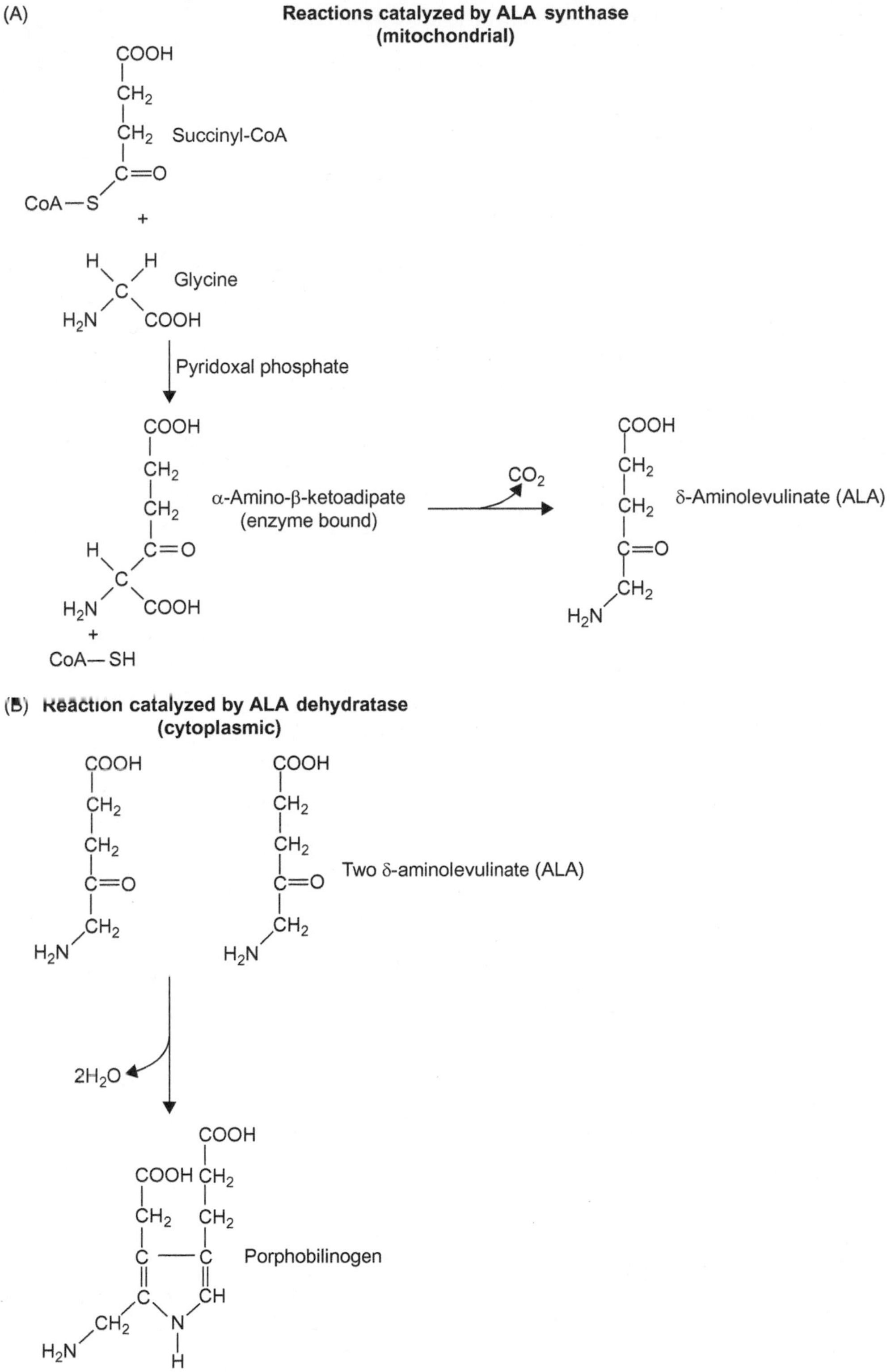

**FIGURE 19.13**   Chemical reactions in heme biosynthesis.

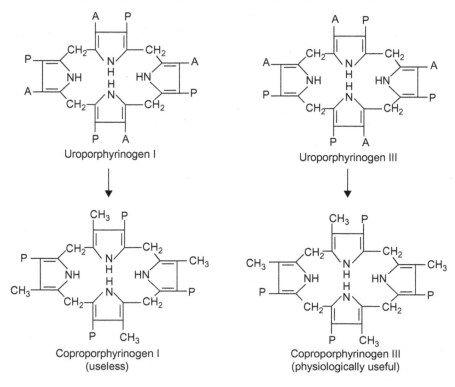

**FIGURE 19.13** (Continued.)

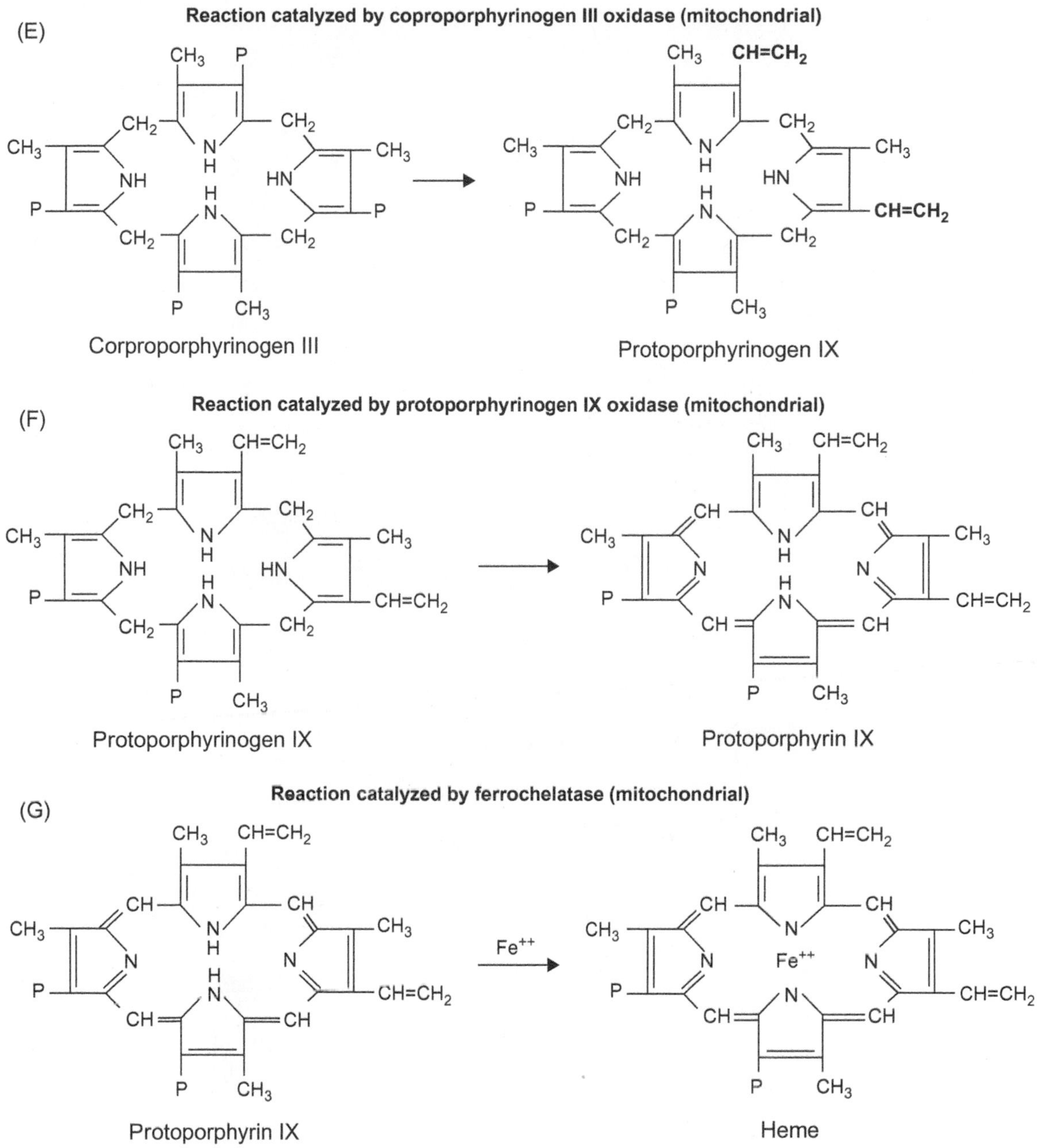

**Reaction catalyzed by coproporphyrinogen III oxidase (mitochondrial)**

(E)

Corproporphyrinogen III → Protoporphyrinogen IX

**Reaction catalyzed by protoporphyrinogen IX oxidase (mitochondrial)**

(F)

Protoporphyrinogen IX → Protoporphyrin IX

**Reaction catalyzed by ferrochelatase (mitochondrial)**

(G)

Protoporphyrin IX → Heme

**FIGURE 19.13** (Continued.)

The final step in the synthesis of heme is the insertion of iron by ferrochetalase (Fig. 19.13 **part G**) into **protoheme IX**. Because all of the intermediates in the biosynthesis of heme are **tetrapyrroles** or **porphyrins**, the overall reactions can be classified as **porphyrin synthesis**, somewhat analogous to the synthesis of vitamin B12 in certain other (than mammalian) species. The enzyme (**ALA synthase**) catalyzing the initial step of **heme synthesis** is regulated by the intracellular concentrations of both heme and **iron**. When the cell concentration of iron is low from a deficient diet, especially in bone marrow cells and liver cells, there is a decrease in the synthesis of porphyrin in order to eliminate its toxicity if it accumulates.

Mutations in the genes for ALA synthase or for other enzymes in the heme biosynthetic pathway cause inability to form heme (**genetic porphyria**). A list of genetic porphyrias appears in Table 19.4.

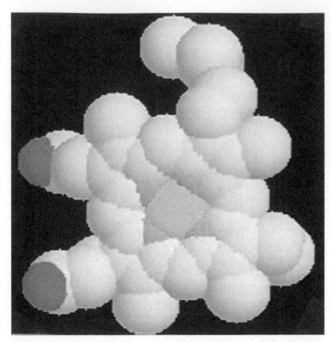

**FIGURE 19.14**   Space-filling model of heme. The iron atom is in *yellow* at the center. *Reproduced from http://course1.winona.edu/sberg/308s08/ Lec-note/EnzymesA.htm.*

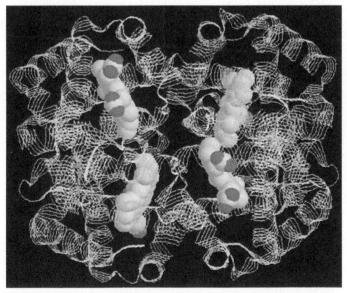

**FIGURE 19.15**   Space-filling model of hemoglobin. Hemoglobin contains four globins and four hemes. Where visible, the iron atom of heme is in *yellow. Reproduced from http://faculty.stcc.edu/AandP/AP/AP2pages/Units18to20/blood/hemoglob.htm.*

If the single gene for **ferrochetalase**, is mutated, this can lead to **erythropoietic protoporphyria**, a disease that would be carried in the germ line (X chromosome). Porphyrias also may be acquired, for example, in liver damage. **Hematin** or **heme arginate** can be administered intravenously to limit a severe attack. Other medications to limit pain and other symptoms are in use.

Ferrochetalase, the final enzyme in heme synthesis, is found in the inner mitochondrial membrane in such a position that its catalytic center faces the space of the **mitochondrial matrix**. The prosthetic group of the enzyme is an **iron-sulfur cluster** (2Fe2S) in the form of a chelate of iron with cystine residues. The binding of iron to ferrochetalase involves a histidine group (H207) and a glutamate residue (E87) functional in the catalytic center. The enzymatic

**TABLE 19.4 Subtypes of Inherited Porphyrias Based on the Enzyme that is Mutated**

| Deficient Enzyme | Associated Porphyria | Type of Porphyria | Inheritance | Symptoms | Prevalence |
|---|---|---|---|---|---|
| δ-Aminolevulinate (ALA) synthase | X-linked sideroblastic anemia (XLSA) | Erythropoietic | X-linked | | |
| δ-Aminolevulinate dehydratase (ALAD) | Doss porphyria/ ALA dehydratase deficiency | Hepatic | Autosomal recessive | Abdominal pain, neuropathy | Extremely rare (fewer than 10 cases ever reported) |
| Hydroxymethylbilane (HMB) synthase (or PBG deaminase) | Acute intermittent porphyria (AIP) | Hepatic | Autosomal dominant | Periodic abdominal pain, peripheral neuropathy, psychiatric disorders, tachycardia | 1 in 10,000 20,000 |
| Uroporphyrinogen (URO) synthase | Congenital erythropoietic porphyria (CEP) | Erythropoietic | Autosomal recessive | Severe photosensitivity with erythema, swelling and blistering. Hemolytic anemia, splenomegaly | 1 in 1000,000 or less |
| Uroporphyrinogen (URO) decarboxylase | Porphyria cutanea tarda (PCT) | Hepatic | Autosomal dominant | Photosensitivity with vesicles and bullae | 1 in 10,000 |
| Coproporphyrinogcn (COPRO) oxidase | Hereditary coproporphyria (HCP) | Hepatic | Autosomal dominant | Photosensitivity, ncurologic symptoms, colic | 1 in 500,000 |
| Protoporphyrinogen (PROTO) oxidase | Variegate porphyria (VP) | Mixed | Autosomal dominant | Photosensitivity, neurologic symptoms, developmental delay | 1 in 300 in South Africa 1 in 75,000 in Finland |
| Ferrochelatase | Erythropoietic protoporphyria (EPP) | Erythropoietic | Autosomal dominant | Photosensitivity with skin lesions. Gallstones, mild liver dysfunction | 1 in 75,000 200,000 |
| | Transient erythroporphyria of infancy | | | Purpuric skin lesions | |

Reproduced from http://en.wikipedia.org/wiki/Porphyria.

reaction at the inner mitochondrial membrane involves another protein, **frataxin**, which acts as the iron donor for the iron-sulfur cluster. Holofrataxin (without its binding to the iron-sulfur cluster) binds to ferrochetalase on the inner mitochondrial membrane. It also binds to the **iron-sulfur cluster synthetic unit (ISU)**. The dimer of ferrochetalase (active form) contains a **frataxin-binding site** on its matrix side. Holofrataxin is a high affinity-binding partner for holoferrochetalase. Holofrataxin can deliver iron to ferrochetalase and is the mediator for the final step in the synthesis of heme. Frataxin is involved in the biosyntheses of both heme and iron-sulfur clusters. Shown in Fig. 19.16 is the **iron-binding site of frataxin** where the iron atom is bound through coordination with residues of histidine and aspartate.

In Fig. 19.17, the function of frataxin within **mitochondria** is presented.

**Holofrataxin** (frataxin bound with iron) is used as an iron donor for the biosynthetic pathways of both heme and iron-sulfur cluster. Fig. 19.18 shows models for the regulation of frataxin in the cell when the holofrataxin (Hftx) level is low (*top figure*) or when the holofrataxin level is high (*bottom figure*).

Heme synthesis responds to oxygen deficiency (hypoxia) by an increase in ferrochetalase mRNA. There exists a hypoxia-inducible transcription factor, **HIF-1**. Two HIF-1-binding motifs have been found in the gene promoter of the ferrochetalase gene. When cellular oxygen levels are low HIF-1 (HIF-1α, hypoxia-inducible factor-1) binds to the **ferrochetalase gene promoter** resulting in increased transcription of the enzyme. **Reactive oxygen species (ROSs)** negatively affect HIF-1, as might be expected.

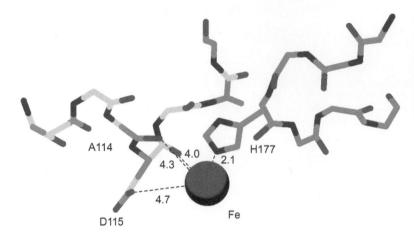

A114    4.0
     4.3    2.1
        4.7    H177

D115         Fe

**FIGURE 19.16** Iron binding to **frataxin**. Frataxin molecules are colored in *green* and *yellow*. **Iron** is depicted as a *red* sphere coordinated to His-177. Distances in Å (*dotted lines*) are between the nucleus of iron and its closest neighbors. *Reproduced from Figure 5 of S. Dhe-Paganon et al., "Crystal structure of human frataxin", J. Biol. Chem., 275: 30753–30756, 2000 and http://www.jbc.org/content/vol275/issue40/images/large/bc3608148005.jpeg.*

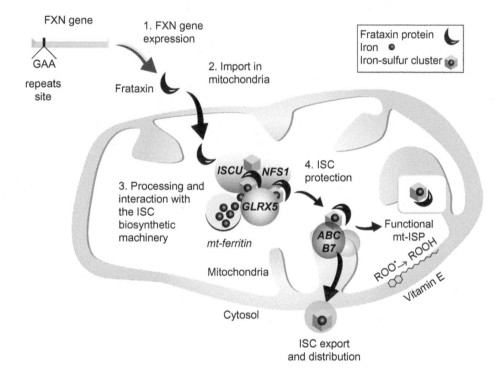

FXN gene

1. FXN gene expression

GAA
repeats site

Frataxin

2. Import in mitochondria

Frataxin protein
Iron
Iron-sulfur cluster

ISCU    NFS1

GLRX5

*mt-ferritin*

4. ISC protection

3. Processing and interaction with the ISC biosynthetic machinery

Mitochondria

ABC B7

Functional mt-ISP

ROO˙ → ROOH

Vitamin E

Cytosol

ISC export and distribution

**FIGURE 19.17** Frataxin function in the mitochondria. The scheme illustrates the **iron-sulfur cluster** (**ISC**) biosynthesis machinery present in the mitochondrial matrix encompassing the **ISCU-NFS1 protein complex** associating glutaredoxin 5 (**GLRX5**) with the frataxin protein. It makes use of iron, possibly delivered by mitochondrial **ferritin** to synthesize ISC, also distributed among several of the mitochondrial proteins (including several membrane-bound respiratory chain components: complexes I, II, and III and the matrix-soluble **aconitase**). In addition to its role in the biogenesis of ISC, the frataxin protein might be associated with ISC after its synthesis. The detoxifying role of **vitamin E** in the mitochondrial inner membrane is also indicated. ISP, ISC-containing protein; *mt*, mitochondria. *Reproduced from http://www.biomedcentral.com/1741-7015/9/112/figure/F1.*

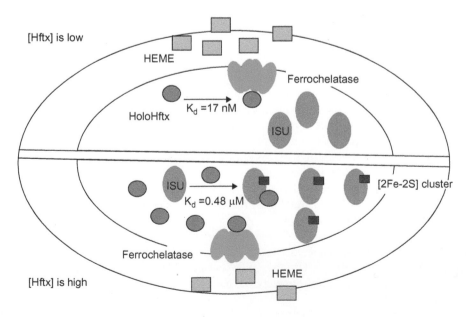

[Hftx] is low

HEME

Ferrochelatase

HoloHftx    $K_d$ =17 nM

ISU

ISU    $K_d$ =0.48 μM

[2Fe-2S] cluster

Ferrochelatase

HEME

[Hftx] is high

**FIGURE 19.18** Cellular models for the regulation of frataxin chemistry in iron-sulfur cluster and heme biosynthesis. Holofrataxin (Hftx) is used as an iron donor for both heme and iron-sulfur cluster biosynthetic pathways. Under normal cell growth conditions, the frataxin concentration is sufficient for both heme and iron-sulfur cluster syntheses. The level of frataxin is downregulated in erythroid differentiation, as is the iron-sulfur cluster biosynthesis pathway. However, heme biosynthesis remains essentially normal as a consequence of the distinct binding affinities of frataxin to ISU and to ferrochelatase. *Reproduced from Figure 3 of T. Yoon and J.A. Cowan, "Frataxin-mediated iron delivery to ferrochetalase in the final step of heme biosynthesis", J. Biol. Chem., 279: 25943–25946, 2004.*

# HEMOGLOBIN FORMATION

**Heme** interacts directly with **globin** to form a specific **heme-globin complex**. Then, the heme pocket collapses around the porphyrin, and a bond is formed between the proximal histidine residue (of globin) and the heme iron atom. This interaction is (nonenzymatically) governed by the concentrations of the reactants, heme and globin. Hemoglobin is a heterotetramer containing two $\alpha$-globin molecules and two $\beta$-globin molecules, each of which has a heme associated with it (four hemes as shown in Fig. 19.15). The globins are encoded by different genes: the $\alpha$-globin gene is located on chromosome 16, while the $\beta$-globin gene is located on chromosome 11. $\alpha$-Globin is a protein of 141 amino acids, while $\beta$-globin is a protein of 146 amino acids. Other globin variants also exist, and there is one gene for each of the other globin structures. The $\alpha$-globin gene promoter functions independently, while the $\beta$-**globin gene promoter** is dependent on an enhancer.

Hemoglobin synthesis takes place as **erythroid cells** are differentiating from immature red blood cells to mature red blood cells. When the red cells mature, the synthesis of hemoglobin stops and the hemoglobin already formed must survive for 120 days, the lifetime of the erythrocyte. When the formation of $\beta$-globin is insufficient, there is a significant excess of $\alpha$-globin over $\beta$-globin generating a group of thalessemias called $\beta$-**thalessemias**. $\alpha$-Globin, by itself, is unstable and forms aggregates on the membranes of red blood cells, damaging them and causing anemia. Under normal conditions, there is a small excess of $\alpha$-globin over $\beta$-globin without generating a thalessemia. This small excess of $\alpha$-globin becomes bound to a chaperone-like protein, the $\alpha$-**hemoglobin stabilizing protein** (**AHSP**). The interaction between $\alpha$-globin and AHSP (association constant $= 10^7\,M^{-1}$) is weaker than the interaction between $\alpha$-globin and $\beta$-globin (association constant $= 10^{10}\,M^{-1}$). Thus, when a molecule of $\beta$-globin appears, it can displace the AHSP from the $\alpha$-globin-AHSP complex, and it can interact with the released $\alpha$-globin. The two globins interact about 1000 times more avidly (comparing association constants) than the interaction between $\alpha$-globin and AHSP. AHSP forms a triple helix bundle, and $\alpha$-globin has a surface region for the binding of AHSP, whereas $\beta$-Globin does not have a surface domain that interacts with AHSP. Thus, the appearance of $\beta$-globin molecules causes the dissociation of the $\alpha$-globin-AHSP dimer by competition, and the hemoglobin tetramer can be formed (1 dimer of $\alpha$-globin + 2 hemes; 1 dimer of $\beta$-globin plus 2 hemes). In the hemoglobin tetramer, the AHSP-binding surface (of the $\alpha$-globin component) is buried, and since there is no such surface on the $\beta$-globin component, **AHSP** cannot bind.

The normal **red blood cell** contains about 4 mM hemoglobin with a normal excess of $\alpha$-**globin** that is about 10%–20% (about 0.4 to 0.8 mM of the hemoglobin concentration). The cellular concentration of AHSP is about the same, 0.4 mM, so the excess of $\alpha$-globin is in a complex with AHSP. In $\beta$-**thalassemia**, the large excess of $\alpha$-globin forms aggregates and causes the release of the iron atom from **heme**. The free iron atom can generate toxic **oxygen free radicals** (ROS), and the free $Fe^{2+}$ can interact with peroxide (ROS) to form $Fe^{3+}$ and hydroxyl ions ($OH + OH^-$). $Fe^{3+}$ can interact with $H_2O_2$ to form $Fe^{2+}$ and another radical ($OOH + H^+$). Experimentally, the addition of AHSP can reduce free radical formation generated by the excess of $\alpha$-globin. It does this by trapping the iron associated with $\alpha$-globin in the $Fe^{3+}$ state so that it does not interact with peroxide and cycle back to $Fe^{2+}$ (left side of Fig. 19.19, part A).

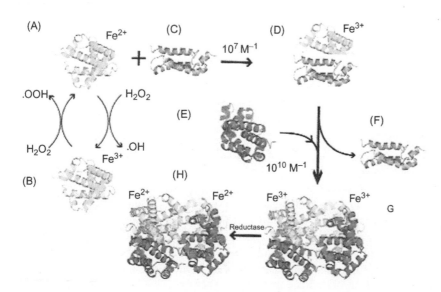

FIGURE 19.19 The hypothetical mechanism of AHSP. (A). Excess of $\alpha$-globin associated with heme whose iron is in the ferrous form; it can interact with ROS to produce $\alpha$-globin $Fe^{3+}$. $\alpha$-Globin $Fe^{3+}$, in turn, can interact with $H_2O_2$ to form $\alpha$-globin $Fe^{2+}$ and •OOH (B). (C) AHSP interacts with $\alpha$-globin (A) to form $\alpha$-**globin-AHSP complex** (D). This reacts with a molecule of $\beta$-globin (with its heme) (E) causing the dissociation of AHSP (F) and forming the tetrameric hemoglobin (1 dimer of $\alpha$-globin-$Fe^{3+}$ and 1 dimer of $\beta$-globin-$Fe^{3+}$) (G). The $Fe^{3+}$ iron atoms of hemoglobin are reduced to the ferrous form ($Fe^{2+}$) by a reductase to form the *functional hemoglobin tetramer with 4 hemes, all of which contain $Fe^{2+}$* (H). *Reproduced from http://www.science.org.au/sats2004/images/mackay23d.jpg (no longer available on the Internet) Australian Academy of Science.* AHSP, $\alpha$-hemoglobin stabilizing protein.

**TABLE 19.5 Characteristics of Thalessemias**

| Category | Anemia | MCV | % Hb A$_2$ | % HB F |
|---|---|---|---|---|
| β-Thalassemia | | | | |
| Heterozygous | Mild | ↓ | ↑ | Variable |
| Homozygous | Severe | ↓ | Variable | ↑ up to 90% |
| β-δ-Thalassemia | | | | |
| Heterozygous | Mild | ↓ | N in↓ | >5% |
| Homozygous | Moderate to severe | ↓ | Absent | 100% |
| α-Thalassemia | | | | |
| Single-gene defect | None | N to ↓ | N | N |
| Double-gene defect | Mild | ↓ | N to ↓ | <5% |
| Triple-gene defect | Moderate | ↓ | N to ↓ (Hb H or Bart's present) | Variable |

*HbH* (or Hb Bart) is caused by a deletion removing both α-globin genes (chromosome 16) plus a deletion removing only a single gene (16pter-p13.3). The β-globin gene cluster is located on chromosome 11 p15.5. The consequent excess of β-globin chains within red blood cells will form a β4 tetramer characteristic of HbH, generating moderate anemia; *HbF* is fetal hemoglobin consisting of α2γ2 globins. *HbA2* consists of 2 α-globins and 2 δ-globins (α2δ2) that is found in low levels in normal red blood cells; an elevated level of HbA2 is characteristic of β-thalassemia carriers; *MCV*, mean corpuscular volume; *down arrow*, decreased; *up arrow*, increased, *N*, normal.
Reproduced from G. Litwack, *Human Biochemistry & Disease*, Academic Press/Elsevier, page 763, San Diego.

Table 19.5 summarizes the characteristics of thalessemias.

**α-Thalessemias** also can occur when there is a defective synthesis of **α-globin**. In **β-thalessemia,** there is a decreased production of **β-globin** chains. β-Thalessemia is an autosomal recessive disease. There is a condition called thalessemia minor in which heterozygotes are carriers with only mild to moderate **microcytic anemia**. Only a single gene is involved in β-thalessemia, however, two genes are involved in α-thalessemia. Heterozygotes with a defect in a single gene do not express symptoms (**α-thalessemia-2**). Heterozygotes with defects in two genes express anemia (**α-thalessemia-1**), and the clinical symptoms are similar to β-thalessemia.

In the normal red blood cell, **hemoglobin** is formed with four reduced iron atoms in the hemes, and hemes with oxidized iron (Fig. 19.19B) will not be produced nor should there be an appearance of ROS (e.g., •OOH). Thus, the reaction of **A** and **C**, in Fig. 19.19, will form **D** followed by the additional reactions to form the normal hemoglobin (following the reductase action). When β-globin is formed, it binds heme immediately and forms hemoglobin with the slight excess of α-globin. β-Globin synthesis, therefore, seems to be the rate-limiting event in the normal formation of hemoglobin in the immature red blood cell.

# TRACE ELEMENTS

## Trace Metals

Trace elements include a variety of substances in very small amounts, probably in the μg levels or even smaller amounts, usually taken into the body through the diet or by breathing air. One group within the trace elements is the trace metals and many of these are of great importance in the body because they are formed into the cofactors of bodily proteins and enzymes. Besides calcium and iron (already discussed as a component of heme), there are other important metals, such as copper, zinc, selenium, magnesium, manganese, cobalt, and molybdenum plus the nonmetal halogen, iodine. In addition to these, other metals that can be taken into the body are: lithium, chromium, nickel, vanadium, tungsten, arsenic, and some others. Zinc is a cofactor in more than one hundred enzymatic reactions. Some examples of the enzymatic reactions that incorporate trace metals will be discussed. An excess intake of these metals can cause problems (excessive intake of iron can cause **hemochromatosis**, for example, either due to excessive intake or to genetics). Table 19.6 lists examples of proteins or enzymes that incorporate trace metals as cofactors.

**TABLE 19.6** Certain Trace Metals and Examples of Proteins or Enzymes that Incorporate Them as Cofactors

| Trace Metal | Protein or Enzyme Examples |
|---|---|
| Copper ($Cu^{2+}$) | Ceruloplasmin, superoxide dismutase, lysyl oxidase, cytochrome c oxidase, dopamine β-hydroxylase |
| Selenium ($Se^{2-}$) | Selenodeiodinases, iodothyronine deiodinase, selenophosphate synthetase, glutathione peroxidase, thioredoxin reductase |
| Zinc ($Zn^{2+}$) | Carbonic anhydrase, cytosine deaminase, zinc family transporters (Zip), zinc finger motifs (e.g., glucocorticoid receptor) |
| Magnesium ($Mg^{2+}$) | ATP synthase, glutathione synthesis, many enzymes involving MgATP, hexokinase, phosphofructokinase, pyruvate kinase, aldolase, phosphoglycerate kinase |
| Manganese ($Mn^{2+}$) | Pyruvate carboxylase, Phosphoenolpyruvate carboxykinase, catalase, superoxide dismutase, arginase |
| Calcium ($Ca^{2+}$) | CaT1, calcium channel (TRPV family), calbindin, calcium-ATPase, 1,4-lactonase, 3-ketovalidoxylamine A C-N lyase |
| Molybdenum ($Mo^{2+}$) | Sulfite oxidase, xanthine dehydrogenase, aldehyde oxidase, nitrate reductase, ethylbenzene dehydrogenase glyceraldehyde 3-phosphate ferredoxin oxidoreductase, respiratory-arsenate reductase |

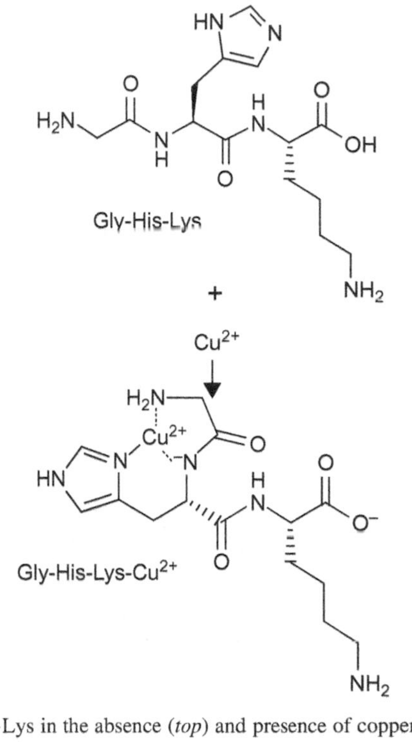

**FIGURE 19.20** Chemical structures of Gly−His−Lys in the absence (*top*) and presence of copper ($Cu^{2+}$) at *bottom*. Histidine residues are important for copper-binding.

# Copper

## Ceruloplasmin

Copper is stored mainly in the liver but is contained in most tissues. The total amount of copper in the body is less than 100 mg. In addition to the content of copper in enzymes (Table 19.6) and in tissues, 90%−95% of the copper in the bloodstream is incorporated into the circulating protein **ceruloplasmin (Cp)**. Cp functions as a copper transporter, in copper mobilization and in copper homeostasis. In addition to its transporter function is has the enzymatic activity of **ferroxidase (iron II: oxygen oxidoreductase)** catalyzing the oxidation of iron (and other metals); ferroxidase activity of ceruloplasmin converts $Fe^{2+}$ to $Fe^{3+}$ in the reaction: $4Fe^{2+} + 4H^+ + O_2 \rightleftharpoons 4\ Fe^{3+} + 2H_2O$. In this process (reaction to the right), Cp-$Cu^{2+}$ is converted to Cp-$Cu^{1+}$. Cp contains six bound copper ions.

A tripeptide isolated from human plasma albumin has been shown to bind copper; it is **GHK (Gly−His−Lys)** as shown in Fig. 19.20 indicating that copper-binding sites contain histidine residues.

(A)

(B)

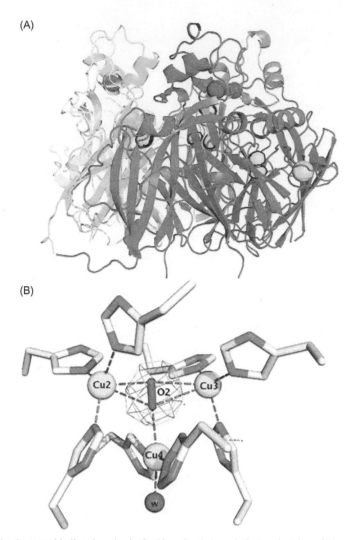

**FIGURE 19.21** (A) Shown are the 6 copper-binding domains in Cp (domains 1 through 6: domain 1 in *red*; domain 2 in *orange*; domain 3 in *yellow*; domain 4 in *green*; domain 5 in *blue*; and domain 6 in *purple*). The locations of bound metals in Cp are: $Cu^{2+}$ as *blue spheres* outlined in *black* for clarity; $Ca^{2+}$ as *olive-green spheres* and $Na^+$ as *red spheres*. (B) The trinuclear cluster, accounting for 3 of the 6 copper ions, is located between domains 1 and 6 and has a water molecule attached to copper ions. There is a dioxygen within the type-3 cluster. Note the participation of **histidine residues**. *Reproduced from http://www.ncbi.nlm.nih.gov/pmc/articles/PMC2483498/#!po = 95.8333.*

Cp is a single polypeptide chain of 1046 amino acid residues and a molecular weight of 132,000. It contains 3 contiguous homologous units of 350 amino acid residues, and each homologous unit is labeled $A_1$, $A_2$, and B. There are attachment sites for 4 glucosamine oligosaccharides. In Fig. 19.21A is shown 6 **copper-binding domains** in Cp.

Thus, 3 copper ions are attached in the **trinuclear cluster** (interface of domains 1 and 6), while the remaining 3 copper ions are in **mononuclear sites** located each in domains 2, 4, and 6. Each of the mononuclear copper ions is coordinated with 2 His residues and 1 free cysteine residue, and the copper ions in domains 4 and 6 also coordinate weakly to a Met residue, while the copper ion in domain 2 is coordinated again to 2 His residues and to a Leu residue in place of the Met residue. The structural arrangements of the copper ions of the trinuclear center and the mononuclear copper ion in domain 6 are identical to the copper structures in **ascorbate oxidase**. The biogenic amines (epinephrine, norepinephrine, serotonin, and DOPA) all bind in domain 6 near the site of cation oxidation. There are binding sites for $Ca^{2+}$ and $Na^+$ in domain 1 near the trinuclear center of copper binding. Since Cp is the major storage and transporting protein for copper in the plasma, it must be able to provide copper ions to tissue cells when needed by the cells. One suggestion for this transport involves ascorbate that becomes oxidized to dehydroascorbate when $Cp$-$Cu^{2+}$ is reduced to $Cu^{1+}$. In this case, presumably, $Cu^{1+}$ should have a much lower association constant for Cp than $Cu^{2+}$. The $Cu^{1+}$ and dehydroascorbate are both thought to cross the cell membrane and enter the cell. Conversely, the copper ion could be transported into the cell by the hCTR1 transporter (see below).

A **deficiency of copper** in plasma leads to diseases, such as **Wilson's disease, aceruloplasminemia,** and **anemia**. The condition is relatively rare (1/40,000). Since Cp is the major storage protein circulating in the plasma, its loss through a mutation in the gene encoding it means that the plasma storage is low or nonexistent but the tissue levels of copper are very high. In aceruloplasminemia, copper content of the liver, brain, and other organs increases (copper overloading), and eventually tissue damage occurs (there are some drugs available that can chelate copper and may be useful as treatments).

In the experimental absence of copper, Cp has been shown to misfold irreversibly and can no longer bind copper ion. Thus, in order to become functional as a copper-binding protein, *Cp must bind copper when it is first synthesized*. It seems possible that Cp in some of these diseases is irreversibly misfolded for some reason and becomes nonfunctional. Perhaps, a mutation of the gene-encoding information for Cp causes this kind of result.

## Wilson's Disease

**Wilson's disease** is an autosomal recessive condition wherein mutated **gene *ATP7B*** occurs in each parent. This disease is one in which it becomes difficult to excrete copper from the body, a process normally carried out by the liver where copper is released into the bile and excreted through the GI tract. The ***ATP7B* gene** encodes information for a transmembrane **copper-transporting P-type ATPase** (transporter using the energy of ATP to drive the transport of an ion across a membrane).

## Human Copper Transporter 1

This is a high affinity copper transporter moving copper, against a concentration gradient, from the extracellular space to the intracellular space of mammalian cells. In addition to the ability to transport copper, this transporter also can take up the anticancer agent **cisplatin** that contains an atom of platinum coordinated to 2 chlorines and 2 amino groups. A general model for the mechanism of the hCTR1 transporter is shown in Figs. 19.22 and 19.23.

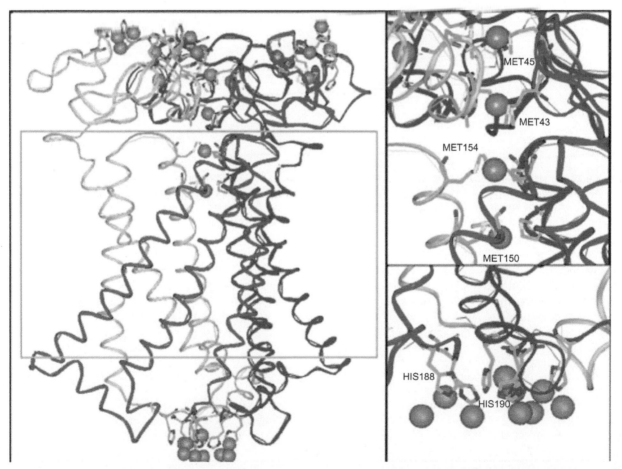

**FIGURE 19.22** All atom model of the hTCR$_1$ trimer and predicted copper binding sites (magenta spheres). Copper binding sites are found (1) at each of the 4 stacked methionine triads (upper right panel), (2) at the lower end of the vestibule on the cytoplasmic side of the protein (lower right panel), and (3) at various Met-rich sites in the N-terminal domain. The predicted transmembrane region is delimited by the green box.

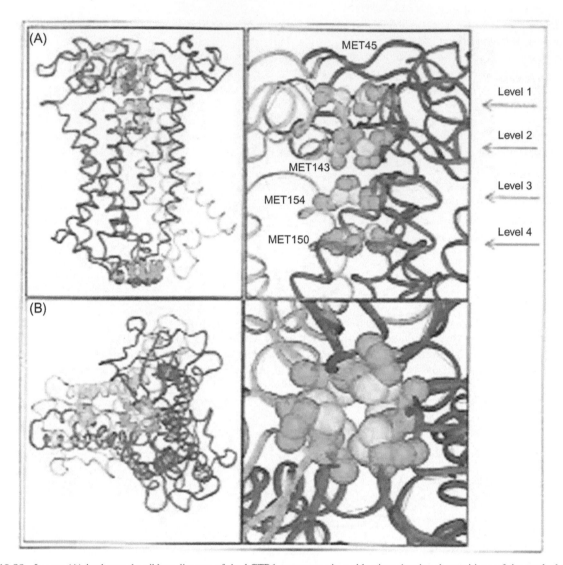

**FIGURE 19.23** In part (A) is shown the ribbon diagram of the hCTR1 transporter in a side view showing the positions of the stacked methionine triads of Met 45, Met 43, Met 154, and Met 150. Also shown are His 188 and His 190 residues at the C-terminus. An enlargement of the four levels containing the methionine residues is shown in the *top right*. In (B) shown is a view from the top of the transporter (the sulfur groups of methionine are in *yellow*). At the *right* a portion of the top view is enlarged. *Reproduced from http://www.ncbi.nlm.nih.gov/pmc/articles/PMC3590913/#! po = 7.14286.*

In this model, two triads of methionine delineate the regions between the two membranes of the transporter, and there are two additional methionine triads that are located in the extracellular N-terminus. This series of four triads creates four levels of ion transport in a stepwise fashion. The metal ions, then, are transported into and then through the channel's intramembranous regions by stepwise interactions with the thioether bonds of the methionine residues. The methionine residues, arranged in a stepwise fashion (in 4 levels), are shown in the model in Fig. 19.23.

## Cellular Prion Protein

The specific function of the normal cellular prion protein is not yet understood; however, it is clear that there are four **copper-binding sites** in the N-terminal domain between amino acids 60 and 92 (**Fig. 4.4**). In the N-terminal region of the normal cellular prion protein (PrP$^c$), the repeating copper-binding sites have the sequence HGGGW, again, as in the copper-binding region of albumin (Fig. 19.20), the presence of His—Gly is indicated. The location of the copper-binding repeats is shown in Fig. 19.24.

Copper bound to the prion protein influences certain interactions between the N-terminal and the C-terminal regions. One theory is that PrP$^c$ can act as a recycling receptor for the uptake of copper from the extracellular space. In this

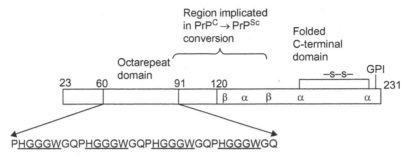

FIGURE 19.24 Organization of the PrP^c and the octarepeat domain. The schematic of the PrPc locates the globular C-terminal domain, the **glycosyl-phosphatidylinositol (GPI) membrane anchor** and the octarepeat domain. Also shown is a flexible region implicated in multimerization (aggregation) that accompanies the conversion PrP^c → PrP^sc (Prion scrapie). Copper binding within the octarepeats involves the specific residues HGGGW (underlined). *Reproduced from http://www.ncbi.nlm.nih.gov/pubmed/14967054.*

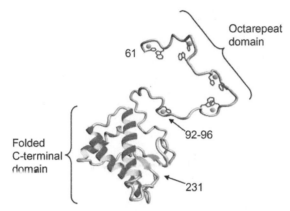

FIGURE 19.25 Three-dimensional rendering of PrPc 61-231 amino acid residues with copper ions included. Crystal structure coordinates were used for the octarepeat-binding units **HGGGW**. The PrP (92−96) segment was based on the model shown in **Fig. 18.26**. The intervening regions were built in a relaxed conformation. Copper ions are shown as *blue* spheres. *Reproduced from http://www.ncbi.nlm.nih.gov/pubmed/14967054.*

regard, copper rapidly stimulates the **endocytosis of PrP^c** from the cellular surface. Normally, the PrP^c can be broken down by ADAM (**a d**isintegrin **a**nd **m**etalloproteinase) protease, and the nature of the fragmentation is influenced both by copper and zinc. However, the scrapie form of the protein (PrP^sc) is an aggregate and is protease-resistant. The scrapie form of the prion protein aggregates leads to neurodegenerative disease in animals and humans (**Creutzfeldt−Jakob disease**). For this conversion to the pathogenic form to take place, a person having only PrP^c must ingest animal protein containing PrP^sc (see **Chapter 4** on Proteins). In Fig. 19.25 is shown a model of the PrP^c protein with the C-terminal domain on the *left* and the N-terminal domain, containing the copper-binding sites, on the *right*.

The chemical structures of the two types of copper-binding sites in PrP^c are shown in Fig. 19.26. The structure on the *left* represents the binding of **copper** to the four binding sites (HGGGW) in the region of 60−92 amino acid residues in the N-terminal domain, and the structure on the *right* shows the single binding site for copper in the region of amino acids 91−96 (**GGGTH**). The single site, GGGTH, may have a tighter association between copper and the protein than the four sites (HGGGW) in the N-terminal domain.

A molecular model based on the crystal structure is shown for the **HGGGW-Cu^{2+}complex** in Fig. 19.27.

## Copper−Zinc Superoxide Dismutase (SOD)

There are at least four different superoxide dismutases: an iron-SOD, a mitochondrial Mn-SOD, a Ni-SOD, a prokaryotic Fe-SOD, and a eukaryotic Cu,Zn-SOD. Here, we discuss the last in the context of copper ion. CuZnSOD is a homodimer of 16 kDa subunits each of which contains 151 amino acids. The **SOD enzyme** catalyzes the conversion of a **reactive oxygen specie (ROS)** that is $O_2^{-\bullet}$, a **superoxide radical** formed in the mitochondria as a byproduct of electron transport chain activity. These radicals poach electrons from molecules in proximity and generate a cascade of electron-poaching that results in cellular damage. The SOD attacks this damaging process by converting the superoxide radicals to hydrogen peroxide ($H_2O_2$), a more benign ROS that can be neutralized subsequently by **catalase** in which $H_2O_2$ is converted to water and oxygen, or, in a lipid environment, $H_2O_2$ is degraded to water by **glutathione**

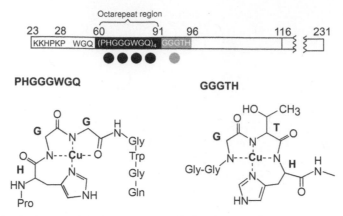

**FIGURE 19.26**  Locations and molecular features of the five main copper-binding sites in PrPᶜ. Bond line models of the equatorial $Cu^{2+}$ coordination spheres for the two different types of binding are shown. The structure for the single HGGGW-$Cu^{2+}$ unit was determined from crystallographic and spectroscopic data. The structure is maintained for each HGGGW unit in the full octarepeat region. The coordination sphere depicted for GGGTH-$Cu^{2+}$ is a model based on spectroscopic data. *Reproduced from http://www.ncbi.nlm.nih.gov/pubmed/14967054.*

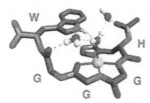

**FIGURE 19.27**  Crystal structure of the $Cu^{2+}$-HGGGW complex. Equatorial $Cu^{2+}$ coordination is from the histidine imidazole, the deprotonated amides from the next two glycines, and the amide carbonyl of the second glycine. In addition, the NH of the indole is within hydrogen bonding distance to the oxygen of the axial water. Two additional intramolecular-ordered water molecules are also shown. *Reproduced from http://www.ncbi.nlm.nih.gov/pubmed/149670545.*

**peroxidase**. The set of reactions catalyzed by Cu,Zn-SOD is as follows (note that in this reaction Cu II is converted to Cu I and then back to Cu II):

$$Cu^{2+} + O_2^{-\bullet} \rightarrow Cu^{+} + O_2$$
$$Cu^{+} + O_2 + 2H^{+} \rightarrow Cu^{2+} + H_2O_2$$
$$\text{Net reaction: } 2O_2^{-\bullet} + 2H^{+} \rightarrow H_2O_2 + O_2$$

The ROS, $H_2O_2$, is then converted by catalase to harmless products:

$$2H_2O_2 \rightarrow 2H_2O + O_2$$

or, $H_2O_2$ can be rendered harmless by the reaction catalyzed by glutathione peroxidase if the environment is lipid in nature:

$$H_2O_2 + 2H^{+} \rightarrow 2H_2O$$

In Fig. 19.28 is pictured the **copper- and zinc-binding domains** in the active center of human Cu, Zn-SOD. Note that Cu is coordinated by four His residues.

The Zn ion is coordinated by three His residues and one Asp residue. The Cu ion and the Zn ion are connected through a **histidine bridge**. The substrate, superoxide, lodges near the active site by electrostatic attraction through a "cationic funnel." When a superoxide radical approaches the enzyme, the superoxide donates an electron to the active site causing the histidine bridge to break. The copper ion moves away and the rings on the copper rotate about 20° in the direction of the Zn ion. As the second superoxide radical approaches the active site, reduced copper is reoxidized, and the original electron moves over to the second superoxide molecule with two protons forming one molecule of **hydrogen peroxide** (see chemical reactions above). The catalytic reaction occurs quickly because the superoxide does not actually bind to the catalytic center but is actively drawn into the "funnel."

*SOD1*, the gene for CuZnSOD, is on chromosome 21 (21q22.1). Mutations in this gene cause **neurodegenerative diseases** including **amyotrophic lateral sclerosis** (**ALS**). ALS due to mutation of the *SOD1* gene leads to little or no SOD activity resulting in oxidative damage to the motor neurons leading to cell death.

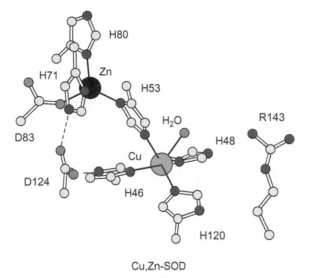

Cu,Zn-SOD

**FIGURE 19.28** Active site of Cu,Zn-SOD. The *green* sphere depicts the $Cu^{2+}$ ion and the *black* sphere depicts the $Zn^{2+}$ion. The amino acid residues are numbered according to human cytoplasmic Cu,Zn-SOD (SOD1), although the coordinates used were those of the bovine enzyme (2SOD.pdb). The water molecule above and to the right of the $Cu^{2+}$ion represents the presumed substrate-binding site (outer-sphere binding via R143 and H63 is envisioned for the reduced state). *Solid lines* indicate bonds, and *dashed lines* indicate hydrogen bonds. *Reproduced from A.-F. Miller,* Current Opinion in Chemical Biology, *8: 162−168, 2004.*

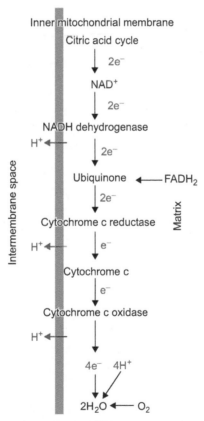

**FIGURE 19.29** The respiratory chain, located on the inner mitochondrial membrane consists of: NADH dehydrogenase, cytochrome c reductase, and cytochrome c oxidase, catalyze the stepwise transfer of electrons from NADH (and FADH) to oxygen molecules to form water (with protons). *Redrawn from http://users.rcn.com/jkimball.ma.ultranet/BiologyPages/C/CellularRespiration.html.*

Copper is involved in other enzymes as well, such as **lysyl oxidase**, an important enzyme in the synthesis of collagen and elastin, **dopamine beta-hydroxylase**, the enzyme converting dopamine to norepinephrine, and **cytochrome c oxidase**, an enzyme in the **electron transfer respiratory chain** Figs. 19.29–19.31.

The copper ion in **cytochrome c oxidase** is coordinated by three histidine residues.

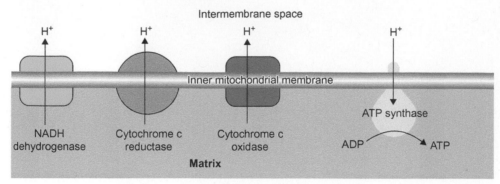

**FIGURE 19.30**  Components of the chemiosmosis process in mitochondria. *Redrawn from http://users.rcn.com/jkimball,ma.ultranet/BiologyPages/C/CellularRespiration.html.*

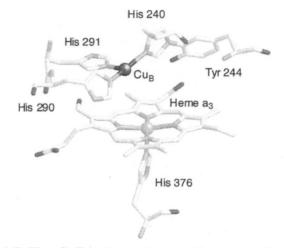

**FIGURE 19.31**  Structure of fully reduced (Fe III . . . Cu II) bovine cytochrome c oxidase structure, Cu . . . Fe = 5.1 Å. The diagram was assembled by using PDB ID 1OCR coordinates and the program RASMOL. *Reproduced from Kim, E., et al., PNAS, **100**: 3623-3628, 2003 or http://www.pnas.org/content/100/7/3623/F1.large.jpg.*

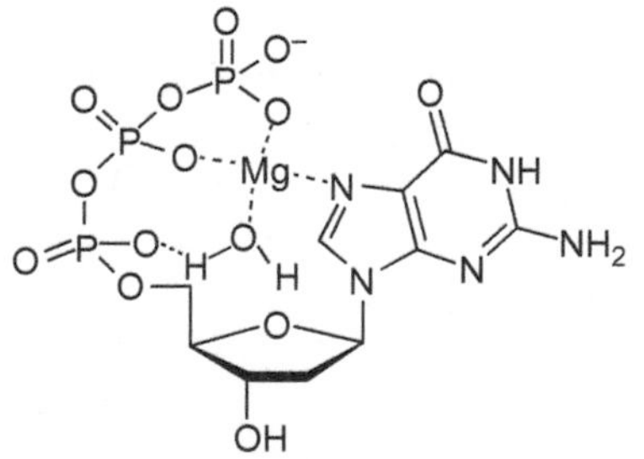

**FIGURE 19.32**  Magnesium is bound to ATP mainly through hydrogen bonding to the three phosphates of ATP and a nitrogen on the purine ring. *Reproduced from http://homepages.strath.ac.uk/~bas96104/Metals/Metals.htm.*

# Magnesium ($Mg^{2+}$)

Magnesium is involved in more than 300 metabolic reactions, including nucleic acid synthesis in that magnesium forms a complex with most nucleotides; **MgATP** is most significant (Fig. 19.32).

It is a cofactor for many enzymes involved in the synthesis of carbohydrates, lipids, and the synthesis of **glutathione**. It is a component of muscular contraction and relaxation where $Mg^{2+}$ is bound to ATP and to two amino acid residues of **myosin** (Fig. 19.33).

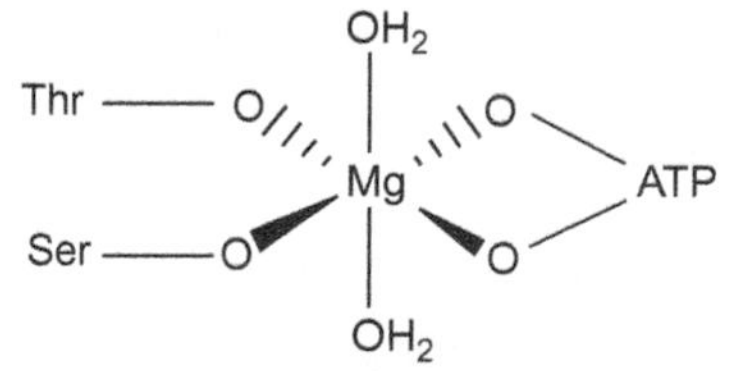

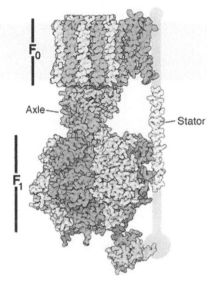

**FIGURE 19.33** Binding of magnesium to ATP and myosin. Magnesium is coordinated between two terminal phosphate groups of ATP and two amino acid residues of the myosin molecule, threonine and serine. *Reproduced from http://homepages.strath.ac,uk/~bas96104/Metals/Metals.htm.*

**FIGURE 19.34** ATP synthase is an enzyme, a molecular motor, an ion pump, and another molecular motor all in one nanoscale machine. There are two rotary motors. The motor at the top ($F_0$) is embedded in a membrane (*gray bar*) and is powered by the flow of hydrogen ions across the membrane (protomotive force). As the protons flow through the rotor, they turn a circular rotor (*blue*). This rotor is connected to a second motor ($F_1$), a chemical motor. The motors are connected by a stator (*right*) so that when $F_0$ turns, $F_1$ also turns. Thus, the turning of the first rotor by the passage of protons causes the second motor to turn (F1 becomes a generator) and generates ATP. *Reproduced from http://www.rcb.org/pdb/education_discussion/molecule_of_the_month/images/ATPsynthase.gif.*

Although many of the sites occupied by magnesium often can be replaced by manganese, magnesium works better. Magnesium transport is of interest, and some of these topics will be discussed in this section.

Magnesium is critical for the synthesis of ATP by **ATP synthase**. Energy is produced by the transportation of protons down their concentration gradient (Figs. 19.29 and 19.30), and this energy is used to couple ADP and inorganic phosphate to form ATP, the primary energy source of a cell. ATP synthase is an amazing molecular motor whose structure is shown in Fig. 19.34.

There are many types of transporters that move magnesium ions across a variety of membranes. The structures of these various types of **magnesium transporters** have been summarized in **Fig. 18.15**. Aspects of magnesium transport in the vascular smooth muscle cell are shown in Fig. 19.35.

Increased transmembrane $Mg^{2+}$ transport through the channel domain occurs through activation of the transporter TRPM7 and results in the activation of the downstream targets, **annexin-1, calpain** and **myosin IIA** heavy chain through the **alpha-kinase domain**. In addition to the $Mg^{2+}$ channel, this transporter has attached to its cytoplasmic side **PKC** and alpha-kinase. The intracellular $Mg^{2+}$ itself is a negative regulator of TRPM7 activity (Fig. 19.35).

As for the distribution of the approximate amount of 25 g, or so, of $Mg^{2+}$ in the body, 60% or more is located in the skeleton, 25% is in muscle, 7% is located in other tissue cells, and about 1% is in the extracellular spaces. Since $Zn^{2+}$ will utilize the same transporter as $Mg^{2+}$, high dietary intake of $Zn^{2+}$ will interfere with the uptake of $Mg^{2+}$ by the intestine. Tables 19.7 and 19.8 list recommended daily allowances for dietary magnesium and food sources of magnesium, respectively.

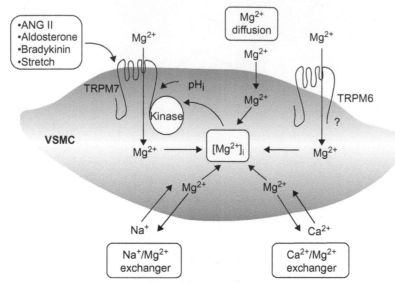

**FIGURE 19.35** Transport mechanisms regulating intracellular free $Mg^{2+}$ concentration ($[Mg^{2+}]$) in vascular smooth muscle cells (VSMC). Magnesium influx and efflux are modulated by passive diffusion and transient receptor potential, **melastatin 7** (TRPM7 and possibly TRPM6) through the $Na^{+}/Mg^{2+}$ exchanger and via the $Ca^{2+}/Mg^{2+}$ exchanger. -, Inhibitory effect;?, unknown effect; *pH*, intracellular pH; *ANG II*, angiotensin II; subscript *i*, intracellular. *Reproduced from http://ajpheart.physiology.org/content/294/3/H1103.figures-only.*

**TABLE 19.7 Recommended Daily Allowances for Dietary Magnesium**

| Life Stage | Age | Males (mg/day) | Females (mg/day) |
|---|---|---|---|
| Infants | 0−6 months | 30 (AI) | 30 (AI) |
| Infants | 7−12 months | 75 (AI) | 75 (AI) |
| Children | 1−3 years | 80 | 80 |
| Children | 4−8 years | 130 | 130 |
| Children | 9−13 years | 240 | 240 |
| Adolescents | 14−18 years | 410 | 360 |
| Adults | 19−30 years | 400 | 310 |
| Adults | 31 years and older | 420 | 320 |
| Pregnancy | 18 years and younger | − | 400 |
| Pregnancy | 19−30 years | − | 350 |
| Pregnancy | 31 years and older | − | 360 |
| Breastfeeding | 18 years and younger | − | 360 |
| Breastfeeding | 19−30 years | − | 310 |
| Breastfeeding | 31 years and older | − | 320 |

*AI*, adequate intake, which is established when an RDA cannot be determined.
Reproduced from http://lpi.oregonstate.edu/infocenter/minerals/magnesium/.

# Zinc ($Zn^{2+}$)

As indicated in Table 19.6, zinc is a component of **carbonic anhydrase**, **cytosine deaminase**, **zinc family transporters** (**Zip**), **zinc finger motifs**, and **ATP synthase**, to name a few proteins. Some proteins contain zinc coordinated by three histidine residues and water or hydroxyl groups ($Zn^{2+}(N\text{-His})_3/OH_2$). In certain zinc finger motifs, zinc is coordinated by four cysteine residues and, in other cases, zinc can be coordinated by a combination of histidine and cysteine residues.

**TABLE 19.8** Food Sources of Magnesium

| Food | Serving | Magnesium (mg) |
| --- | --- | --- |
| 100% bran cereal (e.g., All Bran) | 1/2 cup | 128.7 |
| Oat bran | 1/2 cup dry | 96.4 |
| Shredded wheat | 2 biscuits | 54.3 |
| Brown rice | 1 cup cooked | 83.8 |
| Almonds | 1 ounce [22 almonds] | 81.1 |
| Hazelnuts | 1 ounce | 49.0 |
| Peanuts | 1 ounce | 49.8 |
| Lima beans | 1/2 cup cooked | 62.9 |
| Black-eyed peas | 1/2 cup cooked | 42.8 |
| Spinach, chopped | 1/2 cup cooked | 78.3 |
| Swiss chard, chopped | 1/2 cup cooked | 75.2 |
| Okra, sliced | 1/2 cup cooked | 45.6 |
| Molasses, blackstrap | 1 tablespoon | 43.0 |
| Banana | 1 medium | 34.2 |
| Milk 1% fat | 8 fluid ounces | 33.7 |

Reproduced from http://lpi.oregonstate.edu/infocenter/minerals/magnesium/.

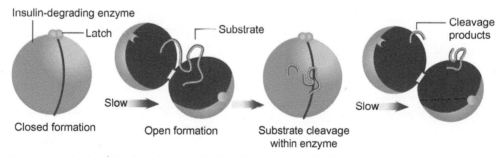

**FIGURE 19.36** Schematic mechanism of action of the insulin-degrading enzyme (IDE) illustrating the closed and opened conformations of the enzyme during the catalytic reaction. The latch mechanism (*green*) holds the enzyme in a closed state, delaying entry of the substrate or exit of the cleavage products. Mutations that disrupt the latch promote the open conformation of the enzyme so that mutants accept substrates and release products more readily than naturally occurring IDE and are, thus, more active. *Reproduced from http://www.nature.com/nature/journal/v443/n7113/fig_-tab/nature05210_F1.html.*

In addition to those proteins mentioned, zinc is a cofactor for the metalloprotease, the **insulin-degrading enzyme (IDE)**. In addition to the B chain of insulin, this enzyme also degrades amyloidβ, glucagon, amylin, TGFα, and β-endorphin. Its mass is 113 kDa, and it is located on the cell membrane surface, in the soluble cytoplasm and in the extracellular matrix. IDE cleaves the **B chain of insulin** at several locations and has a complex reaction mechanism both chemically and physically. Physically it is reported to have a "latch" mechanism with a closed and open conformation. When the latch of the enzyme is open, it binds substrate, after which it closes and cleaves the substrate within the enzyme and then opens to release the products (Fig. 19.36).

IDE has a complex reaction mechanism with transitional intermediates and more stable intermediates in which coordinated zinc ion plays a key role in the catalytic center. The structure of the coordinated zinc ion involves two histidine residues and the oxygens of a glutamate residue. In addition, a water molecule is complexed both to zinc and to another glutamate residue of the enzyme as shown in Fig. 19.37.

In addition to the distorted tetrahedral formed by the interactions of the water molecule, glutamate 111 and the two histidines, the enzyme has another site that interacts with the N-terminus of the **insulin B chain** (consisting of 3 amino acid residues: Phe141, Tyr16 and Leu17) and IDE cleaves the B chain at several points.

Another zinc enzyme is **carbonic anhydrase**, an extremely fast enzyme that catalyzes the conversion of carbon dioxide to carbonic acid through an intermediate conversion to **bicarbonate ion** as shown in Fig. 19.38 (*top*).

The active site of carbonic anhydrase is shown in Fig. 19.39.

The **zinc ion** in **carbonic anhydrase** is coordinated by three histidine side chains and a molecule of water and this causes the polarization of the hydrogen to oxygen bond resulting in increased negativity of the oxygen and weakening the bond. There is a fourth histidine residue, close to the water molecule, that accepts a proton, leaving the hydroxide

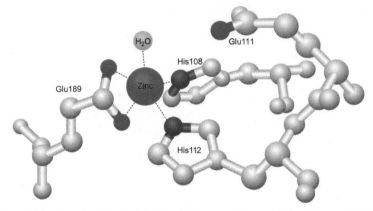

**FIGURE 19.37** The zinc complex within the active site of IDE. The zinc ion complexes with the nitrogens of histidine 108 and histidine 112 and the oxygens of glutamate 189. The water molecule (*cyan*) is also complexed with zinc as well as glutamate 111. Interactions between the water and Glu111 are responsible for peptide hydrolysis. *Reproduced from http://maptest.rutgers.edu/drupal/?q = node/247.*

$$CO_2 + H_2O \rightleftharpoons H_2CO_3 \rightleftharpoons H^+ + HCO_3^-$$

$$HCO_3^- \rightleftharpoons H^+ + CO_3^{2-}$$

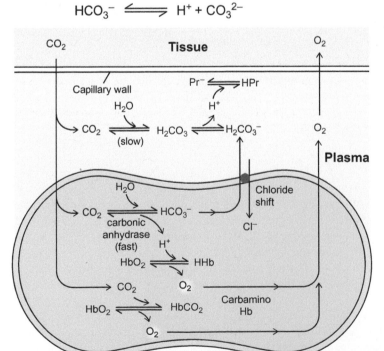

**FIGURE 19.38** The chemical reactions involved in the conversions of carbon dioxide to carbonic acid to bicarbonate ion catalyzed by carbonate anhydrase (*top*). The illustration at the *bottom* shows the action of carbonic anhydrase in the **red blood cell** where incoming $CO_2$ from plasma is converted by the enzyme to bicarbonate ion and a proton. The proton reacts with heme-bearing oxygen to form reduced heme and free oxygen which then leaves the red blood cell and gains access to the extracellular space and finally oxygenates the tissues. *Reproduced from http://www.bio.miami.edu/tom/courses/protected/ECK/CH13/figure-13-10a.jpg.*

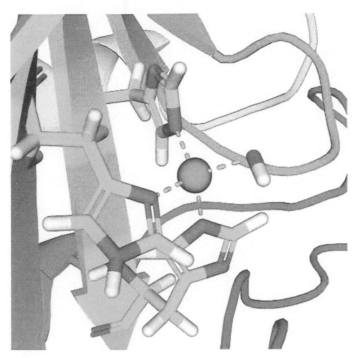

**FIGURE 19.39**    Illustration of the **active site of human carbonic anhydrase II**, showing three histidine residues and a hydroxide group (of a water molecule) coordinating (*dashed lines*) the zinc at the center. From PDB1CA2. *Reproduced from http://upload.wikimedia.org/wikipedia/commons/7/7f/ Carbonic_anhydrase_1CA2_active_site.png.*

coordinated to zinc. Additionally, a pocket in the enzyme close to the hydroxyl brings in $CO_2$, allowing the electron-rich hydroxide to attack the $CO_2$ and form bicarbonate. *The carbonic anhydrase reaction is key to the elimination of $CO_2$ produced by metabolism.*

There are several carbonic anhydrases located in the soluble cytoplasm, in the mitochondria, associated with membranes and carbonic anhydrases that are secreted. These four groups also generate isoforms so that there is a large number of carbonic anhydrase enzymes, and they are listed in Table 19.9.

The product of the carbonic anhydrase reaction, bicarbonate ion, is transported across the red blood cell membrane into the plasma; at the same time, chloride ion enters the red blood cell from the plasma, and this exchange is known as the **chloride shift**.

Zinc in carbonic anhydrase can be replaced by **cadmium** or **mercury** (mercury can be toxic) to produce an enzyme that is more stable than when it is complexed with zinc and such replacement results in an enzyme that has decreased activity.

Another zinc-containing enzyme (metalloenzyme) is **cytosine deaminase** that catalyzes the conversion of cytosine to uracil. The mechanism of this reaction, showing the participation of the zinc cofactor, is shown in Fig. 19.40.

Zinc is an important ingredient in the interaction of certain proteins and DNA. The metal is part of a structure known as a **zinc finger,** and in several proteins, especially steroid receptors, there are two zinc fingers as shown for the mineralocorticoid receptor, the glucocorticoid receptor and the androgen receptor in **Fig. 16.13A**. In these zinc fingers, the zinc is coordinated by four cysteine residues that is also the case for the **GATA3 protein**, another example of a DNA-binding protein.

Absorption of zinc from the diet in the small intestine is accomplished by the **zinc family transporters (Zip)**. **Zip4** is located in the apical membrane and moves zinc from the intestinal lumen into the intestinal cell while **Zip5**, located in the basolateral membrane, moves zinc from the intestinal cell into the extracellular space and into the blood. **Albumin** is the major transporting protein of zinc in blood (although zinc seems to be tightly bound to α2-macroglobu-lin; it is probably not the main transporter in plasma). Since fatty acids also bind to albumin, presumably at the same site as zinc, they may cause the dissociation of zinc from albumin, allowing zinc to enter tissue cells. The Zip transporters in cells of widely distributed tissues are 8-membrane spanning proteins (Fig. 19.41).

The recommended daily dose of zinc is 11 mg for adult males and 8 mg for adult females. During pregnancy, 11 mg/day is required and during lactation, 12 mg/day. For children, depending on age, the daily requirement is 3 to

**TABLE 19.9 Comparison of Mammalian Carbonic Anhydrases**

| Isoform | Gene | Molecular mass | Location (cell) | Location (tissue) | Specific activity of human enzymes (except for Mouse CA-XV) ($s^{-1}$) | Sensitivity to sulfonamides (acetazolamide in this table) $K_I$ (nM) |
|---|---|---|---|---|---|---|
| CA-I | CA1 (http://www.genenames.org/data/hgnc_data.php?hgnc_id = 1368) | 29 kDa | Cytosol | Red blood cell and GI tract | $2.0 \times 10^5$ | 250 |
| CA-II | CA2 (http://www.genenames.org/data/hgnc_data.php?hgnc_id = 1373) | 29 kDa | Cytosol | Almost ubiquitous | $1.4 \times 10^6$ | 12 |
| CA-III | CA3 (http://wwww.genenames.org/data/hgnc_data.php?hgnc_id = 1374) | 29 kDa | Cytosol | 8% of soluble protein in Type I muscle | $1.3 \times 10^4$ | 240,000 |
| CA-IV | CA4 (http://www.genenames.org/data/hgnc_data.php?hgnc_id = 1375) | 35 kDa | Extracellular GPI linked | GI tract, kidney, endothelium | $1.1 \times 10^6$ | 74 |
| CA-VA | CA5A (http://www.genenames.org/data/hgnc_data.php?match=CA5A) | 34.7 kDa (predicted) | Mitochondria | Liver | $2.9 \times 10^5$ | 63 |
| CA-VB | CA5B (http://www.genenames.org/data/hgnc_data.php?match = CA5B) | 36.4 kDa (predicted) | Mitochondria | Widely distributed | $9.5 \times 10^5$ | 54 |
| CA-VI | CA6 (http://www.genenames.org/data/hgnc_data.php?hgnc_id = 1380) | 39–42 kDa | Secretory | Saliva and milk | $3.4 \times 10^5$ | 11 |
| CA-VII | CA7 (http://www.genenames.org/data/hgnc_data.php?hgnc_id = 1381) | 29 kDa | Cytosol | Widely distributed | $9.5 \times 10^5$ | 2.5 |
| CA-IX | CA9 (http://www.genenames.org/data/hgnc_data.php?hgnc_id = 1383) | 54, 58 kDa | Cell membrane associated | Normal GI tract, several cancers | $1.1 \times 10^6$ | 16 |
| CA-XII | CA12 (http://www.genenames.org/data/hgnc_data.php?hgnc_id = 1371) | 44 kDa | Extracellularly located active site | Kidney, certain cancers | $4.2 \times 10^5$ | 5.7 |
| CA-XIII | CA13 (http://www.genenames.org/data/hgnc_data.php?hgnc_id = 14914) | 29 kDa | Cytosol | Widely distributed | $1.5 \times 10^5$ | 16 |
| CA-XIV | CA14 (http://www.genenames.org/data/hgnc_data.php?hgnc_id = 1372) | 54 kDa | Extracellularily located active site | Kidney, heart, skeletal muscle, brain | $3.1 \times 10^5$ | 41 |
| CA-XV | CA15 (http://www.genenames.org/data/hgnc_data.php?hgnc_id = 80733) | 34–36 kDa | Extracellular GPI linked | Kidney, not expressed in human tissues | $4.7 \times 10^5$ | 72 |

Reproduced from http://en.wikipedia.org/wiki/Carbonic_anhydrase.

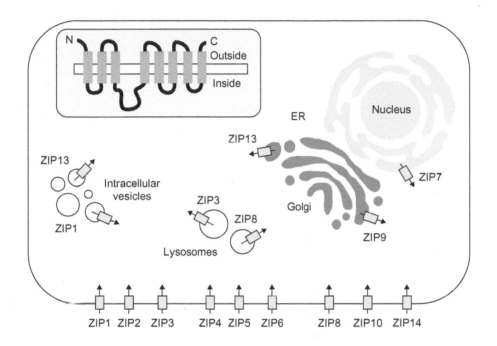

**FIGURE 19.40** A proposed catalytic mechanism for cytosine deaminase from yeast. The Glu64 residue is part of the enzyme.

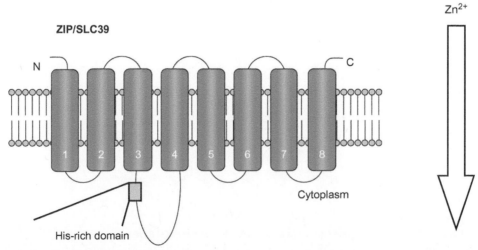

**FIGURE 19.41** Upper panel shows the eight membrane typical Zip transporters. Lower panel shows an enlargement. In the upper panel, the various Zip transporter family members are arrayed according to their locations in the cell. *Reproduced from J. Jeong and D.J.Eide, "The SLC39 family of zinc transporters", Mol. Aspects. Med., 34: 612−619, 2013.*

8 mg. There may be competition between zinc and folic acid at the intestinal level since zinc and folic acid can form a complex.

## Molybdenum (Mo$^{2+}$)

The coenzyme form of molybdenum is a **pterin**. Mo$^{2+}$ is a coenzyme or cofactor for several enzymes, a few of which are listed in Table 19.6. The formation of the **molybdenum cofactor** is shown in Fig. 19.42.

The uptake of Mo$^{2+}$ from the intestine (the molybdenum content in foods depends on the amount of molybdenum in the soil in which the food is grown) is not well understood, although experimental studies indicate that it is primarily absorbed in the proximal small intestine without the agency of a protein, is nonsaturable, does not require energy and

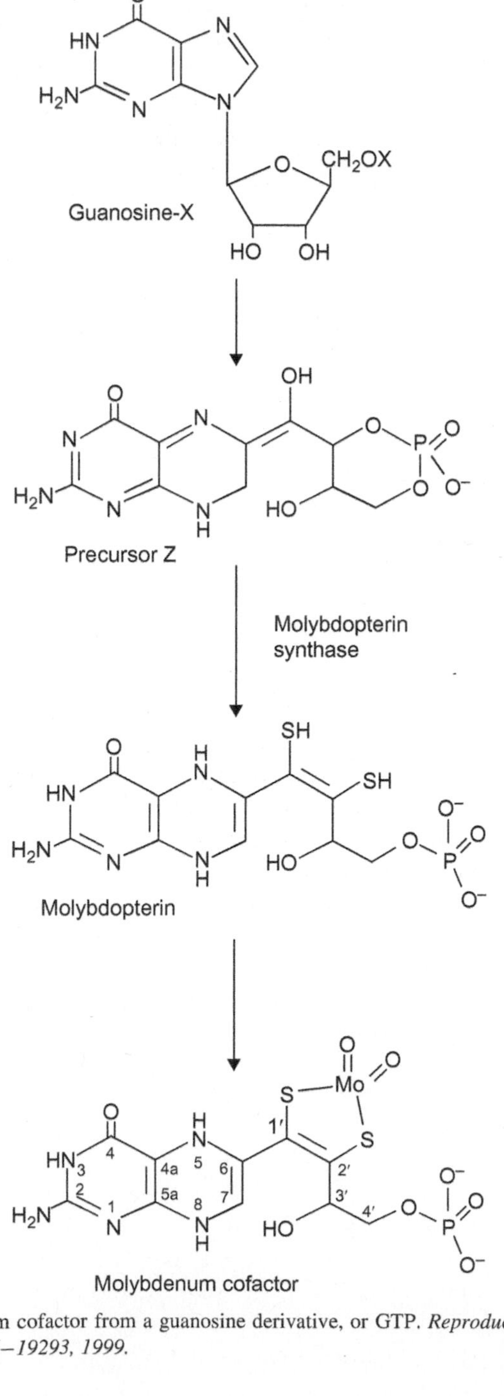

**FIGURE 19.42** Synthesis of the molybdenum cofactor from a guanosine derivative, or GTP. *Reproduced from S. Unkles et al., "Eukaryotic molybdopterin synthase", J. Biol. Chem. **274**: 19286–19293, 1999.*

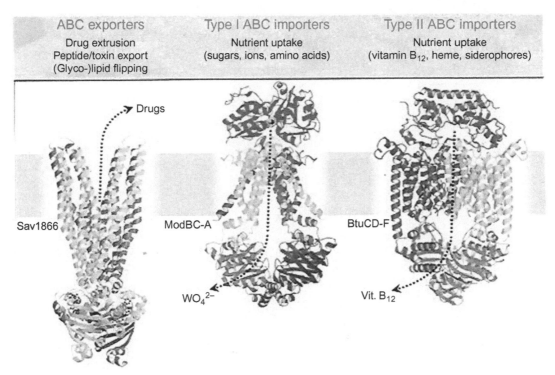

**FIGURE 19.43** Types of ABC transporters for uptake into cells and for export from cells. *Reproduced from http://www.mol.biol.ethz.ch/groups/ locher_group/Research.*

can enter the intestinal cell as a complex with ascorbic acid. Interestingly, in bacteria, $Mo^{2+}$ is absorbed by an ABC-like transporter. The $Cl^-$ transporter involved in cystic fibrosis is an example of an **ABC transporter (Fig. 18.1A–18.1C)**. ABC transporters (**ATP-b**inding **c**assette transporters) are plentiful in human cells and are used for a variety of transport activities (Fig. 19.43) including the export of drugs from cells, the uptake of nutrients, such as sugars and ions (tungstate ion ($WO_4^{2-}$) is an example) and vitamin B12.

A human protein called **gephyrin** (93 kDa) facilitates the clustering of inhibitory neuroreceptors in the postsynaptic membrane, and it catalyzes the insertion of $Mo^{2+}$ into the final step in the synthesis of the molybdenum cofactor (Fig. 19.42). A lack of gephyrin due to mutations of the gephyrin gene, causes stiff muscles, a serious disease that can result in death. Gephyrin forms oligomers to generate a postsynaptic platform to bring various proteins (e.g., receptors, cytoskeletal proteins, and downstream signaling proteins) into proximity (Fig. 19.44).

Human tissues contain less than 1 ppm of $Mo^{2+}$, and the greater concentrations are found in liver, adrenal glands, kidneys, and bone, and smaller amounts are found in muscle, lungs, brain spleen, and small intestine. Molybdenum deficiency is a rare occurrence as are genetic diseases known as sulfite oxidase deficiency or molybdenum cofactor deficiency. Loss of function of gephyrin can result in severe brain damage. The recommended daily allowances for molybdenum are listed in Table 19.10.

Good **sources of dietary molybdenum** are beans, lentils, peas, grains, and nuts.

Tetrathiomolybdate has been used in humans to treat **Wilson's disease** in which the accumulation of $Cu^{2+}$ in liver and brain leads to damage.

## Selenium ($Se^{2-}$)

The selenium-containing enzymes are listed in Table 19.6. **Thioredoxin reductase**, in a conserved C-terminal sequence, contains a **selenocysteine** residue (Gly−Cys−**SeCys**−Gly). The enzyme loses activity if SeCys is deleted through a mutation in the gene for this enzyme. The enzyme uses the coenzymes, FAD and NADPH, in the reaction it catalyzes. The domains of the enzyme and the sequence containing the SeCys residue are shown in Fig. 19.45.

The reaction catalyzed by thioredoxin reductase is shown in Fig. 19.46.

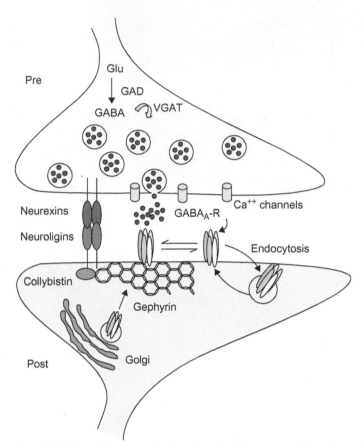

**FIGURE 19.44** Structural organization of **GABAergic** (gamma-aminobutyric acidergic) **synapses**. The postsynaptic organization comprises a large number of proteins that allow the correct targeting, clustering, and stabilization of **GABA$_A$ receptors**. Among them, gephyrin forms hexagonal lattices that trap GABA$_A$ receptors in precise apposition to presynaptic release sites. Cell adhesion molecules of the neuroligin-neurexin families bridge the cleft and ensure **transynaptic signaling**, essential for the maintenance of a proper E/I (excitatory/inhibitory) balance. *Red dots* represent GABA; *GAD*, glutamic acid decarboxylase; *VGAT*, vesicular glutamic acid transporter. *Reproduced from http://www.hindawi.com/journals/np/2011/297153/fig3/.*

**TABLE 19.10 Recommended Daily Allowances for Molybdenum**

| Life Stage | Age | Males (µg/day) | Females (µg/day) |
|---|---|---|---|
| Infants | 0−6 months | 2 (AI) | 2 (AI) |
| Infants | 7−12 months | 3 (AI) | 3 (AI) |
| Children | 1−3 years | 17 | 17 |
| Children | 4−8 years | 22 | 22 |
| Children | 9−13 years | 34 | 34 |
| Adolescents | 14−18 years | 43 | 43 |
| Adults | 19 years and older | 45 | 45 |
| Pregnancy | All ages | — | 50 |
| Breastfeeding | All ages | — | 50 |

*AI*, adequate intake, which is established when an RDA cannot be determined.
Reproduced from http://lpi.oregonstate.edu/infocenter/minerals/magnesium/.

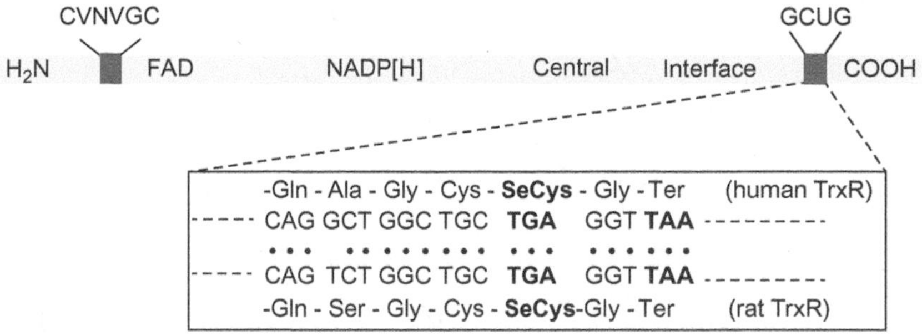

FIGURE 19.45 Domain structure of human thioredoxin reductase. *Ter*, termination (stop) codon for mitochondria. Redrawn from L. Zhong and A. Holmgren, *J. Biol. Chem.*, **275**: 18121–18128, 2000.

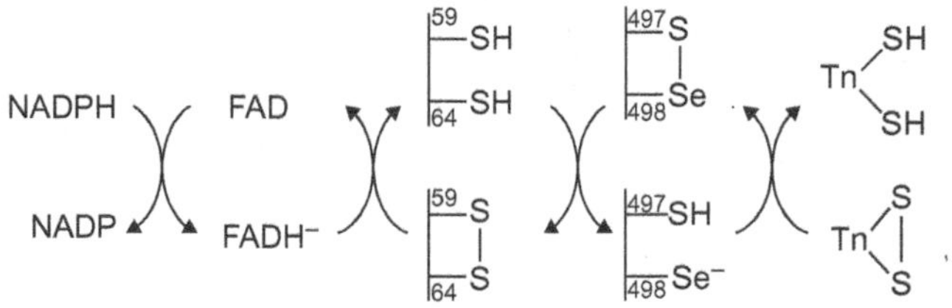

FIGURE 19.46 Reaction catalyzed by human thioredoxin reductase. *Tn*, thioredoxin. *Reproduced from http://biochem.uvm.edu/research/research-interests/.*

## Selenodeiodinases

These enzymes (D1, D2, and D3) either deiodinate **thyroxine** (**T4**) to the biologically active form of **triiodothyronine** (**T3**), or they inactivate the thyroid hormone by removing essential iodides from the inner or outer rings. The activation of thyroxine to triiodothyronine (T3), that binds to the **thyroid hormone receptor**, is carried out by D2 and D1 and the inactivation of T3 to form **3,3′-triiodothyronine** and the inactivation of **thyroxine** to form **3,3,5′-triiodothyronine** (**reverse T3**) is carried out by D3. D1 and D2 can further convert the inactive reverse T3 to **3,3′-diiodothyronine**. These reactions are shown in Fig. 19.47.

**D1** can deiodinate both rings of the iodinated thyronines, and it is located in the plasma membrane where it is positioned so that its catalytic domain faces the soluble cytoplasm. This enzyme is located primarily in liver and kidney. **D2** is the major activating enzyme and attacks the outer ring of the active hormone precursor, thyroxine (T4), and has its transmembrane domain anchored to the endoplasmic reticulum so that its active site faces the perinuclear cytosol. D2 is located in heart, skeletal muscle, central nervous system, adipose, thyroid, and pituitary tissues. **D3** is the major inactivating enzyme, and it deiodinates the inner ring of T4 or T3. It is found anchored to the plasma membrane, and its active site faces the extracellular space. D3 can recycle bidirectionally between the plasma membrane and early endosomes. It is located primarily in fetal tissues, placenta and perinatal pancreatic β-cells. In adult tissues, D1 also can perform the functions of D3. Information about the three deiodinases, and an illustration representing the **active site of D2 deiodinase** is shown in Fig. 19.48.

Fig. 19.49 shows the enzymatic mechanism of **D1 deiodinase** catalyzing the conversion of **reverse T3** to T2 (3,3'-diiodothyronine).

## Selenophosphate Synthase (SePS)

This enzyme catalyzes the reaction between **selenide** (**HSe⁻**) and ATP to produce **selenophosphate** (**SeP**). SeP is a selenium donor for the synthesis of **selenocysteine**. The structures of cysteine and selenocysteine are compared in Fig. 19.50.

SePS catalyzes the phosphorylation of selenide with magnesium ATP to form selenium phosphate (SeP), AMP, and Pi. SeP can donate its selenium to tRNAs (e.g., Ser-tRNA$^{Sec}$) that subsequently incorporate selenium into specific

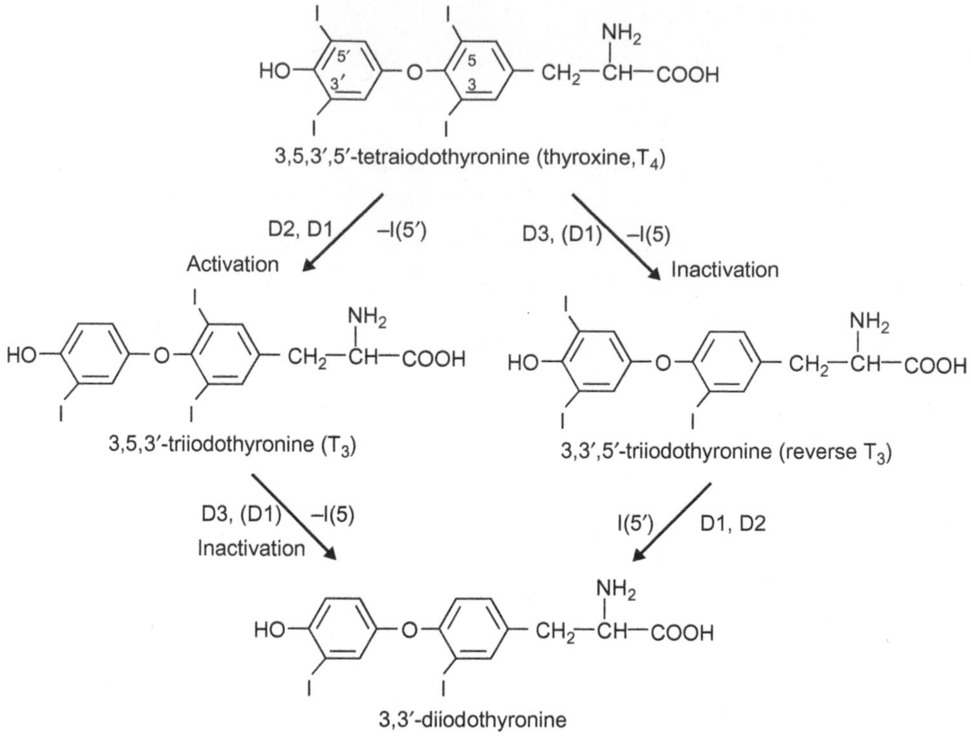

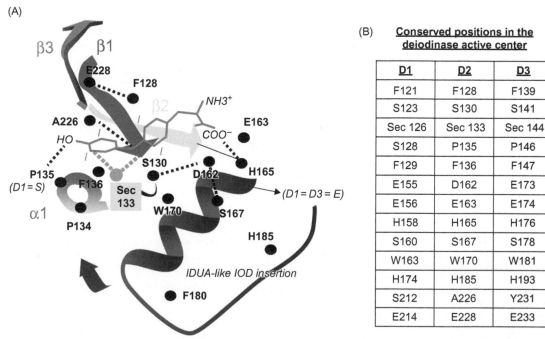

**FIGURE 19.47** Principal iodothyronines activated and inactivated by removal of iodide by selenodeiodinases. *D1*, *D2*, and *D3* refer to isoforms of the same enzyme.

(A)

(B) <u>Conserved positions in the deiodinase active center</u>

| D1 | D2 | D3 |
|---|---|---|
| F121 | F128 | F139 |
| S123 | S130 | S141 |
| Sec 126 | Sec 133 | Sec 144 |
| S128 | P135 | P146 |
| F129 | F136 | F147 |
| E155 | D162 | E173 |
| E156 | E163 | E174 |
| H158 | H165 | H176 |
| S160 | S167 | S178 |
| W163 | W170 | W181 |
| H174 | H185 | H193 |
| S212 | A226 | Y231 |
| E214 | E228 | E233 |

**FIGURE 19.48** Schematic representation of the putative active site of deiodinases deduced from sequence alignment and from the associated modeling. The positions shown are those of D2 and the table contains the corresponding positions and residues in D1 and D3. According to this model, H-bonds/ion pairs between His-165 and the carboxyl group stabilize the iodothyronine (thyroxine structure is shown in the *green stick* model). A similar interaction might take place between the hydroxyl group and Ser-128 in D1 but not in D2 and D3 where a Pro (*P135*) replaces Ser. Assuming these anchors, the essential **Sec133** lies between the inner and outer rings of the iodothyronine where it might interact with iodine atoms (*orange dashed lines*). The specificity of D1, D2, and D3 is further tuned by several amino acids in close proximity (e.g., Ser-130, Asp-162, and Ser-167). Two other groups of amino acids centered on the conserved Glu-228/Phe-128 and Trp-170/His-185 positions are in the vicinity of the active center and modulate its functional properties. After His-165, the IDUA-like IOD (*IDUA*, iduronidase) insertion has a speculative character, and alternatively, the highly conserved Trp-170 may directly interact with T4 molecules as well as His-185. The IUDA-like IOD insertion probably constitutes a cap that may cover the active site upon ligand binding. *Reproduced from http://www.jbc.org/content/278/38/36887/F1.expansion.html.*

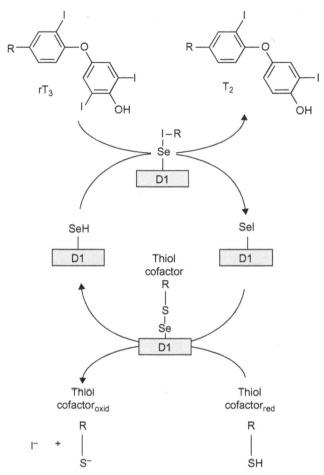

**FIGURE 19.49** Schematic model of thyroid hormone catalysis by type I deiodinase. D1-catalyzed reaction follows ping-pong kinetics (similar to many aminotransferases) with two substrates, the first being iodothyronine and the second being an as yet unidentified endogenous intracellular thiol cofactor. *rT3*, 3,3′,5′-triiodothyronine; *T2*, diiodothyronine; *D1*, deiodinase type I; *Se*, selenium; *I*, iodide. *Reproduced from http://joe.endocrinology-journals.org/content/209/3/283/F1.expansion.html.*

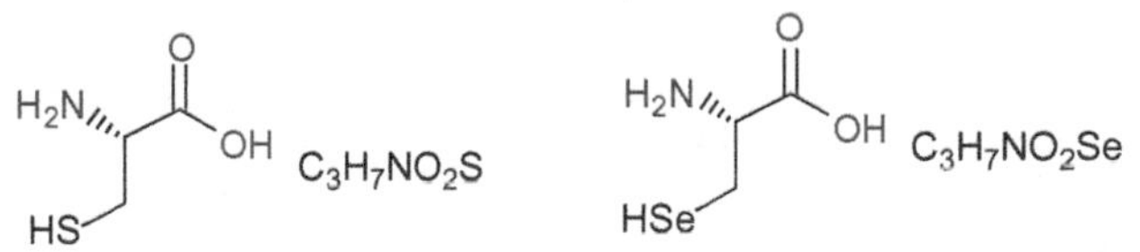

**FIGURE 19.50** Structures of cysteine (*left*) and selenocysteine (*right*).

proteins. **Selenium cysteine synthase (Sec synthase)** catalyzes the conversion of Ser-tRNA[Sec] to Sec-tRNA[Sec] (selenocysteine is a *bona fide* amino acid with all of the components of protein synthesis specific to it) by the addition of Se from SeP. Sec-tRNA is recognized by a specific elongation factor and is brought to Se-protein mRNA for the synthesis of **selenoprotein**. This set of reactions using SeP is shown in Fig. 19.51.

The active site of **selenophosphate synthase (SePS)** is shown in Fig. 19.52.

Good sources of **dietary selenium** are oysters, Brazil nuts, beef, shrimp, crab-meat, salmon, halibut, pork, liver, brown rice, cereals, and eggs. Adults should consume $5-20$ μg selenium daily.

# Calcium ($Ca^{2+}$)

Calcium is important in the body both structurally ($>99\%$ in bones and teeth; bone remodeling is a continuous process in which the formation and breakdown of bone vary with age) and as an important **second messenger** in cells. Most of the residual calcium is located in blood, muscle, and interstitial fluid. In addition to its role in signaling systems in cells,

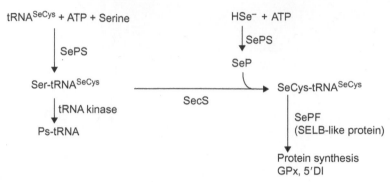

**FIGURE 19.51**    A schematic model of mammalian Sec synthesis.

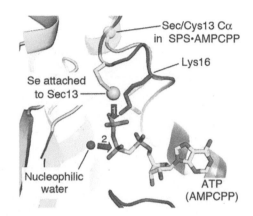

**FIGURE 19.52**    Models of the active N-terminal loop conformation in selenophosphate synthase (SePS) with selenium attached to Sec13. The coordinates of the N-terminal mobile loop (residues 9-17 in chain A of the enzyme) were modified so that Sec13-Se⁻ may reside in proximity of the ATP γ-phosphate group while maintaining the Lys16-ATP interaction. The model was energy-minimized with the CNS program. The modeled loop is shown in *magenta,* and the attached Se atom is shown as a *gold sphere.* The original loop trace in SPS-AMPCPP (chain A) is represented as a transparent *light pink* tube, and the Sec/Cys13 Cα atom is shown as a *yellow-green sphere.* The putative nucleophilic water for ADP hydrolysis is represented as a *magenta sphere. Reproduced from http://www.sciencedirect.com/science/article/pii/S0022283608010486.*

calcium is important in muscle contraction, in expansion of blood vessels and in nervous system functions. Normal plasma levels of total calcium vary between 9 and 10.5 mg/dL and the levels of ionized calcium between 4.5 and 5.6 mg/dL. It is the ionized calcium ($Ca^{2+}$) that determines its biological effects. The level of circulating **albumin** to which calcium is bound in the plasma has an effect on the total calcium level but does not affect the ionized portion.

Calcium and phosphate are interdependent, and they regulate many biological processes. The balance between calcium and phosphate is related to dietary intake and subsequent intestinal absorption (duodenum and jejunum). The levels of calcium (high or low) determine the mode of transport from the intestinal lumen into the intestinal cell. At high levels of calcium (and phosphate), uptake occurs by a passive transport mechanism, possibly through tight junctions between cells. When the diet is low in calcium, active transport is required and this involves the **TRPV6 channel**, **vitamin D**, the **vitamin D receptor (VDR)**, and the product of its transcriptional action, the protein **calbindin**, a calcium-binding protein. Calbindin binds calcium in the cytosol and transports it across the cell to the basolateral side where calcium is transported to the extracellular space (and ultimately into the bloodstream) by the energy-dependent (using ATP) transporter PMCA. The summary of intestinal transport of calcium is shown in Fig. 19.53.

The genes controlling the intestinal apical entry of calcium and the basolateral expulsion and entry into the bloodstream are tightly linked. In the small intestine, calcium, when it is low in the diet, is absorbed through the calcium transport channel (CaT), also known as TRPV6 that is activated by calcium storage depletion. When dietary calcium is high, the ion moves through the intestinal brush border cells into the extracellular space by passive transport. Thus, low dietary $Ca^{2+}$ requires an adaptation involving the apical channel **TRPV6**, the active form of vitamin D (1,25-dihydroxyvitamin D3), its activation of the **VDR,** and the production of the protein calbindin that transports the $Ca^{2+}$ across the

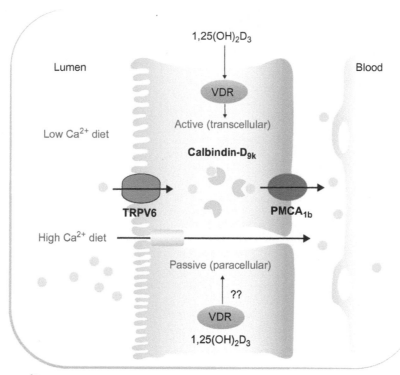

$1,25(OH)_2D_3$

Lumen

Blood

VDR

Low $Ca^{2+}$ diet

Active (transcellular)

**Calbindin-$D_{9k}$**

TRPV6

$PMCA_{1b}$

High $Ca^{2+}$ diet

Passive (paracellular)

??

VDR

$1,25(OH)_2D_3$

**FIGURE 19.53** Transport of $Ca^{2+}$ from the intestinal lumen into the enterocyte, across the cytoplasm through the basolateral membrane under conditions of low or high levels of $Ca^{2+}$ in the diet. When the amount of calcium in the diet is low, $Ca^{2+}$ is transported into the intestinal cell through the TRPV6 apical channel moving the $Ca^{2+}$ into the cytosol. In the cytosol $Ca^{2+}$ is bound to the transport protein calbindin whose synthesis results from the binding of **1,25-dihydroxyvitamin D3** (the active form) to its receptor (VDR) that activates the calbindin gene to express the mRNA and then the protein. Calbindin-$Ca^{2+}$ moves to the basolateral side of the cell where $Ca^{2+}$ is released and energetically (ATP) transported to the extracellular space through the **PMCA channel** in the basolateral membrane. The $Ca^{2+}$ will gain access to the bloodstream. It is unclear whether the action of vitamin D3 is required for passive transport when the diet contains high levels of calcium. *Reproduced from http://nature.com/bonekeyreports/2014/140205/bonekey20133230.html.*

cytoplasm to the basolateral side of the enterocyte. At the basolateral side of the cell, $Ca^{2+}$ is pumped energetically, using ATP, into the extracellular space through the PMCA channel, and $Ca^{2+}$ enters the bloodstream.

If the calcium from the intestine is not sufficient to meet bodily needs, calcium will be sequestered through the breakdown of bone, regulated by **parathyroid hormone**, and through decreased mineralization of new bone. The normal diet provides about 25 mmol of calcium and of this about 10 mmol calcium is absorbed through the small intestine, while 5 mmol is excreted so that 5 mmol calcium is netted into the body daily. Parathyroid hormone stimulates the adaptive process (induced by low dietary calcium) of intestinal calcium absorption by generating, in the kidney, the active form of **vitamin D** and by facilitating the reabsorption (TRPV5 or ECaC) of calcium in the kidney. Calcium absorption and reabsorption takes place in epithelial cells that include the intestine, kidney, mammary glands, pancreas, prostate, salivary gland, and placenta. The **TRPV6 intestinal channel** has 6 transmembrane domains and an anchored N-terminus to an **ankyrin repeat** on the cytoplasmic side. An aspartate residue (D542) is involved in the permeation of the calcium cation. There is speculation as to whether there is a constitutively active TRPV6 and one that is inactive but activated by depletion of the intracellular calcium store. Another version concerns an inactivated transporter housed inside the cell and moved to its functional position in the membrane after an appropriate signal.

The kidney version of the calcium uptake transporter is **ECaC1** that acts similarly to CaT or TRPV6. Calcium transport takes place in the **distal nephron** in a three-step process. Calbindin plays a role in controlling intracellular concentrations of calcium, and $Ca^{2+}$ is pumped out of the cell through a **calcium-ATPase (PMCA1b)** in the basolateral membrane and by a **$Na^+/Ca^{2+}$ antiporter**. The activated VDR stimulates the expression of ECaC1 and calbindin. ECaC1 is a 6-transmembrane spanning transporter with a short hydrophobic stretch between transmembrane domains 5 and 6. Both the C-terminal and N-terminal tails contain potential regulatory sites for phosphorylation by PKC and also ankyrin repeats and a **PDZ domain** (80−90 stretches of amino acids that help anchor transmembrane proteins to the cytoskeleton and bind to a short region of the C-terminus of other specific proteins). The reabsorption of $Ca^{2+}$ along the kidney tubule is illustrated in Fig. 19.54.

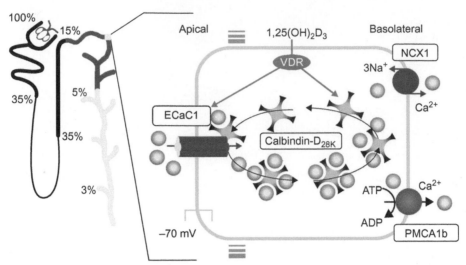

**FIGURE 19.54** $Ca^{2+}$reabsorption along the **kidney tubule**. The remaining $Ca^{2+}$at different sites of the nephron is indicated in percentages (*left figure*). Transcellular $Ca^{2+}$transport takes place only in the distal nephron (*pink*) and is carried out as a 3-step process. Following entry of $Ca^{2+}$through the epithelial $Ca^{2+}$channel, ECaC1, cytosolic $Ca^{2+}$is buffered by calbindin-$D_{28k}$. At the basolateral membrane, $Ca^{2+}$is extruded by a $Ca^{2+}$-**ATPase** (**PMCA1b**) and by a **sodium-calcium exchanger** (**NCX1**). $1,25(OH)_2D_3$ regulates this process by stimulating the expression of ECaC1 and calbindin-$D_{28k}$. Reproduced with permission from Fig. 1 of D. Muller *et al.*, "The epithelial calcium channel, ECaC1, molecular details of a novel player in renal calcium handling," *Nephrol. Dial. Transplant.* **16**: 1329−1335, 2001.

**TABLE 19.11 Recommended Daily Intakes of Calcium**

| Male and Female Age | Calcium (mg/day)[a] | Pregnancy & Lactation |
|---|---|---|
| 0−6 months | 210 | N/A |
| 7−12 months | 270 | N/A |
| 1−3 years | 500 | N/A |
| 4 years | 800 | N/A |
| 9−13 years | 1300 | N/A |
| 14−18 years | 1300 | 1300 |
| 19−50 years | 1000 | 1000 |
| 51 + years | 1200 | N/A |

[a]*mg, milligrams.*
Reproduced from http://ods.od.nih.gov/factsheets/calcium.asp.

In this figure, the role of **calbindin** is clearly shown to ferry, as many as four atoms of $Ca^{2+}$per molecule of calbindin, across the cellular cytoplasm. The reabsorption process involves the uptake of $Ca^{2+}$ through the ECaC1 channel, binding of $Ca^{2+}$ to calbindin, ferry the calbindin-$Ca^{2+}$ complex across the cytoplasm with release of $Ca^{2+}$ in the basolateral region and extrusion of $Ca^{2+}$ to the extracellular space and, ultimately, to the bloodstream via two transporters, the NCX1 antiport, exchanging three atoms of $Na^+$ for one atom of $Ca^{2+}$and calcium ATPase (PMCA1b).

The recommended daily intake of calcium for various ages is shown in Table 19.11.

Yogurt and dairy products are rich in calcium as well as other sources, such as sardines, tofu, salmon, spinach, turnip greens, kale, whole wheat, and white bread as well as commercial products fortified with calcium.

Regarding the **second messenger activity of $Ca^{2+}$**, tissue cells exhibit a variety of mechanisms for the entry of calcium across the plasma membrane. There are at least three major types of calcium uptake mechanisms: **capacitative calcium entry (CCE)**, **receptor-operated channels (ROC)**, and **second messenger-operated channels (SMOC)**. These types are illustrated in Fig. 19.55.

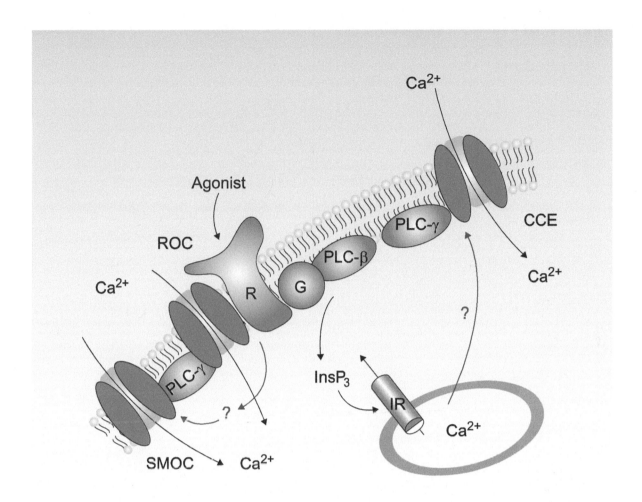

**FIGURE 19.55** Activation of calcium entry into cells across the plasma membrane can occur through a variety of mechanisms. These include capacitative calcium entry (CCE), receptor-operated channels (ROC), and second messenger-operated channels (SMOC). CCE is initiated by mechanisms that bring about depletion of intracellular calcium stores, such as G-protein (G) dependent PLC-β activation and formation of the calcium-mobilizing messenger, InsP3. InsP3 causes discharge of stored calcium through the calcium permeable InsP3 receptor (IR). The depletion of specialized intracellular calcium stores signals the activation of store-operated or CCE channels in the plasma membrane. ROC activation may involve more direct interaction of receptors with the channels, or, alternatively, the channels may function as the receptors for extracellular ligands. In the case of SMOC activation, second messengers, such as InsP3 or diacylglycerol (DAG), interact with and initiate activation of the channels. CCE and SMOC depend, in some way, on PLCγ. *InsP3*, IP3, or inositol *tris*phosphate. *Reproduced from http://www.nature.com/ncb/journal/v4/n12/fig_tab/ncb1202-e280_F1.html.*

Ligands for the hormone receptor activated channels include catecholamines, TRH, ADH (AVP), LHRH, and oxytocin. Once the ROC is activated by the binding of ligand, associated G-protein bound with GDP, exchanges GDP for GTP which binds in its place and the activated G-protein activates **PLCγ** to produce (**PIP2 [phosphatidylinositol *bis*phosphate]**) and **DAG (diacylglycerol)**. CCE becomes activated admitting extracellular $Ca^{2+}$ along with the activated ROC channel. $PIP_2$ is phosphorylated to form **$IP_3$ (inositol *tris*phosphate)**, a second messenger that releases $Ca^{2+}$ from its ER store through the agency of the **$IP_3$ receptor** on the ER membrane. At a certain level of cytosolic $Ca^{2+}$, it will bind to calmodulin forming a third messenger, **$Ca^{2+}$-calmodulin (CaCM)**. Calmodulin is a small protein (148 amino acids; molecular weight, 15,706) that can be modified by phosphorylation, acetylation, methylation, or proteolysis to tailor its actions. Calmodulin exists in one conformation in the absence of bound $Ca^{2+}$, but when $Ca^{2+}$ binds (up to 4 atoms/molecule), the conformation changes to one which avails two separate grooves and exposure of nonpolar residues (generated by Met residues), and these nonpolar domains allow interactions with the nonpolar regions of many other target proteins. The motif in calbindin responsible for $Ca^{2+}$-binding is called the **EF hand,** and there are four such motifs (Fig. 19.54) in the calbindin molecule (Fig. 19.56).

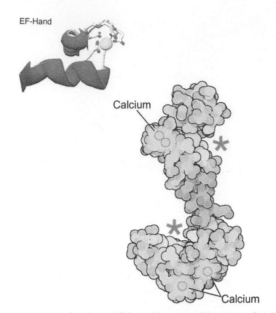

**FIGURE 19.56    Calmodulin (CM)** has many target proteins that exhibit a wide array of functions. Calcium-calmodulin (CaCM) is a key subunit for **phosphorylase kinase**, a regulator of glycogen breakdown. CM binds and activates other kinases and phosphatases that play important roles in cell signaling, in transport and in cell death. Nonpolar interactions occur between CaCM and its target proteins through the abundant Met residues in CM. $Ca^{2+}$-binding exposes the nonpolar surfaces of CM that bind to nonpolar regions of target proteins. The structure at the *upper left* shows one of the four EF hand motifs (a helix-loop-helix structure) in CM that binds $Ca^{2+}$(*yellow sphere*). The structure on the right is CaCM that contains two clefts (*red stars*) exposed after a conformational change due to $Ca^{2+}$-binding and also shows the nonpolar surfaces containing Met residues (*yellow*). *Reproduced (EF hand motif) from http://swissmodel.espasy.org/course/gifs/efhand.gif and the structure of CaCM, on the* right, *is reproduced from ftp://resources.rcsb.org/motm/tiff/44-Calmodulin-1cfd-1cll.tif.*

CM is expressed by all eukaryotic cells and gives rise to CaCM in the presence of sufficient calcium rendering CaCM a multifunctional messenger protein that can interact with a variety of target proteins, such as **calcium-calmodulin kinase (CaCM kinase)**. In general, the CaCM target proteins are unable to bind $Ca^{2+}$ themselves, allowing CaCM to fulfill this requirement and such proteins use CaCM as a $Ca^{2+}$ sensor and signal transducer. CaCM is a mediator in metabolism (e.g., glycogen breakdown), inflammation, muscle contraction, intracellular movements, memory (short- and long-term), apoptosis, and in the immune response. $Ca^{2+}$, a specific signal, itself, plays multiple roles in the depolarization of cells that require **membrane depolarization** in order to release materials (e.g., hormones, such as insulin) stored in vesicles (in the pancreatic β-cells, in the case of insulin).

Concerning the potential role of $Ca^{2+}$ in **apoptosis** (programmed cell death), recent research has shown that the **voltage-dependent anion channel (VDAC)**, located in the outer mitochondrial membrane, mediates interactions between the mitochondria and other locations in the cell by ability of VDAC to transport anions, cations, ATP, metabolites, and $Ca^{2+}$(VDAC has specific $Ca^{2+}$-binding sites) across the outer mitochondrial membrane. Because VDAC regulates the release of apoptotic proteins (specifically cytochrome c) from the mitochondria, VDAC transport of **cytochrome c** is a critical step in the **intrinsic pathway of apoptosis**. $Ca^{2+}$(either directly or indirectly) promotes the **oligomerization of VDAC**, essential to its function in moving proteins from the mitochondria to the cytoplasm in the process of apoptosis. The oligomerization of VDAC is related directly to calcium ion concentration ([$Ca^{2+}$]) because increasing mitochondrial [$Ca^{2+}$] generates both VDAC oligomerization and apoptosis in the absence of other apoptotic stimuli, whereas a reduction in mitochondrial [$Ca^{2+}$] reduces VDAC oligomerization and inhibits apoptosis. Clearly, $Ca^{2+}$ is being recognized as an important factor in mediating intrinsic apoptosis.

**Muscle contraction** (Fig. 19.57) takes place when $Ca^{2+}$ binds to **troponin** to remove the inhibited state of contraction and also $Ca^{2+}$ removes the triose phosphate isomerase subunit from the myosin head that has ATPase activity. ATPase action removes the energy source for muscular contraction so that its removal preserves ATP that drives contraction.

Skeletal muscle contracts in response to a neurotransmitter that generates an **action potential** that causes $Ca^{2+}$ to flow into the sarcoplasm from the sarcoplasmic reticulum (ER store). $Ca^{2+}$ binds to troponin-tropomyosin molecules causing troponin to change its shape to expose actin-myosin binding sites. Myosin and actin interact and the muscle shortens (contracts). After the action potential passes, cytoplasmic $Ca^{2+}$ returns to the ER store, and calcium ions

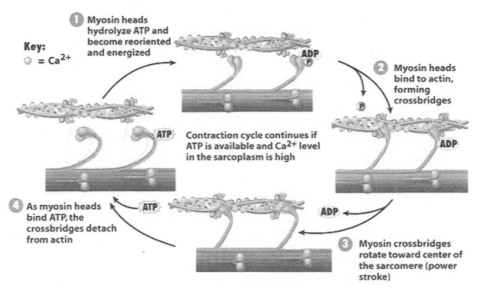

**Key:**
◯ = Ca²⁺

1. Myosin heads hydrolyze ATP and become reoriented and energized

2. Myosin heads bind to actin, forming crossbridges

Contraction cycle continues if ATP is available and Ca²⁺ level in the sarcoplasm is high

4. As myosin heads bind ATP, the crossbridges detach from actin

3. Myosin crossbridges rotate toward center of the sarcomere (power stroke)

**FIGURE 19.57** The role of Ca²⁺ in muscle contraction. *Reproduced with permission (required) from http://classconnection.s3.amazonaws.com/ 1517/flashcards/715536/jpg/picture1.jpg.*

dissociate from troponin so that it returns to its normal shape. This allows tropomyosin to mask the actin-myosin binding sites on the actin filament and causes muscle relaxation. Thus, the level of [Ca²⁺] in the soluble sarcoplasm (cytosol) controls muscle contraction. ATP (generated from creatine phosphate stores) resets the myosin head and releases the actin filament.

Calcium is important when the contents of vesicular stores need to be discharged from the cell. An example is the **secretion of insulin** from the **pancreatic β-cell**. In this cell, the elevation of plasma glucose causes glucose to be transported into the β-cell through the **GLUT2 transporter**. The incoming glucose transits glycolysis, generating ATP that increases the **ATP/ADP ratio** in the cell. The rise in ATP inhibits the **ATP-sensitive potassium channel** (K⁺$_{ATP}$ channel) causing a buildup of potassium ions in the cell and the depolarization of the cell membrane. This causes the **voltage-gated calcium channel** to open and admit Ca²⁺into the cell from the extracellular space. Elevated calcium ions in the cytosol lead to release of **insulin**, stored in secretory vesicles, into the extracellular space and into the bloodstream. The mechanism by which Ca²⁺ acts in the exocytosis process is not completely understood although it may involve **SNARE** (soluble N-maleimide-sensitive factor attachment protein receptor) proteins participating in fusion (SNARE proteins in the membranes of the secretory vesicle and the plasma membrane may form a bridge that mediates the fusion process) of the vesicular and plasma membranes and possibly **myosin-like proteins** with which Ca²⁺could interact to enhance the ejection process. SNARE could activate the release of vesicular luminal Ca²⁺ that could bind to and activate the membrane fusion apparatus. The overall activities involved in the entry of glucose into the β-cell culminating in the release of stored insulin are shown in Fig. 19.58.

## Iodine (I)

Iodine is another critical micronutrient because it is essential for the formation of the thyroid hormone in the thyroid gland and functions to maintain the bodily metabolic rate. Thyroid hormone synthesis takes place in the follicular cells of the thyroid gland, and it is regulated by the **thyroid-stimulating hormone** (TSH, also **thyrotropin**) that is signaled to be released from the anterior pituitary thyrotrope by the hypothalamic hormone, **thyrotropin-releasing hormone** (TRH). Under specific regulation, the hypothalamus releases **TRH** which courses through the closed portal system to reach the TRH receptor on the cell membrane of the **thyrotrope** that causes TSH to be released into the general circulation. TSH binds to its receptor on the membrane of the thyroid follicular cell (thyrocyte) causing a stimulation of adenylate cyclase and the production of **cAMP** that culminates in the synthesis of **thyroxine** (T4) and a much smaller amount of **triiodothyronine** (T3), the biologically active form of the hormone. The action of TSH also accounts for the active level of the **deiodinase** enzyme that is responsible for the formation of T3 from T4. Finally, both T4 and a lesser amount of T3 are secreted from the thyroid gland into the bloodstream where the quantity of T3 represents about 20% of the total formed, and the remaining T3 (~80%) is formed (mainly by deiodination of T4) in the tissue cells,

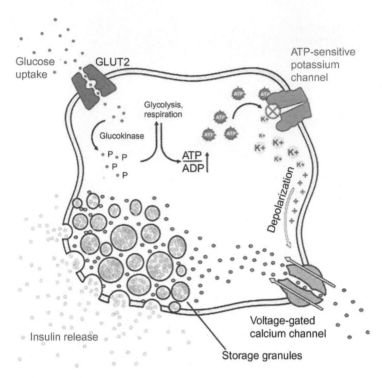

**FIGURE 19.58** Insulin secretion in β-cells is triggered by rising blood glucose levels. Uptake of glucose by GLUT2 transporters results in the glycolytic phosphorylation of glucose and an increase in ATP causing a rise in the ATP:ADP ratio. The rise in [ATP] inactivates the ATP-sensitive K$^+$ channel that depolarizes the membrane causing the calcium channel to open and allow Ca$^{2+}$ to flow into the cell. The elevation of Ca$^{2+}$ leads to exocytotic release of insulin from its storage granule. *Reproduced from http://www.betacell.org/content/articlepanelview/article_id/1/panel_id/2.*

primarily the liver hepatocytes. The control of thyroid hormone secretion under conditions of normal availability of dietary iodine or insufficient dietary iodine is shown in Fig. 19.59.

Under the influence of TSH, iodide enters the cell at the basolateral side through the 13 transmembrane Na$^+$/I$^-$ symporter (NIS) where two atoms of Na$^+$ are transported with one atom of iodide. Na$^+$ also can be transported, against a concentration gradient, out of the cell into the extracellular space (basolateral side of the thyrocyte) by a Na$^+$/K$^+$ATPase. In consequence, the thyroid gland can concentrate iodide up to 50-fold the concentration in the blood. TSH operates by binding to the TSH receptor on the plasma membrane of the thyrocyte (thyroid follicle cell) leading to the activation of adenylate cyclase and the production of cAMP. cAMP, through **PKA**, probably activates (phosphorylates) a transcription factor for the gene encoding mRNA for the NIS transporter resulting in the stimulation of its synthesis. Consequently, more iodide ion is taken into the thyrocyte. Iodide crosses the cytoplasm to the apical side of the thyrocyte where it is ejected into the follicle through the **iodide/chloride transporter, pendrin**. Through the availability of H$_2$O$_2$, generated by a **NADPH oxidase**, iodide is oxidized to iodine (I$_2$) that reacts with the tyrosine residues of **thyroglobulin (Tg)** and is converted, through several steps, to the thyroid hormones (T4 and T3) catalyzed by the enzyme **thyroid peroxidase (TPO)**. The reactions are as follows:

$$2I^- + H_2O_2 \rightarrow I_2$$

$$I_2 + tyrosine(Tg) \rightarrow Tg\text{-monoiodotyrosine (MIT) or Tg-diiodotyrosine (DIT)}$$

$$Tg\text{-DIT} + Tg\text{-DIT} \rightarrow Tg\text{-}3, 5, 3', 5'\text{-tetraiodothyronine (thyroxine or T4)}$$

$$Tg\text{-DIT} + Tg\text{-MIT} \rightarrow Tg\text{-}3, 5, 3'\text{-triiodothyronine (T3)}$$

Then:

$$Tg\text{-(MIT, DIT, T4, T3)} + lysosomal\ proteases \rightarrow MIT + DIT + T4 + T3 + Tg$$

$$MIT + DIT + deiodinases \rightarrow I^- + tyrosine$$

Both TPO and NADPH oxidase are located in the apical membrane of the thyrocyte. These reactions derivatize the tyrosine residues in thyroglobulin (in the colloid) to thyroglobulin bound with MIT, DIT, T4, and T3. Substituted (with

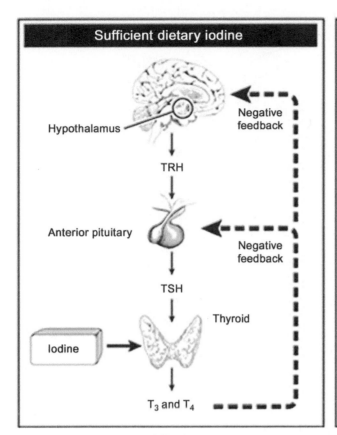

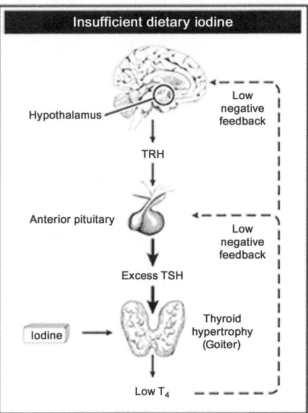

**FIGURE 19.59** Iodine intake and thyroid function. In response to thyrotropin-releasing hormone (TRH) secretion by the hypothalamus, the pituitary gland secretes thyroid-stimulating hormone (TSH), which stimulates iodine trapping, thyroid hormone synthesis, and release of T3 (triiodothyronine) and T4 (thyroxine) by the thyroid gland. When dietary iodine uptake is sufficient, the presence of adequate circulating T4 and T3 feeds back at the level of both the hypothalamus and pituitary, decreasing TRH and TSH production. When circulating T4 levels decrease (lowering the negative feedback), the pituitary increases its secretion of TSH, resulting in increased iodine trapping as well as increased production and release of both T3 and T4. Dietary iodine deficiency results in inadequate production of T4. In response to decreased blood levels of T4, the pituitary gland increases its output of TSH. Persistently elevated TSH levels may lead to hypertrophy of the thyroid gland, also known as **goiter**. *Reproduced from http://lpi.oregonstate.edu/infocenter/minerals/iodine/thyroid.html.*

MIT, DIT, T4, T3) thyroglobulin reenters the apical thyrocyte by endocytosis with **megalin**, a member of the LDL receptor family in the apical membrane of the thyrocyte. The substituted Tg is taken up by the **lysosome** where the hormones, T4 and T3, and the iodinated tyrosines, MIT and DIT are split off from thyroglobulin by proteolytic action. MIT and DIT are deiodinated and the released iodide ions join the cascade of iodide ions moving, again through **pendrin** for further production of the thyroid hormones. These steps are summarized in Fig. 19.60.

T3 and T4 are partially hydrophobic and may dissolve into the basolateral membrane for passive transport to the bloodstream, or there may be a transport mechanism to achieve this step. T4 is the predominant secretory product and T3, the biologically active hormone, is present in much lower concentration. Both T4 and T3 in the blood feedback negatively on the synthesis and release of TSH in the thyrotrope of the anterior pituitary and on the hypothalamus to inhibit the release of TRH.

In the target tissues, such as liver, T4 is deiodinated to T3, the active form of the hormone, that enters the nucleus and activates thyroid hormone specific genes. An overall summary is shown in Fig. 19.61.

**Thyroxine** is mainly carried in the plasma as bound forms with the **thyroid-binding globulin (TBG)** that binds 75% of T4 or with **transthyretin**, a prealbumin and **albumin**. There is some T4 in the free form that is not bound to plasma proteins and the free form is able to enter target cells. The normal total T4 (bound + free form) in blood is in the range of 4.5−11.5 μg/dL. The amount of free T4 in normal blood is 0.8−2.8 ng/dL (∼0.03%), less than 1/1000 of the bound hormone, reflecting the equilibria between the bound forms and free T4. About 20% of circulating T3 is produced in the thyroid gland, and the remaining 80% of T3 is produced in target cells (e.g., hepatocyte) by 5′-deiodinase of the outer ring of T4. Thyroid hormones enter target cells facilitated by the human **monocarboxylate transporter 8**

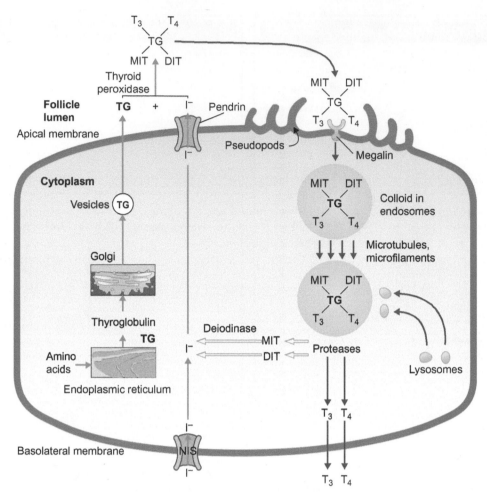

**FIGURE 19.60** Summary of the steps from the import of iodide ion through the **2Na$^+$/I$^-$ symporter** (**NIS**) on the basolateral side of the thyrocyte, through the derivatization of thyroglobulin forming the thyroid hormones and intermediates in the follicle, to the entry through the apical membrane of derivatized thyroglobulin through megalin and the release of thyroid hormones from thyroglobulin in the lysosome and their exit from the thyrocyte through the basolateral membrane. *Reproduced from http://users.atw.hu/blp6/BLP6/HTML/common/M97803233045827-041-f005.jpg.*

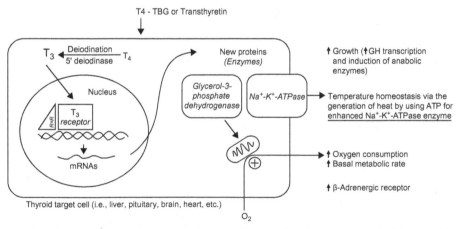

**FIGURE 19.61** Overview of the fate of thyroid hormone in a target cell, such as the liver (also kidney and brain). *Reproduced from http://www.nbs. csudh.edu/chemistry/faculty/nsturm/CHE452/images/T3T4Receptors.gif.*

(**MCT8**). Once in the cytoplasm of the target cell, T4 is deiodinated by hormone deiodinase (D2) to T3. T3 enters the nucleus and binds to the heterodimeric **thyroid hormone receptor**. TR then interacts with thyroid-specific gene promoters to activate transcription. A similar situation occurs in brain where T4 enters the **astrocyte** through the **OATp1c1 transporter** (**organic anion transporter polypeptide**) where it is converted to T3 by **D2 deiodinase,** and T3 leaves the astrocyte through a transporter and enters a neuron through the MCT8 transporter where it carries out its gene activation functions.

## SUMMARY

Iron deficiency anemia, the most common form of anemia, exemplifies the information in this chapter as it is related to micronutrients. About 2.5 g of iron are incorporated into hemoglobin heme, and iron is needed for myoglobin and proteins in the electron transport chain. Ferritin and hemosiderin are iron storage proteins. Various enzymes require nonheme iron as cofactors. Ferric iron ($Fe^{3+}$) in the diet (in the form of heme and nonheme iron-containing proteins) is reduced to ferrous iron ($Fe^{2+}$) by ascorbic acid (vitamin C), and the resulting complex enhances the absorption of nonheme iron from the intestine ($Fe^{3+}$ is converted to $Fe^{2+}$ to be absorbed). The divalent metal ion transporter (DMT1) also is involved in the transport of dietary iron across intestinal epithelial cells. Heme, itself, is toxic and intestinal ER heme oxygenase converts heme to biliverdin and $Fe^{2+}$ that is incorporated into ferritin and exported from the cell. Many proteins and certain transporters are involved in the metabolism of iron. Phytic acid is a component of many plant foods and, although they may be rich in iron, phytic acid forms a complex with iron and prevents intestinal absorption. Ferrous iron can be given orally to correct dietary iron deficiency.

An excess of iron over the immediate iron stores will bind to a specific aconitase enzyme, known as the IRE (iron-responsive element) binding protein (IREBP). This binding causes a change in the conformation of IREBP so that it no longer binds to transferrin receptor mRNA resulting in increased production of ferritin which increases the cellular capacity to store the excess iron.

The synthesis of heme starts with δ-aminolevulinic acid derived from succinyl-CoA from the mitochondrial TCA cycle and glycine from the cytoplasm. The several intermediate reactions are presented that consummate in the deposition of iron into the heme molecule by the terminal enzymatic reaction of ferrochetalase. Mutation of the gene for ferrochetalase leads to erythropoietic protoporphyria.

Copper is stored mainly in the liver although it occurs in most tissues. Copper is transported in the bloodstream by ceruloplasmin that also has the activity of ferroxidase (converting $Fe^{2+}$ to $Fe^{3+}$ and oxidizes other metals, as well). Histidine residues are usually involved in copper-binding domains of copper-binding proteins. A deficiency of copper can lead to Wilson's disease, aceruloplasminemia and anemia. Wilson's disease involves the mutation of the *ATP7B* gene that encodes a transmembrane copper-transporting P-type ATPase. Human copper transporter 1 (hCTR1) is a high-affinity transporter of copper moving it against a concentration gradient from the extracellular to intracellular space. The mechanism of this transport is described. Another copper-binding protein is the cellular prion protein that theoretically could be a recycling receptor for copper uptake from the extracellular space. The lower affinity copper-binding sites in cellular prion have the motif HGGGW, whereas the single high-affinity site has the sequence GGGTH. Ingestion of the prion scrapie form leads to Creutzfeldt−Jakob disease.

There are four different superoxide dismutases (SODs): iron-SOD, Mn-SOD, Ni-SOD, CuZnSOD and a prokaryotic Fe-SOD. SOD converts $O_2^{-\bullet}$ (a superoxide radical, a ROS) formed in the mitochondria as a byproduct of electron transport chain activity generating $H_2O_2$. $H_2O_2$ is converted to water by lysosomal catalase or, in a lipid environment, by glutathione peroxidase. Mutations in the *SOD1* gene for CuZnSOD can cause neurodegenerative diseases, including amyotrophic lateral sclerosis.

Magnesium is a component of several enzymes and forms a complex with ATP. Magnesium is critical for the synthesis of ATP by ATP synthase (terminal protein in the electron transport chain). Melastatin 7 (TRPM7) is a magnesium channel (transporter) and, in addition, the metal also can enter and leave cells by passive diffusion. Intracellular magnesium is a negative regulator of TRPM7.

Zinc is a component of carbonic anhydrase, cytosine deaminase, zinc family transporters (Zip), zinc finger motifs, and ATP synthase as well as other proteins. In certain zinc finger motifs, zinc is coordinated by four cysteine residues as well as by a combination of cysteine and histidine residues. In some proteins, zinc is coordinated by three histidines and a molecule of water. Zinc is a cofactor of the insulin-degrading metalloprotease (IDE) that cleaves the insulin B chain at several locations. In the active site of this enzyme, zinc is coordinated with two histidines and the oxygens of two glutamates and a water molecule.

Molybdenum is a cofactor for several enzymes. The coenzyme form of molybdenum is a pterin. The molybdenum cofactor is formed from GTP. Molybdenum can enter the intestinal cell as a complex with ascorbic acid or may involve a transporter as it does in bacteria, although this aspect is not well known. The human protein gephyrin catalyzes the insertion of molybdenum in the final step of the synthesis of the molybdenum cofactor. Mutations in the gene for gephyrin cause stiff muscles, a disease that can result in death. Loss of function of gephyrin also can result in severe brain damage.

Selenium is a component of several enzymes. For example, thioredoxin reductase contains a selenocysteine residue in the sequence GlyCys*SeCys*Gly. The metal is also a component in selenodeiodinases (D1, D2, D3). These enzymes either deiodinate thyroxine (D2, D1) to the biologically active form, triiodothyronine (T3), or they remove essential iodides (D3) from the inner or outer rings to inactivate the thyroid hormone. Selenophosphate synthase (SePS) catalyzes the reaction between selenide (HSe) and ATP to produce selenocysteine. The enzyme also catalyzes the phosphorylation of selenide with MgATP to form selenium phosphate (SeP). SeP can donate Se to a tRNA that incorporates Se into specific proteins. Selenium cysteine synthase (Sec synthase) causes the conversion of Ser-tRNA$^{Sec}$ by the addition of Se from SeP. Sec-tRNA is recognized by a specific elongation factor and is brought to Se-protein mRNA for the synthesis of selenoprotein.

Calcium is a structural component of bone and teeth and also is an important second messenger. It is involved in muscle contraction (where $Ca^{2+}$ binds to troponin), expansion of blood vessels and in nervous system functions. Calcium circulates in the blood bound to albumin and in the free ionized form, the latter of which determines its biological effects. High levels of dietary calcium are absorbed by passive transport through tight junctions between cells but low dietary levels of calcium are transported through the TRPV6 channel. Intracellular calcium ion is bound to calbindin, a transport protein induced through the action of the active form of vitamin D (1,25-dihydroxyvitamin D3) bound to the vitamin D receptor (a member of the steroid nuclear receptor gene family). Calbindin (containing 4 EF hand motifs that bind $Ca^{2+}$) transports calcium ion across the intestinal enterocyte to the basolateral side where it is transported to the extracellular space (and ultimately to the bloodstream) by the PMCA transporter. Calcium levels increase in the cellular cytoplasm to depolarize certain cells allowing the secretion of stored (in vesicles) secretory products, such as insulin from the pancreatic beta cell.

Dietary iodine is essential for the formation of thyroid hormones in the thyroid follicle cell. Iodide ion is transported into the cell through the membrane iodide transporter (NIS). Inside the cell iodide crosses the cytoplasm to the apical side of the cell where it is ejected into the follicle through the transporter, pendrin (iodide/chloride transporter). NADPH oxidase avails $H_2O_2$ that oxidizes iodide to iodine that reacts (catalyzed by thyroid peroxidase) with the tyrosine residues in thyroglobulin (in the follicle store) to generate the thyroid hormones (T4 and T3) along with iodinated tyrosine derivatives. Iodinated thyroglobulin enters the thyrocyte (by endocytosis with megalin receptor) where it is taken up by lysosomes where T4, T3, MIT, and DIT are split out from iodinated thyroglobulin. The hormones, T4 and the biologically active form, T3, are released into the bloodstream and are taken up by target cells where T3 activates the nuclear thyroid hormone receptor inducing gene expression of ultimate proteins that affect the oxidation rates in target cells.

# SUGGESTED READING

### Literature

Anderson, G.J., Frazer, D.M., 2006. Iron metabolism meets signal transduction. Nat. Genet. 38, 503−504.

Barnes, C.M., Theil, E.C., Raymond, K.N., 2002. Iron uptake in ferritin is blocked by binding of [Cr(TREN)(H2O)(OH)]$^{2+}$, a slow dissociating model for [Fe(H2O)6]$^{2+}$. Proc. Natl. Acad. Sci. U.S.A. 99, 5195−5200.

Bianco, A.C., et al., 2002. Biochemistry, cellular and molecular biology, and physiological roles of the iodothyronine selenodeiodinases. Endocr. Rev. 23, 38−89.

Brent, G.A., 2012. Mechanisms of thyroid hormone action. J. Clin. Invest. 122, 3035−3043.

Callebaut, I., et al., 2003. The iodothyronine selenodeiodinases are thioredoxin-fold family proteins containing a glycoside hydrolase clan GH-A-like structure. J. Biol. Chem. 278, 36887−36896.

Cheng, K.T., Ong, H.L., Liu, X., Ambudkar, I.S., 2013. Contribution and regulation of TRPC channels in store-operated $Ca^{2+}$ entry. Curr. Top. Membr. 71, 149−179.

Dailey, H.A., et al., 2000. Ferrochetalase at the millennium: structures, mechanisms and [2Fe-2S] clusters. Cell. Mol. Life Sci. 13−14, 1909−1926.

Delmondes de Carvalho, F., Quick, M., 2011. Surprising substrate versatility in SLC5A6: Na$^+$-coupled I⁻ transport by the human Na$^+$/multivitamin transporter (hSMVT). J. Biol. Chem. 286, 131−137.

Dhe-Paganon, S., et al., 2000. Crystal structure of human frataxin. J. Biol. Chem. 275, 30753−30756.

Jeong, J., Eide, D.J., 2013. The SLC39 family of zinc transporters. Mol. Aspects Med. 34, 612−619.

Kim, E., et al., 2003. Superoxo, μ-peroxo, and μ-oxo complexes from heme/O$_2$ and heme-Cu/O$_2$ reactivity: copper ligand influences in cytochrome c oxidase models. Proc. Natl. Acad. Sci. 100, 3623–3628.

Maia, A.L., Goemann, I.M., Meyer, E.L., Wajner, S.M., 2011. Deiodinases: the balance of thyroid hormone; type 1 iodothyronine deiodinase in human physiology and disease. J. Endocrinol. 209, 283–297.

Maia, A.L., Goemann, I.M., Meyer, E.L.S., Wajner, S.M., 2011. Type 1 iodothyronine deiodinase in human physiology and disease. J. Endocrinology. 209, 283–297.

Martin, A., Thompson, A.A., 2013. Thalessemias. Pediatr. Clin. North Am. 60, 1383–1391.

McDermid, J.M., 2012. Iron. Adv. Nutr. 3, 532–533.

Mendel, R.R., 2013. The molybdenum cofactor. J. Biol. Chem. 288, 13165–13172.

Miller, A.-F., 2004. Superoxide dismutases: active sites that save, but a protein that kills. Curr. Opin. Chem. Biol. 8, 162–168.

Miller, D.S., 2010. Regulation of P-glycoprotein and other ABC drug transporters at the blood-brain barrier. Trends Pharmacol. Sci. 31, 246–254.

Muller, D., et al., 2001. The epithelium calcium channel, ECaC1, molecular details of a novel player in renal calcium handling. Nephrol. Dial. Transplant. 16, 1329–1335.

Millhauser, G.L., 2004. Copper binding in the prion protein. Acc. Chem. Res. 37, 79–85.

Sanyo, Y., et al., 2013. Hepcidin bound to α$_2$-macroglobulin reduces ferroportin-1 expression and enhances its activity at reducing serum iron levels. J. Biol. Chem. 288, 25450–25465.

Shen, Y., Joachimiak, A., Rosner, M.R., Tang, W.-J., 2006. Structures of human insulin-degrading enzyme reveal a new substrate recognition mechanism. Nature. 443, 870–874.

Touyz, R.M., 2008. Transient receptor potential melastatin 6 and 7 channels, magnesium transport, and vascular biology: implications in hypertension. Am. J. Physiol. Heart Physiol. 294, H1103–H1118.

Unkles, S., et al., 1999. Eukaryotic molybdopterin synthase. J. Biol. Chem. 274, 19286–19293.

Vashchenko, G., MacGillivray, R.T., 2013. Multi-copper oxidases and human iron metabolism. Nutrients. 5, 2289–2313.

Watt, N.T., Griffiths, H.H., Hooper, N.M., 2013. Neuronal zinc regulation and the prion protein. Prion. 7, 203–208.

Wilks, A., Heinzl, G., 2014. Heme oxygenation and the widening paradigm of heme degradation. Arch. Biochem. Biophys. 544, 87–95.

Yi, L., et al., 2009. Heme regulatory motifs in heme oxygenase-2 form a thiol/disulfide redox switch that responds to the cellular redox state. J. Biol. Chem. 284, 20556–20561.

Yoon, T., Cowan, J.A., 2004. Frataxin-mediated iron delivery to ferrochetalase in the final step of heme biosynthesis. J. Biol. Chem. 279, 25943–25946.

Zhong, L., Holmgren, A., 2000. Essential role of selenium in the catalytic activities of mammalian thioredoxin reductase revealed by characterization of recombinant enzymes with selenocysteine mutations. J. Biol. Chem. 275, 18121–18128.

## Books

Prasad, K.N., 2011. Micronutrients in Health and Disease. CRC Press, Taylor and Francis Group, Boca Raton, FL, 357 pages.

Uthman, E., 1998. Understanding Anemia. University Press of Mississippi, Jackson, MI, (available on Kindle).

Chapter 20

# Vitamins and Nutrition

## VITAMIN D DEFICIENCY

There are many locations in North America and around the world where sunlight is limited. As many as half or more patients seen in clinical practice may be deficient in vitamin D, indicating the seriousness of this problem. Children growing up in a region with limited sunlight and remaining there as adults often prove to be vitamin D deficient and at risk for developing colon or other cancers because this vitamin is needed by many biological systems, particularly by the immune system. The immune system is essentially the surveillance system against the development of cancer. Adults having suffered from vitamin D deficiency may be at risk for cancers of the colon, breast, or prostate. Moreover, the elderly produce less vitamin D through exposure to sunlight than do younger persons. In addition, those with certain intestinal diseases, such as Crohn's disease, Whipple's disease, or sprue are unable to absorb dietary vitamin D. Table 20.1 describes the prevalence of vitamin D deficiency in clinical patient populations.

The clinical risk factors for vitamin D deficiency are reviewed in Table 20.2.

In Table 20.3 are shown the laboratory findings that suggest possible vitamin D deficiency and ranges of total serum 25-hydroxyvitamin D.

Information on the biosynthesis of the active form of vitamin D following sunlight exposure and the characteristics of the vitamin D receptor and its actions is reported in Chapter 16, Steroid Hormones and in a discussion of fat-soluble vitamins below. Although some vitamin D is excreted through its excretory product calcitroic acid, considerable amounts are retained in storage forms. In general, the fat-soluble vitamins A, D, E, and K are stored, whereas the water-soluble B vitamins and ascorbic acid, etc. can be excreted except when coenzyme forms are bound with proteins; some coenzymes dissociate from the enzyme in time.

The symptoms of vitamin D deficiency can be subtle, or they can express as muscle weakness and bone pain. The vitamin is need for calcium deposition into skeletal collagen. Calcium is required for muscular functions. Head sweating can be indicative. As this vitamin is fat-soluble, it can be stored when fat reserves are great as in overweight or obesity; in these cases, the vitamin D requirement increases owing to the poor availability of vitamin D in fat stores. Dark skin screens out sunlight so the darker the skin, the greater the requirement for sunlight to generate the vitamin. After 50 years of age, sunlight is utilized less efficiently for the formation of active vitamin D precursors, and the aged kidney is less efficient in hydroxylating the intermediate (25-hydroxyvitamin D) at the one position.

**Hypervitaminosis D**, a toxic condition, occurs with a daily intake of more than 10,000 IU (1 IU = 25 ng) per day compared to the normal adult intake of 600–800 IU per day (upper level refers to persons older than 71 years). This is vitamin D toxicity that results in abnormally high blood levels of calcium that can damage kidneys, soft tissues, and bones for extended periods. Toxicity symptoms are constipation, loss of appetite, dehydration, irritability, fatigue, muscle weakness, hypercalciuria, excessive thirst, polyuria, and high blood pressure. Lowering vitamin intake can result in reversal of vitamin D toxicity.

## VITAMINS

### Water-Soluble Vitamins

Vitamins are essential factors, not synthesized in the body, but required in the diet because they are needed mostly for the formation of **coenzymes**. Water-soluble vitamins will be discussed first. The major water-soluble vitamins are as follows: thiamine (vitamin B1, aneurine), riboflavin (vitamin B2), niacin (vitamin B3), pantothenic acid (vitamin B5), pyridoxine, pyridoxal and pyridoxamine (3 forms of vitamin B6), biotin (vitamin H), folic acid (vitamin B9), cobalamin (vitamin B12), and ascorbic acid (vitamin C).

Human Biochemistry. DOI: http://dx.doi.org/10.1016/B978-0-12-383864-3.00020-X

**TABLE 20.1 Prevalence of Vitamin D Deficiency in Commonly Encountered Clinical Patient Populations**

| | |
|---|---|
| Nursing home or housebound residents: mean age, 81 years | 25%–50% |
| Elderly ambulatory women, aged >80 years | 44% |
| Women with osteoporosis, aged 70–79 years | 30% |
| Patients with hip fractures: mean age, 77 years | 23% |
| African–American women, aged 15–49 years | 42% |
| Adult hospitalized patients: mean age, 62 years | 57% |

Reproduced from http://www.ncbi.nlm.nih.gov/pmc/articles/PMC2912737/table/T1/.

**TABLE 20.2 Clinical Risk Factors for Vitamin D Deficiency**

- Decreased intake
  - Inadequate oral intake
  - Malnutrition (poor oral intake)
  - Limited sun exposure
- Gastrointestinal
  - Malabsorption (e.g., short bowel syndrome, pancreatitis, inflammatory bowel disease, amyloidosis, celiac sprue, and malabsorptive bariatric surgery procedures)
- Hepatic
  - Some antiepileptie medications (increased 24-hydroxylase activity)
  - Severe liver disease or failure (decreased 25-hydroxylase activity)
- Renal
  - Aging (decreased 1-α hydroxylase activity)
  - Renal insufficiency, glomerular filtration rate <60% (decreased 1-α hydroxylase activity)
  - Nephrotic syndrome (decreased levels of vitamin D-binding protein)

Reproduced from http://www.ncbi.nlm.nih.gov/pmc/articles/PMC2912737/table/T2/.

**TABLE 20.3 (A) Laboratory and Radiographic Findings that Suggest Possible Vitamin D Deficiency. (B) Reference Ranges for Total Serum 25-hydroxyvitamin D from the Mayo Clinic**

(A)
- Laboratory
  - Low 24-h urine calcium excretion (in the absence of thiazide use)
  - Elevated parathyroid hormone level
  - Elevated total or bone alkaline phosphatase level
  - Low serum calcium and/or serum phosphorus level
- Radiographic
  - Decreased bone mineral density (osteopenia or osteoporosis)
  - Nontraumatic (fragility) fracture
  - Skeletal pseudofracture

(B)
| | |
|---|---|
| Severe deficiency[a,b] | <10 ng/mL |
| Mild-to-moderate deficiency[c] | 10–24 ng/mL |
| Optimal[d] | 25–80 ng/mL |
| Possible toxicity | >80 ng/mL |

[a]SI conversion factor: To convert 25(OH)D values to nmol/L, multiply by 2.496.
[b]Could be associated with osteomalacia or rickets.
[c]May be associated with secondary hyperparathyroidism and/or osteoporosis.
[d]Levels present in healthy populations.
Reproduced from http://www.ncbi.nlm.nih.gov/articles/PMC2912737/table/T3/ for (A) and from http://www.ncbi.nlm.nih.gov/articles/PMC12737/table/T4/ for (B).

# Thiamine (Vitamin B1, Aneurine)

Thiamine is required in the diet and for adults the daily intake is about 1.5 mg/day. Thiamine is needed for the metabolism of carbohydrates and for the functioning of the heart and nervous system. **Thiamine deficiency** (*beriberi*) occurs in populations that depend on polished rice as the major food staple (the vitamin is present in unpolished rice). In the United States, thiamine deficiency can occur in alcoholics who have poor nutrition and whose absorption processes in the gut may be compromised; this condition is known as the **Wernicke–Korsakoff syndrome**. Thiamine deficiency in the United States appears to be one of the most common causes of **dementia**. The best food sources of thiamine are as follows: liver, pork, whole grain cereals, potatoes, and breads.

The active form of thiamine is in the form of **thiamine pyrophosphate** (TPP or TDP) that is a coenzyme for **pyruvate dehydrogenase**, **α-ketoglutarate dehydrogenase**, and **transketolase** (in the pentose phosphate pathway). TPP is formed from thiamine by **thiamine diphosphotransferase** (thiamine diphosphokinase) as shown in Fig. 20.1.

The conversion of pyruvate to acetyl-CoA involves three enzymes (pyruvate dehydrogenase (PDH), **dihydrolipoyl transacetylase (DLT)**, and **dihydrolipoyl dehydrogenase (DLD)**). PDH is a huge complex of about 9 million Daltons consisting of many copies of the three enzymes. PDH is a tetramer of 2 α-subunits and 2 β-subunits. The enzymes are physically arranged so that the products of one enzyme are in close proximity to the next. The overall reactions and the physical relationships of the three enzymes are shown in Fig. 20.2.

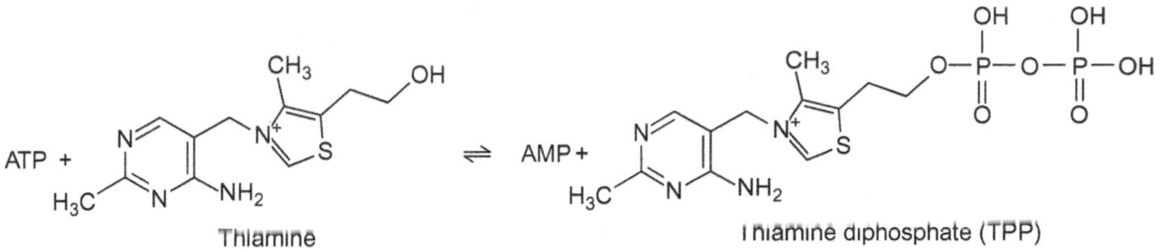

**FIGURE 20.1**   Conversion of thiamine to its coenzyme form, thiamine diphosphate (TPP) by the action of thiamine diphosphotransferase (thiamine diphosphokinase).

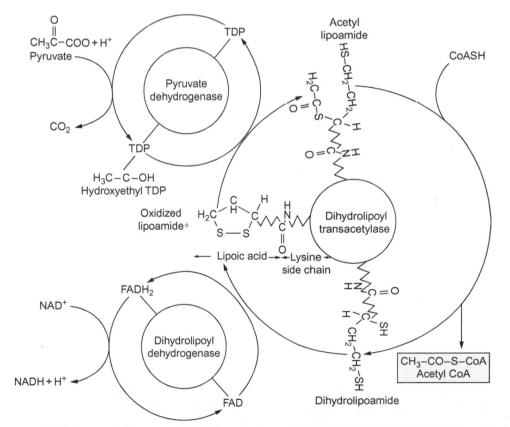

**FIGURE 20.2**   Conversion of pyruvate to acetyl-CoA catalyzed by the pyruvate dehydrogenase complex. **Lipoic acid** is joined by an amide link to a lysine residue of dihydrolipoyl transacetylase forming a long flexible arm that allows the **lipoic acid** prosthetic group to rotate between the active sites of each enzyme in the complex. *TDP*, thiamine diphosphate; *TPP*, thiamine pyrophosphate.

The summary of all of these reactions is:

$$\text{Pyruvate} + \text{CoA} + \text{NAD}^+ \rightarrow \text{Acetyl-CoA} + \text{CO}_2 + \text{NADH} + \text{H}^+$$

Five coenzymes participate in these reactions: thiamine pyrophosphate, lipoic acid and FAD; CoA and $\text{NAD}^+$ are the fourth and fifth coenzymes, and they are the stoichiometric cofactors. The overall reactions are as follows:

$$\text{CH}_3\text{COCOO}^- (\text{decarboxylation}) \rightarrow \text{CO}_2 + \text{CH}_3\text{CO}^- (\text{oxidation}) \rightarrow 2e^- + \text{pyruvate}$$

$$\text{CH}_3\text{CO}^+ + \text{CoA (transfer to CoA)} \rightarrow \text{CH}_3\text{COSCoA} + \text{Acetyl-CoA}$$

The activity of PDH is controlled by the phosphorylation of serine residues (3) by **PDH kinase** (**PDK**) on the PDH $\alpha$-subunit. The four known isoforms of PDK are differently distributed in different tissues. Phosphorylation of PDH by PDK inhibits PDH activity, and the activity of the enzyme is restored by the action of **PDH phosphatase** (**PDHP**). There are two isoforms of PDHP, and they are tissue specific.

Three molecules of NADH (nicotinamide adenine dinucleotide, reduced form) and one molecule of $\text{FADH}_2$ (flavin adenine dinucleotide, reduced form) are produced for each molecule of **acetyl-CoA** metabolized through one turn of the **citric acid cycle**. When NADH is reoxidized through the respiratory chain, three molecules of ATP are formed from three molecules of ADP, and the reoxidation of $\text{FADH}_2$ produces two molecules of ATP. **Succinate thiokinase** generates one molecule of ATP equivalent in the form of GTP when the substrate is phosphorylated by this enzyme. In total, 9 molecules of ATP are generated from $\text{NADH}_2$, and 2 molecules of ATP are generated from $\text{FADH}_2$ plus one molecule of ATP from the succinate thiokinase reaction produces 12 ATP molecules per turn of the cycle from acetyl-CoA that is equivalent to one molecule of pyruvate. Since two molecules of pyruvate are generated from one molecule of glucose, the glucose molecule requires two turns of the citric acid cycle, generating 24 molecules of ATP. Additionally, in glycolysis two NADPH are produced from the glyceraldehyde-3-phosphate dehydrogenase reaction. When these are reoxidized through the respiratory chain, 6 ATP molecules are produced. Added to this, 2 ATPs are produced in the phosphoglycerate kinase reaction, and 2 more ATPs are produced in the pyruvate kinase reaction, adding up to 10 ATPs but actually only 8 ATPs are generated from glycolysis because 2 ATPs are used up in the hexokinase and phosphofructokinase reactions. So, 24 plus 8 gives 32 ATP molecules. The phosphorylation of substrate in the succinate thiokinase reaction gives 2 ATP and 4 more ATP molecules are produced in the respiratory chain oxidation of 2 $\text{FADH}_2$ molecules in the succinate dehydrogenase reaction. Thus, there is a net of 38 ATPs when both cycles are added under aerobic conditions. One ATP is used up in the transport of $\text{H}^+$ into the mitochondrion with pyruvate and malate.

In addition to its role as a coenzyme, thiamine has a role in nerve impulses through the sodium/potassium gradient. Thiamine deficiency causes neurological malfunctioning. Experimental electrical stimulation of nerves causes **thiamine monophosphate** and free **thiamine** to be released into the medium and decreased **TPP** and **thiamine triphosphate** in the cell. The **Wernicke−Korsakoff syndrome** describes the thiamine-deficient diet of chronic alcoholics. Also, there is evidence for an $\text{H}^+$/**thiamine antiporter** in the placental brush border.

## Riboflavin (Vitamin B₂)

The coenzyme forms of riboflavin are **flavin mononucleotide** and **flavin adenine dinucleotide** (Fig. 20.3).

Flavin mononucleotide (FMN) is formed first by the action of **riboflavin kinase** on riboflavin. FMN is then converted to flavin adenine dinucleotide (FAD) by the action of **FAD pyrophosphorylase** with ATP. These reactions are shown in Fig. 20.4.

Many flavoproteins contain $\text{Mg}^{2+}$ or other metals. Examples of such enzymes are **succinate dehydrogenase** and **xanthine dehydrogenase** catalyzing reactions shown in Fig. 20.5.

Succinate dehydrogenase has four domains: (1) located in the mitochondrial matrix containing the binding site for succinate and is the location of the reduction to fumarate and also where **FAD** is converted to $\textbf{FADH}_2$ (histidine is the covalent link between the flavin and the peptide chain); (2) also located in the mitochondrial matrix; (3) located in the inner mitochondrial membrane where **heme** is located and where **ubiquinone** is reduced; and (4) also in the inner mitochondrial membrane adjacent to C containing heme and the reduction of ubiquinone. **Succinate dehydrogenase** constitutes Complex II of the mitochondrial respiratory transport chain.

**Riboflavin** is degraded when exposed to light, generating **lumichrome** as the product as shown in Fig. 20.6.

Riboflavin can become deficient in newborns treated with phototherapy for **hyperbilirubinemia**. In general, **riboflavin deficiency** causes aversion to light (**photophobia**), inflammation of the mouth, face, and tongue (**glossitis**), excessive oiliness of face and scalp (**seborrhea**), and angular **stomatitis** (fissures and inflammation of the lower lip).

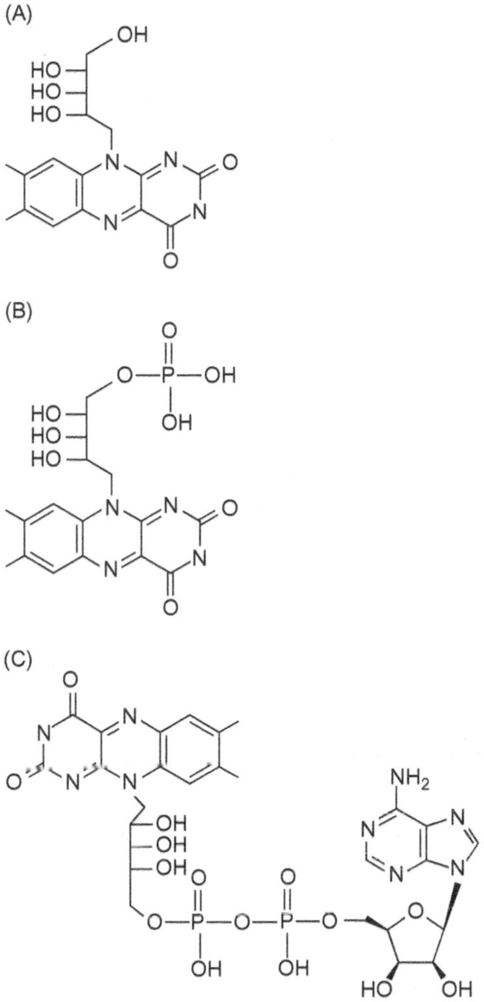

FIGURE 20.3   Structure of riboflavin (A), flavin mononucleotide (B), and flavin adenine dinucleotide (C).

Recommended daily intakes for persons of various ages in mg are as follows: babies (birth−6 months), 0.3 mg; infants (7−12 months), 0.4 mg; children (1−3 years), 0.5 mg; children (4−8 years), 0.6 mg; children (9−13 years), 0.9 mg; boys (14−18 years), 1.3 mg; girls (14−18 years), 1.0 mg; men (19 years plus), 1.3 mg; women (19 years plus), 1.1 mg; pregnant females, 1.4 mg; breastfeeding females, 1.6 mg (data from http://umm.edu/health/medical/altmed/supplement/vitamin-b2-riboflavin). If riboflavin is given as a supplement, the best absorption is between meals.

## Niacin (Vitamin B$_3$)

There are two vitamin forms of niacin: **nicotinic acid** and **nicotinamide** (Fig. 20.7).

Nicotinamide exists as two coenzyme forms: **nicotinamide adenine dinucleotide** (**NAD⁺, NADH**) and **nicotinamide adenine dinucleotide phosphate** (**NADP⁺, NADPH**) whose structures are shown in Fig. 20.8.

Formation of the coenzyme forms of **nicotinic acid** and **nicotinamide** is shown in Fig. 20.9.

There are many enzymes that have NAD⁺ or NADP⁺ as their coenzymes, such as **lactate dehydrogenase** and malate dehydrogenase. Lactate dehydrogenase (**LDH**) is a good example. It catalyzes the freely reversible reaction:

$$\text{lactate} + \text{NAD}^+ \leftrightarrow \text{pyruvate} + \text{NADH} + \text{H}^+$$

as shown in Fig. 20.10.

In Fig. 20.11, the catalytic center of lactate dehydrogenase is pictured. In the enzymatic reaction, the −CH$_3$ group of pyruvate is replaced by −NH$_2$ to form an oxamate (in the reaction proceeding from the left to the right in

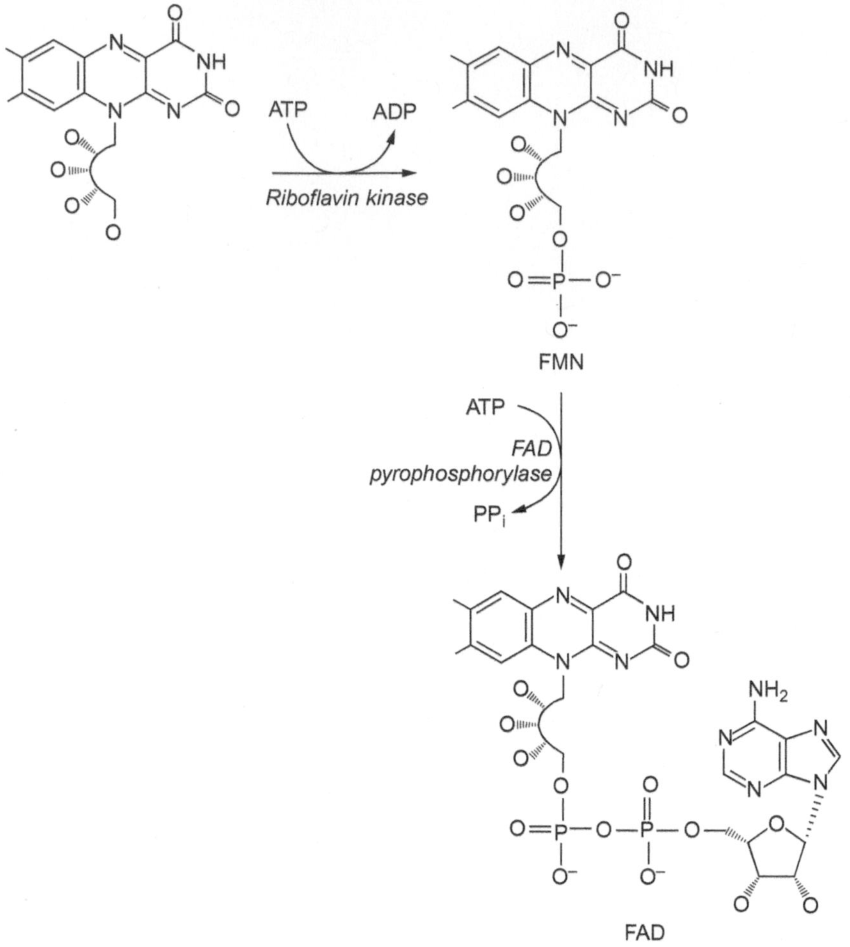

**FIGURE 20.4** Riboflavin is taken into the body in the diet followed by the tissue syntheses of FMN and FAD.

Fig. 20.10). The hydride transfer from NADH takes place on the C2 of pyruvate, and a hydrogen transfer occurs from His 195 of the enzyme protein to the pyruvate C2 oxygen to generate lactate and $NAD^+$.

A small amount of **niacin** can be synthesized from **tryptophan**, but it requires about 60 mg of tryptophan to generate 1 mg of niacin (particularly in the form of NAD) through this highly inefficient process and the niacin generated through this mechanism (Fig. 20.12) cannot satisfy the body's requirements (about 15 mg/day), therefore, sufficient niacin must be obtained in the diet.

Deficiency of niacin in the human (**pellagra**) results in inflammation of the tongue (glossitis), dermatitis, and diarrhea. Niacin deficiency can be the result of the malabsorption of tryptophan in the intestine and kidneys (**Hartnup disease**). Hartnup disease is an autosomal recessive disorder deriving from mutations in the gene for the **$Na^+$-dependent, $Cl^-$-independent neutral amino acid transporter** principally in the apical brush border membrane of the small intestine and the proximal tubule of the kidney. Hartnup disease has the symptoms of niacin deficiency plus cerebellar ataxia and aminoaciduria. The incidence of this disorder in New South Wales is reported to be 1 in 33,000. In **malignant carcinoid syndrome** (carcinoid tumors arise from neuroendocrine cells), there is the synthesis of excess serotonin and niacin deficiency can be a factor. High incidence of carcinoid tumors occurs in the gastrointestinal tract including the small bowel and the appendix. The symptoms include flushing of the face, severe diarrhea, and asthma attacks.

## Pantothenic Acid (Vitamin B$_5$, Pantothenate)

Pantothenic acid is synthesized in microorganisms from **β-alanine** and **pantoic acid**. Pantothenic acid is a component of **Coenzyme A** and also the **acyl carrier protein (ACP)** of **fatty acid synthase** (see Chapter 9: Lipids on lipids for many reactions involving CoA). A great many enzymes (at least 70) require CoA or ACP rendering pantothenate critical in the metabolism of carbohydrates, fats, and proteins and the functioning of the citric acid cycle. The synthesis of

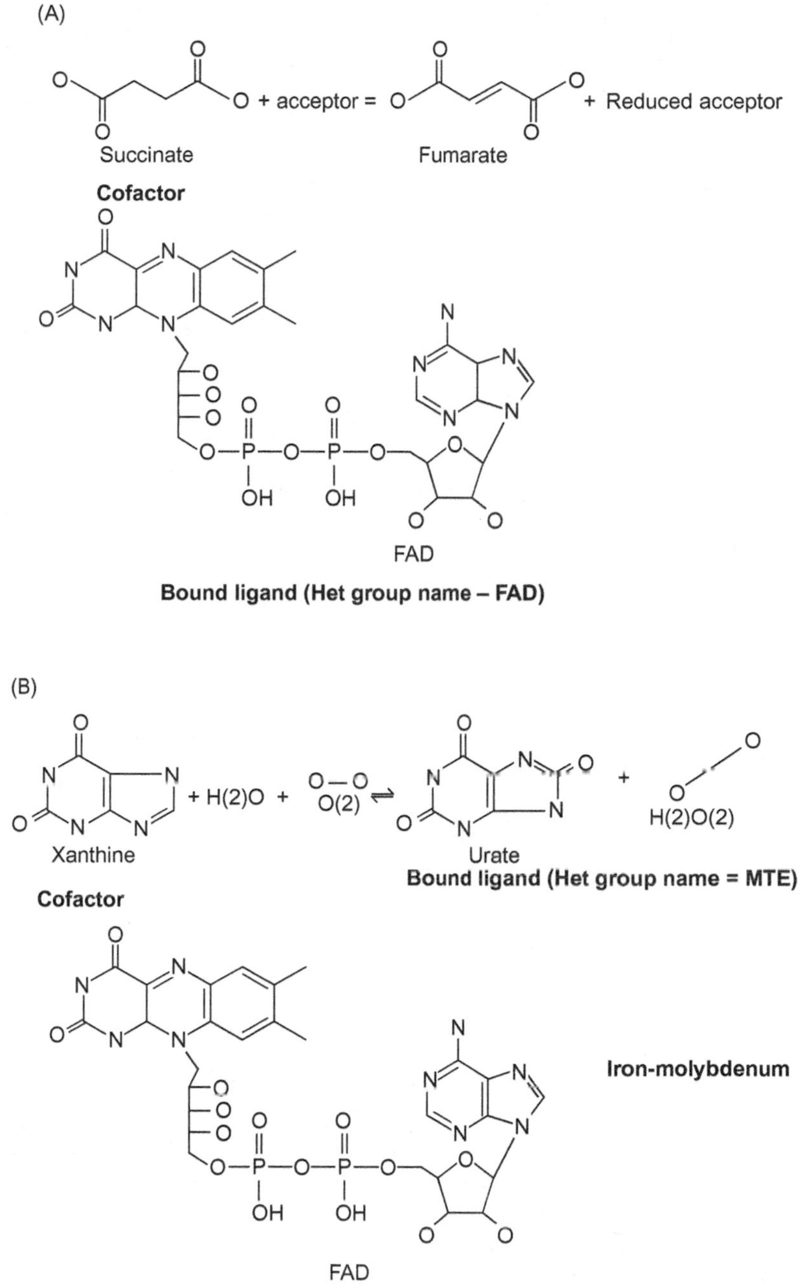

(A)

Succinate + acceptor = Fumarate + Reduced acceptor

**Cofactor**

FAD

**Bound ligand (Het group name – FAD)**

(B)

Xanthine + H(2)O + O(2) ⇌ Urate + H(2)O(2)

**Cofactor**

Bound ligand (Het group name = MTE)

**Iron-molybdenum**

FAD

**FIGURE 20.5** Reactions catalyzed by succinate dehydrogenase (A) and xanthine dehydrogenase (B). FADH$_2$ is a product in these reactions.

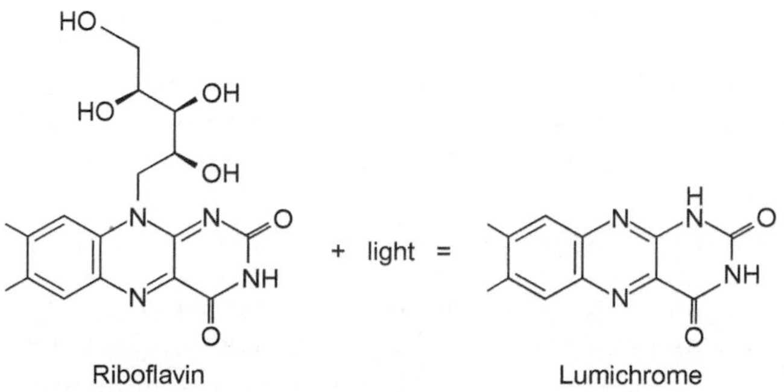

Riboflavin + light = Lumichrome

**FIGURE 20.6** Riboflavin is degraded by visible light to form lumichrome.

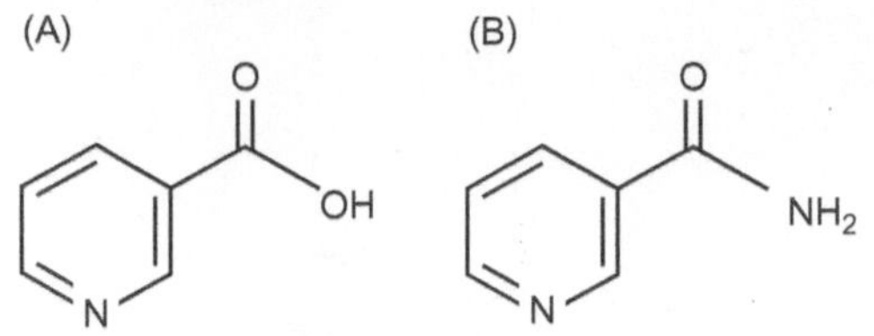

FIGURE 20.7 Structures of nicotinic acid (A) and nicotinamide (B).

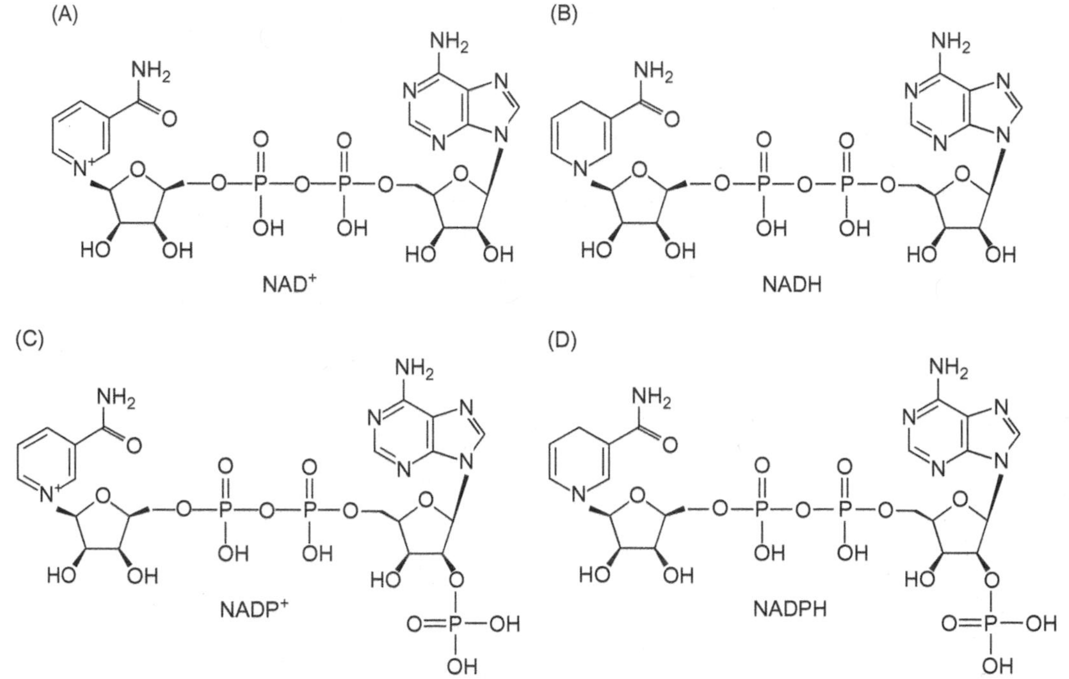

FIGURE 20.8 Structures of NAD$^+$ (A), NADH (B), NADP$^+$ (C), and NADPH (D).

CoA starting with the vitamin, pantothenate (itself synthesized in microorganisms from β-alanine and pantoic acid) is shown in Fig. 20.13.

Adequate intake of pantothenate is 1.7 mg/day in infants to 5 mg/day in adults. Pantothenate is easily obtained in a normal diet. Good sources are avocado, yogurt, cooked chicken, sweet potatoes, cereals, meats and legumes, and many others containing lesser amounts. Pantothenic acid deficiency in humans is rare.

## Pyridoxine, Pyridoxal, and Pyridoxamine (Vitamin B$_6$)

The three forms of the vitamin in the diet are convertible in the body to the coenzyme form, **pyridoxal-5′-phosphate** (**PLP**) as shown in Fig. 20.14.

The kinase for the three vitamin forms phosphorylates them (using ATP) to their respective 5′-phosphate derivatives. **PMP oxidase** converts pyridoxine-5′-phosphate to the coenzyme form, PLP. Pyridoxamine-5′-phosphate is converted by an oxidase and an aminotransferase to **PLP**. The **transaminase mechanism** in which an amino acid is bound to the enzyme and forms an aldimine (**Schiff base**) that is converted to an enzyme ketimine; releases the keto acid product as the coenzyme is converted to pyridoxamine phosphate that becomes the enzyme ketimine, then the aldimine, and finally the original form that binds a second amino acid molecule to start the reaction cycle again (see Fig. 13.6).

The reaction between enzyme and pyridoxal phosphate forming the Schiff base aldimine is shown in Fig. 20.15. This figure also shows the attachment of pyridoxal to pyridoxal kinase through lysine and aspartate residues of the enzyme.

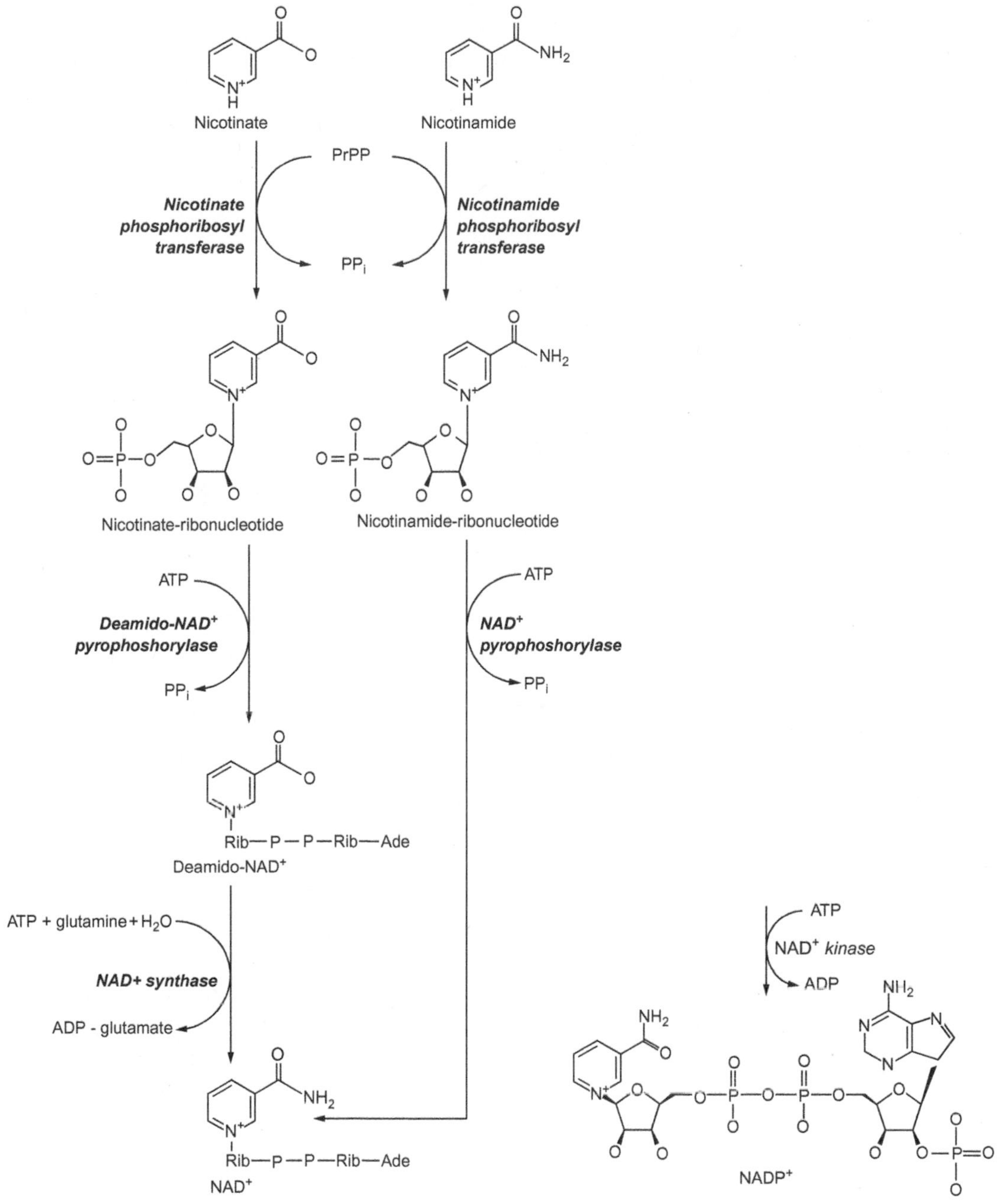

**FIGURE 20.9** Formation of coenzyme forms of nicotinic acid or nicotinamide in human tissues. *Ade*, adenine; *P*, phosphate; *Rib*, ribose.

PLP is a coenzyme for a great many enzymes including aminotransferases, amino acid racemases, amino acid decarboxylases, and others including glycogen phosphorylase.

The daily requirement for **vitamin B$_6$** in adults is in the range of 1.4−2.0 mg/day. A slightly higher amount is required by pregnant or lactating women.

## Biotin (Vitamin H)

Biotin is a cofactor for carboxylase enzymes. There are five human enzymes for which biotin is a coenzyme: **acetyl-CoA carboxylase I** (soluble cytoplasm), **acetyl-CoA carboxylase II** (mitochondrial fatty acid oxidation), **pyruvate**

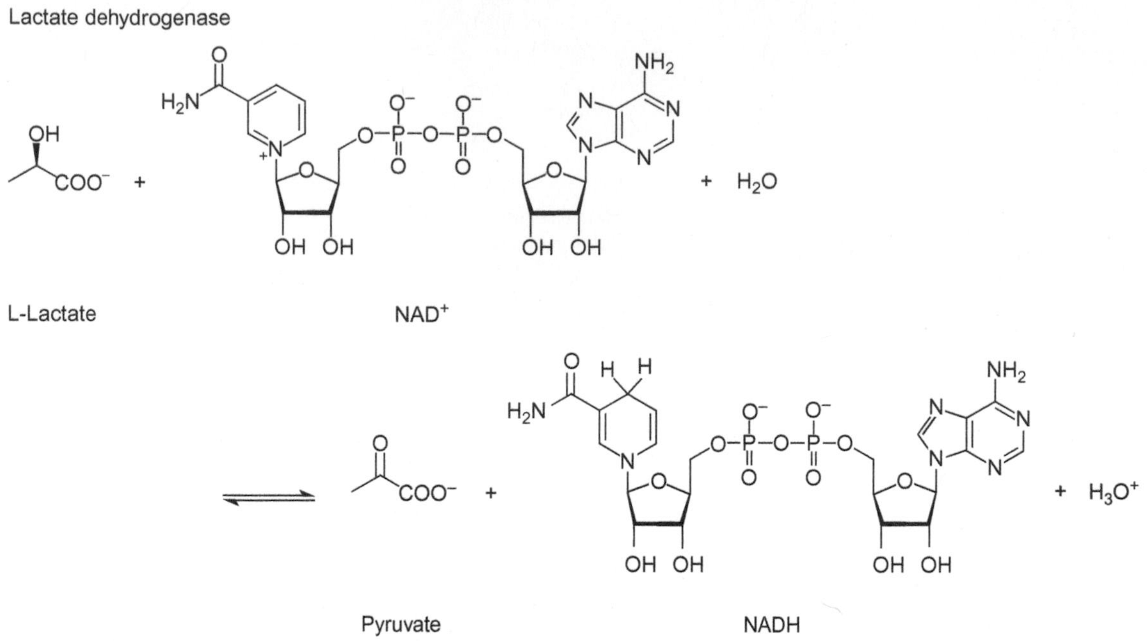

Lactate dehydrogenase

**FIGURE 20.10** Chemical structures of the substrates and products in the lactate dehydrogenase reaction.

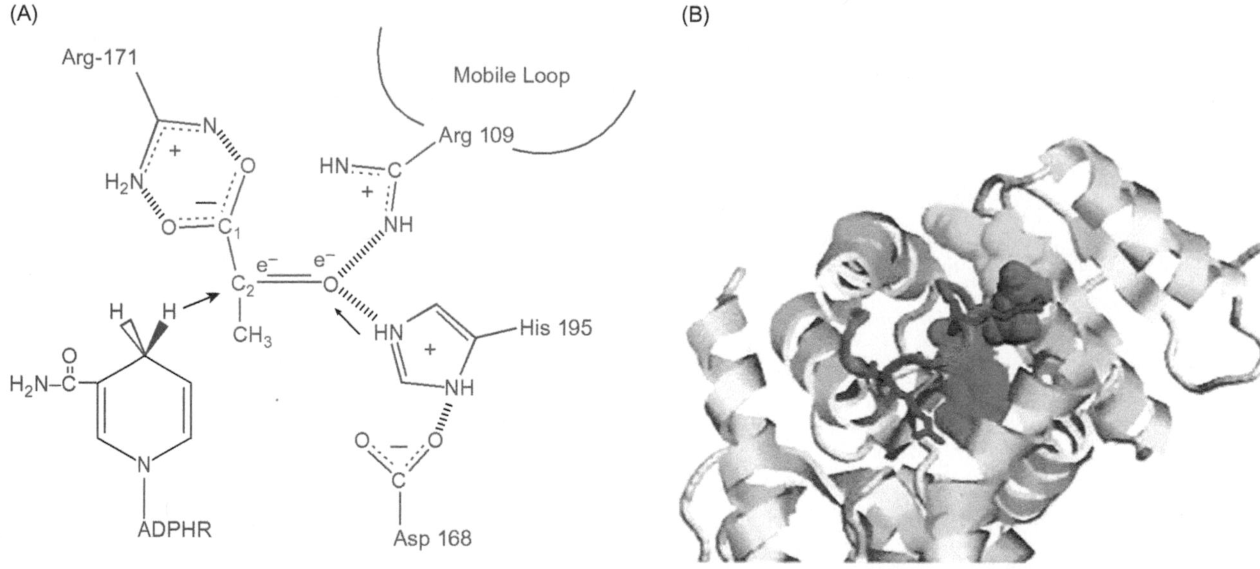

(A)                                    (B)

**FIGURE 20.11** (A) A cartoon of the active site of lactate dehydrogenase showing the relative arrangement of reacting groups. The substrate pyruvate is shown; the $-CH_3$ group is replaced by $-NH_2$ to form oxamate. Hydride transfer is indicated by the *bold arrow*, hydrogen transfer by the *light arrow*. (B) A ribbon diagram of the structure of LDH in the vicinity of the active site. The protein is shown in *gray* with the "mobile" loop shown in *purple* and its two Arg residues shown in *red*; the loop closes over the binding pocket after both NADH and substrate are bound and Arg 109 of the mobile loop comes in contact with bound substrate. NADH is shown with its adenosine (*yellow*), pyrophosphate (*blue*), and nicotinamide (*green*) moieties. Bound substrate is not shown but it is located at the end of the nicotinamide moiety of NADH. *Reproduced from http://www.bioc.aecom.yu.edu/labs/calllab/highlights/LDH.htm.*

**carboxylase**, **methylcrotonyl-CoA carboxylase**, and **propionyl-CoA carboxylase**. Acetyl-CoA carboxylase II catalyzes the reaction:

$$\text{bicarbonate } + \text{ acetyl-CoA} \rightarrow \text{malonyl-CoA}$$

a controlling reaction in the synthesis of fatty acids. Acetyl-CoA carboxylase operates in the synthesis of lipids from acetate. Pyruvate carboxylase is involved in liver gluconeogenesis, particularly from amino acids. Methylcrotonyl-CoA

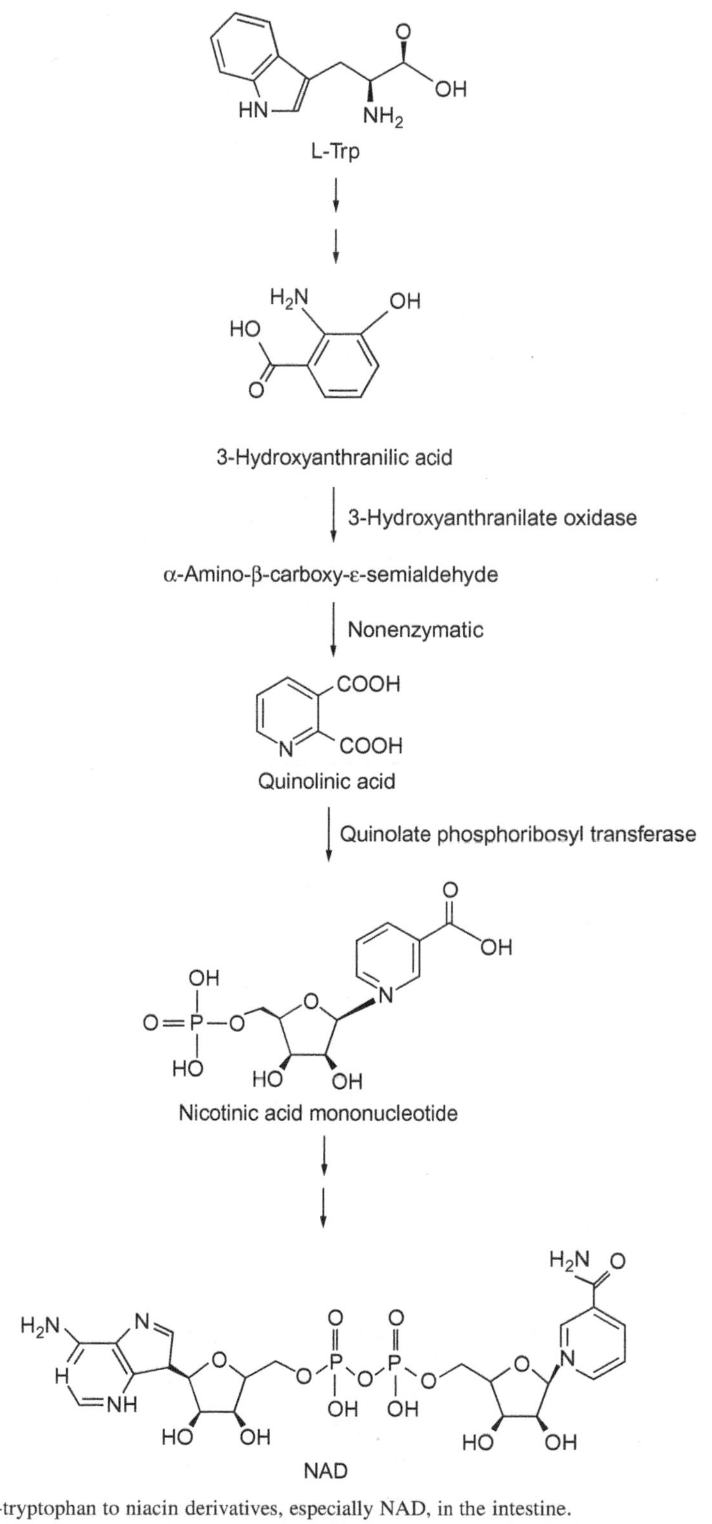

**FIGURE 20.12** Conversion of L-tryptophan to niacin derivatives, especially NAD, in the intestine.

carboxylase is involved in the catabolism of the essential amino acid, leucine. Proprionyl-CoA carboxylase plays a role in the metabolism of certain amino acids, cholesterol, and odd chain fatty acids.

Taking pyruvate carboxylase as an example of this class of enzymes, it catalyzes the reaction:

$$\text{pyruvate} + \text{ATP} + \text{bicarbonate} \rightarrow \text{oxaloacetate} + \text{ADP} + \text{inorganic phosphate}$$

shown in Fig. 20.16.

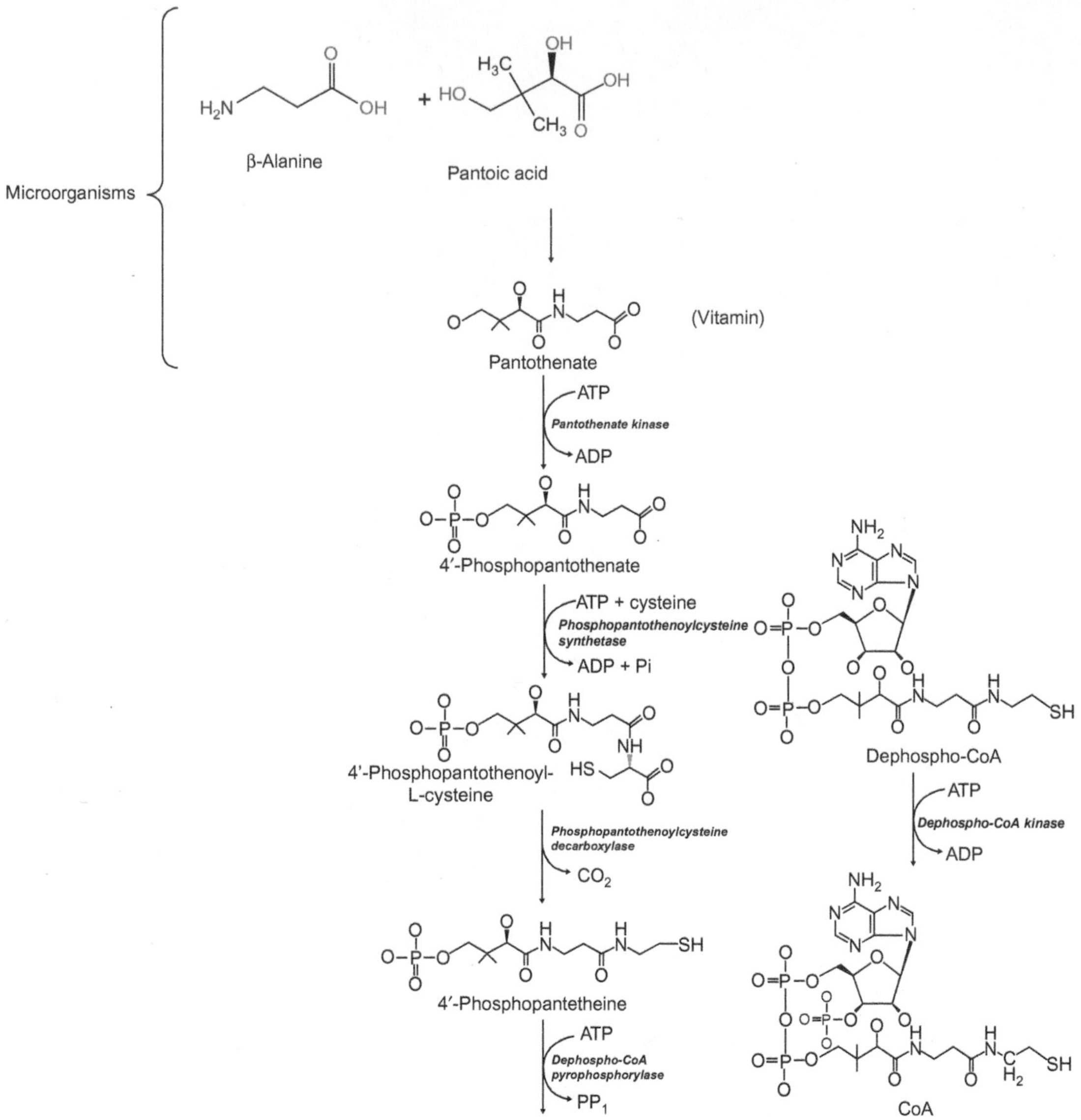

FIGURE 20.13 Synthesis of coenzyme A from the vitamin, pantothenate.

The biotin coenzyme reacts with the enzyme in the first part of the reaction; then it swivels over to the second part of the reaction. Biotin is the carrier of carbon dioxide, and biotin is linked to the enzyme via an ε-amino group of lysine (Fig. 20.17).

When carboxylase holoenzymes are degraded, the biotin is cleaved from the ε-amino group of the active site of the enzyme by another enzyme, **biotinidase**, causing the release of free biotin. The residue of biotin still bound to the enzyme (Fig. 20.17, *middle*) is called **biocytin**. The active site of **pyruvate carboxylase** is pictured diagrammatically in Fig. 20.18.

**Biotin** also can be bound to histones, and the **biotinylated histones** play a role in the regulation of DNA synthesis, transcription, and cell proliferation. The free biotin released by biotinidase from carboxylases or biotinylated histones is available for binding to other apocarboxylases as shown for the "**biotin bicycle**" in Fig. 20.19.

Biotin binds to other proteins and **avidin**, in particular, found in egg white is a strong binder (cooking inactivates the interaction). Other biotin binding proteins are **strepavidin**, **homocitrate synthase,** and **isopropylmalate synthase**. Good sources of dietary biotin are liver, soy flour, egg yolk, cereal, and yeast. Biotin deficiency is rare since intestinal bacteria synthesize biotin. In general, normal biotin intake varies from 5 μg/day in the infant to 30 μg/day in the adult.

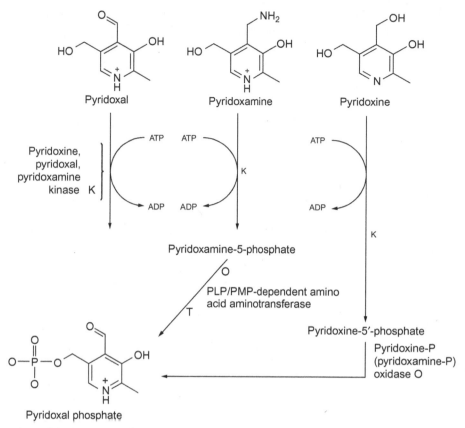

**FIGURE 20.14** Conversions of the three forms of vitamin $B_6$ into the coenzyme form, pyridoxal-5'-phosphate. These reactions occur primarily in the liver. $K$, kinase for the three vitamin forms; $O$, pyridoxine phosphate/pyridoxamine phosphate oxidase; $T$, PLP/PMP-dependent amino acid aminotransferase; $P$, phosphate; $PMP$, pyridoxamine phosphate.

(A)

Free PLP
($A_{max}$ = 388 nm)

Carbinolamine
intermediate
($A_{max}$ ~ 336 nm)

Protonated PLP aldimine
(Schiff base)
($A_{max}$ ~ 420 nm)

(B)

Loop II

Asp233

Lys229

PL

Loop I

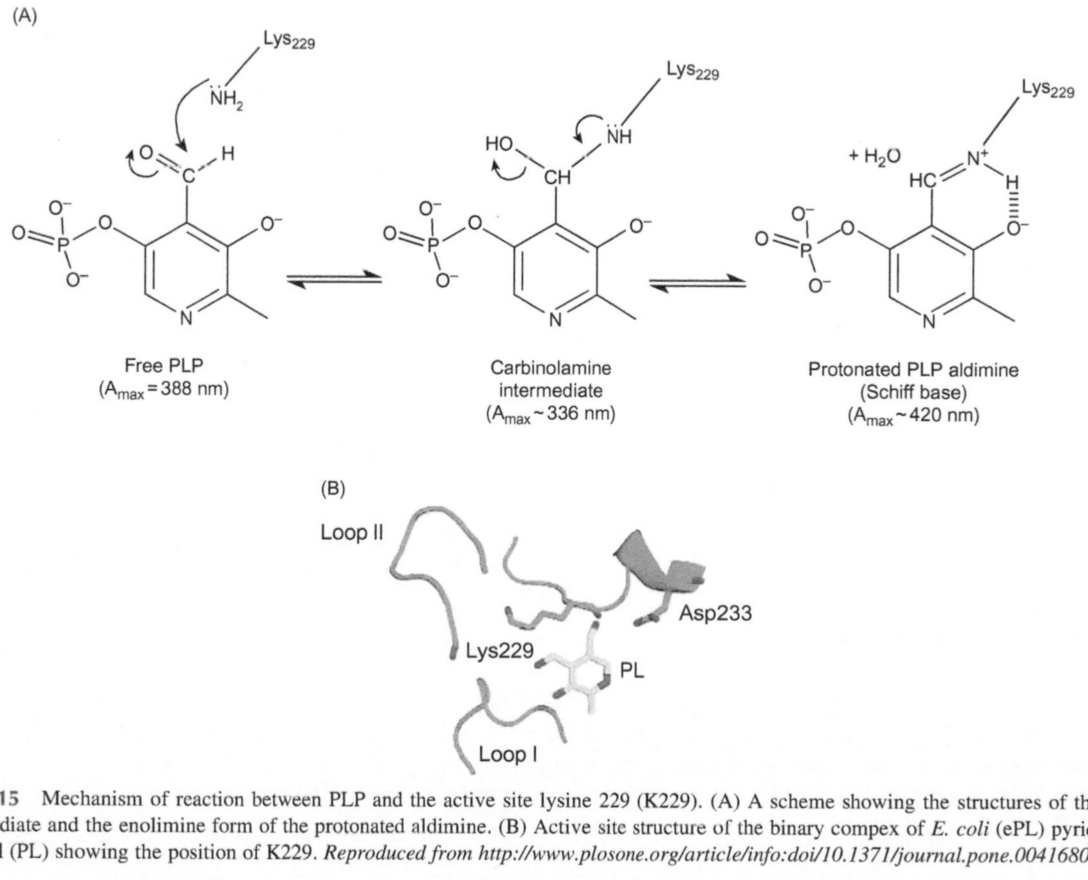

**FIGURE 20.15** Mechanism of reaction between PLP and the active site lysine 229 (K229). (A) A scheme showing the structures of the carbinolamine intermediate and the enolimine form of the protonated aldimine. (B) Active site structure of the binary compex of *E. coli* (ePL) pyridoxal kinase and pyridoxal (PL) showing the position of K229. *Reproduced from http://www.plosone.org/article/info:doi/10.1371/journal.pone.0041680.*

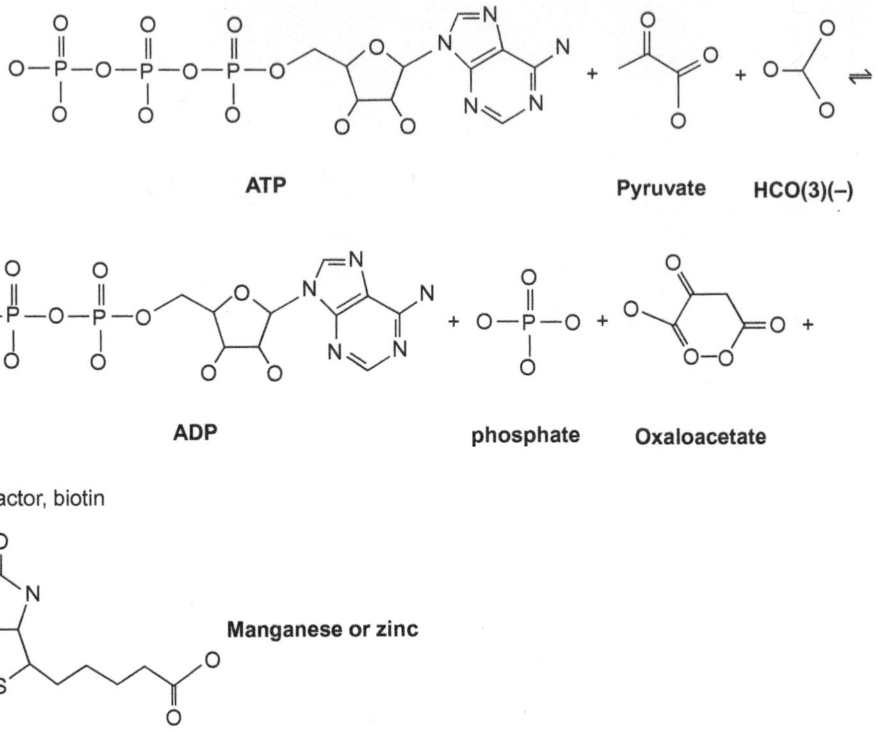

**FIGURE 20.16** The reaction catalyzed by pyruvate carboxylase.

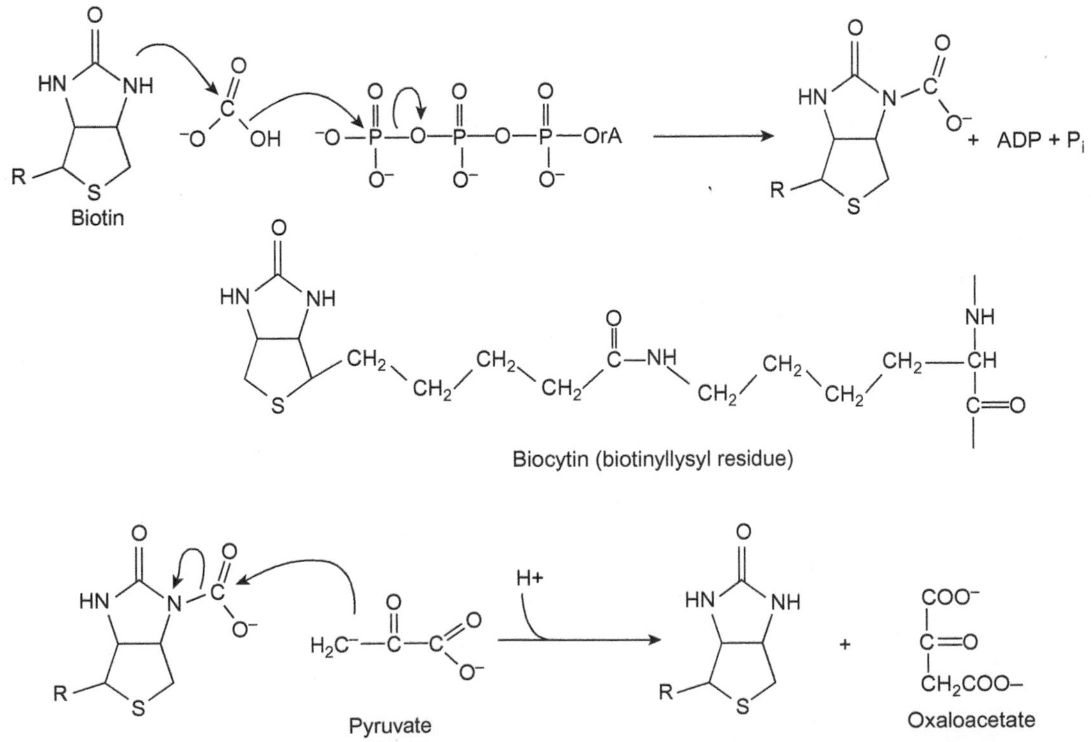

**FIGURE 20.17** The pyruvate carboxylase reaction (pyruvate + $CO_2$ + ATP → oxaloacetate + ADP + $P_i$) in which biotin is coenzyme, and $Mg^{2+}$ is cofactor. Carboxylases are synthesized as apoenzymes without the coenzyme, biotin. The active form (holoenzyme) appears when biotin binds to an ε-amino group of a lysine residue in the apoenzyme. The reaction (above) is catalyzed by the holoenzyme.

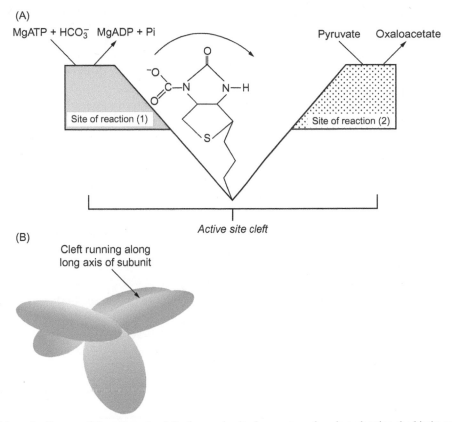

(A)

MgATP + $HCO_3^-$   MgADP + Pi

Site of reaction (1)

Pyruvate    Oxaloacetate

Site of reaction (2)

*Active site cleft*

(B)

Cleft running along
long axis of subunit

**FIGURE 20.18** (A) Schematic diagram of the active site cleft of one subunit of pyruvate carboxylase showing the biotin prosthetic group acting to carry the carboxyl group from the site of reaction (1) to reaction site (2). (B) Diagram showing the tetrahedron-like arrangement of subunits of pyruvate carboxylase and illustrating the cleft that runs along the long axis of each subunit and which has been proposed to be the location of the active site.

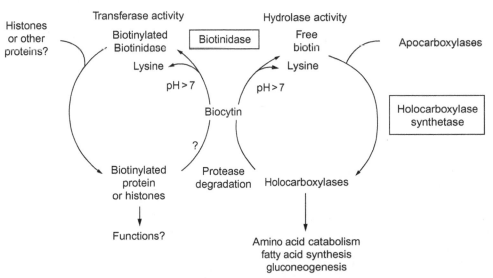

Transferase activity          Hydrolase activity

Histones
or other
proteins?

Biotinylated
Biotinidase

Biotinidase

Free
biotin

Apocarboxylases

Lysine          Lysine

pH > 7          pH > 7

Biocytin

Holocarboxylase
synthetase

?

Biotinylated
protein
or histones

Protease
degradation

Holocarboxylases

Functions?

Amino acid catabolism
fatty acid synthesis
gluconeogenesis

**FIGURE 20.19** The biotin bicycle, which includes biotinyl-hydrolase activity and biotinyl-transferase activity. *Redrawn from Figure 1 of B. Wolf, "Biotinidase: its role in biotinidase deficiency and biotin metabolism"*, J. Nutr. Biochem., *16: 441–445, 2005.*

## Cobalamin (Vitamin B$_{12}$)

Vitamin B$_{12}$ has a complex ring structure (tetrapyrrole) and an atom of **cobalt** in the center as shown in Fig. 20.20.

**Vitamin B$_{12}$** is synthesized by microorganisms. In ingested meat, the vitamin is bound to proteins and is released from the bound form by hydrolysis by stomach acid or by the action of trypsin in the intestine. A protein secreted by **parietal cells** of the stomach is the **intrinsic factor** that binds **cobalamin**. The **cobalamin-intrinsic factor complex** is absorbed in the ileum after attaching to **cubilin**, a receptor for the cobalamin-intrinsic factor complex. Cubilin facilitates the absorption of the complex. Mutation of the gene for cubilin results in impaired absorption of the cobalamin-intrinsic factor complex generating the rare **megaloblastic anemia** (formation of red blood cells requires vitamin B$_{12}$, folic acid and iron; a deficiency in any one of these, causes anemia). **Pernicious anemia,** a form of megaloblastic anemia is an autoimmune disorder in which parietal cells of the stomach are destroyed (by autoantibodies) so that there is insufficient intrinsic factor available to facilitate absorption of vitamin B$_{12}$. Normally, vitamin B$_{12}$ is bound to the transporting protein, **transcobalamin II**, in the intestinal cells, and the vitamin B$_{12}$-transcobalamin II complex is carried in the bloodstream to tissue cells and to the liver where it is stored as a complex of **vitamin B$_{12}$-transcobalamin III** (transcobalamin III is also present in blood). In the liver, the stored form of vitamin B$_{12}$ can be released into the small intestine through the bile; then vitamin B$_{12}$ is released from the complex, becomes bound to intrinsic factor and cycles again through the intrinsic factor-vitamin B$_{12}$-cubilin complex, and eventually reaches the bloodstream bound to the transporting protein.

The **5′-deoxyadenosine derivative** (Fig. 20.21) of cobalamin is the coenzyme for **methylmalonyl-CoA mutase** that catalyzes the conversion of methylmalonyl-CoA to succinyl-CoA in the system catabolizing fatty acids.

This form of cobalamin is involved in **methionine synthase** that catalyzes the conversion of homocysteine to methionine. In this reaction, a methyl group is transferred from N$^5$-methyltetrahydrofolate to hydroxycobalamin. When cobalamin is deficient because the intrinsic factor is absent or inactive, folic acid becomes trapped as the N$^5$-methyltetrahydrofolate due to the inactivity of methionine synthase. Also, other tetrahydrofolate derivatives cannot be formed, and these are required for the synthesis of purines and the pyrimidine, **thymidine**. Additionally, there is an

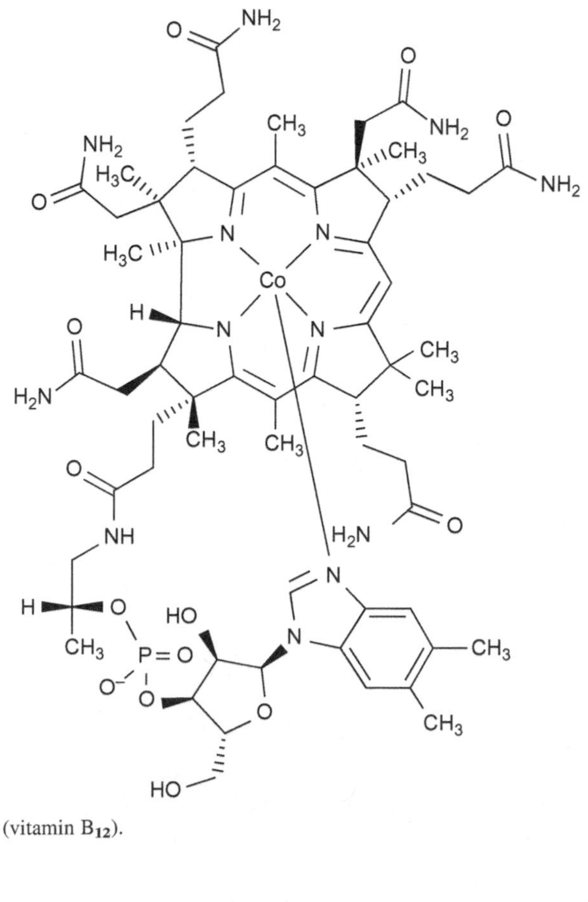

**FIGURE 20.20** Structure of cobalamin (vitamin B$_{12}$).

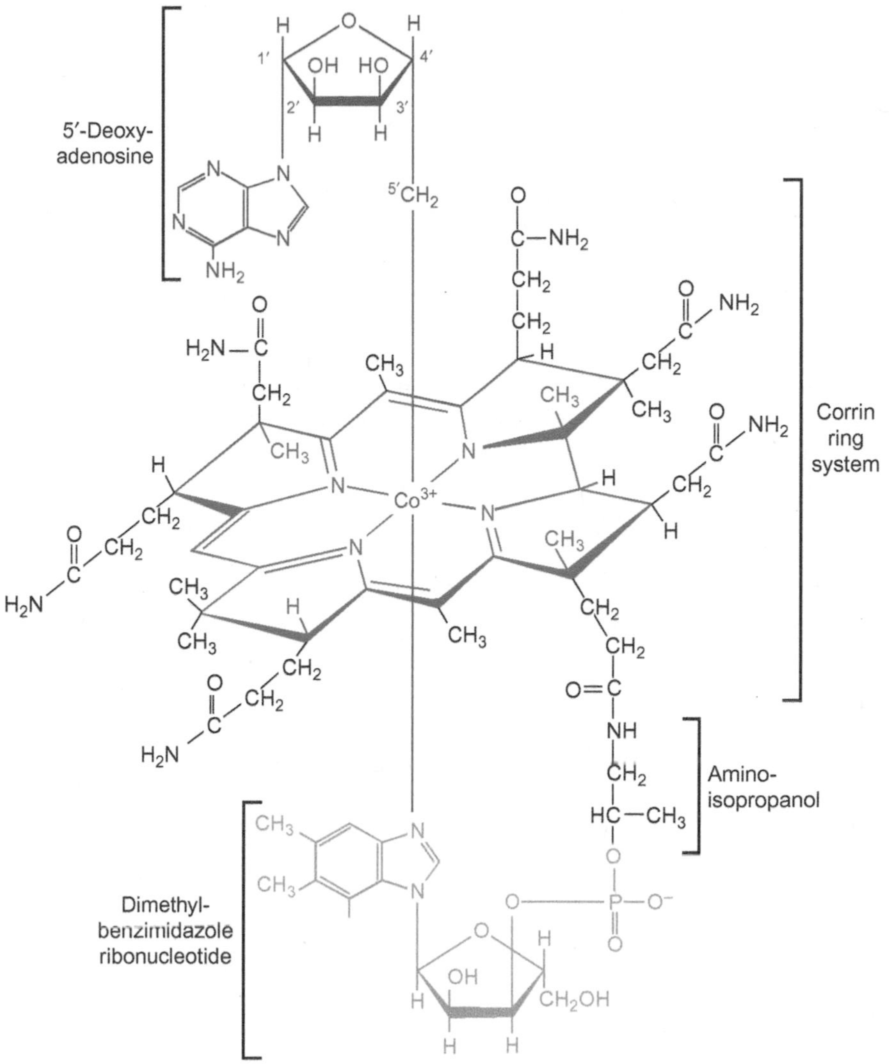

**FIGURE 20.21** Structure of the 5′-deoxyadenosine derivative of cobalamin. *From http://www.bioinfo.org.cn/book/biochemistry/chapt16/494.jpg.*

increase in methylmalonyl-CoA resulting in the inhibition of fatty acid synthesis essential for the turnover of the **myelin sheath** (of neurons), and this results in **progressive demyelination**.

## Folic Acid (Vitamin B₉, Pteroylglutamic Acid)

The structure of folic acid contains **6-methylpteridine** linked to **para-aminobenzoic acid (PABA)** linked to a **glutamate** residue as shown in Fig. 20.22. The human cannot synthesize PABA.

Folic acid in the diet (folic acid is synthesized in plants and yeasts) can occur as mixtures of **polyglutamates** containing as many as seven glutamic acid residues (Fig. 20.23).

The polyglutamates are hydrolyzed by peptidases in the intestine to the natural form of folate with a single or a reduced number of glutamate residues, and these glutamate-degraded forms are recognized by the **folic acid receptor (RFC1 transporter)** in the intestine. The RFC1 transporter is structurally related to the **thiamine transporter**, since both derive from the same gene family; the gene-encoding information for the folate transporter (RFC1) is **SLC19A1**, while the gene-encoding information for the thiamine transporter is **SLC19A2**. Removal of conjugated glutamate residues reduces the negative charges on the folate molecule making it more easily passed through the basolateral membrane of the intestinal cell destined for the bloodstream. There are other receptors in some tissue cells, such as a **proton-coupled folate transporter (PCFT)** and **folate receptor 1(FOLR1)**. PCFT is an intestinal proton-coupled folate transporter that also transports **heme** from the gut lumen into duodenal epithelial cells. Subsequently, iron is

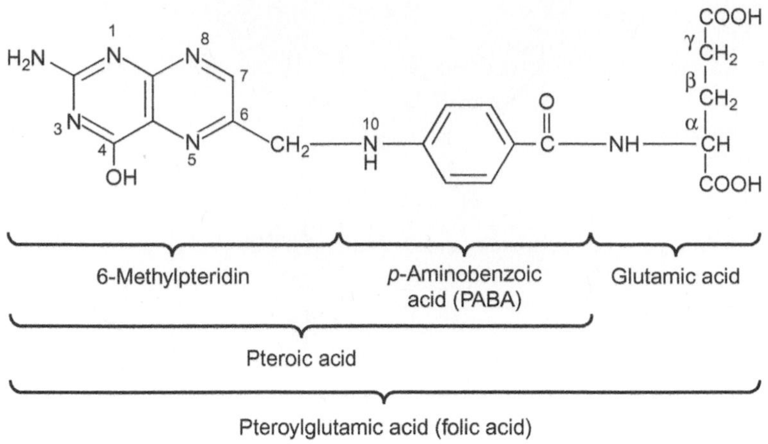

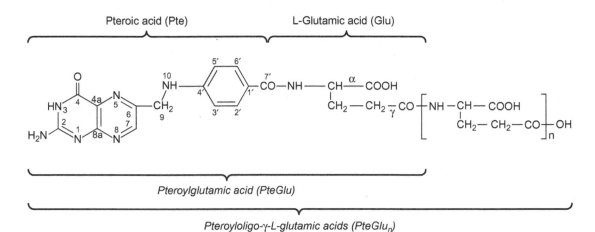

**FIGURE 20.22** Structure of folic acid.

**FIGURE 20.23** Structure of **Pteroyl-γ-L-glutamic acids** where $n$ can be as many as six glutamate residues (Pteroyloligo-γ-L-glutamic acids).

released from heme and moved into the bloodstream. FOLR1 binds folic acid and its reduced derivatives and is the transporter for **5-methyltetrahydrofolate** into cells.

In the hepatocyte, folic acid is reduced to **tetrahydrofolate** (**THF**) by **dihydrofolate reductase** (**DHFR**). During biosynthetic reactions, THF derivatives transfer one-carbon units (methyl, methylene, formyl, or formimino groups; see Fig. 20.24).

There are **one-carbon transfer reactions** in the biosyntheses of serine, methionine, glycine, choline, purine nucleotides, and deoxythymidylate monophosphate (dTMP). *The role of $N^5,N^{10}$-methylene-THF in the regeneration of dTMP may be the most important **role of folate**.* The formation of methionine from homocysteine catalyzed by **methionine synthase** also involves cobalamin with $N^5,N^{10}$-**methylene-THF**. Methionine synthase catalyzes the conversion of homocysteine to methionine where methyltetrahydrofolate is the coenzyme. Fig. 20.25 shows the active center of the enzyme and how the methyltetrahydrofolate is positioned in the Fol barrel of the enzyme structure.

**Folate deficiency** (often occurring in alcoholics with inadequate diets) diminishes **dTMP synthesis** that can lead to cell cycle arrest in the S-phase of rapidly dividing hemopoietic cells. Folate deficiency also can occur when there is inadequate absorption of the vitamin. Anticoagulants and oral contraceptives interfere with folate absorption. As mentioned before, folate deficiency can lead to **megaloblastic anemia**, identical to that induced by deficiency of vitamin $B_{12}$. The recommended daily intake of folic acid in the adult is 400 μg/day. Fortified cereals, orange juice, spinach, asparagus, lentils, and garbanzo beans are good sources of folic acid.

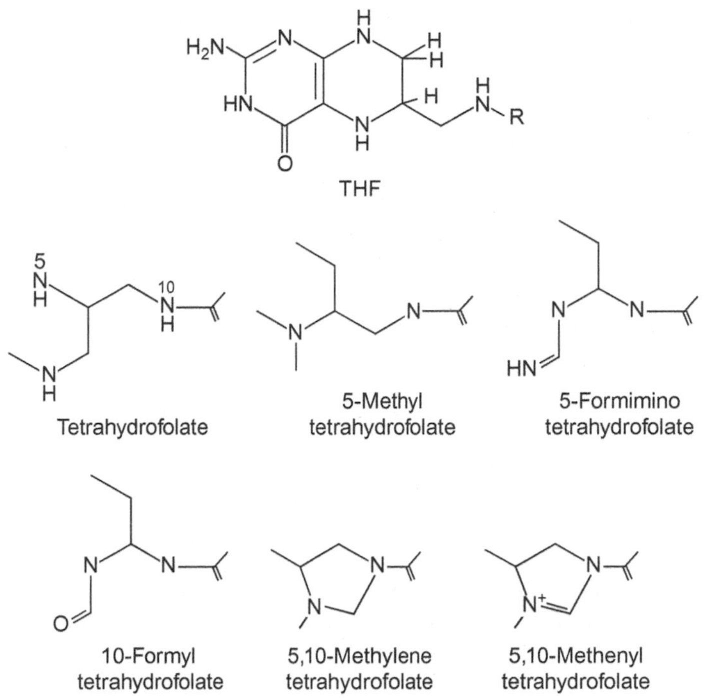

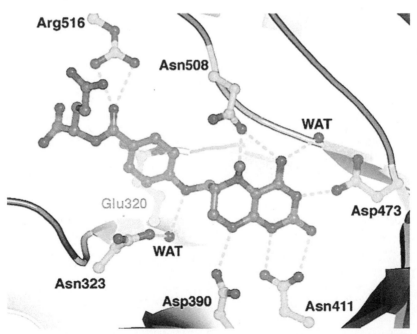

THF

5-Methyl
tetrahydrofolate

5-Formimino
tetrahydrofolate

Tetrahydrofolate

10-Formyl
tetrahydrofolate

5,10-Methylene
tetrahydrofolate

5,10-Methenyl
tetrahydrofolate

**FIGURE 20.24** Structure of tetrahydrofolate (top) and some fractional one-carbon derivatives.

**FIGURE 20.25** The **active site of methionine synthase** showing methyltetrahydrofolate and its interactions with conserved residues from the Fol barrel of the enzyme. The extended arrangement of the folate is determined by the interactions of the **para-aminobenzoic acid (PABA)** side chain: the ring stacks with Glu320, N10 makes a through-water interaction with the conserved Asn323 and Arg516 binds the PABA carbonyl oxygen. In the saturated pyrazine ring of the pterin, the C6 side chain substituent is axial, and the N5-methyl group is equatorial, pointing toward the reader. Hydrogen bonds are indicated for donor−acceptor pairs that are closer than 3.2A˚. The side chain amine of Asn508 may reorient in the ternary complex to allow the oxygen to interact with a protonated pterin N5. The viewpoint is approximately along the expected approach of the corrin ring (of cobalamin). *WAT*, water. *Reproduced from J.C. Evans et al., PNAS, **101**: 3729−3736, 2004.*

## Ascorbic Acid (Vitamin C)

Ascorbic acid is an important **antioxidant** that reduces other compounds (e.g., reactive oxygen species) and is converted to its oxidized form, **dehydroascorbic acid (DHA)**, in the process. Humans are unable to synthesize ascorbic acid, although other species are able to do so. The loss of the ability to synthesize ascorbic acid seems to parallel the inability of humans and certain other species to degrade **uric acid** (loss of the enzyme uricase through mutation) that is also a strong reducing agent. Inability to synthesize ascorbate is due to a defective pseudogene ($\psi GULO$) that encodes information for the enzyme, L-**gulonolactone oxidase** which is the final enzyme required for the synthesis of ascorbate. In the human and in other species that have lost the ability to synthesize ascorbate, there is a compensatory vitamin C recycling mechanism. In this system, after ascorbic acid is used to reduce some oxidized substance and ascorbic acid is, itself, oxidized to dehydroascorbic acid, DHA can again be reduced back to ascorbic acid, either by an **NADH** system or by the action of **glutathione (GSH)** as suggested by Fig. 20.26. This is of particular importance in the red blood cell.

Loss of L-gulonolactone oxidase and uricase through ancient mutations probably benefitted early primates by conferring a survival advantage in maintaining blood pressure when dietary and environmental changes were occurring. However, in modern society having moved to the Western diet and relative physical inactivity, these mutations increase the risk for hypertension and cardiovascular disease.

Ascorbic acid can be imported into cells through a **Na$^+$/ascorbate symporter**, although it is more likely imported as the oxidized form (DHA) down its concentration gradient (DHA is more concentrated outside the cell) through **facilitative glucose transporters**, such as **GLUT 1**. The blood−brain barrier (BBB) is unable to pass ascorbic acid but does transport DHA by way of GLUT 1, and the transported DHA is subsequently reduced back to ascorbic acid by GSH.

One of the main functions of ascorbic acid is in the hydroxylation of proline in collagen, where ascorbic acid is a cofactor for the hydroxylation reaction (Fig. 20.27).

This hydroxylation is required for aggregation of the collagen molecule into a triple helix (Fig. 20.28).

**Ascorbate**, then, is essential for the maintenance of connective tissue and also for wound healing where the synthesis of connective tissue is an early step of the process. Ascorbate is also necessary for bone remodeling because collagen is in the organic matrix of bone. In addition, ascorbate is involved in the synthesis of **epinephrine** from tyrosine (Fig. 20.29).

Ascorbate is needed in severe stress when the adrenal store is rapidly depleted of ascorbate, and consequently it plays a role in the formation of cortisol. Vitamin C also is involved in signal transduction of NFκB. DHA transported into the cell through a glucose transporter inhibits IκBα kinase (IKKβ). Reactive oxygen species (ROS) that are intermediates in the activation of NFκB are quenched by ascorbate; ascorbate becomes oxidized to DHA that inhibits IKKβ and IKKα enzymatic activities. In this way, ascorbate depresses NFkB signaling (Fig. 20.30).

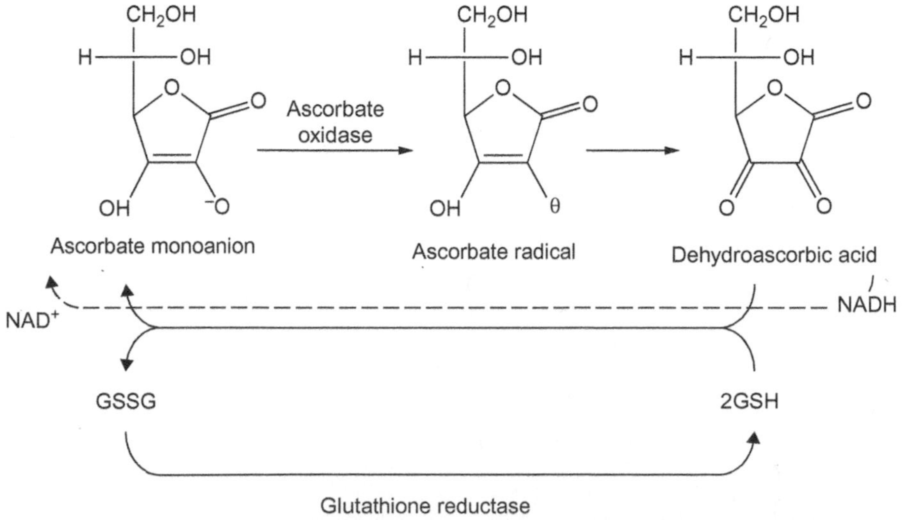

**FIGURE 20.26**   The structures of ascorbic acid and its oxidation product. Ascorbic acid can be oxidized by ascorbic acid oxidase or by its action on other oxidized compounds effecting their reduction. The structures of ascorbic acid (*left*) and its oxidation product, dehydroascorbic acid (DHA, *right*), are shown with the intermediate form. In cells, DHA is recycled back to ascorbic acid by GSH or by an NADH system.

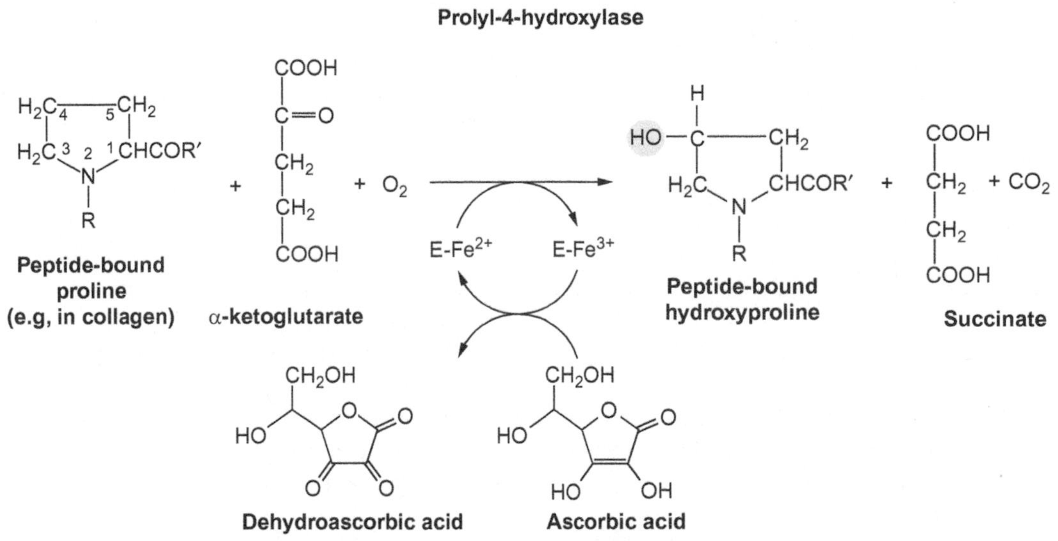

**Prolyl-4-hydroxylase**

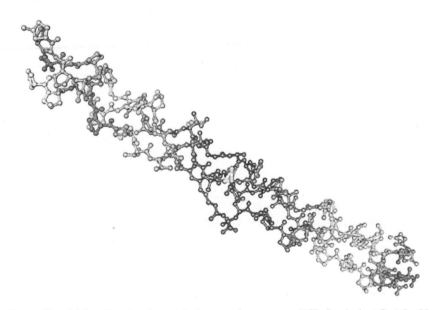

**FIGURE 20.27** Prolyl-4-hydroxylase reaction with substrate prolyl residue in a peptide. The hydroxylation of peptide-bound proline is coupled to the oxidative decarboxylation of $\alpha$-ketoglutarate and enzyme-bound $Fe^{2+}$ is rapidly converted to $Fe^{3+}$. Enzyme-bound $Fe^{3+}$ can be reduced again by ascorbate to reactivate the enzyme. *E*, enzyme; *R*, amino acid residues.

**FIGURE 20.28** **Tropocollagen fiber triplex.** The strands contain the repeating sequence [(Gly-Pro-hydroxyPro)$_n$] with occasional appearance of other amino acids, such as histidine.

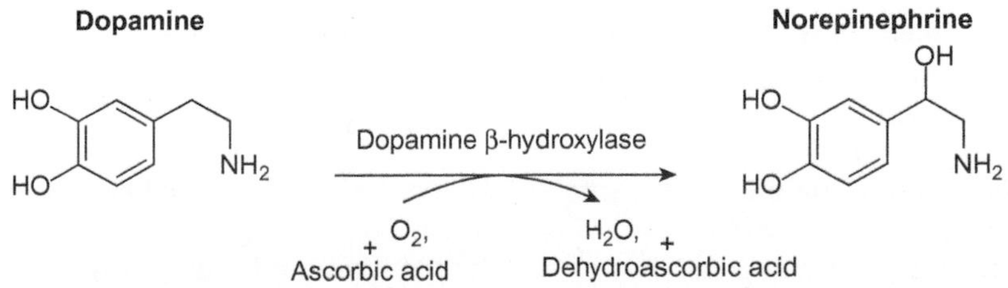

**FIGURE 20.29** Ascorbic acid is a coenzyme for **dopamine β-hydroxylase** that catalyzes the conversion of dopamine to norepinephrine.

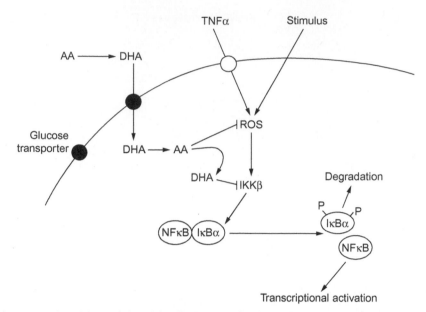

**FIGURE 20.30** Schematic representation of the regulation of signaling responses by vitamin C. Vitamin C, as DHA, enters the cell through glucose transporters and is rapidly reduced to ascorbic acid (AA) by GSH or other reducing systems (e.g., NADH). ROS induces NFκB signaling responses by activating IKKβ and AA quenches ROS, inhibiting the activation of IKKβ. Throughout these processes, AA becomes oxidized to DHA and DHA inhibits IKKβ. *AA*, Ascorbic acid; *DHA*, dehydroascorbic acid; *IκBα*, I kappa B alpha; *IKKβ*, I kappa B kinase; *NFκB*, nuclear factor kappa B; *ROS*, reactive oxygen species.

Ascorbate can be degraded to DHA to diketogulonic acid and then to other acids, such as oxalic acid.

The adult male requires about 90 mg ascorbate per day, and the adult female requires about 75 mg ascorbate per day. Excellent sources of ascorbic acid are as follows: guava, red bell pepper, papaya, and concentrated orange juice.

## FAT-SOLUBLE VITAMINS

### Vitamin A (β-Carotene)

This vitamin is found primarily in plant food, especially in carrots. **β-Carotene** is the ingested provitamin, and it is processed in the body to its biologically active form, **retinoic acid**(s) (there are two biologically active retinoic acids, and there are other metabolites), the ligands for retinoic acid receptors. β-Carotene is made up of two retinal molecules linked head to tail. In the intestine, it is cleaved by **β-carotene dioxygenase** to yield two retinal molecules (Fig. 20.31).

Actually, the biosynthesis of 9-*cis*-retinoic acid is not so well understood. It is likely that for the formation of both retinoic acids, both **retinol dehydrogenase** and **retinal dehydrogenase** are required. Retinol dehydrogenase activity takes place in the endoplasmic reticulum:

retinol (retinol dehydrogenase) → retinal (retinal dehydrogenase) → retinoic acid.
The possible scenario for the synthesis of 9-*cis*-retinoic acid is:
9-*cis*-retinol (9-cis-retinol dehydrogenase) → 9-*cis*-retinaldehyde (oxidase) → 9-*cis*-retinoic acid.

Retinoic acid is the ligand for the retinoic acid receptor, RAR, that heterodimerizes with RXR of the **nuclear receptor superfamily** and also interacts with coactivator and corepressor proteins. On the other hand, 9-cis-retinoic acid is the ligand for the RXR receptor, and the RXR forms many **heterodimers** with other receptors of the nuclear receptor gene family and also interacts with coactivator and corepressor proteins. An overview of the metabolism of β-carotene is presented in Fig. 20.32.

A derivative of retinol, **retinol palmitate**, is incorporated into **chylomicrons** and enters the bloodstream. The liver absorbs chylomicron remnants where retinol is stored as an ester. For transport out of the liver to other tissues, retinol ester is hydrolyzed to free retinol. Retinol can then bind to the **liver retinol-binding protein** that can be transported to the cell surface for secretion. The complex travels to other tissues where free **retinol** enters the cell and becomes bound to a **cellular retinol-binding protein (CRBP)**. On the other hand, **retinoic acid** transport is facilitated by its binding to **albumin**. As mentioned before, retinoic acid and 9-cis-retinoic acid act like hormones as they are ligands for RAR and

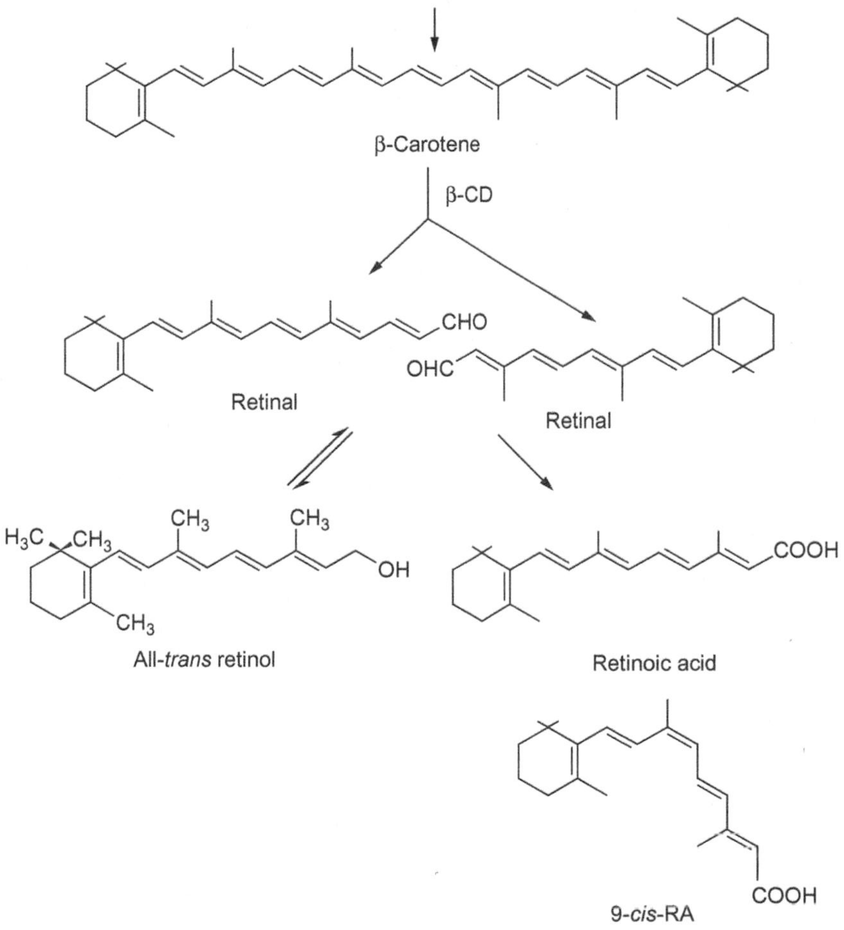

**FIGURE 20.31** Conversion of β-carotene to retinal by β-carotene dioxygenase (β-CD). Retinal can be converted to **all-*trans*-retinoic acid** or to **9-*cis*-retinoic acid** (**RA**) and other metabolites. *All trans-RA is the ligand for the retinoic acid receptor (RAR), and 9-cis-RA is the ligand for the retinoic X receptor (RXR).*

RXR that form heterodimeric receptors (in particular RXR) with other receptors, such as TR (RXR-TR), VDR (RXR-VDR), and RAR (RXR-RAR). Retinoic acid plays a role in early embryonic development in germ layer formation, body axis formation, neurogenesis, cardiogenesis, and development of organs, such as the pancreas, lung, and eye.

Eventual **visual functions** are directly dependent on RA. The photosensitive pigment, **opsin**, is covalently coupled (in a Schiff base linkage to a His residue of opsin) to 11-*cis*-retinal to form **rhodopsin** (**RD**). Rhodopsin is a receptor that is located in the rod cell membrane. Coupling of 11-*cis*-retinal occurs at three transmembrane domains of rhodopsin. The absorption of light by rhodopsin (bleaching) is the primary event of vision in the rod cells of the eye. This occurs by isomerization of the 11-*cis* double bond to the 11-*trans* configuration as shown in Fig. 20.33.

When rhodopsin (RD) encounters light energy it disintegrates into **scotopsin** through a series of transient intermediates (bathoRD, lumiRD, metaRDI, and metaRDII). **Metarhodopsin II** (MetaRDII, an enzyme) causes a change in the rod membrane charge and activates membrane-bound enzyme, **transducin**. **Transducin** activates a **phosphodiesterase** that hydrolyzes **cGMP** that facilitates the opening of sodium channels in the rod membrane so that Na$^+$ can flow in two directions preventing the hyperpolarization of the inner rod segment. When light impacts, the cascade of **metaRDII-transducin-phosphodiesterase** is initiated, causing the degradation of cGMP and the closing of the sodium ion channels in the rod membrane so that it becomes hyperpolarized causing propagation of nerve impulses to the brain. This cascade is reversed by another enzyme of the rod outer segment, **rhodopsin kinase** allowing the opening of the sodium ion channels. The ability to respond to light is restored and **11-*trans*-retinal** is reconverted into **11-*cis*-retinal** for recombination with **scotopsin** to reform **rhodopsin** (Fig. 20.34).

Deficiency of **vitamin A** causes **night blindness** and other problems, such as susceptibility to cancer owing to the loss of the antioxidant activity of β-carotene that reduces free radicals; vitamin A deficiency also causes susceptibility to infections and anemia.

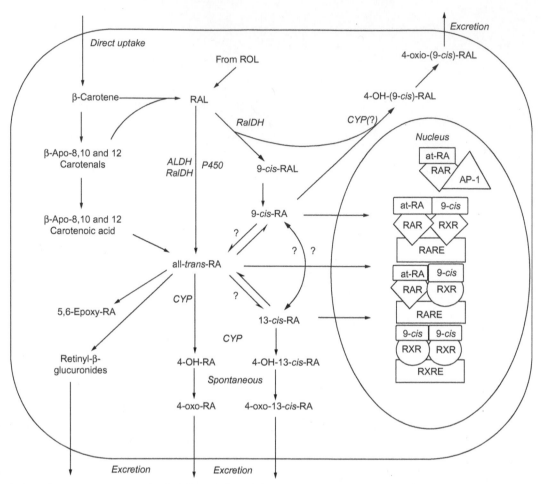

**FIGURE 20.32** Overview of the pathways of **retinoid metabolism**. Within the nucleus are shown homodimers and heterodimers of RAR and RXR and their DNA binding sites for receptor-ligand complexes. *ALDH*, aldehyde dehydrogenase; *at-RA*, all *trans*-retinoic acid; *9-cis*, 9-cis-retinoic acid; *RA*, retinoic acid; *RAL*, retinal; *RalDH*, retinal dehydrogenase; *RAR*, retinoic acid receptor; *ROL*, retinol; *RXR*, retinoic X receptor; *CYP*, cytochrome P450; *RARE*, RAR response element; *RXRE*, RXR response element.

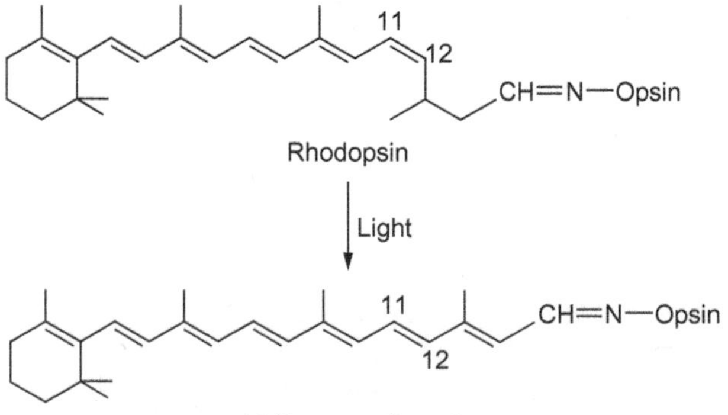

**FIGURE 20.33** A primary event in the visual process in rod cells of the eye is the absorption of light by rhodopsin that isomerizes 11-*cis* double bond to the 11-*trans* configuration.

*Fat-soluble vitamins, such as vitamin A are stored as esters in the liver, and they are not catabolized and excreted like the water-soluble vitamins.* Too great an intake of vitamin A, like the other fat-soluble vitamins, can cause toxicity and toxicity, in this case, can produce bone pain, enlargement of the liver and spleen, nausea, and diarrhea.

Cooked liver and fish oils are the best sources of vitamin A from animal products, and dairy products are good sources. Of plant foods, the best sources are carrots, raw or cooked followed by mango, sweet potato, and spinach. Many other vegetables are good sources of vitamin A.

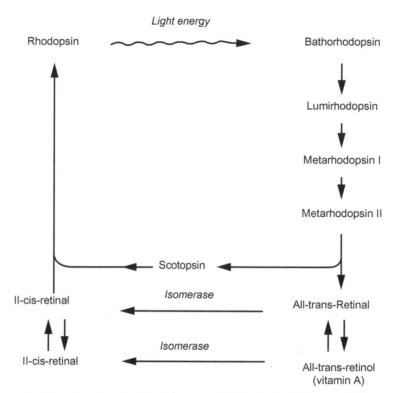

**FIGURE 20.34** The rhodopsin cycle. *Reproduced from http://www.vetmed.vt.edu/education/curriculum/vm8054/eye/rhodopsin.htm.*

Daily requirements in terms of retinol activity equivalents are, for adults (19—50 years), 900 μg/day for the male and 700 μg/day for the female in the same age group; however, female requirements increase during pregnancy and lactation.

## Vitamin D (1,25-Dihydroxyvitamin D₃, Active Form)

Vitamin D can be synthesized in the body by irradiation of the skin, containing **7-dehydrocholesterol**, a precursor of cholesterol, to a precursor of the active hormonal form (**cholecalciferol**). By several metabolic steps involving different organs, the vitamin is converted to a hormone, **1,25-dihydroxyvitamin D₃** (the active form of vitamin D) that regulates homeostasis of calcium and phosphorus.

As acetate is the precursor of steroids, **7-dehydrocholesterol** is formed in the skin from acetate and with **ultraviolet light** (i.e., in sunlight), it is converted to cholecalciferol (vitamin D₃). This enters the bloodstream and is converted in the liver to **25-hydroxyvitamin D₃** by the action of **25-hydroxylase**. 25-Hydroxyvitamin D3 is transported to the kidneys where it is converted to 1,25-dihydroxyvitamin D₃ (by the action of **1α-hydroxylase**), the hormonal form of the vitamin that can activate the **vitamin D receptor** (**VDR**). These steps are outlined in Fig. 16.25. 1,25-**Dihydroxyvitamin D₃** (**calcitriol**) is distributed to the tissues via the bloodstream. A transport protein (**vitamin D binding protein**) has been observed for the transport of **25-hydroxyvitamin D₃** from the liver to the kidney in the bloodstream, and this **precalcitriol** has a relatively long half-life. Calcitriol may be distributed to the tissues via this same transport protein, although there is little information available about the plasma transport of calcitriol; in any case, its half-life is much shorter than that of 25-hydroxyvitamin D₃ being only a few hours. The regulation of the concentration of calcitriol in serum is controlled by **parathyroid hormone** (**PTH**) and by **phosphorus** levels, and some reports indicate that the level of calcitriol in serum may be controlled directly by serum $Ca^{2+}$ independently of PTH and of [phosphorus]. If so, this would represent a feedback mechanism, since the serum level of $Ca^{2+}$ is stimulated by calcitriol (in complex with VDR) through the induction of intestinal **calbindin** that carries $Ca^{2+}$ from the apical side to the basolateral side of the intestinal cell (availing it to the bloodstream). Calcitriol in complex with VDR also controls the formation of **calbindin** and the **kidney epithelial $Ca^{2+}$ channel**.

Once calcitriol has been distributed to tissues, especially the target tissues, intestine, thyroid, and kidney plus about 35 other target tissues that contain (smaller amounts of) the calcitriol receptor, calcitriol binds to VDR in the nucleus, forming the transcriptional complex similar to that shown in Fig. 20.35.

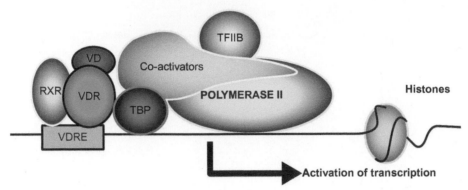

**FIGURE 20.35**   The **vitamin D receptor (VDR) transcriptional complex**. VDR bound to calcitriol (VD) binds to the vitamin D responsive element (VDRE) in the target gene promoter as a heterodimer with the retinoid X receptor (RXR; the ligand is 9-cis retinal). This complex associates with the TATA-binding protein (TBP), coactivators, specific transcription factors (TFIIB), and with acetylated histones to induce transcription of a target gene, such as the gene-encoding information for the protein calbindin in the intestine. *Reproduced from http://www.iontechopen.com/source/html/18632/media/image3.png.*

Detailed actions of the calcitriol-VDR complex are shown in Fig. 20.36.

VDR also acts as a receptor for the secondary bile acid, **lithocholic acid**. Some of the tissues affected by **vitamin D** through the action of the calcitriol-VDR complex and the effects of the vitamin are summarized in Fig. 20.37.

**Parathyroid hormone (PTH)** (also a decreased level of serum phosphorus) induces 1α-hydroxylase in the kidney. VDR action represses 1α-hydroxylase in the kidney and elevates the **calcitriol inactivating enzyme CYP24** (Fig. 20.38).

Of great import for the action of this vitamin are its effects on the stimulation of the immune system and the prevention of colon cancer. Children who grow up in areas where there is relatively little sunshine seem to be more at risk, as adults, for developing colon cancer owing to vitamin D deficiency. Otherwise, in areas of normal weather with adequate sunshine, vitamin D deficiency would be relatively rare. Good dietary sources of vitamin D are canned tuna and salmon, milk and fortified rice.

## Vitamin E (α-Tocopherol)

Vitamin E is an **antioxidant**, and there are several compounds with vitamin E activity, although α-tocopherol is the most active. There are three groups of compounds with vitamin E activity. These are **tocopherols** (with four forms: α-, β-, γ-, and δ-tocopherols); **tocotrienols** (with four forms: α-, β-, γ-, and δ-tocotrienols); and **tocomonoenol**. Structures of the natural (d-forms) of α-tocopherol and α-, β-, γ-, and δ-tocotrienol are shown in Fig. 20.39.

Tocopherols have a saturated phytol side chain, whereas the tocotrienols have three double bonds in the side chain. Different relatives of α-tocopherol may have different unique activities. γ-Tocopherol has been found to inhibit cyclooxygenase activity. In platelets, α-tocopherol inhibited the activity of cyclooxygenase causing a reduction in the formation of thromboxane and prestaglandin D2 (PGD2) and inhibited aggregation. γ-Tocopherol-inhibited **cyclooxygenase (COX-2,** induced form) in macrophages and epithelial cells and inhibited PGE2 and inflammation. α-Tocotrienol [unsaturated isoprenoid (farnesyl) side chain] but not α-tocopherol (saturated side chain) is a modulator of **12-lipoxygenase (12-LOX**). Tocotrienol suppresses the early action of **c-Src kinase** and lowers the tyrosine phosphorylation of 12-LOX, thus protecting neurons from various forms of damage. Tocotrienol appears to interact directly with 12-LOX and docks in two positions forming a complex with the enzyme to inhibit its activity. It has been shown that **α-tocotrienol**, in nanomolar concentrations, inhibited the toxicity induced by glutamate, homocysteine, or L-buthionine sulfoximine essentially by modulating the activity of 12-LOX and is a major neuroprotective form of vitamin E.

Vitamin E in the diet is absorbed in the intestinal enterocytes and incorporated into **chylomicrons** for export to the lymphatic system and eventual delivery to the liver. Under some experimental conditions, vitamin E was exported from enterocytes with **apolipoprotein B-lipoproteins** and **HDL**s. Vitamin E is located in cell membranes, circulating lipoproteins and in adipose tissue, a major storage site. The vitamin, acting as an antioxidant, prevents peroxidation of unsaturated fatty acids of the cell membrane. Following scavenge of a peroxy-free radical, the oxidized radical form of vitamin E can be regenerated by interacting with **ascorbic acid**, retinol or ubiquinol. When not reacting with these agents, **α-tocopherol** can scavenge two peroxy-free radicals and then be excreted as a glucuronate derivative in the bile.

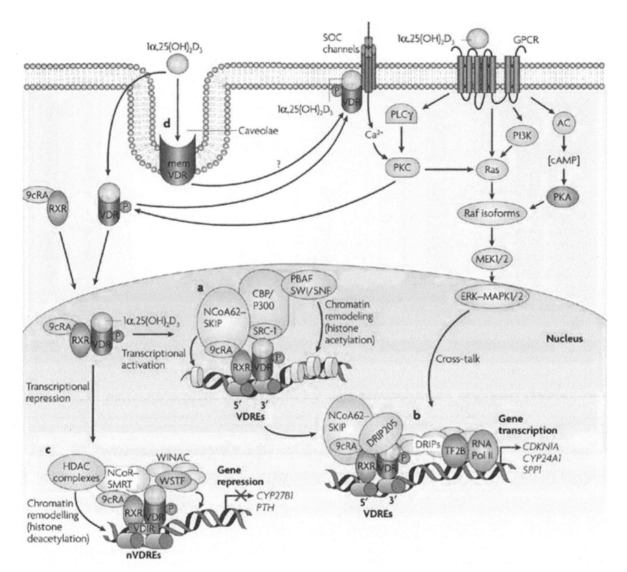

**FIGURE 20.36** Classical action of 1,25-dihydroxyvitamin D$_3$ (calcitriol) is mediated by the binding of **VDR-9-cis-retinoic acid receptor (RXR)** complex at the **vitamin D response elements (VDREs)**. **a** Transcriptional action involves the coactivators, **steroid receptor coactivators (SRCs)** nuclear coactivator-62kDa-Ski-interacting protein (NCaA62-SKIP) and **histone acetyltransferases (HATs)**, **CREB-binding protein (CBP)-p300**, and polybromo-and SWI-2-related gene 1-associated factor (PBAF-SNF) to acetylate histones to derepress chromatin. **b** Binding of the vitamin D receptor-interacting protein 205 (DRIP205) to the **activation function 2 (AF2)** of VDR (+ RXR) attracts a mediator complex containing other **vitamin D receptor-interacting proteins (DRIPs)** that bridge the VDR-RXR-NCoA62-SKIP-DRIP205 complex with **transcription factor 2B (TF2B)** and RNA polymerase II (RNA Pol II) for transcription initiation. The presence of the multiprotein complex facilitates increased transcription of genes, such as *CDKN1A* (which encodes the cyclin-dependent kinase inhibitor p21), *CYP24A1* (which encodes **24-hydroxylase**), and *SPP1* (which encodes **osteopontin**). **c** Calcitriol-mediated transcriptional repression involves VDR-RXR heterodimer associated with VDR-interacting repressor (VDRIR) bound to E-box-type negative VDREs (nVDREs), dissociation of the HAT coactivator and recruitment of **histone deacetylase (HDAC)** corepressor, Williams syndrome transcription factor (WSTF) potentiates transrepression by interacting with a multifunctional ATP-dependent chromatin remodeling complex (WINAC) and chromatin. This leads to the repression of genes, such as *CYP27B1* (which encodes **1α-hydroxylase**) and *PTH* (which encodes **parathyroid hormone**). **d** Nongenomic rapid actions of calcitriol are hypothesized to involve calcitriol binding to cytosolic (VDR) and membrane VDR (memVDR) also found in **caveoli**, and speculated to activate the **mitogen-activated protein kinase (MAPK)-extracellular signal-regulated kinase (ERK)**1 and 2 cascade through the phosphorylation (P) and activation of Raf by **protein kinase C (PKC)** by Ca$^{2+}$ influx through store-operated Ca$^{2+}$ (SOC) channels. Calcitriol stimulates SOC Ca$^{2+}$ influx (in muscle cells) by trafficking of the classic VDR to the plasma membrane, where the VDR interacts with the SOC channel. Ca$^{2+}$ influx activates Ca$^{2+}$ messenger systems, such as PKC. Activated PKC can phosphorylate VDR. Calcitriol-binding to G protein-coupled receptors (GPCRs) activates **phospholipase Cγ (PLCγ)**, Ras, **phosphatidylinositol 3-kinase (PI3K)**, and protein kinase A (PKA) pathways and induces MAPK-ERK1 and 2 signaling. Activated Raf-MAPK-ERK may engage in cross talk with the classical VDR pathway to modulate gene expression. *AC*, adenylate cyclase; *cAMP*, cyclic adenosine monophosphate. *Reproduced from http://www.nature.com/nrc/journal/v7/n9/fig_tab/nrc2196_F2.html.*

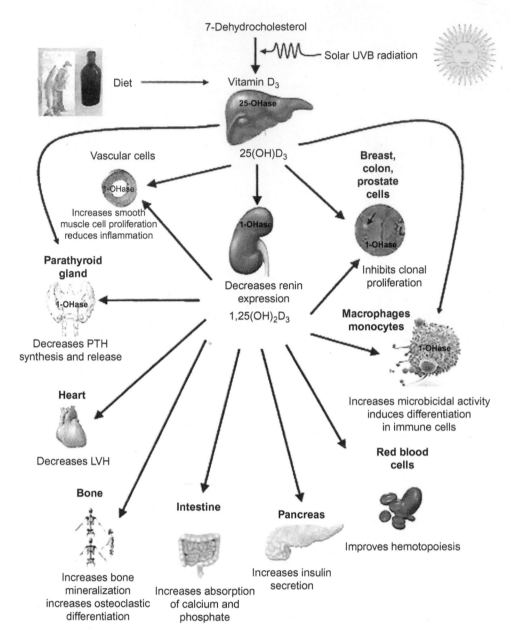

**FIGURE 20.37** Through the effect of UV light in sunlight, 7-dehydrocholesterol (synthesized from acetate) is converted to vitamin $D_3$. (cholecalciferol $D_3$). After transport to the liver which contains **25-hydroxylase (25-OHase)**, the product, **25-hydroxy $D_3$ ($25(OH)D_3$)**, is transported in the serum by the vitamin D transport protein to the kidney where, after the action of **1α-hydroxylase (1-OHase)**, the hormone-active form, 1,25-dihydroxycholecalciferol D3 ($1,25-(OH)_2D_3$) or **calcitriol**, is formed and distributed to the tissues via the bloodstream. The effects in various tissues through VDR are indicated. *LVH*, left ventricular hypertrophy; *PTH*, parathyroid hormone. *Reproduced from http://ajprenal.physiology.org/content/ajprenal/297/6/ F1502/F1.large.jpg.*

Good food sources of vitamin E are wheat germ oil, almonds, sunflower seeds, several kinds of nuts and peanut butter. The average intake of this vitamin for adults is about 15 mg/day. Adults on a normal diet usually do not become deficient in vitamin E. A deficiency, if it should occur, causes increased red blood cell fragility.

## Vitamin K

The two forms of naturally occurring vitamin K are vitamin K1, **phytylmenaquinone** and vitamin K2, **multiprenylmenaquinone** (Fig. 20.40).

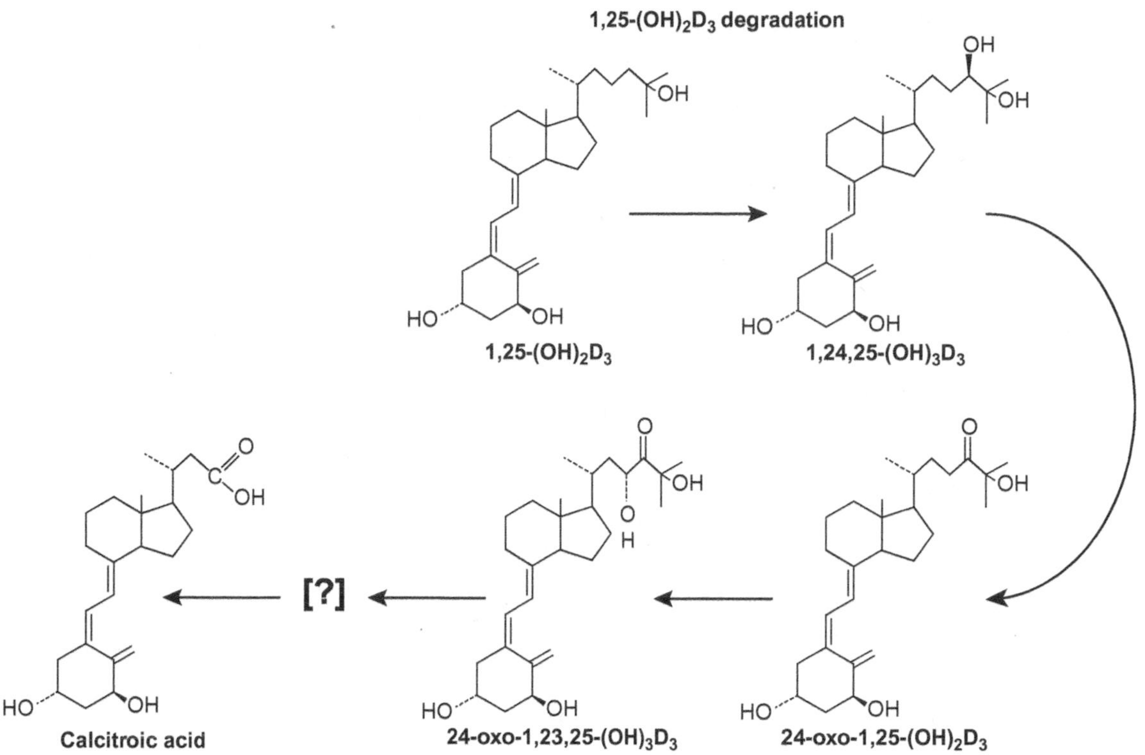

**1,25-(OH)₂D₃ degradation**

**FIGURE 20.38** Degradation of 1,25-(OH)₂D₃. The *CYP24A1* gene is induced by 1,25-(OH)₂D3. The resulting enzyme carries out all the reactions shown to produce the biologically inactive excretion product, **calcitroic acid**. Presumably, a similar set of reactions takes place with 25-OH-D3 as the substrate. Clearly, 1,25-(OH)₂D3 programs its own destruction through the *CYP24A1*. *Reproduced from http://www.nature.com/bonekeyreports/2014/ 140108/bonekey2013213/fig_tab/bonekey2013213_F4.html.*

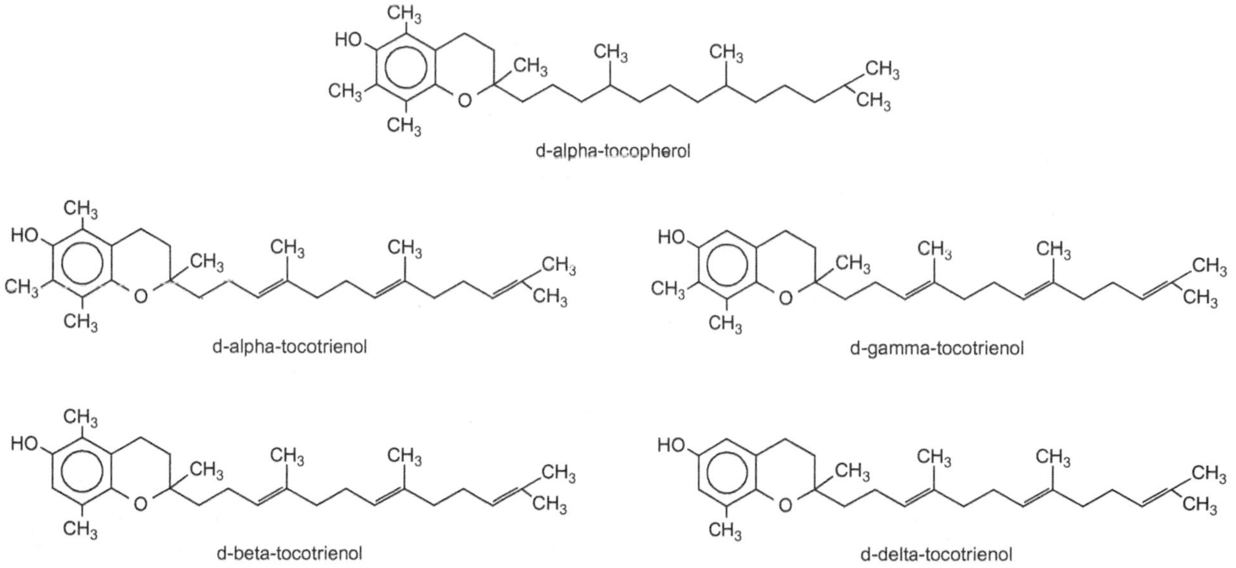

**FIGURE 20.39** Structures of α-tocopherol and four isoforms of tocotrienols. *Reproduced from http://www.bradolinegroup.com/images/VITAL%20E %20tocotrienols01.jpg.*

These compounds have in common a methylated naphthoquinone ring structure with a varying aliphatic side chain at the 3-position. Vitamin K₁ occurs in plants and is therefore a dietary form and vitamin K₂ is synthesized in the intestine by bacteria. Vitamin K₃, **menadione**, is a synthetic molecule with vitamin K activity. Vitamin K stimulates **blood clotting** by the maintenance of normal levels of blood-clotting protein factors II, VII, IX, and X and proteins C and S.

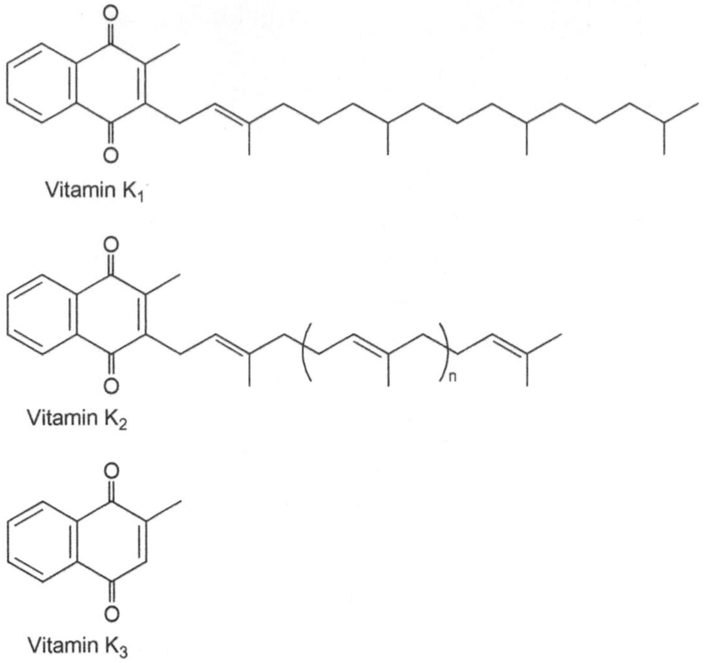

Vitamin K₁

Vitamin K₂

Vitamin K₃

**FIGURE 20.40** Structures of molecules with vitamin K activity. Chemical structures of vitamin $K_1$ (*top*), vitamin $K_2$ (*middle*), and synthetic vitamin $K_3$, menadione (*bottom*).

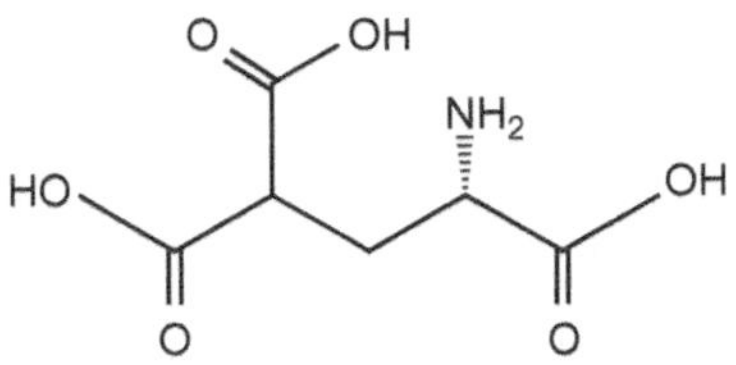

**FIGURE 20.41** Structure of $\gamma$-carboxyglutamic acid.

Clotting factors are modified posttranslationally, becoming active as they course through the endoplasmic reticulum. The modification of these factors is the carboxylation of multiple glutamate residues in the clotting factor protein by a vitamin K-dependent **$\gamma$-carboxylase** that converts clotting factors into **$\gamma$-carboxylglutamate (Gla)-containing proteins**. The structure of Gla is shown in Fig. 20.41.

The modifying system is located on the membrane of the endoplasmic reticulum (ER) and includes the $\gamma$-carboxylase and the enzyme **2,3-epoxide reductase (VKOR)** that converts vitamin K to the 2,3-epoxide form. VKOR provides the reduced vitamin K cofactor for the $\gamma$-carboxylase. A reductase regenerates the hydroquinone form of the vitamin. An ER chaperone protein, **calumenin**, associates with the **$\gamma$-carboxylase** and inhibits it. The system containing $\gamma$-carboxylase, VKOR and **calumenin** is shown in Fig. 20.42.

The formation of several carboxylated glutamate residues in clotting factors results in the carboxylation of the $\gamma$-carboxylase enzyme. Vitamin K-dependent (VKD) proteins bind to a high affinity site on the enzyme, and this interaction increases the affinity of the $\gamma$-carboxylase for vitamin K. Warfarin (Fig. 20.43) is an anticoagulant.

Warfarin is metabolized in the liver by components of the **cytochrome P450** system. In the case of S-warfarin, the metabolizing system is **CYP2C9** and in the case of R-warfarin, the metabolizing system is CYP2C19, CYP1A2, and CYP3A4. VKOR, the vitamin K epoxide reductase is attached to the inner plasma membrane. VKOR reduces the **vitamin K epoxide** to the **vitamin K hydroquinone**. The enzyme has a specific binding site for vitamin K epoxide and an overlapping binding site for warfarin. Thus, when warfarin is bound, vitamin K epoxide cannot attach to the substrate-binding site, and VKOR is inactivated. Vitamin K hydroquinone is the cofactor for $\gamma$-glutamyl carboxylase that carboxylates glutamic acid residues (**Gla**) in clotting factors (II, VII, IX, and X) and in specific proteins (C, S, and Z) that ultimately allows these proteins to attach to membrane phospholipids. In the process of carboxylation of proteins by

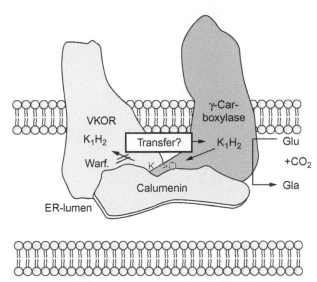

**FIGURE 20.42** A molecular model of the γ-carboxylation system. **VKOR (2,3-epoxide reductase)** and γ-carboxylase are parts of an enzyme complex in the ER lipid bilayer. Calumenin binds to γ-carboxylase as an inhibitory chaperone and also affects the activity and **warfarin** sensitivity of VKOR. *Warf*, warfarin, *Gla*, γ-carboxyglutamic acid, *ER*, endoplasmic reticulum. Redrawn from Fig. 8 of J.-K. Tie et al., "Chemical modification of cysteine residues is a misleading indicator of their status as active site residues in the vitamin K-dependent γ-glutamyl carboxylation reaction," *J. Biol. Chem.*, **279**: 54079−54087, 2004.

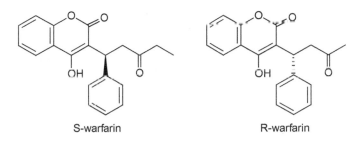

**FIGURE 20.43** Chemical structures of warfarin (*upper*) and dicoumarol, the original compound from which warfarin was derived. As a pharmaceutical, warfarin comes as a racemic mixture of two compounds, S-warfarin (*upper left*) and R-warfarin (*upper right*). S-Warfarin is three−five times more potent than R-warfarin and is, therefore, the active component of the mixture. Note that the difference between the two structures is in the orientation of the benzene substituent (*lower part of structure*); in S-warfarin, the benzene substituent tends to be moved toward the reader, whereas in R-warfarin, the benzene substituent tends to be moved away from the reader (*hashed line*).

**γ-glutamyl carboxylase**, vitamin K hydroquinone is oxidized to vitamin K epoxide. These activities are summarized in Fig. 20.44.

The carboxylase enzyme has a high affinity for vitamin K-dependent (VKD) proteins and the contact between the enzyme and these proteins increases the affinity of the carboxylase for vitamin K hydroquinone. The action of warfarin is to block indirectly the carboxylation of VKD proteins by inhibiting VKOR's ability to reduce the vitamin K epoxide to the required hydroquinone coenzyme form. Consequently, VKD proteins accumulate and these undercarboxylated and inactive clotting proteins prevent normal blood clotting. Prothrombin (factor II) containing Glas, chelates $Ca^{2+}$ and interacts with membrane phospholipids. Activated factor X, a protease, converts prothrombin to thrombin (factor IIa)

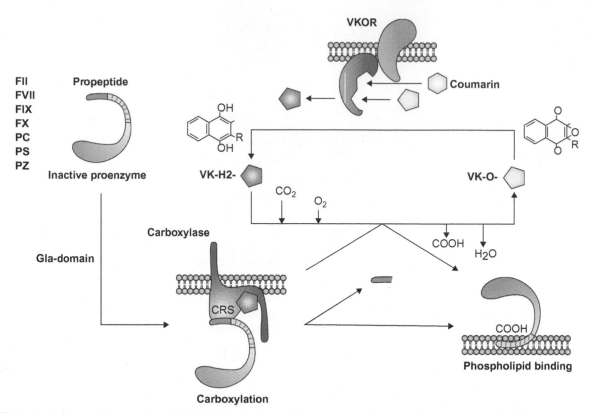

**FIGURE 20.44** Summary of the activities of **VKOR** (**vitamin K epoxide reductase**) and **gamma-glutamate carboxylase** that requires vitamin K hydroquinone as its coenzyme. γ-Glutamate carboxylase catalyzes the **carboxylation of the glutamate residues (Gla)** of specific protein factors (II, VII, IX, and X) in the blood-clotting mechanism as well as other specific proteins involved in blood clotting (C, S, and Z). These are carboxylated by the carboxylase with vitamin K hydroquinone as coenzyme, and the proteins containing Gla become capable of binding to phospholipids in membranes, an important property in clotting. In the catalyzed reaction, **vitamin K hydroquinone** is converted to **vitamin K 2,3-epoxide** and must be converted back to the hydroquinone form by VKOR before it, again, combines with γ-glutamate carboxylase. "*Coumarin,*" warfarin; *CRS*, carboxylation recognition sequence. *Reproduced from http://www.tankonyvtar.hu/hu/tartalom/tamop425/0011_1A_Molelkularis_diagnoszitka_en_book/ch11.html.*

that facilitates the conversion of fibrinogen to a monomer that, in turn, polymerizes to form a cross-linked fibrin polymer. This last process is positively regulated by factor XIIIa. Thrombin also accomplishes the conversion of factor XIII to factor XIIIa.

Vitamin K is absorbed in the intestine in the presence of bile salts and interacts with **chylomicrons**. When the absorption of fat is a problem, as in bile duct obstruction, vitamin K can become deficient. However, a deficiency of vitamin K is rare in view of the fact that intestinal bacteria can synthesize vitamin $K_2$.

The daily intake of **vitamin K** is normally 80 μg for adult males and 60 μg for females. The best food sources of the vitamin are found in leafy vegetables, such as kale, broccoli, parsley, Swiss chard, spinach, and leaf lettuce.

## BALANCED NUTRITION

A balanced diet contains nutrients, vitamins, and calories in optimal proportions as shown in a food guide pyramid (Fig. 20.45). Since this food pyramid was constructed, it is clear that the number of servings of carbohydrates (bottom of pyramid) should be reduced. Lately, there has been an emphasis on vegetarianism (Fig. 20.46). Currently, the discussion on what constitutes an optimal diet is somewhat controversial. Clearly, a reduction in carbohydrate intake and an emphasis on varying the diet seems reasonable.

In Table 20.4 are shown the calculations of calorie requirements for women and men.

### Nutrient Intake

The daily nutrient intakes based on a daily 2000 calorie diet are listed in Table 20.5.

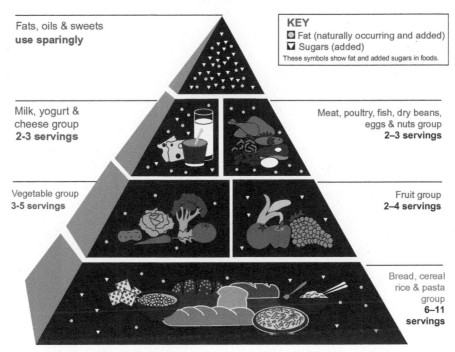

**FIGURE 20.45** Food guide pyramid. A guide to daily food choices.

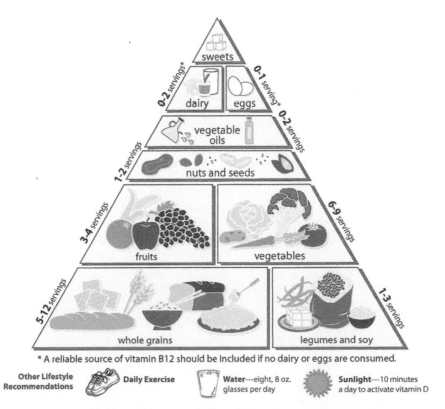

**FIGURE 20.46** Food guide pyramid for vegetarians. *Reproduced from commons.wikimedia.org/wiki/File:Loma_Linda_University_Vegetarian_Food_Pyramid.jpg.*

**TABLE 20.4** Calculations of Calorie Requirements for Men and Women

**Calories for Women**

- **Work out your weight in pounds**
  - Example: Hilary is 150 pounds
- **Multiply your weight according to your lifestyle**
  - **Sedentary:** multiply by 12
  - A very inactive Hilary needs 150 × 12 = 1800 calories
  - **Light exercise:** multiply by 13
  - If Hilary takes light exercise, she needs about 150 × 13 = 1950 calories
  - **Moderate exercise:** multiply by 14
  - If Hilary takes moderate exercise she needs about 150 × 14 = 2100 calories

**Calories for Men**

- **Work out your weight in pounds**
  - Example: Bill is 165 pounds
- **Multiply your weight according to your lifestyle**
  - **Sedentary:** multiply by 13
  - A very inactive Bill needs 165 × 13 = 2145 calories
  - **Light exercise:** multiply by 14
  - If Bill takes light exercise, he needs about 165 × 14 = 2310 calories
  - **Moderate exercise:** multiply by 15.25
  - If Bill takes moderate exercise, he needs about 165 × 15.25 = 2516 calories
  - **Moderate-heavy exercise:** multiply by 16.5
  - If Bill takes moderate-heavy exercise, he needs about 165 × 16.5 = 2722 calories
  - **Regular heavy exercise:** multiply by 18
  - If Bill takes regular heavy exercise, he needs about 165 × 18 = 2970 calories

Please note: Calorie needs vary according to many things, including, gender, age, height, level of activity, body fat percentage and type of food eaten. This calorie guide offers a ballpark estimation only.
Reproduced from http://www.diet-i.com/weight_loss/calories.htm.

## SUMMARY

Vitamin D deficiency is virtually epidemic throughout the world, especially in locations with limited sunlight. Chronically, this deficiency generates a risk of developing certain cancers (colon, breast, and prostate). On the other hand, excess vitamin D intake through oversupplementation can lead to toxicity. Although vitamin D can be metabolized to an excretory form, vitamin D, as well as the other fat-soluble vitamins, has binding proteins that are used for storage. Of the fat-soluble vitamins (A, D, E, K) required in the diet, only vitamin D is generated by exposure to sunlight and is metabolized by the liver and kidneys to a form that is active as a hormone (1,25-dihydroxyvitamin D3) which binds and activates the vitamin D receptor.

The B vitamins represent the water-soluble vitamins. The water-soluble vitamins considered are as follows: thiamine (vitamin B1), riboflavin (vitamin B2), niacin (vitamin B3), pantothenic acid (vitamin B5), pyridoxine, pyridoxal, pyridoxamine (vitamin B6), biotin (vitamin H), cobalamin, (vitamin B12), folic acid (vitamin B9), and ascorbic acid (vitamin C). These vitamins are metabolized to coenzymes or function as redox agents (e.g., ascorbate). Deficiencies of these vitamins are relatively rare and are related to inadequate diets and/or problems of malfunctioning intestinal absorption systems.

A guide to daily food choices from the US Department of Agriculture, US Department of Health and Human Services is presented. Good nutrition coupled with exercise is optimal for good health. Genetic variation, in part, determines the way people eat and exercise, and such variations are not factored in to daily recommendations. Consideration of balanced nutrition is presented with aids for the calculation of appropriate caloric intake and the nutrients for a 2000 calorie per day intake.

**TABLE 20.5** Daily Values of Nutrients Based on a 2000 calorie/day Diet

| Nutrient | Unit of Measure | Daily Values |
|---|---|---|
| Total fat | grams (g) | 65 |
| Saturated fatty acids | grams (g) | 20 |
| Cholesterol | milligrams (mg) | 300 |
| Sodium | milligrams (mg) | 2400 |
| Potassium | milligrams (mg) | 3500 |
| Total carbohydrate | grams (g) | 300 |
| Fiber | grams (g) | 25 |
| Protein | grams (g) | 50 |
| Vitamin A | International Unit(IU) | 5000 |
| Vitamin C | milligrams (mg) | 60 |
| Calcium | milligrams (mg) | 1000 |
| Iron | milligrams (mg) | 18 |
| Vitamin D | International Unit (IU) | 400 |
| Vitamin E | International Unit (IU) | 30 |
| Vitamin K | micrograms (µg) | 80 |
| Thiamin | milligrams (mg) | 1.5 |
| Riboflavin | milligrams (mg) | 1.7 |
| Niacin | milligrams (mg) | 20 |
| Vitamin $B_6$ | milligrams (mg) | 2.0 |
| Folate | micrograms (µg) | 400 |
| Vitamin $B_{12}$ | micrograms (µg) | 6.0 |
| Biotin | micrograms (µg) | 300 |
| Pantothenic acid | milligrams (mg) | 10 |
| Phosphorus | milligrams (mg) | 1000 |
| Iodine | micrograms (µg) | 150 |
| Magnesium | milligrams (mg) | 400 |
| Zinc | milligrams (mg) | 15 |
| Selenium | micrograms (µg) | 70 |
| Copper | milligrams (mg) | 2.0 |
| Manganese | milligrams (mg) | 2.0 |
| Chromium | micrograms (µg) | 120 |
| Molybdenum | micrograms (µg) | 75 |
| Chloride | milligrams (mg) | 3400 |

Based on a 2000-calorie intake, for adults and children 4 or more years of age.
Reproduced from http://www23.netrition.com/rdi_page.html.

# SUGGESTED READING

## Literature

Basit, S., 2013. Vitamin D in health and disease: a literature review. Br. J. Biomed. Sci. 70, 161–172.

Corpe, C.P., et al., 2013. Intestinal dehydroascorbic acid (DHA) transport mediated by the facilitative sugar transporters, GLUT2 and GLUT8. J. Biol. Chem. 288, 9092–9101.

Deeb, K.K., Trump, D.L., Johnson, C.S., 2007. Vitamin D signaling pathways in cancer: potential for anticancer therapeutics. Nat. Rev. Cancer. 7, 684–700.

Evans, J.C., et al., 2004. Structures of the N-terminal modules imply large domain motions during catalysis by methionine synthase. Proc. Natl. Acad. Sci. U.S.A. 101, 3729–3736.

Furger, E., et al., 2013. Structural basis for universal corrinoid recognition by the cobalamin transport protein haptocorrin. J. Biol. Chem. 288, 25466–25476.

Henning, P., Conaway, H.H., Lerner, U.H., 2015. Retinoid receptors in bone and their role in bone remodeling. Front. Endocrinol. 6, 31.

Johnson, R.J., et al., 2008. The planetary biology of ascorbate and uric acid and their relationship with the epidemic of obesity and cardiovascular disease. Med. Hypotheses. 71, 22–31.

Kennel, K.A., Drake, M.T., Hurley, D.L., 2010. Vitamin D deficiency in adults: when to test and how to treat. Mayo Clin. Proc. 85, 752–758.

Krautler, B., 2012. Biochemistry of B12-cofactors in human metabolism. Subcell. Biochem. 56, 323–346.

Losdale, D., 2015. Thiamine and magnesium deficiencies: keys to disease. Med. Hypotheses. 84, 129–134.

Manara, P., Jain, S.K., 2012. Vitamin D up-regulates glucose transporter 4 (GLUT4) translocation and glucose utilization mediated by cystathionine-γ-lyase (CSE) activation and H25 formation in 3T3L1 adipocytes. J. Biol. Chem. 287, 42324–42332.

Nishizuka, Y., Hayaishi, O., 1963. Studies on the biosynthesis of nicotinamide adenine dinucleotide. J. Biol. Chem. 238, 3369–3377.

Reboul, E., Borel, P., 2011. Proteins involved in uptake, intracellular transport and basolateral secretion of fat-soluble vitamins and carotenoids by mammalian enterocytes. Prog. Lipid Res. 50, 388–402.

Said, H.M., 2011. Intestinal absorption of water-soluble vitamins in health and disease. Biochem. J. 437, 357–372.

Shoori, M.A., Saedisomeolia, A., 2014. Riboflavin (vitamin B2) and oxidative stress: a review. Br. J. Nutr. 111, 1985–1981.

Tie, J.-K., Jin, D.-Y., Stafford, D.W., 2014. Conserved loop cysteines of vitamin K epoxide reductase complex subunit 1-like 1 (VKORC1L1) are involved in its active site regeneration. J. Biol. Chem. 289, 9396–9407.

Valdivielso, J.M., Cannata-Andia, J., Coll, B., Fernandez, E., 2009. A new role for vitamin D receptor in chronic kidney disease. Am. J. Physiol. Ren. Physiol. 297, F1502–F1509.

Wolf, B., 2005. Biotinidase: its role in biotinidase deficiency and biotin metabolism. J. Nutr. 16, 441–445.

## Books

Feldman, D., Pike, J.W., Adams, J.S. (Eds.), 2011. Vitamin D. third ed. Academic Press, San Diego.

Litwack, G. (Ed.), 2006. Vitamin K, Vitamins & Hormones. Academic Press, San Diego.

Litwack, G. (Ed.), 2008. Vitamins and the Immune System, Vitamins & Hormones. Academic Press, San Diego.

Litwack, G. (Ed.), 2016. Vitamin D Hormone, Vitamins & Hormones. Academic Press/Elsevier, San Diego.

Litwack, G. (Ed.), 2007. Vitamin E, Vitamins & Hormones. Academic Press, San Diego.

Litwack, G. (Ed.), 2008. Folic Acid and Folates, Vitamins & Hormones. Academic Press, San Diego.

Litwack, G. (Ed.), 2007. Vitamin A, Vitamins & Hormones. Academic Press, San Diego.

Litwack, G. (Ed.), 2012. Adiponectin, Vitamins & Hormones. Academic Press, San Diego, 480 pages.

Chapter 21

# Blood and Lymphatic System

## DEEP VEIN THROMBOSIS

The normal function of a **blood clot** is to increase wound healing. However, when a clot is formed in a major blood vessel (often in the leg) unrelated to wound healing, it can be a major threat to life. Adventitious clot formation can occur when an individual spends long periods of time without movement, such as when sitting in a car or plane or lying in bed without much movement. Older people are especially prone. There are other factors that may contribute to accidental blood clotting: use of oral contraceptives, surgery, childbirth, massive trauma, burns, or fractures of the hips or femur. These clots (deep vein thrombosis, DVT) usually occur in vessels that carry blood to the heart. In 1 year, these types of clots will occur in one person in a thousand. There have been reports of 100,000 to 300,000 deaths per year in the United States and an incidence of DVT overall of up to 900,000 per year. Clots in a deep vein can break apart allowing parts of the clot (**embolus**) to travel in the bloodstream, coming to rest in the lungs or the heart. Clots can occur in the legs, hips, pelvis (iliac or femoral vein), or arms (subclavian vein). The **deep veins** are shown in Fig. 21.1.

In general, the site of a clot can be detected by **Doppler ultrasound scanning**. Sometimes, clots can occur without swelling or pain and **pulmonary emboli** can be detected by **magnetic resonance scanning**, although it may be difficult to diagnose a clot of this type when there are no symptoms except shortness of breath on climbing stairs and weakness due to compromise of the major circulation. Pulmonary emboli may interfere with the flow of blood into the lungs and vessels in the heart can be blocked, sometimes resulting in a heart attack and death. A clot may travel beyond the lungs and the heart to the brain and cause a stroke. Hypercoagulability of the blood comes into question following a blood clot and, if it exists, **hypercoagulability** can result from various factors. Characteristics of deep vein clots are shown in Fig. 21.2.

Proteins that are inhibitors of clotting may be underexpressed. Protein C is one such protein and **protein S** is a cofactor of **protein C. Antithrombin III** could also be limiting. If the levels of these proteins are low in blood, this could be the cause of **hypercoagulability**. There are about two million persons each year, globally, who have been found to be hypercoagulable. In the case of a patient with cancer, there could be adventitious blood clotting, especially if the tumor interferes with blood flow. Usually, **heparin** is injected to thin the blood; it has a short half-life of about 1—2 hours. Then increasing oral doses of a blood thinner, usually **coumadin** (**warfarin**), are taken until the proper zone of prothrombin time of clotting is reached; this is expressed as an **International Normalized Ratio** (**INR**) and the appropriate target is usually an INR of 2.5 in an acceptable range of 2 to 3. The amount of oral coumadin varies from person to person and is the result of many variables. Usually, blood samples are drawn monthly, after a patient has been stabilized, to determine that the clotting time is appropriate. In many cases, the blood thinner must be taken for life; in other cases where hypercoagulability is not a problem, the coumadin regimen may be terminated. As anticoagulation therapy extends the clotting time, excessive bleeding may become a threat. With long-term anticoagulant therapy, there is a 3% chance of a major **hemorrhage** in a year and 20% of these hemorrhages are fatal. If a surgical or dental procedure that may incur bleeding is needed for a patient on coumadin, the drug can be stopped for 4—5 days prior to the procedure during which time the level of the anticoagulant in blood will have been cleared and the anticoagulant can be resumed after the procedure has been completed without danger.

## Factor V Leiden

One inherited or acquired trait that predisposes to venous clots is a deficiency of the physiological anticoagulant **protein C**. This is a rare inherited trait that predisposes to venous clots and habitual spontaneous abortions. The most

Human Biochemistry. DOI: http://dx.doi.org/10.1016/B978-0-12-383864-3.00021-1

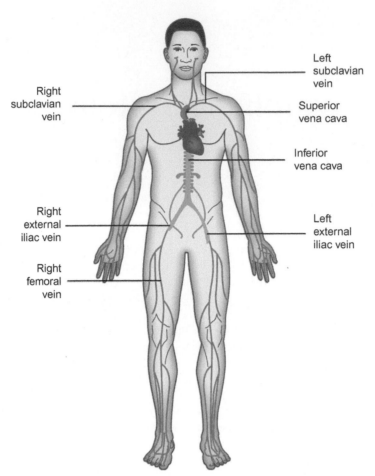

FIGURE 21.1   A deep veins. *Reproduced from Litwack, G.*, Human Biochemistry & Disease, *page 954, Academic Press, 2008.*

common hereditary form of hypercoagulation is the expression of factor V Leiden. Factor V Leiden is an inherited gene encoding a variant of factor V and expression of factor V Leiden makes blood more likely to clot (**thrombophilia**). Acquired forms involve elevated concentrations of **factor VIII**. Protein C cleaves and inactivates factor Va (activated factor V) and factor VIIIa. The activation of protein C results from the activation of **thrombomodulin** by **thrombin**. **Activated protein C (APC)** then combines with its cofactor, **protein S** and the complex binds to a platelet membrane that contains a receptor for APC. Once APC is on the platelet membrane, it cleaves and inactivates factor Va and factor VIIIa. *The activated altered factor V (Leiden) is resistant to inactivation by protein C.* The majority of patients with **activated protein C resistance (APCR)** express **factor V Leiden**. Fig. 21.3 shows the blood-clotting process and the consequence of the expression of factor V Leiden.

The use of **oral contraceptives** increases the chance of venous thrombosis over a normal person not using oral contraceptives by fourfold. These and other conditions are listed in Table 21.1.

There are also cases of **idiopathic** (without known cause) **hypercoagulation** that are not related to expression of factor V Leiden, to deficiencies of protein C or protein S or to other measurable factors in the blood coagulation system. In these unusual cases, a factor could be identified in the future.

**Protein C** is a vitamin K-dependent protease that is activated by thrombin to form **activated protein C (APC)**. It has similarities to other serine proteases. The receptor for APC is the **endothelial protein C receptor (EPCR)** that recognizes the **Gla domain** and a phospholipid domain of protein C. The EPCR accelerates the thrombin- and thrombomodulin-dependent generation of APC. EPCR binds both protein C and APC with equal affinity. In addition to inactivation of factors Va and VIIIa (together with thrombomodulin), APC inhibits inflammation (Fig. 21.4).

The cofactor of protein C is **protein S**. Fig. 21.5 shows an illustration of the structural domains of protein S, protein C, thrombin, and thrombomodulin.

## Deep vein thrombosis (DVT)

(A)

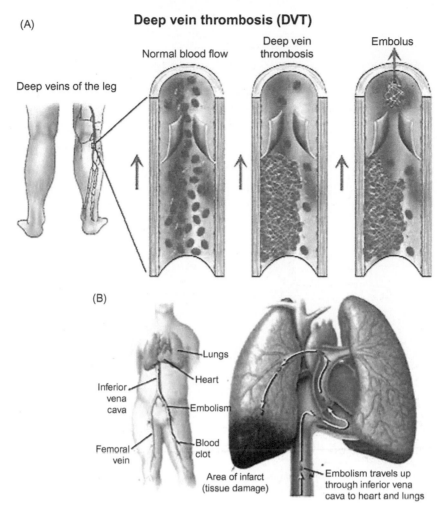

FIGURE 21.2   (A) Deep vein thrombus in the leg. (B) Site of pulmonary embolus. *(A) Reproduced from http://www.sirweb.org/patPub/dvtimages/ DVT_normal_and_embolus.gif. (B) Reproduced from http://www.sinweb.org/patpub/dvtimages/DVT_normal_and_embolus.gif.*

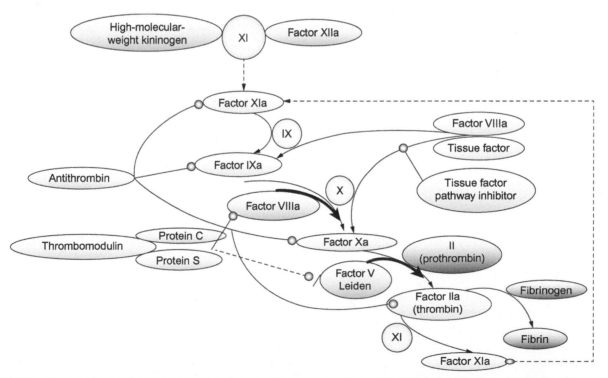

FIGURE 21.3   Blood clotting mechanism showing the consequence of the expression of factor V Leiden. Factor V activates prothrombin to form activated factor II (thrombin) that forms a **fibrin clot** by activation of fibrinogen. Normally, protein C and protein S complexed with thrombomodulin inhibits factor V but factor V Leiden is resistant to this inhibition. Consequently, factor V Leiden, when expressed, produces blood that is more prone to form clots.

**TABLE 21.1** The Role of Factor V Leiden in Venous Thromboembolic Disease. The Risk is Shown Relative to a Normal Person Without Factor V Leiden

| Thrombophilic Status | Relative Risk of Venous Thrombosis |
| --- | --- |
| Normal | 1 |
| OCP[a] use | 4 |
| Factor V Leiden, heterozygous | 5–7 |
| Factor V Leiden, heterozygous + OCP | 30–35 |
| Factor V Leiden, homozygous | 80 |
| Factor V Leiden, homozygous + OCP | ??? > 100 |
| Prothrombin gene mutation, heterozygous | 3 |
| Prothrombin gene mutation, homozygous | ??? possible risk of arterial thrombosis |
| Prothrombin gene mutation, heterozygous + OCP | 16 |
| Protein C deficiency, heterozygous | 7 |
| Protein C deficiency, homozygous | Severe thrombosis at birth |
| Protein S deficiency, heterozygous | 6 |
| Protein S deficiency, homozygous | Severe thrombosis at birth |
| Antithrombin deficiency, heterozygous | 5 |
| Antithrombin deficiency, homozygous | Thought to be lethal before birth |
| Hyperhomocysteinemia | 2–4 |
| Hyperhomocysteinemia combined with factor V Leiden, heterozygous | 20 |

The terms *heterozygous* and *homozygous* are genetic terms. The human genome contains two copies of the information. If the copies are the same, they are homozygous (e.g., AA); if the copies are different, they are heterozygous (e.g., Aa).
[a]*OCP = oral contraceptive pill.*
Reproduced from http://web.archive.org/web20070425232308/http://www-admin.med.uiuc.edu/hematology/PtFacV2.htm (web address no longer available), reproduced from *Human Biochemistry & Disease* by G. Litwack (2008), Table 14–1, page 857.

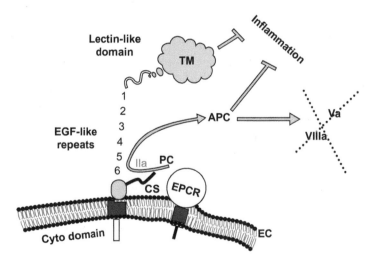

**FIGURE 21.4** Actions of **activated protein C** (**APC**). **Thrombomodulin** (**TM**) is depicted with its five structural domains, including the cytoplasmic (cyto) and transmembrane domains, a serine/threonine rich region with an attached chondroitin sulfate (CS) moiety, 6 EGF-like repeats, and the N-terminal lectin-like domain. EGF-like repeats 4 to 6 of TM provide cofactor function for thrombin (IIa)-mediated activation of PC, a step that is further amplified by the endothelial cell protein C receptor (EPCR). APC cleaves coagulation factors Va and VIIIa, thereby down-regulating thrombin generation, and directly interferes with inflammation. The lectin-like domain of TM also suppresses inflammation. *PC*, protein C; *EC*, vascular endothelial cell; *EGF*, epidermal growth factor. *Reproduced from M. Van de Wouwer, D. Collen and E.M. Conway,* Arterioscler. Thromb. Vasc. Biol., *24: 1374, 2004.*

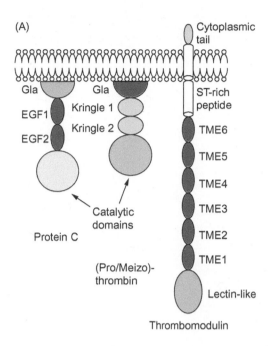

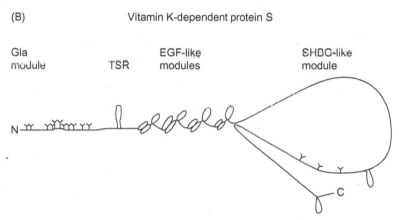

**FIGURE 21.5**   (A) Illustration showing the structural domains of **protein C, thrombin,** and **thrombomodulin**. Protein C, activated protein C, prothrombin, and the active precursor meizothrombin bind to phospholipid membranes through their highly homologous $\gamma$ **carboxyglutamate (Gla)**-rich domains. Des (F1) **meizothrombin** (Xa-Va intermediate product of thrombin generation) and $\alpha$-thrombin lack the Gla-domain and kringle 1, or both kringle domains, respectively. *EGF*, epidermal growth factor-like domain; *Kringle*, a protein domain that folds into a large loop stabilized by three disulfide bridges; *TME*, 6 EGF-like domains of thrombomodulin; *ST*, serine-threonine; *SHBG*, sex hormone-binding globulin. (B) Illustration showing the structural domains of protein S, the cofactor of protein C. *TSR*, thrombin-sensitive region; *SHBG*, sex hormone binding globulin. *(A) Reproduced from part a of http://www.nature.com/nature/journal/v404/n6777/fig_tab/404518a)_F1.html. (B) Reproduced from part a of Figure 14–11 of* Human Biochemistry & Disease, *G. Litwack, Academic Press/Elsevier, 2008, page 862.*

## BLOOD-CLOTTING MECHANISM

The blood-clotting system consists of a series of precursor proteins that become activated by hydrolysis through a series of proteases. A serine protease becomes activated and then activates the next protein in the cascade and this process continues until **fibrinogen** becomes activated forming the long and stringy **fibrin** that forms a clot. The formation of a blood clot is schematically illustrated in Fig. 21.6A and a scanning electron micrograph of a clot formed in a blood vessel is shown in Fig. 21.6B.

Some of the proteins in the clotting cascade are named for the original investigators of those proteins. For example, **factor XII** is also the **Hageman factor, factor IX** is also the **Christmas factor,** and **factor X** is also the **Stuart factor**. The blood-clotting cascade is shown in Fig. 21.7.

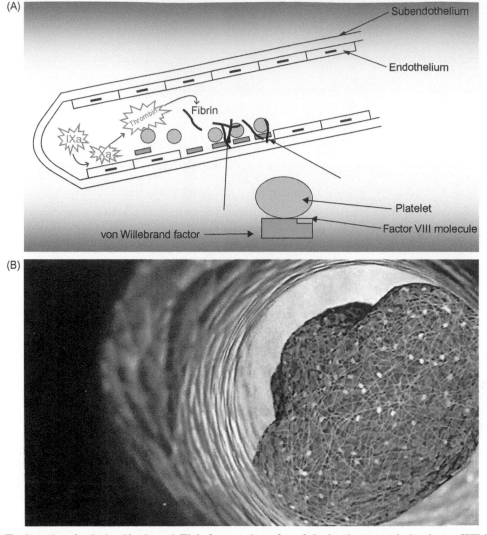

**FIGURE 21.6** (A) The formation of a clot in a blood vessel. Fibrin forms on the surface of platelets that are attached to the **von Willebrand factor** (blood vessel wall endothelial cell-produced glycoprotein that binds to specific proteins and facilitates platelets sticking together) complexed with factor VIII attached to the blood vessel wall. (B) Scanning electron micrograph of a clot in a blood vessel. *(A) Reproduced from http://focus.hms.harvard.edu/2002/Nov8_2002/hematology.html. (B) Reproduced from http://media-cache-ak0.pinimg.com/736x/a7/64/d8/a764d82e8f349304d43275a739bbbd4b.jpg.*

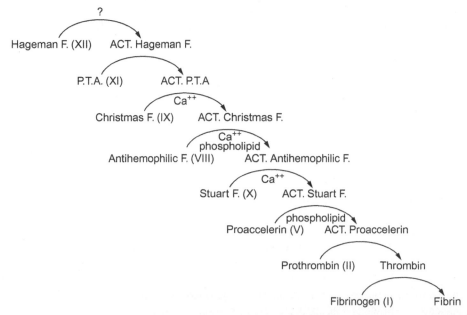

**FIGURE 21.7** The blood-clotting cascade. The question mark at the top of the cascade indicates an unknown event involved in the beginning of the clotting mechanism. *ACT*, activated; *F*, factor; *P.T.A.*, predicted transmitting ability. *Reproduced from Figure 5 of E. W. Davie, "A brief historical review of the waterfall/cascade of blood coagulation", J. Biol. Chem. **278**: 50819–50832, 2003.*

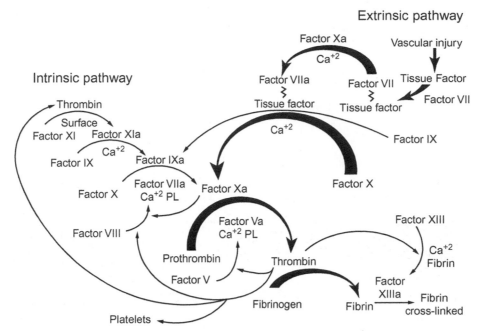

**FIGURE 21.8** Intrinsic and extrinsic pathways of **blood coagulation**. *Reproduced from http://www.jbc.org.content/vol278/issue51/images/large/bc5139454005.jpeg.*

Two types of events can signal the clotting pathway. In the case of a deep vein thrombus initiated by an undetermined cause within the body, the **intrinsic pathway** of blood-clotting is activated, whereas signals from outside the body, such as trauma trigger the **extrinsic pathway** of blood clotting. These pathways are shown in Fig. 21.8.

At the bottom of both pathways, **fibrinogen** (about 3% of the blood proteins) is soluble and in its center it has a sticky region covered by short amino acid chains that are negatively charged. **Thrombin**, activated from prothrombin by **factor X**, is a serine protease that hydrolyzes the short charged amino acid chains to expose the sticky region and forms **fibrin**. Fibrin molecules, so formed, stick together to form the clot. In the normal situation, clots do not form adventitiously because of the presence of α-**1-antitrypsin**, a serine protease inhibitor produced by the tissues and another inhibitor, **antithrombin**, that blocks the action of thrombin. In the normal situation, when a clot is to be dissolved in the process of wound healing, the protease, **plasmin**, a blood enzyme, is released by the activation of plasminogen, and it breaks down fibrin and dissolves the blood clot.

A deficiency of **factor VIII**, the **antihemophilic factor**, leads to a familial bleeding tendency in males. This disease is **hemophilia** and it affects about 1 in 5000 males. Factor VIII is the cofactor of activated **factor IX** in the factor X-activating complex in the blood-clotting intrinsic pathway (Figs. 21.7 and 21.8). The formation, life cycle, and degradation of factor VIII are shown in Fig. 21.9.

# BLOOD

Blood in the **circulatory system** transports oxygen to the tissues and carbon dioxide back from the tissues for expiration in the lungs. In addition to the population of cells in the blood, it carries hormones, enzymes, various nutrients, and plasma proteins. In order to maintain a critical pH range of 6.8 to 7.4, it contains a pH buffering system. Blood carries water and toxic urea to the kidneys for clearance. The nutrients in blood are amino acids, glucose, vitamins, minerals, fatty acids, and glycerol. The total volume of blood is about 5 L, and it circulates through the kidneys to remove toxins from the body into the urine. Forty-five percent of the blood volume is occupied by cells of which 99% are red blood cells (erythrocytes) and the remaining 1% is made up of leukocytes and platelets. The noncell volume (55%) is plasma that is 92% water and contains various ions (sodium, chlorine potassium, manganese, and calcium) and blood plasma proteins [albumin, globulins, fibrinogen, and various hormones (proteins, peptides, and steroids)]. A diagram of the human systemic circulation is shown in Fig. 21.10.

Blood is oxygenated by the lungs and enters the left auricle through the pulmonary vein and then it is pumped to the left ventricle of the heart and then out through the aorta (reverse arrow in Fig. 21.10). The **aorta** is the major artery

*Biosynthesis & secretion*

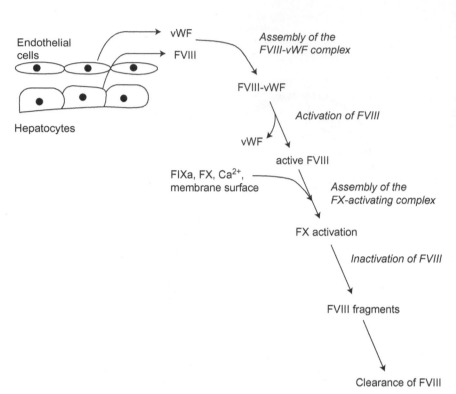

**FIGURE 21.9** The lifespan of factor VIII. Factor VIII is synthesized by various tissues, including liver, kidney, and spleen as an inactive single chain protein. After extensive post-translational processing, factor VIII is released into the circulation as a set of heterodimeric proteins. This heterogeneous population of factor VIII molecules readily interacts with **vWF (von Willebrand factor)** that is produced and secreted by vascular endothelial cells. Upon triggering of the coagulation cascade and subsequent generation of serine proteases, factor VIII is subject to multiple proteolytic cleavages. These cleavages are associated with dramatic changes of the molecular properties of factor VIII, including dissociation of vWF and development of biological activity. After conversion into its active conformation and participation in the factor VIII-activating complex, activated factor VIII rapidly loses its activity. This process is governed by both enzymatic degradation and subunit dissociation. *Reproduced from P.J. Lenting, J.A. van Mourik and K. Mertens, "The life cycle of coagulation factor VIII in view of its structure and function", Blood, **92**: 3983–3996, 1998.*

leading away from the heart to the liver by way of the **hepatic artery**, to the small intestine by way of the **mesenteric artery**, to the kidneys by way of the **renal artery**, and to the legs by way of the **iliac artery**. As for the upper body, oxygenated blood is carried to the arms through the **subclavian artery** and to the head by the **carotid artery**. After the tissues have been oxygenated, the deoxygenated blood travels back to the right ventricle of the heart from the legs via the **iliac vein**, from the kidneys via the **renal vein**, and from the small intestine via **hepatic portal vein** to the liver. The liver empties the deoxygenated blood into the **hepatic vein** to the right ventricle of the heart via the **inferior vena cava**. Deoxygenated blood from the head and from the arms via the **subclavian vein** flows into the **superior vena cava** to the right ventricle of the heart. The deoxygenated blood is then pumped from the heart into the lungs through the **pulmonary artery** and in the lungs $CO_2$ from the blood is expired and the newly oxygenated blood finally enters the aorta and is recycled. In 1 day, the heart beats about 100,000 times and the approximately 5 L of blood circulates three times each minute so that in 1 day the blood travels a total of 12,000 miles.

## Transport of Oxygen

**Hemoglobin** in red blood cells is the oxygen transporter in blood. The counterpart to hemoglobin in muscle is **myoglobin**, a small globular monomeric protein, like hemoglobin (heterodimeric tetramer), that binds oxygen. The kinetics of oxygen binding by the two proteins is different, as shown in Fig. 21.11.

Myoglobin is a monomeric globulin (binds 1 heme) and binds oxygen in a linear fashion, whereas hemoglobin is a tetramer (2 α-globins + 2 β-globins + 4 hemes) (Fig. 21.12) and displays allosteric kinetics in which an initial lag occurs as the oxygen concentration is increased.

When about 10 Torr (10 mmHg) of oxygen pressure is reached, the rate of oxygen binding increases until saturation is reached. The iron in heme must be in the ferrous state ($Fe^{2+}$) in order to bind oxygen. The allosteric kinetics of oxygen binding to hemoglobin can be explained in the following way. The first molecule of the oxygen ligand binds with a low affinity; however, the binding of oxygen alters the conformation of hemoglobin such that the succeeding molecules of oxygen bind with increased affinity reflecting a conformational change in the other hemoglobin subunits. Therefore, there could be two forms of hemoglobin, one with low affinity for oxygen and the other with higher affinity for oxygen.

Head
Arms
Lungs
Jugular vein
Subclavian vein
Pulmonary artery
Superior vena cava
Inferior vena cava
Hepatic vein
Carotid artery
Subclavian artery
Pulmonary vein
Aorta
LA
RA
LV
RV
Heart
Liver
Hepatic artery
Hepatic portal vein
Small intestine
Mesenteric artery
Renel vein
Kidneys
Renal artery
Iliac artery
Iliac vein
Legs
Oxygenated blood
Deoxygenated blood
RA - Right auricle/atrium
LA - Left auricle/atrium
RV - Right ventricle
LV - Left ventricle
Arrows indicate direction of blood flow


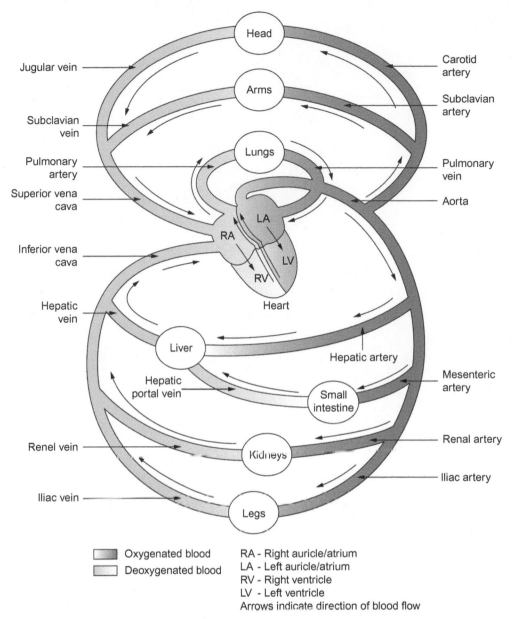

**FIGURE 21.10** Diagram of the human systemic circulation.

Venous blood O2 pressure
Arterial blood O2 pressure
100
50
Fractional saturation
Mb
Hb
Resting muscle
Working muscle
10
20
100
O2 partial pressure (Torr) mm Hg


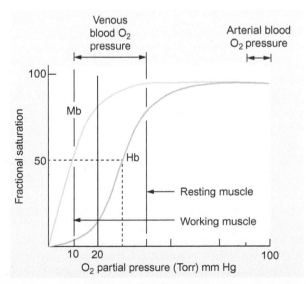

**FIGURE 21.11** Different oxygen binding characteristics between hemoglobin and myoglobin. Myoglobin (Mb) (less than 20 Torr in muscle) binds oxygen under conditions in which hemoglobin releases oxygen. At this pressure hemoglobin (Hb) releases almost all of its oxygen while myoglobin binds over 90% of the released oxygen. The hyperbolic curve of Mb binding is typical of a noncooperative process, whereas the sigmoidal curve of Hb binding is typical of cooperativity. 1 Torr is the pressure to support a column of 1 mmHg @ 0°C and standard gravity.

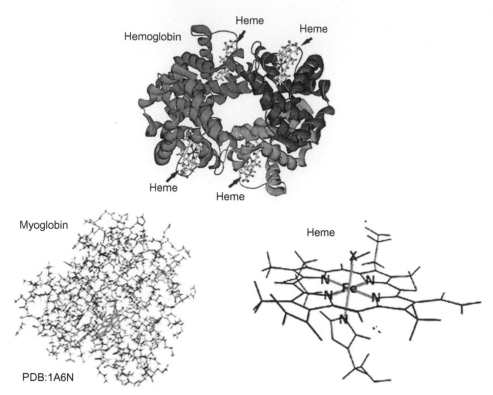

**FIGURE 21.12** *Upper figure* is a model of hemoglobin containing two molecules of α-globin and two molecules of β-globin and four hemes (*ball-stick structures*). Myoglobin (*lower left figure*) is a monomeric globulin containing one molecule of bound heme. The lower right figure enlarges the structure of heme that is linked to the protein through a histidine. *Upper figure is reproduced from http://www.spring8.or.jp/en/news_publications/press_release/2009/090408-2_fig/fig1.png. Bottom two figures are reproduced from http://en.wikibooks.org/wiki/File:Heme2.jpg.*

In the higher affinity conformation of hemoglobin, oxygen would bind with high affinity to all four subunits. Fig. 21.13 shows two possible explanations. In the **symmetry model**, the two forms of **hemoglobin** exist in equilibrium and the oxygen ligands all bind to the four subunits with equally high affinity to the high affinity conformation. In the **sequential model**, the first oxygen ligand binds to one subunit and alters its conformation to the high affinity form. The succeeding oxygen molecules perform similarly until the four hemoglobin subunits are all in the high affinity conformation. The purpose of cooperativity (allosterism) between the four oxygen binding sites is that 1.7 times as much oxygen can be delivered to the tissues compared to a protein in which the oxygen binding sites were independent of each other.

The differences between the three-dimensional structures of the fully oxygenated form (**R form**; for "relaxed") of hemoglobin compared to deoxyhemoglobin (**T form**; for "taut") have been solved. This difference results from a rotation of about 15° between the two α–β dimers (α-globin dimer and β-globin dimer). This rotation alters the bonds between the side chains of the two dimers, causing the heme molecule to alter its position. In deoxyhemoglobin (T structure), the iron atom is displaced from the plane of the porphyrin ring. This movement of the iron atom makes it harder for oxygen to bind to it and, thus, the affinity of oxygen for the T form is reduced. Conversely, in the R form, the iron atom is directly in the plane of the porphyrin ring and, in this position, is more able to bind oxygen; thus, the R form has an increased affinity for oxygen. The high oxygen pressure of the environment in the lungs causes the shift from the T to the R form. In the low oxygen environment of the tissues, on the other hand, the transformation of the R form to the T form occurs. *Hemoglobin, therefore, is able to adjust its structure in the presence or relative absence of oxygen.*

The situation in uncontrolled **diabetes**, for example, can produce another alteration of the hemoglobin molecule. In this case, blood glucose circulates at a high level and, as a result, *hemoglobin can become glycosylated.* Residues of glucose bind to amino acid residues of hemoglobin. Elevated levels of glycated hemoglobin ($HbA_{1c}$) are diagnostic of diabetes. Levels above 10% of hemoglobin, HbA0, that are glycated are indicative of poor metabolic control of carbohydrates and the level of glycated hemoglobin can be used to monitor treatment. In the situation where there high glucose levels are maintained in blood, other proteins as well as hemoglobin can be glycated, including **insulin**. Normally,

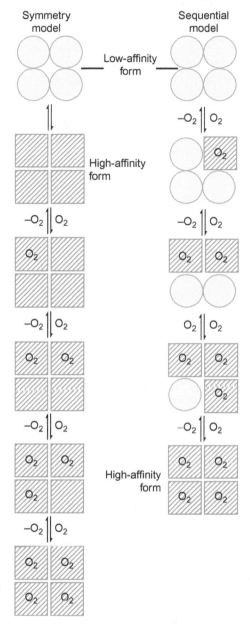

**FIGURE 21.13**   Models to explain the molecular mechanism of positive cooperativity when hemoglobin loads oxygen. In these models, hemoglobin exists in two forms: low affinity and high affinity. However, the difference is that in the Symmetry model, all subunits of hemoglobin exist in either the low affinity or high affinity form. In the Sequential model, the binding of an oxygen molecule leads to the conversion of that subunit to the high affinity form, and so as each molecule of oxygen binds, it converts its subunit to the high affinity form.

there may be some glycated insulin in blood but in diabetes, the level of **glycated insulin** rises by two to threefold. This alteration may affect 10% of the circulating insulin. In the glucose tolerance test where insulin lowers the level of blood glucose over time, glycated insulin has this activity but decreased by a factor of 20%−40%. However, glycated insulin binds to the insulin receptor normally, so that its negative affect may occur at some step other than receptor binding. Elevated levels of **HbA1c** can indicate kidney problems, as well. Experimentally, hemoglobin is highly sensitive to damage by glucose that can result in destruction of the heme and the release of the iron atom. Hemoglobin is primarily glycated on valine and lysine residues *in vivo*. A representation of the reaction of glucose with the terminal amino group of a lysine residue (of hemoglobin) is shown in Fig. 21.14.

HbA1c demonstrates an altered oxygen saturation curve, indicating that glycation can modify the conversion of oxyhemoglobin to deoxyhemoglobin and glycated hemoglobin has a greater affinity for oxygen than HbA0 and a decreased tendency to dissociate oxygen. This is represented by a shift in the oxygenation curve (Fig. 21.11) to the left. The result

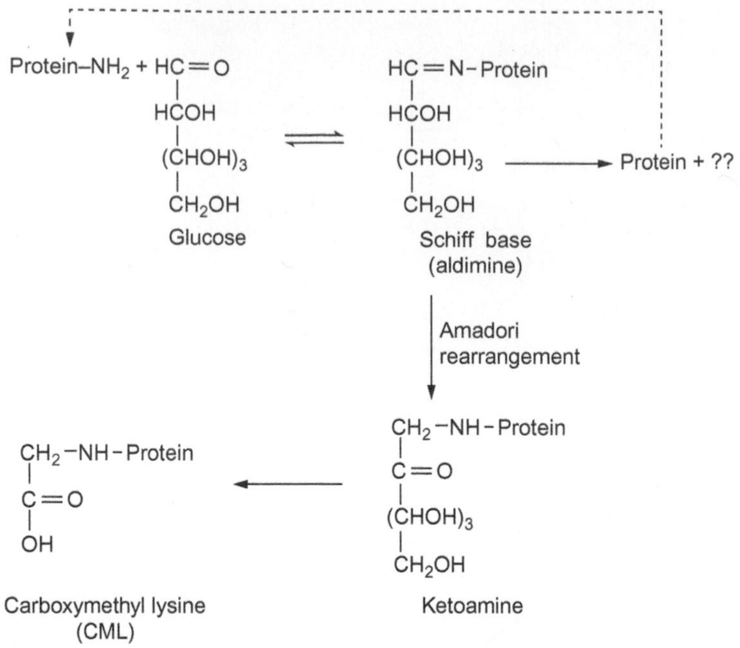

**FIGURE 21.14** A schematic representation of the reactions involved in the glycation of proteins. The open chain form of D-glucose reacts with the ε-amino group of a lysine residue to form a **Schiff base** (a functional imine group formed by the condensation of an amine in the carbonyl of an aldehyde or ketone) which undergoes an Amadori rearrangement to form a ketoamine. This ketoamine is subject to a series of reactions that result in AGEs (advanced glycosylation end products), such as **carboxymethyl lysine (CML)**.

is that the affinity of HbA1c for oxygen is increased as represented by tighter binding and consequently, there is a decrease in the delivery of oxygen to the tissues, all of which increases the risk of coronary artery disease in diabetics and overall poor outcome. Normally, red blood cells have the capacity to deglycosylate proteins but this activity is overrun by glucose when it is at sustained elevated levels in the circulation.

In addition, there are several possible genetic mutations in the hemoglobin molecule that result in serious consequences (Table 21.2).

## Carbon Dioxide

In addition to its ability to bind oxygen, **hemoglobin** can bind carbon dioxide less avidly than it binds oxygen. In the presence of $CO_2$, oxygen is released from hemoglobin, a reaction known as the **Bohr effect**. The oxygen binding curves of hemoglobin in the presence or absence of $CO_2$ are shown in Fig. 21.15.

In the red blood cell with the presence of **carbonic anhydrase**, most of the $CO_2$ is converted to carbonic acid:

$$CO_2 + H_2O \rightleftharpoons H_2CO_3$$

Carbonic acid dissociates into protons and hydrogen carbonate ions:

$$H_2CO_3 \rightleftharpoons H^+ + HCO_3^-$$

In order to maintain the balance of negative charges inside and outside of the red blood cell, $HCO_3^-$ diffuses out from the erythrocyte cytoplasm into the plasma in exchange for chloride ions ($Cl^-$), known as the **chloride shift**. The dissociation of carbonic acid into ions releases protons ($H^+$) that decrease the pH of blood outside the cell and that react with oxyhemoglobin to release oxygen:

$$Hb \cdot 4O_2 + H^+ \rightleftharpoons HHb^+ + 4O_2$$

The release of oxygen from HbAO in the presence of $CO_2$, the Bohr effect, is described by this reaction. Tissue respiration results in the presence of local $CO_2$, causing release of oxygen from oxyhemoglobin from the red blood cell to oxygenate the tissue cell. The increased level of $CO_2$ in plasma evokes alterations within the red blood cell that result in release of oxygen from hemoglobin as shown in Fig. 21.16.

**TABLE 21.2** Some Missense Point Mutations in Human Hemoglobin

| Effect | Residue Changed | Change | Name | Consequence of Mutation | Explanation |
|---|---|---|---|---|---|
| Sickling | β 6 (A3) | Glu → Val | S | Sickling | Val fits into EF pocket in chain of another hemoglobin molecule |
| | β 6 (A3) | Glu → Ala | G Makassar | Not significant | Ala probably does not fit the pocket as well |
| | β 121 (GH4) | Glu → Lys | O Arab, Egypt | Enhances sickling in S/O heteroygotes | β 121 lies close to residue β 6; Lys increases interaction between molecules |
| Change in O2 affinity | α 87 (F8) | His → Tyr | M Iwate | Forms methemoglobin, decreases $O_2$ affinity | The His normally ligated to Fe has been replaced by Tyr |
| | α 141 (HC3) | Arg → His | Suresnes | Increases $O_2$ affinity by favoring R state | Replacement eliminates bond between Arg141 and Asn126 in deoxy state |
| | β 74 (E18) | Gly → Asp | Shepherds Bush | Increases $O_2$ affinity by decreasing in BPG binding | The negative charge at this point decreases BPG binding |
| | β 146 (HC3) | His → Asp | Hiroshima | Increases $O_2$ affinity, reduced Bohr effect | Disrupts salt bridge in deoxy state and removes His that binds a Bohr effect proton |
| | β 92 (F8) | His → Gln | St. Etienne | Loss of heme | The normal bond from FS to Fe is lost, and the polar glutamine tends to open the heme pocket |
| Heme loss | β 42 (CD1) | Phe → Ser | Hammersmith | Unstable, loses heme | Replacement of hydrophobic Phe with Ser attracts water into the heme pocket |
| Dissociation of tetramer | α 95 (G2) | Pro →Arg | St. Lukes | Dissociation | Chain geometry is altered in subunit contact region |
| | α 136 (H19) | Leu → Pro | Bibba | Dissociation | Pro disrupts helix H |

*BPG*, 2,3-bisphosphoglycerate; *EF pocket*, acceptor site in hemoglobin; *S/O heterozygotes*, a rare compound heterozygous hemoglobinopathy.
Reproduced from http://www.chemsoc.org/exemplarchem/entries/2004/durham_mcdowall/prot-evo.html.

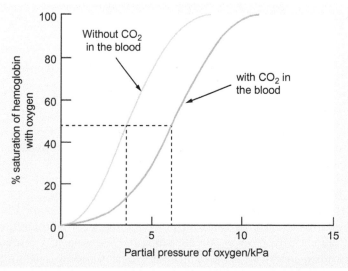

**FIGURE 21.15** Oxygen dissociation curves of hemoglobin in the absence or presence of carbon dioxide in the blood. When $CO_2$ is present in the blood, the binding curve is displaced to the right indicating that more oxygen is required to saturate hemoglobin to form oxyhemoglobin (HbAO). *kPa*, kilopascals; *1 kPa*, 14.7 lbs/in.[2]. *Redrawn from http://www.chemsoc.org/networks/learnet/cfb/transport.htm.*

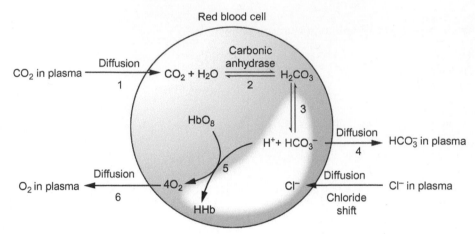

**FIGURE 21.16** Reactions occurring in and around the red blood cell when carbon dioxide (*upper left*) increases in plasma. *Redrawn from http://www.chemsoc.org/networks/learnnet/cfb/transport.htm.*

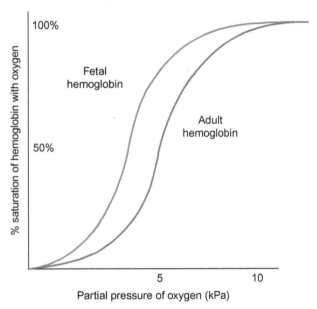

**FIGURE 21.17** Comparison of the oxygen saturation curves for fetal hemoglobin and adult hemoglobin. *Reproduced from http://click4biology.info/c4b/h/images/H6/foetal.gif.*

Carbon dioxide can react directly with hemoglobin by forming a carbamate with an $\alpha$-amino group of an amino acid residue:

$$HbAAresidue-NH2 + CO_2 \rightleftharpoons HbAAresidue-NH-C=O(O-)$$

In this way, hemoglobin carries $CO_2$ from the tissues to the plasma to the lungs where it is expired. In the fetus, there is a different form of hemoglobin, **fetal hemoglobin** (**HbF**, $\alpha_2\gamma_2$). HbF has a higher affinity for oxygen than adult hemoglobin (HbA0) as shown in Fig. 21.17.

The fetal subunit, $\gamma$-globin, has 42 amino acids whereas the adult replacement subunit, $\beta$-globin, has 147 amino acid residues. The $\alpha$-subunit of HbF is the same as the adult $\alpha$-globin. The switch between $Hb\alpha_2\gamma_2$ and $Hb\alpha_2\beta_2$ takes place at birth.

**Carbon monoxide** (CO) poisoning results from the fact that hemoglobin has about 250 fold higher affinity for CO than for $O_2$. When CO binds Hb, oxygen can no longer compete for binding and respiration fails. Carbon monoxide is not bound so strongly that it cannot be reversed by administration of pure oxygen. Thus, pure oxygen greatly increases the mass of ligand over the carbon monoxide (unless pure carbon monoxide is consistently present) indicating that while the binding of CO is strong, it is not irreversible. Accordingly, a person exposed to CO transiently, can be treated with pure oxygen to reverse the effect of CO.

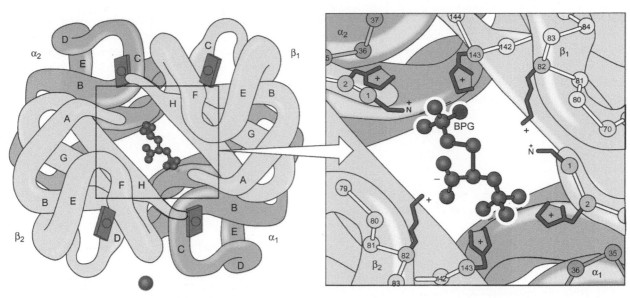

**FIGURE 21.18** Binding of **2,3-*bis*phosphoglycerate (2,3-BPG)** to adult hemoglobin (Hbα₂β₂). 2,3-BPG binds in the cavity between the two β-globin molecules (*left-hand figure*) and because BPG is negatively charged (phosphate groups), it binds electrostatically with positively charged groups in the region of the cavity, shown on the *right-hand figure*. In the *left-hand figure*, the heme groups are indicated by *red rectangles with a circle in the middle. Reproduced from http://www.pearsonhighered.com/mathews/ch07/fi7p18.htm.*

**2,3-Bisphosphoglycerate (2,3-BPG)** is a negative allosteric regulator of the oxygen-binding function of hemoglobin. 2,3-BPG is derived from 1,3-bisphosphoglycerate in the glycolytic pathway. In glycolysis, glucose is converted to two molecules of glyceraldehyde-3-phosphate that is subsequently converted to 1,3-BPG by **glyceraldehyde-3-phosphate dehydrogenase**. 1.3-BPG can be converted to 2.3-BPG by **BPG mutase**. Both 1,3-BPG and 2,3-BPG can be converted to 3-phosphoglycerate, the normal pathway intermediate, and it becomes metabolized through several steps to pyruvate. In the red blood cell, the synthesis of 2,3-BPG is a major pathway for the utilization of glucose and 2,3-BPG is a means of controlling the affinity of hemoglobin for oxygen. 2,3-BPG binds to deoxyhemoglobin (T state), limiting the binding of oxygen to hemoglobin. When 2,3-BPG is low in concentration in the red blood cell hemoglobin is converted to the oxygenated form (R state) more readily. As HbF (fetal hemoglobin) has a lower affinity for 2,3-BPG, the fetus has greater access to oxygenated hemoglobin. This is because there are γ-globins in place of β-globins in the adult form of hemoglobin and 2,3-BPG binds in the cavity between the β-globin chains and it creates electrostatic interactions with positively charged groups that surround this space as shown in Fig. 21.18.

This opening between the β-subunits is only large enough to accommodate BPG in the T state (deoxyhemoglobin). When the T form is converted to the R form (oxygenated hemoglobin), the bonds (positive charges) that hold 2,3-BPG in place are broken and the BPG is released. Hemoglobin will remain in the T state as long as the oxygen concentration is low and when 2,3-BPG is present, there must be more oxygen binding sites filled in order to accomplish the conversion from T state to R state. In the fetus, **fetal hemoglobin (HbF)** has a lower affinity for 2,3-BPG than adult Hb, thus providing it greater access to oxygen.

## Degradation of the Red Blood Cell

After the lifetime of the erythrocyte of 120 days, 90% of the red blood cells are removed from the bloodstream by cells that have engulfing activity: liver macrophages, spleen, and lymph nodes. The remaining 10% of dying red blood cells hemolyze in the bloodstream. *The elements of the red blood cells are degraded by lysosomal enzymes in the macrophage cells.* Hemoglobin is dissociated to heme and globins. Globins are either degraded to free amino acids or reused for new hemoglobin molecules. Iron is removed from the heme molecule in the phagocyte and it is either stored or released into the bloodstream. In the blood, iron binds plasma transferrin and it can be taken up by bone marrow where it can be converted into heme again and form hemoglobin.

## Bilirubin

Heme is converted to **biliverdin** by **heme oxygenase** (with $NADPH + H^+ + O_2$) releasing $Fe^{3+}$ and CO. Biliverdin (green pigment) is converted to yellow **bilirubin** by **biliverdin reductase** (with $NADPH + H^+$). These reactions occur in the **reticuloendothelial system** (liver, spleen, and lymph nodes). Bilirubin moves into the bloodstream in combination with albumin and then moves into the liver with the release of albumin. There, bilirubin is converted to cholebilirubin (bilirubin diglucuronide) and 2UDP by **bilirubin-glucuronyl transferase** (with 2 UDPGlucose). **Cholebilirubin** is secreted by the hepatocyte into the biliary canaliculi with the bile into the small intestine where it is converted to free **bilirubin**. In the large intestine bilirubin is converted to **urobilinogen** (yellow−brown) by the intestinal bacteria that convert urobilinogen to **urobilin** (brown). Some urobilinogen reenters the blood and is removed by the kidney to the urine where it gives urine its yellow color. These pigments give feces the characteristic color.

## Blood Cells

Blood cells occupy about 40% of the blood volume and the plasma occupies the rest. When the clotted blood and the cells are removed, the remainder is the serum. The cells consist of the red blood cells (erythrocytes), white cells (leukocytes) and platelets. The blood cells are derived from the bone marrow containing **pleuripotent stem cells**, either myeloid or lymphoid. The generation of blood cells derived from pleuripotent stem cells is illustrated in Fig. 21.19.

**Hemoglobin** in the red blood cell carries oxygen to the tissues and $CO_2$ from the tissues to the lungs for expiration, as discussed previously. The red blood cell does not have a nucleus and, in its place, is seen as having a central faded spot that varies in size depending on the amount of hemoglobin in the cell. The red cell is round (Fig. 21.19) and is 6−8 μm in diameter.

**Polymorphonuclear leukocytes**, or **segmented neutrophils** are mature **phagocytes**. They migrate through tissues to engulf and destroy microorganisms. They also respond to inflammatory stimuli (e.g., to interleukin-1, released from macrophages as a result of infection, injury to tissues or to histamines released by circulating basophils, mast cells or tissues, and blood platelets). Of the peripheral leukocytes, neutrophils occupy 40%−75%. They are 9−16 μm in diameter. Neutrophils in the normal individual have 3−4 nuclear lobes (segments) with connecting filaments. However, neutrophils can be hypersegmented with six or more lobes and this condition is associated with iron-deficient **anemia**. Hypersegmented neutrophils also may signal the onset of other types of anemia. Neutrophils may display coarse or clumped chromatin with a large cytoplasmic space.

The **eosinophil** is a mature granulocyte that responds to parasitic infections or allergic reactions. Eosinophils occupy 1%−4% of peripheral leukocytes and have diameters of 9−15 μm. The eosinophil may contain 1−3 lobes in the nucleus and, like the neutrophil, they may exhibit coarse or clumped chromatin and a large cyto space. **Eosinophils** are capable of releasing proteins and enzymes from granules that can destroy a parasitic membrane. Granular substances that can be released are: major basic protein, cationic proteins, peroxidase, arylsufatase B, phospholipase D, and histaminase.

**Basophils** are granulocytes that contain granules of **heparin** and vasoactive substances that are released upon activation. They occupy about 0.5% of the total leukocyte population. Basophils are involved in immediate hypersensitivity reactions (type I, e.g., a bee sting) and in some delayed hypersensitivity reactions. Their diameters are 10−15 μm, making them the smallest of the circulating granulocytes. They have coarse chromatin and the nucleus and cytoplasm occupy similar volumes. The homogeneous cytoplasm contains large granules.

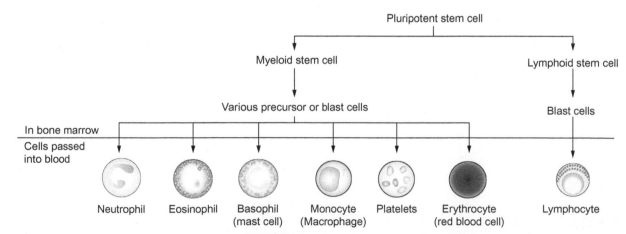

FIGURE 21.19   Production of blood cells from pleuripotent stem cells.

The **monocyte** (large mononuclear phagocyte) is an immature stage of the macrophage. The diameter of a monocyte is from 10 to 30 μm. Its nucleus can be shaped like a horseshoe folding over on itself and can be twice the size of the vacuolated cytoplasm. Chromatin is arranged in strands. Before emigrating into the tissues, monocytes circulate in the peripheral blood. In the liver, they are **Kupfer cells**; in the brain, **microglia**; in the kidney, **mesangial cells**, and in the bone, they are **osteoclasts**.

**Platelets** are cyto fragments of megakaryocytes circulating as small discs in the peripheral blood. They are instrumental in forming blood clots (Fig. 21.6) and in the maintenance of the lining of the blood vessel. Platelets have no nucleus and have diameters of 14 μm. Normal blood contains 130,000 to 450,000 platelets per microliter.

**Lymphocytes** vary in size and in the number of granules in their cytoplasm. Common lymphocytes in peripheral blood are 7–10 μm in diameter containing nuclei that are about 7 μm in diameter. Much of the nuclear chromatin is condensed. The nucleus almost completely fills the cytoplasm and may be 3–5 times larger than the cytoplasm. If they achieve immune competence within the bone marrow, they are **B cells** or if they become immune competent in the thymus, they are **T cells**. Secondary lymphoid tissue includes: lymph nodes, adenoids, tonsils, and **mucosa-associated lymphoid tissue** (**MALT**). MALT includes bronchus-associated lymphoid tissue, gut-associated lymphoid tissue, nasopharyngeal-associated lymphoid tissue, and urogenital-associated lymphoid tissue. B cells respond to antigens by producing antibodies or lymphokines.

**Band neutrophils** (young neutrophils) occupy 1%–3% of peripheral leukocytes. Before segmentation, the nucleus appears as a U-shaped rod with coarse and clumped chromatin.

A patient's status usually requires a **complete blood cell count** (**CBC**). Table 21.3 reviews the components of the complete blood count (CBC) and information obtained.

## Blood Proteins

Major proteins are serum albumins (55% of plasma proteins), globulins, especially γ-globulins (mainly antibodies), fibrinogens, and hemoglobin. Other proteins are as follows: serum haptoglobin (hemoglobin-binding protein in the blood), protein C (a vitamin K-dependent plasma protein that cleaves activated factors V and VIII and inhibits blood coagulation and clot formation shown in Fig. 21.4). Thromboplastin (factor III, platelet tissue factor, and thrombokinase) is a plasma protein (enzyme complex) in tissues, platelets, and white blood cells; it catalyzes the conversion of prothrombin to thrombin in the presence of $Ca^{2+}$ in the extrinsic pathway of blood coagulation (Fig. 21.8) and appears to result from an interaction of coagulation factors V, VIII, IX and X, and factor IV. There can exist abnormal proteins in blood, such as macroglobulin, cryoglobulin, or myeloma protein. Regulatory peptides and fragments of proteins are also found in serum (Table 21.4).

## Blood Type and Rh Factor

There are four human **blood types**: A (42% in the United States), B (10% in the United States), AB (3% in the United States), and O (45% in the United States). These types derive from codominant alleles (genotype). There are three different alleles: $I^A$, $I^B$, and i (recessive). Type A can be homozygous ($I^A I^A$) or heterozygous ($I^A i$) and type B can be either homozygous ($I^B I^B$) or heterozygous ($I^B i$). Type AB has codominant alleles ($I^A I^B$). Type O is homozygous recessive (ii).

If a mother had a type A blood type ($I^A I^A$ or $I^A i$) and the father had a type B blood type ($I^B I^B$ or $I^B i$), there would be four genetic crosses possible for the determination of the blood type of their child: $I^A I^A$ X $I^B i$, $I^A I^A$ X $I^B I^B$, $I^A i$ X $I^B I^B$, and $I^A i$ X $I^B i$. If the father was type AB (genotype $I^A I^B$) and the mother was type O (genotype ii), half of the offspring would be type A ($I^A i$) and half would be type B ($I^B i$) but they would not be AB or O.

Macromolecules on the surface of **red blood cells** determine a blood type. Different blood types could be incompatible when they are mixed together and could coagulate in a blood vessel. **Blood type A** has antigens on the surface of the red blood cell and antibodies to type B in the plasma. **Blood type B** has B antigens on the erythrocyte surface and the plasma contains antibodies to blood type A. **Blood type AB** has both the A and B antigens on the surface of the red blood cell and no antibodies to either type in the plasma. **Blood type O** has no A or B antigens on the surface of the red blood cell but has antibodies to both group A and B antigens. Considerations of antigens and antibodies in blood are vital when transfusions are needed. Type O used to be considered a universal donor; however, there can be complications because of the content of antibodies. The best choice is to use the identical type blood that characterizes the recipient. Table 21.5 summarizes the blood types and the types that can be given or received in transfusions.

**TABLE 21.3 Components of the Complete Blood Count (CBC)**

| Test | Name | Measuring | Use | Increased/Decreased |
|---|---|---|---|---|
| WBC | White blood cell | Total number of WBCs per volume of blood (sum of all types of WBCs) | The body uses WBCs to fight infection. Each type has a slightly different job. WBC is measured to make sure there are a sufficient number and to help detect and monitor conditions that lead to increases or decreases in total WBCs and/or to increases in one or more types of WBCs | May be increased with infections, inflammation, cancer, leukemia; decreased with some medications (such as methotrexate), some auto-immune conditions, some severe infections, bone marrow failure, and congenital marrow aplasia (marrow does not develop normally) |
| % Neutrophil | Neutrophil/band/seg | Measures the percentage of each of five types of WBC, compared to total WBC count | | This is a dynamic population that varies somewhat from day to day depending on what is going on in the body. Significant increases in particular types are associated with different temporary and acute and/or chronic conditions. An example of this is the increased number of lymphocytes seen with lymphocytic leukemia |
| % Lymphs | Lymphocyte | | | |
| % Mono | Monocyte | | | |
| % Eos | Eosinophil | | | |
| % Baso | Basophil | | | |
| Neutrophil | Neutrophil/band/seg | Measures the actual number of each type of WBC per volume of blood | | |
| Lymphs | Lymphocyte | | | |
| Mono | Monocyte | | | |
| Eos | Eosinophil | | | |
| Baso | Basophil | | | |
| RBC | Red blood cell | Total number of RBCs per volume of blood | Primarily measured to detect decreased production, increased loss, or increased destruction of RBCs, to detect anemia, and sometimes to help detect erythrocytosis (too many RBCs) | Decreased with anemia; increased when too many made and with fluid loss due to diarrhea, dehydration, burns |
| Hgb | Hemoglobin | Total amount of oxygen-carrying protein inside RBCs | | Mirrors RBC results |
| Hct | Hematocrit | Percentage of blood volume made up of RBCs (solid versus liquid portion of blood) | | Mirrors RBC results |
| MCV | Mean corpuscular volume | Average size of RBCs | The size of RBCs and the average amount of hemoglobin inside them can help classify different types of anemia | Increased with $B_{12}$ and folate deficiency, decreased with iron deficiency and thalassemia |

| | | | Mirrors MCV results |
|---|---|---|---|
| MCH | Mean corpuscular hemoglobin | Average amount (weight) of hemoglobin inside each RBC | |
| MCHC | Mean corpuscular hemoglobin concentration | Average concentration (%) of hemoglobin inside each RBC | May be decreased when MCV is decreased; increases limited to amount of Hgb that will fit inside a RBC |
| RDW | RBC distribution width | Measures variation in size of RBCS, most normal RBCs are the same size | Help classify anemia | Increased RDW indicates mixed population of RBCs. Immature RBCs tend to be larger |
| Platelet | | Total number of platelets per volume of blood; platelets are special cell fragments that are important in blood clotting | Determine whether number is adequate to control bleeding | Decreased or increased with conditions that affect platelet production; decreased when greater numbers are used, as with bleeding; decreased with some inherited disorders (such as Wiskott-Aldrich and Bernard Soulier), with systemic lupus erythematosus, pernicious anemia, hypersplenism, spleen takes too many out of circulation), leukemia, and chemotherapy |
| MPV | Mean platelet volume | Average size of platelets | Help evaluate decreased platelets | Vary with platelet production. Younger platelets are larger than older ones |

**TABLE 21.4  Proteins, Regulatory Peptides and Protein Fragments Found in Plasma**

| | |
|---|---|
| Peptide hormones | Angiotensin I, Guanylin (22–115), Uroguanylin (89–112), Atrial natriuretic factor (CDD/ANP99-126), GLP-1 |
| Cytokines, growth factors | HCC-1, IGF-1, IGF-2, osteoinductive factor, PDGF, CTAP-III, pigment endothelium-derived factor |
| Defensins | β-Defension 1, propeptides of neutrophil defensins 1 to 3 |
| Plasma proteins | Albumin, fibrinogen A, fibrinogen B, β-2-microglobulin, zinc-α-2-glycoprotein, α-2-HS-glycoprotein [fetuin], serum amyloid protein A, haptoglobin, profilin, desmocollin, thymosin-β-4 and β-10, apolipoprotein C-III, uteroglobin, ubiquitin, gelsolin |
| Transport proteins | Retinol binding protein, α-1-microglobulin, transferrin, transthyretin, TGF β-binding protein, IGF-binding protein 2 and 3 |
| Complement factors | Factor C3, factor D, factor C4A (anaphylatoxin) |
| enzymes, inhibitors | Lysozyme, cystatin C, α-1-antitrypsin, pancreatic trypsin inhibitor, plasminogen, α-2-antiplasmin, carboxypeptidase N, inter-α-trypsin inhibitor component II, somatomedin B, vitronectin |
| Matrix proteins | Collagens α-I-[I], α-2-[I], α-3-[IV], α-1-[XVIII], and osteopontin |

*Human Biochemistry & Disease*, by G. Litwack (2008), Table 14-4, page 889.
Reproduced from Table 2 of http://www.abrt.org/jbt/1998/December98/dec98rcjurpens.html.

**TABLE 21.5  The Blood Groups, Antigens on the Surface of Red Blood Cells, Antibodies in Plasma, the Blood Groups That Can be Donated to and the Blood Groups That Can be Received in Transfusion**

| Blood Group | Antigens | Antibodies | Can Give Blood to | Can Receive Blood From |
|---|---|---|---|---|
| AB | A and B | None | AB | AB, A, B, O |
| A | A | B | A and AB | A and O |
| B | B | A | B and AB | B and O |
| O | None | A and B | AB, A, B, O | O |

HDB, Table 14-6, Page 890.
Reproduced from http://nobelprize.org/medicine/educational/landsteiner/readmore.html.

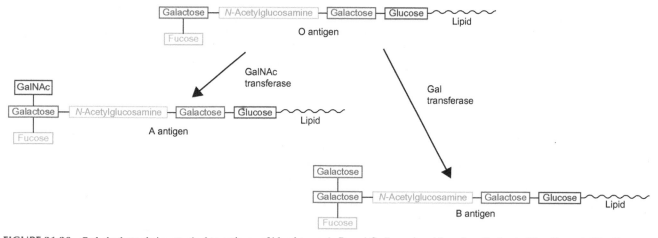

**FIGURE 21.20**  Carbohydrate chains attached to antigens of blood types A, B, and O. *Reproduced from http://upload.wikimedia.org/wikipedia /commons//f/f5/ABO_antigens.JPG.*

## Origin of Nonself Blood Type Antibodies

As mentioned above, blood type A has blood type B antibodies and blood type B has blood type A antibodies while blood type AB has neither and blood type O has both antibodies. Carbohydrate groups are antigenic and most strong antigens contain sugars or substituted sugars. In the case of blood type O, there are no carbohydrates added to the antigen chain that is recognized as self, whereas in blood type A, there is a galactosyl-*N*-acetyl group added to the galactose terminus (Fig. 21.20). In the case of blood type B, there is an additional galactose added to the type O chain. Thus, antibodies are formed to these added carbohydrates because they are recognized as nonself, whereas the chain of type O antigen is recognized as self. Both of these substances (galactose- and galactosyl-*N*-acetyl-) will have been consumed through the diet and **isogglutinin antibodies** will be formed. Isogglutinin antibodies will not cross the placenta and are less dangerous to the fetus than antibodies to the Rh factor discussed below.

## Rh Antigen

In addition to the ABO system of antigens on the red blood cell surface, there is the Rh antigen that may be present or absent. Those who do have the Rh antigen on the erythrocyte surface are Rh positive and those who lack the Rh antigen on the erythrocyte surface are Rh negative. The presence or absence of the Rh antigen further defines the blood type. Thus, blood type O positive is 37.4% of the American population, O negative, 6.6%, AB positive 3.4%, AB negative, 0.6%, B positive 8.5%, B negative 1.5%, A positive 35.7%, and A negative 6.3%.

An Rh negative mother and an Rh positive father will have a child that has a 60% chance of being Rh positive. If the blood of the baby mixes with the mother's blood, the mother would make IgG antibodies against the fetus' red blood cells that could cross the placenta and cause their destruction. If the mother has not already made antibodies to the fetus' Rh antigen, there does exist an Rh immunoglobulin (RhoGAM or RhIG, $Rh_0(D)$ immune globulin) that can prevent the mother's generation of antibodies. This is a solution, injected intramuscularly, of IgG anti-D antibodies that suppress the mother's immune system from attacking Rh positive blood cells. There are three Rh (Rhesus) antigens (RhAG, RhCE, RhD) and their predicted structures are shown in Fig. 21.21.

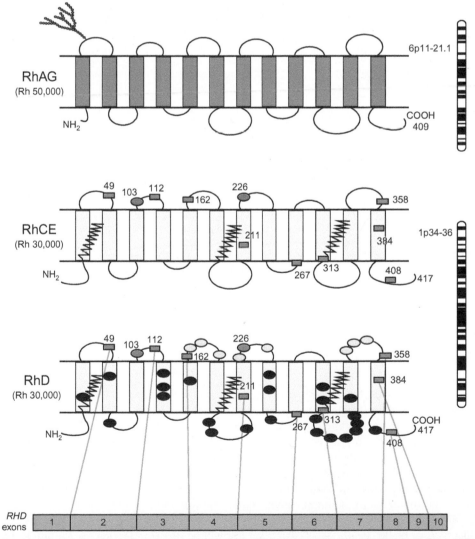

**FIGURE 21.21** Predicted structure of three Rh antigens. Model of topology for RhAG (50,000 Da, 409 amino acids; RhCE and RhD (also called $Rh_0D$, 30,000 Da) have similar topology. Eight D-specific amino acids are located on the exofacial surface (*yellow ovals*) and 24 of these reside in the transmembrane and cytoplasmic domains (*black ovals*). *Red ovals* represent critical amino acids for C and E antigens; *purple ovals* represent Ser103 and Ala226 on RhD. *Zigzag lines* represent the Cys-Leu-Pro motifs that may be involved in the palmitoylation sites. The *N*-glycan on the first loop of RhAG is indicated by the branched structure of *red circles*. *Reproduced from Figure 1 of N.D. Avent and M.E. Reid, "The Rh blood group system: a review", Blood, 95: 375–387, 2000.*

**TABLE 21.6 Transfusability of Various Blood Types Based on the ABO System and Rh Antigen**

**If Your Blood Type Is . . .**

| Type | You Can Give Blood To | You Can Receive Blood From |
|---|---|---|
| A+ | A+, AB + | A+, A−, O+, O− |
| O+ | O+, A+, B+, AB+ | O+, O− |
| B+ | B+, AB+ | B+, B−, O+, O− |
| AB+ | AB+ | Everyone |
| A− | A+, A−, AB+, AB− | A−, O− |
| O− | Everyone | O− |
| B− | B+, B−, AB+, AB− | B−, O− |
| AB− | AB+, AB− | AB−, A−, B−, O− |

**Out of 100 Donors . . .**

| | |
|---|---|
| 84 donors are RH+ | 16 donors are RH− |
| 38 are O+ | 7 are O− |
| 34 are A+ | 6 are A− |
| 9 are B+ | 2 are B− |
| 3 are AB+ | 1 is AB− |

Source: AABB.ORG. http://chapters/redcross/org/br/northernohio/INFO/bloodtype.html.

**TABLE 21.7 Inheritance of Blood Types and Rh Factor**

| Blood Type | | Mother's Type | | | |
|---|---|---|---|---|---|
| | | O | A | B | AB |
| Father's Type | O | O | O, A | O, B | A, B |
| | A | O, A | O, A | O, A, B, AB | A, B, AB |
| | B | O, B | O, A, B, AB | O, B | A, B, AB |
| | AB | A,B | A, B, AB | A, B, AB | A, B, AB |

**Inheritance of Rh Factor**

| Rh Factor | | Mother's Type | | |
|---|---|---|---|---|
| | | $Rh^+$ | | $Rh^-$ |
| Father's Type | $Rh^+$ | $Rh^+$, $Rh^-$ | | $Rh^+$, $Rh^-$ |
| | $Rh^+$ | $Rh^+$, $Rh^-$ | | $Rh^-$ |

HDB, Tables 14.8 & 14.9 combined, page 893.
Reproduced from http://www.mascupid.com/health/bloodinherit.htm.

From the figure, it can be seen that the Rh antigen is a 12-transmembrane protein. The Rh antigen exists on the red blood cell membrane in a complex of proteins. They are glycoproteins **Rh50**, **CD47**, and **glycophorin B**. They need to interact with the red blood cell membrane in order to form a multisubunit complex. It has been shown that a deficiency of the Rh50 protein (consists of 26 amino acids) can prevent the assembly or transport of the Rh membrane complex to the surface of the red blood cell. Furthermore, this protein is located in the contractile vacuole that is the organelle responsible for maintenance of osmotic equilibrium within the cell. However, when the gene encoding information for Rh50 is disrupted, there seems to be little effect on cellular osmotic equilibrium, suggesting that although this protein is located in a particle that functions to maintain osmotic equilibrium in the cell, Rh50 may have a different function.

A person with **Rh negative blood** does not have Rh antibodies but if that person is transfused with $Rh^+$ **blood**, antibodies will develop. About 84% of blood donors will be $Rh^+$; the rest will be Rh−. Considering the Rh factor, the ABO blood typing system becomes more complicated and allowable transfusions from donors and to recipients, are listed in Table 21.6.

Inheritance of blood types and Rh factor are shown in Table 21.7.

## LYMPHATIC SYSTEM

The lymphatic system consists of a network of tissues and organs consisting partly of lymph vessels, lymph nodes, and lymph fluid. Organs that are part of the lymphatic system are tonsils, adenoids, spleen, and thymus gland. In the human, there are up to 700 lymph nodes responsible for filtration of the lymph before it is returned to the circulatory system. The thymus is the storage location for immature lymphocytes and the subsequent generation of active T cells.

The lymphatic system contains thin-walled vessels and valves that promote flow in one direction. Movement of the lymph fluid through valves is caused by skeletal muscle contraction. Lymphatic vessels join with two ducts, the **thoracic duct** and the **right lymph duct** that empty near the heart. An overview of the lymphatic system is shown in Fig. 21.22.

The lymphatic system collects water lost from the blood capillaries and tissues and this water is emptied into the blood circulation. Proteins and lipids that are too large to be taken up by the blood capillaries are transported by the lymph system. Also, some of the extracellular fluid is absorbed by lymph capillaries and drained into the larger vessels of the lymphatic system. Lymph capillaries that interface with tissue cells and blood capillaries are shown in Fig. 21.23.

**Lymph** is collected from the left side of the body, the digestive tract, and the right side of the lower body, all of which flows into the **thoracic duct**. In 1 hour, about 100 mL of lymph from the thoracic duct is emptied into the **left subclavian vein** and the lymph from the right side of the head, neck, and chest is collected by the **right subclavian vein**. The increased production of lymph generates from increased capillary blood pressure and decreased concentration of plasma proteins. If the lymphatic system is unable to accommodate an overproduction of lymph, it can accumulate and distend the tissues with the production of edema. Antibodies can be produced owing to the concentration of white

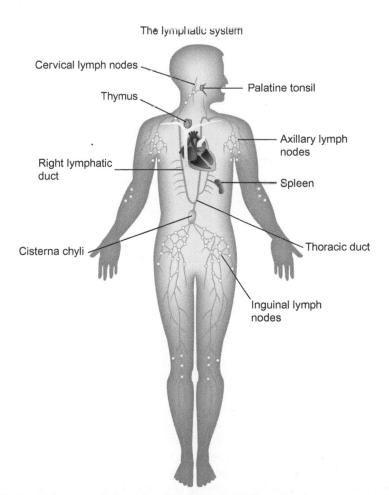

The lymphatic system

Cervical lymph nodes

Palatine tonsil

Thymus

Axillary lymph nodes

Right lymphatic duct

Spleen

Cisterna chyli

Thoracic duct

Inguinal lymph nodes

**FIGURE 21.22** Overview of the lymphatic system. *Redrawn from http://www.paradoja7.com/lymphatic-system-pictures-2/.*

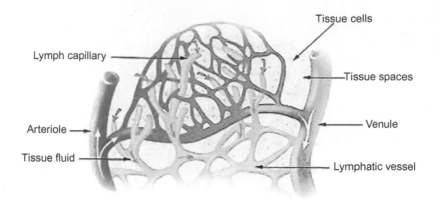

**FIGURE 21.23**    Lymph capillaries in tissue spaces in contact with tissue cells. *Reproduced from http://training.seer.cancer.gov/anatomy.lymphatic/ components/.*

blood cells in the lymph nodes that act as part of the **immune system**. Antibodies and lymphocytes are able to enter the bloodstream through the subclavian veins.

In its composition, lymph fluid is similar to blood plasma. As fluids accumulate and pass through the arterial capillary walls, it is removed by small lymphatic vessels and returned to the bloodstream that prevents edema and aids in the maintenance of normal blood volume and pressure. In contrast to blood vessels, lymphatic vessels carry fluid away from the tissues. Lymph capillaries are found everywhere in the body except for bone marrow, the central nervous systems, and tissues that lack blood vessels, such as the epidermis. Fluid can enter the lymph capillary but not leave it because within the lymph capillary wall endothelium, squamous cells overlap to form a one-way valve. The very small lymph capillaries join to form vessels and the vessels join to form larger lymphatic trunks capable of draining large regions. These trunks merge to the point where the lymph drains into the right lymphatic duct from the upper right quadrant of the body and the thoracic duct drains the rest (Fig. 21.22). Although there does not exist a pump in the lymphatic system, pressure gradients move lymph through the vessels by the actions of skeletal muscles, respiratory movement, and vessel wall smooth muscle.

The lymphatic system is also the **immune system** consisting of a network of specialized organs (digestive tract linings, respiratory tract linings, and in skin) and cells (lymphocytes, macrophages, etc.) that protect the body from invasion by foreign bacteria or viruses. Immune/lymphatic (I/L) cells are immunocompetent and can distinguish between "self" and "nonself." Neutrophils and macrophages act nonspecifically within hours. Specifically acting adaptive components respond more slowly within 4−7 days and can detect changes in cells, such as the development of cancer cells that have altered molecules on their cell surfaces. Cells of the I/L system are derived from precursor bone marrow cells and circulate in both blood and lymph and migrating into connective tissue or to immune organs. **Macrophages** identify bacteria or viruses as nonself, engulf, and destroy them. Antibodies bind to the foreign bacterial or viral particle and the antibody−bacterium complex binds to an antibody receptor on the macrophage membrane and allows the engulfment of the foreign particle by the macrophage cell (Fig. 21.24A and B).

The engulfed bacterium is digested by the macrophage lysosomes.

## Thymosin

Thymosin is a 5-Da polypeptide hormone secreted by the thymus gland. Thymosin α1 stimulates the development of precursor T cells in the thymus to mature T cells. Of the thymosin peptide family, **thymosin β4**, is the most abundant member and is also expressed in many cell types. Thymosin β4 is the principal G-actin sequestering molecule in mammalian cells, playing an important role in the organization of the cytoskeleton. The structures of thymosins and the complex between thymosin β4 and actin are shown in Fig. 21.25.

The activities of **thymosin β4** are many and include: promotion of cell migration, blood vessel formation, survival of cells, differentiation of stem cells, modulation of cytokines, chemokines, and certain proteases as well as up-regulation of matrix molecules and gene expression. Thymosin β4 is being used in clinical applications. One example is its use in treating patients with ST elevation myocardial infarction in combination with other methods of treatment.

Notably, thymosin β4 is secreted from **platelets** and aids in the formation of crosslinks with fibrin in a time- and calcium-dependent manner in the process of clot formation. This crosslinking is mediated by **factor XIIIa**, a

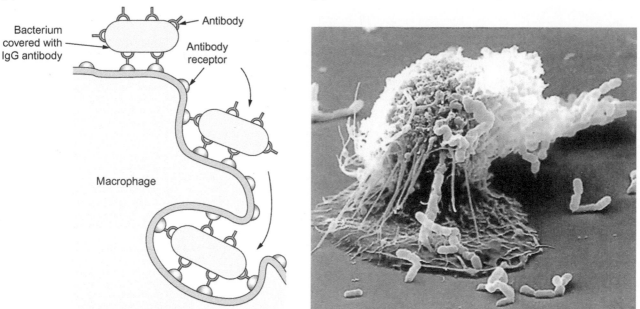

**FIGURE 21.24** (A) A macrophage engulfs a bacterium coated with antibodies. Cell surface receptors against antibodies on the cell surface bind the antibody−bacterium complex facilitating engulfment and internal destruction of the bacterium. (B) A macrophage in the process of ingesting bacteria. Projections from the cell are pseudopodia that are extensions of the cell membrane to pick up and ingest bacteria. *(A) Redrawn from W.K. Purves et al.,* Life: The Science of Biology, *fourth ed., Sinauer Associates and WH Freeman, Garlansville, VA, 2002. (B) Reproduced from http://smg.photobucket.com/user/Aegeri/media/Macrophage.jpg,html.*

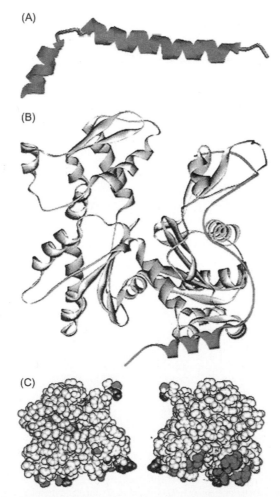

**FIGURE 21.25** (A) Structure of one of the thymosin family members, thymosin β9. (B) Complex of thymosin β4 (*red*) and actin (*gray*) monomer. (C) Molecular contact sites between actin and thymosin β4 (*dark green*). Image on the *left* is a "front" view and image on the *right* is back view. Part (A) is reproduced from Stoll, I.R., Voelter, W., & Holak, T.A., "Conformation of thymosin b9 in water/Fluoroalcohol solution determined by NMR spectroscopy", *Biopolymers*, **41**: 673 (1979); parts (B) and (C) were reproduced from Dos Remedios, C.G., et al., "Actin binding proteins: regulation of cytoskeletal microfilaments", *Physiol. Rev.*, **83**: 433-473 (2003).

**transglutaminase** that is released with thymosin β4 from stimulated platelets. In this way, thymosin β4 stabilizes the blood clot (Fig. 21.26).

The maturation of the **T cell** from the bone marrow to the thymus to the peripheral blood, lymph nodes, spleen, and skin is pictured in Fig. 21.27.

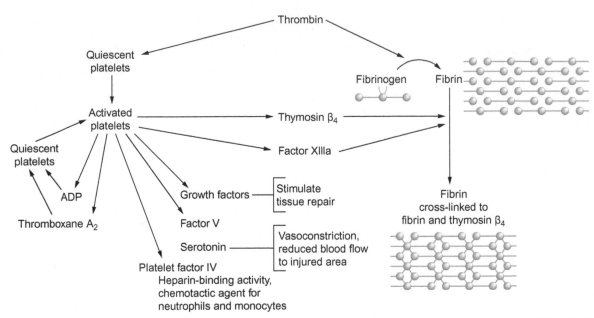

**FIGURE 21.26**   Thymosin β4 is coreleased from activated platelets together with factor XIIIa (transglutaminase) to form crosslinks with fibrin during clot formation. *Redrawn from http://www.fasebj.org/content/vol16/issue7/images/large/380421351s01.jpeg.*

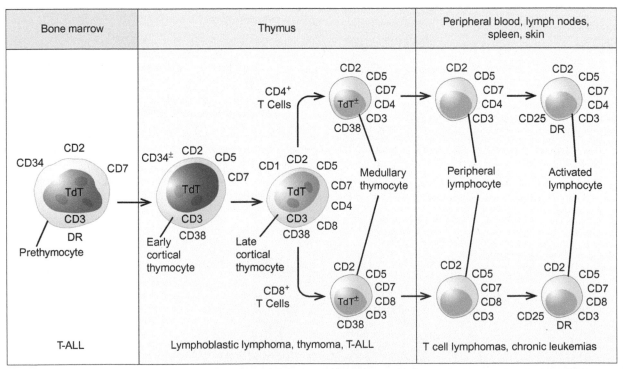

**FIGURE 21.27**   Sequential expression of selected antigens during T cell development. Depicted is the pathway of T cell differentiation from the bone marrow stem cell to thymocyte maturation in the thymus, leading to the diversion of the CD4$^+$ helper and CD8 cytotoxic suppressor cell sublineages, each of which undergoes further differentiation in peripheral lymphoid tissues. Certain disorders are shown under the associated phenotypes (*Dr*, HLA class II; *T-ALL*, T cell acute lymphoblastic leukemia; *TdT*, terminal deoxynucleotidyl transferase). *Reproduced from http://what-when-how.com/wp-content/uploads/2012/04/tmpDC36.jpg.*

# SUMMARY

Deep vein thrombosis is a clinical example of the blood circulatory system. There are a great many factors that are involved in **blood clotting** and in the regulation of the blood clotting mechanism. The many factors in this mechanism are considered. The circulatory system is responsible for carrying oxygen to the tissues and carrying carbon dioxide from the tissues to the lungs for expiration. The formation of **hemoglobin** and its functions are reviewed. Two possible kinetic mechanisms of oxygen binding are presented. Myoglobin, the oxygen-carrying protein in skeletal muscle is reviewed.

In **diabetes**, the blood level of glucose is increased leading to the **glycation of proteins**, including insulin. The mechanism of glycation of proteins is presented.

The reaction between hemoglobin and carbon dioxide is discussed as well as the interaction of hemoglobin and oxygen in the presence or absence of carbon dioxide. The reactions within the red blood cell are reviewed. The saturation curves for hemoglobin binding of oxygen in the adult and fetus are presented.

**2,3-Bisphosphoglycerate (2,3-BPG)** is a regulator of hemoglobin. The mechanism of its binding to hemoglobin and its effects are presented.

The lifetime of the **red blood cell** is 120 days after which 90% of the red blood cells are removed by **macrophages** in the liver, spleen, and lymph nodes. The rest are hemolyzed in the bloodstream. The contents of the erythrocytes are degraded by lysosomal enzymes of macrophages. Globins are broken down to free amino acids or reused for assembly of new hemoglobin molecules. Heme is subjected to a number of reactions that ultimately lead to the formation of **urobilinogen** and **urobilin** that provide the color to feces and urine.

Blood cells are derived from **pleuripotent stem cells** (myeloid or lymphoid) in bone marrow. The lymphoid precursors are destined for maturity in the thymus gland.

The **cell types** that represent 40% of the blood volume are polymorphonuclear leukocytes, eosinophils, basophils, monocytes, platelets, lymphocytes, and band neutrophils; these are considered with their characteristics. The proteins that occur in blood are: serum albumins, globulins, fibrinogens, hemoglobin, serum haptoglobin, protein C, and thromboplastin. Abnormal disease proteins are macroglobin, cryoglobin, and myeloma protein.

The basis of **blood typing** using the ABO system is presented along with consideration of the **Rh factor**. Prediction of paternity based on blood type is presented and the allowable transfusions for donor and recipients are described.

The **lymphatic system** consists in part as a network of lymph vessels, lymph nodes, and lymph fluid as well as the organs: tonsils, adenoids, spleen, and **thymus** gland. Lymph is collected from the left side of the body, the digestive tract and the right side of the lower body and all these sources flow into the **thoracic duct**, one of two main ducts in the system. The other main duct is the **right lymphatic duct**. Lymph from the thoracic duct empties into the left subclavian vein. Lymph from the right side of the head, neck, and chest is collected by the right lymphatic duct and into the subclavian vein. White cells in the lymph nodes can produce **antibodies** that enter the circulation through the subclavian veins. The lymphatic system, including specialized organs in the GI tract, lungs, and skin represent the **immune system** owing to the content of lymphocytes.

**Macrphages** are part of the lymphatic system and the mechanism of engulfment of foreign particles is shown.

The **thymus** is part of the lymphatic system and it produces a family of polypeptide hormones called **thymosins** of which **thymosin β4** is a major member. The structure of thymosins is presented as well as the interaction of thymosin β4 with **actin**, which is the principal activity of the hormone in the organization of the cytoskeleton. Together with **factor XIIIa**, thymosin stimulates and stabilizes **fibrin crosslinks** in clot formation. In the thymus, the T-cell precursor is matured into an active T cell.

# SUGGESTED READING

## Literature

App, C., Knop, J., Huff, T., Sticht, H., Hanappel, E., 2013. Thymosin β4 and tissue transglutaminase. Molecular characterization of cyclic thymosin β4. Protein J. 32, 484−492.

Assman, N., Finlay, D.K., 2016. Metabolic regulation of immune responses: therapeutic opportunities. J. Clin. Invest. 126, 2031−2039.

Avent, N.D., Reid, M.E., 2000. The Rh blood group system: a review. Blood. 95, 375−387.

Burstyn-Cohen, T., Heeb, M.J., Lemke, G., 2009. Lack of protein S in mice causes embryonic lethal coagulopathy and vascular dysgenesis. J. Clin. Invest. 119, 2942−2953.

Cueni, L.N., Detmar, M., 2008. The lymphatic system in health and disease. Lymphat. Res. Biol. 6, 109−122.

Davie, E.W., 2003. "A brief historical review of the waterfall/cascade of blood coagulation". J. Biol. Chem. 278, 50819−50832.

Dos Remedios, C.G., et al., 2003. "Actin binding proteins: regulation of cytoskeletal microfilaments". Physiol. Rev. 83, 433−473.

Fluentes, R.F., et al., 2016. A chimeric platelet-targeted urokinase prodrug selectively blocks new thrombus formation. J. Clin. Invest. 126, 483−494.

Gordon, S., Taylor, P.R., 2005. Monocyte and macrophage heterogeneity. Nat. Rev. Immunol. 5, 953–964.

Johnsen, J.M., et al., 2012. Discrepancies between ABO genotype and ABO glycan phenotypes. Blood. 120, Abstract 274.

Kane, W.H., Mruk, J.S., Majerus, P.I., 1982. Activation of coagulation factor V by a platelet protease. J. Clin. Invest. 70, 1092–1100.

Lenting, J.A., van Mourik, J.A., Mertens, K., 1998. The life cycle of coagulation factor VIII in view of its structure and function. Blood. 92, 3983–3996.

Levantino, M., et al., 2012. The Monod–Wyman–Changeux allosteric model accounts for the quaternary transition dynamics in wild type and recombinant mutant human hemoglobin. Proc. Natl. Acad. Sci. U.S.A. 109, 14894–14899.

Ruf, W., 2010. New players in the sepsis-protective activated protein C. J. Clin. Invest. 120, 3084–3087.

Spronk, H.M., Borissoff, J.L., Cate, H., 2013. New insights into modulation of thrombin formation. Curr. Atherscler. Rep. 15, 363.

Stoll, I.R., Voelter, W., Holak, T.A., 1979. "Conformation of thymosin b9 in water/fluoroalcohol solution determined by NMR spectroscopy". Biopolymers. 41, 673.

Tam, M.P., et al., 2013. Autoxidation of oxygen binding properties of recombinant hemoglobins with substitutions at the Val-62 or βVal-67 position of the distal heme pocket. J. Biol. Chem. 288, 25512–25521.

Van de Wouwer, M., Collen, D., Conway, E.M., 2004. "Thrombomodulin-protein C-EPCR system: integrated to regulate coagulation and inflammation". Arterioscler. Thromb. Vasc. Biol. 14, 1374–1383.

Wells, P., Anderson, D., 2013. The diagnosis and treatment of venous thromboembolism. Hematology. 203, 457–463.

Yu, J., et al., 2009. Human induced pleuripotent stem cells free of vector and transgene sequences. Science. 324, 797–801.

## Books

Elgert, K.D., 2009. Immunology: Understanding the Immune System. 726 pages. John Wiley & Sons, Hoboken, NJ.

Hemker, H.C., Loelinger, E.A., Veltkamp, J.J., 2012. Human Blood Coagulation: Biochemistry, Clinical Investigation and Therapy. Springer Science & Business Media, New York, NY.

Litwack, G. (Ed.), 2008. *Vitamin K*, Volume 78 of *Vitamins & Hormones*. Academic Press/Elsevier, Amsterdam, NL, Cambridge, MA, San Diego, CA.

Noordegraaf, A., 2011. Blood in Motion. 323 pages. Springer, New York, NY.

Purves, W.K., et al., 2002. Life: The Science of Biology, fourth ed. Sinauer Associates and W.H. Freeman, Garlansville, VA.

# List of Abbreviations

| | |
|---|---|
| **AAT** | alanine aminotransferase |
| **Aβ** | amyloid beta |
| **ABC transporter** | ATP binding cassette transporter (family) |
| **AC** | adenylate cyclase |
| **ACAT** | acylCoA cholesterol acyltransferase |
| **ACC** | acetylCoA carboxylase |
| **ACDP2** | ancient conserved domain protein 2 |
| **ACE** | angiotensin converting enzyme |
| **Ach** | acetylcholine |
| **ACP** | acyl carrier protein |
| **ACTH** | adrenocorticotrophic hormone (also: corticotropin) |
| **AD** | Alzheimer's disease |
| **ADH** | antidiuretic hormone (same as AVP) |
| **AdipoR1** | adiponectin receptor 1 |
| **ADRP** | adipophilin |
| **AHSP** | alpha-hemoglobin stabilizing protein |
| **AICD** | amyloid precursor intracellular domain |
| **AKU** | alkaponuria |
| **ALAS** | 5-aminolevulinic acid synthase |
| **ALDO** | aldosterone |
| **Alpha MSH** | melanocortin (melanocyte-stimulating hormone) |
| **ALS** | amyotrophic lateral sclerosis |
| **ALT** | alanine aminotransferase |
| **AMPA** | alpha-amino-3-hydroxy-5-methyl-4-isoxazole propionic acid |
| **AMPK** | $5'$-adenosine monophosphate-activated protein kinase |
| **ANF** | atrionatriuretic factor |
| **AngII** | angiotensin II |
| **APC** | activated protein C |
| **APCR** | activated protein C resistance |
| **API-RE** | fos/jun response element |
| **Apo-B100** | apolipoprotein B100 |
| **APP** | amyloid precursor protein |
| **APPL1** | adaptor protein with phosphotyrosine interaction |
| **APRT** | phosphoribosyltransferase |
| **AQ** | aquaporin |
| **AR** | androgen receptor |
| **ARH** | autosomal recessive hypercholesterolemia |
| **AST** | aspartate aminotransferase |
| **AVC** | atrioventricular canal |
| **AVP** | arginine vasopressin (same as ADH) |
| **BAC** | bacterial artificial chromosome |
| **BAG1** | BCL2-associated athanogene-1 |
| **BCKAD** | branched chain alpha-keto acid dehydrogenase |

| | |
|---|---|
| **BDNF** | brain-derived neurotrophic factor |
| **beta-END** | beta-endorphin |
| **11beta-HSD2** | 11beta-hydroxysteroid dehydrogenase |
| **BMP** | bone morphogenic (or morphogenetic) protein |
| **Bohr effect** | When oxygen is released from hemoglobin in the presence of carbon dioxide |
| **1,3-BPG** | 2,3-*bis*phosphoglycerate |
| **BRE** | TFIIB recognition element |
| **BXR** | benzoate X receptor |
| **CAD** | coronary artery disease |
| **CADD** | computer-assisted drug design |
| **CADPR** | cyclic ADP ribose |
| **CAM** | calcium/cadmodulin |
| **CaMKK** | calcium-cadmodulin-dependent protein kinase |
| **CARbeta** | constitutive androstane receptor |
| **CAT** | carnitine translocase |
| **Cav 1** | caveolin 1 |
| **CBC** | complete blood cell count |
| **CBG** | corticosteroid-binding globulin (transcortin) |
| **CBP** | CREB-binding protein |
| **CCE** | capacitative calcium entry |
| **CCK2R** | gastrin receptor (cholecystokinin kinase 2 receptor) |
| **CD** | collecting duct |
| **CD37** | HSP90 co-chaperone |
| **CD38** | cyclicADP ribose hydrolase |
| **CD47** | red blood cell membrane glycoprotein |
| **CdK** | cyclin-dependent kinase |
| **CE** | cholesteryl ester |
| **Cer** | ceramide |
| **CHIF** | corticosteroid hormone-induced factor |
| **Chloride shift** | bicarbonate ion to red blood cell from plasma; chloride ion enters red blood cell from plasma |
| **CJD** | Creutzfeld-Jakob disease |
| **CK2** | casein kinase 2 |
| **CM** | cadmodulin |
| **CML** | carboxymethyl lysine |
| **CNS** | central nervous system |
| **hCTR1** | human copper transporter 1 |
| **Coumadin** | (also Warfarin) compound that prevents blood clotting |
| **Cox-1** | constitutive cyclooxygenase |
| **Cox-2** | induced form of cyclooxygenase |
| **CP** | choroid plexus |
| **CPK** | creatine phosphokinase |
| **CPSII** | carbamoylphosphate synthase II |
| **CPT1** | carnitylpalmitoyltransferase |
| **CRBP** | cellular retinol-binding protein |
| **CREB** | cAMP-responsive element binding protein |
| **CS** | chondroitin sulfate |
| **CSF** | colony stimulating factor |
| **CTFR** | cystic fibrosis transmembrane conductance regulator |
| *CYP7A1* | cholesterol 7alpha-hydroxylase gene |
| **DAG** | diacylglycerol |
| **DBD** | DNA-binding domain |
| **DD** | death domain |

| | |
|---|---|
| **DHA** | dehydroascorbic acid |
| **DHFR** | dihydrofolate reductase |
| **DHAP** | dihydroxyacetone phosphate |
| **DHEA** | dehydroepiandrosterone |
| **DHFR** | dihydrofolate reductase |
| **1,25-dihydroxyvitamin D3** | active form of vitamin D (also calcitriol) |
| **DMT** | divalent metal ion transporter |
| **DOPA** | 3,4-dihydroxyphenylalanine |
| **DOR delta** | delta receptor |
| **DPP-4** | dipeptidylpeptidase 4 |
| **DRIP** | vitamin D receptor-interacting protein |
| **DVT** | deep vein thrombosis |
| **Ea** | energy of activation |
| **EAAT** | excitatory amino acid transporter |
| **EC** | vascular endothelial cell |
| **ECL** | enterochromaffin-like |
| **EF** | elongation factor |
| **EGF** | epidermal growth factor |
| **EGFR** | epidermal growth factor receptor |
| **EGFR-TK** | epidermal growth factor receptor-tyrosine kinase |
| **ElK/ERKR** | ETS domain-containing protein/extracellular regulated kinase |
| **EPB50** | ezrin-radixin-moesin-binding phosphoprotein |
| **Eph5** | human Ephrin type-A receptor 5 |
| **EPCR** | endothelial cell protein C receptor |
| **EPO** | erythropoietin |
| **EPOR** | erythropoietin receptor |
| **ER** | estrogen receptor |
| **ERAP140** | tissue specific nuclear coactivator |
| **ERE** | estrogen response element |
| **ERK** | extracellular signal-regulated kinase (also MAPK) |
| **ERp57** | a disulfide isomerase |
| **ES** | enzyme-substrate complex |
| **etc** | electron transport chain |
| **FABP** | fatty acid binding protein |
| **FACS** | fatty acid CoA synthase |
| **fAd** | full-length adiponectin |
| **FADD** | *fas*-associated death domain |
| **FAK** | focal adhesion kinase |
| **FAS** | fatty acid synthase |
| **FATP** | fatty acid transport protein |
| **FDB** | familial defective apolipoprotein B-100 |
| **FFA** | free fatty acid |
| **FHBL** | familial hypobetalipoproteinemia |
| **FISH** | fluorescence *in situ* hybridization |
| **FKBP52** | immunophilin 52 kDa FK506-binding protein |
| **FLICE** | FADD-like ICE |
| **FLIP** | FLICE-like inhibitory protein/caspase 8 inhibitor |
| **FOLR1** | folate receptor 1 |
| **FoxA1** | Forkhead box A1 |
| **FSH** | follicle-stimulating hormone |
| **5-FU** | 5-fluorouracil |
| **FXR** | farnesoid X receptor |
| $\mathbf{\Delta G^0}$ | Gibbs free energy |

| | |
|---|---|
| **GABA** | gamma-aminobutyric acid (also 4-amino butanoic acid) |
| **gAd** | globular adiponectin |
| **GAP** | GTPase-activating protein |
| **GATA4** | Zn finger DNA-binding protein |
| **GCase** | glucocerebrosidase or glucosyl ceramidase |
| **G-CSF** | granulocyte stimulating factor |
| **GDH** | glutamate dehydrogenase |
| **GHRE** | growth hormone response element |
| **GHRH** | growth hormone releasing hormone |
| **GHSR1** | growth hormone secretagogue receptor |
| **GIP** | general import pore |
| **GIP** | glucose-dependent insulinotropic polypeptide |
| **Gla** | gamma-carboxyglutamate |
| **GlcCer** | glucosylceramide |
| **GLP** | glucagon-like peptide |
| **GLP-1** | glucagon-like peptide 1 |
| **Glycophorin B** | red blood cell membrane glycoprotein |
| **GnRH** | gonadotropin releasing hormone |
| **G3P** | glycerol-3-phosphate |
| **G6PD** | glucose-6-phosphate dehydrogenase |
| **GPD1** | glycerol-3-phosphate dehydrogenase |
| **GPI** | glycosylphosphatidylinositol |
| **GPR30** | G protein-coupled receptor 30 |
| **GPT** | glutamate-pyruvate transaminase |
| **GRalpha** | full-length classical glucocorticoid receptor |
| **GRbeta** | short isoform of glucocorticoid receptor |
| **GRE** | glucocorticoid response element |
| **GRIP** | glucocorticoid receptor interacting protein |
| **GSAP** | gamma-secretase activating protein |
| **GSD** | glycogen storage disease |
| **GSH** | glutathione |
| *pseudoGULO* | defective pseudogene |
| **HAT** | histone acetyltransferase |
| **HbA$_{1c}$** | glycated hemoglobin |
| **HbAO** | adult hemoglobin |
| **HbF** | fetal hemoglobin |
| **HbH** | hemoglobin Bert |
| **HCG** | human chorionic gonadotropin |
| **HDAC** | histone deacetylase |
| **HDL** | high density lipoprotein |
| **HER-2** | human epidermal growth factor receptor-2 |
| **HETE** | hydroxyeicosatetraenoic acid |
| ***HFE*** | hemochromatosis gene |
| **HGPRT (also HPRT)** | hypoxanthine-guanine phosphoribosyltransferase |
| **HHCE** | hyperhomocysteinemia |
| **HIF-1** | hypoxia-inducible factor |
| **HIOMT** | hydroxyindole-O-methyltransferase |
| **HMGCoA** | 3-hydroxy-3-methylglutarylCoA |
| **HRE** | hormone responsive element |
| **HSC** | heat shock cognate |
| **HSP** | heat shock protein |
| **HSL** | hormone-sensitive lipase |
| **HT3R** | hydroxytryptamine 3 receptor |

| | |
|---|---|
| **ICAM** | intercellular adhesion molecule |
| **IDA** | iron deficiency anemia |
| **IDL** | intermediary density lipoprotein |
| **IEF** | isoelectric focusing |
| **IFN** | interferon |
| **IGF-1** | insulin-like growth factor |
| **IGFBP3** | insulin-like growth factor binding protein-3 |
| **IKKbeta** | IkBalpha kinase |
| **IL** | interleukin |
| **IMC** | inter-digestive migrating contractions |
| **INR** | International Normalized Ratio (measures blood coagulation time) |
| **INR** | initiator element |
| **IP3** | inositol-1,4,5-*tris*phosphate |
| **IRE** | iron responsive element |
| **IREBP** | IRE binding protein |
| **IRF7** | interferon regulatory factor 7 |
| **IRS** | insulin receptor substrate |
| **ISC** | iron-sulfur cluster |
| **ISRE** | IFN-stimulated response element |
| **ISU** | iron-sulfur cluster synthetic unit |
| **JAK** | Janus tyrosine kinase |
| **Ki-*ras*** | Kirsten *ras* |
| **KOR** | kappa receptor |
| **Kringle** | protein domain that folds into a large loop stabilized by 3 disulfide bridges |
| **LacCer** | lactosylceramide |
| **LANR** | low affinity neurotrophin receptor |
| **LAT1** | tyrosine transporter |
| **LBD** | ligand binding domain |
| **LCAT** | lecithin-cholesterol acyltransferase (or phosphatidylcholine-sterol O-acyltransferase) |
| **LD** | lipid droplet |
| **LDH** | lactate dehydrogenase |
| **LDL** | low density lipoprotein |
| **LDLR** | LDL receptor |
| **LETM1** | inner membrane protein, an anchor protein, forming a complex with mitochondrial ribosome |
| **LH** | luteinizing hormone |
| **12-LOX** | 12-lipoxygenase |
| **LRP** | low density lipoprotein (LDL) receptor-related protein |
| **Lsm** | like Sm proteins |
| **LXR** | liver X receptor |
| **M** | Methionine |
| **MALT** | mucosa-associated lymphoid tissue |
| **MAPK** | *Ras*/mitogen-activated protein kinase (also ERK) |
| **MasR** | Mas receptor |
| **Mb** | myoglobin |
| **MCT** | monocarboxylate symporter |
| **MDR** | multi-drug resistance transporter |
| **MEK** | mitogen-activated kinase |
| **MELAS** | Mf encephalomyopathy with lactic acid acidosis and stroke-like episodes |
| **MFS** | major facultative transporter superfamily |
| **miRNA** | microRNA |
| ***MITTL1*** | gene whose mutations cause >80% of MELAS cases |
| **MNCX** | $Na^+/Ca^{2+}$ release exchanger |
| **MOR** | μ mu receptor |

| | |
|---|---|
| **MPP** | mitochondrial processing peptidase |
| **M6PR** | mannose-6-phosphate receptor |
| **Mr** | mineralocorticoid receptor |
| **MSUD** | maple syrup urine disease |
| **MTX** | methotrexate |
| **MUCX** | $Na^+/Ca^{2+}$ exchanger |
| **MVB** | multivesicular body |
| **MX** | transglutaminase |
| **MyD88** | dimeric myeloid differentiation protein 88 |
| **MyoD** | myogenic regulating factor involved in muscle differentiation |
| **NANA** | N-acetylneuraminic acid |
| **NBD** | nucleotide-binding domain |
| **NCOR** | major nuclear corepressor |
| **NCT** | nicastrin |
| **NCX** | low affinity/high capacity $Na^+/Ca^{2+}$ exchanger |
| **Nedd4-2** | ubiquitin ligase |
| **NEP** | neprilysin |
| **NEP** | norepinephrine |
| **Neuropeptide 1** | nocistatin |
| **Neuropeptide 2** | Nocil |
| **NFκB** | nuclear factor kappa B |
| **NFT** | intraneuronal neurofibrillary tangle |
| **NGF** | nerve growth factor |
| **NHE** | sodium-proton exchanger (antiport) |
| **NIS** | $2Na^+/I^-$ symporter |
| **NK** | natural killer |
| **NLS** | nuclear localization signal |
| **NMDA** | N-methyl-D-aspartate |
| **NMDAR** | N-methyl-D-aspartate receptor |
| **NMR** | nuclear magnetic resonance |
| **Nociceptin** | (also Orphanin FQ) 17 amino acid peptide anxiolytic |
| **NOD** | nucleotide-binding and oligomerization domain |
| **NOP1** | nociceptin receptor |
| **eNOS** | endothelial nitric oxide synthase |
| **iNOS** | induced nitric oxide synthase |
| **NP** | neurophysin |
| **NPII** | neurophysin II |
| **NPD** | Niemann-Pick disease |
| **NPY** | neuropeptide Y |
| **NSAID** | non-steroidal anti-inflammatory drug |
| **dNTP** | deoxynucleotide triphosphate |
| **Nup** | nucleoporin |
| **OA** | osteoarthritis |
| **OFT** | outflow tract |
| **o-Has** | hyaluronan oligomers |
| **OPG** | osteoprotegerin |
| **ORCC** | outwardly rectifying chloride channel |
| **OsRE** | osmolar response element |
| **OT** | oxytocin |
| **p53** | tumor suppressor DNA-binding protein |
| **PABA** | para-amino benzoic acid |
| **P1** | bacteriophage-derived vector |
| **PAGE** | polyacrylamide gel electrophoresis |

| | |
|---|---|
| **PAM** | peptidylglycine alpha-amidating monooxygenase |
| **Pant** | phosphopantetheine |
| **PAP** | phosphatidic acid phosphatase |
| **PAR** | pregnane receptor |
| **pCAF** | p300/CBP-associated protein |
| **PCFT** | proton-coupled folate transporter |
| **PCR** | polymerase chain reaction |
| **PDGF** | platelet-derived growth factor |
| **PDHP** | pyruvate dehydrogenase phosphatase |
| **PDK** | phosphoinositide-dependent kinase |
| **PDK** | pyruvate dehydrogenase kinase |
| **PDZ domain** | derived from 3 proteins: PSD95, Dig1 & zo-1 and contains amino acid sequence GLGF |
| **PEDF** | pigment epithelium-derived factor |
| **PEG** | polyethylene glycol |
| **PEP** | mitochondrial processing enhancing protein |
| **PEPCK** | phosphoenolpyruvate carboxykinase |
| **PEPT1** | proton-dependent dipeptide transporter |
| **PHD** | prolyl hydroxylase |
| **pI** | isoelectric focusing |
| **PI3K** | phosphoinositol-3 kinase |
| **PIP2** | phosphoinositol-*bis*phosphate |
| **PKA** | protein kinase A |
| **PKC** | protein kinase C |
| **PKU** | phenylketonuria |
| **PLA2** | phospholipase A2 |
| **PLC$\gamma$** | phospholipase C gamma |
| **PLD** | phospholipase D |
| **PLIN** | perilipin |
| **PLP** | pyridoxal-5′-phosphate |
| **PMCA** | low capacity plasma membrane $Ca2^+$ ATPase |
| **PNMT** | phenylethanolamine N-methyltransferase |
| **PNS** | peripheral nervous system |
| **POMC** | proopiomelanocortin |
| **POT** | proton-dependent oligopeptide transporter |
| **PPAR** | peroxisome proliferator-activated receptor |
| **PPase2** | protein phosphatase 2 |
| **PPP** | pentose phosphate pathway |
| **PR** | progesterone receptor |
| **PRF** | prolactin releasing factor |
| **PRL** | prolactin |
| **Protein C** | blood clotting inhibitor |
| **Protein S** | blood clotting inhibitor |
| **PrP$^c$** | normal cellular prion protein |
| **PRPP** | 5-phospho-D-ribosyl-1-pyrophosphate |
| **PrP$^{sc}$** | diseased form (scrapie) of prion protein |
| **P.T.A.** | predicted transmitting ability |
| **PTF** | peptide transport family |
| **PTH** | parathyroid hormone |
| **n-3PUFA** | long chain n-3 polyunsaturated fatty acid |
| **QSAR** | quantitative structure-activity relationships |
| **RA** | retinoic acid |
| **RANKL** | receptor activator of NFκB ligand |
| **RAR** | retinoic acid receptor |

| | |
|---|---|
| **Rb** | retinoblastoma protein |
| **RD** | rhodopsin |
| **RDS** | respiratory distress syndrome |
| **RER** | rough endoplasmic reticulum |
| **RFC1** | folate transporter |
| **Rh50** | red blood cell membrane glycoprotein |
| **RISC** | RNA-induced silencing complex |
| **RNAi** | RNA interference |
| **siRNA** | small interfering RNA |
| **snoRNA** | small nucleolar RNA |
| **ROC** | receptor-operated channels |
| **ROCK** | Rho-associated kinase |
| **ROMK** | inwardly rectifying $K+$ channel |
| **ROS** | reactive oxygen species |
| *RRND* | gene for prion protein (also "Doppel") |
| **RXR** | retinoic X receptor |
| **SCAP** | sterol-regulated escort protein |
| **Schiff base** | functional imine formed by condensation of an amine and the carbonyl of an aldehyde or ketone |
| **Sec synthase** | selenium cysteine synthase |
| **SeP** | selenophosphate |
| **SePS** | selenophosphate synthase |
| **SER** | smooth endoplasmic reticulum |
| **SERT** | serotonin transporter |
| **SgK** | serum and glucocorticoid hormone-induced kinase |
| **SGLT** | sodium glucose transporter (symporter) |
| **SH2** | Src homology domain |
| **SHBG** | sex hormone-binding globulin |
| **SHIP** | SH2-containing inositol 5-phosphatase |
| **SINE** | short interspersed elements |
| **SMOC** | second messenger-operated channel |
| **SMRT** | major nuclear corepressor |
| **Smurf** | Smad ubiquitin regulatory factor |
| **SNARE** | soluble NSF [N-ethylmaleimide sensitive factor] attachment protein receptor |
| **SOD** | superoxide dismutase |
| **S1P** | sphingosine-1-phosphate |
| **SRE** | steroid response element |
| **SREBP** | sterol regulatory element-binding protein |
| **SS** | somatostatin |
| **SSBP** | single-stranded binding protein |
| **Star** | steroid acute response |
| **STAT** | transcriptional factor and signal transducer |
| **STRP** | short tandem repeat polymorphism |
| **SWI/SNF** | switch/sucrose non-fermentable |
| **SXR** | steroid and xenobiotic receptor |
| **T3** | triiodothyronine |
| **T4** | thyroxine |
| **TAB2** | TAK-binding protein 2 |
| **TAF** | transcriptional activator factor |
| **TBG** | thyroid binding globulin |
| **TBP** | TATA box-binding protein |
| **TdT** | terminal deoxynucleotide transferase |
| **Tbx5** | T-box transcription factor |

| | |
|---|---|
| TFIII | transactivation factor III |
| TFR | transferrin receptor |
| Tg | thyroglobulin |
| TGF | transforming growth factor |
| TGN | trans Golgi network |
| THF | tetrahydrofolate |
| TIM | translocase of inner mitochondrial membrane |
| TM | thrombomodulin |
| TNF | tumor necrosis factor |
| TNFRSF | tumor necrosis factor receptor superfamily |
| TOM | translocase of outer mitochondrial membrane |
| TPO | thyroid peroxidase |
| TR | thyroid hormone receptor |
| TRAF6 | TNF receptor-associated factor 6 |
| TRAIL | tumor necrosis factor-related apoptosis-inducing ligand |
| transcortin | corticosteroid binding globulin |
| TRAP | thyroid hormone-associated protein |
| TRH | thyrotropin-releasing hormone |
| TrkR | tropomyosin receptor kinase receptor |
| TSH | thyroid stimulating hormone (also thyrotropin) |
| TSR | thrombin-sensitive region |
| Ub | ubiquitin |
| UDPG | UDP glucose |
| UNC-45 | HSP90 cochaperone |
| $v_0$ | initial velocity |
| $V_{max}$ | maximal velocity |
| VCAM | vesicular cell adhesion molecule |
| VDAC | voltage dependent anion channel |
| VDR | vitamin D receptor |
| VDRE | vitamin D responsive element |
| v-erbA | viral erythroblastoma A (related protein) |
| pVHL | Von Hippel Landau protein |
| VLDL | very low density lipoprotein |
| VIP | vasoactive intestinal peptide |
| Vitamin B9 | folic acid (pteroylglutamic acid) |
| VKD | vitamin K dependent |
| VKOR | 2,3-epoxide reductase |
| VMAT | vesicular monoamine transporter |
| VSD | ventricular septal defect |
| VSMC | vascular smooth muscle cell |
| vWF | von Willebrand factor substrate |
| YAC | yeast artificial chromosome |
| Zip | zinc family transporters |

# Appendix 1

# Abbreviations of the Common Amino Acids

| Amino Acid | Molecular Weight | 3-Letter | 1-Letter | Character |
|---|---|---|---|---|
| Alanine | 89.09 | Ala | A | Neutral; aliphatic |
| Arginine | 174.20 | Arg | R | Basic |
| Asparagine | 132.12 | Asn | N | Neutral; amide |
| Aspartic acid | 133.10 | Asp | D | Acidic |
| Cysteine | 121.16 | Cys | C | Neutral; SH group |
| Glutamic acid | 147.13 | Glu | E | Acidic |
| Glutamine | 146.15 | Gln | Q | Neutral; amide |
| Glycine | 75.07 | Gly | G | Neutral; aliphatic |
| Histidine | 155.16 | His | H | Aromatic ring; basic |
| Isoleucine | 131.17 | Ile | I | Neutral; aliphatic; hydrophobic; *essential* |
| Leucine | 131.17 | Leu | L | Neutral; aliphatic hydrophobic; *essential* |
| Lysine | 146.19 | Lys | K | Basic; *essential* |
| Methionine | 149.21 | Met | M | Neutral; SH group; *essential* |
| Phenylalanine | 165.19 | Phe | F | Neutral; contains aromatic benzene ring; *essential* |
| Proline | 115.13 | Pro | P | Contains ring; imino acid |
| Serine | 105.09 | Ser | S | Contains OH; neutral |
| Threonine | 119.12 | Thr | T | Neutral; contains OH; *essential* |
| Tryptophan | 204.23 | Trp | W | Aromatic indole ring; neutral; *essential* |
| Tyrosine | 181.19 | Tyr | Y | Aromatic benzene ring; OH |
| Valine | 117.15 | Val | V | Neutral; aliphatic *essential* |

# Appendix 2

# The Genetic Code

*Base pairing in DNA*: A (adenine) forms two hydrogen bonds with T (thymine), and G (guanine) forms three hydrogen bonds with C (cytosine). In *RNA*, the same base pairing occurs except that T is replaced by U (uracil). The DNA strand begins with the 5′-hydroxyl (or 5′-phospho) group of the beginning nucleotide and stretches to the final nucleotide's 3′-hydroxyl group. Thus, the upstream to downstream direction is from 5′ to 3′. In double-stranded DNA, a complementary strand binds to the first strand in an antiparallel mode, such that the first strand runs in the 5′ to 3′ direction, while the antiparallel strand runs in the opposite direction: 3′ to 5′, creating a double helical structure. Each nucleotide residue of double-stranded DNA is opposed by a complementary nucleotide (A:T and G:C). The sense strand of DNA contains the codons while the antisense strand (template strand) contains anticodons. mRNA is synthesized from the antisense strand and therefore contains the same information as the sense template strand. A sample of this process is shown below.

| | |
|---|---|
| Coding sense strand (5′ to 3′): | ACGTTACGCAAGGCCAGA |
| Template antisense strand (3′ to 5′): | TGCAATGCGTTCCGGTCT |
| mRNA (5′ to 3′) from template strand: | AUGUUACGCAAGGCCAGA |
| Protein | met-leu-arg-lys-ala-arg |

The amino acids are specified in triplets as shown for DNA in the next table and for RNA is the subsequent table.

In the preceding table, the *ter* or end codons are equivalent to a stop codon. The preceding table is reproduced from page 1 of http://psyche.uthct.edu/shaun/SBlack/genetics.html.

The same table for mRNA (where U replaces T) is shown here and is reproduced from http://users.rcn.com/jkimball.ma.ultranet/BiologyPages/C/Codons.html.

The codons in mRNA are bonded to anticodons in tRNAs, each of which carries the specific amino acid to the ribosomal site of protein synthesis, as illustrated in Figure 2-33.

| | | Second Position of Codon | | | | | |
|---|---|---|---|---|---|---|---|
| | | T | C | A | G | | |
| | | TTT Phe [F] | TCT Ser [S] | TAT Tyr [Y] | TGT Cys [C] | T | |
| F | T | TTC Phe [F] | TCC Ser [S] | TAC Tyr [Y] | TGC Cys [C] | C | T |
| i | | TTA Leu [L] | TCA Ser [S] | TAA *Ter* [end] | TGA *Ter* [end] | A | h |
| r | | TTG Leu [L] | TCG Ser [S] | TAG *Ter* [end] | TGG Trp [W] | G | i |
| s | | CTT Leu [L] | CCT Pro [P] | CAT His [H] | CGT Arg [R] | T | r |
| t | C | CTC Leu [L] | CCC Pro [P] | CAC His [H] | CGC Arg [R] | C | d |
| | | CTA Leu [L] | CCA Pro [P] | CAA Gln [Q] | CGA Arg [R] | A | |
| P | | CTG Leu [L] | CCG Pro [P] | CAG Gln [Q] | CGG Arg [R] | G | P |
| o | | ATT Ile [I] | ACT Thr [T] | AAT Asn [N] | AGT Ser [S] | T | o |

| | | Second Position of Codon | | | | | |
|---|---|---|---|---|---|---|---|
| | | T | C | A | G | | |
| s | A | ATC Ile [I] | ACC Thr [T] | AAC Asn [N] | AGC Ser [S] | C | s |
| i | | ATA Ile [I] | ACA Thr [T] | AAA Lys [K] | AGA Arg [R] | A | i |
| t | | ATG Met [M] | ACG Thr [T] | AAG Lys [K] | AGG Arg [R] | G | t |
| i | | GTT Val [V] | GCT Ala [A] | GAT Asp [D] | GGT Gly [G] | T | i |
| o | | GTC Val [V] | GCC Ala [A] | GAC Asp [D] | GGC Gly [G] | C | o |
| n | G | GTA Val [V] | GCA Ala [A] | GAA Glu [E] | GGA Gly [G] | A | n |
| | | GTG Val [V] | GCG Ala [A] | GAG Glu [E] | GGG Gly [G] | G | |

| Second nucleotide | | | | |
|---|---|---|---|---|
| U | C | A | G | |
| UUU **Phenylalanine** (Phe) | UCU **Serine** (Ser) | UAU **Tyrosine** (Tyr) | UGU **Cysteine** (CyS) | U |
| UUC Phe | UCC Ser | UAC Tyr | UGC Cys | C |
| UUA **Leucine** (Leu) | UCA Ser | UAA **STOP** | UGA **STOP** | A |
| UUG Leu | UCG Ser | UAG **STOP** | UGG **Tryptophan** (Trp) | G |
| CUU **Leucine** (Leu) | CCU **Proline** (Pro) | CAU **Histidine** (His) | CGU **Arginine** (Arg) | U |
| CUC Leu | CCC Pro | CAC His | CGC Arg | C |
| CUA Leu | CCA Pro | CAA **Glutamine** (Gln) | CGA Arg | A |
| CUG Leu | CCG Pro | CAG Gln | CGG Arg | G |
| AUU **Isoleucine** (Ile) | ACU **Threonine** (Thr) | AAU **Asparagine** (Asn) | AGU **Serine** (Ser) | U |
| AUC Ile | ACC Thr | AAC Asn | AGC Ser | C |
| AUA Ile | ACA Thr | AAA **Lysine** (Lys) | AGA **Arginine** (Arg) | A |
| AUG **Methionine** (Met) or **START** | ACG Thr | AAG Lys | AGG Arg | G |
| GUU **Valine** Val | GCU **Alanine** (Ala) | GAU **Aspartic acid** (Asp) | GGU **Glycine** (Gly) | U |
| GUC (Val) | GCC Ala | GAC Asp | GGC Gly | C |
| GUA Val | GCA Ala | GAA **Glutamic acid** (Glu) | GGA Gly | A |
| GUG Val | GCG Ala | GAG Glu | GGG Gly | G |

The mRNA code for amino acids in mitochondria differs in some respects from the nuclear directed processes already discussed. UGA encodes Trp (tryptophan) rather than being a stop codon. AUA specifies Met (methionine) as opposed to isoleucine in the nuclear process and AGA as well as AGG are the stop codons.

A rare amino acid, selenocysteine, is encoded by UGA where the translation machinery is able to distinguish the use of UGA for the seleno-amino acid from the stop codon, although UGA is also used for the stop codon.

# Appendix 3

# Weights and Measures

| | |
|---|---|
| **M** | Molar; molecular weight in grams per liter |
| **mM** | millimolar; Molar/1000; $10^{-3}$ M; 0.001 M |
| **μM** | micromolar; Molar/1,000,000; $10^{-6}$ M; 0.000001 M |
| **nM** | nanomolar; Molar/1,000,000,000; $10^{-9}$ M; 0.000000001 M |
| **fM** | femptomolar; Molar/1,000,000,000,000,000; $10^{-15}$ M; 0.000000000000001 M |
| **m** | meter |
| **cm** | centimeter; meter/100; $10^{-2}$ m; 0.01 m |
| **mm** | millimeter; meter/1000; $10^{-3}$ m; 0.001 m |
| **μm** | micrometer; meter/1,000,000; $10^{-6}$ m; 0.000001 m |
| **Angstrom** | meter/10,000,000,000; $10^{-10}$ m; 0.0000000001 m |
| **g** | gram |
| **mg** | milligram; gram/1000; $10^{-3}$ g; 0.001 g |
| **μg** | microgram; gram/1,000,000; $10^{-6}$ g; 0.000001 g |
| **ng** | nanogram; gram/1,000,000,000; $10^{-9}$ g; 0.000000001 g |
| **kg** | kilogram; gram $\times$ 1000; $10^{3}$ g |
| **L** | liter; 1000 cm$^3$; 1000 mL |
| **dL** | deciliter; liter/10; 100 cm$^3$; 100 mL |
| **mL** | milliliter; liter/1000; 1 cm$^3$; $10^{-3}$ L |
| **μL** | microliter; liter/1,000,000; $10^{-6}$ L |

# Index

*Note*: Page numbers followed by "*f*" and "*t*" refer to figures and tables, respectively.